BIOPHYSICO-CHEMICAL PROCESSES INVOLVING NATURAL NONLIVING ORGANIC MATTER IN ENVIRONMENTAL SYSTEMS

BIOPHYSICO-CHEMICAL PROCESSES INVOLVING NATURAL NONLIVING ORGANIC MATTER IN ENVIRONMENTAL SYSTEMS

Edited by

NICOLA SENESI
BAOSHAN XING
PAN MING HUANG

WILEY

A JOHN WILEY & SONS, INC., PUBLICATION

Published by John Wiley & Sons, Inc., Hoboken, New Jersey
Published simultaneously in Canada

For general information on our other products and services or for technical support, please contact our Customer Care Department within the United States at (800) 762-2974, outside the United States at (317) 572-3993 or fax (317) 572-4002.

Wiley also publishes its books in a variety of electronic formats. Some content that appears in print may not be available in electronic formats. For more information about Wiley products, visit our web site at www.wiley.com.

Library of Congress Cataloging-in-Publication Data:

Biophysico-chemical processes involving natural nonliving organic matter in environmental systems / edited by Nicola Senesi, Baoshan Xing, Pan Ming Huang.
p. cm.—(Wiley-IUPAC series in biophysico-chemical processes in environmental systems)
Includes index.
ISBN 978-0-470-41300-5 (cloth)
1. Environmental chemistry. 2. Bioorganic chemistry. 3. Soil biochemistry. 4. Humus.
I. Senesi, N. (Nicola) II. Xing, Baoshan. III. Huang, P. M.
TD193.B547 2009
577′.14—dc22

2008055879

Printed in the United States of America

10 9 8 7 6 5 4 3 2 1

CONTENTS

SERIES PREFACE

Scientific progress is based ultimately on unification rather than fragmentation of knowledge. Environmental science is the fusion of physical and life sciences. Physical, chemical, and biological processes in the environment are not independent but rather interactive processes. Therefore, it is essential to address physical, chemical, and biological interfacial interactions to understand the composition, complexity, and dynamics of ecosystems. Keeping these domains separate, no matter how fruitful, one cannot hope to deliver on the full promise of modern environmental science. The time is upon us to recognize that the new frontier in environmental science is the interface, wherever it remains unexplored.

The Division of Chemistry and the Environment of the International Union of Pure and Applied Chemistry (IUPAC) has recently approved the creation of an IUPAC-sponsored book series entitled *Biophysico-Chemical Processes in Environmental Systems* to be published by John Wiley & Sons, Hoboken, NJ. This series addresses the fundamentals of physical–chemical–biological interfacial interactions in the environment and the impacts on (1) the transformation, transport and fate of nutrients and pollutants, (2) food chain contamination and food quality and safety, and (3) ecosystem health, including human health. In contrast to classical books that focus largely on separate physical, chemical, and biological processes, this book series is unique in integrating the frontiers of knowledge on both fundamentals and impacts on interfacial interactions of these processes in the global environment.

With the rapid developments in environmental physics, chemistry, and biology, it is becoming much harder, if not impossible, for scientists to follow new developments outside their immediate area of research by reading the primary research literature. This book series will capture pertinent research topics of significant current interest and will present to the environmental science community a distilled and integrated version of new developments in biophysico-chemical processes in environmental systems.

This book is Volume 2 of this series. It can be used as an advanced reference book on biophysico-chemical processes involving natural nonliving organic matter in the global environment for senior, undergraduate, and graduate students in environmental sciences and engineering. It represents an instrumental reference for chemists and biologists studying environmental systems as well as for geochemists, environmental engineers, and soil, water, and atmosphere scientists. It will serve as a useful resource book for professors, instructors, research scientists, professional consultants, and other persons working on environmental and ecological sciences.

P. M. Huang
N. Senesi

PREFACE

A large body of scientific literature is available on the fundamentals and analytical methods for investigation of physico-chemical and biological interfacial reactions and their impacts on nonliving natural organic matter (NOM) in nature, which is currently an area of very active research. However, to advance the frontiers of knowledge on the subject matter in a comprehensive manner would require a concerted effort of scientists in relevant physical and life sciences such as chemistry, mineralogy, geochemistry, microbiology, ecology, and soil, sediment, atmospheric, and aquatic sciences. Environmental science is indeed the fusion of physical and life sciences. Scientific progress in advancing the understanding of NOM in the environment is based ultimately on unification rather than fragmentation of knowledge.

In recognition of the above, the book *Biophysico-Chemical Processes Involving Natural Nonliving Organic Matter in Environmental Systems* is Volume 2 of the newly created Wiley-IUPAC series and consists of 18 chapters organized in two parts: (1) Fundamentals and Impact of Mineral–Organic–Biota Interactions on the Formation, Transformation, Turnover, and Storage of Natural Nonliving Organic Matter and (2) Analytical Methods for Investigation of Natural Nonliving Organic Matter. The formation, transformation, turnover, and storage of natural nonliving organic matter is influenced markedly by mineral–organic–biota interfacial interactions. The overall goal of this book is to provide the scientific and professional communities with an up-to-date and critical evaluation by world-leading scientists on biophysico-chemical processes of NOM in various environmental compartments. The specific objectives of this book are to address (1) the fundamentals and the impact of mineral–organic matter–biota interactions on the formation, nature and properties, transformation, turnover, and storage of NOM in various environmental systems and (2) the state-of-the-art analytical methods for investigating the biophysico-chemical processes involving NOM in nature. The book also identifies the gaps in knowledge on the subject matter and as such provides future directions to stimulate scientific research to advance the chemical science on biophysico-chemical interfacial reactions of natural nonliving organic matter in natural habitats, leading to the subsequent development of innovative management strategies to sustain environmental quality and ecosystem health on a global scale.

In contrast to the classic books that largely focus on separate physico-chemical and biological aspects, this book aims to integrate the frontiers of knowledge on NOM in soil, sediment, water, and air. This book, contributed by a multidisciplinary group of soil, water, sediment, atmosphere, and environmental scientists, along with renowned experts in analytical chemistry, provides the scientific community with a critical evaluation of the state of the art on (1) the fundamentals of reactions and processes of natural nonliving organic matter in the global environment and (2) the most modern and advanced analytical methods and techniques used for their investigation.

The book is an important guide to scientists interested in environmental chemistry and engineering, and it represents an important addition to the scientific literature and a valuable source of reference for students, professors, scientists and engineers. The latest advances in physico-chemical methods and techniques to study various aspects of natural nonliving organic matter are also reviewed critically and addressed clearly. The chapter authors are scientists who are internationally renowned experts in their fields, and all the chapters have been reviewed by at least two external referees.

We wish to thank all of the authors and referees who generously contributed their time and knowledge to ensure the high quality of this volume. We also express our gratitude to the staff of IUPAC and John Wiley & Sons for their invaluable support and cooperation in the publication of the book.

N. Senesi
B. Xing
P. M. Huang

ABOUT THE EDITORS

NICOLA SENESI

Nicola Senesi is Professor of Soil Chemistry and Head of the Department of Agroforestry and Environmental Biology and Chemistry at the University of Bari, Bari, Italy, where he has been actively involved in research and teaching since 1969. He has taught courses in soil chemistry, soil science, agricultural chemistry, wood chemistry and technology, organic chemistry, and general and inorganic chemistry. He has been a visiting professor and/or scientist for various periods at universities in Canada, the US, Somalia, Indonesia, Switzerland, Argentina, Brazil, Venezuela, and Colombia. Dr. Senesi is Fellow of the American Society of Agronomy (since 1995) and the Soil Science Society of America (since 1996), and a recipient of the Golden Medal of the Polish Soil Science Society (1994). He has been conferred a Doctorate Honoris Causa by the Institute National Polytechnique de Toulouse, France, in 2000. Dr. Senesi is currently President of the Division VI-Chemistry and the Environment and Chair of the Subcommittee Biophysico-Chemical Processes in Environmental Systems of the International Union of Pure and Applied Chemists, President of the Mediterranean Scientific Association for Environmental Protection, and President of the Italian Soil Science Society, and was formerly President of the International Humic Substances Society and Chairman of Division II-Soil Properties and Processes of the International Union of Soil Science. He is currently an Associate Editor of *Geoderma*, *Soil Science*, *Pedosphere*, *Pure and Applied Chemistry*, and *CLEAN—Soil, Air, Water*. Dr. Senesi's research is focused on fundamental and applied aspects of chemistry and biochemistry of organic matter from soils and other systems and materials, and its interactions with soil-applied organic chemicals and trace metals, by the use of advanced physico-chemical techniques and biochemical tools. Specific topics of research include the abiotic interactions of herbicides and endocrine disruptors with humic substances, the complexation mechanisms between trace metals of agricultural and environmental importance and natural and artificial humic materials, the physiological and antimutagenic effects of humic substances on plants, and the implications of recycling organic wastes on soil fertility and crop production. He has also pioneered the application of fractal geometry to the study of molecular conformation and aggregation processes of natural soil organic colloids. The results of his research are documented in some 300 scientific and technical papers and some 60 book chapters and invited reviews. Dr. Senesi has also co-edited 12 books and proceedings volumes.

BAOSHAN XING

Baoshan Xing is Professor of Environmental and Soil Chemistry (since 2004) in the Department of Plant, Soil and Insect Sciences, University of Massachusetts, Amherst, where he has been actively involved in teaching and research since 1996. Dr. Xing received his Ph.D. degree from University of Alberta, Canada, in 1994. Dr. Xing's current research includes sorption and fate of organic chemicals in soils and sediments, environmental behavior and toxicity of engineered nanomaterials, interfacial processes in the environment, natural organic matter characterization and chemistry, and application of spectroscopic and microscopic techniques in environmental and soil sciences. Along with students and colleagues, he has published over 150 refereed articles; his research work is ranked in the top 1% of cited authors for journals in environmental sciences and ecology. Currently, Dr. Xing is teaching environmental soil chemistry, advanced soil chemistry, environmental impact of manufactured nanomaterials, and inorganic contaminants in soil, water, and sediment. He has been invited to present his research results at many universities and institutions. Dr. Xing is a technical editor of *Journal of Environmental Quality* and an editorial board member of *Environmental Pollution*, *Chemosphere*, *International Journal of Phytoremediation*, *Communications in Soil Science and Plant Analysis*, and *Pedosphere*. Dr. Xing won the outstanding research award of his college (Natural Resources and the Environment) in 2003 and of the Northeast Branch of Soil Science Society of America and American Society of Agronomy in 2006. He was selected as a Cheung Kong Scholar by the Education Ministry of China in 2007. He was also promoted early to Associate Professor with tenure in 2000 and to Full Professor in 2004. His research program is recognized internationally.

PAN MING HUANG

Pan Ming Huang received his Ph.D. degree in soil science at the University of Wisconsin, Madison, in 1966. He is Professor Emeritus of Soil Science at the University of Saskatchewan, Saskatoon, Canada. His research work has significantly advanced the frontiers of knowledge on the formation chemistry and nature and surface reactivity of mineral colloids, organic matter, and organomineral complexes of soils and sediments and their role in the dynamics, transformations, and fate of nutrients, toxic metals, and xenobiotics in terrestrial and aquatic environments. His research findings, embodied in well over 300 refereed scientific publications, including nine research papers published in *Nature* and others in leading journals, book chapters, and books, are fundamental to the development of sound strategies for managing land and water resources in the Earth's critical zone.

He has developed and taught courses in soil physical chemistry and mineralogy, soil analytical chemistry, and ecological toxicology. He has trained and inspired M.Sc. and Ph.D. students and postdoctoral fellows, and has received visiting scientists from all over the world. He has served on numerous national and international scientific and academic committees. He has also served as a member of many editorial boards, such as *Soil Science Society of America Journal*, *Geoderma*, *Chemosphere*, *Water, Air and Soil Pollution*, *Soil Science and Plant Nutrition*, and *Pedosphere*. He has served as a titular member of the Division of Chemistry and the Environment of the International Union of Pure and Applied Chemistry and is the founding chairman of the Working Group MO, "Interactions of Soil Minerals with Organic Components and Microorganisms," and the founding chair of Commission 2.5, "Soil Physical/Chemical/Biological Interfacial Reactions" of the International Union of Soil Sciences. He received the Distinguished Researcher Award from the University of Saskatchewan, the Soil Science Research Award from the Soil Science Society of America, the Distinguished Alumnus Award and the Chair Professorship Award of National Chung Hsing University, and the Y.Q. Tang Chair Professorship Award from Zhejiang University. He is a fellow of the Canadian Society of Soil Science, the Soil Science Society of America, the American Society of Agronomy, the American Association for the Advancement of Science, and the World Innovation Foundation.

LIST OF CONTRIBUTORS

Abbt-Braun, G., University of Karlsruhe, Germany

Benner, R., University of South Carolina, USA

Brandes, J., University of Georgia, USA

Carletti, P., University of Padova, Italy

Castanha, C., Lawrence Berkeley National Laboratory, USA

Cesco, S., University of Udine, Italy

Chen, Y., The Hebrew University of Jerusalem, Israel

Chilom, G., South Dakota State University, USA

Contin, M., University of Udine, Italy

da Costa, Duarte, A., University of Aveiro, Portugal

Da Silva, W. T. L., Brazilian Agricultural Research Corporation (EMBRAPA), Brazil

De Nobili, M., University of Udine, Italy

Eckhardt, K.-U., University of Rostock, Germany

Fleckenstein, H., SUNY Stony Brook, USA

Frimmel, F. H., University of Karlsruhe, Germany

Gaspar, A., GSF, National Research Center for Environment and Health, Germany

Hardie, A. G., Stellenbosch University, South Africa

Hatfield, K., University of Florida, USA

Hayes, M. H. B., University of Limerick, Ireland

Hertkorn, N., GSF, National Research Center for Environment and Health, Germany

Hofmann, D., Research Centre Jülich, Germany

Huang, P. M., University of Saskatchewan, Canada

Jacobson, C., SUNY Stony Brook, USA

Jandl, G., University of Rostock, Germany

Knicker, H., Instituto de Recursos Naturales y Agrobiologia de Sevilla, Seville, Spain

Konstantinov, A. I., Lomonosov Moscow State University, Russia

Kulikova, N. A., Lomonosov Moscow State University, Russia

Kunenkov, E. V., Lomonosov Moscow State University, Russia

Leboeuf, E. J., Vanderbilt University, USA

Lehmann, J., Cornell University, USA

Leinweber, P., University of Rostock, Germany

Martin-Neto, L., Brazilian Agricultural Research Corporation (EMBRAPA), Brazil

Milori, D. M. B. P., Brazilian Agricultural Research Corporation (EMBRAPA), Brazil

Muscolo, A., University of Reggio Calabria, Italy

Nardi, S., University of Padova, Italy

Oliveira Duarte, R. M. B., University of Aveiro, Portugal

Perdue, E. M., Georgia Institute of Technology, USA

Perminova, I. V., Lomonosov Moscow State University, Russia

Pinton, R., University of Udine, Italy

Pizzeghello, D., University of Padova, Italy

Plaza, C., Centro de Ciencias Medioambientales (CSIC), Spain

Rice, J. A., South Dakota State University, USA

Schlichting, A., Steinbeis-Transferzentrum Soil Biotechnology, Germany

Schmitt-Kopplin, P., GSF, National Research Center for Environment and Health, Germany

Schulten, H.-R., University of Rostock, Germany

Senesi, N., University of Bari, Italy

Simões, M. L., Brazilian Agricultural Research Corporation (EMBRAPA), Brazil

Simpson, A. J., University of Toronto, Canada

Simpson, M. J., University of Toronto, Canada

Solomon, D., Cornell University, USA

Swanston, C. W., USDA Forest Service, USA

Thieme, J., University of Göttingen, Germany

Torn, M. S., Lawrence Berkeley National Laboratory and University of California, USA

Trumbore, S. E., University of California, USA

Varanini, Z., University of Verona, Italy

Xing, B., University of Massachusetts, USA

Zhang, L., Vanderbilt University, USA

1

EVOLUTION OF CONCEPTS OF ENVIRONMENTAL NATURAL NONLIVING ORGANIC MATTER

M. H. B. Hayes

Chemical and Environmental Sciences, University of Limerick, Limerick, Ireland

Biophysico-Chemical Processes Involving Natural Nonliving Organic Matter in Environmental Systems, Edited by Nicola Senesi, Baoshan Xing, and Pan Ming Huang

1.1. INTRODUCTION

The importance of soil organic matter (SOM) has been recognized since the dawn of agriculture. There were extensive writings about agriculture in Roman times, and the literature that had accumulated was assembled in one volume by Petrus Crescentius (ca. 1240), a senator in Bologna (Russell, 1973). Applications to soil of by-products of vegetable, animal, and human origins have always been made. Even the process of maturing, prior to soil applications, of what we would now regard as composts has been known for over 2000 years. Columella (Lucius Iunius Moderatus Columella, Cadice, Spain, 1st century A.D.) described how organic wastes had to be processed before their use as soil amenders.

Russell (1973) referred to a communication by Palissy around 1563 that states: "You admit that when you bring dung into a field it is to return to the soil something that has been taken away." In reference to straw he writes "being burnt on the ground it serves as a manure because it returns to the soil those substances that had been taken away." It was known then that the application of char, or the residue from the pyrolysis or burning of organic materials and wastes under restricted air conditions, enhanced soil fertility greatly. For several millenia in pre-Columbian times the Indians of the Amazon region had burned their organic refuse slowly in restricted air supplies and returned the char to the soil. That char gave rise to an amazing fertility that is still very evident in comparison to the infertile unamended neighboring oxisol soils (Woods and Glaser, 2004). These chars in the Amazonian Dark Earth (ADE), or the Terra Preta de Indio soils, will inevitably have some mineral elements, and the slow oxidation of the chars has given rise to humic-type materials. However, the amazing fertility of these soils cannot be attributed to the elementary compositions of the chars. The char, as well as trapping and holding nutrients and growth promoting substances that could be released in the pyrolysis processes, is considered to provide refuge for such soil microorganisms as arbuscular mycorrhizal fungi (Rillig et al., 2001). These form symbiotic relationships with plant roots. The fungal hyphae effectively extend the root system and transport nutrients from the soil to the plant roots. The hyphae are considered to liberate and to make available for the plant phosphate complexed in the iron oxides associated with oxisols, and such fungi are known to secrete mucopolysaccharides that strongly adsorb to mineral surfaces and in this way help to promote and to stabilize soil aggregates. In return the fungi receive their organic sustenance from the plants of the Amazon region. The modern interest in the pyrolysis and the gasification of organic substrates and the residuals from second-generation biorefining processes (which give bio-oil and biochar products) has awakened interest in the benefits of biochar as soil amenders (Marris, 2006; Hayes, 2006a).

The early statement that "corruption is the mother of vegetation" doubtless arose from the observation that manures, composts, dead animal bodies, and parts thereof such as blood, hair, hoofs, and so on, increased plant growth. John Woodward (cited by Russell, 1973), in a paper published in the *Philosophical Transactions of the Royal Society* (Vol. 21, p. 382), observed that the falloff in yields of crops grown in successive years on unmanured land could be rectified when "supplied with a new fund of matter, of like sort with that it first contained; which supply is made in several ways, either by the ground's being fallow some time, until the rain has poured down a fresh stock upon it; or by tiller's care in manuring it." He considered that the best manures were parts either of vegetables or of animals, which ultimately are "derived from vegetables."

Applications of the sciences to the study of soils began slowly around the middle of the 18th century. Because there was not the instrumentation needed for studies of complex mixtures, such as SOM, progress was slow until the second half of the 20th century, and the most striking advances have been made in the present generation.

1.2. ORGANIC MATTER AT THE "DAWN" OF SOIL RESEARCH

In Chapter 1 of her book *Soil Organic Matter*, Kononova (1966) has provided a good treatise on the evolution of the chemical and biological approaches to the study of soil organic matter. The mixture of organic materials in soils presents problems in separation and in compositional and structural studies. Thus significant verifiable advances in aspects of the compositions and structures of the organic components in soils had to await the development of separation and analytical procedures and instrumentation. In the same way, awareness of the processes of organic matter transformations in soil was retarded because the development of the science of soil microbiology lagged behind that of the chemistry that had allowed some significant advances to be made. By the middle of the 18th century, there was appropriate awareness that humus has an important bearing on soil fertility. Lomonosov (1763) recognized that soil humus had its origins in the rotting with time of plants and animal bodies, and Komov (1789) associated the desirable agronomic properties and fertility to the presence of humus. He stressed the importance of applying farmyard manure and recommended the sowing of perennial grasses.

The book by Wallerius (1761) is regarded as the first scientific guide to agricultural chemistry. In it he recognized the formation of humus during the decomposition of plants; and he was aware of some of its important properties, such as its capacity to absorb water and retain nutrients. However, at that time humus was considered to be a plant food. deSaussure (1804) was first to show that humus had more C, H, and O than the original plant residues. However, he took the view that plants take their carbon chiefly from carbonic acid in the air. He considered that it was taken directly into plants as a food. The experimental data of von Thaer (1809) led him to consider that soil humus was a direct source of plant food. The book by Sir Humphrey Davy (Davy, 1813), who studied the hypotheses of other writers, set out in the new chemical language the accepted knowledge of the time. He did not totally accept de Saussure's concept of carbonic acid from the air, though he did concede that some plants could derive their C from that source. In general he considered that the carbon was taken in through the roots. German (1836, 1837) was among the last to accept the concept of humus as a source of plant nutrients. However, Liebig (1846) cast scorn in the theory of humus as a plant nutrient. Russell (1973) describes how Liebig, "with polished invective," scorned the plant physiologists of his day for their widely held view that plants derive their carbon from the soil. Russell quotes Liebig as follows: "All explanations of chemists must remain without fruit, and useless, because, even to the great leaders in physiology carbonic acid, ammonia, acids, and bases are sounds without meaning, words without sense, in terms of an unknown language, which awake no thoughts and no associations."

The experiments quoted by the physiologists were considered by Liebig "to be valueless for the decision of any question. These experiments are considered by them as convincing proofs, whilst they are fitted only to awake pity." Liebig's ridicule killed the theory of humus as a nutrient.

1.3. SCIENTIFIC APPROACHES TO STUDIES OF SOIL HUMIC SUBSTANCES

1.3.1. The Emergence of Procedures for the Isolation of Soil Humic Components

Chemical knowledge advanced significantly during the 18th and 19th centuries, and these advances had a significant bearing on studies of the chemistry of SOM. It was recognized that the isolation and fractionation of components of SOM was a prerequisite for studies of their compositions.

Achard (1786) is regarded as the first to isolate and to make a fractionation of humic substances (HS). He treated peat with potassium hydroxide, added acid to the dark solution, and obtained an amorphous dark precipitate. More precipitate was obtained from the darker (more humified) layers than from the "less rotted" residues in the upper layers. Later, Vauquelin (1797, 1798) isolated in alkali solution humic-type substances from elm wood infected by fungi, and Thomson (1807) proposed the term *ulmin* for the isolates.

Toward the middle of the 20th century, it was considered generally that HS were polymeric or at least macromolecular, as well as being polyelectrolytes. It was realized that at the pH of most fertile agricultural soils the polyelectrolytes were rendered insoluble through ion-exchange reactions with divalent and polyvalent metal cations. Mineral acids exchanged the metal cations for H^+ to give rise to H^+-exchanged HS, allowing the molecules to remain associated through hydrogen bonding and van der Waals forces mechanisms. In this state the humic materials would have the properties of H^+-exchanged polyelectrolytes. Raising the pH would give rise to dissociation of the acidic functionalities, allowing the conjugate bases to solvate in aqueous media.

Bremner (1950) showed that significant oxidation of organic (humic) matter took place in basic media. Earlier, Bremner and Lees (1949) had shown that a sodium pyrophosphate solution (0.1 M), neutralized to pH 7, was an effective solvent for humified material. The pyrophosphate forms a complex with the charge-neutralizing divalent and polyvalent cations and the freed conjugate bases dissolve in the aqueous medium. This system is successful in dissolving the more highly oxidized humic components (with greater carboxyl functionalities), but it will not dissolve the less humified materials in which phenolic groups (which dissociate at higher pH values) contribute significantly to the charge density.

Hayes et al. (2008) described the uses of 0.1 M sodium pyrophosphate (Pyro) solutions for exhaustive extractions of soil organic matter at pH 7, pH 10.6, and at pH 12.6 (Pyro + 0.1 M NaOH). They showed that the fractions were compositionally different, with the most transformed (oxidized) fractions isolated at the lower pH value. However, only about 26% of the organic matter was isolated in the sequential process.

When the International Humic Substances Society (IHSS) was founded at a meting in the US Geological Survey in September 1981, it was decided to hold symposia at which designated persons would present state-of-the-art information about various aspects of the humic sciences. This new society aimed to unite the soil and water humic scientists. Intensive studies on aquatic HS had arisen from the observation by Rook (1977) that coloured waters when chlorinated gave rise to mutagenic chlorinated hydrocarbons. This caused the US Geological Survey, led by Drs. Bob Avert and Ron Malcolm, to initiate intensive studies into all aspects of aquatic HS. At the International Soil Science Society meetings in Edmonton in 1978, Dr. Malcolm and Dr. Patrick MacCarthy (Colorado School of Mines) introduced soil humic scientists to the initiatives being taken by the USGS, and that led to the eventual formation of the IHSS in 1981. It was determined at the founding meeting (which the writer attended) that funding from the USGS would be used to provide a set of IHSS Soil and Water Standards, and procedures were agreed for the preparation of these standards. The writer, then at the University of Birmingham, England, was given the task of isolating humic and fulvic acids from a Florida Sapric Histosol; and Professor R. S. Swift, then at Lincoln College of the University of Canterbury, NZ, agreed to isolate these fractions from an Elliott Mollisol from Illinois. Dilute NaOH was the solvent of choice for the extraction of the IHSS Soil Standards, and Swift (1996) has outlined the procedures used. Invited contributions at the first meeting of the IHSS at Estes Park in 1983 dealt with the geochemistry, isolation and fractionation, and characterization of HS (Aiken et al., 1985a). Hayes (1985) presented the information that was available then about the isolation of HS from soil, and Aiken (1985) did likewise for their isolation from waters.

Hayes (1985) drew up a set of criteria that he considered to be important for good organic solvents for HS. Earlier, Whitehead and Tinsley (1964) had outlined criteria that they considered to be important for effective solvents for HS. These were:

1. A high polarity and a high dielectric constant to assist the dispersion of charged humic molecules
2. A small molecular size to penetrate through the humus matrix
3. The ability to disrupt existing hydrogen bonds, and to provide alternative groups to form humic–hydrogen bonds
4. The ability to immobilize metallic cations.

Hayes (1985, 2006b) has listed the properties of organic solvents that might be considered for the isolation of soil humic components. He checked the extents to which selected organic solvents dissolved H^+-exchanged humic acids (HAs), and he concluded that good organic solvents have electrostatic factor (the product of relative permittivity and dipole moment) values greater than 140 and have pK_{HB} (the measure of the strength of a solvent as an acceptor in hydrogen bonding) values greater than 2. Dimethylformamide (DMF) and dimethylsulfoxide (DMSO) meet these requirements; both of these were shown to be good solvents for the HAs, with DMSO being the better of the two. Hayes also discussed applications of solubility parameter data. The best of the organic solvents tested had δ_p (dispersion force), δ_h (hydrogen bonding), and δ_b (proton acceptor) parameters greater than 6, 5, and 5, respectively. Solvation is greatest when the product of δ_a (solvent) × δ_b (solute), or vice versa, is maximum (Hayes, 1985).

The major solvent systems still involve base, and 0.1 M NaOH is the solvent of choice of the IHSS (Swift, 1996). Song et al. (2008) used a modification of the IHSS procedure and extracted a Mollisol soil exhaustively in 0.1 M NaOH adjusted to pH 7, then at pH 10.6, and then with the unadjusted solution (pH 12.6). Subsequently the residual soil material was exhaustively extracted with 0.1 M NaOH + 6 M urea (see also Hayes, 2006b). The NMR data show significant differences between the humic components isolated at the different pH values, but the extract in the NaOH/urea solvent (which would be humin in the classical definitions) was similar to that isolated at pH 12.6. This would-be humin (Section 1.3.3) material in the classical definitions was in fact composed of HAs and fulvic acids (FAs) trapped within the humin matrix (Song et al., 2008). Subsequently the residual organic matter associated with the fine clay (humin material) was exhaustively extracted with DMSO + 6% concentrated H_2SO_4, and 93% of the humin residue was solvated and recovered. The remaining clay–humin associations can be released by dissolving the silicates in HCl/HF (Preston and Newman, 1992).

1.3.2. The Emergence of Procedures for the Fractionation of Soil Humic Components

Berzelius (1806) was the first to consider the humic fractions that are still extensively worked with. His *humic acids* were soluble in aqueous base and precipitated upon acidification of the media. He regarded as humin the inert material that was not dissolved in base. Light yellow materials were left in solution following the precipitation of the humic acids, and Berzelius called these *crenic* and *apocrenic acids* (he considered the latter to be an oxidation product of crenic acid), components that effectively complexed ammonia and various metals to give these elements greater mobility (compared with the salts of humic acids). Crenic and apocrenic acids would be covered by the term *fulvic acids* introduced subsequently. The mobility in the soil profile of salts of crenic and apocrenic acids was later used by Sibirtsev (1900, 1901) to explain aspects of podzolization. Berzelius ascribed to the thesis that soil fertility and plant nutrition were associated with the presence of humus and that, because crops deplete soil humus, it is necessary to apply organic manures. The various views held at the time are incorporated in his textbook of chemistry (Berzelius, 1839).

Fractionation on the basis of solubilities at different pH values has always been a major procedure for the fractionation of HS, and thus there arose the primary fractionations into HAs and FAs. Further fractionations on the basis of different solubilities in alcohol gave rise to additional components, as outlined in Section 1.3.3.

The development of electrophoretic techniques afforded possibilities for fractionations based on charge density differences. Duxbury (1989) has reviewed applications of different electrophoretic separation methods, including zone electrophoresis, moving boundary electrophoresis, isotachophoresis, and isoelectric focusing (IEF). Preparative column electrophoresis (Clapp, 1957) and continuous flow paper electrophoresis (Hayes, 1960; summarized by Hayes et al., 1985) methods have been used to separate components isolated from sapric histosol soils. These techniques allowed separation of polysaccharides from the colored components; the electrophoretograms of the colored components were diffuse, showing a continuum of components of different charge densities.

The availability of gel filtration techniques during the 1960s allowed fractionation to be achieved on the basis of size differences. The most noteworthy work using these techniques is attributed to Cameron et al. (1972) (see Swift, 1985). They, using gel filtration and discrete pore size membranes, fractionated a HA extract into 11 different size fractions and determined the molecular weight of the fractions using ultracentrifugation techniques (see Section 1.4.7).

Leenheer (1985) has reviewed procedures used by water scientists for the fractionation of aquatic HS. Water scientists introduced the Rohm and Haas resins XAD-8 [(poly)methylmethacrylate] and XAD-4 (styrenedivinly benzene) for the separation and isolation of HAs, FAs, and XAD-4 acids. The less polar HA and FA components sorb on XAD-8, and the polar components elute through the resin but are held by XAD-4. The HAs and FAs are recovered during back elution in dilute base, and the HAs are then precipitated at pH 2. The XAD-4 acids are also back-eluted in base, H^+-exchanged using IR-120 H^+-exchanged resin, and freeze-dried. The resin techniques are applicable to soil extracts, and they have been used successfully by Hayes et al. (2008) for the fractionation of extracts from soils and their drainage waters.

Techniques for the isolation and fractionation of carbohydrate and peptide components of SOM are discussed in Sections 1.5.2 and 1.6.

1.3.3. Definitions of Soil Humic Components

Mulder (1861–1862), who had been a student of Berzelius, classified the HS he isolated as:

Ulmin and *Humin*, the components insoluble in alkali
Ulmic acid (brown) and *Humic acid* (black), the components soluble in alkali
Crenic acid and *Apocrenic acid*, the components soluble in water

These definitions were essentially the same as those put forward by Berzelius. Mulder considered, however, that, besides humus substances, products from the decomposition of organic residues, such as leucine, butyric acid, valeric acid, and formic and ethanoic acids, could exist in soil. These observations are of interest because of the information that has emerged in the past half-century about growth inhibitors and stimulators from low-molecular-weight extracts from SOM and composts.

Mulder considered that the different isolates were chemically individual compounds, and on the basis of elemental analyses data he assigned to the different fractions the empirical folmulae:

Humin, $C_{10}H_{30}O_{15}$	Ulmin, $C_{10}H_{32}O_{11}$
Humic acid, $C_{10}H_{24}O_{12}$ (or $C_{10}H_{30}O_{15}$)	Ulmic acid, $C_{10}H_{30}O_{15}$
Crenic acid, $C_{10}H_{21}O_{16}$	Apocrenic acid, $C_{21}H_{12}O_{12}$

The numbers of humic fractions continued to increase, and terms such as *"mucic acid," lignoic acid*, and *hymatomelanic* acid were introduced; all of these were considered to represent chemically individual compounds, which of course they were not. Sprengel (1826) promoted fractionation on the basis of solubility characteristics in aqueous media, and then the terms *humic acids*, *fulvic acids*, and *humin* became generally recognized.

Definitions (based on the solubility criteria) of soil humic components have not changed much in the last 200 years. In the definitions of the International Humic Substances Society, as stated by Aiken et al. (1985b), *humic substances* are "a general category of naturally occurring, biogenic, heterogeneous organic substances that can generally be characterized as being yellow to black in color, of high molecular weight, and refractory." They classified formally the three major fractions as *humin*, "that fraction of humic substances that is not soluble in water at any pH value"; *humic acid*, that fraction of humic substances that is not soluble under acid conditions (below pH 2), but becomes soluble at greater pH"; and *fulvic acid*, "that fraction of HS that is soluble under all pH conditions." In the classification of Kononova (1966, 1975) adapted by Hayes and Swift (1978), SOM is grouped into:

1. Unaltered materials, which include fresh debris and nontransformed components of older debris;
2. Transformed products, or humus, bearing no morphological resemblances to the structures from which they were derived.

The transformed, or humified, components consist of humic and nonhumic substances. The humic substances are defined by Aiken et al. (1985b), as described above. The nonhumic substances belong to recognizable classes, such as polysaccharides, polypeptides, and so on. These can be synthesized by microorganisms or can arise from modifications of similar compounds in the original debris. It is questionable, on the basis of emerging information, that HS can be considered to be of high molecular weight (Piccolo, 2001; Simpson, 2002). Inevitably, these substances will have high molecular weight components, but there is support for the concept of molecular associations that give rise to pseudo-macromolecular properties.

The major solvent systems still involves base, and 0.1 M NaOH is the solvent of choice of the IHSS (Swift, 1996). Hayes (1985, 2006b) reviewed the principles and the procedures for the isolation of HS, and the more recent publication refers to solvent systems that isolate additional HA and FA materials using exhaustive extractions at increasing pH values, followed by exhaustive extractions with 0.1 M NaOH + 6 M urea (see Section 1.3.1). The components in intimate associations with the clays, isolated in DMSO/H_2SO_4 in the solvent sequence, were largely biological molecules (see Sections 1.4.7 and 1.7) and would not, in the classical definitions, be HS.

1.3.4. The Need to Redefine the Fractions of Soil Humic Substances

It would be pointless to draw up a classification system that takes account of several fractions based on charge density differences, or even differences in solubilities in organic solvent systems. Consideration might be given to the hymatomelanic acid, or the alcohol-soluble component described by Hoppe-Seyler (1889). It would be important to distinguish between the FA fraction (or the material that is soluble in acidic and basic media) and the FAs as defined by the IHSS (or the fractions recovered when the FA fraction in solution at pH 2 is passed on to XAD-8 resin). The XAD-4 acids are not true humic components. These are rich in carbohydrate and peptide biological molecules, which are, of course, components of SOM but should be considered to be outside of the definitions of HS as referred to above. Similarly, nonhumic components associated with the HAs can be recovered by dissolving the HAs in dilute base, diluting the solution to <20 ppm and passing on to XAD-8 resin. Again the polar components wash from the resin (Hayes, 2006b). Thus it would be appropriate to consider as the HA fraction the materials precipitated at pH 1 or 2 and to be aware that polar nonhumic substances will be associated in the precipitates. Consideration should be given to the humin fraction. On the basis of the recent compositional studies referred to above, humin materials will, for the most part, be biological molecules in association with other organic components and with the mineral colloids.

The terms HAs, FAs, and humin are general terms that refer to broad differences between soil organic components. In order to provide more discrete definitions, careful considerations need to be given to additional fractionation procedures that can give more discrete and distinct fractions.

1.4. EMERGENCE OF CONCEPTS OF COMPOSITIONS AND STRUCTURES OF SOIL HUMIC COMPONENTS

1.4.1. Pre- and Some Early 20th-Century Concepts of Soil Humic Components and Functions

As pointed out in Section 1.3.3, the organic fractions isolated in and fractionated from aqueous media were considered by the early workers to be individual compounds. The influences of microorganisms in the genesis were not recognized until the end of the 19th century, largely because the science of microbiology had not been developed. Thus the emphasis was on chemical synthesis. The prevailing concepts at the time considered that humus materials were formed from oxidation products of plant materials. Detmer (1871) considered that the oxidation of cellulose according to Eq. (1.1) gives rise to humic materials.

$$13C_6H_{10}O_5 + 36[O] \rightarrow C_{60}H_{54}O_{27} + 18CO_2 + 38H_2O \qquad (1.1)$$

van Bemmelen (1888) failed to isolate compounds that could be considered to be pure, and he concluded that crenic acids, apocrenic acids, ulmic acids, HAs, and humin were not homogeneous materials. Because of the developments in colloid chemistry toward the end of the 19th century, he was able to conclude that such materials were amorphous and colloidal and that the formulae suggested for these had no significance.

The emphasis on microbiology inspired by Pasteur focused interest on the roles that microorganisms have in the transformations of organic debris in the soil environment. Several studies—for example, those by vonPost (1862), Darwin (1881), Müller (1887), and Ramann (1888)—indicated that the genesis of humus is a biological, and not a chemical, process. That initiated the biological and chemical studies based on the release of building blocks from the degradations of plant components. Perhaps the definitive influence can be attributed to Dokuchaev (1883), who defined soil as a natural body formed through the combined action of natural factors and, in particular, the biological factors contributing to soil formation such as vegetation cover and the activities of living organisms. Humus was considered to have an important role in soil formation and in soil fertility. These concepts introduced a new era of soil humus studies.

At the end of the century there was general acceptance that HS are complex compounds of a synthetic nature formed as the result of decomposition involving two or more plant-derived materials. For example, Dehérain (1902) considered that HS synthesis involves interactions between proteins and "encrusting substances," mainly lignin. This concept was later developed as the "ligno-protein complex" of Waksman and Iyer (1932, 1933) (see Section 1.4.5) and as described by Waksman (1936).

Schreiner and Shorey (1909, 1910) of the US Department of Agriculture regarded soil humus as a complex mixture of organic substances arising from the decomposition of materials of plant and of animal origins. Because they considered HS to be artificial products formed in the processes of extractions with alkali, they focused on the isolation of organic substances using the techniques of organic chemistry. They used the acid filtrates after the precipitation of HAs, along with the ethanol extracts of the precipitated materials. In this way they isolated and identified more than 40 compounds that included hydrocarbons, sterols, fats, organic acids, carbaldehydes, organophosphorus, and N-containing compounds.

In a review of the work of Schreiner and Shorey, Shmuk (1924) considered that their approach tended to divide the humus concept into small groups of peripheral units and overlooked the major reserve of organic substances in the soil.

Trusov (1914) carried out a systematic study of humus formation. Initially he subjected plant components—proteins, cellulose, plant oils, and tannins—to treatment with strong acids. Later, recognizing the importance of biological processes, he (Trusov, 1915) studied the humification of plant components under normal biological conditions and then (Trusov, 1916) studied the transformations of plant residues, leaves, grass, and woody species under similar conditions. He concluded that plant components most readily utilized by microorganisms are first converted to microbial plasma and this then participates to give rise to humus. The plant residues not utilized by microorganisms (such as lignin, tannins, etc.) were considered to be direct sources of HS. These concepts have relevance at the present time.

Trusov died prematurely, but his contemporary, Shmuk, advanced the approaches initiated by him. He (Shmuk, 1924) was first to establish that soil HAs contained benzenoid structures, although Hoppe-Seyler (1889) had shown that peats and coals had aromatic units in their compositons. By esterification procedures, Shmuk showed that humic substances had hydroxyl (of phenolic origins) as well as carboxylic functional groups. He proposed that two components were contained in the HA molecule; one of these was an organic N-containing compound (Shmuk, 1914) of microbial origin, and the second one was the benzene ring. He regarded these components to be linked and not present as a mixture.

In a period of investigation and teaching, starting in 1902 and culminating in 1939, the studies of Williams (1939) led to conclusions for the era, as listed by Kononova (1966):

1. Humus substances exist in soil as a natural body.
2. Various plant materials that undergo complex biochemical transformations serve as sources of humus substances.
3. Plant materials decompose to more simple products from which the complex humic substances are synthesized.
4. Microbial enzymatic processes are involved in the decomposition and in the synthesis processes.

Thus it was accepted at the beginning of the 20th century that microorganisms had an important role in the synthesis of humus substances. Some considered the synthesis to be wholly biological, but there was a growing concept suggesting that the compounds released in the microbial breakdown of organic substrates could condense to give products from chemical synthesis processes.

Concepts of the compositions of coals influenced many in considerations of the structures of soil HS. For example, the proposal of Fuchs (1931) for structures of coal HAs (Figure 1.1) influenced soil humic scientists. The proposed structure is composed of heterocyclic aliphatic functionalities, some phenol-derived units, and considerable amounts of carboxylic and hydroxyl acidic functionalities. It may be possible that such structures could arise under conditions of elevated temperature and pressure, with oxidation taking place subsequently. Whereas such conditions might prevail during the synthesis of coals, they would be most unlikely to take place during the transformations of organic materials in the soil environment.

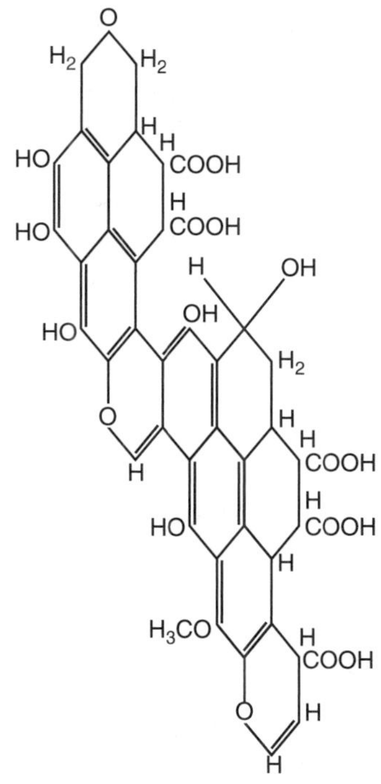

Figure 1.1. Structure of humic acid as proposed by Fuchs (1931).

1.4.2. The Maillard (Melanoidin) Concept of Humic Substances

The Maillard reaction is discussed in considerable detail Section 2.5.2 of Chapter 2 of this book. This section deals only with relevant work involving this reaction that was carried out prior to 1960.

Ellis (1959) has defined the Maillard reaction as the "reaction of the amino group of amino acids, peptides, or proteins with the glycosidic hydroxyl group of sugars," and he and Hodge (1953) presented excellent reviews of the chemistry of the Maillard–Browning reaction. The initial reaction of glucose with glycine is followed by other more complex changes that result eventually in the formation of brown pigments and polymers (see Section 2.5.2, Chapter 2). The formation of the brown pigmented products is generally regarded as the Browning reaction. It was argued that browning is brought about by the effects of pH on sugars, and that this can take place over a wide range of pH values, whereas the Maillard reaction requires alkaline media (Maillard, 1912, 1916). He (Maillard, 1912) deduced that the CO_2 evolved was from the glycine; and Wolfrom et al. (1953), using ^{14}C-labeled glucose, deduced that 90–100% of the CO_2 was from the glycine. Maillard (1917) showed that products of the reaction of glucose with glycine gave materials that resembled in many respects those of natural humic materials from soil. That stimulated interest in the abiotic synthesis of HS from sugars and amino acids liberated from the hydrolysis of polysaccharide and peptide materials.

Burdon (2001) has raised questions that would suggest that the Maillard reaction is not a significant contributor to the synthesis of soil HS. He has pointed out that there are not sufficient concentrations of reducing sugars or of amino acids in the soil solution to allow the reaction to take place to any great extent; and since the Maillard reaction proceeds best under alkaline conditions (Ellis, 1959), there should be more humic substances in alkaline soils, which there are not. Burdon referred to the spectroscopic data which show that lignin is a major contributor to the aromaticity of humic materials, whereas the aromatic components in Maillard reaction products are largely composed of heterocyclic N-containing substances.

Hayes (1960) observed that products from heating glucose with glycine in aqueous media under reflux conditions had some compositional properties and reactivities similar to those of the HAs isolated in aqueous base from a sapric histosol. The techniques used for the comparisons included differential thermal analysis (DTA) (see Figure 1.2). However, that work was done before the introduction of NMR to such studies.

1.4.3. The Enders Concept of Humic Compositions

Enders and Fries (1936) observed what they considered to be a relationship between melanoidins (or browning products) and humic acids. Later, he and Marquardt (Enders and Marquardt, 1941) showed that methylglyoxal [$CH_3C(O)C(O)H$] could give rise to melanoidins and caramel products. Then, based on his experiments to establish mechanisms of formation of methylglyoxal from hexose sugars, Enders (1942, 1943a) showed that glucose in aqueous solution was in equilibrium with the triose, and the triose could give rise to methylglyoxal. Enders and Sigurdsson (1944) found this equilibrium to be pH- and temperature-dependent under test tube conditions.

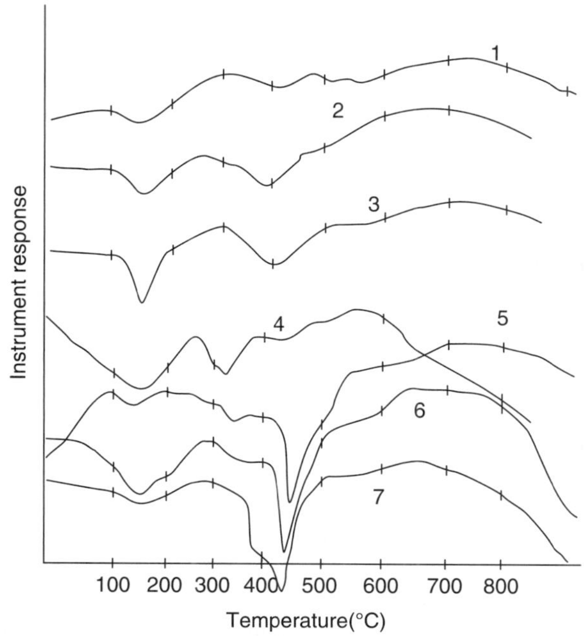

Figure 1.2. Differential thermal analysis for the humic acid fraction isolated in NaOH from a sapric histosol (1), from the acid precipitate isolated from products of the reaction of methylglyoxal with glycine (2), and from the acid precipitate formed from the reaction of glucose with glycine (3), alkali lignin (4), casein (5), lignin–casein 3:1 complex (6), and lignin–casein 6:1 complex (7).

Enders and Sigurdsson (1943) postulated a pathway from methylglyoxal to acetaldehyde (via pyruvic acid) and acetaldehyde could polymerize via the aldol condensation mechanism. In order to show how HAs could be formed under physiological conditions, Enders (1943b) postulated that methylglyoxal could be released by soil microorganisms under conditions unfavorable for microbial growth (lack of substrate, low or high temperature, etc.) and then polymerize rapidly in the presence of amino acids. Later, Enders and Sigurdsson (1947) showed the presence of methylglyoxal in 10 of the 16 soils tested, and Enders et al. (1948) showed that the products of this reaction had many properties similar to those of soil HS.

Schuffelen and Bolt (1950) found that the C/N ratio in the product from the reaction of methylglyoxal and glycine varied with the concentrations of the reactants used. The product with a C/N ratio of 9.2 had titration curves and exchange capacity values similar to those for a Dalgrund peat. Hayes (1960) reacted equimolar (1.25 molar) concentrations of methylglyoxal and glycine at 97 °C in an atmosphere of O_2 and under N_2. In the course of synthesis of the browning product from the methylglyoxal/glycine system, 72% decarboxylation had taken place in the aerobic system in 40 h, and the HA-type product formed had 60% C, 3.3% N, and a C/N ratio of 18.2. The comparable data for the HA isolated from a sapric histosol were 59%, 3.0%, and 19.7%, respectively. The brown components in both products had the same electrophoretic mobility and relatively comparable differential thermal analysis (DTA) thermograms (Figure 1.2). The products formed from similar reactions of glucose and glycine, sodium glucuronate and glycine, and xylose and glycine had lesser similarities to the histosol HA.

Based on the reasoning of Burdon (2001), the extents to which methylglyoxal polymerization products and the methylglyoxal/glycine reaction products are synthesized will await NMR evidence for the nature of the aromatic components in the synthetic structures.

1.4.4. Phenols and the Synthesis of Humic Substances

The polyphenol pathway for the synthesis of HS is discussed in Section 2.5.1 of Chapter 2.

Trusov (1915, 1916), as indicated in Section 1.4.1, introduced the concept that polyphenols and quinones contributed to the synthesis of HS. Subsequently, as the result of enzymatic oxidation (with oxidase enzymes from microorganisms), the phenols oxidize to quinones; and these, through further condensation, are converted into dark-colored humic-type substances.

The writer, when an undergraduate in University College, Dublin, happened, purely by chance, to enter the Physics lecture theatre when Wolfgang Flaig (1960) was addressing attendees at the International Peat Conference in 1952. There the writer was introduced to the concept of synthesis from polyphenol and quinone structures. He was not ready at that time to appreciate the mechanistic inferences but was in a better position to do so six years later when Professor Flaig visited The Ohio State University. The concepts involved in the synthesis processes are outlined in the chapter by Flaig et al. (1975).

Martin et al. (1967) cultured the fungus *Epicoccum nigrum* on a glucose/asparagine medium containing yeast extract and inorganic salts. After five weeks of incubation at pH 2, they isolated mycelium-free "HA-type" substances. Some of these were similar to leonardite HAs in terms of elementary composition, total acidity, cation-exchange capacity, and carboxyl, phenolic, and hydroxyl contents, as well as in terms of molecular weight distributions. Haider and Martin (1967) illustrated a plausible scheme for the genesis from nonaromatic precursors of phenols identified from the *E. nigrum* culture. There is strong evidence from that era to indicate that fungi contribute significantly to the genesis of soil humus. Intracellular enzymatic synthesis would seem to be likely, but extracellular enzymatic catalytic synthesis cannot be ruled out (Hayes and Swift, 1978). Burdon (2001) has posed a very relevant question when he asked "Why should a micro-organism expend energy and resources making a material that it has no use for? Any organism that did this would become extinct because of competition by organisms that did not waste energy and resources in this way." His succeeding arguments leave room to consider that the "humic-type" materials are formed by the enzymatic catalysis that render waste products innocuous to the organism and are stored inside the fungal mycelium.

The structural proposal for humic acids by Stevenson (1982) had significant logic, based on the state of information at that time. The structure (Figure 1.3) has compositional aspects of phenols derived from lignin and from tannins, and paper chemistry exercises would allow reactions and interactions needed to give structures of the type predicted. The structure shown is highly aromatic, and the proposed aliphatic moieties are saccharide- and peptide-derived. Modern analytical procedures invariably show peptide and saccharide components to be associated with

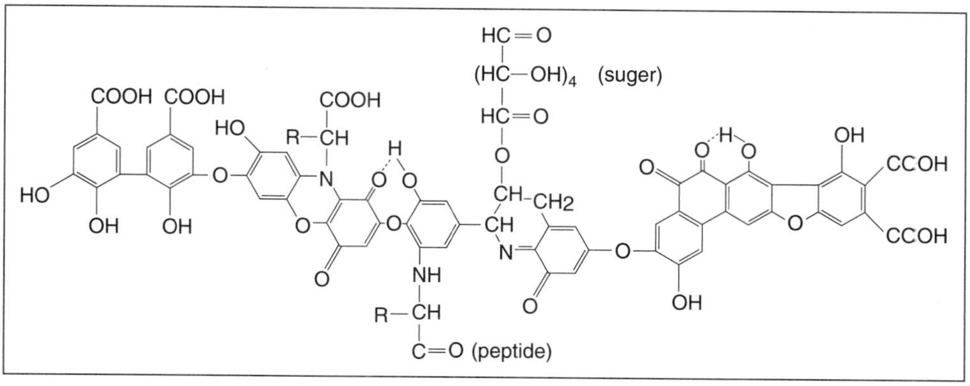

Figure 1.3. Humic acid-type structure, as postulated by Stevenson (1982).

HAs, but there has not been convincing evidence so far for covalent linkages between these and the lignin-derived components that are accepted as major contributors to soil HA structures.

1.4.5. The "Ligno-Protein" Concept of Humic Acids

As seen in Section 1.4.1, various authors had considered the involvement of lignin and of proteins in the genesis of HS. Specifically, Dehérain (1902) saw HS as products of interactions between proteins and "encrusting substances," mainly lignin. The ligno-protein theory that evolved is generally attributed to Waksman and his colleagues (Waksman and Iyer, 1932, 1933; Waksman, 1936). Their work indicated that oxidized lignin, when reacted with protein (casein), gave a product similar to HAs; and based on their work, the concept of HAs as complexes of oxidized lignin with protein predominated for a generation.

In the latter part of the 1950s, this author (Hayes, 1960) attempted to repeat the experimentation used by Waksman and Iyer (1932, 1933). He exhaustively washed powdered wheat straw with boiling water, then with hot dilute hydrochloric acid, and extracted twice for 5 h in an autoclave at 120 °C, each time with a 4% sodium hydroxide solution. The combined filtrates were acidified to pH 4 with hydrochloric acid and the precipitate formed was washed free of chloride and freeze-dried. A "ligno-casein complex" was formed by reacting three parts of the lignin extract and one part casein in a 0.1 M solution of sodium hydroxide and collecting the precipitate formed when the pH was adjusted to 4. This "complex" was washed free of chloride and freeze-dried. A 6:1 lignin–protein complex was formed in the same way.

The DTA thermograms shown in Figure 1.2 are for the HAs isolated in base from a sapric histosol, from the product from the reaction of methylglyoxal with glycine, for that for glucose with glycine, for alkali lignin, for casein, and for "lignin–casein 3:1 and 6:1 complexes." The thermograms show a degree of similarity between the sapric histosol HA and the methylglycol/glycine and glucose/glycine products, but definite differences are evident between these and the thermograms for the alkali lignin and casein. The thermograms for the lignin–casein complexes are composites of those for lignin and for casein and are very different from the HAs. That ended any adherence to the "ligno-protein complex" theory on the part of this author. However, on the basis of the information that was available at the time, the ligno-protein complex theory was logical and was based on the evidence we have today that lignin derivatives and peptides are important components of SOM, and they may well be present in associations, though not as materials linked covalently to significant extents (see Sections 1.4.7 and 1.7).

1.4.6. The Haworth Concept of Humic Structures

On the basis of the fused aromatic structures identified in the digests of the zinc dust distillation and fusion reductions of "acid boiled" HA, the Haworth Group (Haworth, 1971; Cheshire et al., 1967) proposed that humic molecules have a polycyclic aromatic core to which polysaccharides, simple phenols, proteins or peptides, and metals are attached by chemical or physical bonding processes, as summarized in Figure 1.4. The zinc dust distillation and fusion procedures are very harsh, and the yields are always low. It has become accepted that the procedures can lead to excessive bond breaking, and the recombination of fragments can give fused aromatic structures (see Clapp et al., 2005; Hayes and Swift, 1978). The polycyclic aromatic core thesis was based on the premise that yields of these aromatic materials were greater from the HAs than from nonhumic precursors. However, the formation of artifacts under these conditions is inevitable (Burdon, 2001). Polyhydroxyaromatic compounds, quinones, and furfurals, for example, would give rise to fused aromatic structures under the conditions applied. Cheshire et al. (1968) found that such compounds gave polycyclic aromatic structures from zinc-dust distillation at 500–550 °C, but at 400 °C only small yields of anthracene were obtained from hydroxybenzenes, and no aromatic structures were detected in the distillates from furfural and polymers of *ortho*- and *para*-benzoquinone. Because the same products were detected in the same proportions in the digests of the HAs at the higher and lower temperatures, it was concluded that the products released were largely from the humic materials. Long-chain hydrocarbons were identified in the digests, and these can be considered to have survived the distillation process.

The Haworth model (Figure 1.4) has been referenced widely, and in many instances it has been offered as an example of humic structures. Humic acid-type components containing fused aromatic structures can occur in soils. There is increasing awareness of char, black carbon, or charcoal products in soils that have been

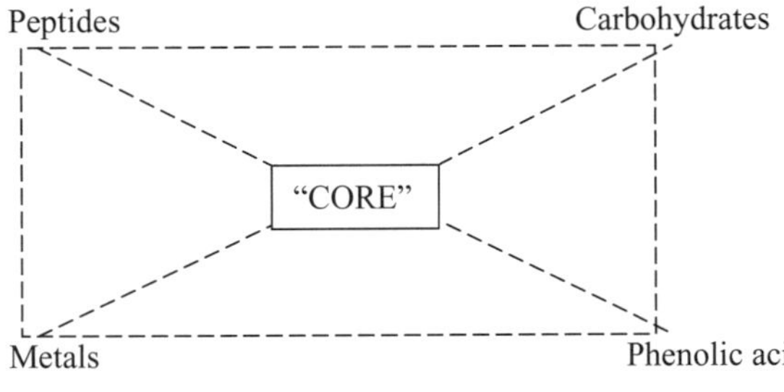

Figure 1.4. The Haworth concept of humic acid structure.

subjected to vegetation burning over time (Skjemstad et al., 2002). The char materials will, of course, be composed of fused aromatic structures, and there will be peripheral oxidation of the aromatic structures to give the acidic properties that could classify these under the operational definitions of HAs. However, in the cases of most soils, the char/oxidized char contents will be relatively small. Hence there is no convincing evidence to support the Haworth structural hypothesis.

1.4.7. Concepts of the Compositions of Humic Substances in the Modern Era

The availability of modern fractionation and spectroscopic instrumentation has allowed rapid advances to be made in our understanding of the compositions and aspects of the structures of HS. During the last half-century, Morris Schnitzer (Schnitzer, 2000) contributed greatly to our awareness of the compositions of HS. He commenced his degradation studies in the 1950s, "in the days of sail" insofar as modern instrumentation is concerned, with the oxidation of the HAs from the Ao and Bh horizons of a podzol from Prince Edward Island, Canada (Schnitzer and Wright, 1960). In terms of compositions, the greatest initial advance was made when GCMS instrumentation became available. Through extensive investigations, using a variety of chemical degradation procedures and GCMS instrumentation, Schnitzer and his colleagues provided indications of the types of molecules that could compose humic structures (reviewed by Schnitzer and Khan, 1972; Schnitzer, 1978, 2000). However, since many of the procedures used in degrading humic molecules are harsh, often involving vigorous oxidation and reduction procedures, the digest products in many instances may best be described as derivatives of the structures that compose the HS. Hayes and Swift (1978, 1990), repeated by Clapp et al. (2005), have reviewed the mechanisms involved in the different degradation reactions, and mechanistic considerations allow conclusions to be drawn about the types of structures that could give rise to the products identified in the degradation digests. Thus, through mechanistic awareness it has been possible to get indications of the types of structures that compose humic molecules.

The founding of the International Humic Substances Society (IHSS) in 1981 (as referred to in Section 1.3.1) has given a considerable boost to the humic sciences. Before that time there was little communication between soil and water scientists with interests in dissolved (DOM) or particulate (POM) organic matter in waters. The response of the US Geological Survey [Water Resources Division (USGS WRD, Denver, Colorado)] and the EPA (Cincinnati, Ohio) to the discovery by Rook (1977) of mutagenicity generated by the chlorination of FAs in natural waters brought about a major research effort in the WRD, led initially by Drs. Bob Avert and Ronald Malcolm. Following the first international meeting of the IHSS at Estes Park, Colorado (referred to in Section 1.3.1), a second meeting was convened in the University of Birmingham a year later. It was clear in Birmingham that soil and water humic scientists had learned from their associations of the previous year, and the book of invited contributions, *Humic Substances II: In Search of Structure* (Hayes et al., 1989a), covered in considerable detail the various degradative processes used in humic studies, listed the digest products identified, and referred to mechanistic interpretations that would suggest origins for the digest products. There were extended treatises on the different spectroscopic techniques used at that time for the characterization of humic molecules, and attention was also focused on the physico-chemical properties of HS, including studies of molecular sizes, shapes, and charges. In a chapter written after the meeting, consideration was given to the emergence of structural forms as seen on the basis of the data and concepts presented (Hayes et al., 1989b).

Two major IHSS publications (Aiken et al., 1985a; Hayes et al., 1989b) gave a comprehensive account of the states of the art in the humic sciences in the mid-1980s. However, very significant advances have been made in the past 20 years.

Awareness of sizes and of shapes is of major importance for considerations of the reactivities and interactions of humic molecules. Major advances were made when appropriate fractionation procedures became available. Advances in isolation, fractionation, and instrumental analytical methods have led to our current awareness of the compositions and aspects of the structures (including shapes and sizes) of humus materials. Thus, the development in the 1960s of gels of relatively discrete pore sizes allowed Cameron et al. (1972) to isolate 11 humic fractions from a sapric histosol. They considered these fractions to be reasonably homogeneous with respect to size. Molecular weight (MW) and frictional ratio values were determined from ultracentrifugation data. The plot of frictional ratio versus MW (Figure 1.5) gave a linear relationship for fractions with MW values up to ~400,000 Da, and the nonlinearity for the samples with the higher MW values was attributed to branching of the structures or to silicate contamination [see review by Swift (1989)]. The model based on the linearity of the MW versus frictional ratio plot suggested a random coil conformation for the HAs. That concept was remarkably convenient for explaining many of the interactions of HS. However, in the past 10 years there has been considerable emphasis on concepts of molecular associations that give rise to pseudo-macromolecular-type structures. The shift from concepts of macromolecular structures has been aided by the work of von Wandruszka (1998), which suggests the formation of intramolecular micelles, and by that of Wershaw (1999), which suggests molecular aggregation. Piccolo (2001) has provided evidence to

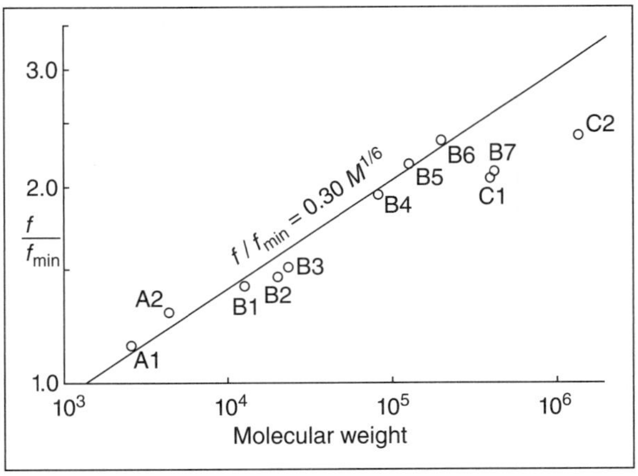

Figure 1.5. Relationship between the frictional ratio (f/f_{min}) and the molecular weight for different humic acid fractions isolated from a sapric histosol. The line is the theoretically derived relationship between frictional ratio and molecular weight for a randomly coiled polymer. After Cameron et al. (1972).

indicate how molecular associations can be disrupted to release from pseudo-macromolecular structures molecules of much lower sizes. Kenworthy and Hayes (1997) furthered that concept by showing that pyrene sorbed by what might be considered as HA aggregates or molecular associations was released by additions of very small amounts of ethanoic acid. Measurements of diffusion by Simpson (2002) and by Simpson et al. (2002), using diffusion-ordered spectroscopy (DOSY), have shown that additions of trace amounts of ethanoic acid (known to disaggregate proteins) gave different diffusivities for the lignin-derived, the carbohydrate, and the peptide components of an HA. These components gave a single diffusion coefficient in the absence of the ethanoic acid (see Figure 15.13, Chapter 15).

Senesi (1999) introduced and reviewed fractal geometry for the quantitative study of the morphology developed by HS under different experimental conditions, and he has discussed the underlying aggregation processes. The fractal approach may have limitations with regard to gaining direct evidence for the nature of the associations that are important for organic matter in the soil environment, but it can have applications for modeling the extents of the associations and the morphologies of the products.

Electron spin resonance (ESR) spectroscopy, as reviewed by Senesi and Steelink (1989), has given an awareness of the abundances of free radicals in humic fractions and of the possibilities for interactions that can arise from these.

Infrared spectroscopy was widely used in the second half of the 20th century, and this technique has allowed some advances to be made in awareness of functionalities in, and of complexes formed by, humic molecules. However, the greatest advances in determinations of functionalities, in aspects of compositions and structures, and now in aspects of humic interactions have been made since the introduction of solid-state ^{13}C NMR spectroscopy (Wilson, 1987; Malcolm, 1989). Chapter 15 in this book (by Simpson and Simpson) has reviewed in detail the applications of NMR in the solid and liquid states to studies of compositions and interactions of NOM. We now have a good indication of the types of functionalities that compose HS, and combinations of modern NMR technologies and principal component analysis (PCA) techniques allow us to deduce the origins of some of the functionalities (Novotny et al., 2007).

There is abundant evidence from NMR for the contributions of lignin to the compositions of the HA and FA fractions in the NOM of soils and waters. This is evident from the methoxyl resonance (56 ppm) and the O-aromatic resonance at 140–150 ppm. As humification proceeds, evidence for methoxyl diminishes, but the aromatic functionalities will still be derived mainly from altered lignin. The contributions of tannins to humic compositions have been recognised relatively recently, and tannins from decaying leaves and vegetation can be significant contributors to the HAs and FAs of soils and waters. Tannins come under the general category of phenolic extractives. These are a heterogeneous class of compounds consisting of: *hydrolyzable tannins*, which upon hydrolysis mainly yield glucose and gallic and ellagic acids; *flavonoids*, or polyphenols with a $C_6C_3C_6$ carbon skeleton, and their polymers are called *condensed tannins*; *lignans* formed from the oxidative coupling of two phenylpropane (C_6C_3) units; *stilbene* (1,2-diphenylethylene) derivatives which, because of their conjugated double bonds, are very reactive; and *tropolones*, characterized by an unsaturated seven-membered carbon ring (with a high resistance to biological degradation).

Tannins will be associated with lignin derivatives in the NMR spectra of HS. There can be significant resonance in the 100- to 110-ppm resonances, which hence can overlap with the anomeric carbon resonance (centered at 105 ppm) of saccharides. However, the uses of chemical shift anisotropy (CSA) and dipolar dephasing (DD) NMR techniques (Mao and Schmidt-Rohr, 2004) allow the distinction to be made between tannin and anomeric carbon resonances.

Humin has been regarded as the most intractable component of SOM. It must be considered to be a very important component, however, because typically it represents more than 50% of the organic carbon in a soil (Kononova, 1966; Stevenson, 1982, 1994) and more than 70% of the organic carbon in unlithified sediments (Durand and Nicaise, 1980; Rice, 2001). The definition of humin (Section 1.3.3) is similar to that of a protokerogen (Calvin and Philip, 1976; Rice, 2001), which is often used in petroleum geochemistry to describe insoluble organic matter in unlithified sediments.

Humin is considered by some to be at the stage in the generalised biogeochemical carbon cycle where organic carbon in the biosphere begins to become part of the geosphere (Rice, 2001). A part of this process would involve the removal of the more labile components of the organic carbon soil inputs. Consequently, humin would be regarded as the oldest of the three humic fractions (Rice, 2001). It was suggested that the mechanism by which nature slows the mineralisation of organic matter might involve the accumulation of an ill-defined, amorphous and heterogenous mixture of organic molecules (Swaby and Ladd, 1966; MacCarthy and Rice, 1991; Rice, 2001). This mixture would require either a very large assemblage of enzymes or an uncharacteristically versatile enzyme to effect its rapid mineralisation (Rice, 2001).

When considered from this perspective, humin has a significant role as a sink for carbon.

Solid-state NMR has done much to dispel the mysteries of humin compositions, and significant advances have recently been made using proton NMR in the liquid state (see Section 15.3.3 of Chapter 15). Based on solid-state ^{13}C NMR spectra, Hatcher et al. (1980) concluded that a repeating aliphatic structural unit, possibly attributable to branched and cross-linked algal or microbial lipids, is common to both soil and sediment humin samples. Hatcher et al. (1983) viewed the increase in humin relative to the other humic fractions as a "selective preservation" of the aliphatic compounds of the sediments and did not support condensation theories.

Derenne and Lageau (2001) have favored the concept of "selective preservation" of plant residues with a degree of resistance to microbial degradation. Studies of soils, sediments, and fossil fuels have shown that the organic components are rich in long-chain polymethylene functional groups (Deshmukh et al., 2005). These components are considered to be derived in terrestrial systems from plant cuticles containing the biopolymers cutin and cutan, as well as from suberized parts of plant organs containing suberin (Nierop, 1998). It has been shown that the aliphatic carbon content increases with increasing SOM decomposition (Baldock et al., 1997), which would indicate that these components are diagenetically resistant and survive unaltered over time (Deshmukh et al., 2005).

The two major polymeric lipid components found in plant cuticles are cutin and cutan. Whereas cutin is the polyester biopolymer that is solubilized upon saponification treatment, cutan is a nonsaponifiable and nonextractable polymeric substance found in certain cuticles. These molecules provide a protective barrier between the plant and its external environment and between different organs of the plant (Kolattukudy, 1980). Holloway (1984, 1994) has reviewed the structure of cutin.

Depolymerization of cutin using reactions such as alkaline hydrolysis, transesterification, and hydrogenolysis with $LiAlH_4$ in tetrahydrofuran (THF) gives monomers that are identified by gas chromatography–mass spectrometry (GC/MS) after derivatization (Kolattukudy, 1980). Such degradative studies suggest that cutin has a primary architecture comprised of C_{16} and C_{18} fatty acids, hydroxy fatty acids, and epoxy fatty acids (Kolattukudy, 1980). Deshmukh et al. (2005) have listed and given relevant ^1H (ppm) and ^{13}C (ppm) shifts for straight-chain and branched long-chain acids, alcohols, aliphatic hydrocarbons, epoxides, esters, and lesser amounts of aromatic esters in the cutan and cutin/cutan mixture in *Agave americana*. These components can be expected to persist in humin materials.

Suberins or polyestolides are related to cutins. These are complex polymers composed of ω-hydroxy monobasic acids linked by ester bonds. They also contain α,β-dibasic acids esterified with diols, as well as ferulic and sinapic acid moieties. Suberins are enriched with molecules having 16 and 18 carbon atoms. They also have ethylenic and hydroxyl functionalities, and ester and ether cross-linking can occur.

Although there is mostly indirect evidence, extensive ester cross-linking of these monomeric species gives cutins and suberins a three-dimensional architecture (Kolattukudy, 1984; Deshmukh et al., 2005).

Humin can no longer be considered to be the most intractable of the components of HS. As will be seen in Chapter 15, huge advances have been made, using proton NMR, in understanding the compositions of humin materials dissolved in DMSO/H_2SO_4. Evidence from solid-state (^{13}C) and liquid-state (proton) NMR has shown that humin material extracted from fine clays in DMSO + 6% H_2SO_4, after prior exhaustive exhaustive isolations at pH 7.0, 10.6, and 12.6 and in 0.1 M NaOH + 6 M urea, was largely composed of small amounts of oxidized lignin residues (trapped in the humin matrix), significant amounts of proteins/peptides and carbohydrate materials, peptidoglycan and lipoprotein structures, and large amounts of aliphatic waxes, cuticlar materials, and lipids (Simpson et al., 2007). The inference from that information is that the high affinities for the inorganic colloids of the biological molecules provides protection from microbial attack. Such biological molecules would mean that humin should not be classified as a humic substance, based on the classical definitions outlined in Section 1.3.3.

1.5. EVOLUTION OF STUDIES OF SOIL SACCHARIDES

The book by Cheshire (1979) has reviewed studies of soil polysaccharides, and he and his colleagues have contributed significantly to our awareness of the subject. Chapters by Hayes and Swift (1978), Cheshire and Hayes (1990), and Clapp et al. (2005) have reviewed relevant aspects of the chemistry and compositions of polysaccharides and have discussed the reactivities of saccharides in the soil environment.

Interest in soil saccharides, and especially in soil polysaccharides, is relatively recent, and their studies may be considered to be in the modern era of soil organic matter research. Martin (1945, 1946) established that the "slimy" bacterial products shown by Waksman and Martin (1939) to aggregate sand–clay mixtures were polysaccharides. Stacey and colleagues at the University of Birmingham, England (Haworth et al., 1946), and Geoghegan and Brian (1946, 1948) at the ICI Research Station, Jealotts Hill, Berkshire, England, were, at about that time, investigating the uses of bacterial dextran and levan polysaccharides for the improvement of soil structure and for the retention of water in soil. The Haworth Group, Geoghegan and Brian, and Martin, though working on a similar topic at about the same time, were unknown to each other.

The research that followed focused on statistical analyses (Rennie et al., 1954; Chesters et al., 1957) that correlated good soil structure with microbial gums and polysaccharide substances. Confidence in the roles of polysaccharides in soil stabilization was questioned when Mehta et al. (1960) showed that soil aggregate structures were preserved when a Swiss Braunerde was treated with periodate ($NaIO_4$). Periodate degrades polysaccharides. Hence it could be argued that soil crumbs stabilized by polysaccharides would disintegrate upon treatment with periodate. However, Greenland et al. (1961, 1962) showed that the periodate-treated soil crumbs were stabilized by fungal hyphae and myceliae (which resisted the periodate treatment). Also, crumbs in soils with free $CaCO_3$ were shown by Clapp and Emerson (1965) to resist degradation after treatment with periodate.

1.5.1. Origins of Soil Saccharides

Cheshire and his colleagues at the Macaulay Institute, Aberdeen, Scotland, carried out extensive studies on the origins of soil saccharides. When readily degradable ^{14}C-labeled carbohydrate substrates, such as glucose and starch, were incubated under field conditions, the labels were scrambled and the strongest labels in the newly synthesized sugars were in the hexoses glucose, galactose, and mannose and in the deoxyhexoses rhamnose and fucose, and there was much less labeling in the pentoses arabinose and xylose (Cheshire et al., 1969, 1971). That suggested that microorganisms were largely responsible for the hexoses and deoxyhexoses in soil. When the experiments were extended for up to two years, high levels of activity developed in the deoxyhexoses and especially in rhamnose. Because the compositions of the labeled sugars bore only a superficial resemblance to those in the whole soil, or in the soil extracts, the authors considered that sufficient time may not have elapsed to allow for differential degradation rates for the newly synthesized carbohydrates. The data would support the view that microbial synthesis is only a partial contributor to the origins of soil carbohydrates (Cheshire and Hayes, 1990).

In other soil incubation studies Cheshire et al. (1974, 1979) found that some sugars in straw in labeled plant material decomposed rapidly, but others decomposed relatively slowly, and about 15% of these remained after five years. There was little indication that xylose or cellulose was synthesized by soil microorganisms. However, studies by Cheshire and Anderson (1975) showed that plant residues are essential to maintain the soil carbohydrate levels. Total carbohydrate contents of soils fallowed for 10 years fell by as much as 50%. Because there was no significant change in the compositions of the soil carbohydrate as the result of the fallow, it was concluded that the sugars were equally susceptible to metabolism regardless of the management system.

Cheshire and Hayes (1990) have pointed out that although a particular sugar might be only a small proportion of the sugars in plants, this sugar could with time become predominant in a residue if it should have even a slightly greater resistance to degradation than the other sugars in the medium. This point was demonstrated earlier when Bacon and Cheshire (1971) showed that 2-*O*-methylfucose and 2-*O*-methylxylose are prominent among the methylated soil saccharides, despite the fact that these sugars are minor components of the hemicelluloses of some plants. The *O*-methyl functionality is considered to block enzymatic processes, leading to degradation of the sugars.

Sugar abundances ratios, such as (Mannose + Galactose)/(Xylose + Arabinose) and (Rhamnose + Fucose)/(Xylose + Arabinose), have been used to indicate the origins, plant or microbial, of sugars in soil saccharides (Oades, 1984; Murayama, 1984). The logic is based on the types of data, referred to above [and referenced in Clapp et al. (2005)], that indicate that xylose and arabinose are largely found in plants and that mannose, galactose, rhamnose, and fucose are more likely to be found in microbial cells. In the case of (Mannose + Galactose)/(Xylose + Arabinose), ratio values <0.5 would suggest origins in plants and values on the order of 2 would suggest microbial sources. It could also be meaningful to compare the ratios Mannose/Rhamnose and Galactose/Fucose for each sample. These and other ratio values have been used recently by Hayes et al. (2008) to show that the saccharides in fractions of HAs and FAs from the same soil have different origins (plant or microbial).

1.5.2. Isolation and Fractionation of Soil Saccharides

Procedures for the isolation of polysaccharides from soil and their fractionation have been reviewed by Mehta et al. (1961), Swincer et al. (1968), Greenland and Oades (1975), Hayes and Swift (1978), Cheshire (1979), Cheshire and Hayes (1990), Stevenson (1994), and Clapp et al. (2005). Sodium hydroxide is probably the best of the aqueous solvents used. Its efficiency might well be related to charges, arising predominantly from uronic acids. These are characteristic of many soil polysaccharides, though sulfonated polysaccharides (typically with SO_3H functionalities on C-6 of the hexose sugars) may also contribute to charge. Some of the dipolar aprotic solvents, especially DMSO, can be expected to be good solvents for soil saccharides. DMSO has the ability to break hydrogen bonds and can even dissolve cellulose. Häusler and Hayes (1996) have shown that considerable amounts of saccharides were removed from the HA fraction from a sapric histosol when that fraction was dissolved in DMSO/HCl and passed on to XAD-8 resin. The HA was sorbed by the resin, and saccharides were washed through. Thus the use of DMSO/HCl (or DMSO/H_2SO_4) and of XAD-8 and XAD-4 resins in tandem may well provide a useful method for the isolation of soil saccharides. Saccharides would be retained by the XAD-4 resin.

The bulk of the saccharides in soil aqueous extracts are contained in the FA fraction, or the organic fraction that remains in solution on acidification of soil extracts. The introduction of XAD resin technology (see Section 1.3.2) for the fractionation of soil organic extracts has allowed some separation of saccharides from what are considered to be the true FAs. Swincer et al. (1968) deserve to be credited with that concept. They used Polyclar-AT [a (poly)vinylpyrrolidone resin used for clarifying wine] to remove color (HS) from soil saccharide extracts, but XAD-8 has been found to be superior for that purpose. Barker et al. (1965, 1967) and Hayes et al. (1975) used gel chromatography and ion-exchange chromatography techniques to remove color successfully from their saccharide extracts.

Attempts to fractionate soil polysaccharides into components that would satisfy the criteria of purity have failed. That is not surprising because a vast, inhomogeneous mixture of polysaccharides are likely to exist because of the countless microbial species in soil. It has been possible to isolate components that were relatively homogeneous with regard to charge, and possibly size, but generally these had four or more component sugars and were likely to be mixtures. Clapp (1957), Finch et al. (1967), Barker et al. (1967), Clapp and Davis (1970), and Clapp et al. (1979) achieved a considerable degree of physicochemical homogeneity in their isolates, but all contained four or more sugars and must be regarded as mixtures (see Clapp et al., 2005). Hayes et al. (1975) isolated a polysaccharide material from a histosol that contained about 70% glucose, but it had six other sugars in concentrations ranging from 2% to 9% and hence must be considered to be impure.

1.5.3. Polysaccharides and Soil Aggregates

Pagliai et al. (1979) have shown that the stabilities of aggregates incorporating β-glycosidic-linked (poly)glucose (which gives a linear helical structure) are proportional to the molecular weights (MWs) of the polymers. Also, the stabilities could be related to the viscosities and the extents of adsorption. These authors have also shown that the relationship between MW and biodegradation is not straightforward. The first polymer segments to contact the clays would orientate flat on the sorbent. As the surface is covered, loops would extend away from the surface (Burchill et al., 1981), and these loops would not be protected from microbial attack if microorganisms or the enzymes they secrete gain access to the loops.

To stabilize preformed aggregates, the polysaccharide would need to diffuse to the adsorption site in the internal surfaces. Cellulose [β-(1→4) linked (poly)glucose] is not an effective binding agent when added to soil because the strong hydrogen bonds between the β-linked strands make the polymer insoluble in water. The polymer would dissolve, should these bonds be broken, and could then be expected to be a good aggregate former and stabilizer. Page (1980) demonstrated that cellulose xanthate (in which the –OH group on C-6 is replaced by $-OCS_2$), a water-soluble cellulosic polymer, is an excellent stabilizer of soil aggregates. Harrison (1982) showed that when the xanthate was introduced to clay media, CS_2 was given off and the cellulose remained sorbed to the clay.

Clapp and Emerson (1972) used a series of homo- and hetero-polysaccharides of varying intrinsic viscosities and charges (containing uronic acids) in studies of binding and the stabilization of aggregates of Ca^{2+}-montmorillonite preparations [see also Clapp et al. (1991)]. Sorption of the neutral polysaccharides increased with their intrinsic viscosity (η) values. When Na^+-exchange was used to disperse the aggregates, samples that adsorbed high levels of polysaccharide were dispersed only after prolonged treatments with 0.05 M periodate. Anionic polysaccharides adsorbed least, and it was clear that stabilization by the polyanions involved complexation with the polyvalent metals neutralizing the charges on the clay surfaces. Polysaccharide mixtures extracted from soils adsorbed to varying extents, yet failed to stabilize the aggregates. That was ascribed to their lower (than required) η and MW values and hence their inabilities to bridge between clay particles.

Finch et al. (1967) calculated that the plateau adsorption by Georgia Na^+-exchanged kaolinite of a soil polysaccharide corresponded to a surface coverage of $80\,m^2\,g^{-1}$. Surface area measurements by standard techniques gave a value of $96\,m^2\,g^{-1}$ for the clay, and this would suggest contamination by a clay of higher surface area (montmorillonite). Periodate oxidation of the adsorbed polysaccharide was greatly retarded, suggesting that the adsorbed material was held close to the surface. Less protection would be afforded to loops extending away from the surface. The adsorption data are in line with those of Moavad et al. (1974) for their Na^+- and K^+-kaolinites. Coordination provides one explanation for the vast increases in adsorption, especially by the Fe-exchanged clay, but the extents to which (hydr)oxides were involved in the adsorption mechanisms is not known.

Aggregate formation may be visualized as the interaction of mobile clay particles with mucigel from plant roots and extracellular polysaccharides from bacteria [see Oades (1990)]. Many plants during growth appear to give rise to a breakdown of microaggregates in the rhizosphere. This could result from a priming action of the indigenous organic matter.

It is inevitable that polysaccharides are important for the formation of microaggregates, but there is enough evidence to indicate that fungal hyphae and plant rootlets are important for the stabilization of macroaggregates. Interest is being focused on the symbiotic relationship which arbuscular mycorrhizal fungi have with plants and on the nature of the secretions by such fungi (Rillig et al., 2001). Peptidoglycan, mucopolysaccharide, and such secretions by these and other fungi will have close contact with the soil inorganic colloids and can be expected to aid in the formation and stabilization of soil aggregates.

Hayes and Swift (1978), Theng (1979), and Clapp et al. (1991, 2005) have dealt in depth with the adsorption of neutral and charged polysaccharides by clays.

1.5.4. General Conclusions from Studies of Soil Saccharides

The fact that polysaccharides are major aggregating agents in soils is well established, and there is a reasonable awareness of the compositions and aspects of the structures of the kinds of polysaccharides that can give rise to aggregate formation and stabilization. However, there is only a limited awareness of the compositions and the structures of the indigenous soil polysaccharides that promote the formation of soil aggregates. The study by Finch et al. (1967) has shown ways by which polysaccharides that interact with clays can be isolated. However, there must be the will to persist in extended studies, such as those that engaged Cheshire and his colleagues at the Macaulay Institute, if this important area of SOM studies is to be advanced. There was considerable emphasis on soil polysaccharides during the period between the late 1940s and the mid-1980s, but interest has not been maintained. These studies were carried out in an era when the instrumentation needed for rapid advances was limited.

Soil polysaccharides can amount to as much as 20% of the humic fractions isolated in aqueous media, and identification of the classes of components in humin materials in associations with the soil clays indicates that carbohydrates contribute significantly to those isolated in the DMSO/H_2SO_4 medium (see Sections 1.4.7 and

1.7). Qualitative and quantitative identification of the sugars in the hydrolysates of these humin isolates may give indications about their origins (plant or microbial), and identification of the configurations of the sugar linkages could lead to deductions about the sorption mechanisms. To achieve the latter objective will require isolation of the polysaccharide and/or mucopolysaccharide components. That will not be an easy task, but it is doable.

1.6. SOIL PEPTIDES

Amino acids, amino groups, amino sugars, and nucleic acid derivatives usually account for >95% of the organic N in soils (Anderson et al., 1989), and many other N-containing compounds have been reported in trace amounts (Stevenson, 1994). Anderson et al. (1989) have found traces of L-phosphatidic acid, choline, ethanolamine, and uric acid (the end product of N metabolism of many animals), which can be oxidized to allantoin, cyanuric acid, and urea.

It is inevitable that there will be considerable amounts of peptide and amino acids in soils at any time because of the inputs from plant roots or from plant materials that are directly added to the soil or that enter the soil as senescent matter from vegetative cover (introduced by organisms such as earthworms). Additions can also arise from the microfauna, composed of bacteria, fungi, and viruses, all of which can number in the region of 1×10^7 to $1 \times 10^{10}\,g^{-1}$ of dry weight of soil (Burns, 1990), and estimates of living tissue microbial biomass range from 17 to $22\,g\,m^{-2}$ of soil (Jenkinson and Ladd, 1981). Significant contributions also arise from protozoa, eelworms, earthworms, soil insects, burrowing animals, and so on. If not protected, protein and peptide materials in dead organisms are rapidly recycled by the soil biota. When conditions are right, some will interact with other organic species to give products such as melanoidins (referred to in Section 1.4.2). These can have a degree of resistance to microbial decomposition. Interactions with clays, as well as steric constraints provided by microaggregates, can also provide protection. McLaren (1954) and McLaren et al. (1958) were first to show how enzymes could interact with clays, and enter the interlamellar spaces. McLaren et al. (1975) also showed how the activity of enzymes is preserved in associations with humus.

Knicker (2000) has referred to her NMR evidence that shows that >80% of the organic N in soils is in peptide-like structures. Knicker et al. (2002) did detect a clear shoulder in the chemical shift region for pyrrole- or indole-heteroaromatic N (−145 to −220 ppm) in the ^{15}N-NMR spectrum of the deepest layer of a peat that was at least 10,000 years old. However, this peat could be considered to be at the beginning of the coalification stage, and there is abundant evidence for heterocyclic N in coal.

There are no accurate determinations of the overall contributions of peptide components to the compositions of HS and of SOM. The difficulties arise from the problems of separating these from other components of the SOM mixtures. Use of DMSO + 1% 12 M HCl and XAD-8 resin [(poly)methylmethacrylate] technology enabled Appelqvist et al. (1996) to decrease the amino acid content of a HA by 23%, and the decrease was uniform for the different amino acid groupings. This still left considerable amounts of peptide material in association with the HAs. Some may well have been in association with the humic matrix, and some may have been covalently linked to the humic core through, for example, the formation of Schiff base structures through interactions between free amino groups in argenine and lysine with carbaldehyde and keto groups in the humic core. Such amino functional groups can also form covalent linkages with carbon α to the keto group in the quinones in humic structures, as indicated by Hayes and Swift (1978).

The possibilities for the identification of forms of N, other than peptide N, in soil fractions have been greatly advanced by recent developments in solid-state NMR pulse sequences. Carbon directly bonded to N can now be observed through saturation-pulse induced dipolar exchange with recoupling (SPIDER), as described by Schmidt-Rohr and Mao (2002). That technique allows detection of N-substituents on aromatic nuclei. This has led to increasing evidence for organically immobilized N in SOM. For example, yields of lowland rice (*Oryza sativa* L.) have been shown to decrease by more than 35% during 20–30 years of double and triple cropping. The total soil N did not decrease. The SPIDER evidence suggested that the unavailable organic N was bonded to lignin residues that accumulate in the anaerobic conditions (Olk et al., 2000; Schmidt-Rohr and Mao, 2002).

The same amino acids are found in plants, microorganisms, and animals, and so it is difficult to assign origins to peptide materials on the basis of the amino acid compositions of hydrolysis digests. Beavis and Mott (1996, 1999) claimed that amino acid fingerprints could distinguish between different sources of amino acids. They studied the amino acid profiles of the Rothamsted Wilderness soils, of manured and unmanured soils from the Park Grass experiment, and from the manured and unmanured Broadbalk continuous wheat plots at the Rothamsted Experimental Station, Harpenden, England, and their data indicated that amino acid fingerprints could distinguish between grass and arable experiments, and the amino acid profile for the Wilderness wooded soil was distinguishable from those for the grassland and arable soils.

Hayes et al. (2008) identified the amino acids in a variety of fractions from grassland soils and their drainage waters and segregated these into acidic, basic, neutral hydrophilic, and neutral hydrophobic groups. They then calculated the ratios of the abundances of the different groups in the different fractions. By comparing the distributions of the ratios with those of the sugars (Section 1.5.1), it was possible to get some indications about the likely origins (plant or microbial) of the amino acids.

1.7. THE FUTURE FOR STUDIES OF ENVIRONMENTAL NATURAL ORGANIC MATTER

The advances that are being made this century in the studies of the compositions and of some aspects of the structures of components of the NOM of soils and of waters far outstrip those made in any previous relatively short period of time. This is largely attributable to the advances in instrumentation and in particular in NMR technology, as outlined in Chapter 15 of this book. NMR techniques such as dipolar dephasing (DD), chemical shift anisotropy (CSA), diffusion-ordered spectroscopy (DOSY), diffusion editing, and 2-D NMR techniques have given significant information about aspects of compositions, sizes, and associations of components of NOM materials. The uses of ultrahigh-resolution Fourier transform ion cyclotron mass spectrometry (FT–ICR MS), using electrospray ionization (ESI) is capable of resolving individual species in complex DOM mixtures with masses as high as

~1000 Da (Stenson et al., 2003). Because *m/z* values can be determined with very high resolution and precision, usually to the fifth decimal place, accurate *m/z* values can be calculated for each peak, allowing the determination of elemental formulae that can be assigned to within 1 ppm error (Sleighter and Hatcher, 2007). This technique will be especially applicable initially to DOM materials.

There is a good, but not detailed, awareness of aspects of the compositions of various components of NOM. Lignin would appear to be the source of the major components of transforming organic matter isolated from soils in aqueous basic media. There is clear evidence for lignin functionalities in the NMR spectra (O-aromatic and methoxyl substituents) of these. There is evidence also for significant amounts of carbohydrate and peptide materials, and in general there is less convincing evidence for contributions from aliphatic hydrocarbon moieties. The contributions of tannins to the humic components has not been widely appreciated, but information from dipolar dephasing (DD) and chemical shift anisotropy (CSA) NMR procedures has resolved the overlaps that occur in the 100- to 110-ppm region of the spectrum where tannins and anomeric carbon (from carbohydrates) resonances can overlap (see Section 1.4.7). As humification progresses, the definite evidence for lignin residues diminishes; and there are some indications of increasing inputs from microbial sources, especially to the carbohydrate and peptide components.

On the basis of data from DOSY and diffusion editing, we now know that the components soluble in aqueous base tend not to be macromolecular (Simpson et al., 2002), at least not to the extents considered previously. More detail needs to be known about the nature of the associations that give pseudo-macromolecular properties.

Humin (see Section 1.4.7) has been regarded as the most recalcitrant component of SOM. Until recently, it was considered to be an intractable mixture of materials that were outside the capabilities of methods used for the isolation and fractionation (without significant alteration) of components of SOM. The recent work by Simpson et al. (2007) and by Song et al. (2008) has shown that humin can indeed be isolated and, to some extent, fractionated. The exhaustive aqueous extraction processes used by these authors isolated classical HA and FA fractions, including those released by urea/NaOH systems, before humin components were isolated in DMSO + 6% concentrated H_2SO_4 [see also Hayes (2006b)]. Recent studies in the author's laboratory have shown that >90% of SOM can be solubilized using aqueous and nonaqueous media. Also, the residual materials in association with the soil inorganic colloids have been shown to be similar to the major component isolated in the DMSO/H_2SO_4 medium (Song et al., 2008).

Solid-state, liquid-state, and 2-D NMR have given good indications of the compositions of the humin materials, and these have been found to be rich in carbohydrate, protein/peptide, peptidoglycan, waxes, lipids, and aliphatic hydrocarbon components (see Chapter 15). The aliphatic hydrocarbon contributions from waxes, lipids, long-chain hydrocarbons, acids and esters, and cutins/cutans/suberins predominate. Cutins/cutans and suberins are relatively recalcitrant components of plants (Deshmukh et al., 2005) and are likely to be significant contributors to the aliphatic materials. The lignin-derived species in humin may be regarded as components trapped in the humin matrix. Humin materials are surprisingly low in aromatic components, and the major contributors to the aromaticity would seem to be from small amounts of lignin-derived residues, some tannin materials, and aromatic amino acids (Simpson et al., 2007; Song et al., 2008).

The humin components isolated in DMSO/H_2SO_4 media contain macromolecular components, such as peptides, peptidoglycans, and possibly mucopolysaccharides (Simpson et al., 2007), and these would appear to be strongly sorbed to the soil inorganic colloids. Such species can therefore be considered to have importance in soil particle interactions.

It can be said that we know in broad terms the aspects of the compositions and the origins of at least some of the components of SOM, but we lack detailed awareness of the ways in which the components are associated, how they interrelate and interact in providing degrees of resistance to decomposition, how they interact with the mineral colloids to provide the basis of soil structure, and how they interact with anthropogenic chemicals that enter the soil environment. There is much work to be done, and this work will still involve getting more complete information about compositions, aspects (though not necessarily details) of structure, and associations. That approach is more important than concerns about details of structures. It is likely that the compositions and awareness of structures of humin molecules will become known in the not-too-distant future because these are largely biological molecules. However, it will be highly challenging to resolve the structures of the biologically transformed molecules because of the difficulties faced in isolating materials of sufficient purity for structural studies, at least in the cases of soil HAs.

There is a need to resume studies of soil saccharides and peptides. These can compose as much as 30–40% (when account is taken of the compositions of humin materials). Much is known about how polysaccharides of known structures interact with soil colloids, but it has not been possible as yet to know in sufficient detail the structures of the polysaccharides that persist in the soil. Hence we do not know the mechanisms of their binding to soil mineral colloids. The same applies for the peptide materials, though it is clear that polysaccharides and peptides have important roles in soil structure formation and stabilization.

The need to place an urgent emphasis of NOM cannot be overemphasized. The study of NOM has never been given the priority emphasis that it deserves. Political and popular interest has been raised from time to time—as, for example, by (a) the dust bowls in the 1930s in the United States and (b) the realization in the 1970s that chlorinated aquatic humic substances can be carcinogenic. However, the level of interest that would lead to substantial support for NOM (and especially SOM) research has not been sustained. Essential lessons from the past are not heeded. It is known, for example, that the great civilisations in the plains of the Tigris and Euphrates did not perish or disband directly as the result of invasions by vandals. A thriving population relied on conservational agriculture in which the value of organic amendments was appreciated. When the irrigation systems were damaged and conservational emphasis was lost, the organic reserves became depleted and the population disbanded. Similar happenings are taking place throughout the world at this time.

There is now much emphasis on carbon sequestration, carbon sinks, and carbon "footprints," but there is only peripheral awareness of the vast and vital sink that the soil provides for carbon. Few seem to realize that, worldwide, there is more than three times the amount of carbon in SOM than there is in all living matter on the surface of the earth. Depletion of that reserve is more serious than is appreciated because vast amounts of carbon are released into the atmosphere as the result of soil mismanagement. In order to emphasize the vast reserves of carbon in Irish soils (where 90% of the arable soils are in grassland), this author has pointed out that the amount of organic matter in one hectare of local Limerick grassland soils is equivalent to the mass of 6000 Kerry Gaelic Football players (the current All-Ireland champions), each weighing ~85 kg. After 30 years of continuous nonconservational cultivation, the organic matter loss would be equivalent to the mass of ~2000 of these athletes. This loss must be regarded as fossil carbon, though it could be replaced over time should the land use be returned to grassland. The degradation of soil structure resulting from the biological oxidation of the organic matter is more serious than the release of the CO_2.

The current controversy surrounding the uses of food crops as source materials for fuel additives and substitutes, and the increasing demands for food for an increasing world population will inevitably focus interest on agriculture and eventually on soils. More pressure will be placed on soil resources, and it will be evident that soil degradation will have serious consequences for water quality. This should lead to an awareness of the essential role that NOM has on the quality of life and on the need to conserve SOM. It will also be realized that carbon lost from soils as the result of mismanagement is fossil carbon. Therefore, it is logical to expect that it will be seen that a comprehensive awareness is needed of the transformations that lead to the genesis and the losses of SOM, of the compositions and aspects of the structures of components of SOM, and of the mechanisms by which SOM influence vital reactions and interactions that take place in the soil.

Studies at the frontiers of NOM research in the present era require state-of-the-art instrumentation, and it will be very difficult for any one laboratory to have all the equipment needed to make comprehensive advances in certain aspects of NOM studies. Thus future frontiers research may involve a limited number of laboratories fully equipped with state-of-the-art equipment operated by experts. To obtain optimum advances and instrumentation, scientists might be encouraged to take their samples, prepared by state-of-the-art procedures, to Centres of Excellence equipped with the state-of-the-art equipment and operators.

REFERENCES

Achard, F. K. (1786). Chemische Untersuchung des Torfs. *Crell's Chem. Ann.* **2**, 391–403.

Aiken, G. R. (1985). Isolation and concentration techniques for aquatic humic substances. In *Humic substances in Soil, Sediment, and Water: Geochemistry, Isolation and Characterization*, ed. Aiken, G. R., McKnight, D. M., Wershaw, R. L., and MacCarthy, P., eds., John Wiley & Sons, New York, pp. 363–385.

Aiken, G. R., McKnight, D. M., Wershaw, R. L., and MacCarthy, P., eds. (1985a). *Humic Substances in Soil, Sediment, and Water: Geochemistry, Isolation and Characterization*, John Wiley & Sons, New York.

Aiken, G. R., McKnight, D. M., Wershaw, R. L., and MacCarthy, P. (1985b). An introduction to humic substances in soil, sediment, and water. In *Humic Substances in Soil, Sediment, and Water: Geochemistry, Isolation and Characterization*, Aiken, G. R., McKnight, D. M., Wershaw, R. L., and MacCarthy, P., eds., John Wiley & Sons, New York, pp. 1–9.

Anderson, H. A., Bick, W., Hepburn, A., and Stewart, M. (1989). Nitrogen in humic substances. In *Humic Substances II: In Search of Structure*, Hayes, M. H. B., MacCarthy, P., Malcolm, R. L., and Swift, R. S., eds., John Wiley & Sons, Chichester, pp. 223–253.

Appelqvist, I. A. M, Graham, C. L., and Hayes, M. H. B. (1996). Isolation, fractionation, and characterization of humic acids from a sapric histosol. In *Humic Substances and Organic Matter in Soil and Water Environments*, Clapp, C. E., Hayes, M. H. B., Senesi, N., and Griffith, S. M., eds., IHSS, University of Minnesota, St. Paul, pp. 33–39.

Bacon, J. S. D., and Cheshire, M. V. (1971). Apiose and mono-ethyl sugars as minor constituents of the leaves of deciduous trees and various other species. *Biochem. J.* **124**, 555–562.

Baldock, J. A., Oades, J. M., Nelson, P. N., Skene, T. M., Golchin, A., and Clark, P. (1997). Assessing the extent of decomposition of natural organic materials using solid-state ^{13}C NMR spectroscopy. *Aust. J. Soil Rese.* **35**, 1061–1083.

Barker, S. A., Finch, P., Hayes, M. H. B., Simmonds, R. G., and Stacey, M. (1965). Isolation and preliminary characterization of soil polysaccharides. *Nature (London)* **205**, 68–69.

Barker, S. A., Hayes, M. H. B., Simmonds, R. G., and Stacey, M. (1967). Studies on soil polysaccharides. 1. *Carbohydrate Res.* **5**, 13–24.

Beavis, J., and Mott, C. J. B. (1996). Effects of land use on the amino acid composition of soils: 1. Manured and unmanured soils from the Broadbalk continuous wheat experiment, Rothamsted, England. *Geoderma* **72**, 259–270.

Beavis, J., and Mott, C. J. B. (1999). Effects of land use on the amino acid composition of soils: 2. Soils from the Park Grass experiment and Broadbalk Wilderness, Rothamsted, England. *Geoderma* **91**, 173–190.

Bemmelen, J. van (1888). Die Absorptionsverbindungen und das Absorptionsvermogen der der Ackererde. *Landw. Vers St.*, **35**, 69.

Berzelius, J. (1806). Undersökning af Adolfsbergs Brunnsvatten. Undersökning af Porlakällvatten Berzelius' och Hisinger's Afhandlingar i physik kemi och mineralogi. Stockholm. See Istoriya agrikul'tury (1940). *Sb. Izd. Akad. Nauk SSSR.*

Berzelius, J. (1839). *Lehrbuch der Chemie* (3rd edition; translator Wöhler). Dresden and Leipzig.

Bremner, J. M. (1950). Some observations on the oxidation of soil organic matter in the presence of alkali. *J. Soil Sci.* **1**, 198–204.

Bremner, J. M., and Lees, H. (1949). Studies of soil organic matter: II. The extraction of organic matter from soil by neutral reagents. *J. Agric. Sci.* **39**, 274–279.

Burchill, S., Hayes, M. H. B., and Greenland, D. J. (1981). Adsorption. In *The Chemistry of Soil Processes*, Greenland, D. J., and Hayes, M. H. B., eds., John Wiley & Sons, Chichester, pp. 221–400.

Burdon, J. (2001). Are the traditional concepts of the structures of humic substances realistic? *Soil Sci.* **166**, 752–769.

Burns, R. G. (1990). Microorganisms, enzymes and soil colloid surfaces. In *Soil Colloids and Their Associations in Aggregates*, De Boodt, M. F., Hayes, M. H. B., and Herbillon, A., eds., Plenum, New York, pp. 337–361.

Calvin, M., and Philip, R. P. (1976). Possible origin for insoluble organic (kerogen) debris in sediments from insoluble cell-wall material of algae and bacteria. *Nature* **232**, 134–136.

Cameron, R. S., Thornton, B. K., Swift, R. S., and Posner, A. M. (1972). Molecular weight and shape of humic acid from sedimentation and diffusion measurements on fractionated extracts. *J. Soil Sci.* **23**, 394–408.

Cheshire, M. V. 1979. *Nature and Origin of Carbohydrates in Soils*. Academic Press, London.

Cheshire, M. V., and Anderson, G. (1975). Soil polysaccharides and carbohydrate phosphates. *Soil Sci.* **119**, 356–372.

Cheshire, M. V., Cranwell, P. A., Falshaw, C. P., Floyd, A. J., and Haworth, R. D. (1967). Himic acid II. Structure of humic acids. *Tetrahedron* **23**, 1669–1682.

Cheshire, M. V., Cranwell, P. A., and Haworth, R. D. (1968). Humic acid III. *Tetrahedron* **24**, 5155–5167.

Cheshire, M. V., and Hayes, M. H. B. (1990). Composition, origins, structures, and reactivities of soil polysaccharides. In *Soil Colloids and Their Associations in Aggregates*, De Boodt, M. F., Hayes, M. H. B., and Herbillon, A., eds., Plenum, New York, pp. 307–336.

Cheshire, M. V., Mundie, C. M., and Shepherd, H. (1969). Transformation of (^{14}C) glucose and starch in soil. *Soil Biol. Biochem.* **1**, 117–130.

Cheshire, M. V., Mundie, C. M., and Shepherd, H. (1971). The origin of the pentose fraction of soil polysaccharide. *J. Soil Sci.* **22**, 222–236.

Cheshire, M. V., Sparling, G. P., and Inkson, R. W. E. (1979). The decomposition of straw in soil. In *Straw Decay and Its Effect on Disposal and Utilization*, Grossard, E., ed., pp. 65–71.

Cheshire, M. V., Sparling, G. P., and Shepherd, H. (1974). Transformation of sugars when rye hemicellulose labelled with (^{14}C) decomposes in soil. *J. Soil Sci.* **25**, 90–98.

Chesters, G., Attoe, O. J., and Allen, O. N. (1957). Soil aggregation in relation to various soil constituents. *Soil Sci. Soc. Am. Proc.* **21**, 272–277.

Clapp, C. E. (1957). *High Molecular Weight Water-Soluble Muck: Isolation and Determination of Constituent Sugars of a Borate Complex-Forming Polysaccharide Employing Electrophoretic Techniques*. Ph.D. dissertation, Cornell University.

Clapp, C. E., and Davis, R. J. (1970). Properties of extracellular polysaccharides from Rhizobium. *Soil Biol. Biochem.* **2**, 109–117.

Clapp, C. E., Dawson, J. E., and Hayes, M. H. B. (1979). Composition and properties of a purified polysaccharide isolated from an organic soil. In *Proceedings, International Symposium on Peat in Agriculture and Horticulture*, Schallinger, K. M., ed., Special Publication 205, Agriculture Research Organization, Bet Dagan, Israel, pp. 153–167.

Clapp, C. E., and Emerson, W. W. (1965). The effect of periodate oxidation on the strength of soil crumbs: I. Qualitative studies. II. Quantitative studies. *Soil Sci. Soc. Am. Proc.* **29**, 127–134.

Clapp, C. E., and Emerson, W. W. (1972). Reactions between Ca-montmorillonite and polysaccharides. *Soil Sci.* **114**, 210–216.

Clapp, C. E., Harrison, R., and Hayes, M. H. B. (1991). Interactions between organic macromolecules and soil inorganic colloids and soils. In *Interactions at the Soil Colloid–Soil Solution Interface*, Bolt, G. H., De Boodt, M. F., Hayes, M. H. B., and McBride, M. B., eds., Kluwer, Dordrecht, pp. 409–468.

Clapp, C. E., Hayes, M. H. B., Simpson, A. J., and Kingery, W. L. (2005). The chemistry of soil organic matter. In *Chemical Processes in Soils*, Tabatabai, A., and Sparks, D. L., eds., Soil Science Society of America, Book Series no. 8, Madison, WI, pp. 1–150.

Darwin, C. (1881). *The Formation of Vegetable Mould, Through the Action of Worms, with Observations on Their Habits*, John Murray, London.

Davy, H. (1813). *Elements of Agricultural Chemistry*, London.

Dehérain, P. (1902). *Traité de chimieagricole*, 2nd ed., Paris.

Derenne, S., and Lageau, C. (2001). A review of some important families of refractory macromolecules: Composition, origin, and fate in soils and sediments. *Soil Science* **166**, 833–847.

Deshmukh, A. P., Simpson, A., Hadad, C. M., and Hatcher, P. G. (2005). Insights into the structure of cutin and cutan from *Agave americanas* leaf cuticle using HRMAS NMR spectroscopy. *Org. Geochem.* **36**, 1072–1085.

Detmer, V. (1871). Natural humic substances of soil and their importance in agriculture. See Istoriya agrikul'tury (1940). *Izd. Akad. Nauk SSSR.*

Dokuchaev, V. V. (1883). *Russian Chernozem (Russkii chernozem)*; see also Soch. 3 (1949). *Izd. Akad. Nauk SSSR.*

Durand, B., and Nicaise, G. (1980). Procedures for kerogen isolation. In *Kerogen*, Durand, B., ed., Technip, Paris, pp. 35–53.

Duxbury, J. M. (1989). Studies of the molecular size and charge of humic stubstances by electrophoresis. In *Humic Substances II: In Search of Structure*, Hayes, M. H. B., MacCarthy, P., Malcolm, R. L., and Swift, R. S., eds., John Wiley & Sons, Chichester, pp. 593–620.

Ellis, G. P. (1959). The Maillard reaction. *Adv. Carbohydrate Chem.* **14**, 63–124.

Enders, C. (1942). pH-Abhangigkeit und Gleichgewichtsreaktion der Aufspaltung der Zucker zu Methylglyoxal. *Biochem. Z.* **312**, 349–353.

Enders, C. (1943a). pH-Abhangigkeit und Gleichgewichtsreaktion der Aufspaltung der Zucker zu Methylglyoxal. II. Mitteilung. *Biochem. Z.* 265–269.

Enders, C. (1943b). Uber den Chemismus der Huminsauerbildung unter Physiologischen Bedingungen. IV. Mitteilung. Die Rolle der Mikroorganismen bei den humifizierungsvorgangen. *Biochem. Z.* **315**, 259–292.

Enders, C., and Fries, G. (1936). Zur Analogie von Melanoidinen und Huminsauren. *Kolloid Z.* **76**, 289–291.

Enders, C., and Marquardt, R. (1941). Methylglyoxal als Zwischenprodukt der Melanoidin und Karmelbildung. *Naturwissenschaften* **29**, 46–47.

Enders, C., and Sigurdsson, S. (1943). Uber den Chemismus der Huminsauerbildung unter physiologischen Bedingungen. V. Mitteilung. Die einleitende Phase der Huminsaurebildung. Eine Aldolkondensation van methylglyoxal. *Ber. Deutsche Chem. Ges.* **76**, 560–563.

Enders, C., and Sigurdsson, S. (1944). Zur Existenz und Bedentung der Zuker-Trioseglichgewichte. V. Mitteilung: Untersuchungen uber Temperaturabhangigkeit, Reversibilitat und pH-Abhangigkeit der Zucker-Triosgleichgewichte mit Hilfe von Ariyam als Reagens. *Biochem. Z.* **316**, 303.

Enders, C., and Sigurdsson, S. (1947). Uber den Chemismus den Huminsaurenbildung unter physiologischen Bedingungen. VII. Mitteilung. *Biochem. Z.* **318**, 44–46.

Enders, C., Tschapek, M., and Glane, R. (1948). Vergleichende Untersuchungen einiger kolloider Eigenschaften von naturlichen Huminsauren und synthetischen Melanoidinen. *Kolloid Z.* **110**, 240–244.

Finch, P., Hayes, M. H. B., and Stacy, M. (1967). Studies of soil polysaccharides and on their interactions with clay preparations. *Int. Soc. Soil Sci. Trans., Comm. IV and VI* (Aberdeen, 1966), Aberdeen University Press, Aberdeen, pp. 19–32.

Flaig, W. (1960). Comparative chemical investigations on natural humic compounds and their model substances. *Sci. Proc. Royal Dublin Soc.* **4**, 49–62.

Flaig, W., Beutelspacher, H., and Reitz, E. (1975). Chemical composition and physical properties of humic substances. In *Soil Components*, Vol. **1**, Gieseking, J. E., ed., Springer-Verlag, Berlin, pp. 1–219.

Fuchs, W. (1931). *Die Chemie der Kohle*, Springer, Berlin.

Geoghegan, M. J., and Brian, R. C. (1946). Influence of bacterial polysaccharides on aggregate formation in soils. *Nature (London)* **158**, 837.

Geoghegan, M. J., and Brian, R. C. (1948). Aggregate formation in soil. 2. Influence of various carbohydrates and proteins in aggregation of soil particles. *Biochem. J.* **43**, 14–14.

German, R. O. (1836). A chemical investigation of chernozem soils in our southern districts for determining their various properties. *Zemled. Zh. Moskov. Obshch. Sel. Khoz.* **5**.

German, R. O. (1837). Chemical investigation on chernozem occurring in southern districts of Russia. *Zemled. Zh. Moskov. Obshch. Sel. Khoz.* **1**.

Greenland, D. J., Lindstrom, G. R., and Quirk, J. P. (1961). Role of polysaccharides in stabilisation of natural soil aggregates. *Nature (London)* **191**, 1283–1284.

Greenland, D. J., Lindstrom, G. R., and Quirk, J. P. (1962). Organic materials which stabilize natural soil aggregates. *Soil Sci. Soc. Am. Proc.* **26**, 366–371.

Greenland, D. J., and Oades, J. M. (1975). Saccharides. In *Soil Components, Vol. 1: Organic Components*, Gieseking, J. E., ed., Springer-Verlag, Berlin, pp. 213–257.

Haider, K., and Martin, J. P. (1967). Synthesis and transformation of phenolic compounds by *Epicoccum nigrum* in relation to humic acid formation. *Soil Sci. Soc. Am. Proc.* **31**, 766–772.

Harrison, R. (1982). *A Study of Some Montmorillonite–Organic Complexes*. Ph.D. dissertation, University of Birmingham.

Hatcher, P. G. D., VanderHart, L., and Earl, W. L. (1980). Use of solid-state ^{13}C NMR in structural studies of humic acids and humin from Holocene sediments. *Org. Geochem.* **2**, 87–92.

Hatcher, P. G., Breger, I. A., Dennis, L. W., and Maciel, G. E. (1983). Solid-state ^{13}C-NMR of sedimentary humic substances: New revelations on their chemical compostion. In *Aquatic and Terrestrial Humic Materials*, Christman, R. F., and Gjessing, E. T., eds., Ann Arbor Science, Ann Harbor, MI, pp. 37–81.

Häusler, M. J., and Hayes, M. H. B. (1996). Uses of the XAD-8 resin and acidified dimethylsulfoxide in studies of humic acids. In *Humic Substances and Organic Matter in Soil and Water Environments: Characterization, Transformations and Interactions*, Clapp, C. E., Hayes, M. H. B., Senesi, N., and Griffith, S. M., eds., IHSS, University of Minnesota, St. Paul, 25–32.

Haworth, R. D. (1971). The chemical nature of humic acids. *Soil Sci.* **111**, 71–79.

Haworth, W. N., Pinkard, F. W., and Stacey, M. (1946). Function of bacterial polysaccharides in soil. *Nature (London)* **158**, 836–837.

Hayes, M. H. B. (1960). *Subsidence and Humification in Peats*. Ph.D. dissertation, The Ohio State University.

Hayes, M. H. B. (1985). Extraction of humic substances from soil. In *HumicSsubstances in Soil, Sediment, and Water: Geochemistry, Isolation and Characterization*, Aiken, G. R., McKnight, D. M., Wershaw, R. L., and MacCarthy, P., eds., John Wiley & Sons, New York, pp. 329–362.

Hayes, M. H. B. (2006a). Biochar and biofuels for a brighter future. *Nature* **443**, Letters to the Editor, 144.

Hayes, M. H. B. (2006b). Solvent systems for the isolation of organic components from soils. *Soil Sci. Soc. Am. J.* **70**, 986–994.

Hayes, M. H. B., Dawson, J. E., Mortensen, J. L., Clapp, C. E., and Hausler, M. J. (1985). Comparisons of synthetic "humic-like" substances with soil humic acids. In *Volunteered Papers, 2nd International Conference, International Humic Substances Society* (Birmingham, 1984), Hayes, M. H. B., and Swift, R. S., eds., IHSS, University of Minnesota, St. Paul, pp. 157–165.

Hayes, M. H. B., MacCarthy, P., Malcolm, R. L., and Swift. R. S., eds. (1989a). *Humic Substances II. In Search of Structure*, John Wiley & Sons, Chichester.

Hayes, M. H. B., MacCarthy, P., Malcolm, R. L., and Swift, R. S. (1989b). Structures of humic substances: The emergence of "forms." In *Humic Substances II. In Search of Structure*, Hayes, M. H. B., MacCarthy, P., Malcolm, R. L., and Swift, R. S., eds., John Wiley & Sons, Chichester, pp. 689–733.

Hayes, M. H. B., and Swift, R. S. (1978). The chemistry of soil organic colloids. In *The Chemistry of Soil Constituents*, Greenland, D. J., and Hayes, M. H. B., eds., John Wiley & Sons, Chichester, pp. 179–320.

Hayes, M. H. B., and Swift, R. S. (1990). Genesis, isolation, composition and structures of soil humic substances. In *Soil Colloids and Their Associations in Aggregates*, DeBoodt, M. F., Hayes, M. H. B., and Herbillon, A., eds., Plenum, New York, pp. 245–305.

Hayes, M. H. B., Stacey, M., and Swift, R. S. (1975). Techniques for fractionating soil polysaccharides. *Trans. 10th Intern. Congr. Soil Sci. (Moscow)*, Suppl. Vol., 75–81.

Hayes, T. M., Hayes, M. H. B., Skjemstad, J. O., and Swift, R. S. (2008). Studies of compositional relationships between organic matter in a grassland soils and its drainage waters. *Eur. J. Soil Sci.* **59**, 603–616.

Hodge, J. E. (1953). Chemistry of Browning reactions in model systems. *Agric. Food Chem.* **1**, 928–943.

Holloway, P. J. (1984). Cutins and suberins, the polymeric plant lipids. In *CRC Handbook of Chromatography: Lipids*, Vol. **1**, Mangold, H. K., Zweig, G., and Sherma, J., eds., CRC Press, Boca Raton, FL, pp. 321–345.

Holloway, P. J. (1994). Plant cuticles: Physicochemical characteristics and biosynthesis. *NATO ASI Ser., Ser. G: Ecol. Sci.* **36** (Air Pollutants and the Leaf Cuticle), 1–13.

Hoppe-Seyler, F. (1889). Uber Huminsubstanzen, ihre Emstenung and ihre Eigenschaften. 2. *Physiol. Chem.* **13**, 66–121.

Jenkinson, D. S., and Ladd, J. N. (1981). Microbial biomass in soil: measurement and turnover. In *Soil Biochemistry*, Vol. **5**, Paul, E. A., and Ladd, J. N., eds., Marcel Dekker, New York, pp. 415–417.

Kenworthy, I. P., and Hayes, M. H. B. (1997). Investigations of some structural properties of humic substances by fluorescence quenching, In *Humic Substances, Peats, and Sludges. Health and Environmental Aspects*, Hayes, M. H. B., and Wilson, W. S., eds., The Royal Society of Chemistry, Cambridge, pp. 39–45.

Knicker, H. (2000). Double cross polarization magic angle spinning ^{15}N ^{13}C NMR spectroscopic studies for characterization of immobilized nitrogen in soils. *Proceedings, 10th IHSS International Conference* (Toulouse), France, pp. 1105–1108.

Knicker, H., Hatcher, P. G., and Gonzales-Vila, F. J. (2002). Formation of heteroaromatic nitrogen after prolonged humification of vascular plant remains as revealed by nuclear resonance spectroscopy. *J. Environ. Qual.* **31**, 444–449.

Kolattukudy, P. E. (1980). Biopolyester membranes of plants: Cutin and suberin. *Science* **208**, 990–1000.

Kolattukudy, P. E. (1984). Biochemistry and function of cutin and suberin. *Can. J. Bot.* **62**, 2918–2933.

Komov, I. I. (1789). Agriculture (O zemledelii). Moscow.

Kononova, M. M. (1966). *Soil Organic Matter: Its Nature, Its Role in Soil Formation and in Soil Fertility*, 2nd English edition, Pergamon Press, Oxford.

Kononova, M. M. (1975). Humus of virgin and cultivated soils. In *Soil Components*, Vol. **1**, Gieseking, J. E., ed., Springer-Verlag, Berlin, pp. 475–526.

Leenheer, J. A. (1985). Fractionation techniques for aquatic humic substances. In *Humic substances in Soil, Sediment, and Water: Geochemistry, Isolation and Characterization*,

Aiken, G. R., McKnight, D. M., Wershaw, R. L., and MacCarthy, P., eds., John Wiley & Sons, New York, pp. 409–429.

Liebig, J. von (1846). *Die Chemie in ihrer Anwendung auf Agrikultur und Physiologie*, 5th edition, Vieweg, Braunschweig.

Lomonosov, M. (1763). *The First Principles of Metallurgy or Ore-mining. Humus (Second Supplement Soil Layers), St. Petersburg*. See also Istoriya agrikul'tury (1940). *Izd. Akad. Nauk SSSR*.

MacCarthy, P., and Rice, J. A. (1991). An ecological rationale for the heterogeneous nature of humic substances. In *Scientists on Gaia*, Schneider, S., and Boston, P. J., eds., MIT Press, Cambridge, MA, pp. 339–345.

Maillard, L. C. (1912). Action des acides amines sur les sucres; formation des melanoidines par voie methodique. *Acad. Sci. Compt. Rend.* **154**, 66–68.

Maillard, L. C. (1916). Synthese des matieres humiques par action des acides sur les sucres reducteurs. *Ann. Chim.* **5**, 258–317.

Maillard, L. C. (1917). Identite des matieres humiques de synthese avec les matieres humiques naturelles. *Ann. Chim.* **7**, 113–152.

Malcolm, R. L. (1989). Application of solid-state ^{13}C N. M. R. spectroscopy to geochemical studies of humic substances. In *Humic Substances II. In Search of Structure*, Hayes, M. H. B., MacCarthy, P., Malcolm, R. L., and Swift, R. S., eds., John Wiley & Sons, Chichester, pp. 339–372.

Mao, J. D., and Schmidt-Rohr, K. (2004). Separation of aromatic-carbon ^{13}C NMR signals from di-oxygenated alkyl bands by a chemical-shift-anisotropy filter. *Solid State Nucl. Magn. Reson.* **26**, 36–45.

Marris, E. (2006). Black is the new green. *Nature* **442**, 624–626.

Martin, J. P. (1945). Microorganisms and soil aggregation. I. Origin and nature of some of the aggregating substances. *Soil Sci.* **59**, 163–174.

Martin, J. P. (1946). Microorganisms and soil aggregation. II. Influence of bacterial polysaccharides on soil structure. *Soil Sci.* **61**, 157–166.

Martin, J. P., Richards, S. J., and Haider, K. (1967). Properties and decomposition and binding action in soil "humic acid" synthesized by *Epicoccum nigrum*. *Soil Sci. Soc. Am. Proc.* **31**, 657–662.

McLaren, A. D. (1954). The adsorption and reactions of enzymes and proteins on kaolinite. *J. Phys. Chem.* **58**, 129–137.

McLaren, A. D., Peterson, G. H., and Barshad, I. (1958). The adsorption reactions of enzymes and proteins on clay minerals. IV. Kaolinite and montmorillonite. *Soil Sci. Soc. Am. Proc.* **22**, 239–244.

McLaren, A. D., Puktie, A. H., and Barshad, I. (1975). Isolation of humus with enzymatic activity from soil. *Soil Sci.* **119**, 178–180.

Mehta, N. C., Dubach, P., and Deuel, H. (1961). Carbohydrates in the soil. *Adv. Carbohydr. Chem.* **16**, 335–355.

Mehta, N. C., Streuli, H., Müller, M., and Deuel, H. (1960). Role of polysaccharides in soil aggregation. *J. Sci. Food Agric.* **11**, 40–47.

Moavad, H., Guzev, V. S., Babyeva, I. P., and Zuyagintsev, D. G. (1974). Adsorption of the extracellular polysaccharide of the yeast *Lipomyces lipofer* on kaolinite. *Pochvovedenie* **11**, 79–84.

Mulder, G. J. (1861–1862). *Die Chemie der Ackerkrume*. Berlin.

Müller, P. (1887). *Studien uber die naturlichen Humusformen*.

Murayama, S. (1984). Changes in the monosaccharide composition during the decomposition of straws under field conditions. *Soil Sci. Plant Nutr.* **30**, 367–381.

Nierop, K. G. J. (1998). Origin of aliphatic compounds in a forest soil. *Org. Geochem.* **29**, 1009–1016.

Novotny, E. H., deAzevedo, E. R., Bonagamba, T. J., Cunha, T. J. F., Madari, B. E., Benites, V. deM., and Hayes, M. H. B. (2007). Studies of the compositions of humic acids from Amazonian Dark Earth soils. *Environ. Sci. Technol.* **41**, 400–405.

Oades, J. M. (1984). Soil organic matter and structural stability: Mechanisms and implications for management. *Plant and Soil* **76**, 319–337.

Oades, J. M. (1990). Associations of colloids in soil aggregates. In *Soil Colloids and Their Associations in Aggregates*, DeBoodt, M. F., Hayes, M. H. B., and Herbillon, A., eds., Plenum, New York, pp. 463–483.

Olk, D. C., Brunetti, G., and Senesi, N. (2000). Decrease in humification of organic matter with intensified lowland rice cropping: A wet chemical and spectroscopic investigation. *Soil Sci. Soc. Am. J.* **64**, 1337–1347.

Page, E. R. (1980). Cellulose xanthate as a soil conditioner: Laboratory experiments. *J. Sci. Food Agric.* **31**, 1–6.

Pagliai, H., Guidi, G., and Petruzzelli, G. (1979). Effect of molecular weight on dextran–soil interactions. In *Modifications of Soil Structure*, Emerson, W. W., Bond, A. D., and Dexter, A. R., eds., John Wiley & Sons, New York, pp. 175–180.

Piccolo, A. (2001). The supramolecular structure of humic substances. *Soil Sci.* **166**, 810–832.

Post, H. von (1862). Studien uber die koprogenen Erdbildungen unser Tage. Cited by Ramann (1888), *Landw. Jb*, **17**, 405.

Preston, C. M., and Newman, R. H. (1992). Demonstration of spatial heterogeneity in the organic-matter of de-ashed humin samples by solid-state C-13 Cpmas Nmr. *Can. J. Soil Sci.* **72**, 13–19.

Ramann, E. (1888). Die Arbeiten von Post uber Schlamm. Moor, Torf, und Humus. *Landw. Jb.* **17**, 405.

Rennie, D. A., Truog, E., and Allen, O. N. (1954). Soil aggregation as influenced by microbial gums, level of fertility, and kind of crop. *Soil Sci. Soc. Am. Proc.* **18**, 399–403.

Rice, J. A. 2001. Humin. *Soil Sci.* **166**, 848–857.

Rillig, M. C., Wright, S. F., Nichols, K. A., Schmidt, W. F., and Torn, M. S. (2001). Large contribution of arbuscular mycorrhizal fungi to soil carbon pools in tropical forest soils. *Plant Soil* **233**, 166–177.

Rook, J. J. (1977). Chlorination reactionsof fulvic acids in natural waters. *Env. Sci and Technol.* **11**, 478–482.

Russell, E. W. (1973). *Soil Conditions and Plant Growth*, 10th edition, Longmans, London, Chapter 1.

Saussure, Th. de (1804). Recherches chimiques sur la vegetation. *Paris, Ann.* **12**, 162.

Schmidt-Rohr, K., and Mao, J.-D. (2002). Selective observation of nitrogen-bonded carbons in solid-state NMR by saturation-pulse induced dipolar exchange with recoupling. *Chem. Phys. Lett.* **359**, 403–411.

Schnitzer, M. (1978). Humic substances: Chemistry and reactions. In *Soil Organic Matter*, Schnitzer, M., and Khan, S. U., eds., Elsevier North-Holland, New York, pp. 1–64.

Schnitzer, M. (2000). A lifetime perspective on the chemistry of soil organic matter. *Adv. Agron.* **68**, 1–58.

Schnitzer, M., and Khan, S. U. (1972). *Humic Substances in the Environment*, Marcel Dekker, New York.

Schnitzer, M., and Wright, J. R. (1960). Studies of the oxidation of organic matter of the Ao and Bh horizons of a podzol. *Trans. 7th Intern. Congr. Soil Sci. (Madison)*, 112–119.

Schreiner, O., and Shorey, E. C. (1909). *The Isolation of Harmful Substances from Soils*. USDA Bureau of Soils, Bulletin 53.

Schreiner, O., and Shorey, E. C. (1910). *Chemical Nature of Soil Organic Matter*. USDA Bureau of Soils, Bulletin 74.

Schuffelen, A. C., and Bolt, G. H. (1950). Some notes on the synthesis of humic compounds. Overfruk Uit Het. Landbouwkundig Titzschrift 62 ste Jaargang No. 4/5.

Senesi, N. (1999). Aggregation patterns and macromolecular morphology of humic substances: A fractal approach. *Soil Sci.* **164**, 841–856.

Senesi, N., and Steelink, C. (1989). Application of ESR spectroscopy to the study of humic substances. In *Humic Substances II*. In *Search of Structure*, Hayes, M. H. B., MacCarthy, P., Malcolm, R. L., and Swift, R. S., eds., John Wiley & Sons, Chichester, pp. 373–408.

Shmuk, A. A. (1914). Some data on the forms of nitrogen in soils. *Zh. opyt. Agron.* **15**, 139.

Shmuk, A. A. (1924). The chemical nature of soil organic matter. *Byull. Pochvoveda*, 5–7.

Sibirtsev, N. M. (1900, 1901). *Soil Science. 2. Teaching of Soil as a Mass (Uchenie o pochve kak o masse)*. St. Petersburg.

Simpson, A. J. (2002). Determining the molecular weight, aggregation, structures and interactions of natural organic matter using diffusion ordered spectroscopy. *Magn. Reson. Chem.* **40**, S72–S82.

Simpson, A. J., Kingery, W. L., Hayes, M. H. B., Spraul, M., Humpfer, E., Dvortsak, P., Kerssebaum, R., Godejohann, M., and Hofmann, M. (2002). The structures and associations of organic molecules in the terrestrial environment. *Naturwissenschaften* **89**, 84–88.

Simpson, A. J., Song, G., Smith, E., Lam, B., Novotony, E. H., and Hayes, M. H. B. (2007). Unraveling the structural components of soil humin using solution state nuclear magnetic resonance spectroscopy. *Environ. Sci. Technol.* **41**, 876–883.

Skjemstad, J. O., Reicosky, D. C., Wilts, A. R., and McGowan, J. A. (2002). Charcoal carbon in U. S. agricultural soils. *Soil Sci. Soc. Am. J.* **66**, 1249–1255.

Sleighter, R. L., and Hatcher, P. G. (2007). The application of electrospray ionisation coupled to ultrahigh resolution mass spectrometry for the molecular characterization of natural organic matter. *J. Mass Spectrom.* **42**, 559–574.

Song, G., Novotny, E. H., Simpson, A. J., Clapp, C. E., and Hayes, M. H. B. (2008). Sequential exhaustive extractions, and characterisations using solid and solution state NMR, of the humic, including humin, components in a Mollisol soil. *Eur. J. Soil Sci.* **59**, 505–516.

Sprengel, C. (1826). Über Pflanenzhumus, Humussäure and Humussaure Slaze. *Kastnes Arch. Ges. Naturlehre* **8**, 145–220.

Stenson, A. C., Landing, W. M., Marshall, A. G., and Cooper, W. T. (2003). Exact masses and chemical formulas of individual Suwannee River fulvic acids from ultrahigh resolution electrospray ionisation fourier transform ion cyclotron resonance mass spectrometry. *Anal. Chem.* **75**, 1275–1284.

Stevenson, F. J. (1982, 1994). *Humus Chemistry. Genesis, Composition, Reaction*, John Wiley & Sons, New York.

Swaby, R. J., and Ladd, J. N. (1966). Stability and origin of soil humus. In *The Use of Isotopes in Soil Organic Matter Studies*, Int. Atomic Energy Agency, Pergamon Press, Oxford, pp. 153–159.

Swift, R. S. (1985). Fractionation of soil humic substances. In *Humic substances in Soil, Sediment, and Water: Geochemistry, Isolation and Characterization*, Aiken, G. R., McKnight, D. M., Wershaw, R. L., and MacCarthy, P., eds., John Wiley & Sons, New York, 387–408.

Swift, R. S. (1989). Molecular weight, shape, and size of humic substances by ultracentrifugation. In *Humic Substances II. In Search of Structure*, Hayes, M. H. B., MacCarthy, P., Malcolm, R. L., and Swift, R. S., eds., John Wiley & Sons, Chichester, 467–495.

Swift, R. S. (1996). Organic matter characterization. In *Methods of Soil Analysis. Part 3. Chemical Methods*, Sparks, D. L., ed., Soil Science Society of America, Book Series no. 5, Madison, WI., pp. 1011–1069.

Swincer, G. D., Oades, J. M., and Greenland, D. J. (1968). Studies on soil polysaccharides. I. The isolation of polysaccharides from soil. *Aust. J. Soil Res.* **6**, 211–224.

Thaer, A. von (1809). *The Basis of Rational Agriculture.*

Theng, B. K. G. (1979). *Formation and Properties of clay–Polymer Complexes. Developments in Soil Science 9*, Elsevier, Amsterdam.

Thomson, T. (1807). *A System of Chemistry*, Edinburgh. See also Istoriya agrikul'tury (1940). *Izd. Akad. Nauk SSSR.*

Trusov, A. G. (1914, 1915). The humification of compounds which are constituents of plants. *Sel Khoz. Lesovod.,* October, November; (1915) ibid., April, July, November; 1916, *ibid.,* March, September, October, November.

Trusov, A. G. (1916). Some chemical–biological processes occurring during the humification of plant residues. *Zh. Opyt. Agron.* **17**.

Vauquelin, C. (1797, 1798). Sur une maladie des arbres que attaque spécialment l'orme et que est analogue à un ulcére. *Ann. Chim.* **21**, 39. See also Istoriya agrikul'tury (1940). *Izd. Akad. Nauk SSSR.*

Waksman, S. A. (1936). *Humus, Origin, Chemical Composition, and Importance in Nature*, Tindall and Cox, London.

Waksman, S. A., and Iyer, K. R. N. (1932). Contribution to our knowledge of the chemical nature and origin of humus: I. On the synthesis of the "humus nucleus." *Soil Sci.* **34**, 43–69.

Waksman, S. A., and Iyer, K. R. N. (1933). Contribution to our knowledge of the chemical nature and origin of humus: IV. Fixation of proteins by lignins and formation of complexes resistant to microbial decomposition. *Soil Sci.* **36**, 69–82.

Waksman, S. A., and Martin, J. P. (1939). The role of microorganisms in the conservation of the soil. *Science* **90**, 304–305.

Wallerius, J. (1761). Agriculturae Fundamenta Chemica. Upsala. See also Istoriya agrikul'tury (1940). *Izd. Akad. Nauk SSSR.*

Wandruszka, R. von (1998). The micellar model of humic acid: Evidence from pyrene fluorescence measurements. *Soil Sci.* **163**, 921–930.

Wershaw, R. L. (1999). Molecular aggregation of humic substances. *Soil Sci.* **164**, 803–813.

Whitehead, D. C., and Tinsley, J. (1964). Extraction of soil organic matter with dimethylformamide. *Soil Sci.* **97**, 34–42.

Williams, V. R. (1939). *Soil Science and the Basis of Agriculture (Pochvovedenie s osnovami zemledeliya).* Sel'Khozgiz.

Wilson, M. A. (1987). *NMR Techniques and Applications in Geochemistry and Soil Chemistry*, Pergamon Press, Oxford.

Wolfrom, M. L., Schlicht, R. C., Langer, A. W., Jr., and Rooney, C. S. (1953). Chemical interactions of amino compounds and sugars: VI. The repeating unit in Browning polymers. *J. Am. Chem. Soc.* **75**, 1013.

Woods, W. I., and Glaser, B. (2004). Towards an understanding of Amazonian Dark Earths. In *Amazonian Dark Earths: Explorations in Space and Time*, Glaser, B., and Woods, W. I., eds., Springer-Verlag, Berlin.

2

FORMATION MECHANISMS OF HUMIC SUBSTANCES IN THE ENVIRONMENT

P. M. HUANG AND A. G. HARDIE

Department of Soil Science, University of Saskatchewan, Saskatoon, Canada

Biophysico-Chemical Processes Involving Natural Nonliving Organic Matter in Environmental Systems, Edited by Nicola Senesi, Baoshan Xing, and Pan Ming Huang

2.1. INTRODUCTION

Soil and related environments are both an important natural habitat of biota and a natural reservoir of biotic debris consisting of plant remains and dead animals and microorganisms. With time, dead remains are subject to continuous turnover, either mineralized or transformed to diverse organic components which are termed *humus*. This process is referred to as humification. Humus is composed of *humic substances* plus *nonhumic substances* that have become stabilized and are thus an integral part of soil and related environments (Table 2.1).

The stocks of organic matter in soils results from the balance between inputs and outputs of organic C within the below-ground environment (Figure 2.1). Inputs are primarily controlled by net primary productivity; outputs are dominated by the efflux of carbon dioxide (CO_2) from the soil surface, although methane (CH_4) efflux and hydrologic leaching of dissolved organic and inorganic and particulate organic C compounds can also be important (Davidson and Janssens, 2006). During the turnover process of organic C, organic matter may become physically protected in the interior of soil aggregates (Oades, 1988; Six et al., 2002), where microorganisms and their enzymes may have limited access and where O_2 concentration may also be low. Similarly, organic compounds can be physically protected from degradation by water-soluble enzymes if they have low water solubility or if they occur in hydrophobic domains of humified organic matter (Spaccini et al., 2002). Furthermore, organic matter may become adsorbed onto surfaces of minerals, especially short-range ordered Al and Fe (oxy)hydroxides through complexation, thus chemically protecting it from decomposition (Oades, 1988; Torn et al., 1997; Huang, 2004; Rasmussen et al., 2005).

The transformation of biotic debris to humus proceeds in two stages (Hayes, 1991; Stevenson, 1994; Bollag et al., 1998). The first stage involves degradation processes that lead to the formation of biological residues, their "partial decomposition" products, and substrates (simpler structural units), which are primarily mediated by microorganisms and free enzymes (Haider et al., 1975; Bollag et al., 1998). The prevailing vegetation greatly affects the amount and type of biomolecules released during decomposition processes, which in turn will affect the nature of substances and the amount of CO_2 released. Climate and microbial populations affect the rate

TABLE 2.1. Definitions of Environmental Organic Matter and Humic Substances

Term	Definition
Litter	Macroorganic matter (e.g., plant residues) that lies on the soil surface.
Light fraction	Undecayed plant and animal tissues and their partial decomposition products that occur within soil proper and that can be recovered by flotation with a liquid of high density.
Biomass	Organic matter present as organisms.
Humus[a]	Total of the diverse organic components in the environment exclusive of undecayed plant and animal tissues, their "partial decomposition" products, and the soil biomass.
Soil organic matter	Same as humus.
Humic substances	A series of relatively high-molecular-weight, yellow to black colored substances formed by secondary synthesis reactions. The term is used as a generic name to describe the colored material or its fractions obtained on the basis of solubility characteristics. These materials are distinctive to the soil (or sediment) environment in that they are dissimilar to the biopolymers of microorganisms and higher plants (including lignin).
Nonhumic substances	Compounds belonging to known classes of biochemistry, such as amino acids, carbohydrates, fats, waxes, resins, organic acids, etc. Humus probably contains most, if not all, of the biochemical compounds synthesized by living organisms.
Humin	The alkali insoluble fraction of soil organic matter.
Humic acid	The dark-colored organic material that can be extracted from soil by dilute alkali and other reagents and is insoluble in dilute acid.
Hymatomelanic acid	Alcohol-soluble portion of humic acid.
Fulvic acid	Fraction of soil organic matter that is soluble in both alkali and acid.
Generic fulvic acid	Pigmented material in the fulvic acid fraction.

[a]The term humus is generally used synonymously with soil organic matter and refers to those organic substances that do not occur in the form of plant residues or their decay products (i.e., the "light fraction") (Waksman, 1936; Stevenson, 1994). The "light fraction" is sometimes included with the definition of "soil organic matter," in which case the term "humus" has a restricted meaning and refers to humic substances plus resynthesis products that have become stabilized, and is thus an integral part of soil and related environments. However, the absolute demarcation is blurred, and it should be noted that strict adherence to the definitions will not always be possible. Furthermore, humic polymers may anchor and encapsulate unstable biomolecules by various adsorption forces or chemical binding (Bollag et al., 1998). Any biomolecules intimately associated with humic substances that cannot be separated effectively by chemical and physical methods may be considered as humic components (Sutton and Sposito, 2005). Therefore, many unstable biological constituents may survive in humus in the environment for a significant length of time in the humification process (Bollag et al., 1998).

Source: Adapted with permission from Stevenson, F. J. (1994). *Humus Chemistry: Genesis, Composition, Reactions*, 2nd ed., John Wiley and Sons, New York.

of decomposition of biological residues. Recent studies show that (1) the biotic community is able to disintegrate any organic matter of natural origin, (2) molecular recalcitrance of organic matter is relative rather than absolute, (3) recalcitrance is only important during early decomposition and in active surface soils, and (4) during late decomposition and in subsoils, the relevance of spatial inaccessibility and

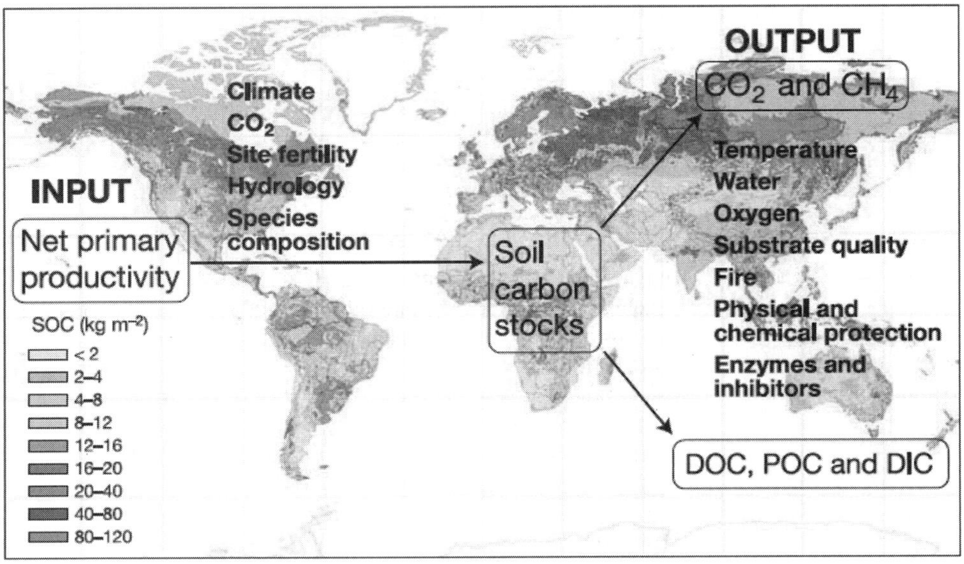

Figure 2.1. Diagram of factors controlling the main inputs and outputs of soil carbon, superimposed over a global map of soil organic carbon stocks. DOC, POC, and DIC stand for dissolved organic C, particulate organic C, and dissolved inorganic C, respectively. The background soil organic carbon (SOC) map (Miller Projection; 1:100,000,000). See color insert. Reprinted from Davidson, E. A., and Janssens, I. A. (2006). Temperature sensitivity of soil carbon decomposition and feedbacks to climate change. *Nature* **440**, 165–173, with permission from Macmillan.

organo-mineral interactions for soil organic matter increases (von Lützow et al., 2006). In the second stage of the transformation, the substrates are further transformed by synthetic processes catalyzed by enzymes (biotic catalysts) (Stevenson, 1994; Bollag et al., 1998) and mineral particles (abiotic catalysts) (Shindo and Huang, 1982, 1984b; Wang and Huang, 1986; Wang et al., 1986; Huang, 1990, 2004; Bollag et al., 1998; Jokic et al., 2004b). Environmental mineralogy and surface chemistry greatly influence the turnover and storage of organic matter (Torn et al., 1997; Guggenberger and Haider, 2002; Huang et al., 2005). Biotic and abiotic catalysts in humification all have significant roles to play; one cannot be considered more important than the other, because they interact with each other to influence humification which is one of the most important processes in the C cycle.

In the environment, humification is pivotal in transforming biomolecules originating from organized structures typical of organisms to randomly polymerized, heterogeneous humic substances characteristic of biogeochemical systems. The objective of this chapter is to integrate the existing information on our understanding of the mechanisms of the transformation of biological debris and the resultant formation of humic substances in soils and related environments. It is hoped that this review would provide a well-balanced view on this subject matter and lead the way to further advancement on our knowledge on this very important and exciting area of science.

2.2. CURRENT CONCEPTS OF THE NATURE OF HUMIC SUBSTANCES

Traditionally, humic substances (HS) were viewed as heterogeneous, high-molecular-weight polymers—as reflected, for example, in the widely accepted definition according to Aiken et al. (1985): "HS are a category of naturally occurring, biogenic, heterogeneous organic substances that can be generally characterized as being yellow-to-black in colour, of high molecular weight, and refractory." MacCarthy (2001) provided a broader definition for the term: "HS refers to a category of naturally occurring materials found in or extracted from soils, sediments, and natural waters. They result from the decomposition of plant and animal residues." Despite the important role of HS in the sustainability of life, their chemical nature and reactivities still remain poorly understood and there remains some contention with regard to their molecular structures.

One school of thought states that HS are collections of diverse, relatively low molecular mass organic components forming dynamic supramolecular associations stabilized by hydrophobic interactions and hydrogen bonds (Burdon, 2001; Piccolo, 2001; Simpson, 2002; Diallo et al., 2003; Sutton and Sposito, 2005; Kelleher and Simpson, 2006). Another widely known school of thought states that HS are formed by polymerization and polycondensation of simple biomolecules derived from the degradation of biological residues (Schnitzer, 1986; Stevenson, 1994; Huang, 2004; Jokic et al., 2004b; Allison, 2006a). However, there is no conclusive evidence to disprove either view (Clapp et al., 2005; Schaumann, 2006a). A number of reports support the polymer-sorption model for HAs (Xia and Pignatello, 2001; Xing, 2001; Lu and Pignatello, 2002, 2004). Sorption nonlinearity in the undissolved phase is attributed to polymer properties of the sorbent; hysteresis and conditional effects can up to now only be explained with the polymer analogy (Schaumann, 2006a). According to the chemical terminology of the International Union of Pure and Applied Chemistry (IUPAC), a macromolecule (polymer molecule) is a molecule of high relative molecular mass, the structure of which essentially comprises the multiple repetition of units derived, actually or conceptually, from molecules of low relative molecular mass (McNaught and Wilkinson, 1997). Polymer molecules do not have a definite formula since they consist of chains of different lengths (Daintith, 1990). The IUPAC definition of a supramolecule is a system of two or more molecular entities held together and organized by means of intermolecular (noncovalent) binding interactions. Macromolecules as well as small molecules tend to form supramolecular structures, the properties of which largely determine the chemical and physical nature of the whole material (Steed and Atwood, 2000). Although the supramolecular model has not explicitly been shown for unfractionated dissolved organic matter (DOM) and unaltered humic substances including humin, the combination of all studies suggests supramolecular as well as macromolecular characteristics of natural organic matter (NOM) (Schaumann, 2006a). Neither macromolecules nor supramolecules can be excluded in solid and dissolved NOM.

Recent studies have shown that the soil biotic community is able to disintegrate any organic matter of natural origin including black C unless it is physically inaccessible and/or chemically protected (von Lützow et al., 2006). Therefore, mere associations of biological residues and metabolites would not be able to make up the stable environmental HS and to account for their darkness in color. Mere hydrophobic interactions and hydrogen bonding between colorless biomolecules such as lipids, proteins, and polysaccharides cannot provide a logical explanation of the browning reaction and the resultant dark color of HS. Browning of biomolecules during oxidative degradative processes is produced by five known biochemical pathways in nature: (i) the Maillard reaction (condensation reaction of reducing sugars and amino compounds to produce melanoidins), (ii) oxidation of polyphenols to produce quinones and subsequent polymerization reactions of quinones (enzymatic and nonenzymatic reactions), (iii) quinone–amine polycondensation reactions (iv) ascorbic acid browning, and (v) oxidative lipid–protein polycondensation reactions (Rouet-Mayer et al., 1990; Hidalgo and Zamora, 2000; Zamora and Hidalgo, 2005; Bittner, 2006). Most of these browning reactions involve carbonyl–amine reactions that result in the formation of highly colored high- and low-molecular-weight polymers (Hidalgo and Zamora, 2000). Quinones are produced during the decomposition of lignin and by the oxidation of polyphenols; they are highly unstable and very reactive in aqueous media (Stevenson, 1994; Filley et al., 2002; Telysheva et al., 2007). They readily combine with amine, sulfydryl, phenol, indole and imidazole groups of amino acids, peptides, and proteins to give even more intensely colored products than simple quinones or phenol polymers (Bittner, 2006). Abundant research evidence at the molecular level shows that biomolecules such as amino acids, sugars, and polyphenols, derived from the breakdown of biological residues and from biological metabolites, undergo polymerization and/or polycondensation, especially by catalysis of enzymes and mineral particles (clay minerals, short-range ordered Mn, Fe, and Al oxides and (oxy)hydroxides, and primary minerals) (e.g., Stevenson, 1994; Bollag et al., 1998; Huang, 2000; Jokic et al., 2001b, 2004a, 2004b; Wang and Huang, 2003, 2005). These reactions evidently account for the browning reaction and the darkness in color of HS. Furthermore, HS are known to contain free radicals, as shown by electron paramagnetic resonance studies (Schnitzer, 1978). Free radicals drive polymerization reactions with other biomolecules and organic pollutants. Black C is another possible contributor to the color of HS in soils, which forms as a result of condensation of aromatic structures of organic residues during the burning process.

Most biomolecules, such as polysaccharides, simple sugars, lipids, and proteins, are crystalline (International Centre for Diffraction Data, 2006). If HS consist merely of associations of biological residues, they should have characteristic crystal structures that can be rigorously studied and identified by X-ray diffraction analysis. However, the research evidence clearly shows that environmental organic matter has to be considered as highly amorphous material, which additionally contains microcrystalline regions like polymethylene crystallite (Hu et al., 2000; Schaumann, 2006b).

Environmental organic matter is a composite of humic and nonhumic substances, which is formed through operation of various biotic and abiotic mechanisms, with differing importance. The relative importance of these mechanisms and the nature and properties of the resultant organic matter vary with natural vegetation, microbial populations and activities, enzymatic activities, mineralogical composition and surface chemistry, management practices, and the environment. Therefore, the formation of environmental organic matter is a result of concerted reactions of various biotic and abiotic processes.

2.3. DECOMPOSITION OF ORGANIC RESIDUES IN THE ENVIRONMENT

Plants are the primary source of organic matter in soil and related environments, whereas microorganisms and fauna, which facilitate plant residue decomposition, are considered secondary sources of degradable organic matter. Kassim et al. (1981) estimated that microbial biomass contributed 1–4% of the soil organic carbon, while the total edaphon only contributed about 10%. Only a small portion of the organic residues entering the soil is finally transformed into stable humic substances. Schlesinger (1990) estimated that only 0.7% of the annual terrestrial net primary production is transformed into refractory humic substances. Organic matter entering the soil can be divided into four major groups of biomolecules, namely, polysaccharides (e.g., cellulose, hemicellulose, chitin), proteins, lipids/aliphatic materials (e.g., waxes, cutin, suberin), and lignins. The relative amounts of these biomolecules vary greatly between plant species (Kögel-Knabner, 2002). Biomolecules, such as lipids and lignins (recalcitrant fractions), take longer to be degraded than polysaccharides, sugars, and proteins (labile fractions) and can accumulate during the initial phase of organic residue decomposition (Kalbitz et al., 2003). However, soil microbial communities can and will degrade any type of organic residues entering the soil, even black carbon (Bird et al., 1999; Hamer et al., 2004), provided that it is physically accessible to them and that there is enough oxygen and moisture present (von Lützow et al., 2006).

2.3.1. Organisms Involved in Degradation Processes

Microbes and fauna are primarily responsible for the decomposition of organic residues in the environment. Megafauna (e.g., rodents), macrofauna (e.g., earthworms, beetle larvae, termites), and mesofauna (e.g., collemboles and mites) are known as primary decomposers. They are responsible for physically breaking the plant litter and other organic residues into smaller pieces, redistributing it in the soil profile, enzymatically altering plant tissues in the gut and ultimately exposing the larger biomolecules in the residues to further chemical attack by the secondary decomposers (Wolters, 2000). Large amounts of litter have to be consumed by invertebrates because of its low nutritional value. Earthworms have been observed to consume up to 90% of the annual leaf fall litter in a single month of spring (Knollenberg et al., 1985). In general, litter that contains high amounts of carbohydrates and N-containing biomolecules is preferentially ingested, while litter containing large amounts of lignin polyphenols and tannins is avoided; thus, this can lead to an enrichment of recalcitrant biomolecules (Wolters, 2000). Ingested soil, organic residues, and stabilizing compounds are thoroughly mixed in invertebrate digestive tracts; this can lead to the exposure of lignocellulose to microflora in the gut (Hammel, 1997), as well as to the formation of stable organo-mineral complexes (Barois et al., 1993). In general, soil fauna are not able to digest lignin, cellulose, and other large highly aliphatic or aromatic compounds. Some of them do possess gut microflora that are able to partially degrade resistant organic matter, but even these species primarily assimilate C from less recalcitrant forms of organic matter (Wolters, 2000). Some saprophytic fungi and protozoa are also important in the primary decomposition process because these organisms produce extracellular enzymes that catalyze the dissolution of outer protective tissues.

Microorganisms are known as secondary decomposers. Secondary decomposition is performed exclusively by microorganisms, which include most of the saprophytic bacteria and fungi (Ross, 1989), by means of extracellular enzymes. These microorganisms are able to further degrade and utilize the decomposition products of the primary decomposers. They are specialized to facilitate the chemical degradation of large biomolecules, such as lignin and cellulose, into their constituent biochemical subunits which can provide substrates for further reaction with soil enzymes or minerals to form humic polymers (Stevenson, 1994). Although soil fauna are primary decomposers, decomposition of organic matter in soils is predominantly mediated by microorganisms, with only about 10–15% of organic C energy utilized by soil fauna (Wolters, 2000). Plants, however, serve as the major source of biomass in the humification process. All other organisms contribute as a minor source of biomass after death.

Extracellular enzymes play a fundamental role in the global carbon cycle. Microorganisms must produce extracellular enzymes of the correct structural specificity to hydrolyze the high-molecular-weight substrates, such as polysaccharides and lignin, so that they are small enough (generally less than 600 Da) to be taken up and metabolized by their cells (Weiss et al., 1991). Fungi are the most efficient degraders of lignin and cellulose, in particular white rot and brown rot species from the basidiomycetes and ascomycetes groups. These fungi are able to break down the large recalcitrant biomolecules by producing extracellular oxidoreductive enzymes, such as laccase and peroxidase, and biochelators containing redox reactive metals (ten Have and Theunissen, 2001; Xu and Goodell, 2001). White rot fungi are the only species capable of completely degrading lignin to CO_2 and water (Kirk and Farrell, 1987), whereas brown rot fungi are only able to modify lignin through demethylation and demethoxylation (Eriksson et al., 1990). The microorganisms responsible for the breakdown of organic residues are virtually found in all soils, differing only in numbers and proportions, even in the deepest layers up to 500 m below the surface (Bollag et al., 1998).

2.3.2. Degradation Processes in the Formation of Substrates and Preservation Products

The rate of organic residue decomposition in soils and related environments is ultimately controlled by its biological stability, which is a function of the following four main factors, namely, (i) its biochemical recalcitrance, (ii) the biological capability and capacity of the environment, (iii) decomposition rate modifiers (e.g., temperature, moisture, exposure time) and (iv) physical protection mechanisms (Baldock et al., 2004). Recent studies have shown that the physical protection mechanisms, such as the spatial inaccessibility of organic matter in soil micropores, are the most important factors in controlling the stability of organic matter in soils (Mikutta et al., 2006; von Lützow et al., 2006).

Biochemical recalcitrance of biomolecules is related to their molecular weight and complexity, as well as, to the presence of ether-bridges, quaternary and tertiary C-atoms, amide groups, phenyl- and heterocyclic N-groups, long-chain hydrocarbons and polymerized aromatic groups (Haider and Martin, 1981; von Lützow et al.,

TABLE 2.2. Mechanisms of Chemical Recalcitrance of Primary and Secondary Sources of Organic Matter

Specific Mechanism	Cause	Compounds and Precursors
Primary recalcitrance Plant litter and rhizodepostion	C–C bondings	*n*-fatty acids, n-alkanes, branched alkanes, *n*-alkenes, *n*-alcohols, sterines, mono-, di-, triesters
	C–O–C–, C–C, R–C–R bonding, aromatic polymers structure	Lignin
	Aromatic polymer structure	Tannins
Secondary recalcitrance Microbial and faunal products	Macromolecular structure	Chitin (*N*-acetyl-D-glucosamine in β-(1–4)-glycosidic bonds)
	Aromatic polymer structure	Melanin
	Macromolecular structure	Murein (peptidoglucan)
	C–C bonding	Phospholipids (*n*-C4:0 to *n*-C26:0 fatty acids)
	Macromolecular structure	Ceratin (scleroprotein)

Source: Adapted from von Lützow, M., Kögel-Knabner, I., Ekschmitt, K., et al. (2006). Stabilization of organic matter in temperate soils: mechanisms and their relevance under different soil conditions—a review. *Eur. J. Soil Sci.* **57**, 426–445, with permission from Wiley-Blackwell.

2006). Table 2.2, adapted from von Lützow et al. (2006), summarizes the mechanisms of chemical recalcitrance from primary and secondary sources of organic matter. The compounds most resistant to degradation are those containing polymerized aromatic rings, such as lignin, and compounds containing polymethylenic structures, such as lipids and waxes (Derenne and Largeau, 2001).

2.3.2.1. Decomposition Phases. Organic matter entering the soil and related environments goes through a number of stages of degradation. Baldock and Skjemstad (2000) studied the changes that occur in organic matter during the decomposition process using physical fractionation and solid-state ^{13}C NMR spectroscopy (Figure 2.2). Initially the organic residues have a chemical structure (denoted by chemical shift in ppm) and C/N ratio similar to that of the materials from which they were derived and a particle size >20 μm. The NMR spectrum reveals that the residues are rich in *O*-alkyl groups (50–100 ppm), which is typical for material rich in polysaccharides, such as fresh plant tissues. The first phase of decomposition involves a decrease in the particle size (2–20 μm) of the residues and a rapid and preferential consumption and degradation of the labile fraction (proteins, sugars, polysaccharides) by fauna and microorganisms. This relatively rapid process (weeks to months) involves assimilation of the labile fraction's C, N, P, and S (about 5–10%) by the fauna and microbes and results in the release of CO_2 and other inorganic species such as ammonium, phosphates, and sulfates (about 70%) (Haider, 1992). It also results in the accumulation of recalcitrant forms of organic matter, such as lignin and alkyl structures. Figure 2.2 shows the accumulation of aromatic lignin (100–

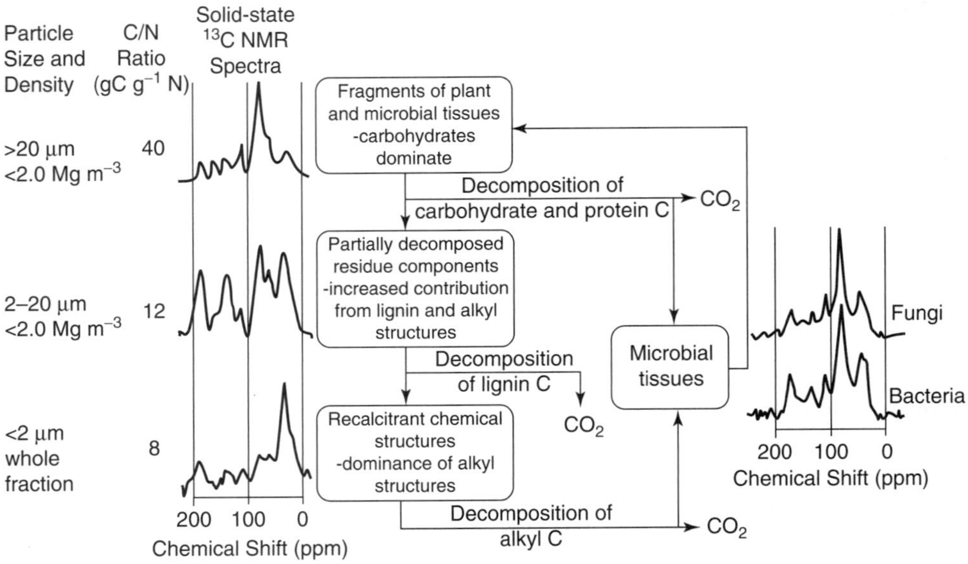

Figure 2.2. Changes in particle size, C/N ratio, and chemical composition of organic matter in mineral soil with increasing extent of oxidative decomposition. Reprinted from Baldock, J. A., and Skjemstad, J. O. (2000). The role of soil mineral matrix in protecting natural organic materials against biological attack. *Org. Geochem.* **31**, 697–710, with permission from Elsevier.

200 ppm) and alkyl structures (below 50 ppm) in the residues from the first decomposition stage (2–20 µm). At the same time the *O*-alkyl C peaks (50–100 ppm) remain dominant because of the assimilation of plant C into *O*-alkyl microbial structures. In the second phase of decomposition the accumulated lignin is slowly decomposed, primarily by white-rot fungi. These organisms do not actually derive energy or assimilate C from the lignin degradation but benefit from an exposure of labile *O*-alkyl C (holocellulose) buried in the lignin structures (Haider, 1992; ten Have and Theunissen, 2001). The decomposition of lignin results in a decrease in the amount of aromatic C, which is seen in Figure 2.2 in the NMR spectrum of the final decomposition product, where the aromatic features have significantly decreased (100–200 ppm). In Figure 2.2 it can also be seen that in the final decomposition stage, the alkyl-C fraction has become the most prominent fraction (0–50 ppm), which Baldock and Skjemstad (2000) speculated was due to its highly recalcitrant nature. However, recent NMR studies have shown that a significant fraction of the stable lipids (alkyl-C) found in the soil actually originate from the microbial biomass rather than plant derived carbon (Poirer et al., 2006; Quénéa et al., 2006b).

2.3.2.2. Breakdown Processes

Polysaccharides, Proteins, and Simple Monomers. Sugar and amino acid monomers can be rapidly degraded, within hours, by fauna or microorganisms that use these compounds as their primary energy source. Even though polysaccharides and proteins are large heteropolymers, they are also easily degraded even within weeks or months. This is because they contain hydrolytic bonds that are decomposed by a ubiquitous group of enzymes known as hydrolases (e.g., glucosidase, amidase, pectinase, xylanase, proteases, chitinase) which are able to hydrolyze ester, glycoside, ether, peptide, and other C–N bonds (von Lützow et al., 2006).

Cellulose and Hemicellulose. Cellulose is a linear and highly ordered polymer of cellobiose (D-glucopyranosyl-β-1,4–D-glucopyranose) and is the most abundant biopolymer found in plant residues. Hemicellulose is the second most abundant biopolymer and is a complex polysaccharide with a lower molecular weight than cellulose. It consists of D-xylose, D-mannose, D-galactose, D-glucose, L-arabinose, 4-*O*-methyl-glucuronic, D-galacturonic, and D-glucuronic acids, linked together by β-1,4- and occasionally β-1,3-glycosidic bonds (Cowling and Brown, 1969). Cellulose and hemicellulose are often found together strongly bonded with lignin in woody tissues and is thus known as lignocellulose (Pérez et al., 2002). A host of fungi—in particular, white-rot, brown-rot, and soft-rot fungi—and selected bacteria are able to break down cellulose and hemicellulose using a variety of hydrolytic and oxidoreductive enzymes. All the enzymatic decomposition reactions occurs exocellularly because the biopolymers are insoluble (Pérez et al., 2002).

Microorganisms that degrade cellulose produce a variety of hydrolytic enzymes that work together to break down the polymers. This includes cellulases, which are responsible for hydrolyzing the β-1,4-glycosidic linkages of cellulose. Two classes of cellulases exist, namely endoglucanases and cellobiohydrolases. Endoglucanases can hydrolyze internal bonds in amorphous regions while cellobiohydrolases act on existing chain ends. Once the β-1,4–glycosidic linkages are broken, cellobiose molecules are released, which are further broken down by β-glucosidases into two glucose molecules (Pérez et al., 2002). The released glucose can then be utilized by the cellulolytic microorganisms or other microbes as sources of energy. Thus, the final products of decomposition under aerobic conditions are CO_2 and water, or methane and water under anaerobic conditions (Béguin and Aubert, 1994; Leschine, 1995).

Brown-rot basidiomycetes fungi are one of the most common degraders of cellulose and hemicellulose in the Northern Hemisphere, being the dominant wood decay fungi in coniferous forests (Xu and Goodell, 2001; Filley et al., 2002). The brown-rot fungi are only able to demethylate and slightly depolymerize lignin, as well as being able to oxidize its side chains—unlike white-rot fungi, which are able to completely degrade lignin and cellulose at the same time (Filley et al., 2002). In the past it was thought that the degradation of cellulose by brown-rot fungi was merely an enzymatic process. It was later realized that the degradation occurs not only in the vicinity of the hyphae but also deeper into the wood cells, indicating a nonenzymatic process because enzymes are too large to penetrate unmodified wood cells (Xu and Goodell, 2001). The exact mechanism of brown-rot decay has not yet been established, but it is known that iron, hydrogen peroxide, biochelators, and oxalate each play an important role in the process. The initial stage of decomposition is thought to involve Fenton chemistry [Eq. (2.1)] because the fungi are known to produce extracellular H_2O_2 and the wood substrate contains iron, which after reaction result in the production of hydroxyl anions and free radicals (Koenigs, 1974; Xu and Goodell, 2001).

$$Fe^{2+} + H_2O_2 \rightarrow Fe^{3+} + OH^- + HO\cdot \qquad (2.1)$$

The resultant hydroxyl radicals are able to cleave long-chain cellulose polymers and thus help to initiate and accelerate decay. The fungi also produce low-molecular-weight (1500–5000 Da) *ortho*-dihydroxy phenolic compounds (catecholates) which act as Fe(III) iron chelators and as a source of electrons for iron reduction (Goodell et al., 1997; Paszczynski et al., 1999). These biochelators are small enough to penetrate the unmodified wood's lignocellulose tissue and to initiate cellulose decomposition. It has also been shown that the brown *ortho*-dihydroxy-rich lignin residues produced by the fungi can be enriched in metals, particularly calcium, iron, manganese, and magnesium (Ostrofsky et al., 1997). This is of particular relevance to synthetic humification reactions because polyphenols, particularly *ortho*-polyphenols, are very reactive in participating in polymerization and polycondensation reactions with amino compounds, especially in the presence of redox reactive metals (Huang, 2000).

Hemicellulose degradation requires a host of enzymes working together, mainly from the glycosidase family of enzymes, due to its complex and heterogeneous structure. These include β-mannases, β-xylanases, β-mannosidase, β-xylosidase, β-glycosidase, α-galactosidases, and accessory hemicellulases, xylan esterases, ferulic and *p*-courmaric esterases, α-l-arabinofuranosidases, and α-4-*O*-methyl glucuronosidases. The degradation of a common hemicellulose, *O*-acetyl-4-*O*-methylglucuronxylan, requires the synergistic action of endoxylanase, acetyl esterase, α-glucuronidase, and β-xylosidase. The products of hemicellulose decomposition are monomeric sugars and acetic acid (Pérez et al., 2002).

Lignin. Lignin is a large insoluble, amorphous heteropolymer that consists of phenylpropane units—namely, coniferyl alcohol (guaiacyl propanol), coumaryl alcohol (*p*-hydroxyphenylpropanol), and sinapyl alcohol (syringyl propanol)—that are randomly joined together by C–C and aryl–ether linkages. Lignin contains no hydrolytic bonds and is highly resistant to biological and chemical degradation (Martínez et al., 2005).

The first stage in lignin degradation involves the production of nonspecific, extracellular oxidoreductive enzymes by basidiomycetes fungi—in particular, white-rot fungi. The microorganisms do not actually metabolize the lignin C, but rather utilize holocellulose that is exposed during the decomposition process (Haider, 1992). There are two major classes of oxidoreductase enzymes that are involved in lignin degradation, namely, peroxidases (EC 1.11.1) and laccases (EC 1.10.3.2). The peroxidases are considered true lignases because they have a high redox potential and are able to degrade phenolic as well as nonphenolic lignin units, whereas the laccases have a low redox potential and are only able to degrade phenolic lignin units, which often constitute less than 10% of the whole lignin polymer (Martínez et al., 2005). Aryl alcohol oxidase and glyoxal oxidase are also considered important oxidoreductase enzymes in lignin degradation because they generate H_2O_2 for the peroxidases and have also been shown to inhibit polymerization reactions of the phenolic lignin fragments (Galliano et al., 1991; Ander and Marzullo, 1997).

White-rot fungi produce three peroxidases that are involved in lignin degradation, namely, lignin peroxidase (LiP) and manganese peroxidase (MnP), first discovered in *Phanerochaete chrysosporium* in the 1980s, and versatile peroxidase (VP), recently discovered in *Pleurotus* and *Bjerkandera* species (Martínez, 2002). Lignin peroxidase has a heme group in its active center, and it catalyzes a variety of oxidations, all of which are dependent on H_2O_2. These include Cα-Cβ cleavage of the propyl side chains of lignin, hydroxylation of benzylic methylene groups, oxidation of benzyl alcohols to the corresponding aldehydes or ketones, oxidation of aromatic ethers, phenol oxidation, and even aromatic ring cleavage of nonphenolic lignin model compounds (Cullen, 1997; Pérez et al., 2002). Manganese peroxidase is also a heme protein like LiP, except it attacks aromatic lignin structure indirectly using Mn(III) chelates and the associated free radicals that form during the process. It functions by oxidizing Mn^{2+} to Mn^{3+}. White-rot fungi secrete oxalic and other organic acids that form Mn(III) chelates which stabilize the Mn(III) and allow it to diffuse in solution so that it can oxidize phenolic compounds. Low-molecular-weight redox mediators play an important role during the initial steps of lignin degradation because the compact molecular architecture of the intact plant cell-wall prevents the penetration of enzymes to be in direct contact with lignin (Martínez, 2002). The recently discovered versatile peroxidase (VP) has the combined catalytic properties of LiP, MnP, and plant/microbial peroxidases oxidizing phenolic compounds.

Laccases are blue-copper phenoloxidases that catalyze one-electron oxidation reactions, mainly the oxidation of phenolic lignin structures to phenoxy radicals in the presence of mediators (Leonowicz et al., 2001). They also catalyze the demethylation of lignin, methoxyphenol acids, and methoxyaromatics (Burton, 2003). Laccases are also able to catalyze the polymerization of lignin monomers and *o*- and *p*-polyphenols, and thus they could also play an important role in synthetic humification reactions (Bollag et al., 1998). Laccases have been found in mainly wood rotting fungi, but also in *Aspergillus*, thermophilic fungi, and even bacteria (Pérez et al., 2002).

Since lignin is such a large and complex heteropolymer, it is reasonable that it would require the action of many enzymes working together to break it down, rather than just one enzyme, such as LiP. Leonowicz et al. (2001) proposed an integrated lignocellulose degradation mechanism (Figure 2.3), whereby the degradation pathway of polysaccharides, cellulose, and hemicellulose is incorporated into the degradation pathway of lignin. In their proposed mechanism, glucose oxidase operates as a feedback system where lignin and manganese peroxidase function as the initial lignin attacking enzymes, with laccase acting as the demethylating factor. Glucose, produced by cellulose and hemicellulose hydrolysis, becomes the substrate for glucose oxidase. Quinones, produced by laccase from lignin oligomers, can serve as a replacement for the O_2 required by glucose oxidase. D-Gluconolactone is formed as a result of glucose oxidation, which is used as a fungal metabolite. Hydrogen peroxide produced in the reaction catalyzed by glucose oxidase, in turn, activates LiP and MnP. Lignin exposed to the activated peroxidases undergoes decomposition into lower-molecular-weight fragments containing methoxyl groups. Laccase demethylates these oligomers while peroxidases degrade them further into even smaller fragments. Demethylation by laccase after the initial activity of the peroxidases is an important step, because it prevents the inhibition of the depolymerizing activity of LiP by the phenolic groups of native lignin (Ander and Marzullo, 1997). Generation of excess quinones and resulting repolymerization could be counterbalanced by glucose oxidase, which reduces them to the respective phenols. The phenols could then be utilized as substrates by microbes to form

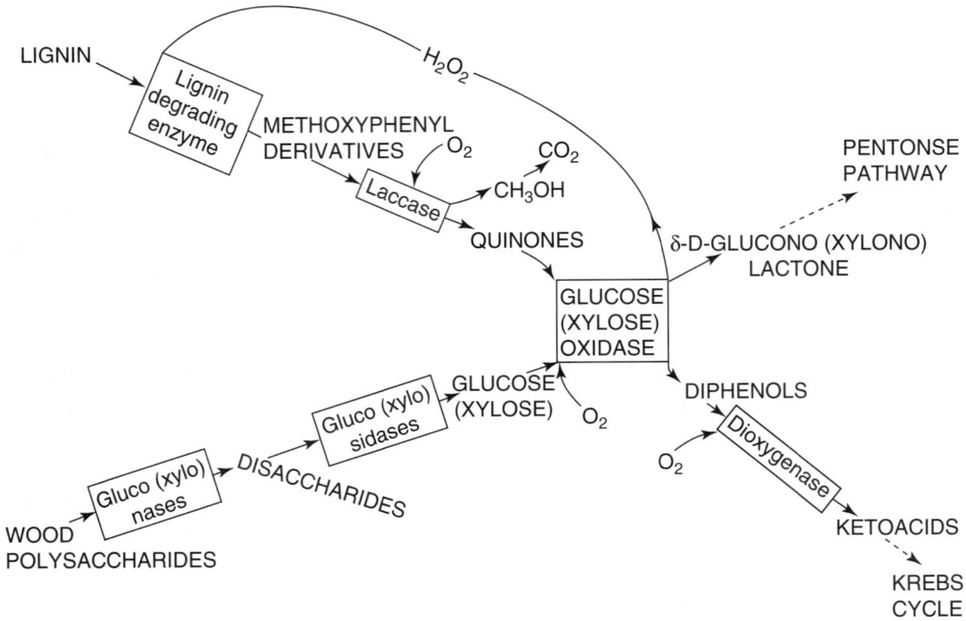

Figure 2.3. Hypothetical mechanism of ligninocellulose transformation by enzymes of white-rot fungi. The initial products of partial wood hydrolysis distinctly induce enzymatic systems accelerating the degradation processes. Reprinted from Leonowicz, A., Cho, N.-S., Luterek, J., et al. (2001). Fungal laccase: properties and activity on lignin. *J. Basic Microbiol.* **41**, 185–227, with permission from Wiley-VCH.

ketoacids via deoxygenases; however, it has been shown that less than 1% lignin carbon is incorporated into new biomass (Haider and Martin, 1981). The phenols can also become substrates for synthetic humification reactions (Bollag et al., 1998; Yavmetidinov et al., 2003; Telysheva et al., 2007). Polyphenols are the most reactive biomolecules in participating in polymerization and polycondensation humification reactions, especially in the presence of enzymes such as laccase and abiotic catalysts, such as Fe and Mn oxides (Bollag et al., 1998). Telysheva et al. (2007) studied the degradation of lignin in planted soils using a range of chemical and spectroscopic analytical techniques. They found that initially the carbohydrate component of the lignin was preferentially degraded, followed by the oxidative decomposition of the lignin moieties which involved removal of side chains (demethylation and demethoxylation). They also observed a progressive decrease in the acidity of lignin phenolic groups which was directly linked to increasing quinone and phenoxy radical formation and subsequent aromatization and condensation (polymerization) of the lignin structures during the degradation processes.

2.3.2.3. Physical and Chemical Protection. Soil mineral colloids, especially short-range ordered (SRO) Al and Fe (oxy)hydroxides, which are virtually noncrystalline, have the ability to complex with organic matter and control the turnover rate and storage of organic C in the soil and related environments (Huang, 1990; Torn et al., 1997; Huang et al., 2002; Rasmussen et al., 2005). Figure 2.4 (Torn et al., 1997)

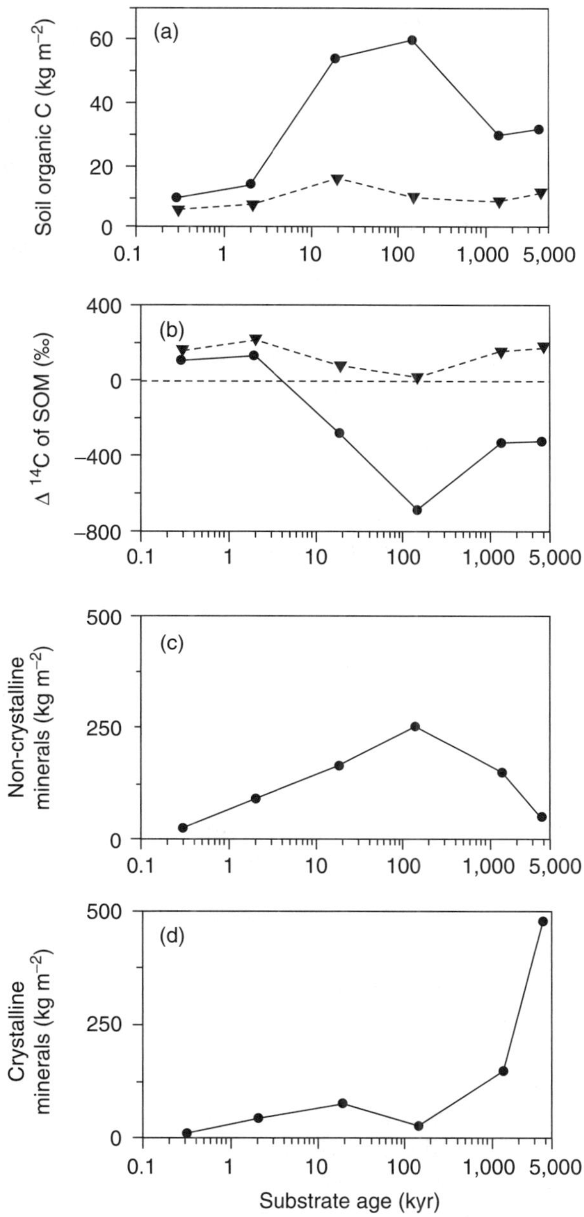

Figure 2.4. Soil inventory carbon in soil organic matter (SOM) **(a)**, $\Delta^{14}C$ of SOM **(b)**, noncrystalline minerals **(c)**, and crystalline minerals **(d)** versus age of soil substrate. Filled circles, total profile; filled triangles, surface (O and A) horizons. Reprinted from Torn, M. S., Trumbore, S. E., Chadwick, O. A., et al. (1997). Mineral control of soil organic carbon storage and turnover. *Nature* **289**, 170–173, with permission from Macmillan.

TABLE 2.3. Regression Analyses for Combined Andesite-Granite (AN) and Granite (GR) Data Explaining Total C Content (n = 6)[a]

Horizon	Equation	Adjusted R^2
A1[b]	—	
A2	$C = 6.83\text{Al}_p{}^c - 1.73$	0.80[d]
Bt1	$C = 1.54\text{Al}_o{}^e - 1.73$	0.89[f]
Bt2	$C = 1.64\text{Al}_o - 0.56$	0.998[f]
BC	$C = 1.66\text{Al}_p + 0.27$	0.99[f]

[a]All values are kg m^{-2}.
[b]No significant regression found with mineral variables.
[c]Al_p stands for pyrophosphate extractable Al.
[d]Significant at the probability level < 0.01.
[e]Al_o stands for oxalate extractable Al.
[f]Significant at the probability level < 0.001.

Source: Reprinted from Rasmussen, C., Torn, M. S., and Southard, R. J. (2005). Mineral assemblage and aggregation control carbon dynamics in a Californian conifer forest. *Soil Sci. Soc. Am. J.* **69**, 1711–1721, with permission from the Soil Science Society of America.

demonstrates the positive relationship between noncrystalline minerals and the turnover rate [$\Delta\,^{14}$C of SOM (‰)] and soil organic C storage. Rasmussen et al. (2005) reported that SRO Al mineral species variation explains nearly all of the variation in C contents in Bt horizons in a California Conifer forest soil (Table 2.3). This SRO Al consists of the precipitated Al–humus complexes and SRO–Al–OH species within interlayers and on edges and external planar surfaces of soil mineral colloids. Short-range ordered Al species possess considerable reactive surface area and microporosity that may contribute to adsorption of organic matter and physical protection within micropore structures (Huang et al., 2002; Yu et al., 2006) and also promote aggregation and occlusion of organic matter in aggregate structures (Baldock, 2002). The results of Rasmussen et al. (2005) are similar to those of Percival et al. (2000), who found that the pyrophosphate-extractable Al (Alp) is the best predictor of New Zealand C stocks in grassland soils and Veldkamp (1994), who found a high correlation between soil C mean residence time (MRT) and Alp and SRO aluminosilicates. Therefore, chemical protection of organic materials and physical protection of plant-like material within aggregates merit close attention in understanding the degradation of biological residues.

Lignin is relatively resistant to biodegradation (Rasse et al., 2006). Thus, it may stabilize reactive components such as cellulose, hemicellulose, or proteins by forming chemical or physical linkage to these molecules (Alexander, 1997). Soil humus polymers may anchor unstable plant constituents by various adsorptive forces or chemical binding (Allison, 1973; Bollag et al., 1998). Highly degradable proteins, for example, may be protected against rapid biodegradation by their nucleophilic addition to aromatic polymers through free NH_2 or SH groups and/or by adsorption on humus and mineral colloids (Huang, 1990; Bollag et al., 1998; Adani et al., 2006). As a result of the acquired stability, the plant components may not be completely mineralized during the initial rapid decomposition phase. Therefore, many relatively unstable plant constituents, such as polysaccharides, proteins, soluble sugars, and amino acids, can survive in the soil and related environments for a sufficient length of time in the humification process.

2.3.3. Decomposition of Organic Material by Fire and Charcoal Formation

Fire plays an important role in C cycling in many regions, particularly in the Boreal forest and peatlands (Preston and Schmidt, 2006). Fire converts biomass and detrital C mainly to gaseous C (predominantly CO_2), and converts 1–3% pyrogenic C, which is characterized by condensed aromatic structures. Fire affects soil organic matter (SOM) by decreasing the proportion of HA and increasing the insoluble humin fraction. There is loss of carbohydrates and O-containing functional groups and an increase in the aromatic C. The proportion of alkyl C may increase and is often associated with the formation of hydrophobicity in the upper soil (Preston and Schmidt, 2006). Fire also leads to a change in the N speciation of SOM, most notably a decrease in amide N (dominant species in fire unaffected soils) and an increase in heterocyclic N (Knicker et al., 2005). Generally speaking, fire causes an increase in the recalcitrant forms of C and N in soils.

Black C is defined as the finer fraction of pyrogenic C that exhibits chemical resistance to oxidation, because it is the most condensed (has relatively little or no O- and H-containing functional groups) (Preston and Schmidt, 2006). Black C can also be extracted by conventional methods used to isolate humic and fulvic acids from soils (Skjemstad et al., 1996). Ponomarenko and Anderson (2001) found that as much as 60% of the organic C fraction in Black Chernozem soils from Saskatchewan, Canada, was resistant to UV oxidation, which indicated a significant presence of black C. Their study highlighted the important contribution of black C to the formation of Black Chernozem soils and a need for a better understanding of the role of black C in carbon sequestration in these soils. Rumpel et al. (2006) showed that the black C content of tropical soils subjected to slash and burn treatment was positively correlated with the organic C contents, suggesting that black C is a component strongly influencing organic carbon sequestration in these types of tropical soils.

Black C has generally been considered inert because of its resistance to oxidation; however, it does eventually degrade, but much more slowly than other plant residues. It has been shown that with aging (decades to millennia), the surfaces of black C become increasingly oxidized, which increases potential for degradation by microorganisms (Schmidt et al., 2002). Bird et al. (1999) found that black C could be degraded in the range of tens to hundreds of years. A number of incubation studies have shown that black C can be microbially degraded at a relatively slow rate. Baldock and Smernick (2002) showed that 2% of wood char was mineralized after 120 days. Brodowski (2004) found that 5–50% of the black carbon in wheat and rice char was degraded within the first 6 months of incubation. Hamer et al. (2004) showed that the presence of a readily available source of C (glucose) significantly increased the rate of microbial degradation of maize char.

Shindo et al. (2004, 2005) recently suggested that charred plant remains serve as an important source of carbon for the formation of humic substances through oxidative degradative processes in Japanese volcanic ash soils, based on evidence of $\delta^{13}C$ values (Table 2.4). The formation of humic substances from the degradation of charred plant residues would be of particular significance in regions where annual

TABLE 2.4. Correlation of $\delta^{13}C$ Values of Charred Plant Fragments, Humic and Fulvic Acids, and Whole Soils

Whole Soil and Constituent	Charred Plant Fragments	Humic Acid	Fulvic Acid	Whole Soil
Charred plant fragments	1			
Humic acid	0.951[a]	1		
Fulvic acid	0.792[b]	0.740[c]	1	
Whole soil	0.968[a]	0.947[a]	0.900[a]	1

[a]Significant at 0.1% level.
[b]Significant at 1% level.
[c]Significant at 5% level.

Source: Reprinted from Shindo, H., Yoshida, M., Yamamoto, A., et al. (2005). $\delta^{13}C$ values of organic constituents in Japanese volcanic ash soils. *Soil Sci.* **170**, 175–182, with permission from Lippincott Williams & Wilkins.

burning of crop residues is practiced, or where frequent fires or volcanic activity occur.

2.4. PATHWAYS OF HUMIC SUBSTANCE FORMATION

2.4.1. Selective Preservation Pathways of Humification

The modified lignin pathway was one of the first theories of humus formation. Waksman (1936) popularized this theory, which proposes that humic acids resulted from the condensation of lignin with microbially produced protein (Figure 2.5). Waksman's theory was primarily based on the observation that (i) lignin is a recalcitrant fraction of plant residues that accumulates during the first stages of organic residue decomposition and (ii) lignin decomposition (as opposed to cellulose decomposition) gives rise to aromatic products, and humic acids also contain aromatic compounds. Hatcher and Spiker (1988) proposed a modified version of Waksman's lignin–protein theory, which included other refractory macromolecules such as cutin, suberin and microbial melanins. They also proposed the degradative formation of humic and fulvic acids from refractory biopolymers (humin), whereby increasing degradation of biopolymers leads to the formation of macromolecules enriched in carboxylic and phenolic functional groups. This increase in acidic functional groups promotes increased solubility in alkali and thus the development of humic acids and then fulvic acids. Bacteria and fungi, in particular white-rot fungi, are known to be able to effectively degrade humic substances (Haider and Martin, 1988; Gramss et al., 1999; Steffen et al., 2002; Granit et al., 2007). Gramss et al. (1999) found that large humic acid molecules were more readily degraded than the smaller fulvic acid molecules by the fungal species and eubacteria studied and that the humic substances served as a sole source of carbon and energy for the microorganisms in the systems they studied.

Selective preservation theories have been questioned by several authors (O'Brien and Stout, 1978; Nadelhoffer and Fry, 1988; Melillo et al., 1989) based on the fact that $\delta^{13}C$ values generally increase with depth in soils, whereas lignin and fatty acids

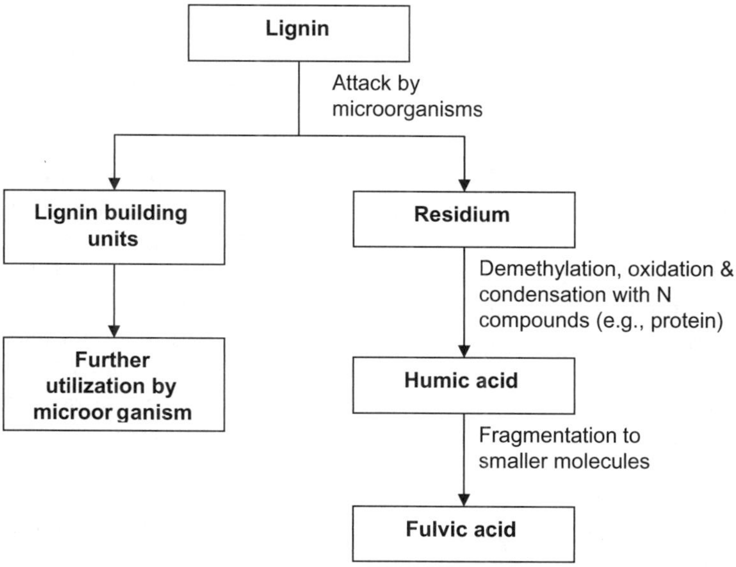

Figure 2.5. Schematic representation of the modified lignin theory of humus formation. Adapted with permission from Stevenson, F. J. (1994). *Humus Chemistry: Genesis, Composition, Reactions*, 2nd ed. John Wiley and Sons, New York.

and waxes have depleted $\delta^{13}C$ values relative to other plant components. Preston et al. (2006) further confirmed this hypothesis by demonstrating that increasing amounts of lignin-derived residues resulted in lower $\delta^{13}C$ values, by studying the lignin-rich residues produced by brown-rot fungi acting on wood debris. Although lignin is less easily attacked by microorganisms than other plant components, mechanisms do exist in nature for its complete aerobic breakdown; otherwise the earth would be deeply buried in undecomposed plant residues (Stevenson, 1994). Recent studies have shown that lignin is quickly degraded in the soil (Rasse et al., 2006). Plant lipids have also been shown to be rapidly degraded in cultivated soil; a significant portion of the stable lipids found in the soil originate from the microbial biomass rather than from plant-derived carbon (Poirer et al., 2006; Quénéa et al., 2006b). It has been suggested that the selective preservation pathway would play a more important role in humification processes in poorly drained soils and lake sediments, because of the lack of the fungi that are responsible for lignin degradation and the lack of oxygen which is necessary for this process (Stevenson, 1994).

2.4.1.1. The Lignin Theory Pathway. Lignin is one of the most abundant constituents of vascular plant tissues and has long been considered to be a major source of stable C in soils. As discussed above, Waksman's modified lignin pathway was one of the earliest theories of humic substance formation. Recent studies using pyrolysis- and TMAH thermochemolysis-GC/MS and multidimensional liquid NMR techniques have confirmed the presence of plant-derived lignin structures in bulk soils and humic acid fractions from agricultural and forest soils (Chefetz et al., 2002; Simpson, 2002; Simpson et al., 2003; Kelleher and Simpson, 2006). A number of studies have shown that lignin residues accumulate and are stabilized by covalent linking to humic substances during mineralization processes in soils (Haider et al.,

1977; Almendros et al., 2000; Tuomela et al., 2002). The polymerization of lignin residues to the existing humic acid fraction is promoted by oxidoreductase enzymes such as laccase, which are secreted by white-rot fungi when decomposing lignin (Dec and Bollag, 1990; Tuomela et al., 2002). It has also been suggested that humic polymers can physically encapsulate partially degraded biological residues (Bollag et al., 1998; Sutton and Sposito, 2005). The physical inaccessibility of the lignin residue accompanied by the inactivating effect of the presence of humic substances and minerals on extracellular enzymes can slow the rate of lignin degradation in soils (Tuomela et al., 2002; Allison, 2006b). Tuomela et al. (2002) observed that the presence of soil significantly decreased the ability of white-rot fungi species to degrade lignin. However, they found that the white-rot fungi were nonetheless able to degrade the ^{14}C-labeled lignin that was bound to the humic acid fraction of a soil. It is well established that bacteria and fungi can degrade humic substances (Gramss et al., 1999).

Adani et al. (2007) investigated the contribution of lignin to the formation of humic acid from Maize plants using pyrolysis GC/MS. They found that there was a substantial preservation of lignin in plant residues incubated with soil in the short term. However, they also observed the modification of the syringyl/guaiacyl ratio and oxidation of the side chains of lignin, which suggested a turnover of lignin-derived molecules in the soil humic acid fraction. Other studies have shown that lignin is degraded relatively quickly in the soil and does not appear to be stabilized in the soil in the long term (Baldock and Nelson, 2000; Kögel-Knabner, 2002; Kiem and Kögel-Knabner, 2003; Rasse et al., 2006). Rasse et al. (2006) estimated that the turnover rate for lignin in a temperate loamy soil was $1.9\,yr^{-1}$ (the turnover rate obeys the first-order rate equation). They also found that about 92% of the lignin was mineralized to CO_2 or assimilated to microbial C, thus only 8% was incorporated in the stabile fraction.

2.4.1.2. Preservation of Other Refractory Biologically Derived Polymers. Plant-derived polymers with polymethylenic structures such as lipids, waxes, cutin, suberin, and other microbially derived lipids are considered the fraction of organic C that is most resistant to degradation (Derenne and Largeau, 2001). It has been hypothesized that the hydrophobicity of these components prevents enzymes from directly interacting with them. Furthermore, clay stabilization of alkyl C components by either surface association or intercalation has also been suggested as a reason for the accumulation of alkyl C in finer fractions (von Lützow et al., 2006). The humin fraction of soil C, which usually accounts for more than 50%, is known to be rich in alkyl C (Rice, 2001). Quénéa et al. (2006b) studied the lipids in particle size fractions of previously forested soil that had been cultivated with maize for 22 years and compared it to a similar continuously forested soil from the same region. They found that microbial reworking of lipid components increased with decreasing particle size in the cultivated soil. They also found that the lipids of the cultivated soil were dominated by microbial lipids while the forest soil contained more plant derived lipids. This indicates that the persistence of plant lipids is also related to type of vegetation as well as management practices. In a related study of the same soils, Quénéa et al. (2006a) studied the so-called "refractory organic macromolecular" fraction of the soils, which is defined as the fraction which is resistant to drastic laboratory hydrolysis. They found that this fraction (20% of TOC) consisted of heterogeneous macromolecules that contained altered lignin, polysaccharides, suberans, bacterial and plant fatty acids, suberin and melanoidin-type components. Feng and Simpson (2007) studied the distribution of organic C in subsoil horizons in Alberta grasslands and found that aliphatic molecules derived from cutin and suberin were preferentially preserved in deeper horizons in comparison to lignin; they also found that trehalose, a fungal polysaccharide, was detected in significant abundance in the subsoil horizons. They concluded that non-plant biomass and eluviation strongly contributed to the composition of organic C in these subsoils. Similarly, Kiem and Kögel-Knabner (2003) showed that microbially produced polysaccharides were stabilized over the long-term within the fine fraction of arable soils, while lignin was not. It would appear that microbial resynthesis products persist for longer periods because they are closely associated with the fine fraction of soils and hence are physically protected.

2.4.2. Synthesis Pathways of Humification

There is a large volume of work documenting the polycondensation and polymerization of simple biomolecules (e.g., polyphenols, amino acids, and sugars), as catalyzed by enzymes and soil minerals, leading to the formation of humified substances (Bollag et al., 1998; Huang, 2000, 2004). Since these catalysts are ubiquitous in soil environments, and the substrate biomolecules are readily available from the continuous decomposition of organic residues as discussed in Section 2.3, it is certain that these reactions occur in the natural environment. Furthermore, these reactions are responsible for browning reaction of biomolecules observed during oxidative decomposition, and they provide an explanation for the dark color of humic substances in soils. The following section reviews the major synthetic pathways that have been investigated.

2.4.2.1. Polyphenol Pathway. According to polyphenol pathway humification theory, quinones of lignin and microbial origin are the major building blocks of humic substances. In this model, the first step is the breakdown of all plant biopolymers into their structural units, some of which polymerize enzymatically or by means of mineral colloid catalysis to form humic molecules of various complexities (Shindo and Huang, 1982; Wang et al., 1986; Stevenson, 1994; Bollag et al., 1998). For both biotic and abiotic catalysts, polyvalent metals such as Cu, Mn, and Fe facilitate the transformation of phenolic compounds by acting as electron acceptors. The order of formation of humic substances by the polyphenol pathway is: fulvic acid → humic acid → components of humin (Stevenson, 1994). The formation of brown-colored polymers by reactions involving quinones and amino acids and proteins is a well-known phenomenon that occurs in plants following mechanical injury or during disintegration of cells (Bittner, 2006) and is also observed when brown-rot fungi decompose wood and leave behind aromatic dihydroxy-rich residues (Filley et al., 2002). Polyphenols, originating from lignin, tannins, microorganisms, root exudates, glycosides, and anthropogenic contaminants, can be oxidized, by means of oxidoreductase enzymes or mineral colloids, to form highly reactive semiquinone free radicals and/or quinones, which readily participate in polymerization and/or polycondensation reactions with amino compounds to form humic substances (Wang et al., 1986; Stevenson, 1994).

Dec et al. (2003) investigated oxidative coupling (polymerization) reactions of a number of polyphenolic compounds, including vanillic acid, as catalyzed by the enzymes laccase and peroxidase, and birnessite (δ-MnO_2). The first step of the oxidative coupling reactions involves oxidation of susceptible polyphenols to form unstable free radicals. This involves the removal of a hydrogen ion and an electron from the hydroxyl group, generating an alkoxy free radical. The free radical intermediate then reacts with positions *ortho* and *para* to the hydroxyl group to form a dimer, and ultimately polymers. This reaction results in the formation of C–C and C–O bonds between phenolic species (Bollag et al., 1995). If the potential of the oxidant is high enough, C–C coupled dimers can be oxidized to form extended quinones. In the presence of certain enzymes and particularly strong oxidants such as Mn oxides, phenolic compounds can also be oxidized to the extent that ring cleavage occurs, allowing the cleavage products to be further degraded by enzymatic activity or mineral catalysis to CO_2 or incorporated into humic substance structures (Wang and Huang, 1992, 2000b, 2005; Majecher et al., 2000). Another effect of oxidative coupling is dehalogenation, decarboxylation, or demethoxylation of the phenolic substrates (Dec et al., 2003). This only occurs if the substituent is attached to a C atom involved in coupling. Electron-withdrawing substituents, such as –COOH and –Cl, are more susceptible to release than electron-donating ones such as OCH_3 and CH_3. The release of organic substituents during oxidative coupling reactions leads to the production of CO_2. Oxidative coupling and degradation are considered environmentally beneficial pathways, because they lead to detoxification of hazardous xenobiotic substrates.

Although self-condensation of quinones or free radicals can occur under soil conditions, these types of reactions are greatly enhanced in the presence of amino compounds, such as amino acids or amino sugars (Bittner, 2006). The condensation reaction between catechol and glycine is shown in Figure 2.6. The reaction of glycine with a quinone C=O group leads to the degradation of glycine and the formation

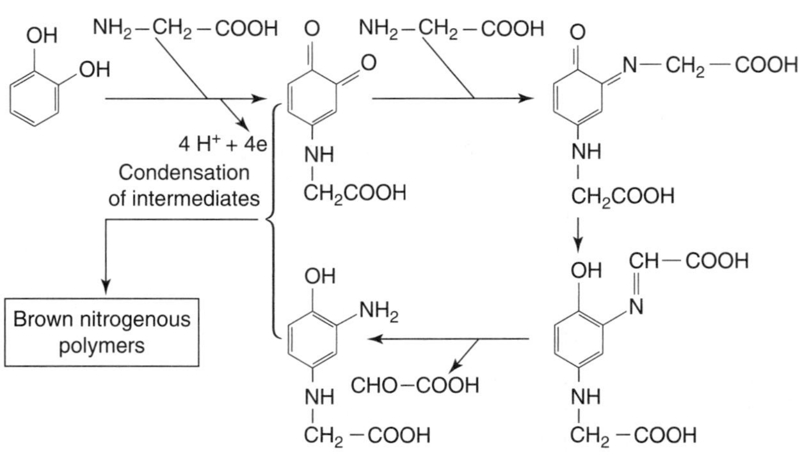

Figure 2.6. Formation of humic substances from quinones and amino acids, as illustrated by the reaction between catechol and glycine. Reprinted with permission from Stevenson, F. J. (1994). *Humus Chemistry: Genesis, Composition, Reactions*, 2nd ed., John Wiley and Sons, New York.

of an aryl amine. Thus, part of the amino acid–N becomes incorporated into the polymer. The condensation of hydroquinone and glycine to form nitrogen complexes is substantially enhanced by mineral colloids such as δ-MnO_2 in the pH range commonly found in the environment (Shindo and Huang, 1984b). The findings indicate that Mn oxides merit close attention in the abiotic formation of organic N complexes from nitrogenous substances and polyphenols and the subsequent turnover of N in the environment.

2.4.2.2. Maillard Reaction Pathway. The Maillard reaction (Maillard, 1913), involving condensation reactions between reducing sugars and amino acids, is considered to be an important pathway in natural humification (Ikan et al., 1996). Sugars and amino acids are among the most abundant constituents of terrestrial and aquatic environments (Anderson et al., 1989). The Maillard reaction consists of a cascade of complex pathways involving interactions between degradation products of the precursor sugars and amino acids, and reaction intermediates, known as Amadori and Heyns compounds (Yaylayan, 1997).

The initial reaction in the Maillard reaction involves condensation between the α-hydroxy carbonyl group of a reducing sugar and the amino group from an amino acid with the formation of a Schiff base which rearranges to form either Amadori or Heyns compounds (Figure 2.7). Aldohexoses generate Amadori compounds whereas ketohexoses generate Heyns products (Yaylayan, 1997). These compounds can undergo retroaldolization reactions forming α-dicarbonyl and α-hydroxyketone compounds (Figure 2.7). All of these compounds are highly reactive and readily

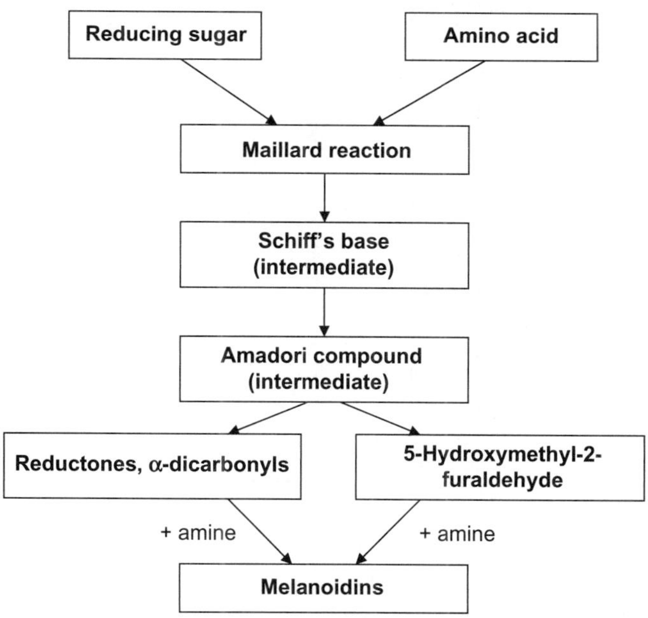

Figure 2.7. A schematic representation of the Maillard reaction. Adapted with permission from Ikan, R. Y., Rubinsztain, Y., Nissenbaum, A., and Kaplan, I. R. (1996). Geochemical aspects of the Maillard reaction. In *The Maillard Reaction: Consequences for the Chemical and Life Sciences*, Ikan, R., ed., John Wiley and Sons, Chichester, UK, 1–25.

polymerize in the presence of amino compounds to form brown-colored melanoidins. Amino acids can react with the α-dicarbonyl compounds, undergoing the Strecker degradation and then forming α-amino ketones. The α-amino ketones may then condense, resulting in the formation of pyrazines (Ho, 1996; Yaylayan, 1997). In addition, the Maillard reaction can result in the formation of polyphenols, such as catechol, resorcinol, hydroquinone, and pyrogallol, which can enhance the degree of browning during the Maillard reaction by affecting the redox potential of the system (Haffenden and Yaylayan, 2005) and undergoing polymerization and polycondensation (Wang and Huang, 2005).

There has been some recent criticism of the Maillard reaction as a possible humification pathway (Burdon, 2001; Sutton and Sposito, 2005; von Lützow et al., 2006). First, the critics argue that the Maillard reaction results in the formation of heterocyclic N, whereas soil N consists primarily of amide N based on ^{15}N CPMAS NMR (Knicker and Lüdemann, 1995; Knicker, 2004) studies. However, Jokic et al. (2004a) clearly showed, using N K-edge XANES, that the Maillard reaction catalyzed by birnessite under ambient temperature conditions and environmentally relevant pH not only produces heterocyclic N but also a significant amount of amide N (Figure 2.8).

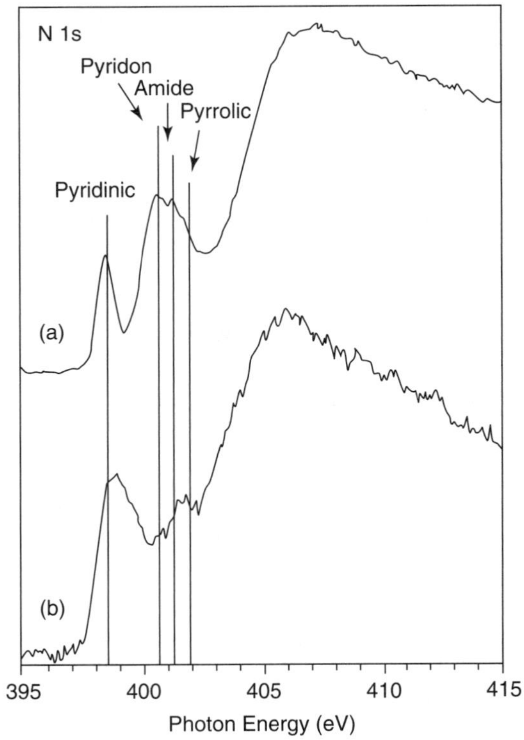

Figure 2.8. N 1s XANES spectra of (a) fulvic acid isolated from a glucose–glycine–δ-MnO_2 system and (b) the lyophilized solid phase. The peaks are assigned to pyridinic (398.6 eV), pyridone (400.7 eV), amide (401.3 eV), and pyrrolic (402.0 eV) moieties. Reprinted from Jokic, A., Schulten, H.-R., Cutler, J. N., et al. (2004). A significant abiotic pathway for the formation of unknown nitrogen in nature. *Geophys. Res. Lett.* **31**, L05502, with permission from the American Geophysical Union.

The spectrum of the fulvic acid is similar to that of asphaltene (Mitra-Kirtley et al., 1993), which has implications with regard to the formation of sedimentary organic matter such as HS, fossil fuels, and kerogen. Pyrrolic N and pyrinilic N are the dominant fractions in fossil fuels (Gorbaty and Kelemen, 2001). The action of birnessite on glucose should promote the autoxidation of glucose which results in the generation of reactive dicarbonyl compounds, hydrogen peroxide and hydroxylating agents (Wolff, 1996) (Figure 2.9). These reactive dicarbonyl compounds would then react with ammonia formed by the known deamination of glycine catalyzed by birnessite (Wang and Huang, 1987), as proposed by Vairavamurthy and Wang (2002), or with glycine (undergoing the Strecker degradation), resulting in either case in the formation of heterocyclic N compounds (Wong and Shibamoto, 1996), as illustrated in Figure 2.10. Ammonium ions are known to react with compounds containing reactive carboxyl groups to form pyridinic structures (Steelink, 1994). Carboxylic acids are produced during the Strecker degradation of amino acids (Wong and Shibamoto, 1996), or by the action of manganese dioxides on simple carbohydrates (Bose et al., 1959). When heated, carboxylic acids react with ammonia to form amides (Smith and March, 2001). The scheme for possible amide formation is shown in Figure 2.11.

The study of Jokic et al. (2004a) reveals that the Maillard (sugar–amino acid condensation) reaction, catalyzed by birnessite, is an abiotic pathway for the forma-

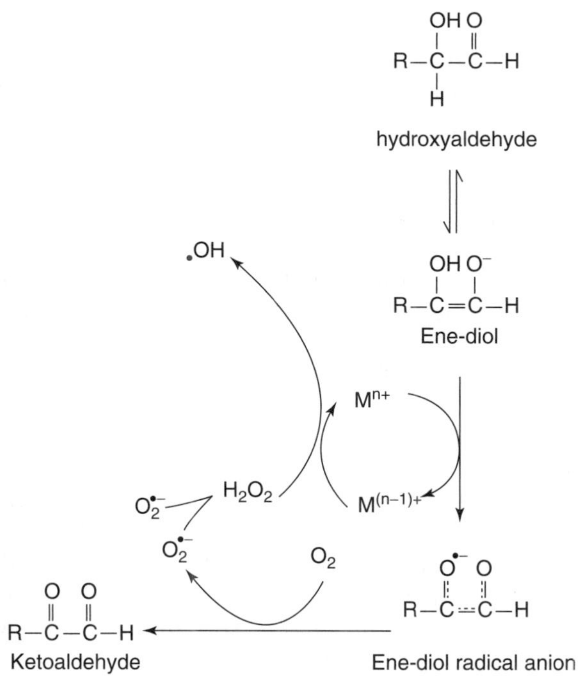

Figure 2.9. Glucose can enolize and reduce transition metals thereby generating superoxide free radicals ($O_2^{\bullet-}$), hydroxyl radicals (•OH), hydrogen peroxide (H_2O_2) and reactive dicarbonyl compounds. Adapted with permission from Wolff, S. P. (1996). Free radicals and glycation theory. In *The Maillard Reaction. Consequences for the Chemical and Life Sciences*, Ikan, R., ed., John Wiley & Sons, Chichester, UK, 73–88.

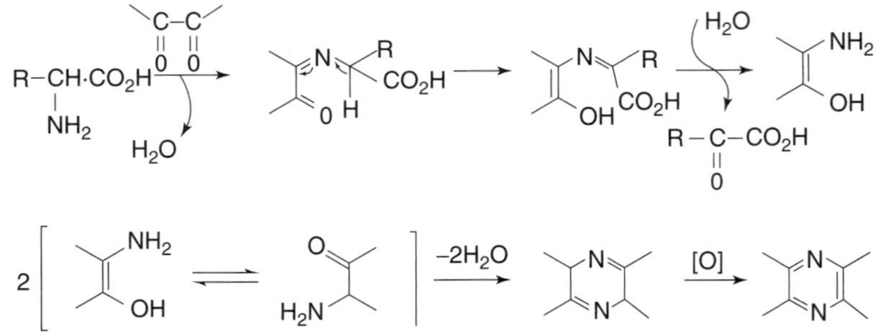

Figure 2.10. Strecker degradation of amino acids and α-dicarbonyls to form heterocyclic compounds. For glycine, R = H. Reprinted with permission from Wong, J. W., and Shibamoto, T. (1996). Geotoxicity of the Maillard reaction products. In *The Maillard Reaction. Consequences for the Chemical and Life Sciences*, Ikan, R., ed., John Wiley & Sons, Chichester, UK, 129–159.

(1) Oxidation of carbohydrate or Strecker aldehyde by δ-MnO_2 to form carboxylic acid, or formation of carboxylic acid during Strecker degradation

(2) Deamination of amino acid by δ-MnO_2

$$R-\overset{\displaystyle R}{\underset{\displaystyle NH_2}{\overset{|}{\underset{|}{C}}}}-COOH \xrightarrow{\delta\text{-}MnO_2} CO_2\uparrow + NH_3 + \text{other products}$$

(3) Amide formation

$$R\underset{\displaystyle O}{\underset{\|}{C}}OH + NH_3 \longrightarrow R\underset{\displaystyle O}{\underset{\|}{C}}NH_2 + H_2O$$

Figure 2.11. Possible amide formation pathway including the key role of MnO_2. Reprinted with permission from Jokic, A., Schulten, H.-R., Cutler, J. N., et al. (2005). Catalysis of the Maillard reaction by δ-MnO_2: A significant abiotic sorptive condensation pathway for the formation of refractory N-containing biogeomacromolecules in nature. In *Soil Abiotic and Botic Interactions and Impact on the Ecosystem and Human Welfare*, Huang, P. M., Violante, A., Bollag, J.-M., and Vityakon, P., eds., Science Publishers, Enfield, NJ, 127–152.

tion of biogeomacromolecules containing organic N originally derived from amino acids under ambient environmental conditions. Their data provided for the first time unequivocal evidence that the Maillard reaction catalyzed by birnessite, which is common in soil and sediment environments, produces (1) amides that are the dominant N types in humic substances, soils, and sediments and (2) heterocyclic N compounds that are often referred to as *unknown* N. In fact, the N XANES study by Vairavamurthy and Wang (2002) showed that heterocyclic N accounted for at least 20–30% of the total N in the humic substances investigated. Furthermore, it has been shown that ^{15}N CPMAS NMR is insensitive to detecting heterocyclic N. Smernik and Baldock (2005) showed that 20–50% of the soil N was not detected by this technique, which they stated belonged to unprotonated, heterocyclic N functional groups.

A second criticism of the Maillard reaction as a potential humification pathway is that it requires a high temperature and pH to proceed because it is highly unfavorable under ambient soil conditions. In order to elucidate some details of this process, Jokic et al. (2001c) applied molecular shape analysis to investigate the initial reaction between D-glucose and glycine to form the Amadori compound fructosylglycine. This initial part of the Maillard reaction is a complex one that involves an intermediate phase resulting in the formation of the Amadori compound and the splitting off of a molecule of water. The structure of the Amadori compound was optimized at a quantum mechanical level and its ground-state electron energy was calculated. Molecular isodensity contours (MIDCOs) and electron density contour surfaces of constant electron density were constructed for D-glucose, glycine, and fructosylglycine in order to study the steric conditions for the reaction. The calculations of Jokic et al. (2001c) showed that the Amadori compound and water on one hand and the separate entities D-glucose and glycine on the other hand are very similar to each other in terms of their ground-state energy. Their results indicated that the potential energy barrier of this reaction is high and therefore the reaction between D-glucose and glycine alone to form fructosylglycine is very slow at room temperature, which is in accord with the experimental observation (Jokic et al., 2001b). In order to decrease the energy barrier, Jokic et al. (2001b) used birnessite, which is commonly present in the environment, as catalyst to enhance the reaction rate. Their data showed that the presence of a redox reactive mineral, in this case birnessite, significantly accelerates the reaction by one to two orders of magnitude under environmentally relevant temperatures (25 °C and 45 °C) and a neutral pH (7.00).

A number of studies have detected Maillard reaction products (melanoidins) in refractory organic matter from natural environments—for example, from sediments from a west African upwelling (Zegouagh et al., 1999) and archeological plant remains (Evershed et al., 1997). Poirer et al. (2000, 2002) and Quénéa et al. (2006a) showed that the refractory organic matter isolated from different soils consists in part of cross-linked melanoidins poorly resolved by ^{13}C NMR spectroscopy. The presence of amide structures in soil as elucidated by ^{15}N NMR is ascribed to either preserved proteinacious structures or melanoidin-type macromolecules (Derenne and Largeau, 2001).

2.4.2.3. Integrated Polyphenol–Maillard Reaction Pathway. Jokic et al. (2004b) were the first to study an integrated Maillard reaction and polyphenol (glucose, glycine, and catechol) pathway of humification, using birnessite as catalyst. Their data showed that the ubiquitous soil mineral, birnessite, significantly accelerates humification processes in an integrated polyphenol–Maillard reaction system under ambient conditions. In nature it is unlikely that the Maillard reaction and polyphenol pathways occur separately, but rather interact closely, since sugars, amino acids, and polyphenols all coexist in soil solutions and natural waters. Jokic et al. (2004b) also showed that the integrated polyphenol–Maillard reaction system is more effective in generating humic polymers than the Maillard reaction alone. Hardie et al. (2007) studied the birnessite-catalyzed polyphenol–Maillard reaction pathway using two structurally different polyphenols, namely pyrogallol and resorcinol, and the Maillard reagents, glucose and glycine. They characterized the resultant reaction products using C K-edge and Mn L-edge NEXAFS spectroscopy. Their results clearly demonstrated that the structure of pyrogallol and resorcinol and their molar ratio to Maillard reagents in the polyphenol–Maillard reaction system significantly affect not only the humification processes but also the biomolecule-induced formation of inorganic C, namely rhodochrosite ($MnCO_3$). The integrated pyrogallol–Maillard reaction systems form the least $MnCO_3$ and are more enriched in Mn(II)-coprecipitated organic components in the solid phase than are the integrated resorcinol–Maillard reaction systems. They concluded that polyphenols with *ortho*-OH groups (pyrogallol) participate in direct electron transfer reactions more readily than polyphenols with *meta*-OH groups (resorcinol) and subsequently undergo oxidative polymerization and ring cleavage reactions to a greater extent, which results in the formation of polymers with a greater aliphatic character (Figure 2.12) and the suppression of $MnCO_3$ formation. They also compared (a) the humic acid fraction formed in the presence of birnessite from the Maillard reaction and (b) two polyphenol–Maillard reaction systems with natural soil and peat humic acid using C K-edge NEXAFS (Figure 2.12), and they found that these humic acids are basically similar, especially the humic acid from the Maillard reaction and pyrogallol–Maillard systems.

2.5. BIOTIC CATALYSIS OF SYNTHETIC HUMIFICATION PATHWAYS

2.5.1. Enzymes

Common extracellular oxidoreductase enzymes in soils, which catalyze the oxidative coupling of phenolic compounds derived from lignins, tannins, and plant and microbial metabolites, include (a) the phenoloxidases tyrosinase (*o*-diphenoloxidase) and laccase (*p*-diphenoloxidase) and (b) peroxidases (Sjoblad and Bollag, 1981). Phenoloxidase enzymes catalyze phenol oxidative coupling reactions in the presence of O_2 by radical formation. Peroxidases catalyze oxidative reactions in the presence of hydrogen peroxide. The oxidized products (quinones) can undergo nucleophilic addition with other quinones or free-NH_2 groups with the eventual production of humic acid-like polymers (Martin and Haider, 1980). Many researchers have studied the formation of humic-like substances from the enzymatic polymerization of phenolic compounds, such as catechol, pyrogallol, or ferulic acid (Martin and Haider, 1971; Ladd and Butler, 1975; Marthur and Schnitzer, 1978; Martin and Haider, 1980; Dec et al., 2001, 2003; Ahn et al., 2006). There has been a lot of recent interest in developing techniques for using these oxidoreductase enzymes as biocatalysts to aid in the breakdown of toxic anthropogenic phenolic compounds and incorporation of the reaction products into existing humus fractions of the soil, thus rendering them harmless (Durán and Esposito, 2000; Burton, 2003; Torres et al., 2003; Gianfreda and Rao, 2004).

Laccases (EC 1.10.3.2). Laccases are cuproproteins belonging to the small group of enzymes named blue oxidase enzymes. They possess four neighbor Cu atoms, which are distributed among different binding sites and are involved in either (a) electron capture or (b) binding with oxygen (Burton, 2003). They catalyze the oxidation of many organic substances, including phenols, diphenols, aminophenols, polyphenols, polyamines, and lignin-related molecules, with concomitant reduction of oxygen to

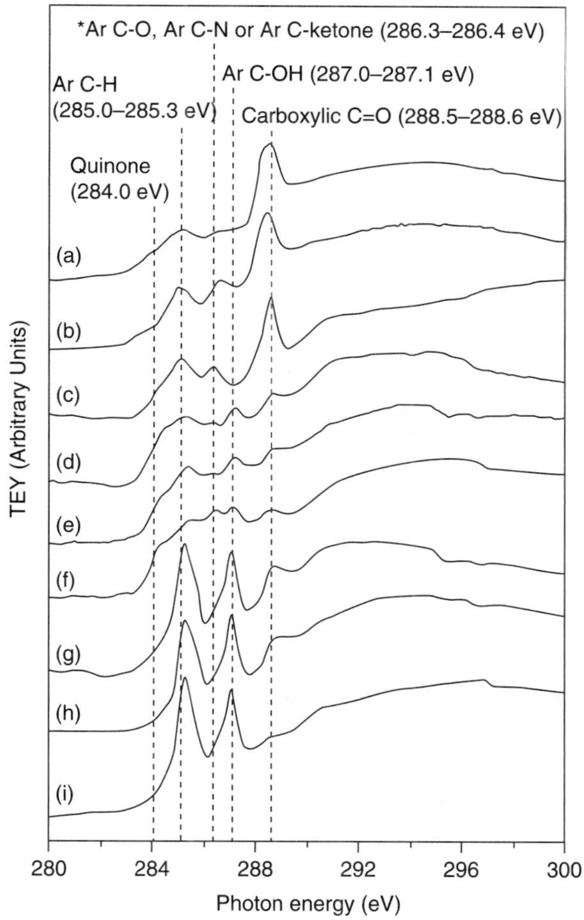

Figure 2.12. Carbon K-edge NEXAFS spectra of the IHSS **(a)** soil and **(b)** peat humic acids, and the humic acids extracted from the supernatants of the reaction systems catalyzed by birnessite: **(c)** Maillard reaction (50 mmole glucose + 50 mmole glycine); pyrogallol–Maillard reaction with **(d)** 50 mmole pyrogallol; **(e)** 100 mmole pyrogallol; **(f)** 50 mmole pyrogallol only; resorcinol–Maillard reaction with **(g)** 50 mmole resorcinol and **(h)** 100 mmole resorcinol; and **(i)** 50 mmole resorcinol only system. *Ar = aromatic. Reprinted from Hardie, A. G., Dynes, J. J., Kozak, L. M., and Huang, P. M. (2007). Influence of polyphenols on the integrated polyphenol-Maillard reaction humification pathway as catalyzed by birnessite. *Annals Env. Sci.* **1**, 91–110, with permission from Northeastern University, Boston, MA.

water. They also catalyze the demethylation of lignin, methoxyphenol acids, and methoxyaromatics, the polymerization of lignin monomers, and the copolymerization of lignin with phenols and acrylamide (Claus, 2004). It is produced by both plants and microorganisms, in particular by fungi. Their molecular masses typically vary between 40,000 and 140,000 Da, with laccases of fungal origin often being lower in molecular mass than those from plants. Laccases are capable of catalyzing the oxidation of *p*-diphenols as well as *o*-diphenols; this distinguishes them from tyrosinase, which can only react with *o*-diphenols (Burton, 2003; Claus, 2004; Baldrian, 2006). Chefetz et al. (1998) studied the interaction of laccase produced by thermophilic fungi (*C. thermophilium*) with a wide range of phenolic substrates and concluded that it could be involved in polymerization reactions that yield humic macromolecules. They hypothesized that during composting reactions, natural and xenobiotic phenols are oxidized by laccase present in the compost environment, resulting in the formation of free radicals that could be spontaneously bound to soluble high-molecular-weight compounds and results in the formation of humic macromolecules.

Tyrosinases (EC1.10.3.1). Tyrosinases are also known as polyphenol oxidases or catecholases. They have a coupled binuclear copper active site. They can catalyze the hydroxylation of monophenols with molecular oxygen to form *o*-biphenols (cresolase activity), as well as the oxidation of *o*-diphenols to *o*-quinones (catecholase activity) (Figure 2.13). These reactions may be separate or sequential. The highly reactive quinones generally undergo further non-enzymatic reactions resulting in the polymerization, and the subsequent formation of melanins. Tyrosinase is produced by plants, animals and microorganisms, in particular by fungi and bacteria (Durán and Esposito, 2000; Burton, 2003; Claus and Decker, 2006).

Peroxidases (EC 1.11.1.7). Peroxidases are hemoproteins, produced mainly by microorganisms and plants, which catalyze oxidation of the recalcitrant nonphenolic lignin units in the presence of hydrogen peroxide (Durán and Esposito, 2000). This is possible because of the formation of a high redox potential oxo-ferryl intermediate during the reaction of the heme cofactor with H_2O_2 (Martínez et al., 2005). Dubey et al. (1998) studied the polymerization of catechol by plant peroxidases and found that the resultant polymers consisted of phenylene and oxyphenylene units (Figure 2.14).

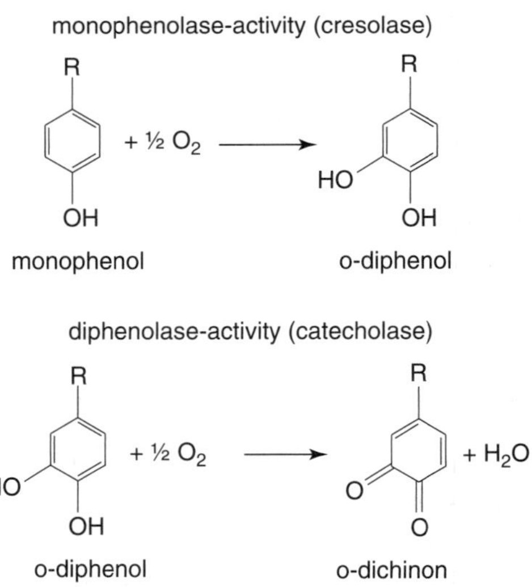

Figure 2.13. Enzymatic activities of tyrosinases. Reprinted from Claus, H., and Decker, H. (2006). Bacterial tyrosinases. *System. Appl. Microbiol.* **29**, 3–14, with permission from Elsevier.

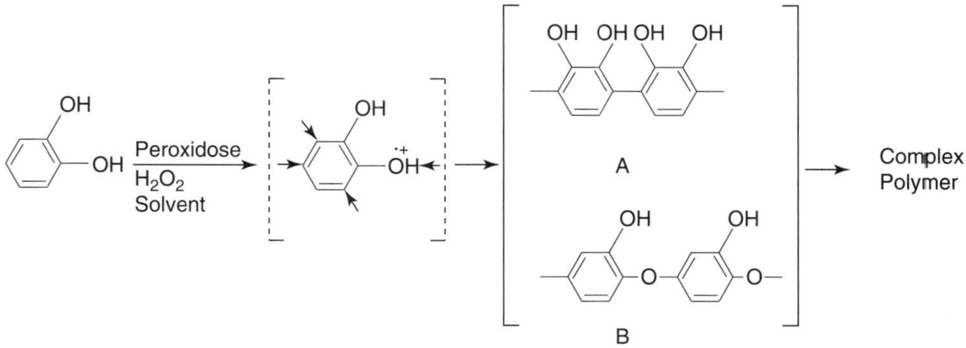

Figure 2.14. Possible reaction sequence of polymer formation from catechol by peroxidase. Reprinted from Dubey, S., Singh, D., and Misra, R. A. (1998). Enzymatic synthesis and various properties of poly(catechol). *Enzyme Microb. Technol.* **23**, 432–437, with permission from Elsevier.

2.5.2. Microorganisms

The production of humic substances by microorganisms is an extracellular process, because the enzymes are secreted into the external solution that contains the phenolic compounds derived from lignin and tannic acid degradation and microbial and plant metabolites. These phenolic compounds can then be enzymatically oxidized to quinones, which can undergo further polymerization or polycondensation reactions with other biomolecules (e.g., amino acids) to form humic polymers (Stevenson, 1994; Bollag et al., 1998; Burton, 2003).

Fungi are the most important group of organisms responsible for the cleavage of lignin. The majority of studies have focused on the basidiomycetes known as "white-rot fungi." White-rot fungi such as *Phanerochaete chrysosporium* and *Coriolus versicolor* are the most efficient ligninolytic organisms that have been found to date. T heir ability to degrade lignin and a wide variety of aromatic compounds is due to a nonspecific extracellular enzyme system, which involves lignin peroxidases, laccases, and manganese-dependent peroxidases, as well as hydrogen-producing oxidases which are also able to catalyze the oxidative polymerization of phenolic compounds (Sjoblad and Bollag, 1981; Lopez et al., 2006) and the polycondensation of phenolic compounds and amino acids (Martin and Haider, 1980). Yavmetidinov et al. (2003) studied the formation of humic-like substances produced by the white-rot fungi, *Coriolus hirsutus* and *Cerrena maxima*, grown on oat straw. They analyzed the humic substances formed using liquid-state ^{13}C NMR and FTIR and found that these humic substances closely resembled the spectra obtained from soil humic acids (Figure 2.15). Both these strains of fungi produce high quantities of the enzyme laccase; therefore, they concluded that these fungi play an important role in not only the degradation of lignin but also the formation of high-molecular-weight humic substances.

Martin and Haider (1971) concluded that microscopic fungi of the *Imperfecti* group play a significant role in the synthesis of humic substances in soil. Their studies showed that fungi such as *Hendersonula toruloidea*, *Epicoccum nigrum*, *Stachybotrys atra*, *Stachysbotrys chartarum*, and *aspergillus sydowi* degrade lignin, as well as cellulose and other organic plant constituents and in the process synthesize

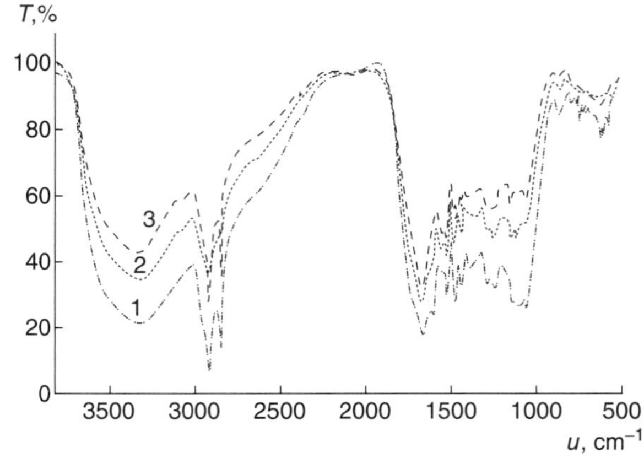

Figure 2.15. FTIR spectra of the humic-like substances produced by (1) *C. maxima*, (2) *C. maxima* + *C. hirsutus*, and (3) *C. hirsutus*. Reprinted from Yavmetidinov, I. S., Stepnova, E. V., Gavrilova, V. P., et al. (2003). Isolation and characterization of humin-like substances produced by wood-degrading white rot fungi. *Appl. Biochem. Microbiol.* **39**, 257–264, with permission from Springer.

humic-acid like polymers. The initial degradation of lignin by fungi involves the release of the primary phenylpropane lignin structural units, such as coniferaldehyde, *p*-hydroxybenzaldehyde and synapylaldehyde. These phenylpropane units are then further degraded into polyphenols by the fungal enzymes (Stevenson, 1994).

Phenoloxidase activity has recently been found in lichenized ascomycetes belonging to a variety of taxonomic groups (Zavarzina and Zavarzin, 2006). The researchers concluded that the oxidases discovered may play an important role in the phenolic metabolism of lichens and be involved in the formation of humus during primary soil formation processes, which may be a previously unknown geochemical function of lichens.

Microorganisms can also synthesize polyphenols which can contribute to humus formation in natural environments, as lignin is not present in all environments. Numerous phenolic and hydroxyl aromatic acids are synthesized by microorganisms from nonaromatic C sources, in particular by actinomycetes and fungi (Stevenson, 1994). Martin and Haider (1971) reviewed the studies which investigated the synthesis of humic acid-like substances by fungi of the *Imperfecti* group. These microscopic fungi can degrade cellulose and other organic constituents besides lignin and in the process synthesize dark-colored melanins from the phenols synthesized by these fungi. One interesting feature is the occurrence of resorcinol and resorcinol-type constituents which are not found in lignin transformation products.

2.6. ABIOTIC CATALYSIS OF SYNTHETIC HUMIFICATION PATHWAYS

The classical definition of a catalyst, according to Oswald, is a substance that increases the rate of a chemical reaction without undergoing any permanent chemical alteration itself. In reality the properties of catalysts do change with use (Twigg, 1989). Actually many substances are destroyed either as a result of the process that gives them their catalytic activity or because of subsequent combination with the products (Moore and Pearson, 1981). From a practical point of view, a catalyst is a substance that changes the rate of the desired reaction, regardless of the fate of the catalyst itself.

Metal oxides, clay minerals and dissolved metals have been shown to catalyze the transformations of natural and synthetic organic compounds, and inorganic substances (Huang, 2000). Many metal ions, especially transition metals, have several oxidation states which enables them to act as catalysts in certain redox reactions. Iron oxides and especially Mn oxides are the most reactive in facilitating transformations of organic compounds. This includes catalysis of the ring cleavage of polyphenols, the deamination, decarboxylation, and dealkylation of amino acids, the polymerization of phenolic compounds and their polycondensation with amino acids, and the Maillard reaction (Shindo and Huang, 1982, 1984a, 1984b; Huang, 2000; Wang and Huang, 2000b, 2003, 2005; Jokic et al., 2004b). Table 2.5 provides a complete summary of all the work that has been conducted on mineral catalysis of humification reactions.

TABLE 2.5. Summary of Research Conducted on Mineral Catalysis of Abiotic Humification Reactions

Inorganic Components as Catalysts	Observations	References
	Abiotic Polymerization of Polyphenols	
Oxides, Oxyhydroxides And Short-Range Ordered Minerals		
Hydrohematite, goethite, hematite, and lepidocrocite	Polymerization of hydroquinone (measurement of optical density)	Scheffer et al. (1959)
Silica gel and disturbed quartz surface	Polymerization of hydroquinone, catechol and pyrogallol (measurement of optical density)	Ziechmann (1959)
Allophane	IR spectra of humified polyphenols; measurement of O_2 uptake of pyrogallol solution	Kyuma and Kawaguchi (1964)
SiO_2, Al_2O_3, and silicoalumina	IR spectra resembling natural humic substances; measurement of O_2 uptake	Wang et al. (1983b)
Short-range ordered Fe(III) oxides	Polymerization of phenolic compounds (measurement of optical density); yields of humic polymers	Shindo and Huang (1984a)
Birnessite (δ-MnO_2)	Polymerization of phenolic compounds (measurement of optical density); Δ log K and RF values; yields of humic polymers	Shindo and Huang (1982, 1984a)
Birnessite (δ-MnO_2) and Fe(III) oxides	Measurement of O_2 uptake; ESR determination of free radical content; FTIR study of mineral/organic polymer complexes	McBride (1987)

TABLE 2.5. *Continued*

Inorganic Components as Catalysts	Observations	References
Birnessite (δ-MnO_2)	Polymerization of pyrogallol (measurement of optical density); ring cleavage of pyrogallol and catechol (measurement of CO_2 release); yields of humic polymers; IR and ^{13}C CPMAS NMR spectra resembling natural HAs	Wang and Huang (1992); Majecher et al. (2000)
Birnessite (δ-MnO_2)	Comparison of reaction mechanisms and products of transformation of catechol by biotic (tyrosinase) and abiotic (birnessite) catalysts	Pal et al. (1994); Nadja et al. (1998, 1999)
Birnessite (δ-MnO_2)	Investigation of the effect of light on ring cleavage of phenolics (measurement of CO_2 release) and abiotic humification	Lee and Huang (1995)
Short-range ordered Mn(IV), Fe(III), Al and Si oxides	Polymerization of pyrogallol (measurement of optical density); ring cleavage of pyrogallol (measurement of CO_2 release); yields of humic polymers; X-ray diffractograms, IR and ^{13}C CPMAS NMR spectra resembling natural HAs	Shindo (1992); Wang and Huang (2000a, 2000b)
Silicic acid, hydroxy-Al ions, and hydroxy-aluminosilicate ions	Polymerization of catechol (measurement of optical density); X-ray powder diffractograms, electron micrographs and FTIR and ^{13}C CPMAS NMR spectra resembling natural HAs	Liu and Huang (2000, 2002)
Clay-Size Layer Silicates		
Halloysite, Al-vermiculite, montmorillonite, illite, and kaolinite	Polymerization of pyrogallol (measurement of optical density)	Kumada and Kato (1970)
Smectites	Polymerization of hydroquinone	Thompson and Moll (1973)
Montmorillonite, illite, and kaolinite	Polymerization of phenolic compounds (measurement of optical density); IR spectra resembling natural humic substances; yields of humic polymers; measurement of O_2 uptake	Filip et al. (1977); Wang and Li (1977); Wang et al. (1978a, 1978b)
Nontronite, bentonite, kaolinite and quartz	Polymerization of hydroquinone and pyrogallol (measurement of optical density); intercalation of humic macromolecules in nontronite (XRD investigation into change in basal d-spacings); IR, ESR and ^{13}C CPMAS NMR spectra resembling natural humic substances	Wang and Huang (1986, 1989b, 1989c)
Montmorillonite	Oxidation and polymerization of catechol, pyrogallol and 2,6-dimethylphenol (FTIR spectra); SEM coupled with energy dispersive X-ray spectrometry investigation of reaction products on surface of clay minerals, ^{13}C NMR, MALDI MS study of reaction products	Birkel et al. (2002)

TABLE 2.5. *Continued*

Inorganic Components as Catalysts	Observations	References
Primary Minerals		
Olivines, pyroxenes, amphiboles, micas and feldspars	Polymerization of phenolic compounds (measurement of optical density); SEM micrographs and IR spectra resembling natural HAs; yields of humic acids; measurement of O_2 uptake	Shindo and Huang (1985a)
Natural Soils		
Oxisol, Inceptisol and silt fraction of Mollisol	IR spectra resembling natural humic substances; yields of humic polymers; ESR spectra which indicate presence of stabilized free radicals (semiquinones) and resemble those of natural HAs; measurement of O_2 uptake	Wang et al. (1978b, 1978c, 1983a)
Forest soils (Alfisol and Inceptisol)	Polymerization of phenolic acids (measurement of optical density); investigated rate of Mn(II) and Al dissolution	Pohlman and McColl (1989)
Mollisol	Polymerization of pyrogallol (measurement of optical density); IR and ESR spectra of humic macromolecules resembling natural substances; measurement of CO_2 release	Wang and Huang (1989a)
	Abiotic Copolymerization of Amino Acids and Polyphenols	
Birnessite (δ-MnO_2)	Polycondensation of glycine and pyrogallol/hydroquinone (measurement of optical density); IR and ESR spectra of N-polymers resembling natural humic substances; yields of humic substances; measurement of O_2 uptake; decarboxylation and dealkylation of glycine	Shindo and Huang (1984b); Wang and Huang (1987, 2005); Wang and Lin (1993)
Nontronite	Polycondensation of glycine and pyrogallol (measurement of optical density); IR and ESR spectra of N-polymers resembling natural humic substances; yields of humic substances	Wang and Huang (1991)
Oxisol and Mollisol clay fractions	Polycondensation of glycine and pyrogallol (measurement of optical density); IR and ESR spectra of N-polymers resembling natural humic substances; yields of humic substances; measurement of CO_2 and NH_3 released	Wang and Huang (2003)
	Maillard Reaction	
Absence of catalysis	Pioneering work on condensation reactions between sugars and amino acids to form melanoidins in the absence of catalysts	Maillard (1913)
Absence of catalysis	Condensation reaction between basic amino acids and sugars in the absence of catalysts results in N-rich polymers similar to humic substances in marine environments	Hedges (1978)

TABLE 2.5. ***Continued***

Inorganic Components as Catalysts	Observations	References
Oxides, Oxyhydroxides and Short-Range Ordered Minerals		
Birnessite (δ-MnO_2)	Condensation of glucose and glycine under soil ambient conditions (measurement of optical density); yields of humic substances; XANES study of change in speciation of Mn; ESR study of Mn speciation in solution; ^{13}C CPMAS NMR spectra of FA fraction resembling spectra of natural FAs	Jokic et al. (2001b)
Birnessite (δ-MnO_2)	Investigation of the effect of light on Maillard reaction—condensation of glucose and glycine (measurement of optical density); yields of humic substances; EPR and XANES study of change in speciation of Mn; FTIR study of solid residues	Jokic et al. (2001a)
Birnessite (δ-MnO_2)	Condensation of glucose and glycine (measurement of optical density); yields of humic substances; XANES, Py-FIMS, and Curie-point Py-GC/MS studies of solid residue show that amide and heterocyclic N compounds are formed (dominant N forms in natural humic substances)	Jokic et al. (2004a)
Goethites	Condensation of arginine and glucose at 37 °C; C and N fractions yields; FTIR spectra of solid residue and supernatant humic substances	Gonzalez and Laird (2004)
Clay-Size Layer Silicates		
Montmorillonite, nontronite, kaolinite and quartz (saturated with Cu^{2+}, Ca^{2+}, Al^{3+})	Condensation of glucose and tyrosine/glycine/ tryptophan at 70 °C: FTIR spectra resembling natural humic substances; yields of humic substances; XRD investigation into change in d-spacings	Arafaioli et al. (1997, 1999); Bosetto et al. (1995, 1997, 2002)
Smectites	Condensation of arginine and glucose at 37 °C; C and N fractions yields; FTIR spectra of solid residue and supernatant humic substances; XRD investigation into change in d-spacings	Gonzalez and Laird (2004)
	Integrated Abiotic Humification Pathway	
Birnessite (δ-MnO_2)	Integration of abiotic polyphenol and Maillard reaction pathways—polycondensation of catechol/pyrogallol/ resorcinol-glucose-glycine (measurement of optical density); yields of HA; XANES study of change in speciation of Mn; ^{1}H NMR and ^{13}C CPMAS NMR spectra of HA fraction; atomic force micrographs resembling natural HAs; C K-edge NEXAFS study of HA	Jokic et al. (2004b); Hardie et al. (2007)

Source: Modified with permission from Wang, T. S. C., Huang, P. M., Chou, C.-H., and Chen, J.-H. (1986). The role of soil minerals in abiotic polymerization of phenolic compounds and formation of humic substances. In *Interactions of Soil Minerals with Natural Organics and Microbes,* Huang, P. M., and Schnitzer, M., eds., SSSA Special Publ. No. 17, Soil Science Society of America, Madison, WI, 251–281.

2.6.1. Oxides, Oxyhydroxides, and Short-Range Ordered Minerals

The surfaces of metal oxides are very reactive in promoting the polymerization of phenolic compounds. They can act as Lewis acids by accepting electrons from hydroxyphenolics, leading to the formation of highly reactive semiquinone radicals that readily undergo coupling reactions with other semiquinones, phenolics, or existing humus (Dec and Bollag, 2000; Huang, 2000). Shindo and Huang (1984a) investigated the catalytic ability of short-ranged ordered Mn, Fe, Al, and Si oxides in the oxidative polymerization of catechol, resorcinol, and hydroquinone. They determined the degree of polymerization (browning) by measuring the visible absorbances (at 600 nm) of the supernatants from the various reactions systems. They showed that the visible absorbance was directly correlated with the yield of humic polymers. They found that Mn oxides (birnessite, cryptomelane, pyrolusite) were the most powerful catalysts of oxidative polymerization reactions, compared to Fe, Al, and Si oxides (Figure 2.16). Wang and Huang (2000a) characterized the pyrogallol-derived polymers formed by the catalysis of short-ranged ordered Mn, Fe, Al, and Si oxides. They showed that the infrared spectra of the FA fraction from the Mn oxide–pyrogallol system closely resembled that of the FA fraction extracted from a natural Borosaprist soil (Figure 2.17).

The catalytic effectiveness of a metal ion depends on its ability to complex with ligands and shift electron density and molecular conformation in ways favorable for the reaction (Hoffmann, 1980; Huang, 2000). The Mn oxide has the highest catalytic power in promoting catechol humification compared with Fe and Al oxides (Liu and Huang, 2001). This is, in part, attributable to the lower electronegativity of Mn. The electronegativity values of Mn, Fe, Al, H, and O are, respectively, 1.55, 1.83, 1.61, 2.20, and 3.44 (Porterfield, 1983). Catechol acts as a hard Lewis base and Al, Fe, and Mn are hard Lewis acids. When Al, Fe, or Mn replaces H in catechol to form metal–catechol complexes, the electron cloud delocalizes from phenolic oxygen into the π-orbital formed from overlap of the $2p$ orbitals of the aromatic C atoms, thus accelerating the formation of semiquinone free radicals and their coupling to polycondensates. The electron cloud around the Mn–O bond in the Mn oxide–phenolic complex should be more delocalized than that around the Al–O bond in the Al–catechol complex and especially the Fe–O bond in the Fe oxide–phenolic complex due to the lower electronegativity of Mn than those of Al and Fe. This partially explains the greater accelerating effect of Mn oxide on the humification of catechol than Fe and Al oxides. Redox reactions also play an important role in many abiotic catalyses (Huang, 2000). Aluminum oxide is not subject to redox reaction. The standard electrode potential ($E°$) values of the overall redox reaction of the Fe(III) oxide–catechol and Mn(IV) oxide–catechol systems are +0.071 V and +0.509 V, respectively, as indicated by the following reactions (Shindo and Huang, 1984a):

$$Fe^{3+} + e^- = Fe^{2+} \qquad E° = 0.770\ V \qquad (2.2)$$

$$MnO_2 + 4H^+ + 2e^- = Mn^{2+} + 2H_2O \qquad E° = 1.208\ V \qquad (2.3)$$

$$C_6H_4(OH)_2 = C_6H_4O_2 + 2H^+ + 2e^- \qquad E° = -0.6992 \qquad (2.4)$$

The positive $E°$ values of the overall redox reactions indicate that the reactions are thermodynamically feasible, and catechol oxidation can thus be accelerated by Fe

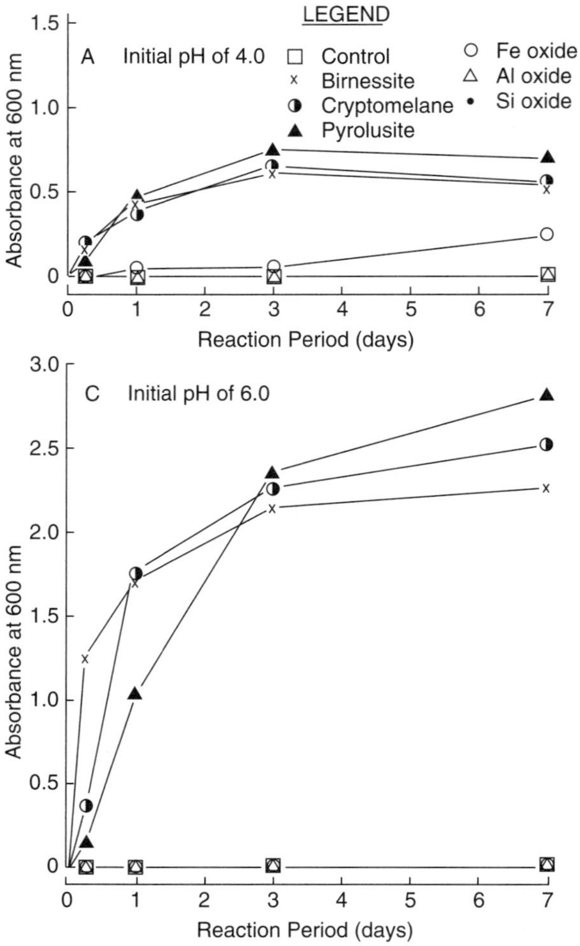

Figure 2.16. Changes in the degree of darkening of hydroquinone solution at pH 4.0 and 6.0 as influenced by various oxides as a function of time Reprinted from Shindo, H., and Huang, P. M. (1984). Catalytic effects of manganese(IV), iron(III), aluminum, and silicon oxides on the formation of phenolic polymers. *Soil Sci. Soc. Am. J.* **48**, 927–934, with permission from the Soil Science Society of America.

oxide and especially Mn oxide. This also explains the stronger catalytic ability of the Mn oxide, when compared to that of Fe and Al oxides, in accelerating catechol oxidation. In addition, the lower point of zero salt effect (PZSE) and more negative charges of the Mn oxide than the Fe and Al oxides could also enhance the oxidation of catechol (Liu and Huang, 2001). More negative charges of the Mn oxide may favor the binding of protons released from catechol [Eq. (2.4)] and subsequently increase the catalytic reaction rate. Therefore, the catalytic ability of a metal oxide in polyphenol transformation depends on the E° value of the overall redox reaction and the ability of the metal ions to complex with ligands, to shift electron density and molecular confrontation in the way conducive to the reaction, and to favor the binding of protons to the metal oxide.

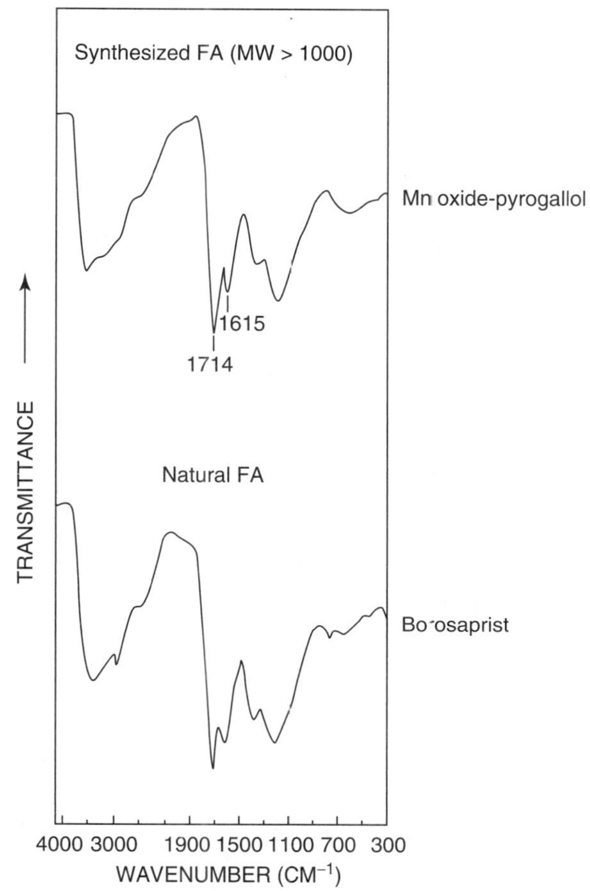

Figure 2.17. Infrared spectra of the synthesized FA (MW > 1000 Da) in the Mn(IV) oxide-pyrogallol system and the FA extracted from a Borosaprist (Terric Humisol). Reprinted from Wang, M. C., and Huang, P. M. (2000). Characteristics of pyrogallol-derived polymers formed by catalysis of oxides. *Soil Sci.* **165**, 737–747, with permission from Lippincott Williams & Wilkins.

It has also been shown that there is a concomitant mechanism resulting in aromatic ring cleavage and the release of CO_2 during metal oxide catalysis of polyphenol polymerization reactions (Wang and Huang, 1994, 2000b; Lee and Huang, 1995; Majecher et al., 2000). Wang and Huang (2000b) studied the ring cleavage and oxidative polymerization of pyrogallol by short-range ordered Mn, Fe, Al, and Si oxides under aerobic and anaerobic conditions. They found that the presence of oxygen significantly enhanced oxide-catalyzed ring cleavage of pyrogallol and the resultant release of CO_2 (Table 2.6). The semiquinone free radicals formed appear to be partially transformed, through ring cleavage, to aliphatic fragments, resulting in the development of carboxyl groups and subsequent decarboxylation and CO_2 release (Liu and Huang, 2001). The ability of short-range ordered oxides of Al and Si, and especially Fe and Mn, in enhancing the formation of aliphatic and carboxylic groups in humic polymers derived from polyphenol humification may, in part, contribute to the high aliphaticity of humic substances in soils (Schnitzer, 1977; Wilson

TABLE 2.6. Release of CO_2 in Oxide–Pyrogallol Systems in Air or N_2 Atmosphere at the End of 90-h Reaction Period

	CO_2 release (μmol)[a]		
Oxide	Air	N_2	$LSD_{0.05}$
Manganese	463.8 ± 0.3[b]	13.7 ± 0.1	1.4
Iron	124 ± 2	21.4 ± 0.1	8.6
Aluminum	52.7 ± 0.4	18.4 ± 0.3	2.2
Silicon	28.4 ± 0.0	2.7 ± 0.0	4.3
No catalyst	24.5 ± 0.3	0	1.3
$LSD_{0.05}$	3.4	0.5	

[a]The amounts of CO_2 released in the systems, which contained 100 mg of oxide (0.2–2 μm) and 0.5 mmol of pyrogallol in 30 mL of sterilized aqueous solution, adjusted to pH 6.00.
[b]The average deviation from the mean.

Source: Reprinted from Wang, M. C., and Huang, P. M. (2000). Ring cleavage and oxidative transformation of pyrogallol catalyzed by Mn, Fe, Al, and Si, oxides. *Soil Sci.* **165**, 934–942, with permission from Lippincott Williams & Wilkins.

and Goh, 1977; Hatcher et al., 1981; Preston et al., 1982) and the variation in their characteristics, including carboxyl group content, with soils from different climatic zones (Schnitzer, 1977).

Liu and Huang (2000) reported that silicic acid and especially hydroxy-Al ions substantially enhance oxidative polymerization of catechol. Liu and Huang (2002) showed that hydroxy-aluminosilicate ions, which are precursors to noncrystalline aluminosilicates, are also effective in promoting the oxidative polymerization of catechol. A proposed mechanism for the catalytic effect of hydroxy-aluminosilicate ion on the humification of catechol is depicted in Figure 2.18. Catechol acts as a hard Lewis base, and Al and Si of hydroxy-aluminosilicate ions are hard Lewis acids. Hydroxy-aluminosilicate ions can complex with catechol through Al–O and Si–O bond formation. The electronegativity values of Al, Si, H, and O are, respectively, 1.61, 1.90, 2.20, and 3.44 (Hueey, 1983). Therefore, when Al or Si replaces H in catechol, the electron cloud delocalizes around the Al–O and Si–O bonds from phenolic oxygen into the π-orbital bonding formed from the overlaps between the 2*p* orbitals of the carbon atoms of the aromatic ring, thus apparently accelerating the formation of semiquinone free radicals and their coupling to form polycondensates. The semiquinone free radicals formed appear to be partially transformed, through ring cleavage, to aliphatic fragments, resulting in the development of carboxyl groups and subsequent decarboxylation and CO_2 release.

The structure and functionality of polyphenolics also affect the oxidative polymerization catalytic ability of soil mineral oxides. Pohlman and McColl (1989) demonstrated that polyhydroxyphenolic acids with *p*- or *o*-OH groups are more readily oxidized to polymeric humic products by Mn oxides than are *m*-polyhydroxyphenolic acids. Similarly, Shindo and Huang (1992) investigated the oxidative polymerization of various diphenols by Mn oxide and found that HA yields increase in the following order: hydroquinone (*p*-OH group) > catechol (*o*-OH group) > resorcinol (*m*-OH group). Birnessite (δ-MnO_2) has been shown to catalyze polycondensation reactions between amino acids and phenolic compounds in the abiotic formation of organic N complexes (Shindo and Huang, 1984b). Birnessite is able to promote the

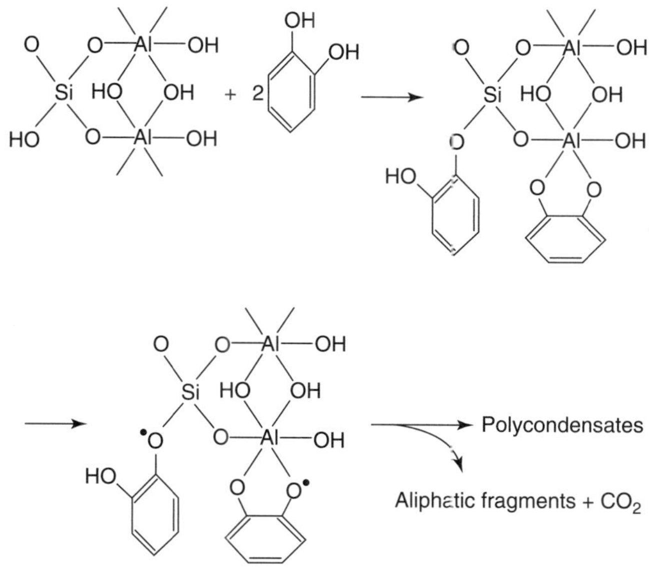

Figure 2.18. Proposed mechanism for the catalysis of hydroxylaluminosilicate ions in catechol humification. Reprinted from Liu, C., and Huang, P. M. (2002). Role of hydroxylaluminosilicate ions (proto-imogolite soil) in the formation of humic substances. *Org. Geochem.* **33**, 295–305, with permission from Elsevier.

deamination and decarboxylation (Wang and Huang, 1987) and dealkylation (Wang and Huang, 1997) of glycine. Wang and Huang (2005) showed that birnessite promotes the incorporation of carboxyl, and especially alkyl C of glycine, into the polycondensates formed with pyrogallol.

Commonly found soil metal oxides, birnessite (Jokic et al., 2001b) and goethite (Gonzalez and Laird, 2004), have also been shown to catalyze the Maillard reaction, under typical pH and temperature ranges found in the natural environment. Jokic et al. (2001b) were the first to report that birnessite catalyzes the Maillard reaction between glucose and glycine. This reaction is kinetically sluggish under ambient temperatures (Jokic et al., 2001c), but the presence of birnessite significantly catalyzes the reaction by decreasing the activation energy required. Jokic et al. (2001a) investigated the effect of light on birnessite catalysis of the Maillard reaction, and they showed that the reaction is promoted by light but also readily occurs in the absence of light. This means that the reaction could readily occur in the subsoil in the presence of a mineral catalyst such as birnessite. Jokic et al. (2004a) showed that birnessite catalyzes the Maillard reaction between glucose and glycine, resulting in the formation of humic substances that contain significant amounts of heterocyclic and amide N, which provides an explanation for one of the pathways for the formation of heterocyclic and amide N found in humic substances in the environment. Jokic et al. (2004b) were the first to investigate an integrated polyphenol–Maillard reaction humification pathway as catalyzed by birnessite, by studying the reaction between glucose, glycine, and catechol. They found that the presence of birnessite significantly accelerates this integrated humification pathway under temperatures and a neutral pH typical of natural environments.

2.6.2. Clay Size Layer Silicates

Besides metal oxides, clay size layer silicates have the ability to catalyze the oxidative transformation of biomolecules (Wang et al., 1978a; Wang et al., 1986; Bollag et al., 1998; Huang, 2000, 2004). Before the pioneering work on the catalytic role of clay size layer silicates in oxidative polymerization of phenolic compounds and the subsequent formation of humic substances (Kumada and Kato, 1970; Filip et al., 1977; Wang and Li, 1977), the conversion of many aromatic amines into their colored derivatives by clay minerals had been investigated (Faust, 1940; Hauser and Legget, 1940). Solomon (1968) reported that, except for talc, a large number of representative clay minerals produce a blue color of varying intensity when brought in contact with a saturated solution of benzidine hydrochloride. The active sites for the oxidation of benzidine are located on the crystal edges and on transition metal atoms in the higher oxidation state that occupy octahedral sites in the silicate layers. Thompson and Moll (1973) measured the oxidative power of smectites by oxidation of hydroquinone to *p*-benzoquinone in a clay slurry. Oxidation occurs in the presence of O_2 (air), but not of N_2 unless Fe^{3+} or Cu^{2+} are the exchangeable cations. Adsorbed O_2 molecules or radicals on the clay surface are apparently responsible for the oxidation.

Montmorillonite, vermiculite, illite, and kaolinite accelerate the formation of HAs to varying degrees (Table 2.7) (Shindo and Huang, 1985b). The promoting effect of 2:1 layer silicates is higher than that of 1:1 layer silicates because of the larger specific surface area and lattice imperfection which favor the adsorption of O_2 molecules or radicals. Many studies focused on the oxidative catalytic ability of transition metal-saturated smectites to transform aromatic molecules (Pinnavaia et al., 1974; Mortland and Halloran, 1976). However, the most common exchangeable cations found on smectites in soils are alkaline earth metals such as Ca. One of the well-identified precursors (Flaig et al., 1975; Hayes, 1991) for the formation of humic substances, hydroquinone, can be transformed in aqueous solution at near neutral pH (6.5) to humic macromolecules and deposited in the interlayers of nontronite saturated with Ca, which is the most common and most abundant exchangeable

TABLE 2.7. Effects of Clay Minerals on the Synthesis of Humic Acids (HA) at an Initial pH of 5.5 at the End of 7 Days

	Yield of HA (g HA-carbon kg^{-1} inorganic material)[a]		
System	Soluble Fraction	Precipitated Fraction	Total
Control[b]	0.68 (100)[c]	0.30 (100)	0.98 (100)
Montmorillonite	1.25 (184)	0.31 (103)	1.56 (159)
Vermiculite	0.98 (144)	0.31 (103)	1.29 (132)
Illite	0.77 (113)	0.32 (107)	1.09 (111)
Kaolinite	0.68 (100)	0.31 (103)	0.99 (101)

[a]1 mL of 0.02 M $KMnO_4$ consumed was calculated as corresponding to 0.45 mg carbon.

[b]In the absence of inorganic material.

[c]The index of the yield of HA in the control system is assigned 100 as the basis for comparison.

Source: Reprinted from Shindo, H., and Huang, P. M. (1985b). The catalytic power of inorganic components in the abiotic synthesis of hydroquinone-derived humic polymers. *Appl. Clay Sci.* **1**, 71–81, with permission from Elsevier.

cation in soils and sediments (Wang and Huang, 1986). Most of the interlayer humic macromolecules are highly resistant to alkaline extraction and are, thus, humin-type materials. Therefore, besides Al interlayering of clays (Barnhisel and Bertsch, 1989), the formation of humic substance interlayers in 2:1 layer silicate, through polymerization of phenol monomers and the associated reactions in soils and sediments, merits attention. The catalytic sequence of selected minerals in the smectite group is: nontronite > montmorillonite > hectorite, which can be related to the dominant structural metals of each mineral. Nontronite contains Fe(III) as the major structural cation in the octahedral sheet, while montmorillonite contains predominantly Al and hectorite Mg and Al (Wang and Huang, 1986).

Wang and Huang (1989b) showed that the edge-sites of kaolinite provided all of its catalytic power whereas in the case of nontronite its edge-sites only partially accounted for its catalytic power. They also found that edge-site adsorption of humic polymers only accounted for a small fraction of the total amount of humic substances sorbed by the minerals. Nontronite also has the ability to cleave the aromatic ring of pyrogallol, catechol and hydroquinone. The ability of nontronite to promote ring cleavage of polyphenols is related to the structure and functionality of the polyphenols (Wang and Huang, 1994). Catechol, with two hydroxyl groups in the *ortho* positions, is more easily cleaved than hydroquinone, which has two hydroxyl groups in the *para* positions (Table 2.8). Pyrogallol, which has three hydroxyls all *ortho* to one another, is by far the most easily cleaved of the three polyphenols.

Clay size layer silicates also have the ability to catalyze the polycondensation of phenolic compounds and amino acids. Wang et al. (1985) examined the catalytic effect of Ca-illite on the formation of N-containing humic polymers in systems containing various phenolic compounds and amino acids. The yields and N contents

TABLE 2.8. Release of Carbon Dioxide in the Nontronite–Polyphenol Systems at the End of a 90-h Reaction Period

Reaction Condition		
Nontronite	Polyphenol	CO_2 Release (μmol[a])
+[b]	Pyrogallol	263
–[c]	Pyrogallol	54
+	Catechol	88
–	Catechol	34
+	Hydroquinone	49
–	Hydroquinone	21

[a]Amount of CO_2 released in the systems containing 1 g of Ca-nontronite (0.2–2 μm) and 5 mmol of pyrogallol, catechol, or hydroquinone in 30 ml of aqueous solution adjusted to pH 6.00.
[b]In the presence.
[c]In the absence.

Source: Reprinted with permission from Wang, M. C., and Huang, P. M. (1994). Structural role of polyphenols in influencing the ring cleavage and related chemical reactions as catalyzed by nontronite. In *Humic Substances in the Global Environment and Implications on Human Health*, Senesi, N., and Miano, T. M., eds., Elsevier, Amsterdam, The Netherlands, 173–180.

of the resultant HAs were dependent on the nature of the amino acids. Nontronite can also catalyze the polycondensation of glycine and pyrogallol (Wang and Huang, 1991). Molecular O_2 chemisorbed on silicates, such as nontronite, has a strong oxidative power (Solomon and Hawthorne, 1983) and seems to be responsible for the ring cleavage of pyrogallol and for the decarboxylation and deamination of glycine (Wang and Huang, 1991). Gonzalez and Laird (2004) showed that four different smectites saturated with various metal cations could catalyze the Maillard reaction between arginine and glucose at an environmentally relevant temperature (37 °C). Of the saturating cations investigated (Ca, Na, Cu(II), and Al), only Cu (II), significantly altered the amount of humic substance produced. They also observed that some of the adsorbed humic substances were intercalated into the smectites.

2.6.3. Primary Minerals

Primary minerals are the rock-forming minerals in the earth's crust and are present in soils and aquatic sediments (Dixon and Weed, 1989; Dixon and Schulze, 2002). These minerals differ in their abilities to accelerate the abiotic polymerization of hydroquinone (Table 2.9). The sequence of the catalytic power of the primary minerals is: tephroite > actinolite > hornblende, fayalite > augite > biotite > muscovite ≅ albite ≅ orthoclase ≅ microcline ≅ quartz (Shindo and Huang, 1985a). The degree of acceleration of the oxidative polymerization of hydroquinone is greatest in the tephroite system which increases the total HA yield more than nine-fold. This is attributable to (1) tephroite (ideal chemical formula, $MnSiO_4$) is a Mn-bearing silicate, (2) part of the Mn in tephroite is present in the higher valence states, and (3) the oxidation of diphenols $[C_6H_4(OH)_2]$ by Mn(III) and Mn(IV) is thermodynamically favorable (Weast, 1978).

The hydroquinone-derived polymers formed in the presence of the tephroite system (Shindo and Huang, 1985a) have similar IR absorption bands to those of soil humic substances (Schnitzer, 1978). The surface features of these polymers

TABLE 2.9. Effects of Primary Minerals on the Synthesis of HA at an Initial pH of 5.5 at the End of 7 Days

	Yield of HA (g HA-carbon kg^{-1} inorganic material)[a]		
System	Soluble Fraction	Precipitated Fraction	Total
Control[b]	0.68 (100)[c]	0.30 (100)	0.98 (100)
Tephroite	6.60 (971)	1.90 (633)	8.50 (867)
Hornblende	3.24 (476)	1.05 (350)	4.29 (438)
Augite	2.90 (426)	0.67 (223)	3.57 (364)
Biotite	2.03 (299)	0.55 (183)	2.58 (263)
Quartz	1.15 (169)	0.32 (107)	1.47 (150)
Microcline	0.98 (144)	0.32 (107)	1.30 (133)

[a]1 mL of 0.02 M $KMnO_4$ consumed was calculated as corresponding to 0.45 mg carbon.
[b]In the absence of inorganic material.
[c]The index of the yield of HA in the control system is assigned 100 as the basis for comparison.

Source: Reprinted from Shindo, H., and Huang, P. M. (1985a). Catalytic polymerization of hydroquinone by primary minerals. *Soil Sci.* **139**, 505–511, with permission from Lippincott Williams & Wilkins.

(Figure 2.19) are similar to those of soil HA and FA (Stevenson and Schnitzer, 1982) with the smallest discrete particles being spheroids with diameters of 0.1 to 0.2 μm (Figure 2.19a) and some aggregation of individual spheroids (Figure 2.19b and 2.19c). Small aggregates resemble moss, whereas the large aggregates are nodule-like (1- to 5-μm diameter) and doughnut like (6- to 8-μm diameter) (Figures 2.19a and 2.19b). The polymers do not appear to be associated with the surfaces of tephroite particles (Figure 2.19d). The role of primary minerals in the oxidative polymerization of polyphenols and the subsequent formation of humic substances in soils and sediments should not be overlooked.

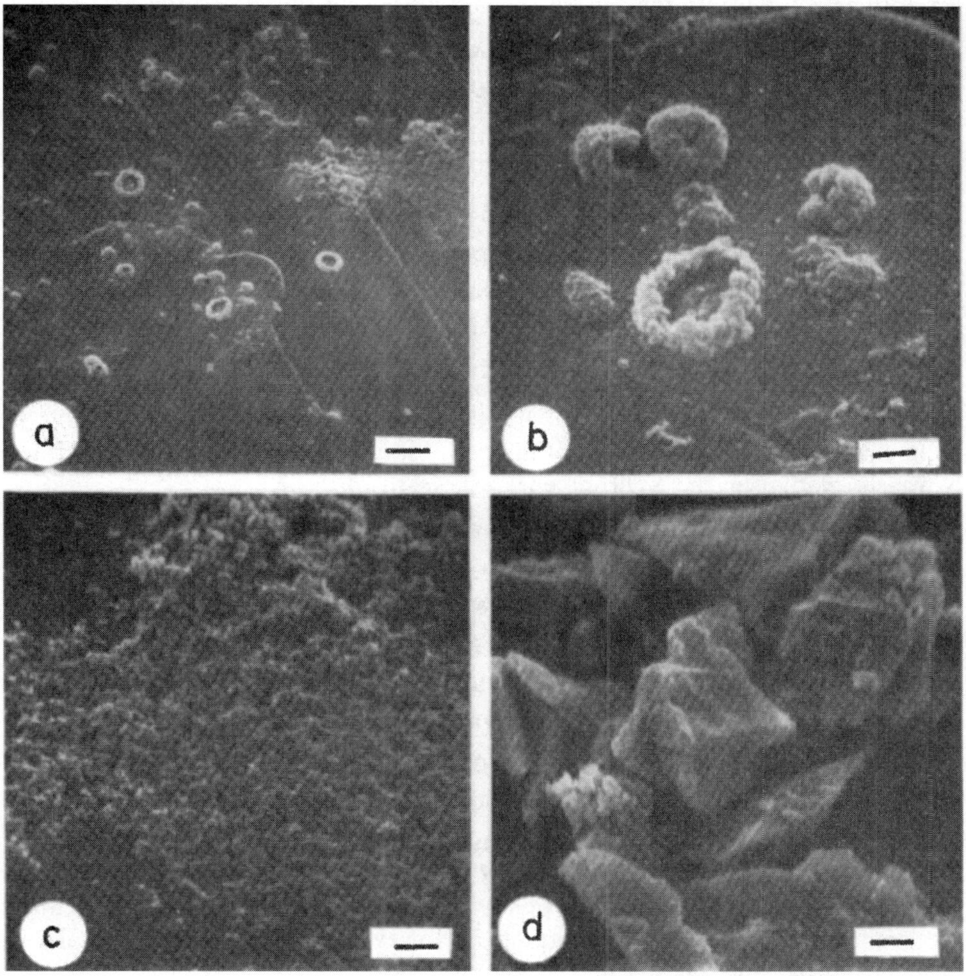

Figure 2.19. SEM micrographs of hydroquinone polymers in the supernatant and mineral particles settled in the tephroite system at the ratio of mineral to hydroquinone solution of 0.01 at the initial pH of 6.0 at the end of 7 days. **(a–c)**: hydroquinone polymers; **(d)** tephroite particles after reaction with hydroquinone. Bar in Figure 2.19a = 10 μm; bars in Figure 2.19b–d = 2 μm. Reprinted from Shindo, H., and Huang, P. M. (1985a). Catalytic polymerization of hydroquinone by primary minerals. *Soil Sci.* **139**, 505–511, with permission from Lippincott Williams & Wilkins.

2.6.4. Natural Soils

A number of studies have investigated the abiotic catalytic ability of natural soils (whole soils and clay fractions) as catalysts for the polymerization of polyphenols (Wang et al., 1978a, 1978c, 1983a; Wang and Huang, 1989a) and phenolic acids (Pohlman and McColl, 1989) and the polycondensation of polyphenols and amino acids (Wang and Huang, 2003). Wang and Huang (1989a) examined the catalytic ability of the sterilized Ap horizon from a Mollisol from Saskatchewan, Canada, with regard to the ring cleavage and polymerization of pyrogallol. They reported that abiotic processes evidently cause the polymerization of pyrogallol as well as its ring cleavage and the formation of its fragments. The IR and ESR spectra of the HS formed closely resemble those of naturally occurring HS. Wang and Huang (2003) showed that the clay fractions from a tropical Oxisol and temperate Mollisol catalyze the abiotic polycondensation of pyrogallol and glycine and the subsequent formation of humic substances. The polymerization of pyrogallol and glycine, the abiotic ring cleavage of pyrogallol, and the deamination of glycine are greatly enhanced by the presence of the sterilized soil clays, all of which increased with increasing temperature (Wang and Huang, 2003). The abiotic catalytic ability of soils in the formation of humic substances is due to the reactive components, namely, Mn, Fe, and Al oxides and (oxy)hydroxides, SRO mineral colloids, and clay-size layer silicates, as well as some reactive primary minerals that have been extensively investigated as discussed in Sections 2.6.1 to 2.6.3.

Little is known on the catalysis of the Maillard reaction and especially the integrated polyphenol–Maillard reaction by natural soils and sediments. Further work is warranted on this subject matter to advance our understanding of the role of abiotic catalysis in the formation of humic substances and related C turnover and N transformations in the environment.

2.7. COMPARISON OF THE MECHANISMS AND SIGNIFICANCE OF BIOTIC AND ABIOTIC CATALYSES OF HUMIFICATION REACTIONS IN NATURAL ENVIRONMENTS

2.7.1. Comparison of the Mechanisms of Biotic and Abiotic Catalyses of Synthetic Humification Reactions

There have been numerous studies in recent years investigating the differences in oxidative coupling reactions of phenols catalyzed by enzymes or mineral colloids. Both mineral colloids and oxidoreductive enzymes contain metals that can act as electron acceptors to catalyze the oxidative transformation of organics; hence there are similarities in their reaction products. However, there are differences in the mechanisms by which these catalysts operate.

Pal et al. (1994) compared the catalysis of oxidative coupling reactions of various phenolic compounds by the enzymes, laccase and tyrosinase, and mineral catalyst, birnessite. Birnessite acts as a heterogeneous catalyst whereas laccase and tyrosinase function as homogeneous catalysts. Laccase and tyrosinase continue to oxidize catechol after repeated additions of the chemical, while birnessite lost its oxidizing activity after the first addition of catechol (Figure 2.20). In the case of birnessite,

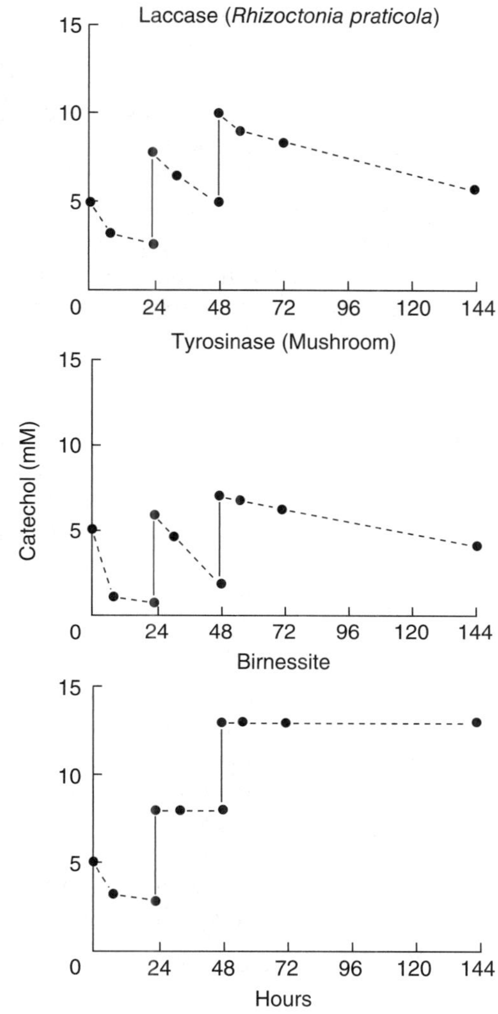

Figure 2.20. Transformation of catechol by laccase (0.4 units ml^{-1}), tyrosinase (0.4 units ml^{-1}) and birnessite (600 ug ml^{-1}) after repeated addition of substrate. Reprinted from Pal, S., Bollag, J.-M., and Huang, P. M. (1994). Role of abiotic and biotic catalysts in the transformation of phenolic compounds through oxidative coupling reactions. *Soil Biol. Biochem.* **26**, 813–820, with permission from Elsevier.

Mn serves as a terminal electron acceptor during the oxidative coupling reactions and thus is altered or consumed. The enzymes on the other hand are able to successfully mediate the transfer of an electron to an electron acceptor (O_2) in a cyclic manner. They concluded that enzymes as homogeneous catalysts appeared to be more effective oxidative agents compared to abiotic agents as heterogeneous catalysts. In the natural environment, however, the Mn^{2+} that is released during birnessite-induced oxidation of phenolic compounds such as catechol could be reoxidized to higher-valency oxides depending on the redox potential of the environment and could thus regain its catalytic potential. This has been observed for Fe oxides

(McBride, 1987). Furthermore, redox transformations between reduced and oxidized Mn forms are strongly catalyzed by bacteria. Microbial Mn(II) oxidation is a major process that can produce Mn oxide coatings in soils and marine sediments (McLean et al., 2002). Based on electron spin resonance studies, the oxidative polymerization of polyphenols (such as catechol) by birnessite (δ-MnO_2) proceeds through the formation of semiquinone radicals (McBride, 1989), whereas with tyrosinase, it proceeds without the semiquinone radical formation as electron transfer is controlled by the binuclear copper active site (Himmelwright et al., 1980). Thus, the formation of humic polymers (melanins) as mediated by tyrosinase or birnessite has been reported to be controlled by different mechanisms (Figure 2.21). Park et al. (1999) investigated the effect of a variety of humic constituents on the transformation of chlorinated phenols and anilines in the presence of peroxidase, tyrosinase, laccase and birnessite. They found that the addition of catechol resulted in a significant reduction of transformation in the peroxidase, laccase and birnessite systems, whereas it enhanced transformation in the tyrosinase systems. They suggested that the varying effect of catechol could be explained by different transformation mechanisms involving either *o*-quinone coupling (with tyrosinase) or free radical coupling (with peroxidase, laccase, or birnessite).

Naidja and Huang (2002) showed that the Henri–Michaelis–Menten theory, which is usually applied to the kinetics of homogeneous enzymatic catalysis reactions, can also be applied to the abiotic catalysis of catechol oxidation by birnessite (Figure 2.22). Naidja et al. (1999) compared the kinetics of catechol oxidation by birnessite and tyrosinase. They found that tyrosinase has a slightly lower activation energy that causes the reaction rate to proceed at a rate three times faster than that of birnessite. However, unlike birnessite, tyrosinase is inactivated at temperatures above 30 °C because of denaturation of the protein molecules; birnessite oxidation activity continues to increase with increasing temperature (Figure 2.23). Therefore, the data suggest that, compared with enzymes such as tyrosinase, metal oxides such as birnessite would play a more important role in the transformation of phenolic compounds in warmer regions.

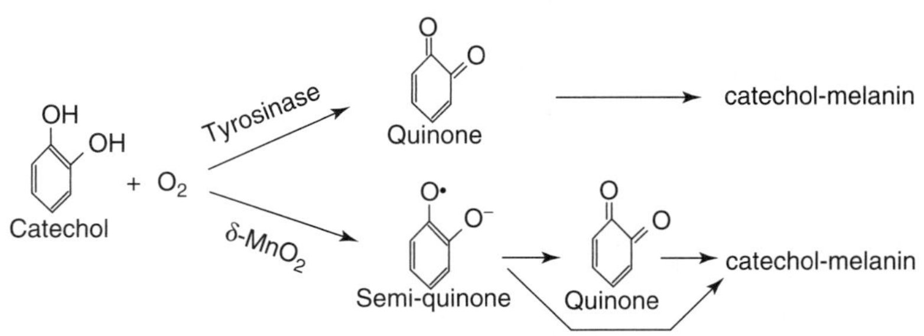

Figure 2.21. Mechanisms of the oxidative polymerization of catechol to melanins (humic polymers) in the presence of tyrosinase or birnessite. Reprinted with permission from Naidja, A., Huang, P. M., Dec, J., and Bollag, J.-M. (1999). Kinetics of catechol oxidation catalyzed by tyrosinase or δ-MnO_2. In *Effect of Mineral-Organic-Microorganism Interactions on Soil and Freshwater Environments*, Berthelin, J., Huang, P. M., Bollag, J.-M., and Andreux, F., eds., Kluwer Academic/Plenum Publishers, New York, 181–188.

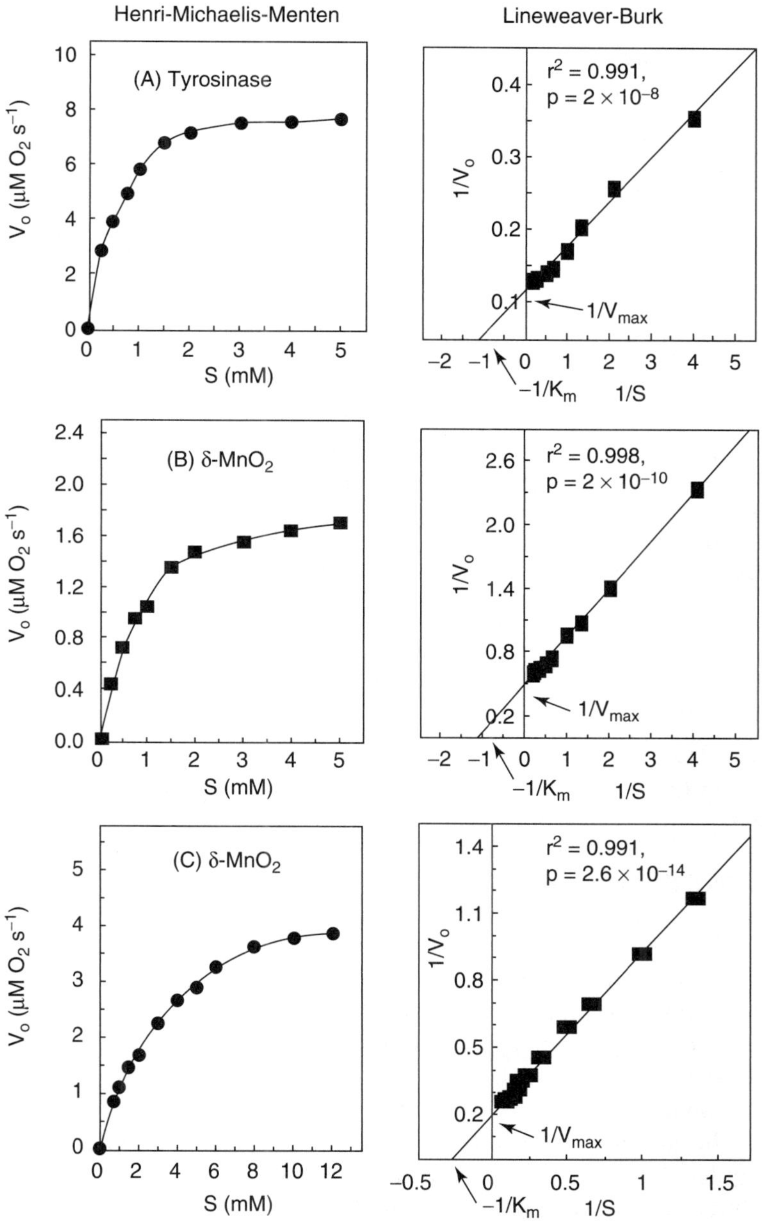

Figure 2.22. Initial velocity of oxygen consumption as a function of the substrate (catechol) concentration in the presence of 0.074 mg (7.11×10^{-9} M) tyrosinase **(A)**, 2.0 mg (2.8×10^{-4} M with a corresponding concentration of the mineral active sites, $[M_0^+]1.71 \times 10^{-6}$) of δ-$MnO_2$ **(B)** and 10.0 mg (1.40×10^{-3} M with a corresponding concentration of the mineral active sites, $[M_0^+]8.54 \times 10^{-6}$) of δ-$MnO_2$ **(C)**. Reprinted from Naidja, A., Liu, C., and Huang, P. M. (2002). Formation of protein-birnessite complex: XRD, FTIR, and AFM analysis. *J. Coll. Interface Sci.* **251**, 46–56, with permission from Elsevier.

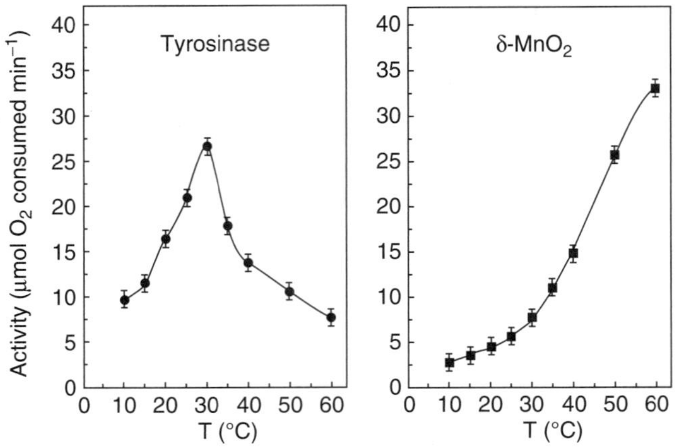

Figure 2.23. Effect of temperature on the activity of tyrosinase (0.148 mg) and δ-MnO_2 (2.0 mg) at an initial pH of 6.0. Reprinted with permission from Naidja, A., Huang, P. M., Dec, J., and Bollag, J.-M. (1999). Kinetics of catechol oxidation catalyzed by tyrosinase or δ-MnO_2. In *Effect of Mineral-Organic-Microorganism Interactions on Soil and Freshwater Environments*, Berthelin, J., Huang, P. M., Bollag, J.-M., and Andreux, F., eds., Kluwer Academic/Plenum Publishers, New York, 181–188.

2.7.2. Comparison of the Products of Biotic and Abiotic Catalyses of Synthetic Humification Reactions

Shindo and Huang (1992) compared the catalytic effects of Mn(IV) oxide and tyrosinase on the oxidative polymerization of diphenols (hydroquinone, catechol, and resorcinol) in the pH range of 4–8. Mn oxide influences the darkening of hydroquinone and resorcinol to a larger extent than does tyrosinase, while the reverse is true for catechol. The yields of humic acids are also significantly influenced by the kind of catalyst and diphenol used. Their findings indicated that the relative catalytic effects of Mn(IV) oxides and tyrosinase in promoting the formation of diphenol-derived humic substances would vary with the type of diphenols in natural systems. Naidja et al. (1998) studied the difference in the reaction products from the transformation of catechol catalyzed by birnessite or tyrosinase. They found that the polymers formed in the tyrosinase–catechol system have a higher degree of aromatic ring condensation than in the birnessite–catechol system. In addition, they found that the products derived from birnessite catalysis contain a greater fraction of lower-molecular-weight fragments and aliphatic components than that of tyrosinase catalysis. Dec et al. (2001) investigated oxidative coupling, decarboxylation, and demethylation of a number of natural phenolic compounds by the phenoloxidase enzymes, peroxidase, laccase, and tyrosinase, and mineral catalyst, birnessite. They observed that birnessite is able to catalyze the ring cleavage of catechol and the decarboxylation of *p*-hydroxybenzoic acid, vanillic acid, *p*-coumaric acid, and ferulic acid to a much greater extent than the enzymes.

2.7.3. The Effect of Environmental Particles on Activity of Biotic Catalysts

A number of environmental particles have been shown to alter the activity of phenoloxidase and other extracellular enzymes associated with the decomposition of organic residues. Mineral colloids influence not only the activity, but also the long-term stability of extracellular enzymes in soils (Naidja et al., 2002) and aquatic environments (Tietjen and Wetzel, 2003).

Ruggiero et al. (1989) investigated the ability of a natural silt loam soil and the clay minerals, montmorillonite (Mte) and kaolinite (Kte), to immobilize laccase. They compared the catalytic abilities of the soil–enzyme and clay–enzyme complexes to degrade 2,4-dichlorophenol. They found that the immobilized laccase remains active in removing the substrate even after 15 repeated cycles of substrate addition (Figure 2.24). However, Claus and Filip (1988) found that the activity of tyrosinase, laccase, and peroxidase is inhibited by immobilization on bentonite. The type of saturating cations on clay surfaces also substantially influences enzymatic activity (Claus and Filip, 1990).

Gianfreda and Bollag (1994) investigated the behavior of laccase and peroxidase in the presence of a montmorillonite, a kaolinite, and a silt loam soil. They observed considerable variation in the retained activities of the two enzymes immobilized on the different supports as well as variation in the amount of each enzyme sorbed (Table 2.10). Interestingly enough, laccase immobilized on montmorillonite showed a higher specific activity (118%) than that of the free enzyme. This may be attributed to the steric modification of the immobilized enzyme or possibly due to the catalytic ability of montmorillonite itself. Their studies showed that the performance of these enzymes is significantly affected by soil mineral colloids.

Naidja et al. (1997) showed that tyrosinase immobilized on montmorillonite coated with Al hydroxide polymers retains a higher specific activity than the free

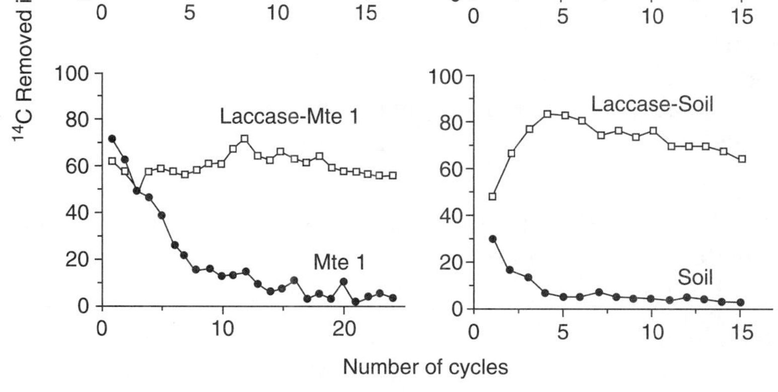

Figure 2.24. Removal of ^{14}C-2,4-dichlorophenol by laccase immobilized on clays and soil. Reprinted from Ruggiero, P., Sarkar, J. M., and Bollag, J.-M. (1989). Detoxification of 2,4-dichlorophenol by a laccase immobilized on soil and clay. *Soil Sci.* **147**, 361–370, with permission from Lippincott Williams & Wilkins.

TABLE 2.10. Immobilization of Laccase and Horseradish Peroxidase on Different Supports

	Enzymatic Activity		
Enzyme and Support	Units Adsorbed[a]	Specific Activity[b]	Residual Specific Activity[c] (%)
Laccase			
Glass beads	28.8	63.7	236.0
Montmorillonite	19.8	31.8	118.0
Kaolinite	13.1	23.1	85.5
Soil	15.7	24.4	90.4
Peroxidase			
Glass beads	8.4	91.6	93.8
Montmorillonite	23.0	102.8	105.2
Kaolinite	9.5	78.9	80.7
Soil	15.0	92.6	94.8

[a]Expressed as μmol O_2 consumed min^{-1} for laccase and μmol guaiacol transformed min^{-1} for peroxidase.

[b]Units adsorbed/protein adsorbed.

[c]Calculated as percentage of the specific activity of the free enzyme (laccase = 27 μmol min^{-1}; peroxidase = 97.7 μmol min^{-1}).

Source: Reprinted from Gianfreda, L., and Bollag, J.-M. (1994). Effects of soils on the behavior of immobilized enzymes. *Soil Sci. Soc. Am. J.* **58**, 1672–1681, with permission from the Soil Science Society of America.

enzyme after 30 days at 25 °C. Naidja et al. (2002) investigated the immobilization of tyrosinase by birnessite. Birnessite was found to have a high affinity for adsorbing tyrosinase and significantly altered its molecular conformation. Ahn et al. (2006) investigated the effect of the presence of birnessite on the catechol oxidative coupling activity of laccase. Birnessite was shown to have an inhibitory effect on catechol oxidation by laccase (Figure 2.25), which was attributed to the formation of humic polymers by catalysis of birnessite and the Mn^{2+} ions released from the mineral.

Humic acids have been shown to slightly inhibit tyrosinase activity by complexing the enzyme (Ruggiero and Radogna, 1988). Allison (2006b) also demonstrated that the addition of humic acid to a soil significantly decreased the polyphenoloxidase activity of the soil.

2.7.4. The Significance of Biotic and Abiotic Catalysts in Synthetic Humification Reactions in Natural Environments

In the past the mineral matrix was considered as inert, only providing stabilization support for enzymes and humic substances; however, due to the overwhelming amount of evidence at the molecular level, there is no doubt that minerals participate in abiotic catalysis of humification reactions in soils. Naidja et al. (2000) referred to mineral particles as the *Hidden Half* of enzyme–clay complexes, which not only prolong the activity of immobilized enzymes but also are readily able to participate in electron transfer reactions. Many environmental factors can negatively affect the

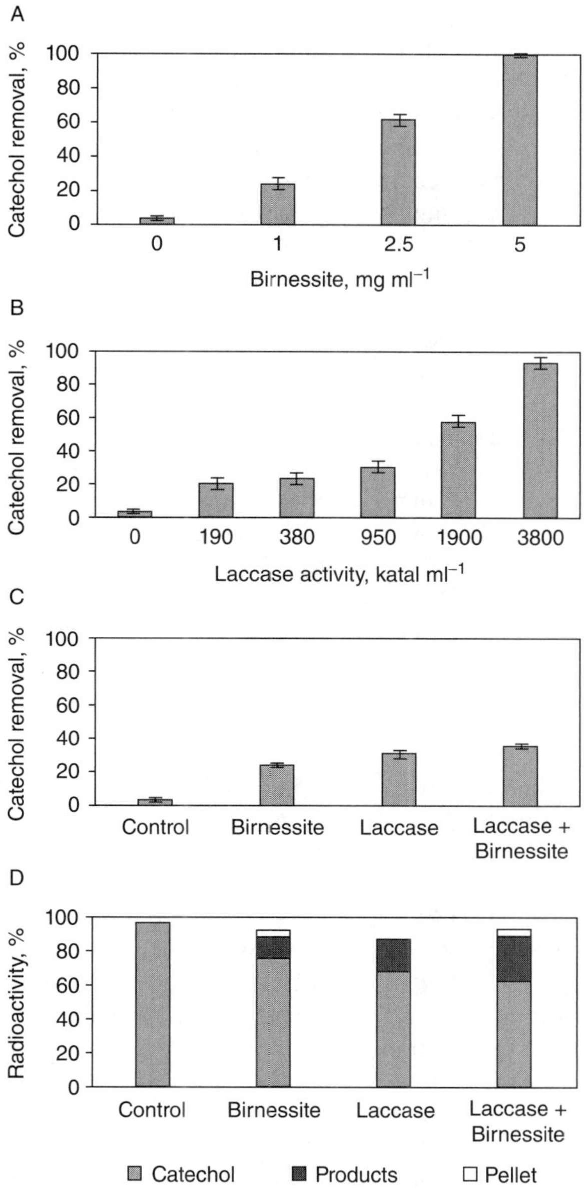

Figure 2.25. Transformation of catechol (0.1 M) in binary and ternary systems: **(A)** Catechol removal by increasing concentrations of birnessite; **(B)** catechol removal by increasing activities of *Trametes villosa* laccase; **(C)** catechol removal by *T. villosa* laccase (950 katal ml^{-1}) and birnessite (1 mg ml^{-1}) applied together; **(D)** distribution of radioactivity after the incubation of ^{14}C-labeled catechol with *T. villosa* laccase (950 katal ml^{-1}) and birnessite (1 mg ml^{-1}). The reactions were carried out in 0.5% NaCl for 24 h at 25 °C. Reprinted from Ahn, M.-Y., Martínez, C. E., Archibald, D. D., Zimmerman, A. R., Bollag, J.-M., and Dec, J. (2006). Transformation of catechol in the presence of a laccase and birnessite. *Soil Biol. Biochem.* **38**, 1015–1020, with permission from Elsevier.

activity of biotic catalysts, such as temperature (Naidja et al., 1999), presence of humic substances (Allison, 2006b), or presence of certain mineral particles (Claus and Filip, 1988, 1990; Ahn et al., 2006). The problem with abiotic catalysts is degeneration of the catalyst, which might be slow depending on the redox conditions in the environment as well as the nature of the mineral catalyst (McBride, 1987). However, microbial oxidation of metals can regenerate oxides (McLean et al., 2002). Furthermore, minerals are highly abundant in soils and sediments and typically make up about 45% of the total soil volume in a loam soil, whereas organic matter contributes to about 5% (Sparks, 2003). Microorganisms, which are the major source of extracellular enzymes in soils, make up about 1–4% of the total organic matter in a soil (Stevenson, 1994). Therefore, besides enzymes, mineral particles should play a significant role in the humification process in the environment, especially under warmer conditions.

Another important consideration is the relative size of enzymes versus that of micropores (less than 2 nm) in environmental particles which are especially abundant in noncrystalline mineral phases. Simple biomolecules (e.g., glucose, MW = 180 Da) can readily enter into micropores and can become stabilized by reacting with the mineral surface, whereas large macromolecules such as enzymes [e.g., laccase, MW = 60,000 Da and diameter = 5 nm (Andersen et al., 1996)] cannot enter and react with the trapped biomolecules.

2.8. CONCLUSIONS AND FUTURE RESEARCH PROSPECTS

Environmental organic matter is a composite of humic and nonhumic substances, which is formed through operation and interactions of various biotic and abiotic processes. Humic substances are formed through both selected preservation (residue) and catalytic synthesis mechanisms. Both enzymatic and mineral catalyses contribute to the formation of humic substances in the environment. The relative importance of these catalytic reactions would depend on vegetation, microbial population and activity, enzymatic activity, mineralogical composition and surface chemistry of environmental particles, management practices, and environmental conditions. Selective preservation pathways would play a more important role in humification processes in poorly drained soils and lake sediments, compared with more aerated environmental conditions.

The existing research data indicate that humic substances have both macromolecular and supramolecular characteristics. The origin of environmental macromolecules (polymers) may include biomolecules from the selective preservation pathway and humification products from catalytic synthesis mechanisms. A supramolecule is a system of two or more molecular entities held together and organized by means of intermolecular (noncovalent) binding interactions. Macromolecules as well as small molecules may form supramolecular structures, the properties of which largely determine the reactivity of the material. Humic polymers may encapsulate or anchor unstable biological constituents by hydrophobic and hydrogen bonding forces and/or chemical binding. Any biomolecules intimately associated with humic polymers may, thus, not be separated effectively by chemical and physical methods and are by operational definition humic components. Therefore, many relatively unstable biological constituents may survive in environmental humus for a significant length of time in the humification process. Furthermore, chemical protection of organic materials by mineral colloids, especially short-range ordered Al and Fe (oxy)hydroxides, and physical protection of plant-like materials within aggregates deserve close attention in understanding the degradation of biological residues, the formation of humic substances, and global C cycling and climate change.

Our knowledge on the intrinsic mechanisms of environmental processes pertaining to the genesis of humic substances in nature remains to be advanced. Previous studies on abiotic and biotic catalyses have focused on polyphenols, amino acids, and sugars, while no work has been done on lipids, one of the most refractory components of plant materials and a significant contributor to the humin fraction of soil organic matter. The interactions of lipids, proteins, and polyphenols in the presence of these catalysts remain to be studied. Our understanding of the influence of pedogenic factors and anthropogenic activities on the transformation of biological constituents to humic substances and the nature and properties of the resultant mineral–humus complexes are still very limited. A vast majority of environmental organic matter is associated with mineral particles. More research should be conducted to uncover the impact of physical–chemical–biological interfacial reactions on biogeochemical reactions, which, in turn, govern the humification processes and the formation of mineral–humus complexes. Also, additional research about the type of organisms responding to the transformation of humic substances as well as their role is needed.

Use of advanced analytical instrumentation—that is, synchrotron-based X-ray absorption spectroscopy, spectromicroscopy and infrared spectroscopy, atomic force microscopy, multidimensional nuclear magnetic resonance spectroscopy, and so on—should shed light on the mystery of environmental humic substances and their complexes with mineral particles. Fundamental understanding of this subject matter at the molecular level and the impacts on the ecosystem would facilitate our development of innovative management strategies to regulate the behaviour of the ecosystem on a global scale. Future research on this extremely important and exciting area of science should be stimulated to restore as well as sustain ecosystem integrity.

ACKNOWLEDGMENT

We acknowledge the funding from Discovery Grant 2383-2008-Huang of the Natural Sciences and Engineering Research Council of Canada.

REFERENCES

Adani, F., Ricca, G., Tambone, F., and Genevini, P. (2006). Isolation of the stable fraction (the core) of the humic acid. *Chemosphere* **62**, 1300–1307.

Adani, F., Spagnol, M., and Nierop, K. G. J. (2007). Biochemical origin and refractory properties of humic acid extracted from maize plants: the contribution of lignin. *Biogeochem.* **82**, 55–65.

Ahn, M.-Y., Martínez, C. E., Archibald, D. D., Zimmerman, A. R., Bollag, J.-M., and Dec, J. (2006). Transformation of catechol in the presence of a laccase and birnessite. *Soil Biol. Biochem.* **38**, 1015–1020.

Aiken, G. R., McKnight, D. M., Wershaw, R. L., and MacCarthy, P. (1985). *Humic Substances in Soil, Sediments and Water: Geochemistry, Isolation and Characterization*. Wiley Interscience, New York, pp. 1–9.

Alexander, M. (1997). *Introduction to Soil Microbiology*, 2nd edition, John Wiley & Sons, New York.

Allison, F. E. (1973). *Soil Organic Matter and Its Role in Crop Production*. Elsevier, Amsterdam, Netherlands.

Allison, S. D. (2006a). Brown ground: A soil carbon analogue for the green world hypothesis? *Am. Nat.* **167**, 619–627.

Allison, S. D. (2006b). Soil minerals and humic acids alter enzyme stability: Implications for ecosystem processes. *Biogeochemistry.* **81**, 361–373.

Almendros, G., Dorado, J., Gonzâlez-Vila, F. J., Blanco, M. J., and Lankes, U. (2000). ^{13}C NMR assessment of decomposition patterns during composting of forest and shrub biomass. *Soil Biol. Biochem.* **32**, 793–804.

Ander, P., and Marzullo, L. (1997). Sugar oxidoreductases and veratryl alcohol oxidase as related to lignin degradation. *J. Biotechnol.* **53**, 115–131.

Andersen, J. E. T., Jensen, M. H., Moller, P., and Ulstrup, J. (1996). Perspectives for in situ scanning tunnel microscopic imaging of metalloproteins at HOPG surfaces. *Electrochim. Acta* **41**, 2005–2010.

Anderson, H. A., Bick, W. Hepburn, A., and Stewart, M. (1989). Nitrogen in humic substances. In *Humic Substances II. In Search of Structure*, Hayes, M. H. B., MacCarthy, P., Malcolm, R. L., and Swift, R. S., eds., Wiley-Interscience, Chichester, UK, pp. 223–253.

Arafaioli, P., Ristori, G. G., Bosetto, M., and Fusi, P. (1997). Humic-like compounds formed from L-tryptophan and D-glucose in the presence of Cu(II). *Chemosphere* **35**, 575–584.

Arafaioli, P., Pantani, O. L., Bosetto, M., and Ristori, G. G. (1999). Influence of clay minerals and exchangeable cations on the formation of humic-like substances (melanoidins) from D-glucose and L-tyrosine. *Clay Miner.* **34**, 487–497.

Baldock, J. A. (2002). Interactions of organic materials and microorganisms with minerals in stabilization of structure. In *Interactions Between Soil Particles and Microorganisms. Impact on the Terrestrial Ecosystem*, Huang, P. M., Bollag, J.-M., and Senesi, N., eds., IUPAC Series on Analytical and Physical of Environmental Systems. Vol. **8**, John Wiley & Sons, Chichester, UK, pp. 85–131.

Baldock, J. A., and Nelson, P. N. (2000). Soil organic matter. In *Handbook of Soil Science*, Sumner, M., ed., CRC Press, Boca Raton, FL, B25–B84.

Baldock, J. A., and Skjemstad, J. O. (2000). The role of soil mineral matrix in protecting natural organic materials against biological attack. *Org. Geochem.* **31**, 697–710.

Baldock, J. A., and Smernik, R. J. (2002). Chemical composition and bioavailability of thermally altered *Pinus resinosa* (Red pine) wood. *Org. Geochem.* **33**, 1093–1109.

Baldock, J. A., Masiello, C. A., Gelinas, Y., and Hedges, J. I. (2004). Cycling and composition of organic matter in terrestrial and marine ecosystems. *Marine Chem.* **92**, 39–64.

Baldrian, P. (2006). Fungal laccases—Occurrence and properties. *FEMS Microbiol. Rev.* **30**, 215–242.

Barnhisel, R. I., and Bertsch, P. M. (1989). Chlorites and hydroxyl-interlayered vermiculite and smectite. In *Minerals in Soil Environments*, Dixon, J. B., and Weed, S. B., eds., Soil Science Society of America, Madison, WI, pp. 729–788.

Barois I, Villemin, G., Lavelle, P., and Toutain, F. (1993). Transformation of the soil structure through *Pontoscolex corethrurus* (Oligochaeta) intestinal tract. *Geoderma* **56**, 57–66.

Béguin, P., and Aubert, J. P. (1994). The biological degradation of cellulose. *FEMS Microbiol. Rev.* **13**, 25–58.

Bird, M. I., Veenendaal, E., Moyo, C., Lloyd, J., and Frost, P. (1999). Stability of elemental carbon in a savanna soil. *Global Biogeochem. Cycles* **13**, 933–942.

Birkel, U., Gerold, G., and Niemeyer, J. (2002). Abiotic reactions of organics on clay mineral surfaces. In *Soil Mineral–Organic Matter–Microorganism Interactions and Ecosystem Health: Dynamics, Mobility and Transformation of Pollutants and Nutrients.* Violante. A., Huang, P. M., Bollag, J.-M. and Gianfreda, L., eds., Elsevier Science B.V., Amsterdam, The Netherlands, pp. 437–447.

Bittner, S. (2006). When quinones meet amino acids: Chemical, physical and biological consequences. *Amino Acids* **30**, 205–224.

Bollag, J. M., Dec, J., and Huang, P. M. (1998). Formation mechanisms of complex organic structures in soil habitats. *Adv. Agron.* **63**, 237–266.

Bollag, J.-M., Myers, C., Pal, S., and Huang, P. M. (1995). The role of abiotic and biotic catalysts in the transformation of phenolic compounds. In *Environmental Impact of Soil Component Interactions,* Vol. 1. *Natural and Anthropogenic Organics*, Huang, P. M., Berthelin, J., Bollag, J.-M., McGill, W. B., and Page, A. L., eds., Lewis Publishers, Cherry Hill, NJ, pp. 299–310.

Bose, J. L., Foster, A. B., Stacey, M., and Webber, J. M. (1959). Action of manganese dioxide on simple carbohydrates. *Nature* **184**, 1301–1302.

Bosetto, M., Arafaioli, P., Ristori, G. G., and Fusi, P. (1995). Formation of melanin-type compounds from L-tryptophan on Ca-saturated and Al-saturated clays. *Fresenius Environ. Bull.* **4**, 369–374.

Bosetto, M., Arafaioli, P., Pantani, O. L., and Ristori, G. G. (1997). Study of the humic-like compounds formed from L-tyrosine on homoionic clays. *Clay Miner.* **32**, 341–349.

Bosetto, M., Arafaioli, P., and Pantani, O. L. (2002). Study of the Maillard reaction products formed by glycine and D-glucose on different mineral substrates. *Clay Miner.* **37**, 195–204.

Brodowski, S. B. (2004). Origin, function, and reactivity of black carbon in the arable soil environment. Ph.D. thesis, University of Bayreuth, Bayreuth, Germany.

Burdon, J. (2001). Are the traditional concepts of the structures of humic substances realistic? *Soil Sci.* **166**, 752–769.

Burton, S. G. (2003). Laccases and phenol oxidases in organic synthesis—A review. *Curr. Org. Geochem.* **7**, 1317–1331.

Chefetz, B., Chen, Y., and Hadar, Y. (1998). Purification and characterization of laccase from *Chaetomium thermophilium* and its role in humification. *Appl. Environ. Microbiol.* **64**, 3175–3179.

Chefetz, B., Tarchitzky, J., Desmukh, A. P., Hatcher, P. G., and Chen, Y. (2002). Structural characterization of soil organic matter and humic acids in particle-size fractions of an agricultural soil. *Soil Sci. Soc. Am. J.* **66**, 129–141.

Clapp, C. E., Hayes, M. H. B., Simpson, A. J., and Kingery, W. L. (2005). Chemistry of organic matter. In *Chemical processes in Soils*, Tabatabai, M. A., and Sparks, D. L., eds., SSSA Book Series 8, Soil Science Society of America., Madison, WI, pp. 1–150.

Claus, H. (2004). Laccases: structure, reactions and distribution, *Micron.* **35**, 93–96.

Claus, H., and Decker, H. (2006). Bacterial tyrosinases. *System. Appl. Microbiol.* **29**, 3–14.

Claus, H., and Filip, Z. (1988). Behavior of phenoloxidases in the presence of clays and other soil-related adsorbents. *Appl. Microbiol. Biotechnol.* **28**, 506–511.

Claus, H., and Filip, Z. (1990). Effects of clays and other solids on the activity of phenoloxidases produced by some fungi and actinomycetes. *Soil Biol. Biochem.* **22**, 483–488.

Cowling, E. B., and Brown, W. (1969). Structural features of cellulosic materials in relation to enzymatic hydrolysis. *Adv. Chem. Ser.* **95**, 152–187.

Cullen, D. (1997). Recent advances on the molecular genetics of ligninolytic fungi. *J. Biotechnol.* **53**, 273–289.

Daintith, J. (1990). *A Concise Dictionary of Chemistry*, Oxford University Press, Oxford, UK.

Davidson, E. A., and Janssens, I. A. (2006). Temperature sensitivity of soil carbon decomposition and feedbacks to climate change. *Nature.* **440**, 165–173.

Dec, J., and Bollag, J.-M. (1990). Detoxification of substituted phenols by oxidoreductive enzymes through polymerization reactions. *Archives Environ. Contamin. Toxicol.* **19**, 543–550.

Dec, J., and Bollag, J.-M. (2000). Phenoloxidase-mediated interactions of phenols and anilines with humic materials. *J. Environ. Qual.* **29**, 655–676.

Dec, J., Haider, K., and Bollag, J.-M. (2001). Decarboxylation and demethoxylation of naturally occurring phenols during coupling reactions and polymerization. *Soil Sci.* **166**, 660–671.

Dec, J., Haider, K., and Bollag, J.-M. (2003). Release of substituents from phenolic compounds during oxidative coupling reactions. *Chemosphere* **52**, 549–556.

Derenne, S., and Largeau, C. (2001). A review of some important families of refractory macromolecules: composition, origin, and fate in soils and sediments. *Soil Sci.* **166**, 833–847.

Diallo, M. S., Simpson, A., Gassman, P., Faulon, J. L., Johnson, J. H., Jr., Godard, W. A., III, and Hatcher, P. G. (2003). 3-D structural modeling of humic acids through experimental characterization, computer assisted structure elucidation and atomic simulations. I. Chelsea soil humic acid. *Environ. Sci. Technol.* **37**, 1783–1793.

Dixon, J. B., and Schulze, D. G. (2002). *Soil Mineralogy with Environmental Applications*, Soil Science Society of America, Madison, WI.

Dixon, J. B., and Weed, S. B. (1989). *Minerals in Soil Environments*. Soil Science Society of America, Madison, WI.

Dubey, S., Singh, D., and Misra, R. A. (1998). Enzymatic synthesis and various properties of poly(catechol). *Enzyme Microb. Technol.* **23**, 432–437.

Durán, N., and Esposito, E. (2000). Potential applications of oxidative enzymes and phenoloxidase-like compounds in wastewater and soil treatment: a review. *Appl. Catal. B.* **28**, 83–99.

Eriksson, K.-E., Blanchette, R. A., and Ander, P. (1990). *Microbial and Enzymatic Degradation of Wood and Wood Components. Springer Series in Wood Science*, Springer Verlag, Berlin, Germany.

Evershed, R. P., Bland, H. A., van Bergen, P. F., Carter, J. F., Horton, M. C., and Rowley-Conwy, R. P. (1997). Volatile compounds in archeological plant remains and the Maillard reaction during the decay of organic matter. *Science* **278**, 432–433.

Faust, G. T. (1940). *Staining of Clay Minerals as a Rapid Means of Identification in Natural and Beneficiation Products*. Report of Investigation 3522. Bureau Mines, Washington, DC.

Feng, X., and Simpson, M. J. (2007). The distribution and degradation of biomarkers in Alberta grassland soil profiles. *Org. Geochem.* **38**, 1558–1570.

Filip, Z., Flaig, W., and Ritz, E. (1977). Oxidation of some phenolic substances as influenced by clay minerals. *Soil Org. Matter Stud. Proc. Symp.*, **2**, 91–96.

Filley, T. R., Cody, G. D., Goodell, B., Jellison, J., Noser, C., and Ostrofsky. A. (2002). Lignin demethylation and polysaccharide decomposition in spruce sapwood degraded by brown rot fungi. *Org. Geochem.* **33**, 111–124.

Flaig, W., Beutelspacher, H., and Rietz, E. (1975). Chemical composition and physical properties of humic substances. In *Soil Components,* Vol. 1: *Organic Components*, Gieseking, J. E., ed., Springer-Verlag, New York, 1–211.

Galliano, H., Gas, G., Seris, J. L., and Boudet, A. M., (1991). Lignin degradation by *Rigidoporus lignosus* involves synergistic action of two oxidizing enzymes: Mn peroxidase and laccase. *Enzyme Microb. Technol.* **13**, 478–482.

Gianfreda, L., and Bollag, J.-M. (1994). Effects of soils on the behavior of immobilized enzymes. *Soil Sci. Soc. Am. J.* **58**, 1672–1681.

Gianfreda, L., and Rao, M. A. (2004). Potential of extracellular enzymes in remediation of polluted soils: a review. *Enzyme Microb. Technol.* **35**, 339–354.

Gonzalez, J. M., and Laird, D. A. (2004). Role of smectites and Al-substituted goethites in the catalytic condensation of arginine and glucose. *Clays Clay Miner.* **52**, 443–450.

Goodell, B., Jellison, J., Liu, J., Daniel, G., Paszczynski, A., Fekete, F., Krishnamurthy, S., Lu, L., and Xu, G. (1997). Low molecular weight chelators and phenolic compounds isolated from wood decay fungi and their role in the fungal biodegradation of wood. *J. Biotechnol.* **53**, 133–162.

Gorbaty, M. L., and Kelemen, S. R. (2001). Characterization and reactivity of organically bound sulfur and nitrogen fossil fuels. *Fuel Process. Technol.* **71**, 71–78.

Gramss, G., Ziegenhagen, C., and Sorge, S. (1999). Degradation of soil humic extract by wood- and soil-associated fungi, bacteria and commercial enzymes. *Microbial Ecol.* **37**, 140–151.

Granit, T., Chen, Y., and Hadar, Y. (2007). Humic acids bleaching by white-rot fungi isolated from biosolids compost. *Soil Biol. Biochem.* **36**, 1040–1046.

Guggenberger, G., and Haider, K. M. (2002). Effects of mineral colloids on biogeochemical cycling of C, N, P and S in soil. In *Interactions between Soil Particles and Microorganisms. Impact on the Terrestrial Ecosystem*, Huang, P. M., Bollag, J.-M., and Senesi, N., eds., IUPAC Series on Analytical and Physical Chemistry of Environmental Systems. Vol. 8, John Wiley & Sons, Chichester, UK, pp. 267–322.

Haffenden, L. J. W., and Yaylayan, V. A. (2005). Mechanism of formation of redox-reactive hydroxylated benzenes and pyrazine in ^{13}C-labled glycine/D-glucose model systems. *J. Agric. Food Chem.* **53**, 9742–9746.

Haider, K. (1992). Problems related to the humification processes in soils of temperate climates. In *Soil Biochemistry*, Vol. 7, Stozky, G., and Bollag, J.-M., eds., Marcel Dekker, New York, pp. 55–94.

Haider, K., and Martin, J. P. (1981). Decomposition in soil of specifically ^{14}C-labled model and cornstalk lignins and coniferylalcohol over two years as influenced by drying, rewetting, and additions of an available C substrate. *Soil Biol. Biochem.* **13**, 447–452.

Haider, K., and Martin, J. P. (1988). Mineralization of ^{14}C-labelled humic acids and of humic acid-bound ^{14}C-xenobiotics by *Phanerochaete chryosporium. Soil Biol. Biochem.* **20**, 425–429.

Haider, K. M., Martin, J. P., and Filip, Z. (1975). Humus biochemistry. In *Soil Biochemistry*, Vol. **4**, Paul, E. A., and McLaren, A. D., eds., Marcel Dekker, New York, pp. 195–244.

Haider, K., Martin, J. P., and Rietz, E. (1977). Decomposition of ^{14}C-labelled coumaryl alcohols; free and linked to dehydropolymer and plant lignins and model humic acids. *Soil. Sci. Soc. Am. J.* **41**, 556–562.

Hamer, U., Marschner, B., Brodowski, S., and Amelung, W. (2004). Interactive priming of black carbon and glucose mineralization. *Org. Geochem.* **35**, 823–830.

Hammel, K. E. (1997). Fungal degradation of lignin. In *Driven by nature—Plant Litter Quality and Decomposition*, Cadisch, G., and Giller, K. E., eds., CAB International, Wallingford, UK, pp. 33–45.

Hardie, A. G., Dynes, J. J., Kozak, L. M., and P. M. Huang. (2007). Influence of polyphenols on the integrated polyphenol–Maillard reaction humification pathway as catalyzed by birnessite. *Annals Environ. Sci.* **1**, 91–110.

Hatcher, P. G., Schnitzer, M., Dennis, L. W., and Maciel, G. E. (1981). Aromaticity of humic substances in soils. *Soil Sci. Soc. Am. J.* **45**, 1089–1094.

Hatcher, P. G., and Spiker, E. C. (1988). Selective degradation of plant biomolecules. In *Humic Substances and Their Role in the Environment*, Frimmel, F. H., and Christman, R. F., eds., Wiley-Interscience, New York, pp. 45–58.

Hauser, E. A., and Legget, M. B. (1940). Color reactions between clays and amines. *J. Am. Chem. Soc.* **62**, 1811–1814.

Hayes, M. H. B. (1991). Concept of the origin, composition, and structure of humic substances. In *Advances in Soil Organic Matter Research: The Impact on Agriculture and the Environment*, Wilson, W. S., ed., Royal Society of Chemistry, Cambridge, UK, pp. 3–22.

Hedges, J. I. (1978). The formation of clay mineral reactions of melanoidins. *Geochim. Cosmochim. Acta* **42**, 69–79.

Hidalgo, F. J., and Zamora, R. (2000). The role of lipids in nonenzymatic browning. *Grasa Aceites* **51**, 35–49.

Himmelwright, R. S., Eickman, N. C., LuBien, C. D., Lerch, K., and Solomon, E. I. (1980). Chemical and spectroscopic studies of the binuclear copper active site of *Neurospora tyrosinase*: Comparison to hemocyanins. *J. Am. Chem. Soc.* **102**, 7339–7344.

Ho, C.-T. (1996). Thermal generation of Maillard aromas. In *The Maillard Reaction. Consequences for the Chemical and Life Sciences*, Ikan, R., ed., John Wiley & Sons, Chichester, UK, pp. 27–35.

Hoffmann, M. R. (1980). Trace metal catalysis in aquatic environments. *Environ. Sci. Technol.* **14**, 1061–1066.

Hu, W.-G., Mao, J., Xing, B., Schmitt-Rohr, K. (2000). Poly(methylene) crystallites in humic substances detected by nuclear magnetic resonance. *Environ. Sci. Technol.* **34**, 530–534.

Huang, P. M. (1990). Role of soil minerals in transformation of natural organics and xenobiotics in soil. In *Soil Biochemistry* Vol. **6**, Bollag, J.-M., and Stotzky, G., eds., Marcel Dekker, New York, pp. 29–115.

Huang, P. M. (2000). Abiotic catalysis. In *Handbook of Soil Science*, Sumner, M. E., ed., CRC Press, Boca Raton, FL, pp. B302–B332.

Huang, P. M. (2004). Soil mineral–organic matter–microorganism interactions: Fundamentals and impacts. *Adv. Agron.* **82**, 391–472.

Huang, P. M., Wang, M. K., Kämpf, N., and Schulze, D. G. (2002). Aluminum hydroxides. In *Soil Mineralogy with Environmental Applications*, Dixon, J. B., and Schulze, D. G., eds., Soil Science Society of America, Madison, WI, pp. 261–289.

Huang, P. M., Wang, M. K., and Chiu, C. Y. (2005). Soil mineral–organic matter–microbe interactions: Impacts on biogeochemical processes and biodiversity in soils. *Pedobiologia* **49**, 609–635.

Hueey, J. E. (1983). *Inorganic Chemistry. Principles of Structure and Reactivity*. Harper International SI Edition, London, UK.

Ikan, R. Y., Rubinsztain, Y., Nissenbaum, A., and Kaplan, I. R. (1996). Geochemical aspects of the Maillard reaction. In *The Maillard Reaction: Consequences for the Chemical and Life Sciences*, Ikan, R., ed., John Wiley & Sons, Chichester, UK, pp. 1–25.

International Centre for Diffraction Data (ICDD) (2006). Powder diffraction file: Organic and organometallic phases search manual for experimental patterns. International Centre for Diffraction Data, Newton Square, PA.

Jokic, A., Frenkel, A. I., and Huang, P. M. (2001a). Effect of light on birnessite catalysis of the Maillard reaction and its implication in humification. *Can. J. Soil Sci.* **81**, 277–283.

Jokic, A., Frenkel, A. I., Vairavamurthy, M. A., and Huang, P. M. (2001b). Birnessite catalysis of the Maillard reaction: Its significance in natural humification. *Geophys. Res. Lett.* **28**, 3899–3902.

Jokic, A., Zimpel, Z., Huang, P. M, and Mezey, P. G. (2001c). Molecular shape analysis of a Maillard reaction intermediate. *SAR and QSAR Environ. Res.* **12**, 297–307.

Jokic, A., Schulten, H.-R., Cutler, J. N., Schnitzer, M., and Huang P. M. (2004a). A significant abiotic pathway for the formation of *unknown* nitrogen in nature. *Geophys. Res. Lett.* **31**, L05502.

Jokic, A., Wang, M. C., Liu, C. Frenkel, A. I., and Huang, P. M. (2004b). Integration of the polyphenol and Maillard reactions into a unified abiotic pathway for humification in nature: the role of δ-MnO_2. *Org. Geochem.* **35**, 747–762.

Jokic, A., Schulten, H.-R., Cutler, J. N., Schnitzer, M., and Huang, P. M. (2005). Catalysis of the Maillard reaction by δ-MnO_2: A significant abiotic sorptive condensation pathway for the formation of refractory N-containing biogeomacromolecules in nature. In *Soil Abiotic and Biotic Interactions and Impact on the Ecosystem and Human Welfare*, Huang, P. M., Violante, A., Bollag, J.-M., and Vityakon, P., eds., Science Publishers., Enfield, NJ, pp. 127–152.

Kalbitz, K., Schmerwitz, J., Schwesig, D., and Matzner, E. (2003). Biodegradation of soil-derived dissolved organic matter as related to its properties. *Geoderma* **113**, 273–291.

Kassim, G., Martin J. P., and Haider, K. (1981). Incorporation of a wide variety of organic substrate carbons into soil biomass as estimated by the fumigation procedure. *Soil Sci. Soc. Am. J.* **45**, 1106–1112.

Kelleher, B. P., and Simpson, A. J. (2006). Humic substances in soils: Are they really chemically distinct? *Environ. Sci. Technol.* **40**, 4605–4611.

Kiem, R., and Kögel-Knabner, I. (2003). Contribution of lignin and polysaccharides to the refractory carbon pool in C-depleted arable soils. *Soil Biol. Biochem.* **35**, 101–118.

Kirk, T. K., and Farrell, R. L. (1987). Enzymatic combustion—The microbial-degradation of lignin. *Annu. Rev. Microbiol.* **41**, 465–505.

Knicker, H. (2004). Stabilization of N-compounds in soil and organic-matter-rich sediments—what is the difference? *Marine Chem.* **92**, 167–195.

Knicker, H., and Lüdemann, H.-D. (1995). N-15 and C-13 CPMAS and solution NMR of N-15 enriched plant material during 600 days of microbial degradation. *Org. Geochem.* **23**, 329–341.

Knicker, H., González-Vila, F. J., Polvillo, O., González, J. A., and Almendros, G. (2005). Fire-induced transformation of C- and N-forms in different organic soil fractions from a Dystric Cambisol under a Mediterranean pine forest (*Pinus pinaster*). *Soil Biol. Biochem.* **37**, 701–718.

Knollenberg, R. W., Merritt, R. W., and Lawson, D. L. (1985). Consumption of leaf litter by *Lumbricus terrestris* (Oligochaeta) in a Michigan woodland floodplain. *Am. Midl. Nat.* **113**, 1–6.

Koenigs, J. W. (1974). Fungal production of hydrogen peroxide in wood. *Arch. Microbiol.* **99**, 129–145.

Kögel-Knabner, I. (2002). The macromolecular organic composition of plant and microbial residues as inputs to soil organic matter. *Soil Biol. Biochem.* **34**, 139–162.

Kumada, K., and Kato, H. (1970). Browning of pyrogallol as affected by clay minerals. *Soil Sci. Plant Nutr.* **16**, 195–200.

Kyuma, K., and Kawaguchi, K. (1964). Oxidative changes of polyphenols as influenced by allophone. *Soil Sci. Soc. Am. Proc.* **28**, 371–374.

Ladd, J. N., and Butler, J. H. A. (1975). Humus–enzyme systems and synthetic, organic polymer–enzyme analogs. In *Soil Biochemistry*, Vol. **4**, Paul, E. A., and McLaren, A. D., eds., Marcel Dekker, NewYork, 143–194.

Lee, J. S. K., and Huang, P. M. (1995). Photochemical effect on the abiotic transformation of polyphenols as catalyzed by Mn(IV) oxide. In *Environmental Impact of Soil Component Interactions,* Vol. 1: *Natural and Anthropogenic Organics*, Huang, P. M., Berthelin, J., Bollag, J.-M., McGill, W. B., and Page, A. L., eds., Lewis Publishers, Boca Raton, FL, pp. 177–189.

Leonowicz, A., Cho, N.-S., Luterek, J., Wilkolazka, A., Wojtas-Wasilewska, M., Matuszewska, A., Hofrichter, M., Wesenberg, D., and Rogalski. J. (2001). Fungal laccase: properties and activity on lignin. *J. Basic Microbiol.* **41**, 185–227.

Leschine, S. B. (1995). Cellulose degradation in anaerobic environments. *Annu. Rev. Microbiol.* **49**, 399–426.

Liu, C., and Huang, P. M. (2000). Catalytic effects of hydroxyl-aluminum and silicic acid on catechol humification. In: *Humic Substances: Versatile Components of Plants, Soil and Water*, Ghabbour, E., and Davies, G., eds., Royal Society of Chemistry, Cambridge, UK, pp. 37–51.

Liu, C., and Huang, P. M. (2001). The influence of catechol humification on surface properties of metal oxides. In *Humic Substances: Structure, Models and Functions*, Ghabbour, E. A. and Davies, G., eds., Royal Society of Chemistry, Cambridge, UK, pp. 253–270.

Liu, C., and Huang, P. M. (2002). Role of hydroxyl-aluminosilicate ions (proto-imogolite soil) in the formation of humic substances. *Org. Geochem.* **33**, 295–305.

Lopez, M. J., Vargas-Garcia, M. C., Suárez-Estrella, F., and Moreno, J. (2006). Biodelignification and humification of horticultural plant residues by fungi. *Int. Biodeter. Biodegrad.* **57**, 24–30.

Lu, Y., and Pignatello, J. J. (2002). Determination of the "conditioning effect" in soil organic matter in support of a pore deformation mechanism for sorption hysteresis. *Environ. Sci. Technol.* **36**, 4553–4561.

Lu, Y., and Pignatello, J. J. (2004). Sorption of apolar aromatic compounds to soil humic acid particles affected by aluminum (III) ion cross-linking. *J. Environ. Qual.* **33**, 1314–1321.

MacCarthy, P. (2001). The principles of humic substances, *Soil Sci.* **166**, 738–751.

Maillard, L. C. (1913). Formation de matières humiques par action de polypeptides sur sucres, *C.R. Acad. Sci.* **156**, 148–149.

Majecher, E. M., Chorover, J., Bollag, J.-M., and Huang, P. M. (2000). Evolution of CO_2 during birnessite-induced oxidation of ^{14}C-labeled catechol. *Soil Sci. Soc. Am. J.* **64**, 157–163.

Martin, J. P., and Haider, K. (1971). Microbial activity in relation to soil humus formation. *Soil Sci.* **111**, 54–63.

Martin, J. P, and Haider, K. (1980). A comparison of the use of phenolase and peroxidase for the synthesis of model humic acid-type polymers. *Soil Sci. Soc. Am. J.* **44**, 983–988.

Martínez, A. T. (2002). Molecular biology and structure-function of lignin-degrading heme peroxidases. *Enz. Microb. Technol.* **30**, 425–444.

Martínez, A. T., Speranza, M., Ruiz-Dueñas, F. J., Ferreira, P., Camarero, S., Guillén, F., Martínez, M. J., Gutiérrez, A., and del Río, J. C. (2005). Biodegradation of lignocellulosics: Microbial, chemical, and enzymatic aspects of the fungal attack of lignin. *Int. Microbiol.* **8**, 195–204.

Marthur, S. P., and Schnitzer, M. (1978). A chemical and spectroscopic characterization of some synthetic analogues of humic acid. *Soil Sci. Soc. Am. J.* **42**, 591–596.

McBride, M. B. (1987). Adsorption and oxidation of phenolic compounds by iron and manganese oxides. *Soil Sci. Soc. Am. J.* **51**, 1466–1472.

McBride, M. B. (1989). Oxidation of 1,2- and 1,4-dihydroxybenzenes in aerated aqueous suspensions of birnessite. *Clays Clay Miner.* **37**, 470–486.

McLean, J. S., Lee, J.-U., and Beverige, T. L. (2002). Interactions of bacteria and environmental metals, fine-grained mineral development and bioremediation strategies. In *Interactions Between Soil Particles and Microorganisms. Impact on the Terrestrial Ecosystem*, Huang, P. M., Bollag, J.-M., and Senesi, N., eds., IUPAC Series on Analytical and Physical Chemistry of Environmental Systems, Vol. **8**, John Wiley & Sons, Chichester, UK, pp. 227–261.

McNaught, A. D., and Wilkison, A. (1997). *Compendium of Chemical Terminology*, 2nd edition, IUPAC Recommendation. Blackwell Science, Oxford, UK.

Melillo, J. M., Aber, J. D., Linkins, A. E., Ricca, A., Fry, B., and Nadelhoffer, K. J. (1989). Carbon and nitrogen dynamics along the decay continuum: plant litter to soil organic matter. *Plant Soil* **115**, 189–198.

Mikutta, R., Kleber, M., Torn, M. S., and Jahn, R. (2006). Stabilization of soil organic matter: association with minerals or chemical recalcitrance? *Biogeochem.* **77**, 25–56.

Mitra-Kirtley, S., Mullins, O. C., Branthauer, J. F., and Cramer, S. P. (1993). Determination of the nitrogen chemical structures in petroleum asphaltenes using XANES spectroscopy. *J. Am. Chem. Soc.* **115**, 252–258.

Moore, J. W., and Pearson, R. G. (1981). *Kinetics and Mechanisms*, 3rd edition, John Wiley & Sons, New York.

Mortland, M. M., and Halloran, L. J. (1976). Polymerization of aromatic molecules on smectites. *Soil Sci. Soc. Am. J.* **40**, 367–370.

Nadelhoffer, K. J., and Fry, B. (1988). Controls on natural N-15 and C-13 abundances in forest soil organic matter. *Soil Sci Soc. Am. J.* **52**:1633–1640.

Naidja, A., and Huang, P. M. (2002). Significance of the Henri–Michaelis–Menten theory in abiotic catalysis: catechol oxidation by δ-MnO_2. *Surface Sci.* **506**, L243–L249.

Naidja, A., Huang, P. M., and Bollag, J.-M. (1997). Activity of tyrosinase immobilized on hydroxyaluminum–montmorillonite complexes. *J. Mol. Catal. A: Chemical* **115**, 305–316.

Naidja, A., Huang, P. M., and Bollag, J.-M. (1998). Comparison of the reaction products from the transformation of catechol catalyzed by birnessite or tyrosinase. *Soil Sci. Soc. Am. J.* **62**, 188–195.

Naidja, A., Huang, P. M., Dec, J., and Bollag, J.-M. (1999). Kinetics of catechol oxidation catalyzed by tyrosinase or δ-MnO_2. In *Effect of Mineral–Organic–Microorganism Interactions on Soil and Freshwater Environments*, ed. Berthelin, J., Huang, P. M., Bollag, J.-M., and Andreux, F., eds., Kluwer Academic/Plenum Publishers, New York, pp. 181–188.

Naidja, A., Huang, P. M., and Bollag, J.-M. (2000). Enzyme–clay interactions and their impact on transformations of natural and anthropogenic organic compounds in soil. *J. Environ. Qual.* **29**, 677–691.

Naidja, A., Liu, C., and Huang, P. M. (2002). Formation of protein-birnessite complex: XRD, FTIR, and AFM analysis. *J. Coll. Interface Sci.* **251**, 46–56.

O'Brien, B. J., and Stout, J. D. (1978). Movement and turnover of soil organic matter as indicated by carbon isotope measurements. *Soil Biol. Biochem.* **10**, 309–317.

Oades, J. M. (1988). The retention of organic matter in soils. *Biogeochemistry.* **5**, 35–70.

Ostrofsky, A., Jellison, J., Smith, K. T., and Shortle, W. C. (1997). Changes in cation concentrations in red spruce wood decayed by brown rot and white rot fungi. *Can. J. Forest Res.* **27**, 567–571.

Pal, S., Bollag, J.-M., and Huang, P. M. (1994). Role of abiotic and biotic catalysts in the transformation of phenolic compounds through oxidative coupling reactions. *Soil Biol. Biochem.* **26**, 813–820.

Park, J.-W., Dec. J., Kim, J.-E., and Bollag, J.-M.. (1999). Effect of humic constituents on the transformation of chlorinated phenols and anilines in the presence of oxidoreductive enzymes or birnessite. *Environ. Sci. Technol.* **33**, 2028–2034.

Paszczynski, A., Crawford, R., Funk, D., and Goodell, B., (1999). *De novo* synthesis of 4,5-dimethoxycatechol and 2,5-dimethoxyhydroquinone by the brown rot fungus *Gloeophyllum trabeum. Appl. Environ. Microbiol.* **65**, 674–679.

Percival, H. J., Parfitt, R. L., and Scott, N. A. (2000). Factors controlling soil carbon levels in New Zealand grasslands: Is clay content important? *Soil Sci. Soc. Am. J.* **64**, 1623–1630.

Pérez, J., Muňoz-Dorado, J., de la Rubia, T., and Martínez, J. (2002). Biodegradation and biological treatments of cellulose, hemicellulose and lignin: An overview. *Int. Microbiol.* **5**, 53–63.

Piccolo, A. (2001). The supramolecular structure of humic substances. *Soil Sci.* **166**, 810–832.

Pinnavaia, T. J., Hall, P. L., Cady, S. S., and Mortland, M. M. (1974). Aromatic radical cation formation on the intracrystal surfaces of transition metal layer silicates. *J. Phys. Chem.* **78**, 994–999.

Pohlman, A. A., and McColl, J. G. (1989). Organic oxidation and manganese and aluminum mobilization in forest soils. *Soil Sci. Soc. Am. J.* **53**, 686–690.

Poirer, N., Derenne, S., Balesdent, J., Rouzaud, J.-N., Mariotti, A., and Largeau, C. (2000). Chemical structure and sources of the macromolecular, resistant, organic fraction isolated from a forest soil (Lacadec, Southwest France). *Org. Geochem.* **31**, 813–827.

Poirer, N., Derenne, S., Balesdent, J., Rouzaud, J.-N., Mariotti, A., and Largeau, C. (2002). Abundance and composition of the refractory organic fraction of an ancient, tropical soil (Poine Noire, Congo). *Org. Geochem.* **33**, 383–391.

Poirer, N., Derenne, S., Balesdent, J., Chenu, C., Bardoux, G., Mariotti, A., and Largeau, C. (2006). Dynamics and origin of the non-hydrolysable organic fraction in a forest and a cultivated temperate soil, as determined by isotopic and microscopic studies. *Eur. J. Soil Sci.* **57**, 719–730.

Ponomarenko, E. V., and Anderson, D. W. (2001). Importance of charred organic matter in Black Chernozem soils of Saskatchewan. *Can. J. Soil Sci.* **81**, 281–297.

Porterfield, W. W. (1983). *Inorganic Chemistry. A Unified Approach.* Harper International SI Edition, London, UK.

Preston, C. M., Rauthan, B. S., Rodger, C., and Ripmeester, J. A. (1982). A hydrogen-1, carbon-13, and nitrogen-15 nuclear magnetic resonance study of *p*-benzoquinone polymers incorporating amino nitrogen compounds ("synthetic humic acid"). *Soil Sci.* **134**, 277–293.

Preston, C. M., and Schmidt, M. W. I. (2006). Black (pyrogenic) carbon: A synthesis of current knowledge and uncertainties with special consideration of boreal regions. *Biogeosciences* **3**, 397–420.

Preston, C. M., Trofymow, J. A., and Flanagan, L. B. (2006). Decomposition, ^{13}C, and the "lignin paradox." *Can. J. Soil Sci.* **86**, 235–245.

Quénéa, K., Largeau, C., Derenne, S., Spaccini, R., Bardoux, G., and Mariotti, A. (2006a). Molecular and isotopic study of lipids in particle size fractions of a sandy cultivated soil (Cestas cultivation sequence, southwest France): Sources, degradation, and comparison with Cestas forest soil. *Org. Geochem.* **37**, 20–44.

Quénéa, K., Derenne, S., Largeau, C., Rumpel, C., and Mariotti, A. (2006b). Influence of change in land use on the refractory organic macromolecular fraction of a sandy spodosol (Landes de Gascogne, France). *Geoderma* **136**, 136–151.

Rasmussen, C., Torn, M. S., and Southard, R. J. (2005). Mineral assemblage and aggregation control carbon dynamics in a Californian conifer forest. *Soil Sci. Soc. Am. J.* **69**, 1711–1721.

Rasse, D. P., Dignac, M.-F., Bahri, H., Rumpel, C., Mariotti, A., and Chenu, C. (2006). Lignin turnover in an agricultural field: from plant residues to soil-protected fractions. *Eur. J. Soil Sci.* **57**, 530–538.

Rice, J. A. (2001). Humin. *Soil Sci.* **166**, 848–857.

Ross, S. (1989). *Soil Processes. A Systematic Approach*. Routledge, London, UK.

Rouet-Mayer, M.-S., Ralambosoa, J., and Philippon, J. (1990). Roles of *o*-quinones and their polymers in the enzymic browning of apples. *Phytochem.* **29**, 435–440.

Ruggiero, P., and Radogna, V. M. (1988). Humic acids–tyrosinase interactions as a model of humic–enzyme complexes. *Soil Biol. Biochem.* **20**, 353–359.

Ruggiero, P., Sarkar, J. M., and Bollag, J.-M. (1989). Detoxification of 2,4-dichlorophenol by a laccase immobilized on soil and clay. *Soil Sci.* **147**, 361–370.

Rumpel, C., Alexis, M., Chabbi, A., Chaplot, V., Rasse, D. P., Valentin, C., and Mariotti, A. (2006). Black carbon contribution to soil organic matter composition in tropical sloping land under slash and burn agriculture. *Geoderma* **130**, 35–46.

Schaumann, G. E. (2006a). Review articles. Soil organic matter beyond molecular structure. Part I: Macromolecular and supramolecular characteristics. *J. Plant Nutr. Soil Sci.* **169**. 145–156.

Schaumann, G. E. (2006b). Review article. Soil organic matter beyond molecular structure. Part II: Amorphous nature and physical aging. *J. Plant Nutr. Soil Sci.* **169**, 157–167.

Scheffer, F., Meyer, B., and Niederbudde, E. A. (1959). Huminstoffbildung unter katalytischer Einwirkung natürlich vorkommender Eisenverbindungen im Modelversuch. *Zeit. Pflanzenern. Dueng. Bodenk.* **87**, 26–44.

Schlesinger, W. H. (1990). Evidence from chronosequence studies for a low carbon-storage potential of soils, *Nature* **348**, 232–234.

Schmidt, M. W. I., Skjemstad, J. O., and Jäger, C. (2002). Carbon isotope geochemistry and nanomorphology of soil black carbon: Black chernozemic soils in central Europe originate from ancient biomass burning, *Global Biogeochem. Cycles* **16**, Art. No. 1123.

Schnitzer, M., (1977). Recent findings on the characterization of humic substances extracted from soils from widely differing climatic zones. In *Proceedings, Symposium on Soil Organic Matter Studies II*, IAEA Bulletin, Vienna, Austria, 77–101.

Schnitzer, M. (1978). Humic substances: Chemistry and reactions. In *Soil Organic Matter*, Schnitzer, M., and Khan, S. U., eds., Elsevier, Amsterdam, The Netherlands, pp. 1–64.

Schnitzer, M. (1986). Binding of humic substances by soil minerals. In *Interactions of Soil Minerals with Natural Organics and Microbes*, Huang, P. M., and Schnitzer, M., eds., SSSA Special Publication No. 17, Soil Science Society of America, Madison, WI, 77–101.

Shindo, H. (1992). Relative effectiveness of short-range ordered Mn(IV), Fe(III), Al, and Si oxides in the synthesis of humic acids from phenolic compounds. *Soil Sci. Plant Nutr.* **38**, 459–465.

Shindo, H., and Huang, P. M. (1982). Role of Mn(IV) oxide in abiotic formation of humic substances in the environment. *Nature* **298**, 363–365.

Shindo, H., and Huang, P. M. (1984a). Catalytic effects of manganese(IV), iron(III), aluminum, and silicon oxides on the formation of phenolic polymers. *Soil Sci. Soc. Am. J* **48**, 927–934.

Shindo, H., and Huang, P. M. (1984b). Significance of Mn(IV) oxide in abiotic formation of organic nitrogen complexes in natural environments. *Nature* **308**, 57–58.

Shindo, H., and Huang, P. M. (1985a). Catalytic polymerization of hydroquinone by primary minerals. *Soil Sci.* **139**, 505–511.

Shindo, H., and Huang, P. M. (1985b). The catalytic power of inorganic components in the abiotic synthesis of hydroquinone-derived humic polymers. *Appl. Clay Sci.* **1**, 71–81.

Shindo, H., and Huang, P. M. (1992). Comparison of the influence of Mn(IV) oxide and tyrosinase on the formation of humic substances in the environment. *Sci. Total Environ.* **117/118**, 103–110.

Shindo, H., Honna, T., Yamamoto, S., and Honma, H. (2004). Contribution of charred plant fragments to soil organic carbon in Japanese volcanic ash soils containing black humic acids. *Org. Geochem.*, **35**, 235–241.

Shindo, H., Yoshida, M., Yamamoto, A., Honma, H., and S. Hiradate. (2005). $\delta^{13}C$ values of organic constituents in Japanese volcanic ash soils. *Soil Sci.* **170**, 175–182.

Simpson, A. J. (2002). Determining the molecular weight, aggregation, structures and interactions of natural organic matter using diffusion ordered spectroscopy. *Magnet. Resonan. Chem.* **40**, S72-S82.

Simpson, A. J., Kingery, W. L., and Hatcher, P. G. (2003). The identification of plant derived structures in humic materials using 3-dimensional NMR spectroscopy. *Environ. Sci. Technol.* **37**, 337–342.

Six, J., Callewaert, P., Lenders, S., De Gryze, S., Morris, S. J., Gregorich, E. G., and Paul, E. A. (2002). Measuring and understanding carbon storage in afforested soils by physical fractionation. *Soil Sci. Soc. Am. J.* **66**, 1981–1987.

Sjoblad, R. D., and Bollag, J.-M. (1981). Oxidative coupling of aromatic compounds by enzymes from soil microorganisms. In *Soil Biochemistry,* Vol. 5, Paul, E. A., and Ladd, J. N., eds., Marcel Dekker, NewYork, 113–125.

Skjemstad, J. O., Clarke, P., Taylor, J. A., Oades, J. M., and McClure, S. G., (1996). The chemistry and nature of protected carbon in soil. *Aust. J. Soil Res.* **34**, 251–271.

Smernik, R. J., and Baldock, J. A. (2005). Does solid-state ^{15}N NMR spectroscopy detect all soil organic matter nitrogen? *Biogeochem.* **75**, 507–528.

Smith, M. B., and March, J. (2001). *March's Advanced Organic Chemistry Reactions, Mechanisms, and Structure*, 5th edition, John Wiley & Sons, New York.

Solomon, D. H. (1968). Clay minerals as electron acceptors and/or electron donors in organic reactions. *Clays Clay Miner.* **16**, 31–39.

Solomon, D. H., and Hawthorne, D. G. (1983). *Chemistry of Pigments and Fillers*, John Wiley & Sons, New York.

Spaccini, R., Piccolo, A., Conte, P., Haberhauer, G., and Gerzabek, M. H. (2002). Increased soil carbon sequestration through hydrophobic protection by humic substances. *Soil Biol. Biochem.* **34**, 1839–1851.

Sparks, D. L. (2003). *Environmental Soil Chemistry*, 2nd edition Academic Press, San Diego, CA.

Steed, J. W., and Atwood, J. L. (2000). *Supramolecular Chemistry: A Concise Introduction*, John Wiley & Sons, Chichester, UK.

Steelink, C. (1994). Application of N-15 NMR spectroscopy to the study of organic nitrogen and humic substances in the soi*l*. In *Humic Substances in the Global Environment*, Senesi, N. and Miano, D., eds., Elsevier, Amsterdam, the Netherlands, 405–427.

Steffen, K. T., Hattaka, A., and Hofrichter, M. (2002). Degradation of humic acids by the litter-decomposing basidomycete *Collybia dryophila. Appl. Environ. Microbiol.* **68**, 3442–3448.

Stevenson, F. J. (1994). *Humus Chemistry: Genesis, Composition, Reactions*, 2nd edition, John Wiley & Sons, New York.

Stevenson, I. L., and Schnitzer, M. (1982). Transmission electron microscopy of extracted fulvic and humic acids. *Soil Sci.* **133**, 179–185.

Sutton, R., and Sposito, G. (2005). Molecular structure in soil humic substances: the new view. *Environ. Sci. Technol.* **39**, 9009–9014.

Telysheva, G., Dobele, G., Meier, D., Dizhbite, T., Rossinka, G., and Jurkjane, V. (2007). Characterization of the transformation of lignocellulosic structures upon degradation in planted soil. *J. Anal. Appl. Pyrolysis* **79**, 52–60.

ten Have, R., and Theunissen, T. M. J. (2001). Oxidative mechanisms involved in lignin degradation by white-rot fungi. *Chem. Rev.* **101**, 3397–3413.

Thompson, T. D., and Moll, J. F. (1973). Oxidative power of smectites measured by hydroquinone. *Clays Clay Miner.* **21**, 337–350.

Tietjen, T., and Wetzel, R. G. (2003). Extracellular enzyme–clay mineral complexes: Enzyme adsorption, alteration of enzyme activity, and protection from photodegradation. *Aq. Ecology* **37**, 331–339.

Torn, M. S., Trumbore, S. E., Chadwick, O. A., Vitousek, P. M., and Hendriks, D. M. (1997). Mineral control of soil organic carbon storage and turnover. *Nature* **289**, 170–173.

Torres, E., Bustos-Jaimes, I., and Le Borgne, S. (2003). Potential use of oxidative enzymes for the detoxification of organic pollutants. *Appl. Catalys. B.* **46**, 1–15.

Tuomela, M., Oivanen, P., and Hatakka, A. (2002). Degradation of synthetic ^{14}C-lignin by various white-rot fungi in soil. *Soil Biol. Biochem.* **34**, 1613–1620.

Twigg, M. V. (1989). *Catalyst Handbook*, 2nd edition, Wolfe Publishing, London, UK.

Vairavamurthy, A., and Wang, S. (2002). Organic nitrogen in geomacromolecules: insights on speciation and transformation with K-edge XANES spectroscopy. *Environ. Sci. Technol.* **36**, 3050–3056.

Veldkamp, E. (1994). Organic-carbon turnover in 3 tropical soils under pasture after deforestation. *Soil Sci. Soc. Am. J.* **58**, 175–180.

von Lützow, M., Kögel-Knabner, I., Ekschmitt, K., Matzner, E., Guggenberger, G., Marschner, B., and Flessa, H. (2006). Stabilization of organic matter in temperate soils: mechanisms and their relevance under different soil conditions—A review. *Eur. J. Soil Sci.* **57**, 426–445.

Waksman, S. A. (1936). *Humus*. Williams & Wilkins, Baltimore.

Wang, M. C., and Huang, P. M. (1986). Humic macromolecule interlayering in nontronite through interaction with phenol monomers. *Nature* **323**, 529–531.

Wang, M. C., and Huang, P. M. (1987). Polycondensation of pyrogallol and glycine and the associated reactions as catalyzed by birnessite. *Sci. Total Environ.* **62**, 435–442.

Wang, M. C., and Huang, P. M. (1989a). Abiotic ring cleavage of pyrogallol and the associated reactions as catalyzed by a natural soil. *Sci. Total Environ.* **81/82**, 501–510.

Wang, M. C., and Huang, P. M. (1989b). Catalytic power of nontronite, kaolinite and quartz and their reaction sites in the formation of hydroquinone-derived polymers. *Appl. Clay Sci.* **4**, 43–57.

Wang, M. C., and Huang, P. M. (1989c). Pyrogallol transformations as catalyzed by nontronite, bentonite and kaolinite. *Clays Clay Miner.* **37**, 525–531.

Wang, M. C., and Huang, P. M. (1991). Nontronite catalysis in polycondensation of pyrogallol and glycine and the associated reactions. *Soil Sci. Soc. Am. J.* **55**, 1156–1161.

Wang, M. C., and Huang, P. M. (1992). Significance of Mn(IV) oxide in the abiotic ring cleavage of pyrogallol in natural environments. *Sci. Total Environ.* **113**, 147–157.

Wang, M. C., and Huang, P. M. (1994). Structural role of polyphenols in influencing the ring cleavage and related chemical reactions as catalyzed by nontronite. In *Humic Substances in the Global Environment and Implications on Human Health*, Senesi, N., and Miano, T. M., eds., Elsevier, Amsterdam, The Netherlands, pp. 173–180.

Wang, M. C., and Huang, P. M. (1997). Catalytic power of birnessite in abiotic formation of humic polycondensates from glycine and pyrogallol. In *The Role of Humic Substances in the Ecosystems and in Environmental Protection*, Proceedings, 8th Conference of the International Humic Substances Society, Drozl, J., Gonet, S. S., Senesi, N., and Webber, J., eds., Wroclaw, Poland, pp. 59–65.

Wang, M. C., and Huang, P. M. (2000a). Characteristics of pyrogallol-derived polymers formed by catalysis of oxides. *Soil Sci.* **165**, 737–747.

Wang, M. C., and Huang, P. M. (2000b). Ring cleavage and oxidative transformation of pyrogallol catalyzed by Mn, Fe, Al, and Si, oxides. *Soil Sci.* **165**, 934–942.

Wang, M. C., and Huang, P. M. (2003). Cleavage and polycondensation of pyrogallol and glycine catalyzed by natural soil clay. *Geoderma* **112**, 31–50.

Wang, M. C., and Huang, P. M. (2005). Cleavage of ^{14}C-labled glycine and its polycondensation with pyrogallol as catalyzed by birnessite. *Geoderma* **124**, 415–426.

Wang, M. C., and Lin, C. H. (1993). Enhanced mineralization of amino-acids by birnessite as influenced by pyrogallol. *Soil Sci. Soc. Am. J.* **57**, 88–93.

Wang, T. S. C., and Li, S. W. (1977). Clay minerals as heterogeneous catalysts in preparation of model humic substances. *Zeit. Pflanzenern. Dueng. Bodenk.* **140**, 669–676.

Wang, T. S. C., Kao, M.-M., and Li, S. W. (1978a). A new proposed mechanism of formation of soil humic substance. In: *Studies and Essays in Commemoration of the Golden Jubilee of Academia Sinica*. Academia Sinica, Tapei. Taiwan, pp. 357–372.

Wang, T. S. C., Kao, M.-M., and Ferng, Y. L. (1978b). Catalytic polymerization of phenolic compounds by clay minerals. *Soil Sci.* **126**, 15–21.

Wang, T. S. C., Li, S. W., and Huang, P. M. (1978c). Catalytic polymerization of phenolic compounds by a latosol. *Soil Sci.* **126**, 81–86.

Wang, T. S. C., Wang, M. C., Ferng, Y. L., and Huang, P. M. (1983a). Catalytic synthesis of humic substances by natural clays, silts and soils. *Soil Sci.* **135**, 350–360.

Wang, T. S. C., Wang, M. C., and Huang, P. M. (1983b). Catalytic synthesis of humic substances by using aluminas as catalysts. *Soil Sci.* **136**, 226–246.

Wang, T. S. C., Chen, J.-H., and Hsiang, W.-M. (1985). Catalytic synthesis of humic acids containing various amino acids and dipeptides. *Soil Sci.* **140**, 3–10.

Wang, T. S. C., Huang, P. M., Chou, C.-H., and Chen, J.-H. (1986). The role of soil minerals in abiotic polymerization of phenolic compounds and formation of humic substances. In *Interactions of Soil Minerals with Natural Organics and Microbes*, Huang, P. M., and Schnitzer, M., eds., SSSA Special Publication No. 17, Soil Science Society of America, Madison, WI, pp. 251–281.

Weast, R. C., ed. (1978). *CRC Handbook of Chemistry and Physics*, 58th edition, CRC Press, Boca Raton, FL.

Weiss, M. S., Abele, U., Weckesser, J., Welte, W., Schiltz, E., and Schulz, G. E. (1991). Molecular architecture and electrostatic properties of a bacterial porin. *Science* **254**, 1627–1630.

Wilson, M. A., and Goh, K. M. (1977). Proton-decoupled pulse Fourier-transform ^{13}C nuclear magnetic resonance of soil organic matter. *J. Soil Sci.* **28**, 645–652.

Wolff, S. P. (1996). Free radicals and glycation theory. In *The Maillard Reaction. Consequences for the Chemical and Life Sciences*, Ikan, R., ed., John Wiley & Sons, Chichester, UK, pp. 73–88.

Wolters, V. (2000). Invertebrate control of soil organic matter stability. *Biol. Fertil Soils*, **31**, 1–19.

Wong, J. W., and Shibamoto, T. (1996). Genotoxicity of the Maillard reaction products. In *The Maillard Reaction. Consequences for the Chemical and Life Sciences*, Ikan, R., ed., John Wiley & Sons, Chichester, UK, pp. 129–159.

Xia, G., and Pignatello, J. J. (2001). Detailed sorption isotherms of polar and apolar compounds in a high-organic soil. *Environ. Sci. Technol.* **35**, 84–94.

Xing, B. (2001). Sorption of naphthalene and phenanthrene by soil humic acids. *Environ. Pollut.* **111**, 303–309.

Xu, G., and Goodell, B. (2001). Mechanisms of wood degradation by brown-rot fungi: Chelator-mediated cellulose degradation and binding of iron by cellulose. *J. Biotech.* **87**, 43–57.

Yavmetidinov, I. S., Stepnova, E. V., Gavrilova, V. P., Lokshin, B. V., Perminova, L. V., and Koroleva, O. V. (2003). Isolation and characterization of humin-like substances produced by wood-degrading white rot fungi. *Appl. Biochem. Microbiol.* **39**, 257–264.

Yaylayan, V. A. (1997). Classification of the Maillard reaction: A conceptual approach. *Trends Food Sci. Technol.* **8**, 13–18.

Yu, G., Saha, U. K., Kozak, L. M., and Huang, P. M. (2006). Kinetics of cadmium adsorption on aluminum precipitation products formed under the influence of tannate. *Geochim. Cosmochim. Acta* **70**, 5134–5145.

Zamora, R., and Hidalgo, F. J. (2005). Coordinate contribution of lipid oxidation and Maillard reaction to the nonenzymatic food browning. *Crit. Rev. Food Sci. Nutr.* **45**, 49–59.

Zavarzina, A. G., and Zavarzin, A. A. (2006). Laccase and tyrosinase activities in lichens. *Microbiology.* **75**, 546–556.

Zegouagh, Y., Derenne, S., Largeau, C., Betrand, P., Sicre, M-.A., Sliot, A., and Rousseau, B. (1999). Refractory organic matter in sediments from the north-west African upwelling system: Abundance, chemical structure and origin. *Org. Geochem.* **30**, 83–99.

Ziechmann, W. (1959). Die Darstellung von Huminsauren im heterogenen System mit neutraler Reaktion. *Zeit. Pflanzenern. Dueng. Bodenk.* **84**, 155–159.

3

ORGANO-CLAY COMPLEXES IN SOILS AND SEDIMENTS

G. Chilom and J. A. Rice

Department of Chemistry and Biochemistry, South Dakota State University, Brookings, South Dakota

Biophysico-Chemical Processes Involving Natural Nonliving Organic Matter in Environmental Systems, Edited by Nicola Senesi, Baoshan Xing, and Pan Ming Huang

3.1. THE COMPONENTS OF SOILS AND SEDIMENTS

3.1.1. Natural Organic Matter

Natural organic matter–mineral complexes typically represent more than 50% of the organic carbon in a soil (Kononova, 1966; Stevenson, 1994) and typically more than 70% of the OC in unlithified sediments (Durand and Nicaise, 1980; Peters et al., 1981; Hatcher et al., 1985; Ishiwatari, 1985; Vanderbrouke et al., 1985; Mayer, 1994, 1995; Keil and Hedges, 1995). While interest in carbon storage and carbon sequestration during the past decade have increased the interest in mineral-bound organic matter (see Chapters 5 and 6 of this book), a unique niche in the biogeochemical carbon cycle and a large capacity for binding anthropogenic organic compounds introduced into a natural environment, it is surprising that 150 years after the first studies there is still no general consensus on the fundamental nature of these organo-mineral nanocomposites (OMN).

Soil scientists have generally thought of OMN primarily as natural organic matter (NOM) complexed to inorganic colloids or clays (e.g., Shah et al., 1975a, 1975b; Banerjee, 1979; Theng, 1979; Cloos et al., 1981), probably because of the generally similar elemental composition of (Rice and MacCarthy, 1991) and similar functional group contents (Stevenson, 1994). In soils, OMN is usually referred to as the "humin" fraction of soil organic matter. Its chemistry has been reviewed by Rice (2001). The term protokerogen is often used in organic geochemistry to describe insoluble organic matter in unlithified sediments (Gillet, 1957; Breger 1960; Huc and Durand, 1973, 1977; Cane, 1976; Steurmer et al., 1978; Tissot and Welte, 1978; Ishiwatari, 1985; Vandenbroucke et al., 1985; Vandenbroucke and Largeau, 2007). The definitions of the terms humin in soil and protokerogen (Calvin and Philip, 1976; Stuermer et al., 1978; Durand and Nicaise, 1980; Peters et al., 1981, Reuter and Perdue, 1984; Taylor et al., 1984; Vandenbroucke and Largeau, 2007) in sediments are essentially equivalent operational definitions; the difference between them is that in the isolation of protokerogen the acid- and alkali-insoluble organic matter fraction (i.e., humin) is subsequently treated with a mixture of HF and HC1 to dissolve mineral matter and produce a concentrated organic isolate (i.e., protokerogen; Durand and Nicaise, 1980; Hatcher et al., 1985; Ishiwatari, 1985).

It is only recently that OMN's structural aspects, along with its role in C cycling and sequestration, have begun to be extensively explored. The biogeochemical carbon cycle places OMN at the point in the cycle where organic carbon produced in the biosphere crosses over to be processed as a part of the geosphere (Rice, 2001). The amount of organic carbon transferred between the spheres as soils and sediments are buried and lithified is small, representing 0.001% to 0.1% of the total organic carbon (TOC) on the earth (Tissot and Welte, 1978; Schlesinger, 1991).

There are many biotic and abiotic processes that occur during early diagenesis that serve to transform organic tissue produced by photosynthesis in the biosphere into organic substances that ultimately become part of the processes operating in the geosphere during later stages of organic metamorphism (Tissot and Welte, 1978; Engel and Macko, 1993; Killops and Killops, 2005). A part of this process is the removal of more labile components of the organic C input into soils or sediments. Consequently, OMN appears to be the oldest of the three humic fractions. Carbon-14 dating typically indicates that OMN organic carbon is ~1000 years old (Goh and

Malloy, 1978; Skujins and Klubeck, 1982; Stevenson, 1994; Anderson and Paul, 1984; Balesdent, 1987; Boulet et al., 1995; Lichtfouse et al., 1995; Pessenda et al., 1998; Gouveia et al., 1999). It has been suggested that the mechanism by which nature slows mineralization of OM may be by the accumulation of an ill-defined, amorphous, and heterogeneous mixture of organic molecules (Swaby and Ladd, 1962, 1966; MacCarthy and Rice, 1991). This mixture would require either a very large assemblage of enzymes or an uncharacteristically versatile enzyme to effect its rapid mineralization.

Numerous studies in the literature demonstrate that a substantial amount (typically >50%) of essentially any organic contaminant introduced into a soil or sediment system is ultimately bound to OMN. Herbicides, insecticides, fungicides, PCBs, and PAHS are bound rapidly and irreversibly to humin, forming what are referred to as bound residues (e.g., Kloskowski and Führ, 1985, 1987; Kloskowski et al., 1986a, 1986b; Xie et al., 1997; Kohl and Rice, 1998). When considered from either an organic or environmental geochemical perspective, it is clear that the contributions of OMN are significant in understanding the processes involved.

We have relied on the NMR spectrum of NOM in organizing this chapter. The NMR spectrum can be conveniently separated into chemical shift regions that are attributed to the major organic carbon component classes present in NOM: aliphatic, protein, carbohydrate, and aromatic (i.e., lignin-derived) carbon types (Figure 3.1). While it is beyond the scope and requirements of this chapter to provide a detailed review of the nature and chemistry of each of these organic materials and the underlying mineral substrates in OMN, the following sections will present brief introductions to each of them in order to provide some background and the interested reader with a starting point in the literature.

3.1.1.1. Lipids. In a geochemical context, lipids are defined as organic compounds that can be extracted with nonpolar organic solvents (e.g., hexane, chloroform, ether) or solvent mixtures (e.g., benzene/methanol) (Breger, 1960). This definition accommodates a diverse group of compounds that includes saturated, unsaturated,

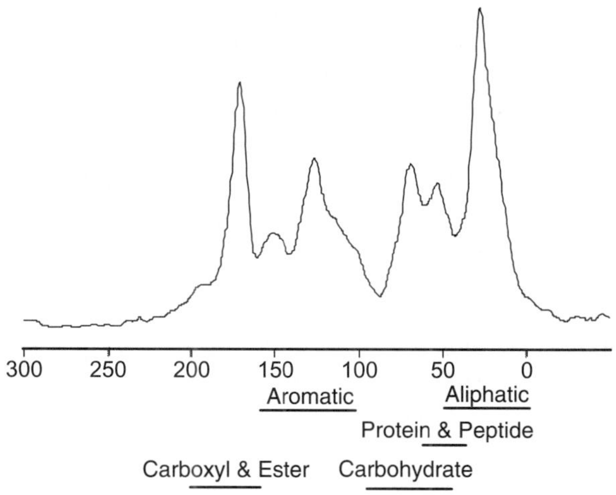

Figure 3.1. Solid-state ^{13}C NMR DPMAS spectrum of a peat humic acid showing chemical shift regions of typically observed organic matter components.

cyclic and aromatic hydrocarbons, fatty acids, fatty alcohols, and so on, and more complex materials such as phospholipids, suberans, cutans, waxes, and resins. These materials may or may not contain heteroatoms (O, N, S, P) and can be functionalized. In soils, lipids represent a relatively small proportion of the total organic carbon (TOC), typically 2–6% (Stevenson, 1994). In sediments, lipids typically represent less than 5% of the TOC and are comprised of compounds such as fatty acids, alkyl monoesters, alkanes, and sterols (Wakeham et al., 1997; Colombo et al., 1997; Rice and MacCarthy, 1989). Soil lipids have been the least investigated; far more work has been done with sediment lipids because of their role in kerogen and petroleum formation. Given the extent of this literature, a convenient starting point is the reviews of soil lipids by Stevenson (1966, 1994), Morrison (1969), Bacon (1969), Braids and Miller (1975), Derenne and Largeau, (2001) and sediment lipids (Eglinton, 1969; Eglinton and Barnes, 1978; Derenne and Largeau, 2001; Killops and Killops, 2005). The common characteristic that is relevant to this chapter is the hydrophobicity of lipids that is a consequence of their operational definition.

3.1.1.2. Proteins. The nature of organic nitrogen compounds in NOM remains somewhat of an enigma (Flaig, 1971; Parsons and Tinsley, 1975; Stevenson, 1994; Schulten and Schnitzer, 1998) despite estimates that ~40% of all soil nitrogen is protein N (Schulten and Schnitzer, 1998). There appears to be agreement that most, if not all, of the noncellular protein present in soils is associated with clay minerals or humic substances. The persistence of proteins in soil and sediment environments is usually attributed to these associations. Proteins in soil (Parsons and Tinsley, 1975; Loll and Bollag, 1983; Boyd and Mortland, 1990; Stevenson, 1994; Schulten and Schnitzer, 1998; Quiquampoix, 2000; Gianfreda et al., 2002; Quiquampoix and Burns, 2007) and sediment environments (Knicker and Hatcher, 1997; Meyers and Ishiwatari, 1993; Nguyen and Harvey, 2003) have been discussed by a number of investigators. Wright and Upadhyaya (1996) described a soil organic material believed to be an iron-containing glycoprotein produced on the hyphae of arbuscular mycorrhizal fungi (phylum *Glomaleromycota*) that they named glomalin. Nichols and Wright (2006) reported glomalin-related proteins are present in soils in concentrations up to nine times as great as humic acid concentrations; furthermore, glomalin is persistent, and it is associated with the insoluble NOM or mineral fractions after treating soils with sodium hydroxide. It contains 3–5% N with a mean C content of nearly 37% (Lovelock et al., 2004). Subsequent work indicates that much of the material present in a glomalin extract appears to be humic acid, an observation that is consistent with the similar operational definitions of the two materials (Schindler et al., 2007).

Proteins contain a variety of functional groups that can bind them to mineral surfaces: carbonyl, alcoholic, carboxylic acid, and amine. Studies have shown that protein adsorption to clays is rapid at a pH below the isoelectric point of the protein (e.g., McLaren, 1954; Armstrong and Chesters, 1964). Conversely, then, protein should be extracted by a solvent system with a pH above the protein's isoelectric point. There are also hydrophobic regions on some proteins that create the possibility for hydrophobic interactions between the sorbed protein and the mineral surface (Quiquampoix, 2000).

3.1.1.3. Carbohydrates. Most of the carbohydrate added to soils and sediments is in the form of cellulose with a smaller amount of hemicellulose and other polysaccharides such as pectin. Cellulose is a β-D-glucose polysaccharide. Hemicellulose does not refer to a specific polysaccharide polymer but to a material associated with cellulose that can be separated using a dilute alkaline solution.

Though plants are the predominant source of most of this material, microbial inputs are important as well (Cheshire, 1977). Carbohydrate extracts from soils and sediments probably contain only a small fraction of the total carbohydrate concentration since much of the carbohydrate content appears to be polysaccharides bound to the humic components (Lowe, 1978), and probably to clay minerals as well. Its presence is recognized by the release of simpler sugars upon hydrolysis of the sample (Greenland and Oades, 1975; Lowe, 1978). Isolated polysaccharide fractions are generally polydisperse mixtures.

3.1.1.4. Lignin. Lignin is a hydrophobic aromatic macromolecule that is an integral component of plant cell walls. After cellulose it is the most abundant biopolymer on the earth's surface. It is perhaps best characterized by the lack of a regular structure that is the result of the irregular cross-linking of *p*-coumaryl alcohol, coniferyl alcohol, and sinapyl alcohol (Sarkanen and Ludwig, 1971; Sjöström, 1993; Boerjan et al., 2003).

Lignin's decomposition in soil and sediment is accomplished primarily by white-rot and brown-rot fungi through enzymatically mediated mechanisms. Its decomposition produces an extremely diverse assemblage of phenols, polyphenols, and aromatic acids that are very reactive toward other NOM components and mineral surfaces (Nord, 1964; Martin and Haider, 1980; Wang et al., 1986).

3.1.1.5. Humic Materials. Humic materials are divided into three fractions based on their solubility in aqueous solutions as a function of pH; humic acid, which is soluble in an alkaline aqueous solution; fulvic acid, which is soluble in an aqueous solution regardless of pH; and humin, which is insoluble in water at any pH value (and contains the OMN in soil organic matter). The chemical characteristics of humic acid and fulvic acid (e.g., Stevenson, 1994; Orlov, 1985; Rashid, 1985; Aiken et al., 1987; Hayes et al., 1987) and humin (Hatcher et al., 1985; Rice, 2001) are described in numerous reviews.

Humic acid is composed of aromatic, aliphatic and carbohydrate carbon compounds. An average humic acid's elemental composition is 55.1% C, 5.0% H, 3.5% N, 35.6% O, and 1.8% S (Rice and MacCarthy, 1991). Its molecular weight distribution is typically broad, and it is a relatively high-molecular-weight material relative to the fulvic acid isolated from the same soil or sediment. It's predominantly functionalized by carboxylic acid and phenolic groups. At least some components of humic acid are surface-active, and these components have been shown to form micelles in concentrated, alkaline aqueous solutions (Piret et al., 1960; Visser, 1964; Wershaw et al., 1969; Tschapek and Wasowski, 1976; Chen et al., 1978; Rochus and Sipos, 1978; Hayano et al., 1982; Hayase and Tsubota, 1984; Guetzloff and Rice, 1994).

Soil- and sediment-derived fulvic acid is also composed of aromatic, aliphatic, and carbohydrate carbon components, though it is generally believed to be more aromatic than the humic acid from that same environment. A typical fulvic acid's elemental composition is 46.2% C, 4.9% H, 2.5% N, 45.6% O, and 1.2% S (Rice and MacCarthy, 1991). The carboxyl group is the predominant functional group in fulvic acid; and it is a stronger acid than humic acid, with average total acidity values approximately twice that of humic acid isolated from the same sample (Orlov, 1985; Rashid, 1985; Stevenson, 1994). Some fulvic acids have been shown to be surface active, though they do not appear to generally be capable of forming micelles (Guetzloff and Rice, 1996).

Both these humic materials are heterogeneous mixtures whose characteristics are perhaps best described by distributions of chemical characteristics. They seem to possess a high degree of disorder to accompany their heterogeneity. Fractal geometry can be used to describe this disorder using noninteger dimensions that are scale invariant (Baveye et al., 2008). Each humic material has been shown to be fractal in the solid-state and in solution (Rice and Lin, 1993) and to aggregate via reaction-limited or diffusion-limited aggregation mechanisms that can be related to their fractal dimensions (Rice, 2008). When bound to minerals, the resulting composite is also fractal (Malekani and Rice, 1997; Malekani et al., 1997). Rice (2008) has reviewed the application of fractal geometry to humic materials.

3.1.2. Clays

3.1.2.1. Clay Minerals and Clay Colloids. The literature on clays and clay colloids is expansive, but there remains a degree of uncertainty in many areas of their study due to their inherent heterogeneity. Descriptions of the structures and properties of clay minerals can be found in Grim (1968), Brindley and Brown (1980), Newman and Brown (1987), Sposito et al. (1999), and Giese and van Oss (2002).

Clay minerals and clay colloids are the products of the advanced weathering of primary silicates. They are comprised mainly of silica and alumina, often with appreciable amounts of alkali and alkaline earth metals and iron. Most also have varying amounts of water bound to their surfaces and can take on a variety of different chemical and physical properties depending on the amount of water adsorbed. They have the ability to exchange or bind cations and anions and are capable of complex formation with a wide variety of organic molecules.

Clays are comprised of the two structural units depicted in Figure 3.2. One unit consists of two planes of oxygens or hydroxyls between which aluminum, magnesium, or iron atoms are octahedrally coordinated. When Al^{3+} or Fe^{3+} are present in this plane, only two-thirds of the available cation positions are occupied, and the unit geometry is dioctahedral. When Mg^{2+} is present, all the available positions are filled and the unit is trioctahedral. The second main structural unit consists of silica tetrahedra that are arranged in a hexagonal formation that repeats indefinitely. All tetrahedra share three oxygens, and they are arranged such that the bases are in the same plane forming a sheet with hexagonal perforations. In clay minerals, these sheets combine to form different structural units that repeat throughout the clay. In certain clay minerals, one tetrahedral sheet combines with one octahedral sheet to form what is called a 1:1 layer clay. In these clays, the silica tetrahedra and octahedral sheets are bonded to form a single structural unit. There are also 2:1 layer clays, where an octahedral sheet is sandwiched between two tetrahedral sheets. These 1:1 or 2:1 layers repeat indefinitely in the clay mineral, separated by an interlayer spacing that is of variable thickness in some clays. The interlayer space is often occupied by cations, organic molecules, or water, all of which can affect the properties of the clay.

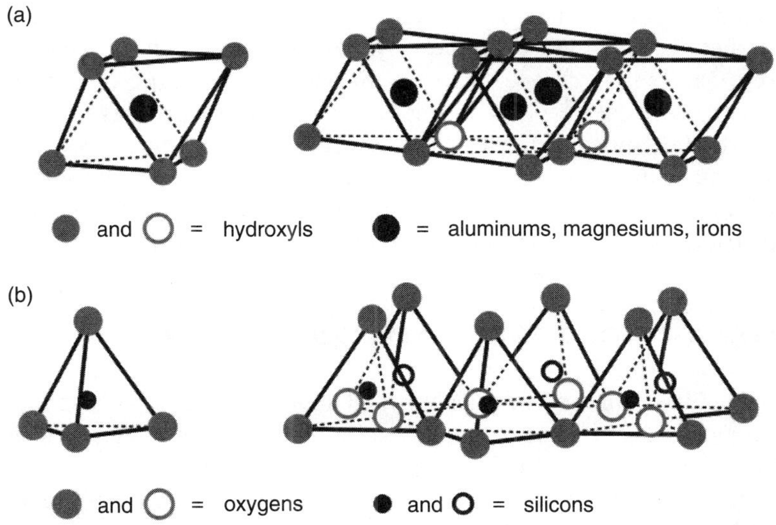

Figure 3.2. Structural units of clay minerals and clay colloids: **(a)** Octahedral sheet, **(b)** tetrahedral sheet. Reprinted with permission from Grim, R. E. (1968). *Clay Mineralogy*, 2nd edition, McGraw-Hill, New York.

3.1.2.2. Surface and Interfacial Chemistry of Clays. The variety of possible structural variations in clay minerals lead to a surface chemistry that is highly variable. As described by Sposito (1984), Bolt and van Riemsdijk (1987), Stumm (1992), and Sposito et al. (1999), the siloxane surface of clay particles contains four types of features that may be relevant to adsorption of organic matter: charge sites due to ditrigonal cavities; a surface charge that is the result of isomorphic substitution of Al and/or Si within the clay mineral structure; charges on the edges of clay particles; and hydrophobic regions on the siloxane surface that are believed to exist in the absence of extensive isomorphic substitution. The result is a material that generally possesses a negative charge at environmental pH values.

Cation exchange by clays is a process by which the negative charges present on the clay surface are balanced through cation adsorption. These negative charges are the result of isomorphic atomic substitutions, broken bonds, and proton exchange involving surface hydroxyls (Grim, 1968), and the relative importance of each varies with the clay. The amount of a cation sorbed is defined by the cation exchange capacity (CEC) and is dependent on the clay. The CEC is typically expressed in units of cmol kg^{-1} of clay. Typical values of the CEC of 2:1 layer clays such as montmorillonite are 80–150 cmol kg^{-1}; 1:1 layer clays such as kaolinite have CECs in the range of 3–15 cmol kg^{-1} (Grim, 1968).

Clays are known to form complexes with natural and anthropogenic organic molecules found in soils and sediments (Theng, 1976; Sposito, 1984; Lagaly, 1987; Sposito et al., 1999). Because many of these molecules are negatively charged at environmental pH values, it is believed that complexation with the negatively charged clay surface is facilitated through what is known as "cation bridging." The biological activity of these organic compounds can be significantly altered by adsorption onto clay minerals, creating great interest in the nature of the complexes. Studies have shown that different exchangeable cations result in varying amounts of an organic ligand sorbed by a clay (Greenland, 1971; Zielke et al., 1989; Varadachari et al., 1991). For example, clays exchanged with di- or trivalent cations sorb more organics than clays exchanged with monovalent cations, thereby leaving little doubt that the cation plays an important role in the process (Varadachari et al., 1991). All clay minerals possess a high degree of heterogeneity as a result of random structural distortions. For this reason, it is suspected that the sorbed species actually experience a variety of slightly differing local chemical environments resulting in a distribution of characteristics in the composite material.

3.2. ADSORPTION OF ORGANIC MATTER TO CLAYS

Organic substances can bind to clays through a variety of mechanisms that depend on the properties of the organic compound and the mineral surface. This section will restrict itself to mechanisms most relevant to NOM. Mortland (1986) has reviewed interaction mechanisms in broader detail. Scheidegger and Sparks (1996) have published a detailed review of adsorption–desorption mechanisms.

3.2.1. Adsorption

Adsorption isotherms are used to quantitatively describe adsorption at the solid/liquid interface (Hinz, 2001). They represent the distribution of the solute species between the liquid solvent phase and solid sorbent phase at a constant temperature under equilibrium conditions. While adsorbed amounts as a function of equilibrium solute concentration quantify the process, the shape of the isotherm can provide qualitative information on the nature of solute–surface interactions. Giles et al. (1974) distinguished four types of isotherms: high affinity (H), Langmuir (L), constant partition (C), and sigmoidal-shaped (S); they are represented schematically in Figure 3.3.

The S-shaped isotherm has an initial slope that increases with increasing equilibrium solute concentration and has two causes. Giles et al. (1974) attributed the S-shape to cooperative adsorption due to solute–solute interactions. These interactions stabilized the solute at the solid surface, and therefore the first adsorbed molecules enhance the adsorption of the next solute molecules. At high concentration, when the sites of the solid surface are saturated with solute the slope of adsorption isotherm start to decrease again. Sposito (1984) explained the S-shaped isotherm by a competing reaction within the solution. Solution ligands compete with surface

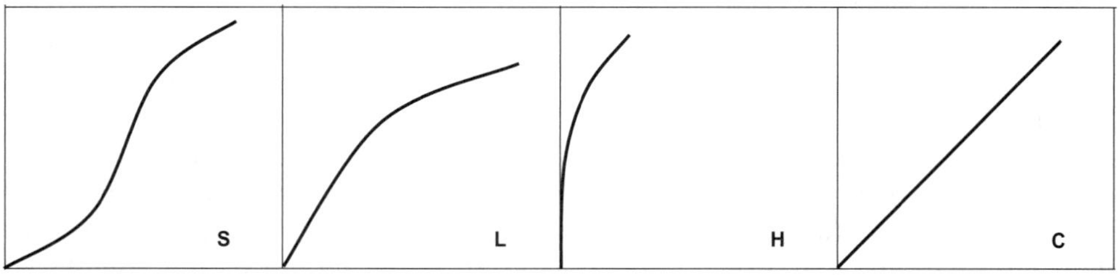

Figure 3.3. Types of solute adsorption isotherms.

sites for the solute molecules until the ligands are complexed with the solutes, and then any additional solute molecules are free to adsorb on the solid surface. In an L-shaped isotherm, the initial slope decreases steadily with increasing solute concentration, indicating that it is becoming increasingly difficult for the solute molecules to find vacant adsorption sites. A H-type isotherm has a very steep initial slope, indicating that the affinity of the solute for the solid surface is very high. This type of isotherm indicates highly specific interactions or significant van der Waals interactions. The C-type isotherm is linear with a constant slope that is independent of solute concentration. This isotherm type may be the result of either a constant partitioning of the solute between the interfacial region and external solution or the creation of more sites as the adsorption process progresses (Sposito, 1984).

3.2.1.1. Lipids. The adsorption of fatty acids onto clays gives L-shape adsorption isotherms and is characterized by weak interactions such as van der Waals forces, hydrogen bonds, and hydrophobic interactions (Meyers and Quinn, 1973; Bayrak, 2006). Ulrich et al. (1988) showed that long-chain fatty acids sorb onto oxide surfaces by two mechanisms, by surface complex formation, and by the hydrophobic effect, which becomes significant for molecules with a carbon chain of eight or more carbons (Evanko and Dzombak, 1998). Hedges (1977) concluded that the degree of fatty acid–mineral interaction in aqueous system is controlled by the aqueous solubility of the fatty acids. Up to 60% of the initial concentration of stearic acid was sorbed independently of the type or quantity of the clay. Increasing fatty acid molecular weight and solution salt concentrations also increased fatty acid adsorption. There are also reports of intercalation of fatty acids into clays, but the characteristics of the resulting composites were very sensitive to the method of preparation (Brindley and Moll, 1965; Weiss and Roloff, 1965; Meyers and Quinn, 1973).

3.2.1.2. Proteins. Studies on the adsorption of proteins by clays were reported beginning in the 1940s and early 1950s (Ensminger and Gieseking, 1941; Talibudeen, 1950). Large quantities of protein can be adsorbed by clays, reaching a maximum value at or near the protein isoelectric point, and the adsorption isotherms are generally of type L (Greenland, 1965; Fusi et al., 1989; Boyd and Mortland, 1990; Gianfreda et al., 1992). There are still questions about the mechanism of adsorption as well as about the structure of the adsorbed protein layer (Quiquampoix et al., 1995). Sinegani et al. (2005) reported that adsorption of cellulase (MW 30,000–100,000) on montmorillonite did not result in expansion of mineral structure and concluded that the adsorption is entirely external and not in the interlayer. Similar results were obtained for the adsorption of catalase (MW 238,000) onto Ca-montmorillonite (Harter and Stotzky, 1973), invertase (MW 270,000), and urease (MW 480,000) onto montmorillonite (Gianfreda et al., 1991, 1992). On the other hand, Naidja and Huang (1996; Naidja et al., 1995) reported that molecules of aspartase (MW 180,000) and tyrosinase (MW 120,000) were intercalated between montmorillonite layers. A detectable intercalation was also reported with acid phosphatase (MW 100,000) and glucose oxidase (MW 153,000) adsorption onto montmorillonite (Garwood et al., 1983; Rao et al., 1996). There are also contrasting reports about structural changes induced by adsorption of proteins on clays. Proteins such as α-chymotrypsin (Baron et al., 1999) and bovine pancreas ribonuclease (Haynes and Norde, 1994) are considered "hard" proteins because they show minimal modification of secondary structure when adsorbed on solid surfaces. Bovine serum albumin (Haynes and Norde, 1994; Quiquampoix et al., 1995; Servagent-Noinville et al., 2000) and some phytases (Yoon and Lenhoff, 1992) are "soft" proteins that show large conformational changes upon adsorption. Both enthalpic (electrostatic and van der Waals forces) and entropic effects (hydrophobic interactions and molecular structural rearrangements) contribute to adsorption mechanisms, depending on the nature of the substrate surface (Quiquampoix et al., 1995, 2002). It is hypothesized that "hard" proteins sorb onto hydrophobic surfaces under all charge conditions and onto hydrophilic surfaces by electrostatic interactions. The "soft" proteins are believed to sorb onto all surfaces because the increase in entropy outweighs the unfavorable enthalpy of adsorption (Forciniti and Hamilton, 2005).

3.2.1.3. Carbohydrates. There are few studies on the adsorption of simple sugars onto clays. Those that have been reported found that they sorb in small amounts; for example, less than 1% of initial concentration of glucose was adsorbed by either montmorillonite or kaolinite (Hedges, 1977). Higher adsorption capacities have been reported for oligosaccharides than for monosaccharides, along with higher values reported for methylated sugars than for nonmethylated ones (Greenland, 1956). Adsorption of polysaccharides has received considerable attention due to their role in the soil aggregate formation. The adsorption of polysaccharides onto clay generally gives an H-type isotherm (Clapp et al., 1991) that depends on the structure and charge of the polysaccharide. Chenu et al. (1987) showed that the adsorption of uncharged polysaccharides by clays occurs via weak interactions (van der Waals forces and H-bonding) that depend on structural factors such as the polysaccharide tertiary and quaternary structure (i.e., conformation), molecular weight, and solubility. Positively charged polysaccharides adsorb due to interactions between the cationic components and the negatively charged surfaces of the clays (Clapp et al., 1991). Uncharged and positively charged polysaccharides adsorb onto the external clay particle surface (Chenu et al., 1987) but also adsorb in the interlamellar spaces (Olness and Clapp, 1973). Clays generally sorb anionic polysaccharides on their external surfaces in a process that is dependent on environmental conditions such as pH, on the type and concentrations of cations (Dontsova and Bigham, 2005), and on the charge density and spatial conformation of the macromolecule (Labille et al., 2005).

3.2.1.4. Lignin. The adsorption of lignin-based copolymers on different single-cation forms of kaolinite and montmorillonite gives L-type isotherms (Prikhod'ko, 1982) and depends on the nature of the mineral and exchangeable cations. Tadjerpisheh and Ziechmann (1994) reported that clays can modify and partially sorb difficult-to-decompose lignin and the degree to which lignin is modified depended on the clay mineral structure and its degree of cation-saturation. For example, montmorillonite produces a stronger modification of lignin than kaolinite. As a result of its reaction with clay, the proportion of phenolic OH, carboxyl, and carbonyl groups in lignin decreases and the proportion of β-O-4 ether linkages increase.

3.2.1.5. Humic Materials. Adsorption studies of model compounds have been primarily directed toward single compound classes rather than mixtures, even simple mixtures, which might be used to approximate the adsorption behavior of NOM. Compared to protein adsorption to clays, which has been more extensively studied, studies of lipids and lignin are scarce because of the experimental difficulties related to the low aqueous solubility of these materials. Sequential adsorption of model compounds might be more appropriate and closely approximate OMN as it exists in nature, but such studies appear to be nonexistent. It is possible to have competitive adsorption in sequential adsorption because some compounds will compete for the same sites. It is also possible to have sorbed organics that will enhance the adsorption of further organic components. There are also very few studies describing sequential adsorption, probably due to the experimental difficulties in quantifying the extent of adsorption.

Adsorption of humic substances by clays has been extensively investigated, and the qualitative aspects of adsorption of particular clay–humic compound pairs has been the topic of a substantial number of scientific papers (for example, Schnitzer and Kodama, 1966; Schnitzer and Khan, 1972; Theng, 1976, 1979; Chassin et al., 1977; Schnitzer, 1986; Chaney and Swift, 1986; Hayes and Himes, 1989; Rebhun et al., 1992; Baham and Sposito, 1994; Wershaw et al., 1996a,b; Chandrakanth and Amy, 1996; Vermeer et al., 1998; Arnarson and Keil, 2000; Schulten and Leinweber, 2000; Specht et al., 2000). Early work by Evans and Russell (1959) on the adsorption of soil humic and fulvic acids onto H^+/Al^{3+}- and Ca^{2+}-exchanged montmorillonite and kaolinite clays found that the adsorption isotherms were C-type. The same isotherm shape was also observed by Theng and Scharpenseel (1975) for the adsorption of humic acids onto various homoionic-exchanged clays. Chassin et al. (1977) studied the adsorption of Na^+-humates and Na^+-fulvates by Al^{3+}- and Ca^{2+}-montmorillonite and found L-shaped isotherms that reached adsorption saturation at equilibrium solution concentrations between 0.8 and 1.2 mg/cm^3. This same study found S-shaped isotherms for the adsorption of the mixtures of humates and fulvates that indicated that each humic substance was not sorbed in the same ratio at all points in the isotherm. A more recent study (Kumar et al., 2001) showed that at low concentrations of humic acid the isotherms are C-type, and with increasing concentration the isotherm reaches a plateau producing an L-type isotherm. Majzik and Tombácz (2007a,b) concluded that Ca^{2+}-bridging is a dominant process in driving humic acid accumulation on clay surfaces particularly under environmentally realistic conditions.

Factors that influence the adsorption of humic materials onto mineral surfaces are listed in Table 3.1.

TABLE 3.1. Chemical Characteristics that Influence the Formation of OMN Complexes in Natural Systems

Mineral	NOM	Solution Properties
Chemical properties (oxide content, structure)	Chemical properties (aromaticity, aliphaticity, hydrophobicity, polarity)	pH
Particle size	Aqueous solubility	Ionic strength
Surface area	Molecular weight	Specific cations (bridging, bi- or multivalent)
Exchangeable cations		Temperature

The affinity of humic substances for mineral surfaces varies with the chemical composition of the mineral surface because different mechanisms of interaction are involved (Schulthess and Huang, 1991). Under conditions typical of natural waters, Davis (1982) observed that adsorption occurred readily for alumina and iron oxide but not for silica. This observation was explained in terms of the acidity of surface hydroxyl groups, but in actuality this contribution to the free energy of adsorption is relatively small. Alumina and iron oxides sorb humic substances mainly through ligand exchange (Zhou et al., 1994; Chorover and Amistadi, 2001) as does kaolinite (Shen, 1999). In contrast, ligand exchange mechanisms make relatively small contributions to the adsorption of humic material by montmorillonite (Chorover and Amistadi, 2001; Feng et al., 2005) because the mechanism is based on cation bridging, entropy-driven processes, and hydrophobic effects. Saturating the clay with polyvalent cations can enhance adsorption of humic substances onto mineral surfaces (Theng and Scharpenseel, 1975).

Various studies have suggested that hydrophobic organic components with a high molecular weight may preferentially adsorb to mineral surfaces (e.g., Namjesnik-Dejanovic et al., 2000). Adsorption of humic acid to kaolinite and montmorillonite performed with the same solution conditions indicated that adsorption onto kaolinite was greater than onto montmorillonite even though the latter has a much larger surface area and higher cation exchange capacity (Zhou et al., 1994; Feng et al., 2005). They suggested that the kaolinite surface prefers high-molecular-weight components compared to the montmorillonite surface which adsorbs a wider range of molecular weight components. Fulvic acid and aquatic NOM have a higher adsorption affinity for goethite than for kaolinite (Meier et al., 1999; Namjesnik-Dejanovic et al., 2000; Wang and Xing, 2005).

Preferential uptake of high-molecular-weight organic matter was observed for both goethite and kaolinite, but a smaller decrease in the weight-average molecular weight of solution-phase NOM occurred upon adsorption to kaolinite (Figure 3.4). In addition, the NOM components remaining in solution were also less aromatic suggesting that selective adsorption may also influences the distribution of organic components in natural systems. Balcke et al. (2002) analyzed the relationship between the adsorption–desorption of NOM onto clay surfaces and structural characteristics for 11 different humic substances sorbed to kaolin clay. The authors determined that the adsorption affinity correlates directly with the aromaticity and inversely with the polarity. They also observed a strong correlation between the molecular weight and the negative charge state of humic material. In a study of the interactions between synthetic goethite and 18 different soil organic matter samples, Kaiser (2003) also observed that the degree of adsorption correlated to the degree of aromaticity but that the strongest correlation was with the content of acidic functional groups, specifically the total carboxyl content. This author found that aromatic and aliphatic structures alone had only a small effect on NOM adsorption to synthetic goethite, and instead, the number and position of acidic groups attached to aromatic NOM components appeared to control adsorption. Recent studies focusing on the properties of adsorbed NOM have concluded that it is the aliphatic components that are preferentially sorbed to the clay surface (Feng et al., 2005; Wang and Xing, 2005; Simpson et al., 2006). These studies have also shown that different clays exhibit selective adsorption for different components of a humic substance. The choice of organic matter samples in studying the organo-mineral complex

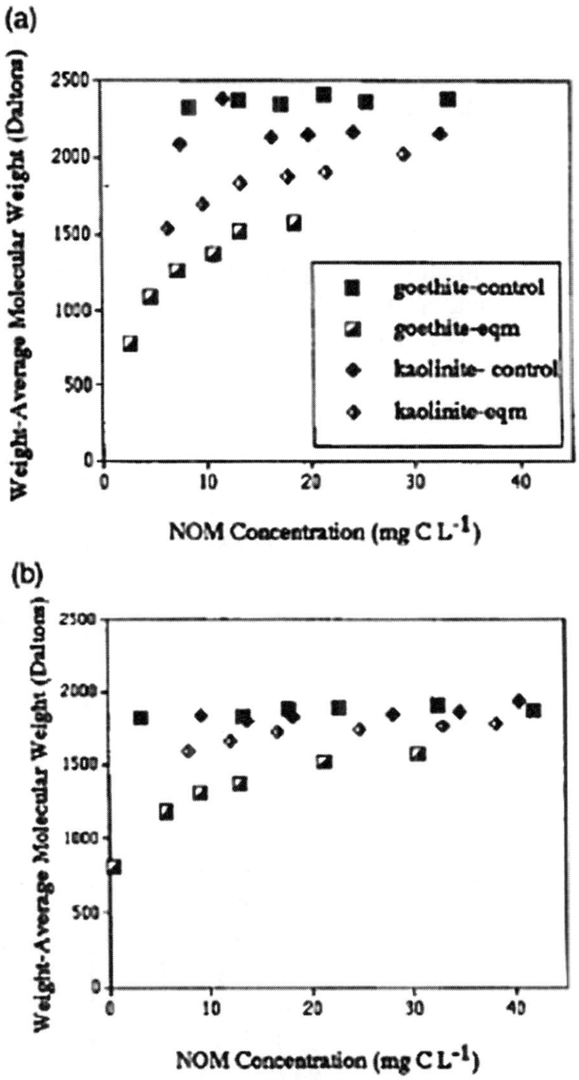

Figure 3.4. Change in weight-average molecular weight of NOM samples from the Suwannee River **(a)** and Great Dismal Swamp **(b)** by goethite and kaolinite. The term "eqm" refers to the concentration after 24-h equilibration time. Reprinted from Meier, M., Namjesnik-Dejanovic, K., Maurice, P. A., Aiken, G. R., Chin, Y. P., and Cabaniss, S. (1999). Fractionation of aquatic natural organic matter upon sorption to goethite and kaolinite. *Chem. Geol.* **157**, 275–284, with permission from Elsevier.

properties has to cover a range of chemical properties in order to address the importance of the nature of organic matter. Chorover and Amistadi (2001) found that aromatic components were not preferentially adsorbed to montmorillonite surface but rather appear to be adsorbed in direct proportion to their presence in the bulk organic matter in solution.

Wang and Xing (2005) observed that the fractions sorbed by kaolinite are more aliphatic than those sorbed by montmorillonite. Feng et al. (2005) showed that montmorillonite displayed a higher uptake of aromatics and proteins while components with CH_2 groups were preferentially sorbed to kaolinite surfaces. It should be apparent that while the process of selective adsorption of humic substances onto clay surfaces is generally accepted, there is considerable variation in the nature of the fractions that are reported to be preferentially sorbed. Simpson et al. (2006) used ^{1}H high-resolution magic-angle spinning NMR to study the chemical characteristics of clay-associated organic matter (Figure 3.5). Using model compound mixtures and soil extracts, they showed that the organo-clay complexes formed were primarily aliphatic. The amount of carbohydrate, peptide, and aromatic compounds sorbed was considered small. Wang and Xing (2005) used solid-state ^{13}C RAMP CP/MAS (ramped amplitude cross-polarization/magic-angle spinning) to reveal a predominantly aliphatic character for the organo-mineral complexes formed despite

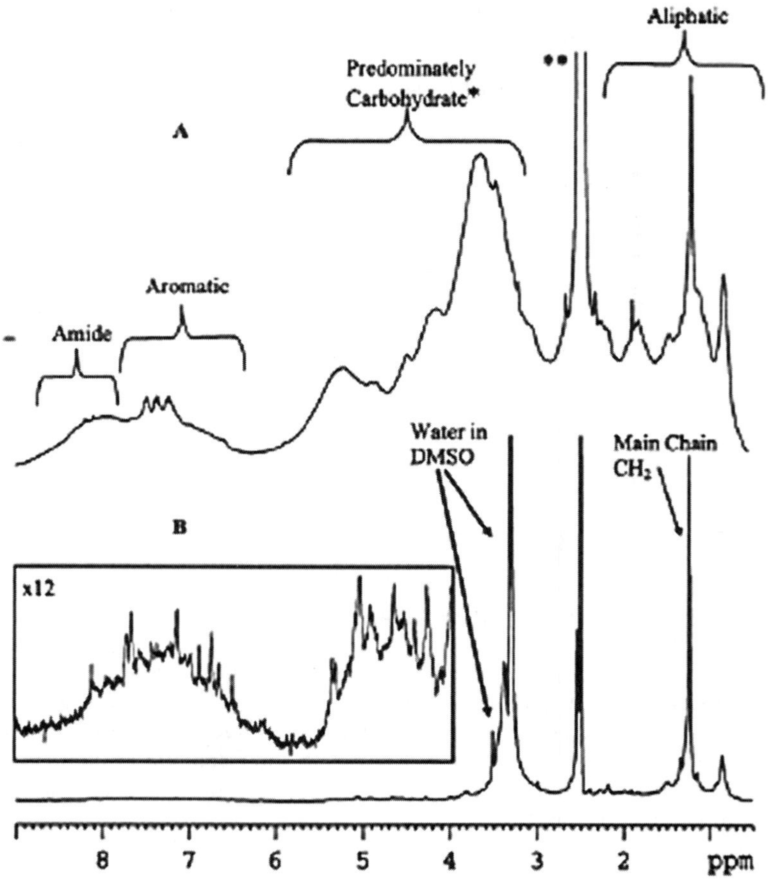

Figure 3.5. (A) ^{1}H NMR spectrum of the Brooksville fulvic acid (BFA) dissolved in d_6-DMSO and **(B)** HR-MAS NMR spectrum of the BFA–clay complex swollen in d_6-DMSO. Inset shows that lower abundance aromatic species are present in the spectrum in part B. Reprinted from Simpson, A. J., Simpson, M. J., Kingery, W. L., Lefebvre, B. A., Moser, A., Williams, A. J., Kvasha, M., and Kelleher, B. P. (2006). The application of 1H high-resolution magic-angle spinning NMR for the study of clay–organic associations in natural and synthetic complexes. *Langmuir* **22**, 4498–4503, with permission from the American Chemical Society.

the predominant aromatic nature of the NOM sample used. Both of these studies provide direct evidence of the aliphatic character of organo-mineral complexes using organic samples that have different characteristics.

Solution properties such as concentration, pH, ionic strength, relative concentrations of mono- and multivalent cations, and temperature have an impact on the adsorption of humic substances. Kumar et al. (2001) found that irrespective of the type of clay mineral, humic material adsorption increases linearly with the concentration of humic acid. Adsorption is pH-dependent with lower uptake at higher pH values due to increased ionization and greater solubility of humic substances (Schnitzer and Kodama, 1966; Rashid et al., 1972; Karickhoff and Brown, 1979; Theng, 1979; Zhou et al., 1994). At environmental pH values bridging between multivalent cations and the clay surface is a dominant adsorption mechanism (Majzik and Tombácz, 2007a). An increase in ionic strength is accompanied by an increase in the humic adsorption due to charge neutralization on mineral surfaces and the compression of the diffuse layer of cations associated with basal surfaces of layer silicates (Tombácz et al., 1988, 1990; Feng et al., 2005). The solubility of humic substances also decreases with increasing ionic strength (Kipton et al., 1992), which will favor their transfer from the solution phase to the mineral surface. There are few studies on the influence of temperature on the adsorption of humic substances onto mineral surfaces. Adsorption of humic substances onto clays and other minerals can be endothermic or exothermic. Zhou et al. (1994) reported that adsorption of Aldrich humic acid and a river-water humic acid increased with increasing temperature while the adsorption of water-derived fulvic acid and hydrophilic fractions decreased with increasing temperature in artificial seawater at pH 8. They suggested that the temperature dependence of the adsorption distinguishes between chemisorption and physical adsorption. Chemisorption requires activation energy and is favored by increasing temperature, while physical adsorption increases with decreasing temperature. Ghabbour et al. (2004) reported that the adsorption of an aqueous humic acid onto kaolinite is endothermic and results in an entropy increase attributed to dehydration of the kaolinite by the humic acid.

While important for the quantitative characterization of adsorption processes, adsorption studies provide only a phenomenological description of the process and do not generally reveal much in the way of a detailed molecular picture of the resulting organo-mineral composite.

3.2.2. Characterization of Organo-Mineral Complexes

Adsorption of NOM onto mineral surfaces produces a composite that possesses physical and chemical properties distinct from either of its constituent components. The ill-defined, heterogeneous nature of NOM makes the interpretation of data from the characterization of naturally occurring OMN complexes problematic. In this respect, studies involving NOM- component classes (e.g., lipids, proteins, etc.) and reference minerals may offer insights. The characterization of model NOM–mineral composites provides the opportunity to employ techniques specific to the interaction of interest.

The surface chemistry of OMN may be dominated by adsorbed organic matter that masks the properties of the supporting mineral to varying extents (Davis, 1982;

Beckett and Le, 1989; Tombácz, 2003). The presence of as little as 1 wt% NOM-modified mineral chemistry surface to the extent that it was concluded to be the controlling factor (Bertsch and Seaman, 1999).

Measurements of the surface charge of OMN by either electrophoretic mobility experiments or conventional potentiometric acid–base titration suggest that almost all particles in natural systems are negatively charged due to NOM surface coatings (Tombácz, 2003). Kretzschmar et al. (1997) showed that the surface charge density continuously shifted to more negative values when increasing amounts of humic acid were adsorbed to the kaolinite surface.

Nitrogen adsorption measurements of soil samples upon removal of the natural organic matter showed that the specific surface area and pore volume of the OMN are smaller than those of the mineral constituents (Tombácz et al., 1998). Different explanations have been provided to explain these observations. Tombácz et al. (1998) and Kaiser and Guggenberger (2003) presumed that the organic matter plugs the clay pores, whereas others suggested (Pennell et al., 1995; Pachepsky et al., 1995) that clay aggregates previously bound together by organic matter have disaggregated.

Both scanning electron microscopy (Heil and Sposito, 1995; Laird, 2001) and scanning force microscopy (Heil and Sposito, 1995) suggest that organo-mineral colloids exhibit an amorphous, roughened surface. Heil and Sposito (1995) reported that illitic soil colloids possessed a rough irregular surface in contrast to a smooth, flat surface observed for illite. The surface fractal dimensions determined by small-angle X-ray scattering (Malekani et al., 1997) decreased upon removal of organic matter from soil or humin samples. Malekani et al. (1997) showed that mineral components of humin have smooth surfaces over length scales of approximately 1–15 nm, and it is the organic matter coatings that are responsible for their surface roughness (Table 3.2).

A NOM coating on mineral surfaces enhances the stability of colloidal mineral particles by providing electrostatic and steric stabilization (Kretzschmar et al., 1997; Tombácz, 2003). Optical density measurements of kaolinite in the absence and presence of various amounts of humic acid showed that small addition of humic acid resulted in a large increase in colloidal stability (Kretzschmar et al., 1997). Tombácz (2003) reached the same conclusion using dynamic light scattering to measure coagulation kinetics in single and composite systems of clay minerals and aluminum and iron oxides in the presence of humic substances. At low loadings, humic materials affect the colloidal stability of the metal oxides and silicate clay minerals differently. The aggregation of a metal oxide is enhanced by sorption of humic materials under acidic conditions but under the same conditions silicate clay minerals are dispersed in the presence of humic materials. At some NOM surface loading, the differences between the charge properties of oxide and clay mineral surfaces are minimized and the colloidal stability of either dispersion in generally enhanced.

Thermal analysis of humic–mineral complexes has shown there is an overall reduction in the decomposition temperatures of humic acid that has been complexed to a mineral surface. Changes in the exothermic peak temperatures of humic substances in the free and complexed state are well-documented for synthetic mineral complexes with humic and fulvic acid (Schnitzer and Kodama, 1972; Tan, 1977; Schnitzer and Ghosh, 1982) as well as for authentic soil complexes.

TABLE 3.2. Surface Characteristics of Whole Soil, Humin, Humin with No Lipids (Humin-L), and Oxidized Humin (Humin-O.M) Resulting from Sequential Removal of Organic Matter

Soil Fraction	% Organic Matter	PZC	Pore Diameter (nm)	% Surface Area in <10-nm Pores	Fractal Dimension[a,b]	
					D_3	D_α
PSL Soil	4.15 ± 0.53	5.95	5.6 ± 0.1	74	2.9 (1–10)	2.9 (0.4–4)
PSL Humin	11.70 ± 2.45	6.30	7.1 ± 0.1	88	2.7 (1–10)	2.9 (0.4–2)
PSL Humin-L	7.88 ± 1.10	6.91	5.8 ± 0.4	99	2.3 (2–8)	3.0 (0.4–3)
PSL Humin-O.M	1.98 ± 0.72	7.20	4.5 ± 0.2	85	2.9 (1–9)	3.0 (0.3–4)
SCL Soil	4.60 ± 0.10	4.70	4.3 ± 0.1	63	2.9 (1–9)	
SCL Humin	12.49 ± 0.55	4.83	4.7 ± 0.1	63	2.9 (1–10)	2.9 (0.4–4)
SCL Humin-L	6.59 ± 0.91	4.91	5.1 ± 0.1	64	2.5 (1–10)	2.9 (0.4–4)
SCL Humin-O.M	1.25 ± 0.19	5.02	5.3 ± 0.5	65	2.7 (1–12)	
CLS Soil	4.54 ± 0.48	4.61	13.5 ± 0.1	81	2.5 (1–11)	
CLS Humin	11.69 ± 0.37	4.57	11.3 ± 0.9	95	2.4 (1–11)	
CLS Humin-L	8.66 ± 0.54	4.78	10.5 ± 0.2	85	2.3 (1–15)	
CLS Humin-O.M	10.6 ± 0.98	4.97	7.5 ± 0.7	80	2.2 (1–15)	
GPP Soil	68.69 ± 0.22	3.75	5.8 ± 0.3	73	2.5 (1–11)	2.5 (0.4–4)
GPP Humin	55.30 ± 2.87	4.10	20.8 ± 2.2	75	2.2 (2–8)	2.2 (0.5–1)
GPP Humin-L	47.09 ± 2.20	4.62	14.3 ± 1.4	76	2.0 (2–8)	2.6 (0.4–2)
GPP Humin-O.M	22.12 ± 2.34	4.90	6.6 ± 0.4	90	20. (2–8)	2.6 (0.4–1)

[a]Values in parentheses represent length scale in millimeters.
[b]Absolute uncertainty associated with each D value is ±0.1.

Source: Reprinted from Malekani, K., Rice, J. A., and Lin, J.-S. (1997). The effect of sequential removal of organic matter on the surface morphology of humin. *Soil Sci.* **162**, 333–342, with permission from the World Scientific Publishing Company.

Rashid et al. (1972) reported that decarboxylation of humic acid in the free state and in physical mixture with illite occurred at 300 °C, while in a humic acid–illite complex it occurred at 275 °C. Satoh (1984a,b) found that the humic acid extracted from the clay fraction of a volcanic ash soil gave exothermic peaks at 320 °C and 480 °C, whereas for the humic–clay complex the exotherms occurred at 295 °C and 390 °C. The overall reduction in the thermal stability of humic acid in clay–humic complexes has been attributed to ring strain caused by multiple attachments of humics to the mineral surface. Yet recent studies have reported that a small fraction of humic acid was thermally stabilized in the clay-humic complex. Ahmed et al. (2002) analyzed the thermal stability of natural clay–humic complexes isolated from five soils. They showed that while an overall reduction in the decomposition temperature of humic acid in the clay complexes occurred for all the samples, minor exothermic peaks also occurred at higher temperatures for most of the samples indicating that a fraction of humic acid was thermally stabilized. The authors suggested the existence of two types of humic–clay bonding in these complexes, with the majority of humic substances being destabilized due to an increase in the molecular strain and a minor fraction being stabilized due to a reduction in molecular strain.

3.2.3. Nature and Structure of Organo-Mineral Complexes

NOM–mineral interactions involve a number of processes such as cation and water bridging, cation and anion exchange, ligand exchange, van der Waals interactions, and hydrophobic bonding (Mortland, 1970; Greenland, 1971; Theng, 1979; Sposito, 1984).

Bridging mechanisms occur at the adsorption of anionic, carboxylate, and uncharged polar functional groups such as amino, carbonyl, carboxyl, and hydroxyl by mineral surfaces bearing solvated exchangeable cations (Figure 3.6). Cation bridge formation occurs when the anionic group of the organic matter replaces a water molecule in the primary shell of the exchangeable cation and is directly coordinated to the exchangeable cation. When the cation has high solvation energy, it is less likely that solvating water molecules will be displaced. Anionic or polar organic groups may also form hydrogen bonds with the water molecules from the primary shell of the exchangeable cation and therefore the organic groups are indirectly coordinated to exchangeable cations through solvating water molecules forming the water bridge.

Cation exchange occurs at the adsorption of organic cations onto mineral surfaces. Organic cations may exchange with alkaline earth and alkali metal cations that neutralize the negative electrical charges responsible for the cation exchange capacity of the mineral. Typical examples of organic cations present in NOM are protonated amino groups as in the case of alkylamines and amino acids. Clay minerals have selective affinity for organic cations, with higher values for large organic cations over smaller ones. Theng (1979) showed that the affinity of clay for the organic cation within a group of primary, secondary, and tertiary amines is linearly

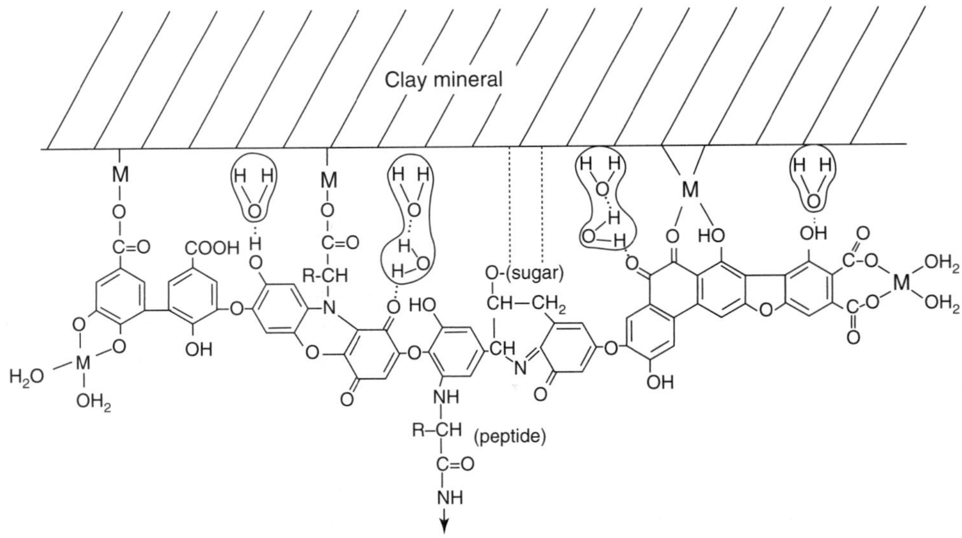

Figure 3.6. Modes of interaction of natural organic matter with a clay mineral surface. The figure depicts cation (M) and water bridging, along with van der Waals interactions (through the sugar moiety). Reprinted with permission from Stevenson, F. J., and Ardakani, M. S. (1972). Organic matter reactions involving micronutrients in soils. In *Micronutrients in Agriculture*, Mortvedt, J. J., Giordano, P. M., and Lindsay, W. L., eds., Soil Science Society of America, Madison, WI, pp. 79–114.

related to the molecular weight and to the size and shape of the ions. Recently, Teppen and Aggarwal (2007) found that the selectivity of smectites for organic cations is due to pure hydrophobic expulsion of the larger cation from water in case of organic cations with identical head-groups, to a stronger preference of aqueous-phase for the cation with smaller head-group than the clay preference in case of two organic cations with different head-groups and to van der Waals forces in case of clay already substantially loaded with organic cations. As exchange takes place, cations are not uniformly distributed across all clay mineral surfaces, resulting in two types of cation sites in the clay layers. Segregation is assumed to limit the extent of conformational changes of organocations upon adsorption (Sposito, 1984).

Electrostatic anion exchange does not often occur; and when it does, it is under acidic conditions when clay exchange sites are occupied by iron or aluminum ions. NOM anions, functionalized primarily by carboxylate groups, exchange with inorganic anions to form inner-sphere complexes with protonated surface hydroxyl groups (Sposito, 1984). Besides the simple coulombic attraction operating in anion exchange, organic anions can interact with mineral surfaces through specific or ligand exchange mechanisms. Ligand exchange is a stronger interaction than bridging or electrostatic ion exchange mechanisms and occurs when the anionic group penetrates the coordination shell of aluminum or iron ions and becomes incorporated in place of surface hydroxyl or $-OH_2$.

Van der Waals interactions between two molecules are very weak, but their contributions can be of considerable importance in the adsorption of neutral polar and nonpolar molecules, particularly those with a high molecular weight.

A hydrophobic bonding mechanism is the result of an overall entropy gain of the system due to the displacement of water molecules from the clay surface by sorbed organic compounds. The contributions of this mechanism will be more pronounced in the adsorption of high-molecular-weight organic matter (Parfitt and Greenland, 1970; Mortland, 1970).

There is experimental as well as theoretical evidence for the existence of all these mechanisms. Indirect evidence is provided by batch adsorption experiments performed under systematically controlled conditions such as pH, temperature, and ionic composition. Although these experiments do not offer definitive evidence for the presence or absence of a particular mechanism, they are often used to characterize the interactions responsible for the adsorption of organic matter at the mineral surface. For example, at pH values less than the point of zero charge of sesquioxides, adsorption to those minerals is mainly due to a ligand exchange mechanism (Gu et al., 1994; Murphy and Zachara, 1995; Shen, 1999; Feng et al., 2005). However, there are electrostatic as well as hydrophobic contributions that cannot be distinguished (Balcke et al., 2002). Increasing adsorption with increasing ionic strength is attributed to van der Waals interactions and cation and water bridge formation. The contribution of cation bridges can be evaluated from experiments with polyvalent cations. Electrostatic interactions are assumed to play a major role when adsorption increases with decreasing ionic strength. Ligand exchange requires overcoming the process activation energy and therefore increases with increasing temperature. Arnarson and Keil (2000) estimated the relative contribution of van der Waals interaction, ligand exchange, and cation bridging to be 60%, 35%, and 5%, respectively, for adsorption of NOM to montmorillonite in a $CaCl_2$ solution. Analyzing adsorption of humic acid to kaolinite and montmorillonite in a $CaCl_2$ solution, Feng et al. (2005) estimated that van der Waals interactions accounted for 22% of the overall adsorption mechanism, ligand exchange for 32%, and cation bridging for 41%. Majzik and Tombácz (2007a) observed that cation bridging is highly influenced by the Ca/humic acid ratio, increasing significantly as the ratio increases.

Direct evidence of specific ligand exchange interactions is provided by spectroscopic analysis (Parfitt et al., 1977; Yost et al., 1990; Gu et al., 1994; Yoon et al., 2005). Parfitt et al. (1977) used infrared spectroscopy to demonstrate that the adsorption of a fulvic acid onto goethite involves complexation between fulvic acid $-COO^-$ and the hydroxyl group of goethite. Evidence of preferential adsorption of high-molecular-weight and hydrophobic components from size exclusion chromatography (Specht et al., 2000) and NMR spectroscopy (Wershaw et al., 1996a; Xing, 2001; Simpson et al., 2006) has been used to support a hydrophobic adsorption mechanism.

Sutton and Sposito (2006) applied molecular modeling to study the interactions of a model humic substance in the protonated and Ca-saturated forms with Ca-montmorillonite. The authors concluded that protonated humic materials were involved in direct hydrophobic and hydrogen bonding interactions with the clay mineral. In addition, a few polar organic groups adsorbed via water bridging interactions. They also suggested that the Ca-saturated humic substance adsorbed via extensive cation bridging, a less important water-bridging contribution, and H-bonding interactions mediated by water molecules.

The particular contribution of each adsorption mechanism depends on the specific mineral–humic system under investigation and the location and conformation of adsorbed organic matter. Several authors have reported the existence of organic coatings on the mineral surface as bilayers (Wershaw et al., 1996a,b; Evanko and Dzombak, 1998; Namjesnik-Dejanovic and Maurice, 2001; Zavarzina, 2001; Ghabbour et al., 2004). Ghabbour et al. (2004) showed that adsorption of humic acid on clays and minerals occurs in two consecutive steps and suggested a different molecular conformation for humic acid on kaolinite. The first step is monolayer formation and the second is an association of "free" humic acid with adsorbed humic acid. Vermeer et al. (1998) measured the thickness of the humic acid adsorbed layer onto mineral particles and concluded that the humics adsorbed to suspended particles can be described as a dynamic layer that extends into the bulk solution. Using atomic force microscopy, Liu et al. (2000) also measured the thickness of humic acid layers adsorbed on mica and suggested that humic acid molecules sorb at discrete sites to form monolayer islands at the mica/water interface. Using the same technique, Namjesnik-Dejanovic and Maurice (2001) have observed NOM spheres and bilayers on mineral particles and concluded that the conformation of sorbed humics represents complex structures and aggregates. While many of the studies report that adsorption of humics is restricted to the external surfaces of minerals (Baham and Sposito, 1994), there are studies that suggest the existence of organic matter in interlamellar adsorption sites. An increase in the interlayer spacing of montmorillonite upon adsorption of humic/fulvic acid at $pH < 4$ was taken as evidence for the intercalation of the organic molecules in the interlayer space (Schnitzer and Kodama, 1966; Martinez and Rodriguez, 1969). X-ray diffraction as well as thermal analysis investigations found evidence of humic acid combined with aluminum hydroxyls between the montmorillonite layers (Moinereau, 1977; Singer and Huang, 1988; Violante et al., 1999).

3.2.4. Geochemistry of Organo-Mineral Complexes

The reactivity of OMN toward heavy metals and small organic compounds is affected by the interaction between the components of the OMN complex and cannot be extrapolated simply by considering just the reactivity of the components alone (Piccolo, 1989; Azimonti et al., 1994; Robertson and Leckie, 1994). Vermeer et al. (1999) observed that the overall adsorption of a specific metal ion to the organo-mineral complex was smaller than predicted by the additivity rule when this metal ion has a more pronounced affinity for the humic acid than for the mineral oxide, whereas it is larger than predicted by the additivity rule when the metal ion has a higher affinity for the oxide than for the humic acid. They concluded that this deviation from the expected behavior was due to the interaction between the negatively charged humic acid and the positively charge oxide surface.

Several studies have examined the properties and behavior of OMN complexes in comparison to those of the organic or mineral components individually. Spark et al. (1997) showed that adsorption of heavy metals in mineral–humic acid systems is dependent on adsorption of humic acid by the mineral and on the solubility of the metal–humic acid complex. They compared the adsorption of heavy metals such as copper, zinc, cobalt, and cadmium by mineral, humic acid, and combined mineral–humic acid systems. Enhanced adsorption occurred in the combined systems when goethite and silica were present. It was explained by secondary reactions in which metal–humic acid complexes were adsorbed by the minerals. No enhancement occurred for combined systems with alumina and kaolinite, which the authors postulated was due to competing reactions between humic acid and metals for surface sites on these minerals that resulted in an inhibition of cation adsorption. There are reports that have found that adsorption affinity and isotherm nonlinearity of clays increased in the presence of the adsorbed humic materials (Wang and Xing, 2005; Wang et al., 2005). Wang and Xing (2005) reported that phenanthrene adsorption to humic acid–montmorillonite and humic acid–kaolinite complexes increases 18 and 35 times, respectively, compared to the pure mineral (Figure 3.7). These results could be expected since studies of the adsorption on pure minerals and humic substances showed that the latter materials often have an adsorption affinity for phenanthrene that is orders of magnitude higher than the pure minerals (Boyd et al., 2001; Xing, 2001). When comparisons are made between humic acid–clay complexes and the original humic acid, there are reports of increased affinity (Wang and Xing, 2005) as well as decreased affinity (Wang et al., 2005) for hydrophobic anthropogenic organic compounds. The variation in the results reported by these studies can be attributed to the differences in the organic matter components adsorbed to the mineral surface in each particular system. A more meaningful comparison can be made if this contribution is subtracted, in effect normalizing the adsorption phenomenon between the sorbed and organic matter that remains in the solution phase.

A quantitative assessment of OMN's impact in natural systems is hard to achieve because it would require the quantitative determination of these complexes in a natural soil or sediment sample. Given the operational nature of current methods for isolation and purification of naturally occurring OMN complexes, it is not possible to prevent the possibility of significant alteration of the organic matter or co-extraction of other uncomplexed organic or inorganic soil or sediment components. Yet qualitative aspects of the "architecture" of these nanocomposites are

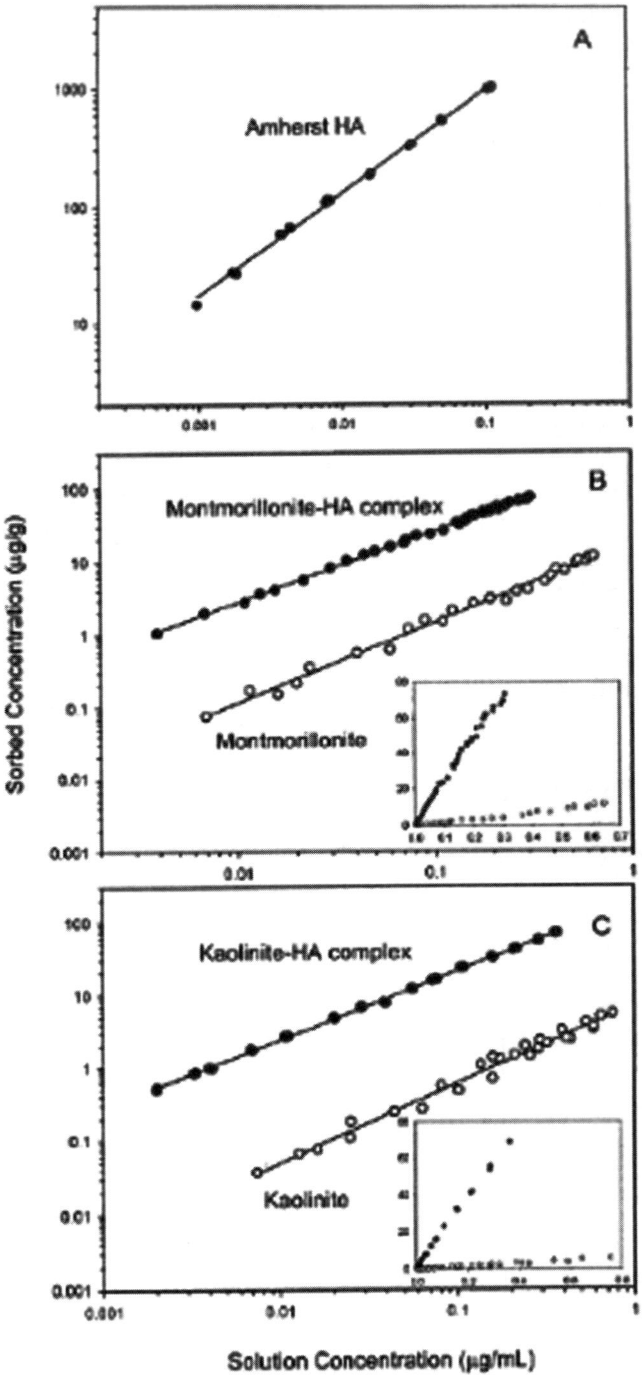

Figure 3.7. Phenanthrene sorption isotherms on **(A)** the whole Amherst peat soil humic acid, **(B)** montmorillonite and a montmorillonite–humic acid complex (5:1 ratio), and **(C)** kaolinite and kaolinite–humic acid complex (5:1 ratio). Insets in parts B and C are the respective isotherms presented on a linear scale. Reprinted from Wang, K., and Xing, B. (2005). Structural and sorption characteristics of adsorbed humic acid on clay minerals. *J. Environ. Qual.* **34**, 342–349, with permission from the Soil Science Society of America.

accessible. For example, most studies seem to indicate that NOM coats mineral particles, though there appears to be little direct evidence to that effect. AFM studies actually seem to indicate that sorbed organic matter is "clumped" on to mineral surfaces (e.g., Liu et al., 2000). Complex formation then defines the NOM–mineral bond, which implies that sorbed NOM has different conformational characteristics than the bulk NOM, and it may have different chemical properties if preferential adsorption has occurred. In effect, the clay surface may serve as a template in forming OMN complexes, which in turn may template the sorptive characteristics of the resulting material.

3.3. FUTURE RESEARCH OPPORTUNITIES

This review has shown that there is considerable depth to the current understanding of NOM–mineral interactions. Much of this understanding is based on adsorption experiments. While the work to date has established many aspects of the nature of the binding of the OM to mineral matter, with few exceptions it has not shown how the organic components interact with each other. What is lacking is a molecular-level understanding of what might be referred to as the "design principles" affecting the assembly of composite structures formed by protein, lipid, lignin, carbohydrate, and humic material–mineral interactions.

Three specific areas can be identified to serve as foci for expanding the research on this material: (i) The nature of the organic components' interactions need to be ascertained. Do the lipids (whose chemistry is dominated by aliphatic components) and humic (whose chemistry is dominated by aromatic, carboxyl, and carbohydrate components) actually exist as distinct domains in organo-mineral complexes? (ii) What is the effect of the mineral surface on adsorbed macromolecule conformation? How does conformation impact the adsorption of additional NOM components? (iii) Finally, a better understanding of the interfacial chemistry of these organo-mineral composites needs to be developed in order to understand the fate of many organic contaminants introduced into natural systems.

Given the role of these composites in both organic and environmental geochemical issues of recognized importance, developing such an understanding is vital.

REFERENCES

Ahmed, N., Varadachari, C., and Ghosh, K. (2002). Soil clay–humus complexes. II. Bridging cations and DTA studies. *Aust. J. Soil Res.* **40**, 705–713.

Aiken, G. R., McKnight, D. M., Wershaw, R. L., and MacCarthy, P. (1987). *Humic Substances in Soil, Sediment, and Water: Geochemistry, Isolation, and Characterization*, John Wiley & Sons, New York.

Anderson, D. W., and Paul, E. A. (1984). Organomineral complexes and their study by radiocarbon dating. *Soil. Sci. Soc. Am. J.* **48**, 298–301.

Armstrong, D. E., and Chesters, G. (1964). Properties of protein–bentonite complexes as influenced by equilibrium conditions. *Soil Sci.* **98**, 39–52.

Arnarson, T. S., and Keil, R. G. (2000). Mechanisms of pore water organic matter adsorption to montmorillonite. *Mar. Chem.* **71**, 309–320.

Azimonti, G., Bidoglio, G., Righetto, L., and Bellobono, I. R. (1994). Cobalt and mercury adsorption on alumina colloids in the presence of humic substances. In *Humic Substances in the Global Environment and Implications to Human Health*, Senesi, N., and Miano, T. M., eds., Elsevier, Amsterdam, 1079–1084.

Bacon, J. S. D. (1969). Soil Lipids. In *Organic Geochemistry. Methods and Results*, Murphy M. T. J., and Eglinton, G., eds., Springer-Verlag, New York, 358–375.

Baham, J., and Sposito, G. J. (1994). Adsorption of dissolved organic carbon extracted from sewage sludge on montmorillonite and kaolinite in the presence of metal ions. *J. Environ. Qual.* **23**, 147–153.

Balcke, G. U., Kulikova, N. A., Hesse, S., Kopinke, F. D., Perminova I. V., and Frimmel, F. H. (2002). Adsorption of humic substances onto kaolin clay related to their structural features. *Soil Sci. Soc. Am. J.* **66**, 1805–1812.

Balesdent, J. (1987). The turnover of soil organic fractions estimated by radiocarbon dating. *Sci. Total Environ.* **62**, 405–408.

Banerjee, S. K. (1979). Humin fraction of soil humus. *J. Indian Chem. Soc.* **56**, 1094–1097.

Baron, M. H., Revault, M., Servagent-Noinville, S., Abadie, J., and Quiquampoix, H. (1999). Chymotrypsin adsorption on montmorillonite: Enzymatic activity and kinetic FTIR structural analysis. *J. Coll. Interface Sci.* **214**, 319–332.

Baveye, P., Boast, C. W., Gaspard, S., and Tarquis, A. M. (2008). Introduction to fractal geometry, fragmentation processes and multifractal measures: Theory and operational aspects of their application to natural systems In *Bio-Physical Chemistry of Fractal Structures and Process in Environmental Systems*, Senesi, N., and Wilkinson, K., eds. IUPAC Series on Analytical and Physical Chemistry of Environmental Systems. Vol. **11**, John Wiley & Sons, Chichester, pp. 11–67.

Bayrak, Y. (2006). Application of Langmuir isotherm to saturated fatty acid adsorption. *Micropor. Mesopor. Mater.* **87**, 203–206.

Beckett, R., and Le, N. P. (1989). The role of organic matter and ionic composition in determining the surface charge of suspended particles in natural waters. *Coll. Surf.* **44**, 34–59.

Bertsch, P. M., and Seaman, J. C. (1999). Characterization of complex mineral assemblages: Implications for contaminant transport and environmental remediation. *Proc. Natl. Acad. Sci.* **96**, 3350–3357.

Boerjan, W., Ralph, J., and Baucher, M. (2003). Lignin biosynthesis. *Annu. Rev. Plant Biol.* **54**, 519–549.

Bolt, G. H., and van Riemsdijk, W. H. (1987). Surface Chemical Processes in Soil. In *Aquatic Surface Chemistry: Chemical Processes at the Particle–Water Interface*, Stumm, W., ed., John Wiley and Sons, New York, pp. 127–164.

Boulet, R. L., Pessenda, C. R., Everaldo, E. C. C., and Melfi, A. J. (1995). Evaluation of the rate of surface biological accumulation of soil organic matter using radiocarbon dating on charcoal and soil humin fraction. Case study of Latosols from the basin of Lagoa Campestrre, Salitre, Minas Gerais, Brazil. *C. R. Acad. Sci., Ser. II: Sci. Terre Planetes* **320**, 287–294.

Boyd, S. A., and Mortland, M. M. (1990). Enzyme interactions with clays and clay–organic matter complexes. In *Soil Biochem. 6*, Bollag, J.-M., and Stotzky, G., eds., Marcel Dekker, New York, pp. 1–28.

Boyd, S. A., Sheng, G., Teppen, B. J., and Johnston, C. J. (2001). Mechanisms for the adsorption of substituted nitrobenzenes by smectite clays. *Environ. Sci. Technol.* **35**, 4227–4234.

Braids, Q. C., and Miller, R. H. (1975). Fats, waxes, and resins in soil. In *Soil Components.* Vol. I.: *Organic Compounds*, Geiseking, J. E., ed., Springer, New York, pp. 343–368.

Breger, I. A. (1960). Diagenesis of metabolites and a discussion of the origin of petroleum hydrocarbons. *Geochim. Cosmochim. Acta* **19**, 297–308.

Brindley, G. W., and Brown, G. (1980). *Crystal Structures of Clay Minerals and Their X-Ray Identification*. Mineralogical Society Monograph No. 5. The Mineralogical Society, London.

Brindley, G. W., and Moll, W. F., Jr. (1965). Clay-organic studies. IX. Complexes of natural and synthetic Ca-montmorillonites with fatty acids. *Am. Min.* **50**, 1355–1370.

Calvin, M., and Philip, R. P. (1976). Possible origin for insoluble organic (kerogen) debris in sediments from insoluble cell-wall material of algae and bacteria. *Nature* **232**, 134–136.

Cane, R. F. (1976). The origin and formation of oil shale. In *Oil Shale*, Yen, T. F., and Chilinger, G. V., eds., Elsevier, New York, pp. 25–60.

Chandrakanth, M. S., and Amy, G. (1996). Effect of ozone on the colloidal stability and aggregation of particles coated with natural organic matter. *Environ. Sci. Technol.* **30**, 431–443.

Chaney, K., and Swift, R. S. (1986). Studies on aggregate stability. I. Re-formation of soil aggregates. *J. Soil Sci.* **37**, 329–335.

Chassin, P., LeBerre, B., and Nakaya, N. (1977). Influence des substances humiques sur les propriétés des argiles. II. Adsorption des acides humiques et fulviques par la montmorillonite. *Clay Min.* **12**, 261–271.

Chen, Y., Senesi, N., and Schnitzer, M. (1978). Chemical and physical characteristics of humic acid and fulvic acid extracted from soils of the Mediterranean region. *Geoderma* **26**, 87–104.

Chenu, C., Pons, C. H., and Robert, M. (1987). Interaction of kaolinite and montmorillonite with neutral polysaccharides. In *Proceedings, International Clay Conference, Denver 1985*, Schultz, U. G., van Olphen, H., and Mumpton, F. A., eds., Clay Minerals Society, Bloomington, IN, pp. 375–381.

Cheshire, M. V. (1977). Origins and stability of soil polysaccharide. *Eur. J. Soil Sci.* **28**, 1–10.

Chorover, J., and Amistadi, M. K. (2001). Reaction of forest floor organic matter at goethite, birnessite and smectite surfaces. *Geochim. Cosmochim. Acta* **65**, 95–109.

Clapp, C. E., Harrison, R., and Hayes, M. H. B. (1991). Interactions between organic macromolecules and soil inorganic colloids and soils. In *Interactions at the Soil Colloid–Soil Solution Interface*, ed. Bolt, G. H., 409–468.

Cloos, P., Badot, C., and Herbillon, A. (1981). Interlayer formation of humin in smectites. *Nature* **289**, 391–393.

Colombo, J. C., Silverberg, N., and Gearing, J. N. (1997). Lipid biogeochemistry in the Laurentian Trough II. Changes in composition of fatty acids, sterols and aliphatic hydrocarbons during early diagenesis. *Org. Geochem.* **26**, 257–274.

Davis, J. A. (1982). Adsorption of natural dissolved organic matter at the oxide/water interface. *Geochim. Cosmochim. Acta* **46**, 2381–2393.

Derenne, S., and Largeau, C. (2001). A review of some important families of refractory macromolecules: Composition, origin, and fate in soils and sediments, *Soil. Sci.* **166**, 833–847.

Dontsova, K. M., and Bigham, J. M. (2005). Anionic polysaccharide sorption by clay minerals. *Soil Sci. Soc. Am. J.* **69**, 1026–1035.

Durand, B., and Nicaise, G. (1980). Procedures for kerogen isolation. In *Kerogen*. ed. Durand, *Editors Technip*, Paris, pp. 35–53.

Eglinton, G. (1969). Organic geochemistry. The organic chemist's approach. In *Organic Geochemistry. Methods and Results*, M. T. J., Murphy, and G., Eglinton, eds., Springer-Verlag, New York, pp. 20–73.

Eglinton, G., and Barnes, P. J. (1978). Organic matter in aquatic sediments. In *Environmental Biogeochemistry and Geomicrobiology,* Vol. 1: *The Aquatic Environment*, Krumbein, W. E., ed., Ann Arbor Science, Ann Arbor, MI, pp. 25–46.

Engel, M. H., and Macko, S. A. (1993). *Organic Geochemistry: Principles and Applications.* Plenum Press, New York.

Ensminger, L. E., and Gieseking, J. E. (1941). The adsorption of proteins by montmorillonitic clays and its effect on base exchange capacity. *Soil Sci.* **51**, 125–132.

Evanko, C. R., and Dzombak, D. A. (1998). Influence of structural features on sorption of NOM-analogue organic acids to goethite. *Environ. Sci. Technol.* **32**, 2846–2855.

Evans, L. T., and Russell, E. W. (1959). The adsorption of humic and fulvic acids by clays. *J. Soil Sci.* **10**, 119–132.

Feng, X., Simpson, A. J., and Simpson, M. J. (2005). Chemical and mineralogical controls on humic acid sorption to clay mineral surfaces. *Org. Geochem.* **36**, 1553–1566.

Flaig, W. (1971). Organic compounds in soil. *Soil Sci.* **111**, 19–33.

Forciniti, D., and Hamilton, W. A. (2005). Surface enrichment of proteins at quartz/water interfaces: a neutron reflectivity study. *J. Col. Interface Sci.* **285**, 458–468.

Fusi, P., Ristori, G. G., and Calamai, L., Stotzky G. (1989). Adsorption and binding of protein on "clean" (homoionic) and "dirty" (coated with Fe oxyhydroxides) montmorillonite, illite and kaolinite. *Soil Biol. Biochem.* **21**, 911–920.

Garwood, G. A., Mortland, M. M., and Pinnavaia, T. J. (1983). Immobilization of glucose oxidase on montmorillonite clay: Hydrophobic and ionic modes of binding. *J. Mol. Catal.* **22**, 153–163.

Ghabbour, E. A., Davies, G., Goodwillie, M. E., O'Donaughy, K., and Smith T. L. (2004). Thermodynamics of peat-, plant-, and soil-derived humic acid sorption on kaolinite. *Environ. Sci. Technol.* **38**, 3338–3342.

Gianfreda, L., Rao, M. A., and Violante, A. (1991). Invertase (β-fructosidase): Effects of montmorillonite, Al–hydroxide and $Al(OH)_x$–montmorillonite complexes. *Soil Biol. Biochem.* **23**, 581–587.

Gianfreda, L., Rao, M. A., and Violante, A. (1992). Adsorption, activity and kinetic properties of urease on montmorillonite, aluminium hydroxide and $Al(OH)_x$–montmorillonite complexes. *Soil Biol Biochem* **24**, 51–58.

Gianfreda, L., Rao, M. A., Sannino, F., Saccomandi, F., and Violante, A. (2002). Enzymes in soil: Properties behavior and potential applications. In *Developments in Soil Science*, Vol. **28B**, Huang, P. M., and Bollag, J.-M., eds., Elsevier, Amsterdam, pp. 301–327.

Giese R. F., and van Oss, C. J. (2002). *Colloid and Surface Properties of Clays and Related Minerals*, Marcel Dekker, New York.

Giles, C. H, Smith, D., and Huitson, A. (1974). A general treatment and classification of the solute adsorption isotherms. I. Theoretical. *J. Coll. Interface Sci.* **47**, 755–765.

Gillet, A. (1957). Von der Cellulose zum Anthrazit. Die Entwicklung der Humine aus holzifer Materie IV. Die Kohlebildende Entwinklung durch oxycarbonischen Abbau. *Brennstoff-Chem.* **38**, 42–49.

Goh, K. M., and Malloy, B. P. J. (1978). Radiocarbon dating of paleosols using soil organic matter components. *J. Soil Sci.* **29**, 567–573.

Gouveia, S. E. M., Pessenda, L. C. R., and Aravena, R. (1999). Dating of the humin fraction of soil organic matter and its comparison with ^{14}C ages of fossil charcoal, *QJIM Nova* **22**, 810–814.

Greenland, D. J. (1956). The adsorption of sugars by montmorillonite. I. X-ray studies. *J. Soil Sci.* **7**, 319–328.

Greenland, D. J. (1965). Interaction between clays and organic compounds in soils. Part 1. Mechanisms of interaction between clays and defined organic compounds. *Soils Fertil.* **28**, 415–425.

Greenland, D. J. (1971). Interaction between humic and fulvic acids and clays. *Soil Sci.* **111**, 34–41.

Greenland, D. J., and Oades, D. M. (1975). Saccharides. In *Soil Components,* Vol. I.: *Organic Compounds*, Geiseking, J. E., ed., Springer, New York, pp. 213–261.

Grim, R. E. (1968). *Clay Mineralogy*, 2nd edition, McGraw-Hill, New York.

Gu, B., Schmitt, J., Chen, Z., Liang, L., and McCarthy, F. (1994). Adsorption and desorption of natural organic matter on iron oxide: Mechanisms and models. *Environ. Sci. Technol.* **28**, 38–46.

Guetzloff, T. F., and Rice, J. A. (1994). Does humic acid form a micelle? *Sci. Total Environ.* **152**, 31–35.

Guetzloff, T. F., and Rice, J. A. (1996). Micellar nature of humic colloids. In *Humic and Fulvic Acids: Isolation, Structure, and Environmental Role*, Gaffney, S., Marley, A., and Clark, S. B., eds., ACS Symposium Series, No. 651, American Chemical Society, Washington, D.C., pp. 18–25.

Harter, R. D., and Stotzky, G. (1973). X-ray diffraction, electron microscopy, eletrophoretic mobility, and pH of some stable smectite-protein complexes. *Soil Sci. Soc. Amer. Proc* **37**, 116–123.

Hatcher, P. G., Breger, I. A., Maciel, G. E., and Szeverenyi, N. M. (1985). Geochemistry of humin. In *Humic Substances in Soil, Sediment, and Water: Geochemistry, Isolation, and Characterization*, Aiken, G. R., McKnight, D. M., Wershaw, R. L., and MacCarthy, P. J., eds., John Wiley & Sons, New York, pp. 275–302.

Haynes, C. A., and Norde, W. (1994). Globular proteins at solid/liquid interfaces. *Coll. Surf. B Biointerf.* **2**, 517–566.

Hayano, H., Shinozuka, N., and Hyakutake, M. (1982). Surface active properties of marine humic acids. *Yukagaku* **31**, 357–362.

Hayase, K., and Tsubota, H. (1983). Sedimentary humic acid and fulvic acid as surface active substances. *Geochim. Cosmochim. Acta* **47**, 947–952.

Hayase, K., and Tsubota, H. (1984). Reversed-phase liquid chromatography of molecular weight-fractionated sedimentary fulvic acid. *J. Chromatogr.* **295**, 530–532.

Hayes, M. H. B., and Himes, F. L. (1989). Nature and properties of humus-mineral complexes. In *Interactions of Soil Minerals with Natural Organics and Microbes*, Huang, P. M., and Schnitzer, M., eds., Soil Science Society of America, Special Publication No. 17, Madison, WI, pp. 103–158.

Hayes, M. H. B., MacCarthy, P., Malcolm, R. L., and Swift, R. S. (1987). *Humic Substances II, In Search of Structure*, John Wiley & Sons, Chichester.

Hedges, J. I. (1977). The association of organic molecules with clay minerals in aqueous solutions. *Geochim. Cosmochim. Acta* **41**, 111–1123.

Heil, D., and Sposito, G. (1995). Organic matter role in illitic soil colloids flocculation: III Scanning force microscopy. *Soil Sci. Soc. Am. J.* **59**, 266–269.

Hinz, C. (2001). Description of sorption data with isotherm equations. *Geoderma* **99**, 225–243.

Huc, A. Y., and Durand, B. (1973). Etudes des acides humiques et de L'humine de sediments. Recents consideres comme precurseurs des kerogenes. In *Advances in Organic Geochemistry 1973*, Tissot, B., and Bienner, E, eds., Editions Technip, Paris, pp. 53–72.

Huc, A. Y., and Durand, B. M. (1977). Occurrence and significance of hurnic acids in ancient sediments. *Fuel* **56**, 73–80.

Ilés, E., and Tombácz, E. (2003). The role of variable surface charge and surface complexation in the adsorption of humic acid on magnetite. *Colloids Surf. A: Physicochem. Eng. Aspects* **230**, 99–109.

Ishiwatari, R. (1985). Geochemistry of humic substances in lake waters. In *Humic Substances in Soil, Sediment, and Water: Geochemistry, Isolation, and Characterization*, Aiken, G. R., McKnight, D. M., Wershaw, R. L., and MacCarthy, P. J., eds., John Wiley & Sons, New York, pp. 147–180.

Kaiser, K. (2003). Sorption of natural organic matter fractions to goethite (α-FeOOH): effect of chemical composition as revealed by liquid-state ^{13}C NMR and wet-chemical analysis. *Org. Geochem.* **34**, 1569–1579.

Kaiser, K., and Guggenberger, G. (2003). Mineral surfaces and soil organic matter. *Eur. J. Soil Sci.* **54**, 219–236.

Karickhoff, S. W., and Brown, D. S. (1979). Determination of octanol/water distribution coefficients, water solubilities, and sediment/water partition coefficients for hydrophobic organic pollutants. EPA/600/4-79/032 (Order No. PB80-103591).

Keil, R. G., and Hedges, J. I. (1995). Sedimentary organic matter preservation: An assessment and speculative synthesis. *Mar. Chem.* **49**, 81–115.

Killops, S., and Killops, V. (2005). *Introduction to Organic Geochemistry*, 2nd edition, Blackwell, Malden, MA.

Kipton, H., Powell, J., and Town, R. M. (1992). Solubility and fractionation of humic acid; effect of pH and ionic medium. *Anal. Chim. Acta* **267**, 47–54.

Kloskowski, R., and Führ, F. (1985). Behaviour of [ring14C] hydroxy-simazine in a parabraun soil. *Chemosphere* **14**, 1913–1920.

Kloskowski, R., and Führ, F. (1987). Aged and bound herbicide residues in soil and their bioavailability. *J. Environ. Sci. Health* **B22**, 623–642.

Kloskowski, R., Führ, F., and Mittelstaedt, W. (1986a). Formation of bound residues of [benzene ring-U-14C]- anilazine and [triazine-U-14C]-anilazine in parabraunerde (Alfisol soil) and their availibility to maize. In *Quantification, Nature, and Bioavailability of Bound 14C-Pesticide Residues in Soil, Plants, and Food*. Int. Atomic Energy Agency, Vienna, 65–70.

Kloskowski, R., Führ, F., and Mittelstaedt, W. (1986b). Plant availability of bound anilazine residues in a degraded loess soil. *J. Environ. Sci. Health* **B21**, 487–505.

Knicker, H., and Hatcher, P. G. (1997). Survival of protein in an organic-rich sediment: Possible protection by encapsulation in organic matter. *Naturwiss.* **84**, 231–234.

Kohl, S. D., and Rice, J. A. (1998). The binding of contaminants to humin: A mass balance. *Chemosphere* **36**, 251–261.

Kononova, M. M. (1966). *Soil Organic Matter*, Nowakowski, T. A., and Newman, A. C. D., translators, Pergamon Press, Oxford.

Kretzschmar, R., Hesterberg, D., and Sticher, H. (1997). Effects of adsorbed humic acid on surface charge and flocculation of kaolinite. *Soil Sci. Soc. Am. J.* **61**, 101–108.

Kumar, M., Ghosh, S. K., and Manjaiah, K. M. (2001). Adsorption of humic acid on homoionic clays. *Clay Res.* **20**, 1–10.

Labille, J., Thomas, F., Milas, M., and Vanhaverbeke, C. (2005). Flocculation of colloidal clay by bacterial polysaccharides: Effect of macromolecule charge and structure. *J. Coll. Interface Sci.* **284**, 149–156.

Lagaly, G. (1987). Clay-organic interactions: Problems and recent results. In Proceedings of the International Conference, Denver 1985, The Clay Minerals Society, pp. 343–351.

Laird, D. (2001). Nature of clay–humic complexes in an agricultural soil: II. Scanning electron microscopy analysis. *Soil Sci. Soc. Am. J.* **65**, 1419–1425.

Lichtfouse, E., Sen, O., Houot, S., and Barriuso, E. (1995). Isotope evidence for soil organic matter carbon pools with distinct turnover rate-II. Humic substances. *Org. Geochem.* **23**, 845–847.

Liu, A., Wu, R. C., Eschenazi, E., and Papadopoulos, K. (2000). AFM on humic acid adsorption on mica. *Coll. Surf. A* **174**, 245–252.

Loll, M. J., and Bollag, J.-M. (1983). Protein transformation in soil. *Adv. Agron.* **36**, 351–382.

Lovelock, C. E., Wright, S. F., Clark, D. A., and Ruess, R. W. (2004). Soil stocks of glomalin produced by arbuscular mycorrhizal fungi across a tropical rain forest landscape. *J. Ecol.* **92**, 278–287.

Lowe, L. E. (1978) Carbohydrates in soil. In *Soil Organic Matter*. Schnitzer, M., and Khan, S. U., eds., Elsevier, Amsterdam, pp. 65–93.

MacCarthy, P., and Rice, J. A. (1991). An ecological rationale for the heterogeneous nature of humic substances. In *Scientists on Gaia*, Schneider, S., and Boston, P. J., eds., MIT Press, Cambridge, MA, pp. 339–345.

Majzik, A., and Tombácz, E. (2007a). Interaction between humic acid and montmorillonite in the presence of calcium ions I. Interfacial and aqueous phase equilibria: Adsorption and complexation. *Org. Geochem.* **38**, 1319–1329.

Majzik, A., and Tombácz, E. (2007b). Interaction between humic acid and montmorillonite in the presence of calcium ions II. Colloidal interactions: Charge state, dispersing and/or aggregation of particles in suspension. *Org. Geochem.* **38**, 1330–1340.

Malekani, K., and Rice, J. A. (1997). Fractal character of humin and its components. *Fractals* **5**, 83–100.

Malekani, K., Rice, J. A., and Lin, J.-S. (1997). The effect of sequential removal of organic matter on the surface morphology of humin. *Soil Sci.* **162**, 333–342.

Martin, J. P., and Haider, K. (1980). Microbial degradation and stabilization of ^{14}C-labeled lignins, phenols, and phenolic polymers in relation to soil humus formation In *Lignin Biodegradation: Microbiology Chemistry and Potential Applications*, Vol. **II**. Kent-Kirk, T., Higuchi, T., and Chang, H., eds., CRC Press, Boca Raton, FL, pp. 77–100.

Martinez, F. M., and Rodriguez, J. L. P. (1969). Interlamellar adsorption of a blackearth humic acid on Na-montmorillonite. *Z. Pflanzen. Bodenk.* **124**, 52–57.

Mayer, L. M. (1994). Surface area control of organic carbon accumulation in continental shelf sediments, *Geochim. Cosmochim. Acta* **58**, 1271–1284.

Mayer, L. M. (1995). Relationships between mineral surfaces and organic carbon concentrations in soils and sediments. *Chem. Geol.* **114**, 347–363.

McLaren, A. D. (1954). The adsorption and reactions of enzymes on kaolinite. I. *J. Phys. Chem.* **58**, 129–137.

Meyers, P. A., and Ishiwatari, R. (1993). Lacustrine organic geochemistry—An overview of indicators of organic matter sources and diagenesis in lake sediments. *Org. Geochem.* **20**, 867–900.

Meyers, P. A., and Quinn, J. G. (1973). Factors affecting the association of fatty acids with mineral particles in sea water. *Geochim. Cosmochim. Acta* **37**, 1745–1759.

Meier, M., Namjesnik-Dejanovic, K., Maurice, P. A., Aiken, G. R., Chin, Y. P., and Cabaniss, S. (1999). Fractionation of aquatic natural organic matter upon sorption to goethite and kaolinite. *Chem. Geol.* **157**, 275–284.

Moinereau, J. (1977). Absorption of humic compounds by a hydrogen(+), aluminum(3+) montmorillonite. Presence of smectite in the interlayer space of organomineral complexes of the andosols. *Clay Min.* **12**, 75–82.

Morrison, R. A. (1969). Soil Lipids. In *Organic Geochemistry—Methods and Results*, Eglinton, G., and Murphy, M. T. J., eds., Springer Verlag, New York, pp. 558–575.

Mortland, M. M. (1970). Clay–organic complexes and interactions. *Adv. Agron.* **22**, 75–117.

Mortland, M. M. (1986). Mechanisms of adsorption of nonhumic organic species by clays. In *Interactions of Soil Minerals with Natural Organics and Microbes*, Huang, P. M., and Schnitzer, M., eds., Soil Science Society of America, Special Publication No. 17, Madison, WI, pp. 59–76.

Murphy, E. M., and Zachara, J. M. (1995). The role of sorbed humic substances on the distribution of organic and inorganic contaminants in groundwater. *Geoderma* **67**, 103–124.

Naidja, A., and Huang, P. M. (1996). Deamination of aspartic acid by aspartase–Ca–montmorillonite complex. *J. Mol. Catal. A: Chem.* **106**, 255–265.

Naidja, A., Violante, A., and Huang, P. M. (1995). Adsorption of tyrosinase onto montmorillonite as influenced by hydroxyaluminum coatings. *Clays Clay Min.* **43**, 647–655.

Namjesnik-Dejanovic, K., and Maurice, P. A. (2001). Conformations and aggregate structures of sorbed natural organic matter on muscovite and hematite. *Geochim. Cosmochim. Acta* **65**, 1047–1057.

Namjesnik-Dejanovic, K., Maurice, P. A., Aiken, G. R., Cabaniss, S., Chin, Y. P., and Pullin M. J. (2000). Adsorption and fractionation of a muck fulvic acid on kaolinite and goethite at pH 3.7, 6 and 8. *Soil Sci.* **165**, 545–559.

Newman, A. C. D., and Brown, G. (1987). The chemical constitution of clays. In *Chemistry of Clays and Clay Minerals*, Newman, A. C. D., ed., Mineralogical Society Monograph No. 6, pp. 1–128.

Nichols, K. A., and Wright, S. F. (2006). Carbon and nitrogen in operationally defined soil organic matter pools. *Biol. Fertil. Soils* **43**, 215–220.

Nord, F. F. (1964). The formation of lignin and its biochemical degradation. *Geochim. Cosmochim. Acta* **28**, 1507–1521.

Nguyen, R., and Harvey, R. (2003). Preservation via macromolecular associations during *Botryococcus braunii* decay: proteins in the Pula Kerogen. *Org. Geochem.* **34**, 1391–1403.

Olness, A. E., and Clapp, C. E. (1973). Occurrence of collapsed and expanded crystals in montmorillonite–dextran complexes. *Clays Clay Min.* **7**, 113–118.

Orlov, D. S. (1985). *Humus Acids of Soils*. tranls. Kothekar, V., A. A. Balkema, Rotterdam.

Pachepsky, Y. A., Polubesova, T. A., Hajos, M., Jozefaciuk, G., and Sokolowska, Z. (1995). Parameters of surface heterogeneity from laboratory experiments on soil degradation. *Soil Sci. Soc. Am. J.* **59**, 410–417.

Parfitt, R. L., and Greenland, D. J. (1970). Adsorption of polysaccharides by montmorillonite. *Soil Sci. Soc. Am. Proc.* **34**, 862–866.

Parfitt, R.L., Fraser, A. R., and Farmer, V.C. (1977). Adsorption on hydrous oxides. III. fulvic acid and humic acid on goethite, gibbsite and imogolite. *J. Soil Sci.* **28**, 289–296.

Parsons, J. W.; Tinsley, J. (1975). Nitrogenous substances. In *Soil Components,* Vol. I.: *Organic Compounds*, Geiseking, J. E., ed., Springer, New York, pp. 263–304.

Pennell, K. D., Boyd, S. A., and Abriola, L. M. (1995). Surface area of soil organic matter reexamined. *Soil Sci. Soc. Am. J.* **59**, 1012–1018.

Pessenda, L. C. R., Gouveia, S. E. M., Aravena, R., Gomez, B. M., Boulet, R., and Ribeiro, A. S. (1998). ^{14}C dating and stable carbon isotopes of soil organic matter in forest savanna boundary areas in the southern Brazilian Amazon region. *Radiocarbon* **40**, 1013–1022.

Peters, K., Rohrback, B., and Kaplan, I. (1981). Geochemistry of artificially heated humic and sapropelic sediments-I: Protokerogen. *Am. Assoc. Pet. Geol. Bull.* **65**, 688–705.

Piccolo, A. (1989). Reactivity of added humic substances towards plant-available heavy metals in soils. *Sci. Total Environ.* **81–82**, 607–614.

Piret, E., White, R., Walther, H., and Madden, A., Jr. (1960). Some physico-chemical properties of peat humic acids. *Sci. Proc. R. Dublin Soc.* **A1**, 69–79.

Prikhod'ko, V. E. (1982). Effect of exchangeable cations on the interactions between minerals and lignin-based copolymers. *Pochvovedenie* **4**, 57–63.

Quiquampoix, H. (2000). Mechanisms of protein adsorption on surfaces and consequences for extracellular enzyme activity in soil. In *Soil Biochemistry 10*, Bollag, J.-M., and Stotzky, G., eds., Marcel Dekker, New York, pp. 171–206.

Quiquampoix, H., and Burns, R. G. (2007). Interactions between proteins and soil mineral surfaces: Environmental and health consequences. *Elements* **3**, 401–406.

Quiquampoix, H., Servagent-Noinville, S., and Baron, M. H. (2002). Enzyme adsorption on soil mineral surfaces and consequences for the catalytic activity. In *Enzymes in the Environment*, Burns, R. G., and Dick, R. P., eds., Marcel Dekker, New York, 285–306.

Quiquampoix, H., Abadie, J., Baron, M. H., Leprince, F., Matumoto-Pintro, P. T., Ratcliffe, R. G., and Staunton, S. (1995). Mechanisms and consequences of protein adsorption on soil mineral surfaces. In *Proteins at Interfaces. Protein Adsorption on Soil Mineral Surfaces*. Horbett, T. A., and Brash, J. L., eds., American Chemical Society Symposium Series 602, Washington, D.C., pp. 321–333.

Rao, M. A., Palmiero, F., Gianfreda, L., and Violante, A. (1996). Interaction of acid phosphatase with clays, organic molecules and organo-mineral complexes. *Soil Sci.* **161**, 751–760.

Rashid, M. A. (1985). *Geochemistry of Marine Humic Compounds*, Springer-Verlag, New York.

Rashid, M. A., Buckley, D. E., and Robertson, K. R. (1972). Interactions of a marine humic acid with clay minerals and a natural sediment. *Geoderma* **8**, 11–27.

Rebhun, M., Kalabo, R., Grossman, L., Manka, J., and Rav-Acha, C. (1992). Sorption of organics on clay and synthetic humic–clay complexes simulating aquifer processes. *Water Res.* **26**, 79–84.

Reuter, J. H., and Perdue, E. M. (1984). A chemical structural model of early diagenesis of sedimentary humus/proto-kerogens. *Mitt. Geol.-Palaon. Inst.* **56**, 249–262.

Rice, J. A. (2001). Humin, *Soil Sci.* **166**, 848–857.

Rice, J. A. (2008). Applications of fractals in the study of humic materials. In *Bio-Physical Chemistry of Fractal Structures and Process in Environmental Systems*, Senesi, N., and Wilkinson, K., eds., IUPAC Series on Analytical and Physical Chemistry of Environmental Systems, Vol. **11**, John Wiley & Sons, Chichester, pp. 221–238.

Rice, J. A., and Lin, J. S. (1993). Fractal nature of humic materials. *Environ. Sci. Technol.* **27**, 413–414.

Rice, J. A., and MacCarthy, P. (1989). Characterization of a stream sediment humin. In *Influence of Aquatic Humic Substances on Fate and Treatment of Pollutants*, Suffet, I. H., and MacCarthy, P., eds., ACS Advances Chemistry, Series, No. 219, American Chemical Society, Washington, D.C., pp. 41–54.

Rice, J. A., and MacCarthy, P. (1991). Statistical evaluation of the elemental composition of humic substances, *Org. Geochem.* **17**, 635–648.

Robertson, A. P., and Leckie, J. O. (1994). Humic acid/goethite interactions and their effect on copper binding. In *Humic substances in the Global Environment and Implications to Human Health*, Senesi, N., and Miano, T. M., eds., Elsevier, Amsterdam, pp. 487–492.

Rochus, W., and Sipos, S. (1978). Dir Micellbilding bei Huminstoffen. *Agrochim.* **22**, 446–454.

Sarkanen, K. V., and Ludwig, C. H., eds. (1971). *Lignins: Occurence, Formation, Structure, and Reactions*, Wiley Interscience, New York.

Satoh, T. (1984a). Organo-mineral complex status in soils. I. Thermal analytical characteristics of humus in the soils. *Soil Sci. Plant Nutr.* **30**, 1–12.

Satoh, T. (1984b). Organo-mineral complex status in soils. II. Thermal nature of organo-mineral complex particles and their humic substances. *Soil Sci. Plant Nutr.* **30**, 95–104.

Scheidegger, A. M., and Sparks, D. L. (1996). A critical assessment of sorption–desorption mechanisms at the soil mineral/water interface. *Soil Sci.* **161**, 813–831.

Schindler, F. V., Mercer, E., and Rice, J. A. (2007). Chemical characteristics of glomalin extracted from soil. *Soil Biol. Biochem.* **39**, 320–329.

Schlesinger, W. H. (1991). *Biogeochemistry: An Analysis of Global Change*, Academic Press, San Diego, CA.

Schnitzer, M. (1986). Binding of humic substances by soil mineral colloids. In *Interactions of Soil Minerals with Natural Organics and Microbes*, Huang, P. M., and Schnitzer, M., eds., Soil Science Society of America, Special Publication No. 17, Madison, WI, pp. 77–101.

Schnitzer, M., and Ghosh, K. (1982). Characterization of water soluble fulvic acid–copper and fulvic acid–iron complexes. *Soil Sci.* **134**, 354–363.

Schnitzer, M., and Khan, S. U. (1972). *Humic Substances in the Environment*. Marcel Dekker, New York.

Schnitzer, M., and Kodama, H. (1966). Montmorillonite: Effect of pH on its adsorption of a soil humic compound. *Science* **153**, 70–71.

Schnitzer, M., and Kodama, H. (1972). Differential thermal analysis of metal–fulvic acid salts and complexes. *Geoderma* **7**, 93–103.

Schulten, H.-R., and Leinweber, P. (2000). New insights into organic-mineral particles: composition, properties and models of molecular structure. *Biol. Fertil. Soils* **30**, 399–432.

Schulten, H.-R., and Schnitzer, M. (1998). The chemistry of soil organic nitrogen: A review. *Biol. Fertil. Soils* **26**, 1–15.

Schulthess, C. P., and Huang, C. P. (1991). Humic and fulvic acid adsorption by silicon and aluminum oxide surfaces on clay minerals. *Soil Sci. Soc. Am. J.* **55**, 34–42.

Servagent-Noinville, S., Revault, M., Quiquampoix, H., and Baron, M.-H. (2000). Conformational changes of bovine serum albumin induced by adsorption on different clay surfaces: FTIR analysis. *J. Coll. Interface Sci.* **221**, 273–283.

Shah, R. K., Choski, M. R., and Joshi, B. C. (1975a). Development studies on soil organic matter: Humus. *Chem. Era* **4**, 31–34.

Shah, R. K., Choski, M. R., and Joshi, B. C. (1975b). Development studies on soil organic matter: Humin. *Chem. Era* **6**, 1–3.

Shen, Y. H. (1999). Sorption of humic acid to soil: The role of soil mineral composition. *Chemosphere* **38**, 2489–2499.

Simpson, A. J., Simpson, M. J., Kingery, W. L., Lefebvre, B. A., Moser, A., Williams, A. J., Kvasha, M., and Kelleher, B. P. (2006). The application of ^{1}H high-resolution magic-angle spinning NMR for the study of clay–organic associations in natural and synthetic complexes. *Langmuir* **22**, 4498–4503.

Sinegani, A. A. S., Emtiazi, G., and Shariatmadari, H. (2005). Sorption and immobilization of cellulase on silicate clay minerals. *J. Coll. Interface Sci.* **290**, 39–44.

Singer, A., and Huang, P. M. (1988). Thermal analysis of AlOH polymer/montmorillonite/humic acid complexes. *Thermochim. Acta* **135**, 307–312.

Sjöström, E. (1993). *Wood Chemistry: Fundamentals and Applications*, Academic Press, New York.

Skujins, J., and Klubeck, B. (1982). Soil biological properties of a montane forest sere: Corroboration of Odum's postulates. *Soil Biol. Biochem.* **14**, 505–513.

Spark, K. M., Wells, J. D., and Johnson, B. B. (1997). Sorption of heavy metals by mineral–humic acid substrates. *Aust. J. Soil. Res.* **35**, 113–122.

Specht, C. H., Kumke, M. U., and Frimmel, F. H. (2000). Characterization of NOM adsorption to clay minerals by size exclusion chromatography. *Water Res.* **34**, 4063–4069.

Sposito, G. (1984). *The Surface Chemistry of Soils*, Oxford University Press, New York.

Sposito, G., Skipper, N. T., Sutton, R., Park, S.-H., Soper, A. K., and Greathouse, J. A. (1999). Surface geochemistry of the clay minerals. *Proc. Natl. Acad. Sci. USA* **96**, 3358–3364.

Steurmer, D. H., Peters, K. E., and Kaplan, I. R. (1978). Source indicators of humic substances and protokerogen. Stable isotope ratios, elemental compositions and electron spin resonance spectra. *Geochim. Cosmochim. Acta.* **42**, 989–997.

Stevenson, F. J. (1966). Lipids in soil. *J. Am. Oil Chem. Soc.* **43**, 203–210.

Stevenson, F. J. (1994). *Humus Chemistry: Genesis, Composition, Reactions*, John Wiley & Sons, New York.

Stevenson, F. J., and Ardakani, M. S. (1972). Organic matter reactions involving micronutrients in soils. In *Micronutrients in Agriculture*, Mortvedt, J. J., Giordano, P. M., and Lindsay, W. L., eds., Soil Science Society of America, Madison, WI, pp. 79–114.

Stumm, W. (1992). *Chemistry of the Solid–Water Interface*. John Wiley & Sons, New York.

Sutton, R., and Sposito, G. (2006). Molecular structures in soil humic substances: A new view. *Environ. Sci. Technol.* **39**, 9009–9015.

Swaby, R. J., and Ladd, J. N. (1962). Chemical nature, microbial resistance, and origin of soil humus. In *International Society of Soil Science Transactions, IV and V, Communications*, Neale, G. J., ed., Palmerston North, New Zealand, pp. 197–202.

Swaby, R. J., and Ladd, J. N. (1966). Stability and origin of soil humus. III. In *The Use of Isotopes in Soil Organic Matter Studies*. International Atomic Energy Agency, Pergamon Press, Oxford, pp. 153–159.

Tadjerpisheh, N., and Ziechmann, W. (1994). Reaction of clay minerals with lignin and analysis of the reaction products. *Chem. Erde* **54**, 199–212.

Talibudeen, O. (1950). Interlamellar adsorption of protein monolayers on pure montmorillonoid clays. *Nature* **163**, 236–237.

Tan, K. H. (1977). Formation of metal–humic acid complexes by titration and their characterization by differential thermal analysis and infrared spectroscopy. *Soil Biol. Biochem.* **10**, 123–129.

Taylor, J., Young, C., Parkes, R., Eglinton, T., and Douglas, A. (1984). Structural relationships in protokerogens and other geopolymers from oxic and anoxic sediments. *Org. Geochem.* **6**, 279–286.

Teppen, B. J., and Aggarwal, V. (2007). Thermodynamics of organic cation exchange selectivity in smectites. *Clays Clay Min.* **55**, 119–130.

Theng, B. K. G. (1976). Interactions between montmorillonite and fulvic acid. *Geoderma* **15**, 243–251.

Theng, B. K. G. (1979). *Formation and Properties of Clay–Polymer Complexes*. Elsevier, Amsterdam.

Theng, B. K. G., and Scharpenseel, H. W. (1975). The adsorption of ^{14}C-labeled humic acid by montmorillonite. Proceedings, International Clay Society, Mexico City, pp. 643–653.

Tissot, B. P., and Welte, D. H. (1978). *Petroleum Formation and Occurrence*. Springer-Verlag, Berlin.

Tombácz, E. (2003). Effect of environmental relevant organic complexants on the surface charge and the interaction of clay mineral and metal oxide particles. In *Role of Interfaces in Environmental Protection*, Barany, S., ed., Kluwer, Dordrecht, 397–424.

Tombácz, E., Gilde, M., Ábrahám, I. and Szántó, F. (1988). Effect of electrolyte concentration on the interaction of humic acid and humate with montmorillonite, *Appl. Clay Sci.* **3**, 31–52.

Tombácz, E., Gilde, M., Ábrahám, I. and Szántó, F. (1990). The possibility of heterocoagulation between montmorillonite and humic substances, *Appl. Clay Sci.* **5**, 265–270.

Tombácz, E., Szekeres, M., Barany, L., and Michéli, E. (1998). Surface modification of clay minerals by organic polyions. *Coll. Surf. A* **141**, 379–384.

Tschapek, M., and Wasowski, C. (1976). The surface activity of humic acid. *Geochim. Cosmochim. Acta* **40**, 1343–1345.

Ulrich, H. J., Stumm, W., and Cosovic, B. (1988). Adsorption of aliphatic fatty acids on aquatic interfaces. Comparison between two model surfaces: The mercury electrode and δ-AI_2O_3, colloids. *Environ. Sci. Technol.* **22**, 37–41.

Vandenbroucke, M., Pelet, R., and Debyser, Y. (1985). Geochemistry of humic substances in marine environments. In *Humic Substances in Soil, Sediment, and Water: Geochemistry, Isolation, and Characterization*, Aiken, G. R., McKnight, D. M., Wershaw, R. L., and MacCarthy, P. eds., John Wiley & Sons, New York, pp. 249–273.

Vandenbroucke, M., and Largeau, C. (2007). Kerogen origin, evolution and structure. *Org. Geochem.* **38**, 719–833.

Varadachari, C., Mondal, A. H., and Ghosh, K. (1991). Some aspects of clay–humus complexation: Effect of exchangeable cations and lattice charge. *Soil Sci.* **151**, 220–227.

Vermeer, A. W. P., McCulloch, J. K., van Riemsdijk, W. H., and Koopal, L. K. (1999). Metal ion adsorption to complexes of humic acid and metal oxides: Deviations from the additivity rule. *Environ. Sci. Technol.* **33**, 3892–3897.

Vermeer, A. W. P., van Riemsdijk, W. H., and Koopal, L. K. (1998). Adsorption of humic acid to mineral particles. 1. Specific and electrostatic interactions. *Langmuir* **14**, 2810–2819.

Violante, A., Arienzo, M., Sannino, F., Colombo, C., Piccolo, A., and Gianfreda., L. (1999). Formation and characterization of OH–Al–humate–montmorillonite complexes. *Org. Geochem.* **30**, 461–468.

Visser, S. (1964). Oxidation–reduction potentials and capillary activities of humic acids. *Nature* **204**, 581.

Wakeham, S. G., Lee, C., Hedges, J. I., Hernes, P. J., and Peterson, M. L. (1997). Molecular indicators of diagenetic status in marine organic matter. *Geochim. Cosmochim. Acta* **61**, 5363–5369.

Wang, K., and Xing, B. (2005). Structural and sorption characteristics of adsorbed humic acid on clay minerals. *J. Environ. Qual.* **34**, 342–349.

Wang, T. S. C., Huang, P. M., Chou, C.-H., and Chen, J.-H. (1986). The role of soil minerals in the abiotic polymerization of phenolic compounds and formation of humic substances. In *Interactions of Soil Minerals with Natural Organics and Microbes*, Huang, P. M., and Schnitzer, M., eds., Soil Science Society of America, Special Publication No. 17, Madison, WI, pp. 251–281.

Wang, X. P., Shan, X. Q., Luo, L., Zhang, S. Z., and Wen, B. (2005). Sorption of 2,4,6-trichlorophenol in model humic acid–clay systems. *J. Ag. Food Chem.* **53**, 3548–3555.

Weiss, A., and Roloff, G. (1965). The role of organic derivatives of mica-like layer silicates during the formation of petroleum. *Intern. Clay Conf., Proc. Conf., Stockholm* **2**, 373–378.

Wershaw, R., Burcar, P., and Goldberg, M. (1969). Interaction of pesticides with natural organic material. *Environ. Sci. Technol.* **3**, 271–273.

Wershaw, R. L, Llaguno, E. C., and Leenheer, J. A. (1996a). Mechanism of formation of humus coatings on mineral surfaces. 3. Composition of adsorbed organic acids from compost leachate on alumina by solid-state ^{13}C NMR. *Coll. Surf. A* **108**, 213–223.

Wershaw, R. L., Llaguno, E. C., Leenheer, J. A., Sperline, R. P., and Song, Y. (1996b). Mechanism of formation of humus coatings on mineral surfaces. 2. Attenuated total reflectance spectra of hydrophobic and hydrophilic fractions of organic acids from compost leachate on alumina. *Coll. Surf. A* **108**, 199–211.

Wright, S., and Upadhyaya, A. (1996). Extraction of an abundant and unusual protein from soil and comparison with hyphal protein of arbuscular mycorrhizal fungi. *Soil Sci.* **161**, 575–586.

Xie, H., Guetzloff, T. F., and Rice, J. A. (1997). Fractionation of pesticide residues bound to humin. *Soil Sci.* **162**, 421–429.

Xing, B. (2001). Sorption of naphthalene and phenanthrene by soil humic acids. *Environ. Pollution* **111**, 303–309.

Yoon, T. H., Johnson, S. B., and Brown, G. E., Jr. (2005). Adsorption of organic matter at mineral/water interfaces: 4. Adsorption of humic substances at boehmite–water interfaces and impact on boehmite dissolution. *Langmuir* **21**, 5002–5012.

Yoon, L., and Lenhoff, A. (1992). Computation of the electrostatic interaction energy between a protein and a charged suface. *J. Phys. Chem.* **96**, 3130–3134.

Yost, E. C., Tejedor-Tejedor, M. I., and Anderson, M. A. (1990). *In situ* CIR-FTIR characterization of salicylate complexes at the goethite/aqueous solution interface. *Environ. Sci. Technol.* **24**, 822–828.

Zavarzina, A. G. (2001). Sorption of soil-originated humic acids on clay minerals, *Proceedings, Eleventh Annual V. M. Goldschmidt Conference*, May 20–24, Hot Springs, Va.

Zhou, J. L., Rowland, S., Mantoura, R. F. C., and Braven, J. (1994). The formation of humic coatings on mineral particles under simulated estuarine conditions—A mechanistic study. *Water Res.* **28**, 571–579.

Zielke, R., Pinnavaia, T. J., and Mortland, M. M. (1989). Adsorption and reactions of selected organic molecules on clay mineral surfaces. In *Reactions and Movement of Organic Chemicals in Soils*, Sawhney, B. L., and Brown, K., eds., SSSA Special Publication No. 22, Madison, WI, pp. 81–97.

4

THE EFFECT OF ORGANIC MATTER AMENDMENT ON NATIVE SOIL HUMIC SUBSTANCES

C. Plaza

Centro de Ciencias Medioambientales, CSIC, Madrid, Spain

N. Senesi

Dipartimento di Biologica e Chimica Agroforestale ed Ambientale, University of Bari, Bari, Italy

4.1. INTRODUCTION

The natural sources of soil organic matter (SOM) are indigenous plant and animal debris that are decomposed and partly mineralized by soil biota. Any factor that

Biophysico-Chemical Processes Involving Natural Nonliving Organic Matter in Environmental Systems, Edited by Nicola Senesi, Baoshan Xing, and Pan Ming Huang

decreases C inputs and/or increases soil biological activity, or destabilizes SOM physical protection, such as changes in land use (e.g., deforestation), inappropriate crop cultivation and harvesting, and erosion, favor the loss of SOM. On the other hand, any soil management practice that increases C inputs, reduces biological activity, and/or enhances SOM physical stabilization, such as crop rotation, reduced tillage, intensive use of cover crops, and organic amendment, promotes the accumulation of SOM (Figure 4.1) (Gregorich et al., 1998; Lal, 2001, 2004; Siemens and Janssens, 2003; Worrall et al., 2003; Smith, 2004; Schulze and Freibauer, 2005). Nowadays, organic amendment represents the most common, efficient, and cheap practice for restoring, maintaining, and/or improving SOM content and its physical, chemical, and biological functions in soil (Senesi et al., 1996, 2007; Senesi and Plaza, 2007).

Traditional organic amendments include crop residues and animal manures. However, locally these amendments may be scarce or other organic residues may be abundant; thus a wide variety of other organic materials of natural or artificial origin are currently used with the aim of restoring organic fertility of organic-depleted soils, maintaining it in intensively cropped soils and enhancing it in intrinsically organic-poor soils. These materials include organic wastes generated in large amount from agriculture and forestry activities (e.g., crop residues, animal slurry and manures, forest maintenance wastes, wood chips, and saw mill wastes), municipalities (e.g., municipal sewage sludges and urban solid wastes), and light industries (e.g., olive-oil processing residues, winery and distillery wastes, paper mill sludges, and leather wastes) (Brebbia et al., 2004; Nortcliff, 2005). Soil application of organic wastes is generally preceded by treatments having the aim of lowering or eliminating a number of possible negative effects and hazards that raw materials may have on soil physical, chemical, and biological properties. The most common treatments

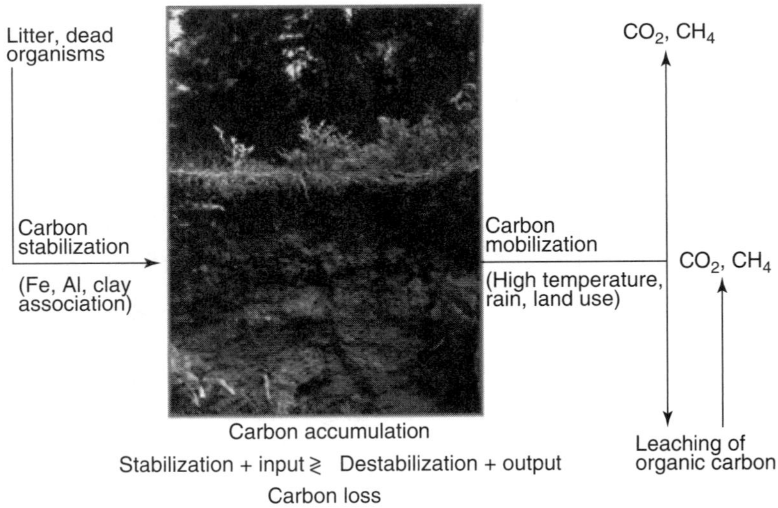

Figure 4.1. Balance of organic matter in soils. Reprinted from Schulze, E. D., and Freibauer, A. (2005). Carbon unlocked from soils. *Nature* **437**, 205–206, with permission from Macmillan Publishers Ltd.

of organic wastes include composting, vermicomposting, and anaerobic and aerobic digestions (De Bertoldi et al., 1996; Edwards, 2004; Williams, 2005).

Organic amendments are rich in organic compounds including humic substances (HS); therefore, their application to soil affects to various extent the content, composition, and properties of native SOM. Since soil HS have a major role in a wide number of agronomic, environmental, and geochemical processes, the knowledge of the short-term and/or long-term effects of organic amendments on the status, quality, and reactivity of indigenous soil HS is of critical importance (Senesi et al., 1996, 2007; Senesi and Plaza, 2007).

4.2. MINERALIZATION AND HUMIFICATION PROCESSES IN AMENDED SOILS

Organic residues in soil are subjected to decay in several, mainly microbial-controlled stages of decomposition, utilization, and transformation leading to more refractory organic substances (Stevenson, 1994). In the initial phase, readily decomposable organic compounds, such as sugars, starches, hemicelluloses, and amino acids, and some of the more resistant materials, such as cellulose, are rapidly decomposed with production of CO_2 and other volatile compounds, organic acids, and several incompletely oxidized compounds. This phase is followed by a stage in which organic intermediates and newly formed biomass tissues are utilized together with the remainder of cellulose and part of the lignin, with production of new biomass and further loss of C as CO_2. The final stage is characterized by a gradual decomposition of more resistant plant components, such as lignin, and the formation of newly synthesized, more stable products, the HS, which may persist for thousand of years (Stevenson, 1994).

Soil HS feature a colloidal, polydispersed, and polyelectrolytic character, a mixed aliphatic and aromatic nature, and the presence of various chemically reactive functional groups, including carboxyls and phenolic and alcoholic hydroxyls (Stevenson, 1994). Several pathways have been proposed for the formation of soil HS, the major ones being the lignin–protein theory, the sugar–amine theory, and the polyphenol theory (Stevenson, 1994; Senesi and Loffredo, 1999). The classical lignin–protein theory considers plant lignin as the main source of soil HS. According to this theory, lignin is incompletely utilized by soil microorganisms and undergoes a preliminary series of modifications, including loss of methoxyl groups, generation of *o*-hydroxyphenols, and oxidation of terminal aliphatic side chains to form carboxylic groups. The *o*-dihydroxybenzene units resulting from demethylation of lignin would further oxidize to quinines capable of undergoing condensation reactions with amino compounds produced by microbial synthesis. This process would yield first humin, then humic acids (HAs), and finally fulvic acids (FAs). In the sugar–amine theory, nonenzymatic condensation of reducing sugars and amino acids formed as by-products of microbial metabolism and further polymerization reactions, are postulated to play an important role in the formation of soil HS. According to the polyphenol theory, polyphenols of lignin origin or synthesized by microorganisms are the major building blocks from which soil HS are formed. In this theory, polyphenols are converted by polyphenoloxidase enzymes to quinones, which react with N-containing compounds and polymerize to produce first FAs, then HAs, and finally humin.

The accumulation and mineralization of SOM is greatly influenced not only by the amount but also by the origin and nature of the organic material added to soil (Figure 4.2) (Levi-Minzi et al., 1990; García et al., 1992a; Bernal et al., 1998; Flavel and Daniel, 2006; Pedra et al., 2007). Treatment processes such as composting or vermicomposting of raw organic wastes before application to soil lead to an extensive mineralization and partial humification—that is, a wide conversion of easily degradable organic matter to refractory organic compounds that resemble native soil HS (Senesi, 1989; Senesi et al., 1996; Paré et al., 1998; Plaza et al., 2005a, 2007; Romero et al., 2007; Senesi and Plaza, 2007). In principle, these treatments are accelerated versions of processes involved in the natural decomposition of organic debris in soil, achieved through the provision of the most favorable conditions for microbial activity. Because of the larger proportion of stable compounds, the application of treated amendments as compared to raw amendments to soil generally represents a more effective means of increasing SOM content. For example, Eghball (2002) reported that after 4 years of amendment, 36% of the C added as composted manure was retained as soil organic C compared with 14–25% of the C added as uncomposted manure. Similarly, in an experiment conducted in Spain, five annual applications of 24 t ha^{-1} of municipal solid waste compost resulted in more soil organic C accumulated than that occurring when fresh cow manure or fresh sewage sludge was applied at the same rate of dry matter (Albiach et al., 2001).

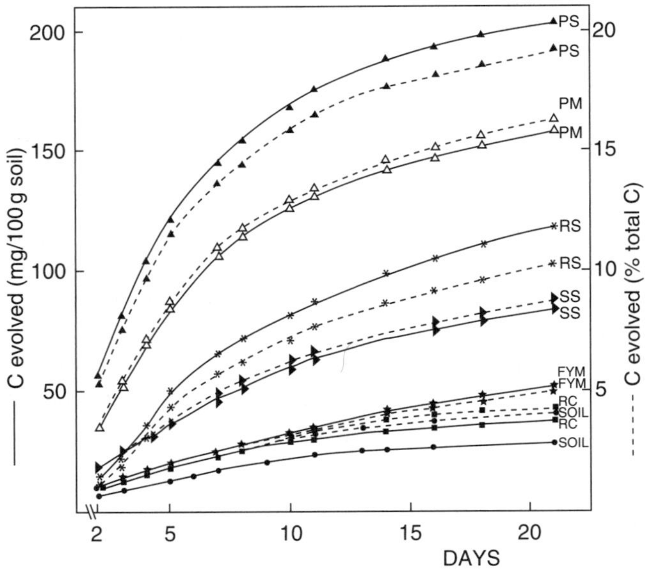

Figure 4.2. Carbon released as CO_2 from unamended soil and soils amended with pig slurry (PS), poultry manure (PM), cattle farmyard manure (FYM), aerobic sewage sludge (SS), municipal solid waste fuse compost (RC), and rye straw (RS) at a rate of 10 g kg^{-1} during incubation at 22 °C. Reprinted from Levi-Minzi, R., Riffaldi, R., and Saviozzi, A. (1990). Carbon mineralization in soil amended with different organic materials. *Agric. Ecosyst. Environ.* **31**, 325–335, with permission from Elsevier.

4.3. HUMIFICATION PARAMETERS

A number of indexes based on the distribution of organic C in humic and nonhumic fractions are used for the evaluation of the humification level in organic amendments and amended soils (Senesi, 1989). Among others these include: the degree of humification (DH), which is calculated as DH% = 100 × (HAC + FAC)/TEC; the humification rate (HR), which is calculated as HR% = 100 × (HAC + FAC)/TOC; and the humification index (HI), which is calculated as HI = NHC/(HAC + FAC) (Sequi et al., 1986; Ciavatta et al., 1988). TOC represents the total organic carbon in the sample; TEC represents the total extractable C by an alkaline solution (NaOH and/or $Na_4P_2O_7$); HAC and FAC represent the C content in HA and FA fractions, respectively; and NHC represents the nonhumified C content calculated by difference, that is, NHC = TEC − (HAC + FAC).

During the treatment processes of organic materials aimed at producing high-quality soil amendments, such as composting or vermicomposting, nonhumic components, such as cellulose, hemicellulose, and lignin, are degraded at different rates (Sánchez-Monedero et al., 1999; Romero et al., 2007). Water-soluble carbohydrates and phenols present within the NHC fraction of organic wastes are shown to play a very important role in organic matter degradation and humification. In particular, water-soluble carbohydrates constitute the principal C source for microorganisms responsible for organic matter degradation. Furthermore, a significant negative correlation is found to exist between the phenol content in aqueous extracts and several humification indexes, such as the HAC/TOC, HAC/TEC, and HAC/FAC ratios, measured for organic materials during common treatment processes (Sánchez-Monedero et al., 1999). These results strongly suggest that phenols act as major precursors in the formation of HS.

Wide evidence exists in the literature that soil amendment with organic materials not only increases the content of TOC, TEC, HAC, FAC, and NHC but also causes significant changes in the organic C distribution (Table 4.1). In particular, a significant decrease of DH and HR and an increase of HI are commonly measured in organically amended soils, especially in those that have received high amendment rates of untreated materials (Adani and Tambone, 2005; Brunetti et al., 2005, 2007a; Adani et al., 2007). These results indicate that application of organic amendments leads to a decrease of native SOM stability by increasing the nonhumified SOM fraction more than the humic fractions.

4.4. COMPOSITIONAL AND STRUCTURAL FEATURES OF HUMIC SUBSTANCES IN AMENDED SOILS

Numerous chemical, physico-chemical, and spectroscopic methods and techniques have been applied to study the following on a molecular scale: (a) the compositional and structural properties of HS in organic amendments of various origins and nature and (b) the effects of added organic materials on indigenous soil HS. These include, among others, elemental and acidic functional group analyses, gel-filtration chromatography, potentiometry, and ultraviolet–visible (UV–Vis), fluorescence, Fourier transform infrared (FT-IR), nuclear magnetic resonance (NMR), and electron spin resonance (ESR) spectroscopies (Senesi et al., 1996, 2007; Chen, 2003; Senesi and

TABLE 4.1. Total Organic C (TOC), Total Extractable C (TEC), Humic and Fulvic Acid C (HAC + FAC), and Nonhumified C (NHC) Contents, Degree of Humification (DH), Humification Rate (HR), and Humification Index (HI) of Soils Amended with $10\,t\,ha^{-1}\,yr^{-1}$ of Sewage Sludge for 10 Years (SO + SS10), 36 and $72\,t\,ha^{-1}\,yr^{-1}$ of Composted Food and Gardening Wastes for 4 Years (SO + CW36 and SO + CW72, Respectively), 10 and $20\,t\,ha^{-1}\,yr^{-1}$ of Crude or Exhausted Olive Oil Mill Pomace for 1 Year (SO + CP10 and SO + CP20, and SO + EP10 and SO + EP20, Respectively), and 300 and $600\,m^3\,ha^{-1}\,yr^{-1}$ of Lagooned or Catalytically Digested Olive Oil Mill Wastewater for 1 Year (SO + LWW300 and SO + LWW600, and SO + CWW300 and SO + CWW600, Respectively), with the Corresponding Unamended Soils (SO)

Soil	TOC ($g\,kg^{-1}$)	TEC ($g\,kg^{-1}$)	HAC + FAC ($g\,kg^{-1}$)	NHC ($g\,kg^{-1}$)	DH (%)	HR (%)	HI
SO[a]	17.7	9.1	6.5	2.5	72	37	0.39
SO + SS10[a]	16.8	11.3	6.3	5.0	56	38	0.79
SO[b]	12.0	6.6	3.3	3.3	50	28	1.01
SO + CW36[b]	11.0	5.4	3.2	2.2	59	29	0.70
SO + CW72[b]	15.0	6.5	3.9	2.6	60	26	0.65
SO[c]	10.3	7.9	7.0	0.9	89	68	0.13
SO + CP10[c]	11.5	8.8	7.3	1.5	83	63	0.21
SO + CP20[c]	12.4	9.6	7.8	1.9	81	63	0.24
SO + EP10[c]	11.7	9.1	7.5	1.5	83	64	0.20
SO + EP20[c]	13.0	10.2	8.3	1.9	81	64	0.23
SO[d]	10.3	7.9	7.0	0.9	89	68	0.13
SO + LWW300[d]	11.8	9.3	7.7	1.6	83	65	0.21
SO + LWW600[d]	13.8	10.2	8.2	2.1	80	59	0.25
SO + CWW300[d]	12.4	9.4	8.1	1.3	87	66	0.15
SO + CWW600[d]	14.5	10.9	9.0	1.9	83	62	0.21

[a]From Adani and Tambone (2005).
[b]From Adani et al. (2007).
[c]From Brunetti et al. (2005).
[d]From Brunetti et al. (2007a).

Plaza, 2007). The chemical and physico-chemical data available in the literature obtained by application of these methods and techniques will be discussed separately in the following text.

4.4.1. Elemental Composition

Elemental analysis is a common tool used for the characterization and differentiation of HS isolated from organic amendments and unamended and amended soils. It provides information on the distribution of major elements, typically C, H, N, S, and O, in HS, thus setting limits for HS possible molecular composition. The atomic ratios C/N, C/H, and O/C are also useful in identifying types of HS, monitoring their structural changes, and devising HS structural formulas (Stevenson, 1994; Senesi and Loffredo, 1999).

Important differences are apparent in the elemental composition of HAs and FAs from untreated and treated materials. In particular, HAs and FAs from mature composts and vermicomposts are generally characterized by C, H, and N contents smaller than, and O content and C/N and C/H ratios larger than, those measured in untreated substrates (Miikki et al., 1997; Veeken et al., 2000; Amir et al., 2004; Plaza et al., 2005a, 2007; Huang et al., 2006; Romero et al., 2007). These results indicate a partial loss of N-containing groups and aliphatic side chains, possibly due to intense mineralization achieved by microbial activity during the treatment processes, and formation of more oxidized, aromatic, condensed, and polymerized structures comparable with those of native soil HAs and FAs (Schnitzer, 1978).

The elemental composition and related atomic ratios of HAs and FAs isolated from some representative organic amendments of various nature and source, and unamended and amended soils are shown in Table 4.2. Although the elemental composition of HAs and FAs show significant variability on dependence on the nature, origin, and treatment of the amendment, they generally feature larger H, N,

TABLE 4.2. Elemental Composition (Moisture- and Ash-Free) and Atomic Ratios of Humic Acids (HAs) and Fulvic Acids (FAs) Isolated from Liquid Swine Manure (LSM), Sewage Sludge (SS), Composted Food and Gardening Wastes (CW), Soils Amended with 90 and 150 $m^3\,ha^{-1}\,yr^{-1}$ of LSM for 7 Years (SO + LSM90 and SO + LSM150, Respectively), 10 and 25 $t\,ha^{-1}\,yr^{-1}$ of SS for 20 Years (SO + SS10 and SO + SS25, Respectively), 36 and 72 $t\,ha^{-1}\,yr^{-1}$ of CM for 4 Years (SO + CW36 and SO + CW72, Respectively), 90 $t\,ha^{-1}\,yr^{-1}$ of SS for 3 Year (SO + SS90), with the Corresponding Unamended Soils (SO)

Sample and Origin	C ($g\,kg^{-1}$)	H ($g\,kg^{-1}$)	N ($g\,kg^{-1}$)	S ($g\,kg^{-1}$)	O ($g\,kg^{-1}$)	Atomic Ratios		
						C/N	C/H	O/C
HAs								
LSM[a]	655	90	54	14	188	14.3	0.6	0.2
SO[a]	566	49	48	3	334	13.8	1.0	0.4
SO + LSM90[a]	570	51	47	5	328	14.1	0.9	0.4
SO + LSM150[a]	571	52	47	5	325	14.2	0.9	0.4
SS[b]	535	63	100	19	283	6.2	0.7	0.4
SO[b]	538	39	37	2	384	17.2	1.2	0.5
SO + SS10[b]	539	41	38	3	379	16.4	1.1	0.5
SO + SS25[b]	543	50	47	4	355	13.4	0.9	0.5
CW[c]	547	51	76	10	316	8.4	0.9	0.4
SO[c]	523	50	58	7	362	10.5	0.9	0.5
SO + CW36[c]	533	51	63	9	344	9.9	0.9	0.5
SO + CW72[c]	542	50	65	9	334	9.7	0.9	0.5
FAs								
LSM[d]	585	110	44	35	226	15.5	0.4	0.3
SO[d]	474	77	37	7	405	14.9	0.5	0.6
SO + LSM90[d]	501	80	39	8	372	15.0	0.5	0.6
SO + LSM150[d]	507	80	39	9	365	15.1	0.5	0.5
SS[e]	408	66	28	82	416	17.0	0.5	0.8
SO[e]	457	54	21	19	448	25.4	0.7	0.7
SO + SS90[e]	360	79	47	60	455	8.9	0.4	0.9

[a]From Hernández et al. (2007).
[b]From Brunetti et al. (2007b).
[c]From Adani et al. (2007).
[d]From Hernández et al. (2006).
[e]From Sposito et al. (1982).

and S contents and smaller O content and C/N, C/H, and O/C ratios than do native soil HAs and FAs, while C content fluctuates around the mean C value in soil HAs and FAs (Sposito et al., 1982; Hernández et al., 2006, 2007; Brunetti et al., 2007b; Adani et al., 2007). The high levels of N and S in amendment HAs and FAs may be ascribed to the presence of protein decomposition products and S-containing residues. The low O content and C/N, C/H, and O/C suggest that amendment HAs and FAs are freshly formed, low-humified materials rich in saturated aliphatic structures (with a low C/H ratios) as compared to native soil HAs and FAs.

In general, HA and FA isolated from amended soils have elemental compositions that are intermediate between those of the unamended soil HA and FA and those of amendment HA and FA. In particular, organic amendment affects soil HA and FA composition by generally increasing N and S contents, and decreasing C/H ratio (Sposito et al., 1982; Piccolo et al., 1992; Hernández et al., 2006, 2007; Brunetti et al., 2007b; Adani et al., 2007). These effects, however, become less and less apparent with time. For example, in agreement with previous findings on similar systems (Boyd et al., 1980), García-Gil et al. (2004a) found greater modifications for the HA extracted 9 months than 36 months after sludge application from the same soil.

4.4.2. Molecular Weight Distribution

Gel chromatography has been extensively used to fractionate HS on the basis of molecular sizes and determine molecular-weight (MW) distribution by calibration with homologous compounds of known molecular weight (Stevenson, 1994). The gel chromatography curves of HAs and FAs isolated from various organic amendments, including sewage sludge, animal manure, and composts obtained from various mixtures, are dominated by a peak of a high-MW organic fraction, differently from gas chromatography curves commonly obtained for native soil HAs and FAs (Almendros et al., 1983a,b; Senesi, 1989; Piccolo et al., 1992). These results may be ascribed to the relevant presence in the amendments, especially in the uncomposted ones, of large-size lignin constituents with adsorbed peripheral lipidic and/or peptidic chains, and not to lignin degradation products typical of native soil HS (Almendros et al., 1983a,b; Senesi, 1989; Piccolo et al., 1992). As a consequence, HS isolated from amended soils generally show a slight enhancement of the high-MW fractions as compared to the MW distribution of native soil HS (Figure 4.3) (Piccolo et al., 1992).

4.4.3. Acid-Base Properties

The presence of acidic functional groups, mostly carboxyl and phenolic OH groups, in the molecular structure of soil HS renders them major players in the acid–base buffering capacity of soils and in the fate, bioavailability, and physico-chemical behavior of macro- and micronutrients, toxic metal ions, and several xenobiotic organic compounds in soil (Ritchie and Perdue, 2003; Senesi and Loffredo, 2005). Consequently, the effects of amendment on the acid–base properties of soil HAs and FAs is a subject of considerable interest.

During composting and vermicomposting of organic materials, the total acidity and especially the carboxyl group content of HA and FA fractions generally increase (Plaza et al., 2005a, 2007; Romero et al., 2007). However, the acidic functional group

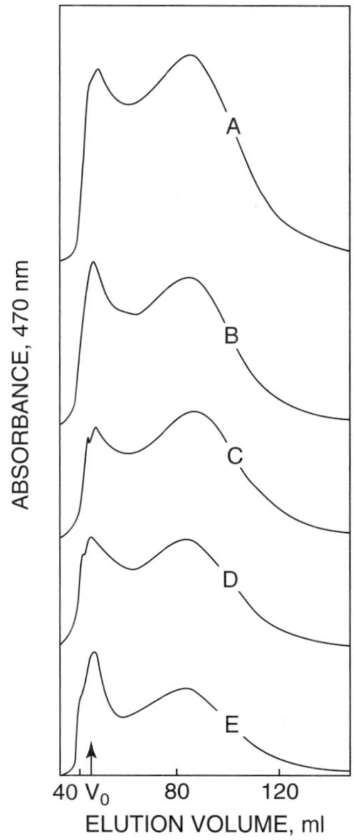

Figure 4.3. Gel permeation chromatograms of humic acids isolated from a soil either unamended (**A**) or amended with 25 t ha^{-1} yr^{-1} of cattle manure for 4 years (**B**) and 25, 50, and 100 t ha^{-1} yr^{-1} of sewage sludge for 4 years (**C**, **D**, and **E**, respectively). Reprinted from Piccolo, A., Zaccheo, P., and Genevini, P. G. (1992). Chemical characterization of humic substances extracted from organic-waste-amended soils. *Bioresource Technol.* **40**, 275–282, with permission from Elsevier.

content of organic amendment HAs and FAs are generally smaller than those of native soil HAs and FAs (Table 4.3) (Sposito et al., 1982; Senesi, 1989; Piccolo et al., 1992; Campitelli et al., 2006; Brunetti et al., 2007b; Hernández et al., 2006, 2007). Similar to the elemental composition, the acidic functional group composition of HAs and FAs from amended soils is generally intermediate between those of the unamended soil and amendment HAs and FAs (Table 4.3) (Sposito et al., 1982; Senesi, 1989; Piccolo et al., 1992; Campitelli et al., 2006; Brunetti et al., 2007b; Hernández et al., 2006, 2007).

The complexity of structures and behavior of HS has led to a variety of models attempting to describe their interactions with protons (Tipping, 2002; Dudal and Gérard, 2004). Of particular interest is the nonideal competitive adsorption (NICA)–Donnan model, which stands out from the others in terms of physicochemical realism, accuracy, number of applications, and possibility of incorporation in existing speciation programs (e.g., ECOSAT and Visual MINTEQ). This model addresses the binding site heterogeneity of HS by assuming a continuous distribution of proton binding sites, discriminates between chemical and electrostatic interactions,

TABLE 4.3. Acidic Functional Group Contents of Humic Acids (HAs) and Fulvic Acids (FAs) Isolated from Cattle Manure (CM), Sewage Sludge (SS), Municipal Solid Waste Compost (MSWC), Liquid Swine Manure (LSM), Soils Amended with 25 t ha^{-1} yr^{-1} of CM for 4 Years (SO + CM25), 25, 50, and 100 t ha^{-1} yr^{-1} of SS for 4 Years (SO + SS25, SO + SS50, and SO + SS100, Respectively), 40 t ha^{-1} yr^{-1} of MSWC for 3 Years (SO + MSWC40), 90 and 150 m^3 ha^{-1} yr^{-1} of LSM for 7 Years (SO + LSM90 and SO + LSM150, Respectively), and 90 t ha^{-1} yr^{-1} of SS for 3 Years (SO + SS90), with the Corresponding Unamended Soils (SO)

Sample and Origin	Total acidity	COOH (g kg^{-1})	Phenolic OH
HAs			
CM[a]	5.3	1.7	3.6
SS[a]	3.3	1.3	2.0
SO[a]	6.3	2.9	3.4
SO + CM25[a]	5.9	2.7	3.2
SO + SS25[a]	5.4	2.5	2.9
SO + SS50[a]	5.3	2.4	2.9
SO + SS100[a]	5.2	2.2	3.0
MSWC[b]	3.7	0.5	3.2
SO[b]	5.3	3.8	1.5
SO + MSWC40[b]	5.3	3.6	1.7
FAs			
LSM[c]	7.5	6.2	1.3
SO[c]	8.8	7.1	1.7
SO + LSM90[c]	8.4	6.8	1.6
SO + LSM150[c]	8.2	6.7	1.5
SS[d]	—	0.7	—
SO[d]	—	8.2	—
SO + SS90[d]	—	7.9	—

[a]From Piccolo et al. (1992).
[b]From Brunetti et al. (2007b).
[c]From Hernández et al. (2007).
[d]From Sposito et al. (1982).

and takes into account ionic strength effects in a generic way by means of an electrostatic Donnan gel model (Koopal et al., 1994, 2001, 2005; Benedetti et al., 1995, 1996a,b; Kinniburgh et al., 1996, 1999; Milne et al., 2001). By fitting the NICA–Donnan model to potentiometric titration data obtained at different ionic strengths, intrinsic (i.e., independent of pH, salt concentration, or metal concentration) acid–base properties of HAs and FAs can be probed. The main fitting parameters of the NICA–Donnan model includes site densities, median affinity constants, and widths of affinity distributions for proton binding to low- and high-affinity sites, which are assumed to be, respectively, carboxylic- and phenolic-type groups. However, it should be noted that the model parameter values obtained by fitting may not be unique and their physical meaning may not be fully retained in practice (Koopal et al., 2005).

Recent studies have shown that the NICA–Donnan model fits very well to the acid–base titration data of HAs and FAs form organic amendments and of unamended and amended soils, pointing out substantial differences in site densities, proton-binding affinities, and widths of affinity distributions (Plaza et al., 2005b,c, 2006a). In particular, the results obtained from these studies indicate that the composting process increases the content of carboxylic- and phenolic-type groups of HAs and FAs, decreases their proton affinity, and increases the heterogeneity of phenolic-type groups of HAs (Plaza et al., 2005b). With respect to unamended soil HAs and FAs, amendment HAs and FAs are generally characterized by smaller acidic functional group contents, larger proton binding affinities of both carboxylic- and phenolic-type groups, and smaller heterogeneity of carboxylic and phenolic-type groups. Amendment with organic materials determines a decrease of acidic functional group contents and an increase of proton binding affinities of carboxylic- and phenolic-type groups of soil HAs and FAs. These effects are more evident in the HA and FA fractions from soils amended with untreated materials than in those from soils amended with composts (Table 4.4) (Plaza et al., 2005c, 2006a). As a whole, composting produces HA and FA fractions with acid–base properties that

TABLE 4.4. Fitting Parameters of the NICA–Donnan Model for Proton Binding to Humic Acids (HAs) and Fulvic Acids (FAs) Isolated from Liquid Swine Manure (LSM), Sewage Sludge (SS), Municipal Solid Waste Compost, Soils Amended with 90 and 150 $m^3 ha^{-1} yr^{-1}$ of LSM for 7 Years (SO + LSM90 and SO + LSM150, Respectively), 40 $t ha^{-1} yr^{-1}$ of SS for 1 Year (SO + SS40), and 40 $t ha^{-1} yr^{-1}$ of MSWC for 3 Years (SO + MSWC40), with the Corresponding Unamended Soils (SO)

Sample and Origin	b[a]	$Q_{max,1}$[b]	$\log\tilde{K}_{H,1}$[c]	m_1[d]	$Q_{max,2}$[e]	$\log\tilde{K}_{H,2}$[f]	m_2[g]	$Q_{max,1}$ + $Q_{max,2}$
HAs								
LSM[h]	0.70	1.36	4.44	0.82	1.58	8.11	0.41	2.93
SO[h]	0.56	3.59	3.06	0.63	2.28	7.58	0.31	5.88
SO + LSM90[h]	0.59	2.75	3.34	0.64	1.89	7.87	0.31	4.64
SO + LSM150[h]	0.63	2.23	3.52	0.77	1.86	7.88	0.33	4.10
SS[i]	0.57	2.17	4.22	0.67	1.21	7.57	0.52	3.38
SO1[i]	0.45	4.42	2.91	0.54	2.20	8.14	0.39	6.62
SO1 + SS40[i]	0.49	3.65	3.25	0.58	1.89	7.81	0.48	5.54
MSWC4[i]	0.50	2.59	3.41	0.64	1.66	7.46	0.40	4.25
SO2[i]	0.46	4.08	2.92	0.52	2.52	8.06	0.36	6.60
SO2 + MSWC40[i]	0.47	3.84	2.99	0.53	2.22	7.82	0.36	6.05
FAs								
LSM[h]	0.76	3.94	3.33	0.96	1.46	8.13	0.45	5.40
SO[h]	0.63	4.89	2.88	0.60	1.91	7.80	0.45	6.80
SO + LSM90[h]	0.66	4.75	2.96	0.77	1.87	7.82	0.43	6.62
SO + LSM150[h]	0.66	4.27	3.23	0.82	1.85	7.95	0.46	6.12

[a]Empirical parameter describing how the Donnan volume varies with ionic strength.
[b]Carboxyl group content (mmol g^{-1} on moisture- and ash-free basis).
[c]Median value of affinity distribution for proton binding by carboxyl groups.
[d]Width of proton-affinity distribution of carboxyl groups.
[e]Phenolic OH group content (mmol g^{-1} on moisture- and ash-free basis).
[f]Median value of affinity distribution for proton binding by phenolic OH groups.
[g]Width of proton-affinity distribution of phenolic OH groups.
[h]From Plaza et al. (2006a).
[i]From Plaza et al. (2005c).

resemble those typical of native soil HAs and FAs and, therefore, induces fewer modifications in amended soil HAs and FAs.

4.4.4. Utraviolet–Visible Spectra

The ultraviolet–visible (UV–Vis) spectra of HAs and FAs are somewhat featureless in showing a continuous increase of absorbance with decreasing wavelength. The absence of any well-defined UV–vis maxima and minima feasibly results from extended overlap of absorbances of a wide variety of chromophores affected by various substitutions (Stevenson, 1994; Senesi and Loffredo, 1999). Despite these limitations, the ratio of absorbances at 465 nm and 665 nm, referred to as the E_4/E_6 ratio, has been found to vary with the nature of HS and has been widely used for characterization purposes (Stevenson, 1994; Senesi and Loffredo, 1999). In particular, the E_4/E_6 ratio appears to be inversely related to the MW and the degree of condensation of aromatic constituents of HS and is considered as an index of humification (Kononova, 1966; Chen et al., 1977).

In general, composting of organic materials causes an increase of the E_4/E_6 ratio of HA and FA fractions (Inbar et al., 1992; Jerzykiewicz et al., 1999; Plaza et al., 2005a, 2007; Fuentes et al., 2006). With some exceptions (Piccolo et al., 1992; García-Gil et al., 2004a; Hernández et al., 2006), the E_4/E_6 ratio of HAs from organic amendments are larger than the corresponding mean values reported for unamended soil HAs (González-Vila and Martin, 1987; García-Gil et al., 2004b; Plaza et al., 2005a; Campitelli et al., 2006). In contrast, the E_4/E_6 ratios of organic amendment FAs are generally smaller than those of soil FAs (Riffaldi et al., 1983; Plaza et al., 2007; Hernández et al., 2007). As a consequence, with some exceptions, organic amendment generally produces an increase of the E_4/E_6 ratio of soil HAs (González-Vila and Martin, 1987; Piccolo et al., 1992; García-Gil et al., 2004a,b; Rivero et al., 2004; Campitelli et al., 2006) and a decrease of that of soil FAs (Hernández et al., 2007).

4.4.5. Fluorescence Spectra

Bidimensional fluorescence spectra are commonly obtained in the three modes of emission, excitation, and synchronous-scan excitation, whereas tridimensional fluorescence (or total luminescence) spectra are obtained in the form of excitation–emission matrix (EEM) plots by measuring the fluorescence intensity emitted as a function of the wavelength over a range of excitation wavelengths. This technique allows to obtain more detailed information than that obtained by using conventional monodimensional fluorescence (Mobed et al., 1996). Fluorescence spectroscopy has provided valuable information on the molecular structure, functionalities, conformation, and intramolecular and intermolecular interactions of HS from organic amendments and unamended and amended soils (Senesi et al., 1990, 1996, 2007; Mobed et al., 1996; Chen et al., 2003; Senesi and Plaza, 2007).

The fluorescence emission spectra of native soil HAs and FAs generally consist of a unique broad band with a maximum wavelength which ranges from 500 to 520 nm for HAs and from 445 to 465 nm for FAs (Senesi et al., 1990). Fluorescence excitation spectra of most soil HAs feature two closely spaced major peaks in the long-wavelength region (around 465 and 450 nm), often accompanied by a minor peak or shoulder in the intermediate wavelength range (at 395–390 nm) (Senesi et al., 1990). Differently, soil FAs generally feature one main excitation peak in the intermediate region of the spectrum (around 390 nm) with additional minor peaks and shoulders at longer and shorter wavelengths (Senesi et al., 1990). Synchronous-scan spectra of soil HAs generally feature only one major peak in the long-wavelength region, often accompanied by faint shoulders at longer and shorter wavelengths, whereas soil FAs generally exhibit two main synchronous-scan peaks at long (450–460 nm) and intermediate (390–400 nm) wavelengths, often with some less intense peaks and shoulders at both sides (Senesi et al., 1990). Fluorescence EEM spectra generally consist of a unique broad band centered at an excitation/emission wavelength pair that is much longer for soil HAs (430–470/500–550 nm) than for soil FAs (320–340/420–440 nm) (Mobed et al., 1996; Bertoncini et al., 2005).

With respect to those of native soil HAs and FAs, fluorescence spectra of HAs and FAs from organic amendments generally show (a) an emission maximum at a much shorter wavelength, (b) more intense excitation peaks at short and intermediate wavelengths and less intense peaks at long wavelength, (c) several synchronous-scan peaks and shoulders with a relative intensity decreasing with increasing wavelength, and (d) fluorescence EEM spectra with the main fluorophore centered at shorter excitation/emission wavelength pairs (Figures 4.4 and 4.5) (Soler-Rovira et al., 2002; Bertoncini et al., 2005; Hernández et al., 2006, 2007; Plaza et al., 2006b,c; Brunetti et al., 2007a,b). These results would suggest the presence of simpler structural components with a smaller degree of aromatic polycondensation, smaller level of conjugated chromophores, and smaller humification degree in HAs and FAs from organic amendments, especially untreated ones, as compared to soil HAs and FAs (Senesi et al., 1990). In general, however, with progressing composting and vermicomposting, the fluorescence features of HAs and FAs tend to approach those typical of native soil HAs and FAs (Figures 4.4 and 4.5) (Plaza et al., 2005a, 2007; Fuentes et al., 2006; Romero et al., 2007).

Fluorescence spectra of HAs and FAs isolated from amended soils, with respect to those of the corresponding unamended soil HAs and FAs, show (a) a shift to a lower wavelength of the fluorescence emission maximum, (b) a relative increase of the intensity in the short and intermediate excitation and emission wavelength regions, with respect to that in the long wavelength regions, and (c) a shift to shorter excitation/emission wavelength pairs of the main peak in the fluorescence EEM spectra (Figures 4.4 and 4.5) (Soler-Rovira et al., 2002; Bertoncini et al., 2005; Hernández et al., 2006, 2007; Plaza et al., 2006b,c; Brunetti et al., 2007a,b). These data indicate a partial incorporation of simple and low humified components typical of amendment HAs and FAs into amended soil HAs and FAs.

4.4.6. Infrared Spectra

Infrared (IR) spectroscopy and Fourier transform IR (FT-IR) have been very useful in studying the molecular structure of HS from organic amendments and unamended soils, as well as the effects of organic amendment on native soil HS (Senesi et al., 1996, 2007; Senesi and Plaza, 2007).

The FT-IR spectra of HAs and FAs from organic amendments, especially uncomposted materials, differ markedly from those of native soil HAs and FAs (Figure 4.6) (Boyd et al., 1980; Piccolo et al., 1992; García-Gil et al., 2004a; Adani and

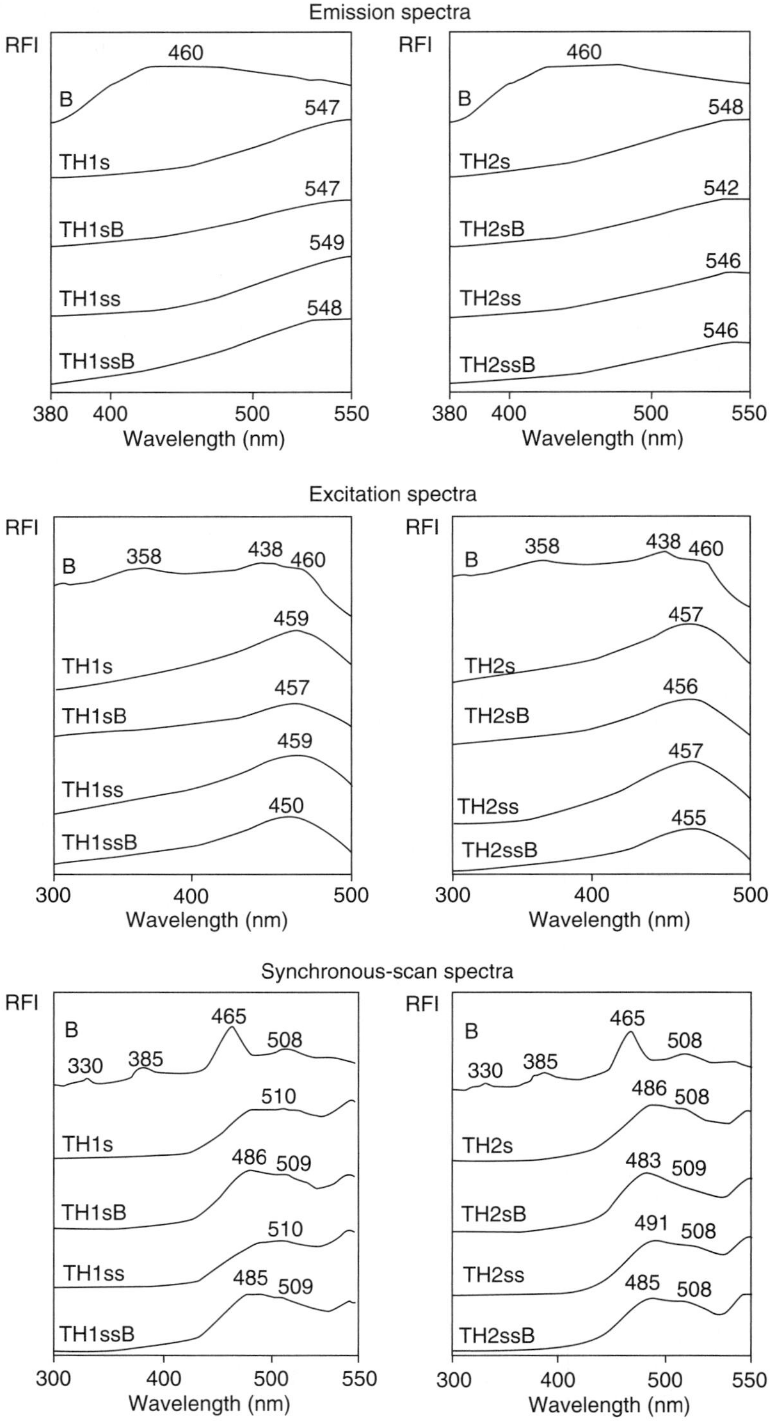

Figure 4.4. Fluorescence emission, excitation, and synchronous-scan spectra of humic acids (HAs) isolated from sewage sludge (B) and surface (s) and subsurface (ss) horizons of two soils either unamended (TH1s and TH1ss, and TH2s and TH2ss, respectively) or amended with 390 t ha^{-1} of sewage sludge (THB1s and THB1ss, and THB2s and THB2ss, respectively). Reprinted from Bertoncini, E. I., D'Orazio, V., Senesi, N., and Mattiazzo, M. E. (2005). Fluorescence analysis of humic and fulvic acids from two Brazilian oxisols as affected by biosolid amendment. *Anal. Bioanal. Chem.* **381**, 1281–1288, with permission from Springer.

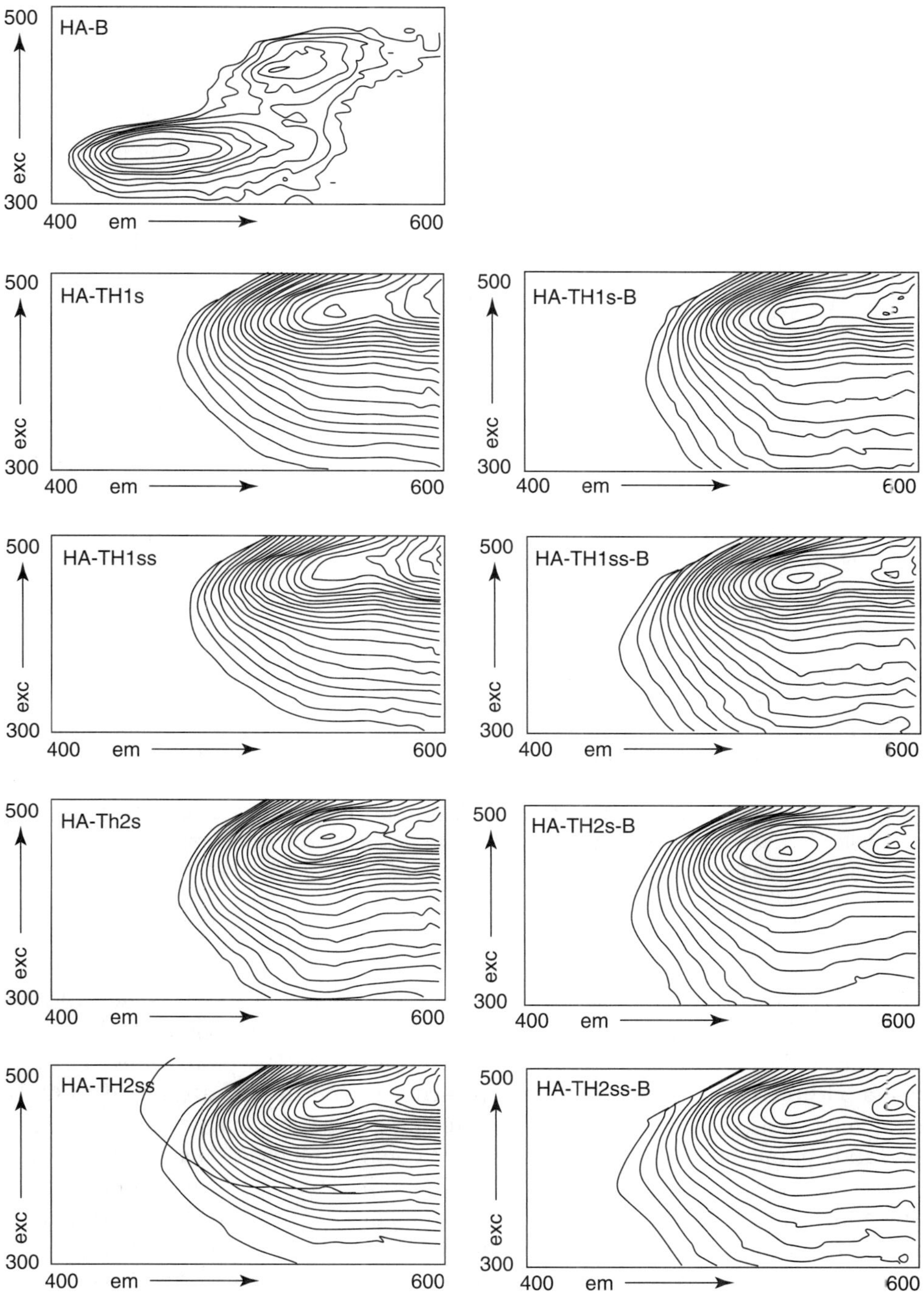

Figure 4.5. Fluorescence excitation–emission matrix spectra of humic acids (HAs) isolated from sewage sludge (B) and two soils either unamended (TH1 and TH2, respectively) or amended with 390 t ha^{-1} of sewage sludge (THB1 and THB2, respectively) sampled from the surface (s, 0–25 cm) and subsurface (ss, 25–50 cm) layers. Reprinted from Bertoncini, E. I., D'Orazio, V., Senesi, N., and Mattiazzo, M. E. (2005). Fluorescence analysis of humic and fulvic acids from two Brazilian oxisols as affected by biosolid amendment. *Anal. Bioanal. Chem.* **381**, 1281–1288, with permission from Springer.

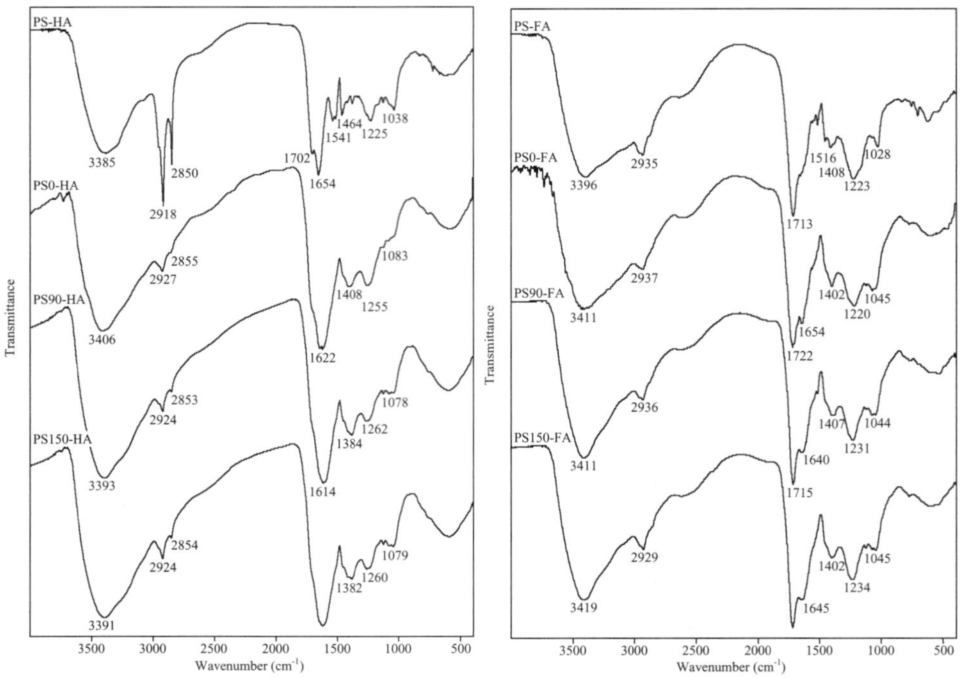

Figure 4.6. Fourier transform infrared spectra of humic acids (HAs) and fulvic acids (FAs) isolated from pig slurry (PS), unamended soil (PS0, and soils amended with 90 and $150\,m^3\,ha^{-1}\,yr^{-1}$ of PS for 7 years (PS90 and PS150, respectively). Reprinted from Hernández, D., Plaza, C., Senesi, N., and Polo, A. (2006). Detection of copper(II) and zinc(II) binding to humic acids from pig slurry and amended soils by fluorescence spectroscopy. *Environ. Pollut.* **143**, 212–220, with permission from Elsevier, and from Hernández, D., Plaza, C., Senesi, N., and Polo, A. (2007). Fluorescence analysis of copper(II) and zinc(II) binding behavior of fulvic acids from pig slurry and amended soils. *Eur. J. Soil Sci.* **58**, 900–908, with permission from Blackwell Publishing.

Tambone, 2005; Brunetti et al., 2005, 2007a,b; Hernández et al., 2006, 2007; Adani et al., 2007). Typical FT-IR features and their corresponding assignments (based on Bellamy, 1975; MacCarthy and Rice, 1985; Stevenson, 1994) are the following: (a) a common broad band in the 3450- to 3300-cm^{-1} region usually attributed to O–H stretching and, secondarily, to N–H stretching of various functional groups; (b) two absorption bands in the 2900-cm^{-1} region due to aliphatic C–H group stretching, whose relative intensity is generally stronger in HAs and FAs from organic amendments, especially uncomposted ones, than in soil HAs and FAs; (c) a band at 1725–1710 cm^{-1} due to C=O stretching of COOH and other carbonyl groups, whose intensity is stronger in FAs than in HAs and stronger in soil HAs and FAs than in organic amendment HAs and FAs; (d) a broad band in the region 1660–1600 cm^{-1} generally considered an envelope of unresolved absorptions mainly due to aromatic C=C, C=O stretching of amide groups (amide I band), quinonic C=O, and/or C=O of H-bonded conjugated ketones, which is generally less broad and intense in organic amendment HAs and FAs than in soil HAs and FAs; (e) a band in the region 1540–1510 cm^{-1} preferentially ascribed to N–H deformation and C=N stretching of amides (amide II band), which is often much more evident in HAs and FAs from organic amendments than in soil HAs and FAs; (f) a band at 1460–1440 cm^{-1} attributed to aliphatic C–H, which is often sharp in HAs and FAs from organic amendments, but only a weak shoulder in soil HAs and FAs; (g) a broad band in the region 1400–1380 cm^{-1}, which is preferentially assigned to O–H deformation and C–O stretching of phenolic OH, as well as C–H deformation of CH_2 and CH_3 groups and/or antisymmetric stretching of COO^- groups, and is generally weaker and narrower in HAs and FAs from organic amendments than in soil HAs and FAs; (h) a broad band in the region 1260–1200 cm^{-1}, which is generally ascribed to C–O stretching and O–H deformation of carboxyls and C–O stretching of aryl ethers; and (i) an absorption in the region 1080–1030 cm^{-1}, generally attributed to C–O stretching of polysaccharides or polysaccharide-like substances, which is often more evident in HAs and FAs from organic amendments than in soil HAs and FAs (Boyd et al., 1980; Piccolo et al., 1992; García-Gil et al., 2004a; Adani and Tambone, 2005; Brunetti et al., 2005, 2007a,b; Hernández et al., 2006, 2007; Adani et al., 2007).

The FT-IR spectra of HAs and FAs isolated from amended soils are generally more similar to those of the corresponding unamended soil HAs and FAs than to those of HAs and FAs from organic amendments. In particular, amended soil HAs and FAs generally show an increased relative intensity of the bands ascribed to amide I and amide II (at about 1650 and 1520 cm^{-1}), aliphatic C–H stretching (2900-cm^{-1} region) and polysaccharide-like-structures (1080–1030 cm^{-1}), and a decrease of the relative intensity of the bands attributed to C=O stretching of carboxyl groups (1725–1710 cm^{-1}) and O–H deformation and C–O stretching of phenolic OH (1400–1380 cm^{-1}) (Boyd et al., 1980; Piccolo et al., 1992; García-Gil et al., 2004a; Adani and Tambone, 2005; Brunetti et al., 2005, 2007a,b; Hernández et al., 2006, 2007; Adani et al., 2007). In general, FT-IR features of amended soil HAs and FAs tend to approach those of amendment HAs and FAs (Hernández et al., 2006, 2007; Adani et al., 2007; Brunetti et al., 2007a,b) with increasing amendment rate and number, whereas with increasing time after application, they resemble more and more those of unamended soil HAs and FAs (Boyd et al., 1980; García-Gil et al., 2004a).

4.4.7. Nuclear Magnetic Resonance Spectra

^{13}C- and ^{1}H-nuclear magnetic resonance (NMR) spectroscopies are among the most powerful tools currently available for the study of HS (Wilson, 1987; Preston, 1996; Kögel-Knabner, 1997, 2000; Hatcher et al., 2001). The ^{1}H- and ^{13}C-NMR spectra confirm the higher aliphatic character of HAs and FAs from organic amendments, with respect to native soil HAs and FAs (Sposito et al., 1978; González-Vila and Martin, 1985, 1987; Inbar et al., 1990, 1991; García et al., 1992b; Giusquiani et al., 1994; González-Vila et al., 1999; Adani and Tambone, 2005; Polak et al., 2005; Adani et al., 2006; Brunetti et al., 2007b). In particular, ^{1}H-NMR spectra of FAs from organic amendments, compared to soil FAs, typically show the following: (a) lower contents of H of terminal methyl groups of methylene chains (0.8–1.0 ppm), attached to C and/or to O (3.3–5.0 ppm), and aromatic (6.1–8.1 ppm); and (b) a higher content of H of methylen chains (1.0–1.4 ppm), alicyclic structures (1.4–1.7 ppm), and methyl and methylene groups onto aromatic rings and carbonyl groups (2.0–3.3 ppm) (Giusquiani et al., 1994). Similar to FAs, HAs from organic amendments show a lower intensity in the range 3.3–5.0 ppm and a higher intensity in the range

1.0–1.4 ppm, with respect to soil HA (Figure 4.7) (Giusquiani et al., 1994; Adani and Tambone, 2005; Polak et al., 2005; Adani et al., 2006).

With respect to native soil HAs and FAs, ^{13}C-NMR spectra of HAs and FAs from organic amendments typically feature the following: (a) more intense signals of terminal methyl groups, methylene C in aliphatic rings, and methylene C in alkyl chains (0–50 ppm); and methoxyl C (at about 55 ppm), O- and N-substituted aromatic C (at about 145 and 155 ppm), and anomeric C of polysaccharide structures (103 ppm); (b) similar or less intense signals of ring C of carbohydrates (at about 70 ppm); and (c) less intense signals of alkyl-substituted aromatic C (at about 130 ppm) and carboxyl and amidic C (at about 170 ppm) (Figure 4.8) (González-Vila and Martin, 1985, 1987; Inbar et al., 1990, 1991; García et al., 1992b; Giusquiani et al., 1994; González-Vila et al., 1999; Adani and Tambone, 2005; Adani et al., 2006; Brunetti et al., 2007b). These data suggest that HA and FA fractions of organic amendments are richer in aliphatic, N-containing, phenolic, methoxyl, and polysaccharide-like groups than are soil HAs and FAs. However, a decrease in aliphatic-C and an increase in aromatic C are apparent in HAs from composted materials (Inbar et al., 1990, 1991; García et al., 1992b).

The ^{13}C-NMR spectra of amended soil HAs differ from those of the corresponding unamended soil HAs in that the former exhibit more pronounced signals assigned to methoxyl C (55 ppm) and to O- and N-substituted aromatic C (145 and 155 ppm) (Figure 4.8), which suggests that HA fractions of organic amendments are partially incorporated into native soil HAs (González-Vila et al., 1999; Adani and Tambone, 2005; Adani et al., 2006; Brunetti et al., 2007b). However, recent studies have shown the existence of no significant differences between (a) the ^{1}H-NMR spectra of HA fractions from unamended soils and (b) the ^{1}H-NMR spectra of the same soils amended with either sewage sludge or municipal solid waste compost (Figure 4.7) (Adani and Tambone, 2005; Adani et al., 2006).

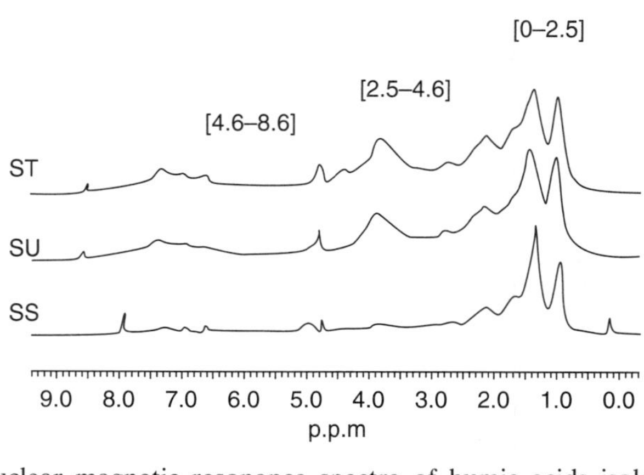

Figure 4.7. ^{1}H nuclear magnetic resonance spectra of humic acids isolated from sewage sludge (SS), unamended soil (SU), and soil amended with sewage sludge at a rate of 10 t ha^{-1} yr^{-1} for 10 years (ST). Reprinted from Adani, F., and Tambone, F. (2005). Long-term effect of sewage sludge application on soil humic acids. *Chemosphere* **60**, 1214–1221, with permission from Elsevier.

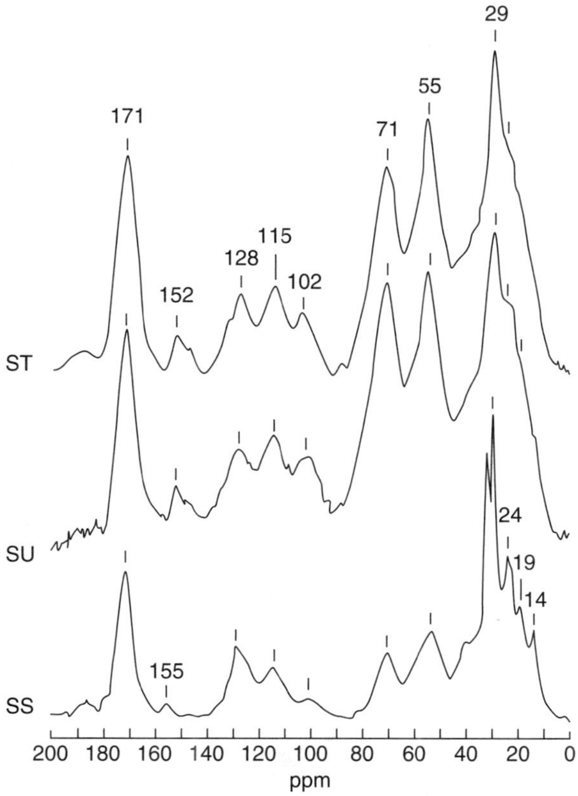

Figure 4.8. ^{13}C nuclear magnetic resonance spectra of humic acids isolated from sewage sludge (SS), unamended soil (SU), and soil amended with sewage sludge at a rate of $10\,t\,ha^{-1}\,yr^{-1}$ for 10 years (ST). Reprinted from Adani, F., and Tambone, F. (2005). Long-term effect of sewage sludge application on soil humic acids. *Chemosphere* **60**, 1214–1221, with permission from Elsevier.

4.4.8. Electron Spin Resonance Spectra

Electron spin (or paramagnetic) resonance (ESR or EPR) spectroscopy is a highly sensitive and accurate analytical technique that can detect and characterize species containing unpaired electrons, including organic free radicals and paramagnetic transition metal ions in free or complexed forms (Poole, 1997). The ESR spectra can provide four types of information: (a) the spectroscopic splitting factor, that is, the *g* value, which provides insight into the chemical nature of the radical; (b) the width of the absorption line, that is, the peak-to-peak separation of the first derivative line, which is influenced by such factors as free radical concentration, temperature, and state of aggregation; (c) the hyperfine splitting, which is measured as the separation between the hyperfine lines and can provide information on the chemical structure of the free radical; and (d) the concentration of unpaired electrons (free radicals).

The ESR spectra of HAs and FAs of any nature and origin, including native soils, organic amendments, and amended soils, show a sharp and narrow resonance characterized by a *g* value at about 2.0040 and by a line width ranging from 0.60 to 0.80 mT, which is attributed to indigenous organic free radicals of semiquinonic nature conjugated with an extended aromatic network (Senesi, 1990a). Generally, this resonance is accompanied by a number of other signals of various complexity and intensity possibly arising from paramagnetic metal ions present as organic complexes and/or mineral impurities (Figure 4.9) (Hervas et al., 1989; Senesi et al., 1989; Jerzykiewicz et al., 1999; Plaza et al., 2002, 2003; Soler-Rovira et al., 2002, 2003; González-Pérez et al., 2006).

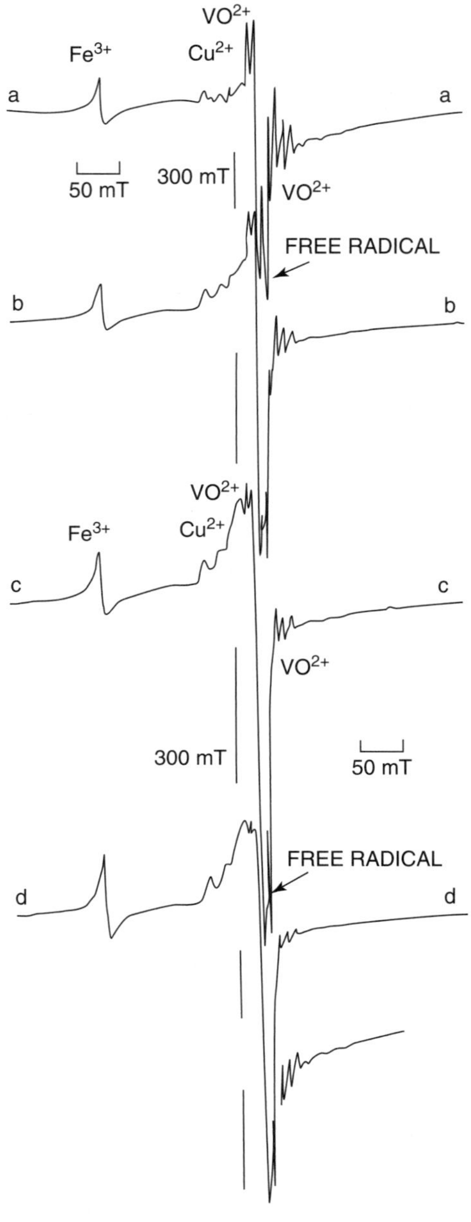

Figure 4.9. Electron spin resonance spectra of humic acids isolated from unamended soil (**a**) and soil amended with 90, 225, and 630 t ha^{-1} of sewage sludge (**b**, **c**, and **d**, respectively). Reprinted from Senesi, N. (1989). Composted materials as organic fertilizers. *Sci. Total Environ.* **81/82**, 521–542, with permission from Elsevier.

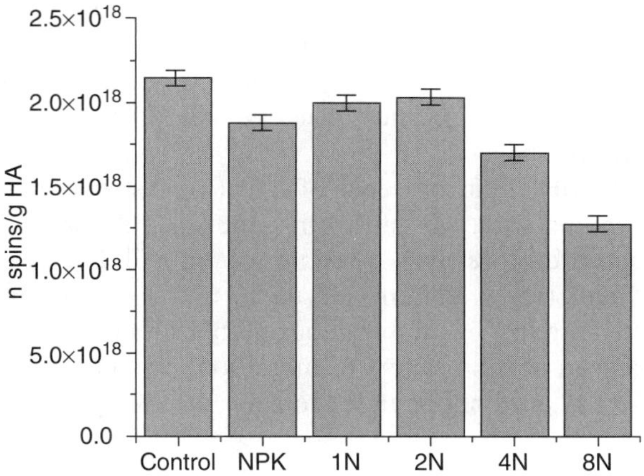

Figure 4.10. Organic free radical concentrations of humic acids isolated from unamended soil (control) and soils amended with mineral fertilizer (NPK) and 3.5, 7, 14, and 28 t ha^{-1} of sewage sludge (1N, 2N, 4N, and 8N, respectively). Reprinted from González-Pérez, M., Martin-Neto, L., Colnago, L. A., Milori, D. M. B. P., De Camargo, O. A., Berton, R., and Bettiol, W. (2006). Characterization of humic acids extracted from sewage sludge-amended oxisols by electron paramagnetic resonance. *Soil Till. Res.* **91**, 95–100, with permission from Elsevier.

The free radical concentrations of HAs and FAs from organic amendments are generally much smaller than those of native soil HAs and FAs (Figure 4.10) (Hervas et al., 1989; Senesi et al., 1989; Jerzykiewicz et al., 1999; Plaza et al., 2002, 2003; Soler-Rovira et al., 2002, 2003; González-Pérez et al., 2006). With increasing amendment rate and number, free radical concentrations of amended soil HAs and FAs decrease with respect to those of the corresponding unamended soil HAs and FAs (Plaza et al., 2002, 2003; González-Pérez et al., 2006). The concentration of organic free radicals in HS is generally related positively to the aromatic polycondensation, polymerization, and humification degree and may influence (a) their reactivity with metal ions and organic chemicals and (b) their physiological activity (Steelink, 1987; Senesi and Steelink, 1989; Senesi, 1990a,b, 1996; Cheshire and Senesi, 1998).

Thus, ESR data confirm (a) the lower-ring polycondensation and polymerization degree of HAs and FAs from organic amendments with respect to soil HAs and FAs and (b) the partial incorporation of the amendment on these structural properties into native soil HAs and FAs.

4.5. REACTIVITY OF HUMIC SUBSTANCES IN AMENDED SOILS

4.5.1. Interaction with Metal Ions

The capacity of binding metal ions is one of the most important physicochemical properties of HS, which is mainly attributed to their large content of oxygenated reactive functional groups that include carboxylic, phenolic, alcoholic and enolic hydroxyl groups, and carbonyl functionalities of various types. Nitrogen-, S-, and P-containing functional groups may also be involved in metal ion binding by HS (Stevenson, 1994; Tipping, 2002). Most processes in which metal ions are involved in soils, including mobility and transport, fixation and accumulation, and chemical reactivity and bioavailability, are affected by their interaction with HS (Senesi, 1992a; Senesi and Loffredo, 2005). For these reasons, the effects of organic amendment on metal ion binding properties of native soil HS is of great environmental and agronomic importance.

Depending on the pH value, presence of salts (ionic strength effect), and degree of saturation of binding sites, HS can form either soluble or insoluble complexes with metal ions and therefore play a double role in soil. Because of their smaller molecular weight and larger contents of acidic functional groups, FAs can form metal complexes that are more soluble, bioavailable, and mobile than those formed by HAs. Thus, FAs can act as carriers of metal ions in soil solution, whereas HAs tend to immobilize and accumulate metal ions in soil solid phases (Stevenson, 1994; Tipping, 2002).

Potentiometric and fluorescence titration studies indicate that HAs from liquid swine manure (LSM), LSM-amended soils, and unamended soils exhibit metal ion binding capacities and stability constants larger than their FA counterparts (Table 4.5) (Plaza et al., 2005d; Hernández et al., 2006, 2007). Apparently, not only the total amount of acidic functional groups, but also the overall chemical structure and/or sterical hindrance may affect the metal ion binding behavior of HAs and FAs. In this respect, aromaticity and humification degree are believed to be strongly related to metal ion binding of HS (Stevenson and Chen, 1991; Kaschl et al., 2002). Furthermore, physical data obtained by several authors (Buffle, 1988; Clapp et al., 1989; Tipping, 2002) suggest that dissolved FAs are approximately spherical, whereas HAs would present more open structures, with greater availability of sites "active" for metal binding.

Fluorescence quenching titration data fitted to a single-site model indicate that, with respect to unamended soil HAs and FAs, HA and FA fractions from organic amendments, especially if uncomposted, are generally characterized by much smaller metal ion binding capacities and stability constants (Table 4.5) (Provenzano et al., 2004; Plaza et al., 2005d, 2006b,c; Hernández et al., 2006, 2007). In general, organic amendment decreases metal ion complexing capacities and binding affinities of soil HAs and FAs (Table 4.5) (Plaza et al., 2005d, 2006b,c; Hernández et al., 2006, 2007).

Extended IR evidence exists of the involvement of carboxylate and phenolic OH groups of HAs and FAs from organic amendments in the formation of mixed electrovalent–covalent coordination bindings with several metal ions (Tan et al., 1971; Sposito et al., 1976; Senesi et al., 1992). Results of IR analysis combined with gel-filtration separation of trace metal-sludge FA solution indicate the involvement of COOH and HSO_3 functional groups in metal complexation (Baham et al., 1978). Direct IR evidence is also obtained for metal binding sites in sludge HS fractions involving either (a) amide N and possibly amide O (Boyd et al., 1979) or (b) amide N and COO^- groups (Hernández et al., 1993; Pignalosa et al., 1994).

The ESR spectroscopy has been used to characterize (a) indigenous Fe(III), Cu(II), and VO(II) ion complexes with anaerobic sludge HAs and FAs and (b) HAs from vermicomposted sludges, municipal solid wastes, and animal manures (Senesi and Sposito, 1984; Hervas et al., 1989; Senesi, 1990a,b; Senesi et al., 1992). The ESR spectra of organic amendment HAs show intense resonances arising from inner-sphere complexes of Cu(II) and Fe(III) and an anisotropic pattern, partially superimposed on the Cu(II) resonance, typical of VO(II)–HA complexes (Senesi, 1990a;

TABLE 4.5. Stability Constants (log K_M) and Complexing Capacities (CC_M) for Cu(II), Zn(II), Cd(II), and Pb(II) Binding to Humic Acids (HAs) and Fulvic Acids (FAs) Isolated from Liquid Swine Manure (LSM), Sewage Sludge (SS), Municipal Solid Waste Compost, Soils Amended with 90 and 150 $m^3 ha^{-1} yr^{-1}$ of LSM for 7 Years (SO + LSM90 and SO + LSM150, Respectively), 40 $t ha^{-1} yr^{-1}$ of SS for 1 Year (SO + SS40), and 40 $t ha^{-1} yr^{-1}$ of MSWC for 3 Years (SO + MSWC40), with the Corresponding Unamended Soils (SO), as Measured by Fluorescence Quenching Titrations

Samples and Origin	$\log K_{Cu}$	$\log K_{Zn}$	$\log K_{Cd}$	$\log K_{Pb}$	CC_{Cu} ($mmol\,g^{-1}$)[d]	CC_{Zn} ($mmol\,g^{-1}$)[d]	CC_{Cd} ($mmol\,g^{-1}$)[d]	CC_{Pb} ($mmol\,g^{-1}$)[d]
HAs								
LSM[a]	4.71	4.24	—	—	1.01	0.66	—	—
SO[a]	5.25	4.49	—	—	2.84	1.86	—	—
SO + LSM90[a]	5.13	4.37	—	—	2.20	1.72	—	—
SO + LSM150[a]	5.02	4.36	—	—	2.04	1.40	—	—
SS[b]	4.65	4.08	4.24	4.95	0.68	0.39	0.50	0.83
SO[b]	5.55	4.43	4.63	5.81	1.25	0.67	0.77	1.95
SO + SS40[b]	5.36	4.31	4.47	5.53	0.93	0.60	0.68	1.51
MSWC[c]	4.97	4.23	4.35	5.09	0.87	0.53	0.59	1.10
SO[c]	5.53	4.44	4.60	5.76	1.21	0.68	0.78	1.80
SO + MSWC40[c]	5.43	4.37	4.49	5.56	1.04	0.65	0.70	1.59
FAs								
LSM[d]	4.45	3.96	—	—	0.67	0.62	—	—
SO[d]	4.94	4.35	—	—	1.90	1.33	—	—
SO + LSM90[d]	4.80	4.31	—	—	1.27	1.19	—	—
SO + LSM150[d]	4.74	4.29	—	—	1.22	1.00	—	—

[a]From Hernández et al. (2006).
[b]From Plaza et al. (2006b).
[c]From Plaza et al. (2006c).
[d]From Hernández et al. (2007).

Senesi et al., 1992). The ESR spectra of sludge FAs also show intense resonances of Fe(III) and Cu(II) complexes, typical of soil FAs, but more than one order of magnitude less intense (Senesi and Sposito, 1984; Senesi, 1990a,b). Three types of binding sites, which involve more N than O ligand atoms, are detected for Cu(II) ions in sludge FAs (Senesi and Sposito, 1984). This case is different from soil FAs, in which one or two mostly oxygenated binding sites are observed for Cu(II) complexes.

The ESR spectrum of sludge-amended soil HA shows important differences from that of unamended soil HA (Figure 4.9) (Senesi et al., 1989). Besides the signal assigned to inner-sphere Fe(III) ions in octahedral or tetrahedral sites, an additional weak resonance is observed at low field in the spectrum of sludge-amended soil HAs, which is also attributed to Fe(III) ions possibly bound to four N atoms in a planar configuration (Senesi, 1990a). The ESR parameters of the inner-sphere complexes of Cu(II) ions in amended soil HAs indicate a large participation of N-containing functional groups, with respect to Cu(II) complexes in the unamended soil HA (Senesi et al., 1989; Senesi, 1990a). Similar to unamended soil HAs, the sludge-amended soil HAs feature an anisotropic ESR pattern typical of inner-sphere complexes of VO(II) ions held by four O ligand atoms in square-planar coordination (Senesi et al., 1989; Senesi, 1990a). Evidence also exists of a progressive increase in the intensity of the Cu(II) resonances—and, to a lesser extent, Fe(III)–HA resonances—and a decreasing intensity and loss of resolution of the VO(II)–HA resonances, with increasing sludge application to soil (Senesi et al., 1989).

4.5.2. Adsorption of Organic Xenobiotics

The important and multiple role played by HS in the behavior, performance, and fate of pesticides and other organic chemicals in soil has been extensively documented (Senesi and Chen, 1989; Senesi, 1992b, 1993a,b; Senesi and Miano, 1995). The HS are able to interact in several ways with organic xenobiotics, resulting in adsorption, catalytic activities, solubility effects, and photosensitization. Among these, adsorption is the most important process that consistently affects the bioavailability and toxicity, degradability, persistence and mobility, transport and accumulation, volatilization, and leaching of pesticides and other organic chemicals in soil systems.

The mechanism and extent of adsorption are ascertained to depend on several factors, which include (a) the physical and chemical nature and properties of both HS and organic xenobiotic and (b) the conditions of the medium. Organic matter in organic amendments is relatively "fresh" or little humified, has composition and properties that differ substantially from native soil HS, and affects the composition and structure of native soil HS. Therefore, organic xenobiotics added to soils interact with a complex mixture of applied and native HS, which is expected to affect both quantitative and mechanistic aspects of adsorption phenomena.

In a recent study (Senesi et al., 2001), the adsorption isotherms of the pesticides triallate, *trans*- and *cis*-chlordane, alachlor, imazethapyr, and rimsulfuron to HAs from sewage sludges, pig slurry, and amended and unamended soils obtained by using a batch equilibrium method coupled with high-preasure liquid or gas chromatographic analysis are discussed. In general, alachlor and triallate exhibit linear or C-type isotherms, imazethapyr nonlinear Freundlich isotherms of the L-type, and rimsulfurom and chlordane Langmuir-type isotherms in their absorption to HAs.

TABLE 4.6. Freundlich Adsorption Constants (1/*n* and *K*) and Distribution Coefficients (K_d) for Adsorption of Alachlor (Generally Linear Isotherms), Imazethapyr (Nonlinear Freundlich Isotherms), and Rimsulfuron (Langmuir Isotherms) on Humic Acids (HAs) Isolated from Two Sewage Sludges (SS1 and SS2), a Soil Amended with 10 t $ha^{-1} yr^{-1}$ of SS1 for 2 Years (SO1 + SS1), and a Soil Amended with 40 t $ha^{-1} yr^{-1}$ of SS2 for 2 Years (SO2 + SS2), with the Corresponding Unamended Soils (SO1 and SO2, Respectively) (from Senesi et al., 2001)

Origin of HA	Alachlor		Imazethapyr		Rimsulfuron	
Samples	K ($l\,kg^{-1}$)	K_d ($l\,kg^{-1}$)	$1/n$ ($l\,kg^{-1}$)	K ($l\,kg^{-1}$)	K_d ($l\,kg^{-1}$)	K_d ($l\,kg^{-1}$)
SS1	205.7	205.6	0.48	247.0	44.6	1102.9
SO1	161.2	177.4	0.64	332.5	95.0	2205.6
SO1 + SS1	147.1	147.4	0.75	122.6	49.6	1907.5
SS2	47.7	169.6	0.82	31.3	16.1	1221.1
SO2	140.8	152.4	0.77	181.6	79.7	2262.7
SO2 + SS2	115.1	116.1	0.55	179.7	36.3	2107.1

Linear isotherms, such as those obtained for alachlor and triallate, indicate that a constant partition of the pesticide occurs between the solution and the substrate HA; that is, adsorption is directly proportional to the pesticide concentration in solution over the whole concentration range tested. Nonlinear L-shaped isotherms, such as those obtained for imazethapyr, indicate that the adsorbent HA has a moderately high affinity for the pesticide molecule in the initial stages of adsorption, which occurs with increasing difficulty as adsorption sites are filled. Finally, Langmuir isotherms, such as those measured for rimsulfuron and *trans*- and *cis*-chlordane adsorption, suggest that these molecules have a moderately high affinity for the substrate HA in the initial stage of adsorption, whereas they have increasing difficulty in finding vacant sites as they are filled, finally reaching a maximum of adsorption.

Freundlich adsorption coefficients (K) and distribution coefficients (K_d), which are indexes of the adsorption capacity of the various HAs for the pesticides, are reported in Tables 4.6 and 4.7. Regardless of the nature of the HA, rimsulfuron is adsorbed in amounts about 10 times higher than those of imazethapyr, which in turn is adsorbed in amounts about two times higher than those of alachlor (Table 4.6). *Trans*- and *cis*-chlordane are absorbed in amounts more than one order of magnitude than those of triallate (Table 4.7). Furthermore, *trans*-chlordane appears to be more adsorbed than *cis*-chlordane by any HA examined.

The origin, composition, and chemical properties of HAs appear to have a smaller effect than the pesticide type on the extent of adsorption. However, the extent of pesticide adsorption by organic amendment HAs differs from that of soil HAs, and it varies depending on the type of pesticide. The same is true for amended soil HAs when compared to unamended soil HAs. In particular, HAs from sewage sludges show a smaller capacity to adsorb imazethapyr and rimsulfuron than do unamended soil HAs, whereas contrasting results are obtained for alachlor (Table 4.6). Pig slurry HA is less effective than any soil HA in adsorption of *trans*- and *cis*-chlordane, whereas the adsorption capacity of this HA for triallate is from about two to four times larger than that of soil HAs (Table 4.7). Similarly, results of Simpson et al.

TABLE 4.7. Adsorption Coefficients (*K*) and Distribution Coefficients (K_d) for Adsorption of Triallate (Linear Isotherms) and *trans*- and *cis*-Chlordane (Langmuir Isotherms) on Humic Acids (HAs) Isolated from Pig Slurry (PS) and Surface (1) and Whole (2) Horizons of Two Soils (UK and PO)

Origin of HA Samples	Triallate		*Trans*-Chlordane	*Cis*-Chlordane
	K ($l kg^{-1}$)	K_d ($l kg^{-1}$)	K_d ($l kg^{-1}$)	K_d ($l kg^{-1}$)
PS	13,573	12,677	101,918	80,096
UK1	7,907	7,283	119,932	119,356
UK2	7,438	6,865	154,916	100,309
PO1	6,247	6,405	154,381	154,608
PO2	3,051	2,782	110,609	87,341

Source: Senesi et al. (2001).

(2003) show that soil HA exhibits a higher affinity for phenanthrene than do peat and compost HAs. Unamended soil HAs generally exhibit a higher adsorption capacity for alachlor, imazethapyr, and rimsulfuron than do the corresponding amended soil HAs (Table 4.6). These trends are apparently related to the molecular structure and chemical properties of the various HAs.

Information on the binding mechanisms involved in the interaction of various pesticides with HAs from organic amendments and amended and unamended soils has been obtained from the comparative analysis of FT IR, fluorescence, and ESR spectroscopic data of laboratory-prepared pesticide–HA interaction products, with respect to the corresponding unreacted HA and the pure pesticide molecule (Senesi et al., 2001). Spectroscopic data suggest that moderately water-soluble, polar pesticides (e.g., alachlor, imazethapyr, and rimsulfuron) are mainly adsorbed to HAs by multiple binding mechanisms, including H-bonding and ionic and charge-transfer processes, whereas low water-soluble, nonpolar pesticides (e.g., triallate and chlordane) preferentially bind to HAs by hydrophobic bonding. The highly aliphatic, low humified HAs from organic amendments tend to bind pesticides by hydrophobic bonding, whereas the well-humified soil HAs appear to prefer chemical binding forms.

4.6. CONCLUSIONS

Chemical and physico-chemical data available in the literature suggest that the composition, structure, properties, and reactivity of HS fractions isolated from organic wastes used as soil amendments are in any case markedly different from those of native soil HS. In general, the former feature larger C, H, N, and S contents and smaller O, organic free radical and acidic functional group contents than does native soil HS. Furthermore, HS fractions in organic amendments are typically characterized by a relatively larger presence of aliphatic, amide, and polysaccharide structures, simple structural components of wide molecular heterogeneity, low degree of aromatic polymerization, low level of conjugated chromophores, and low humification degree. Composting, vermicomposting, and other treatments are proven to be able to induce a loss of aliphatic materials and carbohydrates, a decrease of molecular heterogeneity, and an increase in oxygenation, acidic functional group contents, aromatic polycondensation and polymerization, and humification degree in the HS fractions. These changes lead to the positive result that HS components in processed organic amendments chemically and physico-chemically resemble native soil HS.

The compositional, functional, and structural features of amended soil HS are affected in different ways and at various extent as a function of the nature, origin, and application rate of the amendment and are generally intermediate between those of unamended soil HS and amendment HS, but closer to the former. In general, organic amendment determines an increase of aliphatic, amide, and polysaccharide components, along with a decrease of acidic functional group and organic free radical contents. These effects suggest a partial incorporation of HS fractions of organic amendments into native soil HS, and they are more evident when untreated organic materials are used.

In general, HS fractions from organic soil amendments exhibit smaller binding capacities and affinities for metal ions and organic xenobiotics than do native soil HS, and their application to soil determines, as it may be expected, a decrease of the reactivity of amended soil HS. The relatively small binding capacities and affinities of organic amendment HS may be ascribed to their ascertained typical aliphatic character and small degree of aromatic polycondensation and humification. On the other hand, the relatively large binding capacities and affinities of native soil HS may be related to (a) their typical high content of acidic functional groups and other O-containing ligands groups and (b) the marked aromatic character and high humification degree. The intermediate reactivity generally measured for amended soil HS confirm the partial incorporation of low-humified HS structures from organic amendments into native soil HS.

Despite the extensive research performed, further research is needed to know better the molecular structure of HS in organic amendments and amended soils as well as the mechanisms of HS formation and transformations in order to understand better the biogeochemistry of these materials and their interactions with surrounding environments. Although recent advances in analytical chemistry have allowed great progress in the understanding of HS chemistry, new analytical methods and experimental strategies are still needed. Innovative research is also needed to be targeted to the direct effects that HS fractions in organic amendments may exert on the biochemical, physiological, and genetic processes in plants. Furthermore, investigations of the mechanisms of interaction of HS with inorganic and organic xenobiotics in organically amended soils are also expected in order to better understand the fate, performance, and behavior of these contaminants in soil.

LIST OF ABBREVIATIONS

SOM, soil organic matter; HS, humic substances; DH, degree of humification; HAC, humic acid C; FAC, fulvic acid C; TEC, total extractable C; HR, humification rate; HI, humification index; NHC, nonhumified C; TOC, total organic C; HA, humic acid; FA, fulvic acids; UV–Vis, ultraviolet-visible; FT-IR, Fourier transform infrared; NMR, nuclear magnetic resonance; ESR, electron spin resonance; EEM, excitation–emission matrix.

REFERENCES

Adani, F., and Tambone, F. (2005). Long-term effect of sewage sludge application on soil humic acids. *Chemosphere* **60**, 1214–1221.

Adani, F., Genevini, P., Tambone, F., and Montoneri, E. (2006). Compost effect on soil humic acid: A NMR study. *Chemosphere* **65**, 1414–1418.

Adani, F., Genevini, P., Ricca, G., Tambone, F., and Montoneri, E. (2007), Modification of soil humic matter after 4 years of compost application. *Waste Manage.* **27**, 319–324.

Albiach, R., Canet, R., Pomares, F., and Ingelmo, F. (2001). Organic matter components and aggregate stability after the application of different amendments to a horticultural soil. *Bioresource Technol.* **76**, 125–129.

Almendros, G., Polo, A., and Dorado, E. (1983a). Estudio de los compuestos húmicos en diversos tipos de compost preparados a partir de paja de trigo. II. Caracterización fisicoquímica de los compuestos húmicos. *Agrochimica* **27**, 310–325.

Almendros, G., Lobo, M. C., Polo, A., and Dorado, E. (1983b). Naturaleza y propiedades de la materia orgánica de dos tipos de compost de paja de trigo. *An. Edafol. Agrobiol.* **42**, 2083–2093.

Amir, S., Hafidi, M., Merlina, G., Hamdi, H., and Revel, J. C. (2004). Elemental analysis, FTIR, 13C-NMR of humic acids from sewage sludge composting. *Agronomie* **24**, 13–18.

Baham, J., Ball, N. B., and Sposito, G. (1978). Gel filtrations studies of trace metal–fulvic acid solutions extracted from sewage sludges. *J. Environ. Qual.* **7**, 181–188.

Bellamy, L. J. (1975). *The Infrared Spectra of Complex Molecules*, Chapman & Hall, London, UK.

Benedetti, M. F., Milne, C. J., Kinniburgh, D. G., Van Riemsdijk, W. H., and Koopal, L. K. (1995). Metal ion binding to humic substances: application of the non-ideal competitive adsorption model. *Environ. Sci. Technol.* **29**, 446–457.

Benedetti, M. F., Van Riemsdijk, W. H., and Koopal, L. K. (1996a). Humic substances considered as a heterogeneous gel phase. *Environ. Sci. Technol.* **30**, 1805–1813.

Benedetti, M. F., Van Riemsdijk, W. H., Koopal, L. K., Kinniburgh, D. G., Gooddy, D. C., and Milne, C. J (1996b). Metal ion binding by natural organic matter: from the model to the field. *Geochim. Cosmochim. Acta* **60**, 2503–2513.

Bernal, M. P., Sánchez-Monedero, M. A., Paredes, C., and Roig, A. (1998). Carbon mineralization from organic wastes at different composting stages during their incubation with soil. *Agric. Ecosyst. Environ.* **69**, 175–189.

Bertoncini, E. I., D'Orazio, V., Senesi, N., and Mattiazzo, M. E. (2005). Fluorescence analysis of humic and fulvic acids from two Brazilian oxisols as affected by biosolid amendment. *Anal. Bioanal. Chem.* **381**, 1281–1288.

Boyd, S. A., Sommers, L. E., and Nelson, D. W. (1979). Infrared spectra of sewage sludge fractions: Evidence for an amide metal binding site. *Soil Sci. Soc. Am. J.* **43**, 893–899.

Boyd, S. A., Sommers, L. E., and Nelson, D. W. (1980). Changes in the humic acid fraction of soil resulting from sludge application. *Soil Sci. Soc. Am. J.* **44**, 1179–1186.

Brebbia, C. A., Kungolos, S., Popov, V., and Itoh, H. (2004). *Waste Management and the Environment II*, WIT Press, Southampton, UK.

Brunetti, G., Plaza, C., and Senesi, N. (2005). Olive pomace amendment in Mediterranean conditions: effect on soil and humic acid properties and wheat (*Triticum turgidum* L.) yield. *J. Agric. Food Chem.* **53**, 6730–6737.

Brunetti, G., Senesi, N., and Plaza, C. (2007a). Effects of amendment with treated and untreated olive oil mill wastewaters on soil properties, soil humic substances and wheat yield. *Geoderma* **138**, 144–152.

Brunetti, G., Plaza, C., Clapp, C. E., and Senesi, N. (2007b). Compositional and functional features of humic acids from organic amendments and amended soils in Minnesota, USA. *Soil Biol. Biochem.*, doi:10.1016/j.soilbio.2006.12.012.

Buffle, J. (1988). *Complexation Reactions in Aquatic Systems: An Analytical Approach.* Ellis-Horwood, Chichester, UK.

Campitelli, P. A., Velasco, M. I., and Ceppi, S. B. (2006). Chemical and physicochemical characteristics of humic acids extracted from compost, soil and amended soil. *Talanta* **69**, 1234–1239.

Chen, Y. (2003). Nuclear magnetic resonance, infra-red and pyrolysis: Application of spectroscopic methodologies to maturity determination of composts. *Compost Sci. Util.* **11**, 152–168.

Chen, J., LeBoeuf, E. J., Dai, S., and Gu, B. (2003). Fluorescence spectroscopic studies of natural organic matter fractions. *Chemosphere* **50**, 639–647.

Chen, Y., Senesi, N., and Schnitzer, M. (1977). Information provided on humic substances by E_4/E_6 ratios. *Soil. Sci. Soc. Am. J.* **41**, 352–358.

Cheshire, M. V., and N. Senesi (1998). Electron spin resonance spectroscopy of organic and mineral soil particles. In *Structure and Surface Reactions of Soil Particles*, Huang, P. M., Senesi, N., and Buffle, J., eds., IUPAC Book Series on Analytical and Physical Chemistry of Environmental Systems, John Wiley & Sons, New York, pp. 325–373.

Ciavatta, C., Vittori Antisari, L., and Sequi, P. (1988). A first approach to the characterization of the presence of humified materials in organic fertilizers. *Agrochimica* **32**, 510–517.

Clapp, C. E., Emerson, W. W., and Olness, A. E. (1989). Sizes and shapes of humic substances by viscosity measurements. In *Humic Substances II. In Search of Structure*, Hayes, M. H. B., MacCarthy, P., Malcolm, R. L., and Swift, R. S., eds., John Wiley & Sons, Chichester, UK, pp. 497–514.

De Bertoldi, M., Sequi, P., Lemmes, B., and Papi T. (1996). *The Science of Composting.* Chapman & Hall, London, UK.

Dudal, Y., and Gérard, F. (2004). Accounting for natural organic matter in aqueous chemical equilibrium models: A review of the theories and applications. *Earth-Sci. Rev.* **66**, 199–216.

Edwards, C. A. (2004). *Earthworm Ecology*, CRC Press, Boca Raton, FL.

Eghball, B. (2002). Soil properties as influenced by phosphorus- and nitrogen-based manure and compost applications. *Agron. J.* **94**, 128–135.

Flavel, T. C., and Daniel, V. (2006). Carbon and nitrogen mineralization rates after application of organic amendments to soil. *J. Environ. Qual.* **35**,183–193.

Fuentes, M., González-Gaitano, G., and García-Mina, J. M. (2006). The usefulness of UV–visible and fluorescence spectroscopies to study the chemical nature of humic substances from soils and composts. *Org. Geochem.* **37**, 1949–1959.

García, C., Hernández, T., and Costa, F. (1992a). Mineralization in a calcareous soil of a sewage sludge composted with different organic residues. *Waste Manage. Res.* **10**, 445–452.

García, C., Hernández, T., and Costa, F. (1992b). Characterization of humic acids from uncomposted and composted sewage sludge by degradative and non-degradative techniques. *Bioresource Technol.* **41**, 53–57.

García-Gil, J. C., Plaza, C., Senesi, N., Brunetti, G., and Polo, A. (2004a). Effects of sewage sludge amendment on humic acids and microbiological properties of a semiarid Mediterranean soil. *Biol. Fertil. Soils* **39**, 320–328.

García-Gil, J. C., Ceppi, S. B., Velasco, M. I., Polo, A., and Senesi, N. (2004b). Long-term effects of amendment with municipal solid waste compost on the elemental and acidic functional group composition and pH-buffer capacity of soil humic acids. *Geoderma* **121**, 135–142.

Giusquiani, P. L., Gigliotti, G., Businelli, D., and Macchioni, A. (1994). Spectroscopic comparison between humic and fulvic acids from urban waste compost and soil. In *Humic Substances in the Global Environment and Implications on Human Health*, Senesi, N., and Miano, T. M., eds., Elsevier, Amsterdam, The Netherlands, pp. 1303–1310.

González-Pérez, M., Martin-Neto, L., Colnago, L. A., Milori, D. M. B. P., De Camargo, O. A., Berton, R., and Bettiol, W. (2006). Characterization of humic acids extracted from sewage sludge-amended oxisols by electron paramagnetic resonance. *Soil Till. Res.* **91**, 95–100.

González-Vila, F. J., and Martin, F. (1985). Cehmical structural characteristics of humic acids extracted from composted municipal refuse. *Agric. Ecosyst Environ.* **14**, 267–278.

González-Vila, F. J., and Martin, F. (1987). Modifications of the humic acid fraction in a soil treated with composted municipal refuse. *Sci. Total Environ.* **62**, 459–466.

González-Vila, F. J., Almendros, G., and Madrid, F. (1999). Molecular alterations of organic fractions from urban waste in the course of composting and their further transformation in amended soil. *Sci. Total Environ.* **236**, 215–229.

Gregorich, E. G., Greer, K. J., Anderson, D. W., and Liang, B. C. (1998). Carbon distribution and losses: Erosion and deposition effects. *Soil Till. Res.* **47**, 291–302.

Hatcher, P., Dria, K. J., Kim, S., and Frazier, S. W. (2001). Modern analytical studies of humic substances. *Soil Sci.* **166**, 770–794.

Hernández, T., Moreno, J. L., and Costa, F. (1993). Infrared spectroscopic characterization of sewage sludge humic acids. Evidence of sludge organic matter-metal interactions. *Agrochimica* **37**, 12–17.

Hernández, D., Plaza, C., Senesi, N., and Polo, A. (2006). Detection of copper (II) and zinc(II) binding to humic acids from pig slurry and amended soils by fluorescence spectroscopy. *Environ. Pollut.* **143**, 212–220.

Hernández, D., Plaza, C., Senesi, N., and Polo, A. (2007). Fluorescence analysis of copper(II) and zinc(II) binding behavior of fulvic acids from pig slurry and amended soils. *Eur. J. Soil Sci.* **58**, 900–908.

Hervas, L., Mazuelos, C., Senesi, N., and Saiz-Jiménez, C. (1989). Chemical and physicochemical characterization of vermicomposts and of their humic acid components. *Sci. Total Environ.* **81/82**, 543–550.

Huang, G. F., Wu, Q. T., Wong, J. W. C., and Nagar, B. B. (2006). Transformation of organic matter during co-composting of pig manure with sawdust. *Bioresource Technol.* **97**, 1834–1842.

Inbar, Y., Chen, Y., and Hadar, Y. (1990). Humic substances formed during the composting of organic matter. *Soil Sci. Soc. Am J.* **54**, 1316–1323.

Inbar, Y., Chen, Y., and Hadar, Y. (1991). Carbon-13 CPMAS NMR and FTIR spectroscopic analysis of organic matter transformations during composting of solid wastes from wineries. *Soil Sci.* **152**, 272–282.

Inbar, Y., Hadar, Y., and Chen, Y. (1992). Characterization of humic substances formed during the composting of solid wastes from wineries. *Sci. Total Environ.* **113**, 35–48.

Jerzykiewicz, M., Drozd, J., and Jezierski, A. (1999). Organic radicals and paramagnetic metal complexes in municipal solid waste composts. An EPR and chemical study. *Chemosphere* **39**, 253–268.

Kaschl, A., Römheld, V., and Chen, Y. (2002). Cadmium binding by fractions of dissolved organic matter and humic substances from municipal solid waste compost. *J. Environ. Qual.* **31**, 1885–1892.

Kinniburgh, D. G., Milne, C. J., Benedetti, M. F., Pinheiro, J. P., Filius, J. D., Koopal, L. K., and Van Riemsdijk, W. H. (1996). Metal ion binding by humic acid: Application of the NICA–Donnan model. *Environ. Sci. Technol.* **30**, 1687–1698.

Kinniburgh, D. G., Van Riemsdijk, W. H., Koopal, L. K., Borkovec, M., Benedetti, M. F., and Avena, M. J. (1999). Ion binding to natural organic matter: Competition, heterogeneity, stoichiometry and thermodynamic consistency. *Colloid Surf. A* **151**, 147–166.

Kögel-Knabner, I. (1997). 13C and 15N NMR spectroscopy as a tool in soil organic matter studies. *Geoderma* **80**, 243–270.

Kögel-Knabner, I. (2000). Analytical approaches for characterizing soil organic matter. *Org. Geochem.* **31**, 609–625.

Kononova, M. M. (1966). *Soil Organic Matter*, Pergamon Press, Oxford, UK.

Koopal, L. K., Van Riemsdijk, W. H., De Wit, J. C. M., and Benedetti, M. F. (1994). Analytical isotherm equations for multicomponent adsorption to heterogeneous surfaces. *J. Colloid Interface Sci.* **166**, 51–60.

Koopal, L. K., Van Riemsdijk, W. H., and Kinniburgh, D. G. (2001). Humic matter and contaminants. General aspects and modeling ion binding. *Pure Appl. Chem.* **73**, 2005–2016.

Koopal, L. K., Saito, T., Pinheiro, J. P., and Van Riemsdijk, W. H. (2005). Ion binding to natural organic matter: General considerations and the NICA–Donnan model. *Colloids Surf. A* **265**, 40–54.

Lal, R. (2001). Managing world soils for food security and environmental quality. *Adv. Agron.* **74**, 155–192.

Lal, R. (2004). Soil carbon sequestration to mitigate climate change. *Geoderma* **123**, 1–22.

Levi-Minzi, R., Riffaldi, R., and Saviozzi, A. (1990). Carbon mineralization in soil amended with different organic materials. *Agric. Ecosyst. Environ.* **31**, 325–335.

MacCarthy, P., and Rice, J. A. (1985). Spectroscopic methods (other than NMR) for determining functionality in humic substances. In *Humic Substances in Soil, Sediment and Water. Geochemistry, Isolation, and Characterization*, Aiken, J. R., McKnight, D. M., Wershaw, R. L., and MacCarthy, P., eds., Wiley-Interscience, New York, pp. 527–559.

Miikki, V., Senesi, N., and Hänninen, K. (1997). Characterization of humic material formed by composting of domestic and industrial biowastes, Part 2: Spectroscopic evaluation of humic acid structures. *Chemosphere* **34**, 1639–1651.

Milne, C. J., Kinniburgh, D. G., and Tipping, E. (2001). Generic NICA–Donnan model parameters for proton binding by humic substances. *Environ. Sci. Technol.* **35**, 2049–2059.

Mobed, J. J., Hemmingsen, S. L., Autry, J. L., and McGown, L. B. (1996). Fluorescence Characterization of IHSS humic substances: total luminescence spectra with absorbance correction. *Environ. Sci. Technol.* **30**, 3061–3065.

Nortcliff, S. (2005). Soil organic matter—The potential benefits and necessary precautions required in the use of composts and biosolids as soil amendments in agricultural ecosystems. In *Application of the Emerging Soil Researches to the Conservation of Agricultural Ecosystems*, Yang, J. E., Sa, T. M., and Kim, J. J., eds., KSSSF, KSAE, RDA, Korea, pp. 81–93.

Paré, T., Dinel, H., Schnitzer M., and Dumontet, S. (1998). Transformations of carbon and nitrogen during composting of animal manure and shredded paper. *Biol. Fertil. Soils* **26**, 173–178.

Pedra, F, Polo, A., Ribeiro, A., and Domingues, H. (2007). Effects of municipal solid waste compost and sewage sludge on mineralization of soil organic matter. *Soil Biol. Biochem.*, doi:10.1016/j.soilbio.2006.12.014.

Piccolo, A., Zaccheo, P., and Genevini, P. G. (1992). Chemical characterization of humic substances extracted from organic-waste-amended soils. *Bioresource Technol.* **40**, 275–282.

Pignalosa, V., Amalfitano, C., and Ramunni, A. (1994). Alternative use of sewage sludges in agriculture. *Agrochimica* **38**, 91–96.

Plaza, C., Senesi, N., García-Gil, J. C., Brunetti, G., D'Orazio, V., and Polo, A. (2002). Effects of pig slurry application on soils and soil humic acids. *J. Agric. Food Chem.* **50**, 4867–4874.

Plaza, C., Senesi, N., Polo, A., Brunetti, G., García-Gil, J. C., and D'Orazio, V. (2003). Soil fulvic acid properties as a means to assess the use of pig slurry amendment. *Soil Till. Res.* **74**, 179–190.

Plaza, C., Senesi, N., Brunetti, G., and Mondelli, D. (2005a). Co-composting of sludge from olive oil mill wastewater mixed with tree cuttings. *Compost Sci. Util.* **13**, 217–226.

Plaza, C., Senesi, N., Polo, A., and Brunetti, G. (2005b). Acid–base properties of humic and fulvic acids formed during composting by the NICA–Donnan model. *Environ. Sci. Technol.* **39**, 7141–7146.

Plaza, C., Brunetti, G., Senesi, N., and Polo, A. (2005c). Proton binding to humic acids from organic amendments and amended soils by the NICA–Donnan model. *Environ. Sci. Technol.* **39**, 6692–6697.

Plaza, C., García-Gil, J. C., Polo, A., and Senesi N. (2005d) Copper (II) complexation by humic and fulvic acids from pig slurry and amended and non-amended soils. *Chemosphere* **61**, 711–716.

Plaza, C., Hernández, D., Fernández, J. M., and Polo, A. (2006a). Long-term effects of amendment with liquid swine manure on proton binding behavior of soil humic substances. *Chemosphere* **65**, 1321–1329.

Plaza, C., Brunetti, G., Senesi, N., and Polo, A. (2006b). Molecular and quantitative analysis of metal ion binding to humic acids from sewage sludge and sludge-amended soils by fluorescence spectroscopy. *Environ. Sci. Technol.* **40**, 917–923.

Plaza, C., Brunetti, G., Senesi, N., and Polo, A. (2006c). Fluorescence characterization of metal ion–humic acid interactions in soils amended with composted municipal solid wastes. *Anal. Bioanal. Chem.* **386**, 2133–2140.

Plaza, C., Senesi, N., Brunetti, G., and Mondelli, D., (2007). Evolution of the fulvic acid fractions during co-composting of olive oil mill wastewater sludge and tree cuttings. *Bioresource Technol.* **98**, 1964–1971.

Polak, J., Sułkowski, W. W., Bartoszek, M., and Papiez, W. (2005). Spectroscopic studies of the progress of humification processes in humic acid extracted from sewage sludge. *J. Mol. Struct.* **744–747**, 983–989.

Poole, C. P. (1997). *Electron Spin Resonance: A Comprehensive Treatise on Experimental Techniques*, Dover Publications, Mineola, NY.

Preston, C. (1996). Applications of NMR to soil organic matter analysis: history and prospects. *Soil Sci.* **161**, 144–166.

Provenzano, M. R., D'Orazio, V., Jerzykiewicz, M., and Senesi, N. (2004). Fluorescence behaviour of Zn and Ni complexes of humic acids from different sources. *Chemosphere* **55**, 885–892.

Riffaldi, R., Levi-Minzi, R., and Saviozzi, A. (1983). Humic fractions of organic wastes. *Agric. Ecosyst. Environ.* **10**, 353–359.

Ritchie, J. D., and Perdue, E. M. (2003). Proton-binding study of standard and reference fulvic acids, humic acids, and natural organic matter. *Geochim. Cosmochim. Acta* **67**, 85–96.

Rivero, C., Chirenje, T., Ma, L. Q., and Martinez, G. (2004). Influence of compost on soil organic matter quality under tropical conditions. *Geoderma* **123**, 355–361.

Romero, E., Plaza, C., Senesi, N., Nogales, R., and Polo, A. (2007). Humic acid-like fractions in raw and vermicomposted winery and distillery wastes. *Geoderma* **139**, 397–406.

Sánchez-Monedero, M. A., Roig, A., Cegarra, J., and Bernal, M. P. (1999). Relationships between water-soluble carbohydrate and phenol fractions and the humification indices of different organic wastes during composting. *Bioresource Technol.* **70**, 193–201.

Schnitzer, M. (1978). Humic substances: Chemistry and reactions. In *Soil Organic Matter*, Schnizer, M., and Khan, S. U., eds., Elsevier, Amsterdam, The Netherlands, pp. 1–64.

Schulze E. D., and Freibauer A. (2005). Carbon unlocked from soils. *Nature* **437**, 205–206.

Senesi, N. (1989). Composted materials as organic fertilizers. *Sci. Total Environ.* **81/82**, 521–542.

Senesi, N. (1990a). Application of electron spin resonance (ESR) spectroscopy in soil chemistry. *Adv. Soil Sci.* **14**, 77–130.

Senesi, N. (1990b). Molecular and quantitative aspects of the chemistry of fulvic acid and its interactions with metal ions and organic chemicals. Part. I. The electron spin resonance approach. *Anal. Chim. Acta* **232**, 51–75.

Senesi, N. (1992a). Metal-humic substance complexes in the environment. Molecular and mechanistic aspects by multiple spectroscopic approach. In *Biogeochemistry of Trace Metals*, Adriano, D. C., ed., Lewis Publishers, Boca Raton, FL, pp. 425–491.

Senesi N. (1992b). Binding mechanisms of pesticides to soil humic substances. *Sci. Total Environ.* **123/124**, 63–76.

Senesi, N. (1993a). Nature of interactions between organic chemicals and dissolved humic substances and the influence of environmental factors. In *Organic Substances in Soil and Water: Natural Constituents and Their Influence on Contaminant Behaviour*, Beck, A. J., Jones, K. C., Hayes, M. H. B., and Mingelgrin, U., eds., Royal Society of Chemistry, London, UK, pp. 73–101.

Senesi, N. (1993b). Organic pollutant migration in soils as affected by soil organic matter. Molecular and mechanistic aspects. In *Migration and Fate of Pollutants in Soils and Subsoils*, Petruzzelli, D., and Helfferich, F. G., eds., NATO-ASI Series, Vol. G32, Springer-Verlag, Berlin, Germany, pp. 47–74.

Senesi, N. (1996). Electron spin (or paramagnetic) resonance spectroscopy. In *Methods of Soil Analysis: Chemical Methods*, Sparks, D. L., ed., ASA-CSSA-SSSA, Madison, WI, pp. 323–356.

Senesi, N., and Chen, Y. (1989). Interactions of toxic organic chemicals with humic substances. In *Toxic Organic Chemicals in Porous Media*, ed. Gerstl, Z., Chen, Y., Mingelgrin, V., and Yaron, B., eds., *Ecological Studies*, Vol. **73**, Springer-Verlag, Berlin, Germany, pp. 37–90.

Senesi, N., and Loffredo, E. (1999). The chemistry of soil organic matter. In *Soil Physical Chemistry*, Sparks, D. L., ed., CRC Press, Boca Raton, FL, pp. 239–370.

Senesi, N., and Loffredo, E. (2005). Metal ion complexation by soil humic substances. In *Chemical Processes in Soils*, Tabatabai, M. A., and Sparks, D. L., eds., SSSA, Madison. WI, pp. 563–617.

Senesi, N., and Miano, T. M. (1995). The role of abiotic interactions with humic substances on the environmental impact of organic pollutants. In *Environmental Impact of Soil Component Interactions: Land Quality, Natural and Anthropogenic Organics*, Vol. **1**, Huang, P. M., Berthelin, J., Bollag, J. M., and McGill, W. B., eds., CRC-Lewis, Boca Raton, FL, pp. 311–335.

Senesi, N., and Plaza, C. (2007). Role of humification processes in recycling organic wastes of various nature and sources as soil amendments. *Clean Soil Air Water* **35**, 26–41.

Senesi, N., and Sposito, G. (1984). Residual copper (II) complexes in purified soil and sewage sludge fulvic acids: an electron spin resonance study. *Soil Sci. Soc. Am. J.* **48**, 1247–1253.

Senesi, N., and Steelink, C. (1989). Application of ESR spectroscopy to the study of humic substance structure. In *Humic Substances II. In Search of Structure*, Hayes, M. B. H., MacCarthy, P., Malcolm, R., and Swift, R., eds., John Wiley & Sons, New York, pp. 373–407.

Senesi, N., Sposito, G., Holtzclaw, K. M., and Bradford G. R. (1989). Chemical properties of metal–humic acid fractions of a sewage sludge-amended aridisol. *J. Environ. Qual.* **18**, 186–194.

Senesi, N., Miano, T. M., Provenzano, M. R., and Brunetti, G. (1990). Characterization, differentiation and classification of humic substances by fluorescence spectroscopy. *Soil Sci.* **152**, 259–271.

Senesi, N., Saiz-Jiménez, C., and Miano, T. M. (1992). Spectroscopic characterizaton of metal-humic acid-like complexes of earthworm-composted organic wastes. *Sci. Total Environ.* **117/118**, 111–120.

Senesi, N., Miano, T. M., and Brunetti, G. (1996). Humic-like substances in organic amendments and effects on native soil humic substances. In *Humic Substances in Terrestrial Ecosystems*, Piccolo, A., ed., Elsevier, New York, pp. 531–593.

Senesi, N., Loffredo, E., D'Orazio, V., Brunetti, G., Miano, T. M., and La Cava, P. (2001). Adsorption of pesticides by humic acids from organic amendments and soils. In *Humic Substances and Chemical Contaminants*, Clapp, C. E., Hayes, M. H. B., Senesi, N., Bloom, P. R., and Jardine, P. M., eds., SSSA, Madison, WI, pp. 129–153.

Senesi, N., Plaza, C., Brunetti, G., and Polo, A. (2007). A comparative survey of recent results on humic-like fractions in organic amendments and effects on native soil humic substances. *Soil Biol. Biochem.*, doi:10.1016/j.soilbio.2006.12.

Sequi, P., De Nobili, M., Leita, L., and Cercignani, G. (1986). A new index of humification. *Agrochimica* **30**, 175–179.

Siemens, J., and Janssens, I. A. (2003). The European carbon budget: A gap. *Science* **302**, 1681.

Simpson, M. J., Chefetz, B., and Hatcher, P. G. (2003). Phenanthrene sorption to structurally modified humic acids. *J. Environ. Qual.* **32**,1750–1758.

Soler-Rovira, P. A., Brunetti, G., Polo, A., and Senesi, N. (2002). Comparative chemical and spectroscopic characterization of humic acids from sewage sludges and sludge-amended soils. *Soil Sci.* **167**, 235–245.

Soler-Rovira, P. A., Brunetti, G., Polo, A., and Senesi, N. (2003). Effects of amendment with composted sludge on soil humic acid properties. *Compost Sci. Util.* **11**, 176–184.

Smith, P. (2004). Carbon sequestration in croplands: The potential in Europe and the global context. *Eur. J. Agron.* **20**, 229–236.

Sposito, G., Holtzclaw, K. M., and Baham, J. (1976). Analytical properties of the soluble, metal-complexing fractions in sludge–soil mixtures: II. Comparative structural chemistry of fulvic acid. *Soil Sci. Soc. Am. J.* **40**, 691–697.

Sposito, G., Schaumberg, G. D., Perkins, T. G., and Holtzclaw, K. M. (1978). Investigation of fulvic acid extracted from sewage sludge using carbon-13 and proton NMR spectroscopy. *Environ. Sci. Technol* **12**, 931–934.

Sposito, G., Holtzclaw, K. M., LeVesque, C. S., and Johnston, C. T. (1982). Trace metal chemistry in arid-zone field soils amended with sewage sludge: II. Comparative study of the fulvic acid fraction. *Soil Sci. Soc. Am. J.* **46**, 265–270.

Stevenson, F. J. (1994). *Humus Chemistry: Genesis, Composition, Reactions*. John Wiley & Sons, New York.

Stevenson, F. J., and Chen, Y. (1991). Stability constants of copper(II)-humate complexes determined by modified potentiometric titration. *Soil Sci. Soc. Am. J.* **55**, 1586–1591.

Steelink, C. (1987). Review of ESR spectroscopy of humic substances. In *NMR of Humic Substances and Coal. Techniques, Problems and Solutions*, Wershaw, R. L., and Mikita, M. A., Lewis, eds., Chelsea, MI, pp. 47–72.

Tan, K. H., King, L. D., and Morris, H. D. (1971). Complex reactions of zinc with organic matter extracted from sewage sludge. *Soil Sci. Soc. Am. Proc.* **35**, 748–752.

Tipping, E. (2002). *Cation Binding by Humic Substances*, Cambridge University Press, New York.

Veeken, A., Nierop, K., De Wilde, V., and Hamelers, B. (2000). Characterisation of NaOH-extracted humic acids during composting of a biowaste. *Bioresource Technol.* **72**, 33–41.

Wilson, M. A. (1987). *NMR Techniques and Applications in Geochemistry and Soil Chemistry*. Pergamon Press, Oxford, UK.

Williams, P. T. (2005). *Waste Treatment and Disposal*, John Wiley & Sons, West Sussex, UK.

Worrall, F., Reed, M. S., Warburton, J., and Burt T. (2003). Carbon budget for a British upland peat catchment. *Sci. Total Environ.* **312**, 133–146.

5

CARBON SEQUESTRATION IN SOIL

M. De Nobili and M. Contin

Dipartimento di Scienze Agrarie e Ambientali, University of Udine, Udine, Italy

Y. Chen

Department of Soil and Water Sciences, Faculty of Agricultural, Food, and Environmental Quality Sciences, The Hebrew University of Jerusalem, Rehovot, Israel

5.1. INTRODUCTION

Panels on climate change have underscored the need for drastically improving the management of our agricultural resources to address potential impacts around the globe. While the impact of climate change will be positive in some areas, such as those that will gain longer growing seasons, other areas will be adversely impacted and will required adoption of improved soil and water management practices.

Biophysico-Chemical Processes Involving Natural Nonliving Organic Matter in Environmental Systems, Edited by Nicola Senesi, Baoshan Xing, and Pan Ming Huang

Globally, soil and water scientists should be encouraged to conduct further research into how we should adapt to effectively manage plant, soil, and water resources. It is a critical task of scientists in these disciplines to employ their knowledge to develop methods for reducing the negative impacts of climate change on the soil resulting from climate change.

Deforestation, drainage of wetlands, and, in general, conversion of natural ecosystems to agricultural use have contributed over the last 250 years to about 30% of the total anthropogenic emissions of C to the atmosphere. Soil organic C (SOC) pool (2500 Pg) is the second largest global C pool after the oceanic pool (38,000 Pg), and it stocks more than three times the amount of atmospheric C (750 Pg) and about 5 times the C stored in living biomass. The SOC pool is relatively low in arid sandy soils (30 Mg ha^{-1}) but generally ranges from 50 to 150 Mg ha^{-1} (Lal et al., 2004).

A large fraction of the CO_2 emitted from soil is derived directly from mineralization of stocked soil organic C (SOC) and can be attributed to agricultural management practices.

Dynamics of the SOC pool are not completely understood, yet they are key to understanding why accumulation of CO_2 in the atmosphere is actually proceeding at a much slower rate than predicted by models on the basis of fossil fuel burning and deforestation (IPCC, 2001). Estimates of the current net uptake of C by the terrestrial biosphere in the northern hemisphere have identified the existence of a large (1–2 Pg C yr^{-1}) terrestrial C sink (IPCC, 2001; Nabuurs, 2004; Ciais et al., 1995). For North America and Europe, the terrestrial C sink has been estimated to amount, respectively, to 0.3–0.6 Pg C yr^{-1} (Pacala et al., 2001) and 0.1–0.2 Pg C yr^{-1} (Janssens et al., 2005). If Europe were to maintain its current forest and grassland sink and stop all C losses from arable and peat soils, the terrestrial SOC sink alone would absorb 16% of the European C emissions from fossil fuel consumption (Freibauer et al., 2004).

Soil organic matter (SOM) decomposition could also be the agent of a feedback mechanism that could further enhance the warming trend of the planet (Cox et al., 2000). Under a warmer climate, thawing of high-latitude permafrost regions may result in large releases of CO_2 to the atmosphere (Goulden et al., 1998; Oelke et al., 2004). Furthermore, changes to massive soil drainage due to permafrost melting may have a large impact on the C stored in high-latitude peatlands (Bubier et al., 2003; Lafleur et al., 2003) and may significantly contribute to the climate–carbon cycle feedback (Schimel et al., 1994).

The additional release of CO_2 from SOM mineralization from 1991 to 2051, calculated on the basis of a 0.003 °C yr^{-1} increase in temperature, amounts to 61 Pg C and is equivalent to 19% of that released from fossil fuel combustion assuming unabated use. It is therefore important to quantify precisely this contribution. Uncertainties in estimation of the contribution of this feedback mechanism depend on changes in the distribution pattern and intensity of precipitation, but also on the behavior of the more recalcitrant fractions of SOM (Jenkinson et al., 1991). Better knowledge of the factors that affect decomposition of organic matter (OM) in soil and eventually control rates at which different fractions decompose is urgently required and would be of immediate practical importance.

According to the U.N. Framework Convention on Climatic Change, total worldwide CO_2 emission amounts to 428,941 Gg yr^{-1} and the 10% reduction required by the Kyoto Protocol would correspond to 11,698 Gg C yr^{-1}. Lal (2002b) estimated the global potential for SOC sequestration by adopting recommended agricultural practices on croplands and restoring desertified and degraded ecosystems to be

- 0.7–0.9 Pg C yr^{-1} in cropland soils (Lal and Bruce, 1999)
- 0.9–1.9 Pg C yr^{-1} in desertified lands (Lal et al., 1999)
- 3.0 Pg C yr^{-1} in degraded lands (Lal, 1997)

The required reduction therefore amounts to only a tiny fraction of the theoretical sequestration potential of the world soils. Although evaluation and certification of emission credits for sequestration of C in the terrestrial biosphere is certainly difficult (Marland et al., 2001), it is worthwhile to consider C sequestration in developing possible mitigation plans.

An example of a European country as a case study is given below: Cultivated land in Italy is about 15×10^6 ha; because agricultural soils in this country contain about 1.5% C (7.5×10^7 g C ha^{-1}), the organic C stored as SOM in agricultural soils corresponds to 450,000 Gg for the whole country. Thus an annual increase in SOC storage 0.011% would account for all the required emission reductions for the country. It is obvious that this cannot be the only approach for addressing emission reduction targets, yet such calculations help to point out that C sequestration in soil and climate change feedback mechanisms affecting SOM decomposition are worth increased attention by scientists and decision makers. It is remarkable that, in its present form, the Kyoto protocol does not offer sufficient protection to the large terrestrial C pools.

Soil C sequestration can operationally be defined as the result of the combination of biotic and abiotic natural processes that transfer atmospheric C, first by way of photosynthetic fixation of CO_2 into plant or autotrophic microbial biomass and then into SOM through complex immobilization mechanisms acting on the products of heterotrophic decomposition of this biomass in soil. These processes are the ultimate result of the activity of soil biota, a large and well-adapted biological community, ranging from small mammals and arthropods to microbial predators and microflora. Steady-state levels of C sequestration in soil result from the dynamic balance between the soil C inputs and the mineralization rate supported by the soil biota. The SOC status depends on climatic factors such as precipitation and temperature, oxygen availability, and so on, that regulate both net primary production and activity of soil organisms. Subordinately, pedogenic factors, such as the nature and content of clay minerals, also affect OM stabilization in soil (Figure 5.1). All of these processes control not only the quantity but also the quality of SOM and its potential resistance to decomposition.

The term SOM generally encompasses all the organic components present in the soil including living organisms (Vaughan and Ord, 1985). This broad definition causes a number of difficulties, but more restrictive definitions are not devoid of problems. Stevenson (1994) defines SOM in a way that is similar to a definition suggested earlier by Waksman (1938) for "humus." This definition excludes undecayed or only partially decomposed plant material and tissue, as well as living organisms. Although this definition may seem more rational, in practice SOM defined in this manner is very difficult to analyze either quantitatively or qualitatively because soil microbial biomass, microscopic plant, and root debris cannot be reliably separated from the soil and are currently "analyzed" as total organic C.

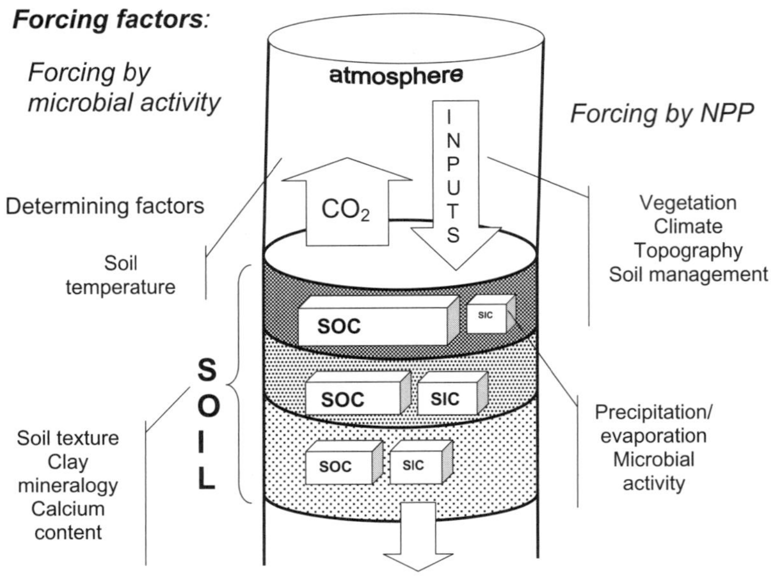

Figure 5.1. Fluxes of C in and out of soil and their forcing factors (SOC, soil organic carbon; SIC, soil inorganic carbon; NPP, net primary production).

The old word "humus" itself has been used in soil science in an often arbitrary and poorly defined way. It is generally agreed that SOM can be divided into nonhumic and humic substances (HS) (Stevenson, 1994). The nonhumic materials comprise organic substances that have defined chemical structures, such as carbohydrates, hydrocarbons, alcohols, aldehydes, resins, and amino acids as well as aliphatic and aromatic acids. Humic substances are largely heterogeneous, and their chemical structure is not sufficiently known. They are comprised of yellow- to black-colored polyphenolic polycarboxilic acids exhibiting a multidispersive array of molecular weights. Yet, their functional groups and reactivity were described in great detail.

Abiotic processes have an important role in SOC sequestration, yet their impact is either limited or dependent on the mechanical action of detritivores (see Section 5.1.2). The total soil C pool contains soil inorganic C (SIC) present as primary and secondary carbonates. The latter are formed by the dissolution of CO_2 in the soil solution and its reaction with dissolved Ca^{2+} and Mg^{2+} (Lal and Kimble, 2000). This process leads to accumulation of inorganic C only in soils of arid and semiarid regions; and the rate of SIC sequestration is low, ranging from 5 to 15 kg C ha^{-1} yr^{-1}. However, where precipitation exceeds the soil's water holding capacity, inorganic carbonates can be leached to groundwater and eventually transferred into the relatively inert geological pool.

The soil atmosphere usually contains relatively high CO_2 concentrations, often reaching 100 times that present in the air above the soil. These high levels result from respiration by plant roots and heterotrophic organisms, and they greatly increase the concentration of CO_2 of the soil solution. This biologically mediated C sequestration in an inorganic pool is likely to be more pronounced in cool rainy climates, yet it has never been thoroughly investigated. These processes deserve more research attention by scientists.

5.1.1. Potential and Attainable Carbon Sequestration

Ingram and Fernandes (2001) defined potential soil C sequestration as the theoretical maximum C storage capacity of a given soil and attainable C sequestration and as the level of sequestration that can actually be achieved. The latter is determined by environmental and pedogenic factors limiting soil C inputs. Attainable C sequestration, however, does not just depend on input levels. As shown in Table 5.1, the dependence of the SOC pool on net primary production (NPP) is bimodal: Attainable sequestration is the combined result of the contrasting effects of factors which control organic C inputs on one side and C decomposition and mineralization on the other. The role of climatic differences in SOC dynamics can be recognized only for relatively homogeneous climatic regions. For example, in the temperate forest soils of Minnesota, Wisconsin and Michigan, SOC increases with mean annual precipitation (Grigal and Ohmann, 1992), and across the Great Plains grassland SOC is positively correlated with annual precipitation and negatively with mean annual temperature (Burke et al., 1989).

Site variables such as topography, soil texture, drainage and slope are non-climate factors considered to be responsible for about 50% of the variation in SOC in grassland and cropland soils (Burke et al., 1989) and for up to 65% of the variation in upland forest soils (Grigal and Ohmann, 1992). In a catena, SOC accumulation can be higher at the summit and footslope positions compared to soils in the backslope and shoulder positions which can be strongly eroded. Drainage affects SOC accumulation by determining the persistence of anaerobic conditions, which in turn slow SOM decomposition and virtually stop decomposition of lignin. Organic soils, formed under anoxic conditions, can attain a SOC pool of 800 Mg ha^{-1} (Lal, 2004) even in warm climates, because of year-round saturation.

Soil texture, especially clay content, has a significant influence on C sequestration (Parton et al., 1987; Burke et al., 1989; Beker-Heidmann and Scharpenseel, 1992; Schimel et al., 1994) by promoting the formation of physically stabilized and chemically stabilized SOC (Parton et al., 1994) and by controlling soil hydrologic properties (Schimel et al., 1994). The significance of the effects of individual site variables is in the order of soil taxon>drainage>texture>slope>elevation (Tan et al., 2004).

TABLE 5.1. Soil C Balance at Equilibrium in Different Ecosystems

Ecosystem	Net Primary Production ($tC ha^{-1} yr^{-1}$)	Annual Soil C Input ($tC ha^{-1} yr^{-1}$)	Soil Organic C ($tC ha^{-1}$)
Continuous wheat, unfertilized	2.6	1.2	26
Continuous wheat, fertilized	5.1	1.9	30
Continuous hay, unfertilized	2.7–3.2	2.0–2.5	77
Native prairie	2.8	1.7	52
Humid savannah	5.0	1.5	56
Sub-humid savannah	1.4	0.5	17
Moist tropical forest	9–10	4.9	44
Cold temperate beech forest	7.1	2.4	72

Source: Reprinted with permission from Jenkinson, D. S. (1981). *Chemistry of Soil Processes*, John Wiley & Sons Ltd.

5.1.2. Organic Matter Decomposition in Soil: The Forcing Factors

In natural environments, organic C inputs enter the soil through leaching of soluble components of litter residues, by rhizodeposition and by the mixing action of invertebrates such as earthworms, insects at various stages of their life cycle, and other arthropods that promote interaction of decaying organic materials with mineral constituents. Soil fauna has an important role in enhancing the contact of organic residues and their decay products with inorganic and organic soil colloids, and therefore it helps to physically stabilize SOM. The activity of detritivores, in particular, is important to the formation of organo-mineral complexes as ingested soil undergoes many alterations including physical realignment of clay particles (Wolters, 2000). Earthworms play an important role in protecting organic C from decay by helping the formation of stable soil aggregates that can contain particulate OM (POM) derived from freshly incorporated plant residue (Bossuyt et al., 2004).

Mineralization of organic residues in soil is mainly carried out by an extremely diverse heterotrophic community referred to as the soil microbial biomass. The soil environment is a rather peculiar natural environment for the growth of microorganisms, in that they have had to adapt to quite extreme growth-limiting factors: (a) discontinuous availability of substrates and water and (b) high variability of soil chemical properties (pH, temperature, oxygen supply) that can vary in the soil environment on both the micro and macro scales (Jenkinson and Ladd, 1981).

The surprising feature of the soil microbial biomass is that its characteristics and general behavior are remarkably similar over widely different pedo-climatic environments. For example, decomposition of ^{14}C-labeled rye grass in soil and the consequent formation of ^{14}C-labeled soil microbial biomass showed no differences between an English soil and a tropical rain forest soil from Nigeria when incubated at optimum moisture and temperature conditions (Jenkinson and Anayaba, 1977).

Gunapala et al. (1998) found minimal differences in the ability of the organisms in soils under long-term conventional or organic management to decompose organic residues. This can be explained considering that the soil microbial biomass maintains in all soils an ATP concentration of 10–12 μmol ATP g^{-1} biomass C (Jenkinson, 1988; Contin et al., 2001) and a high adenylate energy charge (AEC) (0.8–0.95) that are typical of exponentially growing microrganisms in vitro (Brookes et al., 1983). It is therefore immediately able to activate itself and decompose substrates as soon as they become available. Usually only a small fraction of the soil microbial biomass will actually be active at any time, and biomass turnover times are very slow, approximately 1.5 years (Anderson and Domsch, 1985; Jenkinson and Ladd, 1981). These characteristics are almost certainly an evolutionary response to the relatively small annual (typically slightly more than twice the amount of biomass C per year) and discontinuous substrate inputs to most soils. An elevated AEC allows microorganisms to readily activate transport mechanisms whenever substrates become available. This explains the rapid response of the soil microbial biomass to soil rewetting or disturbance by tillage, which results in immediate large bursts of CO_2 emissions.

Mineralization of SOM can be accelerated or retarded by the addition of organic substrates to soil, an effect known as "priming" due to exploitation by microorganisms of SOM otherwise not available or to changes in community composition. The activation of microrganisms through easily available substrates is considered to be the main reason for the occurrence of positive priming effects in soils. One possible mechanism, as reviewed by Kuzyakov et al. (2000), is co-metabolism—that is, enhanced SOM degradation due to microbial growth and the accompanying increased enzyme production. Another explanation is the "trigger molecules hypothesis" (De Nobili et al., 2001). According to their hypothesis, a shift from a dormant to an active state is initiated in microrganisms by cells sensing molecular signals or "trigger molecules." These are probably low-molecular-weight soluble compounds diffusing from substrates entering the soil: Additions of trace amounts of sugars and amino acids have been found to increase rates of SOM mineralization (De Nobili et al., 2001; Hamer and Marshner, 2005). CO_2 itself could also be a possible trigger substance (Insam et al., 1999).

An increased influx of labile carbon to soil may stimulate microbial degradation of SOM. Carney et al. (2007) showed that, in a scrub-oak ecosystem, 6 years of experimental CO_2 doubling reduced soil carbon despite higher plant growth, offsetting 52% of the additional carbon that had accumulated under the elevated CO_2 treatment in aboveground and coarse root biomass. The decline in soil carbon was driven by changes in soil microbial composition and activity. A substantial portion of the "extra" carbon fixed by plants at elevated CO_2 and deposited to soils through increased leaf litter, root exudates, or root turnover is labile and rapidly metabolized by microbial communities (Pendall et al., 2004). Soils exposed to elevated CO_2 had higher relative abundances of fungi and higher activities of a carbon degrading enzyme, which led to more rapid rates of soil organic matter degradation than in soils exposed to ambient CO_2. This points out to a dangerous possible feedback mechanism, forced by the present and future increase trend in atmospheric CO_2 concentrations that could lead to enhanced SOM mineralization and CO_2 emission from soil.

In temperate climates, the microbial biomass C is normally 1–3% of SOC, ranging on average from 180 kg C ha^{-1} in arable soils to 2200 kg C ha^{-1} in woodland and grassland soils, and decreases with mean annual soil temperature. Insam (1990) reported that the microbial C-to-SOC ratio is largest in arid soils and decreases with increasing precipitation, reaching a minimum in soils of balanced precipitation and evaporation. In general, the microbial biomass C in soils under similar geographic conditions is larger in soils of larger SOC content. The microbial C-to-SOC ratio increases with C inputs, so that this parameter can be considered an index of C accumulation (Powlson et al., 1987) as the increase is detectable years before any measurable increase in SOC. Any increase in soil microbial biomass is obviously accompanied by a proportional increase in SOM decomposition and CO_2 emission, so that the rate of SOC accumulation decreases with time (Dilly et al., 2005). Therefore, soils cannot accumulate organic C indefinitely, but they will eventually reach equilibrium conditions when the annual mineralization rate equals the amount of organic C entering the soil each year.

5.2. PROCESSES ENHANCING CARBON SEQUESTRATION IN SOIL

One can hardly report on C sequestration in soils without a mention of Brazil as a source of CO_2 to the atmosphere due to extensive transformations of forests to cropland, which began in the 1970s and still continues. The C stocks have been reduced in Brazilian soils exposed to intensive tillage and reduced inputs. Recently, however, farmers have rapidly adopted no-till practices, reaching 80% of total cropland last year, which have partially restored soil C levels and reduced fuel consumption. Long-term experiments conducted by Amado et al. (2006) to assess the potential of C accumulation in these croplands found that summer legume cover crops, such as pigeon pea and velvet bean in maize cropping systems, showed the highest C accumulation rates (0.38–0.59 $Mg\,ha^{-1}\,yr^{-1}$). The inclusion of these intensive cropping systems also increased the C accumulation rates in no-till soils (0.25–0.34 $Mg\,ha^{-1}\,yr^{-1}$) when compared to the double-crop system used by farmers. Overall, these results are encouraging since they show the results of adoption of conservation management practices in countries with a huge soil C-sequestration potential.

Reforestation has often been indicated as the only effective way to increase terrestrial C sequestration due to the large contribution of the standing wooden biomass (Fan et al., 1998) and the attention of researchers on soil/forest interactions has increased (Evans and Ehleringer, 1994). The attainable C sequestration potential of forest soils depends on the complex interactions between the vegetation and the soil on which reforestation takes place. The appropriate match of tree type and site is obviously the first condition for successful reforestation and its importance has been recognized since the beginning of forestry. Several other aspects of forestry management, such as sustainability of yields under monoculture, adverse effects of clear-felling, and replanting, are also essential.

A recent study by Woodbury et al. (2006) dealt with the conversion of forest into cropland and vice versa. Their basic hypothesis was that converting forests into cropland causes a rapid loss of C from the soil and forest floor, and converting cropland into forest causes a slow gain of C. These investigators developed a model aimed at the evaluation of soil C changes throughout the southern United States from 1900 until 2050. From 1990 to 2004, they found that afforestation caused sequestration of 88 Tg C in the soil and forest floor, and deforestation caused emissions of 49 Tg C. The net effect of previous land-use change on C stocks in soil and forest floor during this period was about six times smaller than the net change in C stocks in trees on all forestland. Thus, they concluded that land-use change effects and forest C cycling during this period were dominated by changes in tree stocks.

Afforestation of up to 30% of present arable land would, over the next century, increase soil C stocks by about 8%, yet would contribute to C mitigation only for 0.8% of annual global anthropogenic CO_2–C emissions (Smith et al., 1997). The potential is therefore apparently small, as compared to a direct reduction in anthropogenic emissions and fossil fuel burning. However, considering the overall costs and benefits of environmental services of sequestering C and N to mitigate air and water pollution, Sparling et al. (2006) demonstrated that the net present value of SOM calculated over recovery periods of 36–125 years was 42–73 times higher than the costs associated with lower productivity. The same authors and others (Pretty et al., 2001) suggest that if additional direct and/or indirect effects of SOM retention, such as erosion control and flood prevention, are to be included in the calculations, the environmental value of SOM will be much greater than that presented by Sparling et al. (2006).

Organic C inputs to soils mainly consist of plant residues that all contain the same classes of organic compounds such as cellulose, hemi-cellulose, starch, proteins, lipids, and polyphenols. Their proportions, which depend on the species and maturity stage of plants, may influence the rate of decomposition. Martens (2000) found greater losses of C from residues with lower amounts of phenolic acids and less C loss from residues higher in phenolic acids. The organic C remaining in the treated soils after about three months was significantly correlated with the phenolic acid content at day 29 and the phenolic acid content and C/N ratio. This indicates that residues with higher amounts of phenolic acids result in higher levels of C retained in the soil.

Sparingly soluble polymethylenic molecules, such as lipids and waxes and polymers such as cutin and suberin (Derenne and Largeau, 2001), are generally considered among the most recalcitrant components, but the most abundant recalcitrant compound in plants is certainly lignin. Lignin contains no hydrolytic bonds but only aliphatic, alcylaryl, and biaryl bonds as well as aromatic rings; and, due to its relative structural complexity, it is not easily degradable. Therefore, it accumulates during the initial phase of degradation of plant residues (Kalbitz et al., 2003a,b). Waksman (1938) concluded that stable humus compounds are formed predominately from partially decomposed lignin fragments. This selective preservation concept was questioned by several authors (O'Brien and Stout, 1978; Volkoff and Cerri, 1987; Nadelhoffer and Fry, 1994; Melillo et al., 1989) on the basis that ^{13}C values normally increase with soil depth compared with the litter. Lignin components and also fats and waxes, are depleted in ^{13}C relative to bulk plant tissues. Selective preservation of these components should thus cause a decrease in ^{13}C as the residue degrades rather than the observed increase.

Although the rate at which components of plant and animal residues are decomposed by the soil microbial biomass varies widely (Stout et al., 1981), none of the classes of naturally produced organic compounds persist in the soil indefinitely as there are always species or a succession of species that can degrade them. Jenkinson and Ladd (1981) pointed out that if it were not so, the completely recalcitrant SOM fractions would accumulate indefinitely in the soil and by now would cover the surface of the earth.

Black C, produced by wild fires and humic substances (HS), the natural by products of SOM decomposition in soil and water systems, are certainly the classes of organic compounds that most closely approximate this recalcitrant behavior. HS occur widely, being found in large amounts not only in the soil and sediments but also in lakes, rivers, ground waters, and even the open ocean (Stevenson, 1994). Besides these relatively refractory substances, more labile compounds can persist in soil for a much longer time than would be predicted from their inherent recalcitrance to decomposition. SOM stabilization (Figure 5.2) is generally considered to occur by three main mechanisms: (i) physical protection, (ii) chemical stabilization, and (iii) biochemical stabilization (Six et al., 2002).

Physical protection is exerted by occlusion of particulate organic matter (POM) inside aggregates. It is responsible for the physical separation of organisms active in decomposition and substrates, reduced oxygen availability in the substrate compartment, and reduced biomass turnover through protection from microbial grazers (Mamilov and Dilly, 2002).

Chemical stabilization of SOM occurs as a result of chemical or physico-chemical binding to soil mineral surfaces (Polubesova et al., 2008). Reversible sorption of labile substrates decreases their concentration in the soil solution and slows

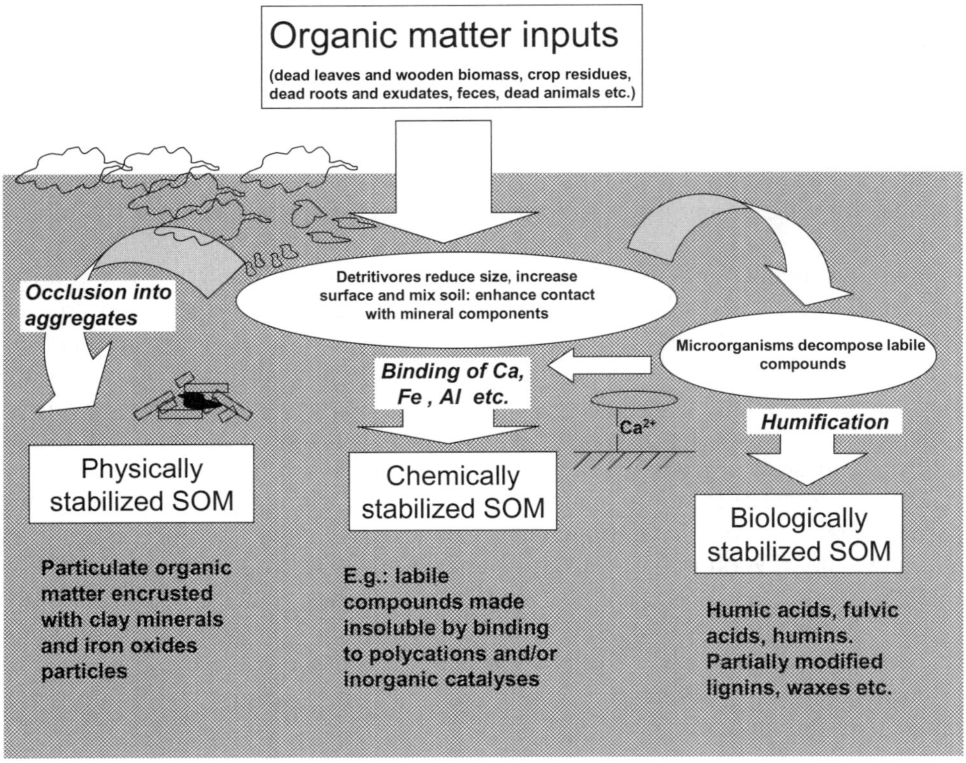

Figure 5.2. Mechanisms of C sequestration in soil.

decomposition. This mechanism accounts for the direct relationship often observed between soil silt plus clay content and amount of silt plus clay protected soil C (Six et al., 2002; Hassink et al., 1997), and for the lower CO_2 evolution observed in clayey soils compared with sandy soils after addition of substrates (Feller and Beare, 1997).

The term biochemical stabilization refers to the biotic or abiotic production of organic substances that are refractory to decomposition by microorganisms and contribute, through condensation and complex formation, to the stabilization of otherwise easily decomposable substrates such as enzymes. This stabilization process coincides with the process of humification.

5.2.1. Physical Protection

Physical stabilization of SOM has been extensively investigated and several exhaustive reviews can be found in the literature concerning its role in C sequestration (Oades and Waters, 1991; Angers and Carter, 1996; Christensen, 1996; Baldock and Skjemstad, 2000) and is also demonstrated in Figure 5.3.

Formation of aggregates, which allows inclusion of particulate organic matter (POM), thereby making it inaccessible to decomposing microorganisms, is a fundamental process in C sequestration. Besides the action of the soil macrofauna (already mentioned in Section 5.2), which aids in aggregate formation by reducing the size

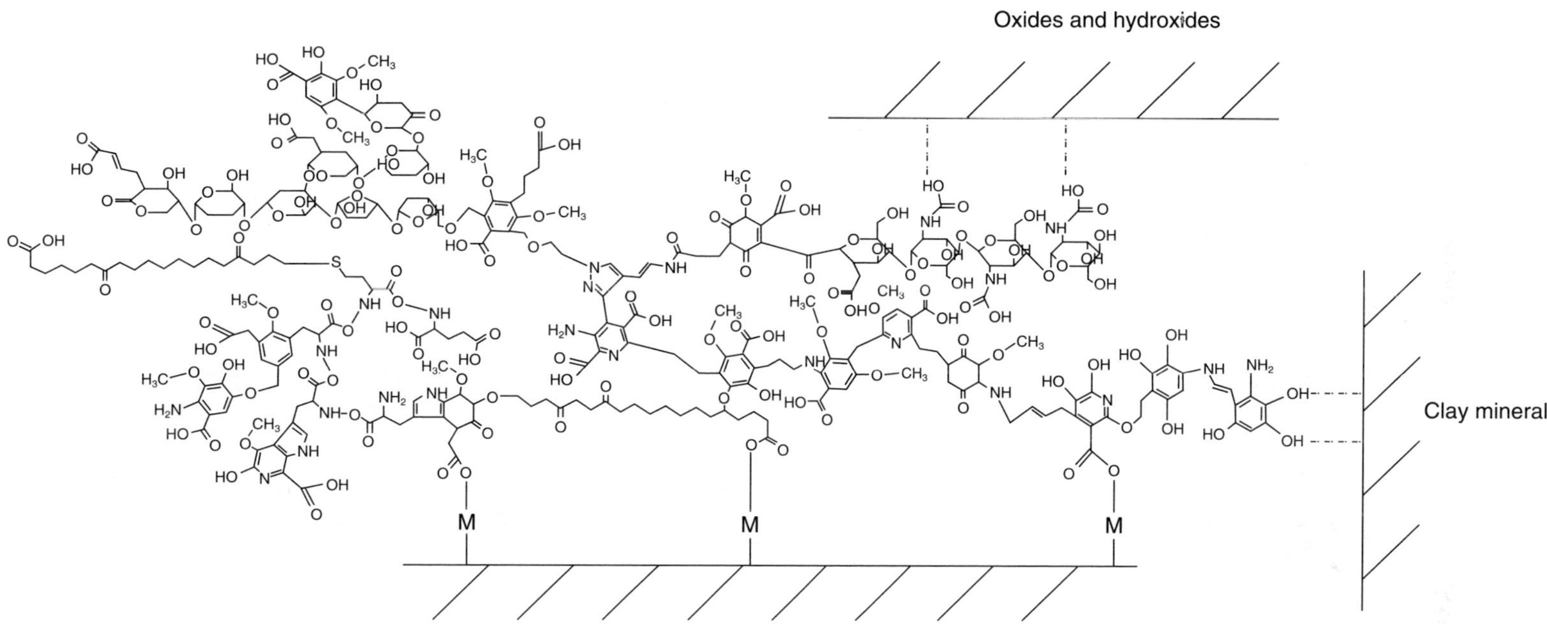

Figure 5.3. A humic acid macromolecule interacting with a surface of a clay mineral. The proposed macromolecular structure of the soil humic acid (HA) is based on the following common average characteristics of humic acids: MW: 6386 Da; elemental analysis (%): C, 53.9; N, 5.0; H, 5.8; O, 35.1; S, 0.5; C/N, 10.7; NMR information (%): aliphatic C, 18.1; aromatic C, 20.9; carbohydrate C, 23.7; metoxy C, 4.9; carboxylic C, 8.4; ketone C, 4.5; phenolic C, 4.2; functional groups (cmol/g): carboxyl, 376; phenol, 188; total acidity, 564. The structure was created using the ACD/ChemSketch program. [HA–clay complex: Chen's group, unpublished (2008). Individual HA molecule: Grinhut et al., 2007.]

of POM and mixing it with inorganic components (Bossuyt et al., 2004), the soil microbial biomass assists in the formation of smaller aggregates (2–20 μm) producing exopolysaccharides (EPS) that bind clay platelets together.

A high clay content inhibits soil respiration (Schimel et al., 1994; Telles et al., 2003). Ladd et al. (1985) monitored for 8 years the mineralization of ^{14}C-labeled plant residues added to four cultivated soils having similar mineralogies but different clay contents (5–42%). The amounts of residual labeled plant C and residual native soil organic C, remaining at the end of the study, were proportional to soil clay content. The reason for this is still largely unexplained, but two main causes contribute to the larger C sequestration potential of clayey soils: One is the pore size distribution and the other is the large specific surface area of clays. The pore size distribution of a soil affects the possibility of decomposer organisms to reach potential organic substrates. Bacteria can only enter pores >3 μm (Kilbertus, 1980). Within pore sizes less than this lower limit, decomposition of SOM can only occur by the action of extracellular enzymes, followed by diffusion of the products of enzyme reactions out of the pores. With increasing clay content, the proportion of small pores out of the total porosity increases, and therefore the potential stabilization of OM against biological attack due to the exclusion of decomposer organisms, increases. Predation of microorganisms by soil fauna is also pore size limited: van der Linden et al. (1989) showed that protozoa and nematodes are, respectively, excluded from pores <5 and <30 μm. Thus, SOM residing in pores smaller than these diameters in the form of molecules, small particles, or bacterial or fungal tissues will not be susceptible to decomposition or predation by soil fauna. Killham et al. (1983) demonstrated that when glucose was placed in pores <6 μm, its turnover was slower than when placed in pores <30 μm.

In a study aimed to quantify the relationship between soil texture and C content of temperate and tropical mineral soils, and working on the hypothesis that the amounts of C that can be associated with clay and silt is limited, Hassink et al. (1997) observed a close relationship between the proportion of silt and clay particles (<20 μm) and the SOC associated with this fraction in the top 10 cm of soil. Cultivation decreased the amount of SOC in the >20-μm fraction more than in the <20-μm fraction, indicating SOC associated with the smaller particles is better protected against decomposition. Hassink et al. (1997) also reported that, as the upper limit for the adsorption of organic inputs to clay and silt is reached, increasing C inputs did not lead to any further increase of association with mineral particles. Once the microaggregates are saturated with SOM, further additions are found mainly in the sand-sized macroorganic matter fraction (Carter, 2002).

Rühlmann (1999) pointed out that the organic C content of long-term bare fallow soils could be used as an indicator of the size of the stable C pool, as the active pool is by force reduced to a minimum. He found a strong relationship between %C of long-term bare fallow soils and the percentage of soil particles below 20 μm and argued that the amount of C in the stable pool could be defined as the capacity of soils to sorb C. Binding of a model humic acid molecule to a clay particle, along with its stabilization against decomposition due to sorption, is demonstrated in Figure 5.3 The amount of C associated with mineral particles showed both a lower and an upper limit, which is in agreement with the observed increase in free HA during C accumulation.

5.2.2. Chemicophysical Stabilization

Chemical stabilization of SOM against microbial decomposition in the presence of Ca^{2+} ions was demonstrated long ago by Sokoloff (1938), who compared the mineralization of soil organic C in soil after the addition of different Ca^{2+} or Na^+ salts. His experiments showed that soluble soil C and mineralizable soil C strongly decreased upon addition of $CaSO_4$ and $CaCl_2$, whereas Na_2SO_4 and NaCl slightly enhanced both C solubilization and C mineralization. Later studies showed that the lower CO_2 evolution observed after addition of Ca^{2+} ions depends on the formation of either inner-sphere or outer-sphere complexes with ionizable decomposition products. The formation of $CaCO_3$ precipitates could also reduce CO_2 evolution. Although ^{14}C-labeled glucose disappeared at about the same rate from a $CaSO_4$ amended soil, a larger amount of glucose-derived ^{14}C persisted in the $CaSO_4$-amended soil for more than 3 months (Baldock and Oades, 1989).

Polyvalent cations such as Ca^{2+}, Al^{3+}, or Fe^{3+} display stabilizing effects in soil (Blaser et al., 1997; Lundstrom et al., 2000), but relatively little is known about the actual mechanism of stabilization, apart from reduced availability attributed to reduction in solubility. Contrary to Ca^{2+}, Al^{3+} and Fe^{3+} can form strong coordination complexes with SOM, and in particular with HS. Long-term incubation experiments showed the Al/C ratio of dissolved organic matter (DOM) is correlated with the half-life of natural DOM (Schwesig et al., 2003). In the case of Al^{3+}, the lower decomposition rates measured may also reflect the effect of a direct toxicity of Al^{3+} to the decomposing microorganisms. A toxic effect is not likely to take place, however, in the case of Fe^{3+}. Pollution by toxic metals can indeed cause reduced foliar litter decomposition and SOM accumulation (Johnson and Hale, 2004), although chronic stress due to heavy metal exposure at lower levels can cause enhanced CO_2 evolution by soil microorganisms (Chander and Brookes, 1991). Divalent cations such as Cu^{2+}, Zn^{2+}, Pb^{2+}, Ni^{2+}, and so on, can cause a reduction in the availability of DOM acting similarly to Ca^{2+} and other polycations (Martin et al., 1966). However, if present in large concentrations, these elements are certainly toxic for the soil microbial biomass.

Soil acidity negatively affects SOM decomposition in several ways. In a study on C transformations during decomposition of plant leaves in soil, Webster et al. (2000) investigated effects of lime additions. They found that soils from limed and unlimed plots contained similar amounts of C (47% and 48% by weight, respectively) and the C/N ratios were 28 and 23, respectively. The respiration rate over 28 d and biomass-C were significantly greater ($P < 0.01$) for the limed than for the unlimed soil. The increase in respiration rate due to liming was smaller than that for biomass C, implying that the respiration rate per unit of biomass (qCO_2) was substantially smaller in the limed (0.22 mmol $CO_2\,mg^{-1}$ biomass C h^{-1}) than in the unlimed soil (2.8 mmol $CO_2\,mg^{-1}$ biomass C h^{-1}). Liming increased the size and activity of the microbial community, and this effect remained detectable 15 years after the amendment. The fact that qCO_2 was smaller in the limed soil suggests that the microbial community used less C catabolically. Consequently, in the limed soil the microorganisms were better adapted to convert a larger proportion of C to biomass.

Kalbitz et al. (2005) measured decomposition of DOM extracted from soil and litter and observed that the organic C mineralized during incubation of sorbed compounds was only about 15–30% of that mineralized in solution. They estimated that sorption increased the mean residence time of the most stable DOM sample from 28 years in solution to 91 years. For highly degradable samples, the portion of residual C nearly doubled upon sorption while with less degradable DOM the stability increased by only 20%. Therefore, the increase in stability due to sorption is larger for labile DOM high in carbohydrates and relatively small for stable DOM high in aromatic and complex molecules. Nevertheless, resistance to decomposition after sorption followed the same order as in solution, and the extent of sorption of recalcitrant compounds was much larger than sorption of labile compounds. The UV, fluorescence, and ^{13}C measurements indicated that aromatic and complex compounds, probably derived from lignin, were preferentially stabilized by sorption of DOM. Thus, the overall sorptive stabilization of stable DOM was four times larger than for the labile DOM. Kalbitz et al. (2005) concluded that stabilization of DOM by sorption depends on the intrinsic stability of organic compounds sorbed.

A particular case of chemico-physical stabilization is the occlusion of C in phytoliths. Phytoliths, also referred to as plant opal, are formed by silica deposition within plant tissues: in cell walls, often replicating the morphology of the living cells, as infillings of the cell lumen, and inside intercellular spaces of the cortex (Piperno, 1988). The C occluded in phytoliths is highly resistant to oxidation (Wilding et al., 1967). Herbaceous plants are generally considered the most prolific producers of phytoliths (Krishnan et al., 2000; Parr et al., 2001). Long-term phytolith accumulation rates under grasslands are commonly 5–10 times greater than under forests (Drees et al., 1989). The rate of phytolith production in a given soil is affected by the monosilicic acid concentration in the soil solution as well as by climate and geomorphology (Drees et al., 1989). These investigators found that the concentration of phytoliths in soil varied by several orders of magnitude: from 8 to 10 $kg\,ha^{-1}\,yr^{-1}$ in New Mexico (Pease and Anderson, 1969) to 300 $kg\,ha^{-1}\,yr^{-1}$ in Oregon (Norgren, 1973).

Although the concentration of phytoliths in soils is generally below 3% on a total soil basis (Drees et al., 1989), some soil horizons are almost completely composed of phytoliths (Riquier, 1960). For example, the phytolyth C produced by a sugarcane crop is comparable (i.e., 18.1 $g\,C\,m^{-2}\,yr^{-1}$) to the short-term rates of C sequestration achievable by land use or tillage changes (Post and Kwon, 2000; West and Post, 2002). These data clearly demonstrate the option of enhancing both short- and long-term C sequestration by cultivation of high phytolith yielding plant species (Parr and Sullivan, 2005).

Stabilization of biomolecules can also result from abiotic humification catalyzed by mineral colloids (see Chapter 2).

5.2.3. Biochemical Stabilization

The humic constituents of SOM are usually regarded as the primary resistant compounds (Stout et al., 1981). In spite of the fact that the accumulation of C in soil is not indefinite even in natural ecosystems, it is certainly true that HS have been accumulating on the surface of the earth since the appearance of life. They now make up a considerable fraction of the soil organic C pool: The amount of C stored as HS is 60×10^{17} g, and it exceeds that which occurs in living organisms (Stevenson, 1994). ^{14}C dating combined with isotope enrichment techniques have been used to measure the relative age and stability of humic fractions (Goh and Pullar, 1977). The range of radiocarbon ages of SOM fractions is a reflection of their relative stability. Campbell et al. (1967) showed total ^{14}C dates of SOM varied widely not only from soil to soil, but also in the SOM fractions from the same surface soil, with HAs being the oldest extractable fraction. The age of the HA fraction increases if SOM is hydrolyzed with 6 M HCl to separate hydrolysable sugars and peptide residues included in the HA molecule. However, in spite of being the most widespread and abundant organic C compounds on earth, the structures of HS are still largely unknown. An advanced attempt to present a logical model of a HA molecule is shown in Figure 5.3. Most of the known chemical and physicochemical properties of HA were considered when this model was proposed (Grinhut et al., 2007). HS are only operationally defined as "a category of naturally occurring, biogenic, heterogeneous organic substances that can generally be characterized as being yellow to black in colour, of high molecular weight and refractory" (Aiken, 1985). This statement is still very far from a proper definition and reflects, in its generality, the nonspecificity that has affected the study of HS: Even now, there is still uncertainty and argument concerning the actual molecular weight range of HS (Swift, 1999). Also the processes involved in the conversion of SOM to HS is still very poorly understood. No single theory is adequate to describe the complex reactions leading to accumulation of HS, which are evidently the product of mixed biological and random chemical condensation reactions of a wide range of plant, animal, and microbial components and of their intermediate decomposition products (Tate, 1992; Stevenson, 1994).

Humic and fulvic acids (HA and FA, respectively) are differentiated on the basis of operational definitions: Both fractions include not only the aromatic components but also a variety of plant components. Thus, in reality, HA and FA are extremely heterogeneous fractions.

FA can either be precursors of HA or be formed during their partial decomposition. Many papers can be cited to support either possibility. This points out the fact that more than one of several possible pathways take place concomitantly, although different conditions may lead one or the other to predominate.

According to the classic lignin theory, which was thought for a long time to be the main pathway of HS formation in soil, HS form from the condensation of proteins with modified lignin residues, probably via the formation of a Schiff base through the reaction of an amino group of the protein with an aldhehyde group of the modified lignin. During decomposition, lignin undergoes structural changes that progressively cause resemblance of HS to partially oxidized lignin. The first step of degradation is demethylation of methoxyl groups, then the oxidation of side chains. This leads to enrichment of the product in acidic functional groups (COOH and phenolic OH). Oxidation to quinones of orthohydroxy-benzene moieties resulting from demethylation favors condensation with NH_3 and amino compounds. Lignin, however, is also depolymerized by fungi, which release dilignol components that are further transformed into coniferyl alcohol, coniferaldehyde, ferulic acid, syringaldehyde, syringic acid, vanillin, vanillic acid, and so on. Lignin is therefore a source of polyphenols that are readily oxidized to quinones, which can either (a) directly condense with amino compounds or (b) polymerize enzymatically to produce HS of increasing complexity. In this hypothesis, FA would therefore be formed first and their subsequent condensation would produce HA.

Many cellulose decomposing microorganisms (myxobacteria) can synthesize polyphenols. During decomposition of plant remains, these compounds would be released into the soil even before the degradation of lignin began. They subsequently would undergo the same reactions of lignin-derived polyphenols to form HS. Structural units typical of both lignin (guaiacyl, coniferyl alcohol, *p*-hydroxycinnamyl alcohol, and synapyl alcohol derivatives) and microbial polyphenols (flavonoids) are produced by oxidative and reductive chemical degradation of HS.

The less probable pathway of formation of HS is through the sugar–amine condensation (Maillard reaction): The precursor molecules (sugar, amino acids, etc.) are continuously released into the soil solution by microorganisms, but are also quickly decomposed and their concentration never builds up. The Maillard reaction occurs at an appreciable rate only at extreme pH values and at elevated temperatures, conditions that are never encountered in soil. Jokic et al. (2001) reported that mineral colloids such as Mn(IV) oxide (birnessite) markedly accelerate the Maillard reaction between glucose and glycine in the ranges of temperatures and pH typical of natural environments. Furthermore, the significance of mineral-catalyzed Maillard reaction has been recently addressed by Horwath (2007). Furthermore, in nature, the Maillard reaction and polyphenol pathway may not occur separately but rather interact with each other, since sugars, amino acids, and polyphenols coexist in soil solutions and natural waters. The literature indicates the significance of linking the polyphenols and Maillard reactions as catalyzed by mineral colloids into an integrated humification pathway (Jokic et al., 2004; Hardie et al., 2007) (see Chapter 2).

The synthesis of HA and FA, by three out of the four main formation pathways proposed, involves either the synthesis ex novo or the liberation of polyphenolic substances. HS are polyphenol polycarboxilic substances themselves, and it is therefore reasonable to assume that an increasing quantity of phenolic substances should be produced or released during humification. During composting (Sequi et al., 1985; De Nobili and Petrussi, 1988), the ratio of nonhumic to humic extractable C decreases linearly with time during SOM transformation and therefore has been used as an index of humification (HI). Due to the complexity of the humification process, which involves condensation and the incorporation of pre-existing organic compounds into HS, the determination of humic C by classical extraction–fractionation procedures fails to give an unequivocable trend during the process. To make this trend visible, it is necessary to separate from true FA the nonphenolic substances that are co-extracted with them and that constitute a large part of the extractable organic C. This approach was introduced by Lowe (1969) for the characterization of soils, but so far has not been applied to the study of humification processes in SOM; it should have received more attention.

One of the major reasons for the relatively high resistance to biodegradation of HS, as compared to easily decomposed plant constituents, is that they are not comprised of repeating subunits; in other words, they are macromolecules rather than biopolymers. The large variability in structure of HS means that a variety of enzymes are needed to catabolize or depolymerize them. The synthesis of this large array of enzymes is extremely unfavorable in terms of energy expenditure, and it is therefore energetically advantageous for a population of microorganisms to catabolize simple polymers rather than develop the complex enzyme system required to mineralize HS. Under these circumstances, any catabolism of the aromatic moieties of the HS can only be the result of co-metabolic processes.

In an investigation by Martens (2000), alkaline extraction of HA was employed on a control soil, a corn-amended soil (high phenolic acids content), and canola residue amended soil (low phenolic acid content) after incubation for 29 and 84 days. It was found that the extractable HA in the nonamended soil maintained $2.8\,g\,kg^{-1}$ soil during incubation, while in the corn-amended soil the HA increased to $3.3\,g\,kg^{-1}$ soil (day 29) and $3.6\,g\,kg^{-1}$ soil (day 84). Incubation of the canola residue resulted in $2.8\,g\,HA\,kg^{-1}$ (day 29); but a decrease of HA to $2.0\,g\,kg^{-1}$ soil was observed after 84 days of incubation, suggesting that the canola residue addition resulted in an increased decomposition of native SOM.

5.2.4. Charred Carbon Storage in Soils

During natural or man-managed fires, incomplete combustion of standing biomass and litter results in the storage of highly refractory black C in soils (Schmidt and Noack, 2000). The global production of black C has been estimated to range from 50 to $270\,Tg\,yr^{-1}$ with as much as 80% of this remaining as residues in soil (Kuhlbusch, 1998; Suman et al., 1997).

Black C is produced by an incomplete combustion of fossil fuels and biomass (Schmidt and Noack, 2000) and includes combustion residues, such as char and charcoal, as well as combustion condensates such as soot (Kuhlbusch, 1998; Hedges et al., 2000). Black C is mainly composed of elemental C and has low O/C ratios, with soot consisting of molecules of the lowest O/C ratios, below 0.2, (Hedges et al., 2000). The size of charcoal pieces ranges from micrometers to meters (Kuhlbusch, 1998), with either visible plant structures preserved during charring (Goldberg, 1985) or homogenized cell walls leading to a fibrous texture of charcoal (Jones and Chaloner, 1991). In contrast, soot particles are smaller (submicrometer scale) with spherical shapes (Akhter et al., 1985).

Since black C has a highly aromatic structure with a low level of substitution with functional groups, it is highly recalcitrant and therefore contributes to the stable fraction of soil C. At the global scale, formation of black C rapidly transfers fast-cyclable C from the biosphere to much slower-cyclable forms that may persist in the soil for millennia. It therefore represents an effective pathway for C sequestration.

However, Daia et al. (2005) found that repeated savannah fires increased black C only slightly compared to the unburned controls, and the effects were not statistically significant. Results of this study provide estimates of black C concentrations for native, uncultivated mixed-grass savanna and indicate that 2–3 fires have little effect on the size of the soil black C pool (Daia et al., 2005).

Black C has been reported to represent as much as 10–45% of the total soil organic C (Glaser et al., 1998; Schmidt et al., 1999; Skjemstad et al., 1996, 2002) and 15–65% of marine sedimentary organic C (Lim and Cachier, 1996; Masiello and Druffel, 1998; Middelburg et al., 1999). Schmidt et al. (2001) tested several forms of thermal oxidation, chemical oxidation by photooxidation, and a chemical oxidation/molecular marker method on Australian soil samples. The resulting black C values for individual samples varied over two orders of magnitude, indicating great disparity between individual methods.

The determination of black C in natural SOM is a difficult task primarily due to the complexity of the SOM structure and the consortium of compounds that fall within the broad definition of black C. Simpson and Hatcher (2004) applied a chemical oxidation technique to remove lignin, such that the amount of black C could be quantified by integrating the aromatic region of a solid-state CP MAS ^{13}C NMR spectrum. In addition, the TMAH GC–MS method was used to evaluate the removal efficiency of lignin through the identification of lignin specific biomarkers. They demonstrated that the chemical oxidation procedure does not result in the transformation of nonpyrogenic C to a pyrogenic form, and the procedure may be more suitable for samples that contain a range of labile and refractory forms of C such as soil and sedimentary SOM samples.

Cheng et al. (2006) incubated for four months black C and black C–soil mixtures at 30 °C and 70 °C, with and without microbial inoculation, nutrient addition, or manure amendment. Incubation caused a decrease in pH from 5.4 to 5.2 and 3.4, as well as an increase in cation exchange capacity by 53% and 538%, respectively. Surface formation of carboxylic functional groups was the reason for the enhanced CEC during oxidation.

Hamer et al. (2004) found that addition of glucose enhanced black C mineralization, therefore its great stability in soils may not be solely attributable to its refractory structure, but also to poor accessibility when physically enveloped by soil particles. For example, Brodowski et al. (2006) found the greatest black C concentrations (7.2% of organic C) in the <53 μm aggregates, whereas the smallest black C concentrations occurred in large macroaggregates (>2 mm). The C-normalized black C concentrations were significantly greater ($P < 0.05$) in the occluded POM than in the free POM within the mineral fractions. This enrichment of black C amounted to factors of 1.5–2.7. Hence, black C was embedded within microaggregates in preference to other organic C compounds.

5.3. STUDIES EMPLOYING ISOTOPES

The simplest way to calculate turnover times for SOM is by dividing the total C stock in a soil by the average CO_2 evolved corrected for root respiration. Raich and Schlesinger (1992) found that the global turnover time for SOM ranges from 14 to 400 years, in accordance with different ecosystems, with an average of 32 years. SOM must, however, contain components that turn over much faster or slower than this average, because the average age of organic C in soil falls in a range of several hundreds of years, as shown by radiocarbon measurements (Campbell et al., 1967). Planetary pollution of the atmosphere caused by surface thermonuclear tests carried out for several decades during and after World War II had at least one unforseen useful consequence: a marked increase of ^{14}C concentration of the atmospheric CO_2 which followed with an enriched ^{14}C labeling of plant inputs. This offered scientists the possibility to calculate (a) the annual input of organic C to the soil and (b) the turnover time of SOM by measurements of total organic C and radiocarbon C in soils sampled before and after the thermonuclear tests. Also, it allowed a more precise calibration of C cycling models, provided that pre-bomb samples of soil were available for background comparison. Jenkinson and Coleman (1994) measured apparent radiocarbon ages for SOM in six experimental sites in southern England and found that turnover times ranged from 685 to 2395 years.

The ^{14}C content of SOM decreases with depth in the soil profile (Martin et al., 1990). An estimate of the passive pool could be obtained by measurements of the ^{14}C of SOM in deeper layers (Harrison and Broecker, 1993). Physical fractionation and sequential extraction have also been used, and they have shown progressively lower $^{14}C/^{12}C$ ratios in decomposing residues.

Campbell et al. (1967) applied sequential extraction to characterize FA, HA, and humin from gray podzolic and chernozemic soils. The fractions of FA and HA extracted by 0.5 M NaOH without acid pretreatment, which they called "mobile humates" (since the researchers assumed that they are not bound to minerals), had a lower mean residence time (ranging from 85 to 785 for HA, respectively, in the chernozemic and gray podzolic soils) as compared to Ca-humates extracted from humin (1410 years in the chernozemic soil) and to the total FA and HA extracted after acid pretreatment (195–1235 years for HA). This study showed that in the chernozemic soil, Ca-humates and clays play an equally important role in the stabilization of HS, whereas in the podzolic soil the oldest fraction was associated with clays.

Information on long-term C exchange between the soil and the atmosphere can be obtained also from ^{13}C abundance measurements where there has been a switch from C-3 to C-4 crop (or vice versa). The isotopic signature, or ratio of $^{13}C/^{12}C$, in the tissues of C-3 plants is different from that of C-4 plants. For example, the natural vegetation in the Canadian prairies includes C-4 plants; and after conversion to arable land and planted with C-3 crops, the SOM shows a distinct decrease in ^{13}C abundance after the first decades of cultivation (Ellert and Janzen, 2006). The same happens in savannah soils in the humid tropics where vegetation can shift from grasses to bush or dense woodland after protection from fire (Martin et al., 1990). After 25 years, colonization by trees had modified the isotopic signature of SOM throughout the soil profile and indicated that only 30–45% of C-4 plant-derived SOM remained in the upper 10 cm of soil whereas it made up to 80% SOM in the 10- to 25-cm layer. While changes in ^{13}C after conversion of a C-3 to a C-4 dominated crop or vegetation may be used to determine the amount and average turnover of the nonpassive fractions, it is not possible to use this concept to distinguish between the active and intermediate pools (Harkness et al., 1991) or to isolate the contribution of the passive fraction.

Measurements of stable isotopic signatures confirm the existence of a relatively passive soil C pool, by the fact that some C-3 plant-derived SOM persists in the soil even after more than a century of cultivation with C-4 plants. Stable isotope studies can also be used to determine whether compost application is effective for increasing C storage in soils. Lynch et al. (2006) found that up to 89% of corn silage compost, 65% of sewage sludge compost, and 42% of dairy manure compost remained in the soil after 2 years. It is possible that these differences are related not only to the degree of stabilization achieved through composting, but also to the lignin content of the various composts since lignin is only slightly degraded during composting (Inbar et al., 1989). Mineralization of lignin by soil microorganisms can reach up to 41% after 2 years as determined from studies of the decay of ^{14}C-labeled synthetic lignin (Martin and Haider, 1979). However, there is no experimental evidence for long-term storage of lignin in soil (Amelung, 1999; Lobe et al., 2002; Rumpel et al., 2002). Dignaca et al. (2005) found that 47% of the initial wheat-derived lignin had been replaced after 9 years and only 9% of SOM had been renewed in the same time. Their results confirm that there is no long-term storage of untransformed plant molecules and that the long turnover times are probably associated with a complex in situ transformation pattern, namely the humification process.

Doane et al. (2003) used natural abundance ^{13}C measurements to identify changes in soil HS. They found incorporation of corn-derived C during the winter into both FA and HA as well as into humin. Corn residue decomposition (28% of incorporated C) contributed to the observed increase in the HA pool (1200 kg C ha^{-1}). During the same period, 6–9% of incorporated C contributed to the FA fraction (up to 370 kg C ha^{-1}). The FA fraction was considerably more enriched in ^{13}C than the HA, and the humin was even more so. Other authors have described a similarity in ^{13}C content of the HS fractions. This, however, does not agree with the hypothesis of transformation of one fraction into another, but indicates the existence of all HS fractions from the first stages of decomposition. Humin, the non-alkali extractable fraction of SOM, also showed significant seasonal changes and exhibited an 8% turnover of C. Moreover, the different HS were affected at different times, indicating that they may result from different pathways.

5.4. EFFECTS OF INCREASING CARBON INPUTS TO SOILS

Both quality and quantity of SOM in soils affect plant growth and health and therefore affect attainable C inputs levels. The effects of SOM decomposition level in the soil on interactions of beneficial microorganisms and pathogens and on plant growth have been largely overlooked (Grebus et al., 1994; Hoitink and Fahy, 1986). In addition, HS have both direct and indirect effects on plant growth (Chen and Aviad, 1990; Chen et al., 1994; Chen et al., 2004).

SOM contributes to soil fertility both directly, by releasing major inorganic nutrients as well as trace elements during its decomposition, and indirectly, by increasing soil cation exchange capacity, by improving soil structure and by increasing soil water holding capacity. Although its nutritional role has been obscured in modern agriculture by the effectiveness of mineral fertilizers, the benefits of SOM became apparent again along with the introduction of high yielding cultivars and improved control of pathogens. SOM is always the main factor in sustaining soil fertility in low-input systems, particularly under tropical climates (Tiessen et al., 1994). In these systems, SOM accumulation has a strong positive feedback effect on net primary production (NPP) and ultimately on soil C inputs.

Manures and composts have been used as a means for increasing soil fertility and crop production, all through the history of farming. Organic residues served as the only means for adding nitrogen (N) and very important means for adding other nutrients until the development of mineral fertilizer production and distribution systems. At present, the chemical industry provides concentrated mineral fertilizers that are easily handled and that can supply the need for any nutrient element. This development offsets the use of organic residues as a sole source for nutrients and, in some instances, eliminates the use of manures and compost to a point where these materials are becoming more of a problem rather than an asset even though laws have been established to deal with the responsibility for the re-utilization and/or disposal of organic wastes. A modern agricultural system should strive for a sustainable approach in which mineral fertilizers and organic wastes are applied in tandem.

SOM and its management, while aiming to promote soil health, go hand-in-hand with the emergence of an ecologically based approach to soil and crop management that stresses prevention of imbalances that otherwise can lead to soil, crop, and environmental problems. Machado et al. (2006) reported that increasing cropping frequency and eliminating tillage can increase the potential for soil C sequestration. SOC content was highest in grass pasture, where there was no tillage, and was the lowest in conventional tillage winter wheat–summer fallow, which involved intensive tillage and less intensive cropping. Continuous cropping, even under conventional tillage, increased the amount of SOC in the soil, particularly in the plow layer. However, mixing surface with subsurface soil during conventional tillage reduces organic C near the surface, making the soil more susceptible to wind and water erosion.

Growth of trees on grassland and arable land modifies soil forming processes and, in general, has two main macroscopic consequences, namely, accumulation of SOM in the upper soil layers and soil acidification. Several other subtle, yet durable, and sometimes irreversible changes have been observed, most associated more with soil acidification and quality of plant residues (conifers) than with the quantity of the latter. In the Broadbalk Classical Experiment at Rothamsted (De Nobili et al., 2008) the soils of the two small areas of the old farmland field, which were fenced off before the harvest of winter wheat in 1882 and were never cultivated again, do not display strong quantitative or qualitative differences in SOM. The pH of these soils remained neutral, in spite of the fact that one section was occasionally cut (stubbed section) so that trees were not allowed to grow and the other was left to revert to woodland. They have, since the time they were fenced off, received similar C inputs ($4\,Mg\,C\,ha^{-1}\,yr^{-1}$) and now sustain about the same soil microbial biomass and have accumulated very similar amounts of SOC. The ^{13}C-CPMAS-TOSS-NMR spectra of HA extracted from the woodland section of this experiment (Figure 5.4) show relatively small differences apart from an increase in the *O*-alkyl signal in free FA and in that of alkyls in free HA from the corresponding spectra of the HA extracted from the stubbed section. Much larger differences were noted between free and bound FA and HA, indicating that these correspond to different humic C pools.

Denef et al. (2002) observed that additions of organic C (OC) and nutrients caused significant increases in unstable and stable macroaggregation in soils of different mineralogy. Moreover, in a treatment without nutrients and with low OC inputs, stable macroaggregation decreased after 14 days of incubation in soils with 2:1 clay minerals, due to exhaustion of available OC and subsequent decrease of microbial activity and production of binding agents. Only in the soil dominated by 1:1 clay minerals, where mineral surfaces are directly bound by electrostatic interactions, no changes in macroaggregation were noticed over time. In soils where SOM is the major binding agent for aggregates, additions of organic substrates resulted in a destabilization of the macroaggregates (Six et al., 2001).

According to a review by Magdoff and Weil (2004), increased SOM can counteract the ill effects of too much clay or too much sand. Increasing the SOM content usually increases total porosity and therefore decreases bulk density. Within a limited range of SOM contents, the relationship for a given soil is nearly linear (Weil and Kroontje, 1979). However, across a wider range of SOM levels, the relationship

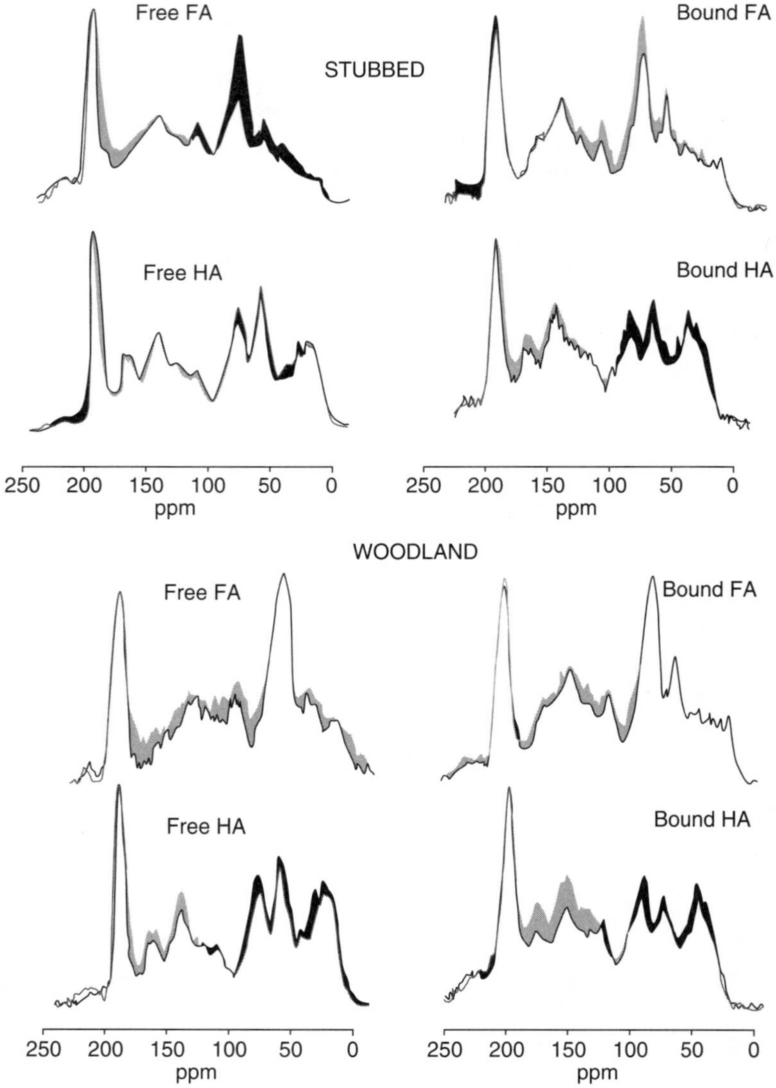

Figure 5.4. ^{13}C CPMAS TOSS NMR spectra of free and bound fulvic and humic acids extracted from the woodland and stubbed sections of the Broadbalk experiment before and after incubation for about 5 months at 25 °C. Areas in black and gray indicate, respectively, a decrease and an increase of the corresponding signals after incubation. Reprinted from De Nobili, M., Contin, M., Mahieu, N., Randall, E. W., and Brookes, P. C. (2008). Assessment of chemical stabilization of organic C in soils from the long-term experiment at Rothamsted (UK). *Waste Management* **28**, 723–733, with permission from Elsevier.

between these two variables is likely to be curvilinear, because at very high levels of SOM, additional OC has little further effect on soil aggregation and influences bulk density mainly because of its low particle density (Franzluebbers et al., 2001). Increased SOM levels are therefore associated with lower energy requirements for soil tillage and represents an additional bonus as it implies a lower emission of CO_2 from fossil fuel usage in agriculture.

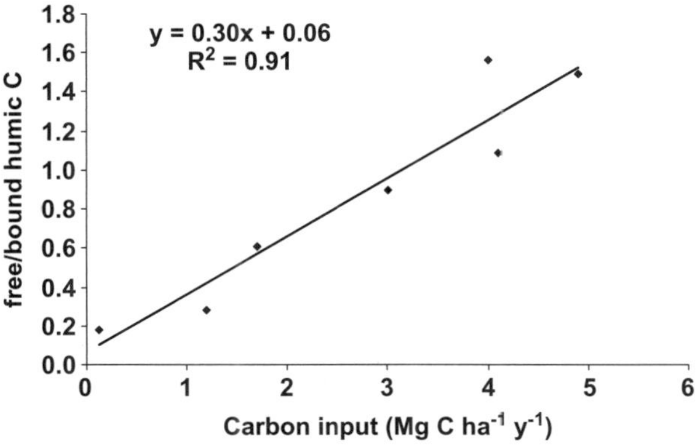

Figure 5.5. Trend of the free to bound humic C ratios in extracts of soils of different C inputs in the Broadbalk experiment at Rothamsted. Reprinted from De Nobili, M., Contin, M., Mahieu, N., Randall, E. W., and Brookes, P. C. (2008). Assessment of chemical stabilization of organic C in soils from the long-term experiment at Rothamsted (UK). *Waste Management* **28**, 723–733, with permission from Elsevier.

De Nobili et al. (1999) showed that the mineral-bound pyrophosphate extractable fraction of HS (chemically stabilized fraction) predominates in mineral soils of low SOM content that had been arable for the last few centuries. In contrast, NaOH-extractable OM, the so-called free or mobile humic C, is more abundant in soils that are currently accumulating SOM. In the Broadbalk plot, which has received only mineral fertilizer for the last 153 years, the ratio of free to mineral-bound extractable C was 0.61. In the same soil, for HAs the ratio of free to bound C was 0.35 and for FAs 0.80. This soil, having been arable land since the 16th century, is considered to be now at equilibrium and is neither losing nor accumulating C. The amount of extractable mineral-bound C of the wooded and stubbed plots soils has, on the contrary, increased about fourfold due to a corresponding increase in SOM. However, an even much larger increase in the NaOH extractable C caused the ratio of free to bound extractable C of plots under continuous wheat to increase above unity. This was a consistent trend in soils which were accumulating organic C. It is also consistent with a "saturation" of the capability of a soil to stabilize SOM by chemico-physical and physical mechanisms (Six et al., 2002). The assumption that soils do not accumulate C beyond their capacity to physically stabilize SOM is therefore not true: Biochemical stabilization also contributes to C sequestration as shown by data of Figures 5.5 and 5.6.

5.5. EFFECTS OF REDUCING CARBON INPUTS TO SOIL

All through the 20th century, and especially ever since the advent of the Green Revolution, modern agriculture has been striving to feed and clothe the ever-increasing population through improved technology, relying heavily on inputs of fertilizers, pesticides, and various other agrochemicals. Undoubtedly, this has been a great blessing to mankind, and enormous strides have indeed been made in the never-ending struggle against starvation, but these have been achieved at a very

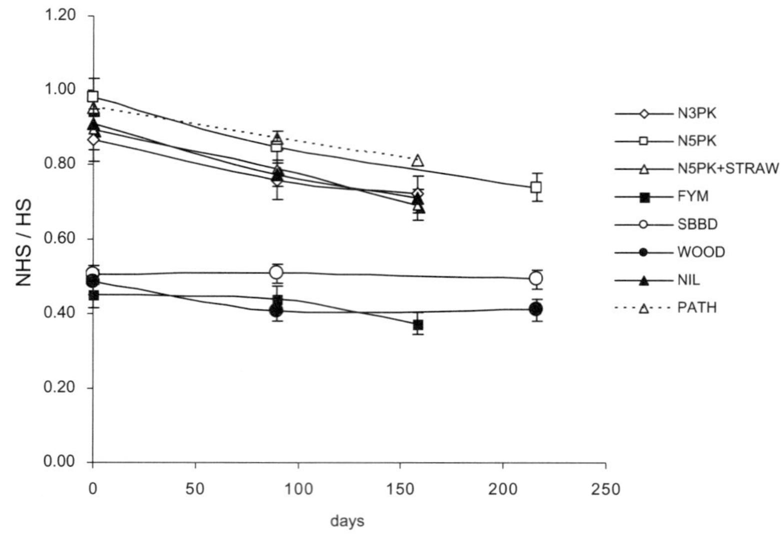

Figure 5.6. Trend of the nonhumic to humic C ratios (NHS/HS) in alkaline pyrophosphate extracts of soils from the Broadbalk experiment at Rothamsted during incubation at 25 °C. Error bars represent the standard deviation. Plot acronyms: N3PK, mineral fertilizers, 144 kg N ha^{-1}; N5PK, mineral fertilizers, 240 kg N ha^{-1}; N5PK+straw, mineral fertilizers, 240 N kg ha^{-1} and wheat straw incorporated; FYM, farmyard manure (35 Mg ha^{-1} yr^{-1}); SBBD, stubbed wilderness section; WOOD, woodland wilderness section; NIL, without fertilizers; PATH, bare strips separating plots without fertilizers. Reprinted from De Nobili, M., Contin, M., Mahieu, N., Randall, E. W., and Brookes, P. C. (2008). Assessment of chemical stabilization of organic C in soils from the long-term experiment at Rothamsted (UK). *Waste Management* **28**, 723–733, with permission from Elsevier.

steep price of increased environmental deterioration. Modern agriculture has become one of the major factors contributing to the degradation of the world's fragile biosphere. This has been manifested in mounting rates of soil, water, and air pollution, increased emissions of greenhouse gases, depletion of the ozone layer in the stratosphere, accumulation of manure and other solid wastes, excessive exploitation of forests and open lands, elimination of the natural habitats of many plants and animals, an almost unprecedented mass extinction of living species, and an alarming destruction of biodiversity. Obviously, this should not take place in the future, and urgent steps have to be taken, on a regional, national, and global level, to stop, avoid, and mitigate these detrimental processes.

Carbon loss rates from terrestrial ecosystems are an order of magnitude faster than that of C sequestration (Körner, 2003), so an effective protection of the already existing C stocks is essential. At present in Europe arable soils are losing C at a rate equivalent to 10% of total fossil fuel emissions (Janssens et al., 2005). Considering that management changes have turned arable soils in North America into large C sinks (Pacala et al., 2001), Smith (2004) estimated that, using biological, social, and economical constraints, current C losses in continental Europe could be reduced by 46 Tg yr^{-1} by the year 2010.

A number of years after a soil is made arable, it reaches a new steady-state condition that is normally characterized by a lower SOM level. The time needed to reach the new equilibrium with respect to C cycling depends on climatic conditions, whereas the new equilibrium level of SOM is principally a function of both C inputs levels and of soil texture. Depletion of SOM upon cultivation is generally attributed to oxidation or mineralization, due to breakdown of aggregates leading to exposure to oxygen of POM, changes in temperature and moisture regimes favoring microbial decay processes, leaching, and accelerated erosion by water runoff or wind (Lal, 2002a; Ingram and Fernandes, 2001). However, although crop harvesting and bare fallowing periods typically reduce C inputs to arable land, whenever C inputs to soil have increased SOM levels increased after cultivation.

Figure 5.6 shows that the proportion of nonhumic substances (NHS) is actually larger in $Na_4P_2O_7$ extracts of monocultured soils than in soils under crop rotation, grassland or woodland. The unexpected lower proportion of NHS in the extracts of soils that are accumulating C implies that HS formation is favored by a surplus of C inputs and that at least some are mineralized when soil microbial biomass is subjected to an insufficient supply of more labile substrates. Soils of different SOM contents from the long term Broadbalk experiment incubated at 25 °C for up to 250 days (De Nobili et al., 2008) showed a slow, but measurable decrease of the NHS/HS ratio only in soils of C inputs equivalent or lower to 4 t ha^{-1} yr^{-1} (Figure 5.6). HS were therefore utilized by the soil microbial biomass to survive at no C inputs and were actually decomposed, in the high C input soils, at a rate comparable to that of nonhumic C (hydrolyzed polysaccharides, proteins, peptides, etc.), which is normally considered much more labile. A possible explanation is that HS formed in these soils have a lower degree of chemical and biochemical stabilization than those found in arable soils of low C input. The biodegradation of HS in the field is supported by the fact that in soils that have reached equilibrium, accumulation of HS ceases and their content remains steady and can even decrease if soil organic C inputs decrease. Nevertheless, information on HS decomposition in soil is lacking. *In vitro*, a large variety of microorganisms have been shown to be able to decompose HS; to some degree (Tate, 1992; Grinhut et al., 2007) and are for the most part stimulated by amendment of cultures with an easily metabolizable C and energy source such as glucose.

During incubation of the Broadbalk soils, the ratio between free and bound C in HS decreased (De Nobili et al., 2008). The largest decreases of free C were measured in soils that contained more HS; however, bound C fractions also decreased after incubation. This suggests that both biochemical and chemical stabilization are not effective for C sequestration against changes in management causing strong reductions of C inputs to the soil.

Changes in $\delta^{13}C$ after incubation were confined to the free FA fractions, which showed an increase of 1.48‰ in the stubbed soil and indicated progressing biological transformation. In contrast, a decrease was observed for the bound FA of both stubbed and woodland soil samples.

^{13}CPMAS-TOSS-NMR spectra of free and bound FA and HA of the two soils clearly showed (Figure 5.4) that these fractions correspond to well-defined, chemically different fractions of HS. All ^{13}C-NMR spectra of free FA display a strong signal in the *O*-alkyl C region and a lower proportion of *N*-alkyl and alkyl C than HA. Particularly intense signals in the *O*-alkyl region are typical of NaOH (free) extractable HS, whereas a larger proportion of aromatic structures were extracted by a second extraction with $Na_4P_2O_7$ (bound HS) from the clay fraction of various soils (Wattel-Koekkoek et al., 2001). The alkyl C to *O*-alkyl ratio has been suggested as an indicator of the decomposability of SOM (Baldock et al., 1997; Webster et al., 2001), high alkyl C to *O*-alkyl C ratios indicating a lower resource quality of SOM. This ratio is the lowest in the free FA fractions and increases in free HA and bound HA.

Humic fractions extracted after 215 days of incubation maintained the characteristic features observed in the ^{13}CPMAS-TOSS-NMR spectra of fractions extracted before incubation. The observed changes (Figure 5.4) were a decrease in the areas of signals corresponding to *O*-alkyls C (65–95 ppm), *N*-alkyl C (45–65 ppm), methoxy C (45–65 ppm), and carbonyl C (185–225 ppm) plus amide and ester C (160–185 ppm) but also alkyl C (0–45 ppm). They also showed an increase in aromatic C (108–145 ppm) and phenolic C (145–160 ppm) areas, which were particularly evident for bound HA.

In summary, although HS are undoubtedly the main chemical form under which C is stored in soils of large C inputs, when the capacity for physical stabilization has been saturated, a decrease in C inputs can cause qualitative as well as quantitative changes in the soil HS stock.

5.6. CONCLUSIONS

Warming of the earth is unequivocal, and a worldwide counter action is immediately needed. The cost of not acting, most economists agree, will exceed the costs of acting now by orders of magnitude. Carbon trading is one weapon in our arsenal. New technologies for energy generation, energy conservation, forestry projects and renewable fuels, as well as private markets, must all be part of a long-term strategy (Rice, 2006).

Co-benefits of increasing soil C stocks must never be forgotten or underestimated both in terms of economics as well as the overall environmental revenue. Increased C sequestration in soil means reduced costs and fossil fuel consumption for soil tillage due to improved soil structure and increased water availability and reduced costs for land protection and water quality maintenance. Additional important effects are reduced erosion, better fertilizer management, and increased retention of pollutants, as well as uncountable indirect benefits related to management of organic wastes, biodiversity, desertification control, and land remediation. Therefore, C sequestration in soils is a winning holistic strategy that should be pursued and encouraged throughout our planet.

REFERENCES

Aiken, G. R. (1985). Isolation and concentration techniques for aquatic humic substances. In *Humic Substances in Soil, Sediment and Water: Geochemistry, Isolation and Characterization*, Aiken, G. R., McNight, D., Wershaw, R. L., and McCarthy, P., eds., Wiley Interscience, New York, pp. 363–385.

Akhter, M. S., Chughtai, A. R., and Smith, D. M. (1985). The structure of hexane soot: I. Spectroscopic studies. *Appl. Spectrosc.* **39**, 143–153.

Amado T. J. C., Bayer, C., Conceição, P. C., Spagnollo, E., Costa de Campos, B.-H., and da Veiga, M. (2006). Potential of carbon accumulation in no-till soils with intensive use and cover crops in Southern Brazil. *J. Environ. Qual.* **35**, 1599–1607.

Amelung, W., Flach, K. W., and Zech, W. (1999). Lignin in particle-size fractions of native grassland soils as influenced by climate. *Soil Sci. Soc. Am. J.* **63**, 1222–1228.

Anderson, T.-H., and Domsch, K. H. (1985). Determination of ecophysical maintenance carbon requirements of soil microorganisms in a dormant state. *Biol. Fertil. Soils* **1**, 81–89.

Angers, D. A., and Carter, M. R. (1996). Aggregation and organic matter storage in cool, humid agricultural soils. In *Advances in Soil Science—Structure and Organic Matter Storage in Agricultural Soils*, Carter, M. R., and Stewart, B. A., eds., Lewis/CRC Press, Boca Raton, FL, pp. 193–211.

Baldock, J. A., and Oades, J. M. (1989). Effect of electrolyte concentration on glucose decomposition in soil. *Aust. J. Soil Res.* **27**, 433–438.

Baldock, J. A., Sewell, T., and Hatcher, P. G. (1997). Decomposition induced changes in the chemical structure of fallen red pine, white spruce and tamarack logs. In *Driven by Nature: Plant Litter Quality and Decomposition*, Cadisch, G., and Giller, K. E., eds., CAB International, Wallingford, UK, pp. 75–83.

Baldock, J. A., and Skjemstad, O. J. (2000). Role of the soil matrix and minerals in protecting natural organic materials against biological attack. *Org. Geochem.* **31**, 697–710.

Beker-Heidmann, P., and Scharpenseel, H. W. (1992). The use of natural ^{14}C and ^{13}C in soils for studies on global climate change. *Radiocarbon* **34**, 535–540.

Blaser, P., Kernebeek, P., Tebbens, L., Breemen, N., and van Luster, J. (1997). Cryptopodzolic soils in Switzerland. *Eur. J. Soil Sci.* **48**, 411–423.

Bossuyt, H., Six, J., and Hendrix, P. F. (2004). Rapid incorporation of carbon from fresh residues into newly formed stable microaggregates within earthworm casts. *Eur. J. Soil Sci.* **55**, 393–399.

Brodowski, S., John, B., Flessa, H., and Amelung, W. (2006). Aggregate-occluded black carbon in soil. *Eur. J. Soil Sci.* **57**, 539–546.

Brookes, P. C., Tate, K. R., and Jenkinson, D. S. (1983). The adenylate energy charge of the soil microbial biomass. *Soil Biol. Biochem.* **15**, 9–16.

Bubier, J., Crill, P., Mosedale, A., Frolking, S., and Linder, E. (2003). Peatland responses to varying interannual moisture conditions as measured by automatic CO_2 chambers. *Global Biogeochem. Cycles* **17**, 1066.

Burke, I. C., Yonker, C. M., Parton, W. J., Cole, C. V., Flach, K., and Schimel, D. S. (1989). Texture, climate, and cultivation effects on soil organic matter content in US grassland soils. *Soil Sci. Soc. Am. J.* **53**, 800–805.

Campbell, C. A., Paul, E. A., Rennie, D. A., and McCallum, K. J. (1967). Applicability of the carbon dating method of analysis to soil humus studies. *Soil Sci.* **140**, 217–223.

Carney, K. M., Hungate, B. A., Drake, B. G., and Megonigal, J. P. (2007). Altered soil microbial community at elevated CO_2 leads to loss of soil carbon. *Proc. Natl. Acad. Sci.* **104**, 4990–4995.

Carter, M. R. (2002). Soil quality for sustainable land management: Organic matter and aggregation interactions that maintain soil functions. *Agron. J.* **94**, 38–47.

Chander, K., and Brookes, P. C. (1991). Microbial biomass dynamics during the decomposition of glucose and maize in metal-contaminated and non-contaminated soils. *Soil Biol. Biochem.* **23**, 917–925.

Chen, Y., and Aviad, T. (1990). Effects of humic substances on plant growth. In *Humic Substances in Soil and Crop Science*, MacCarthy P., ed., ASA, SSA, Madison, WI, pp. 161–186.

Chen, Y., Magen, H., and Riov, J. (1994). Humic substances originating from rapidly decomposing organic matter: Properties and effects on plant growth. In *Humic Substances in the Global Environment and Implications on Human Health*, Senesi, N., and Miano, T. M., eds., *Proceedings of the 6th International Meeting of the International Humic Substances Society*, Monopoli (Bari), Italy, September 20–25, 1992, pp. 427–443,

Chen, Y., De Nobili, M., and Aviad, T. (2004). Stimulatory effects of humic substances on plant growth. In *Soil Organic Matter in Sustainable Agriculture*, Magdoff, F. R., and Weil, R. R., eds., CRC Press, New York, pp. 103–129.

Cheng, C. H., Lehmann, J., Thies, J. E., Burton, S. D., and Engelhard, M. H. (2006). Oxidation of black carbon by biotic and abiotic processes. *Org. Geochem.* **37**, 1477–1488.

Christensen, B. T. (1996). Carbon in primary and secondary organomineral complexes. In: *Structure and Organic Matter Storage in Agricultural Soils*, Carter, M. R., and Stewart, B. A., eds., Advances in Soil Science, CRC Press, Boca Raton, FL, pp. 97–165.

Ciais, P., Tans, P. P., Trolier, M., White, J. W. C., and Francey, R. J. (1995). A large northern hemisphere terrestrial CO_2 sink indicated by the 13C/12C ratio of atmospheric CO_2. *Science* **269**, 1098–1102.

Contin, M., Todd, A., and Brookes, P. C. (2001). The ATP concentration of the soil microbial biomass. *Soil Biol. Biochem.* **33**, 701–704.

Cox, P. M., Betts, R. A., Jones, C. D., Spall, S. A., and Totterdell, I. J. (2000). Acceleration of global warming due to carbon-cycle feedbacks in a coupled climate model. *Nature* **408**, 184–187.

Daia, X., Bouttona, T. W., Glaserb, B., Ansleyc, R. J., and Zech, W. (2005). Black carbon in a temperate mixed-grass savannah. *Soil Biol. Biochem.* **37**, 1879–1881.

De Nobili, M., and Petrussi, F. (1988). Humification index (HI) as evaluation of the stabilization degree during composting. *J. Ferment. Technol.* **66**, 577–583.

De Nobili, M., Contin, M., Mondini, C., and Brookes, P. C. (2001). Soil microbial biomass is triggered into activity by trace amounts of substrate. *Soil Biol. Biochem.* **33**, 1163–1170.

De Nobili, M., Brookes, P. C., Contin, M., Mahieu, N., and Randall, E. W. (1999). Qualitative and quantitative changes in free and mineral bound humic and fulvic acids in the Broadbalk Classical Experiment at Rothamsted. *Humic Substances Environ.* **1**, 17–21.

De Nobili, M., Contin, M., Mahieu, N., Randall, E. W., and Brookes, P. C. (2008). Assessment of chemical stabilization of organic C in soils from the long-term experiment at Rothamsted (UK). *Waste Management* **28**, 723–733.

Denef K., Six, J., Merckx, R., and Paustian, K. (2002). Short-term effects of biological and physical forces on aggregate formation in soils with different clay mineralogy. *Plant and Soil* **246**, 185–200.

Derenne, S., and Largeau, C. (2001). A review of some important families of refractory macromolecules: Composition, origin, and fate in soils and sediments. *Soil Sci.* **166**, 833–847.

Dignaca, M. F., Bahria, H., Rumpela, C., Rassea, D. P., Bardouxa, G., Balesdentb, J., Girardina, C., Chenuc, C., and Mariottia, A. (2005). Carbon-13 natural abundance as a tool to study the dynamics of lignin monomers in soil: An appraisal at the Closeaux experimental field (France). *Geoderma.* **128**, 3–17.

Dilly, O., Gnaß, A., and Pfeiffer, E. M. (2005). Humus accumulation and microbial activities in calcari-epigleyic fluvisols under grassland and forest diked in for 30 years. *Soil Biol. Biochem.* **37**, 2163–2166.

Doane, T. A., Devêvre, O. C., and Horwáth, W. R. (2003). Short-term soil carbon dynamics of humic fractions in low-input and organic cropping systems. *Geoderma* **114**, 319–331.

Drees, L. R., Wilding, L. P., Smeck, N. E., and Senkayi, A. L. (1989). Silica in soils: Quartz and disordered silica polymorphs. In *Minerals in Soil Environments*, Dixon, J. B., and Weed, S. B., eds., SSSA Book Series, no. 1, pp. 913–974.

Ellert, B. H., and Janzen, H. H. (2006). Long term biogeochemical cycling in agroecosystems inferred from ^{13}C and ^{15}N. *J. Geochem. Explor.* **88**, 198–201.

Evans, R. D., and Ehleringer, J. R. (1994). Water and nitrogen dynamics in an arid woodland. *Oecologia* **99**, 1432–1939.

Fan, S., Gloor, M., Mahlman, J., Pacala, S., Sarmiento, J., Takahashi T., and Tans, P. (1998) A large terrestrial carbon sink in North America implied by atmospheric and oceanic carbon dioxide data and models. *Science* **282**, 442–446.

Feller, C., and Beare, M. H. (1997). Physical control of soil organic matter dynamics in the tropics. *Geoderma* **79**, 69–116.

Franzluebbers, A. J., Stuedemann, J. A., and Wilkinson, S. R. (2001). Bermudagrass management in the Southern Piedmont of the USA: I. Soil and surface residue carbon and sulfur. *Soil Sci. Soc. Am. J.* **65**, 8334–8341.

Freibauer, A., Rounsevell, M. D. A., Smith, P., and Verhagen, J. (2004). Carbon sequestration in the agricultural soils of Europe. *Geoderma* **122**, 1–23.

Glaser, B., Haumaier, L., Guggenberger, G., and Zech, W. (1998). Black carbon in soils: The use of benzenecarboxylic acids as specific markers. *Org. Geochem.* **29**, 811–819.

Goh, K. M., and Pullar, W. A. (1977). Radiocarbon dating techniques for tephras in central North Island, New Zealand. *Geoderma* **18**, 265–278.

Goldberg, E. D. (1985). *Black Carbon in the Environment: Properties and Distribution*, John Wiley & Sons, New York.

Goulden, M. L., Wofsy, S. C., Harden, J. W., Trumbore, S. E., Crill, P. M., Gower, S. T., Fries, T., Daube, B. C., Fan, S. M., Sutton, D. J., Bazzaz, F. A., and Munger, J. W. (1998). Sensitivity of boreal forest carbon balance to soil thaw. *Science.* **279**, 214–217.

Grebus, M. E., Watson, M. E., and Hoitink, H. A. J. (1994). Biological, chemical and physical properties of composted yard trimmings as indicators of maturity and plant disease suppression. *Compost Sci. Util.* **2**, 57–71.

Grigal, D. G., and Ohmann, L. F. (1992). Carbon storage in upland forestland of the Lake States. *Soil Sci. Soc. Am. J.* **56**, 935–943.

Grinhut, T., Hadar, Y., and Chen, Y. (2007). Degradation and transformation of humic substances by saprotrophic fungi: Processes and mechanisms. *Fungal Biol. Rev.* **21**, 179–189.

Gunapala, N., Venette, R. C., Ferris, H., and Scow, K. M. (1998). Effects of soil management history on the rate of organic matter decomposition. *Soil Biol. Biochem.* **30**, 1917–1927.

Hamer, U., Marschner, B., Brodowski, S., and Amelung, W. (2004). Interactive priming of black carbon and glucose mineralisation. *Org. Geochem.* **35**, 823–830.

Hamer, U., and Marschner, B. (2005). Priming effects in different soil types induced by fructose, alanine, oxalic acid and catechol additions. *Soil Biol. Biochem.* **37**, 445–454.

Hardie, A. G., Dynes, J. J., Kozak1, L. M., and Huang, P. M. (2007). Influence of polyphenols on the integrated polyphenol-Maillard reaction humification pathway as catalyzed by birnessite. *Ann. Environ. Sci.* **1**, 91–110.

Harkness, D. D., Harrison, A. F., and Bacon, P. J. (1991). The potential of bomb-^{14}C measurements for estimating soil organic matter turnover. In *Advances in Soil Organic Matter Research: The Impact on Agriculture and the Environment*, Wilson W. S., ed., Proceedings of a Symposium, Colchester, 3–4 September, 1990, pp. 240–251.

Harrison, K., and Broecker, W. (1993). A strategy for estimating the impact of CO_2 fertilization on soil carbon storage. *Global Biogeochem. Cycles.* **7**, 69–80.

Hassink, J., Whitmore, A. P., and Kubát, J. (1997). Size and density fractionation of soil organic matter and the physical capacity of soils to protect organic matter. *Eur. J. Agron.* **7**, 189–199.

Hedges, J. I., Eglinton, G., Hatcher, P. G., Kirchman, D. L., Arnosti, C., Derenne, S., Evershed, R. P., Kögel-Knabner, I., de Leeuw, J. W., Littke, R., Michaelis, W., and Rullkötter, J. (2000). The molecularly-uncharacterized component of nonliving organic matter in natural environments. *Org. Geochem.* **31**, 945–958.

Hoitink, H. A. J., and Fahy, P. C. (1986). Basis for the control of soilborne plant pathogens with composts. *Ann. Rev. Phytopathol.* **24**, 93–114.

Horwath (2007). Carbon cycling and formation of soil organic matter. In *Soil Microbiology, Ecology, and Biochemistry*, Paul, E. A., ed., Academic Press, Elsevier, Amsterdam, pp. 331.

Inbar, Y., Chen, Y., and Hadar, Y. (1989). Solid-state carbon-13 nuclear magnetic resonance and infrared spectroscopy of composted organic matter. *Soil Sci. Soc. Am. J.* **53**, 1695–1701.

Ingram, J. S. I., and Fernandes, E. C. M. (2001). Managing carbon sequestration in soils: Concepts and terminology. *Agric. Ecosyst. Environ.* **87**, 111–117.

Intergovernmental Panel on Climate Change (IPCC). (2001). *Climate Change: The Scientific Basis*, Cambridge University Press, Cambridge, UK.

Insam, H. (1990). Are the soil microbial biomass and basal respiration governed by the climatic regime? *Soil Biol. Biochem.* **22**, 525–533.

Insam, H., Bååth, E., Berreck, M., Frostegård, A., Gerzabek, M. H., Kraft, A., Schinner, F., Schweiger, P., and Tschuggnall, G. (1999). Responses of the soil microbiota to elevated CO_2 in an artificial tropical ecosystem. *J. Microbiol. Methods.* **36**, 45–54.

Janssens, I. A., Freibauer, A., Schlamadinger, B., Ceulemans, R., Ciais, P., Dolman, A. J., Heimann, M., Nabuurs, G. J., Smith, P., Valentini, R., and Schulze, E. D. (2005). The carbon budget of terrestrial ecosystems at country-scale—A European case study. *Biogeosciences* **2**, 15–26.

Jenkinson, D. S. (1981). The fate of plant and animal residues in soil. In *Chemistry of Soil Processes*, Greenland, D. J., and Hayes, M. H. B., eds., John Wiley & Sons, New York, pp. 505–561.

Jenkinson, D. S. (1988). Determination of microbial biomass carbon and nitrogen in soil. In *Advances in Nitrogen cycling in Agricultural Ecosystems*. Commonwealth Agricultural Bureau International, Wallingford, UK, pp. 368–386.

Jenkinson, D. S., and Anayaba, A. (1977). Decomposition of carbon-14 labelled plant material under tropical conditions. *Soil Sci. Soc. Am. J.* **41**, 912–915.

Jenkinson, D. S., and Ladd, J. N. (1981). Microbial biomass in soil: Measurement and Turnover. In *Soil Biochemistry*, Vol. **5**, Paul, E.A., and Ladd, J. N., eds., Marcel Dekker, New York, pp. 415–471.

Jenkinson, D. S., Adams, D. E., and Wild, A. (1991). Model estimates of CO_2 emissions from soil in response to global warming. *Nature* **351**, 304–306.

Jenkinson, D. S., and Coleman, K. (1994). Calculating the annual input of organic matter to soil from measurements of total organic carbon and radiocarbon. *Eur. J. Soil Sci.* **45**, 167–174.

Johnson, D., and Hale, B. (2004). White birch (*Betula papyrifera* Marshall) foliar litter decomposition in relation to trace metal atmospheric inputs at metal-contaminated and uncontaminated sites near Sudbury, Ontario and Rouyn-Noranda, Quebec, Canada. *Environ. Poll.* **127**, 65–72.

Jokic, A., Frenkel, A. I., Vairavamurthy, M. A., and Huang, P. M. (2001). Birnessite catalysis of the Maillard reaction: Its significance in natural humification. *Geophysical Res. Lett.* **28**, 3899–3902.

Jokic, A, Wang, M. C., Liu, C, Frenkel, A. I., and Huang, P. M. (2004). Integration of the polyphenol and Maillard reactions into a unified abiotic pathway for humification in nature: The role of δ-MnO_2. *Org. Geochem.* **35**, 747–762.

Jones, T. P., and Chaloner, W. G. (1991). Fossil charcoal, its recognition and palaeoatmospheric significance. *Palaeogeogr. Palaeoclimatol. Palaeoecol.* **97**, 39–50.

Kalbitz, K., Schmerwitz, J., Schwesig, D., and Matzner, E. (2003a). Biodegradation of soil-derived dissolved organic matter as related to its properties. *Geoderma* **113**, 273–291.

Kalbitz, K., Schwesig, D., Schmerwitz, J., Kaiser, K., Haumaier, L., Glaser, B., Ellerbrock, R., and Leinweber, P. (2003b). Changes in properties of soil-derived dissolved organic matter induced by biodegradation. *Soil Biol. Biochem.* **35**, 1129–1142.

Kalbitz, K., Schwesig, D., Rethemeyer, J., and Matzner, E. (2005). Stabilization of dissolved organic matter by sorption to the mineral soil. *Soil Biol. Biochem.* **37**, 1319–1331.

Kilbertus, G. (1980). Microhabitats in soil aggregates. Their relationship with bacterial biomass and the size of the procaryotes present. *Rev. Ecol. Biol. Sol.* **17**, 543–557.

Killham, K., Firestone, M. K., and McColl, J. G. (1983). Acid rain and soil microbial activity: Effects and their mechanisms. *J. Environ. Qual.* **12**, 133–137.

Körner, C. (2003). Slow in, Rapid out—Carbon flux studies and Kyoto targets. *Science.* **300**, 1242–1243.

Krishnan, S., Samson, N. P., Ravichandran, P., Narasimhan, D., and Dayanandan, P. (2000). Phytoliths of Indian grasses and their potential use in identification. *Bot. J. Linnean Soc.* **132**, 241–252.

Kuhlbusch, T. A. J. (1998). Black carbon in soils, sediments, and ice cores. In: *Environmental Analysis and Remediation*, Meyers, R. A., ed., John Wiley & Sons, Toronto, pp. 813–823.

Kuzyakov, Y., Friedel, J. K., and Stahr, K. (2000). Review of mechanisms and quantification of priming effects. *Soil Biol. Biochem.* **32**, 1485–1498.

Ladd, J. N., Amato, M., and Oades, J. M. (1985). Decomposition of plant material in Australian soils. III. Residual organic and microbial biomass C and N from isotope-labelled legume material and soil organic matter, decomposing under field conditions. *Aust. J. Soil Res.* **23**, 603–611.

Lafleur, P. M., Roulet, N. T., Bubier, J. L., Frolking, S., and Moore, T. R. (2003). Interannual variability in the peatland-atmosphere carbon dioxide exchange at an ombrotrophic bog. *Global Biogeochem. Cycles* **17**, 1036.

Lal, R. (1997). Residue management, conservation tillage and soil restoration for mitigating greenhouse effect by CO_2-enrichment. *Soil Tillage Res.* **43**, 81–107.

Lal, R. (2002a). Soil carbon dynamics in cropland and rangeland. *Environ. Poll.* **116**, 353–362.

Lal, R. (2002b). The potential of soils of the tropics to sequester carbon and mitigate the greenhouse effect. *Adv. Agron.* **74**, 155–192.

Lal, R. (2004). Soil carbon sequestration to mitigate climate change. *Geoderma.* **123**, 1–22.

Lal, R., and Bruce, J. P. (1999). The potential of world cropland soils to sequester C and mitigate the greenhouse effect. *Environ. Sci. Pollut.* **2**, 177–185.

Lal, R., and Kimble, J. M. (2000). Importance of soil bulk density and methods of its measurements. In *Assessment Methods for Soil Carbon*, Lal, R., Kimble, J. M., Follett, R. F., and Stewart, B. A., eds., Lewis Publishers, Boca Raton, FL, pp. 31–44.

Lal, R., Follett, R. F., Kimble, J., and Cole, C. V. (1999). Managing U.S. cropland to sequester carbon in soil. *J. Soil Water Cons.* **54**, 374–381.

Lal, R., Griffin, M., Apt, J., Lave, L., and Morgan, M. G. (2004). Managing soil carbon. *Science.* **304**, 393.

Lim, B., and Cachier, H. (1996). Determination of black carbon by oxidation and thermal treatment in recent marine and lake sediments and Cretaceous–Tertiary clays. *Chem. Geol.* **131**, 143–154.

Lobe, I, Du Preez, C. C., and Amelunga, W. (2002). Influence of prolonged arable cropping on lignin compounds in sandy soils of the South African Highveld. *Eur. J. Soil Sci.* **53**, 553–562.

Lowe, L. E. (1969). Distribution and properties of organic fractions in selected Alberta soils. *Can. J. Soil Sci.* **49**, 129–141.

Lundstrom, U., van Breemen, N., and Bain, D. 2000. The podzolization process. A review. *Geoderma.* **94**, 91–107.

Lynch, D. H., Voroney, R. P., and Warman, P. R. (2006). Use of ^{13}C and ^{15}N natural abundance techniques to characterize carbon and nitrogen dynamics in composting and in compost amended soils. *Soil Biol. Biochem.* **38**, 103–114.

Machado, S., Rhinhart, K., and Petrie, S. (2006). Long-term cropping system effects on carbon sequestration in Eastern Oregon. *J. Environ. Qual.* **35**, 1548–1553.

Magdoff, F., and Weil, R. R. (2004). Soil organic matter management strategies. In *Soil Organic Matter in Sustainable Agriculture*, Magdoff, F. R., and Weil, R. R., eds., CRC Press, New York, pp. 45–129.

Mamilov, A. S., and Dilly, O. M. (2002). Soil microbial eco-physiology as affected by short-term variations in environmental conditions. *Soil Biol. Biochem.* **34**, 1283–1290.

Marland G., Fruit K., and Sedjo, R. (2001). Accounting for sequestered carbon: the question of permanence. *Environ. Sci. Pollut.* **4**, 259–268.

Martens, D. A. (2000). Plant residue biochemistry regulates soil carbon cycling and carbon sequestration. *Soil Biol. Biochem.* **32**, 361–369.

Martin, J. P., Ervin, J. O., and Shepherd, R. A. (1966). Decomposition of the iron, aluminium, zinc and copper salts or complexes of some microbial and plant polysaccharides in soil. *Soil Sci. Soc. Am. Proc.* **30**, 196–200.

Martin, J. P., and Haider, K. (1979). Biodegradation of ^{14}C-labeled model and cornstalk lignins, phenols, model phenolase humic polymers, and fungal melanins as influenced by a readily available carbon source and soil. *Appl. Environ. Microbiol.* **38**, 283–289.

Martin, A., Mariotti, A., Balesdent, J., Lavelle, P., and Vuattoux, R. (1990). Estimate of organic matter turnover rate in a savannas soil by 13C natural abundance measurements. *Soil Biol. Biochem.* **22**, 517–523.

Masiello, C. A., and Druffel, E. R. M. (1998). Black carbon in deep-sea sediments. *Science.* **280**, 1911–1913.

Melillo, J. M., Aber, J. D., Linkins, A. E., Ricca, A., Fry, B., and Nadelhoffer, K. J. (1989). Carbon and nitrogen dynamics along the decay continuum: plant litter to soil organic matter. *Plant and Soil* **115**, 189–198.

Middelburg, J. J., Nieuwenhuize, J., and van Breugel, P. (1999). Black carbon in marine sediments. *Mar. Chem.* **65**, 245–252.

Nabuurs, G. J. (2004). Current consequences of past actions: How to separate direct from indirect. In *The Global Carbon Cycle: Integrating Humans, Climate and the Natural World*, Field, C. B., and Raupach, M. R., eds., Island Press, Washington, D.C., pp. 317–326.

Nadelhoffer, K. J., and Fry, B. (1994). Nitrogen isotope studies in forest ecosystems. In *Stable Isotopes in Ecology and Environmental Science*, Lajtha, K., and Michener, R. H. eds., Blackwell Scientific, London, UK, pp. 22–44.

Norgren, A. (1973). Opal phytoliths as indicators of soil age and vegetative history. Ph.D. dissertation. Oregon State University, Corvallis, OR, Dissertation. *Abstr. Intern.* **33**, 3421B.

Oades, J. M., and Waters, A. G. (1991). Aggregate hierarchy in soils. *Aust. J. Soil Res.* **29**, 815–828.

O'Brien, B. J., and Stout, J. D. (1978). Movement and turnover of soil organic matter as indicated by carbon isotope measurements. *Soil Biol. Biochem.* **10**, 309–317.

Oelke, C., Zhang, T., and Serreze, M. C. (2004). Modelling evidence for recent warming of the Arctic soil thermal regime. *Geophys. Res. Lett.* **31**, 1–4.

Pacala, S. W., Hurtt, G. C., Baker, D., Peylin, P., Houghton, R. A., Birdsey, R. A., Heath, L., Sundquist, E. T., Stallard, R. F., Ciais, P., Moorcroft, P., Caspersen, J. P., Shevliakova, E., Moore, B., Kohlmaier, G., Holland, E., Gloor, M., Harmon, M. E., Fan, S. M., Sarmiento, J. L., Goodale, C. L., Schimel, D., and Field, C. B. (2001). Consistent land- and atmosphere-based U.S. carbon sink estimates. *Science.* **292**, 2316–2320.

Parr, J. F., Dolic, V., Lancaster, G., and Boyd, W. E. (2001). A microwave digestion method for the extraction of phytoliths from herbarium specimens. *Rev. Palaeobot. Palynol.* **116**, 203–212.

Parr, J. F., and Sullivan, L. A. (2005). Soil carbon sequestration in phytoliths. *Soil Biol. Biochem.* **37**, 117–124.

Parton, W. J., Schimel, D. S., Cole, C. V., and Ojima, D. S. (1987). Analysis of factors controlling soil organic matter levels in Great Plains grassland. *Soil Sci. Soc. Am. J.* **51**, 1173–1179.

Parton, W. J., Schimel, D. S., Ojima, D. S., and Cole, C. V. (1994). A general model for soil organic carbon dynamics: Sensitivity to litter chemistry, texture and management. In *Quantitative Modelling of Soil Forming Processes*, Bryant, R. B., ed., Soil Science Society of America, Madison, WI, pp. 147–168.

Pease, D. S., and Anderson, J. U. (1969). Opal phytoliths in *Bouteloua eriopoda* Torr. roots and soils. *Soil Sci. Soc. Am. Proc.* **33**, 321–322.

Pendall, E., Betancourt, J. L., and Leavitt, S. W. (1999). Paleoclimatic significance of dD and d13C values in piñon pine needles from packrat middens spanning the last 40,000 years. *Palaeogeog. Palaeoclimatol. Palaeoecol.* **147**, 53–72.

Pendall, E, Bridgham, S, Hanson, P. J., Hungate, B., Kicklighter, D. W., Johnson, D. W., Law, B. E., Luo, Y., Megonigal, J. P., Olsrud, M., Ryan, M. G., and Wan S. (2004). Below-ground process responses to elevated CO_2 and temperature: A discussion of observations, measurement methods, and models. *New Phytol.* **162**, 311–322.

Piperno, D. R. (1988). *Phytolith Analysis: An Archaeological and Geological Perspective*, Academic Press, London, UK.

Polubesova, T., Chen, Y., Navon, R., and Chefetz, B. (2008). Interactions of hydrophobic fractions of dissolved organic matter with Fe^{3+}– and Cu^{2+}–montmorillonite. *Environ. Sci. Technol.* **42**, 4797–4803.

Post, W. M., and Kwon, K. C. (2000). Soil carbon sequestration and land-use change: Processes and potential. *Global Change Biol.* **6**, 317–327.

Powlson, D. S., Brookes, P. C., and Christensen, B. T. (1987). Measurement of soil microbial biomass provides an early indication of changes in total soil organic matter due to straw incorporation. *Soil Biol. Biochem.* **19**, 159–164.

Pretty, J., Brett, C., Gee, D., Hine, R., Mason, C., Morison, J., Rayment, R., Van der Bijl, G., and Dobbs, T. (2001). Policy challenges and priorities for internalising the externalities of modern agriculture. *J. Environ. Plan. Manag.* **44**, 263–283.

Raich, J. W., and Schlesinger, W. H. (1992). The global carbon dioxide flux in soil respiration and its relationship to vegetation and climate. Tellus. Series B, *Chem. Phys. Meteor.* **44**, 81–99.

Rice, C. W. (2006). Introduction to special section on greenhouse gases and carbon sequestration in agriculture and forestry. *J. Environ. Qual.* **35**, 1338–1340.

Riquier, J. (1960). Les phytoliths de certain sols Tropicaux et des podzols. *Transactions of the Seventh International Congress of Soil Science*, Madison, WI, pp. 425–431.

Rühlmann, J. (1999). A new approach to estimating the pool of stable organic matter in soil using data from long-term field experiments. *Plant Soil* **213**, 149–160.

Rumpel, C., Kögel-Knabner, I., and Bruhn, F. (2002). Vertical distribution, age, and chemical composition of organic carbon in two forest soils of different pedogenesis. *Org. Geochem.* **33**, 1131–1142.

Schwesig, D., Kalbitz, K., and Matzner, E. (2003). Effects of aluminium on the mineralization of dissolved organic carbon derived from forest floors. *Eur. J. Soil Sci.* **54**, 311–322.

Schmidt, I. K., Jonasson, S., and Michelsen, A. (1999). Mineralization and microbial immobilization of N and P in arctic soils in relation to season, temperature and nutrient amendment. *Appl. Soil Ecol.* **11**, 147–160.

Schmidt, M. W. I., and Noack, A. G. (2000). Black carbon in soils and sediments: Analysis, distribution, implications, and current challenges. *Glob. Biogeochem. Cycles* **14**, 777–793.

Schmidt, M. W. I., Skjemstad, J. O., Czimczik, C. I., Glaser, B., Prentice, K. M., Gelinas, Y., and Kuhlbusch, T. A. J. (2001). Comparative analysis of black carbon in soils, *Global Biogeochem. Cycles* **15**, 163–167.

Schimel, D. S., Braswell, B. H., Holland, E. A., McKeown, R., Ojima, D. S., Painter, T. H., Parton, W. J., and Townsend, A. R. (1994). Climatic, edaphic, and biotic controls over storage and turnover of carbon in soils. *Glob. Biogeochem.* **8**, 279–293.

Sequi, P., De Nobili, M., Leita, L., and Cercignani, G. (1985). A new index of humification. *Agrochimica* **30**, 175–178.

Simpson, M. J., and Hatcher, P. G. (2004). Determination of black carbon in natural organic matter by chemical oxidation and solid-state ^{13}C nuclear magnetic resonance spectroscopy. *Org. Geochem.* **35**, 923–935.

Six, J., Carpentier, A., Kessel, C. van, Merckx, R., Harris, D., Horwath, W. R., and Lüscher, A. (2001). Impact of elevated CO_2 on soil organic matter dynamics as related to changes in aggregate turnover and residue quality. *Plant Soil* **234**, 27–36.

Six, J., Conant, R. T., Paul, E. A., and Paustian, K. (2002). Stabilization mechanism of soil organic matter: Implications for C-saturation of soils. *Plant Soil* **241**, 155–176.

Skjemstad, J. O., Clarke, P., Taylor, J. A., Oades, J. M., and McClure, S. G. (1996). The chemistry and nature of protected carbon in soil. *Aust. J. Soil Res.* **34**, 251–271.

Skjemstad, J. O., Reicosky, D. C., Wilts, A. R., and McGowan, J. A. (2002). Charcoal carbon in US Agricultural Soils. *Soil Sci. Soc. Am. J.* **66**, 1249–1255.

Smith, P. (2004). Carbon sequestration in croplands: the potential in Europe and the global context. *Eur. J. Agron.* **20**, 229–236.

Smith, P., Powlson, D., Glendinig, M., and Smith, J. (1997). Potential for carbon sequestration in European soils: Preliminary estimates for five scenarios using results from long-term experiments. *Global Change Biol.* **3**, 67–79.

Sokoloff, V. P. (1938). Effect of neutral salts of sodium and calcium on carbon and nitrogen in soils. *J. Agric. Res.* **57**, 201–216.

Sparling, G. P., Wheeler, D., Vesely, E. T., and Schipper L. A. (2006). What is soil organic matter worth? *J. Environ. Qual.* **35**, 548–557.

Stevenson, F. J. (1994). *Humus Chemistry. Genesis, Composition and Reactions*. John Wiley & Sons, New York.

Stevenson, F. J., and Cole, M. A. (1999). *Cycles in Soils: Carbon, Nitrogen, Phosphorus, Sulfur, Micronutrients*, 2nd edition, John Wiley & Sons, New York.

Stout, J. D., Goh, K. M., and Rafter, T. A. (1981). Chemistry and turnover of natural occurring resistant organic compounds in soil. In *Soil Biochemistry*, Vol. **5**, ed. Paul, E. A., and Ladd, J. N., eds., Marcel Dekker, NY, pp. 19–24.

Suman, D. O., Kuhlbusch, T. A. J., and Lim, B. (1997). Marine sediments: A reservoir for black carbon and their use as spatial and temporal records of combustion. In *Sediment Records of Biomass Burning and Global Change*, Clark, J. S., Cachier, H., Goldammer, J. G., and Stocks, B., eds., NATO ASI Series I: Global Environmental Change, Vol. **51**, Springer-Verlag, Berlin, Heidelberg, pp. 271–293.

Swift, R. S., (1999). Macromolecular properties of humic substances: Fact, fiction, and opinion. *Soil Sci.* **164**, 790–802.

Tan, Z. X., Lal, R., Smeck, N. E., and Calhoun, F. G. (2004). Relationships between surface soil organic carbon pool and site variables. *Geoderma.* **121**, 187–195.

Tate, R. L. (1992). III Humic and fulvic acids: Formation and decomposition. In *Soil Organic Matter Biological and Ecological Effects*, Tate, R. L., ed., Krieger, Melebar, FL, pp. 147–164.

Telles, E. de C. C., de Camargo, P. B., Martinelli, L. A., Trumbore, S. E., da Costa, E. S., Santos, J., Higuchi, N., and Oliveira, R. C. (2003). Influence of soil texture on carbon dynamics and storage potential in tropical forest soils of Amazonia. *Glob. Biogeochem. Cycles* **17**, 1040.

Tiessen, H., Cuevas, E., and Chacon, P. (1994). The role of soil organic matter in sustaining soil fertility. *Nature.* **371**, 783–785.

van der Linden, A. M. A., Jeurisson, L. J. J., Van Veen, J. A., and Schippers, G. (1989). Turnover of soil microbial biomass as influence by soil compaction. In *Nitrogen in Organic Wastes Applied to Soil*, Attansen, J., and Henriksen, K., eds., Academic Press, London, UK, pp. 25–36.

Vaughan, D., and Ord, B. G. (1985). Soil organic matter—A perspective on its nature, extraction, turnover and role in soil fertility. Soil organic matter and biological activity. *Dev. Plant Soil Sci.* **16**, 1–35.

Volkoff, B., and Cerri C. (1987). Carbon isotopic fractionation in subtropical Brazilian grassland soils, Comparison with tropical forest soils. *Plant Soil.* **102**, 27–31.

Waksman, S. A. (1938). *Humus. Origin, Chemical Composition and Importance in Nature*, 2nd edition, Williams and Wilkins, Baltimore.

Wattel-Koekkoek, E. J. W., van Genuchten, P. P. L., Buurman, P., and van Lagen, B. (2001). Amount and composition of clay-associated soil organic matter in a range of kaolinitic and smectitic soils. *Geoderma.* **99**, 27–49.

Webster, E. A., Chudek, J. A., and Hopkins, D. W. (2000). Carbon transformations during decomposition of different components of plant leaves in soil. *Soil Biol. Biochem.* **32**, 301–314.

Webster, E. A., Hopkins, D. W., Chudek, J. A., Haslam, S. F. I., Šimek, M., and Pîcek, T. (2001). The relationship between microbial carbon and the resource quality of soil carbon *J. Environ. Qual.* **30**, 147–150.

Weil, R. R., and Kroontje, W. (1979). Physical condition of a Davidson clay loam after five years of heavy poultry manure application. *J. Environ. Qual.* **8**, 387–392.

West, T. O., and Post, W. M. (2002). Soil organic carbon sequestration rates by tillage and crop rotation: A global data analysis. *Soil Sci. Soc. Am. J.* **66**, 1930–1946.

Wilding, L. P., Brown, R. E., and Holowaychuk, N. (1967). Accessibility and properties of occluded carbon in biogenic opal. *Soil Sci.* **103**, 56–61.

Wolters, V. (2000). Invertebrate control of soil organic matter stability. *Biol. Fertil. Soils* **31**, 1–19.

Woodbury, P. B., Heath, L. S., and Smith, J. E. (2006). Land use change effects on forest carbon cycling throughout the Southern United States. *J. Environ. Qual.* **35**, 1348–1363.

6

STORAGE AND TURNOVER OF ORGANIC MATTER IN SOIL

M. S. TORN

Earth Sciences Division, Lawrence Berkeley National Laboratory and University of California, Berkeley, California

C. W. SWANSTON

USDA Forest Service, Northern Research Station, Houghton, Michigan

C. CASTANHA

Earth Sciences Division, Lawrence Berkeley National Laboratory, Berkeley, California

S. E. TRUMBORE

Department of Earth System Science, Center for Global Environmental Change Research and Institute for Geophysics, University of California, Irvine, California

Biophysico-Chemical Processes Involving Natural Nonliving Organic Matter in Environmental Systems, Edited by Nicola Senesi, Baoshan Xing, and Pan Ming Huang

6.1. INTRODUCTION

Historically, attention on soil organic matter (SOM) has focused on the central role that it plays in ecosystem fertility and soil properties, but in the past two decades the role of soil organic carbon in moderating atmospheric CO_2 concentrations has emerged as a critical research area. This chapter will focus on the storage and turnover of natural organic matter in soil (SOM), in the context of the global carbon cycle.

Organic matter in soils is the largest carbon reservoir in rapid exchange with atmospheric CO_2, and thus it is important as a potential source and sink of greenhouse gases over time scales of human concern (Fischlin and Gyalistras, 1997). SOM is also an important human resource under active management in agricultural and range lands worldwide. Questions driving present research on the soil C cycle include: Are soils now acting as a net source or sink of carbon to the atmosphere? What role will soils play as a natural modulator or amplifier of climatic warming? How is C stabilized and sequestered, and what are effective management techniques to foster these processes? Answering these questions will require a mechanistic understanding of how and where C is stored in soils.

The quantity and composition of organic matter in soil reflect the long-term balance between plant carbon inputs and microbial decomposition, as well as other loss processes such as fire, erosion, and leaching. The processes driving soil carbon storage and turnover are complex and involve influences at molecular to global scales. Moreover, the relative importance of these processes varies according to the temporal and spatial scales being considered; a process that is important at the regional scale may not be critical at the pedon scale. At the regional scale, SOM cycling is influenced by factors such as climate and parent material, which affect plant productivity and soil development. More locally, factors such as plant tissue quality and soil mineralogy affect decomposition pathways and stabilization. These factors influence the stability of SOM in part by shaping its molecular characteristics, which play a fundamental role in nearly all processes governing SOM stability but are not the focus of this chapter.

We review here the most important controls on the distribution and dynamics of SOM at plot to global scales, along with methods used to study them. We also explore the concepts of controls, processes, and mechanisms, as well as how they operate across scales. The concept of SOM turnover, or mean residence time, is central to this chapter and so it is described in some detail. The Appendix details the use of radiocarbon (^{14}C), a powerful isotopic tool for studying SOM dynamics. The genesis of this chapter was a NATO Advanced Study Institute on "Soils and Global Change: Carbon Cycle, Trace Gas Exchange and Hydrology," June 16–27, 1997, Chateau de Bonas, France, and the written material prepared after the workshop by S. Trumbore and M. Torn.

6.2. THE AMOUNT OF ORGANIC CARBON STORED IN SOILS

In this section we summarize current estimates of C stocks in soils and explore the factors predicting broad-scale patterns in soil C storage.

6.2.1. Empirical Estimates of Global Carbon Stocks in Soils

Most assessments of global soil C stocks have included only the top meter of soil, but recent estimates have encompassed lower depths. Historical global estimates for the top meter of soil ranged from 800 to 2400 Pg C, more recently narrowing to a range of 1300–1600 Pg C to 1 m. Batjes (1996) estimated that an additional 900 Pg C is stored between 1 and 2 m depth, and Jobbagy and Jackson (2000) revised that estimate to 500 Pg between 1 and 2 m and another 350 Pg between 2 and 3 m depth. Global organic C stocks to 3 m are currently estimated at 2300 Pg, with an additional 1000 Pg contained in permafrost and peatlands (Jobbagy and Jackson, 2000; Zimov et al., 2006). Soil C distribution with ecosystem type is shown in Figure 6.1.

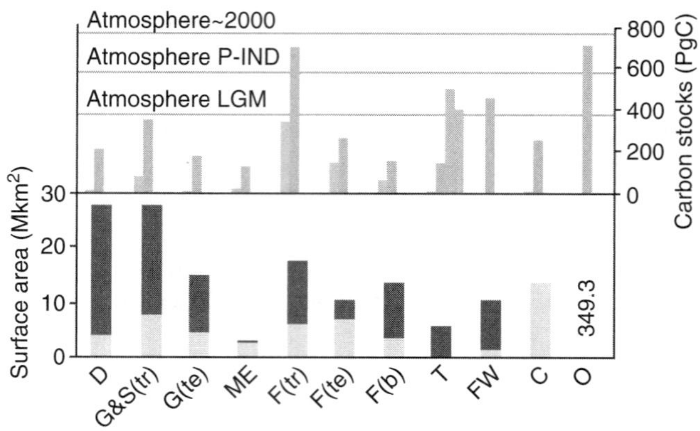

Figure 6.1. Ecosystem area and soil carbon content to 3-m depth. Lower Panel: Global areal extent of major ecosystems, transformed by land use in yellow, untransformed in purple. Data from Hassan et al. (2005) except for Mediterranean-climate ecosystems; transformation impact is from Myers et al. (2000); and ocean surface area is from Hassan et al. (2005). Upper Panel: Total C stores in plant biomass, soil, yedoma/permafrost. D, deserts; G&S(tr), tropical grasslands and savannas; G(te), temperate grasslands; ME, Mediterranean ecosystems; F(tr), tropical forests; F(te), temperate forests; F(b), boreal forests; T, tundra; FW, freshwater lakes and wetlands; C, croplands; O, oceans. Data are from Sabine et al. (2004), except C content of yedoma permafrost and permafrost (light blue columns, left and right, respectively; Zimov et al., 2006), and ocean organic C content (dissolved plus particulate organic; Denman et al., 2007). This figure considers soil C to 3-m depth (Jobbagy and Jackson, 2000). Approximate carbon content of the atmosphere is indicated by the dotted lines for last glacial maximum (LGM), pre-industrial (P-IND) and current (about 2000). Reprinted from Fischlin et al. (2007) in IPCC (2007). See color insert.

Two general approaches have been taken to estimate the global soil C inventory from soil profile data. The first, used by Schlesinger (1977) and Post et al. (1982), relates C storage to climate and vegetation (for example, expressed as Holdridge life zone classifications). For example, Post et al. (1982) generated relationships of climate and vegetation with soil C using 2700 soil profiles, and they used these to calculate a global soil C inventory of 1400 Pg C in the top 1 m. A second approach uses soil mapping units for extrapolation (Eswaran et al., 1993; Batjes, 1996). Eswaran (1993) determined the average C inventory for each soil order, based on data from roughly 1000 pedons from FAO/UNESCO and 15,000 profiles from U.S. Department of Agriculture databases. This soil map-based estimate of soil C inventory globally is 1600 Pg C in the top 1 m. Batjes (1996), using a database of 4353 soil profiles considered to be representative of soil units on the FAO map, estimated 1500 Pg soil organic C to 1-m depth.

Regardless of approach, these global inventories are acknowledged to underestimate the total amount of dead organic matter in ecosystems because they do not include important reservoirs. All omit C stored in surface detritus (including the O horizon), which contains an additional 50–200 Pg C (Matthews, 1997). Moreover, some soils contain significant soil C even below 3 m. For example, some tropical soils and many histosols contain as much C below 1 m as they do above 1 m (Eswaran et al., 1993; Nepstad et al., 1994). In general, highly weathered soils are often very deep (tens of meters) so that even very low C concentrations may add up to a large amount of stored C in the total (Sombroek et al., 1993). Finally, most soil C estimates do not include organic C in permafrost or paleosols, a reservoir containing roughly 900 Pg C in the boreal and arctic regions alone (Figure 6.1).

Not surprisingly, given that they are extrapolations of point observations, global C inventories have large uncertainties. Post et al. (1982) estimate an error of ±200 Pg C (14%), reflecting the variability in soil C inventory within each life zone category. Eswaran et al. (1993) show large coefficients of variation, of 28–70%, for soils within given soil classification categories. Much of the profile data employed in these estimates includes only carbon density and not bulk density, so that the bulk density must be estimated from empirical relationships between C density and bulk density developed from profiles where both were measured (Zinke et al., 1984).

6.2.2. Understanding Variation in Carbon Storage Across the Landscape

Constructing estimates of the amount of C stored in soils requires extrapolating from individual soil profiles to larger regions. Likewise, model simulations of soil C stocks require quantitative relationships with the factors controlling stocks, and distributed data for model testing. For all these reasons, we need to link soil C to factors that are mapped or modeled, globally. Moreover, linking soil C storage to environmental factors at broad spatial and temporal scales will help us gain insight into the large-scale controls on C cycling.

Are there predictable ways in which C storage varies across the landscape? Jenny (1941), expanding on an approach by Dokuchaev (Glinka, 1927; Jenny, 1941) and Rizpolozhenski (Lapenis et al., 2000), suggested that soil properties—including C inventory—may be predicted from soil forming, or "state," factors. This concept is expressed by the "clorpt" equation:

$$\text{Soil property (in this case, SOM inventory)} = f(\text{cl, o, r, p, t})$$

where cl stands for *climate*, o stands for *potential organisms* (vegetation and fauna), r stands for *relief* (aspect and topography), p stands for *parent material*, and t stands for *time*. One of Jenny's important contributions was to develop the experimental approach—carefully selected sets of sites that isolate the state variables of interest—that derives from the clorpt relation (Amundson and Jenny, 1997). In this approach, sites are selected such that the variable (state factor) of interest varies while all other important factors are held relatively constant. For example, to understand the influence of *temperature regimes*, Jenny located a series of sites with the same soil age (*time*), biota, parent material, and precipitation but with different temperature regimes. In contrast, interpreting the influence of, for example, temperature, from gradients in which other state factors vary (such as elevation gradients in which parent material also varies), is difficult.

Note that each state factor can influence soil carbon storage in two ways: by influencing the quantity and quality of plant inputs and by influencing the residence time of organic matter in the soil. Figures 6.2–6.4 show the influence of three state factors on soil C storage. Many of the descriptions of state factors below touch upon the importance of controls on SOM stabilization, and these controls are discussed further in a following section.

6.2.2.1. Climate. Climate has an overriding influence on large-scale patterns in ecosystem properties, including soil C cycling, through its control of plant commu-

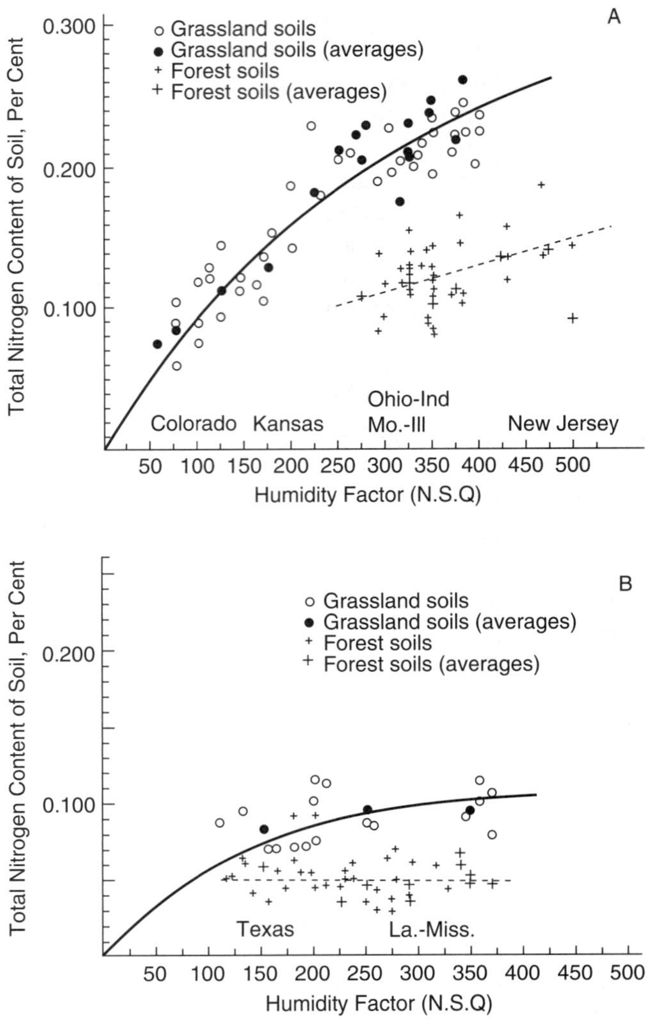

Figure 6.2. (A) Variations in %N (which is proportional to C density) with precipitation along the 11 °C isotherm in the Great Plains of the United States. The humidity factor (NSQ, *Niederschlag-Sattigungsdefizit* from the German, or Meyer's quotient) is the total annual precipitation (mm) divided by the absolute saturation deficit of air (mm mercury). All soils were developed on loess deposits from the last glacial maximum. **(B)** Change in %N with precipitation along the 19 °C isotherm. Note that relative C density (estimated by assuming that the C/N ratio of SOM is fairly constant) is lower at higher mean annual temperature. Reprinted with permission from Jenny, H. (1941). *Factors of Soil Formation*, Dover Publications, New York.

nity composition and productivity (Holdridge, 1947), which affect the quantity and quality of inputs to the SOM pool, as well as of microbial community composition and decomposition activity. Climate acts over a range of time scales as well, influencing which minerals are the stable weathering products, whether physical or chemical erosion processes dominate the landscape, and rates of microbial decay compared to other removal processes.

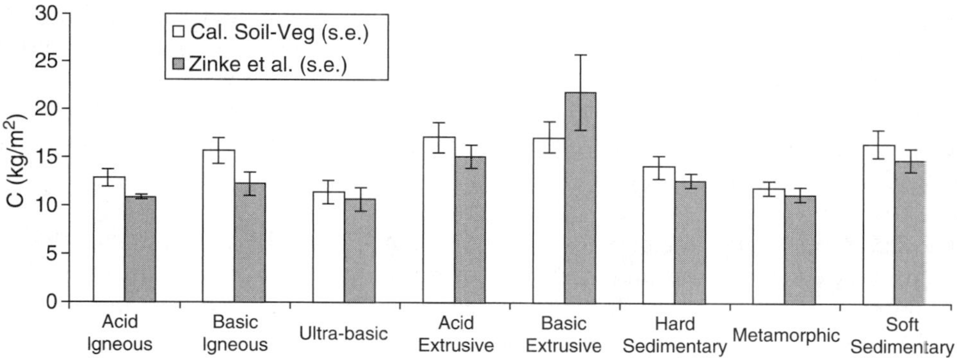

Figure 6.3. Soil organic carbon inventory to 1 m depth by parent material category, for California and globally. Well-drained soils in California (white bars) are from the Soil-Vegetation Survey data set, $n = 568$, well-drained soils only. Worldwide data (gray bars) are from Zinke et al. (1984), $n = 2995$, which includes the California Soil-Vegetation Survey data, all drainage classes. Reprinted from Torn et al. (1997).

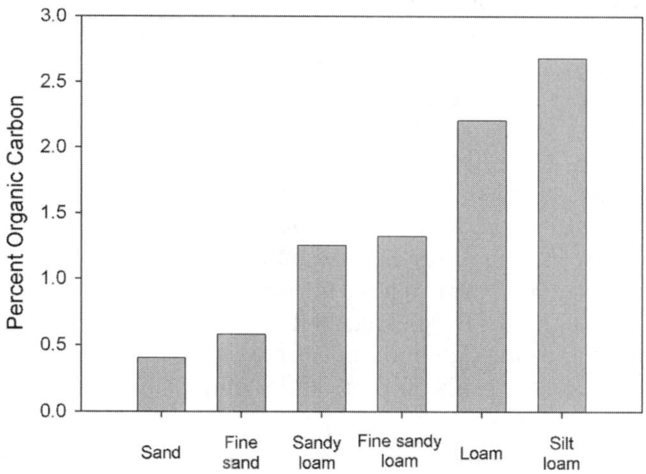

Figure 6.4. Variation of organic carbon density with texture, in surface soils developed on glacial till and loess in Iowa. Soils with more loess have finer texture [data are from Brown (1936), as reported in Jenny (1941)].

Jenny's research is exemplary for both our understanding of climate and soil, and applying the state factor approach. Jenny (1930) measured C storage in soil sampled across the U.S. Great Plains (Figure 6.2), where the parent material consists largely of loess deposited during the last glacial period. In this region, precipitation increases from west to east, while mean annual temperature increases from north to south. By comparing soils at the same longitude but different latitudes, the gradient in temperature can be studied in relative isolation from variation in mean precipitation and parent material. By comparing soils across longitude, the effect of precipitation can be studied with minimal confounding variation in temperature. Jenny used the variation in climate across the Great Plains to explore the influence of temperature and moisture on SOM. Carbon stocks were largest toward the cooler and wetter northeast, and smallest in the hotter and drier southwest (Jenny, 1941). More generally, decomposition is more sensitive to climate than is productivity, so that, all else equal, higher latitudes and elevations tend to have larger C stock than tropical or low elevation ecosystems.

6.2.2.2. Organisms. Organisms—including plants, animals, and soil microbes—control the chemical form and location of organic matter input to soils. In an analysis of more than 2700 soil profiles from three global databases, Jobbagy and Jackson (2000) found that vegetation type was closely correlated with the amount of soil carbon and its distribution with depth. Vegetation controls C cycling in several ways, some of which covary with climate. First, net primary productivity, which depends on plant species and communities, determines C input rates to soil. For soils with similar decomposition rates, those with more productive vegetation will have higher organic C inventories. For example, soil carbon losses after forest conversion to pasture or agriculture are partly attributed to decreases in primary productivity (Trumbore et al., 1995). Second, vegetation type affects tissue chemistry and seasonality of inputs. Finally, plant species differ in the proportion of photosynthate partitioned to roots, shoots, or woody structures. Since roots comprise a large fraction of plant inputs to soil, and soil C decomposition decreases with depth, the depth distribution of root inputs affects soil C storage. All of these attributes have large effects on the transformation and stabilization of organic matter (Steinmann et al., 2004; Bird and Torn, 2006; Zanelli et al., 2006).

While litter chemistry clearly influences initial decomposition rates, the influence of plant–tissue chemistry on the structure and decomposability of SOM is a current and unresolved research question. For example, rates of litter and root decomposition have been correlated with lignin, nitrogen, and nonstructural carbohydrate content (Melillo et al., 1982; Berg et al., 2001; Zhang et al., 2008), but many initially recalcitrant compounds like lignin do not persist in soils (von Lützow et al., 2006). Quideau (1998) found that only 50 years after being planted with different vegetation types, adjacent sites with the same parent material, soil age, and climate had several-fold differences in soil C stock and differences in SOM chemistry, documented by ^{13}C nuclear magnetic resonance, corresponding to the different vegetation inputs.

Fauna also influence soil carbon cycling. Bioturbation mixes and aerates soil, physically breaks down litter, creates flow paths for water in soil, and can reduce surface litter stocks and enhance erosion (Bohlen et al., 2004). For example, along a gradient of European earthworm (*Lumbricus terrestris*) colonization in a deciduous forest of northern Michigan, earthworms are associated with a decrease in litter-layer thickness, apparently mixing some forest floor organic matter into the mineral soil. Thus, fauna can create spatial patterns in SOM stocks.

In addition to being the enzymatic agents of biotic decomposition of SOM, microbial cell by-products are increasingly recognized as major building blocks of SOM. As a result, the microbial community, controlled by climate, vegetation, and soil environment, is a key mediator of organic matter composition and decomposition. The spatial scales of influence, however, have not been well characterized.

6.2.2.3. Relief. Factors like soil slope, drainage, and erosion create significant variations in C stock within and among watersheds. For instance, poorly drained soils tend to have high C stocks due to low availability of O_2 for decomposition. Erosion is a key process underlying landscape patterns. Erosion redistributes nutrient-rich topsoil downhill, thereby increasing fertility in depositional environments and potentially reducing it in eroding environments. The effects of erosion on decomposition are more complicated. Transport of particles breaks down aggregates and increases SOM accessibility for decomposition. On the other hand, burial of eroded C in depositional settings can reduce its decomposition rate (Berhe et al., 2007).

These patterns and processes are particularly important when considering how representative specific soil profiles are of regional soil C storage (Davidson and Lefebvre, 1993), as well as in evaluating the effects of land cover and use change (if it alters erosion rates) on terrestrial C stocks. In terms of spatial patterns within watersheds, C inventories are typically higher at the bottom of slopes for two reasons. Lower slope positions have slower decomposition as a result of fine texture, low O_2, and burial. They also tend to have higher inputs relative to upper slope positions from higher productivity and deposition of eroded material.

In terms of total watershed C stock, recent empirical and modeling studies conclude that erosion—and the balance between its effects on productivity and decomposition in eroding and depositional sites—tends to lead to an *increase* in stock, even if there are local decreases in the eroding sites themselves (Stallard, 1998; Smith et al., 2005; Berhe et al., 2007). The rate at which C accumulates in a watershed due to erosion and deposition depends on the strength of erosion and management of productivity, as well as on the types of depositional settings involved (Berhe et al., 2007; van Oost et al., 2007).

Because so much land area is subject to erosion, this watershed-increase in soil C scales to a large global C sink. Erosion by wind and water affects roughly $10 \times 10^{12}\,m^2$ of land worldwide (Jacinthe and Lal, 2001) and moves 1–5 $Pg\,C\,y^{-1}$, with more than 70% deposited terrestrially (Stallard, 1998). As a result of the erosion effects on decomposition and NPP described above, recent studies have suggested that erosion results in a global terrestrial C sink of 0.25–1 $Pg\,C\,y^{-1}$ (Stallard, 1998; Smith et al., 2005; Berhe et al., 2007).

6.2.2.4. Parent Material. An analysis of worldwide data shows differences in carbon storage among parent materials, in spite of the fact that it does not control for confounding factors such as soil age (Zinke et al., 1984) (Figure 6.3). Parent material—the mineral substrate at the inception of soil development—has a variety of influences over SOM stocks. It affects the chemistry and fertility of soil and thus plant productivity (C inputs to soil); texture, which affects soil moisture retention and thus both productivity and decomposition; and clay content and mineralogy, which affect SOM stabilization. For example, soils developed on volcanic ash or rocks of basic pH often contain more organic carbon than those formed on granitic or acidic parent material (Harradine and Jenny, 1958; Marti and Badia, 1995). Jenny (1980) found increasing C density with increasing loess in soils developed on post-glacial till, a pattern he ascribed to finer texture (Figure 6.4). Indeed, soil texture, particularly clay content, is positively correlated with C storage in many sites and is used as the proxy for mineral control of stabilization in most ecosystem C models.

The reliance on texture as a proxy for mineral stabilization is changing very rapidly, however.

Recent studies suggest that storage and turnover are much more closely related to mineral properties, in particular the poorly crystalline phases, than to texture *per se* (Torn et al., 1997; Masiello et al., 2004; Basile-Doelsch et al., 2005; Kleber et al., 2005). For example, in paired forest soils derived from granitic versus andesitic parent materials, the latter stored almost 50% more C than did the granitic soils despite similar levels of clay and aggregate stability, climate, and vegetation (Rasmussen et al., 2005). Based on these and other recent studies, the amount of mineral-associated C appears correlated with (a) reactive iron and aluminum in short-range-order minerals and humic complexes and (b) their interaction with aggregates.

6.2.2.5. Time. Soils develop over time. Soil profiles thicken and carbon stocks increase with time and soil age, during early-to-mid stages of development, but may decrease as soils become highly weathered. A chronosequence is a series of sites that vary in the period of time over which the soil has developed, or since significant disturbance. In a state factor experiment, all the sites in the chronosequence would have similar climate history, parent material, and, unless included in the study design, vegetation species. Such sequences have been constructed from terraces formed by coastal uplift or rivers, glacier retreat, and volcanic deposits. A chronosequence in Hawaii, developed on ash deposits of different ages (Chadwick et al., 1999), illustrates a trend seen in other, similar studies (Jahn et al., 1992; Percival et al., 2000). Soil organic carbon accumulates for the first several hundred thousand years of soil development, then declines in very old soils (older than a million years). The slow buildup and subsequent decline of carbon have been correlated with changes in the amount and type of soil minerals that can stabilize SOM (e.g., Oades, 1989)—for example, in the amount of reactive iron and aluminum and noncrystalline, secondary minerals like allophane and imogolite, which have large reactive surface areas and stabilize organic compounds (Jahn et al., 1992; Torn et al., 1997; Masiello et al., 2004) (Figure 6.5).

In many cases, the differences in C storage among soils developed on different parent materials decrease as soils reach great age (millions of years). We speculate that over millions of years, the mineral content and the mineral-associated organic C content of different soils tends to converge to a state predicted or constrained by climate.

6.2.2.6. Opportunities and Drawbacks to Gradient Studies and Other Approaches. These state factors provide general rules for predicting how C inventory in soils will vary across large regions and over long time scales. The largest organic C inventory should be in cool, wet climates with high ecosystem productivity, on young volcanic surfaces (as in the Pacific Northwest of the United States). The smallest C stocks should be found in hot, arid regions of low productivity (as in deserts). Another observation, put forth by Jenny, is that the major reservoirs of soil carbon change with latitude: At low latitudes, very little of the total soil C is stored in surface detritus, and most of the C is in the mineral soil. At high latitudes, slow litter decay leads to large accumulations of detrital organic material, and relatively little of the organic C is in the mineral soil.

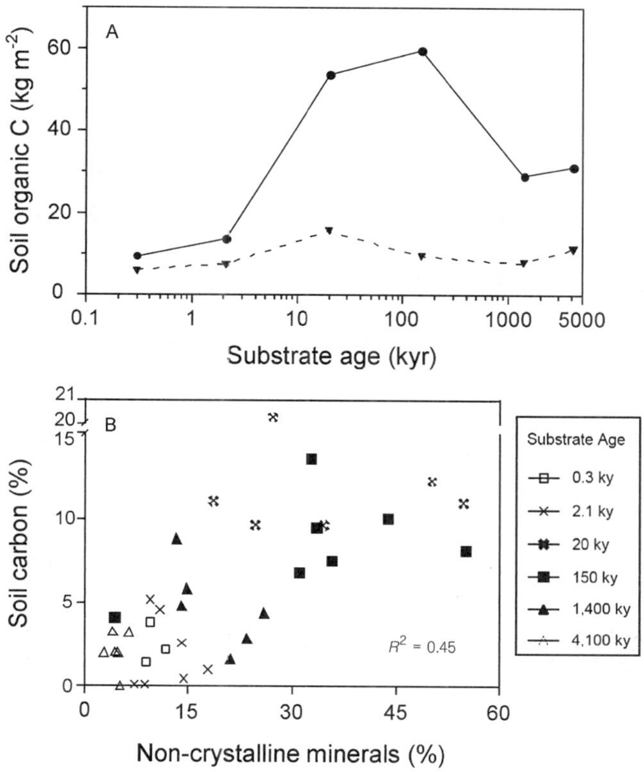

Figure 6.5. Changes in soil organic C storage and mineral content along a chronosequence in Hawaii (Torn et al., 1997). The substrate for soil development are basaltic ash deposits of known age. Climate and vegetation are virtually the same across the sites. **(A)** Soil organic C inventory versus ash substrate age. The solid line is the whole mineral soil to the C horizon, and the dashed line is the top 20 cm. The increase and subsequent decrease in SOM with soil age is mostly due to changes in the subsurface mineral soil. **(B)** The correlation of soil carbon in mineral horizons with the amount of noncrystalline minerals.

The approach of studying gradients is not without drawbacks. It is challenging to locate clean environmental gradients, meaning gradients with a minimum of confounding variability. For example, one of the problems in climate-gradient studies is that climate and vegetation cannot always be separated as independent variables across landscapes. In Hawaii, the same tree species, *Metrosideros polymorpha*, dominates native forests in young to mature and mesic to wet sites, and this is one reason that Hawaii has proven a rich location for gradient studies. For research relevant to anthropogenic environmental change, a fundamental drawback of most gradients is that they reflect gradual or long-term influences rather than rapid or transient responses (Dunne et al., 2004). For example, relationships between soil C stocks and vegetation or climate across natural gradients have taken shape over relatively long time scales. The response of a soil to a rapid change in vegetation or climate may not be of the same magnitude or even in the same direction as that predicted from natural gradients. For example, the difference in soil C stored at two elevations with a 3 °C temperature difference may be quite different from the change in C stocks that would occur if the higher elevation site warmed by 3 °C in a hundred years.

Experimental manipulations are useful not only for controlling variables of concern but also for investigating short-term responses to environmental change. They, too, have drawbacks, however. Field experiments may not run for long enough or encompass a large enough area to predict long-term effects of environmental change. For example, the relationship of C stock to soil temperature was negative during the first nine years of experimental warming in a montane meadow, but positive along a local climate gradient. The C decline observed in the experiment appears to have been a transient, process-rate response because concurrent changes in plant litter quality may lead to increases in soil C storage (Saleska et al., 2002). Another limitation of ecological experiments is that they tend to use step changes in variables rather than matching the gradual rate of expected changes (Shaver et al., 2000). Results from a suite of ecosystem warming manipulations in Europe and North America indicate that (1) the same temperature change can elicit different responses, depending on the initial climate and biogeochemical conditions of the system; (2) temperature affects ecosystems rapidly via process rates and more slowly via species composition and tissue chemistry; and, as a result, (3) the magnitude and direction of the response can change over time.

Finally, processes operating at larger spatial scales may control the storage of C in soils. Fire, for example, is as important a loss mechanism as decomposition for organic C in thick detrital layers in boreal forests (Harden et al., 2000). For fire-prone regions, the net status of the land surface as a C sink or source depends as much on the area burned in a given year as on the responses of decomposition rates to weather variability in unburned areas.

6.3. TURNOVER TIME AND DYNAMICS OF SOIL ORGANIC MATTER

The chapter up to now has focused on the amount of C stored in soils. However, knowing the amount of organic C in soil provides little insight into its roles in ecosystem function or atmospheric feedbacks. For example, a large SOM reservoir that is extremely stable may provide little in the way of plant-available nitrogen and may respond slowly to climate change. It is thus important to understand not only how much C is stored in a reservoir, but also how rapidly the C cycles. This is not a simple proposition, however, because SOM is a complex mixture of compounds that cycle along a continuum of time scales from minutes to tens of thousands of years. Segregating SOM into discrete reservoirs with different turnover times, along with understanding their relationship to biotic and soil conditions, is one of most important challenges for biogeochemical research today. This section summarizes some approaches and observational constraints for characterizing C dynamics in soils.

To lay the foundation for the following section on metrics, consider that the decomposition flux from soil is a function of the soil C stock and its decay rate. More strictly speaking, decomposition of a homogeneous reservoir is treated as a linear, donor-controlled process—meaning that the amount of C decomposed is the product of the C stock (C, $\mathrm{g\,C\,m^{-2}}$), a decomposition-rate constant (k, y), and the time interval (Δt, y). The change in soil C stock between one time point and the next (dC/dt) is the difference between the plant inputs (I) and decomposition outputs

(kC) over that period, or using the symbols in Appendix 2, $dC/dt = I - kC$, where t = time (y) and I = C inputs ($gCm^{-2}y^{-1}$). The other concept we use frequently is turnover time (τ), which is simply the reciprocal of the decomposition rate constant ($\tau = 1/k$). At steady state (i.e., when inputs equal losses) $\tau = I/C$ and the soil C stock is the product of inputs and turnover time. The next section and Appendix 2 expand on these definitions and applications.

6.3.1. Metrics of Carbon Dynamics

It is very useful to think about biogeochemical reservoirs in terms of the time constants describing their dynamics, or their mixing, transport, and age. Several terms are used to describe these time constants. Rodhe (1992) identifies three key terms for expressing the dynamics of cycling for geochemical reservoirs: turnover time, mean residence time, and average age; we have adopted this terminology here. Although under certain conditions these terms may be equivalent, they often differ, and it is important to understand the distinctions among them.

The **turnover time** (τ) of a reservoir is its mixing or refresh rate, and is the time it would take for the reservoir to completely empty if there were no further inputs. For soils, it is a measure of the first-order kinetics for decay ($\tau = 1/k$). At steady state, it is calculated as the inventory divided by the total inputs (or total outputs) to the reservoir. To calculate the turnover time for a soil C reservoir at steady state, we would divide the mass of SOM (C) by the total carbon fluxes (S) from the reservoir or $\tau = C/S$. Fluxes would include decomposition to CO_2 and leaching of dissolved organic.

The **average residence time** (also **mean residence time**, τ_r) of C in the reservoir is the average time spent in the reservoir by individual C atoms when they *leave* the reservoir (as if they were polled on their way out). Finally, the **average age** (τ_a) of C atoms in the reservoir is the average time spent in the reservoir by all the atoms currently *in* the reservoir.

The distinction between these concepts is illustrated by the population of a country: the average **age** of the population might be 40 years while the mean **residence time**, or life expectancy, might be twice that value. Human populations are not, however, a simple case for illustrating the concept of **turnover time** as we described it above, because they are not homogeneous in that all members do not have an equal probability of leaving at any time. If this population were homogeneous with respect to mortality and at steady state, turnover time would be the population divided by the number of members who die each year (the stock divided by the flux out).

In the simple scenario of a homogeneous SOM at steady state, the turnover time, mean residence time, and average age of organic matter in the reservoir are equal ($\tau_0 = \tau_r = \tau_a$). The assumption of steady state is often reasonable for mature, undisturbed ecosystems, and many papers use the terms turnover time and residence time interchangeably. However, this is rarely accurate, not only because there are many studies looking at non-steady-state situations, but also because homogeneity in SOM is not commonly observed. In fact, it is rarely the case that all the C in a bulk soil sample will be homogeneous with respect to turnover (i.e., all C turning over at the same rate). It is thus recommended to divide SOM, physically or virtually, into pools that can be treated as homogeneous. Otherwise, estimates of turnover time may be misleading (Figure 6.6).

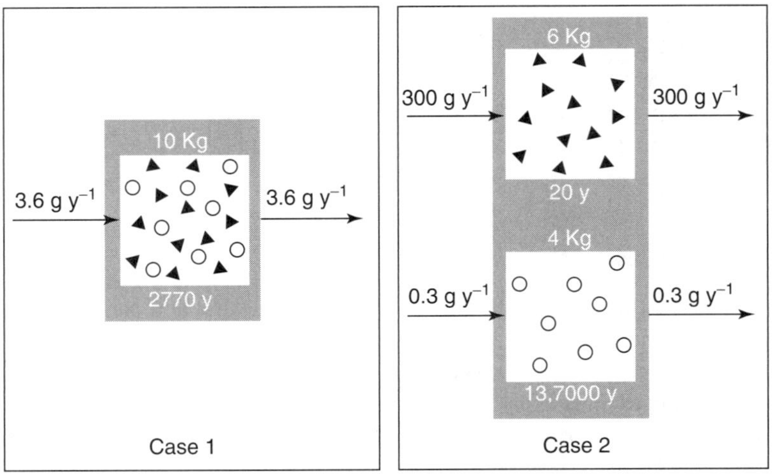

Figure 6.6. Illustration of how turnover time and soil fluxes may be hard to estimate using radiocarbon data. Consider a bulk soil with $10\,kg\,C\,m^{-2}$ with a ^{14}C content equivalent to a turnover time of 2770 yr (0.749 fraction Modern, $F^{14}C$, which gives a conventional ^{14}C age of 2320 yr; see Appendixes 1 and 2). Case 1 models the soil as a single, homogeneous reservoir. The annual flux in or out of the reservoir at steady state is given by flux = $10{,}000\,g\,m^{-2}/2770\,yr = 3.6\,g\,m^{-2}\,yr^{-1}$. In Case 2, the SOM is assumed to be a two-component mixture, with 60% of the C in one pool and 40% in the other. To produce the same total ^{14}C content, the larger pool would have a turnover time of 20 yr, and the smaller pool would have a turnover time of 13,700 yr. The overall ^{14}C content is the same as in Case 1, but the annual flux is now Flu x = $(6000\,g\,m^{-2}/20\,yr) + (4000\,g\,m^{-2}/13{,}700\,yr) = 300.3\,g\,m^{-2}\,yr^{-1}$. The age of respired CO_2 would be ~20 years. Clearly, these two cases have large differences in predicted fluxes and in their implications about how fast the system will respond to changes in inputs or decomposition rates—for example, associated with land use or climate change.

For example, Raich and Schlesinger (1992) calculated the turnover time for C in soils using C inventory (to 1-m depth, and including surface litter) divided by the CO_2 emission observed for the same ecosystem (corrected assuming ~30% was root respiration and ~70% from organic matter decomposition). The turnover times they calculate ranged from 10 y in tropical grasslands to ~500 yr for tundra and wetland environments, with a global average of 32 yr.

Yet, radiocarbon measurements of SOM show that the average age of soil C is several hundreds to thousands of years in temperate and some tropical systems. The apparent contradiction with Raich and Schlesinger's results may be explained if most of the flux of CO_2 from the soil is derived from decomposition of "young" carbon, whereas much of the C residing in the soil is stabilized and decomposing only very slowly. In other words, instead of a large C reservoir with 10-year turnover in tropical grasslands, there is likely a small reservoir of annual-cycling organic matter and a much larger reservoir cycling on time scales of several decades to a century. The distinction is important if we want to predict the rate and magnitude of the response of these grassland soils to disturbances like management or climate change. More generally, quantifying decomposition rates and residence times for C in different compounds and locations in the soils is an important research area.

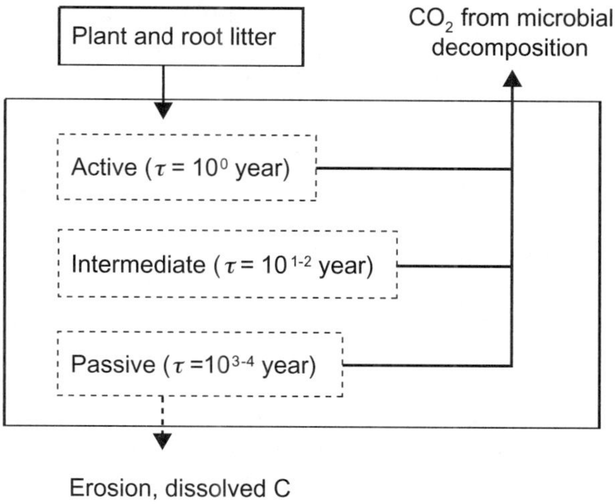

$$C_t/\tau_t = C_a/\tau_a + C_i/\tau_i + C_p/\tau_p \qquad C_a/\tau_a > C_i/\tau_i > C_p/\tau_p$$

Figure 6.7. Simplifed soil carbon cycling scheme. Major inputs (plant litter) to and outputs (respiration and erosion) from the soil carbon reservoir. The observed flux of C out of the soil can be modeled by assuming three pools of carbon: an active pool with a turnover time on the order of years, an intermediate pool with a turnover time on the order of decades to centuries, and a passive pool with a turnover time on the order of millennia. The decomposition constant is $k = 1/\tau$. Subscripts a, i, and p refer to the active, intermediate, and passive C pools, respectively. Adapted with permission from Amundson, R. (2001). The carbon budget in soils. *Annu. Rev. Earth Planet. Sci.* **29**, 535–562.

What are the time scales of soil carbon cycling? As stated above, soil organic matter cycles on a continuum of time scales. A continuous distribution of decomposition rates, however, is difficult to constrain using field or laboratory measurements. There is general agreement that the distribution of SOM decomposition rates tends to cluster at three very different time scales: sub-annual, decadal-century, and longer. Root exudates, microbial cell contents, and some fresh litter compounds decompose on time scales of hours to months to years, and are referred to as the "active pool." Highly stabilized organic matter, typically associated with mineral surfaces or very stable aggregates, persists in soils for thousands of years and is often referred to as the "passive" or "millennial cycling" C pool. The remaining "intermediate" or "slow" C has turnover times in the range of decades to centuries, and it may consist of structural components of plants more resistant to decay, or organic compounds that have been stabilized by their association with soil minerals or aggregate structures. While these pools are broad categories with many exceptions, they have proven useful for many kinds of experimental and modeling studies (Figure 6.7). We will divide our discussion of C dynamics by the time scale involved in decomposition.

6.3.2. Observational Constraints for Determining Soil Carbon Dynamics

No single satisfactory method yet exists by which to separate soil C from the complex soil matrix into discrete components with different turnover times. Instead, soil C dynamics are deduced using many constraints, including physical and chemical fractionation of organic matter, field and laboratory decomposition studies, measures of C fluxes into and out of the soil, measurements of ^{14}C in soils sampled at various times before and after the peak of atmospheric nuclear weapons testing, changes in the ^{13}C content of SOM following a vegetation change from plants with C_3 to C_4 photosynthetic pathways, additions of ^{13}C and ^{14}C tracers, and measurements of changes in the total amount of C in soils of different age or following disturbance. Each of these approaches is suitable for studying different time scales of soil cycling, and combining two or more will often form the most powerful means of elucidating C dynamics. Here, we discuss the use and limitations of some of the most common approaches.

6.3.2.1. Litter Decomposition Experiments. The rate of mass loss of fresh plant litter may be used to estimate litter decomposition rates, assuming first-order kinetics:

$$dM/dt = -kM, \qquad k = -t^{-1}\ln(M_t/M_0)$$

where M_0 is initial litter mass and M_t is mass at time t after deployment. Unless one is tracking isotopically labeled material (see next section), this method is complicated by the need for a litter-containment system that keeps litter fragments in while allowing soil fauna to move in and out (e.g., Harmon et al., 1999), which may create artifacts. Nevertheless, litter bags are a widely accepted method of quantifying and comparing litter decay rates.

Most litter-bag experiments in temperate and tropical ecosystems show relatively rapid initial rates of loss, followed by slower decomposition of the remaining, more recalcitrant compounds (with the absolute rates depending on climate, substrate properties, soil fauna, and soil properties) (Moore et al., 2007; Parton et al., 2007; Zhang et al., 2008). Moreover, experiments using carbon isotopes to follow specific decomposition pathways (Osono et al., 2008) have shown that, while most of the plant litter C decomposes rapidly, a portion is incorporated into components that are stable and persist for many years (Hanson et al., 2005). Therefore, it should not be assumed that labile plant litter, or plant litter with rapid rates of initial degradation, will also produce SOM with rapid turnover rates. Indeed, C dynamics in aboveground litter versus mineral soil may bear little resemblance, reflecting the effects of microbial transformations as well as fundamental differences in biotic and abiotic conditions. In fact, although litter decay rates for different plant functional types have been used to parameterize organic matter turnover in most ecosystem soil C models, the link between litter quality and SOM turnover is not well established.

6.3.2.2. Laboratory Incubations. Laboratory incubations provide a controlled environment for characterizing and comparing C and nutrient dynamics in isolated soils. While subject to artifacts, they do provide one way to quantify the amount of fast-turnover C in soils (Paul et al., 2001). Most often, soils are incubated in jars (Stotzky, 1965; Hart et al., 1994) or microlysimeters (Nadelhoffer, 1990). Incubations have been used to estimate C mean residence times (Torn et al., 2005; Paul et al., 2006) and stability (Whalen et al., 2000; Swanston et al., 2002), interactions of SOM and various nutrients (Zak et al., 1993; Swanston et al., 2004; Torn et al., 2005), and the influence of temperature and moisture on SOM decomposition (Reichstein et al., 2005; Dutta et al., 2006). However, the isolation that allows for controlled conditions also introduces artifacts. Specifically, the altered microclimate, soil disturbance, and lack of continued plant inputs associated with incubations cause changes in substrate quality, microbial communities, and decay rates. We suggest that while incubations are useful for comparative and process-level investigations, it is generally inadvisable to extrapolate rates from the lab to ecological settings.

6.3.2.3. Soil Respiration. Soil respiration, CO_2 flux from soils to atmosphere, is a fundamental flow in the terrestrial carbon cycle and is the primary way that carbon moves from ecosystems back to the atmosphere. Soil respiration is one of the largest fluxes in the global carbon cycle, at 50–80 $\mathrm{Pg\,C\,y^{-1}}$ (Raich and Schlesinger, 1992; Potter et al., 1993; Schimel, 1995). Since the annual exchange of C between the soil and atmosphere is so large (by comparison, fossil fuel use released less than 8 Pg C in 2007), interannual variability in soil respiration is an important source of variation in the rate of increase in atmospheric CO_2 (Trumbore et al., 1995). In most ecosystems, soil respiration makes up >50% of total ecosystem respiration. To estimate global patterns, consider that at steady state, total ecosystem respiration roughly equals gross primary productivity, and heterotrophic respiration roughly equals net primary productivity (because at steady state, the flux of plant inputs is matched by the flux of decomposition). Soil respiration thus varies with latitude, from 80 $\mathrm{g\,C\,m^{-2}\,y^{-1}}$ in deserts to 800–2000 $\mathrm{g\,C\,m^{-2}\,yr^{-1}}$ in tropical forests (Raich and Potter, 1995; Raich and Schlesinger, 1992; Schlesinger, 1977; Trumbore et al., 1995).

The most common method of measuring soil respiration is to place a chamber over the soil and measure the change in headspace CO_2 concentration. This may be done rapidly and accurately by using a chamber connected to a portable infrared gas analyzer, such as that made by LI-COR, Inc. For longer incubation times of 15–60 min, syringe samples of headspace air can be collected and analyzed with a gas chromatograph. For a continuous 24-h measurement, headspace CO_2 can be trapped in soda lime in the chamber. The longer incubation times likely create artifacts in the flux measurements. Eddy covariance methods measure net ecosystem carbon exchange (the difference between all photosynthesis and respiration in their footprint) and provide data to constrain models of soil respiration. If they are located over a bare field or below the plant canopy, they measure soil respiration directly.

Despite its importance in ecosystem C fluxes, soil respiration has limitations as a constraint on SOM turnover, for two main reasons. First, it is difficult to partition soil respiration into its two sources: (1) decomposition of SOM by microbes (heterotrophic respiration) and (2) respiration from live plant roots (autotrophic respiration) (Kuzyakov, 2006). As a result, an increase in soil respiration may indicate not only an increase in SOM decomposition but also an increase in root respiration. Second, it is likely that in most soils only a small fraction of total SOM contributes to heterotrophic respiration. As a result, respiration measurements provide information about the dynamic fraction of SOM (particularly when combined with ^{14}C measurements of respiration) but do not provide information about the large, stable pools unless they are destabilized and contribute to respiration (detectable with $^{14}CO_2$ respiration measurements). Attributing the sources of respiration from different SOM reservoirs, which may respond differently to climatic variables, is not currently attainable through CO_2 flux measurements alone. Changes in atmospheric CO_2 and climate will affect autotrophic and heterotrophic respiration in different ways (Kuzyakov, 2006). To use soil respiration measurements to help understand the effects of climate change on SOM turnover versus net primary productivity (NPP) *in situ*, respiration must be partitioned.

Approaches to partitioning soil respiration typically involve some combination of (1) physically separating respiration sources (e.g., separating roots from soil and measuring respiration of roots and root-free soil), (2) stimulating or suppressing respiration sources (e.g., adding glucose to stimulate microbes or trenching or girdling to reduce root respiration), and/or (3) isotopically labeling respiration sources (e.g., pulse labeling of whole plants or growing C_3 plants in soil produced under C_4 vegetation, combined with measuring the isotopic content of soil respiration) (Kuzyakov and Larionova, 2005). Most of these approaches have been used in laboratory or greenhouse experiments rather than *in situ*. In both the lab and field, it is difficult to avoid disturbance of plant carbon flows or soil structure that unintentionally alter microbial activity and CO_2 flux rates. Nevertheless, these techniques provide valuable constraints on relative fluxes and the effects of environmental variables on them.

6.3.2.4. Isotopic Tools: Tracers. Carbon has three stable or long-lived isotopes: 98.9% of earth's C is ^{12}C, ~1.1% is ^{13}C (a stable isotope), and about one in a trillion (1 in 10^{12}) carbon atoms is ^{14}C. By enriching or depleting the ratios of the rare isotopes in plants, plant litter, or other organic material put in soil, it is possible to follow the pulse of altered isotopic ratios (and the carbon compounds they were associated with) as they move through the system.

Carbon isotopic tracers can be an effective means of characterizing C reservoirs that cycle on sub-daily to decadal time scales. Because of the high cost of isotopically enriched material and the logistical difficulty of labeling large trees or large areas, these studies typically take place in fairly small plots. A notable exception are the free-air CO_2 enrichment (FACE) experiments, which maintain elevated levels of atmospheric CO_2 in open-air sites that are up to 30 m in diameter. If the elevated CO_2 is supplied by a fossil source, the elevated-CO_2 treatment atmosphere is isotopically depleted in both ^{13}C (approximately $\delta^{13}C$ of −21‰ as compared to −8‰) and ^{14}C ($\Delta^{14}C$ of −1000‰ as compared to approximately −60‰). For example, Jastrow et al. (2005) analyzed the depleted ^{13}C pulse in SOM fractions to show accumulation of SOM in elevated CO_2 treatments at several FACE sites. Another exception is the Enriched Background Isotope Study (EBIS) at Oak Ridge National Laboratory in Oak Ridge, Tennessee (Trumbore et al., 2002). EBIS investigators used a combination of a stand-level radiocarbon enrichment and a reciprocal litter transplant from a forest stand with litter that had near-background ^{14}C levels, to partition sources of soil respiration (Cisneros-Dozal et al., 2006) and investigate SOM dynamics (Hanson et al., 2005; Swanston et al., 2005).

6.3.2.5. Natural Abundance Stable Carbon Isotopes (^{13}C). There are trends in ^{13}C of plant, litter, and organic constituents in soil that can be used to investigate carbon cycling, but to date the trends have proven too subtle, the variation too high, or the mechanisms too poorly understood to exploit these patterns definitively. The $\delta^{13}C$ of plant litter and particulate SOM is typically close to that of the plant source, while mineral-associated SOM is often 1–3‰ higher (more enriched). Within the profile, the $\delta^{13}C$ of bulk soil increases by 1–3‰ with increasing depth and decreasing particulate SOM. Comparisons of 100-yr-old and modern soil profiles show that the increase in ^{13}C with depth is not due to fossil fuel effects (Torn et al., 2002), but there is ongoing discussion regarding whether these trends are caused by microbial discrimination or selective preservation of plant compounds (Dijkstra et al., 2006; Mikutta et al., 2006). Resolution of these questions may be difficult using only natural abundance stable isotopes and non-manipulated systems.

Due to differences in their photosynthetic pathways, C_3 and C_4 plant biomass have different $^{13}C/^{12}C$ ratios (C_3:δ13C ≈ −27‰; C_4:δ13C ≈ −13‰; Still et al., 2003). Where a vegetation change from C_3 to C_4 plants (or vice versa) has occurred, the rate of change of $^{13}C/^{12}C$ ratios in SOM will give an idea of the turnover time of SOM (Balesdent et al., 1988; Veldkamp, 1994). This method has been used most commonly in tropical pastures where C_4 grasses have replaced C_3-dominated forest, as well as in the agricultural sites where C_4 maize has replaced native C_3 forest. While generally a very useful method, there are two main limitations to mention: (1) It cannot be used to study soils that have not undergone a vegetation change, and (2) it requires careful measurement of C inventory changes in disturbed versus undisturbed soils. These attempts are complicated by plowing, disking, and/or erosion of the pasture soils, as well as inherent uncertainty in bulk density determinations.

6.3.2.6. Radiocarbon. Radiocarbon (^{14}C) is unstable, with a half-life of 5730 yr, and decays by emission of an electron to form ^{14}N. It is continuously produced in the upper atmosphere by interactions of high-energy cosmic rays with the upper atmosphere. The ^{14}C is oxidized to $^{14}CO_2$ within a few weeks and is then mixed into the troposphere (the lower, well-mixed part of the atmosphere), where it is taken up by plants during photosynthesis and exchanges with the surface waters of the ocean.

If a C reservoir ceases to actively exchange ^{14}C with the atmosphere, the ^{14}C content of the reservoir will begin to decrease because of radioactive decay. This is useful for studying very stable C pools in soils, since they reside long enough for significant decay of ^{14}C to occur. In this case, the more ^{14}C-depleted a soil fraction is, the slower the turnover of the C is [even if the soil fraction also contains some faster-cycling components (Mikutta et al., 2006; Sollins et al., 2006)]. The longest time scales that can be addressed with ^{14}C in this way are on the order of 60,000 yr.

Atmospheric thermonuclear weapons testing, which peaked in 1963, approximately doubled the amount of ^{14}C in the atmosphere (Figure 6.8). Atmospheric $^{14}CO_2$ levels have been decreasing rapidly since then, because of atmospheric exchange with terrestrial and oceanic C reservoirs. This "bomb" ^{14}C spike provides a global isotopic tracer for the C cycle, although still several orders of magnitude below levels of ^{14}C used in most small-scale ^{14}C-tracer studies. The amount of bomb ^{14}C found in SOM provides a direct measure of the amount of fast-cycling (active + slow) SOM. The most straightforward application is to compare the ^{14}C content of SOM sampled prior to 1960 with that of contemporary samples from the same location (Trumbore, 1993, 2000). Where no archived soils are available, however, radiocarbon measurements must be combined with other observational constraints to separate the radiocarbon signature of rapidly cycling from that of

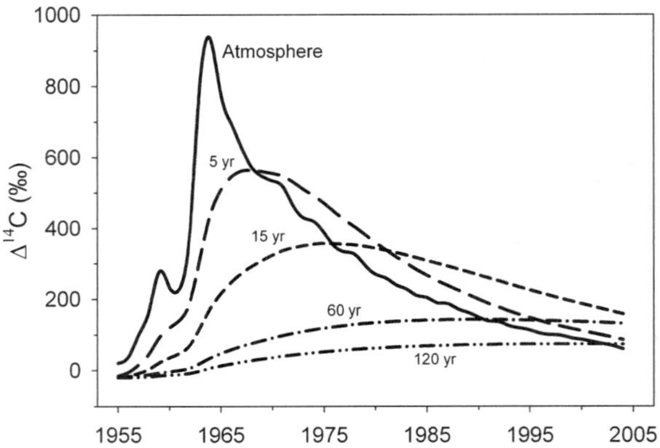

Figure 6.8. Change in atmospheric ^{14}C with time in the Northern Hemisphere (heavy solid line) since 1955. Radiocarbon values are expressed as the per mil variation in $^{14}C/^{12}C$ ratio relative to a standard (see Appendix for definition of units). The lighter lines show the evolution of ^{14}C for homogeneous, steady-state reservoirs with turnover times of 5, 15, 60, and 120 yr.

very refractory organic matter. The time scales of C turnover that may be addressed using bomb ^{14}C range from ~4 to ~100 yr.

The ^{14}C content of soil respiration leaving the soil can be measured using trapped air from the headspace of a chamber (Dörr and Münnich, 1986; Gaudinski et al., 2000). To the extent that ^{14}C reflects recently fixed C versus C fixed years to decades ago, $^{14}CO_2$ measurements provide a useful tool for partitioning the sources of soil respiration (autotrophic plant respiration versus heterotrophic microbial respiration) and the turnover times of the decomposing organic matter that contribute the most to soil respiration (Torn et al., 2005). Working in temperate grassland and forest, Dörr and Münnich (1986) found significant seasonal differences in the ^{14}C content of total soil respiration: Summer emissions were dominated by recently fixed carbon, and winter fluxes were dominated by carbon fixed up to several decades previously (which likely indicates a higher proportion of autotrophic respiration in the growing season as well as seasonal changes in the substrate for decomposition).

In summary, natural abundance radiocarbon is a powerful tool, because it can be used in mature and undisturbed ecosystems (as well as in younger or disturbed ones), and because it can be used to quantify turnover times across a range of time scales. In fact, it may be the only tracer for stable, or slow-cycling, C pools. The radiocarbon content, along with additional constraints regarding, for example, the relative proportions of fast- and slow-cycling SOM, can be used to model turnover times. Appendixes 1 and 2 contain more thorough explorations of radiocarbon methods and applications to SOM studies.

6.3.2.7. Fractionation of Soil Organic Matter. There are numerous approaches to separating SOM pools for analysis, with a corresponding number of underlying conceptual frameworks. The objective of fractionation is usually to reduce the chemical, physical, and/or C-cycling time variation in the fractions compared to the bulk soil. Nonetheless, most isolated SOM fractions are operationally defined and remain a mixture of heterogeneous compounds from multiple sources. The near-term challenge is not to find a single, universal method that will characterize all SOM pools in all soil types, but instead to understand what kind of information is provided by each method for the scientific questions of interest. The most common methods of fractionating SOM are chemical (humic/fulvic separation, acid hydrolysis), density, size, and aggregate.

Humic and fulvic acids are isolated by extracting the soil with alkali, which solubilizes acids from the humin, and then treating the extract with acid to separate the humic and fulvic acids. Ideally, this method separates SOM by chemical characteristics, in which the humic fraction contains compounds with lower oxygen concentration and a higher degree of polymerization, molecular weight, and C concentration compared to those in the fulvic fraction (e.g., Stevenson, 1994). As such, these fractions may provide "signatures" for soils, perhaps varying with factors such as vegetation and management (Miglierina and Rosell, 1995). Chemical separation methods have significant drawbacks. Primarily, the harsher treatments can form new compounds as well as solubilize and extract them. Extraction with NaOH can separate some lignin-derived aromatic C, depositing it into the humic acid while the remainder is left in the humin (Kögel-Knabner et al., 1991). Most polysaccharides would presumably reside in the fulvic fraction. Thus, although these C structures are related spatially and biologically in the soil, chemical fractionation procedures can separate them into several fractions, potentially obscuring their connected roles in soil C cycling. Another common chemical separation is acid hydrolysis, used to isolate N-rich compounds (including proteins and nucleic acids), polysaccharides, and other chemically labile SOM from acid-resistant material such as aromatics and long-chain aliphatics (Paul et al., 2001). More recently, ultraviolet radiation and chemical (H_2O_2 or NaOCl) methods have been used to oxidize some organic matter and leave behind less reactive, and radiocarbon-older, organic residues (Krull et al., 2006). In two comparisons, NaOCl appears to cause less mineral alteration and yield an older OM fraction than does HCl (Mikutta et al., 2006; Zimmermann et al., 2007). These approaches use chemical reactivity of organic matter as a proxy for readiness to microbial degradation, rather than attempting to separate material of different chemistry *per se*.

Physical soil fractionation methods such as density, size, and aggregation aim to isolate pools of SOM based upon their degree of organo-mineral interaction, the extent of protection within aggregates, and the size and location of the aggregates. Density fractionation takes advantage of the differences in density between particulate organic matter and mineral-associated organic matter. The basic approach is that a light fraction is floated on a dense liquid while the denser, or heavy fraction, sinks (Strickland, 1987; Sollins et al., 1999). The light fraction is typically less degraded, more plant-like, and of more recent origin than the C in the mineral-associated heavy fraction (Gregorich et al., 1996; Trumbore and Zheng, 1996). Golchin et al. (1994a) modified this technique to separate the "free," unprotected light fraction first; then they disrupted aggregates to separate the "occluded," aggregate-protected light fraction from the dense fraction. Radiocarbon measurements show that the occluded light fraction can have a slower turnover time than the heavy fraction (Rasmussen et al., 2005; Swanston et al., 2005). The heavy fraction can be further separated by increasing density, generally yielding older, but smaller, organo-mineral pools (Golchin et al., 1994b; Sollins et al., 2006). There is evidence that aggregate dispersion may redistribute C and N between fractions (Cambardella, 1994; Baisden et al., 2002), and some C and N (1–15%) is typically lost during these procedures (Swanston et al., 2004; Crow et al., 2007; Castanha et al., 2008). However, density separation does produce fractions with distinct C, N, and isotopic composition, and it appears to reveal the trends in mineral-associated SOM as well as provide some information about protection of particulate organic matter in macro- and microaggregates (Golchin et al., 1994a,b).

Particle size fractionation is based on the concept that as organic matter is degraded and interacts with minerals, its particle size decreases (Tiessen and Stewart, 1983; Christensen, 1992). Variations in size-fraction methods exist (e.g., Christensen, 1992), but soils are typically dispersed sonically, by shaking with glass beads, or chemically with hexametaphosphate, and the resulting soil slurry is passed through a series of decreasing sieve sizes and centrifuged to isolate fine fractions. Although there are some trends with particle size, it is also evident that C pools separated by size are composed of multiple, chemically protected or biochemically recalcitrant pools with differing residence times that may be classified as a function of origin, chemical composition, and mineralogical interactions (Schmidt and Kogel-Knabner, 2002; Kiem and Kogel-Knabner, 2003). Size separation can be particularly useful in separating organic matter into distinctive chemical pools in organic horizons, in other words where no mineral stabilization is occurring, with material <63 μm diameter of predominantly microbial origin, and larger sizes, made up of plant material (Grandy and Neff, 2008).

The goal of aggregate-based soil fractionation is to isolate C pools according to their location in different soil physical structures, based on a conceptual model of stabilization of C inputs in microaggregates that cycle with macroaggregates (e.g., Oades, 1984; Golchin et al., 1994b; Jastrow et al., 1998; Six et al., 2000a; Six et al., 2000b). Based in part on work by Golchin et al. (1994b) and Cambardella (1994), Six et al. (2000a) developed a fractionation scheme designed to separately isolate SOM found inside and between aggregates of different sizes and stabilities. Physical isolation of intact microaggregates, including those located within macroaggregates, followed by their dispersion provides quantitative information for several process steps related to SOM cycling. Unfortunately, this method does not reduce the problem of losing C and N during the separation and rinsing process (Chan, 2001; Moran et al., 2005).

6.3.2.8. Microbial Fractionations. Microbes ultimately determine what organic compounds will be metabolized in soils. Trumbore (2000) demonstrated that C respired from soils is younger than the mean age of C in organic matter. Measures of radiocarbon in classes of phospholipid fatty acids from microbial cell walls, however, show that microbes consume C substrates with a range of ^{14}C ages. Similar results have been observed with ^{13}C- and ^{15}N-labeled litter added to soil (Bird and Torn, 2006). Combining microbial biomarkers with isotopic analysis is a promising technique but is beyond the scope of this chapter.

6.3.3. Soil Carbon Stock and Bulk Density

The stock of organic matter in soil is one of the most fundamental constraints on estimates of turnover time and tests for models that predict storage or turnover time. The amount of organic carbon stored in a soil profile is calculated from measurements of C density and bulk density (BD) by horizons or depth intervals, i, as follows:

$$\begin{aligned} \text{C inventory}\left(\frac{\text{g C}}{\text{cm}^2}\right) = \sum_{horizon=i}^{n} & \text{Bulk density}_\text{i}\left(\frac{\text{g soil}}{\text{cm}^3}\right) \\ & \times \text{Carbon density}_\text{i}\left(\frac{\text{g C}}{\text{g soil}}\right) \\ & \times \text{depth}_\text{i}\,(\text{cm}) \times (1 - \text{fr gravel})_i \end{aligned}$$

The well-established methods for measuring and reporting C density need no elaboration here. Bulk density, the dry weight of a known volume of soil, including pore space, is a simple concept that is difficult to measure. Bulk density values can vary by a factor of 4 depending, for example, on soil OM content, depth, and compaction. Because it varies and because measurement is difficult to do precisely, the largest uncertainties in determining C stock in a soil profile usually come from estimates of bulk density and the volume of soil that is gravel. The last term in the equation above is a correction for stones and gravel greater than 2 mm in diameter, also called the coarse fraction or gravel fraction. It is difficult to reproducibly collect a precise volume of soil, and in many studies bulk density simply has not been measured at all.

In addition to expressing soil C stocks on an areal basis, it may be preferable, particularly for comparisons of C stocks due to land use changes that alter bulk density, to express soil C stocks on an equivalent mass basis. This approach, nicely explained by Ellert et al. (2001), samples to a depth giving a *constant mass of soil* at all locations or time points, rather than a *constant depth of soil*. Calculating the required depth requires bulk density measurements.

There are several new *in situ* soil carbon measurement techniques undergoing field testing. Some devices, such as those using a neutron generator, measure total C atoms to a known depth, and as a result do not require a bulk density measurement (L. Wielopolski, personal communication, March 2007). In addition, such *in situ* and noninvasive techniques would allow the same location to be measured repeatedly. However, for the time being, and likely for many applications in the future, the importance of carefully measuring bulk density cannot be overstated.

Large plant fragments and organic mats are often excluded from estimates of soil C stocks, either during sample collection or during sieving at 2 mm. The traditional focus in soil science on the "fine soil," however, is not adequate in the context of carbon management, where a complete accounting of organic carbon is desirable. Researchers are adapting to this expanded perspective by developing new protocols and expectations for the reporting of soil characteristics.

6.4. IMPORTANT CONTROLS OF SOIL CARBON DYNAMICS

Definition of three terms that are commonly used interchangeably but nonetheless have distinct meanings will aid our discussion on the controls of soil C dynamics. In the most general sense, a **process** is a series of steps leading to a result; in the context of carbon cycling, the result is the stabilization or destabilization of carbon. A **mechanism** is the crucial step in the process, or the physicochemical condition or transformation that most distinctly results in carbon stabilization or destabilization. A **control** is something that exerts an exceptionally strong influence on a process, rendering mechanisms more or less effective.

Soil C dynamics are the integration of myriad processes leading to stabilization or destabilization of SOM. Sollins et al. (1996) describe SOM cycling within the context of proximal and distal influences—that is, mechanisms that have an immediate influence on C stability and flux (e.g., molecular recalcitrance, mineral and organic interactions, and accessibility) ranging to controls with a distal and more general influence (e.g., the state factors, "cloprt"). This conceptual hierarchy is a useful framework in which to chart the relationships of the numerous influences on soil C turnover. One subtlety is that a proximal influence (mechanism) is not necessarily the dominant factor in C stabilization, nor is a distal influence necessarily a minor factor. We suggest that there is no universally dominant mechanism or control on SOM dynamics. Instead, we consider different controls, as well as what factors make a particular mechanism more important or effective in one place but not another.

6.4.1. Mechanisms of Stabilization

A physical or chemical condition that renders SOM less susceptible to alteration or transport (i.e., more stable) is a mechanism of stabilization. The assumed mode of alteration is often microbial activity, although this is not always explicit. A great deal of thought has gone into (a) defining and comparing the dominant mechanisms that affect C stability and (b) organizing them into a limited number of broad categories (Sollins et al., 1996; Baldock and Skjemstad, 2000; Krull et al., 2003; von Lützow et al., 2006). Additionally, researchers have sought to identify the dominant mechanisms of long-term stabilization, leading some to focus on the inherent molecular recalcitrance of organic molecules (Krull et al., 2003) and leading others to focus on mineral interaction and protection as the fundamental controls (Van Veen and Kuikman, 1990; von Lützow et al., 2006). Here we describe several categories of stabilization mechanisms, largely adapted from Sollins et al. (Sollins et al., 1996), and place them within the context of climate, ecology, and management.

6.4.1.1. Recalcitrance. We use the term "recalcitrance" to refer specifically to the inherent molecular characteristics of SOM that contribute to resistance to microbially mediated degradation within a soil environment (Sollins et al., 1996). Aliphatic (e.g., lipids, waxes) and aryl (e.g., charcoal) compounds tend to have the longest turnover times in many soils (Hamer et al., 2004; Preston and Schmidt, 2006) and are often considered to be more recalcitrant than other organic compounds. Attention to the compounds that have been labeled as recalcitrant in different soils in different ecosystems, however, suggests that molecular characteristics may not convey inherent (i.e., universal) stability, but rather recalcitrance may be more context-specific: (1) Whether a particular turnover time is considered "stable" depends on the cycling rates of other SOM pools in the same soil or region, and (2) the same compounds may be more or less recalcitrant than each other in different environments, depending on controls. A recent review concluded that there is increasing evidence that selective preservation of plant compounds is not important for soil C storage (von Lützow et al., 2006). Black carbon, organic matter transformed by pyrolization, remains a candidate for recalcitrance. Comparing a grassland, woodland, and subtropical rainforest in Australia, Krull et al. (2006) concluded that although the woodland and rainforest had greater aggregate and mineral protective capacity, respectively, the grassland supported a larger reservoir of more stable carbon due to frequent inputs of charcoal from regular fires. In a Russian Steppe soil, on the other hand, black carbon had a profile-total turnover time of less than 300 yr, which was faster than turnover of the bulk SOM (Hammes et al., 2008).

6.4.1.2. Mineral Associations. Direct association between organic C and primary and secondary minerals in soil includes H-bonding, van der Waals forces, ligand exchange, cation bridging, and metal complexation (von Lützow et al., 2006). Multiple layers of organic C may range outward from the mineral surface with decreasing strength of association, and the outer layers may thus be the most actively cycling of the mineral-stabilized C (Sollins et al., 2006; Rillig et al., 2007). The cleavage of the innermost bonds may often prove energetically unfavorable, raising the possibility that it is actually mineral dissolution or evolution that results in destabilization of the SOM at the mineral surface instead of direct degradation. Along a chronosequence in Hawaii, Torn et al. (1997) found that soil C content and radiocarbon abundance were correlated with the changing soil mineral composition, and in particular with Al and Fe (oxy)hydroxide abundance. They concluded that as metastable noncrystalline minerals transitioned into crystalline clays that have lower surface area and charge density, their ability to stabilize SOM was reduced. In this ecosystem, mineral association appears to be the dominant control on long-term SOM stability. The relative effectiveness of Al versus Fe for stabilization was not determined in that study and is likely not generalizable among soils. While both Al and Fe (oxy)hydroxides were correlated with C and ^{14}C content in a California coastal prairie chronosequence, only reactive aluminum species (and not iron) were important in a California montane forest (Rasmussen et al., 2005). For more systematic treatments of mechanisms at the mineral and root interface, see Kleber et al. (2007) and Rillig et al. (2007), respectively.

6.4.1.3. Accessibility. Physical protection that precludes microbial and enzymatic access to SOM may preserve a substrate that would otherwise be rapidly degraded. This type of protection is largely a function of soil structure, occurring primarily within meso- and microaggregates, pores with spaces or entrances too small for soil organisms or enzymes to pass (Oades, 1988; Mayer et al., 2004; Strong et al., 2004). Additionally, highly tortuous diffusional paths may reduce the viability of bacterial "foraging" using enzymes, potentially reducing the likelihood that otherwise degradable SOM is degraded (e.g., Vetter et al., 1998). The influence of aggregate protection can readily be seen in grassland soils under different regimes of physical disturbance; in agroecosystems, aggregate disruption by tillage is usually the foremost cause of soil C loss (e.g., Six et al., 2002). Yet soil structure is intimately related to soil texture and mineralogy. Denef et al. (2004) looked at conventional tillage and no-tillage cropping systems across several soils with differing clay mineralogies. While total SOM storage and aggregate stability appeared to be associated with mineralogy, >90% of the SOM loss related to conventional tillage in all soils was associated with a single size class of microaggregates that were isolated from water-stable macroaggregates. Direct mineral association appeared to be a dominant mechanism of long-term stabilization, but a shorter-term mechanism of stabilization was the limitation of accessibility. In these intensively managed ecosystems, the mode of field preparation can thus become a major control on C stabilization by influencing accessibility to occluded SOM.

6.4.1.4. Biotic Suppression and Climatic Stabilization. Organic C in soils does not simply cycle; rather, *C is cycled by biological activity*. The mechanisms described above ultimately relate to the ability of soil organisms to access and degrade SOM. However, if the organisms themselves are in some way suppressed, the low activity of the soil microbiota becomes the effective mechanism of stabilization. Biotic suppression, and consequent C stabilization, may thus occur through conditions such as O_2 limitation (e.g., flooding), desiccation (e.g., desert environments), extreme or prolonged cold (e.g., boreal and arctic systems), nutrient imbalances (e.g., N concentration; Waldrop and Zak, 2006), and excessively high or low pH (e.g., mine spoils). Some of these ecosystems have high enough NPP to result in significant SOM accumulation, such as in peat bogs (Smith et al., 2004) and boreal forests (Harden et al., 2000). A major concern about climate change is that conditions may become more favorable to microbial activity, possibly leading to destabilization of large quantities of SOM that are currently protected by conditions that suppress biotic activity (Freeman et al., 2001) and fueling positive feedbacks to global warming (Chapin et al., 2000; Kirschbaum, 2000; Davidson and Janssens, 2006).

6.4.2. Mechanisms of Destabilization

Just as mechanisms of stabilization cause greater SOM stability, mechanisms of destabilization render SOM more susceptible to alteration or transport. By definition, they in some way reduce or eliminate the efficacy of the mechanisms of stabilization (Sollins et al., 1996). *The controls on destabilization typically promote disturbances that expose SOM, or otherwise foster a physical environment more advantageous to microbial or faunal degradation of SOM.* In general, the factors that control (enhance) destabilization promote disturbances that expose SOM or otherwise foster a physical environment more advantageous to microbial and mesofaunal degradation of SOM. Examples of natural and anthropogenic soil disturbance include tilling (Six et al., 1999), freeze/thaw and shrink/swell cycles (Denef et al., 2001), erosion and mass wasting (Harden et al., 1999), bioturbation (Stork and Eggleton, 1992), windthrow (Kramer et al., 2004), and fire (Harden et al., 2000). The degradation of a substrate can also act as a mechanism for further destabilization through the production of more labile by-products. Degradation can be from biotic or abiotic sources, such as faunal degradation (Verhoef and Brussaard, 1990), microbial degradation and extracellular enzymatic alteration (Cairney and Burke, 1998), and photo-degradation (Zepp et al., 2003). As a sign of the complexity of soils, destabilization often happens concurrently with stabilization. Even as a compound is degraded or transported, some by-products may be generated that are more stable (or become stabilized more readily) than the original compound (e.g., Wolters, 2000).

6.4.3. Temporal and Spatial Scales of Carbon Cycling

As the scale of analysis moves from centuries to years or hours, and from regions to meters or microns, the relevant controls, processes, and mechanisms also change. A dominant control at a millennial time scale may be largely irrelevant for hourly variation, just as a mechanism that explains patterns at the scale of a field plot may shed no light on processes at the edge of a clay micelle. In this chapter, we have tried to acknowledge the complexity of the spatial and temporal dynamics of carbon cycling by considering controls, processes, and mechanisms at a point in space within the context of time, as well as at a point in time but across spatial scales. Possible examples are numerous and generalizations prone to exception; nonetheless, we offer a few hypotheses to encourage further discussion and debate of the controls on carbon turnover.

Over large spatial scales or among biomes, climate tends to dominate C budgets, particularly at the extremes of temperature and moisture—for example, tundra and deserts—due to direct effects of these extremes on plant production and microbial processes. Within most temperate regions, more complex relationships among state factors, processes, mechanisms, and scale exist that are not easily generalized across the landscape. Seasonal climate extremes, such as summer drought and winter cold, can exert strong controls over C cycling through influence on plant productivity and biotic suppression. At the field scale, vegetation and topography tend to be dominant controls by determining C inputs and strongly influencing hydrology at a given time. Over long time periods, however, mineralogy influences both vegetation and topography through soil development. In the surface soil, which receives most plant C inputs, accessibility and recalcitrance provide greater constraints on surface soil C fluxes than do mineral interactions, even though mineral interactions lead to the most stable C. While there is less C deeper in the soil profile, more of the deep C is stabilized through mineral interactions, which become the dominant mechanism of stabilization in that part of the profile. The primary mechanisms of stabilization in the rooting zone, accessibility and recalcitrance, reflect the dynamic nature of C inputs and soil moisture. The relative importance of these mechanisms will vary greatly with ecosystem properties and management, with accessibility being dominant except in regions with high char inputs. Organo-mineral interactions, especially the "inner layer" molecules, tend to stabilize C for much longer time spans, hundreds to thousands of years, and are associated with the most stable C throughout the profile. At the scale of the mineral surface itself, broad mechanisms like mineral associations take on the nature of processes; and finer-scale mechanisms like charge density of the mineral, polarity and structure of the molecule, and the density of the pore water will most directly result in C stabilization.

6.5. RESPONSES OF SOIL ORGANIC MATTER TO GLOBAL ENVIRONMENTAL CHANGE

Human activities can have profound consequences for soil carbon cycling. Climate change, nitrogen deposition, elevated atmospheric CO_2 concentrations and other atmospheric changes, land use and land cover change, and altered disturbance regimes are all having increasing influence on plant productivity, soil decomposition rates, and soil carbon storage. At the most basic level, changes in soil carbon stocks and net transfer of CO_2 between soil and atmosphere will depend on (a) the balance between plant productivity and SOM decomposition and (b) how global change factors affect these flows.

6.5.1. Productivity and Soil Carbon Storage

Plant productivity is determined by factors such as plant species composition, moisture, soil fertility, growing season length, and solar radiation—many of which are affected by human activities. All else equal, increases in primary productivity and production of plant tissues will lead to increases in soil C stock, while decreases will lead to decreases in soil C stock. The rate of change in soil C stock is determined by the difference between C inputs and outputs, as well as the turnover times of the soil C, which are often not known. Here we review briefly how some environmental factors are expected to alter productivity and explore how the effects on stock depend on the number of soil carbon pools and their turnover times.

Elevated CO_2 can enhance plant growth, albeit with uncertain efficacy and duration, while the accompanying climate change will have variable effects on NPP. Most coupled climate carbon-cycle models predict that terrestrial ecosystem productivity will increase in the first half of this century due to CO_2 fertilization and moderate increases in temperature, but will moderately decline after that due to more severe changes in climate (Sitch, 2003; Gerten, 2004; Friedlingstein et al., 2006; Fischlin et al., 2007). For example, the productivity of intact Amazonian forests has been increasing over recent decades, variously explained by episodic disturbance and recovery dynamics, changing species distribution, CO_2 fertilization, modest warming, reduced tropical cloud cover, and increased radiation (Nemani, 2003; Baker, 2004; Chambers and Silver, 2004; Lewis, 2004; Malhi and Phillips, 2004; Boisvenue and Running, 2006). These C gains are predicted to be transient, however, due to losses associated with escalating heating and drying trends (Malhi and Phillips, 2004). More generally, in the long term, it is unlikely that plant productivity will continue to increase with increasing atmospheric CO_2, due to widespread limitation of NPP by water or nutrients and also because of acclimation of plants to higher CO_2 conditions.

By increasing the amount of N available to plants, nitrogen deposition can contribute to carbon uptake in N-limited (e.g., temperate) ecosystems (Melillo et al., 1995; Schimel, 1995; Trumbore, 2000) but can also lead to changes in plant species, microbial community composition, and soil pH (Boggs, 2005; Silvertown, 2006). Changes in vegetation allocation strategy, litter quality, and soil microbes can lead to large C losses belowground that more than offset C gains associated with increased aboveground productivity (Mack et al., 2004). Any benefits of N deposition are expected to reach a saturation point, after which productivity levels off and eventually diminishes due to other nutrient limitations or increased susceptibility to stresses such as pollution, frost damage, or disease (Agren and Bosatta, 1988; Aber et al., 1989). Other types of atmospheric pollution, such as acid rain, increased tropospheric ozone, and stratospheric ozone depletion, are all predicted to reduce NPP.

In addition to productivity, plant species composition and abundance affect soil C cycling through tissue chemistry and surface energy balance (i.e., the energy balance between land surface and atmosphere). The latter influences soil microclimate, while the former helps determine decomposition pathways and products. Global warming is projected to lead to large-scale vegetation shifts, such as expansion of boreal forests as growing season lengthens, transition from temperate evergreen to deciduous forests due to warming, and from tropical evergreen forest to seasonal forest or to grassland due to drought stress (Fischlin et al., 2007).

If plant productivity increases without a commensurate increase in decomposition rates, more carbon will be sequestered in soil. Since decomposition is proportional to the stock of SOM, stock will build up until the efflux from decomposition reaches a level roughly equal to the higher rate of inputs. While faster-cycling C pools will adjust more rapidly to reach a new steady state, slower cycling pools will build up to a higher stock of C for the same increase in NPP. As an illustrative example, consider the world's shallow carbon stocks (1500 Pg) in equilibrium with global NPP (60 Pg yr^{-1}). The average turnover time of this C is estimated as 25 yr for fast cycling C (Harrison, 1993) or 32 yr for all soil C to 1 m (Raich and Schlesinger, 1992). Now stipulate for this example that CO_2 fertilization and other factors increase NPP by 10% worldwide. The predicted change in global soil stocks will depend on the number of C pools and the turnover time of each pool. For this example, we compare the one-pool scenario considered by Harrison or by Raich and Schlesinger with a two-pool scenario illustrated in Figure 6.9. The initial C inventory, NPP, and bulk turnover time are the same in either case. For the one-pool soil, with $\tau = 25$ yr, increased inputs to soils from CO_2 fertilization would result in a build up of the soil C inventory to 1600 Pg in less than 100 yr. However, if 30% of C inputs have a 2-yr turnover ($\tau = 2$ yr) and 70% of C inputs have a 35 yr turnover ($\tau = 35$ yr), then the new C stock would—over the course of ~150 yr—build up to 2500 Pg C! (For precise estimates on century time scales, leaching of soluble C and erosion should also be considered.) More generally, treating soil C as one pool with

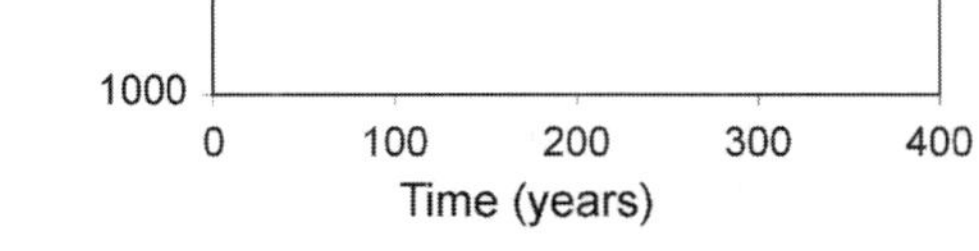

Figure 6.9. Modeled time series of global C stocks assumed to start at 1500 Pg C, following a one-time 10% increase in NPP from 60 to 66 Pg C yr^{-1}. The lower line represents a single pool scenario in which all soil C has a residence time τ of 25 yr. This residence time is derived from the mass-balance equation shown in Section 6.3, dC/dt = NPP − C/τ, and assuming steady state (such that inputs equal outputs). The upper line represents a two-pool scenario in which 70% of the incoming C has a residence time $\tau = 35$ yr and 30% has a $\tau = 2$ yr, such that the overall residence or "bulk" $\tau = 25$ yr. Because it contains a slower cycling pool, the two-pool model accumulates C faster and takes longer to reach a new steady-state equilibrium, with much higher C stocks than the single-pool model.

a simple temperature response function for global change predictions will lead to an underestimate for the short term and an overestimate for the long term.

Detecting whether global soil C stocks have increased in the past decades, and acted as a sink of atmospheric CO_2, is very difficult due to spatial heterogeneity in soils, relatively large analytical uncertainties (especially in bulk density), and the fact that even changes of a small fraction of standing stock are environmentally important. In fact, Post et al. (1995) conclude that a change of global or regional soil C inventory on the order of 1 Pg C would be impossible to measure directly. Even a 1 Pg C yr^{-1} sink continuously for 30 yr would increase the global soil C inventory by only 2%. While changes in C stocks associated with land use change are frequently observed, a recent study documented a *regional* change in carbon stock in unmanaged and managed soils. Bellamy (2005) found that the top 15 cm of soils in Great Britain have lost 2% of their C stocks over the past 25 yr. Because the soil C loss was fairly independent of land use, they conclude that the observed warming in Great Britain is the most likely explanation for the loss.

6.5.2. Climate Change

One of the most important questions regarding SOM is how future climate change will influence decomposition rates, and the flux of CO_2 from soils to the atmosphere, relative to CO_2 uptake by NPP—and thus the potential for positive feedback with climate change. For example, across a gradient of mean annual temperature in intact mature tropical forests, NPP increased, but soil C stocks decreased more steeply, implying a net loss in ecosystem C from faster SOM decomposition (Raich et al., 2006). To some extent, these linkages can be evaluated by land surface models coupled to global climate models. In a recent intercomparison of coupled climate carbon-cycle models (the Coupled Carbon Cycle Climate Model Intercomparison Project, or C^4MIP), all but one of 11 models predicted faster decomposition rates with climate change to 2100 (Friedlingstein et al., 2006). Because modeled NPP did not increase commensurately, most of these simulations predicted decreases or no change in soil carbon stocks and a positive feedback with climate change. Current observations show that terrestrial ecosystems and vegetation—and presumably soil as well—currently act as a large sink for atmospheric CO_2. However, all of the models in the previous study predicted that terrestrial ecosystems will be a less effective sink, and in many simulations become a net C source, after 2070 (Friedlingstein et al., 2006; Fischlin et al., 2007).

The coupling between C and N cycling in soil, in which decomposition also mineralizes nitrogen, means that the net effect of increased decomposition on ecosystem carbon budgets is complex. The direct effect of increased decomposition is the transfer of C from soils to atmospheric CO_2. On the other hand, stimulation of decomposition in relatively undisturbed ecosystems may cause ecosystems to accumulate C as nutrients are transferred from soils (low C/N ratio) to plant reservoirs with higher C/N ratios. Few land surface models simulate these nitrogen transformations in soil and plants, and thus most are likely missing an important modulator of CO_2 and climate impacts. For example, none of the 11 soil modules in the C^4MIP takes into account the release of nitrogen through increased decomposition and its potential to stimulate plant growth. Including this nitrogen feedback in the CENTURY biogeochemical model reduced the loss of soil organic C due to increasing temperature by half compared to simulations without the nitrogen processes (Schimel et al., 1994). Nitrogen cycling in the land surface model CN, coupled with the Community Climate Simulation Model (but not in C4MIP), had a similar effect in maintaining ecosystem C stocks (Thornton and Rosenbloom, 2005). Predicting the ability of coupled C and N cycles to buffer the impacts of climate change requires better understanding and integration of soil decomposition, nutrient cycling, and plant growth.

The magnitude, rate, and duration of ecosystem soil responses to climate change depend on the amount of soil C and the rates at which it cycles. Soils in dry tropical forests tend to have less C per unit area and slower rates of C turnover times than do wetter tropical forests (Raich and Schlesinger, 1992). Modeling and radiocarbon studies show that soil CO_2 fluxes per unit area from tropical forest soils are an order of magnitude greater than those from temperate or boreal forest soils (Trumbore, 2000). These large, fast-cycling C stocks are therefore predicted to dominate short-term, interannual response to climate variations (Townsend et al., 1995; Trumbore et al., 1996). In contrast, the large stocks of SOM in high-latitude tundra, forests, and peatlands cycle very slowly because decomposition is restricted by low temperatures and anoxia from saturation (Carrasco et al., 2006). If these soils are warmed, decomposition and emissions of CO_2 and CH_4 will proceed rapidly, leading to a large positive feedback with climate change. Indeed, year-to-year differences in decomposition of old SOM in some boreal forest soils can determine the status of entire forest stands as net sources or sinks of C (Goulden et al., 1998). Because gross C fluxes at high latitudes are small compared to those in the tropics, it is unlikely that even large interannual variability in those fluxes could be as important in affecting the short-term (annual) C balance of the atmosphere. However, the potential for a large, long-term (decadal to century) response of soil C to climate change is greatest at higher latitudes, because much more organic C is stored there.

In addition to the large C stocks in high-latitude soils, the 900 Pg C in permafrost, about half of which is contained in deep, relict, loess soils from the last glacial period, may also be vulnerable to warming. The radiocarbon content of CO_2 and CH_4 in soil pore spaces, bubbles, and diffusive gas flux in these areas indicates that in many sites this C was fixed 10,000–20,000 years ago; and as the permafrost thaws, C in the ancient soil is being released to the atmosphere (Zimov et al., 2006).

Earth has warmed significantly over the past 150 years. The land areas of earth have warmed 0.27 °C per decade since 1979 and almost 1 °C since 1850 (IPCC, 2007). Considering only the effect of temperature on decomposition, soils should be an increasing source of atmospheric CO_2. However, trends in soil moisture, plant growth, and recovery of C stocks in previously eroded agricultural regions influence C flows in ways that may reverse this pattern in some places. In some regions, the interacting effects of temperature and moisture on plant growth and decomposition, as well as changes in plant litter quality (from changes in species composition and plant partitioning) and the nitrogen interaction described above, may be as important as the direct effects of warming on soil processes. Accurate assessment and prediction require considering microbial, plant, and microclimate influences on carbon flows.

Land surface models do not currently include landscape factors, like soil history and erosion, that can affect whether soils acts as regional net sources or sinks. For example, the retreat of the Laurentide ice sheet at the termination of the last ice age led to a period of terrestrial C accumulation as soils developed (Harden et al., 1992). Erosion and burial of soil in the 19th and early 20th centuries, combined with regrowth of vegetation on eroded soil, likely have led to overall C sinks in soils in the United States and western Europe in the last decades (van Oost et al., 2007). Ignoring such processes may lead to large errors in analyses of land-use change or in coupled carbon-climate models.

6.5.3. Land Use and Land Cover Change

6.5.3.1. Disturbance. Disturbances that affect soil C cycling include fires and floods, deforestation, cultivation, and drainage or fields, bogs, forests, and wetlands. All of these alter C inputs and losses to soil by changing vegetation, soil structure, temperature, water balance, and nutrient availability.Rates of change in organic C stocks in response to disturbance can be an order of magnitude larger than those associated with response to increased productivity or climate variability, because the changes in ecosystem inputs and decomposition rates are more extreme for disturbance.

Wildfires are predicted to get more frequent and severe in many regions due to climate change, particularly in regions that do not practice active fire suppression (Fried et al., 2008). Fire influences ecosystem C cycling by removing biomass and litter and creating black (pyrolyzed) carbon. In boreal forests and Mediterranean chaparral, decomposition is slow enough that it is less important than periodic removal of biomass by fire for returning CO_2 to the atmosphere. In boreal forests, warmer temperatures and more summer drought may be increasing the frequency and severity of fires, which, in turn, eliminate the moss layer that helps insulate the permafrost. Both factors, the loss of permafrost and the more severe burning, are predicted to drive these ecosystems to become net C sources (Harden et al., 2000). As mentioned in the section on recalcitrance, grasslands in Australia with frequent fires have relatively high proportions of chemically recalcitrant black carbon (Krull et al., 2006). Changes in fire frequency linked to climate or land management may ultimately control a region's status as a C source or sink.

6.5.3.2. Land Management. During the past two centuries, agricultural expansion has led to large losses of soil C. Soils may lose a significant portion of their C when native ecosystems are replaced by less productive ones; these changes represent a loss of fast-cycling C rather than passive C pools (Davidson and Ackerman, 1993; Harrison et al., 1993; Trumbore et al., 1995; Stallard, 1998). Tillage leads to substantial losses of old soil C due to physical disruption of soil aggregates and enhanced aeration of the soil that exposes organic matter to microbes and oxidization (Tisdall and Oades, 1982; Tiessen and Stewart, 1983; Baisden et al., 2002; Ewing et al., 2006). Based on a careful assessment of soil C stocks in pairs of uncultivated and cultivated fields, cultivation reduces C stocks by 25–30% within five years in temperate regions and faster (within two years) in the tropics (Davidson and Ackerman, 1993). The fraction of C lost is even higher in the A horizon.

Agricultural management that does not rely on tillage, such as bare-fallow and stubble-mulch practice, can greatly reduce carbon losses (Cambardella and Elliott, 1992). Moreover, converting tilled land to no-till agriculture can lead to rapid increases in soil C stocks, at least near the soil surface (Lal and Bruce, 1999). Consequently, management of cropping systems may have potential for C sequestration and has been promoted as a way to offset anthropogenic C emissions (Kong et al., 2005). Soil C is a resource that is potentially manageable, particularly in agricultural and pasture lands, although it is important to understand that soils have a finite capacity for C sequestration (Six et al., 2002). Predicting these limits requires understanding the mechanisms controlling SOM stabilization.

Globally, land conversion rates are highest in the tropics, and a significant proportion of this zone is in some state of recovery from past disturbance, mainly as forests succeed former pastures and croplands. Where productivity is high, soils under regrowing forests have been identified as another potential C sink, with sequestration capacity contingent on site history and climate. It may be more favorable in (a) wet forests where soil decomposition rates are lower than in dry or moist sites (Silver et al., 2000; Guo and Gifford, 2002) and (b) sites that have not been highly degraded, for example by intensive use of pasture and subsequent compaction, and where preexisting forest root systems were not heavily damaged.

6.5.4. Temporal Dimensions of Soils as Sources or Sinks of Carbon

As soils and ecosystems develop, they gain and lose C. For example, approximately 25% of the world's SOM is stored in soils that began developing after the last major deglaciation (Harden et al., 1992). Based on chronosequence studies, these soils are still functioning as long-term sinks for atmospheric CO_2. Similarly, we predict that older soils may be acting as long-term net sources of CO_2 to the atmosphere because of declines in NPP and weathering of minerals to more stable forms. Export of C from upland soils in dissolved or particulate form ultimately leads to transport into the oceans. While it is unclear whether, globally, soils were at steady state pre-1850, net C fluxes from soils to the atmosphere have been accelerated by large-scale land-cover changes over the past 150 years.

As described above, on decadal-to-century timescales, the net C balance of soils may be dominated by disturbance regimes and frequencies. Disturbance-dominated ecosystems are characterized by short periods of rapid C loss (e.g., from fires, large storms and blow-downs, insect mortality, or floods), followed by longer periods of C accumulation as they recover. Net C accumulation between disturbances may be rapid compared to the long-term rates associated with soil development and the associated alteration of soil minerals. However, when averaged over long times or large spatial scales (which include many stages of disturbance and recovery), the long-term rates should dominate.

Superimposed on the oscillations of disturbance and recovery is interannual variability in C flux from soils driven by variability in productivity and decay. Decay rates are of course directly and rapidly affected by climate anomalies such as droughts and heat waves. However, soil respiration also responds to changes in plant inputs. In addition to the magnitude of variability in these component processes, the net variability will also be determined by the lag time between C uptake by photosynthesis and respiration, which is a function of ecosystem C turnover times (Trumbore, 2000). For example, if most organic C is respired within a year of fixation, enhanced plant productivity in a given year will be offset by increased decomposition in the same year. However, if lag times are longer, higher-than-average productivity in one year would lead to net C gain, with net C loss in following years, as the pulse of high productivity is decomposed slowly over time.

6.6. CONCLUSIONS AND FUTURE PROSPECTS

As the focus on SOM has grown to encompass its pivotal role in the global carbon cycle and climate, the study of organic matter in soil has taken on new importance. Climatic change will cause ecosystems to experience novel and rapidly changing conditions, as well as put new demands on land management for carbon sequestration. Accurately predicting future atmospheric CO_2 concentrations and better managing soil resources will require a clear understanding of the processes and mechanisms controlling SOM storage and turnover.

There is a need for models that can predict ecosystem response to novel or long-term forcing. The numerical models used to simulate soil C cycling largely all share the same rules governing allocation of plant inputs, structure (C pools), and controls of turnover time. They incorporate multiple soil carbon pools, detailed plant growth modules, and temperature response functions (e.g., the CENTURY family, RothC, CASA, IBIS, Orchidee, LPJ, CN). Yet there is much room for improvement, in at least three important ways. First, while temperature and moisture interact to control decomposition and are predicted to change in novel combinations in the future, most models treat their effects as independent. Second, models rely on clay content as a proxy for the host of physical stabilization mechanisms in soil, if they include them at all. Finally, plant tissue chemistry drives model partitioning of inputs into pools of different turnover time, yet intrinsic plant compound recalcitrance is much less important than previously thought. These model simplifications exist in large part because the growing understanding of the processes that influence turnover time has occurred rapidly and has yet to be translated into mathematical functions that operate on an area basis (i.e., per m^2) and depend on variables that are regionally or globally available.

We have in hand sufficient understanding and data to begin development of much-improved model parameterizations, including the influence of plant allocation, soil mineralogy, and climate conditions. However, fundamental research targeting these areas, in parallel with model development, is still needed. There are many fruitful areas for future research; we suggest that priority be given to those processes and ecosystems that are vulnerable to global change, are potentially manageable, represent a large stock of carbon, and could influence atmospheric CO_2 concentrations significantly within the next several decades.

In the last decade, exciting new molecular, genomic, and imaging techniques have emerged for probing SOM at atomic and molecular scales, such as soft-energy X ray (for example, the Advanced Light Source), secondary ion mass spectrometry (SIMS, nano-SIMS), pyrolysis–gas chromatography–mass spectrometry–isotope ratio mass spectrometry (Py-GC-MS-IRMS), and gene-based microarrays. These techniques can be used in combination with isotopic analysis (^{13}C, ^{14}C) to explore detailed characterizations of turnover times. The vanguard, then, is utilizing these techniques in the context of experiments and controlled environmental gradients to gain insight at the landscape scale.

In terms of improving our ability to predict soil C turnover, we identify five priorities for research: (1) The interactive effects of temperature and moisture on microbial decomposition rates, because soils will experience novel and transient conditions; (2) the mechanisms governing protection of OM through interactions with mineral surfaces and due to spatial structure; (3) the mechanisms leading to slower OM turnover times with depth; (4) the potential for nonlinear responses of decomposition to C availability—for example, the role of labile C inputs in stimulating decomposition of less labile OM (i.e., priming) and density-dependent microbial behavior; and (5) how the chemical characteristics of organic compounds, as inputs from different plant species, charred (black) carbon, or microbial cell walls and by-products, influence mechanisms of stabilization and turnover.

6.7. APPENDIX 1. METHODS OF RADIOCARBON (^{14}C) ANALYSIS AND REPORTING OF ^{14}C DATA

6.7.1. Background Information

In this appendix, we briefly describe sample preparation and radiocarbon analysis, as well as the conventions for reporting of ^{14}C data. We have tried to strike a balance between brevity and explanation, addressing common questions we have encountered. Additionally, we urge those interested in using ^{14}C data to read Stuiver and Polach (1977), the paper that established most ^{14}C reporting conventions and from which most of the equations in this appendix were acquired.

6.7.2. Radiocarbon Sample Preparation

There are two methods for measuring radiocarbon: decay counting and accelerator mass spectrometry (AMS). Decay counting measures the electrons emitted during radioactive decay of ^{14}C to ^{14}N, measuring electrical pulses (gas counting) or light pulses (scintillation counting). Samples with a natural abundance of ^{14}C have relatively few decays per gram, because the half-life of ^{14}C is 5730 years. As a result, several grams of carbon and days or weeks of counting are required to observe enough decay events for a precise estimate of the ^{14}C concentration.

AMS directly measures the number of ^{14}C atoms, and the ratio of ^{14}C to ^{13}C and/or ^{12}C, using a high-energy accelerator as an inlet to a mass spectrometer. The key characteristics of ^{14}C-AMS are the electron stripping and ion acceleration, which allow ^{14}C to be distinguished from isobars and molecules that would confuse a standard mass spectrometer. AMS requires only a fairly small sample of 100 μg to 1 mg of C. In addition, the measurement only takes minutes per sample.

To measure ^{14}C in plant tissue or soils with AMS, the organic C must first be completely combusted to CO_2. Enough homogenized sample to provide ~1 mg C is added to a quartz glass tube with CuO and Ag. The tube is then evacuated, sealed, and combusted at ~900 °C (Buchanan and Corcoran, 1959). After combustion, or for gas samples from air or soil gas, the CO_2 in the sample is purified cryogenically and then reduced to graphite on an iron or cobalt catalyst using zinc (Xu et al., 2007) or hydrogen (Vogel et al., 1984). The graphite is pounded or pressed into a small sample holder called a target. In the AMS, the target is bombarded by a Cs beam to deliver a stream of C ions. Samples for decay counting are combusted at similar temperatures in a large vacuum line (Goh, 1991). The resulting CO_2 is cryogenically purified, then counted directly (gas counting) or converted to acetylene or benzene (scintillation counting). For more details on experimental methods, see Goh (1991) and Trumbore (1996).

For both decay counting and AMS, it is critical to prepare standard materials of known $^{14}C/^{12}C$ content. These include the OX1 standard described below and materials relatable to it, as well as materials that are radiocarbon-free. Standards allow assessment of the overall accuracy and the effects of sample pretreatment procedures, and radiocarbon-free samples provide a blank to determine the radiocarbon introduced to the sample during processing.

6.7.3. Reporting of Radiocarbon Data

Both AMS and conventional counting facilities report ^{14}C data as the ratio of ^{14}C activity in the sample to that of a known standard. By convention (Broecker and Olson, 1959; Stuiver and Polach, 1977), the standard is corrected to 0.95 times the activity of an oxalic acid standard (OX1), which is normalized to a $\delta^{13}C$ of −19‰. The sample is also normalized for ^{13}C content as follows: The activity of the sample, A_S, with a $\delta^{13}C$ of δ is corrected to a constant ^{13}C abundance (−25‰) by, using the following equation:

$$A_{SN} = A_S \frac{(1-25/1000)^2}{(1+\delta/1000)^2} \tag{A1.1}$$

where A_{SN} = ^{13}C-corrected sample activity and A_S = un-^{13}C-corrected activity of the sample. These ^{13}C corrections account for mass-dependent isotopic fractionation effects (Stuiver and Polach, 1977) and are a crucial part of the analysis. For example, the $\delta^{13}C$ difference between atmospheric CO_2 and carbon fixed during photosynthesis by C_3 plants is approximately 20‰. Assuming that the fractionation of ^{14}C is roughly twice that of ^{13}C (since the mass difference between 12 and 14 is twice that between 12 and 13; i.e., mass-dependent fractionation), the difference in ^{14}C abundance, if one didn't perform the correction in Eq. (A1.1), between atmospheric CO_2 and photosynthate, will be approximately 40‰ (equivalent to 330 ^{14}C years), even though both CO_2 and photosynthates are the same "age." Reporting radiocarbon data corrected to a common $\delta^{13}C$ value eliminates isotope fractionation effects and allows differences in age to be ascertained directly. It also allows analysis of SOM ^{14}C without having to quantify plant or microbial fractionation.

The standard approach to correcting for ^{13}C that is described above is applicable when fractionation is due to mass-dependent processes. This covers most diffusive and biological processes. One case where fractionation is not mass-dependent is the alteration of $\delta^{13}C$ values by physically mixing CO_2 sources, as is done in elevated CO_2 experiments. For that reason, a different equation should be used for the ^{13}C correction for samples from managed-CO_2 environments or experiments using purposeful C isotope tracer manipulations (Torn and Southon, 2001).

A common term used for reporting ^{14}C data is "fraction Modern" (F or $F^{14}C$; Reimer et al., 2004):

$$F^{14}\mathrm{C} = \frac{A_{\mathrm{SN}}}{A_{\mathrm{ON}}} = \frac{\left(\dfrac{^{14}\mathrm{C}}{^{12}\mathrm{C}+^{13}\mathrm{C}}\right)_{\mathrm{sample}(-25)}}{0.95\left(\dfrac{^{14}\mathrm{C}}{^{12}\mathrm{C}+^{13}\mathrm{C}}\right)_{\mathrm{OX1}(-19)}} \tag{A1.2}$$

where A_{SN} is as defined above for Eq. (A1.1) and A_{ON} is 0.95 times the measured activity of the OX1 standard normalized to a δ^{13}C of −19‰.

The conventional radiocarbon age, reported by AMS labs and used in archaeology, is

$$^{14}\mathrm{C\ age} = -8033 \ln F^{14}\mathrm{C} \tag{A1.3}$$

where 8033 yr is the Libby mean life of radiocarbon. Note that the true mean life of radiocarbon is 8267 yr (the Libby mean life is used for this unit by convention), so the ^{14}C age is not an accurate calculation of true age of the sample. Radiocarbon ages are referenced to 1950 such that 1950 A.D. = 0 B.P. Samples with more ^{14}C than the 1950 atmosphere (i.e., those with $F^{14}\mathrm{C} > 1$) are commonly reported as ">Modern".

Great care must be taken in using the conventional radiocarbon age, which is almost never used directly. In some cases the actual age of an object in soil is required—for example, for determining the age of a seed or a piece of undecomposed sphagnum in a peat bog. Such an age can only be calculated for something that formed in a single year (or short time span) and is presumed not to have exchanged carbon with its surroundings after being added to the soil. In such cases, the Libby age must be converted to a calendar age using appropriate calibration curves—several programs for this are available through the web site of the journal *Radiocarbon*, www.radiocarbon.org. *The age with the Libby half-life is almost never used except as the basis for calculating calibrated ages, and it should not be used to estimate mean residence times of carbon in soil directly.*

The activity of OX1 changes through time as ^{14}C in the standard decays (i.e., A_{ON} measured in 2007 is less than if it were measured in 1950). For dating purposes, both the sample and A_{ON} decrease at the same rate (the radiocarbon decay constant). In other words, F^{14}C is constant with time. However, when considering an open and dynamic system, such as soil, the need arises for a standard that represents a constant value. Stuiver and Polach (1977) thus proposed an *absolute international standard activity* (A_{abs}) that would incorporate a yearly correction for the decay in the OX1 standard:

$$A_{\mathrm{abs}} = A_{\mathrm{ON}} \exp[\lambda(y-1950)] \tag{A1.4}$$

where y is the year of sample collection and $\lambda = 1/8267\,\mathrm{yr}^{-1} = 1.210 \times 10^{-4}\,\mathrm{yr}^{-1}$. (This λ is the true radiodecay constant rather than that derived from the Libby mean life of 8033 yrs.) The ratio A_{SN}/A_{abs} therefore differs from F^{14}C by the factor $\exp(-\lambda(y - 1950))$, and it will decrease with time since the ^{14}C in the sample radiodecays but the amount in the standard stays the same as in 1950.

The most commonly reported ^{14}C unit in biogeochemical studies is Δ^{14}C. This parameter is the deviation in parts per thousand (per mil, ‰) from the absolute standard (A_{abs}):

$$\Delta^{14}C = \left[\frac{A_{SN}}{A_{abs}} - 1\right] \times 1000 = \left[F^{14}C \times \exp\left(\frac{-(y-1950)}{8267}\right) - 1\right] \times 1000 \quad (A1.5)$$

Positive values of $\Delta^{14}C$ indicate the presence of bomb-produced ^{14}C. Conversely, negative values of $\Delta^{14}C$ indicate the predominance of C fixed from the atmosphere long enough ago for significant radioactive decay of ^{14}C to have occurred.

Most radiocarbon measurement facilities provide the analysis results in different formats, depending on the needs of the researchers. Forms in which results are commonly reported include $F^{14}C$, $\Delta^{14}C$, and ^{14}C age. Conversion between the various units can be done using Eqs. (A1.4) and (A1.5). *If only $\Delta^{14}C$ data are reported or published, it is important to state the year of measurement, since values of $\Delta^{14}C$ will be specific to that year.* The analytical precision typically reported with the data is the 1 sigma error, determined from counting statistics and propagating laboratory errors. Typical precision reported for samples with $F^{14}C$ ~1 is ±0.005 (or ±5‰ for $\Delta^{14}C$), and as low as 0.001 (±1‰) for high-precision analyses. Accuracy is usually reported based on the repeated analysis of secondary standards of known $F^{14}C$, or at least materials for which a consensus value exists, and is laboratory-specific.

6.8. APPENDIX 2. MODELING CARBON DYNAMICS USING RADIOCARBON MEASUREMENTS

6.8.1. Background Information

SOM is a heterogeneous reservoir with a variety of turnover times, to which carbon is continuously added (as new plant matter) and lost (as CO_2, leached organic matter, or eroded material). These dynamics preclude using radiocarbon to meaningfully "date" SOM, and at best the ^{14}C-based age of SOM represents the average ^{14}C age of a carbon atom in the soil reservoir. This tells us relatively little about the distribution of C in reservoirs with different turnover times, and can be quite misleading when the SOM has incorporated "bomb ^{14}C" created through atmospheric nuclear weapons testing. In this appendix we describe methods of using ^{14}C to estimate turnover times of soil organic C. These methods differ somewhat when the source is natural radiocarbon ('pre-bomb') or bomb ^{14}C and when the system is assumed to be at steady state or changing.

6.8.2. Steady-State Systems

6.8.2.1. Natural Radiocarbon—For Samples Collected Prior to 1950, or Assumed to Contain No Bomb Radiocarbon. For samples not complicated by the presence of bomb ^{14}C, the ratio of $^{14}C/^{12}C$ measured in a sample represents the rate of decomposition relative to the rate of radiodecay of ^{14}C. This treatment is most useful for very old C found in soils. For a homogeneous carbon-containing reservoir, i, with input rate I_i, first-order decomposition constant k_i, and carbon content C_i, the change in stock over time (balance of inputs and outputs) is

$$dC_i/dt = I_i - k_iC_i \quad (A2.1)$$

For a steady-state reservoir, we have $k_i = I_i/C_i$. Since turnover time (τ) is defined as $1/k_i$, at steady state, τ equals the inventory of carbon divided by the input rate, C_i/I_i.

The balance of ^{14}C atoms in the same reservoir ($^{14}C_i = F_iC_i$) will reflect the rate of loss from decomposition, k_i, as well as the rate constant for radioactive decay of ^{14}C, λ ($\lambda = 1.210 \times 10^{-4}\,yr^{-1}$), and the rate of inputs (in this case, from the atmosphere):

$$dF_i/dt = (1/C_i)(I_iF_{\text{atmosphere}} - (k_i + \lambda)F_iC_i) \tag{A2.2}$$

At steady state and assuming that $F_{\text{atmosphere}}$ before 1959 equals 1, we have

$$F_i = (1/C_i)(I_i/(k_i + \lambda)) \tag{A2.3}$$

Since at steady state, we have $C_i = I_i/k_i$, Eq. (A2.3) may be rewritten as

$$F_i = (k_i/(k_i + \lambda)) \tag{A2.4}$$

For components with short turnover times ($k_i >> \lambda$), a calculated ^{14}C age will approximate the turnover time, τ ($1/k_i$). For components with k_i equal to or less than the decay constant for radiocarbon, the age will be less than the turnover time. For example, the ^{14}C age calculated for a steady-state reservoir with $k_i = 0.01\,yr^{-1}$ ($\tau = 100\,yr$) would be 100 yr, while that for a component with $k_i = 0.0002\,yr^{-1}$ ($\tau = 5000\,yr$) would be 3910 yr.

Note that this approach assumes $F^{14}C = 1.0$ and is constant prior to 1950. Actually the $^{14}C/^{12}C$ of atmospheric CO_2 did vary with time prior to 1900, mostly reflecting changes in the rate of ^{14}C production in the upper atmosphere. During the Holocene, these variations were less than 10%, and they are documented in the calibration data sets based on ^{14}C measured in known-age wood. Between 1900 and 1950, F_{atm} declined due to the addition of ^{14}C-free CO_2 derived from fossil fuels, known as the Suess effect. Modeling of turnover times should use the actual atmospheric ^{14}C inputs to photosynthesis, although it is not as important before 1959 as after.

6.8.2.2. Bomb Radiocarbon. One of the great uses of radiocarbon for SOM studies is the ability to estimate the turnover time of organic carbon based on the degree to which it has incorporated bomb radiocarbon since 1959. This provides one of the only tools to study C dynamics on decadal time scales.

For a steady-state system, a time-dependent model is used because of the irregular shape of the atmospheric $^{14}CO_2$ record. This model accounts for radioactive decay of the ^{14}C since 1950 explicitly, and it requires that we compare measured radiocarbon to a standard with a radiocarbon value that stays constant over time (A_{abs}). For ease, we define F' here as A_{SN}/A_{abs} [see Eq. (A1.4)] for samples measured since 1950; F' equals $\Delta^{14}C/1000 + 1$. For a reservoir at steady state, the balance of radiocarbon entering and leaving the reservoir in year t is given by

$$F'_{C,t} = \frac{[IF'_{\text{atm},t} + C_{t-1}F'_{C,t-1}(1 - k - \lambda)]}{C_t} \tag{A2.5}$$

Since the reservoir is at steady state, we have $C_{(t-1)} = C_{(t)} = I/k$, so Eq. (A2.5) reduces to

$$F'_{C,t} = kF'_{\text{atm},t} + F'_{C,t-1}(1-k-\lambda) \tag{A2.6}$$

Figure 6.8 shows the predicted values of ^{14}C in 1996 for a homogeneous, steady-state reservoir with different turnover times. For turnover times <50 yr, it is clear that two different turnover times may yield the same $\Delta^{14}C$ value. To distinguish which of these two turnover times is correct, we use one of two methods. First, if an archived sample from the same soil is available, radiocarbon measurements may distinguish between the two possibilities. Organic matter with shorter turnover times will have decreased in ^{14}C over the past several decades, while those with longer turnover times will have increased in ^{14}C. If no archived soil is available, knowledge of the C stock and C fluxes into and out of the soil may be used to determine the correct turnover time (since τ = stock/flux), as illustrated here.

The $\Delta^{14}C$ values measured in low-density organic matter isolated from the A horizon of a soil sampled in 1956 and 1992 in the Sierra Nevada were −31‰ and +127‰, respectively (Trumbore et al., 1996). This ^{14}C increase is consistent with either a turnover time of 5 or 57 yr. The total amount of low-density carbon in the A horizon was 6.5 kg C m^{-2}, with low-density carbon accounting for nearly 90% of the carbon in this layer. The 5-year turnover time implies annual C inputs from litter of ~1300 g C m^{-2} yr^{-1}, while the 57-year turnover time implies inputs of only 114 g C m^{-2} yr^{-1}. The measured aboveground litterfall at a nearby site was ~100 g C m^{-2} yr^{-1}. Hence, the most reasonable turnover time is 57 yr for the low-density organic matter in the A horizon.

A potential problem with this approach is the uncertainty as to whether the reservoir under consideration is homogeneous. Bulk SOM is almost certainly heterogeneous, and the bulk ^{14}C value does not give a good idea of SOM dynamics. Even low-density organic matter is made up of relatively fresh litter material (small roots and pieces of leaves) as well as more humified materials that likely have slower turnover. Normally, the soil must be split into components with different turnover times using fractionation methods outlined in the text. For each component, a new measurement constraint (such as total C flux into and out of the soil) must be added to arrive at a unique solution. A second problem is that carbon entering the soil as litter may not have the ^{14}C signature of that year's atmospheric CO_2 (F'_{atm} in the equations above). For example, conifer needles often reside on trees for several years before they fall and are incorporated into soils. Failure to account for these time lags in living vegetation may result in an overestimate of the time required for decomposition (since the turnover time will reflect the time spent in the plant plus soil, rather than the soil alone).

6.8.2.3. Systems That Are Accumulating Soil Carbon. Again, net change in C storage (dC/dt) represents the balance between annual C inputs (I; kg C m^{-2} yr^{-1}) and decomposition (kC, where k is a first-order decomposition rate constant (yr^{-1}), and C_t is the soil layer C inventory in kg C m^{-2}) in year t. The solution to this [integrating Eq. (A2.1) and assuming that C stock = 0 at time t = 0] is

$$C(t) = I/k - (I/k)e^{-kt} = I/k[1-\exp(-kt)] \tag{A2.7}$$

(By assuming that $C = 0$ at $t = 0$, we are modeling only the new C that has accumulated; this is most practical in sites where the new C is layered distinctly from older soil organic matter. If this is not the case, the integration can be done keeping C_0 explicitly.) Using historical site data or radiocarbon data to determine the time of accumulation, t, we can determine the history of C accumulation at a site. A plot of accumulated carbon inventory (C_t) versus the time it took to accumulate (t, from radiocarbon in this case) may be fit with Eq. (A2.7) to derive estimates of I and k describing either decadal (bomb radiocarbon) or millennial (natural radiocarbon) C dynamics (Trumbore and Harden, 1997). An example is shown in Figure 6.10. Alternatively, for a known-age disturbance or soil age, the amount of C accumulated, as well as the amount of radiocarbon accumulated, will uniquely determine I and k.

6.8.2.3.1. Natural Radiocarbon (i.e. Samples not Affected by Bomb-C). Prior to 1950, the ^{14}C content of atmospheric CO_2 was approximately constant relative to the magnitude of the bomb spike or to radiodecay of ^{14}C in SOM cycling on century-millennial timescales. For constant atmospheric ^{14}C content ($F_{atm} = 1.0$ pre-1959), $F_{C,t}$

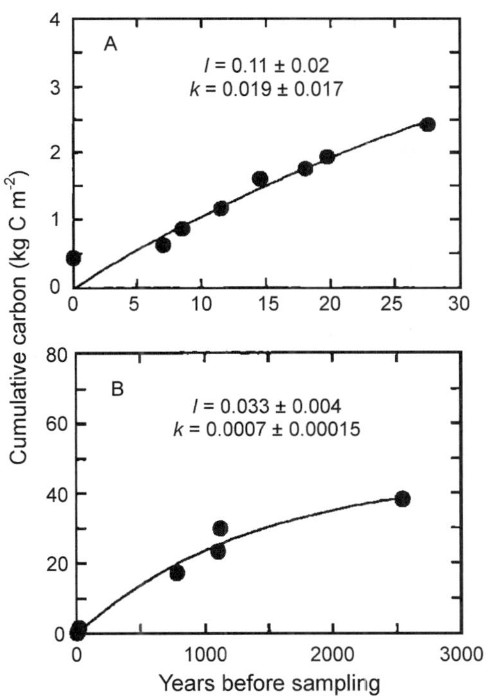

Figure 6.10. Accumulation of C in non-steady-state soils of a mature black spruce/moss forest in central Manitoba, Canada. Data shown are **(A)** for sphagnum moss that has accumulated since the site last burned (~100 yr before sampling), and **(B)** for the humus and charred layer below the regrowing moss and including the A horizon. The soil is developed on the sediments of a lake that dried up ~7000 years ago. The parameters I = plant input ($kg\,C\,m^{-2}\,yr^{-1}$) and k = decomposition constant (yr^{-1}). Reprinted from Trumbore and Harden (1997), with permission from the American Geophysical Union.

may be expressed by including the loss terms for radiocarbon (decomposition plus radiodecay: $k + \lambda$) compared to carbon [decomposition only; Eq. (A2.1)]:

$$F_{C,t} = \frac{\frac{I}{(k+\lambda)}\left[1-e^{(-(k+\lambda)t)}\right]}{C_t} \tag{A2.8}$$

where t is the time since soil (or soil layers) began to form. Substituting for C_t, we obtain

$$F_{C,t} = \frac{k}{(k+\lambda)} * \frac{\left[1-e^{(-(k+\lambda)t)}\right]}{\left[1-e^{(-kt)}\right]} \tag{A2.9}$$

In their study of how different soil minerals affect the long-term turnover rates of carbon in soils of Hawaii, Torn et al. (1997) used this approach, but they incorrectly used F' rather than F and used the Libby half-life of radiocarbon. Although these mistakes had only small effects on the estimates of turnover time reported in that paper, it was this kind of confusion over approaches that led us to write this appendix. Likely far larger (but unquantifiable) errors when using the natural radiocarbon equations given here are (1) the possibility that some bomb C has been incorporated in any sample taken since 1950 (which would lead to underestimation of turnover times) and (2) the assumption that the carbon pools being measured are homogeneous (i.e., all described by a single turnover time).

6.8.2.3.2. Bomb Radiocarbon. To determine the inventory-weighted mean $\Delta^{14}C$ value in 1996, we assume that annual C additions are labeled with the $\Delta^{14}C$ of that year's atmospheric CO_2, and we track the loss of C and ^{14}C with time for each year's C input. We can ignore isotopic fractionation, and we can assume that respired C has the same ^{14}C content as the organic matter in each annual layer because of the ^{13}C correction in the radiocarbon units (see Appendix 1). The equation expressing the inventory-weighted mean ^{14}C content of the soil profile in year t after initiation of accumulation is

$$F'_{C,t} = \frac{\sum_{i=0}^{i=T} C_{i,t} F'_{\text{atm},i}}{\sum_{i=0}^{i=T} C_{i,t}} \tag{A2.10}$$

where T is the total number of years carbon has been accumulating (years since disturbance), and $F'_{\text{atm},i}$ is A_{SN}/A_{abs} for carbon fixed in the year i (assumed to equal that year's atmospheric $\Delta^{14}CO_2/1000 + 1$), and $C_{i,t} = I/k\ (1 - \exp(-kt))$ is the carbon remaining t years after it was fixed in year i ($t = T - i$). For example, consider a layer of moss and detritus sampled in 1994 that began to accumulate 120 years prior, following a fire in a boreal forest. If the rate of C inputs is $120\,g\,C\,m^{-2}\,yr^{-1}$ and the decomposition rate is $0.02\,yr^{-1}$ (turnover time of 50 years), the total amount of carbon accumulated in 120 years will be $5.5\,kg\,C\,m^{-2}$, and the bulk $\Delta^{14}C$ of the moss layer will be +183 ‰. For the same input rate, but with faster decomposition ($0.04\,yr^{-1}$, or turnover time of 25 years), only $3.0\,kg\,C\,m^{-2}$ will have accumulated, with $\Delta^{14}C$ of +231‰. Knowing the bulk amount of C and its inventory-weighted $\Delta^{14}C$ value for a known period of accumulation will uniquely define I and k. Again, complications arise because of the assumption of zero time lag for C storage in vegetation (especially woody biomass that may represent a significant portion of the litter carbon in fire-dominated forest systems), but lag times can be estimated and incorporated.

6.8.3. Converting from k to τ

The relationship between the turnover time, τ, and the decay rate constant, k, used in the appendix, is a simple inverse: $\tau = 1/k$. However, care must be taken when converting between the *mean* value of k and τ, because the inverse of a mean value is not the same as the mean of inverse values (i.e., the mean τ is *not* the inverse of the mean k). To illustrate, consider three samples with estimated k values of 0.1, 0.08, and $0.06\,yr^{-1}$. Mean k is $0.08\,yr^{-1}$. It would be incorrect to calculate the mean τ as $1/0.08 = 12.5\,yr$. Rather, the correct mean τ is the mean of 1/0.1, 1/0.08, and 1/0.06, which is 13 yr. In other words, one must first convert each k value to a τ value and calculate the mean from the replicate τ values. The same is true for converting from mean τ to mean k.

ACKNOWLEDGMENTS

We thank Markus Kleber and Cinzia Fissore for insightful comments. This work was supported by the U.S. Department of Energy under Contract No. DE-AC02-05CH11231 and the Northern Institute of Applied Carbon Science, USDA Forest Service.

REFERENCES

Aber, J. D., Nadelhoffer, K. J., Steudler, P., and Melillo, J. M. (1989). Nitrogen saturation in northern forest ecosystems. *BioScience* **39**(6), 378–386.

Agren, G. I., and Bosatta, E. (1988). Nitrogen saturation of terrestrial ecosystems. *Environ. Pollut.* **54**(3–4), 185–197.

Amundson, R. (2001). The carbon budget in soils. *Annu. Rev. Earth Planet. Sci.* **29**, 535–562.

Amundson, R., and Jenny, H. (1997). On a state factor model of ecosystems. *Bioscience* **47**(8), 536–543.

Baisden, W. T., Amundson, R., Cook, A. C., and Brenner, D. L. (2002). Turnover and storage of C and N in five density fractions from California annual grassland surface soils. *Global Biogeochem. Cycles* **16**(4), 1117.

Baker, T. R. (2004). Increasing biomass in Amazonian forest plots. *Philos. Trans. R. Soc. London Ser. B Biol. Sci.* **359**(1443), 353–365.

Baldock, J. A., and Skjemstad, J. O. (2000). Role of the soil matrix and minerals in protecting natural organic materials against biological attack. *Org. Geochem.* **31**(7–8), 697–710.

Balesdent, J., Wagner, G. H., and Mariotti, A. (1988). Soil organic matter turnover in long-term field experiments as revealed by carbon-13 natural abundance. *Soil Sci. Soc. Am. J.* **52**(1), 118–124.

Basile-Doelsch, I., Amundson, R., Stone, W., Masiello, C. A., Bottero, J.-Y., Colin, F., Masin, F., Borschneck, D., and Meunier, J.-D. (2005). Mineralogical control of organic carbon dynamics in an allophanic soil (La Reunion). *Eur. J. Soil Sci.* **56**, 689–703.

Batjes, N. H. (1996). Total carbon and nitrogen in the soils of the world. *Eur. J. Soil Sci.* **47**(2), 151–163.

Bellamy, P. H. (2005). Carbon losses from all soils across England and Wales 1978–2003. *Nature* **437**(7056), 245–248.

Berg, B., McClaugherty, C., De Santo, A. V., and Johnson, D. (2001). Humus buildup in boreal forests: effects of litter fall and its N concentration. *Can. J. Forest Res. Rev. Can. Recherche Forestiere* **31**(6), 988–998.

Berhe, A. A., Harte, J., Harden, J. W., and Torn, M. S. (2007). The significance of the erosion-induced terrestrial carbon sink. *Bioscience* **57**(4), 337–346.

Bird, J. A., and Torn, M. S. (2006). Fine roots vs. needles: A comparison of ^{13}C and ^{15}N dynamics in a ponderosa pine forest soil. *Biogeochemistry* **79**(3), 361–382.

Boggs, J. L. (2005). Tree growth, foliar chemistry, and nitrogen cycling across a nitrogen deposition gradient in southern Appalachian deciduous forests. *Can. J. Forest Res. Rev. Can. Recherche Forestiere* **35**(8), 1901–1913.

Bohlen, P. J., Groffman, P. M., Fahey, T. J., Fisk, M. C., Suarez, E., Pelletier, D. M., and Fahey R. T. (2004). Ecosystem consequences of exotic earthworm invasion of north temperate forests. *Ecosystems* **7**(1), 1–12.

Boisvenue, C., and Running, S. W. (2006). Impacts of climate change on natural forest productivity—Evidence since the middle of the 20th century. *Global Change Biology* **12**(5), 862–882.

Broecker, W. S., and Olson, E. A. (1959). Lamont radiocarbon measurements VI. *Radiocarbon* **1**, 111–132.

Brown, P. E. (1963). *Soils of Iowa*. Iowa Agr. Expt. Sta., Special Rep. 3.

Buchanan, D., and Corcoran, B. (1959). Sealed tube combustions for the determination of carbon-14 and total carbon. *Anal. Chem.* **31**(10), 1635–1638.

Cairney, J. W. G., and Burke, R. M. (1998). Extracellular enzyme activities of the ericoid mycorrhizal endophyte *Hymenoscyphus ericae* (Read) Korf & Kernan: Their likely roles in decomposition of dead plant tissue in soil. *Plant Soil* **205**(2), 181–192.

Cambardella, C. A. (1994). Carbon and nitrogen dynamics of soil organic matter fractions from cultivated grassland soils. *Soil Sci. Soc. Am. J.* **58**(1), 123–130.

Cambardella, C. A., and Elliott, E. T. (1992). Particulate soil organic-matter changes across a grassland cultivation sequence. *Soil Sci. Soc. Am. J.* **56**(3), 777–783.

Carrasco, J. J., Neff, J. C., and Harden, J. W. (2006). Modeling physical and biogeochemical controls over carbon accumulation in a boreal forest soil. *J. Geophys. Res. Biogeosci.* **111**(G2).

Castanha, C., Trumbore, S., and Amundson, R. (2008). Methods of separating soil carbon pools affect the chemistry and turnover time of isolated fractions. *Radiocarbon.* **50**(1), 83–97.

Chadwick, O. A., Derry, L. A., Vitousek, P. M., Huebert, B. J., and Hedin, L. O. (1999). Changing sources of nutrients during four million years of ecosystem development. *Nature* **397**(6719), 491–497.

Chambers, J. Q., and Silver, W. L. (2004). Some aspects of ecophysiological and biogeochemical responses of tropical forests to atmospheric change. *Philos. Trans. R. Soc. London Ser. B Biol. Sci.* **359**(1443), 463–476.

Chan, K. Y. (2001). Soil particulate organic carbon under different land use and management. *Soil Use Manage.* **17**, 217–221.

Chapin, F. S., Mcguire, A. D., Randerson, J., Pielke, R., Baldocchi, D., Hobbie, S. E., Roulet, N., Eugster, W., Kasischke, E., Rastetter, E. B., Zimov, S. A., and Running, S. W. (2000). Arctic and boreal ecosystems of western North America as components of the climate system. *Global Change Biol.* **6**(s1), 211–223.

Christensen, B. T. (1992). Physical fractionation of soil and organic matter in primary particle size and density separates. *Adv. Soil Sci.* **20**, 1–90.

Cisneros-Dozal, L. M., Trumbore, S., and Hanson, P. J. (2006). Partitioning sources of soil-respired CO_2 and their seasonal variation using a unique radiocarbon tracer. *Global Change Biol.* **12**(2), 194–204.

Crow, S. E., Swanston, C. W., Lajtha, K., Brooks, J. R., and Keirstead, H. (2007). Density fractionation of forest soils: Methodological questions and interpretation of incubation results and turnover time in an ecosystem context. *Biogeochemistry* **85**(1), 69–90.

Davidson, E. A., and Ackerman, I. L. (1993). Changes in soil carbon inventories following cultivation of previously untilled soils. *Biogeochemistry* **20**(3), 161–193.

Davidson, E. A., and Janssens, I. A. (2006). Temperature sensitivity of soil carbon decomposition and feedbacks to climate change. *Nature* **440**(7081), 165–173.

Davidson, E. A., and Lefebvre, P. A. (1993). Estimating regional carbon stocks and spatially covarying edaphic factors using soil maps at three scales. *Biogeochemistry (Dordrecht)* **22**(2), 107–131.

Denef, K., Six, J., Bossuyt, H., Frey, S. D., Elliott, E. T., Merckx, R., and Paustian, K. (2001). Influence of dry-wet cycles on the interrelationship between aggregate, particulate organic matter, and microbial community dynamics. *Soil Biol. Biochem.* **33**(12–13), 1599–1611.

Denef, K., Six, J., Merckx, R., and Paustian, K. (2004). Carbon sequestration in microaggregates of no-tillage soils with different clay mineralogy. *Soil Sci. Soc. Am. J.* **68**(6), 1935–1944.

Denman, K. L., Brasseur, G., Chidthaisong, A., Ciais, P., Cox, P., Dickison, R. E., Hauglustaine, D., Heinze, C., Holland, E., Jacob, D., Lohmann, U., Ramachandran, S., da Silva Dias, P. L., Wofsy, S. C., and Zhang, X. (2007). Couplings between changes in the climate system and biogeochemistry. In *Climate Change 2007: The Physical Science Basis. Contribution of Working Group I to the Fourth Assessment Report of the IPCC*, Soloman, S., Qin, D., Manning, M., Chen, Z., Marquis, M., Averyt, K. B., Tignor, M., and Miller, H. L. Cambridge University Press, Cambridge, United Kingdom, pp. 499–587.

Dijkstra, P., Ishizu, A., Doucett, R., Hart, S. C., Schwartz, E., Menyailo, O. V., and Hungate, B. A. (2006). 13C and 15N natural abundance of the soil microbial biomass. *Soil Biol. Biochem.* **38**(11), 3257–3266.

Dörr, H., and Münnich, K. O. (1986). Annual variations of the ^{14}C content of soil CO_2. *Radiocarbon* **28**(2A), 338–345.

Dunne, J. A., Saleska, S. R., Fischer, M. L., and Harte, J. (2004). Integrating experimental and gradient methods in ecological climate change research. *Ecology* **85**(4), 904–916.

Dutta, K., Schuur, E. A. G., Neff, J. C., and Zimov, S. A. (2006). Potential carbon release from permafrost soils of Northeastern Siberia. *Global Change Biol.* **12**(12), 2336–2351.

Ellert, B. H., Janzen, H. H., and McConkey, B. G. (2001). Measuring and Comparing Soil Carbon Storage. In *Assessment Methods for Soil Carbon*, Lal, R., Kimble, J. M., Follett, R. F., and Stewart, B. A., Lewis Publishers, Boca Raton, FL, pp. 131–146.

Eswaran, H., Van Den Berg, E., and Reich, P. (1993). Organic carbon in soils of the world. *Soil Sci. Soc. Am. J.* **57**(1), 192–194.

Ewing, S. A., Sanderman, J., Baisden, W. T., Wang, Y., and Amundson, R. (2006). Role of large-scale soil structure in organic carbon turnover: Evidence from California grassland soils. *J. Geophysi. Res. Biogeosci.* **111**(G3).

Fischlin, A., and Gyalistras, D. (1997). Assessing impacts of climatic change on forests in the Alps. *Global Ecol. Biogeogr. Lett.* **6**, 19–37.

Fischlin, A., Midgley, G. F., Price, J. T., Leemans, R., Gopal, B., Turley, M. D. A., Rounsevell, M. D. A., Dube, O. P., Tarazona, J., and Velichko, A. A. (2007). Ecosystems, their properties, goods, and services. In *Climate Change 2007: Impacts, Adaptation, and Vulnerability. Contribution of Working Group II to the Fourth Assessment Report of the IPCC*, Parry, M. L., Canziani, O. F., Palutikof, J. P., van der Llinden, P. J., and Hanson, C. E., Cambridge University Press, Cambridge, UK, pp. 211–272.

Freeman, C., Evans, C. D., Monteith, D. T., Reynolds, B., and Fenner, N. (2001). Export of organic carbon from peat soils. *Nature* **412**(6849), 785.

Fried, J. S., Gilless, J. K., Riley, W. J., Moody, T. J., de Blas, C. S., Hayhoe, K., Moritz, M., Stephens, S., and Torn, M. (2008). Predicting the effect of climate change on wildfire behavior and initial attack success. *Climatic Change* **87**, S251–S264.

Friedlingstein, P., Cox, P., Betts, R., Bopp, L., Von Bloh, W., Brovkin, V., Cadule, P., Doney, S., Eby, M., Fung, I., Bala, G., John, J., Jones, C., Joos, F., Kato, T., Kawamiya, M., Knorr, W., Lindsay, K., Matthews, H. D., Raddatz, T., Rayner, P., Reick, C., Roeckner, E., Schnitzler, K. G., Schnur, R., Strassmann, K., Weaver, A. J., Yoshikawa, C., and Zeng, N. (2006). Climate–carbon cycle feedback analysis: Results from the (CMIP)-M-4 model intercomparison. *J. Climate* **19**(14), 3337–3353.

Gaudinski, J. B., Trumbore, S. E., Davidson, E. A., and Zheng, S. H. (2000). Soil carbon cycling in a temperate forest: Radiocarbon-based estimates of residence times, sequestration rates and partitioning of fluxes. *Biogeochemistry* **51**(1), 33–69.

Gerten, D. (2004). Terrestrial vegetation and water balance—Hydrological evaluation of a dynamic global vegetation model. *J. Hydrol.* **286**(1–4), 249–270.

Glinka, K. D. (1927). *Dokuchaiev's Ideas in the Development of Pedology and Cognate Sciences*, USSR Academy of Sciences Russian Pedological Investigations, Leningrad.

Goh, K. M. (1991). Carbon dating. In *Carbon Isotope Techniques*, Coleman, D. C., and Fry, B., eds., Academic Press, San Diego, pp. 125–145.

Golchin, A., Oades, J. M., Skjemstad, J. O., and Clarke, P. (1994a). Soil structure and carbon cycling. *Aust. J. Soil Res.* **32**(5), 1043–1068.

Golchin, A., Oades, J. M., Skjemstad, J. O., and Clarke, P. (1994b). Study of free and occluded particulate organic matter in soils by solid-state ^{13}C CP/MAS NMR spectroscopy and scanning electron microscopy. *Aust. J. Soil Res.* **32**(2), 285–309.

Goulden, M. L., Wofsy, S. C., Harden, J. W., Trumbore, S. E., Crill, P. M., Gower, S. T., Fries, T., Daube, B. C., Fan, S. M., Sutton, D. J., Bazzaz, A., and Munger, J. W. (1998). Sensitivity of boreal forest carbon balance to soil thaw. *Science* **279**(5348), 214–217.

Grandy, A. S., and Neff, J. C. (2008). Molecular C dynamics downstream: The biochemical decomposition sequence and its impact on soil organic matter structure and function. *Sci. Total Environ.* **404**(2/3), 297–303.

Gregorich, E. G., Monreal, C. M., Schnitzer, M., and Schulten, H.-R. (1996). Transformation of plant residues into soil organic matter: chemical characterization of plant tissue, isolated soil fractions, and whole soils. *Soil Sci.* **161**(10), 680–693.

Guo, L. B., and Gifford, R. M. (2002). Soil carbon stocks and land use change: A meta analysis. *Global Change Biology* **8**(4), 345–360.

Hamer, U., Marschner, B., Brodowski, S., and Amelung, W. (2004). Interactive priming of black carbon and glucose mineralisation. *Org. Geochem.* **35**(7), 823–830.

Hammes, K., Torn, M. S., Lapenas, A. G., and Schmidt, M. W. I. (2008). Centennial black carbon turnover observed in a Russia steppe soil. *Eur. Biogeosci. J.* **5**, 1339–1350.

Hanson, P. J., Swanston, C. W., Garten, C. T., Jr., Todd, D. E., and Trumbore, S. E. (2005). Reconciling change in Oi-horizon carbon-14 with mass loss for an oak forest. *Soil Sci. Soc. Am. J.* **69**(5), 1492–1502.

Harden, J. W., Sharpe, J. M., Parton, W. J., Ojima, D. S., Fries, T. L., Huntington, T. G., and Dabney, S. M. (1999). Dynamic replacement and loss of soil carbon on eroding cropland. *Global Biogeochem. Cycles* **13**(4), 885–901.

Harden, J. W., Sundquist, E. T., Stallard, R. F., and Mark, R. K. (1992). Dynamics of soil carbon during deglaciation of the Laurentide ice sheet. *Science* **258**(1921–1924).

Harden, J. W., Trumbore, S. E., Stocks, B. J., Hirsch, A., Gower, S. T., O'Neill, K. P., and Kasischke, E. S. (2000). The role of fire in the boreal carbon budget. *Global Change Biol.* **6**(s1), 174–184.

Harmon, M. E., Nadelhoffer, K. J., and Blair, J. M. (1999). Measuring decomposition, nutrient turnover, and stores in plant litter. In *Standard Soil Methods for Long-Term Ecological Research.* Robertson, G. P., Bledsoe, C. S., Coleman, D. C., and Sollins, P., eds., Oxford University Press, New York, pp. 202–240.

Harradine, F., and Jenny, H. (1958). Influence of parent material and climate on texture and nitrogen and carbon contents of virgin California soils I. Texture and nitrogen contents of soils. *Soil Sci.* **85**(5), 235–243.

Harrison, K. (1993). A strategy for estimating the impact of CO_2 fertilization on soil carbon storage. *Global Biogeochem. Cycles* **7**(1), 69–80.

Harrison, K. G., Broecker, W. S., and Bonani, G. (1993). The effect of changing land use on soil radiocarbon. *Science* **262**(5134), 725–726.

Hart, S. C., Nason, G. E., Myrold, D. D., and Perry, D. A. (1994). Dynamics of gross nitrogen transformations in an old-growth forest: The carbon connection. *Ecology* **75**(4), 880–891.

Hassan, R., Scholes, R., and Ash, N., eds. (2005). *Ecosystems and Human Well-Being*, The Millenium Ecosystem Assessment series, Island Press, Washington, D.C.

Holdridge, L. R. (1947). Determination of World plant formations from simple climatic data. *Science* **105**(2727), 367–368.

IPCC (2007). *Climate Change 2007: The Physical Science Basis*. Contribution of Working Group I to the Fourth Assessment Report of the IPCC, Solomon, S., Qin, D., Manning, M., Chen, Z., Marquis, M., Averyt, K. B., Tignor, M., and Miller, H. L. eds., Cambridge University Press, Cambridge, United Kingdom.

Jacinthe, P. A., and Lal, R. (2001). Mass balance approach to asssess carbon dioxide evolution during erosional events. *Land Deg. and Dev.* **12**, 329–339.

Jahn, R., Zarei, M., and Stahr, K. (1992). Development of andic soil properties and of clay minerals in the semiarid climate of Lanzarote (Spain). *Mineral. Petrograph. Acta* **XXXV-A**, 193–201.

Jastrow, J. D., Miller, R. M., and Lussenhop, J. (1998). Contributions of interacting biological mechanisms to soil aggregate stabilization in restored prairie. *Soil Biol. Biochem.* **30**(7), 905–916.

Jastrow, J. D., Miller, R. M., Matamala, R., Norby, R. J., Boutton, T. W., Rice, C. W., and Owensby, C. E. (2005). Elevated atmospheric carbon dioxide increases soil carbon. *Global Change Biol.* **11**(12), 2057–2064.

Jenny, H. (1930). *A Study on the Influence of Climate Upon the Nitrogen and Organic Matter Content of the Soil*, University of Missouri College of Agriculture, Columbia, MO.

Jenny, H. (1941). *Factors of Soil Formation*, Dover Publications, New York.

Jenny, H. (1980). *The Soil Resource: Origin and Behaviour*, Springer-Verlag, New York.

Jobbagy, E. G., and Jackson, R. B. (2000). The vertical distribution of soil organic carbon and its relation to climate and vegetation. *Ecol. Appl.* **10**(2), 423–436.

Kiem, R., and Kogel-Knabner, I. (2003). Contribution of lignin and polysaccharides to the refractory carbon pool in C-depleted arable soils. *Soil Biol. Biochem.* **35**(1), 101–118.

Kirschbaum, M. U. F. (2000). Will changes in soil organic carbon act as a positive or negative feedback on global warming? *Biogeochemistry* **48**(1), 21–51.

Kleber, M., Mikutta, R., Torn, M. S., and Jahn, R. (2005). Poorly crystalline mineral phases protect organic matter in acid subsoil horizons. *Eur. J. Soil Sci.* **56**(6), 717–725.

Kleber, M., Sollins, P., and Sutton, R. (2007). A conceptual model of organo-mineral interactions in soils: Self-assembly of organic molecular fragments into zonal structures on mineral surfaces. *Biogeochemistry* **85**(1), 9–24.

Kögel-Knabner, I., Zech, W., Hatcher, P. G., and deLeeuw, J. W. (1991). Fate of plant components during biodegradation and humification in forest soils: Evidence from structural characterization of individual biomacromolecules. In *Advances in Soil and Organic Matter Research: The Impact of Agriculture and the Environment*, vol. **90**, Wilson, W. S., eds., Royal Society of Chemistry, London, pp. 61–70.

Kong, A. Y. Y., Six, J., Bryant, D. C., Denison, R. F., and van Kessel, C. (2005). The relationship between carbon input, aggregation, and soil organic carbon stabilization in sustainable cropping systems. *Soil Sci. Soc. Am. J.* **69**(4), 1078–1085.

Kramer, M. G., Sollins, P., and Sletten, R. S. (2004). Soil carbon dynamics across a windthrow disturbance sequence in southeast Alaska. *Ecology* **85**(8), 2230–2244.

Krull, E. S., Baldock, J. A., and Skjemstad, J. O. (2003). Importance of mechanisms and processes of the stabilisation of soil organic matter for modelling carbon turnover. *Funct. Plant Biol.* **30**(2), 207–222.

Krull, E. S., Swanston, C. W., Skjemstad, J. O., and McGowan, J. A. (2006). The importance of charcoal in determining the age and chemistry of organic carbon in surface soils. *J. Geophys. Res. Biogeochem.* **111**(G04001): doi:10.1029/2006JG000194.

Kuzyakov, Y. (2006). Sources of CO_2 efflux from soil and review of partitioning methods. *Soil Biol. Biochem.* **38**(3), 425–448.

Kuzyakov, Y., and Larionova, A. A. (2005). Root and rhizomicrobial respiration: A review of approaches to estimate respiration by autotrophic and heterotrophic organisms in soil. *J. Plant Nutr. Soil Sci.* **168**(4), 503–520.

Lal, R., and Bruce, J. P. (1999). The potential of world cropland soils to sequester C and mitigate the greenhouse effect. *Environ. Sci. Policy* **2**, 177–185.

Lapenis, A. G., Torn, M. S., Harden, J. W., Hollocker, K., Babikov, B. V., Timofeev, A. I., Hornberger, M. I., and Nattis, R. (2000). Scientists unearth clues to soil contamination by comparing old and new soil samples. *EOS* **81**(7), 55–57.

Lewis, S. L. (2004). Concerted changes in tropical forest structure and dynamics: Evidence from 50 South American long-term plots. *Philos. Trans. R. Soc. London Ser. B Biol. Sci.* **359**(1443), 421–436.

Mack, M. C., Schuur, E. A. G., Bret-Harte, M. S., Shaver, G. R., and Chapin, F. S. (2004). Ecosystem carbon storage in arctic tundra reduced by long-term nutrient fertilization. *Nature* **431**(7007), 440–443.

Malhi, Y., and Phillips, O. L. (2004). Tropical forests and global atmospheric change: A synthesis. *Philos. Trans. R. Soc. London B Biol.* **359**(1443), 549–555.

Marti, C., and Badia, D. (1995). Characterization and classification of soil along two altitudinal transects in the eastern Pynenees, Spain. *Arid Soil Res. Rehabil.* **9**, 367–383.

Masiello, C. A., Chadwick, O. A., Southon, J., Torn, M. S., and Harden, J. W. (2004). Weathering controls on mechanisms of carbon storage in grassland soils. *Global Biogeochem. Cycles* **18**(4): GB4023.

Matthews, E. (1997). Global litter production, pools, and turnover times: Estimates from measurement data and regressions models. *J. Geophys. Res.* **102**, 18771–18800.

Mayer, L. M., Schick, L. L., Hardy, K. R., Wagai, R., and McCarthy, J. (2004). Organic matter in small mesopores in sediments and soils. *Geochim. Cosmochim. Acta* **68**(19), 3863–3872.

Melillo, J. M., Aber, J. D., and Muratore, J. F. (1982). Nitrogen and lignin control of hardwood leaf litter decomposition dynamics. *Ecology* **63**(3), 621–626.

Melillo, J. M., Kicklighter, D. W., McGuire, A. D., Peterjohn, W. T., and Newkirk, K. M. (1995). Global changes and its effects on soil organic carbon stocks. In *Environmental Sciences Research Report; The Role of Nonliving Organic Matter in the Earth's Carbon Cycle*, Zepp, R. G., and Sonntag, C., eds., John Wiley & Sons, Chichester, pp. 175–189.

Miglierina, A. M., and Rosell, R. A. (1995). Humus quantity and quality of an entic haplustoll under different soil-crop management systems. *Commun. Soil Sci. Plant Anal.* **26**(19–20), 3343–3355.

Mikutta, R., Kleber, M., Torn, M. S., and Jahn, R. (2006). Stabilization of soil organic matter: Association with minerals or chemical recalcitrance? *Biogeochemistry* **77**(1), 25–56.

Moore, T. R., Bubier, J. L., and Bledzki, L. (2007). Litter decomposition in temperate peatland ecosystems: The effect of substrate and site. *Ecosystems* **10**(6), 949–963.

Moran, K. K., Six, J., Horwath, W. R., and Van Kessel, C. (2005). Role of mineral-nitrogen in residue decomposition and stable soil organic matter formation. *Soil Sci. Soc. Am. J.* **69**(6), 1730–1736.

Myers, N. (2000). Biodiversity hotspots for conservation priorities. *Nature* **403**(6772), 853–858.

Nadelhoffer, K. J. (1990). Microlysimeter for measuring nitrogen mineralization and microbial respiration in aerobic soil incubations. *Soil Sci. Soc. Am. J.* **54**(2), 411–415.

Nemani, R. R. (2003). Climate-driven increases in global terrestrial net primary production from 1982 to 1999. *Science* **300**(5625), 1560–1563.

Nepstad, D. C., de Carvalho, C. R., Davidson, E. A., Jipp, P. H., Lefebvre, P. A., Negreiros, G. H., da Silva, E. D., Stone, T. A., Trumbore, S. E., and Vieira, S. (1994). The role of deep roots in the hydrological and carbon cycles of Amazonian forests and pastures. *Nature* **372**, 666–669.

Oades, J. M. (1984). Soil organic matter and structural stability: Mechanisms and implications for management. *Plant Soil* **76**(1–3), 319–337.

Oades, J. M. (1988). The retention of organic matter in soils. *Biogeochemistry* **5**, 35–70.

Oades, J. M. (1989). An introduction to organic matter in mineral soils. *Minerals in Soil Environments*. Dixon, J. B. and Weed, S. B. Madison, WI, USA, SSSA, 89–153.

Osono, T., Takeda, H., and Azuma, J.-i. (2008). Carbon isotope dynamics during leaf litter decomposition with reference to lignin fractions. *Ecol. Res.* **23**(1), 51–55.

Parton, W., Silver, W. L., Burke, I. C., Grassens, L., Harmon, M. E., Currie, W. S., King, J. Y., Adair, E. C., Brandt, L. A., Hart, S. C., and Fasth, B. (2007). Global-scale similarities in nitrogen release patterns during long-term decomposition. *Science* **315**(5810), 361–364.

Paul, E. A., Collins, H. P., and Leavitt, S. W. (2001). Dynamics of resistant soil carbon of Midwestern agricultural soils measured by naturally occurring ^{14}C abundance. *Geoderma* **104**(3–4), 239–256.

Paul, E. A., Morris, S. J., Conant, R. T., and Plante, A. F. (2006). Does the acid hydrolysis-incubation method measure meaningful soil organic carbon pools? *Soil Sci. Soc. Am. J.* **70**(3), 1023–1035.

Percival, H. J., Parfitt, R. L., and Scott, N. A. (2000). Factors controlling soil carbon levels in New Zealand grasslands: Is clay content important? *Soil Sci. Soc. Am. J.* **64**(5), 1623–1630.

Post, W. M., Emanuel, W. R., Zinke, P. J., and Stangenberger, A. G. (1982). Soil carbon pools and world life zones. *Nature* **298**(5870), 156–159.

Post, W. M., Anderson, D. W., Dahmke, A., Houghton, R. A., Huc, A.-Y., Lassiter, R., Najjar, R. G., Neue, H.-U., Pedersen, T. F., Trumbore, S. E., and Vaikmae, R. (1995). Group report: What is the role of nonliving organic matter cycling on the global scale? *Environmental Sciences Research Report: The role of nonliving organic matter in the Earth's Carbon Cycle*, Zepp, R. G., and Sonntag, C. John Wiley & Sons, Chichester, England, pp. 155–174.

Potter, C. S., Randerson, J. T., Field, C. B., Matson, P. A., Vitousek, P. M., Mooney, H. A., and Klooster, S. A. (1993). Terrestrial ecosystem production: A process model based on global satellite and surface data. *Global Biogeochem. Cycles* **7**(4), 811–841.

Preston, C. M., and Schmidt, M. W. I. (2006). Black (pyrogenic) carbon: A synthesis of current knowledge and uncertainties with special consideration of boreal regions. *Biogeo sciences* **3**(4), 397–420.

Quideau, S. A. (1998). Organic carbon sequestration under chaparral and pine after four decades of soil development. *Geoderma* **83**(3–4), 227–242.

Raich, J. W., and Potter, C. S. (1995). Global patterns of carbon dioxide emissions from soils. *Global Biogeochem. Cycles* **9**(1), 23–36.

Raich, J. W., Russell, A. E., Kitayama, K., Parton, W. J., and Vitousek, P. M. (2006). Temperature influences carbon accumulation in moist tropical forests. *Ecology* **87**(1), 76–87.

Raich, J. W., and Schlesinger, W. H. (1992). The global carbon dioxide flux in soil respiration and its relationship to vegetation and climate. *Tellus Ser. B Chem. Phys. Meteorol.* **44**(2), 81–99.

Rasmussen, C., Torn, M. S., and Southard, R. J. (2005). Mineral assemblage and aggregates control carbon dynamics in a California conifer forest. *Soil Sci. Soc. Am. J.* **69**(6), 1711–1721.

Reichstein, M., Subke, J.-A., Angeli, A. C., and Tenhunen, J. D. (2005). Does the temperature sensitivity of decomposition of soil organic matter depend upon water content, soil horizon, or incubation time? *Global Change Biol.* **11**(10), 1754–1767.

Reimer, P. J., Brown, T. A., and Reimer, R. W. (2004). Discussion: Reporting and calibration of post-bomb ^{14}C data. *Radiocarbon* **46**(3), 1299–1304.

Rillig, M. C., Caldwell, B. A., Wösten, H. A. B., and Sollins, P. (2007). Role of proteins in soil carbon and nitrogen storage: Controls on persistence. *Biogeochemistry* **85**, 25–44.

Rodhe, H. (1992). Modeling biogeochemical cycles. In *Global Biogeochemical Cycles*. Butcher, S. S., Charlson, R. J., Orians, G. H., and Wolfe, G. V., eds., Academic Press, San Diego, pp. 55–72.

Sabine, C. L., Heimann, M., Artaxo, P., Bakker, D. C. E., Chen, C. T. A., Field, C. B., and Gruber, N. (2004). Current status and past trends of the global carbon cycle. In *The Global Carbon Cycle: Integrating Humans, Climate, and the Natural World*, Field, C. B., and Raupach, M. R., eds., Island Press, Washington, D.C.

Saleska, S. R., Shaw, M. R., Fischer, M. L., Dunne, J. A., Still, C. J., Holman, M. L., and Harte, J. (2002). Plant community composition mediates both large transient decline and predicted long term recovery of soil carbon under climate warming. *Global Biogechem. Cycles* **16**(4), 1055.

Schimel, D. S. (1995). Terrestrial ecosystems and the carbon cycle. *Global Change Biol.* **1**, 77–91.

Schimel, D. S., Braswell, B. H., Holland, E. A., McKeown, R., Ojima, D. S., Painter, T. H., Parton, W. J., and Townsend, A. R. (1994). Climatic, edaphic, and biotic controls over storage and turnover of carbon soils. *Global Biogeochem. Cycles* **8**(3), 279–293.

Schlesinger, W. H. (1977). Carbon balance in terrestrial detritus. *Annu. Rev. Ecol. Syste.* **8**, 51–81.

Schmidt, M. W. I., and Kogel-Knabner, I. (2002). Organic matter in particle-size fractions from A and B horizons of a haplic alisol. *Eur. J. Soil Sci.* **53**(3), 383–391.

Shaver, G. R., Canadell, J., Chapin, F. S. I., Gurevitch, J., Harte, J., Henry, G., Ineson, P., Jonasson, S., Melillo, J., Pitelka, L., and Rustad, L. (2000). Global warming and terrestrial ecosystems: A conceptual framework for analysis. *BioScience* **50**(10), 871–882.

Silver, W. L., Ostertag, R., and Lugo, A. E. (2000). The potential for carbon sequestration through reforestation of abandoned tropical agricultural and pasture lands. *Restoration Ecol.* **8**(4), 394–407.

Silvertown, J. (2006). The Park Grass Experiment 1856–2006: Its contribution to ecology. *J. Ecol.* **94**(4), 801–814.

Sitch, S. (2003). Evaluation of ecosystem dynamics, plant geography and terrestrial carbon cycling in the LPJ dynamic global vegetation model. *Global Change Biol.* **9**(2), 161–185.

Six, J., Callewaert, P., Lenders, S., De Gryze, S., Morris, S. J., Gregorich, E. G., Paul, E. A., and Paustian, K. (2002). Measuring and understanding carbon storage in afforested soils by physical fractionation. *Soil Sci. Soc. Am. J.* **66**(6), 1981–1987.

Six, J., Elliott, E. T., and Paustian, K. (1999). Aggregate and soil organic matter dynamics under conventional and no-tillage systems. *Soil Sci. Soc. Am. J.* **63**(5), 1350–1358.

Six, J., Elliott, E. T., and Paustian, K. (2000a). Soil macroaggregate turnover and microaggregate formation: A mechanism for C sequestration under no-tillage agriculture. *Soil Biol. Biochem.* **32**(14), 2099–2103.

Six, J., Merckx, R., Kimpe, K., Paustian, K., and Elliott, E. T. (2000b). A re-evaluation of the enriched labile soil organic matter fraction. *Eur. J. Soil Sci.* **51**(2), 283–293.

Smith, L. C., MacDonald, G. M., Velichko, A. A., Beilman, D. W., Borisova, O. K., Frey, K. E., Kremenetski, V., and Sheng, Y. (2004). Siberian peatlands a net carbon sink and global methane source since the early Holocene. *Science* **303**(5656), 353–356.

Smith, S. V., Sleezer, R. O., Renwick, W. H., and Buddemeier, R. (2005). Fates of eroded soil organic carbon: Mississippi basin case study. *Ecol. Appli.* **15**(6), 1929–1940.

Sollins, P., Glassman, C. A., Paul, E. A., Swanston, C., Lajtha, K., Heil, J. W., and Elliott, E. T. (1999). Soil carbon and nitrogen: pools and fractions. In *Standard Soil Methods for Long-Term Ecological Research*, Robertson, C. S. B. G. P., Coleman, D. C., and Sollins, P. eds., Oxford University Press, New York, pp. 89–105.

Sollins, P., Homann, P., and Caldwell, B. A. (1996). Stabilization and destabilization of soil organic matter: Mechanisms and controls. *Geoderma* **74**(1–2), 65–105.

Sollins, P., Swanston, C., Kleber, M., Filley, T., Kramer, M , Crow, S., Caldwell, B. A., Lajtha, K., and Bowden, R. (2006). Organic C and N stabilization in a forest soil: Evidence from sequential density fractionation. *Soil Biol. Biochem.* **38**(11), 3313–3324.

Sombroek, W. G., Nachtergaele, F. O., and Hebel, A. (1993). Amounts, dynamics and sequestering of carbon in tropical and subtropical soils. *Ambio* **22**, 417–426.

Stallard, R. F. (1998). Terrestrial sedimentation and the carbon cycle: Coupling weathering and erosion to carbon burial. *Global Biogeochem. Cycles* **12**(2), 231–257.

Steinmann, K., Siegwolf, R. T. W., Saurer, M., and Korner, C. (2004). Carbon fluxes to the soil in a mature temperate forest assessed by 13C isotope tracing. *Oecologia* **141**(3), 489–501.

Stevenson, F. J. (1994). Clay-organic complexes and formation of stable aggregates. In *Humus Chemistry: Genesis, Composition Reactions*. 2nd Ed. John Wiley & Sons, New York, pp. 429–452.

Still, C. J., Berry, J. A., Collatz, G. J., and DeFries, R. S. (2003). Global distribution of C-3 and C-4 vegetation: Carbon cycle implications. *Global Biogeochem. Cycles* **17**(1).

Stork, N. E., and Eggleton, P. (1992). Invertebrates as determinants and indicators of soil quality. *Am. J. Altern. Agric.* **7**(1–2), 38–47.

Stotzky, G. (1965). Microbial respiration. *Methods of Soil Analysis. Part 2. Agronomy Monograph No. 9*, Black, C. A., ed., ASA, Madison, WI.

Strickland, T. C. (1987). Improved method for separating light and heavy-fraction organic material from soil. *Soil Sci. Soc. Am. J.* **51**(5), 1391–1392.

Strong, D. T., Wever, H. D., Merckx, R., and Recous, S. (2004). Spatial location of carbon decomposition in the soil pore system. *Eur. J. Soil Sci.* **55**(4), 739–750.

Stuiver, M., and Polach, H. A. (1977). Reporting of ^{14}C data. *Radiocarbon* **19**(3), 355–363.

Swanston, C. W., Caldwell, B. A., Homann, P. S., Ganio, L., and Sollins, P. (2002). Carbon dynamics during a long-term incubation of separate and recombined density fractions from seven forest soils. *Soil Biol. Biochem.* **34**(8), 1121–1130.

Swanston, C., Homann, P. S., Caldwell, B. A., Myrold, D. D., Ganio, L., and Sollins, P. (2004). Long-term effects of elevated nitrogen on forest soil organic matter stability. *Biogeochemistry* **70**(2), 229–252.

Swanston, C. W., Torn, M. S., Hanson, P. J., Southon, J. R., Garten, C. T., Hanlon, E. M., and Ganio, L. (2005). Initial characterization of processes of soil carbon stabilization using forest stand-level radiocarbon enrichment. *Geoderma* **128**(1–2), 52–62.

Thornton, P. E., and Rosenbloom, N. A. (2005). Ecosystem model spin-up: Estimating steady state conditions in a coupled terrestrial carbon and nitrogen cycle model. *Ecol. Modelling* **189**, 25–48.

Tiessen, H., and Stewart, J. W. B. (1983). Particle-size fractions and their use in studies of soil organic matter. II. Cultivation effects on organic matter composition in size fractions. *Soil Sci. Soc. Am. J.* **47**(3), 509–514.

Tisdall, J. M., and Oades, J. M. (1982). Organic matter and water stable aggregates in soils. *J. Soil Sci.* **33**(2), 141–164.

Torn, M. S., Lapenis, A. G., Timofeev, A., Fischer, M. L., Babikov, B. V., and Harden, J. W. (2002). Organic carbon and carbon isotopes in modern and 100-year-old-soil archives of the Russian steppe. *Global Change Biol.* **8**(10), 941–953.

Torn, M. S., and Southon, J. (2001). A new C-13 correction for radiocarbon samples from elevated-CO_2 experiments. *Radiocarbon* **43**(2B), 691–694.

Torn, M. S., Trumbore, S. E., Chadwick, O. A., Vitousek, P. M., and Hendricks, D. M. (1997). Mineral control of soil organic carbon storage and turnover. *Nature (London)* **389**(6647), 170–173.

Torn, M. S., Vitousek, P. M., and Trumbore, S. E. (2005). The influence of nutrient availability on soil organic matter turnover estimated by incubations and radiocarbon modeling. *Ecosystems* **8**(4), 352–372.

Townsend, A. R., Vitousek, P. M., and Trumbore, S. E. (1995). Soil organic matter dynamics along gradients in temperature and land use on the island of Hawaii. *Ecology* **76**(3), 721–733.

Trumbore, S. E. (1993). Comparison of carbon dynamics in tropical and temperate soils using radiocarbon measurements. *Global Biogeochem. Cycles* **7**(2), 275–290.

Trumbore, S. (2000). Age of soil organic matter and soil respiration: Radiocarbon constraints on belowground C dynamics. *Ecol. Appl.* **10**(2), 399–411.

Trumbore, S. E. (1996). Measurement of cosmogenic isotopes by accelerator mass spectrometry: Applications to soil science. In *Applications of Accelerator Mass Spectrometry*, Boutton, T., and Yamasaki, S. eds., Marcel Dekker, New York, pp. 311–340.

Trumbore, S. E., Chadwick, O. A., and Amundson, R. (1996). Rapid exchange between soil carbon and atmospheric carbon dioxide drived by temperature change. *Science* **272**(5260), 393–396.

Trumbore, S. E., Davidson, E. A., Barbosa de Camargo, P., Nepstad, D. C., and Martinelli, L. A. (1995). Belowground cycling of carbon in forests and pastures of Eastern Amazonia. *Global Biogeochem. Cycles* **9**(4), 515–528.

Trumbore, S. E., Gaudinski, J. B., Hanson, P. J., and Southon, J. R. (2002). Quantifying ecosystem-atmosphere carbon exchange with a ^{14}C label. *Eos, Trans. Am. Geophys. Union* **83**, 265–268.

Trumbore, S. E., and Harden, J. W. (1997). Accumulation and turnover of carbon in organic and mineral soils of the BOREAS northern study area. *J. Geophys. Res.* **102**(D24), 28817–28830.

Trumbore, S. E., and Zheng, S. (1996). Comparison of fractionation methods for soil organic matter ^{14}C analysis. *Radiocarbon* **38**(2), 219–229.

van Oost, K., Quine, T. A., Govers, G., Gryze, S. D., Six, J., Harden, J. W., Ritchie, J. C., McCarty, G. W., Heckrath, G., Kosmas, C., Giraldez, J. V., Silva, J. R. M. D., and Merckx, R. (2007). The impact of agricultural soil erosion on the global carbon cycle. *Science* **318**(5850), 626–629.

Van Veen, J. A., and Kuikman, P. J. (1990). Soil structural aspects of decomposition of organic matter by micro-organisms. *Biodegradation* **11**(3), 213–233.

Veldkamp, E. (1994). Organic carbon turnover in three tropical soils under pasture after deforestation. *Soil Sci. Soc. Am. J.* **58**(1), 175–180.

Verhoef, H. A., and Brussaard, L. (1990). Decomposition and nitrogen mineralization in natural and agroecosystems: The contribution of soil animals. *Biodegradation* **11**(3), 175–211.

Vetter, Y. A., Deming, J. W., Jumars, P. A., and Krieger-Brockett, B. B. (1998). A predictive model of bacterial foraging by means of freely released extracellular enzymes. *Microb. Ecol.* **36**(1), 75–92.

Vogel, J. S., Southon, J. R., Nelson, D. E., and Brown, T. A. (1984). Performance of catalytically condensed carbon for use in accelerator mass-spectrometry. *Nucl. Instrum. Methods Phys. Res. Sect. B Beam Interact. Mater. Atoms* **233**, 289–293.

von Lützow, M., Kogel-Knabner, I., Ekschmitt, K., Matzner, E., Guggenberger, G., Marschner, B., and Flessa, H. (2006). Stabilization of organic matter in temperate soils: mechanisms and their relevance under different soil conditions—A review. *Eur. J. Soil Sci.* **57**(4), 426–445.

Waldrop, M. P., and Zak, D. R. (2006). Response of oxidative enzyme activities to nitrogen deposition affects soil concentrations of dissolved organic carbon. *Ecosystems* **9**(6), 921–933.

Whalen, J. K., Bottomley, P. J., and Myrold, D. D. (2000). Carbon and nitrogen mineralization from light-and heavy-fraction additions to soil. *Soil Biol. Biochem.* **32**(10), 1345–1352.

Wolters, V. (2000). Invertebrate control of soil organic matter stability. *Biol. Fertil. Soils* **31**(1), 1–19.

Xu, X. M., Trumbore, S. E., Zheng, S. H., Southon, J. R., McDuffee, K. E., Luttgen, M., and Liu, J. C. (2007). Modifying a sealed tube zinc reduction method for preparation of AMS graphite targets: Reducing background and attaining high precision. *Nucl. Instrum. Methods Phys. Res. Sect. B Beam Interact. Mater. Atoms* **259**(1), 320–329.

Zak, D. R., Grigal, D. F., and Ohmann, L. F. (1993). Kinetics of microbial respiration and nitrogen mineralization in Great Lakes forests. *Soil Sci. Soc. Am. J.* **57**(4), 1100–1106.

Zanelli, R., Egli, M., Mirabella, A., Giaccai, D., and Fitze, P. (2006). Influence of laurophyllous species, *Castanea sativa* and *Quercetum-Betuletum* vegetation on organic matter in soils in southern Switzerland and northern Italy. *Geoderma* **136**(3–4), 723–737.

Zepp, R. G., Callaghan, T. V., and Erickson, D. J., III (2003). Interactive effects of ozone depletion and climate change on biogeochemical cycles. *Photochem. Photobiol. Sci.* **2**(1), 51–61.

Zhang, D., Hui, D., Luo, Y., and Zhou, G. (2008). Rates of litter decomposition in terrestrial ecosystems: global patterns and controlling factors. *J. Plant Ecol.* **1**(2), 85–93.

Zimmermann, M., Leifeld, J., Abiven, S., Schmidt, M. W. I., and Fuhrer, J. (2007). Sodium hypochlorite separates an older soil organic matter fraction than acid hydrolysis. *Geoderma* **139**(1–2), 171–179.

Zimov, S. A., Schuur, E. A. G., and Chapin, F. S. III, (2006). Permafrost and the Global Carbon Budget. *Science* **312**(5780), 1612–1613.

Zinke, P. J., Stangenberger, A. G., Post, W. M., Emmanuel, W. R., and Olson, J. S. (1984). Worldwide organic soil carbon and nitrogen data, Oak Ridge National Laboratory Carbon Dioxide Analysis and Information Center (CDIAC).

7

BLACK CARBON AND THERMALLY ALTERED (PYROGENIC) ORGANIC MATTER: CHEMICAL CHARACTERISTICS AND THE ROLE IN THE ENVIRONMENT

H. Knicker

Lehrstuhl für Bodenkunde, Technische Universität München, Freising-Weihenstephan, Germany

Biophysico-Chemical Processes Involving Natural Nonliving Organic Matter in Environmental Systems, Edited by Nicola Senesi, Baoshan Xing, and Pan Ming Huang

7.1. INTRODUCTION

Nowadays, approximately 10–15 × 10^6 ha of boreal and temperate forest, 20–40 × 10^6 ha of tropical forests, and up to 500 × 10^6 ha of tropical and subtropical savannas, woodlands, and open forests are burnt every year (Goldammer, 1993). Some of those fires may be caused by natural events such as lightning, but most of them are due to human activity. For example, according to surveys, in the United States and Siberia approximately 90% and in Mexico about 97% of all forest fires were reported as human-caused (Zhukov, 1976; Caldararo, 2002).

Burning of biomass, including both natural vegetation fires and fires ignited for agricultural purpose and energy production, releases 2–6 × 10^{15} g C per year (Crutzen and Andreae, 1990; Wittenberg et al., 1998). A part of it will return to the biosphere by rain and photosynthesis (Johnson and Curtis, 2001). Beside volatile products, vegetation fires can transform labile organic components into highly recalcitrant aromatic structures (Almendros et al., 2003). The annual production of this dark-colored combustion residue is estimated to range from 50 to 270 × 10^{12} g (Kuhlbusch and Crutzen, 1995). Because in sediments and soils they can survive for millennia (Hopkins et al., 1990; Masiello and Druffel, 1998) and due to their relative resistance against chemical and thermal oxidation (Skjemstad et al., 1993; Gustafsson and Gschwend, 1998), pyrogenic organic material (PyOM) in soil and sediments is assumed to represent an important sink within the global carbon cycle (Kuhlbusch, 1995; Kuhlbusch and Crutzen, 1995; Kuhlbusch, 1998; Schmidt and Noack, 2000).

One of the terms, commonly used to describe the organic matter that remains after incomplete combustion during vegetation fires, fuel burning, and industrial processes, is black carbon (BC). It embraces a continuum of combustion products ranging from slightly charred degradable biomass to highly condensed, refractory soot (Masiello, 2004) (Figure 7.1). The components within this continuum have high carbon contents and are chemically highly heterogeneous, and their aromaticity increases with heat severity (Freitas et al., 1999). Whereas charcoal particles are the solid combustion residues, soot represents the secondary combustion products formed by condensation of volatiles into polycyclic aromatic hydrocarbons (PAHs), most probably via a radical mechanism. PAHs continue to amass into larger aromatic clusters (Figure 7.2) assembling into concentric shells of graphene stacks (Sergides et al., 1987) with high electrical conductivity. Such soot particles are in submicron order and can be transported for long distances or even remain suspended in the atmosphere for months if not subjected to oxidation and wash-out (Masiello, 2004). Here, they have an effect on the radiation balance. Conversely, charcoal particles will settle within a few to hundreds of meters of the burning source.

The high heterogeneity with respect to the chemical nature and origin of the BC constituents means that various terms are nonspecifically applied for different fractions. For example, the term BC is used not only for the whole continuum (Masiello, 2004) but also more specifically for the more resistant fractions (Kuhlbusch, 1995; Alexis et al., 2006); the expression charcoal or char can be a synonym for the total burnt materials (Knicker et al., 2006) but can also describe more specifically the burnt material identified by visual assessment (Preston and Schmidt, 2006). The expression elemental carbon (EC) is used in association with the oxidation-resistant fraction in analysis of atmospheric aerosols and soot. The fraction remaining after chemical oxidation has been described as chemical-oxidation-resistant elemental carbon (COREC) (Bird and Gröcke, 1997; Rumpel et al., 2006; Knicker et al.,

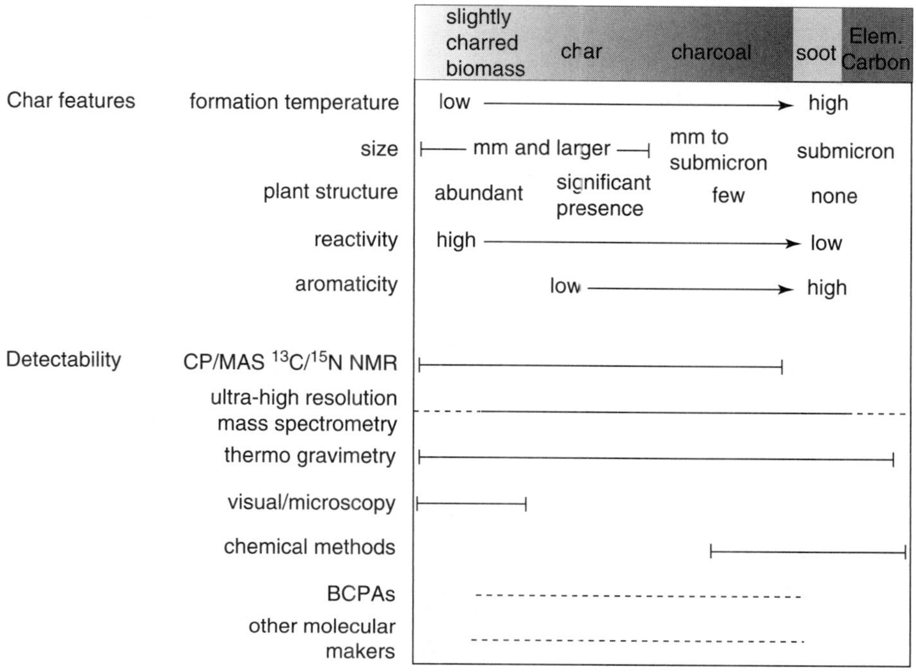

Figure 7.1. The BC continuum, its chemical and physical properties, and the detectability of BC with various techniques. Modified after Masiello (2004).

2007) and the material surviving oxidation in general (chemical, thermal, photo-oxidization) as graphitic BC (GBC) (Preston and Schmidt, 2006). The latter is based on the assumption that the resistance of the remaining materials is caused by their highly condensed polyaromatic nature.

The wide variety of terms often results in problems concerning comparability of BC data, and certainly a better discrimination of the different fractions within the BC continuum is needed. However, this requires a much better understanding of the processes involved in the formation of thermally altered organic material and the chemical nature of the products. Due to this lack of a commonly accepted terminology and to reduce confusion, in the following I use the term "BC" for the whole continuum according to Masiello (2004). "Soot" describes the secondary condensation products, and the expression "charcoal" will be limited to the solid combustion residues. PyOM, on the other hand, will be used to account for the fact that thermally altered material is not limited to a carbon composition but can consist of a considerable proportion of heteroaromatic—in particular, N-containing—constituents.

Deeper insights into the nature of PyOM are needed not only for a better understanding of the role of vegetation fires and burning within the global elemental cycle but also for a better elucidation of the complex and interacting effects of fire on ecosystems. Here, the impact of fire ranges from reduction or elimination of aboveground and topsoil biomass with a subsequent shift in vegetation and microbial populations to alterations of the physical and chemical soil properties. The produced charcoal, accumulating on and in the soil, will finally incorporate into the soil organic matter (SOM) pool and be subjected to humification. This is expected

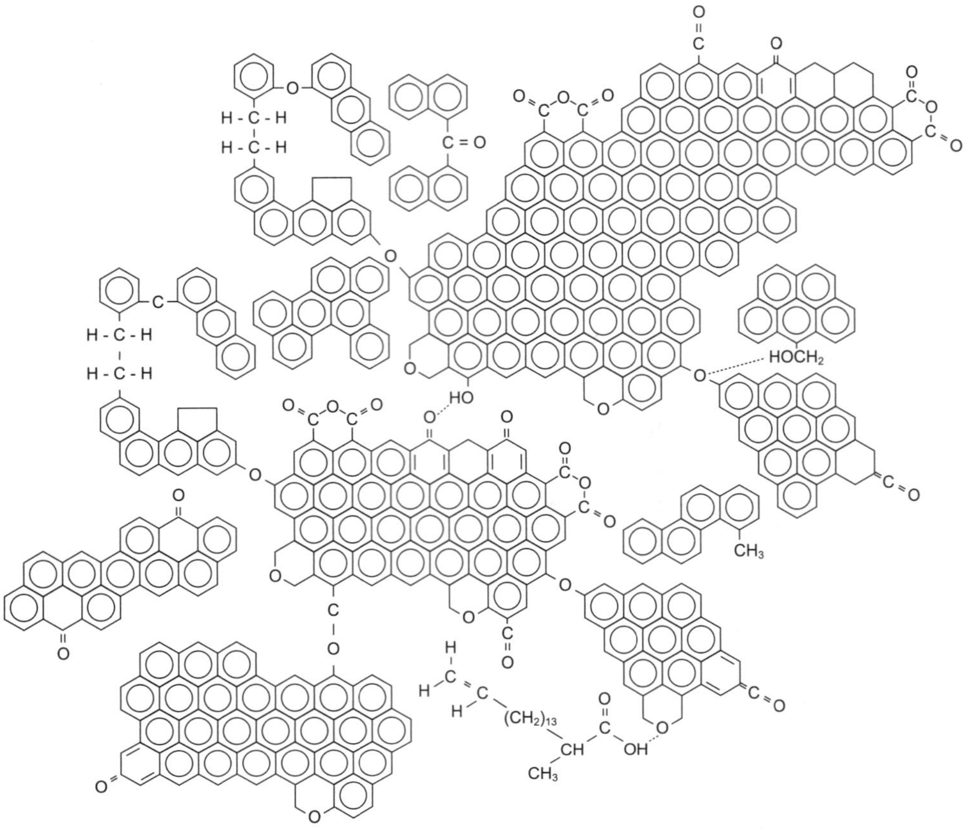

Figure 7.2. Structural model suggested for soot according to Sergides (1987) (redrawn from Schmidt, M. W. I., and Noack, A. G. (2000). Black carbon in soils and sediments: Analysis, distribution, implications, and current challenges. *Global Biogeochem. Cycl.* **14**, 777–793, with permission from the American Geophysical Union.).

to alter humic matter composition and to have long-term effects on microbiologically and biochemically mediated processes.

However, although increasing attention is paid to SOM transformation and stabilization by input of PyOM (Neary et al., 1999; González-Pérez et al., 2004; Certini, 2005), present reports on SOM alteration often lead to divergent conclusions, which is best explained by the heterogeneity of PyOM. In the following review, present knowledge of structural properties of PyOM are summarized and discussed in light of its role as C/N sink and as an important source of organic matter in soils. With respect to its quantitative importance, the impact of structural properties on its detectability with common quantification methods is elucidated.

7.2. TEMPERATURE AND PyOM PRODUCTION

The extent of PyOM productions during organic matter burning depends largely on combustion intensity. With respect to vegetation fires, prescribed (controlled) fires and post-harvest fires are initiated at moderate soil moisture and show generally low severity. Fires in grassland with lower fuel loadings usually have ground-level temperatures of <25 °C, although higher instantaneous temperatures have been measured (Raison, 1979). Wildfires occurring uncontrolled in the presence of an abundant and dry fuel load, on the other hand, can be very severe; but due to high variation in local factors, their severity exhibits large heterogeneity (Kutiel et al., 1995). Temperatures can range from 50 °C to 1500 °C (Neary et al., 1999). During forest fires, maximum ground temperatures are typically in the range of 200–300 °C. In environments with heavy fuel loads such as slash, instantaneous temperatures of at least 1500 °C can occur but soil surface maximum temperatures are usually around 500–700 °C. Rates of spread can vary from 0.5 m $week^{-1}$ in smoldering peat fires to as much as 7 km h^{-1} in large wildfires (Neary et al., 1999). Fast-moving fires on grass may be intense in terms of energy release per unit area, but do not transfer the same amounts of heat to mineral soil or soil organisms as do slow-moving fires in moderate to heavy fuels. The component of fire severity that results in the greatest belowground damage to ecosystems, and hence recovery, is the duration of fire. With depths the severity of heating decreases, depending on factors such as intensity and duration of heat transfer, heat conductivity of the mineral phase, soil porosity, and soil moisture. In general, in dry soils below ground temperatures will rise very slowly because they are a very good insulator (DeBano et al., 1998).

In peat and swamp areas, lowering of the water table has a considerable hazardous potential because drying of the peat creates significant amount of high-energy fuel. Fires that can burn deeply into the peat may even be capable of surviving several centimeters of rainfall. The respective loss of organic material can be as dramatic such that the peat surface sinks clearly with respect to its surroundings (Rollins et al., 1993; Haslam et al., 1998).

At temperatures between 40 °C and 70 °C, biological disintegration starts. Between 70 °C and 90 °C, seed mortality begins and mortality of microbes occurs between 50 °C and 121 °C (Hernández et al., 1997). Although considerable amounts of C can be lost by volatilization already at lower temperatures, organic matter distillation needs temperatures between 200 °C and 315 °C (Lide, 2004). Above 200 °C the carbonization processes start (Freitas et al., 1999), but formation of considerable graphite-like units in a peat sample requires pyrolysis conditions (anoxic) and temperatures above 700 °C. Under oxic conditions, at around 460 °C almost all unprotected SOM will be combusted, leaving nutrient rich ash (Pietikäinen et al., 2000). At 200 °C, N starts to volatilize and above 500 °C half of the N in organic matter is lost to the atmosphere.

7.3. ANALYTICAL CHARACTERIZATION OF PyOM

7.3.1. Elemental Analysis and Van Krevelen Plot

Although several models describing char on a molecular level are reported in the literature (Schmidt and Noack, 2000), a real understanding of the chemical nature of PyOM is still missing, most probably due to the heterogeneous nature of its constituents and the high chemical resistance of some parts that reduces its accessibility to extraction and subsequent analyses.

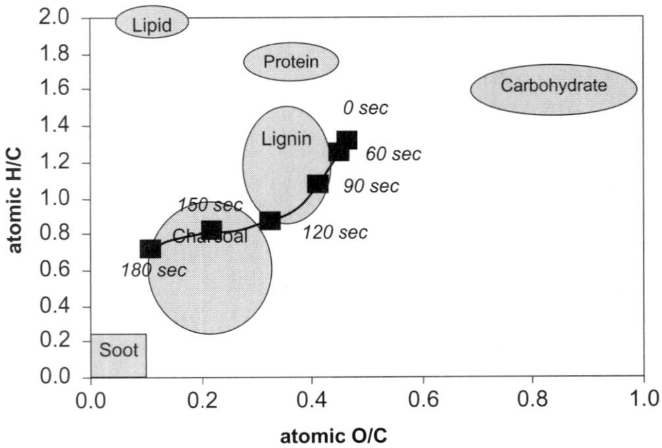

Figure 7.3. Van Krevelen diagram. Black squares plot the changes of the elemental composition during heating of peat material at 350 °C at oxic conditions as reported in Almendros et al. (2003).

First relevant information about the chemical nature of PyOM can be obtained by elemental analysis. Studying fire-induced alteration of SOM of peat during progressive heating under laboratory conditions at 350 °C showed a weight loss of 50% already after 180 s of heating time. The loss of the major elements (C, H, O, N) followed a nonlinear kinetic (Almendros et al., 2003). Compared to C, greater losses were determined for O and H and a relative enrichment was observed for N. These results coincided with the typical behavior of soil humic and fulvic acids, grass residues, sapwood, and wood from *Eucalyptus saligna* and *E. grandis* during laboratory heatings (Almendros et al., 1990; Knicker et al., 1996; Baldock and Smernik, 2002; Trompowsky et al., 2005). In fact, when the changes were plotted in a typical van Krevelen diagram (Figure 7.3), the progressive decrease of the H/C and O/C atomic ratios indicates dehydration reactions in samples subjected to moderate heating, whereas decarboxylation and demethylation were the dominant reactions in the longer-heated samples (Almendros et al., 2003).

7.3.2. NMR Spectroscopy

More detailed information about the chemical properties of a heterogenous and insoluble sample can be obtained with spectroscopic techniques, that is, NMR spectroscopy. Solid-state ^{13}C NMR spectra reveal the presence of PyOM by a higher intensity in the aromatic C region (160 to 110/90 ppm) (Figure 7.4). But because this region overlaps with that of naturally occurring aromatic soil components (e.g., lignin, tannins, and their degradation products), it is difficult to unveil BC structural properties solely by solid-state ^{13}C NMR spectroscopy (Knicker et al., 2005a). On the other hand, if unburned control soils are available, the amount and nature of the PyOM in the fire-affected reference may be discerned by spectral fractionation, as it is proposed in the following. This approach is based on the low thermal stability of carbohydrates. During heating, they are among the first compound classes that are altered into aromatic structures or destroyed, and thus their signal will disappear in the ^{13}C NMR spectrum. Consequently, if a signal of *O*-alkyl C is still visible in a

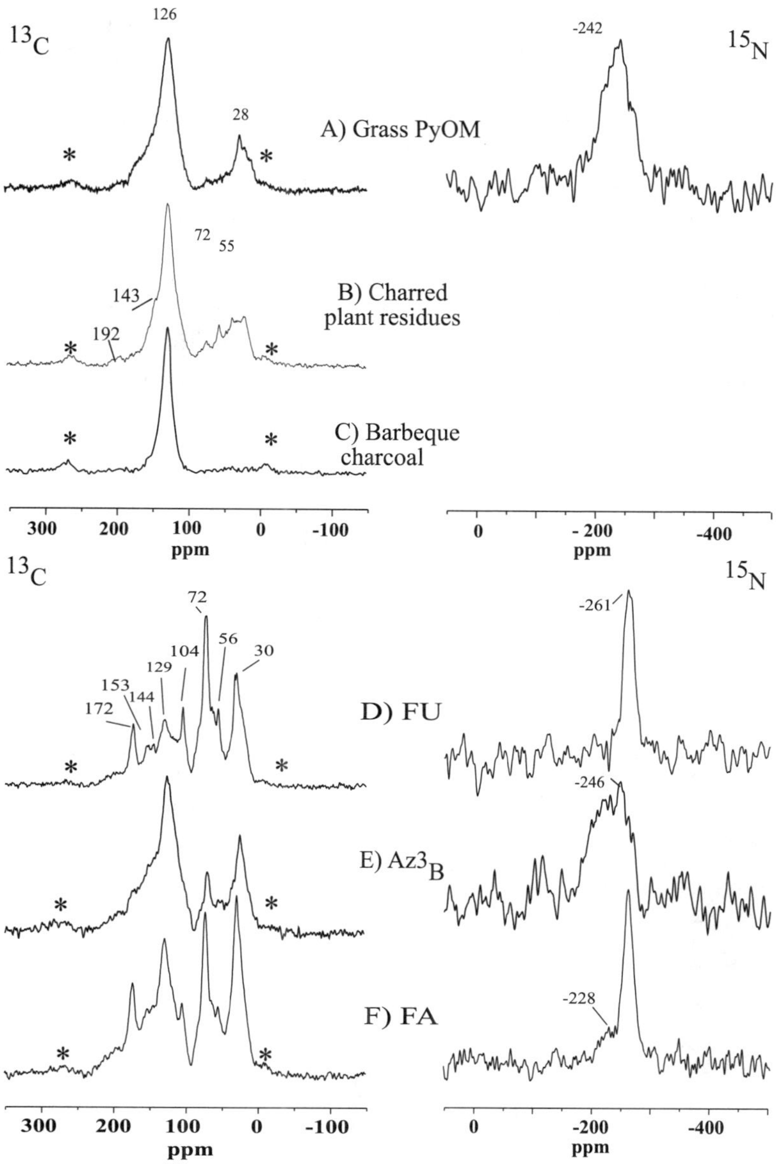

Figure 7.4. Solid-state ^{13}C and ^{15}N NMR spectra of PyOM of plant residues and SOM of A horizons of cambisols from the region around Aznacollár in Southern Spain. Prior to NMR spectroscopy, the soils were demineralized with 10% hydrofluoric acid (HF). **(A)** Grass-PyOM obtained after charring for 4 min at 350 °C under oxic conditions. **(B)** Charcoal found in the fire-affected soil FA. **(C)** Commercially available barbeque charcoal. **(D)** Fire-unaffected soil FU. **(E)** Fire-affected soil Az3, 2 months after an intense fire. **(F)** Soil FA, 5 years after an intense fire.

spectrum of a burnt soil, its intensity must derive from fire-unaffected SOM. Either this material survived in the mineral horizon due to low heat conductivity or it was introduced after the fire as fresh litter that became humified and part of SOM (Knicker et al., 2005a). A second assumption needed is that pre- and post-fire humification processes of the fire-unaffected SOM were comparable at both the burnt and the control site. Then, the relative contribution of unburned material to each chemical shift region can be calculated from the relative intensity distribution of the solid-state ^{13}C NMR spectrum of the control soil. Figure 7.5 gives an example for a control soil and a soil sampled 1 year after a severe fire from an A horizon of a cambisol in Central Spain (Knicker et al., 2006). Subtraction of the relative intensity originating from the fire-unaffected SOM from the total C of the burnt soil yields the proportion that derives from thermally altered organic material. Accord-

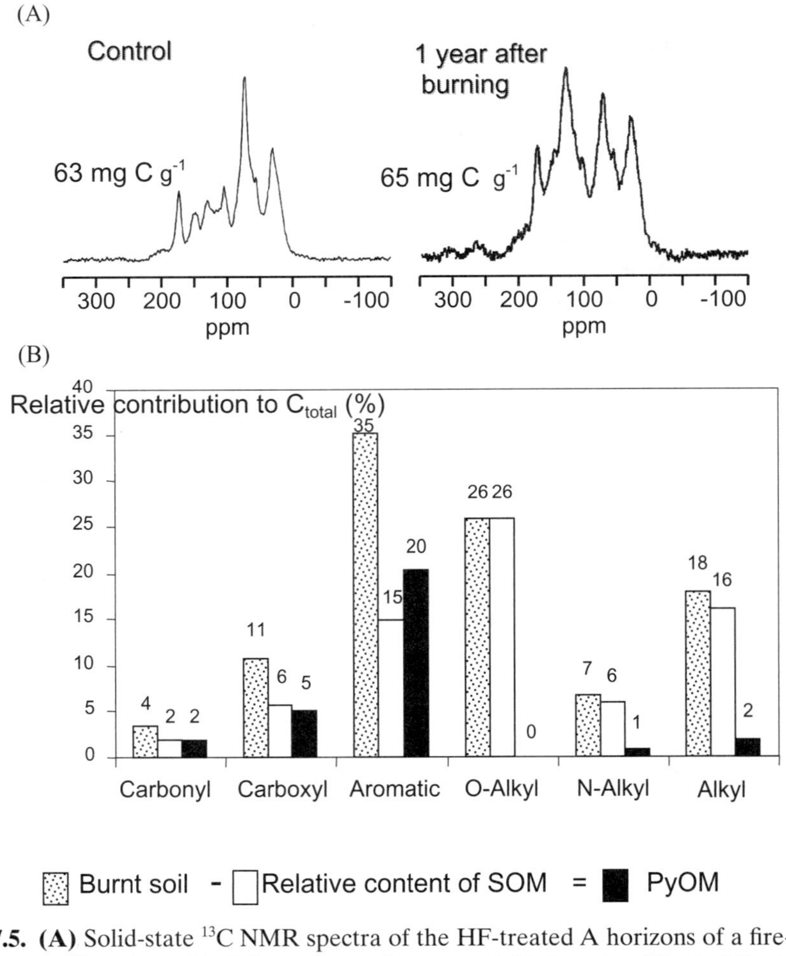

Figure 7.5. (A) Solid-state ^{13}C NMR spectra of the HF-treated A horizons of a fire-affected and a fire-unaffected cambisol from the region around Piedravales (Central Spain), along with their organic C content. **(B)** Relative distribution of the various C groups to the total organic C in the fire-affected soil and the calculated contribution of unburnt SOM. Subtracting the latter from the total organic C yields in the contribution of PyOM.

ingly, for the example in Figure 7.5, a BC content of 30% of the total organic C in the burnt soil was calculated. For the soil FA and Az3 (Figure 7.4), BC contents of 30% and 60% were obtained. It should be noted that here, BC comprises the whole range of slightly to heavily thermally altered organic material.

In geosciences, most solid-state ^{13}C NMR spectra are acquired using the cross-polarization technique. Here, the sensitivity of ^{13}C is increased by magnetization transfer from ^{1}H to ^{13}C spins. The efficiency of this transfer decreases, if the distance between the ^{13}C and the next proton exceeds three bonds and the respective ^{13}C-intensities have been found to be underrepresented (Alemany et al., 1983; Smernik et al., 2002). Based on the assumption that high heat intensities result in aromatic clusters with a considerable proportion of unprotonated core C, the reliability of solid-state NMR spectroscopy in BC research is often questioned (Freitas et al., 1999; Schmidt and Noack, 2000; Baldock and Smernik, 2002). To reveal possible underestimation of such core C, CPMAS NMR spectra are sometimes compared with Bloch decay (BD) MAS NMR spectra that are obtained without cross-polarization by direct excitation of the ^{13}C spins (Smernik et al., 2000). However, in order for this approach to be successful, saturation has to be avoided; and this in particular is difficult if remains of cellulose are still present. For the crystalline domains in cotton, for example, a relaxation time of 266 sec was reported (Teeäär and Lippmaa, 1984), which means that the pulse delay between the single scans has to be in the range of 20 minutes. For soils with C contents below 10%, this requires unfeasible measurement times of weeks in length. On the other hand, comparing CPMAS and BD NMR spectra of peat charred with increasing temperature under a nitrogen atmosphere, no appreciable ^{13}C-intensity loss of the aromatic signal due to inefficient cross-polarization differences was observed until 500 °C (Freitas et al., 1999). After increasing the temperature to 900 °C and 1000 °C, the use of the CP technique was completely unsuccessful. Comparably, no CPMAS ^{13}C NMR signal was obtainable from lignite coke (Abelmann et al., 2005). Therefore, soils containing such material will not yield quantifiable data, and the more time-consuming BD technique has to be applied. On the other hand, charcoal produced under the natural conditions of vegetation fires is unlikely to have experienced conditions that were extreme enough for producing such graphenic structures. In particular, this is true during surface vegetation fires that move fast and require oxic conditions, where all organic material will be volatilized above temperature of 400–500 °C. Recent analysis demonstrated that for such charcoals the protonation is high enough to enable the generation of reliable NMR data (Knicker et al., 2005b). In contrast to former spin counting experiments indicating that only little carbon can be observed with the CPMAS technique (Smernik et al., 2002), new studies in which charcoal was mixed with peat in defined proportions evidenced comparable observability of pyrogenic and non pyrogenic organic matter (Knicker et al., 2005b).

A less common approach for the structural characterization of PyOM utilizes solid-state ^{15}N NMR spectroscopy, revealing pyrogenic N by a clear signal in the chemical shift region of pyrrole/indole N (−150 to −240 ppm) relative to the nitromethane scale (Martin et al., 1981) and a small shoulder assignable to pyridine-type N (−60 to −120 ppm) (Figure 7.4). Both regions exhibited no intensity in spectra of fire-unaffected natural soils, where most of the ^{15}N-signal intensity is detected in the amide N region (−230 to −285 ppm) following a small signal assigned to free amino groups (–325 to –350 ppm). This supports the pyrogenic origin of the heterocyclic N compounds. Thus, their identification in SOM by means of solid-state ^{15}N NMR spectroscopy may be taken as a first indication for the presence of PyOM, which is difficult to detect solely by solid-state ^{13}C NMR spectroscopy, due to overlapping of signals derived from char and from biogenic humic substances.

7.3.3. Thermogravimetric Techniques

Thermal analysis methods (i.e., thermogravimetry-differential scanning calorimetry TG-DSC) is already routinely applied to the characterization of chemical changes in organic matter fractions of soils and sediments, degraded plant tissue, and composts (Dell'Abate et al., 2000; Lopez-Capel et al., 2005b), but was also used to characterize BC forms (Leifeld, 2007). The latter confirmed the idea of BC as a continuum in terms of thermal stability. Addition of charcoal and soot to soils revealed a strong linear relationship between the amount of added and detected BC, from which it was concluded that this method may be suitable to quantify thermally stable BC in soil. Coupling (TG-DSC) to an isotope ratio mass spectrometer (IRMS) enables a better discrimination of the organic matter and BC components of different sources and varying in thermal stability (Lopez-Capel et al., 2005a; De la Rosa et al., 2007).

7.4. STRUCTURAL PROPERTIES OF PyOM

7.4.1. Chemical Alteration of BC During Heating

Following the charring process of plant residues and peat in controlled laboratory heating experiments by solid-state ^{13}C NMR spectroscopy demonstrates that under relatively mild heating conditions, the main processes are dehydration and selective degradation of *O*-alkyl C (Freitas et al., 1999; Baldock and Smernik, 2002; Almendros et al., 2003). For peat heated at 350 °C, the carbohydrate intensity vanished only after 150 s (Almendros et al., 2003). During successive stages of biomass burning, more stable alkyl C and also carboxyl C are removed and only a portion of recalcitrant, cyclic, or branched paraffinic structures remain in a condensed unsaturated matrix. Concomitantly, the aryl C content increases. Based on this observation, the aromatic C/alkyl C ratio was suggested as an index to describe the degree of charring of vegetation residues and SOM (Knicker et al., 2005b).

For soot, atomic H/C ratios of <0.2 were determined which are in accordance with highly condensed aromatic network suggested by Sergides et al. (1987) (Figure 7.1). Graphenic structures are also confirmed for BC yielded under pyrolysis conditions (without oxygen) at temperatures above 700 °C (Freitas et al., 1999). However, for charcoal produced from peat and grass under oxic conditions at 350 °C, simulating more realistically BC production during natural fires and even for commercially available barbeque charcoal atomic H/C ratios >0.48 were yielded (Knicker et al., 1996; Knicker et al., 2005b). Such values demonstrate that on average, every second C is protonated and consequently the chemical structure of charcoal BC must differ greatly from that expected for soot BC. Applying dipolar dephasing solid-state ^{13}C NMR experiments to elucidate interactions between ^{1}H and ^{13}C supported this finding and enabled the conclusion that most of the aromatic C in charcoal BC must occur in relatively small clusters of a maximum of six aromatic rings to be drawn (Knicker et al., 2005b).

The concept of differing chemical structures for soot and charcoal BC is also fostered by the atomic O/C ratios. Whereas soot exhibits typical values of <0.3, charcoal originating from plant and SOM exhibit average atomic O/C ratios of up to 0.5 (Baldock and Smernik, 2002; Almendros et al., 2003; Trompowsky et al., 2005). Such relatively high O concentrations are inconsistent with a large and graphitic network of condensed polyaromatics, but confirm the participation of anhydrosugars, furans, pyranones, anhydrosugars, and 5-hydroxymethylfurfural. These components were reported to derive from cellulose and other carbohydrate components at temperatures of up to 310 °C (Ralph and Hatfield, 1991; Bassilakis et al., 2001; McGrath et al., 2003). Even under anoxic conditions, no PAHs were detected in this temperature range but were formed and released into the gas phase after heating from 300 °C to 600 °C. Addition of oxygen reduced the yields of PAHs. Considering the predominance of carbohydrates in plants, these pyrolysis products are likely to comprise a major proportion in their charcoal BC.

As indicated with model chars, a part of the aromatic C in BC from vegetation residues certainly derives from altered lignin (Knicker et al., 2008a). Pyrolysis of this compound class generally leads to char and volatile products that are mainly substituted methoxyphenols (González-Vila et al., 2001). Pyrolyzing isolated lignin under anoxic conditions, approximately 40% and 60% were volatilized at 450 °C and at 750 °C, respectively (Sharma et al., 2004). Condensation reactions among the volatile products, leading to the formation of coke, were absent under the pyrolysis conditions used. Repeating the experiment under oxic conditions led to comparable char yields at temperature of up to 350 °C, but at 550 °C only 20% was sequestered from complete combustion remaining as char. Elevating the temperature from 250 °C to 400 °C lowered the phenolic C, methoxyl C, and alcoholic C content. Summarizing the chemical alteration of lignin during heating, Sharma et al. (2004) suggest that at low temperatures the predominant reaction is dehydration, followed by dehydration and some decarboxylation at higher temperatures (>350 °C) with the aromatic rings remaining essentially intact.

González-Vila et al. (2001) reported that after progressive charring of grass material at 350 °C, the tetrapyrrole moiety of chlorophylls was rapidly destroyed and the phytol backbone was comparatively more resistant, leading to phytadienes as an intermediate and subsequently to phytenes and pristenes. The alkyl products showed characteristic yields in terms of heating time. The major decreases for the phytadienes and fatty acids were observed after 75 s of heating and for *n*-alkanes after 80 s. Paraffinic structures, waxes, and sterols were comparatively stable.

During biomass and fuel burning, a complex mixture of ill-characterized volatile organic matter are released into the atmosphere (Andreae and Merlet, 2001). It contributes to the formation of aerosols and fine particles of sizes up to 100 μm. After an estimated lifetime of 7.9 days (Cook and Wilson, 1996), they are either degraded or are removed from the atmosphere by precipitation. However, they can be transported a considerable distance. For example, boreal forest fires contribute substantially to atmospheric BC in the Arctic (Cook and Wilson, 1996), and Antarctica receives BC from biomass burning in the tropics (Wolff and Cachier, 1998). On a global scale the amount of atmospheric emission is estimated with 5–6 Tg BC yr^{-1}, which is almost as high as those produced from fossil fuels (Cook and Wilson, 1996; Andreae and Merlet, 2001; Preston and Schmidt, 2006).

Some of the vaporized organic substances formed during combustion by vegetation fires move down the soil profile along steep temperature gradients. On reaching cooler areas they condense, resulting in a water-repellent layer below and parallel to the soil surface (Savage, 1974). In laboratory heating experiments, the water repellency moved down to 10 mm at low fire intensity to up to a depth of 50 mm at high intensity (Robichaud and Hungerford, 2000). Incipient water repellency at different soil depths could also be intensified by heating the organic particles to an extent that they coat and are chemically bonded to mineral soil particles (Giovannini et al., 1983). Little change in water repellency occurs when soils are heated less than 175 °C and strongly intensifies at heatings between 175 °C and 200 °C (DeBano, 2000). Its destruction occurs between 280 °C and 400 °C (Savage, 1974; DeBano et al., 1977). In terms of erosion, the formation of water repellent layers may not be as efficient due to the combustion of overlying vegetation and the litter layer which mitigate the impact of raindrops on soil aggregate disruption (Svenik et al., 1989). It was concluded that fire-induced water repellency produced localized runoff and sediment movement only on hill slopes (Prosser and Williams, 1998). Other plot studies suggested that the hydrological responses depend upon soil dryness and increased runoff was attributed to an increase in the severity of water repellency at lower soil water contents during the dry season (Walsh et al., 1994).

Reported effects of charring on $\delta^{13}C$ are generally small and somewhat erratic (Bird and Gröcke, 1997; Czimczik et al., 2002; Rumpel et al., 2006). During biomass heating the preferential loss of cellulose and selective accumulation of more-depleted lignin is expected to turn the residues isotopically lighter with increasing thermal alteration (Turney et al., 2006). In general, char retains the signature of its C_3 or C_4 plant origin (Hiridate et al., 2004), although much greater depletion has been observed for char derived from some C_4 grasses, especially for natural rather than oxygen-limited laboratory burns. Krull et al. (2003) explained the ^{13}C-depletion in C_4-derived chars from natural burnings by protection of organic matter in silicate structures which were found to be depleted by up to 9%.

7.4.2. Formation of "Black Nitrogen (BN)"

Moderate heating leads to little changes of the C/N ratios, although some increase has been observed indicating a relative enrichment of N-containing structures that are fairly resistant to heating. For burnt woody material, the range can be as wide as 440–630 (atomic ratio) (Trompowsky et al., 2005), although charcoal from young grass material exhibited a narrower range of 6.0–6.9 (w/w) (Knicker et al., 1996). Ranges of C/N ratios between 30–35 (w/w) and 12–14 (w/w) have been reported for charred peat and humic acids (Almendros et al., 1990; Knicker et al., 1996; Almendros et al., 2003). For straw charcoal and vegetation fire residues, C/N ratios of 25 and 40 are reported (Fernandes et al., 2003). Applying solid-state ^{15}N NMR spectroscopy (Figure 7.4) identified the heat-resistant N as pyrrole/indole-type N (−150 to −240 ppm) with a minor contribution of pyridine N (−60 to −120 ppm). They derive mainly from proteins and peptides that thermally decompose via systematic and random depolymerization reactions. Pyrolysis studies have shown that after decarboxylation, deammination, and dehydration of amino acids, one of the most abundant product classes is diketopiperazines (Chiavari and Galletti, 1992; Reeves and Francis, 1998). Further reactions are homolysis of the aliphatic side chains and intramolecular loss of water or ammonium to give cyclic products. At lower temperatures (200–300 °C), low-molecular-weight heterocyclic compounds have been studied; above 500 °C, polynuclear N-containing aromatic structures have been observed (Sharma et al., 2003).

Another important reaction, occurring during charring of biomaterial, is the Maillard reaction (Ikan, 1996). During this process, ammonia or free amino groups of amino acids and amino sugars react with carbonyl groups of sugars to form Schiff bases that subsequently undergo rearrangements via Amadori compounds to produce dark-colored melanoidins. The major compounds found after pyrolysis of four Amadori compounds formed from glucose reacted with asparagine, valine, leucine, and threonine were ketones, aldehydes, pyridines, pyrazines, pyrroles, and carboxylic acids (Coleman and Chung, 2002).

Taking the narrow atomic C/N ratios for pyrrole (C/N = 4), pyridine (C/N = 5), and carbazole (C/N = 12), the approximate contribution of such heterocyclic compounds to BC can be estimated. Whereas for grass PyOM in Figure 7.5, those structures can contribute with up to 50% of its organic C, its contribution to wood BC is negligible. This enables two important conclusions to be drawn. First, considerable differences in the chemical composition have to be encountered for plant BC of varying origin. Second, in environments where N-rich litter is the major source for BC production, N-heteroaromatic C can comprise a fairly important fraction of BC accumulating on and in these soils. Bearing in mind the impact of N availability on biomass production and the relatively high contribution of N-heteroaromatic C to some BC, the heat-resistant and recalcitrant "black nitrogen" (BN) certainly needs to be understood in more detail, if a more complete understanding of the role of charred material as a C-sink in soils is sought. Furthermore, heterocyclic N compounds are assumed to be refractory in nature. Thus, burning may increase the relative contribution of stable organic N to SOM and thus contribute to a decrease of the long-term nitrogen availability for biomass production in soils (Knicker and Skjemstad, 2000).

7.4.3. Conceptual Model for the Chemical Structure of Charcoal

According to results obtained from charring experiments and reports from the pyrolysis literature, clear differences occur between the chemical structure of charcoal remaining as combustion residue and soot being a secondary condensation product of volatiles released during charring. Whereas elemental analysis, pyrolysis, and even NMR spectroscopy can support the graphitic structures proposed by Sergides et al. (1987) for soot, coke, and anthracite coal (Botto and Yuzo, 1993; Abelmann et al., 2003), charcoal reveals much higher chemical heterogeneity. Accordingly, the aromatic skeleton of charcoal accumulating after a vegetation fire must contain remains of the lignin backbone and considerable contributions of thermally altered products such as furans and anhydrosugars from cellulose and heteroaromatic N from peptides (Knicker et al., 2008a). Furthermore, features such as chemical composition or charring temperature are variable but determine contributions of oxygen, nitrogen, and even sulfur functionalities as well as size and orientation of the polycyclic aromatic ring. Although aromatization and condensation are occurring already at charring temperatures below 350 °C, formation of considerable moieties of graphite-like structures require higher temperatures (Freitas et al., 1999). Because under oxic conditions, most of the organic matter will volatilize at temperatures above 350 °C and during vegetation fires, open space enables fast removal of released and volatiles production of soot, the accumulation of the latter is expected to be rare. However, within the soil, some of these products may occur due to precipitation or their evaporation into deeper and cooler regions.

Based on those considerations, a new structural concept for PyOM was proposed (Knicker et al., 2006; Knicker et al., 2008a). According to this concept, charcoal PyOM can be seen as a heterogeneous mixture of thermally altered biomolecules, the degree of alteration being dependent upon the severity of the fire and the chemistry of the original vegetation. Important constituents are furans or pyrroles but also lignin residues and some lipids. According to recent NMR studies (Knicker et al., 2005b), the condensation degree of the aromatic structures is relatively low and does not exceed an average aromatic cluster size of six rings. Whereas intensively charred material possesses an entirely aromatic structure, alkyl-C contents, possibly from thermally altered proteinaceous material, is still present in chars produced at moderate temperature (Knicker et al., 2005b, 2008a). With respect to elemental analysis, N, O, and likely also S substitutions are common features of the charring products. The suggested concept implies that the chemical composition of charcoal can vary to a great extent and that those differences are likely to determine the recalcitrance and behavior of charcoal in the environment. Thus, compared to a more graphitic structural model, degradation of charcoal exhibiting the properties of the concept presented here seems more likely, since small aromatic clusters with considerable substitution with N, O, and S functional groups can be attacked both by biotic and abiotic processes. Biotic oxidation could be performed by lignin-degrading fungi that have been shown to participate in coal degradation (Fakoussa and Hofrichter, 1999). *Phanerochaete chrysosporium*, for example, was found to partially depolymerize coal polymers prepared from nitric acid-treated sub-bituminous German coals (Wondrack et al., 1989). However, if lignin degraders are participating, oxygen is essential for an efficient decomposition of char. Consequently, charcoal accumulated on and within the top horizon is likely to degrade, whereas it will accumulate in buried soils and sediments or at archeological sites due to oxygen depletion. Such a scenario could at least partly explain why the typical pattern of charcoal diminishes from some fire-affected soils with increasing recovery time of the system (Golchin et al., 1997), whereas in sediments char can survive for millennia (Masiello and Druffel, 1998).

7.5. QUANTIFICATION OF PyOM

7.5.1. Common Methods and Their Reliability

Determination of BC concentrations in a National Institute of Standards and Technology reference material (NIST SRM 1649a) by various methods resulted in amounts ranging from 7% to 50%, depending only on the method used (Currie et al., 2002). A newer ring trail study confirmed the high variability (Hammes et al.,

2007). An important reason for the apparent discrepancies of PyOM detection with the various analytical techniques is certainly the high heterogeneity of BC and the fact that each method concentrates on a different region within the combustion continuum (Masiello, 2004). Presently, optical (Bird et al., 2000), thermal (Kuhlbusch, 1995; Gustafsson et al., 2001; Nguyen et al., 2004; Leifeld, 2007), chemical, and photo-oxidation approaches (Masiello et al., 2002; Song et al., 2002), but also spectroscopic (Smernik et al., 2000; Simpson and Hatcher, 2004) and molecular marker methods (Glaser et al., 1998a; Elias et al., 2001), are applied for the identification and quantification of charcoal. Visual or microscopic techniques identify charcoal pieces and quantify their amount by subsequent particle counting or physical separation with weighing. This method, however, misses soot and charcoal degradation products. Alternatively, biomarker essays rely on the identification of typical charring products as molecular biomarker such as benzenepolycarboxylic acids (BPCA) released after oxidation with nitric acid (Glaser et al., 1998b) or levoglucans being a typical pyrolysis product of cellulose (Elias et al., 2001). The latter is used for identification rather than quantification. The BPCA method is also applied for quantification but it is highly sensitive to operating conditions and to the factor used to convert BPCA yields into BC amounts. Furthermore, BPCA are derive from C in larger aromatic clusters which are hardly major constituents in plant-derived charcoal and coals up to the bituminous maturation stage (Abelmann et al., 2003).

Chemical and photo-oxidation as well as thermal methods are based on the refractory nature of PyOM, assuming that the latter is selectively enriched whereas the more labile unburned material is destroyed. With this approach, only the most recalcitrant fractions will be determined. For thermal treatments, assuming char production during the treatment is avoided, this is expected to be soot. However, recent studies have indicated that this technique does not recognize most charcoal produced at temperatures below 850 °C as BC and that even soot does not completely survive thermal treatment at 375 °C (Nguyen et al., 2004). Comparably, chemical oxidation with peroxide also induced substantial degradation of BC (Wolbach and Anders, 1989). Alternatively, oxidation with dichromate showed initial promise in separating labile organic matter and kerogene from highly condensed BC, assuming that the oxidation pattern can be modeled as the sum of three first-order reactions (Wolbach and Anders, 1989). Correspondingly, Rumpel et al. (2006) and Knicker et al. (2007) reported that grass and wood were efficiently degraded already after 4–6 h, whereas BC is expected to survive this treatment. However, applying this technique to fresh pine needle yielded in approximately 12% of the organic carbon (C_{org}) as chemical-oxidation-resistant elemental carbon (COREC), which certainly could not have been derived from BC (Knicker et al., 2007). Based on solid-state ^{13}C NMR spectroscopy, this C_{org} was identified as plant waxes. Comparable results were obtained for other fire-unaffected control soils located under pine and oak forests. Approximately half of the COREC was removed by a subsequent soxhlet extraction, clearly demonstrating that their survival may be explained by its hydrophobic nature rather than an inherent chemical recalcitrance. Survival of paraffinic structures has also been reported in residues after dichromate treatment of some Australian mollisols and oxisols (Krull et al., 2006), in sediment samples (de la Rosa et al., 2007), and in peat humic acid oxidized with nitric acid (Simpson and Hatcher, 2004). Consequently, the resistance of such structures to strong acid can be described as a common feature, which has to be considered in particular if soils with higher lipid contents are subjected to BC determination via chemical oxidation.

Conversely, chemical oxidation of charcoal briquette, beech sawdust BC, and grass BC revealed that even charcoal briquette is attacked by acid and that the amount of determined BC varies greatly with source material and also strongly relies on the applied oxidation time. The oxidation efficiency could not be related to the condensation degree of the aromatic structures (Knicker et al., 2007) and thus is best explained by the differing structure of the different plant BCs. This strongly supports to the questioning of the reliability of BC data obtained by chemical oxidation methods alone. On the other hand, chemical analysis of COREC by solid-state ^{13}C NMR spectroscopy enables the identification of BC by signal intensity assignable to aromatic C, which was absent in the COREC of fire-unaffected samples. However, in using this signal as a means for BC quantification, one has to consider that its intensity is not necessarily proportional to the amount charcoal BC in the bulk sample. In particular, this is of relevance in modeling and global budgeting studies using proxies, such as the estimated BC/CO_2 ratio across fire types (Kuhlbusch and Crutzen, 1995) and literature values from other sites.

7.5.2. Charcoal Yields During Burning

Determining the fire-induced losses to the atmosphere and charcoal production rates in a scrub oak ecosystem of Florida, Alexis et al. (2006) found that post-fire standing dead biomass contained 30% and 12%, while litter contained 64% and 83%, of pre-fire vegetation C and N stocks, respectively. About three-quarters of the fire-induced leaf litter fall was in the form of unburned tissue and the remainder was charcoal, which amounted to 5% of the pre-fire leaf stocks. Charcoal productions ranged between 4% and 6% of the fire-affected biomass. It was observed that these values were below the range of literature values for the transformation of plant tissue in the process of humification (Alexis et al., 2006). This suggests that fire generates a smaller quantity of stable organic C than does humification over decades and potentially centuries.

For savanna fires in South Africa (Kuhlbusch et al., 1996), pre-fire vegetation and litter were approximately 2700 kg ha^{-1}, from which 90% of C was volatilized. A conversion rate of 0.5–2% of BC to pre-fire carbon exposed to fire was determined. Kuhlbusch and Crutzen (1995) found conversion ratios between 0.14% and 2.2% for non-woody biomass substrates and 3.1% for deciduous wood. In boreal forests, charcoal production amounts of 235–932 kg ha^{-1} from experimental fires with crowning or partial crowning in boreal forests were determined (Clark et al., 1998; Ohlsen and Tryterud, 2000; Lynch et al., 2004). Slash-burning after forest clearing produced 7400 kg ha^{-1} of forest floor charcoal at a side in Virginia (Schiffman and Johnson, 1989). For pre-settlement in North America the highest rates of charcoal deposition of 5000 kg ha^{-1} yr^{-1} were estimated for prairie/forest border areas. Such high char deposition for 100 years would result in 500 kg m^{-2} or a 10-cm layer assuming a density of 0.5 g cm^{-3} and no mechanisms for loss (Preston and Schmidt, 2006). However, the reported numbers have to be considered in light of the strong variation of BC quantity determined with the different methods. Thus, presently, the limited data on charcoal and BC production still have to be treated with caution if used for modeling purposes.

7.5.3. Quantity and Distribution of Charcoal in Soils

Estimates of charcoal content in soils range from 2% and 8% in tropical savanna burn sequences (Bird et al., 1999) to 30% of the SOM in U.S. agricultural soils (Skjemstad et al., 2002). Up to 24% was found in a fimic anthrosol from the Amazon region (Glaser et al., 1998b) and 4–17% of soil organic C in native North American prairies soils (Glaser and Amelung, 2003). In a tropical sloping land under slash-and-burn agriculture in Northern Laos, the BC contribution to the total organic C content was 3–5% in the B horizons and between 6% and 7% in the A horizon (Rumpel et al., 2006). For soils along altitudinal transects in the French Alps, the current conifer-dominated forest belt was estimated to contain approximately 10,000–30,000 kg ha^{-1} corresponding to 0.1–20% of SOM (Carcaillet and Talon, 2001). For soils in Corsican forests, Carcaillet and Talon (2001) calculated a charcoal content of 9800–148,000 kg ha^{-1}.

For a number of German chernozems, up to 45% of SOM was identified as charcoal, which mostly occurred in the light fraction (Schmidt et al., 1999). In Japanese andosols the contribution of charred plant fragments in the light fractions ranged from 3.4% to 33% (Shindo et al., 2004). In coastal temperate forest sites of southern Vancouver Island, Canada, BC characteristics were detected in the water-floatable fraction separated from mineral soils (Preston et al., 2002).

In contrast to those studies, most of the charcoal comprising 30% of the SOM in some Australian topsoils was recovered in the <53-μm fractions (Skjemstad et al., 1996; Skjemstad et al., 1999). Comparably, the clay- and silt-size fractions represented also the main charcoal reservoir in native grassland soils from North America (Glaser and Amelung, 2003). In grassland soils with a long fire history, charcoal is associated mostly with the fine fractions, whereas in forest systems and after recent fires, most of the char was found in the light fraction. Possibly this is due to the fact that different plant residues result in different size chars. Additionally, with increasing aging the charcoal particles may suffer ongoing disintegration by physical and chemical processes which leads to smaller particles. Continuing oxidation of such particles forms functional groups (Knicker et al., 2006; Lehmann et al., 2005) that are likely interacting with the mineral phase. Thus, a shift in relative abundance of charcoal is expected since the proportion of char in the light fraction will decrease with time, whereas the disintegrated BC adsorbed to the clay fraction is likely to be selectively enriched. This may explain the finding that in Terra Preta soils, the BC in the heavier fractions was partly embedded within plaques of iron and aluminum oxides on mineral surfaces (Glaser et al., 2000). Enhanced complexation of SOM with metals after burning was also observed by Fernández et al. (1997). The association with Al was favored compared to Fe. Interaction with metal oxides was also suggested to be an important stabilization mechanism for BC occurring in an acid haplic chernozem in Germany (Brodowski et al., 2005).

For some oxisols and ultisols of the Upper Rio Negro region of Colombia and Venezuela, BC enrichment at 30- to 40-cm soil depth is reported (Saldarriaga and West, 1986; Glaser et al., 2000). The latter was attributed to the washing out of light char particles from the surface by selective erosion. A relative enrichment of charcoal in the silt fractions of soil material derived from a depth of between 38 and 48 cm was also evidenced from a Brazilian agriculturally used acrisol (Dieckow et al., 2005) and may have derived from native grassland fires occurring before its conversion for arable cropping. For the dry Alps in France a minimum time for a charcoal particle to migrate down the profile to a depth of 100 cm was calculated to be less than 500 years, but most of the charcoal production of the last two millennia remains in the upper 30 cm (Carcaillet, 2001). Recent studies supported quick transportation into the B horizon of Spanish soils, sampled one year after a severe fire (Knicker et al., 2006). Here, this observation was explained by an increased water solubility of oxidized charcoal constituents. On steep sites, horizontal transport may be more important than vertical transport (Rumpel et al., 2006).

7.6. INTERACTION OF PyOM WITH THE ENVIRONMENT

7.6.1. Impact of Former Vegetation Fires and Charcoal Production on Regional Ecology

Since geological times, natural vegetation fires have participated in the formation of ecological systems. Evidence for this can be obtained from the great coal deposits of Palaeozoic and earlier periods, containing fusain or fusinite which is considered to have originated from burnt plant material (Jones and Chaloner, 1991; Bird, 1995; Schmidt and Noack, 2000). Charcoal particles abundantly found in ancient sediments are further witnesses for fire in the geological past. The presence of a global soot layer at the boundary between the Cretaceous and Tertiary periods has been attributed to a meteorite impact, causing a global wildfire (Wolbach et al., 1988; Venkatesan and Dahl, 1989). Estimates indicated that this event caused the burning of about 25% of the aboveground biomass (Ivany and Salawitch, 1993).

With the development and the spreading of hominids using fire to obtain free land for hunting and agriculture, the importance of anthropogenic fires for changing environments increased continuously (Caldararo, 2002; Carcaillet et al., 2002). Archaeological evidence suggests that modern humans were present on most habitable continents, with "fire-stick in hand," by at least 40,000 years ago and probably earlier (Bird, 1995). Indeed, although grasslands, savannas, and fire-adapted vegetation existed long before the Pleistocene, humans promoted the expansion of savannas by burning the forest margins and by depleting the populations of large herbivorous animals that by grazing contributed to the reduction of large fuel loads. Based on charcoal studies, it has been suggested that in Australia, present fire-adapted species could not have maintained their present extent without human fire intervention occurring since the first arrival of Australian aborigines (Caldararo, 2002). The increasing human influence on European ecosystems during the Neolithic age supported the expansion of evergreen sclerophyllous trees in the Mediterranean basin (Björkman and Bradshaw, 1996; Willcox, 1999) and conifers in cold temperate forests (Björkman and Bradshaw, 1996).

7.6.2. Impact of Charcoal on SOM Quantity and Quality

Compared to SOM formed via biodegradation and humification, PyOM has some important qualitative differences in its molecular structure. Whereas biological processes lead to carboxyl-containing macromolecular products (Knicker and Lüdemann, 1995), thermal treatment removes external O groups, yielding hydrophobic materials with comparatively reduced solubility and colloidal properties (Almendros et al., 1990, 1992). Increased soil hydrophobicity, however, affects soil water distribution. Soils with a frequent input of charred material (e.g., under frequent post-harvest-burning management) may face long-term problems of inhomogeneous distribution of nutrients due to hydrophobic spots that prevent infiltration of water. Additionally, the lack of polar groups is expected to change sorption properties and therefore retention of soil cations, nutrients, or pollutants that can promote their leaching. It can also affect interactions between SOM, clay, and metal oxides and thus the stability and formation of aggregates.

Charcoal produced during vegetation fires derives mostly from the aboveground vegetation and the SOM of the upper few centimeters, since the low heat conductivity of the mineral phase leaves material below the upper few heated centimeters unaffected in its original pre-fire form. Therefore the extent of alteration of the chemical composition and properties by the fire depends on the quality and quantity of the charcoal input and the respective soil mixing mechanisms. Lower charcoal accumulation was observed after intense and very intense fires compared to medium intense burnings; this was attributed to a more complete volatilization during severe heating (Knicker et al., 2006). Stronger fire intensities resulted in wider range of aromatic C to alkyl C ratios of the incorporated charred material, demonstrating that depending on the heat intensity the produced charcoals can vary considerably with respect to their chemical structure. Multiple burning does not necessarily increase the aromaticity of SOM evidenced for a double-burn forest site in Central Spain (Knicker et al., 2006). This may be because a first fire destroys the ground vegetation together with the litter layer and kills most of the trees. During a second fire, following shortly afterwards, the new vegetation cover of shrubs and herbs together with the decaying char remains represent an improved fuel that will be completely combusted without leaving considerable amounts of BC. In addition, within the following post-fire time, fresh litter—rather than charcoal—is the main source for new SOM input, which year by year masks more and more the features imposed by the first BC input. The fresh litter can originate from fire-unaffected or only partly affected litter and roots from the decaying trees. Furthermore, the liming effect of ash and increased nutrient availability (Fernández et al., 1997) can promote a quick development of a new herbaceous layer. Enhanced litter production and decreased microbial activity subsequently leads to an increase of the uncharred SOM pool. Whereas post-fire recovery time of 1–2 years was shown to be too short for sufficient production of biomass and litter to develop a substantial amount of new and uncharred SOM (Knicker et al., 2006) to efficiently mask typical charcoal features, such an effect was observable for a soil sampled 5 years after a forest fire (Knicker et al., 2005a).

On the other hand, several reports have shown that enrichment of PyOM can have a long-term darkening effect on SOM. Relationships between lightness and BC content were observed for a color sequence of German chernozems (Schmidt et al., 1999). In the examination of luvic phaeozems along the Lower Rhine Basin, Germany, clusters of regularly shaped pits were detected that contained charcoal dated from the Mesolithic to the medieval ages. Accordingly, those soils were affected by human activity (Gerlach et al., 2006) and consequently they should be classified as anthrosols rather than as chernozems. Charcoal input was also assumed to play an important factor in the formation of some deep black soils in Russia,

Euroasia, Canada, and California (Schmidt and Noack, 2000) and Australian grass land soils, having been under aboriginal fire management presumably for thousands of years (Skjemstad et al., 1990). The impact of charcoal from burned rice residues on SOM composition was also detected in buried neolithic paddy soils in China (Cao et al., 2006). Considering that formerly, post-harvest burning was a common practice for rice cultivation, stabilized charcoal is likely to be an intergral part of SOM in old paddy field soils. Comparatively, BC can also be considered as a defining property of Terra Preta soils in South America (Glaser et al., 2000). For those soils it was suggested that before the contact with Europeans the native population added charcoal for fertilization purpose. Indeed, compared to surrounding soils, they still show enhanced fertility (Glaser et al., 2002).

7.6.3. Impact of PyOM on the Nature of Extractable SOM

Humic extracts from soot, charred grass, and barley straw treated with HNO_3 and H_2O_2 revealed high aromaticity with considerable contributions of carboxyl C (Haumaier and Zech, 1995). Without any pretreatment, no humic acid was yielded from commercial charcoal and cinder, indicating that depolymerization and carboxylation of the carbonaceous material was required before it became extractable. Comparably, no PyOM was identified in the humic acids of a fire-affected soil sampled 5 years after burning, whereas the bulk soil showed clear indications for charcoal contributions (Knicker et al., 2005a). This points toward the conclusion that also under natural conditions oxidation of charcoal is needed to form humic and fulvic fractions.

Humic acids from volcanic ash soils in Japan that were burned annually contained a greater proportion of aromatic and carbonyl C compared to unburned soils and soils where burning was ceased up to 100 years, ago (Golchin et al., 1997). Based on the observed similarities between the humic acids from volcanic ash soils and those from charred plant residues, it was suggested that charring processes may be one of the mechanisms responsible for humic acid formation in these soils (Shindo et al., 2004). Highly carboxylated humic acids with hydrogen-deficient condensed aromatic structures were also extracted from volcanic ash soils (Kramer et al., 2004) and from 100-year-old char particles found in a forest soil (Hockaday et al., 2006). A significant concentration of heterocyclic N derived from charring processes was identified in type A humic acids extracted from both the subsoil of paddy fields and the surface layer of andosols (Maie et al., 2006). Since burning is not considered to occur in the subsoil of paddy fields, they were supposed to have been transported by leaching.

7.6.4. Stability of PyOM in Soils

As one of the most resistant forms of reduced organic matter, BC is supposed to compose a major proportion of the slow-cycling carbon pools in terrestrial systems and aquatic sediments. But soils under grassland that was invaded by forest after ceasing annual burning demonstrated a comparably fast change of the nature of soil C (Golchin et al., 1997). The greatest changes occurred during the first 20–30 years and were manifested in a decrease in aromatic and an increase in alkyl-C content. It seems that with prolonged recovery time the typical char pattern becomes masked by the input of fresh litter material. However, considering the high degradability of most fresh litter, this either requires a substantially high biomass production or implies that some char is lost. Losses of charcoal can also be the explanation for observations that measurements of BC production and losses are not balanced (Masiello, 2004). Possible mechanisms would be in situ degradation, losses due to surface erosion, or translocation of char into deeper horizons, for example, as oxidized charcoal constituents that are soluble in the soil solution. Such a scenario could explain the increasing aromaticity with depth observed for a Campo soil from Southern Brazil sampled 22 years after ceasing annual burning (Figure 7.6). Subjecting those soils to chemical oxidation with potassium dichromate confirmed relatively low aromatic COREC content of 2% of the total organic C in the top 5 cm, whereas at a depth between 15 and 30 cm this carbon amounted to 4.5% (unpublished data). Considering that this treatment oxidized approximately 60% of

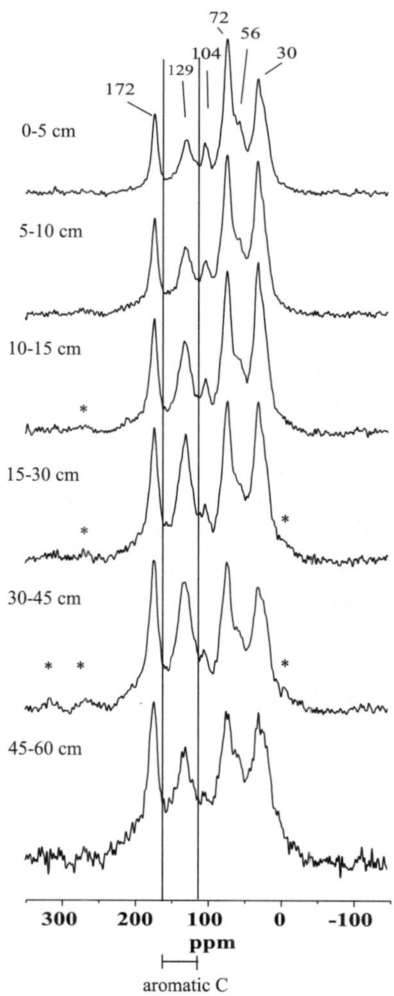

Figure 7.6. Solid-state ^{13}C NMR spectra of HF-treated material obtained from a grassland soils after ceasing annual burning for 22 years. Asterisks indicate spinning side bands.

grass BC (Knicker et al., 2007) and 50% of a model char derived from the Campo vegetation, a correction factor *f* has to be applied to estimate the BC content. The respective factors for those soils were yielded on correlation studies and were calculated to be 2.4 for the topsoil and 3.3 for the subsoil (Knicker et al., 2008b), resulting in a BC contribution to the total organic C of 5% in the topsoil and 14% in the subsoil (unpublished data). This corresponds to 5 and 7 mg BC g^{-1} soil, respectively. Using near-edge-X-ray absorption fine structures spectroscopy, highly oxidized regions were evidenced at the surface of charred material from soils located near Manaus, Brazil (Lehmann et al., 2005). Applying X-ray photoelectron spectroscopy (XPS) to charcoal indicated that oxidation of BC particles was initiated on the surface. Increasing the incubation temperature from 30 °C to 70 °C (abiotic), the oxidiation penetrated into the interior of the particles (Cheng et al., 2006). The presence of carboxyl C formed by the oxidization may also be responsible for a higher microbial degradability. This could explain the observation made by Bird et al. (1999) that BC exhibits less recalcitrance than commonly assumed. In line with this are studies of sedimentary BC accumulated under oxic and anoxic conditions which point to a relatively labile BC fraction. Its decomposition was shown to be on the timescale of thousands of years and to become more efficient under oxic conditions (Middelburg et al., 1999; Masiello and Druffel, 2003). More direct evidence for degradability of charcoal was obtained by demonstrating CO_2 production as a result of microbial degradation of char (Hamer et al., 2004). Addition of glucose resulted in a priming effect and increased mineralization. Adding microbial inoculums to artificially produced charcoal confirmed some microbial degradation (Baldock and Smernik, 2002).

7.7. UNDERSTANDING THE ROLE OF PyOM: WHAT ARE THE MISSING LINKS AND KNOWLEDGE GAPS?

PyOM represents a ubiquitously occurring source material for refractory organic matter in soils and sediments. As an important contributor to the pools of sequestered C and N, its properties greatly affect the ecological functionality of organic matter in soils and sediments as much as its turnover rates. Despite its importance within the global elemental cycles and the recent advances in PyOM research, still very little is known about its structure, distribution, and environmental reactivity. This is in particular true with respect to its chemical structure, where our knowledge is still rather fragmentary, resulting in the assumption that PyOM comprises a highly polycondensated aromatic network. However, as already indicated from elemental analysis, this view can only be valid for carbonized samples or soot that show the corresponding low atomic H/C ratios of smaller than 0.2. Although high aromaticity is certainly a common feature of most charcoals, we cannot necessarily assume that the chemical nature of the aromatic structures of charcoals with different origin is comparable. This conclusion is supported by the high variability of the elemental composition in such materials. Bearing this in mind, those differences in the quality of the aromatic moieties are likely to determine physical and chemical charcoal properties such as porosity, solubility of single components, and accessibility to chemical and biological degradation or resistance to further thermal oxidation. Porosity and solubility are important parameters affecting adsorption of nutrients as well as inorganic and organic pollutants, but also partitioning and aging of the latter. Additionally, the varying chemical recalcitrance has some major impact on the reliability of charcoal data given in the literature. Beside the fact that each method detects only a certain region within the BC continuum, their efficiency also differs strongly for charcoals where chemical differences alter the accessibility to chemical and thermal oxidation. Thus, urgently needed rapid and inexpensive analytical approaches for BC characterization and quantification can only be developed if a more detailed understanding of the charcoal chemistry and its dependence of source and formation conditions has been achieved.

Furthermore, considering that the turnover time of organic soil constituents depends on biological accessibility and utility, this knowledge will guide us to an improved understanding of charcoal degradation and stability. Relating chemical structure to differences in degradation rates of various charcoals would facilitate the determination of the respective turnover times, thereby enabling the inclusion of the charcoal pool into global C cycling models.

REFERENCES

Abelmann, K., Knicker, H., and Kögel-Knabner, I. (2003). Structural differences of organic materials in anthropogenic and natural soils. *Mit. Deutsch. Bodenkl. Ges.* **102**, 157–158.

Abelmann, K., Kleineidam, S., Knicker, H., Grathwohl, P., and Kögel-Knabner, I. (2005). Sorption of HOC in soils with carbonaceous contamination: Influence of organic matter composition. *J. Plant Nutr. Soil Sci.* **168**, 293–306.

Alemany, L. B., Grant, D. M., Pugmire, R. J., Alger, T. D., and Zilm, K. W. (1983). Cross polarization and magic angle sample spinning NMR spectra of model organic compounds. 2. Molecules of low or remote protonation. *J. Am. Chem. Soc.* **105**, 2142–2147.

Alexis, M., Rasse, D. P., Rumpel, C., Bardoux, G., Péchot, N., Schmalzer, P., Drake, B., and Mariotti, A. (2006). Fire impact on C and N losses and charcoal production in a schrub oak ecosystem. *Biogeochemistry* **82**, 201–216.

Almendros, G., González-Vila, F. J., and Martín, F. (1990). Fire-induced transformation of soil organic matter from an oak forest. An experimental approach to the effects of fire on humic substances. *Soil Sci.* **149**, 158–168.

Almendros, G., González-Vila, F. J., Martín, F., Fründ, R., and Lüdemann, H.-D. (1992). Solid state NMR studies of fire-induced changes in the structure of humic substances. *Sci. Tot. Environ.* **117/118**, 63–74.

Almendros, G., Knicker, H., and González-Vila, J. F. (2003). Rearrangement of carbon and nitrogen forms in peat after progressive thermal oxidation as determined by solid-state ^{13}C- and ^{15}N-NMR spectroscopy. *Org. Geochem.* **34**, 1559–1568.

Andreae, M. O., and Merlet, P. (2001). Emission of trace gases and aerosols from biomass burning. *Glob. Biogeochem. Cycl.* **15**, 955–966.

Baldock, J. A., and Smernik, R. J. (2002). Chemical composition and bioavailability of thermally altered Pinus resinosa (Red pine) wood. *Org. Geochem.* **33**, 1093–1109.

Bassilakis, R., Carangelo, R. M., and Wójtowicz, M. A. (2001). TG-FTIR analysis of biomass pyroysis. *Fuel* **80**, 1765–1786.

Bird, M. I. (1995). Fire, prehistoric humanity, and the environment. *Interdisc. Sci. Rev.* **20**, 141–154.

Bird, M. I., and Gröcke, D. R. (1997). Determination of the abundance and carbon isotope composition of elemental carbon in sediments. *Geochim. Cosmochim. Acta* **61**, 3413–3423.

Bird, M. I., Veenendaal, E. M., Moyo, C., Lloyd, J., and Frost, P. (1999). Stability of elemental carbon in a savanna soil. *Glob. Biogeochem. Cycl.* **13**, 923–932.

Bird, M. I., Veenedaal, E. M., Moyo, C., Llyod, J., and Frost, P. (2000). Effect of fire and soil texture on soil carbon in a sub-humid savanna (Matopos Zimbabwe). *Geoderma* **94**, 71–90.

Björkman, L., and Bradshaw, R. H. W. (1996). The immigration of *Fagus sylvatica* L. and *Picea abies* (L.) Karst into a natural forest stand in southern Sweden during the last 2000 years. *J. Biogeograph.* **23**, 235–244.

Botto, R. E., and Yuzo, S., eds. (1993). *Magnetic Resonance of Carbonaceous Solids*, ACS Advances in Chemistry Series No. 229. American Chemical Society, Washington D.C.

Brodowski, S., Amelung, W., Haumaier, L., Abetz, C., and Zech, W. (2005). Morphological and chemical properties of black carbon in physical soil fractions as revealed by scanning electron microscopy and energy-dispersive X-ray spectroscopy. *Geoderma* **128**, 116–129.

Caldararo, N. (2002). Human ecological intervention and the role of forest fires in human ecology. *Sci. Total Environ.* **292**, 141–165.

Cao, Z., Ding, J., Hu, Z., Knicker, H., Kögel-Knabner, I., Yang, L., Yin, R., Lin, X., and Dong, Y. (2006). Ancient paddy soils from the Neolithic age in China's Yangtze River Delta. *Naturwissenschaften* **93**(5), 232–236.

Carcaillet, C. (2001). Are Holocene wood-charcoal fragments strtified in alpine and subalpline soils? Evidence from the Alps based on AMS ^{14}C dates. *Holocene* **11**, 231–242.

Carcaillet, C., and Talon, B. (2001). Soil carbon sequestration by Holocene fires inferred from soil charcoal in the dry French Alps. *Arct. Antarct. Alp. Res.* **33**, 282–288.

Carcaillet, C., Almquist, H., Asong, H., Bradshaw, R. H. W., Carrión, J. S., Gaillard, M.-J., Gajewski, K., Haas, J. N., Haberle, S. G., Hadorn, P., Müller, S. D., Richard, P. J. H., Richoz, I., Rösch, M., Sánchez-Goni, M. F., von Stedingk, H., Stevenson, A. C., Talon, B., Tardy, C., Tinner, W., Tryterud, E., Wick, L., and Willis, K. J. (2002). Holocene biomass burning and global dynamics of the carbon cycle. *Chemosphere* **49**, 845–863.

Certini, G. (2005). Effects of fire on properties of forest soils: A review. *Oecologia* **143**, 1–10.

Cheng, C.-H., Lehmann, J., Thies, J. E., Burton, S. D., and Engelhard, M. H. (2006). Oxidation of black carbon by biotic and abiotic processes. *Org. Geochem.* **37**, 1477–1488.

Chiavari, G., and Galletti, G. C. (1992). Pyrolysis–gas chromatography/mass spectrometry of amino acids. *J. Anal. Appl. Pyrolysis* **123**, 123–137.

Clark, J. S., Lynch, J., Stocks, B. J., and Goldammer, J. G. (1998). Relationships between charcoal particles in air and sediments in west central Siberia. *Holocene* **8**, 19–29.

Coleman, W. M., and Chung, H. L. (2002). Pyrolysis GC–MS analysis of Amadori compounds derived from selected amino acids and glucose. *J. Anal. Appl. Pyrolysis* **62**, 215–223.

Cook, W. F., and Wilson, J. J. N. (1996). A global black carbon aerosol model. *J. Geophys. Res.* **101**, 19395–19409.

Crutzen, P. J., and Andreae, M. O. (1990). Biomass burning in the tropics: Impact on atmospheric chemistry and biogeochemical cycles. *Science* **250**, 1669–1678.

Currie, L. A., Benner, B. A., et al. (2002). A critical evaluation of interlaboratory data on total, elemental, and isotopic carbon in the carbonaceous particle reference material, NIST SRM 1649a. *J. Res. NIST* **107**, 279–298.

Czimczik, C. I., Preston, C. M., Schmidt, M. W. I., Werner, R. A., and Schulze, E. D. (2002). Effects of charring on mass, organic carbon, and stable carbon isotope composition of wood. *Org. Geochem.* **33**, 1207–1223.

De la Rosa, J. M., Knicker, H., López-Capel, E., Manning, D. A. C., González-Pérez, J. A., and González-Vila, F. J. (2008). Direct detection of black carbon in soils by Py-GC/MS, ^{13}C NMR spectroscopy and thermogravimetric techniques. *Soil Sci. Soc. Am. J.* **72**, 258–267.

DeBano, L. F. (2000). The role of fire and soil heating on water repellency in wildland environments: A review. *J. Hydrol.* **231–232**, 195–206.

DeBano, L. F., Mann, L. D., and Hamilton, D. A. (1977). Fire's effects on physical and aggregate stability in hydrophobic soil. In *Proceedings Symposium on Environmental Conservation: Fire and Fuel Management of Mediterranean Ecosystems*, Mooney, H. A., and Conrad, C., eds., Palo Alto, CA, August 1–5, USDA Forest Service General Technical Report, pp. 65–74.

DeBano, L. F., Neary, D. G., and Folliott, P. F. (1998). *Fire's Effect on Ecosystems*, John Wiley & Sons, New York.

Dell'Abate, M. T., Canali, S., Trinchera, A., Benedetti, A., and Sequi, P. (2000). Thermal methods of organic matter maturation monitoring during a composting process. *J. Ther. Anal. Cal.* **61**, 389–396.

Dieckow, J., Mielniczuk, J., Knicker, H., Bayer, C., Dick, P. D., and Kögel-Knabner, I. (2005). Composition of organic matter in a subtropical Acrisol. *Eur. J. Soil Sci.* **56**, 705–715.

Elias, V. O., Simoneit, B. R. T., Cordeiro, R. C., and Turcq, B. (2001). Evaluating levoglucosan as an indicator of biomass burning in Carajás, Amazonia: A comparison to the charcoal record. *Geochim. Cosmochim. Acta* **65**, 267–272.

Fakoussa, R. M., and Hofrichter, M. (1999). Biotechnology and microbiology of coal degradation. *Appl. Microbiol. Biotechnol.* **52**, 25–40.

Fernandes, M. B., Skjemstad, J. O., Johnson, B. B., Wells, J. D., and Brooks, P. (2003). Characterization of carbonaceous combustion residues. I. Morphological, elemental and spectroscopic features. *Chemosphere* **51**, 785–795.

Fernández, I., Cabaneiro, A., and Carballas, T. (1997). Organic matter changes immediately after a wildfire in an Atlantic forest soil and comparison with laboratory soil heating. *Soil Biol. Biochem.* **29**, 1–11.

Freitas, J. C. C., Bonagamba, T. J., and Emmerich, F. G. (1999). ^{13}C high-resolution solid-state NMR study of peat carbonization. *Energy and Fuels* **13**, 53–59.

Gerlach, R., Baumewerd-Schmidt, H., van den Borg, K., Eckmeier, E., and Schmidt, M. W. I. (2006). Prehistoric alteration of soil in the Lower Rhine Basin, Northwest Germany—archaeological ^{14}C and geochemical evidence. *Geoderma* **136**, 38–50.

Giovannini, G., Luchessi, S., and Cerevelli, S. (1983). Water-repellent substances and aggregate stability in hydrophobic soil. *Soil Sci.* **135**, 110–113.

Glaser, B., Guggenberger, G., Haumaier, L., and Zech, W. (1998a). Sustainable soils in the Brazilian Amazon. In *Humic Substances Downunder: Understanding and Managing Organic Matter in Soils, Sediments, and Waters*, 9th IHSS Conference, Adelaide, Australia.

Glaser, B., Haumaier, L., Guggenberger, G., and Zech, W. (1998b). Black carbon in soils: The use of benzenecarboxylic acids as specific markers. *Org. Geochem.* **29**, 811–819.

Glaser, B., Balashov, E., Haumaier, L., Guggenberger, G., and Zech, W. (2000). Black carbon in density fractions of anthropogenic soils of the Brazilian Amazon region. *Org. Geochem.* **31**, 669–678.

Glaser, B., Lehmann, J., and Zech, W. (2002). Ameliorating physical and chemical properties of highly weathered soils in the tropics with charcoal—A review. *Biol. Fertil. Soils* **35**, 219–230.

Glaser, B., and Amelung, W. (2003). Pyrogenic carbon in native grassland soils along a climosequence in North America. *Glob. Biogeochem. Cycl.* **17**, 1064.

Golchin, A., Clarke, P., Baldock, J. A., Higashi, T., Skjemstad, J. O., and Oades, J. M. (1997). The effects of vegetation and burning on the chemical composition of soil organic matter in a volcanic ash soil shown by ^{13}C NMR spectroscopy. I. Whole soil and humic acid fraction. *Geoderma* **76**, 155–174.

Goldammer, J. G. (1993). Historical biogeography of fire: Tropical and subtropical. In *Fire in the Environment, the Ecological, Atmospheric and Climatic Importance of Vegetation Fires*, Crutzen, P. J., and Goldammer, J. G., eds., John Wiley & Sons, Chichester, pp. 297–314.

González-Pérez, J. A., González-Vila, F. J., Almendros, G., and Knicker, H. (2004). The effect of fire on soil organic matter—A review. *Environ. Intern.* **30**, 855–870.

González-Vila, F. J., Tinoco, P., Almendros, G., and Martin, F. (2001). Pyrolysis–GC–MS analysis of the formation and degradation stages of charred residues from lignocellulosic biomass. *J. Agric. Food Chem.* **49**, 1128–1131.

Gustafsson, Ö., and Gschwend, P. M. (1998). The flux of black carbon to surface sediments on the New England continental shelf. *Geochim. Cosmochim. Acta* **62**, 565–472.

Gustafsson, Ö., Bucheli, T. D., Kukulska, Z., Andersson, M., Largeau, C., Rouzaud, J. N., Reddy, C. M., and Eglinton, T. I. (2001). Evaluation of a protocol for the quantification of black carbon in sediments. *Global Biogeochem. Cycl.* **15**, 881–890.

Hamer, U., Marschner, B., Brodowski, S., and Amelung, W. (2004). Interactive priming of black carbon and glucose mineralisation. *Org. Geochem.* **35**, 823–830.

Hammes, K., Schmidt, M. W. I., Smernik, R. J., et al. (2007). Comparison of quantification methods to measure fire-derived (black/elemental) carbon in soils and sediments using reference materials from soil, water, sediment and the atmosphere. *Global Biogeochem. Cycl.* **21**, doi: 10.1029/2006GB002914.

Haslam, S. F. I., Chudek, J. A., Goldspink, C. R., and Hopkins, D. W. (1998). Organic matter accumulation following fires in a moorland soil chronosequence. *Global Change Biol.* **4**, 305–313.

Haumaier, L., and Zech, W. (1995). Black carbon—Possible source of highly aromatic components of soil humic acids. *Org. Geochem.* **23**, 191–196.

Hernández, T., García, C., and Reinhardt, I. (1997). Short-term effect of wildfire on the chemical, biochemical and microbiological properites of Mediterranean pine forest soils. *Biol. Fertil. Soil* **25**, 109–116.

Hiridate, S., Nakadai, T., Shindo, H., and Yoneyama, T. (2004). Carbon source of humic substances in some Japanese volcanic ash soils determined by carbon stable isotope ratio δ^{13}C. *Geoderma* **119**, 133–141.

Hockaday, W. C., Grannas, A. M., Kim, S., and Hatcher, P. G. (2006). Direct molecular evidence for the degradation and mobility of black carbon in soils from ultrahigh-resolution mass spectral analysis of dissolved organic matter from a fire-impacted forest soil. *Org. Geochem.* **37**, 501–510.

Hopkins, M. S., Graham, A. W., Hewett, R., Ash, J., and Head, J. (1990). Evidence of late Pleistocene fires and eucalypt forest from a north Queensland humid tropical reainforest site. *Aust. J. Ecol.* **15**, 345–357.

Ikan, R. (1996). *The Maillard Reaction: Consequences for the Chemical and Life Science*, John Wiley & Sons, Chichester.

Ivany, L. C., and Salawitch, R. J. (1993). Carbon isotopic evidence for biomass burning at the K–T boundary. *Geology* **21**, 487–490.

Johnson, D. W., and Curtis, P. S. (2001). Effect of forest management on soil C and N storage: Meta analysis. *Forest Ecol. Manag.* **140**, 227–238.

Jones, T. P., and Chaloner, W. G. (1991). Fossil charcoal, its recognition and palaeoatmospheric significance. *Paleogeogr. Palaeoclimatol. Palaeoecol.* **97**, 39–50.

Knicker, H., and Lüdemann, H.-D. (1995). N-15 and C-13 CPMAS and solution NMR studies of N-15 enriched plant material during 600 days of microbial degradation. *Org. Geochem.* **23**, 329–341.

Knicker, H., and Skjemstad, J. O. (2000). Nature of organic carbon and nitrogen in physically protected organic matter of some Australian soils as revealed by solid-state ^{13}C and ^{15}N NMR spectroscopy. *Aust. J. Soil Res.* **38**, 113–127.

Knicker, H., Almendros, G., González-Vila, F. J., Martín, F., and Lüdemann, H.-D. (1996). ^{13}C- and ^{15}N-NMR spectroscopic examination of the transformation of organic nitrogen in plant biomass during thermal treatment. *Soil Biol. Biochem.* **28**, 1053–1060.

Knicker, H., González-Vila, F. J., Polvillo, O., González, J. A., and Almendros, G. (2005a). Fire-induced transformation of C- and N-forms in different organic soil fractions from a Dystric Cambisol under a Mediterranean pine forest (*Pinus pinaster*). *Soil Biol. Biochem.* **37**, 701–718.

Knicker, H., Totsche, K.-U., Almendros, G., and González-Vila, F. J. (2005b). Condensation degree of burnt peat and plant residues and the reliability of solid-state VACP MAS ^{13}C NMR spectra obtained from pyrogenic humic material. *Org. Geochem.* **36**, 1359–1377.

Knicker, H., Almendros, G., González-Vila, F. J., González-Pérez, J. A., and Polvillo, O. (2006). Characteristic alterations of quantity and quality of soil organic matter caused by forest fires in continental Mediterranean ecosystems: A solid-state ^{13}C NMR study. *Eur. J. Soil Sci.* **57**, 558–569.

Knicker, H., Hilscher, A., and Müller, P. (2007). How useful is chemical oxidation with dichromate for the determination of "black carbon" in fire-affected soils? *Geoderma* **142**, 178–196.

Knicker, H., Hilscher, A., González-Vila, F. J., and Almendros, G. (2008a). A new conceptual model for the structural properties of char produced during vegetation fires. *Organic Geochem.* **39**, 935–939.

Knicker, H., Wiesmeier, M., and Dick, D. P. (2008b). A simplified method for the quantification of pyrogenic organic matter in grassland soils via chemical oxidation. *Geoderma* **147**, 6974.

Kramer, R. W., Kujawinski, E. B., and Hatcher, P. G. (2004). Identification of black carbon derived structures in a volcanic ash soil humic acid by Fourier transform ion cyclotron resonance mass spectrometry. *Environ. Sci. Technol.* **38**, 3387–3395.

Krull, E. S., Skjemstad, J. O., Graetz, D., Grice, K., Dunning, W., Cook, G., and Parr, J. F. (2003). ^{13}C-depleted charcoal from C4 grasses and the role of occluded carbon in phytoliths. *Org. Geochem.* **34**, 1337–1352.

Krull, E. S., Swanston, C. W., Skjemstad, J. O., and McGowan, J. A. (2006). Importance of charcoal in determining the age and chemisttry of organic carbon in surface soils. *J. Geophys. Res.* **111**, doi: 10.1029/2006JG000194.

Kuhlbusch, T. A. J. (1995). Method for determining black carbon in residues of vegetation fires. *Environ. Sci. Technol.* **29**, 2695–2702.

Kuhlbusch, T. A. J. (1998). Black carbon and the carbon cycle. *Science* **280**, 1903–1904.

Kuhlbusch, T. A. J., and Crutzen, P. J. (1995). Toward a global estimate of black carbon in residues of vegetation fires representing a sink of atmospheric CO_2 and a source of O_2. *Global Biogeochem. Cycl.* **9**, 491–501.

Kuhlbusch, T. A. J., Andreae, M. O., Cachier, H., Goldammer, J. G., Lacaux, J.-P., Shea, R., and Crutzen, P. J. (1996). Black carbon formation by savanna fires: Measurements and implications for global carbon cycle. *J. Geophys. Res.* **101**, 23651–23665.

Kutiel, P., Lavee, H., Segev, M., and Benyamini, Y. (1995). The effect of fire-induced surface heterogeneity on rainfall–runoff–erosion relationships in an eastern Mediterranean ecosystem, Israel. *Catena* **25**, 77–87.

Lehmann, J., Liang, B., Solomon, D., Lerotic, M., Luizao, F., Kinyangi, J., Schäfer, T., Wirick, S., and Jacobsen, C. (2005). Near-edge X-ray absorption fine structure (NEXAFS) spectroscopy for mapping non-scale distribution of organic carbon forms in soil: Application to black carbon particles. *Global Biogeochem. Cycl.* **19**, doi: 10.1029/2004GB002435.

Leifeld, J. (2007). Thermal stability of black carbon characterised by oxidative differential scanning calorimetry. *Org. Geochem.* **38**, 112–127.

Lide, D. R., ed. (2004). *CRC Handbook of Chemistry and Physics*, CRC Press, Boca Raton, FL.

Lopez-Capel, E., Bol, R., and Manning, D. A. C. (2005a). Application of simultaneous thermal analysis mass spectrometry and stable carbon isotope analysis in a carbon sequestration study. *Rapid Comm. Mass. Spect.* **19**, 3192–3198.

Lopez-Capel, E., Sohi, S., Gaunt, J. L., and Manning, D. A. C. (2005b). Use of thermogravimetry–differential scanning calorimetry to characterize modelable soil organic matter fractions. *Am. J. Soil Sci. Soc.* **69**, 3192–3198.

Lynch, J., Clark, J. S., and Stocks, B. J. (2004). Charcoal production, dispersal, and deposition from the Fort Providence experimental fire: Interpretating fire regimes from charcoal records in boreal forests. *Can. J. Forest Res.* **34**, 1642–1656.

Maie, N., Knicker, H., Watanbe, A., and Kimura, M. (2006). Heterocyclic N in the highly humified humic acids extracted from subsoil of paddy fields and surface ando soils. *Org. Geochem.* **37**, 12–19.

Martin, G. J., Martin, M. L., and Gouesnard, J. P., eds. (1981). *^{15}N-NMR Spectroscopy. NMR Basic Principles and Progress*, Vol. **18**, Springer, Heidelberg.

Masiello, C. A. (2004). New directions in black carbon organic geochemistry. *Mar. Chem.* **92**, 201–213.

Masiello, C. A., and Druffel, E. R. M. (1998). Black carbon in deep-sea sediments. *Science* **280**, 1911–1913.

Masiello, C. A., and Druffel, E. R. M. (2003). Organic and black carbon ^{13}C and ^{14}C through the Santa Monica Basin oxic–anoxic transition. *Geophys. Res. Lett.* **30**, 34/1–34/4.

Masiello, C. A., Druffel, E. R. M., and Currie, L. A. (2002). Radiocarbon measurements of black carbon in aerosols and ocean sediments. *Geochim. Cosmochim. Acta* **66**, 1025–1036.

McGrath, T. E., Chan, W. G., and Hajaligol, R. (2003). Low temperature mechanism for the formation of polycyclic aromatic hydrocarbons from the pyrolysis of cellulose. *J. Anal. Appl. Pyrolysis* **66**, 51–70.

Middelburg, J. J., Nieuwenhuize, J., and van Breugel, P. (1999). Black carbon in marine sediments. *Mar. Chem.* **65**, 245–252.

Neary, D. G., Klopatek, C. C., DeBano, L. F., and Fgolliott, P. F. (1999). Fire effects on belowground sustainability: A review and synthesis. *Forest Ecol. Manage.* **122**, 51–71.

Nguyen, T. H., Brown, R. A., and Ball, W. P. (2004). An evaluation of thermal resistance as a measure of black carbon content in diesel soot, wood char, and sediment. *Org. Geochem.* **35**(3), 217–234.

Ohlsen, M., and Tryterud, E. (2000). Interpretation fo the charcoal record in forest soils: Forest fires and their production and deposition of macroscopic charcoal. *Holocene* **10**, 519–525.

Preston, C. M., Trofymow, J. A., Niu, J., and Fyfe, C. A. (2002). Harvesting and climate effects on organic matter characteristics in British Columbia coastal forests. *J. Environ. Qual.* **31**, 402–413.

Preston, C. M., and Schmidt, M. W. I. (2006). Black (pyrogenic) carbon: a synthesis of current knowledge and uncertainties with special consideration of boreal regions. *Biogeosciences* **3**, 397–420.

Prosser, I. P., and Williams, L. (1998). The effect of runoff and erosion in native eucalyptus forest. *Hydrol. Process.* **12**, 251–265.

Raison, R. J. (1979). Modifications of the soil environment by vegetation fires: A review. *Plant and Soil* **51**, 73–108.

Ralph, J., and Hatfield, R. D. (1991). Pyrolysis–GC–MS characterization of forage materials. *J. Agric. Food Chem.* **39**, 1426–1437.

Reeves, J. B., and Francis, B. A. (1998). Pyrolysis–gas chromatography for the analysis of proteins: With emphasis and forages. In *Nitrogen-Containing Macromolecules in the Bio- and Geoshpere*, Stankiewicz, B. A., and van Bergen, P. F., eds., American Chemical Society, Washington, D.C., pp. 47–62.

Robichaud, P. R., and Hungerford, R. D. (2000). Water repellency by laboratory burning of four northern Rocky Mountain forest soils. *J. Hydrol.* **231–232**, 207–219.

Rollins, M. S., Cohen, A. D., and Durig, J. R. (1993). Effects of fires on the chemical and petrographic composition of peat in the Snuggedy Swamp, South Carolina. *Int. J. Coal Geol.* **22**, 101–117.

Rumpel, C., Alexis, M., Chabbi, A. V. C., Rasse, D. P., Valentin, C., and Mariotti, A. (2006). Black carbon contribution to soil organic matter composition in tropical sloping land under slash and burn agriculture. *Geoderma* **130**, 34–45.

Saldarriaga, J. G., and West, D. C. (1986). Holocene fires in the northern Amazon basin. *Quaternary Res.* **26**, 358–366.

Savage, S. M. (1974). Mechanism of fire-induced water repellency in soil. *Soil Sci. Am. Proc.* **38**, 652–657.

Schiffman, P. M., and Johnson, W. C. (1989). Phytomass and detrital carbon storage during forest regrowth in the southeastern United States Piedmont. *Can. J. Forest Res.* **19**, 69–78.

Schmidt, M. W. I., Skjemstad, J. O., Gehrt, E., and Kögel-Knabner, I. (1999). Charred organic carbon in German chernozemic soils. *Eur. J. Soil Sci.* **50**, 351–365.

Schmidt, M. W. I., and Noack, A. G. (2000). Black carbon in soils and sediments: Analysis, distribution, implications, and current challenges. *Global Biogeochem. Cycl.* **14**, 777–793.

Sergides, C. A., Chughtai, A. R., Smith, D. M., et al. (1987). The structure of hexane soot, Part II: Ozonation studies. *Appl. Spectrosc.* **41**, 482–492.

Sharma, R. K., Chan, W. G., Seeman, J. I., and Hajaligol, M. R. (2003). Formation of low molecular weight heterocycles and polycyclic aromatic compounds (PACs) in the pyrolysis of α-amino acids. *J. Anal. Appl. Pyrolysis* **66**, 97–121.

Sharma, R. K., Wooten, J. B., Baliga, V. L., Lin, X., Chan, W. G., and Hajaligol, M. R. (2004). Characterization of chars from pyrolysis of lignin. *Fuel* **83**, 1469–1482.

Shindo, H., Honna, T., Yamamoto, S., and Honma, H. (2004). Contribution of charred plant fragments to soil organic carbon in Japanese volcanic ash soils containing black humic acids. *Org. Geochem.* **35**, 235–241.

Simpson, M. J., and Hatcher, P. G. (2004). Determination of black carbon in natural organic matter by chemical oxidation and solid-state 13C nuclear magnetic resonance spectroscopy. *Org. Geochem.* **35**, 923–935.

Skjemstad, J. O., Le Feuvre, R. P., and Prebble, R. E. (1990). Turnover of soil organic matter under pasture as determined by ^{13}C natural abundance. *Aust. J. Soil Sci. Res.* **28**, 267–276.

Skjemstad, J. O., Janik, L. J., Head, M. J., and McClure, S. G. (1993). High energy ultraviolet photo-oxidation: a novel technique for studying physically protected organic matter in clay- and silt-sized aggregates. *J. Soil Sci.* **44**, 485–499.

Skjemstad, J. O., Clarke, P., Taylor, J. A., Oades, J. M., and McClure, S. G. (1996). The chemistry and nature of protected carbon in soil. *Aust. J. Soil Res.* **34**, 251–271.

Skjemstad, J. O., Taylor, J. A., and Smernik, R. J. (1999). Estimation of charcoal (char) in soils. *Commun. Soil Sci. Plant Anal.* **30**, 2283–2298.

Skjemstad, J. O., Reicosky, D. C., Wilts, A. R., and McGowan, J. A. (2002). Charcoal carbon in U.S. agricultural soils. *Soil Sci. Soc. Am. J.* **66**, 1249–1255.

Smernik, R. J., Skjemstad, J. O., and Oades, J. M. (2000). Virtual fractionation of charcoal from soil organic matter using solid state ^{13}C NMR spectral editing. *Aust. J. Soil Res.* **38**, 665–683.

Smernik, R. J., Baldock, J. A., Oades, J. M., and Wittaker, A. K. (2002). Determination of $T_{1\rho H}$ relaxation rates in charred and uncharred wood and consequences for NMR quantitation. *Solid State Nucl. Magn. Reson.* **22**, 50–70.

Song, J., Peng, P., and Huang, W. (2002). Black carbon and kerogen in soils and sediments. 1. Quantification and characterization. *Environ. Sci. Technol.* **36**, 3960–3967.

Svenik, J., Imeson, A. C., and Verstraten, J. M. (1989). Humus form development and hillslope runoff, and the effects of fire and management under mediterranean forest in NE-Spain. *Catena* **16**, 461–475.

Teeäär, R., and Lippmaa, E. (1984). Solid state ^{13}C NMR of cellulose. A relaxation study. *Polym. Bull.* **12**, 315–318.

Trompowsky, P. M., Benites, V. M., Madari, B. E., Pimenta, S. A., Hockaday, W. C., and Hatcher, P. G. (2005). Characterization of humic like substances obtained by chemical oxidation of eucalyptus charcoal. *Org. Geochem.* **36**, 1480–1489.

Turney, C. S. M., Wheeler, D., and Chivas, A. R. (2006). Carbon isotope fractionating in wood during carbonization. *Geochim. Cosmochim. Acta* **70**, 960–964.

Venkatesan, M. I., and Dahl, J. (1989). Organic geochemical evidence for global fires at the Cretaceous/Tertiary boundary. *Nature* **338**, 57–60.

Walsh, R. P. D., Boakes, D., Coelho, C. O. A., Goncalves, A. J. B., Shakesby, R. A., and Thomas, A. D. (1994). Impact of fire-induced hydrophobicity and postfire forest litter on overland flow in northern and central Portugal. In Viegas, D. X., ed., *Second International Conference on Forest Fire Research*, Coimbra, Portugal, 21–24 November 1994.

Willcox, G. (1999). Charcoal analysis and Holocene vegetation history in southern Syria. *Quat. Sci. Rev.* **18**, 711–716.

Wittenberg, U., Heimann, M., Esser, G., McGuire, A. D., and Sauf, W. (1998). On the influence of biomass burning on the seasonal CO_2 signal as observed at monitoring stations. *Global Biogeochem. Cycl.* **12**, 531–544.

Wolbach, W. S., Gimour, I., Anders, E., Orth, C. J., and Brooks, R. R. (1988). Global fire at the Cretaceous-Tertiary boundary. *Nature* **334**, 665–669.

Wolbach, W. S., and Anders, E. (1989). Elemental carbon in sediments: determination and isotopic analysis in the presence of kerogen. *Geochim. Cosmochim. Acta* **53**, 1637–1647.

Wolff, E. W., and Cachier, H. (1998). Concentrations and seasonal cycle of black carbon in aerosol at a coastal Antarctic station. *J. Geophys. Res.* **103**, 11033–11041.

Wondrack, L., Szanto, M., and Wood, W. A. (1989). Depolymerization of water soluble coal polymer from subbituminous coal and lignite by lignin peroxidase. *Appl. Biochem. Biotechnol.* **20/21**, 765–780.

Zhukov, A. B. (1976). The impact of anthropogenic factors on forest biogeocenosis in Siberia. *Ecol. Bull.* **21**, 41–45.

8

BIOLOGICAL ACTIVITIES OF HUMIC SUBSTANCES

S. NARDI, P. CARLETTI, AND D. PIZZEGHELLO

Dipartimento di Biotecnologie Agrarie, University of Padova, Legnaro, Padova, Italy

A. MUSCOLO

Dipartimento Gestione dei Sistemi Agrari e Forestali, University of Reggio Calabria, Reggio Calabria, Italy

8.1. INTRODUCTION

The presence of humic substances (HS) in the environment has long been recognized (Kononova, 1966; Schnitzer and Khan, 1972; Orlov, 1985; Frimmel and Christman, 1988). The term *humic substances* refers to a category of naturally occurring materials found in, or extracted from, soils, sediments, and natural waters and constitute one of the most abundant forms of organic matter (OM) on the surface of the earth (Woodwell and Houghton, 1977; Woodwell et al., 1978).

Biophysico-Chemical Processes Involving Natural Nonliving Organic Matter in Environmental Systems, Edited by Nicola Senesi, Baoshan Xing, and Pan Ming Huang

Soil fertility depends on the organic matter content, which can vary from less than 1% in young and overworked soils to over 95% in some deep peats. Humic substances influence the chemical and physical properties of the soil and its overall health. In fact these molecules participate in many agronomic, environmental, and geochemical processes (Hayes and Swift, 1978; Stevenson, 1982). Their composition and breakdown rate affect (a) the soil structure and porosity, (b) the water infiltration rate and moisture holding capacity of soils, and (c) the diversity and biological activity of soil organisms. The variety and extent of these reactions and interactions indicate the highly reactive nature of HS, thus highlighting the overall complexity of the plant–soil system.

In addition to the effects exerted by HS on soil properties, HS can influence plant metabolism and morphology by interacting with a variety of biochemical mechanisms and physiological processes, thereby stimulating growth and increasing the total amount of nutrients taken up by the plant (Vaughan and Malcolm, 1985). A considerable amount of literature shows that under laboratory and field conditions, HS can have a beneficial effect on plant growth as measured in terms of increases in lengths, fresh and dry weights of shoots and roots, leaf chlorophyll concentration, number of lateral root initials, and many other biological parameters. Therefore, it has been suggested that these compounds may have a fundamental influence not only on the overall soil fertility and conservation, but also on the physiology of plants (Varanini and Pinton, 2001).

This chapter will approach the subject by introducing a brief overview of the features and definitions of humic substances and then presenting a short review of the history of humus, from the earliest studies to the mid-1980. Then it will focus on the work of Vaughan and Malcom, thereby introducing the modern approaches to the comprehension of the cross-talk between plant and soil. Recent studies will be reviewed with regard to three aspects: the morphological changes in response to humic substances in plant cell cultures and tissues; the modifications induced in macro- and micro-nutrient uptake; and the effects reported on biochemical pathways and processes. To conclude, current knowledge will be summarized and the perspectives for future research will be presented.

8.2. AN OVERVIEW ON DEFINITIONS, FEATURES, PROPERTIES, AND FUNCTIONS OF HUMIC SUBSTANCES

The word *humus* is Latin for "earth, soil, terrain," probably deriving from *humi*, "on the ground," or *humilis*, which means "low, humble." In soil science this term is sometimes used synonymously with soil OM—that is, to denote the organic material in the soil, including humic substances (MacCarthy et al., 1990). In other cases, humus is used to represent only HS (Stevenson, 1982). In order to avoid any misunderstanding in this work, the word *humus* will be avoided or it will be used as a synonym of humic substances.

Humic substances account for various definitions according to different authors, and also its definition changes in literature through the years. These substances are conventionally defined as "a series of relatively high-molecular-weight, yellow-to black-coloured substances, formed by secondary synthesis reactions" (Stevenson, 1982; Soil Science Society of America, 1996) or as "a category of naturally occurring, biogenic, heterogeneous organic substances that can generally be characterized as being yellow-to-black in colour, of high molecular weight (HMW), and refractory" (Aiken et al., 1985). As pointed out by MacCarthy et al. (1990), these statements are really more descriptions of HS rather than definitions, and they teach little about the chemical nature of humic materials. Trying to include the chemistry of these molecules, Andreux (1996) presented HS as "random polymeric, amorphous macromolecules formed by polyaromatic building blocks bridged to each other by ester, ether and C links and carrying variable proportions of carboxyl, hydroxyl, amino and other hydrophilic groups."

Many authors believe that it is essential to have a basic understanding of the compositions and structures of HS in order to make better predictions of their interactions in the soil and water environments such as association/dissociation processes, binding and trapping of biological and anthropogenic materials, and many other biological functions (Schulten and Leinweber, 2000).

Three different viewpoints on the humic substances structural conformation are actually reported in the literature. One suggests that HS are macromolecular and assume random coil conformations in solution (Swift, 1999); a second proposes that HS are molecular associations of relatively small molecules held together by weak interaction forces, thus forming a supramolecular structure (Piccolo and Conte, 1999); a third considers that HS are in solution as micelles or "pseudomicellar" structures (Wershaw, 1999). Viewpoints two and three could be broadly considered to be under the same umbrella (Clapp and Hayes, 1999).

The standpoint of Swift (1989) held assumption that HS are macromolecular polymers in the mode of other macromolecules of nature, such as proteins, polysaccharides, nucleic acids, and lignin. This view has been rationalized by the hypothesis that humus synthesis is based on either the lignin or the polyphenolic theories (Flaig et al., 1975). These theories rely on the supposition that progressive polymerization of humic matter takes place via covalent bonding and that the processes are often mediated by soil enzymes and abiotic catalysts such as metal oxides, primary minerals, and layer silicates (Bollag et al., 1998). This representation has fostered in the scientific community the perception that HS are macromolecular polymers in which simple (though heterogeneous) monomeric units progressively build up into high-molecular-weight polymers by random condensation and oxidation processes. In this model of humus formation, the randomness of the covalent polymerization of monomers accounted for the observed large polydispersity of humic macromolecules. Furthermore, the multiple conformational foldings that a polymeric chain, either linear or branched, would assume in the soil environment provide a plausible explanation for its resistance to microbial degradation and the consequent long residence time observed for soil humic components (Insam, 1996). These conclusions were drawn on the basis of results obtained, among others, through molecular shape and size analyses by ultracentrifugation and computation of frictional ratios of various HS [Cameron et al. (1972a,b), revised in Swift (1999)].

Based on the idea of HS as a macromolecular polymer, some chemical model structures have been proposed combining geochemical, wet-chemical, biochemical, spectroscopic, thermal, agricultural, and ecological data with analytical pyrolysis and modern computational chemistry. An example is shown in Schulten and Leinweber (2000).

The supramolecular model of HS is based on results obtained by low-pressure size exclusion chromatography (SEC), high pressure size exclusion chromatography (HPSEC) and spectroscopic techniques. The supporters of this theory reported that the 280 nm absorption of humic acids (HA) was reversibly shifted from high to low molecular size ranges when organic acids (for example acetic acid) were added to lower the pH of a humic solution from 9.2 to 2 before the elution in a 0.02 mol liter^{-1} alkaline borate buffer (Dell'Agnola and Nardi, 1987; Piccolo et al., 1996a,b). To explain their results, they suggested that, instead of being stable polymers, HS at neutral or alkaline pH values are supramolecular associations of relatively small heterogeneous molecules held together by weak dispersive forces. Moreover, if HS are seen as weakly bound supramolecular associations, their unstable conformations could then be stabilized in real polymeric structures. This could be achieved by increasing the number of intermolecular covalent bonds via an oxidative coupling reaction catalyzed by oxidative enzymes such as the phenoloxidases. Piccolo et al. (2000) turned a loosely bound humic superstructure into a covalently linked polymer. The polymerized HS had high-pressure size exclusion chromatography (HPSEC) absorptions of larger intensities, and the elution volumes were shifted to lower values than the control. Subsequent treatment with acetic acid did not alter significantly the chromatographic appearance of the polymerized HS, whereas it produced disruption of the loosely bound association of the untreated HS resulting from a significant reduction of the intensities of the peaks and their shifts to larger elution volumes. These results were interpreted supporting the supramolecular structure theory. In this concept, one can imagine HS to be relatively small and heterogeneous molecules of various origin that self-organize in supramolecular conformations. Humic superstructures of relatively small molecules are not associated by covalent bonds but are stabilized only by weak forces such as dispersive hydrophobic interactions (van der Waals, π–π, and CH–π bondings) and hydrogen bonds, the latter being progressively more important at low pH values. In humic supramolecular organizations, the intermolecular forces determine the conformational structure of HS, and the complexities of the multiple noncovalent interactions control their environmental reactivity (Piccolo, 2001). Even if still strongly criticized (Swift, 1999; De Nobili and Chen, 1999), this theory is gaining importance in soil science literature (Burdon, 2001; Sutton and Sposito, 2005), and many authors consider that humic substances are both macromolecules and supramolecules.

Operationally, it is common to define HS in terms of the methods used to extract or isolate them from soils, sediments, and natural waters. The classic soil extraction procedure yields three main fractions: humic acid [also defined as high-molecular-weight (HMW) or high-molecular-size (HMS) fraction], fulvic acid (FA) [also defined as low-molecular-weight (LMW) or low-molecular-size (LMS) fraction], and humin. These fractions are defined in terms of their solubility in aqueous media as a function of pH or in terms of their extractability from soils or sediments as a function of the pH of the extracting medium. Humic acid is the fraction of HS that is not soluble in water under acidic conditions, but becomes soluble (or extractable) at higher pH values. Fulvic acid is the fraction that is soluble in aqueous media at all pH values. Humin represents the fraction that is not soluble in an aqueous medium (or is not extractable with an aqueous medium) at any pH value. Actually, humin consists of an aggregate of humic and nonhumic materials (Rice and Mac-

Carthy, 1990); thus, humin is better described as a humic-containing material rather than as a humic substance. The adjective humic is commonly used in a generic way to refer to each of these fractions. It is assumed that free, identifiable constituents such as amino acids, sugars, and polysaccharides, which are co-isolated with the humic material, are removed before the extracted materials are considered to be exclusively humic as distinct from humic-containing. Complete segregations of this type are more readily hoped for than accomplished. Further steps are then taken to wash the HA free of other materials, to separate the FA per se from the other materials in the fulvic acid fraction, to diminish the ash contents of the humic and fulvic extracts, and to fully convert the fractions to their hydrogen forms. Unfortunately, there is no definitive method for absolutely separating all non-humic material from HS. Consequently, some pragmatic compromises must always be made in the extraction and isolation of HS (MacCarthy, 2001).

There are numerous variations of the extraction procedure including: the nature and concentration of the extractant used; the temperature at which the extraction is performed; period of contact with base; steps taken to minimize ash content of extracted products; and the choice of aerobic versus anaerobic conditions during the extraction (MacCarthy, 2001).

All HS are amorphous and consist of complex mixtures. No study has come close to isolating a significant amount of any material that could be referred to as a pure or nearly pure HS. Therefore, most data on HS refer to average properties of a large ensemble of diverse molecules. The precise properties of a given humic extract may depend on the particular substrate chosen and the specific conditions of extraction. Nevertheless, there is a remarkable uniformity in the average properties of all HA, FA, and humins (Schnitzer, 1977; Rice and MacCarthy, 1991). Elemental contents of HA, FA, and humins from all over the world are remarkably consistent (Rice and MacCarthy, 1991). HAs have been reported to have average molecular weights (MW) varying from about 2000 Da for aquatic materials to greater than 10^6 Da for soil-derived materials (Aiken and Wershaw, 1985), and some FAs have a number average MW in the range of about 600–900 Da.

Humic substances have an abundance of oxygen-containing functional groups (carboxyl, phenolic, alcoholic) that dominate their chemical properties. These groups and their relative chemistry are mainly responsible for most of the functions ascribed to HS. The relationship among HS structure, functional groups, and their activity have been previously presented (Nardi et al., 2002), and recent papers on this will be presented in a following paragraph.

The molecular heterogeneity that is characteristic of HS serves a vital role in the ecological system (MacCarthy, 2001). Humic substances constitute the only natural organic material that can survive in bulk and still possess the required chemical reactivity to perform the various functions for sustaining soil quality and promoting plant growth. By virtue of its molecular heterogeneity, this medium is highly biorefractory, but it still possesses the reactive functional groups needed to perform ecologically and environmentally vital tasks. Performing laboratory studies on a group of substances that have been segregated from the other living and nonliving entities in the environment, compared with studying the intact organic/inorganic composites in the presence of microbial communities may appear as an extreme simplification of the rhizosphere environment. However, it seems that such laboratory investigations on isolated humic fractions provide the best hope for unraveling some of the fundamental chemical mysteries regarding the nature of these materials (MacCarthy, 2001).

8.3. HISTORICAL OVERVIEW

For over 8000 years, humans has realized that dark soils are usually productive and that color and productivity are commonly associated with the organic material derived from decaying plant and animal remains (Allison, 1973). Aristotle (384–322 B.C.) is often reported as being the first to suggest that plants absorb their food from soils in an elaborated form by processes similar to those found in animals. Plinio the old (23–79 A.D.) reported in the *Naturalis Historia* that soil produces a kind of manure from itself, thus giving nutrients to plants. The general meaning was also known by the Carthaginian Magone (4th century B.C.), and subsequently by Catone (234–149 B.C.), Varrone (116–27 B.C.) and Columella (1st century A.D.).

These observations later stimulated the general belief that prevailed through the earlier decades of the 19th century—namely, that humus is the only or the major soil product supplying nutrients to plants. The direct utilization of humus by plants was fully developed by Thaer (1808, 1846), who stated that humus comprises a more or less considerable portion of soil; fertility of the soil depends largely upon it since, besides water, humus is the only material that supplies nutrients to plants. This concept was referred to as the humus theory.

A more critical assessment of the role of humus in soil fertility was made by De Saussure (1804), who established that plants could synthesize organic substances using atmospheric CO_2 and water. Successively, Liebig, in a number of publications (Liebig, 1841, 1856), supported other evidence against the humus theory providing fundamental information on the role of minerals in plant nutrition. The controversy between the humus and the mineral theories has continued into the 20th century. Lawes and Gilbert (1905) first demonstrated that soil fertility could be preserved for many years by applying only mineral fertilizers. Meanwhile, Bottomley (1914a,b, 1917, 1920) asserted that small quantities of humic substances favored plant growth. However, Clark and Roller (1924), although working closely with Bottomley, found that humic substances were not essential for plant growth.

In recent years, other researchers have verified a positive effect of HS on plant growth (Niklewski, 1931; Flaig, 1953, 1956; Kononova, 1956; Rerabek, 1960; Chaminade, 1958, 1966). Prát and Pospíšil (1959) demonstrated the accumulation of humic acids in roots of sugar beets and corn by using ^{14}C-labeled materials. However, the amount of radioactivity passing from the roots to the shoots was minimal. Such limited upwards transport has been also confirmed by Führ and Sauerbeck (1967a,b) as well as by Vaughan and Linehan (1976). In 1981, Vaughan and Ord demonstrated that low-molecular-weight fractions are taken up both actively and passively, whereas high-molecular-weight matter is only taken up passively.

8.4. THE VAUGHAN AND MALCOM MILESTONE: TOWARDS A MODERN APPROACH OF HUMIC SUBSTANCES STUDIES

In 1985 the publication of *Soil Organic Matter and Biological Activity*, edited by Vaughan and Malcom, became the basis for modern studies on the interactions between plants and humic substances. The text, in fact, resumes the scientific work carried out in the 19th century and addresses some of the questions that are as yet unresolved today. Among the modifications induced by HS on treated plants, changes in size and development were the first to be studied analytically. The authors state that under certain conditions, HS can stimulate plant growth in terms of increase in plant length, dry or fresh weight (Blanchet, 1958; Gumiński, 1968). These effects depend on (a) the concentration (Elgala et al., 1978) and the source of the substance (Hernando et al., 1977), (b) the plant species (Blanchet, 1958; Gumiński, 1968) and age, and (c) the culture conditions of the trial. It has been also reported that different organs of intact plants respond to humic substances to varying extents.

The composition of the culture medium and the subsequent culture conditions are critical for the extent to which humic substances may influence plant growth. For instance, in some cases HS appeared to be capable of a mitigation of the toxicity effects in case of toxic ion concentrations (Elgala et al., 1978). In addition to increasing the lengths and fresh and dry weights, humic substances can exert a favorable effect on the development of adventitious roots in nutrient culture, thus influencing not only the usual growth of the plant organs but also their morphology.

The positive effects of humic substances often occur when the plants are grown in nutrient solution under anexic conditions in thin film isolators. This shows that the beneficial effects are due to the humic substances *per se*, rather than being mediated by microbial breakdown products.

Moreover, Vaughan and Malcom (1985) also highlighted that, taking all these factors into consideration, it is hardly surprising that there should be wide discrepancy in the literature ranging from no effect, on the one hand, to several hundred percent, on the other.

Another effect described is an increased rate of seed germination, which a member of our group (Visser, 1986) tested and confirmed herself on wheat and pea seeds, unaccompanied by a concomitant increase in the percentage of germination. Moreover, the issue of the uptake of HS into plant tissues was already addressed in the work of Vaughan and Malcom (1985). Some work has been reported (i.e., Prát and Pospíšil, 1959) demonstrating the actual uptake and incorporation of humic substances using ^{14}C-labeled materials. A small fraction of radioactivity has been transported from root to shoot (Vaughan and Linehan, 1976). This research also pointed out that different HS fractions show different behaviors toward their uptake. Humic acids seem to be adsorbed at the root surface or accumulate in the "apparent free space" (Vaughan and Ord, 1981), while at the cellular level it is mainly humic substances of low MW which accumulate (Vaughan et al., 1974b).

A beneficial effect of humic substances on the nutrient uptake and contents of plants has been reported for the major inorganic elements, such as nitrogen, phosphorus, potassium (Mylonas and McCants, 1980), and sulfur (Gumiński, 1968). In addition, the uptake and contents of nutrients such as calcium, magnesium (Mylonas and McCants, 1980), sodium (Vaughan and McDonald, 1976) and copper (Rauthan and Schnitzer, 1981) are also enhanced by humic substances. Most of these reports on the effects of HS on the nutrient contents of plants are purely descriptive, and little attempt has been made to elucidate the mechanisms of the action of the humic material. The authors proposed both an indirect and a direct effect of HS on plant nutrition. In the former case HS may, for example, chelate a cation, thus changing its concentration in growth solution, while in the case of a direct effect, the humic material may influence the permeability of plant membranes or interfere with the active ion uptake carriers and mechanism.

The effects of HS on plant metabolic processes have also been extensively reviewed. For instance, there are many reports that show that humic acids enhance respiration in higher plants (Sladký, 1959; Vaughan, 1967a), and some reports indicate that fulvic acids can sometimes evoke a greater response than humic acids (i.e., Sladký and Tichý, 1959). These results have been interpreted in varying ways. The possibility that HS-induced stimulation could depend on the property of these substances to act as substrates or respiratory chain catalysts is no longer acceptable. In addition, the stimulation of O_2 consumption is only on the order of 25–30% and obtained with intact plants such as tomato (Sladký, 1959) or beet slices (Vaughan, 1967b). The manner in which HS influence respiration has been the subject of considerable speculation and nowadays represents an unresolved controversy.

The effect of HS on photosynthesis suffers from lack of information as the reports focused on the chlorophyll content, which, in turn, does not necessary mean an increase in plant yield (Thomas et al., 1978).

Many reports are presented in the text confirming the effect of HS on both protein synthesis and enzyme activity. HS influence the production of various enzymes—for example, catalase, cytochrome, invertase, and peroxidase (Vaughan et al., 1974a). The use of inhibitors of protein synthesis provides strong evidence in favor of the notion that HA stimulates *de novo* synthesis of invertase and peroxidase (Vaughan and Malcolm, 1979a). Meanwhile, HA were reported to have no effect on the synthesis of phosphatise; thus it influences only some aspects of protein synthesis, and this most likely operates via the formation of new m-RNA.

Most of the reports on enzyme activity displayed analyze the effect of HS on several enzymes present in plant tissue homogenates (Vaughan and Malcolm, 1979b; Malcolm and Vaughan, 1978). Generally, only inhibitions were observed. But these studies may not represent the true nature of what happens when whole plants are grown in solutions of humic substances. Whether the assumption could be made that HA influence enzymes generally in intact tissues awaited further clarification. This aspect has been later explained in many studies, as reported in this text.

The conclusions drawn by Vaughan et al. (1985) highlighted the following:

- An influence on membrane permeability and protein carriers of ions resulting in a more rapid and selective entry of essential elements into the root.
- Activation of respiration and Krebs cycle with a concomitant increase in ATP production.
- An increase in chlorophyll content and photosynthesis giving rise to an enhanced formation of ATP, amino acids, carbohydrates, and proteins.
- An effect on nucleic acid synthesis in which not only the amount of RNA but also the transcription of mRNA is influenced.
- A selective effect on protein synthesis influencing the relative amounts of enzymes, ion carriers, and structural proteins produced.
- An effect on enzyme activity, an inhibition or stimulation depending on the enzyme and its source.

Some of these concepts were the seeds for the development of subsequent research on the biological activity of HS. In particular, recent studies take the effects on plant growth in terms of size for granted, focusing onto the physiological aspect of plant development. Three main strands stand out in the publications of the past 20 years, taking in account of various aspects of the matter. In plant and tissues the morphological changes in response to humic substances have been considered. Another viewpoint concentrated on the macro- and micro-nutrient uptake modifications due to HS treatment. And finally, many authors demonstrated the effects of humic substances on biochemical pathways and processes.

8.5. MORPHOLOGICAL CHANGES IN RESPONSE TO HUMIC SUBSTANCES

The most evident and common effects exerted by HS are the responses on plant growth. Many studies in recent years have confirmed the observations made in this direction by Vaughan and Malcom (1985), who reported data considering various parameters and different treatments with HS extracted from many natural and anthropogenic sources tested against a number of higher plant species.

In Van de Venter et al. (1991), coal-derived sodium humate was found to stimulate primary root growth of seedlings of cantaloupe (*Cucumis melo* L.), lettuce (*Lactuca sativa* L.), and onion (*Allium cepa* L.). Growth enhancement was not due to the release of nutrient elements by the product; and, in the case of lettuce and onion, it was not due to increased availability or uptake of mineral elements. In another paper (Piccolo et al., 1993) lettuce and tomato seeds were treated in anexic conditions with a humic acid derived from an oxidized coal. The fresh weight of total seedlings and per seedling increased in treatments for both species. More recently, a series of papers evaluated the growth responses to vermicomposts and HS extracted from it. In different plant species (Arancon et al., 2004; Atiyeh et al., 2000, 2001, 2002a,b), the results reported support the known activity of humic compounds. Moreover, in maize (*Zea mays* L.) seedlings treated for 7 days with different HA concentrations, elongation of roots was stimulated, resulting in enhanced root surface area (Canellas et al., 2002).

Other evidence on morphological modifications attributed to HS treatment concern root hair development. In this sense, Concheri et al. (1994) identified a strong proliferation of root hairs compared to control plants in wheat root supplemented with HS up to a concentration of 10 mg C liter^{-1}, with best results for HMS and LMS obtained with 5 and 0.1 mg C liter^{-1}, respectively.

Also, Schmidt et al. (2005) found a significant increase in root hair density by working with *Arabidopsis thaliana*, which were treated with water extractable humic substances (WEHS), suggesting that these substances induce a "nutrient acquisition response" that favors the uptake of nutrients via an increase in the absorptive surface area. Furthermore, a phenotypical analysis of an array of mutants harbouring defects in root epidermal patterning revealed that root hair density of the ttg and gl2 mutants, defective in cell specification, was significantly modified, indicating an effect at/or downstream of the determination of the cells.

The elongation/differentiation zone of the principal root includes small, densely meristematic cells that are in continuous metabolic activity and are more susceptible to lateral root formation. In maize the proliferation of the mitotic sites in this zone of roots treated or not treated with 40 mg liter^{-1} earthworm HA has been measured (Canellas et al., 2002). After 3 days of exposure to HA in the growth medium, the treatment clearly stimulated the number of sites of lateral root emergence to a level ranging from 7 to 12 times the control values.

Pinton et al. (1999a) showed the effects of HS on the root apparatus of 4-day-old maize seedlings exposed for 24 h to a nutrient solution, containing 200 μmol liter^{-1} nitrate, in the presence (WEHS) or absence (Control) of the water-extractable humic substances fraction at a final concentration of 5 mg C liter^{-1}. The root apparatus of control plants consisted of a primary and a secondary root. In addition, at the basal zone of the primary root, the presence of lateral roots at the initial phase of development was evident. When WEHS was present together with nitrate in the nutrient solution, a higher proliferation of secondary roots (generally about three per plant) and a higher number of lateral roots more developed in length than those observed in the control plants were evident. A similar effect of WEHS on root morphology was observed when plants were put in contact with 5 mmol liter^{-1} $CaSO_4$ solution for 24 h in the presence of the humic fraction. This behavior has been confirmed in a study in progress based on the analysis of the number of lateral root primordia in Arabidopsis wild-type seedlings treated for 48 hours with a concentration of HS ranging between 0 and 1 mg C liter^{-1}. The results showed an increasing number of lateral root primordia with values fourfold higher in plants treated with 1 mg C liter^{-1} compared to the control (Pizzeghello et al., 2006).

Investigations on root cell morphology highlighted higher differentiation levels in treated roots. In wheat, using both light microscopy (LM) and transmission electron microscopy (TEM) micrographs, cells of the root central cylinder were observed with thicker walls compared to control roots (Nardi et al., 1996). Using epifluorescence micrograph, lignin becomes autofluorescent upon absorbing UV light and the autofluorescent intensity of the cell wall indicates the degree of lignification and differentiation of the xylem vessels (Robert and Roland, 1989). A study on *Pinus sylvestis* (Nardi et al., 2000) revealed that the roots treated with HMS from a rendzic leptosol showed a higher rate of differentiation, measured in terms of autofluorescence (Figure 8.1), compared with the control (Figure 8.2) and with LMS-treated (Figure 8.3), at 1 and 2 mm beyond the root tip.

Several works were carried out to study the morphological changes induced by HS in a simplified system, such as leaves and cells in culture. The use of tissue culture provides a rapid and inexpensive method of screening compounds and has been found to be representative of results obtained in whole-plant experiments studying different aspects (Ehsanpour and Fatahian, 2003; Vidal et al., 2004). In Muscolo et al. (1993), leaf explants of tobacco (*Nicotiana plumbaginifolia*) were compared with cultures supplemented with hormones or HS of various molecular complexity (total, low and high molecular weight) and concentrations (1.0 and 5.0 mg C liter^{-1}). The results showed that all humic fractions at the lowest concentration produced greater leaf explants compared to leaf grown with hormones. Successively, leaf explants of tobacco culture supplemented with either indoleacetic acid (IAA) and/or HEf (humic matter fraction <3500 Da) revealed many morphological changes compared to the control (Nardi et al., 1994). While controls did not have roots, explants grown in the presence of IAA or HEf developed roots. The IAA-treated roots were short and endowed with many root hairs, while the HEf ones were long and with few hairs.

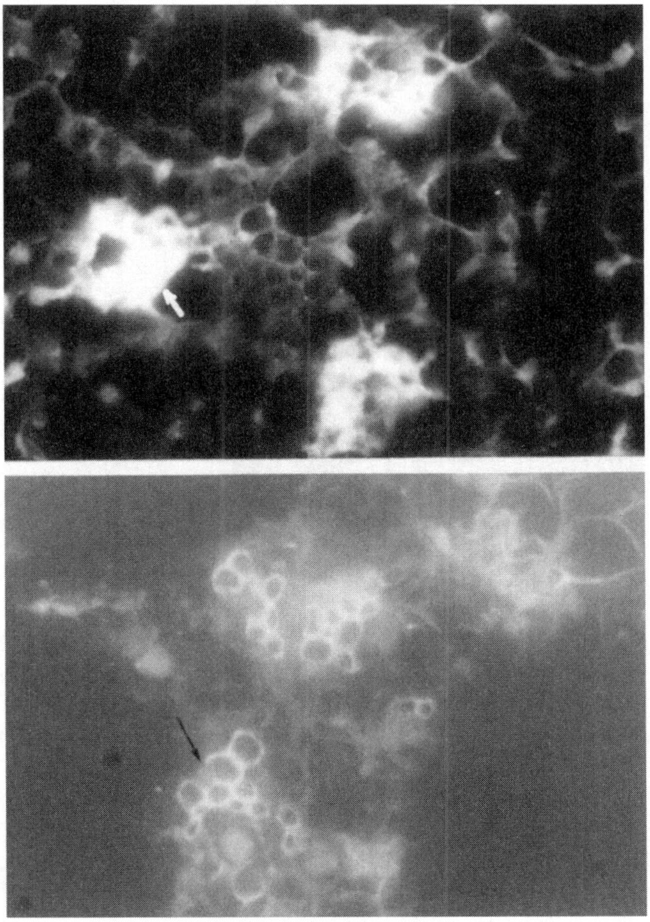

Figure 8.1. Fluorescence micrographs of *Pinus sylvestris* roots at 1 mm (**top**) and 2 mm (**bottom**) behind the root tip after a treatment with HMS (high-molecular-size humic substances). Note the higher rate differentiation of the roots in respect to the control. Reprinted from Muscolo, A., Bovalo, F., Gionfriddo, F., and Nardi, S. (1999). Earthworm humic matter produces auxin-like effects on *Daucus carota* cell growth and nitrate metabolism. *Soil Biol. Biochem.* **31**, 1303–1311, with permission from Elsevier Limited.

Morphological changes induced by HS were also observed in carrot cells culture (Muscolo et al., 1999). Low-molecular-weight humic substances were able to induce growth and shape changes similar to those induced by auxins. 2,4-Dichlorophenoxyacetic acid (2,4-D) induced the best cell growth stimulation, while HEf mimicked 2,4-D action. The presence of IAA, 1-naphthylacetic acid (NAA), or 6-benzylaminopurine (6BAP) stimulated growth, although only to a minor extent compared to HEf. Optical microscopy revealed that the cells grown in the presence of HEf displayed an extended shape similar to those grown in IAA medium. Cells supplemented with 2,4-D and 6BAP, or with 6BAP alone, formed round clumps of cells, while cells cultured only in the presence of 2,4-D formed small round cells. Cells cultured in the presence of NAA exhibited both round clumps and a few extended

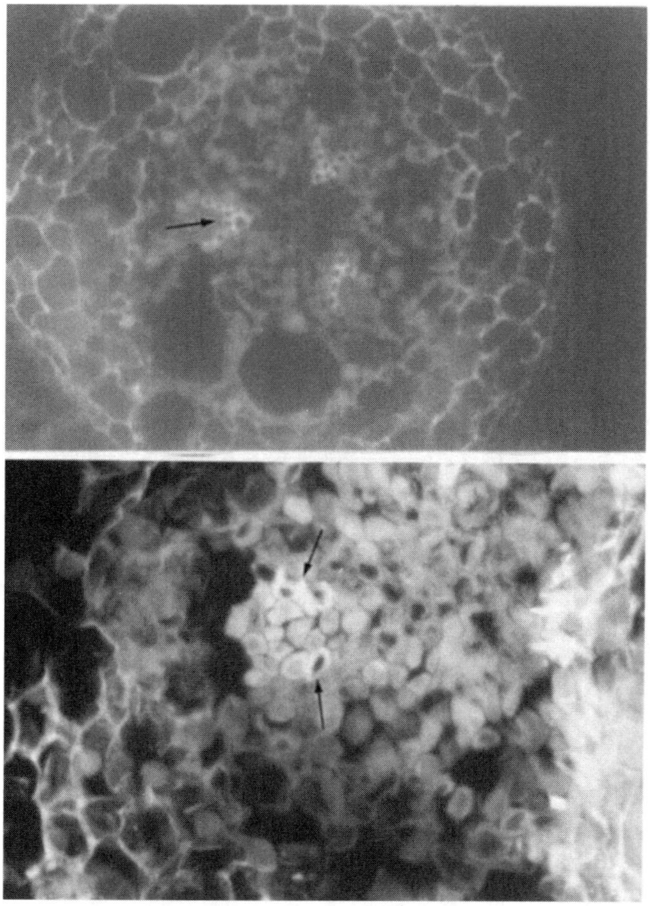

Figure 8.2. Fluorescence micrographs of *Pinus sylvestris* roots at 1 mm (**top**) and 2 mm (**bottom**) behind the root tip (control). Reprinted from Muscolo, A., Bovalo, F., Gionfriddo, F., and Nardi, S. (1999). Earthworm humic matter produces auxin-like effects on *Daucus carota* cell growth and nitrate metabolism. *Soil Biol. Biochem.* **31**, 1303–1311, with permission from Elsevier Limited.

cells. Cells grown in nutrient medium without hormones showed cellular detritus and a few round clumps (Figures 8.4 and 8.5).

Morphological changes were also observed on callus of *Pinus laricio* cultured for 4 weeks with HS extracted from forest soil under *Abies alba* and *Fagus sylvatica* plantation or the hormones 2,4D, IAA, and 6BAP (Muscolo et al., 2005). The results showed that both high- and low-molecular-weight humic fractions inhibited callus growth compared to the control and affected the enzymes involved in carbohydrate metabolism.

8.6. MACRO- AND MICRO-NUTRIENT UPTAKE MODIFICATIONS

One of the unresolved questions raised by the work of Vaughan and Malcom (1985) concerns the mechanisms leading to an altered ion uptake in higher plants, following

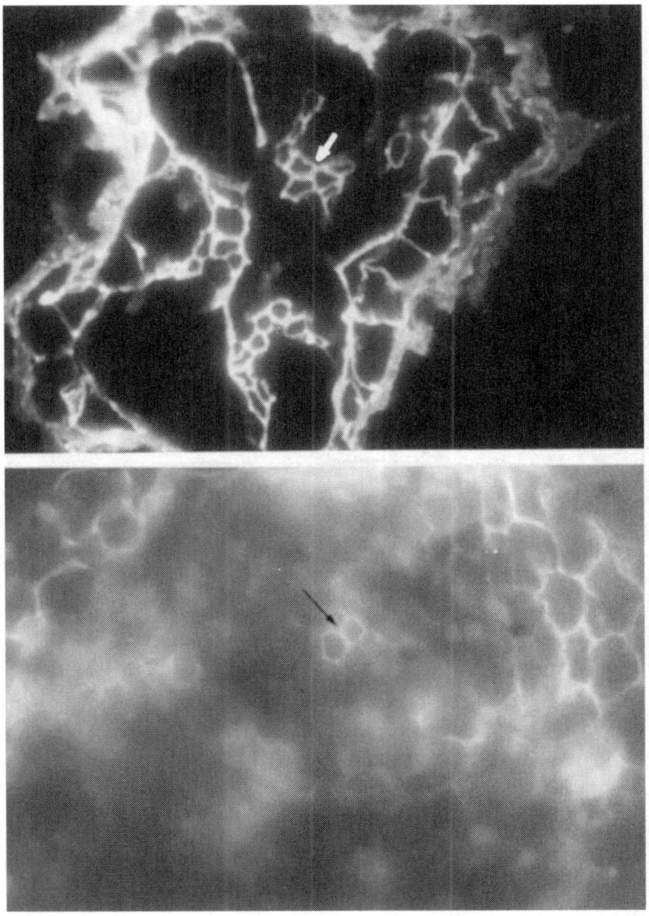

Figure 8.3. Fluorescence micrographs of *Pinus sylvestris* roots at 1 mm (**top**) and 2 mm (**bottom**) behind the root tip after a treatment with LMS (low-molecular-size humic substances). Note the much smaller rate differentiation of the roots treated with LMS than the roots treated with HMS (high-molecular-size humic substances). Reprinted from Nardi, S., Pizzeghello, D., Remiero, F., and Rascio, N. (2000). Chemical and biochemical properties of humic substances isolated from forest soils and plant growth. *Soil Sci. Soc. Am. J.* **64**, 639–645, with permission from the Soil Science Society of America.

HS treatment. The influence of soil humus on ion uptake, and more generally on plant growth, has been examined and reviewed by Chen and Aviad (1990), Varanini and Pinton (1995, 2001), Clapp et al. (2001), Nardi et al. (2002), Tan (2003), and Chen et al. (2004a).

The effects of HS on ion uptake appear to be more or less variable and selective, depending on the HS involved, their concentration, the plant species, and the composition and pH of the medium. Studies on uptake kinetics, use of protein synthesis inhibitors and different experimental conditions suggest that the effect of HS on plant nutrition may be mediated by modulation of the synthesis and functionality of membrane proteins.

Studies have been conducted on the amount and functionality of plasma membrane H^+ATPase. The central role of this enzyme in plant growth and mineral

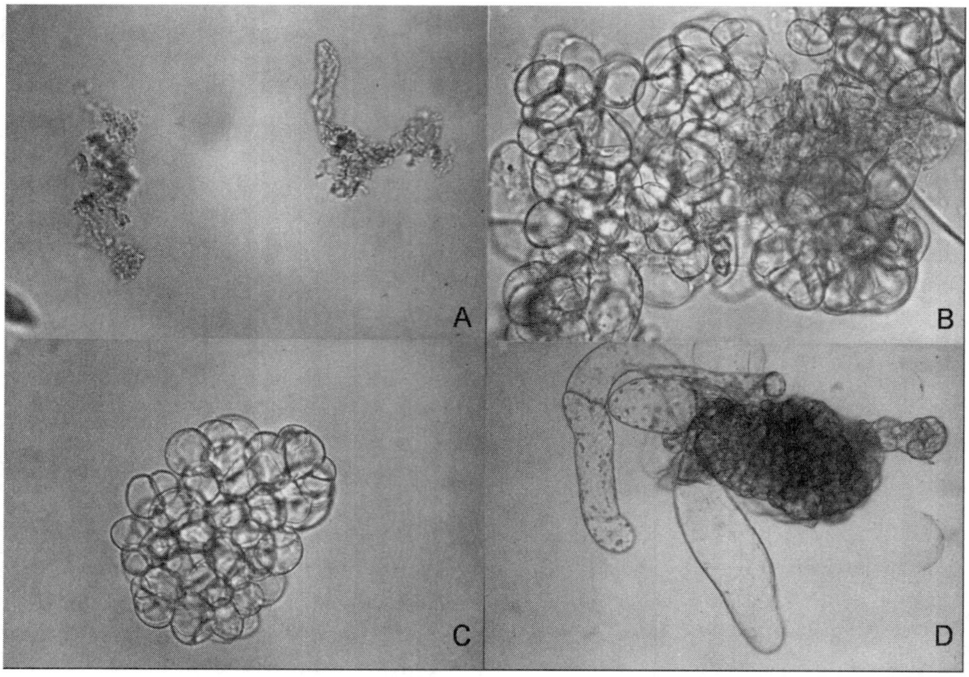

Figure 8.4. Photographs of *Daucus carota* cell suspensions treated with hormones. 2,4-Dichlorophenoxyacetic acid (2,4-D); 6-benzylaminopurine (6BAP). (**A**, control; **B**, control + 2,4-D + 6BAP; **C**, control + 2,4-D; **D**, control + 6BAP). Reprinted from Muscolo, A., Bovalo, F., Gionfriddo, F., and Nardi, S. (1999). Earthworm humic matter produces auxin-like effects on *Daucus carota* cell growth and nitrate metabolism. *Soil Biol. Biochem.* **31**, 1303–1311, with permission from Elsevier Limited.

nutrition has been clearly stated (Palmgren, 1998). The first evidence for an effect of HS on transport proteins concerns the stimulation caused by HMS and LMS fractions on the activity of the K^+-stimulated ATPase (believed to be coincident with the H^+-ATPase of plasma membranes) of microsomal fractions (Maggioni et al., 1987; Nardi et al., 1991; Pinton et al., 1992). Evidence that the plasma membrane proton pump is directly involved in increased nutrient uptake due to the presence of HS is also given by the stimulation of active proton extrusion from intact roots, which was apparent after 2–4 h of incubation (Pinton et al., 1997). This effect has been interpreted as a consequence of a direct stimulation of HS on the proton pump (H^+-ATPase). A further direct proof of an interaction between humic molecules and plasma membrane H^+-ATPase has been given by Varanini et al. (1993), who demonstrated that low-molecular-weight (<5000 Da) humic molecules can stimulate the phosphohydrolitic activity of this enzyme in isolated plasma membrane vesicles, thereby determining an increase of the electrochemical proton gradient which might be, at least in part, responsible for a stimulation of NO_3^- uptake (Pinton et al., 1999a). In fact, transport of this nutrient is a substrate-inducible process and involves proton co-transport and, at higher NO_3^- uptake rates, the quantity and activity of root plasma membrane H^+-ATPase (Santi et al., 1995).

The stimulatory effects of LMS fraction on NO_3^- uptake required long periods of incubation (Albuzio et al., 1986). These mechanisms could be explained through

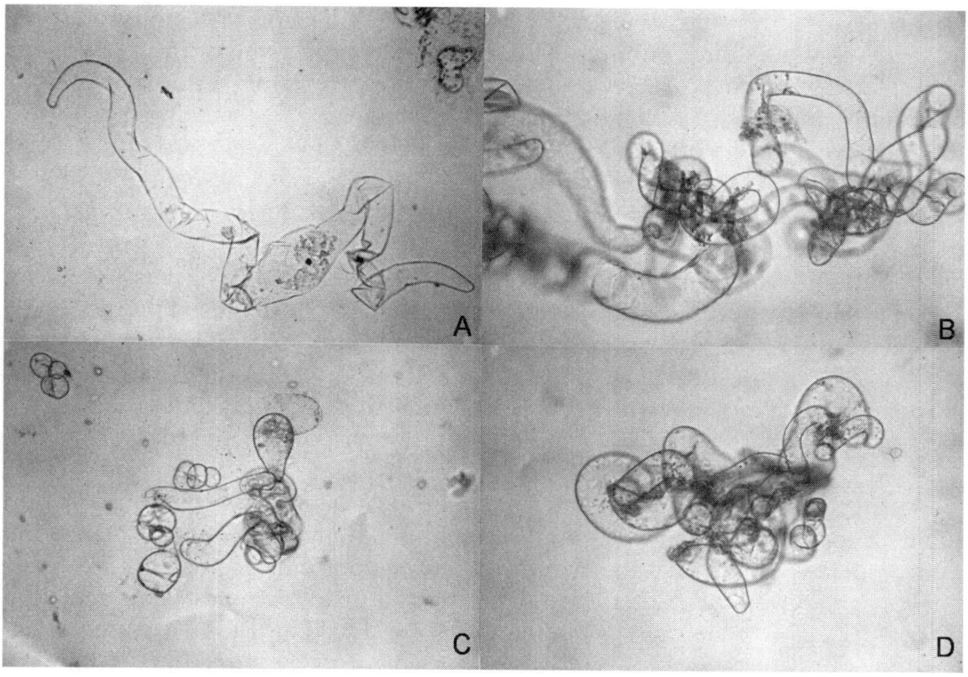

Figure 8.5. Photographs of *Daucus carota* cell suspensions treated with humic substances or hormones. Indoleacetic acid (IAA); 1-naphthylacetic acid (NAA); HEf (humic matter fraction <3500 Da). (**A**, control + IAA; **B**, **D**, control + HEf; **C**, control + NAA). Reprinted from Muscolo, A., Bovalo, F., Gionfriddo, F., and Nardi, S. (1999). Earthworm humic matter produces auxin-like effects on *Daucus carota* cell growth and nitrate metabolism. *Soil Biol. Biochem.* **31**, 1303–1311, with permission from Elsevier Limited.

a regulation of the "coarse" type. In fact, the vanadate-sensitive proton-pumping ATPase (H^+-ATPase) that builds up an electrochemical proton gradient across the plasma membrane (Morsomme and Boutry, 2000) modulates the primary active transport by plant cells. This gradient energizes secondary active transport, accomplished by carrier proteins via symport or antiport. In this context, NO_3^- is taken up by an inducible H^+/NO_3^- symport with a stoichiometry of 2:1 (Miller and Smith, 1996). Another line of evidence supports the hypothesis that LMS fraction could interact with these transport proteins ("fine" regulation), leading to a modulation of NO_3^- uptake. This contention is reinforced by the observation that LMS fraction can reach the apoplast and interact with the plasma membrane of roots (Vaughan, 1986) and cultured carrot cells (Muscolo et al., 2007a).

Other evidence is reported in the paper of Canellas et al. (2002), in which a stimulation of the plasma membrane H^+-ATPase activity took place, apparently associated with an ability to promote expression of this enzyme, as confirmed by western blot analysis. Using antibodies raised against H^+- ATPase PMA2 isoform from *Nicotiana plumbaginifolia* Viv. (Morsomme et al., 1996), it was discovered that the amount of immunoreactive protein at the PMA locus (approximately 96 kD) increased almost threefold in the membrane vesicles isolated from maize roots treated with HA.

Regarding nitrogen assimilation, physiological studies have demonstrated the existence of at least three different NO_3^- uptake systems in plants, with an essentially unregulated low-affinity transport system (LATS) active at high external NO_3^- concentrations and two high-affinity transport systems, one of these being constitutive (cHATS) whereas the other is induced by nitrate (iHATS), which are active at low external concentrations (Glass and Siddiqi, 1995; Forde and Clarkson, 1999). Molecular data have established that multiple gene family members encoding putative nitrate transporters are present for both the high- and low-affinity systems (Glass et al., 2001). Those encoding the HATS transporters, termed Nrt2 genes, have been attributed to the nitrate-nitrite porter family, whereas those encoding the LATS transporters, termed Nrt1 genes, to the peptide transporter family both belong to the major facilitator superfamily of membrane transporters (Forde, 2000). Recently, in roots of maize treated for 48 h with LMS, no increase in terms of transcript accumulation of gene encoding a low-affinity nitrate transporter (ZmNrt1.1), and of a putative high-affinity nitrate transporter (ZmNrt2.1), was found (Quaggiotti et al., 2004). Even though the involvement of other nitrate transporters in the root response to humic acids cannot be ruled out, the absence of stimulation of the transcription of these genes suggests that the increase of nitrate uptake measured after HS treatment may involve post-transcriptional/post-translational mechanisms of regulation of ZmNrt2.1 and/or an indirect modulation of the uptake process. This latter effect may rely on the action of H^+-ATPases responsible for generating the proton electrochemical difference across the plasma membrane essential for nutrient uptake, also considering that nitrate uptake by plants is thought to be a secondary transport driven by the proton electrochemical difference generated by the proton pump (Miller and Smith, 1996; Wang and Crawford, 1996). The analysis of the expression of Mha2, a major maize isoform of H^+- ATPase, showed a strong induction in terms of accumulation of its transcripts in roots of maize seedlings grown in the presence of LMS for 48 h (Quaggiotti et al., 2004).

As far as the shoots are concerned, even though the effects of HS on shoot physiology are mediated by root-shoot signaling events yet to be characterized, the induction of the synthesis of transcripts of ZmNrt2.1, observed in maize seedlings exposed to LMS, together with the absence of significant changes in ZmNrt1.1 and Mha2 transcription and a remarkable downregulation of Mha1 expression, show that LMS also affects gene expression in shoots with striking differences compared with roots (Quaggiotti et al., 2004). ZmNrt2.1 transcription was also shown to be upregulated in maize leaves, as well as in root tissues, in response to nitrate availability, and could, therefore, have been involved in nitrate translocation and compartmentation at the whole plant level. For this reason, it is possible to speculate that the LMS stimulation of the ZmNrt2.1 transcript accumulation in shoots, which is consistent with a higher nitrate content in leaves of LMS-treated plants, may implicate a more efficient translocation of nitrate to its metabolic sinks and reflect a better nitrogen use efficiency, which is also believed to be connected to a more efficient nitrogen remobilization (Hirel et al., 2001).

Finally, owing to their polyanionic (acid) nature, HEf and HSp (humic fraction >3500 Da) do show surfactant-like behavior (Visser 1986; Nardi et al., 1991), as decreased surface tension of water is described when their concentration is increased (Fendler and Fendler, 1975), so HS could simply act as surface-active molecules (Samson and Visser, 1989). By decreasing the pH at the surface of the plasma membranes of root cells, HS may counteract the alkalinization that occurs when NO_3^- is used as a N source, and that is responsible for inhibiting the H^+/NO_3^- symport (Raven and Smith, 1976). Concurrently, when there is a decrease of NO_3^- uptake, at the same time an increase in the NH_4 uptake occurs (Barber, 1984).

The plasma membranes of plant cells possess several redox activities that can be related to both plant nutrition and cell wall formation and lignification (Lüthje et al., 1997; Bérczi and Møller, 2000). In this context, it has been shown that in oat roots, HMS humic fractions inhibited NADH oxidation in either the presence or absence of an artificial electron acceptor (ferricyanide), whereas LMS fractions inhibited this oxidase only if the electron donor (NADH) and acceptor (ferricyanide) were added at the same time (Pinton et al., 1995). While the first effect could be related to the activity of surface peroxidases that can be involved in cell wall formation and thickening (Vianello and Macrì, 1991), the second seems to be exerted on a different redox system with an unknown function (Nardi et al., 2002).

The importance of humic substances in plant nutrition also lies in their ability to chelate metals in soil. The stability order of the complexes formed between metals and humic substances have been determined (Stevenson, 1994; Pandey et al., 2000) and seem to follow the Irwing–Williams series: $Pb^{2+} > Cu^{2+} > Ni^{2+} > Co^{2+} > Zn^{2+} > Cd^{2+} > Fe^{2+} > Mn^{2+} > Mg^{2+}$. It appears evident that due to these properties, humic substances can contribute to the regulation of the chemical balances of metals, thereby influencing their solubility (Stevenson, 1994). With regard to plant availability, solubility and the molecular dimension of humic substances must be taken into account (Varanini and Pinton, 2001). The solubility of the complexes formed by humic acids or fulvic acids with micronutrients depends on the pH, in the presence of salts, and on the saturation degree of binding sites (Varanini and Pinton, 1995). Fractions of higher molecular mass, which may be mostly insoluble, can withhold large amounts of metals which are, consequently, subtracted from precipitation and subsequent crystallization, processes that would decrease their availability (Schwertmann, 1966). Soluble low-molecular-mass humified organic matter that may be present in the soil can help increase metal transport by diffusion to the roots (Pandeya et al., 1998). In fact, thanks to their ability to form complexes with metal cations, it is generally accepted that fulvic acids can mobilize them from soil particles to the root surface, but the quantitative aspects of this process have not yet been elucidated (Varanini and Pinton, 2001). It was recently demonstrated (Leita et al., 2001) that HS interact in solution, not only with metal ions but also with free ligands and complexes of high stability without causing any ligand exchange. The contribution of HS to the migration of micronutrients and toxic elements might, therefore, be much more complex than previously considered (Chen et al., 2004b).

Based on the chelating properties of HS, a hypothesis that is sometimes referred to as the "micronutrient availability hypothesis" (Chen et al., 1994) has been proposed in order to explain the stimulatory effects of HS on plant metabolism. In a number of publications, Chen et al. (2004a) have shown that Fe-enriched organic materials such as peat or manure could serve as a remedy for lime-induced chlorosis. The corrective effect was attributed to the complexation of Fe by HS in these materials. This effect was also analyzed measuring the residual concentrations of Fe, Mn, and Cu in nutrient solutions that were equilibrated at different pH and then thoroughly centrifuged at increasing concentrations of organic matter (Chen et al., 1994). The residual concentrations of Fe increased only at OM concentrations higher than 100 mg liter^{-1} and reached at the highest OM concentration about 7% of the added Fe. More recently, it has been shown (Chen et al., 2004a), working with melon plants and ryegrass, that additions of FA or HA to the nutrient solution not containing Fe does not result in a significant improvement in chlorophyll concentration, whereas adding $FeSO_4$ together with HS enhanced chlorophyll levels, thereby supporting the idea that HS enhance growth due to their complexing properties. In an experiment with Fe-deficient cucumber plants (Pinton et al., 1999b), recovery of dry matter, iron, and chlorophyll contents of plants treated with Fe-HS appeared to be faster or even greater than that observed by supplying other iron sources (Fe-EDTA, Fe-citrate, $FeCl_3$) at the same iron concentration (0.2 μmol liter^{-1}). This endorses the idea that the action exerted by HS on Fe-nutrition may account for more than a mere chelating agent effect. Furthermore, the role of HS on ion absorption by plant roots is not easily explainable, owing to the complex and still unknown nature of these substances. In fact, the effects described in these papers are difficult to compare because HS with different features (due to the origin of the soil and the methods of extraction) were assayed. It is possible that HS may exert several effects on plant functions and that some of these may result, directly or indirectly, in a modulation of ion (thus also Fe) uptake (Nardi et al., 2002).

Varanini and Pinton (2001) proposed an active role of humic substances on iron nutrition and the uptake of ^{59}Fe from soluble ^{59}Fe–humate complexes by cucumber and barley plants has been demonstrated (Cesco et al., 2002). Although abundant in the earth's crust, Fe is mainly present as insoluble Fe(III) precipitates in the soil and is, therefore, largely unavailable to plants, especially at neutral and alkaline pH. Plants are known to possess different mechanisms for responding to limited micronutrient availability. In the case of Fe, two strategies have been observed (Marschner and Römheld, 1994) for dicots and nongraminaceous monocots (strategy I) and for graminaceae (strategy II), respectively.

In a series of works, it has been demonstrated that a water-soluble humic fraction (WEHS) could increase the amount of Fe in the soil solution (Cesco et al., 2000) and that the Fe–WEHS complex could alleviate symptoms of Fe-deficiency in cucumber (Pinton et al., 1999b). More recently, it has been reported that Fe-deficient cucumber plants are able to absorb and translocate to the shoot ^{59}Fe supplied to the nutrient solution as ^{59}Fe–WEHS. In these plants, utilization of ^{59}Fe–WEHS was strongly enhanced by low pH of the root external solution, (Cesco et al., 2002). In the same paper, it has been shown that also a graminaceous plant such as barley was able to use ^{59}Fe from the root extraplasmatic pool, and this utilization was strongly related to the Fe nutritional status.

The results of this work support the view that Fe–WEHS complex may serve as a natural substrate for the inducible plasma-membrane-bound FeIII-chelate reductase in Strategy I plants (Pinton et al., 1999b). In Strategy II plants, an indirect mechanism, conceivably operating via ligand exchange between the humic fraction and the phytosiderophores released, appears to be involved in the use of Fe bound to WEHS.

In conclusion, it appears that the uptake enhancement of macro- and micronutrients due to HS is a synergic sum of various effects exerted by these molecules at a rhizosphere level. Besides a direct source of nutrient subsequent their decomposition and apart from their chelating properties, HS interaction with plant root plasma membrane has been demonstrated in relation to its solubility, its surfactant-like behavior, its involvement in redox activities (in particular, connected to plasmalemma H^+ATPase), and its modulating effect on iron uptake in both Strategy I and Strategy II plants.

8.7. EFFECTS OF HUMIC SUBSTANCES ON BIOCHEMICAL PATHWAYS AND PROCESSES

The uptake of HS into plant tissues was already reported in the text of Vaughan and Malcom (1985). Vaughan (1986), using excised (25–35 mm long) roots from 2-day-old peas (*Pisum sativum*), found that the amount of radioactivity associated with roots increased with the concentration of labeled HA and FA. At all the concentrations used in the incubation media, FA were absorbed more than HA. When pea roots were incubated in labeled humus at different temperatures and in different experimental conditions, two uptake components appeared to be operating: The first was an initial and rapid passive process, while the second was a slower, but continuous, active uptake, dependent on metabolism. Other data indicate that the initial uptake of HS is mainly confined to the cell wall (Vaughan, 1986). In agreement with the latter results, the different treatments [chelation with ethylenediaminetetraacetic acid (EDTA), pronase treatment or NaOH wash] used to remove the bound activity had little effect. This indicates that almost all the labeled HS were tightly bound to the cell wall (Vaughan, 1986). Further investigations on the uptake by plant roots of humic fractions, with different molecular masses, have supplied new insights. When pea roots were cultured at metabolic temperatures in the presence of radioactive LMS fractions, the humic matter was taken up to a greater extent. In addition, it was found that 70% of the radioactivity was present in the supernatant fraction of pea roots. This was in contrast with the value of 25% recovered in the supernatant for the radioactive HMS humic fraction. When pea roots were cultured at low temperatures and in the presence of the two-labeled humic fractions, only the HMS fraction was absorbed by pea roots. These results support the interpretation that HS of all molecular weights can be absorbed and show that the uptake of LMS is dependent on the active component of transport (Vaughan, 1986). The LMS fraction absorbed by roots was then transferred to the shoots; but even in these cases, the amount transferred was not higher than 10–12% (Vaughan, 1986). This pattern has been confirmed by Muscolo et al. (2007a), utilizing LMS and HMS fractions conjugated with fluorescein isothiocianate (FITC). They showed that only the LMS humic fraction was able to interact with the plasma membrane of cultured carrot cells.

These observations, which clearly demonstrate that humic substances are taken up into plant tissues, may sustain the assertion that HS have a direct effect on plant metabolism (Vaughan and Malcolm, 1985).

Direct effects of HS may be reflected in changes in photosynthesis and ATP formation in treated plants. In the case of photosynthesis, our information is still fragmentary and not very recent. Ferretti et al. (1991) showed that HS, applied to the culture medium, increased the activities of the enzymes involved in the photosynthetic sulfate reduction pathway, whereas Merlo et al. (1991) observed in maize leaves a decrease in starch content accompanied by an increase in soluble sugars. These positive biological effects appeared to be mediated by changes in the activity of the main enzymes involved in carbohydrate metabolism. The aspect of ATP metabolism has been addressed by using isolated rat liver mitochondria. It has been shown that HS partially uncoupled oxidative phosphorylation after a short exposure (Visser, 1987). This effect can explain previous results (Visser, 1986), showing that synthetic HS also uncoupled oxidative phosphorylation, thereby decreasing ATP concentration. This partial uncoupling has been confirmed by using isolated higher plant mitochondria (Flaig, 1968), although this effect was also accompanied by an increase in dry matter and sometimes, as in cereals, in yield gain. The apparently contradictory results have been explained by suggesting that the partial uncoupling renders some inorganic phosphate available, without depleting cellular ATP, which is then used in some phosphorylating reactions linked to biosynthetic pathways. However, it has been demonstrated that incubation of mitochondria with HS for a long period resulted in a positive influence on oxidative phosphorylation (Visser, 1987), a result that could explain the finding that HS caused an increase in ATP production (Khristeva et al., 1980). The latter observations are, in any case, difficult to reconcile with the former. In addition, more recent results show that HS determined a decrease (30–40%) of cellular ATP, without affecting O_2 consumption (Nardi et al., 1991). From the above findings and considerations, it is not clear whether HS influence respiration by directly or indirectly interfering with mitochondria, thereby making new experimental work necessary prior to drawing a firmer conclusion.

The mechanism by which humic substances influence enzyme activities is still not completely understood. In recent years this machinery has been closely analyzed in a number of papers considering effects on single enzymes and/or repercussions involving whole metabolic pathways.

Ghorobekova (1987) showed the inhibitory effect of humic matter on protease activity. Inhibition kinetics are of mixed order, and humic acids can be used as a regulator of activity and biosynthesis of proteolytic enzymes.

Passera et al. (1991) studied iron, sulfur, ATP-sulfurylase (ATP-s), and *O*-acetylserine sulfhydrylase (OAS-s) activities in maize leaves treated with 50 mg liter^{-1} of humic matter (>12 kD). Recent evidence suggests that in plants, sulfate assimilation occurs mainly in the leaf and is located in the chloroplast where it utilizes the redox equivalents and the ATP generated during the photosynthetic process. The synthesis of adenosinephosphosulphate, the first step in the reaction chain of sulfate assimilation, is suggested as being catalyzed by ATP-s, while the synthesis of cysteine, catalyzed by OAS-s, appears to be the terminal step. Humic matter positively affected sulfur content and the enzyme activities of sulfur metabolism. The stimulations were different according to the enzyme, leaf age, and the cellular fraction considered.

Concerning nitrate uptake and assimilation (Albuzio et al., 1986) in barley seedlings incubated with HS from a grassland soil and their fractions—nitrate reductase (NR), glutamate dehydrogenase (GDH), and glutamine synthetase (GS) activities—were analyzed together with the rate of nitrate uptake. The enzymatic activities turned out to be stimulated by the treatment of unfractioned humus extract with increases compared to controls of 65%, 35% and 45% respectively.

GDH, GS, and malate dehydrogenase (MDH) activities have been also studied in carrot cells suspensions (Muscolo et al., 1999). All enzymes tested were positively affected by the humic fraction with low molecular weight. The activities of GDH,

GS, and MDH were increased by HEf treatment of 1.79, 1.50, and 1.49-fold, respectively. As a consequence of the increased activity of these enzymes, an increase in the amount of methionine, threonine, isoleucine, and lysine, amino acids derived from the oxalacetate pathway, were found.

The interactions among HS and the enzymes of the nitrogen assimilation pathway were further analyzed in Sessi et al. (2000), considering the repercussions of the treatments on maize seedlings with low (HEf)- and high (HSp)-molecular-weight humic fractions from an agricultural soil and a forest soil. The authors concluded that low- and high-molecular-weight fractions affected the nitrogen metabolism of maize plants differently. HEf was found to interact directly with the systems related to nitrate uptake and also with the activity of the enzymes in nitrogen internal cycling, whereas the effects of HSp was probably limited to cell wall level.

The biological effects of two high-molecular-weight HS extracted from soil with different vegetation cover (*Fagus sylvatica* and *Abies alba*) were assayed by Muscolo et al. (2000) on callus growth of *Pinus laricio*. The results showed that both humic substances strongly inhibited the activity of glucokinase (GK), phosphoglucose isomerase (PGI), aldolase (ALD), and pyruvate kinase (PK), enzymes involved in glucose metabolism. The PGI was the enzyme with the highest percentage of inhibition (96%) compared to the control. The low PGI activity caused significant changes in the levels of aldolase and PK, enzymes belonging to the following steps of glycolytic pathway.

Recently, Nardi et al. (2005) further studied the activities of the enzymes related to the glycolysis pathway in maize seedlings, such as GK, PGI, PPi-dependent phosphofructokinase (PPi-PFK), PK, and those involved in respiration: cytrate synthase, MDH, and the isocitrate dehydrogenase cytosolic form of $NADP^+$-isocitrate dehydrogenase ($NADP^+$-IDH). The results demonstrated enhancement/diminishment effects for the various HS fractions tested, in relation with their molecular size and structural conformations. The low MS fraction affected plant metabolism in the best possible way, thereby confirming its stronger biological activity, compared to HMS, as previously found.

From these results, several observations may be made. On the one hand, there is a seeming unevenness in the enzymatic responses to HS extracted from various sources, tested on different model plant or tissues. This lack of uniformity may be easily explained considering the huge variability in the physiology among the different tissues and species. Moreover, by taking for granted how the responses elicited by HS depend on their structure, it is clear how diversely structured humic extracts from a number of different sources may cause the variability of the enzymatic answers. On the other hand, it appears that the effect exerted by HS, rather then a modulation on a single plant enzyme, seems to be a multi-targeted action involving a whole-system physiologic regulation. In this sense, HS may behave as a signal of the rhizosphere, perhaps containing or eliciting phytohormone production at a plant and/or at a soil biota level. In this context the auxin-like activity of HS is not surprising because it is known that soils vary in their native auxin content (Hamence, 1946) and fertile soils contain greater amounts of auxin than do less fertile ones (Stewart and Anderson, 1942; Dahm et al., 1977). Auxin and gibberellin levels are usually higher in the rhizosphere than in the bulk soil, probably as a consequence of increased microbial populations or of an accelerated metabolism owing to the presence of root exudates. Although numerous soil and rhizosphere microorganisms, as well as the root systems of higher plants, have been reported as producing auxins (Lebuhn and Hartmann, 1993) and gibberellins (Rademacher, 1992), there is little information about their stability and only indirect conclusions have been made about their presence in amounts high enough to be biologically active (Frankenberger and Arshad, 1995). However, Frankenberger and Arshad (1995) found that the active ingredients in humus were not mineral nutrients, but rather organic substances and biologically active metabolites of various microbes. Mineral substances applied in equal amounts to soil had little effects on plant growth. The presence of indoleacetic-acid (IAA) into the humic substances has been tested to verify if HS auxin components are able to justify their biological activity (Muscolo et al., 1998). The concentrations of IAA were estimated to be 0.5% (w/w) by enzyme immunoassay and 3.7% and 2% by radio immunoassay (RIA) in fluid phase using anti-IAA-C and anti-IAA-N antibodies, respectively. Canellas et al. (2002) later reported a detection of IAA in a 30-mg sample of humic acids, although the signal was only about twice the noise level. Unfortunately, the auxin-like activity of HS cannot be fully explained by free IAA in the samples, since the dose response to HS is much broader than for IAA. It is possible that the HS also contain other auxins, such as phenylacetic acid and indole butyric acid in addition to IAA. Later, spectroscopic analyses (Russel et al., 2006) confirmed the presence of low-molecular-weight organic acids in HS. Alternatively, there may be some other unknown component of HS that has auxin-like activity. In fact, all known auxins contain a carboxyl group, in addition to a hydrophobic ring, and the diffuse reflectance infrared spectroscopy (DRIFT) analysis of humic extracts indicated a high content of free carboxyls. Finally, it is possible that IAA may be produced either chemically or enzymatically from HS in contact with the plant cell wall and root plasma membrane.

In an attempt to evaluate the possible interaction of the LMS fraction with plasma membranes (target of IAA) of carrot cells, Muscolo et al. (2007a) labeled IAA, HMS, and LMS with fluorescein isothiocyanate (FITC). The cells in culture were monitored during 10 days of incubation, and the fluorescein staining of carrot cells and the decrease of fluorescein concentration in the culture medium were evaluated. Fluorescent membrane staining was only present in IAA- and LMS-treated cell cultures, and a consequential decrease of fluorescein concentration in their culture media was observed. Pretreatment of carrot cells with unconjugated IAA or LMS humic fraction markedly reduced the fluorescein staining of both FITC-IAA and FITC-LMS humic fraction, giving indirect evidence of the possible binding site of LMS humic fraction to the IAA cell membrane receptors.

Nardi et al. (1994), by using two inhibitors of auxin (TIBA = 2,3,5-triiodobenzoic acid and PCIB = 4-chlorophenoxy-isobutyric acid), demonstrated in *Nicotiana plumbaginifolia* that the LMS component of humic matter is the fraction endowed with auxin-like activity, although the pathways followed by IAA and the LMS fraction in inducing their effects may be somewhat different. Humic matter, IAA, and IAA inhibitors stimulated peroxidase activity in tobacco. When the *Nicotiana* tissues were treated with the humic fraction and IAA, there was a minor polymorphism in the esterase isoenzymes. The presence of both TIBA or PCIB together with LMS or IAA restored the esterase profile obtained from control tissues.

In Muscolo et al. (1996), carrot cells were grown in cultures supplemented with two hormones (2,4-D and 6BAP), and two humic fractions (with high and low molecular weight) extracted from earthworm faeces. The effect of the two fractions have been essayed on the activity of glutamate dehydrogenase, glutamine synthetase, and phosphoenol-pyruvare carboxylase. The LMS fraction was more effective than hormones in stimulating the three enzyme activities, and when this fraction was present with hormones, its action was less efficient. The LMS fraction not only revealed auxin-like activity, but it is was also endowed with a specific mechanism (e.g., protein synthesis) that affects metabolism during plant growth.

A further support to the hypothesis of an IAA-like action of HS is reported in Nardi et al. (1999). In this paper, comparisons were made between the chemical compositions of humic substances extracted from three soils covered by different vegetation and their biological activities assayed using 15- and 30-day-old seedlings of *Pinus sylvestris* and *Picea abies*. In all the cultures tested, the pattern of esterase in Scotch pine and Norway spruce seedlings suggested that IAA and the humic substances caused an increased polymorphism over the control. The appearance of the same band in the electrophoretic pattern of seedlings grown with either humic matter or IAA suggested that humic matter and IAA activity may have a similar effect (Muscolo et al., 1993). These results were in line with the amounts of IAA determined by immunoassay and with the amylase, invertase, and esterase polymorphism.

Similar responses are also exhibited in a work investigating the relationships among litter composition and ^{13}C–nuclear magnetic resonance spectra (^{13}C-NMR) (Nardi et al., 2000). The paper studied $\delta^{13}C$ values and biochemical activities of the humic constituents extracted from the Ah horizons of two undisturbed forest soils located in a unique climatic area under different vegetative covers, *Pinus mugo* T. and *Pinus sylvestris* L. The esterase pattern of *Pinus sylvestris* root seedlings treated with HMS and LMS included an isoenzymatic band present in the electrophoretic pattern of indoleacetic acid that was absent in the control and in other extract patterns. This suggests that humic substances and IAA activity may have similar effects.

Other enzymes showing significant activity modifications following HS treatment are peroxidase, the most common plant scavenging enzyme involved in many metabolic activities, and invertase, hydrolyzing sucrose into hexose substrate available to growing cells and therefore positively correlated to plant growth (Kim and Suzuki, 1989). The modulation on the activities of these enzymes appeared to be very variable in a paper testing HS from 27 *Fagus sylvaticae* forest sites (Pizzeghello et al., 2001) ranging from 16% to 270% of the activity of the untreated control for peroxidase, as well as from 14% to 190% of the activity of the control for invertase. This kind of response has been further indicated comparing the previous data with data obtained testing HS extracted from soils under *Abies alba* forests (Pizzeghello et al., 2002). In all the cultures tested, the pattern of peroxidase activity has been essayed by gel electrophoresis. In the seedlings treated with HS from acid soils the peroxidase pattern included numerous isoenzymatic bands (6–8) that were present in the electrophoretic pattern of IAA treated seedlings and absent in the control plants.

These findings have been recently questioned (Chen et al., 2004a,b) demonstrating that hormone-like activity, measured in terms of invertase and peroxidase activity, is also enhanced by iron and by general healthy plant growth physiology, ignoring the fact that any preparation of HS contains microelements (ash). Although iron certainly plays a signaling role in plant physiology (Schmidt, 2003), this observation overlooks the fact that IAA treatment induced the same responses of HS in comparison to the control (nutrient solution treated) plants. These reactions cannot be attributed either to (a) a high content of micronutrients as ashes or (b) a chelating effect of HS able to improve nutrient assimilation.

In a recent work, Russell et al. (2006) confirmed previous findings, assaying two fractions of earthworm humic substances, differing in nominal molecular weight, on epidermal peels of the Argenteum mutant of pea (*Pisum sativum* L.), where IAA receptors were present. Both of the humic fractions caused stomatal opening in epidermal peels. The response showed a broad biphasic dose dependence, and the effective concentrations were similar for the two fractions. This inducing effect appears to be mediated by phospholipase A_2 (PLA_2) and protein kinase C-like activity (PKC), both enzymes being involved in the signal transduction pathway leading to the response of plants to IAA (Scherer and André, 1989; Nemeth et al., 1998). The maximal stomatal apertures in response to both humic substances were similar to that caused by IAA and somewhat less than the response to white light or fusicoccin. Two inhibitors of phospholipase A_2 selectively blocked the response of stomata to both IAA and humic substances, without affecting the response to light or fusicoccin. The authors concluded that stomatal opening in response to auxin and humic substances involves the activation of a phospholipase A_2 that is not involved in signaling the response to light or fusicoccin.

In Pizzeghello et al. (2006), lateral root formation in *Arabidopsis thaliana*, following humic substance treatment, have been monitored through an established Arabidopsis reporter line containing an auxin synthetic promoter driving GUS (DR5-GUS) as a visual marker for in planta auxin-dependent transductive responses and auxin distribution (Nakamura et al., 2003). Results showed a clear signal corresponding to GUS activity in the sinks of accumulation of auxin in both HS and IAA supplied plant roots. In order to further confirm the link between humic substances and IAA, three specific inhibitors of auxin (1-naphthoxyacetic acid, TIBA and PCIB) were supplied together with the hormone or the humic extract. When the inhibitors were supplied, no GUS staining activity was observed and a very low number of primordia was shown in both HS- and IAA-treated plants.

Schmidt et al. (2005) reported that phenotypes auxin-related mutants of Arabidopsis, all exhibiting a reduced number of root hairs, were not rescued by the application of WEHS, suggesting that functional products of the ethylene and auxin signaling cascade are required for translating the response of root cells to humic molecules. In addition, mutants defective in root hair initiation such as rhd6, known to develop normal hairs in the presence of ethylene or auxin, were not affected by a wide range of applied concentration of WEHS, indicating that HS cannot substitute for these hormones.

The lack of detailed knowledge on the composition of HS makes it very difficult to identify the relationships between the structure and the activity of these substances. The study of these relationships is complicated, as seen above, by the presence of other molecules, such as hormones of microbial origin. Therefore, attempts to relate these two aspects have produced conflicting results. This subject has been already analyzed and reviewed in Piccolo et al. (1992) and Nardi et al. (2002).

More recently, Muscolo et al. (2007b), in order to understand if the biological activity of humic substances may be related to their molecular weight and/or chemical structure, compared the activity of two humic substances derived from an uncultivated couch grass and a forest soil, each separated in fractions with low (<3500 Da) and high (>3500 Da) relative MW on *Pinus laricio* Poiret, spp. *Calabrica callus*. The soluble sugar content, free amino acid pool and the activities of the key enzymes involved in carbon and nitrogen metabolism were investigated. Callus was also treated with different hormones: IAA or 2,4-D and/or 6BAP. A great amount of aliphatic and H-sugar-like component and an intense chemical shift of β-CH_3 region in both grass humic fractions were observed, while high contents of betaine, organic acid, and COOH groups in both forest humic fractions were detected. The grass humic fractions improved the growth of calluses and increased the levels of enzymatic activities, while the forest humic fractions had an inhibitory effect. The data presented may support the view that the biological activity of HS is independent of their molecular weight, since both fractions (HMW and LMW) obtained from the same HS have a similar effect on callus tissues, whereas different HS with different chemical structures exhibited dissimilar responses.

8.8. CONCLUSIONS AND PERSPECTIVES

In the last 40 years of research on HS biological activity, several aspects have been elucidated. The favorable morphological effects of HS on plants regarding growth enhancement have been demonstrated on several plant species under different study conditions. Besides these observations, effects on morphogenesis have also been demonstrated in terms of (a) the induction of lateral root formation and (b) root hair initiation and development in intact plants and stimulation of root and shoot development in treated cell calluses.

HS interact with nutrient assimilation of both macro- and micro-elements, by enhancing the nutrient use efficiency. This capacity is related to both (a) HS chelating properties and (b) an interaction with plasma membrane enzymatic constituents.

Furthermore, various enzymes and biochemical pathways have been found to be influenced by HS treatment with up- and downregulation to different extents. Differences in the physiological responses to HS suggest that the their metabolic targets are not univocal and that the mechanism of action of these substances is not a single widespread common pathway, but instead a result of different signal cascades or biochemical reactions which may change according to the plant species, the phenological condition of the plant, and the kind of humic matter and experimental conditions involved.

These effects are related to the entrance of these substances into plant tissues, reaching both cell wall and cytoplasm. Since 1980s it has been hypothesized that the positive effects of humic compounds on plant metabolism may depend on the uptake of the substance or part of its components. However, the hypothesis that they may act as a signal and interact with the root cells by inducing endogenous activities, independently on their uptake, could not be excluded. In addition, several studies have hypothesized that physiological mechanisms through which the humic substances exert their effects may depend on hormones and, in particular, on an auxin or auxin-like activity. Such hypotheses are based on results demostrating the immunological or spectrometric identification of indol acetic acid (IAA) inside a number of humic substances and on studies carried out using auxin inhibitors. In addition, this hypothesis is supported by reports showing a positive effect of such substances on specific targets of auxin action. However, the measured effects did not always correlate to the amount of IAA detected in the humic acids. For this reason, the presence of different compounds of the auxin family or of molecules that might either mimic the action or stimulate the plant endogenous metabolism of auxin cannot be ruled out.

A number of area of uncertainty remain. The major problem in studying these substances is that HS are very complex mixtures that have no single formula; thus most of the difficulties encountered in chemically defining their structures are reflected into their multifaceted biological effects. New insights into the HS structure may help to identify the correlations existing between the measured effects on plant physiology and their chemical features. Moreover, defining a structural model of HS may also allow the *in silico* testing of the direct interaction between these substances and known plant proteins.

Nowadays the genome of many plant species have been sequenced. This amount of information may be useful in the studies regarding plant responses to HS. For example, mutant or transgenic plant lines may be used in order to discover new aspects on this area.

Finally, the physiological effects of HS have usually been assessed considering a limited amount of enzymatic activities and pathways. In order to evidence the relationships among these useful data, techniques taking into account a wider portion of plant physiology are needed. In this sense, methods of functional genomics, transcriptomics, or proteomics may represent the cutting-edge technology on HS biological activity.

REFERENCES

Aiken, G. R., and Wershaw, R. L. (1985). Molecular size and weight measurements of humic substances. In *Humic Substances in Soil, Sediment, and Water: Geochemistry, Isolation, and Characterization*, Aiken, G. R., McKnight, D. M., Wershaw, R. L., and MacCarthy, P., eds., John Wiley & Sons, New York, pp. 477–492.

Aiken, G. R., McKnight, D. M., Wershaw, R. L., and MacCarthy, P. (1985). An introduction to humic substances in soil, sediment and water. In *Humic Substances in Soil, Sediment, and Water: Geochemistry, Isolation, and Characterization*, Aiken, G. R., McKnight, D. M., Wershaw, R. L., and MacCarthy, P., eds., John Wiley & Sons, New York, pp. 1–9.

Albuzio, A., Ferrari, G., and Nardi, S. (1986). Effects of humic substances on nitrate uptake and assimilation in barley seedlings. *Can. J. Soil Sci.* **66**, 731–736.

Allison, F. E. (1973). *Soil Organic Matter and Its Role in Crop Production*. Developments in Soil Science. Elsevier, New York.

Andreux, F. (1996). Humus in world soils. In *Humic Substances in Terrestrial Ecosystems*, Piccolo, A., ed., Elsevier, Amsterdam, pp. 45–100.

Arancon, N. Q., Edwards, C. A., Atiyeh, R. M., and Metzger, J. D. (2004). Effects of vermicomposts produced from food waste on the growth and yields of greenhouse peppers. *Bioresource Technol.* **93**(2), 139–144.

Atiyeh, R. M., Arancon, N., Edwards, C. A., and Metzger, J. D. (2000). Influence of earthworm-processed pig manure on the growth and yield of greenhouse tomatoes. *Bioresource Technol.* **75**(3), 175–180.

Atiyeh, R. M., Edwards, C. A., Subler, S., and Metzger, J. D. (2001). Pig manure vermicompost as a component of a horticultural bedding plant medium: Effects on physicochemical properties and plant growth. *Bioresource Technol.* **78**(1), 11–20.

Atiyeh, R. M., Arancon, N. Q., Edwards, C. A., and Metzger, J. D. (2002a). The influence of earthworm-processed pig manure on the growth and productivity of marigolds. *Bioresource Technol.* **81**(2), 103–108.

Atiyeh, R. M., Lee, S., Edwards, C. A., Arancon, N. Q., and Metzger, J. D. (2002b). The influence of humic acids derived from earthworm-processed organic wastes on plant growth. *Bioresource Technol.* **84**(1), 7–14.

Barber, S. A. (1984). *Soil Nutrient Bioavailability. A Mechanistic Approach*, Wiley-Interscience, New York.

Bérczi, A., and Møller, I. M. (2000). Redox enzymes in the plant plasma membrane and their possible roles. *Plant Cell Environ.* **23**, 1287–1302.

Blanchet, R. M. (1958). The direct and indirect effect of humified, organic matter on the nutrition of vascular plants. *Ann. Agron.* **9**, 499–532.

Bollag, J.-M., Dec, J., and Huang, P. M. (1998). Formation mechanisms of complex organic structures in soil habitats. *Adv. Agron.* **63**, 237–266.

Bottomley, W. B. (1914a). Some accessory factors in plant growth and nutrition. *Proc. R. Soc. London (Biol.)* **88**, 237–247.

Bottomley, W. B. (1914b). The significance of certain food substances for plant growth. *Ann. Bot. London* **28**, 531–540.

Bottomley, W. B. (1917). Some effects of organic growth-promotion substances (auximones) on the growth of *Lemma minor* in mineral cultural solutions. *Proc. R. Soc. London (Biol.)* **89**, 481–505.

Bottomley, W. B. (1920). The effect of organic matter on the growth of various plants in culture solutions. *Ann. Bot. London* **34**, 353–365.

Burdon, J. (2001). Are the traditional concepts of the structures of humic substances realistic? *Soil Sci.* **166**(11), 752–769.

Cameron, R. S., Swift, R. S., Thornton, K. B., and Posner, A. M. (1972a). Calibration of gel permeation chromatography materials for use with humic acid. *J. Soil Sci.* **23**, 342–349.

Cameron, R. S., Thornton, B. K., Swift, R. S., and Posner, A. M. (1972b). Molecular weight and shape of humic acid from sedimentation and diffusion measurements on fractionated extracts. *J. Soil Sci.* **23**, 394–408.

Canellas, L. P., Olivares, F. L., Okorokova-Façanha, A. L., and Façanha, A. R. (2002). Humic acids isolated from earthworm compost enhance root elongation, lateral root emergence, and plasma membrane H^+-ATPase activity in maize roots. *Plant Physiol.* **130**, 1951–1957.

Cesco, S., Römheld, V., Varanini, Z., and Pinton, R. (2000). Solubilization of iron by water-extractable humic substances. *J. Plant Nutr. Soil Sc.* **163**(3), 285–290.

Cesco, S., Nikolic, M., Römheld, V., Varanini, Z., and Pinton, R. (2002). Uptake of ^{59}Fe- from soluble ^{59}Fe–humate complexes by cucumber and barley plants. *Plant Soil* **241**(1), 121–128.

Chaminade, R. (1958). Influence de la matière organique humifiee sur l'efficacité de l'azote. *Ann. Agron. (Paris)* **9**, 167–192.

Chaminade, R. (1966). Effet physiologique des constituants de la matière organique des sols, sur le métabolisme des plantes, la croissance et le rendement. In *The Use of Isotopes in Soil Organic Matter Studies*, Report FAO/IAEA Technical Meeting Brunswick Volkenrode, 1963, Pergamon, Oxford, pp. 35–47.

Chen, Y., and Aviad, T. (1990). Effects of humic substances on plant growth. In *Humic Substances in Soil and Crop Science: Selected Readings*, MacCarthy, P., Malcolm, R. L., Clapp, C. E., and Bloom, P. R., eds., American Society of Agronomy and Soil Science Society of America, Madison, WI, pp. 161–187.

Chen, Y., Magen, H., and Riov, J. (1994). Humic Substances originating from rapidly decomposing organic matter: Properties and effects on plant growth. In *Humic Substances in the Global Environment and Implications on Human Health*, Senesi, N., and Miano, T. M., eds., Elsevier, Amsterdam, pp. 427–443.

Chen, Y., De Nobili, M., and Aviad, T. (2004a). Stimulatory effects of humic substances on plant growth. In *Soil Organic Matter in Sustainable Agriculture*, Magdoff, F. R., and Weil, R. R., eds., CRC Press, Boca Raton, FL, pp. 103–129.

Chen, Y., Clapp, C. E., and Magen, H. (2004b). Mechanisms of plant growth stimulation by humic substances: the role of organo-iron complexes. *Soil Sci. Plant Nutr.* **50**(7), 1089–1095.

Clapp, C. E., and Hayes, M. H. B. (1999). Sizes and shapes of humic substances. *Soil Sci.* **164**(11), 777–789.

Clapp, C. E., Chen, Y., Hayes, M. H. B., and Cheng, H. H. (2001). Plant growth promoting activity of humic substances. In *Understanding and Managing Organic Matter in Soils, Sediments, and Waters*, Swift, R. S., and Sparks, K. M., eds., International Humic Science Society, St. Paul, MN, pp. 243–255.

Clark, N. A., and Roller, E. M. (1924). "Auximones" and the growth of the green plant. *Soil Sci.* **17**, 193–198.

Concheri, G., Nardi, S., Piccolo, A., Rascio, N., and Dell'Agnola, G. (1994). Effects of humic fractions on morphological changes related to invertase and peroxidase activities in wheat seedlings. In *Humic Substances in the Global Environment and Implications on Human Health*, Senesi, N., and Miano, T. M., eds., Elsevier Science, pp. 257–262.

Dahm, H., Sitek, J. M., and Strzelczyk, E. (1977). Synthesis of auxins by bacterial isolated from the roots of pine seedlings inoculated with rusty forest soil. *Pol. J. Soil Sci.* **10**, 131–137.

Dell'Agnola, G., and Nardi, S. (1987). Hormone-like effect of enhanced nitrate uptake induced by depolycondensed humic fractions obtained from *Allolobophora rosea* and *A. caliginosa* faeces. *Biol. Fertil. Soils* **4,** 115–118.

De Nobili, M., and Chen, Y. (1999). Size Exclusion chromatography of humic substances: Limits, perspectives and prospectives. *Soil Sci.* **164**(11), 825–833.

De Saussure, N. T. (1804). *Recherches chimiques sur la végétation*, Paris.

Ehsanpour, A. A., and Fatahian, N. (2003). Effects of salt and proline on *Medicago sativa* callus. *Plant Cell Tissue Organ Cult.* **73**, 53–56.

Elgala, A. M., Metwally, A. J., and Khalil, R. A. (1978). The effect of humic acid and Na_2 EDDHA on the uptake of Cu, Fe and Zn by barley in sand culture. *Plant Soil* **49**, 41–48.

Fendler, J. H., and Fendler, E. J. (1975). *Catalysis in Micellar and Macromolecular Systems*, Academic Press, New York, pp. 19–41.

Ferretti, M., Ghisi, R., Nardi, S., and Passera, C. (1991). Effect of humic substances on photosynthetic sulphate assimilation in maize seedlings. *Can. J. Soil Sci.* **71**, 239–242.

Flaig, W. (1953). Über den einfluss von spuren von humusstoffen bvw. der modellsubstanzen auf das pflanzenwachstum. *Landwirtsch. Forsch.* **8**, 133–139.

Flaig, W. (1956). Zur chemie der huminsauren und deren modellsubstanzen. In *Transactions 6th International Congress Soil Science*, Paris, France, II, 471–478.

Flaig, W. (1968). Uptake of organic substances from soil organic matter by plants. In *Study Week on Organic Matter and Soil Fertility*, Wiley-Interscience, New York, pp. 723–776.

Flaig, W., Beutelspacher, H., and Rietz, E. (1975). Chemical composition and physical properties of humic substances. In *Soil Components*, Vol **I.**, Gieseking, J. E., ed., Springer-Verlag, New York, pp. 1–219.

Forde, B. G. (2000). Nitrate transport in plants: Structure, function, and regulation. *Biochim. Biophys. Acta* **1469**, 219–235.

Forde, B. G., and Clarkson, D. T. (1999). Nitrate and ammonium nutrition of plants: physiological and molecular perspectives. *Adv. Bot. Res.* **30**, 1–90.

Frankenberger, W. T., and Arshad, M. (1995). *Phytormones in Soils*, Marcel Dekker, New York.

Frimmel, F. H., and Christman, R. F. (1988). *Humic Substances and Their Role in the Environment*, John Wiley & Sons, Chichester.

Führ, F., and Sauerbeck, D. (1967a). The uptake of colloidal organic substances by plant roots as shown by experiments with ^{14}C-labelled humus compounds. In *Report FAO/IAEA Meeting, Vienna*, Pergamon, Oxford, pp. 73–82.

Führ, F., and Sauerbeck, D. (1967b). The uptake of straw decomposition products by plant roots. In *Report FAO/IAEA Meeting, Vienna*, Pergamon, Oxford, pp. 317–327.

Ghorobekova, C. (1987). Humic acids as regulators of activity and biosynthesis of proteolytic enzymes. *FECS Int. 3rd Conf. Chem. Biotechnol. Biol. Act. Natl. Prod., 1985* **5**, 427–430.

Glass, A. D. W., and Siddiqi, M. Y. (1995). Nitrogen absorption by plants roots. In *Nitrogen Nutrition in Higher Plants*, Srivastava, H. S., and Singh, R. P., eds., Associate Publishers, New Delhi, India, pp. 21–56.

Glass, A. D. M., Brito, D., Kronzucker, H. J., Kaiser, B., Okamoto, S., Siddiqi, M. Y., Silim, S., Vidmar, J. J., and Zhuo, D. (2001). Nitrogen transport in plants, with an emphasis on regulation of fluxes to match plant demand. *J. Plant Nutr. Soil Sci.* **164**, 199–207.

Gumiński, S. (1968). Present-day views on physiological effects induced in plant organisms by humic compounds. *Sov. Soil Sci.* 1250–1256.

Hamence, J. H. (1946). The determination of auxins in soils, including a note on synthetic growth substances. *Analyst* **71**, 111–116.

Hayes, M. H. B., and Swift, R. S. (1978). The chemistry of soil organic colloids. In *The Chemistry of Soil Constituents*, Greenland, D. J., and Hayes, M. H. B., eds., John Wiley & Sons, Chichester, pp. 179–320.

Hernando, V., Ortega, B. C., and Fortun, C. (1977). Study of the action of two types of humic acid on the maize plant. In *Soil Organic Matter Studies, Vol. 2. Report of IAEA Meeting Vienna*, Pergamon Press, Oxford.

Hirel, B., Bertin, P., Quillere, Â. I., Bourdoncle, W., Attagnant, C., Dellay, C., Gouy, A., Cadiou, S., and Falque, M. (2001). Towards a better understanding of the genetic and physiological basis for nitrogen use efficiency in maize. *Plant Physiol.* **125**, 1258–1270.

Insam, H. (1996). Microorganisms and soil humus. In *Humic Substances in Terrestrial Ecosystems*, Piccolo A., ed., Elsevier, Amsterdam, pp. 265–292.

Khristeva, L. A., Gallushko, A. M., Gorovaya, A. I., Kolbassin, A. A., Shortshoi, L. P., Tkatshenko, L. K., Fot, L. W., and Luk'Yakenko, N. V. (1980). The main aspects of using physiologically active substances of humus nature. *VI International Peat Congress*, Minnesota, pp. 403–404.

Kim, H. Y., and Suzuki, Y. (1989). Changes in assimilated ^{13}C distribution and soluble acid invertase activity of *Zinnia elegans* induced by Uniconazol an inhibitor of gibberellin biosynthesis. *Plant Physiol.* **90**, 316–321.

Kononova, M. M. (1956). Humus der boden der U.D.S.S.R.; seine nature und rolle in bodenbildungsprozessen. In *Transactions 6th International Congress Soil Science*, Paris, France, II, pp. 557–565.

Kononova, M. M. (1966). *Soil Organic Matter*, Pergamon Press, Elmsford, NY.

Lawes, J. B., and Gilbert, J. H. (1905). Collected papers. In *The Book of the Rothamsted Experiments*, Hall, W. H., ed., London.

Lebuhn, M., and Hartmann, A. (1993). Method for determination of indole-3-acetic acid and related compounds of L-tryptophan catabolism in soils. *J. Chromat.* **629**, 255–266.

Leita, L., De Nobili, M., Catalano, L., Moira, A., Fonda, E., and Vlaic, G. (2001). Complexation of iron-cyanide by humic substances. In *Understanding and Managing Organic Matter in Soils, Sediments and Waters*, Swift, R. S., and Sparks, K. M., eds., International Humic Science Society, St. Paul, MN, pp. 477–482.

Liebig, J. V. (1841). *Organic Chemistry in its Applications to Agriculture and Physiology* (translated by Webster, J. W., and Owen, J.), Cambridge.

Liebig, J. V. (1856). On some points of agricultural chemistry. *J. R. Agric. Soc.* **17**, 284–326.

Lüthje, S., Döring, O., Heuer, S., Lüthen, H., and Böttger, M. (1997). Oxidoreductases in plant plasma membranes. *Biochim. Biophys. Acta* **1331**, 81–102.

MacCarthy, P. (2001). The principles of humic substances. *Soil Sci.* **166**(11), 738–751.

MacCarthy, P., Malcolm, R. L., Clapp, C. E., and Bloom, P. R. (1990). An introduction to soil humic substances. In *Humic Substances in Soil and Crop Science: Selected Readings*, MacCarthy, P., Malcolm, R. L., Clapp, C. E., and Bloom, P. R., eds., American Society of Agronomy and Soil Science Society of America, Madison, WI, pp. 1–12.

Maggioni, A., Varanini, Z., Nardi, S., and Pinton, R. (1987). Action of soil humic matter on plant roots: stimulation of ion uptake and effects on (Mg^{2+} + K^+) ATPase activity. *Sci. Total Environ.* **62**, 355–363.

Malcolm, R. E., and Vaughan, D. (1978). Effects of humic acid fractions on invertase activities in plant tissues. *Soil Biol. Biochem.* **11**, 65–72.

Marschner, H., and Römheld, V. (1994). Strategies of plants for acquisition of iron. *Plant Soil* **165**, 261–274.

Merlo, L., Ghisi, R., Rascio, N., and Passera, C. (1991). Effects of humic substances on carbohydrate metabolism of maize leaves. *Can. J Plant Sci.* **71**, 419–425.

Miller, A. J., and Smith, S. J. (1996). Nitrate transport and compartmentation in cereal root cells. *J. Exp. Bot.* **47**, 843–854.

Morsomme, P., and Boutry, M. (2000). The plant plasma membrane H^+-ATPase: Structure, function and regulation. *Biochim. Biophys. Acta* **1465**, 1–16.

Morsomme, P., d'Exaerde, A. D., DeMeester, S., Thines, D., Goffeau, A., and Boutry, M. (1996). Single point mutations in various domains of a plant plasma membrane H^+-ATPase expressed in *Saccharomyces cerevisiae* increase H^+- pumping and permit yeast growth at low pH. *EMBO J.* **15**, 5513–5526.

Muscolo, A., Felici, M., Concheri, G., and Nardi, S. (1993). Effect of humic substances on peroxidase and esterase patterns during growth of leaf explants of *Nicotiana plumbaginifolia*. *Biol. Fertil. Soils* **15**, 127–131.

Muscolo, A., Panuccio, M. R., Abenavoli, M. R., Concheri, G., and Nardi, S. (1996). Effect of molecular complexity and acidity of earthworm faeces humic fractions on glutamate dehydrogenase, glutamine synthetase and phosphoenolpyruvate carboxylase in *Daucus carota* II cells. *Biol. Fertil. Soils* **22**, 83–88.

Muscolo, A., Cutrupi, S., and Nardi, S. (1998). IAA detection in humic substances. *Soil Biol. Biochem.* **30**, 1199–1201.

Muscolo, A., Bovalo, F., Gionfriddo, F., and Nardi, S. (1999). Earthworm humic matter produces auxin-like effects on *Daucus carota* cell growth and nitrate metabolism. *Soil Biol. Biochem.* **31**, 1303–1311.

Muscolo, A., Panuccio, M. R., Sidari, M., and Nardi, S. (2000). Effect of two different humic substances on some glycolytic enzymes in callus culture of *Pinus Laricio*. *Humic Subst. Environ.* **2**(1/4), 19–24.

Muscolo, A., Panuccio, M. R., Sidari, M., and Nardi, S. (2005). The effects of humic substances on *Pinus callus* are reversed by 2,4-dichlorophenoxy acetic acid. *J. Chem. Ecol.* **31**(3), 577–590.

Muscolo, A., Sidari, M., Francioso, O., Tugnoli, V., and Nardi, S. (2007a). The auxin-like activity of humic substances is related to membrane interactions in carrot cell cultures. *J. Chem. Ecol.* **33**(1), 115–129.

Muscolo, A., Sidari, M., Attinà, E., Francioso, O., and Nardi, S. (2007b). Biological activity of humic substances is related to their chemical structure. *Soil Sci. Soc. Am. J.* doi:10.2136/sssaj2006.0055.

Mylonas, V. A., and McCants, C. B. (1980). Effects of humic and fulvic acids on growth of tobacco. 2. Tobacco growth and ion uptake. *J. Plant Nutr.* **2**, 377–393.

Nakamura, A., Higuchi, K., Goda, H., Fujiwara, M. T., Sawa, S., Koshiba, T., Shimada, Y., and Yoshida, S. (2003). Brassinolide induces IAA5, IAA19, and DR5, a synthetic auxin response element in *Arabidopsis*, implying a cross talk point of brassinosteroid and auxin signalling. *Plant Physiol.* **133**, 1843–1853.

Nardi, S., Concheri, G., Dell'Agnola, G., and Scrimin, P. (1991). Nitrate uptake and ATPase activity in oat seedlings in the presence of two humic fractions. *Soil Biol. Biochem* **23**, 833–836.

Nardi, S., Panuccio, M. R., Abenavoli, M. R., and Muscolo, A. (1994). Auxin-like effect of humic substances extracted from faeces of *Allolobophora caliginosa* and *A. rosea*. *Soil Biol. Biochem.* **26**, 1341–1346.

Nardi, S., Concheri, G., and Dell'Agnola, G. (1996). Biological activity of humus. In *Humic Substances in Terrestrial Ecosystems*, Piccolo, A., ed., Elsevier, The Netherlands, pp. 361–406.

Nardi, S., Pizzeghello, D., Remiero, F., and Muscolo, A. (1999). Biological activity of humic substances extracted from soils under different vegetation cover. *Commun. Soil Sci. Plant Anal.* **30**(5/6), 621–634.

Nardi, S., Pizzeghello, D., Remiero, F., and Rascio, N. (2000). Chemical and biochemical properties of humic substances isolated from forest soils and plant growth. *Soil Sci. Soc. Am. J.* **64**, 639–645.

Nardi, S., Pizzeghello, D., Muscolo, A., and Vianello, A. (2002). Physiological effects of humic substances on higher plants. *Soil Biol. Biochem.* **34**, 1527–1536.

Nardi, S., Vaccaro, S., Pizzeghello, D., Muscolo, A., Baiano, S., and Piccolo, A. (2005). Effetto di differenti frazioni di sostanze umiche sul metabolismo glicolitico e respiratorio in foglie di plantule di mais (*Zea mays* L.). VI Congresso Sezione Italiana dell'International Humic Substances Society, Perugia, Italy, pp. 61–63.

Nemeth, K., Salchert, K., Putnoky, P., Bhalerao, R., Koncz-Kalman, Z., Stankovic-Stangeland, B., Bako, L., Mathur, J., Okresz, L., Stabel, S., Geigenberger, P., Stitt, M., Redei, G. P., Schell, J., and Koncz, C. (1998). Pleiotropic control of glucose and hormone responses by PRL1, a nuclear WD protein, in Arabidopsis. *Gene. Dev.* **12**(19), 3059–3073.

Niklewski, B. (1931). Influence des éléments colloidaux sur la production végétale. *Ann. Agron.* (Paris) **1**, 337–345.

Orlov, D. S. (1985). *Humus Acids of Soils*. Moscow University Press, translated from Russian, Tan, K. H., ed., Amerind Publishers, New Delhi, India.

Palmgren, M. G. (1998). Proton gradients and plant growth: role of the plasma membrane H^+-ATPase. *Adv. Bot. Res.* **28**, 1–70.

Pandey, A. K., Pandey, S. D., and Misra, V. (2000). Stability constants of metal–humic acid complexes and its role in environmental detoxification. *Ecotoxicol. Environ. Saf.* **47**(2), 195–200.

Pandeya, S. B., Singh, A. K., and Dhar, P. (1998). Influence of fulvic acid on transport of iron in soils and uptake by paddy seedlings. *Plant Soil* **198**, 117–125.

Passera, C., Nicolao, L., Ferretti, M., Rascio, N., and Ghisi, R. (1991). Effect of humic substances of enzyme-activities of sulfate assimilation and chloroplast ultrastructure of maize leaves. *Photosynthetica* **25**(1), 39–45.

Piccolo, A. (2001). The supramolecular structure of humic substances. *Soil Sci.* **166**(11), 810–832.

Piccolo, A., and Conte, P. (1999). Molecular size of humic substances. Supramolecular associations versus macromolecular polymers. *Adv. Environ. Res.* **3**(4), 508–521.

Piccolo, A., Nardi, S., and Concheri, G. (1992). Structural characteristics of humic substances as related to nitrate uptake and growth regulation in plant systems. *Soil Biol. Biochem.* **24**, 373–380.

Piccolo, A., Celano, G., and Pietramellara, G. (1993). Effects of fractions of coal-derived humic substances on seed-germination and growth of seedlings (*Lactuga sativa* and *Lycopersicon esculentum*). *Biol. Fertil. Soils* **16**(1), 11–15.

Piccolo, A., Nardi, S., and Concheri, G. (1996a). Macromolecular changes of soil humic substances induced by interactions with organic acids. *Eur. J. Soil Sci.* **47**, 319–328.

Piccolo, A., Nardi, S., and Concheri, G. (1996b). Micelle-like conformation of humic substances as revealed by size-exclusion chromatography. *Chemosphere* **33**, 595–600.

Piccolo, A., Cozzolino, A., Conte, P., and Spaccini, R. (2000). Polymerization of humic substances by an enzyme catalyzed oxidative coupling. *Naturwissenschaften* **87**, 391–394.

Pinton, R., Varanini, Z., Vizzotto, G., and Maggioni, A. (1992). Soil humic substances affect transport properties of tonoplast vesicles isolated from oat roots. *Plant Soil* **142**, 203–210.

Pinton, R., Cesco, S., Santi, S., and Varanini, Z. (1995). Effect of soil humic substances on surface redox activity of oat roots. *J. Plant Nutr.* **18**, 2111–2120.

Pinton, R., Cesco, S., Santi, S., and Varanini, Z. (1997). Soil humic substances stimulate proton release by intact oat seedling roots. *J. Plant Nutr.* **20**, 857–869.

Pinton, R., Cesco, S., Iacoletti, G., Astolfi, S., and Varanini, Z. (1999a). Modulation of NO_3^- uptake by water-extractable humic substances: Involvement of root plasma membrane H^+ATPase. *Plant Soil* **215**, 155–161.

Pinton, R., Cesco, S., Santi, S., Agnolon, F., and Varanini, Z. (1999b). Water extractable humic substances enhance iron deficiency responses by Fe deficient cucumber plants. *Plant Soil* **210**, 145–157.

Pizzeghello, D., Nicolini, G., and Nardi, S. (2001). Hormone-like activity of humic substances in *Fagus sylvaticae* L. forests. *New Phytol.* **151**, 647–657.

Pizzeghello, D., Nicolini, G., and Nardi, S. (2002). Hormone-like activities of humic substances in different forest ecosystems. *New Phytol.* **155**, 393–402.

Pizzeghello, D., Trevisan, S., Quaggiotti, S., Reperti, B., Palme, K., and Nardi, S. (2006). Effetto di sostanze umiche sullo sviluppo della radice in *Arabidopsis thaliana*. XXIV Convegno Nazionale della Società Italiana di Chimica Agraria, Alghero, Italy, p. 92.

Prát, S., and Pospíšil, F. (1959). Humic acids with ^{14}C. *Biol. Plant.* **1**, 71–80.

Quaggiotti, S., Reperti, B., Pizzeghello, D., Francioso, O., Tugnoli, V., and Nardi, S. (2004). Effect of low molecular size humic substances on the expression of genes involved in nitrate transport and reduction in maize (*Zea mays* L.). *J. Exp. Bot.* **55**, 803–813.

Rademacher, W. (1992). Occurrence of gibberellins in different species of the fungal genera *Sphaceloma* and *Elsinoe*. *Phytochemistry* **31**, 4155–4157.

Rauthan, B. S., and Schnitzer, M. (1981). Effect of soil fulvic acid on the growth and nutrient content of cucumber (*Cucumis sativum*) plants. *Plant Soil* **63**, 491–495.

Raven, J. A., and Smith, F. A. (1976). Nitrogen assimilation and transport in vascular land plants in relation to intracellular pH regulation. *New Phytol.* **76**, 415–431.

Rerabek, J. (1960). Humic acid interactions in the growth process. *Biol. Plant.* **2**, 88–97.

Rice, J. A., and MacCarthy, P. (1990). A model of humin. *Environ. Sci. Technol.* **24**, 1875–1877.

Rice, J. A., and MacCarthy, P. (1991). Statistical evaluation of the chemical composition of humic substances. *Org. Geochem.* **17**, 635–648.

Robert, D., and Roland, J. C. (1989). *Biologie Végétale, Organisation Cellulaire*, Doin, Paris.

Russell, L., Stokes, A. R., Macdonald, H., Muscolo, A., and Nardi, S. (2006). Stomatal responses to humic substances and auxin are sensitive to inhibitors of phospholipase A_2. *Plant Soil* **283**, 175–185.

Samson, G., and Visser, S. S. (1989). Surface-active effects of humic acids on potato cell membrane properties. *Soil Biol. Biochem.* **21**, 343–347.

Santi, S., Locci, G., Pinton, R., Cesco, S., and Varanini, Z. (1995). Plasma membrane H^+-ATPase in maize roots induced for NO_3^- uptake. *Plant Physiol.* **109**, 1277–1283.

Scherer, G. F. E., and André, B. (1989). A rapid response to a plant hormone: Auxin stimulates phospholipase A_2 in vivo and in vitro. *Biochem. Biophys. Res. Commun.* **163**, 111–117.

Schmidt, W. (2003). Iron solutions: Acquisition strategies and signalling pathways in plants. *Trends Plant Sci.* **8**(4), 188–193.

Schmidt, W., Cesco, S., Santi, S., Pinton, R., and Varanini, Z. (2005). Water-extractable humic substances as nutrient acquisition signals for root hairs development in *Arabidopsis*. In *Rizosphere 2004—Perspectives and Challenges*, Hartmann, A., Schmid, M., Wenzel, W., and Hinnsinger, P., eds., GSF-Berich, Neuherberg, p. 71.

Schnitzer, M. (1977). Recent findings on the characterization of humic substances extracted from soils from widely differing climatic zones. In *Proceedings of the Symposium on Soil Organic Matter Studies*, Braunsweig, Germany, pp. 117–131.

Schnitzer, M., and Khan, S. U. (1972). *Humic Substances in the Environment*. Marcel Dekker, New York.

Schulten, H. R., and Leinweber, P. (2000). New insights into organic-mineral particles: Composition, properties and models of molecular structure. *Biol. Fert. Soils* **30**(5–6), 399–432.

Schwertmann, U. (1966). Inhibitory effect of soil organic matter on the crystallization of amorphous ferric hydroxide. *Nature* **212**, 645–646.

Sessi, E., Nardi, S., and Gessa, C. (2000). Effects of low and high molecular weight humic substances from two different soils on nitrogen assimilation pathway in maize seedlings. *Humic Subst. Environ.* **2**(1/4), 39–46.

Sladký, Z. (1959). The effect of extracted humus substances on growth of tomato plants. *Biol. Plant.* **1**, 142–150.

Sladký, Z., and Tichý, V. (1959). Applications of humus substances to over ground organs of plants. *Biol. Plant.* **1**, 9–15.

Soil Science Society of America (1996). *Glossary of Soil Science Terms*, SSSA, Madison, WI, p. 134.

Stevenson, F. J. (1982). *Humus Chemistry—Genesis, Composition, Reactions*, John Wiley & Sons, New York, p. 443.

Stevenson, F. J. (1994). *Humus Chemistry: Genesis, Composition, Reactions*, 2nd ed., John Wiley & Sons, New York, p. 496.

Stewart, W. S., and Anderson, M. S. (1942). Auxins in some american soils. *Bot. Gaz.* **103**, 570–575.

Sutton, R., and Sposito, G. (2005). Molecular structure in soil humic substances: The new view. *Environ. Sci. Technol.* **39**(23), 9009–9015.

Swift, R. S. (1989). Molecular weight, size, shape, and charge characteristics of humic substances: some basic considerations. In *Humic Substances II. In Search of a Structure*, Hayes, M. H. B., MacCarthy, P., Malcom, R. L., and Swift, R. S., eds., John Wiley & Sons, Chichester, pp. 449–466.

Swift, R. S. (1999). Macromolecular properties of soil humic substances: Fact, fiction and opinion. *Soil Sci.* **164**(11), 790–802.

Tan, K. H. (2003). *Humic Matter in Soil and the Environment*, Marcel Dekker, New York.

Thaer, A. D. (1808). *Grundriss der Chemie für Landwirte*, Berlin.

Thaer, A. D. (1846). *The Principles of Agriculture*, Shaw, W., and Johnson, C. W., translators, Ridgway, London.

Thomas, S. M., Thorne, G. N., and Pearman, I. (1978). Effect of nitrogen on growth, yield and photorespiratory activity in spring wheat. *Ann. Bot.* **42**, 827–837.

Van de Venter, H. A., Furter, M., Dekker, J., and Cronje, I. J. (1991). Stimulation of seedling root growth by coal-derived sodium humate. *Plant Soil* **138**, 17–21.

Varanini, Z., Pinton, R., De Biasi, M. G., Astolfi, S., and Maggioni, A. (1993). Low molecular weight humic substances stimulated H^+-ATPase activity of plasma membrane vesicles isolated from oat (*Avena sativa* L.) roots. *Plant Soil* **153**, 61–69.

Varanini Z., and Pinton, R. (1995). Humic substances and plant nutrition. In *Progress in Botany*, Vol. **56**, Lüttge, U., ed., Springer, Berlin, pp. 97–117.

Varanini, Z., and Pinton, R. (2001). Direct versus indirect effects of soil humic substances on plant growth and nutrition. In *The Rizosphere*, Pinton, R., Varanini, Z., and Nannipieri, P., eds., Marcel Dekker, Basel, pp. 141–158.

Vaughan, D. (1967a). The stimulation of invertase development in aseptic storage tissue slices by humic acids. *Soil Biol. Biochem.* **1**, 15–28.

Vaughan, D. (1967b). Effect of humic acid on the development of invertase activity in slices of beetroot tissues washed under aseptic conditions. *Humus Planta* **IV**, 268–271.

Vaughan, D. (1986). Effetto delle sostanze umiche sui processi metabolici delle piante. In *Sostanze Umiche Effetti sul Terreno e sulle Piante*, Burns, R. G., Dell'Agnola, G., Miele, S., Nardi, S., Savoini, G., Schnitzer, M., Sequi, P., Vaughan, D., and Visser, S. A., eds., Ramo Editoriale degli Agricoltori, Roma, pp. 59–81.

Vaughan, D., Baker, C. D., and Willoughby, L. G. (1974a). Some aspects of humic acid on two different biological systems. *Plant Soil* **40**, 429–434.

Vaughan, D., Cheshire, M. V., and Mundie, C. M. (1974b). Uptake by beet root tissue and biological activity of ^{14}C-labelled fractions of soil organic matter. *Biochem. Soc. T.* **2**, 126–129.

Vaughan, D., and Linehan, D. J. (1976). The growth of wheat plants in humic acid solutions under axenic conditions. *Plant Soil* **44**, 445–449.

Vaughan, D., and McDonald, I. R. (1976). Some effects of humic acid on the cation uptake by parenchyma tissue. *Soil Biol. Biochem.* **8**, 415–421.

Vaughan, D., and Malcolm, R. E. (1979a). Effect of humic acid on invertase synthesis in roots of higher plants. *Soil Biol. Biochem.* **11**, 247–272.

Vaughan, D., and Malcolm, R. E. (1979b). Effect of soil organic matter in peroxidase activity of wheat roots. *Soil Biol. Biochem.* **11**, 57–63.

Vaughan, D., and Malcolm, R. E. (1985). Influence of humic substances on growth and physiological processes. In *Soil Organic Matter and Biological Activity*, Vaughan, D., and Malcolm, R. E., eds., Martinus Nijhoff/Junk, Dordrecht, The Netherlands, pp. 37–76.

Vaughan, D., Malcolm, R. E., and Ord, B. G. (1985). Influence of humic substances on biochemical processes in plants. In *Soil Organic Matter and Biological Activity*, Vaughan, D., and Malcolm, R. E., eds., Martinus Nijhoff/Junk, Dordrecht, The Netherlands, pp. 77–108.

Vaughan, D., and Ord, B. G. (1981). Uptake and incorporation of ^{14}C-labelled soil organic matter by roots of *Pisum sativum* L. *J. Exp. Bot.* **32**, 679–687.

Vianello, A., and Macrì, F. (1991). Generation of superoxide anion and hydrogen peroxide at the surface of plant cells. *J. Bioenerg. Biomembr.* **23**, 409–423.

Vidal, K., Guermache, F., and Widmer, T. (2004). In vitro culturing of yellow starthistle (*Centaurea solstitialis*) for screening biological control agents. *Biol. Control* **30**, 330–335.

Visser, S. A. (1986). Effetto delle sostanze umiche sulla crescita delle piante. In *Sostanze Umiche. Effetti sul Terreno e sulle Piante*, Burns, R. G., Dell'Agnola, G., Miele, S., Nardi, S., Savoini, G., Schnitzer, M., Sequi, P., Vaughan, D., and Visser, S. A., eds., Ramo Editoriale degli Agricoltori, Roma, pp. 96–143.

Visser, S. A. (1987). Effect of humic substances on mitochondrial respiration and oxidative phosphorylation. *Sci. Total Environ.* **62**, 347–354.

Wang, R. C., and Crawford, N. M. (1996). Genetic identification of a gene involved in constitutive, high-affinity nitrate transport in higher plants. *Proc. Natl. Acad. Sci. USA* **93**, 9297–9301.

Wershaw, R. L. (1999). Molecular aggregation of humic substances. *Soil Sci.* **164**(11), 803–813.

Woodwell, G. M., and Houghton, R. H. (1977). Biotic influences on the world carbon budget. In *Global Chemical Cycles and Their Alterations by Man*, Stumm, W., ed., John Wiley & Sons, New York, pp. 61–72.

Woodwell, G. M., Whittaker, R. H., Reiners, W. A., Likens, G. E., Delwiche, C. C., and Botkin, D. B. (1978). The biota and the world carbon budget. *Science* **199**, 141–146.

9

ROLE OF HUMIC SUBSTANCES IN THE RHIZOSPHERE

R. PINTON AND S. CESCO

Dipartimento di Scienze Agrarie e Ambientali, University of Udine, Udine Italy

Z. VARANINI

Dipartimento di Scienze, Tecnologie e Mercati della Vite e del Vino, University of Verona S. Floriano, Verona, Italy

9.1. INTRODUCTION

Plant survival and crop productivity are strictly dependent on the capability of plants to adapt to different environments. This adaptation is the result of the interaction between roots and biotic and abiotic components of soil, which can determine

Biophysico-Chemical Processes Involving Natural Nonliving Organic Matter in Environmental Systems, Edited by Nicola Senesi, Baoshan Xing, and Pan Ming Huang

changes both in root physiology (Marschner et al., 1986) and anatomy (McCully, 1999) and in the chemical, physical, and microbiological properties of the soil. These phenomena occur in a limited area surrounding the root, the rhizosphere, where nutrient, energy, and signal exchanges make this environment decisively different from bulk soil, from a chemical, a physical, and a microbiological point of view (Pinton et al., 2001). It is well known that movement of water, nutrients, and microbial dynamics are more intricate around the roots than in bulk soil. Changes in pH and redox potential occur frequently. Furthermore, the rhizosphere generally experiences higher mineral weathering rates than bulk soil and is characterized by variable rates of native organic matter turnover. Mineralization and humification processes are likely to be influenced by the unique characteristics of the rhizosphere, thus determining the amount and composition of the organic matter present in this environment. Therefore, dissolved organic matter and humic substances present in the rhizosphere can be considerably different from those of bulk soil due to the plant–microbial–soil particle interactions. Furthermore, the humified organic fractions can affect nutrient dynamics as well as plant growth and nutrition (Varanini and Pinton, 2001).

This chapter will briefly consider the main chemical, biochemical, and biological characteristics of the rhizosphere, with particular emphasis on the presence of gradients (ions, organic molecules, pH, redox, rhizodepositions) determining nutrient cycles and microbial activity (growth and diversity) which in turn may affect the organic matter turnover and humic substances (trans)formation in this environment. The main aim is to describe the role of humic substances on the soil–root relationships, their action as a source of nutrients (including the contribution of associated enzymes), and their complexing and reducing properties (that can affect availability of nutrients and toxic elements). Furthermore, evidence for a more direct action of humic substances on plant nutrition will be presented, such as that exerted on root growth and the mechanisms behind nutrient uptake. With respect to the latter, the possible role of humic substances as rhizospheric signals inducing nutrient acquisition responses, similar to those evoked by fluctuations in the concentration of some nutrients, will be discussed.

9.2. CHEMISTRY AND BIOCHEMISTRY OF THE RHIZOSPHERE

The term *rhizosphere* was first used by Hiltner (1904) to indicate the area of the soil where root exudates released from plant roots can stimulate, inhibit, or have no effect on activities of soil microorganisms; nowadays it is generally used to define "the field of action or influence of a root."

Roots vary enormously in their morphology, longevity, activity, and influence on soil as a result of physiological, environmental, and genetic differences. It can be assumed that a rhizosphere forms around each root as it grows, due to the changes in the physical, chemical, and biological properties of the soil in its immediate vicinity. Specifically, root activities such as growth and water uptake can directly alter physical properties of the rhizosphere. Growth can exert considerable forces which ultimately alter soil density, porosity, and strength (Dexter, 1987). Water uptake can result in substantial and rapid changes in water potential around roots affecting not only the radial transfer of water and solutes into the plant, but also microbial activity in the rhizosphere (Doussan et al., 2003).

It is well known that roots are able to release a wide range of inorganic and organic compounds into the rhizosphere. Soil chemical changes related to the presence of these compounds and their products of microbial degradation are important factors affecting microbial populations, availability of nutrients, solubility of toxic elements, and thereby the ability of plants to cope with adverse soil-chemical conditions (Neumann and Römheld, 2007). In addition, such modifications can determine changes in chemical properties of the soil solution in the rhizosphere, capable in turn of exerting an influence on a wide range of reactions at the soil solid–solution interface. Many of these reactions are involved in the weathering of soil minerals and soil formation processes (pedogenesis) (Hinsinger et al., 1993; Courchesne and Gobran, 1997).

The rhizosphere lacks a precise physical delimitation; rather, it can be described in terms of the longitudinal and radial gradients that develop along the axis of each root as a result of root growth and metabolism, nutrient and water uptake, rhizodeposition, and subsequent microbial growth (Tinker and Nye, 2000). In principle, it can be assumed that there will be gradients with depletion profiles (i.e., the solute concentrations will be lowest at the root surface), as in the case of some plant nutrients (mainly P and K), and accumulation gradients (i.e., solute concentrations are highest at the root surface), as in the case of Ca, Mg and the soluble organic solutes released by the roots (Barber, 1995; Jones and Darrah, 1996). Depending on root type and root portion, root activities change; given the heterogeneous nutrient distribution in soil, the gradients formed in the rhizosphere show a large variability and very distinct microenvironments.

A schematic presentation of the fluxes and gradients occurring in the rhizosphere is reported in Figure 9.1.

9.2.1. Gradients at the Rhizosphere

One of the primary functions of the root is the uptake of nutrients from the soil solution; this activity determines the formation of radial and longitudinal ion gradients in the rhizosphere.

9.2.1.1. Ions. For those nutrients that are present at low concentrations in the soil solution (e.g., K, P), compared with plant's requirement, root uptake results in a decrease in their concentration in the rhizosphere; this phenomenon creates a depletion zone, which is the driving force for the diffusion of those nutrients toward the root surface (Tinker and Nye, 2000; Jungk, 2002). Such depletion zones have been observed in the rhizosphere of many crop plant species for phosphate and for other macronutrient ions such as potassium and, to a lesser extent, nitrate. It has been established that the depletion zone generally extends from <1 mm for phosphate to some millimeters for K, and up to several centimeters for nitrate (Hinsinger, 1998; Jungk, 2002), although it is variable depending on root hair length. Depletion zones of poorly mobile trace pollutants can also develop in the rhizosphere, as shown recently for arsenic (Fitz et al., 2003), nickel (Puschenreiter et al., 2005), and thallium (Al-Najar et al., 2003) in metal hyperaccumulator plants.

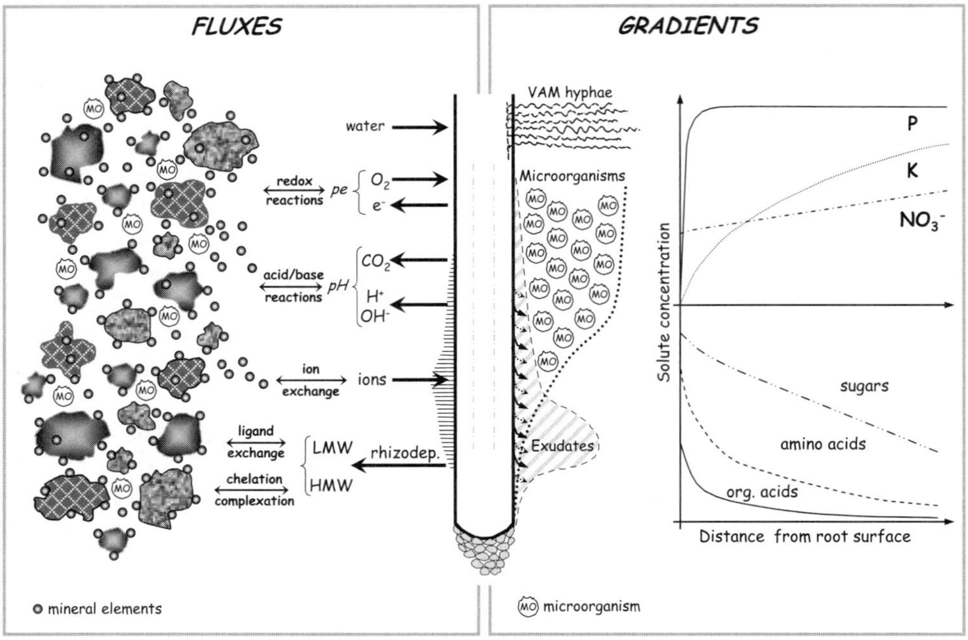

Figure 9.1. Schematic representation of the main fluxes and gradients occurring in the rhizosphere.

The diffusive flux of nutrient also depends on the diffusion coefficient (Jones and Darrah, 1996; Jungk, 2002). The actual diffusion of any nutrient in soils is much slower than in water, because of physical and chemical interactions with the soil solid phase (Tinker and Nye, 2000): The soil water content, tortuosity, and buffer power (ability of soil solid phase to replenish the soil solution) determine the effective diffusion coefficient in soils. Extension of the depletion zone is also related to morphological features of roots. These aspects vary among genotypes (Ge et al., 2000) and are strongly influenced by environmental cues, such as the presence of nutrient-rich patches (Hodge, 2004) and nutrient availability (Lopez-Bucio et al., 2003). It has been shown also that as a consequence of their sink effect and of the depletion of K in solution below a threshold concentration, plant roots could induce the release of interlayer K from K-bearing phyllosilicates such as micas and illites (Hinsinger and Jaillard, 1993).

In addition to this process, the uptake of water by roots results in a movement of water from the bulk soil toward the root surface, along a gradient of water potential. Because the soil solution contains water and dissolved nutrients, the uptake of water by roots generates a convective movement of solutes, that is, mass flow. The corresponding flux of nutrient depends on the flux of water and nutrient concentration in the soil solution (Barber, 1995; Jungk, 2002). For those nutrients that are present in high concentrations in the soil solution, which is typically the case of nitrate in many agricultural soils, mass flow can contribute as a major proportion in the plant acquisition process. For Ca and Mg, it can even exceed the actual flux of nutrients absorbed by the plant, thus resulting in a buildup of their concentration in the rhizosphere. This may ultimately result in the precipitation of Ca salts around roots, such as $CaCO_3$ in calcareous soils (Hinsinger, 1998); this process may severely limit the root-induced acidification of the rhizosphere. In saline soils, accumulation of Na and Cl can occur. Moreover, accumulation of toxic elements (e.g., Al) in acidic soils can be observed in the rhizosphere of plants, which may restrict nutrient uptake.

The transpiration-driven convective movement of water toward the roots may also cause accumulation in the rhizosphere of water-soluble naturally occurring organic molecules (e.g., soluble humic fractions) and/or added into the soil solution (e.g., organic pollutants, Kuiper et al., 2004), leading to (a) changes in biological activities caused by such compounds and (b) possible detoxification processes. Humic fractions have been found in the soil solution, with a highly variable concentration depending on soil type and ranging from 1 up to 400 mg liter^{-1} (Chen and Schnitzer, 1978; Gerke, 1997; Cesco et al., 2000); accumulation of these fractions in the root surroundings might contribute to the chemical complexity of the rhizosphere and affect the metabolism and growth of organisms (roots and microbes) living in that environment.

9.2.1.2. pH and Redox. Root-induced changes in the concentrations of nutrients are not solely due to the diffusion gradient created by root uptake. It is now widely accepted that other root-induced processes occurring in the rhizosphere, such as pH changes, could cause chemical changes affecting nutrient availability (Marschner, 1995).

The changes in pH found in the rhizosphere depend mainly on the ratio between the absorption of anions and cations by roots: in general terms, the preferential uptake of anions causes alkalinization of the rhizosphere, while the preferential uptake of cations causes acidification of the rhizosphere. For example, rhizosphere pH changes are especially pronounced during the uptake of nitrate (pH increase) or ammonium (pH decrease). Other factors that can influence rhizospheric pH in a lower degree include organic acid extrusion, root and microorganism respiration, and redox-coupled pH changes (Hinsinger et al., 2003). In the root cytoplasm, organic acids exist as anions, and it is in this form that they are released into the rhizosphere; thus, even when a massive release occurs (e.g., as in the case of P deficiency—see below), acidification can be observed only when efflux is accompanied by a concomitant proton extrusion, generally due to a strong activation of the root plasma membrane H^+-ATPase, in order to maintain transmembrane charge balance (Ryan et al., 2001).

The contribution of root and microbial respiration to rhizosphere acidification depends on the amount of CO_2 released, the amount of organic material oxidized to CO_2, and the initial pH of the soil (Hinsinger et al., 2003). However, since the first pK_a of H_2CO_3 is 6.36, the contribution of these processes to soil acidification will be significant only in neutral and alkaline soils.

Because of the large demand for O_2 by roots and rhizosphere microorganisms, a large change in the soil's redox potential can be expected. Most redox processes are coupled with the release (oxidation) or consumption (reduction) of protons and therefore may cause pH changes also.

Microbial-driven mineralization of organic matter can also contribute to acidification of the rhizosphere (Badalucco and Nannipieri, 2007). It should be noted that pH variations in the rhizosphere depend also on the soil's buffering capacity

(Hinsinger et al., 2003). This capacity is generally high because of the ability of numerous soil constituents to trap or release protons via (i) cation exchange reactions, (ii) protonation–deprotonation of weak acid groups, and (iii) dissolution or precipitation of soil minerals (reactions which consume or produce protons, respectively). The buffering capacity also varies considerably with soil composition: It is expected to be minimal in sandy soil (which contain little organic matter) and maximal in calcareous soils (because of the consumption of H^+ when calcium carbonates dissolve).

Variation in pH in the rhizosphere can have important effects on the availability of poorly mobile nutrients, such as P, Fe, and Mn, because their solubility depends on pH. Localized release of protons in some root portions, most often in the subapical zone, has frequently been reported to occur as a response to P (Neumann and Römheld, 1999) and Fe (Römheld and Marschner, 1981) shortage.

Alkalinization can be produced by tolerant species, to prevent damages from toxic elements, such as Al (Bagayoko et al., 2000).

Whatever the cause of the changes in rhizosphere pH, the corresponding increase or decrease of proton concentration will promote the dissolution or precipitation of a range of soil minerals; the direct implication of root-induced release of protons in the dissolution of phosphates, silicates, or oxides has been reported (Hinsinger et al., 1993; Hinsinger and Gilkes, 1996; Bertrand and Hinsinger, 2000).

In well-aerated soils, the redox potential varies on average between +500 and +700 mV, but these values can change significantly due to the presence of anoxic microsites (Barber, 1995). The microorganisms and root respiration, as well as the presence of oxido-reduction activity on the surface of the root, may considerably alter the redox potential of the rhizosphere, which is in general characterized by a high frequency of sites with low redox potential. These situations are especially relevant for the absorption of elements such as iron and manganese, which is favored by reducing conditions (Marschner, 1995). A peculiar case is represented by plants grown in flood environment (soils with a negative redox potential), such as rice plants. These are able to maintain an elevated redox potential in their rhizosphere, via a transport through the aerenchyma of oxygen, from the aerial tissues to the roots, where it is released into the rhizosphere. Rhizosphere oxidation is vital to diminish the phytotoxic concentrations of Fe^{2+} and Mn^{2+} found in the flooded environment.

9.2.1.3. Organic Rhizodeposition. Most of the organic compounds present in the rhizosphere are plant-derived organic molecules.

Carbohydrate partitioning from shoot to roots can comprise a substantial proportion of plant-photosynthetic CO_2 fixation (20–70%), and 4–70% of this fraction can be released into the soil as organic rhizodeposition (Lynch and Whipps, 1990; Grayston et al., 1996). This is not only a significant loss of reduced carbon but can also contribute by 30–40% to the total input of soil-organic matter with considerable impact as a source of carbon and nitrogen for soil microorganisms.

According to the mechanisms of release, organic rhizodepositions may be grouped into these major fractions: lysates, leachates from sloughed-off cells, and dead tissues as a consequence of root turnover. In contrast, root exudates (2–10% of translocated carbon) are released from intact root cells either passively as diffusates or actively as excretions or secretions with specific functions (Grayston et al., 1996; Neumann and Römheld, 2007). Rhizodeposition can be influenced in qualitative and quantitative terms by biotic and abiotic factors of plant and soil (Jones et al., 2004); it has been widely reported that rhizodeposition is enhanced in response to high light intensities, temperature extremes, mechanical impedance of the substrate, toxic elements, low-soil pH, nutrient limitation, presence of microorganisms, and plant-developmental stage. The quantity of rhizodeposition can be also influenced by the presence of retrieval mechanisms in plant roots, which are able to reabsorb up to 90% of amino acids and sugars released into the rhizosphere (Jones and Darrah, 1994, 1996).

Organic compounds released from sloughed-off root cells and tissues are a major carbon source for rhizosphere microorganisms but may indirectly have an impact as microbial metabolites on nutrient availability and on exclusion of toxic elements in the rhizosphere (Brimecombe et al., 2007). Continuous root turnover is a general feature of plant development, and insoluble root debris may comprise 50–90% of total rhizodeposition (Darrah, 1991).

High-molecular-weight compounds in rhizodeposition are mainly represented by mucilage and ectoenzymes, which are actively released via exocytosis. Mucilage is released from hypersecretory cells, which subsequently degenerate or are sloughed-off as root border cells.

Mucilage has protective functions for the root meristem and improves root–soil contact by inclusion and aggregation of soil particles. It may also contribute to P desorption and to the exclusion of toxic elements (Al, Cd, Pb) by complexation with galacturonates, mainly in exchange with Ca^{2+} (Neumann and Römheld, 2002). Secreted enzymes contribute to the extracellular enzyme pool; it has been shown that the activity of extracellular enzymes, such as phophatases, proteases, and arylsulfatases, exhibit more activity in the rhizosphere relative to the bulk soil and may have a dramatic effect on the cycling of nutrients such as P, N, and S (Badalucco and Nannipieri, 2007).

Rhizodeposition is dominated by low-molecular-weight compounds, mainly sugars, amino acids and carboxylates, but also phytosiderophores, phenolics, vitamins, and hormones (Uren, 2007). Sugars and amino acids are the largest exudate groups; these compounds may serve as microbial substrates and/or be retrieved by the plant roots (Sacchi et al., 2000; Owen and Jones, 2001). Certain phenolics, carboxylates, and phytosiderophores, which exhibit reducing properties and the capability to complex metals, have been proposed to directly affect the availability of nutrients and of toxic elements in the rhizosphere. It has been shown that their release can be upregulated to help alleviate different kinds of stress (e.g., nutritional shortage and Al toxicity) (Neumann and Römheld, 2007).

Grasses are capable of releasing Fe-chelating compounds called phytosiderophores (Marschner and Römheld, 1994). This process is enhanced under Fe deficiency and is thus part of an efficient strategy to acquire Fe in soils where its availability is minimal, such as calcareous soils. These root exudates can also strongly chelate other metal micronutrients such as Zn and Cu and thus increase their availability (Marschner, 1995; Neumann and Römheld, 2007). Among root exudates, carboxylic anions have also been extensively studied for their impact on the soil biogeochemistry. Their role in the dynamics of nutrients such as P has been clearly demonstrated. This function is based on the complexation of metal cations that play a major role in binding phosphate ions in soils, such as Ca, Fe, and Al, but can also rely on mechanisms (ligand exchange reactions) that promote the desorption of phosphate ions from soil constituents (Jones, 1998; Hinsinger, 2001), or prevent (by competition) the adsorption of phosphate on Fe-oxides (Violante et al., 2003).

Metal complexation by carboxylates is, on the other hand, an important mechanism to exclude uptake of toxic elements, such as Al^{3+}, in plant species and cultivars adapted to acid mineral soils. The release is restricted to the apical root zones, which are most susceptible to the toxic effects of aluminum (Ma, 2000; Brimecombe et al., 2007).

Localized release of root exudates in apical root zones with a low density of microbial colonization and secretion peaks over a limited period of time may counteract rapid microbial degradation and thereby increase the rhizosphere concentrations of exudated compounds for mobilization of nutrients (Römheld, 1991).

Metal complexation by low-molecular-weight organic substances released by plants and microbes have been involved in weathering of minerals and release of nutrients (e.g., Fe) in the rhizosphere (Violante et al., 2003).

9.2.2. Nutrient Cycling and Microbial Activity

It is widely accepted that rhizodeposition supports microbial communities that are more active and abundant in soil around roots than in non-rhizosphere soil. Such microbial communities are crucial to the functioning of the terrestrial ecosystem, not only for their direct effects on plant growth (through, for example, the release of growth factors) but also because they affect the C-flow from plant roots to soil, thus mediating the heterotrophically driven nutrient processes in soil by rhizodeposition (Badalucco and Nannipieri, 2007; Brimecombe et al., 2007; Uren, 2007).

It has been hypothesized that rhizosphere microorganisms may accelerate the decomposition of native soil organic matter and also stimulate the dissolution of insoluble minerals in a way similar to that proposed for plants (pH, redox, and metal-complexation reactions; see above).

This activation of nutrient cycling is based upon a stimulation of microbial activity in the rhizosphere by labile C released by roots (De Nobili et al., 2001).

Falchini et al. (2003) demonstrated that different low-molecular-weight organic compounds, such as glucose, oxalic and glutamic acids, are mineralized at different rates by soil microflora; furthermore, oxalic and glutamic acids detremined a modification in the composition of the bacterial communities.

In the rhizosphere, the bacterial energy circulation follows a loop trajectory (Clarholm, 1994). Nutrients become only temporarily locked up in bacterial biomass surrounding the roots and are later released by microfaunal grazing over bacteria. Rhizodepositions trigger the microbial growth in the rhizosphere with the consequent sequestration of plant available nutrients that would remain locked up into microbial biomass if consumption by protozoa and nematodes would not constantly re-mobilize them for plant uptake (Brimecombe et al., 2007). Using continuous labeling under laboratory conditions, it was demonstrated that decomposition of native organic matter was increased in planted soil under low nutrient conditions (Liljeroth et al., 1994). It was suggested that this was a consequence of the rhizosphere microorganisms being nutrient-limited and that mineralization of native organic matter was required to support assimilation of rhizodeposition, which is typically comprised of low C-to-N ratio compounds. It is likely that this scenario will be particularly evident for ecosystems under elevated CO_2, where C inputs are increased and competition for available nutrients are expected to be intensified (Phillips, 2007). However, it should be kept in mind that the nature of root exudates and native soil organic matter (e.g., C-to-N ratio), as well as the concurrence of demand for mineral N between plant roots and microorganisms, can influence microbial growth and in turn the fate of native organic matter present in the rhizosphere (Kuzyakov, 2002).

Some major nutrients can occur in soil as organically bound nutrients, such as N, P, and S. These organic molecules will require biochemical processes such as enzymatically controlled hydrolysis to evolve into smaller molecules and ultimately release mineral N, P, and S. Extracellular enzymes, such as phosphatases, proteases, and arylsulfatases, exhibit more activity in the rhizosphere relative to the bulk soil. Plant roots can exude some enzymes that will catalyze such breakdown processes, as shown for phosphatases that help plants make use of organic P compounds (Tarafdar and Jungk, 1987). However, soil microorganisms play a major role in these processes, through a variety of enzymatic pathways. Their stimulation by root exudation contributes to an enhanced release of organically bound nutrients in the rhizosphere (Badalucco and Nannipieri, 2007; Martin et al., 2007).

Enzymes released from roots or microorganisms may be adsorbed onto soil organo-mineral colloids and possibly protected against microbial degradation. Extracellular enzymes are essential for the initial degradation of high-molecular-weight substrates like lignin, cellulose, pectins, chitin, or humic molecules, which cannot enter microbial cell envelopes to be processed.

Organic substances released from roots not only act as an energy source for rhizosphere microorganisms, but also promote the chemiotaxis of microorganisms toward roots (de Weert et al., 2002). It is well known that bacteria belonging to *Rhizobium* or *Frankia* genera, as well as mycorrhizal fungi, are able to establish a symbiotic relation with their host plant (Martin et al., 2007; Werner, 2007). These symbioses are relevant to N and P cycling and acquisition by plants.

Rhizobia–legume symbiosis involves a molecular cross-talk between the plant and the microbe; recently, it has been shown that a similar signaling process might be involved in establishing a symbiosis with endomycorrhizal fungi (Perry et al., 2007).

Other rhizobacteria, generally designated as "plant growth-promoting rhizobacteria" (PGPR), stimulate plant growth by producing phytohormones or act indirectly as biocontrol agents through the production of antibiotics and siderophores, and even by inducing plant resistance mechanisms (Brimecombe et al., 2007).

9.3. HUMIC SUBSTANCES IN THE RHIZOSPHERE

As described above, chemical, biochemical, and microbiological conditions in the rhizosphere differ widely from those of bulk soil; this can lead to changes in the dynamics and structure of humified organic matter. On the other hand, little is known about the molecular structure and degree of aggregation of humic molecules in the rhizosphere.

Using FT-IR and TLS-EEM analyses, it was shown that humic acids separated according to conventional methods (alkaline dissolution, acid precipitation) from rhizospheric and bulk soil were characterized by small differences in structural properties (Grasso et al., 2004). Differences in FT-IR spectra and FI values of EEM spectra were interpreted as an indication of the incorporation into humic acids extracted from rhizospheric soils of root exudate components, such as aliphatic and aromatic acids, amino acids, amides, and phenolics.

Chemical characteristics of humic acids isolated from rhizospheric and bulk soil were studied as a function of the plant genotype grown in the soil (Tartaro et al., 1999). Specifically, it was shown that the presence of a strongly acidifying species, such as rape, and a nodulated species, such as field bean, determined greater differences among rhizospheric and non-rhizospheric humic acids, as compared to wheat plants, with respect to total acidity and aromaticity.

Attempts have been made to verify whether organic acids released by plant roots may change the structure of humic molecules (Albuzio and Ferrari, 1989). Analyses by size exclusion chromatography (SEC) have revealed that high-molecular-weight fractions treated with acidic solutions or organic acids release humic molecules of lower molecular weight with a higher biological activities than that of the former molecules. Similar results were also obtained using maize root exudates (Nardi et al., 1997). The disaggregation process can be explained by a micellar behavior of humic substances in solution (Piccolo et al., 1996). This mechanism might be particularly pronounced under environmental stress conditions, such as nutritional shortage (P and Fe), capable of inducing massive release of carboxylates from roots.

Other peculiar conditions of the rhizosphere, such as ion concentration, can influence the molecular complexity of humic substances, due to the possibility of the formation of interchain bonds producing insoluble macromolecules (Sequi et al., 1975). This latter aspect underlines the need of a dynamic evaluation of the processes occurring in the rhizosphere in relation to the activity and the physiological status of the plant.

9.4. ROLE OF HUMIC SUBSTANCES IN THE SOIL–ROOT INTERACTION

One of the most important features in the soil–plant relationship is the rhizosphere extent. This factor is highly variable, ranging from <1 mm to several millimeters and strongly dependent on the gradients that develop in the rhizosphere as a consequence of different processes. In these processes a crucial role is played by the root hairs, tubular outgrowths of root epidermal cells; the development of root hairs is dependent on the genotype and is affected by the environmental conditions (e.g. nutrient availability, abiotic stresses and hormones).

The presence of humic substances interacting with root physiological processes (Varanini and Pinton, 2001) might conceivably affect the extent of rhizosphere—for example, by modulating the release of protons and low-molecular-weight organic root exudates, redox activities present at the root surface, ion uptake rates and pools of available nutrients, and the activity of ectoenzymes. Humic substances may act also through a stimulation of root growth and root hairs proliferation. These effects are particularly important for adaptation of plants to adverse soil conditions and could be useful for the definition of rhizosphere management practices aimed at improving plant growth and nutrient use efficiency (Römheld and Neumann, 2006). Some of these aspects will be described in detail in the following sections.

9.4.1. Source of Nutrients

Humic substances contain N, S, and P. Even though the most important organic source is represented by fresh and easily degradable residues of plants and microorganisms, there is no doubt that humic substances can also be subjected to mineralization, thus releasing inorganic nutrients that can be used by plants. Furthermore, it has to be considered that mineralization is not uniform throughout the soil profile, but is enhanced in the rhizosphere (Badalucco and Nannipieri, 2007), this phenomenon being also influenced by the uneven distribution of gradients (ions, root exudates, microbial growth) along the root axis.

Generally, more than 99% of the total N in the soil is present in the organic fraction if we exclude soil with NH_4^+-fixing clays (Stevenson, 1986); humic substances in particular (which contain 3–6% N) act as a storehouse and supplier of N for plant roots and microorganisms (Schulten and Schnitzer, 1998). A relatively large amount of amine N present in the soil is incorporated in humified matter, up to 50% of which is supposed to be present as peptides and proteins (Mengel, 1996). It is important to emphasize that protease activity is present in the rhizosphere (Badalucco et al., 1997). In addition to a nutritional role linked to the mineralization processes, humic compounds have been hypothesized to directly affect plant nutrition, since it has been suggested that roots may take up low-molecular-weight humic molecules (Popov and Chertov, 1997). Interestingly, plants have been observed to express carriers for amino acids and small peptides at the root level (Varanini and Pinton, 2007). Furthermore, recent studies show that organic nitrogen compounds can be taken up by the roots (Thornton and Robinson, 2005). Certain components of the humic fraction have been found inside root cells and also translocation to the shoots has been reported (Vaughan and Linehan, 1976; Vaughan and Ord, 1981). Experiments performed on rice cells in suspension culture seem to suggest that they may use carbon skeletons from humic molecules to synthesize proteins and DNA (Wang et al., 1999). Although these observations do not allow us to definitely assess if integral humic molecules or their degradation products enter the root cells, on the other hand they confirm that plants can use, at least in part, these sources.

In addition to protein N, peptide N, and amino acid N, humic substances also contain molecules with heterocyclic N, which can account for up to 50% of total N present in humic matter, and are mainly purines, pyrimidines, quinolines, isoquinolines, aminobenzofuranes, piperidine, and pyrrolidine derivatives, which are presumably integrated into the structure of humified substances (Stevenson, 1986). Information relative to the role of the heterocyclic component of humic N in plant nutrition is very scarce. Although they appear to be quite stable, they are not inert and can be microbiologically and chemically converted into inorganic compounds (Schnitzer and Hindle, 1980; Zhuo et al., 1995). The contribution of this fraction to N nutrition in plants is still unknown; and by considering microbial activity, and the particular availability of organic C in the rhizosphere, humic N is likely to be subjected to transformations differing from those in the bulk soil.

In its inorganic form, P is a nutrient showing low solubility and mobility in the soil, because it easily reacts with the soil mineral components (clay, iron and aluminum oxides, and carbonates) (Stevenson, 1986). The P content in humic substances ranges from 0.1% to 1.0% and is particularly abundant in humic acids. By using ^{31}P NMR, it was shown that different forms of P can be associated with humic fractions of different molecular sizes (Bedrock et al., 1994). In fractions of high molecular weight, P is usually found as hexaphosphate inositol, whereas in those of lower molecular weight it is usually present as inorganic orthophosphate. Both types of P have an important nutritional value: The former compound is a substrate for phosphatases, which are particularly abundant in the rhizosphere (Tarafdar and Jungk, 1987), whereas the latter, bound to humic compounds by aluminum and iron bridges, becomes available in the presence of high concentrations of organic acids, usually present in the rhizosphere soil. Indeed, these acids can form complexes with the iron bound to the surface of the humic matter, thus releasing the phosphate (Gerke, 1993; Gerke et al., 1994).

Humic molecules contain S in the form of proteins, amino acidic residues, sulfate esters, and, possibly, stable thiazine rings; inorganic S can be released from these organic compounds following the action of microorganisms using organic carbon as a source of energy. The chemical bonds of the nutrients play an important role in the type of organic matter mineralization, and five classes of bonds (C–C, N–C, S–C, S–O–C, and P–O–C) can be distinguished (Hunt et al., 1986). Microorganisms oxidizing carbon provide energy to mineralize the compounds characterized by the first three types of bonds. On the other hand, compounds with S and P atoms present as esters can be mineralized by the action of extracellular hydrolases such as sulfatases and phosphatases.

9.4.2. Complexing Properties

Humic substances can form complexes with metals, including cationic micronutrients (Linehan, 1985), because of the presence of electron-donor functional groups in these molecules. Therefore, it is obvious that due to these properties humic substances can contribute to the regulation of the availability and solubility of these metals (Stevenson, 1994). Either weak bonds (such as water bridges, or electrostatic attraction due to cation exchange capacity) or strong bonds (involving coordination with single groups or formation of ring structures—chelates—with carboxyl, alcohol, and amino groups) can thereby be formed. The formation of more than one bond between the metal and the organic molecule usually results in a higher stability of the complex.

The stability of the metal chelate complex depends on the number of atoms that form a bond with the metal ion, the number of rings that are formed, the nature and concentration of the metal ion, and the pH (Stevenson, 1994). The stability order of the complexes formed between metals and humic acids has been determined through potentiometric titration and follows the Irwing–Williams series: $Pb^{2+} > Cu^{2+} > Ni^{2+} > Co^{2+} > Zn^{2+} > Cd^{2+} > Fe^{2+} > Mn^{2+} > Mg^{2+}$.

On the other hand, at a pH value of 5.0 there were no large differences in the strength of bonds between humic acids and metals such as Ca, Mg, Mn, Co, Ni, and Zn, whereas bonds with Pb, Cu, and Fe were stronger than with other metals (Schnitzer and Kahn, 1972); this behavior indicates that at different pH values, metal humic substance complexes of different stability are formed in the soil. This aspect is of particular relevance in an environment such as the rhizosphere, where dynamic pH gradients are present mainly due to the availability of nutrients and to their selective uptake by roots. With regard to plant availability, great importance lies in the molecular dimension and solubility of humic substances (Brümmer and Herms,

1983). Fractions of higher molecular mass, which are mostly insoluble, can withhold large amounts of metals, especially in alkaline environments. Metals are thus subtracted from precipitation and subsequent crystallization, processes that would decrease their availability (Schwertmann, 1966), and a reserve of micro-nutrients is created which is in equilibrium with complexing molecules. On the other hand, under conditions of high metal concentrations, complexation by humified organic matter may limit the amount of metals in solution; under these conditions, interchain bonds may form, with possible precipitation of humic molecules. This process can be important for toxic elements, the activity of which can thus be reduced to non-toxic levels (Gerke, 1992).

Soluble humified organic matter of soil (Chen, 1996) may increase metal transport by diffusion to the roots (Pandeya et al., 1998) and favor micro-nutrient uptake by the plants.

Complexing properties of humic substances can also have a great importance in P nutrition (Stevenson, 1991). Indeed Fe^{3+} and Al^{3+} (acidic soils) and Ca^{2+} (calcareous soils) complexed by humic molecules can bound phosphates to humic substances thus making P in a plant-available form (Gerke and Hermann, 1992). Complexation of cations by carboxylates such as those released by plant roots (e.g., citrate) can increase phosphate availability (Gerke, 1993).

In soil–root interactions, an important role of humic substances relies in their capability to affect Fe dynamics in the rhizosphere. Apart from an indirect mechanism involving the stabilization of amorphous Fe oxides by high-molecular-weight humic fractions (Schwertmann, 1991), a direct contribution to Fe availability can derive from the formation of water-soluble Fe-humate complexes, which can move in soil toward the roots (Pandeya et al., 1998) and act as natural Fe-chelates, potentially available for plants. It has been observed that a water-extractable humic substances fraction (WEHS), purified from a water extract of sphagnum peat using XAD-8 amberlite resin, could solubilize Fe present as ferrihydrite and mobilize it in soil, thus making it available for exchange with organic chelating agents such as phytosiderophores released by deficient barley roots (Cesco et al., 2000).

However, the dynamics of Fe mobilization by humic substances depends on the prevailing conditions in the rhizosphere, such as pH and redox potential, and the presence of other types of chelating agents of microbial (siderophores) or plant (organic acids and phytosiderophores) origin. In this context, the different plant strategies in response to limited Fe availability need to be considered. In the case of dicots and non-gramineous monocots (strategy I) the mechanisms are based on an increased reducing capacity of Fe(III) chelates, a necessary step in the uptake process, with a concurrent increase in acidification and release of organic acids into the rhizosphere; in the case of graminaceae (strategy II), molecules having high affinity for Fe (phytosiderophores) are synthesized and released into the rhizosphere when Fe is lacking (Marschner and Römheld, 1994). In this context, it is interesting to observe that response mechanisms to Fe deficiency have been studied almost exclusively using synthetic chelates such as EDTA and EDDHA or, in a few cases, organic acids released by the roots (such as citrate and malate). It is reasonable to suppose, however, that a mixture of natural chelates is present in the soil and in the rhizosphere (Crowley, 2006).

Evidence that humic molecules may play an important role in Fe uptake has been presented (Lobartini and Orioli, 1988; Pinton et al., 1998; Chen et al., 2001). It has been shown that a soluble Fe-humate fraction (WEHS, see above) could also be reduced by intact roots of cucumber plants (Pinton et al., 1999b), supporting the view that soluble Fe-humate can act as one of the naturally occurring substrates for the inducible Fe(III)-chelate reductase.

It is interesting to observe that, unlike other organic molecules of the rhizosphere which can chelate or complex metals (e.g., organic acids, phytosiderophores, microbial siderophores), humic substances are much more stable against microbial degradation. Furthermore, organic acids, which are released in higher amount in some Fe-deficient Strategy I plants, are poor Fe chelators under conditions typical of Fe deficiency (e.g., calcareous soils) (Jones, 1998).

Recent data show that Fe-WEHS could be used by dicots via a reduction-based mechanism more efficiently than other Fe-chelates that may be present in the rhizosphere, such as Fe-phytosiderophores and Fe-citrate (Cesco et al., 2006). It has been shown that Fe-WEHS could also be used by barley plants (strategy II) via a mechanism possibly involving ligand exchange between phytosiderophores and WEHS (Cesco et al., 2002).

Proposed mechanisms for the use of Fe complexed to WEHS-like humic fractions are summarized in Figure 9.2.

Humic substances not only contribute to increase Fe bioavailability through their Fe chelating properties, but are also known to be redox reactive and capable of chemically reducing metals, including Fe^{3+} (Skogerboe and Wilson, 1981; Struyk and Sposito, 2001). Standard redox potentials for fulvic and humic acids have been evaluated to be around 0.5 and 0.7V, respectively. It has been shown that reduction of Fe^{3+} occurs significantly at pH values lower than 4; at higher pH values, reduction is decreased by formation of complexes between Fe^{3+} and humic molecules (Chen et al., 2003).

9.5. DIRECT ACTION OF HUMIC SUBSTANCES ON PLANT NUTRITION AND GROWTH

9.5.1. Effect on Mechanisms of Nutrient Uptake

Humic substances have been shown to stimulate plant growth and nutrient accumulation (for reviews, see Vaughan and Malcolm, 1985; Chen and Aviad, 1990). Various studies performed on excised roots or whole plants show that the uptake of cationic and anionic macronutrients is usually greater when roots are in contact with appropriate concentrations of humic substances (Varanini and Pinton, 1995).

Since the root cell plasma membrane is the main barrier between the cytoplasm and the rhizosphere, it is reasonable to believe that the membrane itself (and associated activities) is one of the primary targets of the effect of humic substances. Among the other plasma membrane (PM) transport proteins H^+-ATPase is acknowledged to play a primary role in plant physiological processes, such as growth and nutrition. In fact, this enzyme is responsible for the electrogenic transport of protons to the cell apoplast and the formation of the consequent electrochemical gradient, which can be used to energize the secondary active transport of nutrients across the plasma membrane or to favor uniport processes according to the potential gradient (Palmgren, 2001).

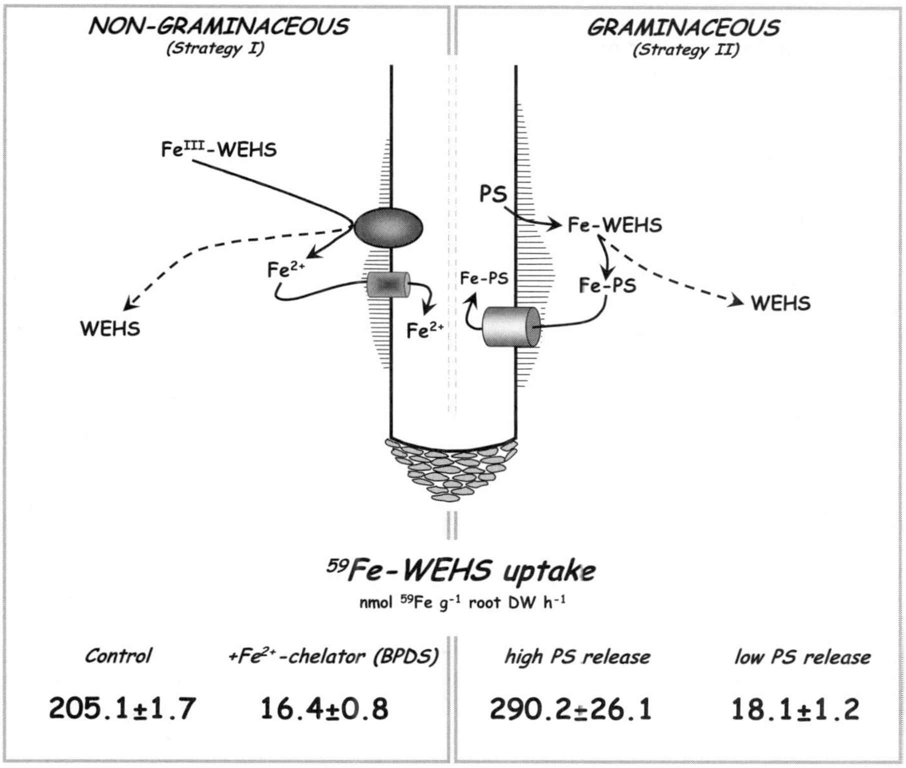

Figure 9.2. Mechanisms for the use of Fe complexed to a water-extractable humic substances fraction (WEHS), separated from a sphagnum peat, by non-graminaceous (strategy I; e.g., cucumber) and graminaceous (strategy II; e.g., barley) Fe-deficient plants. The reduction-based mechanism in strategy I plants is evidenced by the strong inhibition of uptake in the presence of the high-affinity Fe^{2+} chelator BPDS in the assay medium (Cesco et al., 2006). A ligand exchange mechanism in strategy II plants is supported by the highest rates of Fe uptake observed when the assay was run in the period of highest phytosiderophore (PS) release (morning experiment) as compared with the lowest rates obtained during the low PS release period (evening experiment) (Cesco et al., 2002).

Stimulation of active H^+ extrusion from roots (Cesco, 1995; Pinton et al., 1997; Table 9.1) and transmembrane potential hyperpolarization (Slesak and Jurek, 1988) indicated the involvement of the PM H^+-ATPase in the increased nutrient uptake generally observed in the presence of humic substances. Direct proof of an interaction between humic molecules and the PM H^+-ATPase has been obtained by Varanini et al. (1993), who demonstrated that low-molecular-weight (<5 kDa) humic molecules at concentrations compatible with those present in the rhizosphere can stimulate the phospho-hydrolytic activity of this enzyme in isolated PM vesicles (Table 9.1). Further proof of the action of humic molecules on PM H^+-ATPase activity and on nutrient uptake mechanisms was obtained when studying the effect of these molecules on NO_3^- uptake. Transport of this nutrient is a substrate-inducible process and involves H^+ co-transport. At higher uptake rates, the levels and activity of root PM H^+-ATPase increased (Santi et al., 1995). The short-term (4 h) contact

TABLE 9.1. Effect of Different Humic Substances Fractions (HS) on Active Proton Extrusion from Intact Roots[a] and on H^+-ATPase Activty of Plasma Membrane (PM) Vesicles Isolated from Different Plant Roots

	Control	+HS
H^+ release from intact roots[a]	14 ± 1	35 ± 3
(nmol H^+ g^{-1} FW)	(100)	(250)
PM H^+-ATPase activity		
ATP hydrolysis[b]	95 ± 4	122 ± 4
(μmol Pi mg^{-1} prot h^{-1})	(100)	(128)
ATP hydrolysis[c]	74 ± 5	152 ± 8
(μmol Pi mg^{-1} prot h^{-1})	(100)	(205)
ATP-dependent H^+ transport[c]	17.4 ± 1.1	27.0 ± 2.1
(ΔA_{492} mg^{-1} prot h^{-1})	(100)	(155)

Note: Plasma membrane vesicles were either isolated from roots previously treated with HS[b] and then assayed for the ATP-hydrolyzing activity or isolated from untreated roots and then incubated directly in the assay medium with HS[c] for the measurement of the scalar (ATP hydrolysis) and vectorial (ATP-dependent H^+ transport) activity.

[a]Assayed on roots of intact oat plants treated with 5 mg Corg $liter^{-1}$ of a water-extractable humic fraction (WEHS) purified from a sphagnum peat (Cesco, 1995).

[b]Assayed on plasma membrane vesicles isolated from roots of intact maize plants treated for 4 h with 5 mg C org $liter^{-1}$ of WEHS fraction (Pinton et al., 1999a).

[c]Assayed on oat root plasma membrane vesicles incubated with 0.075 mg C org $liter^{-1}$ of a low-molecular-weight (<5 kDa) humic fraction isolated with $Na_4P_2O_7$ from the A_0 horizon of a lythic rendoll (Varanini et al., 1993).

of roots with a low-molecular-weight WEHS in the presence or absence of NO_3^- caused a more rapid development of the NO_3^- uptake capacity and a further increase in PM H^+-ATPase activity, as measured in PM vesicles isolated from maize roots (Pinton et al., 1999a; Table 9.1). Since no increase in protein amount was observed, this effect was attributed to a post-translational regulation of the PM H^+-ATPase. On the other hand, a prolonged treatment with high-molecular-weight humic acids isolated from earthworm compost determined a promoting effect on the activity and amount of PM H^+-ATPase (Canellas et al., 2002), which was attributed to the presence of auxin (IAA) bound in an exchangeable form to the humic molecules. More recently, based on results obtained in a comparative study on the effects on PM and tonoplast H^+-ATPases exerted by IAA and humic acids separated from different sources (soils or organic residues), it was concluded that the presence of humic acids–IAA groups could not completely explain the observed biological effects of humic extracts, thus implying the involvement of other bioactive components or auxin-independent mechanisms of action (Zandonadi et al., 2007). An increase in transcript levels of the PM H^+-ATPase isoform MHA2 in maize roots treated for 48 h with an earthworm low-molecular-weight humic fraction, endowed with IAA, was observed (Quaggiotti et al., 2004). The action of humic molecules on the PM H^+-ATPase can also positively affect the acquisition of sparingly soluble nutrients, such as Fe (Varanini and Pinton, 2006). Increased PM H^+-ATPase activity can contribute to Fe nutrition in several ways: (a) by solubilizing Fe in the apoplast and the rhizosphere; (b) by maintaining favorable conditions for the activity of the Fe(III)-chelate reductase (low apoplast pH and transmembrane electrical potential homeostasis); and (c) by favoring uptake of free Fe^{2+} or Fe(III) complexes (e.g., Fe-phytosiderophores).

9.5.2. Effect on Root Growth

Many authors have observed that plants treated with humic substances have different growth and morphology compared to control plants (Vaughan and Malcolm, 1985; Nardi et al., 2002); these substances were seen to modify root morphology, inducing a proliferation of root hairs in the subapical regions (Concheri et al., 1996) and a higher differentiation rate of root cells (Concheri et al., 1994). Humic acids extracted from different sources were shown to enhance root growth and lateral root proliferation in maize seedlings (Canellas et al., 2002; Zandonadi et al., 2007) and *Arabidopsis* (Dobbss et al., 2007). This kind of root response to the treatment with humic substances has been interpreted in terms of a hormone-like activity of humic substances. Studies have shown that humic substances extracted from different sources could react to antibodies directed against IAA (Muscolo et al., 1998; Pizzeghello et al., 2001); a similar fluorescent staining for IAA and low-molecular-weight humic substances of fluorescein-labeled carrot cellular membranes were also observed (Muscolo et al., 2007). Furthermore, the presence of exchangeable auxin groups in humic substances extracted from earthworm compost was revealed by gas chromatography–mass spectrometry (Canellas et al., 2002).

In an attempt to clarify whether humic substances might have an auxin-like action, the response of *Arabidopsis* roots to the presence of WEHS was studied (Schmidt et al., 2007). Application of WEHS caused an array of changes in root morphology, such as an increase in root hair length and density, formation of ectopic root hairs, and an increase in cell proliferation in the root ground tissue. WEHS affected genes involved in epidermal cell fate specification, suggesting that humic substances can alter developmental programs at an early stage of root cell differentiation. This would in turn determine a remodelling of root morphology leading to an increase in absorptive surface of the root. In this study, no change in auxin-related gene expression was observed after WEHS application to transgenic plants carrying auxin response elements, thus indicating that WEHS do not exert their effect in an auxin-like manner.

9.6. CONCLUSIONS

The evidence presented suggests that humic substances can have a strong influence on the root–soil interaction in the rhizosphere. Particularly promising are the studies aimed at defining structural characteristics of rhizospheric humified fractions of the organic matter, to increase our knowledge on the role of humic substances in the rhizosphere processes. The evaluation of the contribution of humic substances to the plant-soil–microbes interactions is complicated by the dynamic nature of the rhizosphere environment and its spatial variability; and it will require the development of appropriate, although simplified, experimental setups which allow to study the behavior of humic substances in the rhizosphere (e.g., using rhizobox with soils containing defined humic pools and/or treated with well-characterized humic fractions). This task and a better comprehension of the rhizosphere system could only be accomplished by adopting a multidisciplinary approach, which integrates plant physiology, soil chemistry, and microbiology.

Results on the action of humic substances as natural chelates appear to (a) offer new insights for the definition of plant behavior under field conditions and (b) provide agronomical perspectives, such as indication for preparing organic amendments. Improvement of nutrition acquisition, particularly of micronutrients, might be important for favoring plant resistance to biotic and abiotic stresses (Römheld and Neumann, 2006).

Molecular bases of the well-known physiological action of humic substances on membrane activities relevant for root ion acquisition are elucidated (Figure 9.3), and a possible role of some fractions as molecular signals acting in the rhizosphere is emerging (Pinton et al., 2006). An interaction with the PM H^+-ATPase has been observed: Regulatory aspects of this interaction need to be further investigated at the molecular level, considering both transcriptional and post-translational regulation. However, this enzyme may not be the sole molecular target of humic compounds; both lipids and proteins (e.g., carriers) could be involved in the regulation of ion uptake. Effect on root growth and/or root hair proliferation support the view that low-molecular-weight soluble humic fractions induce a "nutrient acquisition response" in the rhizosphere (similar to that evoked by shortage of barely soluble nutrients), which favors nutrient capture via an increase in the absorptive surface area.

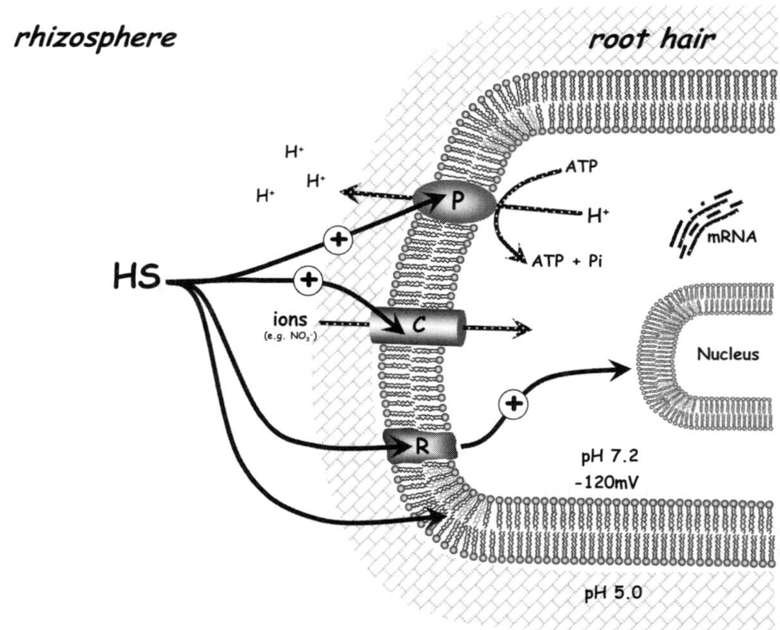

Figure 9.3. Model for the action of humic substances (HS) on plasma membrane-bound targets of a root hair cell. Besides the well-known effects on plasma membrane H^+-ATPase (P) and carriers (C) of mineral nutrients, the envisaged alteration of membrane lipid environment and the possible interaction with an hypothetical membrane receptor (R) for humic molecules which allows transduction of the signal for induction and expression of genes involved in nutrient uptake and root hair development are also presented.

REFERENCES

Albuzio, A., and Ferrari, G. (1989). Modulation of the molecular size of humic substances by organic acids of the root exudates. *Plant Soil* **113**, 237–241.

Al-Najar, H., Schulz, R., and Römheld, V. (2003). Plant availability of thallium in the rhizosphere of hyperaccumulator plants: A key factor for assessment of phytoextraction. *Plant Soil* **249**, 97–105.

Badalucco, L., and Nannipieri, P. (2007). Nutrient transformations in the rhizosphere. In *The Rhizosphere: Biochemistry and Organic Substances at the Soil–Plant Interface*, 2nd edition, Pinton, R., Varanini, Z., and Nannipieri, P., eds., CRC Press, Boca Raton, FL, pp. 111–133.

Badalucco, L., Kuikman, P. J., and Nannipieri, P. (1997). Protease and deaminase activities in wheat rhizosphere and their relation to bacterial and protozoan populations. *Biol. Fertil. Soils* **23**, 99–104.

Bagayoko, M., Alvey, S., Neumann, G., and Buerkert, A. (2000). Root-induced increases in soil pH and nutrient availability to field-grown cereals and legumes on acid sandy soils of Sudano-Sahelian West Africa. *Plant Soil* **225**, 117–127.

Barber, S. A. (1995). *Soil Nutrient Bioavailability: A Mechanistic Approach*, John Wiley & Sons, New York.

Bedrock, C. N., Cheshire, M. W., Chudek, J. A., Goodman, B. A., and Shand, C. A. (1994). Use of ^{31}P-NMR to study the form of phosphorus in peat soils. *Sci. Tot. Environ.* **152**, 1–8.

Bertrand, I., and Hinsinger, P. (2000). Dissolution of an iron oxyhydroxide in the rhizosphere of various crop species. *J. Plant Nutr.* **23**, 1559–1577.

Brimecombe, M. J., De Leij, F. A. A. M., and Lynch, J. M. (2007). Rhizodeposition and microbial populations. In *The Rhizosphere: Biochemistry and Organic Substances at the Soil–Plant Interface*, 2nd edition, Pinton, R., Varanini, Z., and Nannipieri, P., eds., CRC Press, Boca Raton, FL, pp. 73–109.

Brümmer, G., and Herms, U. (1983). Influence of soil reaction and organic matter on the solubility of heavy metals in soils. In *Effects of Accumulation of Air Pollutants in Forest Ecosystems*, Ulrich, B., and Pankrath, eds., Reidel Dordrecht, pp. 233–243.

Canellas, L. P., Olivares, F. L., Okorokova-Façanha, A. L., and Façanha, A. R. (2002). Humic acids isolated from earthworm compost enhance root elongation, lateral root emergence, and plasma membrane H^+-ATPase activity in maize roots. *Plant Physiol.* **130**, 1951–1957.

Cesco, S. (1995). Ruolo di sostanze umiche estratte dal terreno nella regolazione di attività di membrana coinvolte nella nutrizione minerale della pianta. Ph.D. thesis, Dipartimento de Produzione Vegetale e Tecnologie Agrarie, University of Udine, Udine, Italy.

Cesco, S., Nikolic, M., Römheld, V., Varanini, Z., and Pinton, R. (2002). Uptake of ^{59}Fe from soluble Fe–humate complexes by cucumber and barley plants. *Plant Soil* **241**, 121–128.

Cesco, S., Römheld, V., Varanini, Z., and Pinton, R. (2000). Solubilization of iron by a water extractable humic substances fraction. *J. Plant Nutr. Soil Sci.* **163**, 285–290.

Cesco, S., Varanini, Z., and Pinton, R. (2006). Use of Fe complexed to water soluble humic substances by dicotiledonous plants: Mechanisms and efficiency. In *13th International Symposium on Iron Nutrition and Interaction in Plants*, Montpellier, France, S1-O-4, p. 25.

Chen, J., Gu, B. H., Royer, R. A., and Burgos, W. D. (2003). The roles of natural organic matter in chemical and microbial reduction of ferric iron. *Sci. Tot. Environ.* **307**, 167–178.

Chen, Y. (1996). Organic matter reactions involving micronutrients in soils and their effect on plants. In *Humic Substances in Terrestrial Ecosystems*, Piccolo, A., ed., Elsevier Sciences B.V., Amsterdam, pp. 507–529.

Chen, Y., and Aviad, T. (1990). Effects of humic substances on plant growth. In *Humic Substances in Soil and Crop Sciences: Selected Readings*, MacCarthy, P., Clapp, C. E., Malcolm, R. L., and Bloom, P. R., eds., American Society of Agronomy, Soil Science Society of America, Madison, WI, pp. 161–186.

Chen, Y., and Schnitzer, M. (1978). The surface tension of aqueous solutions of soil humic substances. *Soil Sci.* **125**, 7–15.

Chen, Y., Magen, H., and Clapp, C. E. (2001). Plant growth stimulation by humic substances and their complexes with iron. In *Proceedings of the Dalia Greindinger Symposium. The International Fertilizer Society*, Lisbon, No. 470, pp. 1–14.

Clarholm, M. (1994). The microbial loop in soil. In *Beyond the Biomass*, Ritz, K., Dighton, J., and Giller, K. E., eds., Wiley-Sayce, Chichester, pp. 221–230.

Concheri, G., Nardi, S., Piccolo, A., Dell'Agnola, G., and Rascio, N. (1994). Effects of humic fractions on morphological changes related to invertase and peroxidase activities in wheat seedling roots. In *Humic Substances in the Global Environment and Implications on Human Health*, Senesi, N., and Miano, T. M., eds., Elsevier, Amsterdam, pp. 257–262.

Concheri, G., Nardi, S., Reniero, F., and Dell'Agnola, G. (1996). The effects of humic substances within the Ah horizon of a calcic luvisol on morphological changes related to invertase and peroxidase activities in wheat roots. *Plant Soil* **179**, 65–72.

Courchesne, F., and Gobran, G. R. (1997). Mineralogical variation of bulk and rhizosphere soils from a Norway spruce stand. *Soil Sci. Soc. Am. J.* **61**, 1245–1249.

Crowley, D. E. (2006). Microbial siderophores in the plant rhizosphere. In *Iron Nutrition in Plants and Rhizospheric Micreoorganisms*, Barton, L. L., and Abadia, J., eds., Springer-Verlag, Dordrecht, The Netherlands, pp. 169–198.

Darrah, P. R. (1991). Models of the rhizosphere. 1. Microbial dynamics around a root releasing soluble and insoluble carbon. *Plant Soil* **133**, 187–199.

De Nobili, M., Contin, M., Mondini, C., and Brookes, P. C. (2001). Soil microbial biomass is triggered into activity by trace amounts of substrate. *Soil Biol. Biochem.* **33**, 1163–1170.

de Weert, S., Vermeiren, H., Mulders, I. H. M., Kuiper, I., Hendrickx, N., Bloemberg, G. V., Vanderleyden, J., De Mot, R., and Lugtenberg, B. J. J. (2002). Flagella-driven chemotaxis towards exudate components is an important trait for tomato root colonization by *Pseudomonas fluorescens*. *Mol. Plant–Microbe Interact.* **15**, 1173–1180.

Dexter, A. R. (1987). Compression of soil around roots. *Plant Soil* **97**, 401–406.

Dobbss, L. B., Medici, L. O., Peres, L. E. P., Pino-Nunes, L. E., Rumjanek, V. M., Façanha, A. R., and Canellas, L. P. (2007). Changes in root development of *Arabidopsis* promoted by organic matter from oxisols. *Annals Appl. Biol.* **151**, 199–211.

Doussan, C., Pagès, L., and Pierret, A. (2003). Soil exploration and resource acquisition by plant roots: An architectural and modelling point of view. *Agronomie* **23**, 419–431.

Falchini, L., Naumova, N., Kuikman, P. J., Bloemc, J., and Nannipieri, P. (2003). CO_2 evolution and denaturing gradient gel electrophoresis profiles of bacterial communities in soil following addition of low molecular weight substrates to simulate root exudation. *Soil Biol. Biochem.* **35**, 775–782.

Fitz, W. J., Wenzel, W. W., Zhang, H., Nurmi, J., Stipek, K., Fischerova, Z., Schweiger, P., Köllensperger, G., Ma, L. Q., and Stingeder, G. (2003). Rhizosphere characteristics of the arsenic hyperaccumulator *Pteris vittata* L. and monitoring of phytoremoval efficiency. *Environ. Sci. Technol.* **37**, 5008–5014.

Ge, Z., Rubio, G., and Lynch, J. P. (2000). The importance of root gravitropism for inter-root competition and phosphorus acquisition efficiency: Results from a geometric simulation model. *Plant Soil* **218**, 159–171.

Gerke, J. (1992). Orthophosphate and organic phosphate in the soil solution of four sandy soils—Evidence for humic–Fe(Al) phosphate complexes. *Commun. Soil Sci Plant Anal* **23**, 601–612.

Gerke, J. (1993). Solubilization of Fe(III) from humic–Fe complexes, humic/Fe-oxide mixtures and from poorly ordered Fe-oxide by organic acids—Consequences for P adsorption. *Z. Pflanzenernärhr. Bodenk.* **156**, 253–257.

Gerke, J. (1997). Aluminium and iron (III) species in the soil solution including organic complexes with citrate and humic substances. *Z. Pflanzenernärhr. Bodenk.* **160**, 427–432.

Gerke, J., and Hermann, R. (1992). Adsorption of orthophospshate to humic–Fe complexes and to amorphous Fe-oxide. *Z. Pflanzenernähr. Bodenk.* **155**, 233–236.

Gerke, J., Römer, W., and Jungk, A. (1994). The excretion of citric and malic acid by proteoid roots of *Lupinus albus* L.; effects on soil solution concentrations of phosphate, iron, and aluminium in the proteoid rhizosphere in samples of an oxisol and a luvisol. *Z. Pflanzenernärhr. Bodenk.* **157**, 289–294.

Grasso, M., D'Orazio, V., and Senesi, N. (2004). Spectroscopic characterization of soil organic matter fractions from rhizosphere and bulk soils cultivated for tomato and artichoke. In *Proceedings, 12th Meeting of the International Humic Substances Society*, Sao Pedro, Brazil, pp. 368–371.

Grayston, S. J., Vaughan, D., and Jones, D. (1996). Rhizosphere carbon flow in trees, in comparison with annual plants: The importance of root exudation and its impact on microbial activity and nutrient availability. *Appl. Soil Ecol.* **5**, 29–56.

Hiltner, L. (1904). Über neuere Ehrfahrungen und Problem auf dem Gebiet der Bodenbakteriologie unter besonderer Berücksichtigung der Grundüngung und Brache. *Arb. Deut. Landwirtschaft. Gesell.* **98**, 59–78.

Hinsinger, P. (1998). How do plant roots acquire mineral nutrients? Chemical processes involved in the rhizosphere. *Adv. Agron.* **64**, 225–265.

Hinsinger, P. (2001). Bioavailability of soil inorganic P in the rhizosphere as affected by root-induced chemical changes: A review. *Plant Soil* **237**, 173–195.

Hinsinger, P., and Gilkes, R. J. (1996). Mobilization of phosphate from phosphate rock and alumina-sorbed phosphate by the roots of ryegrass and clover as related to rhizosphere pH. *Eur. J. Soil Sci.* **47**, 533–544.

Hinsinger, P., and Jaillard, B. (1993). Root-induced release of interlayer potassium and vermiculitization of phlogopite as related to potassium depletion in the rhizosphere of ryegrass. *J. Soil Sci.* **44**, 525–534.

Hinsinger, P., Elsass, F., Jaillard, B., and Robert, M. (1993). Root-induced irreversible transformation of a trioctahedral mica in the rhizosphere of rape. *J. Soil Sci.* **44**, 535–545.

Hinsinger, P., Plassard, C., Tang, C., and Jaillard, B. (2003). Origins of root-induced pH changes in the rhizosphere and their responses to environmental constraints: A review. *Plant Soil* **248**, 43–59.

Hodge, A. (2004). The plastic plant: Root responses to heterogeneous supplies of nutrients. *New Phytol.* **162**, 9–24.

Hunt, H. W., Stewart, J. A. E., and Cole, C. V. (1986). Concepts of sulphur, carbon and nitrogen transformations in soil: Evaluation of simulating modeling. *Biogeochemistry* **2**, 163–177.

Jones, D. L. (1998). Organic acids in the rhizosphere—A critical review. *Plant Soil* **205**, 25–44.

Jones, D. L., and Darrah, P. R. (1994). Amino-acid influx at the soil–root interface of *Zea mays* L. and its implication in the rhizosphere. *Plant and Soil* **163**, 1–12.

Jones, D. L., and Darrah, P. R. (1996). Re-sorption of organic compounds by roots of *Zea mays* L. and its consequences in the rhizosphere. 3. Characteristics of sugar influx and efflux. *Plant and Soil* **178**, 153–160.

Jones, D. L., Hodge, A., and Kuzyakov, Y. (2004). Plant and mycorrhizal regulation of rhizodeposition. *New Phytol.* **163**, 459–480.

Jungk, A. (2002). Dynamics of nutrient movement at the soil–root interface. In *Plant Roots: The Hidden Half*, 3rd edition, Waisel, Y., Eshel, A., and Kafkafi, U., eds., Marcel Dekker, New York, pp. 587–616.

Kuiper, I., Lagendijk, E. L., Bloemberg, G. V., and Lugtenberg, B. J. J. (2004). Rhizoremediation: A beneficial plant–microbe interaction. *Mol. Plant–Microbe Interact.* **17**, 6–15.

Kuzyakov, Y. (2002). Kreislauf niedermolekularer organischer Substanzen in Böden—Tracerstudien und Theorie. *Habilitationsschrift, Universität Hohenheim, Hohenheimer Bodenkundliche Hefte* **65**, 222 S.

Liljeroth, E., Kuikman, P. J., and Van Veen, J. A. (1994). Carbon transolcation to the rhizosphere of maize and wheat and influence on the turnover of native soil organic matter at different soil nitrogen level. *Plant Soil* **161**, 233–240.

Linehan, D. J. (1985). Organic matter and trace elements. In *Soil Organic Matter and Biological Activity*, Vaughan, D., and Malcolm, R. E., eds., Kluwer Academic Publishers, Dordrecht, pp. 403–421.

Lobartini, J. C., and Orioli, G. A. (1988). Adsorption of iron Fe–humate innutrient solution by plants, *Plant Soil* **106**, 153–157.

Lòpez-Bucio, J., Cruz-Ramìrez, A., and Herrera-Estrella, L. (2003). The role of nutrient availability in regulating root architecture. *Curr. Opin. Plant Biol.* **6**, 280–287.

Lynch, J. M., and Whipps, J. M. (1990). Substrate flow in the rhizosphere. *Plant Soil* **129**, 1–10.

Ma, J. F. (2000). Role of organic acids in detoxification of aluminium in higher plants. *Plant Cell Physiol.* **41**, 383–390.

Marschner, H. (1995). *Mineral Nutrition of Higher Plants*, 2nd edition, Academic Press, London.

Marschner, H., and Römheld, V. (1994). Strategies of plants for acquisition of iron. *Plant Soil* **165**, 261–274.

Marschner, H., Römheld, V., Horst, W. J., and Martin, P. (1986). Root-induced changes in the rhizosphere: Importance for the mineral nutrition of plants. *Z. Pflanzenernährh. Bodenk.* **149**, 441–456.

Martin, F., Perotto, S., and Bonfante, P. (2007). Mycorrhizal fungi: A fungal community at the interface between soil and roots. In *The Rhizosphere: Biochemistry and Organic Substances at the Soil–Plant Interface*, 2nd edition, Pinton, R., Varanini, Z., and Nannipieri, P., eds., CRC Press, Boca Raton, FL, pp. 201–236.

McCully, M. (1999). Roots in soil: Unearthing the complexities of roots and their rhizospheres. *Annu. Rev. Plant Physiol. Mol. Biol.* **50**, 695–718.

Mengel, K. (1996). Turnover of organic nitrogen in soils and its availability to crops. *Plant Soil* **181**, 83–93.

Muscolo, A., Cutrupi, S., and Nardi, S. (1998). IAA detection in humic substances. *Soil Biol. Biochem.* **30**, 199–201.

Muscolo, A., Sidari, M., Francioso, O., Tugnoli, V., and Nardi, S. (2007). The auxin-like activity of humic substances is related to membrane interections in carrot cell cultures. *J. Chem. Ecol.* **33**, 115–129.

Nardi, S., Pizzeghello, D., Muscolo, A., and Vianello, A. (2002). Physiological effects of humic substances on higher plants. *Soil Biol. Biochem.* **34**, 1527–1536.

Nardi, S., Reniero, F., and Concheri, G. (1997). Soil organic matter mobilization by root exudates of three maize hybrids. *Chemosphere* **35**, 2237–2244.

Neumann, G., and Römheld, V. (1999). Root excretion of carboxylic acids and protons in phosphorus-deficient plants. *Plant Soil* **211**, 121–130.

Neumann, G., and Römheld, V. (2002). Root-induced changes in the availability of nutrients in the rhizosphere. In *Plant Roots: The Hidden Half*, 3rd edition, Waisel, Y., Eshel, A., and Kafkafi, U., eds., Marcel Dekker, New York, pp. 617–649.

Neumann, G., and Römheld, V. (2007). The release of root exudates as affected by the plant's physiological status. In *The Rhizosphere: Biochemistry and Organic Substances at the Soil–Plant Interface*, 2nd edition, Pinton, R., Varanini, Z., and Nannipieri, P., eds., CRC Press, Boca Raton, FL, pp. 23–72.

Owen, A. G., and Jones, D. L. (2001). Competition for amino acids between wheat roots and rhizosphere microorganisms and the role of amino acids in plants N acquisition. *Soil Biol. Biochem.* **33**, 651–657.

Palmgren, M. G. (2001). Plant plasma membrane H^+-ATPases: Powerhouses for nutrient uptake. *Annu. Rev. Plant Physiol. Plant Mol. Biol.* **52**, 817–845.

Pandeya, S. B., Singh, A. K., and Dhar, P. (1998). Influence of fulvic acid on transport of iron in soils and uptake by paddy seedlings, *Plant Soil* **198**, 117–125.

Perry, L. G., Alford, É. R., Horiuchi, J., Paschke, M. W., and Vivanco, J. M. (2007). Chemical signals in the rhizosphere: Root–root and root–microbe communication. In *The Rhizosphere: Biochemistry and Organic Substances at the Soil–Plant Interface*, 2nd edition, Pinton, R., Varanini, Z., and Nannipieri, P., eds., CRC Press, Boca Raton, FL, pp. 297–330.

Phillips, R. R. (2007). Towards a rhizo-centric view of plant–microbial feedbacks under elevated atmospheric CO_2. *New Phytol.* **173**, 664–667.

Piccolo, A., Nardi, S., and Concheri G. (1996). Micelle-like conformation of humic substances as revealed by size exclusion chromatography. *Chemosphere* **33**, 595–602.

Pinton, R., Cesco, S., De Nobili, M., Santi, S., and Varanini, Z. (1998). Water- and pyrophosphate-extractable humic substances fractions as a source of iron for Fe-deficient cucumber plants. *Biol. Fert. Soils* **26**, 23–27.

Pinton, R., Cesco, S., Iacolettig, G., Astolfi, S., and Varanini, Z. (1999a). Modulation of NO_3^- uptake by water-extractable humic substances: Involvement of root plasmalemma H^+ ATPase. *Plant Soil* **215**, 155–161.

Pinton, R., Cesco, S., Santi, S., and Varanini Z. (1997). Soil humic substances stimulate proton release by intact oat seedling roots. *J. Plant Nutr.* **20**, 857–869.

Pinton, R., Cesco, S., Santi, S., Agnolon, F., and Varanini, Z. (1999b). Water-extractable humic substances enhance iron deficiency responces by Fe-deficient cucumber plants. *Plant Soil* **210**, 145–157.

Pinton, R., Cesco, S., Schmidt W., and Varanini, Z. (2006). Role of humic substances as rhizospheric signals affecting root growth and mechanisms of nutrient acquisition. In *Proceedings, 13th Meeting of the International Humic Substances Society*, Karlsruhe, vol. 45-1, pp. 45–48.

Pinton, R., Varanini, Z., and Nannipieri, P. (2001). The rhizosphere as a site of biochemical interactions among soil components, plants and microorganisms. In *The Rhizosphere: Biochemistry and Organic Substances at the Soil–Plant Interface*, Pinton, R., Varanini, Z., and Nannipieri, eds., Marcel Dekker, New York, pp. 1–17.

Pizzeghello, D., Nicolini, G., and Nardi, S. (2001). Hormone-like activity of humic substances in *Fagus sylvaticae* forests. *New Phytol.* **151**, 647–657.

Popov, A. I., and Chertov O. G. (1997). On humic substances as a direct nutritive component of plant–soil trophic system. In *The Role of Humic Substances in the Ecosystems and in Environmental Protection*, Drozd, J., Gonet, S. S., Senesi, N., and Weber, J., eds., Polish Society of Humic Substances, Wroclaw, pp. 993–998.

Puschenreiter, M., Schnepf, A., Millán, M. I., Fitz, W. J., Horak, O., Klepp, J., Schrefl, T., Lombi, E., and Wenzel, W. W. (2005). Changes of Ni biogeochemistry in the rhizosphere of the hyperaccumulator *Thlaspi goesingense*. *Plant Soil* **272**, 205–218.

Quaggiotti, S., Ruperti, B., Pizzeghello, D., Francioso, O., Tugnoli, V., and Nardi, S. (2004). Effect of low molecular size humic substances on nitrate uptake and expression of genes involved in nitrate transport in maize (*Zea mays* L.). *J. Exp. Bot.* **55**, 803–813.

Römheld, V. (1991). The role of phytosiderophores in acquisition of iron and other micronutrients in graminaceous species: An ecological approach. *Plant Soil* **130**, 127–134.

Römheld, V., and Marschner, H. (1981). Iron deficiency stress induced morphological and physiological changes in root tips on sunflower. *Physiol. Plant.* **53**, 354–360.

Römheld, V., and Neumann, G. (2006). The rhizsosphere: Contributions of the soil–root interface to sustainable soil Systems. In *Biological Approaches to Sustaible soil Systems*, Uphoff, N., et al., eds., Taylor & Francis, New York, pp. 91–107.

Ryan, P. R., Delhaize, E., and Jones, D. L. (2001). Function and mechanism of organic anion exudation from plant roots. *Annu. Rev. Plant Physiol. Plant Mol. Biol.* **52**, 527–560.

Sacchi, G. A., Abruzzese, A., Lucchini, G., Fiorani, F., and Cocucci, S. (2000). Efflux and active re-absorption of glucose in roots of cotton plants grown under saline conditions. *Plant Soil* **220**, 1–11.

Santi, S., Locci, G, Pinton, R., Cesco, S., and Varanini, Z. (1995). Plasma membrane H^+-ATPase in maize roots induced for NO_3^- uptake. *Plant Physiol.* **109**, 1277–1283.

Schmidt, W., Santi, S., Pinton, R., and Varanini, Z. (2007). Water-extractable humic substances alter root development and epidermal cell pattern in Arabidopsis. *Plant Soil* **300**, 259–267.

Schnitzer, M., and Hindle, D. A. (1980). Effect of peracetic acid oxidation on N-containing components of humic materials. *Can. J. Soil Sci.* **60**, 541–548.

Schnitzer, M., and Kahn, S. U. (1972). *Humic Substances in the Environment*, Marcel Dekker, New York.

Schulten, H. R., and Schnitzer, M. (1998). The chemistry of soil organic nitrogen: A review. *Biol. Fertil. Soils* **26**, 1–15.

Schwertmann, U. (1991). Solubility and dissolution of iron oxides, *Plant Soil* **130**, 1–25.

Schwertmann. U. (1966). Inhibitory effect of soil organic matter on the crystallization of amorphous ferric hydroxide. *Nature* **212**, 645–647.

Sequi, P., Guidi, G., and Petruzzelli, G. (1975). Influence of metals on solubility of soil organic matter. *Geoderma* **13**, 153–161.

Skogerboe, R. K., and Wilson, S. A. (1981). Reduction of ionic species by fulvic acid. *Anal. Chem.* **53**, 228–232.

Slesak, E., and Jurek, J. (1988). Effects of potassium humate on electric potentials of wheat roots. *Acta Univ. Wratisl.* **37**, 13–18.

Stevenson, F. J. (1986). *Cycles of Soil Organic Carbon, Nitrogen, Phosphorous, Sulphur, Micronutrients*. John Wiley & Sons, New York.

Stevenson, F. J. (1991). Organic matter–micronutrient reactions in soil. In *Micronutrients in Agriculture*, Mortvedt, J. J., Cox, F. R., Schuman., L. M., and Welch R. M., Soil Science Society of America, Madison, WI, pp. 145–186.

Stevenson, F. J. (1994). *Humus Chemistry: Genesis, Composition, Reactions*, John Wiley & Sons, New York.

Struyk, Z., and Sposito, G. (2001). Redox properties of standard humic acids. *Geoderma* **102**, 329–346.

Tarafdar, J. C., and Jungk, A. (1987). Phosphatase activity in the rhizosphere and its relation to the depletion of organic phosphorous. *Biol. Fertil. Soils* **3**, 199–204.

Tartaro, E. M., D'Orazio, V., Brunetti, G., Senesi, N., Tartaro M. E., D'Orazio V., Brunetti G., and Senesi N. (1999). Acidi umici del suolo rizosferico a confronto con acidi umici del suolo adiacente. Proceedings IV Congress Italian Chapter IHSS, Alghero, pp. 19–24.

Thornton, B., and Robinson, D. (2005). Uptake and assimilation of nitrogen from solutions containing multiple N sources. *Plant Cell Environ.* **28**, 813–821.

Tinker, P. B., and Nye, P. H. (2000). *Solute Movement in the Rhizosphere*, Oxford University Press, Oxford.

Uren, N. C. (2007). Types, amounts, and possible functions of compounds released into the rhizosphere by soil-grown plants. In *The Rhizosphere: Biochemistry and Organic Substances at the Soil–Plant Interface*, 2nd edition, Pinton, R., Varanini, Z., and Nannipieri, P., eds., CRC Press, Boca Raton, FL, pp. 1–21.

Varanini, Z., and Pinton, R. (1995). Humic substances and plant nutrition. *Prog. Bot.* **56**, Behnke, H. D., Lüttge, U., Esser, K., Kadereit, J. W., and Runge, M., eds., Springer-Verlag, Berlin, pp. 97–117.

Varanini, Z., and Pinton, R. (2001). Direct versus indirect effects of soil humic substances on plant growth and nutrition In *The Rhizosphere: Biochemistry and Organic Substances at the Soil–Plant Interface*, Pinton, R., Varanini, Z., and Nannipieri, eds., Marcel Dekker, New York, pp. 141–157.

Varanini, Z., and Pinton, R. (2006). Plant–soil relationship: Role of humic substances in Iron Nutrition. In *Iron Nutrition in Plants and Rhizospheric Microorganisms*, Barton, L. L., and Abadía, J., eds., Springer-Verlag, Heidelberg, pp. 153–168.

Varanini, Z., and Pinton, R. (2007). Root membrane activities relevant to nutrient acquisition at the plant–soil interface. In *The Rhizosphere: Biochemistry and Organic Substances at the Soil–Plant Interface*, 2nd edition, Pinton, R., Varanini, Z., and Nannipieri, P., eds., CRC Press, Boca Raton, FL, pp. 151–172.

Varanini, Z., Pinton, R., De Biasi, M. G., Astolfi, S., and Maggioni, A. (1993). Low molecular weight humic substances stimulate H^+-ATPase activity of plasma membrane vesicles isolated from oat (*Avena sativa* L.) roots. *Plant Soil* **153**, 61–69.

Vaughan, D., and Linehan, D. J. (1976). The growth of wheat plants in humic acid solutions under axenic conditions. *Plant Soil* **44**, 445–449.

Vaughan, D., and Malcolm, R. E. (1985). Influence of humic substances on growth and physiological processes. In *Soil Organic Matter and Biological Activity*, Vaughan, D., and Malcolm, R. E. eds., Kluwer Academic Publishers, Dordrecht, pp. 37–76.

Vaughan, D., and Ord, B. G. (1981). Uptake and incorporation of ^{14}C-labelled soil organic matter by roots of *Pisum sativum* L. *J. Exp. Bot.* **32**, 679–687.

Violante, A., Barberis, E., Pigna, M., and Boero, V. (2003). Factors affecting the formation, nature and properties of iron precipitation products at the soil–root interface. *J. Plant Nutr.* **26**, 1889–1908.

Wang, W. H., Bray, M. C., and Jones M. N. (1999). The fate of ^{14}C-labelled humic substances in rice cells in culture. *J. Plant Physiol.* **154**, 203–211.

Werner, D. (2007). Molecular biology and ecology of the Rhyzobia–legume symbiosis. In *The Rhizosphere: Biochemistry and Organic Substances at the Soil–Plant Interface*, 2nd edition, Pinton, R., Varanini, Z., and Nannipieri, P., eds., CRC Press, Boca Raton, FL, pp. 237–266.

Zandonadi, D. B., Canellas, L. P., and Façanha, A. R. (2007). Indolacetic and humic acids induce lateral root development through a concerted plasmalemma and tonoplast H^+ pumps activation. *Planta* **225**, 1583–1595.

Zhuo, S., Wen, Q., and Cheng, L. (1995). Availability of non-hydrolyzable soil nitrogen. *Pedosphere* **5**, 183–186.

10

DISSOLVED ORGANIC MATTER (DOM) IN NATURAL ENVIRONMENTS

F. H. FRIMMEL AND G. ABBT-BRAUN
Engler-Bunte-Institut, Wasserchemie, Universität Karlsruhe, Karlsruhe, Germany

10.1. DEFINITION OF DOM

Dissolved organic matter (DOM) consists mostly of a suite of different organic substances which may have a complex structure and manyfold interactions. In order not to get lost in this complexity and to avoid misunderstandings, it is advisable to

Biophysico-Chemical Processes Involving Natural Nonliving Organic Matter in Environmental Systems, Edited by Nicola Senesi, Baoshan Xing, and Pan Ming Huang

define as good as possible the DOM concerned, the time and space frame of the system investigated, and the necessary resolution. In this chapter, DOM is taken as an integrated pool of dissolved organic compounds in aquatic systems, no matter how complex their interactions are. And the dissolved state is understood as part of a dominant liquid phase like rivers, lakes, oceans, groundwater or wastewater, and process water. The authors are well aware that the DOM in other major compartments like the atmosphere, soil, and sediments and their seepage water and porewater have been neglected since they are well represented by other authors and reports. However, most of the applied principles in reaction and transport are transferable for the whole hydrosphere.

It has been widely accepted to define DOM as those organic substances of the TOC (total organic carbon) in aqueous solutions passing through a membrane filter with nominal pore size of 0.45 μm. The analytical methods used determine the carbon content; hence the concentration of DOM and TOC are given either as dissolved organic carbon (DOC) or as TOC concentration. Guidelines for the analytical determination of TOC and DOC are available (e.g., DIN, EN 1484) by using either high-temperature combustion method or chemical oxidation methods followed by the detection of CO_2. In practice, this simple definition has some severe pitfalls: The type of membrane, its surface tension, and the age and state of equilibration can be of significant influence on the results. Therefore a precise description of the kind of filter and of the experimental conditions used is crucial, though often neglected. Another important factor of influence is the pore blocking. It is strongly dependent on the kind of TOC and in particular on the colloidal and particulate matter. As a consequence, the results at the beginning of the filtration can be quite different from those at the end, depending strongly on the amount and kind of material responsible for membrane fouling. Ways around this problem include use of a sequence of membranes with decreasing pore sizes. Determination of the colloidal index (CI, also known as silt density index SI, or fouling index FI) is another possibility to reach operationally defined data on the amount of DOC (ASTM, 2002). The experimental steps of CI determination are given in Figure 10.1. Detailed studies on the interferences from the filter used for filtration and the release of organic compounds were recently published by Khan and Pillai (2007). They showed that the amount and the characteristics of the DOC leached from the filters varies, and they also showed that it depends on the pretreatment of the filters (soaking or discarding initial filtration volume).

Some other common terms are given in Table 10.1. These definitions are mainly due to fractionation procedures, to other classification systems, and to specific research views. In most ecosystems, DOM is derived from natural sources, so it is often referred to as natural organic matter (NOM). Humic substances (HS) can comprise a significant fraction of the DOM. HS can be separated into humic and fulvic acids and nonhumic substances (HA, FA, NHS). The term "refractory" (ROS) commonly reflects a relatively low biodegradability, but it has not been defined properly. The refractory character does not directly show up in most of the classification terms, but it is implicitly present due to the poor biodegradability and generally high molecular masses of the organic matter concerned. It is also reflected in the long-time stability of most of the samples. Some of the terms are synonymous. Since most terms are operationally defined, they need a detailed description of the experimental procedure with regard to their isolation, preparation, and origin. In this

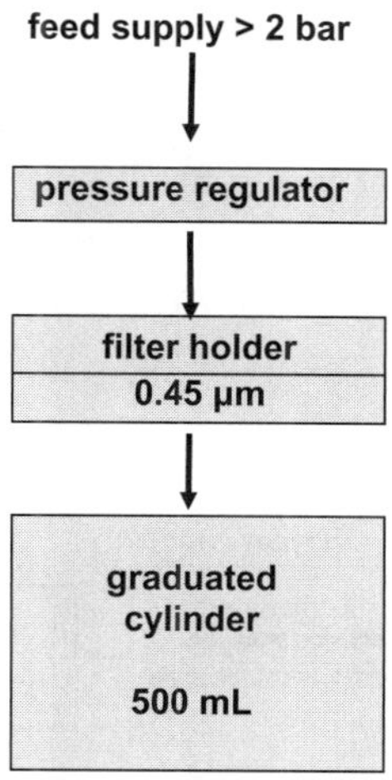

Figure 10.1. Experimental procedure for the determination of the colloidal index (CI). The CI or the silt density index (SDI) test is used to predict and prevent particulate fouling on the membrane surface. It measures the time required to filter a fixed volume of water through a standard 0.45-μm pore-size microfiltration membrane with a pressure of 2.07 bar. The difference between the initial time and the time of a second measurement after normally 15 minutes (after silt was built up) represents the CI or SDI value.

TABLE 10.1. Common Terms Used for DOM and Other Fractions of Organic Matter

Abbreviation	Meaning
AOC	Assimilable organic carbon
BOD	Biological oxygen demand
COD	Chemical oxygen demand
DOC	Dissolved organic carbon (<0.45 μm)
DOM	Dissolved organic matter
FA	Fulvic acid
HA	Humic acid
HS	Humic substances
NHS	Nonhumic substances
NOM	Natural organic matter
OC	Organic matter
POC	Particulate organic carbon (>0.45 μm)
TOC	Total organic carbon
SOM	Soil organic matter
ROM	Recalcitrant organic matter
ROS	Refractory organic substances

chapter for the sake of simplicity and clarity mainly DOC, TOC, and DOM are used, assuming that approximately half of the DOM consists of DOC, which can be measured with relatively high precision.

A way out of the dilemma is the determination of the molecular size distribution of the chromatographable organic carbon (COC) as it is offered from size exclusion chromatography (SEC) equipped with an organic carbon (OC) detection system.

The integrated OC responses can be compared with the DOC concentration of the original sample. In most cases more than 90% of the DOC is chromatographable and can be assigned as DOM. The experimental setup for SEC-DOC characterization is given in Figure 10.2.

In Figure 10.3 a typical size exclusion chromatogram obtained by UV and OC detection is given. Larger molecules show up at shorter retention times than do

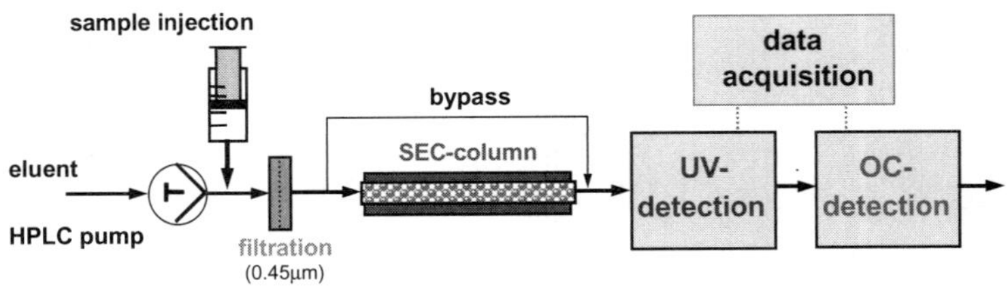

	experimental conditions
eluent	phosphate buffer: 28 mmol/L; pH 6.6 (1.5 g/L $Na_2HPO_4 \cdot 2H_2O$ and 2.5 g/L KH_2PO_4)
flow rate	1.0 mL min^{-1}
SEC columns	TSK HW 50(S) or 40 (S) (250 x 20 mm, Grom Herrenberg)
injection volume	2 mL (column); 0.5 mL (bypass)

Figure 10.2. Experimental setup for the determination of the molecular size distribution of DOM (according to Huber and Frimmel, 1992).

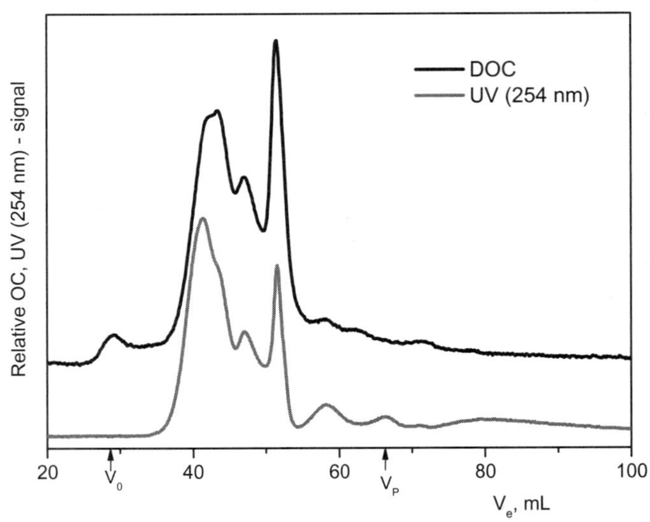

Figure 10.3. Size exclusion chromatogram obtained by UV and OC detection (SEC/UV-DOC) for a river sample (river Neckar, Heidelberg, 22.03.2007, DOC = 2.3 mg liter^{-1}, chromatographable organic carbon (COC = 75%). Column: TSK HW50S (250 × 20 mm). Eluent: phosphate buffer, 26.8 mmol liter^{-1}, UV (254 nm) and organic carbon (OC) detection, exclusion volume V_0, permeation volume V_p).

smaller ones. The UV absorption detection at λ = 254 nm can give valuable additional information on the relative amount of delocalized electrons in the chemical structure. From this, first estimates on the molecular structure can be given (Abbt-Braun et al., 2004).

The application of further element-specific detection systems like ICP-MS (Inductively Coupled Plasma–Mass Spectrometry) can result in the element distribution in the SEC fractions (Schmitt et al., 2003).

Some basic problems, artifacts, and difficulties of the analysis of DOM, mainly obtained from soil seepage water, were discussed by Zsolnay (2003). Methodological aspects concerning the determination of low DOC concentrations as given in marine systems are described by Dafner and Wangersky (2002). Examples for field procedures to collect and preserve freshwater samples for DOC analysis were shown by Kaplan (1994). A simple method to analyze the DOM in aqueous soil samples was suggested by Dilling and Kaiser (2002). They used the light absorption at 260 nm. The method is based on the fact that the hydrophobic fraction of DOM contains almost entirely the aromatic moieties of DOM. They tested the methods for aqueous soil samples from different origin (spruce, pine, and beech litter).

10.2. CHARACTERIZATION METHODS

10.2.1. Isolation Procedures

The concentration of DOM in aquatic systems has a broad range. The same is valid for inorganic constituents. The concentration ranges from <1 mg liter^{-1} DOC for ground- and seawater, up to 40 mg liter^{-1} DOC for brown water in swamps and to several hundred mg liter^{-1} DOC in soil seepage water in the upper layer of the soil horizons. The characteristic DOC concentrations from different aquatic systems are given in Table 10.2 (Thurman, 1985a).

It is most valuable for meaningful results to characterize DOM within the matrix of the sample concerned. However, for most analytical methods, either the DOC concentration is too low or there are interferences with the inorganic constituents. Therefore, a pretreatment of the sample is often needed. Concentration techniques often used are shown in Table 10.3. Some of the methods applied also lead to a concentration of the inorganic water constituents that may interfere with the analytical method and therefore have to be removed. However, these techniques often lead to a fractionation of the original DOM.

Concepts for the isolation of standard and reference material from different origin are recommended from the International Humic Substances Society (IHSS) (IHSS, 2007). The isolation from aqueous solution is based on the ad- and desorption on ionic macroporous resins like XAD-8 resins (acrylic ester, or DAX-8). The XAD material used in the past for the isolation of humic and fulvic acids is not manufactured anymore. The substitute material, DAX-8, has a slightly higher sorption capacity compared to XAD-8, and the content of aliphatics seems to be slightly higher for the XAD-8 product (Peuravuori et al., 2002a,b). According to the XAD method, DOM is fractionated into hydrophobic and hydrophilic fractions (Figure 10.4). The hydrophobic fraction is sorbed onto the resin at low pH value (pH $\leq$ 2) and eluted by an aqueous NaOH solution. The humic acid fraction (HA) is precipi-

TABLE 10.2. Typical DOC Concentrations for Different Aquatic Systems [According to Thurman (1985a)]

Type of Water	DOC (mg $liter^{-1}$) Range (min.–max.)
Ocean	0.5–1
Rivers	5–9
Lakes, oligotrophic	2–3
Lakes, eutrophic	9–16
Soil seepage water	19–31
Groundwater ($CaCO_3$ aquifer)	0–1

TABLE 10.3. Isolation Methods for DOM

Method	Advantages	Disadvantages
Membrane filtration such as ultrafiltration, nanofiltration, and reverse osmosis	Fraction of DOM according to the pore size, high yields, and large volumes possible	Inorganic impurities (ash content), membrane fouling
Nonionic macroporous resins	Broad pH range, simple fractionation, regeneration of resins, large volumes possible, high capacity	High pH values needed for desorption, irreversible sorption possible, resin bleed
Ion exchange resin	Specific ionic interaction, high capacity, large volumes possible	Resin bleed, limited regeneration, irreversible sorption possible
Freeze-drying	Structure protective	Slow, needs pretreatment for good results
Sorbent extraction	Exclusion of salts	Limited solubility of human matter, partly irreversible sorption
Alumina	High sorption capacity, no organic bleed	Irreversible sorption and chemical change of sorptive possible
Activated carbon	High sorption capacity, chemical impurities	Strong irreversible sorption, biofouling

tated at pH = 2, and the soluble fulvic acid fraction (FA) is desalted using a cation exchange resin. This concept or variations using XAD-2 resins (styrene divinylbenzene, or DAX-2) or additional resins like macroporous anion exchange resins to concentrate the hydrophilic fraction are given by Leenheer (1985), Aiken (1988), Naumczyk et al. (1989), Abbt-Braun et al. (1991), and Imai et al. (2001).

In addition, the IHSS recommends a procedure to isolate HA and FA from solid-phase source materials after aqueous extraction (Swift, 1996). Detailed descriptions and critical reviews for the isolation from soil samples are also given by Hayes (1985) and Swift (1985).

Since the last 15 years, membrane techniques like ultrafiltration, nanofiltration, or reverse osmosis have been also applied (Serkiz and Perdue, 1990; Abbt-Braun

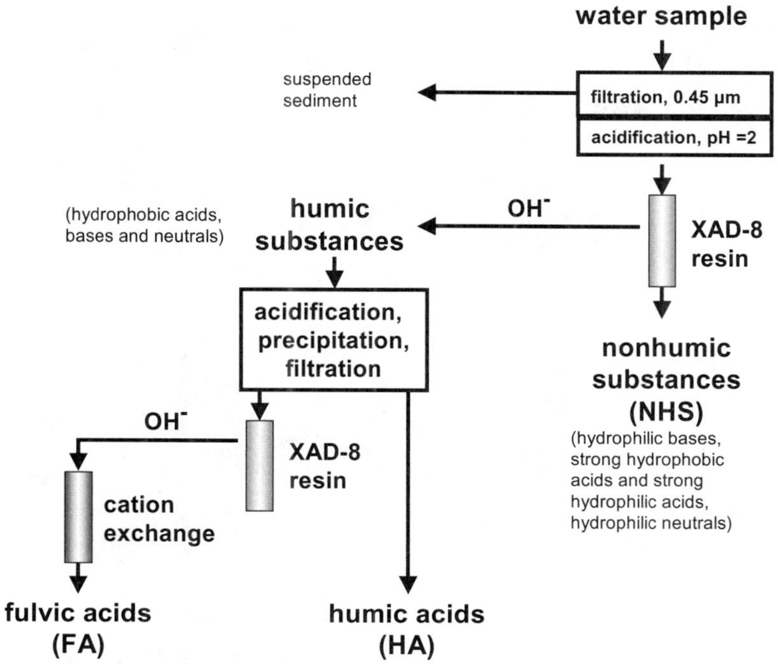

Figure 10.4. XAD-isolation procedure for fulvic and humic acids according to Abbt-Braun and Frimmel (2002).

et al., 1991; Sun et al., 1994; Gjessing et al., 1998, 1999; Kulovaara et al., 1999; Kitis et al., 2001; Abbt-Braun and Frimmel, 2002; Burba et al., 2002; Müller and Frimmel, 2002; Kilduff et al., 2004; Koprivnjak et al., 2006a,b). To reduce the inorganic constituents, cation exchange resins or electrodialysis are used either prior to or after the concentration step.

The fractions obtained after isolation are operationally defined. The yield of the different methods and the relative amounts of the fractions vary according to the reaction conditions applied and to the adsorbents or membranes used. In Figure 10.5 the mass balance for different fractions obtained by the XAD-8 procedure and by ultrafiltration is given for two different samples, a brown water and a wastewater effluent.

According to the operational definition, it is most important that well-defined methods and clear protocols are provided describing the isolation procedure applied. After isolation, freeze-drying is often used to stabilize the final fraction and to prevent chemical and biological reactions during storage. Critical reviews on the isolation and concentration techniques for aquatic substances are given by Aiken (1985), Leenheer (1985), and Abbt-Braun and Frimmel (2002).

10.2.2. Chromatography

High-performance size exclusion chromatography (HPSEC) techniques with multiple online detectors (e.g., UV-, fluorescence-, or element-specific systems, Figure 10.6) enables a better understanding of quantitative and qualitative DOM properties

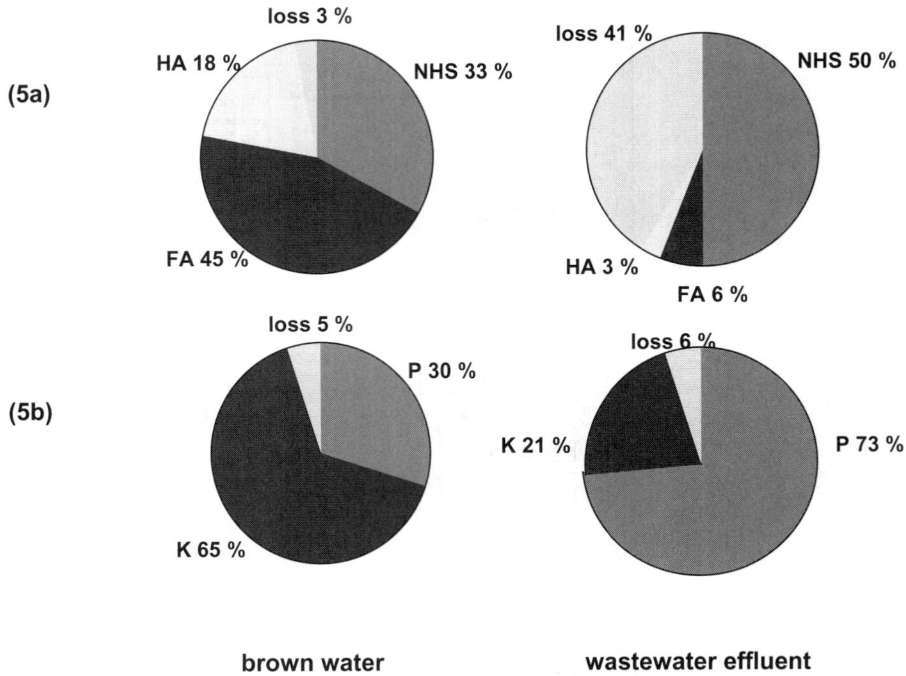

Figure 10.5. Comparison of the mass balance (in % DOC) for different fractions obtained by the XAD-8 procedure (**5a**) (FA, fulvic acid; HA, humic acid; NHS, nonhumic substances) and by ultrafiltration (**5b**) (K, concentrate; P, permeate; cutoff, 4000 g/mol). Abbreviations used: HO, Lake Hohloh, brown water; ABV, wastewater effluent; numbers, sampling occasion.

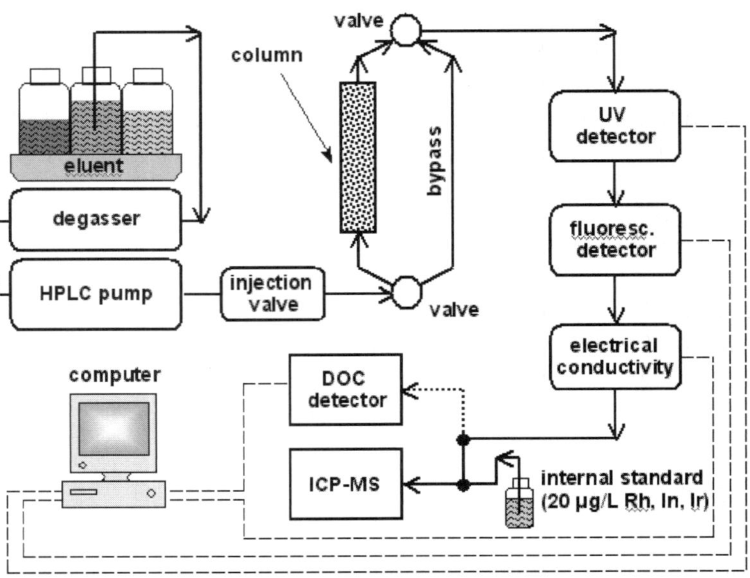

Figure 10.6. Setup for size exclusion chromatography with multidimensional detection.

(Müller et al., 2000; Her et al., 2003). Emphasis has been placed on (a) the use of gel chromatography or gel permeation chromatography for the fractionation of DOM on the basis of molecular size differences and (b) the application of electrophoretic separation methods (Perminova et al., 1998, 2003; Specht and Frimmel, 2000), including electrophoresis, capillary electrophoresis (CE), isotachophoresis, isolelectric focusing, polyacrylamide gel electrophoresis (PAGE), and capillary zone electrophoresis (CEZ) (De Nobili et al., 1989, 1998; Schmitt-Kopplin et al., 1998).

An important advantage of the gel chromatographic approach (size exclusion chromatography) lies in the direct applicability of natural aquatic samples and in the structural information that can be obtained on the fractions from online multidimensional detection of specific elements (e.g., C, N, metals, halogens). In addition, UV (ultraviolet, e.g., at λ = 254 nm) and vis spectral absorbance (visible, e.g., at λ = 436 nm), as well as fluorescence, can be detected (Huber and Frimmel, 1992; Hesse and Frimmel, 1999). Besides the molecular size estimation, gel chromatography is well-suited for a finger printing of the DOM and for the comparison of samples from different sources (Figure 10.7).

Characterization of DOM according to molecular size, polarity, or other structure-based properties include often the demand for element specific information. Questions to be answered might concentrate, for example, on the phosphorous, nitrogen, or sulfur distribution in the different molecular size ranges. From the point of view of transport processes, the heavy metal content in the different fractions might be important for the interactions between inorganic or geogenic colloids and DOM.

In principle, all powerful element-specific methods that are able to monitor continuously the effluents of separation processes commonly in the range of a few ml min^{-1} and in element concentrations of some 10 µg liter^{-1}. A well-suited method is based on modern element-specific quadrupole mass spectrometry (MS) with an inductively coupled plasma (ICP) interface to the separation unit [e.g., liquid chromatography (LC) or field-flow fractionation (FFF)]. The ICP-MS detection can also be used for continuously characterizing the effluent of any kind of packed column (Metreveli and Frimmel, 2007). By this, the transport and elution properties of

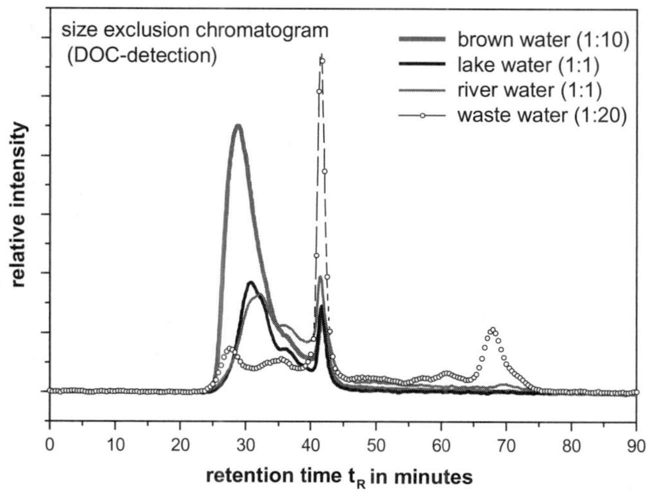

Figure 10.7. Size exclusion chromatogram of DOM samples from different sources, obtained by organic carbon (OC) detection.

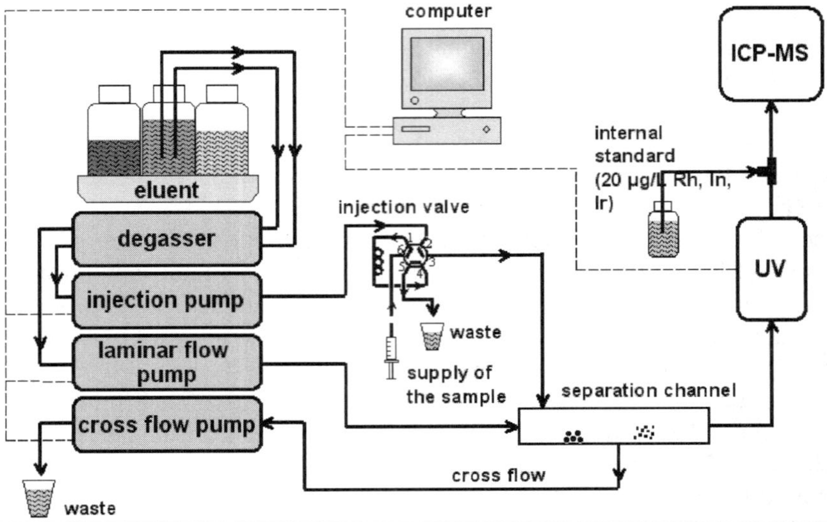

flush time	300 s	injection volume	100 µL
injection time	200 s		
injection flow	0.1 mL/min	cross flow during elution:	
laminar flow during injection	1.8 mL/min	0 - 4 min	1.5 mL/min
focusing time	60 s	> 4 min	exponential decrease
laminar flow during focusing	1.8 mL/min	channel thickness	350 µm
laminar flow at the channel outlet	0.3 mL/min		

Figure 10.8. Experimental setup for element-specific detection of field-flow fractionation.

DOM can be measured under (simulated) natural conditions. Figure 10.8 shows the experimental setup for such investigations.

Separation by flow field-flow fractionation (FFFF) allows the determination of size distribution of molecules and colloids in the submicrometer-diameter range and was first shown by Beckett et al. (1987). Since then, the method has been often used for studies on the mobility of DOM-related nanoparticles. Coupling the FFF with ICP-MS (Exner et al., 2000) leads to useful information on metal speciation.

Isotope dilution mass spectrometry (ICP-IDMS) is a technique sensitive enough for the determination of mass flows of an element in the lowest picogram per second range. Therefore, separated metal/DOM fractions can be investigated (Heumann et al., 2002). Characteristic fingerprints of the element distribution in size exclusion chromatography-separated fractions can be obtained according to the origin of the water samples. Using isotope dilution mass spectrometry (IDMS) kinetic studies can also be conducted.

10.2.3. Spectroscopy

There is a broad variety of spectroscopic methods which can be used for the (partial) characterisation of the molecular structure of the DOM. Many publications have become available on the invasive and noninvasive spectroscopic characterisation of DOM from natural origin (natural organic matter, NOM) (e.g., MacCarthy and Rice, 1985; Cabaniss and Shuman, 1987; Bloom and Leenheer, 1989; Senesi et al., 1989; Hautala et al., 2000; Filippova et al., 2001).

In recent years, those methods that do accept original samples (e.g., from soils or water) without major treatment and concentration have gained importance due to the small degree of denaturing and of added pollutants. This section will focus on the methods that are suited to characterize the organic matter in its dissolved state. Powerful methods are also in use for online detection of chromatographic or otherwise obtained fractions.

Spectroscopic methods like infrared spectroscopy (IR), nuclear magnetic resonance spectroscopy (NMR), and mass spectrometry (MS) are described in other chapters of this textbook.

The spectral absorption in the UV and visible light range belong to the oldest characterization methods for DOM. This is not surprising because of the relatively early availability of spectrometers working in the UV and visible range. On the other hand, it is well known that aquatic systems with high concentrations of DOM (e.g., bog lakes, organic rich aquifers) show typically a yellow to brown color recognizable even with plain human eyes. The old name "Gelbstoff" (Kalle, 1938) for the main DOM points clearly toward the specific color of these substances.

The UV–vis spectra of DOM are poorly resolved with a characteristic strong increase of the absorbance to lower wavelengths (Figure 10.9). This is typical for complex mixtures of substances that have strong intermolecular interactions and a significant amount of unsaturated bonds and lone-pair electrons. Around $\lambda = 254\,\text{nm}$, often a weak shoulder is obvious which is assigned to chromophores with C=C and C=O double bonds that can be conjugated. The spectral absorption exhibits a

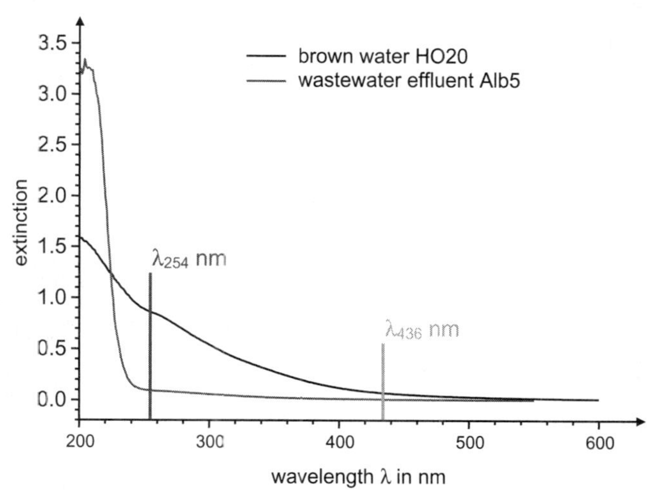

Figure 10.9. UV–vis spectra of DOM from a brown water lake (HO20) and an effluent from a wastewater treatment plant (original water sample, Alb5).

dependence on pH values decreasing specific absorbance as solution pH decreases (Langhals et al., 2000). This reflects the acid–base forms of the chromophores within the molecules; or as suggested by Chen et al. (1977), an increase in particle size due to macromolecular associations. For some studies, differential spectroscopy can be used (Korshin et al., 1999).

Since the UV and visible range absorbance can easily be quantified spectrometrically, this can be used for a fast and straightforward estimation of the amount of DOM. Based on the continuous decrease of absorbance with increasing wavelength which results in a fairly unresolved spectrum, the value of $\lambda = 254\,\text{nm}$ is often used as characteristic information on DOM.

The comparability can be improved by making the spectral information specific, by relating it to the mass concentration. This results is the parameter CbUVA (C-based UV absorbance), which relates the UV absorbance and the DOC concentration of the sample. Typical values are given in Table 10.4.

The values allow a rough characterization of the DOM according to its genesis—for example, whether it is more allochtonous (plant and lignin origin) or more autochtonous (algae- and aquatic microorganisms-derived).

The relatively simple method of UVA determination has led to its broad application as surrogate parameter for the more complicated and less precise determina-

TABLE 10.4. Organic Carbon-Based UV Absorbance (CbUVA) for DOM in Different Samples

Origin/Sample	CbUVA(254) (liter $mg^{-1}\,m^{-1}$)	pH
Brown water		
Lake Hohloh HO10	4.09	4.1
HO14	4.81	3.5
HO16	4.32	4.2
HO14 FA	4.80	2.0
HO14 FA	4.98	7.0
HO14 FA	5.20	11.0
River		
Rio Negro	5.50	5.5
Amazon	2.16	5.2
Rhine	2.66	7.6
Danube	3.05	7.8
Lake, reservoir water		
Kranichsee	2.95	4.1
Lake Constance	2.12	8.2
Kleine Kinzig	2.20	6.2
Wahnbachtalsperre	2.12	7.1
Soil seepage water		
BS1	3.13	4.2
Groundwater		
Fuhrberg FG1	2.92	7.5
Secondary effluent		
ABV2	1.44	7.9
ABV3	1.77	8.0

tion of DOC. The correlation depends on the origin of the water samples but is quite constant for the individual aquatic systems.

A special version of UV–vis spectroscopy is the detection of the sample light emission after irradiative molecular excitation. This phenomenon is called luminescence and specified as fluorescence in case of relatively fast electron relaxation processes without changing spin multiplicity (time scale of microseconds and below).

Fluorescence has turned out to be a fairly typical characteristics of DOM due to numerous double bonds (π electrons) of the molecules involved even though there is partial quenching by functional groups—for example, by nitro, carbonyl, and hydrosulfide structure. Fluorescence spectra show maxima for the excitation and emission wavelength which are source-dependent (Cabaniss and Shuman, 1987; Senesi et al., 1989; Kumke et al., 1998; Kumke and Frimmel, 2002). However, the assignment of the structures responsible for the fluorescence behavior has still not been satisfactory. Three-dimensional images have revealed the entire domain of excitation/emission spectra for DOM (Figure 10.10).

More detailed information about the fluorescence behavior of DOM can be reached by fractionation of the DOM before characterization. Studies from Chen et al. (2003) showed that the polyphenolic-rich DOM fraction exhibited a much more intense fluorescence and a red shift of peak position in comparison with the carbohydrate-rich DOM. Peuravuori et al. (2002b) used different anion exchange resins for the separation of DOM and synchronous fluorescence spectroscopy for characterization. They specified several classes of chromophores by the synchronous techniques ($\lambda_{ex}/\lambda_{em}$ 280/298, 330/348, 355/373, 400/418, 427/445, 460/478, 492/510, and 516/534 nm). Fluorescene spectroscopy is also a sensitive method to distinguish between different sources of DOM. Examples for the different spectroscopic behavior of marine and terrestrial organic material is given by Lombardi and Jardim (1999). For several years, fluorescence spectroscopy has been used for qualitative

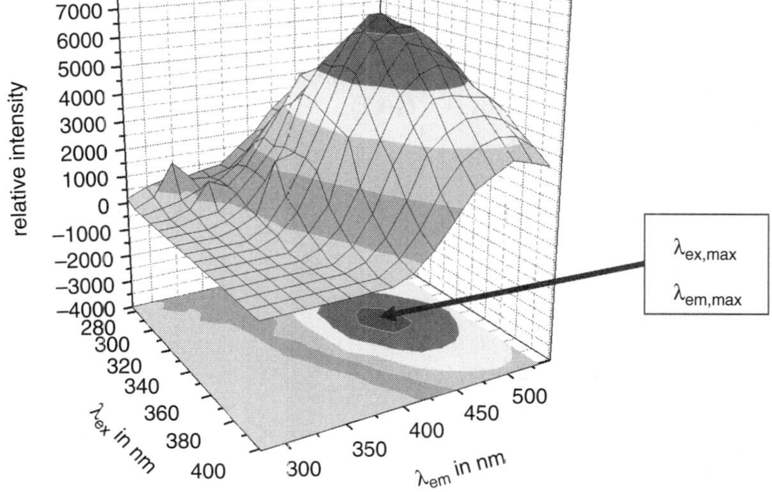

Figure 10.10. Total luminescence spectrum of a water sample (brown water HO19).

identification of the DOM fractions in source water assessment (Marhaba et al., 2000). The sensitivity of fluorescence spectroscopy allows also the investigations in aqueous samples at natural concentrations.

Time-resolved fluorescence spectroscopy shows that there are at least three structural areas, the excitation of which decays at different time windows of the nanosecond range (Kumke and Frimmel, 2002).

Online chromatographic detection by fluorescence can add to the characterization of DOM and help to study complex formation between DOM and paramagnetic metal ions known to be effective fluorescence quenchers (Saar and Weber, 1982; Schmitt et al., 2002).

Due to the high molecular size of the organic molecules in DOM, light scattering is an attractive principle for further characterization. Especially multi-angle laser light scattering (MALLS) can give information on molar masses and radius of the molecules. Manning et al. (2000) show values for the average molar mass of 1.164×10^8 g mol^{-1} and an average root mean square radius (R_z) of 436.0 ± 36 nm at 25 °C for the humic acids investigated. The method is also suited for online detection of chromatographically obtained fractions (Wagoner and Christman, 1999).

10.2.4. Indicator Parameters

Due to the fairly complicated instrumental DOC determination and the relatively high standard deviation of the routinely determined values, there has been an ongoing search for more simple surrogate parameters with meaningful interpretation.

Permanganate consumption belongs to the most traditional wet chemical parameters. It is based on a redox reaction given for a model organic compound $(H_2CO)_n$) given in Eqs. (10.1a) and (10.1b).

$$\overset{+VII}{MnO_4^-} + 5e^- + 8H_3O^+ \rightarrow \overset{+II}{Mn^{2+}} + 12H_2O \tag{10.1a}$$

$$\overset{0}{(H_2CO)_n} + nH_2O \rightarrow n\overset{+IV}{CO_2} + 4ne^- + 4nH^+ \tag{10.1b}$$

The remaining amount of the excessively applied MnO_4^- is quantified by titration with oxalate solution according to Eq. (10.2).

$$2\overset{+VII}{MnO_4^-} + 5\overset{+III}{C_2O_4^{2-}} + 16H_3O^+ \rightarrow 2\overset{+II}{Mn^{2+}} + 10\overset{+IV}{CO_2} + 24H_2O \tag{10.2}$$

There can be interferences from chloride (>300 mg liter^{-1}) and Fe(II), which can be corrected according to:

$$\rho(Fe^{2+}) = 1\,\text{mg liter}^{-1} \equiv \rho(KMnO_4) = 0.57\,\text{mg liter}^{-1}.$$

The oxygen demand can be calculated from the $KMnO_4$ consumption according to:

$$\rho(KMnO_4)\ \text{in mg liter}^{-1} \equiv 3.95 \cdot \rho(O_2)\ \text{in mg liter}^{-1}.$$

Chemical oxygen demand (COD) is another wet chemical oxidation method. The chemical basis is the reaction of dichromate in concentrated sulfuric acid matrix according to Eq. (10.3). COD values are given mg liter^{-1} O_2.

$$\overset{+VI}{CrO_7^{2-}} + 6e^- + 14H_3O^+ \xrightarrow{Ag^+} \overset{+III}{2Cr^{3+}} + 21H_2O \qquad (10.3)$$

The biochemical oxygen demand (BOD) quantifies the biooxidisable water constituents as the oxygen consumption in a mixture with a mixed microbial population. It is mostly determined over a time period of 5 days (BOD_5). The consumed oxygen can be determined by comparing the initial oxygen concentration and the remaining rest concentration. The oxygen concentration can also be followed in air tight reactors over the whole time of the experiment—for example, by means of an oxygen electrode or by measuring the pressure in an air-tight system containing the water to be determined and an oxygen containing gas volume or by electrochemically produced oxygen. From the consumption curve, information on the degradability of DOM can be obtained. The refractory character of most of the DOM leads to relatively small BOD_5 values. The BOD values are given as mg liter^{-1} O_2.

The assimilable amount of DOM (AOC, assimilable organic carbon) can be determined by monitoring the changing turbidity in the sample due to bacterial growth. To reach meaningful results, the aqueous sample to be determined is filtered through a pore membrane, is adjusted with inorganic nutrient solution, and is inoculated with a mixed bacterial population according to a defined protocol (Hambsch et al., 1992). Everything is prepared in a cuvette, and the 12° forward scattered light intensity is measured over a period of 60h (Figure 10.11). From the resulting curve (Figure 10.12), information on growth factors and growth rate can be derived. The higher the amount of growth factor, the higher the availability of assimilable carbon. The higher the growth rate, the faster the assimilation and the greater the degradation of DOM.

The AOC contributes to the microbiological dimension of DOM characterization, the assessment of bacterial growth, and the biological availability of the organic matter. This information is of special value with respect to the refractory character of DOM and its nutritional character for microorganisms.

10.3. STRUCTURE OF DOM

10.3.1. Elemental Composition

The elemental composition represents a fundamental characteristics of DOM, and it is a simple way of characterizing DOM. Despite the broad variety in the origin of DOM, the refractory part shows fairly narrow windows of the elemental content. The elemental analysis includes C, O, H, N, and S, which can be determined by good precision using elemental analysis by specific combustion and the detection of the volatile gases formed (Huffman and Stuber, 1985). In addition, P, halogens, metals, and the ash content can be analyzed. In order to get the content of almost all elements, the organic material has to be isolated from the original sample. Some typical

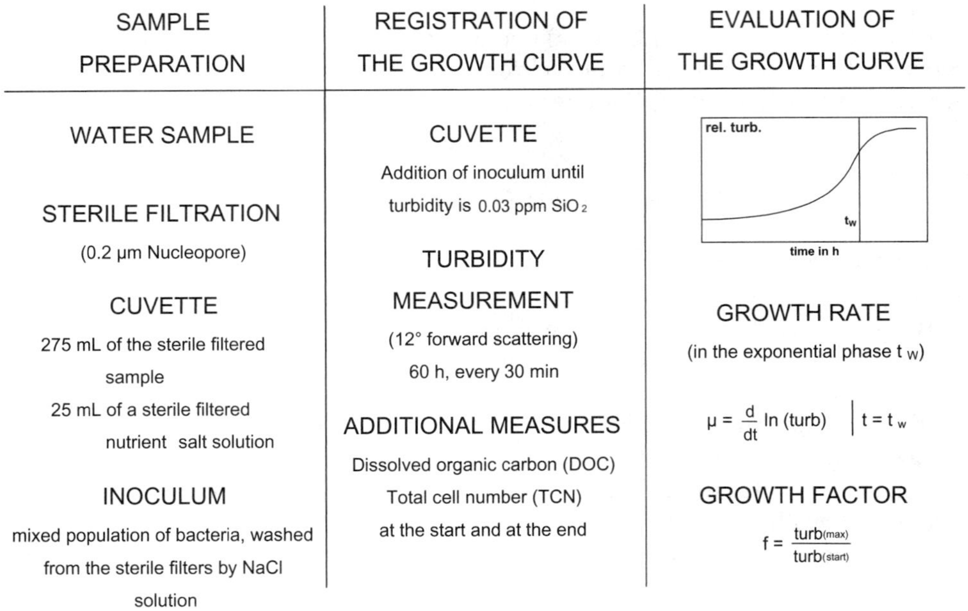

Figure 10.11. Protocol for the determination of the assimilable organic carbon, according to the Werner-AOC determination (Hambsch et al., 1992).

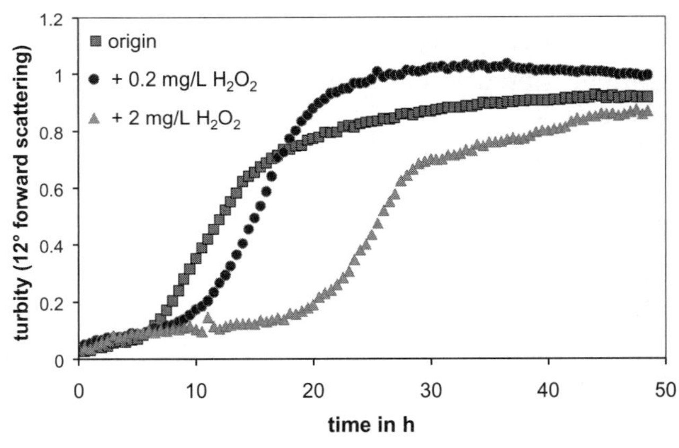

Figure 10.12. Growth curve of DOM (sample: Lake Constance, original sample and after addition of different H_2O_2 concentrations).

data for the elemental composition of isolated DOM are given in Table 10.5. The content of C, H, O, N, and S provides essential information on the origin of the sample (Abbt-Braun and Frimmel, 2002), and it is especially useful to calculate C-mass-specific data of different samples. The H-to-C ratio (H/C) is an indicator for the amount of saturation of the C atoms within the molecule. The O-to-C ratio (O/C) is assumed to be an indicator of the carbohydrate content and the degree of humification. For samples showing high ash contents, N, S, and O values are often questionable and should be used with caution.

TABLE 10.5. Elemental Composition (C, H, O, N, S, Ash) of FA and HA (Isolated by XAD) and of K Samples (Isolated by Ultrafiltration) from Brown Water and Wastewater Effluent Water

Mass Percent, %	C	H	N	O	S	Ash
Fulvic acids (FA), n = 33[a]	36–54	3–5	0.8–3.8	23–40	0.1–1.7	0.1–16
HO10 FA						
Brown water lake	52.7	3.7	0.7	41.5	0.7	0.7
HO10 HA						
Brown water lake	51.8	3.2	1.1	35.3	0.8	7.8
HO12 K						
Brown water lake, concentrate after ultrafiltration	49.1	4.0	1.3	40.9	1.3	3.4
FG1 FA						
Groundwater	55.6	4.9	1.4	31.9	1.6	4.6
FG1 HA						
Groundwater	51.7	4.0	2.9	25.6	—	15.92
BS1 FA						
Soil seepage water	53.3	3.6	0.7	41.8	0.3	0.3
BS1 HA						
Soil seepage water	53.4	3.7	1.1	36.5	0.4	5.1
ABV2 FA						
Wastewater effluent	50.9	5.0	2.9	33.2	2.8	5.2
AB3 K						
Wastewater effluent concentrate after ultrafiltration	15.5	2.1	2.3	29.7	3.6	46.8

[a]Range for FAs isolated from aqueous systems.
Ash: calculated by 100% − Σ(C, H, N, O, S).
Abbreviations used: HO, Lake Hohloh; ABV, wastewater effluent; BS1, soil seepage water; FG1, groundwater; numbers, sampling occasion; FA, fulvic acid; HA, humic acid; K, sample concentrate gained after ultrafiltration.

10.3.2. Functional Groups and Building Blocks

Due to the complexity of DOM fractionation has revealed more detailed information on the structural subunits prior to the application of advanced analytical methods. Most effective is the combination of different spectroscopic methods using UV–vis absorbance, fluorescence, ^{1}H- and ^{13}C-nuclear magnetic resonance, and Fourier transform–infrared (FT-IR) spectroscopy. In some studies, also electron paramagnetic resonance spectroscopy (EPR) is used (e.g., Chen et al., 2002).

The applicability of these methods is, however, often restricted due to the available concentration of the DOM and due to the disturbing inorganic matrix effects. On the other hand, to avoid denaturating, it is attractive to avoid isolation and to use whereever possible the original aqueous samples. This applies mainly to UV–vis, fluorescence-, and FT-IR spectroscopy (Kalbitz et al., 2000).

Since early 1980, degradation methods have been used in DOM research with the aim to produce identifiable subcompounds that can be related to components of the macromolecular structure often referred as building blocks. Oxidative, hydrolytic, and pyrolytic methods were used as degradation reactions in combination with mass spectrometry (MS). The majority of the compounds identified were acids. The aromatic acids range from benzenedi- to benzenehexacarboxylic acids. There is also evidence for substitutents on the aromatic ring like monohydroxy to trihydroxy (or methoxy) groups. However, there is always the question of denaturing reactions and of structural transformations during the degradation procedure. This can lead to different degradation products formed from one single precursor. In addition, only a fraction (usually less then 30%) of the original sample is actually analyzed, whereas the major rest of the analytical sample is lost. All this adds to the limitation of the degradation procedures.

Oxidation reactions included the use of alkaline permanganate, alkaline copper(II)oxide, and aqueous chlorine (Schnitzer and Khan, 1972; Christman et al., 1989). The degradation products consisted of aromatic and aliphatic acids. Aliphatic dicarboxylic acids ranging from ethanedioic to decanedioic acids were identified. Methylation prior to oxidation prevented phenolic groups from degradation and allowed gas chromatography (GC) analysis.

Analytical pyrolysis with field ionization mass spectrometry (online Py-FIMS) or in combination with GC/MS (Curie point Py-GC/MS) led to a significant increased number of identified subunits (e.g., Bracewell et al., 1989; Schulten et al., 2002). In addition, the application of tetramethylammonium hydroxide (TMAH) methylation, followed by GC/MS, was successfully applied. The most abundant pyrolysis products identified are: benzene, phenol and furan derivatives, aliphatic and carboxylic compounds, and indene derivatives (Schulten et al., 2002). New approaches have been used for the quantification of *n*-alkyl fatty acids of DOM and isolated fractions in the form of individual compounds after solvent extraction followed by derivatization with TMAH.

In recent years the application of electrospray ionization (ESI) mass spectrometry, quadrupole time-of-flight (QqTOF) mass spectrometry, and Fourier transform ion cyclotron resonance (FT-ICR) are used for further structural characterization of DOM (Kujawinski et al., 2002; Kim et al., 2003; Stenson et al., 2003; Koch et al., 2005; Tremblay et al., 2007; Reemtsma et al., 2008). MS/MS capabilities provide the screening for selected ions, and FT-ICR allows exact molecular formula determination for selected peaks. In addition, SEC can be coupled to ESI and FTICR-MS to study different DOM fractions. Homologous series of structures can be revealed, and many pairs of peaks differ by the exact masses of $-H_2$, $-O$, or $-CH_2$. Several thousand molecular formulas in the mass range of up to more than 600 Da can be identified and reproduced in element ratio plots (O/C versus H/C plots). Limitations of ESI used by SEC-MS are shown by These and Reemtsma (2003).

The combination of pyrolysis with GC/FTIR was presented by Davies et al. (2002), who showed that gas chromatography coupled with FTIR spectroscopy can be used as a complementary technique to the conventional GC/MS analysis, by an easier determination of structural isomers (e.g., *p*-cresol, *m*-cresols, *p*-cresol).

Hydrolysis by acids, bases, or enzymes opened the door to information on further building units of complex biogenic matter by using powerful HPLC methods to identify amino acids, carbohydrates, and fatty acids in the low concentration range of $\mu g\ liter^{-1}$ (Parson, 1989; Jahnel et al., 1998, 2002). It was shown that monosaccharides, like pentoses, hexoses, deoxycarbohydrates, and amino sugars, can make up to 8 mass percent of the organic carbon (OC) of FA and HA, isolated from DOM. The values for nonhumic substances (NHS) turned out to be significantly higher (Frimmel et al., 2002; Jahnel et al., 2002). Glucose, galactose, mannose, and xylose were predominantly found. Furthermore, amino acids were released by a hydrochloric acid treatment and with the proteolytic enzyme *pronase E* (Jahnel et al., 2002). The total amount of amino acid carbon related to the total DOC is typically quite low (samples investigated in the study: below 5%), and the percentage of amino-acids-derived N related to the total N content may vary in a broad range. It can account from 4% for groundwater FAs up to 30% for soil seepage HAs, or wastewater treatment effluents. By this the organically bound amino nitrogen has turned out to be a valuable indicator for the humification stage and by this the refractory character of the DOM.

There are still many open questions concerning the redox properties of dissolved organic matter. Recently, studies on the role of DOM for the microbial activity as electron acceptor for microbial respiration has emerged as a relevant driver in the reductive degradation of pollutants in soils (Lovely et al., 1996; Kappler and Haderlein, 2003). It has been shown for different DOM sources that the electron-transfer capacities and the reaction kinetics are of great importance for the understanding of the turnover of organic matter in soil habitats (Bauer et al., 2007). Redox-active functional groups have been characterized as quinoid structures by fluoresence spectroscopy (Cory and McKnight, 2005). This is an interesting structural related information to the abiotic and biochemical function of DOM in terrestrial systems.

10.4. INTERACTIONS OF DOM

10.4.1. Metals

Due to the electron donor groups, DOM is capable to function as ligand in metal complexes. This leads to higher values of the dissolved metal concentrations. The total ligand functionality of isolated DOM can be determined by titration with strong acids and bases (e.g., HCl, NaOH) (de Wit et al., 1993; Perdue, 1998; Garnier et al., 2004). The resulting proton capacity (H-CAP) up to a pH value of 7 reflects most of the carboxylic functional groups; above a pH of 7, phenolic groups and other basic functions are represented (Abbt-Braun and Frimmel, 2002). In Table 10.6, some characteristic values are given. It is obvious that the acidic functional groups (≤7) are dominant over the basic ones (≥7).

The lone pairs of heteroatoms in protonable functional groups can also react via coordinative bonds with metals. Several methods have become available to estimate the amount of functional groups capable of complexation (Frimmel and Geywitz, 1983; Weber, 1988). As copper (II) forms fairly stable coordinative bonds with electron donor ligands, the complexation capacity for copper ($CC_{Cu(II)}$) has often been used and interpreted as maximum ligand functionality of DOM versus two valent metals ions (Table 10.5). Details were revealed with polarographic studies (Frimmel et al., 1984) and titrations using ion-sensitive electrodes. For realistic interpretation the competion amongst different metals has to be taken into account (Lu and Allen, 2002).

The well-known metal series of stability for complexes with electron donor sites of the ligands also holds for DOM and was found by Burba et al. (2002) for typical environmental samples to be

TABLE 10.6. DOC Normalized Proton Capacities (H-CAP) and Copper Complexation Capacities ($CC_{Cu(II)}$) of Different FA and Concentrates Gained after Ultrafiltration (K) and Evaporation (G)[a]

Sample	H-CAP Total (μmol mg^{-1} DOC)	H-CAP$_{pH<7}$ pH < 7 (μmol mg^{-1} DOC)	H-CAP$_{pH>7}$ pH > 7 (μmol mg^{-1} DOC)	(H-CAP$_{pH<7}$)/(H-CAP$_{pH>7}$)	$CC_{Cu(II)}$ (μmol mg^{-1} DOC)
Brown water					
HO10 FA	16.0 ± 0.46	11.1 ± 0.16	4.9 ± 0.30	2.3	2.2 ± 0.24
HO16 FA	14.4 ± 0.28	11.1 ± 0.42	3.2 ± 0.15	3.5	3.5 ± 0.47
HO12 K	8.2 ± 0.04	5.2 ± 0.10	2.9 ± 0.14	1.8	0.9 ± 0.02
HO16 G	11.3 ± 0.12	8.4 ± 0.00	2.9 ± 0.12	2.9	1.9 ± 0.10
Soil seepage					
BS1 FA	16.7 ± 0.68	11.1 ± 0.05	5.5 ± 0.64	2.0	2.1 ± 0.13
Groundwater					
FG1 FA	11.4 ± 0.26	10.0 ± 0.04	1.4 ± 0.30	7.2	0.5 ± 0.01
Secondary effluent					
ABV3 FA	12.4 ± 0.10	10.7 ± 0.04	1.7 ± 0.14	6.3	1.3 ± 0.14
ABV3 K	—	—	—	—	1.2 ± 0.21

[a]Mean values and standard deviations were calculated from at least n = 3.

$$Al \approx Fe(II) >> Cu(II) > Zn(II) >> Mn(II) \approx Ca^{2+} \approx Mg^{2+}$$

A limitation of most experimental approaches to determine complexation capacities or stability constants for DOM–metal complexes is the integrative information obtained. There is the assumption of independent ligand sites even though they might be interactive in the macromolecule, and there is mostly no way to specify the individual electron donor/acceptor sites and the resulting ligand field. Ways to get around this pitfall have been taken, for example, in case of the specific complexation of mercury by thiol functional groups (Frimmel et al., 1980). This ecologically significance of mercury complexation has again come into the focus of interest recently (Haitzer et al., 2003; Ravichandra, 2004).

Recent techniques that show a big potential to get detailed information on the functional groups that bind to metals are X-ray absorption near-edge spectroscopy (XANES) and extended X-ray absorption fine structure (EXAFS). Studies on the binding of Co(II) to different soil-derived organic matter were shown by Ghabbour et al. (2007). EXAFS studies with Cu(II) showed the formation of either one or two five-membered chelate rings (Karlsson et al., 2006). Time-resolved luminescence spectroscopy using lanthanide and actinide ions as probes is a powerful tool to study the interaction of DOM and metal ions without any further separation step (Tiseanu et al., 1998; Moulin and Moulin, 2001; Kumke et al., 2005; Planque et al., 2005).

From the point of view of soil and water quality, Al speciation has gained high interest. Donnan membrane technique on gibbsite suspension can be used to study the Al complexation in natural waters (Weng et al., 2002). Metal complexation by DOM has also shown to be of great importance for describing transport processes (Schmitt et al., 2003). Here the sorption of DOM on geosorbents plays an important role (van Riemsdijk et al., 2006; Weng et al., 2006). Practical interest was also given to the leaching of metals in coal mining areas because there is always washing out with soil seepage water (Suteerapataranon et al., 2006). In case of disposal of high-level nuclear waste, the prediction of the behavior of radionuclides in aquifers is needed (e.g., Maes et al., 2003).

In recent years, model calculations based on thermodynamic equilibria for all components present in the considered aqueous system including DOM with an averaged complexation functionality have obtained increasing attention. By this, the relation of free (hydrated) metal ions and their complexed form can be estimated. Mathematical models that are widely used are the *Humic Ion Binding Model VI* (Tipping, 1998, 2002) and the *NICA-Donnan model* (Kinniburgh et al., 1999; Weber et al., 2006). A hysteretic behavior during titration of different humic and fulvic acids was observed by several authors (Cooke et al., 2007). A recently published review summarizes the interaction of several metals and natural organic matter with specific emphasis on models and the prediction on the interaction of metals in the environment (Merdy et al., 2006).

10.4.2. Organic Micropollutants and Xenobiotics

The behavior of organic micropollutants and xenobiotics in the aqueous environment is mainly determined by sorption and by biotic and abiotic degradation processes. The sorption on solid particles is highly influenced by the organic carbon content of the particles (e.g., Kördel, 1997). Many studies have been published for hydrophobic organic compounds like mono- and polyaromatic hydrocarbons, polychlorinated biphenyls, or chlorinated aliphatic hydrocarbons (e.g., Grathwohl, 1990; Chiou et al., 1999; Dewulf et al., 1999; Carmo et al., 2000). Binding or association of DOM with hydrophobic organic contaminants has also been studied (e.g., Carter and Suffet, 1982; Chiou et al., 1986; Suffet and MacCarthy, 1989; Maxim and Kögel-Knabner, 1995). Here the structure of DOM which shows hydrophobic and hydrophilic regions at the same time results in a solubility enhancement for lipohilic xenobiotics in water. Chiou et al. (1986) showed that DOM from different origin have a different influence on the apparent solubility of DDT. HAs isolated from river water caused a stronger effect than FAs isolated from the river water.

From the many papers dealing with the interaction of DOM and xenobiotics, the ones on polycyclic aromatic hydrocarbons (PAH) and phenols are of special ecological interest. The interaction of DOM with pyrene was investigated by Kumke et al. (1994) and several other authors (Kopinke et al., 2000; Löhmannsröben et al., 2002). There was a clear visible effect when the fluorescence of the PAH was decreased in the presence of DOM. The interaction gained strength at lower pH values. A detailed description of the interaction of DOM isolated from freshwater sediments with pyrene was given by Akkanen et al. (2005). They demonstrated the effect of the origin of the sediments, and the influence of the extraction procedure on the functionality of the DOM. In his work, the sorption of pyrene on DOM was used as an indicator for functionality of DOM. The effect of DOM and pH value on the geosorption of phenols was studied by Amirir et al. (2005). The authors used a combined sorption and complex formation model.

The analytical availability of hydrophobic organic micropollutants could be reduced to a high extent in the presence of DOM. Gjessing et al. (2007) showed that the analytical recovers of polychlorinated biphenyls (PCBs) was also affected by the quality and the nature of DOM.

The molecular interaction of xenobiotics and DOM plays also a major role in the photochemical and photocatalytical reactions. Numerous articles on the direct and indirect photolysis of xenobiotics mediated by DOM or HS have been published. For example, Prosen and Zupancic-Kralj (2004) studied the photolysis and hydrolysis of atrazine in the presence of HAs. The rate constants increased up to 10-fold; atrazine, desethylatrazine, and desisopropylatrazine converted to their 2-hydroxy analogs. Doll and Frimmel (2003, 2005a,b) showed that the presence of DOM increased the photochemical degradation by using simulated sunlight of carbamazepine, whereas the degradation of iomeprol was retarded. In the presence of DOM the photocatalysis (TiO_2/UV) of clofibric acid, carbamazepine and iomeprol was retarted. These effects are due to competition for active sites and surface deactivation of the catalyst by adsorption.

A comprehensive review about the effect of DOM on the bioavailabilty of organic xenobiotics is given by Haitzer et al. (1998). Most studies show that DOM concentration of up to 10 mg liter^{-1} decreases the bioavailability of organic chemicals. It is also obvious that the difference in the character of the DOM due to its origin is of vital importance on the results.

10.4.3. Particulate Matter

Most DOMs have medium to high molecular size, causing their refractory character. Due to the operational definition of dissolved state "<0.45 μm," it is interesting to know the molecular size distribution and the molecular structure and elemental composition of the size fractions. Beyond size exclusion chromatography discussed in Section 10.2.2, characterization of DOM interacting with microparticles has emerged recently to describe and understand the particulate matter influenced transport processes. DOM often causes a stabilization effect of colloidal distributions derived from (negative) surface charges. Neutralization by, for example, cations can on the other hand result in enhanced agglomeration and coagulation. The elucidation of these phenomena is important not only for the basic understanding of the interactions of particles in the nano- and microscale but also for optimizing engineered processes like flocculation, sedimentation, and filtration.

Because the chapter is about DOM, detailed information about the role of colloids and the analytical techniques are given elsewhere (e.g., Buffle and Leppard, 1995; Kretzschmar et al., 1999; Frimmel et al., 2007). Different separation techniques, like ultrafiltration, size exclusion chromatography, and flow field-flow fractionation can be coupled with UV–vis absorption and ICP-MS to show the interaction of metals and colloids. Elements like Ni, Cu, Cr, and Co are associated mainly with smaller-size DOM fractions whereas Al, Fe, lanthanides, Sn, and Th are associated with larger-size DOM fractions (Bolea et al., 2006). The laser-induced breakdown detection (LIBD) is a new, sensitive method for the quantification of aquatic colloids of lower-range nanometer size in very low concentration, which cannot be detected by conventional light scattering methods (Bundschuh et al., 2001). Some authors show that the techniques allow us to differ between humic colloids made from Aldrich humic acid and the inorganic components of actinides (Bouby et al., 2002).

10.5. OCCURRENCE OF DOM

Source and ecological matrix are of detrimental influence on the amount and character of DOM. The dissolved status demands a minimum of hydrophilicity. In addition, there are some further facts that play a major role in the occurrence of DOM in the different aqueous phases. Some important points are:

- Plants and other water external residues (allochtonous matter)
- Microbial and other water internal residues (authochtonous matter)
- Physical–chemical separation (hydrophilic/hydrophobic fractions)
- Chemical interactions (complex formation, bound residues)
- Photo(-cata)lytic reactions (formation, transformation/oxidative degradation)
- Microbial reactions (transformation, metabolization)
- Agglomeration and dissolution (particle formation)

The specific aspects dominating in the different aquatic systems lead to a clear differentiation of the DOM regimes. As a consequence, these are not only varying concentrations but also different types of DOM as shown in Table 10.7.

10.6. HUMAN IMPACT

DOM is ubiquitous in rivers, lakes, groundwater, and oceans. It therefore plays a dominant role in the biosphere as well as in treatment of fresh water, for industrial use and human consumption. The main aspects in addition to the function as microbial nutrients are (a) the interactions with other water constituents like metals and xenobiotics and (b) the reactions with chemicals that are used for water disinfection (e.g., chlorine). The latter leads to the problem of disinfection by-product (DBP) formation, which is of toxicological relevance.

10.6.1. Wastewater

Human activities and the connected water usage lead to a variety of anthropogenic influences on DOM. The many years of careless exposure of rivers and lakes to untreated and not sufficiently treated wastewater from production sites and households have led to a severe load of the aquatic systems with inorganic, organic, and industrially synthesized products. Their effective elimination has turned out to be a high challenge and has become the key for sustainable water management.

DOM in effluents of a wastewater treatment plant consist of fairly hydrophilic organic material. It is refractory in character and has been classified as humic-like substances (Frimmel et al., 2005). It is obvious from the data given in Table 10.5 (see

TABLE 10.7. Typical Characteristics for DOM of Different Aquatic Systems

Aquatic System	Average Concentration as DOC mg $liter^{-1}$ (range)	Global Amount of Water ($10^6 km^3$)[a]	Global Amount of Organic Carbon ($10^9 t$)	Dominating Structure	References, for Average DOC Concentration
Ocean	0.5 (0.3–2; 0–300 m) (0.2–0.8; below 300 m)	1370	685	Amino acids, carbohydrates, unsubstituted alkyl carbon	Williams (1971), Duursma and Dawson (1981), Harvey and Boran (1985), Benner (1998), Benner et al. (2005)
Freshwater					
Ice and snow	0.5 (0.1–5.0)	29 (glacier and polar caps)	14.5	Mono- and diacids, phenols	Thurman (1985a), Laird et al. (1988), Grannas et al. (2004)
Rivers	7 (5–9)	0.0017	0.012	Aliphatics substituted by carboxylic, phenolic, and methoxyl groups: derived from plant/soil and microbial/algal organic material	Malcolm (1985), McKnight and Aiken (1998), Aitkenhead-Peterson et al. (2003), Bertilsson and Jones (2003)
Lakes	2.2 (oligotrophic 2–3) 12 (eutrophic 9–16)	0.125	0.28	Aliphatics substituted by carboxylic, phenolic, and methoxyl groups: derived from plant/soil and microbial/algal organic material	Steinberg and Muenster (1985), McKnight and Aiken (1998), Steinberg (2003), Aitkenhead-Peterson et al. (2003), Bertilsson and Jones (2003)
Soil seepage water	25 (19–31)	0.065 soil humidity	1.6	Aromatic groups substituted by carboxylic, methoxyl, and hydroxyl groups, carbohydrates: derived from plant organic material	Abbt-Braun (1992), Frimmel (1992), Frimmel et al. (2002)
Atmosphere	20–90% organics (mass of aerosols, 15–60% out of this HULIS[b])	0.013		Aromatic, phenolic, and acidic functional groups	Dinar et al. (2006), Kalberer et al. (2004), Graber and Rudich (2006)
Groundwater ($CaCO_3$ aquifer)	0.7	5.3 (750 m to 4 km deep) 4.2 (<750 m deep)	3.7 2.9	Derived from soil or kerogen	Thurman (1985b), Abbt-Braun (1992), Frimmel (1992), Frimmel et al. (2002)

[a]Global amount of water, data according to Andrews et al. (1996).
[b]HULIS: humic-like substances, macromolecular organic substances in atmospheric aerosols.

Section 10.3.1) that the elemental composition of the operationally defined isolated FA and HA fractions is in general not much different from the geogenic FA and HA fractions. But there are differences obvious in the C and O content (lower values for wastewater-derived fractions) and in the H, N, and S content (higher values for wastewater-derived fractions). This clearly reflects the recent genesis of the material from microbial degradation reactions in the secondary treatment plant. Relatively low specific UV absorbance and yellow color underline the observations and the interpretation that there is a good part of protein- and polysaccharide (peptidoglycans)-like structure, leftovers from microbiological cells. Solid-state MAS-NMR data (Figure 10.13) are in good agreement with that assumption. This matter of humic-like character will be gradually further degraded in the receiving water, and by this the structure will approach the one of the geogenic organic matter and finally become part of it. Ma et al. (2001) and Imai et al. (2002) isolated different fractions by reverse osmosis or XAD procedure, or both. They found that the fractions from wastewater treatment plant effluents consist of simpler compounds than the complex natural DOM and show high aliphatic character. A detailed characterization of wastewater effluent water and their different fractions is given by Abbt-Braun and Frimmel (2002).

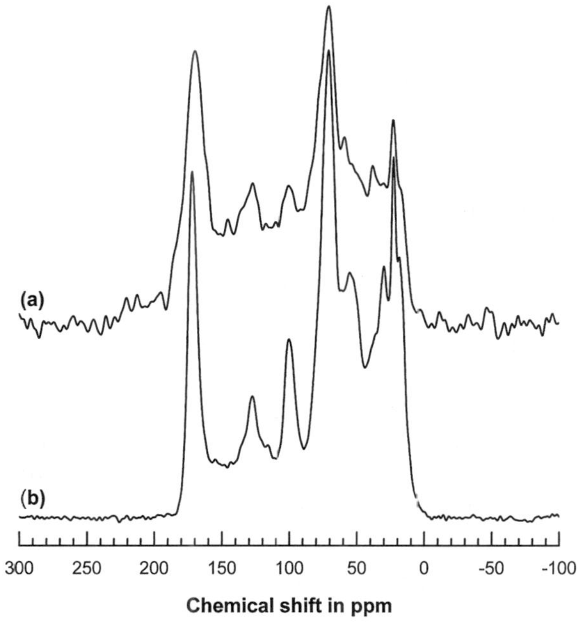

Figure 10.13. ^{13}C CPMAS NMR spectra of the freeze-dried wastewater effluent sample (**a**) and the humic acid fraction (**b**). The signal of inorganic carbon (carbonate at 167 ppm) in the spectrum (**a**) was removed by subtracting a fitted Gaussian function. The total integral of the two spectra (220 ppm – 0 ppm) was scaled to the same area. ^{13}C CPMAS, Bruker MSL-100 spectrometer, ^{13}C resonance frequency: 25.1 MHz. MAS: 5 kHz, chemical shift was calibrated externally to the carboxyl resonance of glycine (176.03 ppm) with respect to the tetramethylsilane (TMS) scale.

10.6.2. Drinking Water

Because DOM in raw waters used for drinking water is mainly derived from natural sources, it is generally referred to as natural organic matter (NOM) in the following paragraph.

Water quality demands in different water usages often ask for a specific treatment. Drinking water in particular has to meet strict quality standards from both the chemical and microbiological point of view to ensure a hygienically and toxicologically safe water supply for the population. The role of NOM in water treatment is many-sided and closely related to the adverse effects of NOM with respect to human consumption. There is

- its aesthetically unwanted yellow or brownish color,
- the toxicity potential derived from its carrier function for heavy metals and organic pollutants as well,
- its formation potential for toxic products from a variety of reactions, and
- its direct and indirect growth potential for microorganisms including pathogens.

In raw water for drinking water supply, DOC concentrations of a few mg liter^{-1} or below are commonly accepted. For higher concentrations or for cases where NOM-related adverse effects might occur, water treatment processes have to be applied.

The common water treatment operations can be classified into

- separation and fractionation processes and
- processes which are based on chemical reactions.

In addition to classical flocculation and sand bed filtration, NOM can efficiently be removed from waters through ultra- and nanofiltration. In a comparative study with three different raw waters, the rejected fraction, in which NOM was relatively enriched, had the highest AOX- and THM-formation potentials (AOX, organically bound halogens absorbable on activated carbon; THM, trihalogen methane) (Gorenflo, 2003). Changes in the molecular structure as well as in the molecular size distribution of the DOM were observed after membrane filtration of the raw waters. Oxidation and photodegradation processes led to the formation of low-molecular-weight organic acids, amino acids, and carbohydrates. Photochemically induced bleaching in surface water, which is an important process in nature, was enhanced in the presence of dissolved iron (Fe(III)), whereas it was hindered in the presence of dissolved copper (Cu(II)) (Brinkmann et al., 2003). In Table 10.8, typical treatment steps and their effects on NOM are given.

It is well known that the prevalence of NOM among the dissolved organic compounds in raw waters—even though they are not toxic *per se*—can cause directly or after reaction with technically introduced reactants (e.g., chlorine, ozone) severe problems that have to be assessed on the basis of an individual, daily intake approach. The consumption of 2 liters of drinking water per capita and day during a lifetime of up to 80 years can be taken as a basis for such an assessment. In addition, the refractory organic matter can be split by oxidation reactions (e.g., ozone, OH radicals, and chlorine) into degradation products with lower molecular weight and by

TABLE 10.8. Water Treatment Steps and Effects on DOM (resp. NOM) in Raw Water

Treatment Step	Effect on DOM, resp. NOM	Resulting Major Effect on Treated Water
Sedimentation	None	Removal of coarse matter
Bank filtration (slow sand filtration)	Adsorption, biodegradation	Decolorization, elimination of organic carbon
Flocculation	Adsorption, precipitation, formation of complexes	Decolorization, elimination of organic carbon
Sand filtration	Retention of colloids	Decolorization, decrease of turbidity
Membrane filtration (nanofiltration)	Partial rejection of DOM	Decolorization, softening
GAC adsorption	Adsorption (selective removal of small molecules)	Decolorization
Oxidation (e.g., O_3, H_2O_2, AOPs)	Attack on double bonds, cleavage of molecules	Bleaching, increase of bioavailability, increased biological regrowth potential
Disinfection (e.g., Cl_2, ClO_2)	Halogenation, AOX and THM formation	Odor, formation of toxic (carcinogenic, mutagenic) compounds

AOP, advanced oxidation processes; AOX, organically bound halogens absorbable on activated carbon; GAC, granular activated carbon; THM, trihalogen methane.

Source: Frimmel (1999).

this better bioavailability. As a consequence, stimulation of adverse bacterial growth or regrowth after disinfection can occur.

Pitfalls of the different water treatment processes are the formation of extensive amounts of sludge, which has to be deposited off, as is the case with flocculation, the formation of fouling layers during membrane filtration, or DBP formation after disinfection of NOM-containing waters.

Flocculation of NOM using Fe(III) or Al(III) salts can reduce the DOC concentration of surface water to a high extent.

$$Me^{3+} + 3OH^- \rightleftarrows Me(OH)_3 \downarrow$$

Adsorption affinity of the NOM on aluminum or iron hydroxide flocs can be different for different NOM fractions. Bose and Reckhow (2007) showed that the adsorption is increased with increasing charge of the NOM fractions. Preozonation increases the efficiency for removal of the DOC concentration.

During ozonation of natural surface water the concentration of oxygen-containing low-molecular-weight compounds is increased. Many of these compounds support microbiological growth and can be analytically quantified as assimilable organic carbon (AOC, see Section 10.2.4). Its technical elimination can be achieved with the help of slow sand filtration. It is evident that phytoplankton can contribute substantially to the organic carbon load and that algae cause a substantial increase in the concentration of DOC and AOC after ozonation (Hammes et al., 2007).

Bank filtration and artificial recharge of groundwater (e.g., along the river Rhine and Elbe) provide important drinking water resources for a variety of cities.

Although the method has been used for decades, there are still open questions concerning the DOM transformation in the underground. Recent studies have shown that the redox conditions and travel time significantly influence the DOM degradation kinetics and by this the efficiency of AOX and trace compound removal (Grünheid et al., 2005).

Activated carbon filtration is a classical method for DOM removal by adsorption and partially biodegradation (Sontheimer et al., 1988). Recent studies showed that the breakthrough of the DOC by using granular activated carbon (GAC) filters are slower at higher water temperatures (best removal at 35 °C). UV–vis spectra and size exclusion chromatograms of water samples treated at different water temperatures indicate that an increase in temperature especially supports the adsorption of small DOM molecules as well as molecules with spectral absorbance at higher wavelengths, specifically aromatic structures of DOM (Schreiber et al., 2005).

High DOM concentrations cause also high disinfection by-products formation when using chlorine-based disinfection as last water treatment step (e.g., Hua and Reckhow, 2007). The kinetics of the formation of trihalomethane at the chlorination of NOM and of the chlorine consumption was shown by Gallard and von Gunten (2002). Far more than 100 disinfection by-products have been identified (Richardson et al., 2002; Krasner et al., 2006). Some of them like 3-chloro-4-(dimethyl)-5-hydroxy-2(5*H*)-furanone (MX) are highly mutagenic (Kronberg et al., 1988; Frimmel and Schmiedel, 1993).

A comprehensive review concerning the influence of natural organic matter on coagulation processes, on the sorption onto activated carbon, and on ion-exchange and membrane filtration, as well as on ozonation and chlorination processes in water treatment, is given in the textbook of Suffet and MacCarthy (1989).

10.6.3. Miscellaneous

Anthropogenic influence on DOM is manyfold. Especially the broad application of organic xenobiotics (i.e., man-made synthetics), and their way into the aquatic environment has led to the appearance of those substances and their metabolites also in the matrix of DOM. Bound residues have been found, for example, as result of agricultural application of diversity of pesticides (Roberts et al., 1984; Gevao et al., 2000; Spiteller et al., 2002; Burauel and Baßmann, 2005). Pharmaceuticals can also be seen as potential anthropogenic sources for their integration into DOM (Halling-Sorensen et al., 1998; Löffler et al., 2005). The merge of the geogenic and anthropogenic compounds is of general interest from the point of view of reaction yield and kinetics, and it is of high ecological relevance how much of the different physiological effects remain active after the DOM integrative reaction and whether there can be a reactivation of effects when xenobiotics or their partial structures get deliberated from the DOM matrix. Isotope ratios of carbon, nitrogen, and other heteroatoms have turned out to be a promising tool for studying the biorelated drivers in these reactions (Gleixner et al., 2002). It can be concluded that the strong and diverse interactions between anthropogenic and geogenic compounds leads to an unresoluable mixture of new structural components that carry the footprint of human activities and industrial development. The needs of sustainability are obvious.

REFERENCES

Abbt-Braun, G. (1992). A: Spectroscopic characterization of humic substances in the ultraviolet and visible region. B: Interpretation of the pyrolysis products of isolated humic and fulvic acid. In *Progress in Hydrogeochemistry*, Matthess, G., Frimmel, F., Hirsch, P., Schulz, H. D., and Usdowski, H.-E., eds., Springer, Berlin, A: pp. 24–36; B: pp. 46–51.

Abbt-Braun, G., and Frimmel, F. H. (2002). The relevance of reference materials—Isolation and general characterisation. In *Refractory Organic Substances (ROS) in the Environment*, Frimmel, F. H., Abbt-Braun, G., Heumann, K. G., Hock, B., Lüdemann, H.-D., and Spiteller, M., eds., Wiley-VCH, Weinheim, pp. 1–37.

Abbt-Braun, G., Frimmel, F. H., and Lipp, P. (1991). Isolation of organic substances from aquatic and terrestrial systems. *Z. Wasser-Abwasser-Forsch.* **24**, 285–292.

Abbt-Braun, G., Lankes, U., and Frimmel, F. H. (2004). Structural characterization of aquatic humic substances—The need for a multiple method approach. *Aquat. Sci.* **66**, 151–170.

Aiken, G. R. (1985). Isolation and concentration techniques for aquatic humic substances. In *Humic Substances in Soil, Sediment and Water*, Aiken, G. R., McKnight, D. M., Wershaw, R. L., and MacCarthy, P., eds., Wiley-Interscience, New York, pp. 363–385.

Aiken, G. R. (1988). Critical evaluation of the use of macroporous resins for the isolation of aquatic humic substances. In *Humic Substances and Their Role in the Environment*, Frimmel, F. H., and Christman, R. F., eds., John Wiley & Sons, Chichester, pp. 15–28.

Aitkenhead-Peterson, J. A., McDowell, W. H., and Neff, J. C. (2003). Sources, production, and regulation of allochthonous dissolved organic matter inputs to surface waters. In *Aquatic Ecosystems—Interactivity of Dissolved Organic Matter*, Findlay, S. E. G., and Sinsabaugh, R. L., eds., Academic Press, Amsterdam, pp. 25–70.

Akkanen, J., Lyytikäinen, M., Tuikka, A., and Kukkonen, J. V. K. (2005). Dissolved organic matter in pore water of freshwater sediments: Effects of separation procedure on quantity, quality and functionality. *Chemosphere* **60**, 1608–1615.

Amirir, F., Börnick, H., and Worch, E. (2005). Sorption of phenols onto aqueous sandy aquifer material: the effect of dissolved organic matter (DOM). *Water Res.* **39**, 933–941.

Andrews, J. E., Brimblecombe, P., Jickells, T. D., and Liss, P. S. (1996). *An Introduction to Environmental Chemistry*, Blackwell Science, London.

ASTM Standard D4189-95 (2002). Standard test method for silt density index (SDI) of water. *The American Standard for Testing Material.*

Bauer, M., Heitmann, T., Macalady, D. L., and Blodau, C. (2007). Electron transfer capacities and reaction kinetics of peat dissolved organic matter. *Environ. Sci. Technol.* **41**, 139–145.

Beckett, R., Jue, Z., and Giddings, J. C. (1987). Determination of molecular weight distributions of fulvic and humic acids using flow field-flow fractionation. *Environ. Sci. Technol.* **21**, 289–295.

Benner, R. H. (1998). Cycling of dissolved organic matter in the ocean. In *Aquatic Humic Substances. Ecology and Biochemistry*, Hessen, D. O., and Tranvik, L. J., eds., Springer, Berlin, pp. 317–342.

Benner, R. H., Louchouarn, P., and Amon, R. M. W. (2005). Terrigeneous dissolved organic matter in the Arctic Ocean and its transport to surface and deep waters of the North Atlantic. *Global Biochem. Cycles* **19**. GB2025, doi:10.1029/2004 GB002398.

Bertilsson, S., and Jones, J. B., Jr. (2003). Supply of dissolved organic matter to aquatic ecosystems: Autochthonous sources. In *Aquatic Ecosystems—Interactivity of Dissolved Organic Matter*, Findlay, S. E. G., and Sinsabaugh, R. L., eds., Academic Press, Amsterdam, pp. 3–24.

Bloom, P., and Leenheer, J. A. (1989). Vibrational, electronic, and high-energy spectroscopic methods for characterizing humic substances. In *Humic Substances II: In Search of Structure*, Hayes, M. H. B., MacCarthy, P., Malcolm, R. L., and Swift, R. S., eds., Wiley-Interscience, Chichester, pp. 409–446.

Bolea, E., Bouby, M., Laborda, F., Castillo, J. R., and Geckeis, H. (2006). Multielement characterization of metal–humic substances complexation by size exclusion chromatography, asymmetrical flow field-flow fractionation, ultrafiltration and inductively coupled plasma–mass spectrometry detection: A comparative approach. *J. Chromatogr. A* **1129**, 236–246.

Bose, P., and Reckhow, D. A. (2007). The effect of ozonation on natural organic matter removal by alum coagulation. *Water Res.* **41**, 1516–1524.

Bouby, M., Manh, T. N., Geckeis, H., Schwerbaum, F., and Kim, J.-I. (2002). Characterization of aquatic colloids by a combination of LIBD and ICP-MS following the size fractionation. *Radiochim. Acta* **90**, 727–732.

Bracewell, J. M., Haider, K., Larter, S. R., and Schulten, H.-R. (1989). Thermal degradation relevant to structural studies of humic substances. In *Humic Substances II: In Search of Structure*, Hayes, M. H. B., MacCarthy, P., Malcolm, R. L., and Swift, R. S., eds., John Wiley & Sons, New York, pp. 181–222.

Brinkmann, T., Sartorius, D., and Frimmel, F. H. (2003). Photobleaching of humic rich dissolved organic matter. *Aquat. Sci.* **65**, 1–10.

Buffle, J., and Leppard, G. G. (1995). Characterization of aquatic colloids and macromolecules. 1. Structure and behavior of colloidal material. *Environ. Sci. Technol.* **29**, 2169–2175.

Bundschuh, T., Knopp, R., and Kim, J. I. (2001). Laser-induced breakdown detection (LIBD) of aquatic colloids with different laser systems. *Colloids Surf. A: Physicochem. Eng. Aspects* **177**, 47–55.

Burauel, P., and Baßmann, F. (2005). Soils as filter and buffer for pesticides—Experimental concepts to understand soil functions. *Environ. Pollut.* **133**, 11–16.

Burba, P., Jakubowski, B., and Van den Bergh, J. (2002). Characterization of refractory organic substances and their metal species by combined analytical procedures. In *Refractory Organic Substances (ROS) in the Environment*, Frimmel, F. H., Abbt-Braun, G., Heumann, K. G., Hock, B., Lüdemann, H.-D., and Spiteller, M., eds., Wiley-VCH, Weinheim, pp. 73–88.

Cabaniss, S. E., and Shuman, M. S. (1987). Synchronous fluorescence spectra of natural waters: Tracing sources of dissolved organic matter. *Marine Chem.* **21**, 37–50.

Carmo, A. M., Hundal, L. S., and Thompson, M. L. (2000). Sorption of hydrophobic organic compounds by soil materials: Application of unit equivalent Freundlich coefficients. *Environ. Sci. Technol.* **34**, 4363–4369.

Carter, C. W., and Suffet, I. H. (1982). Binding of DDT to dissloved humic materials. *Environ. Sci. Technol.* **16**, 735–740.

Chen, J., Gu, B., LeBoeuf, E. J., Pan, H., and Dai, S. (2002). Spectroscopic characterization of the structural and functional properties of natural organic matter fractions. *Chemosphere* **48**, 59–68.

Chen, J., LeBoeuf, E. J., Dai, S., and Gu, B. (2003). Fluorescence spectroscopic studies of natural organic matter fractions. *Chemosphere* **50**, 639–647.

Chen, Y., Senesi, N., and Schnitzer, M. (1977). Information provided on humic substances by E4/E6 ratios. *Soil. Sci. Soc. Am. J.* **41**, 352–358.

Chiou, C. T., Malcolm, R. L., Brinton, T. I., and Kile, D. E. (1986). Water solubility enhancement of some organic pollutants and pesticides by dissolved humic and fulvic acids. *Environ. Sci. Technol.* **29**, 502–508.

Chiou, C. T., McGroddy, S. E., and Kile, D. E. (1999). Partitioning characteristics of polycyclic aromatic hydrocarbons on soil and sediments. *Environ. Sci. Technol.* **33**, 2390–2394.

Christman, R. F., Norwood, D. L., Seo, Y., and Frimmel, F. H. (1989). Oxidative degradation of humic substances from freshwater environments. In *Humic Substances II: In Search of Structure*, Hayes, M. H. B., MacCarthy, P., Malcolm, R. L., and Swift, R. S., eds., John Wiley & Sons, New York, pp. 33–68.

Cooke, J. D., Hamilton-Taylor, J., and Tipping, E. (2007). On the acid–base properties of humic acid in soil. *Environ. Sci. Technol.* **41**, 465–470.

Cory, R. M., and McKnight, D. M. (2005). Fluorescence spectroscopy reveals ubiquitous presence of oxidized and reduced quinones in dissolved organic matter. *Environ. Sci. Technol.* **39**, 8142–8149.

Dafner, E. V., and Wangersky, P. J. (2002). A brief overview of modern directions in marine DOC studies Part I: Methodological aspects. *J. Environ. Monit.* **4**, 48–54.

Davies, N., Hill, W., and Kuckuk, R. (2002). Structural characterization of refractory organic substances by means of pyrolysis-GC/FTIR. In *Refractory Organic Substances (ROS) in the Environment*, Frimmel, F. H., Abbt-Braun, G., Heumann, K. G., Hock, B., Lüdemann, H.-D., and Spiteller, M., eds., Wiley-VCH, Weinheim, pp. 232–238.

De Nobili, M., Gjessing, E., and Sequi, P. (1989). Sizes and shapes of humic substances by gel chromatography. In *Humic Substances II: In Search of Structure*, Hayes, M. H. B., MacCarthy, P., Malcolm, R. L., and Swift, R. S., eds., John Wiley & Sons, New York, pp. 561–591.

De Nobili, M., Bragato, G., and Mori, A. (1998). Combined effects of molecular size and electroendosmotic flow on the capillary electrophoretic behaviour of humic substances. *Acta Hydrochim. Hydrobiol.* **26**, 186–190.

De Wit, J. C. M., van Riemsdijk, W. H., and Koopal, L. K. (1993). Proton binding to humic substances. 1. Electrostatic effects. *Environ. Sci. Technol.* **27**, 2005–2022.

Dewulf, J., Van Langenhove, H., and Grare, S. (1999). Sediment/water and octanol/water equilibrium partitioning of volatile organic compounds: Temperature dependence in the 2–25 C range. *Water Res.* **33**, 2424–2436.

Dilling, J., and Kaiser, K. (2002). Estimation of the hydrophobic fraction of dissolved organic matter in water samples using UV photometry. *Water Res.* **36**, 5037–5044.

DIN, EN 1484, European standard (1997). Water analysis—guidelines for the determination of total organic carbon (TOC) and dissolved organic carbon (DOC). In *Deutsche Einheits-verfahren*, John Wiley & Sons, Lieferung, **1998**, p. 40.

Dinar, E., Taraniuk, I., Graber, E. R., Katsman, S., Moise, T., Anttila, T., Mentel, T. F., and Rudich, Y. (2006). Cloud condensation nuclei properties of model and atmospheric HULIS. *Atmos. Che. Phys.* **6**, 2465–2482.

Doll, T. E., and Frimmel, F. H. (2003). Fate of pharmaceuticals-photodegradation by simulated solar UV-light. *Chemosphere* **52**, 1757–1769.

Doll, T. E., and Frimmel, F. H. (2005a). Removal of selected persistent organic pollutants by heterogeneous photocatalysis in water. *Catal. Today* **101**, 195–202.

Doll, T. E., and Frimmel, F. H. (2005b). Photocatalytic degradation of carbamazepine, clofibric acid and iomeprol with P25 and Hombikat UV100 in presence of natural organic matter (NOM) and of other organic water constituents. *Water Res.* **39**, 403–411.

Duursma, E. K., and Dawson, R. (1981). *Marin Organic Chemistry*, Elsevier, Amsterdam.

Exner, A., Theisen, M., Panne, U., and Niessner, R. (2000). Combination of asymmetric flow field-flow fractionation (AF^4) and total-reflexion X-ray fluorescence analysis (TXRF) for determination of heavy metals associated with colloidal humic substances. *Fresenius J. Anal. Chem.* **366**, 254–259.

Filippova, E. M., Fadeev, V. V., Chubarov, V. V., Dolenko, T. A., and Glushkove, S. M. (2001). Laser fluorescence spectroscopy as a method for studying humic substance. *Appl. Spectrosc. Rev.* **36**, 87–117.

Frimmel, F. H. (1992). Isolation and general characterization of organic acids from pore water. In *Progress in Hydrogeochemistry*, Matthess, G., Frimmel, F., Hirsch, P., Schulz, H. D., and Usdowski, H.-E., eds., Springer, Berlin, pp. 9–13.

Frimmel, F. H., ed. (1999). *Wasser und Gewässer—Ein Handbuch*, Spektrum Akademischer Verlag, Heidelberg.

Frimmel, F. H., and Geywitz, J. (1983). Zur koordinativen Bindung von Metallionen an Gewässerhuminstoffe. *Fresenius Z. Anal. Chem.* **316**, 582–588.

Frimmel, F. H., and Schmiedel, U. (1993). Pyrolysis GC/MS identification of halogenated furandiones and other products from chlorinated aqueous humic solutions. *Fresenius J. Anal. Chem.* **346**, 707–710.

Frimmel, F. H., Sattler, D., and Quentin, K.-E. (1980). Photochemischer Abbau von Quecksilber-Thioverbindungen in sauerstoffhaltigem und sauerstofffreiem Wasser. *Vom Wasser* **55**, 111–120.

Frimmel, F. H., Immerz, A., and Niedermann, H. (1984). Complexation capacities of humic substances isolated from freshwater with respect to copper(II), mercury(II) and iron(II,III). In *Complexation of Trace Metals in Natural Waters*, Kramer, C. J. M., and Diunker, J. C., eds., Martinus Nijhoff/Dr. W. Junk Publishers, The Hague, pp. 329–343.

Frimmel, F. H., Abbt-Braun, G., Heumann, K. G., Hock, B., Lüdemann, H.-D., and Spiteller, M., eds. (2002). *Refractory Organic Substances (ROS) in the Environment*, Wiley-VCH, Weinheim.

Frimmel, F. H., Lankes, U., Lüdemann, H.-D., and Müller, M. B. (2005). Structural characterization of humic substances from waste water. In *Humic Substances—Molecular Details and Applications in Land and Water Conservation*, Ghabbour, E. A., and Davies, G., eds., Taylor and Francis, New York, pp. 3–22.

Frimmel, F. H., von der Kammer, F., and Flemming, H.-C., eds. (2007). *Colloidal Transport in Porous Media*, Springer-Verlag, Berlin Heidelberg.

Gallard, H., and von Gunten, U. (2002). Chlorination of natural organic matter: Kinetics of chlorination and of THM formation. *Water Res.* **36**, 65–74.

Garnier, C., Mounier, S., and Benaim, J. Y. (2004). Influence of dissolved organic carbon content on modelling natural organic matter acid–base properties. *Water Res.* **38**, 3685–3692.

Gevao, B., Semple, K. T., and Jones, K. C. (2000). Bound pesticide residues in soils: a review. *Environ. Pollut.* **108**, 3–14.

Ghabbour, E. A., Scheinost, A. C., and Davies, G. (2007). XAFS studies of cobalt(II) binding by solid peat and soil-derived humic acids and plant-derived humic acid-like substances. *Chemosphere* **67**, 285–291.

Gjessing, E. T., Alberts, J. J., Bruchet, A., Egeberg, P. K., Lydersen, E., McGown, L. B., Mobed, J. J., Münster, U., Pempkowiak, J., Perdue, E. M., Ratnaweera, H., Rybacki, D., Takacs, M., and Abbt-Braun, G. (1998). Multi-method characterization of natural organic matter isolated from water: characterization of reverse-osmosis isolates from water of two semi-identical dystrophic lakes basins in Norway. *Water Res.* **32**, 3108–3124.

Gjessing, E. T., Egeberg, P. K., and Håkedal, J. (1999). Natural organic matter in drinking water—the "NOM-Typing Project," background and basic characteristics of original samples and NOM isolates. *Environ. Int.* **25**, 145–159.

Gjessing, E. T., Steiro, C., Becher, G., and Christy, A. (2007). Reduced analytical availability of polychlorinated biphenyls (PCB's) in colored surface water. *Chemosphere* **66**, 644–649.

Gleixner, G., Kracht, O., Schmidt, H.-L., and Schulze, E.-D. (2002). Isotopic evidence for the origin and formation of refractory organic substances. In *Refractory Organic Substances (ROS) in the Environment*, Frimmel, F. H., Abbt-Braun, G., Heumann, K. G., Hock, B., Lüdemann, H.-D., and Spiteller, M., eds., Wiley-VCH, Weinheim, pp. 146–162.

Gorenflo, A. (2003). Rückhalt und Fouling von natürlichen organischen Substanzen bei der Nano- und Ultrafiltration. Ph.D. thesis, Universität Karlsruhe (TH).

Graber, E. R., and Rudich, Y. (2006). Atmospheric HULIS: How humic-like are they? A comprehensive and critical review. *Atmospheric Chem. A: Physics* **6**, 729–753.

Grannas, A. M., Shewpson, P. B., and Filley, T. R. (2004). Photochemistry and nature of organic matter in arctic and antarctic snow. *Global Biochem. Cycles* **18**, GB1006, doi:10.1029/2003GB002133.

Grathwohl, P. (1990). Influence of organic matter from soils and sediments from various origins on the sorption of some chlorinated aliphatic hydrocarbons: Implications on K_{oc} correlation. *Environ. Sci. Technol.* **27**, 2360–2366.

Grünheid, S., Amy, G., and Jekel, M. (2005). Removal of bulk dissolved organic carbon (DOC) and trace organic compounds by bank filtration and artificial recharge. *Water Res.* **39**, 3219–3228.

Haitzer, M., Höss, S., Traunspurger, W., and Steinberg, C. (1998). Effects of dissolved organic matter (DOM) on the bioconcentration of organic chemicals in aquatic organisms—A review. *Chemosphere* **37**, 1335–1362.

Haitzer, M., Aiken, G. R., and Ryan, J. N. (2003). Binding of Mercury(II) to aquatic humic substances: Influence of pH and source of humic substances. *Environ. Sci. Technol.* **37**, 2436–2441.

Halling-Sorensen, B., Nielsen, S., Lanzky, P., Ingerslev, F., Lutzhoft, H., and Jorgensen, S. (1998). Occurrence, fate and effects of pharmaceutical substances in the environment—A review. *Chemosphere* **36**, 357–393.

Hambsch, B., Werner, P., and Frimmel, F. H. (1992). Bacterial growth measurements in treated waters from different origin. *Acta Hydrochim. Hydrobiol.* **20**, 9–14.

Hammes, F., Meylan, S., Salhi, E., Köster, K., Egli, T., and von Gunten, U. (2007). Formation of assimilable organic carbon (AOC) and specific natural organic matter (NOM) fractions during ozonation of phytoplankton. *Water Res.* **41**, 1447–1454.

Harvey, G. R., and Boran, D. A. (1985). Geochemistry of humic substances in seawater. In *Humic Substances in Soil Sediment and Water*, Aiken, G. R., McKnight, D. M., Wershaw, R. L., and MacCarthy, P., eds., Wiley-Interscience, New York, pp. 233–247.

Hautala, K., Peravuori, J., and Phihlaja, K. (2000). Measurement of aquatic humus content by spectroscopic analyses. *Water Res.* **34**, 2246–2258.

Hayes, M. H. B. (1985). Extraction of humic substances from soil. In *Humic Substances in Soil, Sediment and Wate*, Aiken, G. R., McKnight, D. M., Wershaw, R. L., and MacCarthy, P., eds., Wiley-Interscience, New York, pp. 329–362.

Her, N., Amy, G., McKnight, D., Sohn, J., and Yoon, Y. (2003). Characterization of DOM as a function of MW by fluorescence EEM and HPLC-SEC using UVA, DOC, and fluorescence detection. *Water Res.* **37**, 4295–4303.

Hesse, S., and Frimmel, F. H. (1999). Biochemical characterization of refractory organic substances. *Acta Hydrochim. Hydrobiol.* **27**, 94–97.

Heumann, K. G., Marx, G., Rädlinger, G., and Vogl, J. (2002). Heavy metal and halogen interactions with fractions of refractory organic substances separated by size exclusion chromatography. In *Refractory Organic Substances (ROS) in the Environment*, Frimmel, F. H., Abbt-Braun, G., Heumann, K. G., Hock, B., Lüdemann, H.-D., and Spiteller, M., eds., Wiley-VCH, Weinheim, pp. 55–72.

Hua, G., and Reckhow, D. A. (2007). Comparison of disinfection byproduct formation from chlorine and alternative disinfectants. *Water Res.* **41**, 1667–1678.

Huber, S. A., and Frimmel, F. H. (1992). A new method for the characterization of organic carbon in aquatic systems. *Int. J. Environ. Anal. Chem.* **49**, 49–57.

Huffman, E. W. D., and Stuber, H. A. (1985). Analytical methodology for elemental analysis of humic substances. In *Humic Substances in Soil, Sediment and Water*, Aiken, G. R., McKnight, D. M., Wershaw, R. L., and MacCarthy, P., eds., John Wiley & Sons, New York, pp. 433–455.

IHSS (2007). http://www.ihss.gatech.edu/

Imai, A., Fukushima, T., Matsushige, K., and Kim, Y. H. (2001). Fractionation and characterization of dissolved organic matter in a shallow eutrophic lake, its inflowing rivers, and other organic matter sources. *Water Res.* **35**, 4019–4028.

Imai, A., Fukushima, T., Matsushige, K., Kim, Y.-H., and Choi, K. (2002). Characterization of dissolved organic matter in effluents from wastewater treatment plants 2002. *Water Res.* **36**, 859–870.

Jahnel, J. B., Ilieva, P., Abbt-Braun, G., and Frimmel, F. H. (1998). Amino acids and carbohydrates in refractory organic acids (in German: Aminosäuren und Kohlenhydrate als Strukturbestandteile von refraktären organischen Säuren). *Vom Wasser* **90**, 205–216.

Jahnel, J. B., Brinkmann, T., Abbt-Braun, G., and Frimmel, F. H. (2002). Occurence of amino acids, carbohydrates, and low-molecular-weight organic acids in refractory organic substances. In *Refractory Organic Substances (ROS) in the Environment*, Frimmel, F. H., Abbt-Braun, G., Heumann, K. G., Hock, B., Lüdemann, H.-D., and Spiteller, M., eds., Wiley-VCH, Weinheim, pp. 264–281.

Kalberer, M., Paulsen, D., Sax, M., Steinbacher, M., Dommen, J., Prevot, A. S. H., Fisseha, R., Weingartner, E., Frankevich, V., Zenobi, R., and Baltensberger, U. (2004). Identification of polymers as major components of atmospheric organic aerosols. *Science* **303**, 1659–1662.

Kalbitz, K., Geyer, S., and Geyer, W. (2000). A comparative characterization of dissolved organic matter by means of original aqueous samples and isolated humic substances. *Chemosphere* **40**, 1305–1312.

Kalle, K. (1938). The problem of Gelbstoff in the sea. *Oceangr. Mar. Biol. Annu. Rev.* **4**, 91–104.

Kaplan, L. A. (1994). A field and laboratory procedure to collect, process, and preserve freshwater samples for dissolved carbon analysis. *Limnol. Oceanogr.* **37**, 1470–1476.

Kappler, A., and Haderlein, S. (2003). Natural organic matter as reductant for chlorinated aliphatic pollutants. *Environ. Sci. Technol.* **37**, 2714–2719.

Karlsson, T., Persson, P., and Skyllberg, U. (2006). Complexation of copper(II) in organic soils and in dissolved organic matter—EXAFS evidence for chelate ring structures. *Environ. Sci. Technol.* **40**, 2623–2628.

Khan, E., and Pillai, S. S. (2007). Interferences contributed by leaching from filters on measurements of collective organic constituents. *Water Res.* **41**, 1841–1850.

Kilduff, J. E., Mattaraj, S., Wigton, A., Kitis, M., and Karanfil, T. (2004). Effects of reverse osmosis isolation on reactivity of naturally occurring dissolved organic matter in physicochemical processes. *Water Res.* **38**, 1026–1036.

Kim, S., Simpson, A. J., Kujawinski, E. B., Freitas, M. A., and Hatcher, P. G. (2003). High resolution electrospray ionization mass spectrometry and 2D solution NMR for the analysis of DOM extracted by C18 solid phase disk. *Organic Geochem.* **34**, 1325–1335.

Kinniburgh, D. G., van Riemsdijk, W. H., Koopal, L. K., Borkovec, M., Benedtti, M. F., and Avena, M. J. (1999). Ion binding to natural organic matter: competition, heterogeneity, stoichiometry, and their modynamic consistency. *Colloids Surf. A.* **151**, 147–166.

Kitis, M., Kilduff, J. E., and Karanfil, T. (2001). Isolation of dissolved organic matter (DOM) from surface waters using reverse osmosis and its impact on the reactivity of DOM to formation and speciation of disinfection by-products. *Water Res.* **35**, 2225–2234.

Koch, B. P., Witt, M., Engbrodt, R., Dittmar, T., and Kattner, G. (2005). Molecular formulae of marine and terrigenous dissolved organic matter detected by electrospray ionization Fourier transform ion cyclotron resonance mass spectrometry. *Geochim. Cosmochim. Acta* **69**, 3299–3308.

Kopinke, F.-D., Georgi, A., Mackenzie, K., and Kumke, M. U. (2000). Sorption and chemical reactions of polycyclic aromatic hydrocarbons with dissolved refractory organic subtances and related model compounds. In *Refractory Organic Substances (ROS) in the Environment*, Frimmel, F. H., Abbt-Braun, G., Heumann, K. G., Hock, B., Lüdemann, H.-D., and Spiteller, M., eds., Wiley-VCH, Weinheim, pp. 475–515.

Koprivnjak, J. F., Perdue, E. M., and Pfromm, P. H. (2006a). Isolation of natural organic matter from fresh waters using reverse osmosis and electrodialysis. In *Humic Substances—Linking Structure to Functions*, Frimmel, F. H., and Abbt-Braun, G., eds., Proceedings of the 13th Meeting of the IHSS. Schriftenreihe des Lehrstuhls für Wasserchemie und der DVGW Forschungsstelle am Engler-Bunte-Institut Universität Karlsruhe, Band 45, pp. 29–32.

Koprivnjak, J. F., Perdue, E. M., and Pfromm, P. H. (2006b). Coupling reverse osmosis with electrodialysis to isolate natural organic matter from fresh waters. *Water Res.* **40**, 3385–3392.

Kördel, W. (1997). Fate and effects of contaminants in soil as influenced by natural organic material—Status of information. *Chemosphere* **35**, 405–411.

Korshin, G. V., Benjamin, M. M., and Li, C.-W. (1999). Use of differential spectroscopy to evaluate the structure and reactivity of humics. *Water Sci. Technol.* **40**, 9–16.

Krasner, S. W., Weinberg, H. S., Richardson, S. D., Pastor, S., Chinn, R., Sclimenti, M. J., Onstad, G., and Thruston, A. D., Jr. (2006). The occurrence of a new generation of disinfection byproducts. *Environ. Sci. Technol.* **40**, 7175–7185.

Kretzschmar, R., Borkovec, M., Grolimund, D., and Elimelech, M. (1999). Mobile subsurface colloids and their role in contaminant transport. *Advances in Agronomy* **66**, 121–194.

Kronberg, L., Holmbom, B., Reunanen, M., and Tikkanen, L. (1988). Identification and quantification of the ames mutagenic compound 3-chloro-4-(dichloromethyl)-5-hydroxy-2(5*H*)-furanone and of its geometric isomer (*E*)-2-chloro-3-(dichloromethyl)-4-oxobutenoic acid in chlorine-treated humic water and drinking water extracts. *Environ. Sci. Technol.* **22**, 1097–1103.

Kujawinski, E. B., Freitas, M. A., Zang, X., Hatcher, P. G., Green-Church, K. B., and Jones, R. B. (2002). The application of electrospray ionization mass spectrometry (ESI MS) to the structural characterization of natural organic matter. *Organic Geochem.* **33**, 171–180.

Kulovaara, M., Metsämuuronen, S., and Nyström, M. (1999). Effects of aquatic humic substances on a hydrophobic ultrafiltration membrane. *Chemosphere* **38**, 3485–3496.

Kumke, M. U., and Frimmel, F. H. (2002). Stationary and time-resolved fluorescence for refractory organic substances characterization. In *Refractory Organic Substances (ROS) in the Environment*, Frimmel, F. H., Abbt-Braun, G., Heumann, K. G., Hock, B., Lüdemann, H.-D., and Spiteller, M., eds., Wiley-VCH, Weinheim, pp. 215–231.

Kumke, M. U., Löhmannsröben, H.-G., and Roch, T. (1994). Fluorescence quenching of polycyclic aromatic compounds by humic acid. *Analyst* **119**, 997–1001.

Kumke, M. U., Abbt-Braun, G., and Frimmel, F. H. (1998). Time-resolved fluorescence measurements of aquatic natural organic matter (NOM). *Acta Hydrochim. Hydrobiol.* **26**, 73–81.

Kumke, M. U., Eidner, S., and Krüger, T. (2005). Fluorescence quenching and luminescence sensitisation in complexes of Tb(III) and Eu(III) with humic substances. *Environ. Sci. Technol.* **39**, 9528–9533.

Laird, L. B., Taylor, H. E., and Kennedy, V. C. (1988). Snow chemistry of the Cascade–Sierra Nevada mountains. *Environ. Sci. Technol.* **20**, 275–290.

Langhals, H., Abbt-Braun, G., and Frimmel, F. H. (2000). Association of humic substances: Verification of Lambert–Beer's law. *Acta Hydrochim. Hydrobiol.* **28**, 329–332.

Leenheer, J. A. (1985). Fractionation techniques for aquatic humic substances. In *Humic Substances in Soil, Sediment and Water*, Aiken, G. R., McKnight, D. M., Wershaw, R. L., and MacCarthy, P., eds., Wiley-Interscience, New York, pp. 409–429.

Löffler, D., Römbke, J., Meller, M., and Ternes, T. (2005). Environmental Fate of Pharmaceuticals in Water/Sediment Systems. *Environ. Sci. Technol.* **39**, 5209–5218.

Löhmannsröben, H.-G., Schultze, U., and Skrivanek, T. (2002). Investigations of the interaction between polycyclic aromatic compounds and refractory organic substances with stationary and time-resolved fluorescence and absorption spectroscopy. In *Refractory Organic Substances (ROS) in the Environment*, Frimmel, F. H., Abbt-Braun, G., Heumann, K. G., Hock, B., Lüdemann, H.-D., and Spiteller, M., eds., Wiley-VCH, Weinheim, pp. 516–534.

Lombardi, A. T., and Jardim, W. F. (1999). Fluorescence spectroscopy of high performance liquid chromatography fractionated marine and terrestrial organic materials. *Water Res.* **33**, 512–520.

Lovely, D. R., Blunt-Harris, E. L., Phillips, E. J. P., and Woodward, J. C. (1996). Humic substances as electron acceptors for microbial respiration. *Nature* **382**, 445–448.

Lu, Y., and Allen, H. E. (2002). Characterization of copper complexation with natural dissolved organic matter (DOM)—Link to acidic moieties of DOM and competition by Ca and Mg. *Water Res.* **36**, 5083–5101.

Ma, H., Allen, H., and Yin, Y. (2001). Characterization of isolated fractions of dissolved organic matter from natural waters and a wastewater effluent. *Water Res.* **35**, 985–996.

MacCarthy, P., and Rice, J. A. (1985). Spectroscopic methods (other than NMR) for determining functionality in humic substances. In *Humic Substances in Soil, Sediment and Water*, Aiken, G. R., McKnight, D. M., Wershaw, R. L., and MacCarthy, P., eds., Wiley-Interscience, New York, pp. 527–559.

Maes, A., Bruggeman, C., Geradets, K., and Vancluysen, J. (2003). Quantification of the interaction of Tc with dissolved boom clay humic substances. *Environ. Sci. Technol.* **37**, 747–753.

Malcolm, R. L. (1985). Geochemistry of stream fulvic and humic substances. In *Humic Substances in Soil Sediment and Water*, Aiken, G. R., McKnight, D. M., Wershaw, R. L., and MacCarthy, P., eds., Wiley-Interscience, New York, pp. 181–209.

Manning, T. J., Bennett, T., and Milton, D. (2000). Aggregation studies of humic acid using multiangle laser light scattering. *Sci. Total Environ.* **257**, 171–176.

Marhaba, T. F., Van, D., and Lippincott, R. L. (2000). Rapid identification of dissolved organic matter, fractions in water by spectral fluorescent signatures. *Water Res.* **34**, 3543–3550.

Maxim, C. R., and Kögel-Knabner, I. (1995). Partitioning of polycyclic aromatic hydrocarbons (PAH) to water-soluble soil organic matter. *Eur. J. Soil Sci.* **46**, 193–204.

McKnight, D. M., and Aiken, G. R. (1998). Sources and age of aquatic humus. In *Aquatic Humic Substances*, Hessen, D. O., and Tranvik, L. J., eds., Springer, Berlin, pp. 9–39.

Merdy, P., Hutzlier, S., and Koopal, L. K. (2006). Modeling metal–particle interactions with an emphasis on natural organic matter. *Environ. Sci. Technol.* **40**, 7459–7466.

Metreveli, G., and Frimmel, F. H. (2007). Influence of Na-bentonite colloids on the transport of heavy metals in porous media. In *Colloidal Transport in Porous Media*, Frimmel, F. H., von der Kammer, F., and Flemming, H.-C., eds., Springer-Verlag, Berlin, pp. 29–53.

Moulin, V., and Moulin, C. (2001). Radionuclide speciation in the environment: a review. *Radiochim. Acta* **89**, 773–778.

Müller, M. B., and Frimmel, F. H. (2002). A new concept for the fractionation of DOM as a basis for its combined chemical and biological characterization. *Water Res.* **36**, 2643–2655.

Müller, M. B., Schmitt, D., and Frimmel, F. H. (2000). Fractionation of natural organic matter by size exclusion chromatography—Properties and stability of fractions. *Environ. Sci. Technol.* **34**, 4867–4872.

Naumczyk, J., Szpyrkowicz, L., and Grandi, F. Z. (1989). Organics isolation from fresh and drinking waters by macroporous anion-exchange resins. *Water Res.* **23**, 1593–1597.

Parson, J. W. (1989). Hydrolytic degradations of humic substances. In *Humic Substances II: In Search of Structure*, Hayes, M. H. B., MacCarthy, P., Malcolm, R. L., and Swift, R. S., eds., John Wiley & Sons, New York, pp. 99–120.

Perdue, E. T. (1998). Chemical composition, structure, and metal binding properties. In *Aquatic Humic Substances*, Hessen, D. O., and Tranvik, L. J., eds., Springer-Verlag, Berlin, pp. 41–61.

Perminova, I. V., Frimmel, F. H., Kovalevskii, D. V., Abbt-Braun, G., Kudryavtsev, A. V., and Hesse, S. (1998). Development of predictive model for calculation of molecular weight of humic substances. *Water Res.* **32**, 872–881.

Perminova, I. V., Frimmel, F. H., Kudryavtsev, A. V., Kulikova, N. A., Abbt-Braun, G., Hesse, S., and Petrosyan, V. S. (2003). Molecular weight characteristics of humic substances from different environments as determined by size exclusion chromatography and their statistical evaluation. *Environ. Sci. Technol.* **37**, 2477–2485.

Peuravuori, J., Lehtonen, T., and Pihlaja, K. (2002a). Sorption of aquatic humic matter by DAX-8 and XAD-8 resins Comparative study using pyrolysis gas chromatography. *Anal. Chim. Acta* **471**, 219–226.

Peuravuori, J., Koivikko, R., and Pihlaja, K. (2002b). Characterization, differentiation and classification of aquatic humic matter separated with different sorbents: synchronous scanning fluorescence spectroscopy. *Water Res.* **36**, 4552–4562.

Planque, G., Maurice, Y., Moulin, V., Toulhoat, P., and Moulin, C. (2005). On the use of spectroscopic techniques for interaction studies, part I: Complexation between europium and small organic ligands. *Appl. Spectrosc.* **59**, 432–441.

Prosen, H., and Zupancic-Kralj, L. (2004). Evaluation of photolysis and hydrolysis of atrazine and its first degradation products in the presence of humic acids. *Environ. Pollut.* **133**, 517–529.

Ravichandra, M. (2004). Interactions between mercury and dissolved organic matter—A review. *Chemosphere* **55**, 319–331.

Reemtsma, T., These, A., Springer, A., and Linscheid, M. (2008). Differences in the molecular composition of fulvic acid size fractions detected by size-exclusion chromatography—On line Fourier transform ion cyclotron resonance (FTICR–) mass spectrometry. *Water Res.* **42**, 63–72.

Richardson, S. D., Simmons, J. E., and Rice, G. (2002). DBPs: The next generation. *Environ. Sci. Technol.* **36**, 198A–205A.

Roberts, T. A., Klein, W., Still, G. G., Kearney, P. C., Drescher, N., Desmoras, J., Esser, H. O., Ahronson, N., and Vonl, J. W. (1984). Non-extractable pesticide residues in soil and plants. *Appl. Chem.* **56**, 945–956.

Saar, R. A., and Weber, J. H. (1982). Fulvic acid: modifier of metal–ion chemistry. *Environ. Sci. Technol.* **16**, 510A–517A.

Schmitt, D., Taylor, H. E., Aiken, G. R., Roth, D. A., and Frimmel, F. H. (2002). Influence of natural organic matter on the adsorption of metal ions onto clay minerals. *Environ. Sci. Technol.* **36**, 2932–2938.

Schmitt, D., Saravia, F., Frimmel, F. H., and Schuessler, W. (2003). NOM-facilitated transport of metal ions in aquifers: Importance of complex-dissociation kinetics and colloid formation. *Water Res.* **37**, 3541–3550.

Schmitt-Kopplin, P., Garrison, A. W., Perdue, E. M., Freitag, D., and Kettrup, A. (1998). Capillary electrophoresis of humic substances. Facts and artefacts. *J. Chromatogr. A* **807**, 101–109.

Schnitzer, M., and Khan, S. U. (1972). *Humic Substances in the Environment*, Marcel Dekker, New York.

Schulten, H.-R., Leinweber, P., and Jandl, G. (2002). Analytical pyrolysis of humic substances and dissolved organic matter in water. In *Refractory Organic Substances (ROS) in the Environment*, Frimmel, F. H., Abbt-Braun, G., Heumann, K. G., Hock, B., Lüdemann, H.-D., and Spiteller, M., eds., Wiley-VCH, Weinheim, pp. 163–187.

Schreiber, B., Brinkmann, T., Schmalz, V., and Worch, E. (2005). Adsorption of dissolved organic matter onto activated carbon—The influence of temperature, absorption wavelength, and molecular size. *Water Res.* **39**, 3449–3456.

Senesi, N., Miano, T. M., Provenzano, M. R., and Brunetti, G. (1989). Spectroscopic and compositional comparative characterization of IHSS reference and standard fulvic and humic acids of various origin. *Sci. Total Environ.* **81/82**, 143–156.

Serkiz, S. M., and Perdue, E. M. (1990). Isolation of dissolved organic matter from the Suwannee River using reverse osmosis. *Water Res.* **24**, 911–916.

Sontheimer, H., Crittenden, J. C., and Summers, R. (1988). *Activated Carbon for Water Treatment*, 2nd edition in English of *Adsorptionsverfahren zur Wasserreinigung*. DVGW-Forschungsstelle, Karlsruhe.

Specht, C. H., and Frimmel, F. H. (2000). Specific interactions of organic substances in size exclusion chromatography. *Environ. Sci. Technol.* **34**, 2361–2366.

Spiteller, M., Klaus, U., and Pfeifer, T. (2002). Analysis of the binding of Amitrole and Anilazine to aquatic and terrestrial refractory organic substances. In *Refractory Organic Substances (ROS) in the Environment*, Frimmel, F. H., Abbt-Braun, G., Heumann, K. G., Hock, B., Lüdemann, H.-D., and Spiteller, M., eds., Wiley-VCH, Weinheim, pp. 446–476.

Steinberg, C. E. W. (2003). *Ecology of Humic Substances in Freshwaters*. Springer, Berlin.

Steinberg, C., and Muenster, U. (1985). Geochemistry and ecological role of humic substances in lakewater. In *Humic Substances in Soil Sediment and Water*, Aiken, G. R., McKnight, D. M., Wershaw, R. L., and MacCarthy, P., eds., Wiley-Interscience, New York, pp. 105–145.

Stenson, A. C., Marshall, A. G., and Cooper, W. T. (2003). Exact masses and chemical formulas of individual Suwannee River fulvic acids from ultrahigh resolution electrospray ionisation Fourier transform-ion cyclotron resonance mass spectrometry. *Anal. Chem.* **75**, 1275–1284.

Suffet, I. H., and MacCarthy, P., eds. (1989). *Aquatic Humic Substances—Influence on Fate and Treatment of Pollutants*, American Chemical Society, Washington, D.C.

Sun, L., Perdue, E. M., and McCarthy, J. F. (1994). Using reverse osmosis to obtain organic matter from surface and ground waters. In *Humic Substances in the Global Environment and Implications on Human Health*, Senesi, N., and Miano, T. M., eds., Elsevier, Amsterdam, pp. 901–910.

Suteerapataranon, S., Bouby, M., Geckeis, H., Fanghänel, T., and Grudpan, K. (2006). Interaction of trace elements in acid mine drainage solution with humic acid. *Water Res.* **40**, 2044–2054.

Swift, R. S. (1985). Extraction of humic substances from soil. In *Humic Substances in Soil, Sediment and Wate*, Aiken, G. R., McKnight, D. M., Wershaw, R. L., and MacCarthy, P., eds., Wiley-Interscience, New York, pp. 387–408.

Swift, R. S. (1996). Organic matter characterization. In *Methods of Soil Analysis. Part 3. Chemical Methods*, Sparks, D. L., et al., eds., Soil Science Society of America Book Series no. 5, Madison, WI, pp. 1018–1020.

These, A., and Reemtsma, T. (2003). Limitations of electrospray ionization of fulvic and humic acids as visible from size exclusion chromatography with organic carbon and mass spectrometric detection. *Anal. Chem.* **75**, 6275–6281.

Thurman, E. M. (1985a). *Organic Geochemistry of Natural Waters*, Martinius Nijhoff/Dr. W. Junk Publishers, Lancaster.

Thurman, E. M. (1985b). Humic substances in groundwater. In *Humic Substances in Soil Sediment and Water*, Aiken, G. R., McKnight, D. M., Wershaw, R. L., and MacCarthy. P., eds., Wiley-Interscience, New York, pp. 87–103.

Tiseanu, C.-D., Kumke, M. U., Frimmel, F. H., Klenze, R., and Kim, J.-I. (1998). Time-resolved fluorescence spectroscopy of fulvic acid and fulvic acid complexed with Eu^{3+}—A comparative study. *J. Photochem. Photobiol. A: Chemistry* **117**, 175–184.

Tipping, E. (1998). Humic ion binding model VI: An improved description of the interactions of protons and metals ions with humic substances. *Aquat. Geochem.* **4**, 3–48.

Tipping, E. (2002). *Cation Binding by Humic Substances*, Cambridge University Press, New York.

Tremblay, L. B., Dittmar, T., Marshall, A. G., Cooper, W. J., and Cooper, W. T. (2007). Molecular characterization of dissolved organic matter in a North Brazilian mangrove porewater and mangrove-fringed estuaries by ultrahigh resolution Fourier transform–ion cyclotron resonance mass spectrometry and excitation/emission spectroscopy. *Marine Chem.* **105**, 15–29.

van Riemsdijk, W. H., Koopal, L. K., Kinniburgh, D. G., Benedetti, M. F., and Weng, L. (2006). Modeling the interactions between humics, ions, and mineral surfaces. *Environ. Sci. Technol.* **40**, 7473–7480.

Wagoner, D. B., and Christman, R. F. (1999). Molar mass and size of Norwegian aquatic NOM by light scattering. *Environ. Int.* **25**, 275–284.

Weber, J. W. (1988). Binding and transport of metals by humic materials. In *Humic Substances and Their Role in the Environment*, Frimmel, F. H., and Christman, R. F., eds., John Wiley & Sons, Chichester, pp. 165–178.

Weber, T., Allard, T., Tipping, E., and Benedetti, M. F. (2006). Modeling iron binding to organic matter. *Environ. Sci. Technol.* **40**, 7488–7493.

Weng, L., Temminghoff, E. J. M., and Van Riemsdijk, W. H. (2002). Aluminum speciation in natural waters: measurement using Donnan membrane technique and modeling using NICA-Donnan. *Water Res.* **36**, 4215–4226.

Weng, L., Van Riemsdijk, W. H., Koopal, L. K., and Hiemstra, T. (2006). Adsorption of humic substances on goethite: Comparison between humic acids and fulvic acids. *Environ. Sci. Technol.* **40**, 7494–7500.

Williams, P. M. (1971). The distribution and cycling of organic matter in the ocean. In *Organic Compounds in Aquatic Environments*, Faust, S. D., and Hunter, J. V., eds., Marcel Dekker, New York, pp. 145–164.

Zsolnay, Á. (2003). Dissolved organic matter: artefacts, definitions, and functions. *Geoderma* **113**, 187–209.

11

MARINE ORGANIC MATTER

E. M. PERDUE

School of Earth and Atmospheric Sciences, Georgia Institute of Technology, Atlanta, Georgia

R. BENNER

Department of Biological Sciences, University of South Carolina, Columbia, South Carolina

Biophysico-Chemical Processes Involving Natural Nonliving Organic Matter in Environmental Systems, Edited by Nicola Senesi, Baoshan Xing, and Pan Ming Huang

11.1. INTRODUCTION

Throughout all compartments of the earth system, living organisms are composed of carbon, hydrogen, oxygen, nitrogen, phosphorus, sulfur, and a variety of other essential elements. These elements are typically accumulated in relatively constant proportions during the life span of each living organism and are released at different rates upon its death. Although the rates and mechanisms of accumulation and release of the elements are different, all elements are ultimately restored to their original forms and to the compartments of the earth system from which they were initially assimilated into living organisms. The journey through time and space taken by each element is called its biogeochemical cycle.

The biogeochemical cycle of any essential element is intrinsically worthy of scientific inquiry. Practical concerns, however, tend to motivate much of modern scientific inquiry, and such is the case with biogeochemical cycles. Current concern over the measured accumulation of carbon dioxide in the earth's atmosphere and the observed worldwide retreat of glaciers to higher elevations has focused much scientific effort on the biogeochemical cycle of carbon. Emphasis is placed on understanding linkages between global warming and the accumulation of CO_2, a greenhouse gas, in the earth's atmosphere.

The measured accumulation of CO_2 in the earth's atmosphere is clear evidence that the total flux of CO_2 from all other compartments of the earth system to the atmosphere exceeds the total flux of CO_2 back to those compartments. Most box models of the biogeochemical cycle of carbon consider at least five compartments (e.g., atmosphere, terrestrial biomass, soil, marine biomass, seawater) through which carbon is cycled relatively rapidly and at least two compartments (e.g., marine organic sediments, marine inorganic sediments) through which carbon is cycling much more slowly. The quantity of dissolved inorganic carbon (DIC) in seawater is at least a factor of five greater than the sum of all other forms of rapidly cycled carbon. This vast pool of carbon is a source for aquatic photosynthesis and a sink for aquatic respiration of marine organic matter. Among the five compartments through which carbon is cycled more rapidly, the quantity of nonliving organic carbon in the seawater compartment equals or exceeds the total amount of organic carbon in terrestrial and marine biomass and approximately equals the amount of carbon (as CO_2) in the earth's atmosphere. Any comprehensive assessment of the global carbon cycle must therefore necessarily include a thorough study of the biogeochemistry of marine organic matter.

Within the larger context of the global carbon cycle, marine organic matter may be viewed conceptually as a transitional material that lies somewhere along the diagenetic vector that leads from marine biomass to CO_2. The extent to which a given sample of marine organic matter has been modified through photodegradation and biodegradation as it is transported through space and time in the world's oceans will be manifested in its bulk elemental composition (C, H, O, N, P, S) and in ancillary properties that are derived from bulk elemental composition (unsaturation—the concentration of rings and pi bonds, average oxidation state of organic carbon, etc.). In this review, significant emphasis will thus be placed on methods of isolation of marine organic matter from seawater and on the aforementioned chemical properties. As it turns out, accurate elemental analyses can only be obtained on dry, low-ash samples of marine organic matter, so the two topics are inextricably coupled.

11.1.1. Terminology

The scientific literature regarding marine organic matter is the cumulative result of research in several traditional scientific disciplines, each with its own specialized definitions and acronyms. This review must draw upon all of that literature, so it is beneficial to introduce some of the specialized terminology that will be encountered by the reader. A useful introduction is provided by simply compiling a list of commonly used acronyms, all of which are defined and used in the definitive volume on marine organic matter that was published recently by Hansell and Carlson (2002).

The most basic distinction is between particulate organic matter (POM) and dissolved organic matter (DOM). The operational boundary between these two fractions is established by filtration of a water sample through a "submicron" filter whose nominal pore size varies according to the type of filter. Common examples (pore sizes) include precombusted glass-fiber filters (0.7 μm), silver filters (0.45 μm), preconditioned polycarbonate filters (0.2 μm), and preconditioned mixed-ester filters (0.2 μm). All of these filters will remove most living organisms (except viruses and small bacteria), and all have been shown to be insignificant sources of carbonaceous impurities, if adequately preconditioned. In coastal waters and in other highly productive waters, filtration is often used to remove POM. In samples from depths of 200 m or greater and from relatively nonproductive surface regions of the open ocean, the percentage of POM is negligible, and such waters are seldom filtered. The measured concentrations of organic carbon in unfiltered samples of seawater are reported as total organic carbon (TOC). The small difference between "total" and "dissolved" organic matter in such samples is of much less concern than the possibility of contamination caused by unnecessary handling of samples or by physical breakage of cells. The term DOM is thus broadly applied to samples that contain very few particles, whether or not the samples have actually been filtered.

A very popular method for concentrating and isolating DOM from seawater is tangential-flow ultrafiltration (UF) using a membrane with a nominal molecular weight cutoff of 1000 Da. The fraction of DOM that is retained during this process is called ultrafiltered dissolved organic matter (UDOM), and it is also called high-molecular-weight dissolved organic matter (HMW DOM). The fraction of DOM that passes through the UF membrane and cannot be recovered from seawater is called low-molecular-weight dissolved organic matter (LMW DOM). The remaining acronyms in Table 11.1 refer to properties other than physical size of organic matter. Chromophoric dissolved organic matter (CDOM) is that fraction of DOM which absorbs not only visible light but also ultraviolet light in the UV-A (315–400 nm) and UV-B (280–315 nm) regions of the electromagnetic spectrum. A closely related term is colored detrital materials (CDM), which includes both CDOM and detrital particles whose optical properties are very similar to those of CDOM. This term is needed whenever it is necessary to analyze the optical properties of unfiltered seawater, as is the case when processing data sets that are acquired using satellites. The fluorescent fraction of DOM is called fluorescent dissolved organic matter (FDOM). Finally, the fraction of DOM that most strongly resists biodegradation is called refractory DOM (RDOM).

TABLE 11.1. Common Acronyms Used to Describe Marine Organic Matter

Acronym	Definition
POM	Particulate organic matter
DOM	Dissolved organic matter
UDOM	Ultrafiltered dissolved organic matter
HMW DOM	High-molecular-weight dissolved organic matter
LMW DOM	Low-molecular-weight dissolved organic matter
CDOM	Chromophoric dissolved organic matter
CDM	Colored detrital materials
FDOM	Fluorescent dissolved organic matter
RDOM	Refractory dissolved organic matter

11.1.2. Analytical Approaches

11.1.2.1. Carbon. The remarkable chemical heterogeneity of organic matter in seawater precludes quantification of all individual components. Therefore, the overall concentration of organic matter is expressed in terms of the concentration of organic carbon in the sample. The concentration of organic carbon, in turn, is quantified after conversion of all organic carbon in the sample to CO_2. In the older literature, units of mg C liter^{-1} were used; however, units of μmol C liter^{-1} are much more common in recent literature, especially in the literature of marine chemistry. The analytical measurement is known either as total organic carbon (TOC) or as dissolved organic carbon (DOC), depending on whether or not the sample was filtered prior to the analysis. Particulate organic carbon (POC) is the difference between TOC and DOC.

The conversion of organic carbon to CO_2 may be accomplished using either wet chemical oxidation (WCO) or high temperature combustion (HTC) (Menzel and Vaccaro, 1964; Sharp, 1973). Because the product to be generated and quantified is CO_2, both WCO and HTC methods must be preceded by a step in which inorganic carbon is removed from the sample. This step includes acidification of the sample to pH 2 with a nonvolatile inorganic acid, followed by sparging for around 10 min with a CO_2-free gas to strip out CO_2 as completely as possible. In WCO methods, a sparged sample (5–100 ml) is then reacted with a strong oxidant such as potassium persulfate and/or irradiated with short-wavelength ultraviolet light (e.g., 254 nm). In HTC methods, a sparged sample (50–200 μl) is then injected into a flowing stream of oxygen gas in a furnace that is heated to around 1000 K. Both WCO and HTC methods are known to have system blanks, for which appropriate corrections must be made. The problem is far more challenging for HTC methods, and difficulties in accounting for HTC blanks were the impetus for a substantial amount of research in the late 1980s and early 1990s on the analysis of organic carbon in seawater. Some of the challenges in the analysis of low concentrations of organic carbon, especially in saline waters, are discussed by Hedges and Lee (1993) and Aiken (1992).

The forms of organic carbon that are defined on the basis of optical properties of DOM are never physically separated from the whole DOM, so their concentrations in seawater are expressed indirectly in terms of the corresponding optical property. For example, the concentration of FDOM is generally expressed in terms of fluorescence intensity, which is a function of the wavelengths of excitation and emission. Fluorescence intensity may be measured at a single pair of wavelengths for excitation and emission (e.g., 355 nm and 450 nm). Alternatively, a single wavelength can be used for excitation while emission is monitored over a range of wavelengths (or vice versa). The most comprehensive measurement of fluorescence intensity combines wide ranges of wavelength for both excitation and emission to obtain an excitation-emission matrix spectrum (EEMS).

The concentration of CDOM may be expressed in terms of absorbance at a specified wavelength (A_λ), or absorbance may be normalized to the optical path length (r) to obtain the absorption coefficient at a specified wavelength ($a_\lambda = 2.303A_\lambda/r$), where r and a_λ are in units of m and m^{-1}, respectively. Wavelengths in the approximate range of 280–450 nm have often been used for this purpose. To a good first approximation, absorption coefficients decrease exponentially in this range of wavelength (Hu et al., 2002; Helms et al., 2008).

The concentration of RDOM is simply equal to the measured concentration of DOM in deep waters (>1000 m), where its apparent radiocarbon age of 4000–6000 years is substantially greater than the timescale of thermohaline circulation in the earth's oceans (Druffel et al., 1992). Bioassay experiments have been used to verify the refractory nature of DOM in the deep sea (Barber, 1968).

11.1.2.2. Nitrogen and Phosphorus. Every size fraction (including the particulate fraction) of marine organic matter may contain both inorganic and organic forms of N and P, so a parallel set of analytical methods and acronyms exist for bulk quantification of organic N and organic P. Because organic N and organic P are distributed among an exceedingly complex mixture of organic compounds, they (like organic C) are quantified indirectly after conversion to their respective inorganic forms. Because inorganic forms of N and P cannot be removed from samples prior to analysis for organic N and P, it is necessary to quantify inorganic N and P before the organic forms of N and P are converted into inorganic species.

The most commonly encountered form of inorganic N in particulate matter in seawater is NH_4^+, which can be bound very strongly by illites. In most studies, only total nitrogen (TN) is measured in particulate matter, but some authors have adapted methods from the soils literature to quantify separately total inorganic nitrogen (TIN) and total organic nitrogen (TON) (e.g., Sampei and Matsumoto, 2001; Schubert and Calvert, 2001). In these particular studies, an elemental analyzer was used to measure TN of dried samples of particulate matter before and after exhaustive oxidation and rinsing to remove organic N. The two analytical results were interpreted as TN and as TIN, and the difference between them is TON.

The sum of all dissolved forms of N in seawater, excluding $N_2(aq)$, is called total dissolved nitrogen (TDN), and it consists primarily of NH_4^+, NO_3^-, NO_2^-, and dissolved organic nitrogen (DON). The three inorganic species can be quantified separately, although NO_3^- and NO_2^- are often combined into a single analysis. When seawater samples are oxidized using potassium persulfate (Valderrama, 1981) or strongly irradiated with ultraviolet light (Armstrong et al., 1966; Armstrong and Tibbitts, 1968; Koroleff, 1976; D'Elia et al., 1977; Solórzano and Sharp, 1980a), all forms of TDN are converted to NO_3^- and NO_2^-, which are then commonly quantified using the NO_3^-/NO_2^- colorimetric method. When seawater samples are subjected to high-temperature combustion (Suzuki and Sugimura, 1985), all forms of TDN are converted to NO, which is quantified using a chemiluminescence detector. The concentration of DON is calculated as the difference between TDN and the sum of the three inorganic forms of N. Because the calculated concentration of DON is the result of several separate chemical analyses, the accumulation of analytical errors can introduce considerable uncertainty into this measurement.

Seawater contains both particulate P and dissolved P, and, as was the case for N, these fractions contain both inorganic and organic forms of P. Particulate P is first isolated by filtration of seawater samples, typically using precombusted glass-fiber filters having a nominal pore size of 0.7 µm. Filters are either extracted directly with dilute HCl to solubilize particulate inorganic phosphorus (PIP) or are first combusted using a high-temperature ashing oxidation method (Aspila et al., 1976; Solórzano and Sharp, 1980b) to convert organic P to inorganic P and then extracted with dilute HCl to solubilize total particulate phosphorus (TPP). Particulate organic phosphorus (POP) is the difference between TPP and PIP (e.g., Loh and Bauer, 2000; Benitez-Nelson et al., 2004; 2007; Némery and Garnier, 2007; Yoshimura et al., 2007). All extracts are analyzed for P using the procedures described in the next paragraph.

The analytical chemistry of P relies heavily on the molybdate blue method of Murphy and Riley (1962), which targets primarily ortho-phosphoric acid (H_3PO_4) and its conjugate bases. In this method, reactive forms of P combine with ammonium molybdate to form a molybdophosphoric acid complex, which, upon reduction with ascorbic acid or stannous chloride, yields a deep blue solution whose absorbance is proportional to the concentration of phosphate in the sample. It has been long recognized that the method also detects some hydrolysable forms of both inorganic P and organic P, so the quantity measured by this analytical method is known as soluble reactive phosphorus (SRP). A significant advance in the analytical chemistry of P in seawater is the magnesium-induced coprecipitation (MAGIC) procedure of Karl and Tien (1992) and its more recent modification by Thomson-Bulldis and Karl (1998). This procedure relies upon coprecipitation of inorganic P with $Mg(OH)_2(s)$ that forms upon addition of NaOH to seawater samples to raise pH to around pH 9. If a sample is first pretreated to destroy organic matter and convert all forms of P as completely as possible to ortho-phosphate, the quantity measured by the analytical method is known as total dissolved phosphorus (TDP). The difference between TDP and SRP is very often interpreted as dissolved organic phosphorus (DOP). This analytical methodology and its inherent ambiguities have been reviewed thoroughly and recently by Karl and Björkman (2002).

11.1.2.3. Hydrogen and Oxygen. Organic H equals the quantity of H in $H_2O(g)$ that is generated by combustion of a dry, ash-free sample of organic matter in an excess of $O_2(g)$. Organic O is most often calculated by difference, assuming that the sum of mass percentages of all major elements in a dry, ash-free sample equals 100%. Alternatively, organic O can be measured directly. Huffman and Stuber (1985) have described several of the more common methods of elemental analysis of organic matter. Both moisture and inorganic ash interfere strongly with the measurement of organic H and O, and only a very limited number of such analyses are available for marine organic matter.

11.2. INVENTORIES AND FLUXES

11.2.1. Reservoirs of Organic Matter in the Ocean

Less than 0.2% of the organic carbon in the ocean is contained in the biomass of marine organisms. The vast majority of the organic carbon in the ocean is found in

TABLE 11.2. Reservoirs and Fluxes of Organic Carbon in the Ocean[a]

Reservoirs and Fluxes	Amount	Reference[b]
Reservoirs		
Biomass	3	1
POC	15	2
DOC	685	3
SOC	1000	4
Fluxes		
Marine net primary production	50	1
Riverine POC	0.15	5
Riverine DOC	0.25	5
Atmospheric POC + DOC	0.2	6, 7

[a]Carbon reservoirs and annual fluxes are given in units of Pg C (10^{15} g). POC, particulate organic carbon; DOC, dissolved organic carbon; SOC, sedimentary organic carbon.

[b]1, Denman et al. (2007); 2, Eglinton and Repeta (2004); 3, Hansell and Carlson (1998); 4, Hedges and Oades (1997); 5, Hedges et al. (1997); 6, Romankevich (1984); 7, Willey et al. (2000).

particulate, dissolved, and sedimentary forms of nonliving organic matter (Table 11.2). This is in sharp contrast to the terrestrial environment, where approximately 25% of the organic carbon in the active surface reservoir is found in biomass, mostly as wood (Denman et al., 2007). Microorganisms dominate ocean biota and have very short life spans compared with land plants. Thus, the terrestrial and ocean reservoirs of nonliving organic carbon are more closely linked to photosynthetic production in these systems than to biomass.

Dissolved organic matter is the dominant form of organic carbon in the ocean water column. Concentrations of DOM are highest in surface waters (0–200 m), where most of the biological fixation and remineralization of carbon occur. Concentrations of DOC are typically 60–90 μmol liter^{-1} in surface waters and 35–45 μmol liter^{-1} in deep (>1000 m) waters (Benner, 2002). Particulate organic matter is also most abundant in the surface ocean, accounting for a few percent of the total organic carbon in the water column. Sinking particles play an important role in the ocean carbon cycle by transporting organic matter to the deep ocean and sediments. This process, referred to as the *biological pump*, removes carbon dioxide from the atmosphere and functions as a long-term sink in the global carbon cycle. Submicron particles numerically dominate the reservoir of suspended particulate organic matter (Koike et al., 1990). Sedimentary organic matter (SOM) is the largest reservoir of organic carbon in the ocean. The rain of sinking particles to the sediments supports a highly diverse population of benthic organisms, but most sedimentary organic matter is highly resistant to biodegradation.

11.2.2. Sources and Fluxes of Organic Matter to the Ocean

Photosynthesis in marine surface waters is the dominant source of organic matter to the ocean. Net primary production (50 Pg C) accounts for over 98% of the annual flux of organic carbon to the ocean (Table 11.2) and is similar in magnitude to net primary production on land (Denman et al., 2007). Rates of primary production are typically higher in coastal waters than in open ocean waters due to higher concentrations and availability of essential nutrients, such as N, P, Si, and Fe. Ocean margins account for about 15–30% of the net primary production in the ocean (Gattuso et al., 1998). The reservoirs of particulate organic carbon (POC) and sedimentary organic carbon (SOC) include biomass carbon, but the vast majority of the carbon in these reservoirs is nonliving, as in the DOC reservoir. Continental runoff and atmospheric deposition contribute about 0.6 Pg C as POM and DOM to the ocean annually. Terrestrial and marine organic matter have different chemical compositions and reactivities, and the relative contributions of these sources to the reservoirs of organic carbon in the ocean have been a focus of research for decades (Hedges et al., 1997). Fluxes of marine organic carbon to the ocean are approximately 100-fold greater than fluxes of terrestrial organic carbon (Table 11.2).

Yields of lignin phenols, unique tracers of terrestrial organic matter, and bulk stable carbon isotopic compositions indicate that >90% of the ocean DOC reservoir is of marine origin (Druffel et al., 1992; Opsahl and Benner, 1997). Observed concentrations and distributions of lignin phenols in the ocean are consistent with patterns of global riverine discharge, ocean circulation, and photochemical and microbial degradation as the primary mechanisms for the removal of terrestrial DOC (Hernes and Benner, 2006). Concentrations of terrestrial organic matter are much higher in coastal waters and sediments, particularly in regions with high riverine discharge. On average, DOC and POC account for about 60% and 40%, respectively, of global riverine discharge of organic carbon (Meybeck, 1982). Riverine DOC is rich in aromatic components that absorb ultraviolet light and are major components of the chromophoric DOM in coastal waters (see Section 11.3.2). Riverine POC is largely deposited in deltaic and margin sediments, where most carbon burial in the ocean occurs (Hedges and Keil, 1995; Burdige, 2006). A recent estimate indicates that one-third of the organic carbon buried in marine sediments is of terrestrial origin (Burdige, 2005).

The circulation of seawater through hydrothermal systems of mid-ocean ridges is an important source or sink for many elements in the ocean (Edmond et al., 1979), but little is known about the role of these systems in the organic matter balance of the oceans. A recent study investigated the concentrations of DOC in a range of hydrothermal systems in the deep Pacific Ocean (Lang et al., 2006). Concentrations of DOC are depleted in high-temperature ridge-axis and warm off-axis vent fluids relative to deep Pacific bottom waters, indicating that these systems are a net sink of organic matter. Elevated concentrations of DOC occur in low-temperature diffuse hydrothermal systems, indicating these systems are a net source of organic matter. High DOC concentrations are likely due to release from biota in highly productive ridge-axis diffuse vents. Lang et al. (2006) conclude that total and net hydrothermal fluxes of DOC are small relative to other oceanic sources and sinks; therefore this flux is not included in Table 11.2.

11.3. TRANSFORMATIONS

There are tens of thousands of structural forms of organic matter in nature, and the biological reactivity of each of these compounds is influenced by molecular structure and environmental factors, such as temperature and availability of molecular oxygen. The biological diversity of the decomposer community, which is predominantly microorganisms, is also critical for determining transformation pathways and fates of organic matter, and environmental conditions have an important influence in shaping microbial diversity. The vast majority of organic matter synthesized by living organisms is remineralized to carbon dioxide, water, and various inorganic forms of N, P, and S within an annual cycle, but a very small fraction of this newly synthesized organic matter escapes remineralization processes and survives for centuries or more. The study of both of these processes, remineralization and preservation, is essential for understanding the inner workings of global biogeochemical cycles. The remineralization of organic matter provides biota with the essential elements required by all living organisms, and the preservation of organic matter determines the redox state of the earth's surface and the composition of the atmosphere (Berner, 1982).

Remineralization and preservation processes lead to very different end products but are often influenced by similar factors. The physical shielding of organic matter from enzymes and other hydrolytic and oxidative processes has been shown to inhibit the decomposition and remineralization of marine organic matter (Lee et al., 2004). Organic matter bound within the inorganic matrix of marine organisms, such as siliceous diatom frustules, can be preserved for long periods of time in marine sediments (Ingalls et al., 2003). The sorption of organic matter onto mineral surfaces has also been shown to impede its decomposition, leading to long-term preservation in marine sediments (Mayer, 1994; Hedges and Keil, 1995). Shielding and sorption mechanisms also enhance organic matter preservation by reducing the exposure of organic matter to oxic conditions (Hartnett et al., 1998).

11.3.1. Biotransformations

Microorganisms, principally bacteria, are responsible for the biodegradation of a large fraction of the organic matter in the ocean. All biologically synthesized molecules and some xenobiotic molecules and geomolecules are susceptible to enzymatic transformation and remineralization. Macromolecules are the dominant form of organic matter synthesized by organisms and released into the environment by a variety of mechanisms, including direct cellular release, predation, excretion, and viral lysis (Nagata, 2000). Many of these molecules are too large (≥600 Da) to be transported directly into a cell (Weiss et al., 1991). The hydrolysis and transformation of larger molecules is initiated outside the cell by extracellular enzymes either (a) occurring on the cell surface, (b) occurring in the periplasmic space of bacteria, or (c) being released into the environment (Chróst, 1991). Molecular structure directly and indirectly influences enzymatic transformations. For example, the diversity of enzymes required for substrate utilization is directly dependent upon the heterogeneity of intermonomeric linkages and the tertiary structure of the substrate. Molecular structure determines the aqueous solubility of biomolecules and partitioning between polar and nonpolar phases, thereby indirectly influencing microenvironmental conditions and potential for biotransformations.

Biologists and geochemists have long sought chemical indicators of the biological reactivity and the extent of alteration or diagenetic state of natural organic matter. Numerous studies have investigated various aspects of the chemical nature of organic matter as indicators of reactivity and diagenetic state (Benner, 2003; Keil et al., 2000; Wakeham et al., 1997). Two types of molecular indicators are commonly used to investigate the diagenetic state of organic matter: (a) the fraction of total organic carbon identified as specific molecules and (b) the molecular composition of organic matter. The major classes of biomolecules, such as amino acids, carbohydrates, and lipids, comprise a relatively large fraction of the carbon in living organisms, and the fraction of carbon identified as specific biomolecules decreases with increasing decomposition and diagenesis, as the fraction of carbon that is no longer recognizable as biomolecules increases (Hedges et al., 2000). Most approaches utilizing carbon-normalized yields of specific biomolecules as indicators of the bioavailability of organic matter have been qualitative, but a quantitative approach was recently used to estimate the concentrations of labile (highly bioreactive) and semilabile (moderately bioreactive) DOM in the surface ocean (Davis and Benner, 2007). The major biomolecules that are preferentially utilized during microbial decomposition of organic matter, such as proteins, carbohydrates, and nucleic acids, are relatively oxidized forms of organic matter compared with the aliphatic and alicyclic molecules, such as carboxyl-rich alicyclic molecules (CRAM), that become more prevalent in aged and diagenetically altered DOM (Figure 11.1).

Changes in the molecular composition of organic matter during decomposition are also indicative of reactivity and diagentic alteration. Compositional indicators are often derived from common biochemical classes, such as amino acids, lipids, and carbohydrates. Some molecular indicators occur in a limited group of organisms, such as diatoms, whereas others are found in all organisms. Thus, these compounds

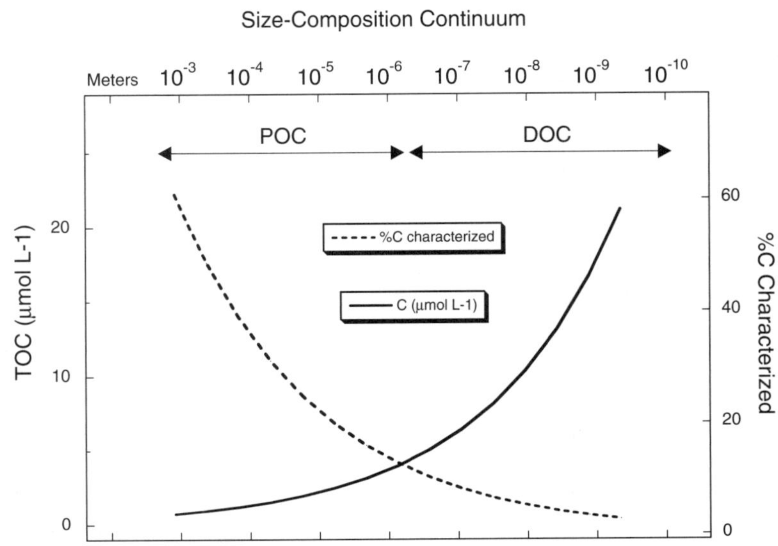

Figure 11.1. A conceptual diagram of the size–composition continuum of organic matter in the ocean. The seawater concentration of total organic carbon (TOC) in various size fractions increases with decreasing size of particles, colloids, and dissolved molecules. The percentages of carbon characterized as specific molecules, such as amino acids and neutral sugars, decreases with decreasing size. Most of the organic carbon resides in the ocean as small molecules that have not been structurally characterized.

can provide information about the reactivity and diagenetic state of a selected subset of organic matter or the bulk reservoir. The composition of amino acids has been most widely used as an indicator of the diagenetic state of bulk organic matter (Keil et al., 2000). Some diagenetic trends in molecular composition are apparent, such as the increase in mole percentage of glycine during decomposition, but others are subtle and are more easily determined using statistical approaches such as principal component analysis (Dauwe and Middelburg, 1998; Dauwe et al., 1999).

11.3.2. Phototransformations

Photochemical transformations of organic molecules in the surface ocean play an important role in the carbon cycle (Mopper et al., 1991). Photochemical transformations of DNA impact organisms and biological processes, such as photosynthesis, and they impact the remineralization and fate of organic matter (Mopper and Kieber, 2002; Tedetti and Sempéré, 2006). The absorption of light is essential for photochemical reactions, and the chromophoric components of DOM (CDOM) play a major role in the absorption of ultraviolet and visible light in the ocean (Nelson et al., 1998). The ultraviolet portion (290–400 nm) of the light spectrum reaching the earth's surface is the most energetic and has the greatest potential for driving photochemical transformations of organic matter inside cells (e.g., DNA dimerization) as well as in seawater. These shorter wavelengths (UV-B and UV-A) are also most efficiently absorbed by CDOM. The quantitative effects of photochemical transformations on the cycles of bioactive elements in the ocean are largely dependent on the penetration of ultraviolet radiation in the surface ocean. The concentrations of CDOM and particles are much higher in the coastal than open ocean, thus UV penetration is typically much greater in the open ocean (up to 40 m) compared with the coastal ocean (up to 10 m) (Tedetti and Sempéré, 2006).

The major carbon-containing photodegradation products are carbon dioxide, carbon monoxide, and a variety of low-molecular-weight carbonyl compounds (Mopper and Kieber, 2002). Direct photochemical transformations of an organic molecule occur through the absorption of light by a chromophore(s) within the same molecule. Molecules that do not directly absorb light can undergo secondary reactions that are photosensitized by light-absorbing molecules (Zepp, 1988). Photomineralization of DOM to carbon dioxide is the dominant process in coastal waters receiving runoff and CDOM from land (Miller and Zepp, 1995), and there is evidence for both direct and photosensitized reactions in this process (Mopper and Kieber, 2002). The DOM in rivers is largely derived from plant litter and soils and is rich in CDOM and aromatic constituents that are much more susceptible to photochemical transformations than algal DOM, which is depleted in CDOM and aromatic constituents (Blough and Del Vecchio, 2002). A relatively large percentage (~45%) of DOM in blackwater rivers is susceptible to photomineralization (Obernosterer and Benner, 2004).

Photochemical transformations of organic matter in natural waters are complex and poorly understood at a mechanistic level. Photodegradation typically leads to an overall reduction in the molecular size and weight distribution of DOM, a major loss or alteration of the CDOM components, and an alteration in the susceptibility of the remaining DOM to microbial decomposition and remineralization. In general, photodegradation of DOM with a substantial CDOM component, as indicated by

UV absorption and fluorescence, leads to an enhancement of microbial decomposition and remineralization, whereas photodegradation of DOM with minimal CDOM has no net effect or reduces subsequent microbial decomposition and remineralization. Continental runoff is rich in CDOM, and photodegradation results in substrates that enhance microbial decomposition and the remineralization of terrestrial DOM (Miller and Moran, 1997; Smith and Benner, 2005). Very little CDOM in the open ocean appears to be of terrestrial origin (Hernes and Benner, 2006), and the photodegradation of surface water DOM results in a reduction in microbial decomposition, whereas photodegradation of deep water DOM results in an increase in microbial decomposition (Benner and Biddanda, 1998).

Surface ocean DOM is largely colorless and "photobleached," and the net production of CDOM in the open ocean is difficult to detect and measure (Nelson and Siegel, 2002). Marine CDOM is largely produced in mesopelagic ocean waters (100–1000 m depth) where sinking particulate material from surface waters is biologically oxidized (Yamashita and Tanoue, 2008). This process is the major source of marine CDOM in the open ocean, whereas continental runoff of terrestrial CDOM is the dominant source of CDOM in the coastal ocean. The chemical compositions of CDOM from continental runoff and the deep open ocean are very different, and this likely affects the photochemical pathways and fates of organic matter and associated bioactive elements as well as the effects on plankton community structure and biological processes, such as photosynthesis. Photochemical transformations of organic matter in the ocean clearly play important roles in biogeochemical cycles, and this area of research is expanding rapidly as understanding of these processes grows.

11.3.3. Size-Composition Continuum

Small molecules are the dominant form of organic matter in the ocean. About 75% of the DOC in the ocean passes an ultrafiltration membrane with a pore size of ~1 nm, corresponding to a nominal molecular weight of 1000 Da (Benner et al., 1992; Ogawa and Ogura, 1992). Thus, the size distribution of organic carbon is heavily skewed to oligomers and other low-molecular-weight compounds that are resistant to microbial degradation and photochemical mineralization. The relationship between the molecular size of DOM and its microbial utilization was investigated in a series of experiments with marine waters, and larger size classes (>1 nm) of DOM were more rapidly utilized by microorganisms than smaller size classes (<1 nm), leading to the concept of the size–reactivity continuum (Amon and Benner, 1994, 1996). The size–reactivity continuum hypothesis states that macromolecules and other high-molecular-weight organic matter are more bioavailable than low-molecular-weight organic matter in the ocean. Several studies have confirmed this relationship between size and reactivity for POM and DOM in the ocean (Hama et al., 2004; Mannino and Harvey, 1999, 2000; Tamburini et al., 2003). The sorption of organic matter to mineral surfaces confounds the relationship between size and reactivity, making it difficult to apply the size–reactivity hypothesis to SOM.

The observation that small molecules are relatively abundant in seawater and resistant to decomposition is puzzling to microbial ecologists and geochemists alike. Small biomolecules are typically utilized rapidly by microorganisms. Clearly, there is something unusual about the chemical composition of the vast reservoir of low-molecular-weight organic matter in the ocean that makes it resistant to decomposition, but characterizing the composition and structure of this material has been extremely challenging. A very small fraction of the low-molecular-weight DOM in the ocean has been characterized as specific molecules (Benner, 2002).

The chemical composition of marine organic matter also varies with size in a similar relationship as observed in the size-reactivity continuum (Figure 11.1). A relatively large fraction of macromolecular organic matter can be characterized at the molecular level. About 60–80% of the carbon in marine plankton and 30–40% of the carbon in sinking particles in the upper ocean can be characterized as specific biomolecules, such as amino acids, neutral sugars, and lipids (Wakeham et al., 1997; Lee et al., 2004). A smaller percentage (15–20%) of the carbon in suspended POM, which is largely submicron particles (Koike et al., 1990), can be characterized as specific biomolecules in the upper ocean. Only 8–14% of the carbon in high-molecular-weight DOM and 2–5% of the carbon in low-molecular-weight DOM can be characterized as specific biomolecules in the upper ocean (Benner, 2002). The percentages of characterized carbon in each of these size fractions of organic matter decreases with increasing depth in the ocean, indicating the increasing diagenetic alteration of organic matter as it is transported in the ocean's abyss. The size–reactivity and size–composition continuum models provide a conceptual framework for understanding diagenetic processes and carbon cycling in the ocean. A size-age continuum based on radiocarbon ages of different size classes of organic matter has also been observed (Loh et al., 2004), providing further support for the concept that diagenetic processes lead to the production of smaller molecules that are eventually lost from the analytical window of many molecular characterizations.

11.4. CHEMICAL PROPERTIES

11.4.1. Concentrations of Dissolved Organic C, N, and P

As indicated in Section 11.2. most nonliving organic carbon in the ocean water column resides in the dissolved phase, and the following sections focus on DOM. There are very many analyses in the literature for the concentrations of C, N, and P in marine DOM. These analyses are generally made on bulk samples of seawater using the methods discussed in Section 11.1.2, rather than on isolated samples of marine DOM. A general summary of bulk chemical measurements on marine DOM was given by Benner (2002) for the surface ocean and deep ocean. Selected ranges of concentration of DOC, DON, and DOP are given in Table 11.3. Bronk (2002) presented a thorough review of the biogeochemistry of DON, in which the bulk analytical measurements for DON were tabulated for a large number of natural waters. Water types and average DON values from that review are also given in Table 11.3. Karl and Björkman (2002) thoroughly described the distribution of DOP in the world's oceans. They summarized 233,118 measurements of DOP (generally obtained as the difference between TDP and SRP; see Section 11.1.2.2) from the world's oceans. This data set included 139,747 measurements from waters at depths exceeding 200 m. The tabulated results for DOP in Karl and Björkman (2002) are less conveniently organized than those for DON in Bronk (2002), and average values are not given for different types of natural waters. The

TABLE 11.3. Dissolved Organic Carbon, Nitrogen, and Phosphorus in Ocean Waters

Type of Sample	DOC (μM)	DON (μM)	DOP (μM)	Source[a]
Surface ocean	60–90	3.5–7.5	0.1–0.4	1
Deep ocean	35–45	1.5–3.0	0.02–0.15	1
Surface ocean		5.8 ± 2.0		2
Deep ocean		4.3 ± 2.1		2
Coastal/continental Shelf		9.9 ± 8.1		2
Estuaries		22.5 ± 17.3		2
Selected rivers		34.7 ± 20.7		2
Open ocean			0.15 ± 0.08	3
Continental shelf			0.20 ± 0.12	3
Coastal/estuarine			0.43 ± 0.24	3

[a]1, Benner (2002); 2, Bronk (2002); 3, Karl and Björkman (2002).

results in Table 11.3 were obtained by lumping together all depths for a selected type of sample, and ranges of values in the original data set were replaced by averages.

Two general trends are evident from the average data in Table 11.3. First, concentrations of DOC, DON, and DOP are greater in surface waters, where photosynthesis and primary production are greatest. Heterotrophic utilization of biologically labile DOM reduces the concentrations of DOC, DON, and DOP in deeper waters, leaving a low and fairly uniform concentration of biologically resistant DOM in deep ocean waters. Upon close examination, there are spatial trends in deep ocean waters. For example, concentrations of DOC are slightly greater in North Atlantic deep waters (48 μM) than in North Pacific deep waters (34 μM). Hansell (2002) attributes the small gradient in the concentration of DOC to microbial mineralization and mixing of DOM during the roughly 1000-year journey of deep waters from the North Atlantic Ocean to the Southern Ocean and finally back to the North Pacific Ocean.

The data in Table 11.3 also illustrate the strong gradients of decreasing DON and DOP from coastal regions to the open ocean. A similar trend is commonly observed for DOC, but Table 11.3 does not contain supporting data. This trend is also attributed, in part, to higher biological productivity in coastal regions, but the input of terrestrially derived DOM through riverine discharge is also an important contributing factor to the higher concentrations of DOC often observed in coastal regions. This general trend of higher concentrations of DOM in coastal regions is not apparent in some regions lacking significant continental runoff, demonstrating the strong influence of terrestrially derived nutrients and organic matter on coastal ocean biogeochemical processes.

A detailed assessment of vertical trends in DOC, DON, and DOP can be obtained using the comprehensive Hawaii Ocean Time-series (HOT) data set of 4953 samples that were taken at Station ALOHA between October 1988 and December 2005 (http://hahana.soest.hawaii.edu/hot/hot-dogs/interface.html). The raw data were coalesced into 25 depth intervals ranging from 0–10 m to 4500–5000 m, and average concentrations of DOC, DON, and DOP were calculated for each depth interval. Average DOC concentrations ranged from 41.2 to 89.4 μmol kg^{-1}, average DON concentrations ranged from 2.1 to 5.8 μmol kg^{-1}, and average DOP concentrations ranged from 0.02 to 0.23 μmol kg^{-1}. All of these results fall within the ranges given by Benner (2002) for surface and deep oceans. The concentrations of DOC, DON, or DOP in each vertical profile were divided by the highest average concentration of that element to yield normalized concentrations that range from zero to one for DOC, DON, and DOP. These normalized results are given in Figure 11.2 to compare the relative rates at which DOC, DON, and DOP decrease with depth.

Below a depth of around 1000 m, there is little variation in the concentration of DOC. This may be true as well for DON and DOP, but the increasingly irregular profiles at lower concentrations of DON and DOP preclude a definitive interpretation. Although concentrations of DOP are rather low in deep ocean waters (≈0.02 μmol kg^{-1}), the irregularity in the vertical profiles of DON and DOP in Figure 11.2 is most likely attributable to the statistically small number of observations—only around two observations per depth interval between 2000 m and 5000 m.

Figure 11.2 clearly illustrates the strongly preferential remineralization in surface waters of DOM that is relatively enriched in P, and to a lesser extent N. At a depth of 1000 m, an average of 50% of DOC, 64% of DON, and 85% of DOP have been consumed by heterotrophic organisms. Clark et al. (1998) also concluded that ~84% of DOP in UDOM was remineralized between surface and deep waters in the Pacific. It follows that N/C, P/C, and P/N molar ratios will decrease sharply from the surface to around 1000 m, after which they appear to remain relatively constant, notwithstanding the larger variance in the data at greater depths. Such depth-related changes in chemical properties are commonly interpreted as being driven by biogeochemical processes, but it is important to recognize that water masses in the

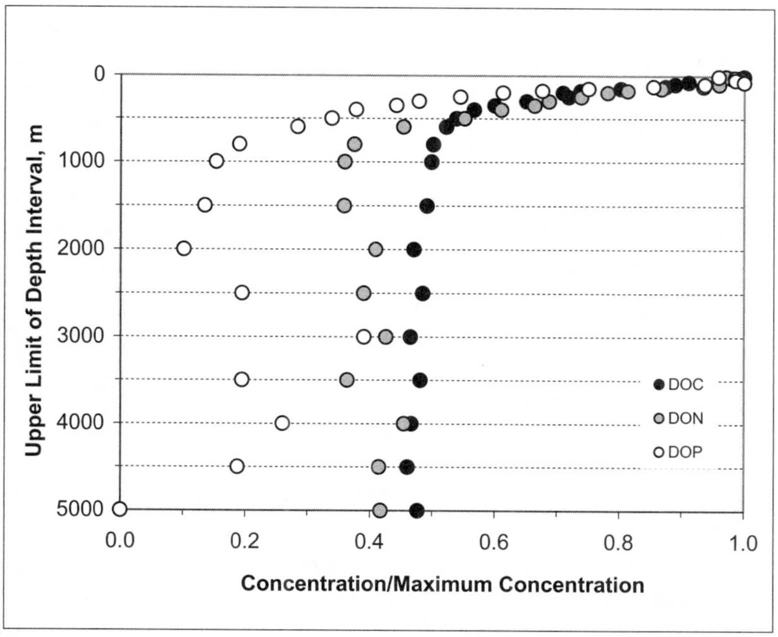

Figure 11.2. Normalized vertical profile of the concentrations of DOC, DON, and DOP at Station ALOHA.

deep ocean were formed in different geographic locations and may not have the same initial chemical properties as surface waters at the sampling location.

Despite the insights provided by measurements of the concentrations of DOC, DON, and DOP and their molar ratios in bulk seawater, these data alone are insufficient to calculate the mass percentage of carbon in marine DOM, the average oxidation state of carbon in marine DOM, or the degree of unsaturation of marine DOM. All such calculations require that the concentrations of organically bound H and O also be measured. Such measurements cannot be made directly on aqueous samples.

11.4.2. Isolation and Fractionation

Isolation of DOM from seawater is a daunting task, both because of its very low concentration (35–85 μmol liter^{-1}, 0.4–1.0 mg liter^{-1}) and, much more importantly, because it is intimately mixed with more than 35,000 mg liter^{-1} of inorganic solutes. Nonetheless, the more detailed chemical and biogeochemical insight into the nature of marine DOM that can only be obtained using representative, concentrated, low-ash samples has provided a continuous impetus for the development of suitable isolation methods. Three strategies are currently being employed:

1. Solid-phase extractions (SPE), in which DOM is selectively concentrated on a solid-phase extractant such as XAD-2 resin or C_{18} adsorbent. In these methods, both inorganic solutes and water are removed concurrently, and a suitable solvent is used to desorb the concentrated, desalted DOM from the solid-phase extractant. In principle, all inorganic solutes may be removed by SPE.
2. Tangential-flow ultrafiltration (UF), in which DOM is concentrated using UF membranes that are highly permeable to both water and inorganic solutes. Some inorganic solutes are also retained by UF membranes and must be washed from the concentrated DOM by diafiltration against pure water. Not all inorganic solutes are removed by this method.
3. Reverse osmosis/electrodialysis (RO/ED), in which inorganic solutes are removed using ED and water is removed using RO. Not all inorganic solutes are removed by this method.

A few representative applications of each method of isolation of DOM from seawater will be discussed. The interested reader is directed to the recent review paper by Mopper et al. (2007) for a more detailed discussion of methods of isolation of DOM from seawater.

11.4.2.1. Solid-Phase Extractions Using XAD Resins. The Amberlite XAD series (Rohm and Haas Co., Philadelphia, PA, USA) have been most often used for isolation of marine DOM by SPE. XAD resins are nonionic macroporous copolymers that differ in pore size, surface area, and polarity. Their generally large specific surface areas and more-or-less reversible adsorption of organic solutes from aqueous solution have made them well-suited for isolation of selected fractions of DOM from natural waters. Even though XAD resins have been used far more often to isolate DOM from fresh waters than from seawater, the first reported use of XAD resins to isolate DOM from natural waters was that of Stuermer and Harvey (1974), who used XAD-2 resin to isolate humic and fulvic acids from seawater. The XAD method was extended to fresh waters by Weber and Wilson (1975), who used XAD-2 resin to isolate humic materials from a blackwater pond and river, and by Mantoura and Riley (1975), who examined the operational characteristics of the adsorption and desorption of humic materials with XAD-2 resin. Leenheer and Huffman (1976) extended the basic isolation method using XAD resins to develop a more comprehensive fractionation scheme for classification of organic solutes in water. Their scheme employed XAD-2 and XAD-8 resins, cation- and anion-exchange resins, and aqueous HCl and NaOH solutions to separate DOM into hydrophilic and hydrophobic fractions, each of which was further separated into acidic, basic, and neutral fractions. Aiken et al. (1979) and Aiken (1988) provide detailed comparisons of several types of XAD resins, along with discussion of (a) mechanisms of adsorption, fractionation, and alteration of DOM by adsorption on XAD resins and (b) effects of several eluents on the properties of isolated DOM.

In the simplest type of protocol (which is most commonly used), water samples are acidified to pH 2 with HCl and passed through a column of XAD resin. Hydrophobic organic acids (which are largely protonated at pH 2) and hydrophobic neutral compounds are adsorbed to the resin. Hydrophobic bases, which are protonated cations at pH 2, and all hydrophilic compounds in DOM pass through the column and are usually discarded. Back-elution with NH_4OH or NaOH is used to desorb the hydrophobic acid fraction of DOM, and this fraction is usually the material that is recovered for laboratory experiments. The use of NH_4OH in this step has yielded samples of marine DOM with elevated levels of N due to irreversible reactions between DOM and NH_4OH (see later discussion). Back-elution of the column with a water-miscible organic solvent may recover some of the adsorbed hydrophobic neutral compounds, but more exhaustive Soxhlet extractions with water-immiscible organic solvents may be required to recover all adsorbed DOM from the XAD resin. This step is usually omitted.

A more complex protocol is sometimes used, in which the acidified solution that has been passed through XAD-2 or XAD-8 resin is then passed through a column of XAD-4 resin. Back-elution of adsorbed DOM from the XAD-4 resin with NaOH yields an additional quantity of DOM that is variably known as hydrophilic acids, XAD-4 acids, transphilic acids, or amphiphilic acids.

Several examples in which DOM was adsorbed to a single type of XAD resin can be used to illustrate some basic trends. Stuermer and Harvey (1977) adsorbed DOM from surface and deep samples in the Sargasso Sea on a column of XAD-2 resin, which was back-eluted with NH_4OH and CH_3CH_2OH to recover the hydrophobic acid fraction (HbA) and hydrophobic neutral fraction (HbN) of marine DOM. HbA and HbN fractions accounted for 4.5% and 3.4%, respectively, of DOM in the surface water sample. In contrast, 22.5% of DOM in the deep water sample was isolated as the HbA fraction, and an additional 8.2% of DOM was isolated in the HbN fraction. Slauenwhite and Wangersky (1996) used XAD-2 resin to adsorb DOM from coastal surface samples in Halifax Harbour. Using both NaOH and CH_3OH as eluents, they were able to recover less than 15% of DOM (HbA and HbN combined). Druffel et al. (1992) used XAD-2 resin for adsorption and NaOH for desorption of DOM to recover 22% ± 2% of marine DOM from four samples in the Sargasso Sea. Using XAD-8 resin for adsorption and NaOH for desorption of DOM, they recovered 19% ± 4% of marine DOM from two samples in the Sargasso Sea. The yield of DOM increased slightly with depth between 50 m and 3237 m. Obernosterer and Herndl (2000) used XAD-8 resin to separate humic and nonhumic DOM in seawater from the northern Adriatic Sea and the coastal North Sea. Humic substances (isolated on XAD-8 resin) accounted for 15% ± 7% of DOC in the northern Adriatic Sea and 43% ± 7% in the coastal North Sea. Esham et al. (2000) used XAD-8 resin to isolate humic substances from the Satilla River estuary in Georgia, and the water they sampled had both a high salinity (30‰) and a high concentration of DOC (1375 μm). The DOM in this sample probably consists mainly of terrestrially derived DOM, and their yield of 63% of DOC is consistent with the average yield (54% ± 14%) of humic substances (as DOC) from freshwaters (Perdue and Ritchie, 2003). Collectively, these studies have shown that XAD-2 and XAD-8 recover a similar percentage of marine DOM, that the yield of DOM is much greater in coastal waters than in the open ocean, and that the yield of DOM is greater in deep waters than in surface waters. The HbN:HbA ratio for DOM isolated from seawater using XAD resins (36:100 in deep water; 76:100 in surface water) is much greater than for freshwater, where the median HbN:HbA ratio is only 11:100 (Perdue and Ritchie, 2003).

Several studies have employed a tandem treatment, in which the sample is passed first through a column of either XAD-2 or XAD-8 resin and then through a column of XAD-4 resin. Various eluents and methods of elution have been used. Bussmann (1999) used XAD-2 and XAD-4 columns in tandem to isolate humic substances from polar seawater. Using NaOH to desorb marine DOM from both the XAD-2 and XAD-4 columns, they recovered 16–27% of DOC as the HbA fraction from surface waters, and 36% of DOC was obtained from a deep water sample. Druffel et al. (1992) also used XAD-4 resin in tandem with XAD-2 and XAD-8 columns to increase overall recoveries of marine DOM (see earlier discussion). When XAD-4 was used in tandem with XAD-2 resin, the average yield of DOM from four samples increased from 22% ± 2% to 26% ± 3%. A much greater effect of the XAD-4 resin was obtained when it was used following XAD-8 resin, where the average yield of DOM from two samples increased from 19% ± 4% to 35% ± 6%. As noted in the preceding paragraph, no significant trends with depth are evident in the data of Druffel et al. (1992). These yields represent only the HbA fraction of marine DOM; and, other than the lack of a strong trend of increasing yield with depth, these results agree rather well with those of Bussmann (1999). Engbrodt and Kattner (2005) also used XAD-2 and XAD-4 resins sequentially to extract DOM from polar seawater, and they used NaOH and CH_3OH to elute HbA and HbN, respectively, from each column. Four fractions were generated from the two columns and two eluents; however, only the combined yield of HbA+HbN was tabulated for each sample. They reported average yields of 45% ± 9% of DOC from surface waters and 60% ± 6% of DOC from deep waters of the Fram Strait and the Greenland Sea. Because the reported yields include both HbA and HbN, yields are higher than those reported by Bussmann (1999) or Druffel et al. (1992). Unlike Druffel et al. (1992), who reported that yields of DOM varied very little with depth, Bussmann (1999) and Engbrodt and Kattner (2005) both reported substantially greater yields from deep ocean waters, which is consistent with studies in which only a single type of XAD resin was used.

For the numerous efforts to isolate DOM from seawater using XAD resins that have been reviewed here, yields range from 5% to 63%, roughly averaging 34% ± 16% and tending to be greater in coastal or deep waters than in surface waters of the open ocean. When XAD resins are used to isolate humic substances from freshwaters, the average yield of humic substances is 54% ± 14% (Perdue and Ritchie, 2003). It follows that roughly one-third of the DOM in seawater and one-half of the DOM in freshwaters is sufficiently hydrophobic to be extracted from acidified water samples by this method. Of all the methods of isolation, only SPE using XAD resins has yielded isolated samples of marine DOM from which nearly all sea salts have been removed (see subsequent discussion of other methods).

11.4.2.2. Solid-Phase Extractions using C_{18} Adsorbents. Another popular adsorbent used to isolate DOM from seawater by SPE is C_{18}, a highly hydrophobic adsorbent in which C_{18} alkyl groups are covalently bonded to silica. Organic acids in seawater are largely protonated in the pH 2–3 range, under which conditions a significant fraction of marine DOM can be isolated on C_{18}. Most often, CH_3OH is used to elute adsorbed DOM from C_{18}, a clear departure from the otherwise similar XAD method in which adsorbed DOM is usually eluted with NaOH. Because the conditions of adsorption are so similar for XAD resins and C_{18} adsorbents, chemically similar fractions of DOM are likely to be isolated by the two types of adsorbents.

Mills and Quinn (1981) used C_{18} to isolate an average of 17% ± 10% of DOC from waters of Narragansett Bay ranging in salinity from 22‰ to 30‰. Amador et al. (1990) reported that 30% of DOC in a 400-ml sample of seawater can be adsorbed by a Sep-Pak C_{18} cartridge. Koch et al. (2005) recovered an average of 24% of DOC using C_{18} SPE to isolate DOM from the Weddell Sea in Antarctica. Simjouw et al. (2005) isolated DOM from brackish and saline waters of the Chesapeake Bay system by SPE, using C_{18} disks. In two series of experiments, they reported average recoveries of 39% and 33%, respectively, of DOC. They also used C_{18} to recover DOM from samples that had previously been ultrafiltered using a 1000-Da membrane, and they recovered an average of 30% of DOC. Dittmar et al. (2008) have used C_{18} cartridges for solid-phase extractions of marine DOM along an 81-km transect along the North Brazil shelf, where they isolated 39% ± 4% of DOM.

In several of the above-cited studies, additional data were presented that confirm the selectivity of C_{18} for chromophores, fluorophores, and aromatic compounds, and its low affinity for nitrogenous DOM. For example, Amador et al. (1990) reported that 42% of UV absorbance and 48% of fluorescence intensity were removed when only 30% of DOM was adsorbed by C_{18}. Likewise, Simjouw et al. (2005), in two series of experiments using brackish and saline waters of the Chesapeake Bay system, reported average yields of 59% and 51% of UV absorbance (integrated from 250 nm to 400 nm), respectively, which is considerably greater than the reported yield of DOM (39% and 33%). Even when the low-molecular-weight fraction of marine DOM in the permeate solution from a 1000-Da UF membrane was passed through C_{18}, the removal of UV absorbance (43%) greatly exceeded the yield of DOM (30%). Louchouarn et al. (2000) reported that lignin-derived phenols are recovered quantitatively from riverine, estuarine, and marine waters by SPE on C_{18}. Given the generally low recoveries of DOM that have been reported (17–39%), aromatic constituents of DOM are preferentially enriched in that fraction of marine DOM which can be adsorbed to the C_{18} phase. Koch et al. (2005) used C_{18} SPE to isolate DOM from the Weddell Sea in Antarctica, where they recovered an average of 24% of DOC, and they found that the isolated DOM was strongly depleted in N (atomic C/N = 35.6) relative to the source water (C/N = 11–18).

Stabenau and Zika (2004) used a hydrophilic–lipophilic copolymer called Nexus (Varian, Inc.) to isolate DOM from seawater. The average recovery of DOC was 30.4%; however, an average of 79% of color (absorbance between 290 and 400 nm) was removed. In comparison with C_{18}, Nexus recovers a similar percentage of DOC but a much higher percentage of color, indicating that the isolated material is even less representative of whole DOM than the fraction that can be isolated using C_{18}. Dittmar et al. (2008) have used PPL (a styrene–divinylbenzene polymer) cartridges for solid-phase extractions of both coastal and noncoastal marine DOM. Along an 81-km transect along the North Brazil shelf, they isolated 62% ± 6% of DOM, which is considerably greater than the average yield they obtained from these sampling sites using C_{18} cartridges. The yield of DOM decreased to 43% ± 4% in noncoastal seawater from the Gulf of Mexico and the Weddell Sea.

In summary, when SPE is conducted using C_{18}, Nexus, or PPL, the recovery of DOM (as DOC) ranges from 17% to 43% in seawater, with higher yields in estuarine and coastal waters. Isolated samples are typically enriched in chromophores, fluorophores, and aromatic compounds, and they are relatively depleted in nitrogen, relative to DOM in unprocessed seawater.

11.4.2.3. Ultrafiltration. In tangential-flow ultrafiltration (UF), DOM is concentrated using UF membranes that are highly permeable to both water and inorganic solutes. Some inorganic solutes are also retained by UF membranes and must be washed from the concentrated DOM by diafiltration against pure water. Not all inorganic solutes are removed by this method. In most studies, the high-molecular-weight fraction of marine DOM (HMW-DOM) is isolated using tangential-flow UF with 1000-Da membranes, although other UF membranes are sometimes used (e.g., Amador et al., 1990). Regardless of the choice of membrane, the separation of DOM from seawater using UF depends more directly on physical than on chemical properties, so chemical fractionation should be less of a problem than is the case with SPE methods.

Benner et al. (1992) used a 1000-Da tangential-flow UF membrane to isolate marine DOM from surface (10 m), oxygen minimum layer (765 m), and deep (4000 m) waters of the Pacific Ocean at Station ALOHA (22°45.0′N, 158°00.0′W), just north of Oahu, Hawaii. Recoveries of DOC decreased with increasing depth, ranging from 33% at 10 m to 22% at 4000 m. The atomic C/N ratios of the three isolated samples ranged from 15.3 to 22.5—well within the normally observed range for bulk DOM in seawater (10–25), so their results suggest that samples isolated by UF do not preferentially exclude N-containing compounds, as is the case for SPE methods (see earlier discussion). The total carbohydrate contents of bulk seawater and isolated DOM samples varied consistently with increasing depth, further supporting the authors' claim that the samples isolated by UF were reasonably representative of the whole DOM, even though the average recovery of DOC was around 25%.

Benner et al. (1997) extended the application of the UF method to an additional 16 samples from the Pacific Ocean, three samples from the Atlantic Ocean, and two samples from the Gulf of Mexico. Including the three samples from Benner et al. (1992), the average recovery of DOC for 24 samples of marine DOM was 26% ± 5%. The marine DOM samples that were isolated using UF had very narrow distributions of atomic C/N (17.1 ± 1.5), $\delta^{13}C$ (−21.7 ± 0.2), and $\delta^{15}N$ (8.1 ± 0.9). Modest spatial trends in some of these properties were identified by the authors; however, the data set of Benner et al. (1997) generally supports the hypothesis that DOM samples that are isolated using UF have bulk properties that are compositionally representative of whole DOM, even though the percent recovery of DOC is rather low. Aluwihare et al. (1997) isolated marine DOM from surface waters (1–15 m) at 12 sampling locations in the Atlantic and Pacific Oceans. Eleven samples were isolated using UF, and one sample was obtained using dialysis with a 1000-Da membrane. They reported yields of 25–35% of DOC in this study.

Several research groups have used UF to isolate DOM from waters having a wide range of salinity. For example, Amador et al. (1990) recovered 39% of DOC from Biscayne Bay and 18–21% of DOC from the Sargasso Sea with either 500-Da or 5000-Da stirred-cell UF membranes. The trend of much greater yield of DOM from coastal waters (Biscayne Bay) than from noncoastal waters (Sargasso Sea) is observed consistently when UF is used to isolate DOM from saline waters. Benner and Opsahl (2001) used UF to isolate DOM from 31 surface water samples collected in four different seasons across a salinity gradient from the Mississippi River to full-strength seawater in the Gulf of Mexico. The percentage of DOC isolated (1000-Da membrane) from river water averaged 46%, and the percentage isolated from seawater (36‰ salinity) averaged 26% (recovered after diafiltration and freeze-drying). The percentages of DOC recovered by UF decreased fairly linearly across the salinity gradient. The average atomic C/N ratios of isolated DOM decreased from 23.6 in the river to 18.5 in surface waters of the Gulf of Mexico. Benner et al. (2005) used UF to isolate 67 DOM samples from river, estuarine, and marine waters in the Arctic. The percentages of DOC isolated from rivers and low-salinity estuarine waters ranged from 52% to 64% and the percentages isolated from marine surface and deep waters ranged from 21% to 32%. The atomic C/N ratios of DOM isolated from river and low-salinity estuarine waters ranged from 38.2 to 47.6 and from 16.2 to 19.7 for DOM isolated from surface and deep marine waters. Guo et al. (1995) used UF to isolate DOM from 18 sampling sites/depths in the Trinity River, Gulf of Mexico, and Middle Atlantic Bight. Salinity ranged from 0‰ to 36.4‰. The reported concentrations of DOC and colloidal organic carbon (COC) can be used to calculate yields of 31–68% for the fraction of marine DOM that was concentrated using a 1000-Da UF membrane. If the data for two samples having salinities of less than 25‰ are omitted, the average calculated yield of DOC is 41% ± 6%. Likewise, Guo et al. (1996) used UF to isolate DOM from 18 sampling sites/depths in the Middle Atlantic Bight. Salinity ranged from 5.0‰ to 36.3‰. The authors reported yields of 28–65% for the fraction of marine DOM that was concentrated using a 1000-Da UF membrane. If the data for five samples having salinities of less than 25‰ are omitted, the average calculated yield of DOC is 36% ± 9%. Guo et al. (2003) used UF to isolate DOM from 29 sampling sites/depths in the Trinity River, Chesapeake Bay, the Middle Atlantic Bight, Galveston Bay, and the Gulf of Mexico. Of the 14 samples for which results had not already been published by Guo et al. (1995) and Guo et al. (1996), 11 samples were collected at salinities ranging from 0‰ to 25‰. This data set reveals very clearly the impact of salinity (or variables that co-vary with salinity) on yields of DOC by the UF method. At low salinity, around 65% of DOC was concentrated by a 1000 Da UF membrane. The yield decreased linearly with increasing salinity up to around 34‰, at which point the yield of DOC was around 52%. Beyond that salinity, the yield of DOC dropped sharply to a minimum value of 34% at the highest salinity. DOC measurements on aliquots of "filtered and ultrafiltered seawater" were used to obtain the yields cited here, so those yields probably do not reflect potentially large losses of DOM that occur during diafiltration of the concentrated ultrafiltrate to remove co-concentrated sea salts. It is perhaps for this reason that the results of Guo and coworkers are systematically higher than the reported yields in other studies cited here. The DOM samples that were isolated using UF by Guo et al. (2003) had narrow distributions of atomic C/N (18.7 ± 3.4), $\delta^{13}C$ (−24.4 ± 1.7), and $\delta^{15}N$ (6.9 ± 1.7).

Minor et al. (2002) used UF to isolate DOM from the lower Chesapeake Bay, USA, and the Oosterschelde estuary, The Netherlands. Although it is unclear what yield of DOC was obtained, the average for 46 samples collected from the Oosterschelde estuary appears to be around 39%. Simjouw et al. (2005) also used both UF and C_{18} SPE (see earlier discussion) to isolate DOM from brackish and saline waters of the Chesapeake Bay. They recovered an average of 51% of DOC and 56% of UV absorbance (integrated from 250 to 400 nm). Relative to their experiments using C_{18} for SPE, there is not much difference between the recoveries of DOC and UV absorbance, indicating that the UF method yields a more representative fraction than C_{18} SPE, at least as far as UV absorbance is concerned. These recoveries are generally greater than achieved by UF in most other studies, probably because the samples were collected in estuarine waters having salinities that are well below that of open-ocean seawater.

In summary, when DOM is isolated from marine waters using UF, higher yields are obtained in coastal waters (50%–60%) than in the open ocean (25%–40%), and there is a slightly decreased yield with increasing depth in the open ocean. Atomic C/N ratios and $\delta^{13}C$ values for isolated DOM are similar to the values in bulk seawater, indicating little or no fractionation of the samples with respect to these properties, even though as little as 25% of DOM is actually isolated. Samples that are isolated by UF typically contain a large percentage of sea salts, even after exhaustive diafiltration against pure water. For example, the percentage of organic carbon in the 24 samples isolated using UF by Benner et al. (1997) is 17% ± 7%. Assuming that marine DOM is around 50% carbon by weight, the isolated samples contain around 34% DOM and 66% inorganic sea salts and water.

11.4.2.4. Reverse Osmosis/Electrodialysis. Very recently, Perdue and co-workers have developed a novel process in which reverse osmosis (RO) is used to remove water from seawater and electrodialysis (ED) is used to desalt seawater. The coupled RO/ED process was developed initially by Koprivnjak et al. (2006) to purify RO-concentrated samples of DOM from freshwaters, which are nearly always contaminated with sulfuric acid (H_2SO_4) and silicic acid (H_4SiO_4). It was possible to remove nearly all H_2SO_4 and H_4SiO_4 from such samples with an average 85% overall recovery of DOC.

Vetter et al. (2007) and Gurtler et al. (2008) have described the essential aspects of the RO/ED process for isolation of DOM from seawater. The authors used the RO/ED process to isolate marine DOM from 16 seawater samples, and the average yield of DOC was 75% ± 12%. Although the method is still very much in the developmental stage and many variations of the experimental protocol have been used to isolate the 16 samples of marine DOM, the general RO/ED process can be considered as three distinct phases of operation. In the first phase, ED is used to lower the conductivity of a sample of seawater from around 50 mS cm^{-1} to around 15 mS cm^{-1}. The removal of 70% of the ionic solutes in the sample during the first phase causes a proportionate decrease in the osmotic pressure of the sample from 2590 kPa to 780 kPa. At that point, water can be removed at a reasonable rate by RO at a conveniently low pressure of around 1380–1725 kPa. In the second phase of the RO/ED process, RO and ED are used simultaneously to remove most of the water in the sample while maintaining the conductivity at 15 mS cm^{-1}. The second phase ends when the final desired volume of the sample is reached. In the third phase of the RO/ED process, ED is used to lower the conductivity of the sample to a final value. In a typical experiment, a 200-liter sample of seawater having a conductivity of 50 mS cm^{-1} is processed in 6–8 h to obtain a 6-liter sample having a conductivity of 50 µS cm^{-1}. A simple analysis of mass balance indicates that 75% of DOC can be recovered using the RO/ED process while removing 97% of water and 99.997% of conductivity. For reference, the conductivities of the final RO/ED samples are one-third (or less) that of the world-average river, based on various published estimates of its chemical composition. At the time of publication of this review, the RO/ED process has achieved the most consistently high yields of marine DOM. Molar C/N ratios have been measured for five of the RO/ED samples, and the average C/N ratio of 17.6 ± 1.3 is very similar to typical C/N ratios in the region of the Atlantic Ocean from which the samples were collected. More advanced methods of characterization (UV/visible spectrophotometry, ^{13}C NMR spectroscopy, ^{1}H NMR spectroscopy, and Fourier transform ion cyclotron resonance spectroscopy) all indicate that the isolated samples of marine DOM are representative of bulk DOM in seawater (Koprivnjak et al., 2009).

Samples isolated using the RO/ED process also contain unacceptably high concentrations of residual sea salts, even though 99.997% of conductivity has been removed from these samples. For five recently isolated samples of marine DOM, the percent carbon in the isolated samples is 12% ± 8%, so only around 25% of the isolated sample is DOM, the remainder being residual sea salts and water. Accurate elemental compositions that include H and O cannot be obtained on such samples.

11.4.2.5. Summary of Methods of Isolation of Marine DOM. In summary, the SPE and UF methods isolate a greater fraction of DOC from river and estuarine waters than from seawater. These results indicate that DOM in river and estuarine waters generally has a more hydrophobic nature and a greater colloidal fraction than DOM in seawater. It is also interesting to compare the DOM isolated from surface and deep ocean waters by these two methods. The SPE method isolates a greater fraction of DOC from deep water than from surface water, whereas the UF method isolates a greater fraction of DOC from surface water than from deep water. These results are consistent with observations about the mechanisms of isolation of DOM by SPE and UF, and they indicate that deep water DOM has a greater hydrophobic nature and a smaller colloidal fraction than surface water DOM.

Approximately one-third of marine DOM is isolated by the chemically based SPE methods, and one-third of marine DOM is isolated by the more physically based UF method. The degree to which these two fractions overlap chemically is unknown, but it is at least mathematically possible that they are entirely different fractions of marine DOM. An average of 75% of marine DOM is recovered by the new RO/ED method, so samples collected by the RO/ED method must necessarily include at least a portion of each of the fractions that can be isolated by the SPE and UF methods. Although samples isolated thus far using the RO/ED method contain too much inorganic residue to permit accurate measurements of elemental composition, the method yields the largest and presumably most representative fraction of marine DOM of any current method of isolation.

11.4.3. Elemental Composition

During routine elemental analyses of humic substances, elemental compositions are usually corrected for the presence of water and inorganic matter in the sample; that is, the elemental composition is presented on a dry, ash-free basis. The correction procedure is based on the assumption that all water can be removed by drying samples to constant weight and that the mass of inorganic residue that remains after an elemental analysis is equal to the mass of inorganic matter in the original sample. Both assumptions are likely to be invalid. Huffman and Stuber (1985) thoroughly discussed the problem of obtaining truly dry samples for elemental analysis. Even carefully dried samples rapidly regain water when exposed to normal laboratory air, even during the time required to weigh a sample and prepare it for elemental analysis. For example, the water content of the standard fulvic acids and humic acids of the International Humic Substances Society (IHSS) averages 11.2% ± 4.0%. Huffman and Stuber (1985) recommended that samples should be allowed to equilibrate with the moisture in laboratory air before making an analysis. The equilibrated sample should then be used for an elemental analysis and a separate analysis for H_2O using a Karl Fischer titration. This recommendation is seldom followed, and it is very common for elemental compositions of humic substances to reflect the presence of residual water. The presence of sea salts, some of which can exist as crystalline hydrated minerals, may exacerbate the problem. Further difficulties are presented if inorganic constituents of a sample react with $O_2(g)$ or, even worse, with the $CO_2(g)$ or $H_2O(g)$ that are formed by combustion of the organic matter. For these reasons, reliable elemental analyses can only be obtained on relatively dry, low-ash samples. For instance, the IHSS standard fulvic acids and humic acids, with an average ash content of 1.6% ± 1.3%, have sufficiently low ash content that reliable elemental analyses can be performed.

The dominant elements in marine DOM are expected to be C, H, and O, with lesser quantities of N, S, and P. Concentrations of organic C, N, and P can be measured directly on seawater without isolating DOM, using methods that were described in Section 11.1.2, and those concentrations can be used to calculate molar ratios of C/N, C/P, and N/P. Concentrations of organic C, N, and P cannot be converted into conventional mass-based elemental compositions (%C, %N, and %P), because the total mass of DOM cannot itself be measured directly in seawater. To be able to calculate the mass percentages of all major elements in marine DOM, samples must be isolated and purified to yield dry, low-ash materials.

The average molar C/N ratio (over all depths) for the HOT data set from Station ALOHA is 16.4 ± 2.5. From the compilation of Bronk (2002), the average molar

C/N ratio for 24 surface ocean samples and 18 deep ocean samples are 13.6 ± 2.8 and 14.7 ± 2.8, respectively. These values serve as a point of reference for evaluating the isolation methods that are described in Section 11.4.2. Samples of DOM that were isolated from seawater by SPE using either XAD-2 or XAD-8 resin (Druffel et al., 1992) have an average molar C/N ratio of 46.7 ± 5.8. Likewise, the DOM that was isolated from seawater by SPE C_{18} (Koch et al., 2005) has an average C/N ratio of 35.6. Both data sets confirm the strong bias of chemically based SPE methods against N-containing constituents of marine DOM. In contrast, samples of marine DOM that are isolated from seawater using UF (Benner et al., 1997) have an average molar C/N ratio of 17.1 ± 1.5, and samples isolated from seawater using RO/ED (Vetter et al., 2007; Gurtler et al., 2008; Koprivnjak et al., 2009) have an average molar C/N ratio of 17.6 ± 1.3, both of which are very close to the average value in the HOT data set and only slightly greater than the data from Bronk (2002). These physically based methods of isolation exhibit little or no bias against N-containing constituents of marine DOM.

11.4.3.1. Elemental Analyses that Include H and O. Unlike C, N, and P, organically bound H and O cannot be measured on aqueous solutions of marine DOM, and accurate elemental analyses that include H and O are only possible if freeze-dried DOM samples contain less than around 5 wt% ash (e.g., residual sea salts). Because of the great difficulty in isolating marine DOM from its seawater matrix, published elemental analyses of marine DOM seldom include H and O. This situation is particularly unfortunate, because the bulk elemental composition of DOM provides substantial insight into its average structural composition (Perdue, 1984), reactivity (Ritchie and Perdue, 2008), bioavailability (Sun et al., 1997), and diagenetic state (Reuter and Perdue, 1984). It is appropriate, therefore, to discuss briefly the important ancillary properties of marine DOM that are derived from bulk elemental compositions that include H and O, if for no other reason than to make the case for the development and implementation of much better methods of isolation of representative, low-ash samples of marine DOM.

The average oxidation state of C in marine DOM (Z_C) is determined primarily by the stoichiometric proportions of organically bound H and O, because those elements are much more abundant than N, P, and S in marine DOM.

Given the relatively low abundance of organic S and P, Z_C is approximately given by

$$Z_C = 2(O/C) - (H/C) + 3(N/C) \tag{11.1}$$

where O/C, H/C, and N/C are molar ratios.

Organic compounds containing rings and/or pi bonds have the capacity, in principle, to accept additional H atoms in their molecular formulae. The deficiency of H in such compounds is known as unsaturation or double-bond equivalents. Perdue (1984) has shown that unsaturation (U_{total}) can be calculated rigorously for a complex mixture such as marine DOM, provided that the concentration of organic H is known. Given the relatively low abundance of organic S and P, U_{total} is approximately given by

$$U_{total} = C_{total} + \frac{1}{2}N_{total} - \frac{1}{2}H_{total} + 1000/M_n \tag{11.2}$$

where all elemental concentrations in Eq. (11.2) are in units of $mmol\,g^{-1}$, and M_n is the number-average molecular weight of the sample. U_{total} can be incorporated into a Monte Carlo analysis that yields probabilistic proportions of aliphatic (sp^3-hybridized) carbon, aromatic carbon, and other sp^2-hybridized carbon in complex mixtures of organic compounds. Through such calculations, U_{total} can be used to validate or invalidate peak assignments and areas that are obtained using ^{13}C NMR spectrometry (Perdue, 1984; Wilson et al., 1987; Ritchie and Perdue, 2008). U_{total} has also been used to interpret diagenetic modification of organic matter in marine sediments (Reuter and Perdue, 1984) and to interpret spatial trends in the bioavailability of DOM in a coastal river (Sun et al., 1997). If biodegradation and photodegradation selectively consume or preserve unsaturated compounds within marine DOM, that result will be reflected in U_{total}.

In short, Z_C and U_{total} provide powerful, rigorous constraints on the interpretation of other analytical data from ^{13}C NMR spectrometry or ultra-high-resolution mass spectrometry, which must ultimately be reconciled with the bulk chemistry of a sample of DOM. Calculations of Z_C and U_{total} for bulk samples of DOM are only possible if elemental analyses for H and O are performed. This is a strong incentive for developing methods to more readily obtain suitable dry, ash-free samples of marine DOM.

In all cases where complete elemental analyses have been found, samples were isolated by SPE using XAD resins. As discussed earlier, this method isolates mainly hydrophobic organic compounds from seawater and is strongly biased against N-containing compounds. Polar ionic solutes have little or no affinity for XAD resins. There are perhaps more published results than have been found in this review, but the data in Table 11.4 are hopefully representative of elemental analyses for isolated samples of marine DOM. For reference, two calculated estimates of the bulk chemical composition of marine phytoplankton are included in Table 11.4. All elemental compositions in Table 11.4 are expressed as molar quantities in Redfield format, using an empirical formula that contains 106 moles of C.

Several observations should be mentioned regarding this small data set. First, four of the samples were eluted from XAD resins using NH_4OH or NH_4OH in CH_3OH. These four samples have higher proportions of both N and H than the other 21 samples, all of which were eluted from the XAD resin using NaOH. Stuermer and Harvey (1977) reported that two subsamples that were separately eluted from XAD-2 resin using NaOH and NH_4OH had atomic C/N ratios of 15.1 and 13.4, respectively. The N content of their final isolated samples of marine fulvic acid was therefore increased 12.7% by exposure to NH_4OH. Thorn and Mikita (1992) carefully studied reactions of NH_4OH with several well-characterized humic substances, using ^{15}N NMR to identify N-containing moieties that were formed by these reactions. They reported that the reaction is favored at alkaline pH and that it is sometimes associated with oxidative uptake of O_2. The degree of incorporation of NH_4OH in their experiments was much greater than that measured by Stuermer and Harvey (1977).

The two estimates of the chemical composition of marine phytoplankton provide a point of reference for the isolated samples of marine DOM. Significant trends can be better observed in this admittedly small data set by considering the plots of molar H/C versus O/C and H/C versus N/C in Figure 11.3. It is strikingly clear in Figure 11.3A that the samples isolated using NH_4OH are distinctly enriched in H, and one

TABLE 11.4. Measured Elemental Compositions of Isolated Samples of Marine DOM[a]

Sample	Eluent	C	H	O	N	S	P	Source[b]
XAD-2 FA	NH_4OH	106	171	58	12	0.4	n.d.	1
XAD-2 FA+HA	NH_4OH/CH_3OH	106	186	76	15	0.0	n.d.	2
XAD-2 FA+HA	NH_4OH/CH_3OH	106	187	71	14	0.0	n.d.	2
XAD-2 FA+HA	NH_4OH/CH_3OH	106	256	115	35	15.1	n.d.	2
XAD-2 FA+HA	NaOH	106	127	74	2	n.d.	n.d.	3
XAD-2 FA+HA	NaOH	106	132	83	2	n.d.	n.d.	3
XAD-2 FA+HA	NaOH	106	131	70	2	n.d.	n.d.	3
XAD-2 FA+HA	NaOH	106	134	70	2	n.d.	n.d.	3
XAD-2 FA+HA	NaOH	106	128	70	3	n.d.	n.d.	3
XAD-2 FA+HA	NaOH	106	102	72	2	n.d.	n.d.	3
XAD-2 FA+HA	NaOH	106	125	69	3	n.d.	n.d.	3
XAD-2 FA+HA	NaOH	106	109	70	2	n.d.	n.d.	3
XAD-8 FA+HA	NaOH	106	135	58	2	n.d.	n.d.	3
XAD-8 FA+HA	NaOH	106	131	59	2	n.d.	n.d.	3
XAD-8 HA	NaOH	106	112	46	8	1.2	n.d.	4
XAD-8 FA	NaOH	106	125	50	3	0.5	n.d.	4
XAD-8 FA	NaOH	106	135	51	2	0.3	n.d.	5
XAD-4 acids	NaOH	106	118	85	4	n.d.	n.d.	3
XAD-4 acids	NaOH	106	109	84	5	n.d.	n.d.	3
XAD-4 acids	NaOH	106	106	95	4	n.d.	n.d.	3
XAD-4 acids	NaOH	106	114	87	5	n.d.	n.d.	3
XAD-4 acids	NaOH	106	137	70	6	n.d.	n.d.	3
XAD-4 acids	NaOH	106	134	71	6	n.d.	n.d.	3
XAD-4 acids	NaOH	106	120	64	7	0.8	n.d.	4
XAD-4 acids	NaOH	106	127	60	6	0.5	n.d.	5
Marine biomass	Calculation	106	175	42	16	0.0	1	6
Marine biomass	Calculation/NMR	106	178	40	17	0.3	1	7

[a]Elemental compositions are expressed in Redfield format, using an empirical formula that contains 106 moles of C.

[b]1, Stuermer and Harvey (1974); 2, Dalvi et al. (2000); 3, Druffel et al. (1992); 4, Esteves et al. (2007); 5, Aiken (2007); 6, Anderson (1995); 7, Hedges et al. (2002); modified to incorporate H_3PO_4 as a phosphate monoester: $R–OH + H_3PO_4 = R–O–PO(OH)_2 + H_2O$.

sample has unreasonably high H/C and O/C ratios. The variation of H/C with O/C for the samples extracted with NH_4OH is approximately linear and closely follows the dashed line in Figure 11.3A, which represents the trend due to gain/loss of H_2O. Humic substances are always rather difficult to dry completely, and the problem may be somehow exacerbated in samples that were isolated using NH_4OH. All of the samples that were isolated using NaOH have chemical compositions that plot well below the biomass triangle that is defined by lipids, proteins, and sugars, and the locus of points is displaced down and to the right of the estimated composition of marine phytoplankton. This pattern is reminiscent of the elemental compositions of humic substances and NOM samples from freshwater (Perdue and Ritchie, 2003), and most of the data for humic substances from freshwater are also obtained on samples that were adsorbed on XAD resins and eluted using NaOH.

Even though terrestrially derived DOM is transported continuously from the land to the oceans, substantial evidence supports the hypothesis that nearly all

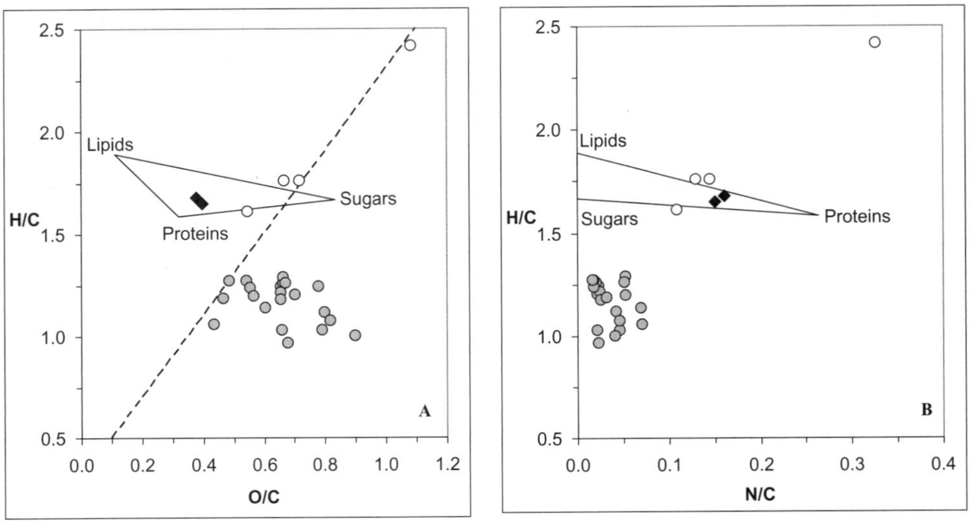

Figure 11.3 Plots of (**A**) molar H/C versus O/C and (**B**) molar H/C versus N/C for major components of biomass (large triangle), marine phytoplankton (black diamonds), marine humic substances eluted with NH_4OH (white circles), marine humic substances eluted with NaOH (gray circles). The dashed line in part A represents the gain/loss of H_2O by a sample.

marine DOM is autochthonous. To quote Hedges et al. (1997), "Either our global budgets and distribution estimates are greatly in error, or both dissolved and particulate organic matter of terrestrial origin suffer rapid and remarkably extensive remineralization at sea." Assuming that this hypothesis is valid, then why should humic substances that have been isolated from freshwater and seawater by SPE using XAD resins have similar H/C and O/C ratios? Shuman (1990) offered a likely explanation: "… that XAD resins select a uniform fraction of the DOM, which gives the illusion of uniform chemical properties, and that the acidic fraction that is not extracted, the so-called 'hydrophilic acid' fraction, carries important and differentiating information that is ignored." As noted earlier, roughly one-third of marine DOM and one-half of freshwater DOM are sufficiently hydrophobic to be isolated using XAD resins. It is chemically reasonable that fractions of DOM from freshwater and seawater that have a comparable degree of hydrophobicity might also have comparable average H/C and O/C molar ratios. It is less clear how humic substances that are formed by diagenesis of terrestrial biomass (which contains a significant percentage of lignin) and marine biomass (which contains no lignin) both yield significant quantities of compositionally similar humic substances. One possible explanation is that these humic substances have a common microbial origin. The chemical imprint of bacteria and fungi is evident in highly decomposed plant litter and soil organic matter (Guggenberger et al., 1999, Amelung et al., 2002; Tremblay and Benner, 2006) as well as in nonliving particulate and dissolved organic matter in the ocean (McCarthy et al., 1998; Ogawa et al., 2001; Benner and Kaiser, 2003; Kaiser and Benner, 2008). It appears that a substantial fraction of the humic substances in disparate environments is derived from the microorganisms that decompose the parent organic matter.

In Figure 11.3B, a strong linear relationship between H/C and N/C is also evident. Again, the four samples isolated using NH_4OH are distinctly enriched in H and N. In addition to the irreversible processes that covalently link some N from NH_4OH to DOM (see earlier discussion), some of the incorporated N may well be the result of inconsistent and incomplete removal of NH_4OH. Removal of all NH_4OH by simply drying or freeze-drying the solution eluted from XAD resin with NH_4OH is only possible if all $NH_4^+ \cdot RCO_2^-$ salts decompose to form R(COOH)(s) and NH_3(g), where R represents any organic moiety to which a carboxyl group is attached. The reaction is thermodynamically less favorable for strongly acidic carboxyl groups, which are relatively abundant in humic substances, so it is possible that some $NH_4^+ \cdot RCO_2^-$ salts remain in the dried samples. One of the authors of this review has used triethylamine ($N(C_2H_5)_3$) to elute humic substances from XAD resins and has found it impossible to remove all of the $N(C_2H_5)_3$ unless cation exchange resins are used to exchange $HN(C_2H_5)_3^+$ for H^+ ion, a step that was not used to remove any exchangeable NH_4^+ from the samples in Table 11.4.

Molar C/N ratios for the humic substances in Table 11.4 that were extracted using NH_4OH range from 3.1 to 9.1, which is well outside the range of measured C/N ratios for DOM in surface (13.6 ± 2.8) and deep (14.7 ± 2.8) seawater (Bronk, 2002). Molar C/N ratios of other types of natural waters (Bronk 2002) are even higher: coastal seawater (17.7 ± 4.3), estuaries (21.1 ± 14.3), and rivers (25.7 ± 12.5). For the remaining samples in Table 11.4, which were isolated using NaOH, 13 samples that were isolated using XAD-2 or XAD-8 resin have an average C/N ratio of 43.9 ± 11.9, which is significantly greater than the values that are usually observed in seawater. Interestingly, the average C/N ratios of samples that did not adsorb on XAD-2 or XAD-8 resin but were isolated subsequently using XAD-4 resin (20.3 ± 3.1) are much closer to C/N values of DOM in seawater.

11.4.3.2. Indirect Estimates of the Elemental Composition of Marine DOM. Fourier transform ion cyclotron resonance (FTICR) mass spectrometry has been used recently to analyze distributions of molecular mass in DOM. The mass resolution of the method is sufficient to assign unique molecular formulae to thousands of peaks in a typical spectrum. Given the molecular formulae of so many peaks, average elemental compositional data can be estimated, although it is firmly established that the method is selective for specific classes of organic compounds, depending on the mode of ionization. Kim et al. (2003) used FTICR-MS (negative ion mode) to characterize a freshwater DOM sample. They calculated an intensity-weighted average H/C of 1.13 and an O/C of 0.34 for the DOM sample, and they noted that the bulk chemical analysis of this sample yielded H/C = 1.19 and O/C = 0.80; thus, although the H/C from FTICR-MS is approximately correct, the O/C is much too low.

This approach has also been attempted by Koch et al. (2005), who used FTICR-MS (positive ion mode) to analyze samples that were isolated by C_{18} SPE from the Weddell Sea in Antarctica. For six samples from depths of 30–4600 m, an average of 1064 ± 245 molecular formulae could be assigned for each sample by assuming that the compounds contained only C, H, and O. Those molecular formulae were used, in turn, to calculate an intensity-weighted average elemental composition for each sample. If the tabulated average elemental compositions and atomic ratios of Koch et al. (2005) for the six samples of marine DOM are themselves averaged, the overall average atomic ratios for their samples of marine DOM are: H/C = 1.27 ± 0.03 and O/C = 0.36 ± 0.01. They also analyzed a standard fulvic acid from the Suwannee River (IHSS ID: 2S101F) and obtained an average H/C = 1.05 and O/C = 0.42. Compared to the bulk average H/C of 0.99 and O/C of 0.62 for this standard material, the weighted-average H/C value from FTICR-MS is close to the expected value; however, the corresponding O/C value is much lower than anticipated. This result is consistent with the observations of Kim et al. (2003) for freshwater DOM. Expressed in the general format of a Redfield formula containing 106 moles of C, the average composition of Koch et al. (2005) is $C_{106}H_{134}O_{38}$. Because molecular formulae for observed peaks were assumed to contain only C, H, and O, the final average chemical composition does not contain N, S, and P.

The fact that the measured average O/C ratios of isolated samples significantly exceed the average O/C ratios that can be calculated from FTICR-MS data leads to two important conclusions. First, bulk average properties of isolated samples of DOM provide a powerful constraint on the interpretation of FTICR-MS data. As noted earlier in Section 11.4.3.1, existing methods of isolation of marine DOM must be improved sufficiently that low-ash samples can be isolated and accurate bulk elemental compositions can be obtained for isolated samples of marine DOM, if for no other reason than to constrain the interpretation of spectroscopic data. Second, even though FTICR-MS provides unprecedented, detailed insight into the chemical compositions of highly complex mixtures such as marine DOM, the results are not quantitative and should not be accepted uncritically.

^{13}C NMR measurements can also be used to estimate the average chemical composition of marine organic matter. Hedges et al. (2002) combined ^{13}C NMR data and molecular formulae of major building blocks of biomass to obtain a more realistic average empirical formula for marine biomass (Redfield formula). First, they determined the linear combination of the ^{13}C NMR spectra of protein, lipid, and carbohydrate that most closely fit the ^{13}C NMR spectrum of a sample of marine plankton. The estimated empirical formulae of proteins ($C_{106}H_{168}O_{34}N_{28}S$), lipids ($C_{18}H_{34}O_2$), and carbohydrates ($C_6H_{10}O_5$) were combined with the NMR-derived proportions of the three endmembers to estimate the bulk composition of marine plankton. The improved Redfield formula is $C_{106}H_{177}O_{37}N_{17}S_{0.3}$. Hedges et al. (2002) deliberately omitted P from the formula. Incorporation of H_3PO_4 through formation of a phosphate monoester with accompanying loss of H_2O yields the result tabulated earlier in Table 11.4 ($C_{106}H_{178}O_{40}N_{17}S_{0.3}P$).

An earlier, independent estimate of the chemical composition of marine biomass was given by Anderson (1995). Marine biomass was modeled as a mixture of four endmembers (proteins, $C_{3.83}H_{6.05}O_{1.25}N$; carbohydrates, $C_6H_{10}O_5$; lipids, $C_{40}H_{74}O_5$; and nucleic acids, $C_{9.625}H_{12}O_{6.5}N_{3.75}P$). Only a narrow range of compositions could be generated that met the following imposed constraints: C/N = 106/16, 1–7% of carbon in nucleic acids, and at least 5% of carbon in each of the other three endmembers. The N/P ratio was simply forced to equal 16/1, without actually solving for P from the compositions and mixing fractions of the endmembers (it was assumed that there are other minor sources of P). The "best guess" of Anderson (1995) for the chemical composition of marine biomass is $C_{106}H_{175}O_{42}N_{16}P$ which is in remarkably good agreement with the subsequent NMR-based results of Hedges et al. (2002).

Sannigrahi et al. (2005) analyzed 11 DOM samples and five samples of particulate organic matter (POM: 0.1–60 μm) that were isolated from seawater using tangential-flow ultrafiltration. Even though their samples were definitely not living biomass, they adopted the method of Hedges et al. (2002) with only minor modifications, using an average lipid composition of $C_{18}H_{36}O_2$ and constraining the proportion of protein ("amino acids") so that the actual N content of the sample was not exceeded. They applied the same method to particulate organic matter (POM) and ultrafiltered DOM. All samples were thus resolved into linear combinations of the three classes of biomolecules. Although not given in their paper, the average empirical formulae for POM ($C_{106}H_{181}O_{50}N_{11}S_{0.4}$) and DOM ($C_{106}H_{182}O_{58}N_7S_{0.3}$) can be calculated from the results of Sannigrahi et al. (2005).

^{1}H NMR spectra have been used to analyze 11 DOM samples that were isolated from seawater using tangential-flow ultrafiltration and one sample that was isolated using dialysis (Aluwihare et al., 1997). A linear combination of three endmembers (carbohydrates, lipids, and acetate) was sufficient to account in some detail for the ^{1}H NMR spectra of the samples. The relative proportions of the endmembers in the 12 samples were reported by Aluwihare et al. (1997). These results can be coupled with assumed chemical compositions of the endmembers to calculate the corresponding average chemical composition of isolated samples of marine DOM. Aluwihare et al. (1997) did not attempt this analysis; however, if the compositions of carbohydrates and lipids used by Hedges et al. (2002) and the composition of acetic acid ($C_2H_4O_2$) are coupled with the relative proportions of endmembers that were given in Aluwihare et al. (1997), the average chemical composition of their 12 samples, expressed in the general format of a Redfield formula containing 106 moles of C, is calculated to be $C_{106}H_{182}O_{83}$. The final average chemical composition does not contain N, S, and P, because their endmembers did not contain any of these elements.

Hertkorn et al. (2006) analyzed DOM that was isolated by ultrafiltration from surface and deep waters. They used ^{1}H and ^{13}C NMR spectra and FTICR mass spectra to resolve DOM into three normative components: heteropolysaccharides, carboxyl-rich alicyclic molecules (CRAM), and peptides (accounting for all N). They reported that surface DOM contains 60.1% heteropolysacharides, 22.9% CRAM, and 23.6% peptides and that deep DOM contains 27.1% heteropolysaccharides, 50.9% CRAM, and 20.7% peptides. Unlike Sannigrahi et al. (2005), who gave empirical formulae of the endmembers of the three-component model that was used to analyze their ^{13}C NMR data, Hertkorn et al. (2006) did not provide empirical formulae for heteropolysaccharides, CRAM, and peptides. It is unclear how to calculate average chemical formulae of surface and deep DOM from their results.

The indirect estimates of chemical composition that have been discussed here are presented in Table 11.5 using the Redfield format (an empirical formula of DOM containing 106 moles of C) to facilitate comparisons among these samples and the samples that were isolated using XAD resins (see Table 11.4). The data are also presented as a van Krevelen plot in Figure 11.4. The samples that were isolated by SPE using C_{18} and analyzed by FTICR-MS fall well outside the biomass triangle in Figure 11.4, reminiscent of the humic substances that were isolated by SPE using XAD resins and eluted from the resins using NaOH (see Figure 11.3A). H/C values are comparable; however, the samples isolated using C_{18} have generally much lower O/C ratios than those samples that were isolated using XAD resins. Although the fractions of DOM that can be recovered from seawater by SPE using both XAD resins and C_{18} adsorbents are expected to be relatively enriched in hydrophobic

TABLE 11.5. Calculated Elemental Compositions of Isolated Samples of Marine DOM[a]

Sample	Analysis	C	H	O	N	S	P	Source[b]
C_{18} DOM	FTICR-MS	106	137	39	n.d.	n.d.	n.d.	1
C_{18} DOM	FTICR-MS	106	137	39	n.d.	n.d.	n.d.	1
C_{18} DOM	FTICR-MS	106	133	39	n.d.	n.d.	n.d.	1
C_{18} DOM	FTICR-MS	106	136	37	n.d.	n.d.	n.d.	1
C_{18} DOM	FTICR-MS	106	128	37	n.d.	n.d.	n.d.	1
C_{18} DOM	FTICR-MS	106	137	37	n.d.	n.d.	n.d.	1
UDOM	^{13}C NMR	106	178	67	7	0.2	n.d.	2
UDOM	^{13}C NMR	106	180	65	7	0.2	n.d.	2
UDOM	^{13}C NMR	106	178	66	7	0.3	n.d.	2
UDOM	^{13}C NMR	106	180	62	7	0.3	n.d.	2
UDOM	^{13}C NMR	106	182	58	7	0.3	n.d.	2
UDOM	^{13}C NMR	106	182	55	8	0.3	n.d.	2
UDOM	^{13}C NMR	106	184	54	7	0.3	n.d.	2
UDOM	^{13}C NMR	106	181	60	7	0.3	n.d.	2
UDOM	^{13}C NMR	106	187	47	7	0.3	n.d.	2
UDOM	^{13}C NMR	106	188	48	6	0.2	n.d.	2
UDOM	^{13}C NMR	106	186	51	6	0.2	n.d.	2
UDOM	^{1}H NMR	106	182	87	n.d.	n.d.	n.d.	3
UDOM	^{1}H NMR	106	183	79	n.d.	n.d.	n.d.	3
UDOM	^{1}H NMR	106	184	78	n.d.	n.d.	n.d.	3
UDOM	^{1}H NMR	106	181	82	n.d.	n.d.	n.d.	3
UDOM	^{1}H NMR	106	184	78	n.d.	n.d.	n.d.	3
UDOM	^{1}H NMR	106	183	79	n.d.	n.d.	n.d.	3
UDOM	^{1}H NMR	106	181	87	n.d.	n.d.	n.d.	3
UDOM	^{1}H NMR	106	184	86	n.d.	n.d.	n.d.	3
UDOM	^{1}H NMR	106	182	87	n.d.	n.d.	n.d.	3
UDOM	^{1}H NMR	106	182	86	n.d.	n.d.	n.d.	3
UDOM	^{1}H NMR	106	181	83	n.d.	n.d.	n.d.	3
Dialysis (1 kDa)	^{1}H NMR	106	183	86	n.d.	n.d.	n.d.	3

[a]Elemental compositions are expressed in Redfield format, using an empirical formula that contains 106 moles of C.
[b]1, Koch et al. (2005); 2, Sannigrahi et al. (2005); 3, Aluwihare et al. (1997).

components of DOM, it does not follow that the measured chemical compositions of the isolated samples would be comparable. The chemical compositions of the XAD isolates were analyzed directly by elemental analysis and are expected to be representative of the isolated samples. In contrast, the average chemical compositions of the C_{18} isolates were calculated by averaging the chemical compositions of all the compounds whose chemical formulae could be assigned using FTICR-MS, and the average chemical compositions thus obtained are, at best, representative of those components of the isolated samples that can be ionized and detected by this method. In the examples that were discussed previously in which H/C and O/C were calculated from FTICR-MS data, calculated average H/C ratios were in good agreement with experimental values obtained by elemental analysis of bulk samples, but calculated average O/C ratios were much too low.

All other data in Table 11.5 are for UDOM samples whose elemental compositions are calculated from a linear combination of three endmembers. The biomass

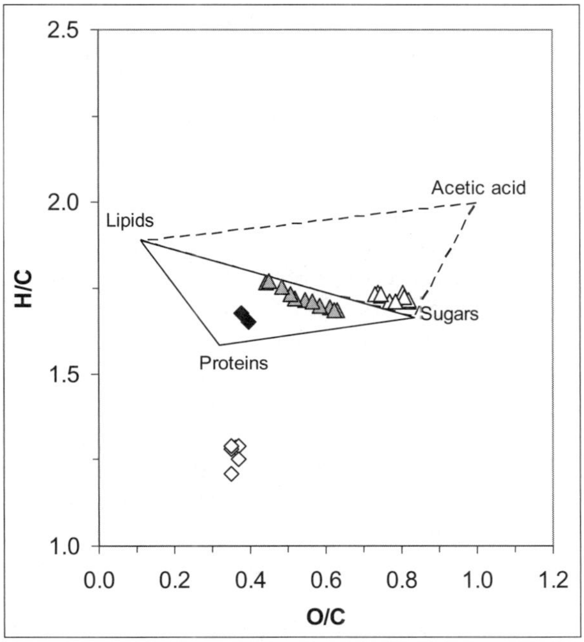

Figure 11.4. Plot of molar H/C versus O/C for major components of biomass (large triangle), marine phytoplankton (black diamonds), and marine DOM that was isolated by SPE using C_{18} and for which chemical composition was calculated from FTICR-MS (white diamonds) and from marine DOM that was isolated by ultrafiltration and for which chemical composition was calculated from ^{13}C NMR spectra (gray triangles) or ^{1}H NMR spectra (white triangles). Chemical compositions that have been obtained from mixing fractions based on ^{1}H NMR spectra (white triangles) were obtained using the assumed compositional triangle given by the dashed lines.

triangle in Figure 11.4 (solid lines) is defined using the compositions for lipids, proteins, and sugars from Hedges et al. (2002). The compositional triangle for the data of Aluwihare et al. (1997) is denoted by the dashed lines in Figure 11.4. The chemical compositions shown in Table 11.5 and Figure 11.4 follow directly from the reported mixing fractions and the assumed chemical compositions of endmembers. In all such calculations, the computed chemical compositions are constrained mathematically to fall inside the triangle that is defined by the endmembers.

Until the problem of isolating in high yield significant quantities of low-ash DOM from seawater is solved, scientists will continue to struggle to obtain accurate bulk chemical compositions for samples of marine DOM. The range of compositions that are presented in Tables 11.4 and 11.5 leads directly to corresponding ranges for the average oxidation state of organic carbon in marine DOM [Z_C in Eq. (11.1)] and the degree of unsaturation of marine DOM [U_{total} in Eq. (11.2)]. Using the data from Tables 11.4 and 11.5, values of Z_C and U_{total} have been calculated for 54 chemical compositions of samples of isolated marine DOM and for two estimated compositions of marine biomass. The results are given in Figure 11.5. The solid black diamonds identify the marine biomass, and the dashed horizontal and vertical lines intersect at the average Z_C and U_{total} of marine biomass.

The data in Figure 11.5 are obviously grouped according to the method of isolation and the method by which the average chemical composition was obtained. In comparison with marine biomass ($Z_C = -0.47\,\text{eq}\,\text{mol}^{-1}$), all samples of Koch et al. (2005) are relatively reduced. As noted earlier, Kim et al. (2003) found that the known average chemical composition of a sample of freshwater DOM was far more oxidized (O/C = 0.80) than the average chemical composition that was derived from FTICR-MS data (O/C = 0.34). The corresponding values of Z_C are +0.41 and −0.45, respectively. Likewise, Koch et al. (2005) found that the known average chemical composition of Suwannee River fulvic acid was far more oxidized (O/C = 0.62) than the average chemical composition that was derived from FTICR-MS data (O/C = 0.36). The corresponding values of Z_C are +0.25 and −0.21, respectively. The six results from Koch et al. (2005) in Table 11.5 are consistent with this pattern, so the apparently reduced state of those samples is probably an artifact. A few samples of Sannigrahi et al. (2005) are also reduced relative to marine biomass. The majority of samples in Tables 11.4 and 11.5 (81%) are oxidized relative to marine biomass. Overall, the data in Figure 11.5 are consistent with a generally oxidative trajectory for diagenesis of marine biomass, although deterministic biases that favor the isolation and/or quantification of more oxidized components of DOM cannot be ruled out.

With the exception of the four samples that were exposed to NH_4OH (see Table 11.4), all samples that were isolated by SPE using either XAD resins or C_{18} adsorbents are more unsaturated (higher U_{total}) than marine biomass ($11.6\,\text{mmol}\,\text{g}^{-1}$). Given that chromophoric and fluorescent components of DOM are preferentially

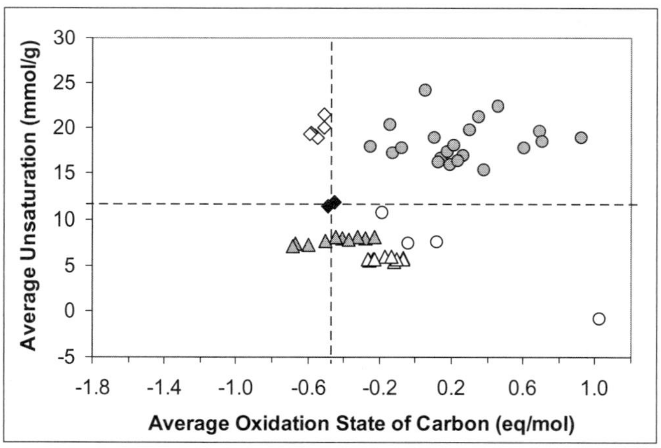

Figure 11.5. The average degree of unsaturation and average oxidation state of carbon in DOM samples that were isolated using XAD resins and eluted using either NH_4OH (white circles) or NaOH (gray circles), samples that were isolated by C_{18} SPE and whose average chemical compositions were calculated from FTICR-MS results (white diamonds), and samples that were isolated by ultrafiltration and whose chemical compositions have been calculated from either ^{13}C NMR results (gray triangles) or ^{1}H NMR results (white triangles). The chemical compositions of marine biomass from Table 11.4 are plotted (black diamonds), and the dashed vertical and horizontal lines intersect at the average chemical composition of marine biomass.

recovered by SPE (see earlier discussion), the greater U_{total} in samples isolated by SPE is probably real. In contrast, all samples that were isolated by UF are less unsaturated (lower U_{total}) than marine biomass, and they are much less unsaturated than samples isolated by SPE. If these samples are actually representative of bulk marine DOM, then unsaturated components of marine biomass are selectively degraded during the production and diagenesis of marine DOM, whether by photodegradation, biodegradation, or some combination of the two (see Section 11.3).

11.5. CONCLUSIONS AND FUTURE DIRECTIONS

After several decades of research, fundamental aspects of the chemical composition and structure of marine organic matter remain elusive. Advances in the chemical characterization of marine organic matter are, in large part, dependent on the development of quantitative methods for its concentration and isolation from seawater. Each of the major methods currently used for the isolation of marine DOM recovers around one-third of the DOM in seawater (solid-phase extractions, using XAD resins or C_{18} adsorbents, and ultrafiltration). A coupled reverse osmosis-electrodialysis method has recently been used to recover an average of 75% ± 12% of marine DOM from 16 seawater samples; however, the method has emerged too recently to have been well tested at this time.

The published literature on soil organic matter and organic matter in freshwaters contains hundreds (or perhaps thousands) of full elemental analyses (CHONSP). Such data provide a direct measure of the average degree of oxidation of organic matter, its average degree of unsaturation (rings and π-bonds), and even the distribution of carbon among aliphatic, aromatic, and other functional classes. The entire data set for marine DOM is derived from a few samples that were isolated from seawater by solid-phase extraction using XAD resins. The dataset can be expanded to some extent using ^{13}C and ^{1}H NMR results and FTICR-MS results, but these approaches are indirect and yield results that cannot be confirmed by direct measurements.

New methods of isolation of marine DOM need to be developed so that significant quantities of low-ash, dry DOM can be routinely isolated in high yield from seawater. Such samples would be suitable for complete elemental analyses, spectroscopic characterization, and quantification of major classes of biomolecules. Even if such methodologies are forthcoming in the very near future, the state-of-the-science regarding marine DOM is not expected to advance beyond the current level in the study of organic matter from soil and fresh waters. In other words, there will be plenty of work remaining for the next generation of scientists.

REFERENCES

Aiken, G. R. (1988). A critical evaluation of the use of macroporous resins for the isolation of aquatic humic substances. In Christman, R., and Gjessing, E. T., eds., *Humic Substances and Their Role in the Environment*, Dahlem Konferenzen, John Wiley & Sons, Chichester, pp. 15–30.

Aiken, G. R. (1992). Chloride interference in the analysis of dissolved organic carbon by the wet oxidation method. *Environ. Sci. Technol.* **26**, 2435–2439.

Aiken, G. R. (2007). Personal communication.

Aiken, G. R., Thurman, E. M., Malcolm, R. L., and Walton, H. F. (1979). Comparison of XAD macroporous resins for the concentration of fulvic acid from aqueous solution. *Anal. Chem.* **51**, 1799–1803.

Aluwihare, L. I., Repeta, D. J., and Chen, R. F. (1997). A major biopolymeric component to dissolved organic carbon in surface sea water. *Nature* **387**, 166–169.

Amador, J. A., Milne, P. J., Moore, C. A., and Zika, R. G. (1990). Extraction of chromophoric substances from seawater. *Mar. Chem.* **29**, 1–17.

Amelung, W., Lobe, I., and DuPreez, C. C. (2002). Fate of microbial residues in sandy soils of the South African Highveld as influenced by prolonged arable cropping. *Eur. J. Soil Sci.* **53**, 29–35.

Amon, R. M. W., and Benner, R. (1994). Rapid cycling of high-molecular-weight dissolved organic matter in the ocean. *Nature* **369**, 549–552.

Amon, R. M. W., and Benner, R. (1996). Bacterial utilization of different size classes of dissolved organic matter. *Limnol. Oceanogr.* **41**, 41–51.

Anderson, L. A. (1995). On the hydrogen and oxygen content of marine plankton. *Deep-Sea Res.* **42**, 1675–1680.

Armstrong, F. A. J., and Tibbitts, S. (1968). Photochemical combustion of organic matter in sea water for nitrogen, phosphorus, and carbon determination. *J. Mar. Biol. Assoc. UK* **48**, 143–152.

Armstrong, F. A. J., Williams, P. M., and Strickland, J. D. H. (1966). Photo-oxidation of organic matter in seawater by ultraviolet radiation, analytical and other applications. *Nature* **211**, 481–483.

Aspila, K. I., Agemian, H., and Chau, A. S. (1976). A semi-automated method for the determination of inorganic, organic and total phosphate in sediments. *Analyst* **101**, 187–197.

Barber, R. T. (1968). Dissolved organic carbon from deep water resists microbial oxidation. *Nature* **220**, 274–275.

Benitez-Nelson, C. R., O'Neill, L., Kolowith, L. C., Pellechia, P. J., and Thunell, R. C. (2004). Phosphonates and particulate organic phosphorus cycling in an anoxic marine basin. *Limnol. Oceanogr.* **49**, 1593–1604.

Benitez-Nelson, C. R., Madden, L. P. O., Styles, R. M., Thunell, R. C., and Astor, Y. (2007). Inorganic and organic sinking particulate phosphorus fluxes across the oxic/anoxic water column of Cariaco Basin, Venezuela. *Mar. Chem.* **105**, 90–100.

Benner, R. (2002). Chemical composition and reactivity, In *Biogeochemistry of Marine Dissolved Organic Matter*, Hansell, D. A., and Carlson, C. A., eds., Academic Press, New York, pp. 59–90.

Benner, R. (2003). Molecular indicators of the bioavailability of dissolved organic matter. In *Aquatic Ecosystems: Interactivity of Dissolved Organic Matter*, Findlay S. E. G., and Sinsabaugh R. L., Eds., Elsevier Science, Amsterdam, pp. 97–119.

Benner, R., and Biddanda, B. (1998). Photochemical transformations of surface and deep marine dissolved organic matter: Effects on bacterial growth. *Limnol. Oceanogr.* **43**, 1373–1378.

Benner, R., and Kaiser, K. (2003). Abundance of amino sugars and peptidoglycan in marine particulate and dissolved organic matter. *Limnol. Oceanogr.* **48**, 118–128.

Benner, R., and Opsahl, S. (2001). Molecular indicators of the sources and transformations of dissolved organic matter in the Mississippi river plume. *Org. Geochem.* **32**, 597–611.

Benner, R., Louchouarn, P., and Amon, R. M. W. (2005). Terrigenous dissolved organic matter in the Arctic Ocean and its transport to surface and deep waters of the North Atlantic. *Global Biogeochem. Cycles* **19**, GB2025, doi:10.1029/2004GB002398.

Benner, R., Pakulski, J. D., McCarthy, M., Hedges, J. I., and Hatcher, P. G. (1992). Bulk chemical characterization of dissolved organic matter in the ocean. *Science* **255**, 1561–1564.

Benner, R., Biddanda, B., Black, B., and McCarthy, M. (1997). Abundance, size distribution, and stable carbon and nitrogen isotopic compositions of marine organic matter isolated by tangential-flow ultrafiltration. *Mar. Chem.* **57**, 243–263.

Berner, R. A. (1982). Burial of organic carbon and pyrite sulfur in the modern ocean: Its geochemical and environmental significance. *Am. J. Sci.* **282**, 451–473.

Blough, N. V., and Del Vecchio, R. (2002). Chromophoric DOM in the coastal environment, In *Biogeochemistry of Marine Dissolved Organic Matter*, Hansell, D. A., and Carlson, C. A., eds., Academic Press, pp. 455–507.

Bronk, D. A. (2002). Dynamics of DON. In *Biogeochemistry of Marine Dissolved Organic Matter*, Hansell, D. A., and Carlson, C. A., eds., Academic Press, New York, pp. 153–247.

Burdige, D. J. (2005). Burial of terrestrial organic matter in marine sediments: A re-assessment. *Global Biogeochem. Cycles* **19**, GB4011, doi:10.1029/2004GB002368.

Burdige, D. J. (2006). *Geochemistry of Marine Sediments*, Princeton University Press, Princeton, NJ, 630 pp.

Bussmann, I. (1999). Bacterial utilization of humic substances from the Arctic Ocean. *Aquat. Microb. Ecol.* **19**, 37–45.

Chróst, R. J. (1991). Environmental control of the synthesis and activity of aquatic microbial ectoenzymes, In *Microbial Enzymes in Aquatic Environments*, Chróst, R. J., ed., Springer-Verlag, New York, pp. 29–59.

Clark, L. L., Ingall, E. D., and Benner, R. (1998). Marine phosphorus is selectively remineralized. *Nature* **393**, 426.

D'Elia, C. F., Steudler, P. A., and Corwin, N. (1977) Determination of total nitrogen in aqueous samples using persulfate digestion. *Limnol. Oceanogr.* **22**, 760–764.

Dalvi, A. G. I., Al-Rasheed, R., and Javeed, M. A. (2000) Studies on organic foulants in the seawater feed of reverse osmosis plants of SWCC. *Desalination* **132**, 217–232.

Dauwe, B., and Middelburg, J. J. (1998). Amino acids and hexosamines as indicators of organic matter degradation state in North Sea sediments. *Limnol. Oceanogr.* **43**, 782–798.

Dauwe, B., Middelburg, J. J., Herman, P. M. J., and Heip, C. H. R. (1999). Linking diagenetic alterantion of amino acids and bulk organic matter reactivity. *Limnol. Oceanogr.* **44**, 1809–1814.

Davis, J., and Benner, R. (2007). Quantitative estimates of labile and semi-labile DOC in the Artic Ocean: A molecular approach. *Limnol. Oceanogr.* **52**, 2434–2444.

Denman, K. L., Brasseur, G., Chidthaisong, A., Ciais, P., Cox, P. M., Dickinson, R. E., Hauglustaine, D., Heinze, C., Holland, E., Jacob, D., Lohmann, U., Ramachandran, S., da Silva Dias, P. L., Wofsy, S. C., and Zhang, X. (2007). Couplings between changes in the climate system and biogeochemistry, In *Climate Change 2007: The Physical Science Basis. Contribution of Working Group I to the Fourth Assessment Report of the Intergovernmental Panel on Climate Change*, Solomon, S., Qin, D., Manning, M., Chen, Z., Marquis, M., Averyt, K. B., Tignor, M., and Miller, H. L. eds., Cambridge University Press, Cambridge, UK, pp. 499–587.

Dittmar, T., Koch, B., Hertkorn, N., and Kattner, G. (2008). A simple and efficient method for the solid-phase extraction of dissolved organic matter (SPE-DOM) from seawater. *Limnol. Oceanogr. Methods* **6**, 230–235.

Druffel, E. R. M., Williams, P. M., Bauer, J. E., and Ertel, J. R. (1992). Cycling of dissolved and particulate organic matter in the open ocean. *J. Geophys. Res. Oceans* **97**, 15639–15659.

Edmond, J. M., Measures, C., McDuff, R. E., Chan, L. H., Collier, R., Grant, B., Gordon, L. I., and Corliss, J. B. (1979). Ridge crest hydrothermal activity and the balance of the major and minor elements in the ocean—Galapagos data. *Earth Planet. Sci. Lett.* **46**, 1–18.

Eglinton, T. I., and Repeta, D. J. (2004). Organic matter in the contemporary ocean. In *Treatise on Geochemistry*, Vol. 6, Holland, H. D., and Turekian, K. K., eds., Elsevier Pergamon, Oxford, UK, pp. 145–180.

Engbrodt, R., and Kattner, G. (2005). On the biogeochemistry of dissolved carbohydrates in the Greenland Sea (Arctic). *Org. Geochem.* **36**, 937–948.

Esham, E. C., Ye, W., and Moran, M. A. (2000). Identification and characterization of humic substances-degrading bacterial isolates from an estuarine environment. *FEMS Microbiol. Ecology* **34**, 103–111.

Esteves, V. I., Otero, M., Santos, E. B. H., and Duarte, A. C. (2007). Stable carbon isotope ratios of tandem fractionated humic substances from different water bodies. *Org. Geochem.* **38**, 957–966.

Gattuso, J.-P., Frankignoulle, M., and Wollast, R. (1998). Carbon and carbonate metabolism in coastal aquatic ecosystems. *Annu. Rev. Ecol. Syst.* **29**, 405–434.

Guggenberger, G., Frey, S. D., Six, J., Paustian, K., and Elliott, E. T. (1999). Bacterial and fungal cell wall residues in conventional and no-tillage agroecosystems. *Soil Sci. Soc. Am. J.* **63**, 1188–1198.

Guo, L., Santschi, P. H., and Warnken, K. W. (1995). Dynamics of dissolved organic carbon (DOC) in oceanic environments. *Limnol. Oceanogr.* **40**, 1392–1403.

Guo, L., Santschi, P. H., Cifuentes, L. A., Trumbore, S. E., and Southon, J. (1996). Cycling of high-molecular-weight dissolved organic matter in the Middle Atlantic Bight as revealed by carbon isotopic (^{13}C and ^{14}C) signatures. *Limnol. Oceanogr.* **41**, 1242–1252.

Guo, L., Tanaka, N., Schell, D. M., and Santschi, P. H. (2003). *Mar. Ecol. Prog. Ser.* **252**, 51–60.

Gurtler, B. K., Vetter, T. A., Perdue, E. M., Ingall, E., Koprivnjak, J.-F., and Prfromm, P. H. (2008). Combining reverse osmosis and pulsed electrical current electrodialysis for improved recovery of dissolved organic matter from seawater. *J. Mem. Sci.* **323**, 328–336.

Hama, T., Yanagi, K., and Hama, J. (2004). Decreases in molecular weight of photosynthetic products of marine phytoplankton during early diagenesis. *Limnol. Oceanogr.* **49**, 471–481.

Hansell, D. A. (2002). DOC in the global ocean carbon cycle. In *Biogeochemistry of Marine Dissolved Organic Matter*, Hansell, D. A., and Carlson, C. A., eds., Academic Press, New York, pp. 685–715.

Hansell, D. A., and Carlson, C. A. (1998). Deep ocean gradients in the concentration of dissolved organic carbon. *Nature* **395**, 263–266.

Hansell, D. A., and Carlson, C. A., eds. (2002). *Biogeochemistry of Marine Dissolved Organic Matter*, Academic Press, New York, 774 pp.

Hartnett, H. E., Keil, R. G., Hedges, J. I., and Devol, A. H. (1998). Influence of oxygen exposure time on organic carbon preservation in continental margin sediments. *Nature* **391**, 572–574.

Hedges, J. I., and Keil, R. G. (1995). Sedimentary organic matter preservation: An assessment and speculative synthesis. *Mar. Chem.* **49**: 81–115.

Hedges, J. I., and Lee, C., eds. (1993) Measurement of dissolved organic carbon and nitrogen in natural waters, *Mar. Chem.* **41**, 290 pp.

Hedges, J. I., and Oades, J. M. (1997). Comparative organic geochemistries of soils and marine sediments. *Org. Geochem.* **27**, 319–361.

Hedges, J. I., Keil, R., and Benner, R. (1997). What happens to terrestrial organic matter in the ocean? *Org. Geochem.* **27**, 195–212.

Hedges, J. I., Eglinton, G., Hatcher, P. G., Kirchman, D. L., Arnosti, C., Derenne, S., Evershed, R. P., Kögel-Knabner, I., de Leeuw, J. W., Littke, R., Michaelis, W., and Rullkötter, J. (2000). The molecularly-uncharacterized component of nonliving organic matter in natural environments. *Org. Geochem.* **31**, 945–958.

Hedges, J. I., Baldock, J. A., Gelinas, Y., Lee, C., Peterson, M. L., and Wakeham, S. G. (2002). The biochemical and elemental compositions of marine plankton: A NMR perspective. *Mar. Chem.* **78**, 47–63.

Helms, J. R., Stubbins, A., Ritchie, J. D., Minor, E. C., Kieber, D. J., and Mopper, K. (2008). Absorption spectral slopes and slope ratios as indicators of molecular weight, source, and photobleaching of chhromophoric dissolved organic matter, *Limnol. Oceanogr.* **53**, 955–969.

Hernes, P. J., and Benner, R. (2006). Sources and fluxes of terrigenous organic matter in the North Atlantic Ocean and North Atlantic Deep Water. *Mar. Chem.* **100**, 66–79.

Hertkorn, N., Benner, R., Frommberger, M., Schmitt-Kopplin, P., Witt, M., Kaiser, K., Kettrup. A., and Hedges, J. I. (2006). Characterization of a major refractory component of marine dissolved organic matter. *Geochim. Cosmochim. Acta* **70**, 2990–3010.

Hu, C., Muller-Karger, F. E., and Zepp, R. G. (2002) Absorbance, absorption coefficient and apparent quantum yield: A comment on common ambiguity in the use of these optical concepts. *Limnol Oceanogr.* **47**, 1261–1267.

Huffman, E. W. D., and Stuber, H. A. (1985). Analytical methodology for elemental analysis of humic substances. In *Humic Substances in Soil, Sediment, and Water. Geochemistry, Isolation, and Characterization*, Aiken, G. R., McKnight, D. M., Wershaw, R. L., and MacCarthy, P., eds., John Wiley & Sons, New York, pp. 433–455.

Ingalls, A. E., Lee, C., Wakeham, S. G., and Hedges, J. I. (2003). The role of biominerals in the sinking flux and preservation of amino acids in the Southern Ocean along 170 °W. *Deep-Sea Res. II* **50**, 709–734.

Kaiser, K., and Benner, R. (2008). Major bacterial contribution to the ocean reservoir of detrital organic carbon and nitrogen. *Limnol. Oceanogr.* **53**, 99–112.

Karl, D. M., and Björkman, K. M. (2002). Dynamics of DOP. In *Biogeochemistry of Marine Dissolved Organic Matter*, Hansell, D. A., and Carlson, C. A., eds., Academic Press, New York, pp. 249–366.

Karl, D. M., and Tien, G. (1992). MAGIC: A sensitive and precise method for measuring dissolved phosphorus in aquatic environments. *Limnol. Oceanogr.* **37**, 105–116.

Keil, R. G., Tsamakis, E., and Hedges, J. I. (2000). Early diagenesis of particulate amino acids in marine systems. In *Perspectives in Amino Acid and Protein Geochemistry*, Goodfriend, G. A., Collins, M. J., Fogel, M. L., Macko, S. A., and Wehmiller, J. F., eds., Oxford University Press, New York, pp. 69–82.

Kim, S., Kramer, R. W., and Hatcher, P. G. (2003). Graphical method for analysis of ultrahigh-resolution broadband mass spectra of natural organic matter, the Van Krevelen diagram. *Anal. Chem.* **75**, 5336–5344.

Koch, B. P., Witt, M., Engbrodt, R., Dittmar, T., and Kattner, G. (2005). Molecular formulae of marine and terrigenous dissolved organic matter detected by electrospray ionization Fourier transform ion cyclotron resonance mass spectrometry. *Geochim. Cosmochim. Acta* **69**, 3299–3308.

Koike I., Hara, S., Terauchi, K., and Kogure, K. (1990). Role of sub-micrometre particles in the ocean. *Nature* **345**, 242–243.

Koprivnjak, J.-F., Perdue, E. M., and Pfromm, P. H. (2006). Coupling reverse osmosis with electrodialysis to isolate natural organic matter from fresh waters. *Water Res.* **40**, 3385–3392.

Koprivnjak, J.-F., Pfromm, P. H., Ingall, E., Vetter, T. A., Schmitt-Kopplin, P., Hertkorn, N., Frommberger, M., Knicker, H., and Perdue, E. M. (2009). Chemical and spectroscopic characterization of marine dissolved organic matter isolated using coupled reverse osmosis-electrodialysis. *Geochim. Cosmochim. Acta*, in press.

Koroleff, F. (1976) Total and organic nitrogen. In *Methods of Seawater Analysis*, Grasshoff, K., ed., Verlag Chemistry, Weinheim, pp. 167–181.

Lang, S. Q., Butterfield, D. A., Lilley, M. D., Johnson, H. D., and Hedges, J. I. (2006). Dissolved organic carbon in ridge-axis and ridge-flank hydrothermal systems. *Geochim. Cosmochim. Acta* **70**, 3830–3842.

Lee, C., Wakeham, S., and Arnosti, C. (2004). Particulate organic matter in the sea: The composition conundrum. *Ambio* **33**, 565–575.

Leenheer J. A., and Huffman, E. W. D., Jr. (1976) Classification of organic solutes in water by using macroreticular resins. *J. Res. U.S. Geol. Survey* **4**, 737–751.

Loh, A. N., and Bauer, J. E. (2000). Distribution, partitioning, and fluxes of dissolved and particulate organic C, N, and P in the eastern North Pacific and Southern Oceans. *Deep-Sea Res. II* **47**, 2287–2316.

Loh, A. N., Bauer, J. E., and Druffel, E. R. M. (2004). Variable ageing and storage of dissolved organic components in the open ocean. *Nature* **430**, 878–881.

Louchouarn, P., Opsahl, S., and Benner, R. (2000). Isolation and quantification of dissolved lignin from natural waters using solid-phase extraction and GC/MS. *Anal. Chem.* **72**, 2780–2787.

Mannino, A., and Harvey, H. R. (1999). Lipid composition in particulate and dissolved organic matter in the Delaware Estuary: Sources and diagenetic patterns. *Limnol. Oceanogr.* **45**, 775–788.

Mannino, A., and Harvey, H. R. (2000). Biochemical composition of particles and dissolved organic matter along an estuarine gradient: Sources and implications for DOM reactivity. *Limnol. Oceanogr.* **45**, 775–788.

Mantoura, R. F. C., and Riley, J. P. (1975) The analytical concentration of humic substances from natural waters. *Anal. Chim. Acta* **76**, 97–106.

Mayer, L. M. (1994). Surface area control of organic carbon accumulation in continental shelf sediments. *Geochim. Cosmochim. Acta* **58**, 1271–1284.

McCarthy, M., Hedges, J. I., and Benner, R. (1998). Major bacterial contribution to marine dissolved organic nitrogen. *Science* **281**, 231–234.

Menzel, D. W., and Vaccaro, R. F. (1964) The measurement of dissolved organic and particulate carbon in seawater. *Limnol. Oceanogr.* **9**, 138–142.

Meybeck, M. (1982). Carbon, nitrogen, and phosphorous transport by world rivers. *Am. J. Sci.* **282**, 401–450.

Miller, W. L., and Moran, M. A. (1997). Interaction of photochemical and microbial processes in the degradation of refractory dissolved organic matter from a coastal marine environment. *Limnol. Oceanogr.* **42**, 1317–1324.

Miller, W. L., and Zepp, R. G. (1995). Photochemical production of dissolved inorganic carbon from terrestrial organic matter: Significance to the oceanic organic carbon cycle. *Geophys. Res. Lett.* **22**, 417–420.

Mills, G. L., and Quinn, J. G. (1981). Isolation of dissolved organic matter and dissolved copper–organic complexes from estuarine waters using reverse-phase liquid chromatography. *Mar. Chem.* **10**, 93–102.

Minor, E. C., Simjouw, J.-P., Boon, J. J., Kerkhoff, A. E., and van der Horst, J. (2002). Estuarine/marine UDOM as characterized by size-exclusion chromatography and organic mass spectrometry. *Mar. Chem.* **78**, 75–102.

Mopper, K., and Kieber, D. J. (2002). Photochemistry and the cycling of carbon, sulfur, nitrogen and phosphorous, In *Biogeochemistry of Marine Dissolved Organic Matter*, Hansell, D. A., and Carlson, C. A., eds., Academic Press, New York, pp. 455–507.

Mopper, K., Zhou, X., Kieber, R. J., Kieber, D. J., Sikorski, R. J., and Jones, R. D. (1991). Photochemical degradation of dissolved organic carbon and its impact on the oceanic carbon cycle. *Nature* **353**, 60–62.

Mopper, K., Stubbins, A., Ritchie, J. D., Bialk, H. M., and Hatcher, P. G. (2007). Advanced instrumental approaches for characterization of marine dissolved organic matter: Extraction techniques, mass spectrometry, and nuclear magnetic resonance spectroscopy. *Chem. Rev.* **107**, 419–442.

Murphy, J., and Riley, J. P. (1962). A modified single solution method for the determination of phosphate in natural waters. *Anal. Chim. Acta* **27**, 31–36.

Nagata, T. (2000). Production mechanisms of dissolved organic matter. In *Microbial Ecology of the Oceans*, Kirchman, D. L., ed., John Wiley & Sons, New York, pp. 121–152.

Nelson, N. B., Siegel, D. A., and Michaels, A. F. (1998). Seasonal dynamics of colored dissolved organic material in the Sargasso Sea. *Deep-Sea Res. I* **45**, 931–957.

Nelson, N. B., and Siegel, D. A. (2002). Chromophoric DOM in the open ocean, In *Biogeochemistry of Marine Dissolved Organic Matter*, Hansell, D. A., and Carlson, C. A., eds., Academic Press, New York, pp. 547–578.

Némery, J., and Garnier, J. (2007). Typical features of particulate phosphorus in the Seine estuary (France). *Hydrobiologia* **588**, 271–290.

Obernosterer, I., and Herndl, G. J. (2000). Differences in the optical and biological reactivity of the humic and nonhumic dissolved organic carbon component in two contrasting coastal marine environments. *Limnol. Oceanogr.* **45**, 1120–1129.

Obernosterer, I., and Benner, R. (2004). Competition between biological and photochemical processes in the mineralization of dissolved organic carbon. *Limnol. Oceanogr.* **49**, 117–124.

Ogawa, H., and Ogura, N. (1992). Comparison of two methods for measuring dissolved organic carbon in sea water. *Nature* **356**, 696–698.

Ogawa, H., Amagai, Y., Koike, I., Kaiser, K., and Benner, R. (2001). Production of refractory dissolved organic matter by bacteria. *Science* **292**, 917–920.

Opsahl, S., and Benner, R. (1997). Distribution and cycling of terrigenous dissolved organic matter in the ocean. *Nature* **386**, 480–482.

Perdue E. M. (1984) Analytical constraints on structural features of humic substances. *Geochim. Cosmochim. Acta* **48**, 1435–1442.

Perdue, E. M., and Ritchie, J. D. (2003). Dissolved organic matter in fresh waters. In *Surface and Ground Water, Weathering, Erosion and Soils*, Drever, J. I., ed., Treatise on Geochemistry, Vol. **5**, Holland, H. D., and Turekian, K. K., eds., Elsevier-Pergamon, Oxford, UK. pp. 273–318.

Reuter, J. H., and Perdue, E. M. (1984). A chemical structural model of early diagenesis of sedimentary humus protokerogens. In *Mitteilungen aus dem Geologisch-Paläontologischen Institut der Universität Hamburg*, Heft 56, Degens, E. T., Krumbein, W. E., and Prashnowsky, A. A., eds., Hamburg, pp. 249–262.

Ritchie, J. D., and Perdue, E. M. (2008). Analytical constraints on acidic functional groups in humic substances. *Org. Geochem.* **39**, 783–799.

Romankevich, E. A. (1984). *Geochemistry of Organic Matter in the Ocean*, Springer-Verlag, Berlin, **344** pp.

Sampei, Y., and Matsumoto, E. (2001). C/N ratios in a sediment core from Nakaumi Lagoon, southwest Japan—Usefulness as an organic source indicator. *Geochem. J.* **35**, 189–205.

Sannigrahi, P., Ingall, E. D., and Benner, R. (2005). Cycling of dissolved and particulate organic matter at station Aloha: Insights from ^{13}C NMR spectroscopy coupled with elemental, isotopic and molecular analyses. *Deep-Sea Res. I* **52**, 1429–1444.

Schubert, C. J., and Calvert, S. E. (2001). Nitrogen and carbon isotopic composition of marine and terrestrial organic matter in Arctic Ocean sediments: Implications for nutrient utilization and organic matter composition. *Deep-Sea Res. I* **48**, 789–810.

Sharp, J. H. (1973) Total organic carbon in seawater—Comparison of measurements using persulfate oxidation and high temperature combustion. *Mar. Chem.* **1**, 211–229.

Shuman, M. S. (1990). Carboxyl acidity of aquatic organic matter: possible systematic errors introduced by XAD extraction. In *Organic Acids in Aquatic Ecosystems*, Perdue, E. M., and Gjessing, E. T., eds., John Wiley & Sons, New York, pp. 97–109.

Simjouw, J.-P., Minor, E. C., and Mopper, K. (2005). Isolation and characterization of estuarine dissolved organic matter: Comparison of ultrafiltration and C_{18} solid-phase extraction techniques. *Mar. Chem.* **96**, 219–235.

Slauenwhite, D. E., and Wangersky, P. J. (1996). Extraction of marine organic matter on XAD-2: Effect of sample acidification and development of an in situ pre-acidification technique. *Mar. Chem.* **54**, 107–117.

Smith, E. M., and Benner, R. (2005). Photochemical transformations of riverine dissolved organic matter: Effects on estuarine bacterial metabolism and nutrient demand. *Aquat. Microb. Ecol.* **40**, 37–50.

Solórzano, L., and Sharp, J. H. (1980a). Determination of total dissolved nitrogen in natural waters. *Limnol. Oceanogr.* **25**, 751–754.

Solórzano, L., and Sharp, J. H. (1980b). Determination of total dissolved phosphorus and particulate phosphorus in natural waters. *Limnol. Oceanogr.* **25**, 754–758.

Stabenau, E. R., and Zika, R. G. (2004). Correlation of the absorption coefficient with a reduction in mean mass for dissolved organic matter in southwest Florida river plumes. *Mar. Chem.* **89**, 55–67.

Stuermer, D. H., and Harvey, G. R. (1974). Humic substances from seawater. *Nature* **250**, 480–481.

Stuermer, D. H., and Harvey, G. R. (1977). The isolation of humic substances and alcohol-soluble organic matter from seawater. *Deep Sea Res.* **24**, 303–309.

Sun, L., Perdue, E. M., Meyer, J. L., and Weis, J. (1997). Use of elemental composition to predict bioavailability of dissolved organic matter in a Georgia river. *Limnol. Oceanogr.* **42**, 714–721.

Suzuki, Y., and Sugimura, Y. (1985) A catalytic oxidation method for the determination of total nitrogen dissolved in seawater. *Mar. Chem.* **16**, 83–97.

Tamburini, C., Garcin, J., and Bianchi, A. (2003). Role of deep-sea bacteria in organic matter mineralization and adaptation to hydrostatic pressure conditions in the NW Mediterranean Sea. *Aquat. Microb. Ecol.* **32**, 209–218.

Tedetti, M., and Sempéré, R. (2006). Penetration of ultraviolet radiation in the marine environment. A review. *Photochem. Photobiol.* **82**, 389–397.

Thomson-Bulldis, A., and Karl, D. M. (1998). Application of a novel method for phosphorus determinations in the oligotrophic North Pacific Ocean. *Limnol. Oceanogr.* **43**, 1565–1577.

Thorn, K. A., and Mikita, M. A. (1992). Ammonia fixation by humic substances: a nitrogen-15 and carbon-13 NMR study. *Sci. Total Environ.* **113**, 67–87.

Tremblay, L., and Benner, R. (2006). Microbial contributions to N-immobilization and organic matter preservation in decaying plant detritus. *Geochim. Cosmochim. Acta* **70**, 133–146.

Valderrama, J. C. (1981). The simultaneous analysis of total nitrogen and total phosphorus in natural waters. *Mar. Chem.* **10**, 109–122.

Vetter, T. A., Perdue, E. M., Ingall, E., Koprivnjak, J.-F., and Pfromm, P. H. (2007). Combining reverse osmosis and electrodialysis for more complete recovery of dissolved organic matter from seawater. *Separation Purif. Technol.* **56**, 383–387.

Wakeham, S. G., Lee, C., Hedges, J. I., Hernes, P. J., and Peterson, M. (1997). Molecular indicators of diagenetic status in marine organic matter. *Geochim. Cosmochim. Acta* **61**, 5363–5369.

Weber J. H., and Wilson S. A. (1975) The isolation and characterization of fulvic acid and humic acid from river water. *Water. Res.* **9**, 1079–1084.

Weiss, M. S., Abele, U., Weckesser, J., Welte, W., Schiltz, E., and Schulz, G. E. (1991). Molecular architecture and electrostatic properties of a bacterial porin. *Science* **254**, 1627–1630.

Willey, J. D., Kieber, R. J., Eyman, M. S., and Avery, G. B., Jr. (2000). Rainwater dissolved organic carbon: Concentrations and global flux. *Global Biogeochem. Cycles* **14**, 139–148.

Wilson, M. A., Vassallo, A. M., Perdue, E. M., and Reuter, J. H. (1987). A compositional and solid state nuclear magnetic resonance study of humic and fulvic acid fractions of soil organic matter. *Anal. Chem.* **59**, 551–558.

Yamashita, Y., and Tanoue, E. (2008). Production of bio-refractory fluorescent dissolved organic matter in the ocean interior. *Nature Geosci.*, doi:10.1038/ngeo279.

Yoshimura, T., Nishioka, J., Saito, H., Takeda, S., Tsuda, A., and Wells, M. (2007). Distributions of particulate and dissolved organic and inorganic phosphorus in North Pacific surface waters. *Mar. Chem.* **103**, 112–121.

Zepp, R. G. (1988). Environmental photoprocesses involving natural organic matter. In *Humic Substances and Their Role in the Environment*, Frimmel, F. H., and Christman, R. F., eds., John Wiley & Sons, New York, pp. 193–214.

12

NATURAL ORGANIC MATTER IN ATMOSPHERIC PARTICLES

A. da Costa Duarte and R. M. B. Oliveira Duarte

Centre for Environmental and Marine Studies (CESAM), Department of Chemistry, University of Aveiro, Aveiro, Portugal

12.1. INTRODUCTION TO ATMOSPHERIC AEROSOLS

Atmospheric aerosols have been studied for a long time. Particularly, and over the last 10–20 years, there has been an increasing interest concerning the role of aerosols on global climate change and public health. Atmospheric aerosols affect the climate by radiative forcing and by altering cloud properties, precipitation efficiency, and ice formation (Penner et al., 2001; Lohmann and Feichter, 2005; Forster et al., 2007; Denman et al., 2007). Aerosol particles may also play a significant role in heteroge-

Biophysico-Chemical Processes Involving Natural Nonliving Organic Matter in Environmental Systems, Edited by Nicola Senesi, Baoshan Xing, and Pan Ming Huang

neous chemical reactions and other multiphase processes, and, therefore, alter the chemical composition of both the gas and aerosol phases in the atmosphere (Andreae and Crutzen, 1997; Raes et al., 2000). Epidemiological studies have also shown that particulate matter can pose adverse health effects, such as respiratory, cardiovascular, allergic and lung cancer diseases (Künzli et al., 2000; Donaldson et al., 2003; Bernstein et al., 2004).

All these climatic and human health effects associated with airborne particulate matter depend on the concentration, size, and chemical composition of particles themselves (Penner et al., 2001; Kaufman et al., 2002). Efforts have been made over the years to elucidate the chemical composition of atmospheric aerosols as a function of size and to achieve a chemical mass closure of aerosols (Krivácsy and Molnár, 1998; Temesi et al., 2001; Maenhaut et al., 2002; Smolík et al., 2003; Sciare et al., 2005). Furthermore, studies have also been conducted on diurnal, seasonal, and geographical variation of the concentrations of main aerosol components (Matsumoto et al., 1998; Molnár et al., 1999; Ruellan and Cachier, 2001; Viidanoja et al., 2002; Putaud et al., 2004; Duarte et al., 2005). While the inorganic component of aerosols has been studied extensively, their organic counterpart is not yet completely understood. This organic component constitutes a major and highly variable fraction of air particulate matter (Andrews et al., 2000; Putaud et al., 2004). However, the multitude of molecular forms, sources, and physical properties makes a complete characterisation of the aerosol organic component extremely difficult, but yet a much more challenging task.

In this chapter, we made an attempt to provide a comprehensive review of the current state-of-the-art on sources, chemical nature, and physical properties of organic aerosols. This review begins with an overview of few basic concepts on atmospheric aerosols, followed by a description of the major constituents of atmospheric aerosols. The sources, transformations, and removal processes of organic aerosols are outlined and followed by an overview of the major environmental and human health issues associated with organic aerosols. The chemical and physical characterization of organic aerosols is then reviewed and is finally followed by a list of uncertainties and suggestions that require further studies.

An aerosol is generally defined as a suspension of solid or liquid particles in a gas, and the most evident example of an aerosol is the air (Horvath, 2000). However, in atmospheric research, the term "aerosol" usually denotes the suspended particles that contain a large proportion of condensed matter, whereas clouds are considered as separate phenomena (Pöschl, 2005).

Depending on their origin, the size range of the atmospheric particles span over five orders of magnitude, from a few nanometers to hundreds of micrometers, and consist of a variety of chemical species (Horvath, 2000; Seinfeld and Pankow, 2003). A few examples are listed in Table 12.1. Several studies covering a wide range of conditions indicate that the airborne particle concentrations are highly variable. The clean Arctic air masses contain 1–1.5 μg of particles per cubic meter (m^3) of air (Hidy and Blanchard, 2005), remote continental and rural air contains up to $20\,\mu g\,m^{-3}$ (Horvath, 2000; Van Dingenen et al., 2004; Hidy and Blanchard, 2005), and in a dust storm event $1500\,\mu g\,m^{-3}$ have been found (Xie et al., 2005). In near-city and urban environments, on average, $20–50\,\mu g\,m^{-3}$ of aerosol mass concentrations have been measured (Van Dingenen et al., 2004), and values higher than $100\,\mu g\,m^{-3}$ can be found in polluted towns (Ruellan and Cachier, 2001; Valavanidis et al., 2006). The

TABLE 12.1. Examples of Particles in the Atmosphere

Diameter	Substance	Origin
~50 μm	Rubber, mineral material, insect fragments	Erosion, tire wear, biologic fragmentation
~10–5 μm	Fog droplets, pollen, bacteria	Condensation of water, biologic origin
	Fly ash, soil material	Combustion, wind erosion, resuspended road dust
~1–0.5 μm	Sea salt, "wet particles"	Sulfate, nitrate grown to solution droplets by uptake of water in a humid environment
	SO_4^{2-}, NO_3^-, organics	Combustion processes, end products of condensation on existing particles
~0.1 μm	Soot, cigarette tar, and ash	Internal combustion engine, smoking
~30 nm	Metallurgic fumes and condensation	Metal processing, primary particles in Diesel engine
~10 nm	Gas to particle conversion	Reactions of precursor gases
~3 nm	SO_2, NO_x; volatile organics, NH_3	Volcanoes, biological processes, anthropogenic activities

Source: Adapted from Horvath (2000).

particle numbers range from $2 \times 10^7\,m^{-3}$ in clean polar air masses to values higher than $10^{11}\,m^{-3}$ in polluted towns (Horvath, 2000).

Besides mass concentration, atmospheric particles are often characterized by their size distribution. Aerosols are typically sized in terms of the *aerodynamic equivalent diameter* (d_{ae}) of the particle, usually expressed in micrometer (μm) or nanometer (nm) (Mark, 1998). Atmospheric particles are usually nonspherical and with unknown density. Therefore, the d_{ae} of a particle is usually defined as the diameter of an equivalent unit density sphere ($\rho = 1\,g\,cm^{-3}$) having the same terminal velocity as the particle in question (Mark, 1998; Seinfeld and Pandis, 1998).

Since atmospheric aerosols comprise particles with a wide range of sizes, it is often convenient to use mathematical models to describe the atmospheric aerosol distribution (Seinfeld and Pandis, 1998). A series of mathematical models have been proposed, of which the *lognormal distribution* has been the most used in atmospheric applications (Seinfeld and Pandis, 1998; Horvath, 2000). Useful discussions of the various aerosol size distribution models are provided by Seinfeld and Pandis (1998) and Jaenicke (1998). In general, atmospheric aerosols size distributions are shown graphically in terms of the volume (or mass) distributions, surface area distributions, or number distributions as a function of particle size (Jaenicke, 1998).

In 1978, Whitby suggested the use of a combination of three lognormal distributions to characterize an urban aerosol (Horvath, 2000). This so-called multimodal size distribution, from which the current nomenclature for aerosol particles has been developed, reflects the diversity of formation mechanisms and the randomness of both particle transformation and removal processes (Seinfeld and Pandis, 1998; Horvath, 2000; Raes et al., 2000). Accordingly, atmospheric particles are classified into three distinct modes:

1. The *nucleation mode* ($d_{ae} < 0.1\,\mu m$) accounts for the majority of particles by number; but because of their small size, these particles rarely account for more than a few percent of the total mass of atmospheric particles. These particles originate from condensation of supersaturated vapors from combustion processes and from the nucleation of atmospheric particles to form fresh particles (Seinfeld and Pandis, 1998; Horvath, 2000).
2. The *accumulation mode* ($0.1 < d_{ae} < 1\,\mu m$); particles included in this mode originate from coagulation of particles in the nucleation mode and from condensation of vapors onto existing particles. These particles usually accounts for a substantial part of the aerosol mass and for most of the aerosol surface area (Seinfeld and Pandis, 1998).
3. The *coarse mode* ($d_{ae} > 1\,\mu m$) include particles generated by mechanical processes and introduced directly into the atmosphere from anthropogenic and natural sources (Horvath, 2000). A few examples include sea spray, erosion, resuspension, and industrial and agricultural activities.

Recently, a fourth mode has been introduced into this nomenclature: It appears that particles with sizes less than $0.1\,\mu m$ consist of two modes, the *nucleation mode*, which includes particles with d_{ae} between 0.01 and $0.03\,\mu m$ representing quite recently formed particles, and the *Aitken mode* containing particles between 0.03 and $0.1\,\mu m$ (Horvath, 2000).

Due to a large range of sizes and different removal processes, airborne particles show different residence times in the lower atmosphere (troposphere). Particles in the nucleation mode ($d_{ae} < 0.03\,\mu m$) require minutes to hours to grow larger than about $0.1\,\mu m$ solely by condensation and coagulation. On the other hand, large particles ($d_{ae} > 100\,\mu m$) only require minutes to hours to settle under gravitational field. Particles with sizes between 0.1 and $1\,\mu m$ have the longest atmospheric lifetimes (days to weeks) since dry removal is very slow, and the condensation and coagulation processes tend to accumulate the aerosol in this size range (Seinfeld and Pandis, 1998; Horvath, 2000; Raes et al., 2000). Accumulation mode aerosols are of utmost importance, not only because they account for a substantial part of the aerosol mass with the longest atmospheric lifetime, but also because they are generally thought to dominate light scattering and constitute the majority of cloud condensation nuclei (CCN) (Penner et al., 2001). It has been suggested, however, that sea salt and soil dust can be important contributors to both light scattering and CCN, thus affecting the radiative balance of the atmosphere (Penner et al., 2001; Forster et al., 2007).

Most of the studies on size-resolved aerosol mass concentrations in areas with different levels of pollution show that particulate matter typically exhibit a bimodal distribution, with most of their mass being found in the submicron size range ($d_{ae} < 1\,\mu m$) and an additional minor mode in the coarse fraction ($1 < d_{ae} < 10\,\mu m$) (Maenhaut et al., 2002; Smolík et al., 2003; Wang et al., 2003; Gajananda et al., 2005; Samara and Voutsa, 2005). However, with instrumentation allowing more precise measurements, the aerosol mass size distribution was found to be multimodal with the preponderance of a fine mode ($d_{ae} \leq 0.2\,\mu m$) and an accumulation mode ($d_{ae} \sim 0.5\,\mu m$), with a minor coarse mode at $d_{ae} \sim 3$–$4\,\mu m$ (Raes et al., 2000; Pillai and Moorthy, 2001; Berner et al., 2004). Traditionally, atmospheric researchers classify airborne particles into three size classes: coarse ($2.5 < d_{ae} < 10\,\mu m$), fine

(d_{ae} < 2.5 µm) and ultrafine (d_{ae} < 0.1 µm), which reflects the concern about the effects of atmospheric particulate matter in public health.

12.2. MAJOR CONSTITUENTS OF ATMOSPHERIC AEROSOLS

Atmospheric particles in the troposphere are composed of a complex mixture of highly water-soluble inorganic salts, insoluble mineral dust, and carbonaceous material (which includes organic compounds plus elemental carbon) (Jacobson et al., 2000). Studies in which the chemical composition has been determined as a function of particle size demonstrate a correlation between the chemical composition and the size mode of atmospheric aerosols (Mészáros et al., 1997; Krivácsy and Molnár, 1998; Alves et al., 2000; Maenhaut et al., 2002; Smolík et al., 2003; Samara and Voutsa, 2005).

A hypothetical aerosol size/composition distribution is shown in Figure 12.1, indicating that crustal materials (e.g., CO_3^{2-}, Si, Al, Fe, Ca, and Mn), sea spray (e.g., Mg, Na, and Cl), and biogenic organic particles (e.g., pollen, spores, and plant fragments) are usually found in the coarse aerosol fraction (2.5 < d_{ae} < 10 µm) (Mészáros et al., 1997; Krivácsy and Molnár, 1998; Matsumoto et al., 1998; Seinfeld and Pandis, 1998; Maenhaut et al., 2002; Smolík et al., 2003). Wind erosion, primary emissions, mechanical disruption, sea spray, and volcanic eruptions all contribute to the concentrations of these species (Seinfeld, 1986; Seinfeld and Pandis, 1998).

Some studies have also shown that carbonaceous material can be responsible for about 1.2–31% of the coarse fraction mass concentration of the atmospheric aerosol (Maenhaut et al., 2002; Hueglin et al., 2005). Nevertheless, the highest concentrations (17–48%) of organic matter and elemental carbon are found predominantly

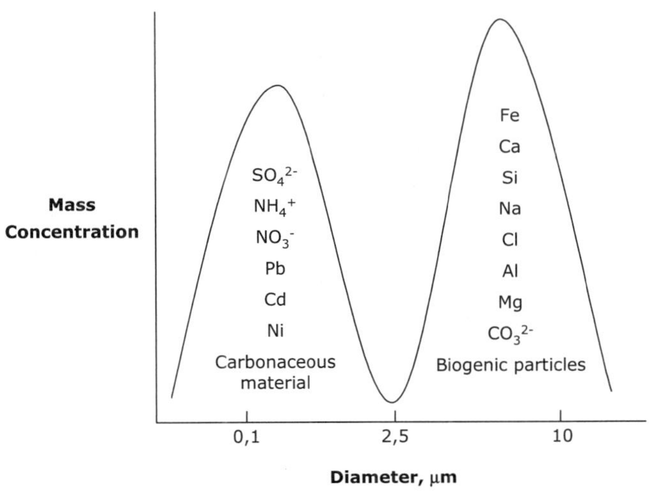

Figure 12.1. Theoretical aerosol mass size distribution profile showing a typical segmentation of chemical species into fine (d_{ae} < 2.5 µm) and coarse (2.5 < d_{ae} < 10 µm) modes. [Adapted from Seinfeld (1986), Seinfeld and Pandis (1998), Krivácsy and Molnár (1998), and Samara and Voutsa (2005).]

in the fine particles ($d_{ae} < 2.5\,\mu m$) (Zappoli et al., 1999; Krivácsy et al., 2001b; Lee and Kang, 2001; Maenhaut et al., 2002; Hueglin et al., 2005).

Other species, present mainly in the fine mode, are sulfate (SO_4^{2-}), ammonium (NH_4^+), and nitrate (NO_3^-) ions. In general, SO_4^{2-} is the main inorganic constituent, accounting for about 4–32% of the fine aerosol mass, while the contribution of NH_4^+ and NO_3^- ions to the total mass of fine particles is only about 2.2–15% and 1.8–16%, respectively (Mészáros et al., 1997; Krivácsy and Molnár, 1998; Zappoli et al., 1999; Lee and Kang, 2001; Krivácsy et al., 2001b; Maenhaut et al., 2002; Wang et al., 2003; Yang et al., 2005). However, the NO_3^- ion can also be found in coarse particles due to the occurrence of chemical reactions between coarse particles and nitric acid, while the presence of NO_3^- in the fine fraction usually results from nitric acid/ammonia reactions leading to the formation of ammonium nitrate (Seinfeld and Pandis, 1998).

Several studies have also shown that an important group of potential toxic elements at trace concentrations, such as Pb, Cd, V, As, Ni, Se, Zn, Ba, and Cr, exhibit mostly a unimodal fine mode size distribution (Mészáros et al., 1997; Maenhaut et al., 2002; Manoli et al., 2002; Hueglin et al., 2005; Lin et al., 2005). In urban areas, these elements arise from anthropogenic sources including traffic, residual oil combustion and industrial processes (Mészáros et al., 1997; Lin et al., 2005). The presence of toxic anthropogenic elements in fine particles enhances the adverse health effects, because smaller particles are inhaled and the deposition efficiency of these particles is highest in the alveolar region (Hughes et al., 1998). On the other hand, a bimodal distribution have been found for K, Zn, V, Ni, and Rb, which reflects both the anthropogenic and natural origin of these elements (Krivácsy and Molnár, 1998; Maenhaut et al., 2002; Smolík et al., 2003). In the particular case of K, its fine mode is most likely predominantly due to emissions from biomass and waste burning (Maenhaut et al., 2002).

Besides different size distributions in the atmospheric aerosol, the spatial and seasonal distributions of trace metals and inorganic ions concentrations can also be highly variable. For example, the seasonal variations due to meteorological conditions are translated into higher concentrations of inorganic ions during colder seasons (Mészáros et al., 1997; Krivácsy and Molnár, 1998). Traffic and industrial-related elements have their highest concentrations near the sources in urban and industrialized areas, which gradually decrease toward clean sites (Hueglin et al., 2005). However, elements with similar concentrations at sites representing different pollution levels indicate that emission sources are either spatially uniformly distributed or no major regional sources exist (Hueglin et al., 2005). Elemental concentrations can also be influenced by major wind patterns (such as speed and direction), suggesting that the travel path of the air mass is also a key factor in the spatial distribution of trace metals concentrations (Schmeling, 2003).

Carbonaceous materials (predominantly found in the fine size mode) and sometimes the dominant fraction of the total fine particle mass (Andrews et al., 2000; Putaud et al., 2004) have been usually classified as: organic carbon (OC), elemental carbon (EC), and inorganic carbon (IC). The latter fraction typically consists of mineral carbonates derived almost exclusively from soil dust (Seinfeld and Pankow, 2003). Since mineral carbonates are commonly discarded from chemical aerosol mass closures, data on total carbon (TC) content of air particulate matter at sites representing different pollution levels refers only to the sum of OC and EC contents. Prior to the determination of TC, carbonates are usually removed by exposing the filters to HCl fumes (Gelencsér, 2004) or by acidification (Zappoli et al., 1999).

A specific subset of aerosol carbon (specifically, the EC) is also sometimes called black carbon (BC). Both EC and BC terms are commonly used interchangeably in the literature to designate the carbon content of the graphite-like material contained in combustion-generated primary carbonaceous aerosols (Pöschl, 2005). However, regarding the OC/EC and OC/BC boundaries, such uniform fraction does not exist, as highlighted in the model of Pöschl (2005) in Figure 12.2. As suggested by the author, there is no real sharp boundary but a continuous decrease of thermochemical refractiveness and specific optical absorption from graphite-like structures to nonrefractive and colorless organic compounds, respectively. Since the EC (or BC) cannot be unambiguously separated from OC by any of the available analytical methods, both OC and EC (or BC) are operationally defined and depend on the analytical technique for the OC/EC (or OC/BC) differentiation (Krivácsy et al., 2001a; Gelencsér, 2004). A very useful and in-depth discussion of the various methods for the determination of the main carbonaceous aerosol components is provided by Gelencsér (2004). At this point, we would like to draw attention to some important remarks when dealing with OC/EC (or OC/BC) differentiation.

The term BC implies that this component is responsible for the absorption of visible light and is used when optical methods are applied for its determination. However, these optical methods are nonspecific and must be calibrated with refractory material of known optical properties (Gelencsér, 2004). When such methods are in use, the OC content is operationally defined as the difference between the TC and BC (OC = TC – BC). The term EC is preferred instead, when the refractory carbon content is determined by thermal and thermo-optical methods (Gelencsér, 2004). In such cases, the OC content is operationally defined as the difference between the TC and EC (OC = TC – EC). Care should be taken, however, when using thermal methods for the OC/EC split. These thermal procedures are prone to positive artifacts by charring (i.e., carbonization) of some organic constituents during heating, which causes a strong bias and overestimation of EC (Gelencsér, 2004). However, thermo-optical methods can correct for the bias by continuous

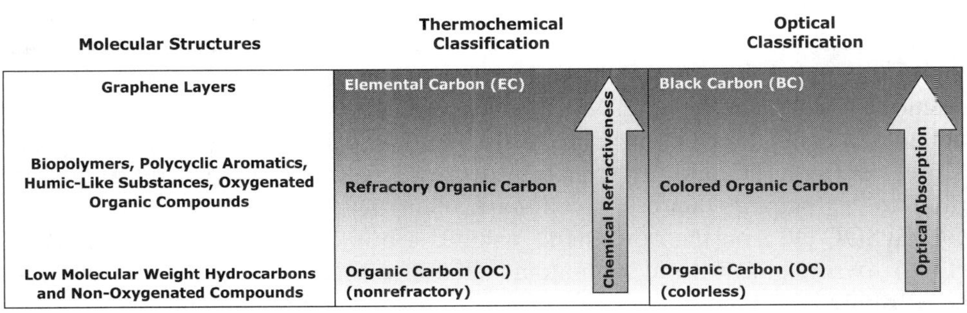

Figure 12.2. Model for molecular structures of black carbon (BC), elemental carbon (EC), and organic carbon (OC) and respective optical and thermochemical classification according to Pöschl (2005).

monitoring the filter blackness with a laser beam (Pio et al., 1993). The correction is accomplished by measuring the fraction of EC oxidized that is necessary to return the reflectance or transmittance of the filter back to the initial value before charring occurs (Gelencsér, 2004).

Nowadays, thermo-optical methods are considered the most reliable measurement techniques for OC/EC split in atmospheric aerosols. Nevertheless, methods for TC/EC/BC analysis in atmospheric particles are still open to debate and their different analytical approaches have been the main cause for performing intercomparison studies (Schmid et al., 2001; ten Brink et al., 2004). The TC measurements showed good agreement, whereas the results of EC/BC determinations were highly variable due to EC overestimation by thermal methods. Furthermore, caution must be taken when using BC as an estimative of the EC content in aerosols, and vice versa: BC and EC measurements are associated to the carbon content of colored and refractory organic compounds, respectively, which can lead to substantially different results between methods (Pöschl, 2005).

From a chemical and morphological point of view, EC/BC fraction can be pictured as more or less disordered stacks of graphene layers or large polycyclic aromatics with a surface coverage by oxygen-containing functional groups and nitrogen species (Smith and Chughtai, 1995; Gelencsér, 2004). In contrast, the OC fraction is composed of thousands of organic compounds, ranging from low-molecular-weight compounds (e.g., malonic and oxalic acids) (Sempéré and Kawamura, 1994) to *n*-alkanes, polycyclic aromatic hydrocarbons, terpens, carbonyls, and *n*-alkanols (Alves et al., 2002). These data have been recently complemented with measurements of water-soluble organic carbon (WSOC), which accounts for a highly variable fraction (12–95%) of the total organics (Sempéré and Kawamura, 1994; Facchini et al., 1999; Zappoli et al., 1999; Decesari et al., 2000, 2001; Krivácsy et al., 2001b; Mayol-Bracero et al., 2002; Kiss et al., 2002; Carvalho et al., 2003; Mader et al., 2004). The WSOC fraction comprises all ionic, polar, and less polar organic compounds which are extractable with water. However, the current knowledge of the chemical nature of this water-soluble organic fraction is far from being complete, as detailed in Section 12.5.2. Nevertheless, the increasing interest on WSOC is fueled by the realization that this fraction may also have and effect on the hygroscopic behavior of atmospheric particles and play an important role in cloud nucleation process (Facchini et al., 1999; Gysel et al., 2004).

Three cases of characteristic concentrations of carbonaceous fractions of fine particulate matter collected in an urban, rural, and remote area are summarized in Table 12.2; more similar data can be found in the literature. The reported data show that, regardless the pollution level, the amount of OC is an important fraction (72–90%) of the aerosol carbonaceous material, which implies the need to obtain a deeper insight into the chemical and physical properties of this fraction. Moreover, both TC and EC contents decrease from urban to rural and from rural to high-alpine samples of air particulate matter. The WSOC fraction of OC (WSOC/OC), on the other hand, exhibits a pronounced increase from urban (13%) to rural (48%) and high-alpine (55%) samples. These observations can be attributed to different aerosol sources (e.g., higher EC content of the urban sample as a result of fossil-fuel combustion or biomass-burning emissions) and also to chemical aging and oxidative transformation of organic aerosol components (Pöschl, 2005).

TABLE 12.2. Average Concentrations of TC, EC, OC, and WSOC in Areas with Different Pollution Levels

Sampling Site	Sampling Time	TC (μg C m^{-3})	EC (μg C m^{-3})	OC (μg C m^{-3})	WSOC (μg C m^{-3})	Reference
Jungfraujoch (high-alpine), $PM_{2.5}$	Jul–Aug 1998	1.3	0.2	1.1	0.6	Krivácsy et al. (2001b)
K-puszta (rural), $PM_{1.5}$	Jun–Aug 1996	5.5	0.6	5.0	2.4	Zappoli et al. (1999)
Paris (urban), $PM_{2.0}$	Aug–Oct 1997	48.2	13.6	34.6	4.5	Ruellan and Cachier (2001)

PM_X: particulate matter with an aerodynamic diameter less than X μm.

12.3. SOURCES, TRANSFORMATION, AND REMOVAL OF ORGANIC AEROSOLS (OAs)

Depending on their sources, organic aerosol (OA) components have been classified as primary or secondary. Primary OAs are directly emitted into the atmosphere in the condensed phase (liquid or solid particles) or as semivolatile vapors, which condense under the atmospheric conditions (Jacobson et al., 2000; Pöschl, 2005). Secondary OAs are developed *in situ* by chemical reactions of gas-phase compounds, or by condensation of gaseous species on existing particles (Jacobson et al., 2000; Seinfeld and Pankow, 2003). Both primary and secondary OAs can be of either natural or anthropogenic origin (Seinfeld and Pankow, 2003). The relative contributions of both primary and secondary sources to ambient OAs depend on the nature and strengths of the local emissions, and on meteorological and atmospheric chemical conditions (Jacobson et al., 2000; Seinfeld and Pankow, 2003). Depending on their origin (which influences their size), atmospheric particles may be limited to the geographic region where they entered the atmosphere or transported over long distances; meanwhile chemical aging can also alter the chemical and physical properties of OAs (Jacobson et al., 2000). EC is generally regarded as coming only from primary emissions, which makes of this aerosol constituent a very suitable tracer for the primary component of atmospheric particulate matter (Seinfeld and Pankow, 2003).

In principle, any process that release particles into the atmosphere is considered as a primary source of OAs. The main primary sources of OAs include open biomass burning and residential wood combustion (Rogge et al., 1998; Simoneit, 2002); fossil fuel combustion (domestic, industry, and traffic) (Hildemann et al., 1991; Rogge et al., 1993e, Penner et al., 2001; Forster et al., 2007); wind-driven or traffic-related suspension of soil and road dust, road, tire, and brake abrasion (Penner et al., 2001; Seinfeld and Pankow, 2003); biogenic materials (plant and animal debris, bacteria, virus, pollen, fungi and spore) (Bauer et al., 2002; Gelencsér, 2004); and wave and bubble breaking in water bodies (Jacobson et al., 2000; Penner et al., 2001).

Secondary OA components are formed from chemical reactions and gas-to-particle conversion of volatile organic compounds (VOCs) in the atmosphere. VOCs are emitted into the atmosphere from natural sources in maritime and terrestrial environments, as well as from anthropogenic sources (Kanakidou et al., 2005). Biogenic secondary OA precursor gases include mono-terpenes (α-pinene, β-pinene, sabinene, and limonene), sesquiterpenes, diterpenes, terpenoid alcohols, *n*-carbonyls, aromatics, and higher-molecular-weight compounds (Kanakidou et al., 2005). Recently, it has been suggested that isoprene, which accounts for about half of all natural VOCs emissions, can contribute to secondary OAs formation (Limbeck et al., 2003; Claeys et al., 2004). Until then, isoprene was generally not considered as a major producer of secondary OAs (Kanakidou et al., 2005). Examples of anthropogenic secondary OA precursor gases include toluene, xylene, trimethylbenzene, and other aromatics (Kanakidou et al., 2005).

A summary of the fine OAs and BC global sources can be derived from the Third Assessment Report of the Intergovernmental Panel on Climate Change (IPCC) (Penner et al., 2001), as shown in Table 12.3. Here we will use the term BC instead of EC since the former is more relevant for climate change studies. Table 12.3 suggests, in spite of the large uncertainties, that biomass burning represents the main source of OAs and that this emission is similar in both hemispheres, while fossil fuel combustion in the Southern Hemisphere is almost negligible. Globally, biogenic VOCs, which are emitted mainly by vegetation, are estimated to be the predominant source of secondary OAs, whereas in urban areas anthropogenic VOCs can be the dominant source. Biomass burning is also a main source of BC in both hemispheres, while emissions by fossil fuel combustion is the main source of BC in the Northern Hemisphere but almost negligible in the Southern Hemisphere. The global annual emission estimates of OC and BC compiled in the 2001 IPCC Report have been recently updated in the inventory paper of Bond et al. (2004). Based on 1996 fuel-use data, the authors suggested that the estimates of global annual emissions of OC

TABLE 12.3. Summary of the Global and Regional Emissions of Fine OAs and BC Expressed as Tg C per Year. Uncertainties, Expressed as Ranges, Are Also Reported for the Global Estimative

Source		Northern Hemisphere	Southern Hemisphere	Global
Organic Aerosols (OAs)	Biomass burning	28	26	54 (45–80)
	Fossil fuel combustion	28	0.4	28 (10–30)
	Biogenic VOCs oxidation	8.2	7.4	16 (8–40)
	Anthropogenic VOCs oxidation	0.45	0.15	0.6 (0.3–1.8)
	Total fine OAs	64	34	98 (60–150)
Black Carbon (BC)	Biomass burning	2.9	2.7	5.7 (5–9)
	Fossil fuel combustion	6.5	0.1	6.6 (6–8)
Total fine aerosol		600	200	800

Adapted from Penner et al., 2001.

and BC from fossil fuels, biofuels, and opening biomass burning are lower than the previous estimates by 25–35%. Nevertheless, this inventory suggests that approximately 74% of OC emissions come from biomass burning sources, whereas fossil fuels and biomass burning sources contribute quite evenly to BC emissions (38% and 42%, respectively) (Bond et al., 2004). In spite of the uncertainties in emission inventories published in 2001 (Table 12.3) and 2004 (Bond et al., 2004), it becomes apparent that OAs play an important role in the atmosphere.

As shown in Figure 12.3 (adapted from Seinfeld and Pankow, 2003), the formation of secondary organic particulate matter may proceed through different pathways. Secondary OAs formation is thought to be initiated with the formation of semivolatile organic compounds (SVOCs), through the oxidation of gaseous species by one of three electrophilic species present in trace amounts in the atmosphere: hydroxyl ($^{\bullet}OH$) and nitrate ($NO_3^{\bullet}$) radicals, and ozone (O_3) (Jacobson et al., 2000; Gelencsér, 2004). The SVOCs may nucleate homogeneously or heterogeneously to form new particles, or condense through adsorption or absorption on preexisting aerosol particles, which generally consists of inorganics, organics, and cloud droplets (Seinfeld and Pankow, 2003; Pöschl, 2005). A third pathway for secondary OAs formation has recently been suggested, consisting of heterogeneous reactions involving the SVOCs or between the SVOCs and the VOCs, as well as oxidation of VOCs in the presence of an acid aerosol catalyst (Limbeck et al., 2003; Gelencsér, 2004; Kanakidou et al., 2005). Regardless of the formation pathway, secondary OAs formation exhibits a strong and nonlinear dependence on temperature, relative humidity, and concentrations of organic and inorganic nucleating and condensing vapors,

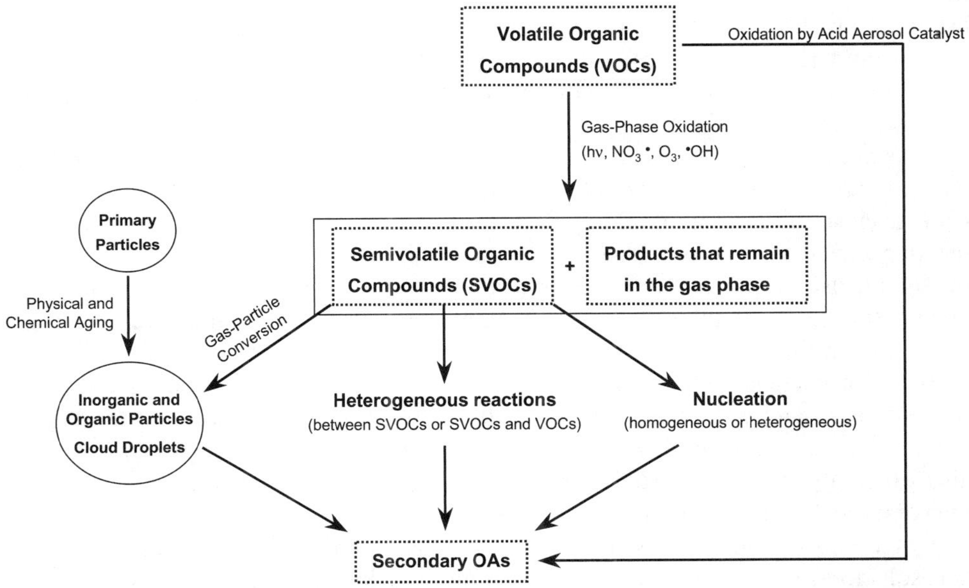

Figure 12.3. Schematics of pathways for the formation of secondary OAs. Adapted from Seinfeld and Pankow (2003).

which depend on atmospheric transport as well as local sources and sinks (Pöschl (2005), and references therein).

Once in the atmosphere, OAs can also undergo a wide range of physical and chemical aging processes under atmospheric conditions. OA components can react with atmospheric photooxidants (e.g., $^{\bullet}OH$, $NO_3^{\bullet}$, and O_3), acids (e.g., H_2SO_4 and HNO_3), water, and UV radiation, forming for instance more polar and hygroscopic products than the precursor material. These atmospheric transformation processes can also occur at the surface layers of BC or EC (Pöschl, 2005).

Chemical aging of OAs are thought to entail oxidation, nitration, hydrolysis, and photolysis transformation of hydrocarbons and derivatives with one or few functional groups into multifunctional hydrocarbons derivatives. The cleavage of organic molecules releases SVOCs, VOCs, CO, and CO_2 from the aerosol phase by volatilization (Pöschl, 2005). On the other hand, oxidative transformations and degradation of biopolymers may lead to the formation of atmospheric polymers with characteristics resembling those of aquatic and soil humic substances. Moreover, condensation and radical-induced oligomer and/or polymer formation following the aerosol-phase photochemistry degradation of organic components can furnish less volatile higher-molecular-weight products and promote the formation of secondary OAs (Pöschl, 2005). However, the exact mechanisms and kinetics of chemical aging processes and their significance for the chemical formation of secondary OAs remain to be understood (Kanakidou et al., 2005; Pöschl, 2005).

Chemical oxidation reactions and radical-induced hydrophobic-to-hydrophilic aging processes tend to increase the water solubility of OAs and, therefore, are thought to enhance the activity of atmospheric OAs as cloud condensation nuclei (CCN). As discussed by Gysel et al. (2004), at 75–90% of relative humidity (RH) the inorganic fraction dominates the water uptake (59–80%). Nevertheless, under the same conditions of RH, between 20% and 40% of total particulate water is associated with water-soluble organic matter. More data concerning the multiphase aerosol and cloud processes, as well as the chemical reactivity of carbonaceous aerosol components, have been compiled in the reviews of Jacobson et al. (2000), Kanakidou et al. (2005), and Pöschl (2005) (and references therein).

OAs in the air are removed from the atmosphere by two main processes: dry and wet (precipitation) deposition. Dry deposition of aerosols is a significant removal process close to the ground (Kanakidou et al., 2005) and the mechanisms involving dry deposition are gravitational sedimentation, impaction on plants, buildings, turbulent diffusion, and diffusion (Horvath, 2000). Gravitational sedimentation has been assumed as the main process that removes coarse OAs (particles larger than 2.5 μm) from the atmosphere, being affected by both particle size and aerosol mass density. For particles smaller than 2.5 μm, where most OA components seem to be present, it is mainly the wet deposition that determines the removal of these particles from the atmosphere (Kanakidou et al., 2005). Falling raindrops, however, can also efficiently remove particles in the range of several micrometers to tens of micrometers by gravitational impaction or washout (Horvath, 2000).

Wet deposition processes depend mostly on microphysical properties of the aerosol, cloud formation, conversion of cloud droplets into rain drops, sedimentation, and evaporation of rain (Kanakidou et al., 2005). Recently, Gysel et al. (2004) discussed the deliquescence properties of water-soluble organic matter and how they affect the hygroscopic growth of atmospheric OAs. Depending on the chemical

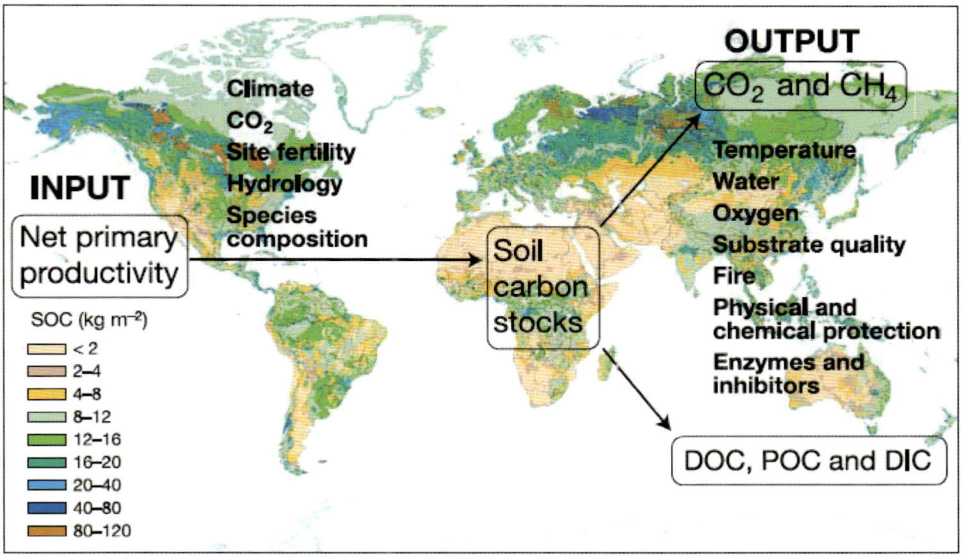

Figure 2.1. Diagram of factors controlling the main inputs and outputs of soil carbon, superimposed over a global map of soil organic carbon stocks DOC, POC, and DIC stand for dissolved organic C, particulate organic C, and dissolved inorganic C, respectively. The background soil organic carbon (SOC) map (Miller Projection; 1:100,000,000). Reprinted from Davidson, E. A., and Janssens, I. A. (2006). Temperature sensitivity of soil carbon decomposition and feedbacks to climate change. *Nature* **440**, 165–173, with permission from Macmillan.

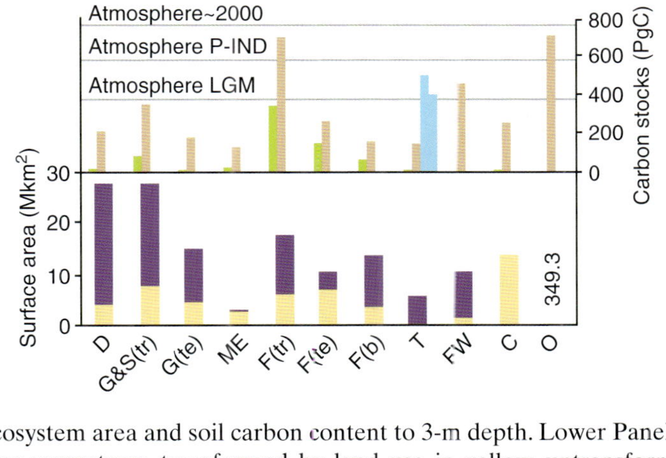

Figure 6.1. Ecosystem area and soil carbon content to 3-m depth. Lower Panel: Global areal extent of major ecosystems, transformed by land use in yellow, untransformed in purple. Data from Hassan et al. (2005) except for Mediterranean-climate ecosystems; transformation impact is from Myers et al. (2000); and ocean surface area is from Hassan et al. (2005). Upper Panel: Total C stores in plant biomass, soil, yedoma/permafrost. D, deserts; G&S(tr), tropical grasslands and savannas; G(te), temperate grasslands; ME, Mediterranean ecosystems; F(tr), tropical forests; F(te), temperate forests; F(b), boreal forests; T, tundra; FW, freshwater lakes and wetlands; C, croplands; O, oceans. Data are from Sabine et al. (2004), except C content of yedoma permafrost and permafrost (light blue columns, left and right, respectively; Zimov et al., 2006), and ocean organic C content (dissolved plus particulate organic; Denman et al., 2007). This figure considers soil C to 3-m depth (Jobbagy and Jackson, 2000). Approximate carbon content of the atmosphere is indicated by the dotted lines for last glacial maximum (LGM), pre-industrial (P-IND) and current (about 2000). Reprinted from Fischlin et al. (2007) in IPCC (2007).

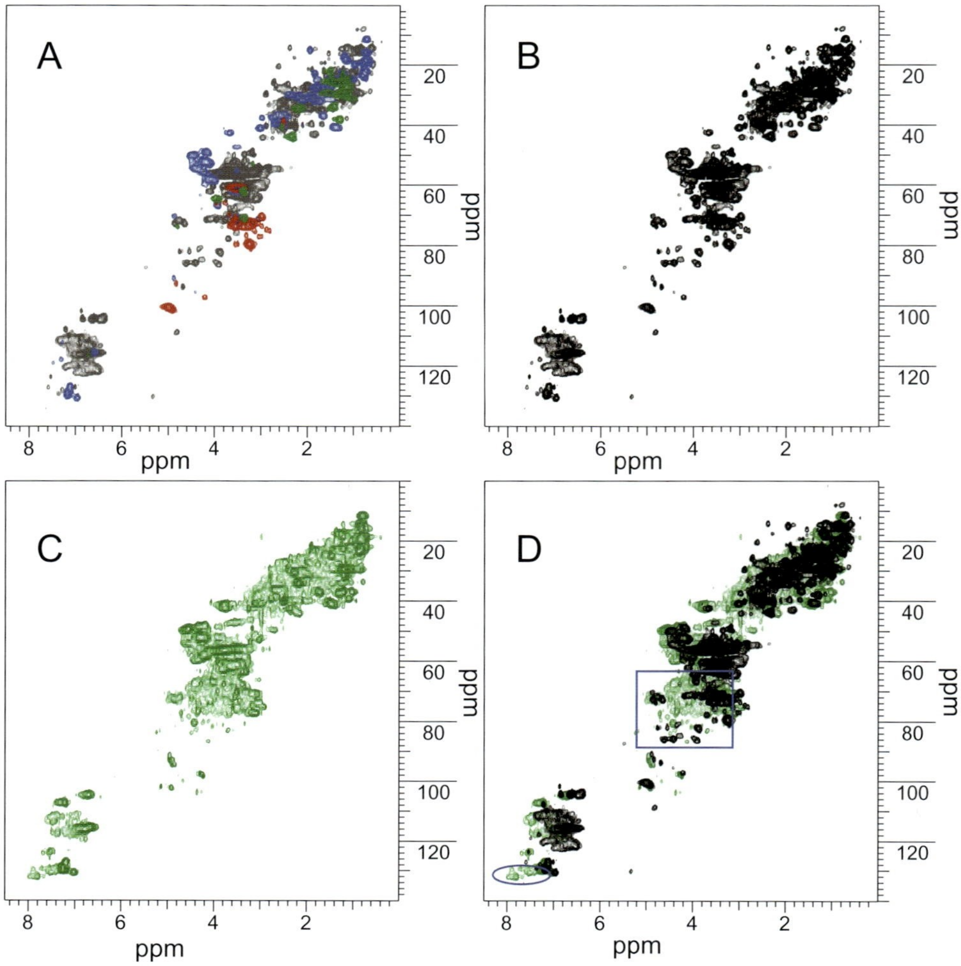

Figure 15.3. Overlaid HSQC spectra of biopolymers on IHSS peat. (**A**) Biopolymers; lignin (gray), amylopectin (red), albumin (blue) and cuticle (green) overlaid on each other. (**B**) All biopolymers are illustrated in black. (**C**) IHSS humic acid extract from peat. (**D**) Biopolymers (black) overlaid on IHSS peat (green). The highlighted areas in 2D are those not well represented by biopolymers in the HA, namely complex carbohydrates and *p*-hydroxybenzoates from lignin [see Kelleher and Simpson (2006) for more details]. Reprinted from Kelleher, B. P., and Simpson, A. J. (2006). Humic substances in soils: Are they really chemically distinct? *Environ. Sci. Technol.* **40**, 4605–4611, with the permission of the American Chemical Society.

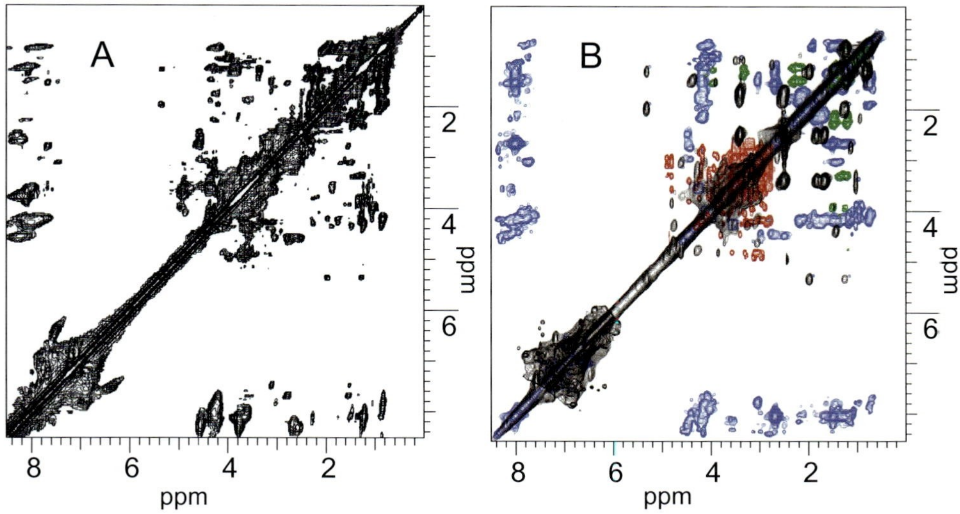

Figure 15.4. TOCSY spectra of the IHSS peat humic acid (**A**) and biopolymers (**B**). The four biopolymers are lignin (gray), albumin (blue), amylopectin (red), and cuticle (green). Reprinted from Kelleher, B. P., and Simpson, A. J. (2006). Humic substances in soils: Are they really chemically distinct? *Environ. Sci. Technol.* **40**, 4605–4611, with the permission of the American Chemical Society.

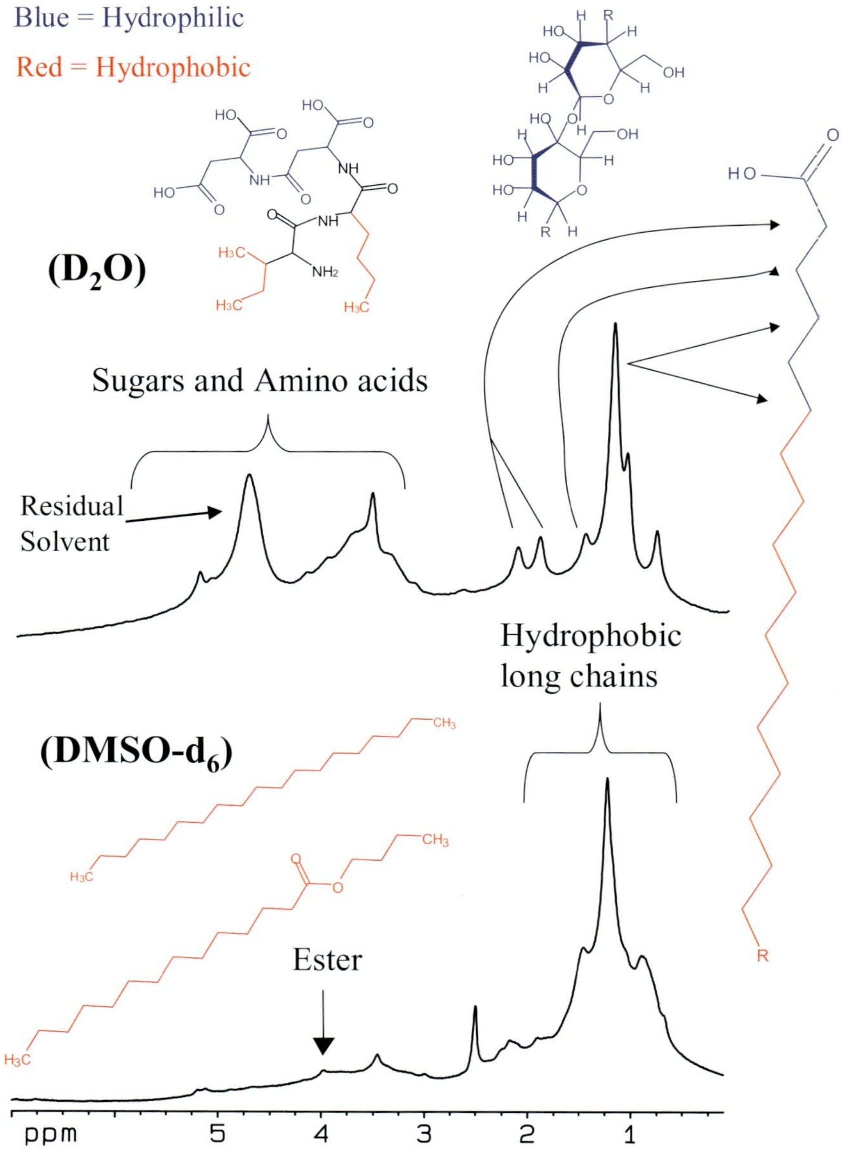

Figure 15.7. ^{1}H HR-MAS NMR of a forest soil. (**Top**) Sampled and analyzed "as is" after the addition of 10 µl of D_2O as a lock signal. Resonances in the top spectrum are those that are in contact with water, and thus at the soil–water interface. (**Bottom**) Same sample as top, but freeze-dried and swollen in DMSO-d_6. Note that DMSO is an excellent swelling solvent and penetrates into both the polar and hydrophobic domains in NOM (Simpson et al., 2001b).

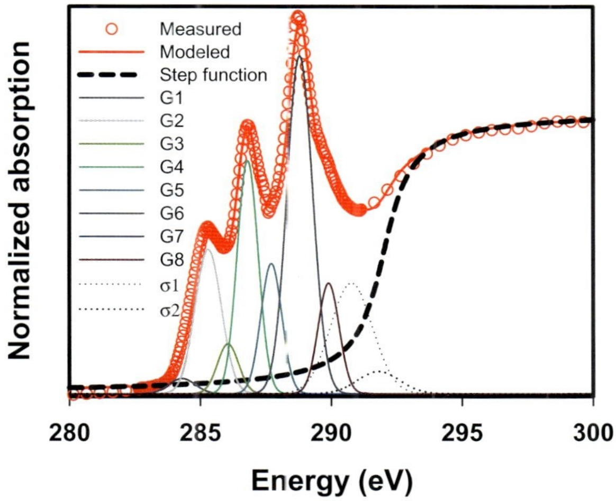

Figure 17.2. Carbon NEXAFS spectrum of NOM from the Suwannee River (IHSS standard humic acid mounted on indium foil; total electron yield using a dwell time of 200 msec and an exit slit of 50 μm, calibrated to CO at 287.38 eV, Canadian Light Source SGM beamline 11-ID.1) to show pre-edge features and the so-called "edge". The spectrum is deconvoluted using a series of Gaussian curves (G) at energy positions of known transitions, along with a step function at the edge as described by Solomon et al. (2005).

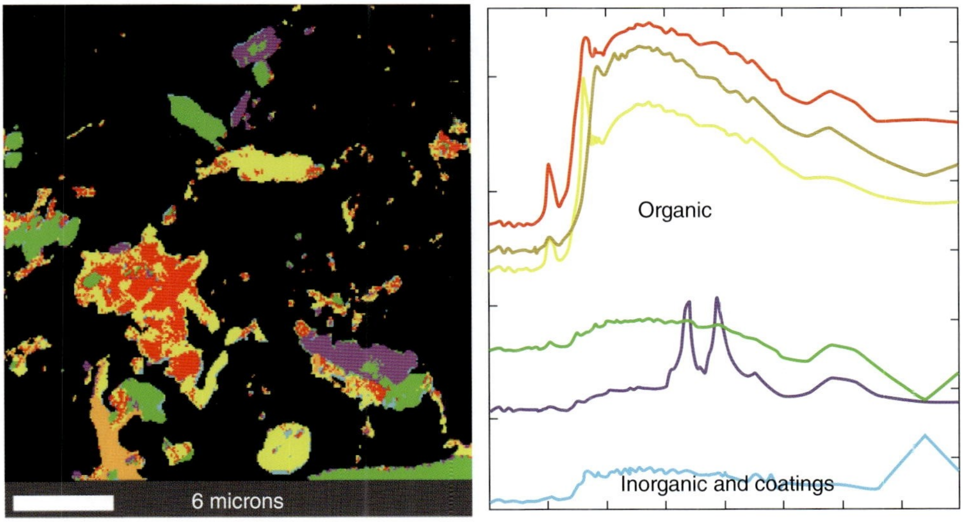

Figure 17.17. Principal component analysis map of sample **(left)** and color-coded spectra **(right)** from a sample of marine suspended particulate matter. The lower three spectra are characteristic of low organic mineral phases, while the upper three organic phases have distinctively different C-NEXAFS spectra. Background regions are shown in black (J. Brandes, unpublished data 2007).

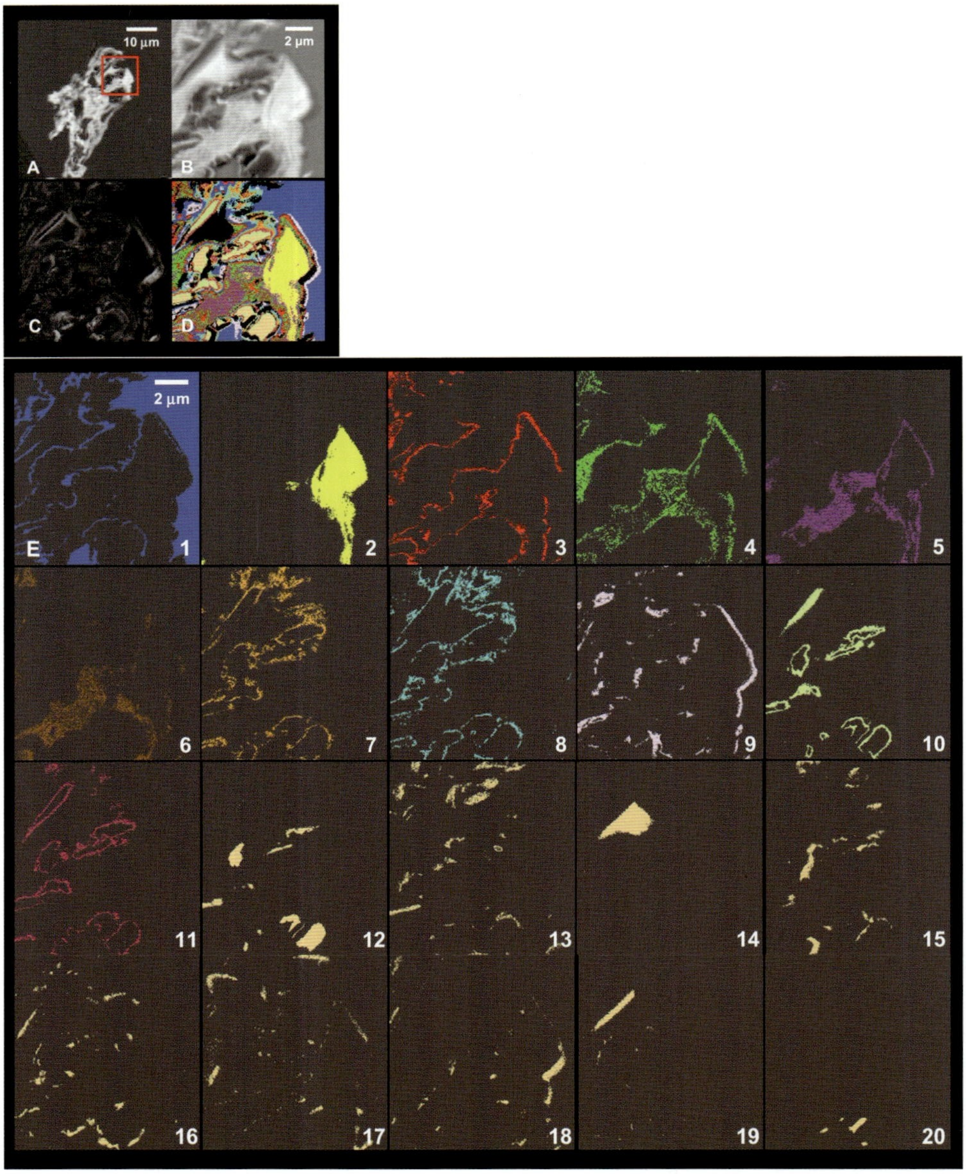

Figure 17.18. **(A)** Carbon contents in an entire microaggregate from an alfisol at Arnot Forest in Upstate New York (500-nm resolution). **(B)** Detail of the microaggregate (red box in A) (50-nm resolution). **(C)** X-ray map of B. **(D)** Cluster map [3 components, 20 clusters, without first component; PCA_GUI 1.0 developed by Lerotic et al. (2004)]. **(E)** Individual clusters from D; numbers in each cluster map correspond to spectra shown in Figure 17.19 (J. Lehmann, unpublished data 2006, measured as described in Lehmann et al., 2007).

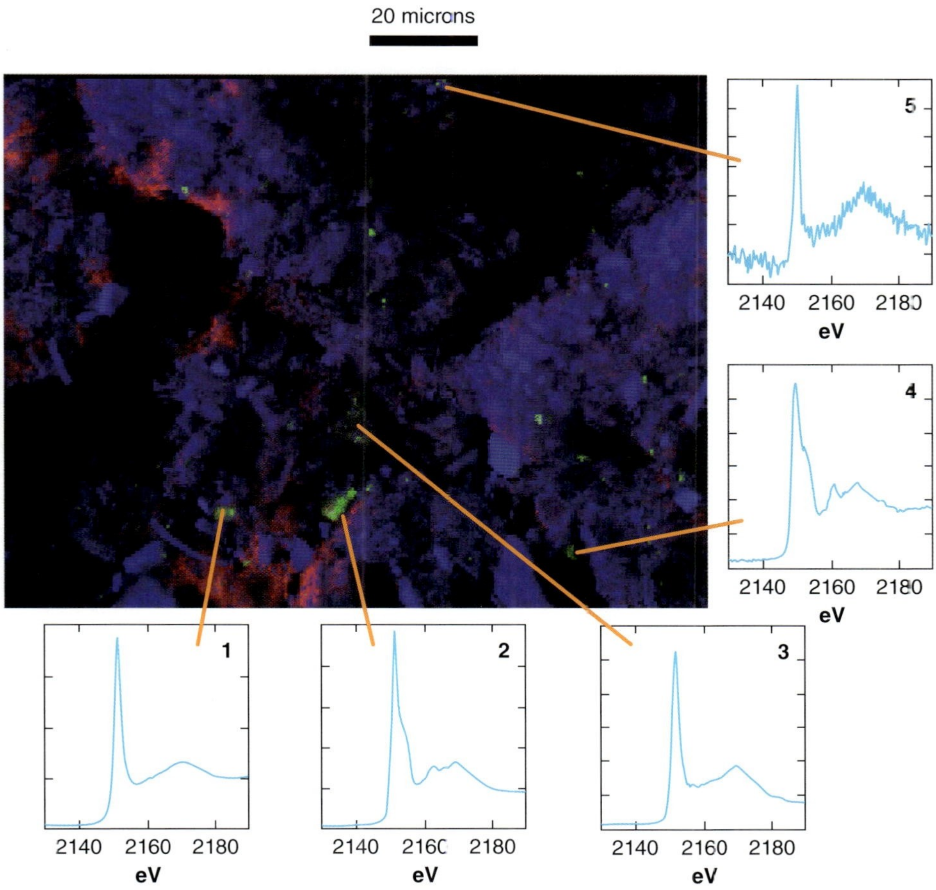

Figure 17.20. Phosphorus covariance map and fluorescence P-NEXAFS spectra collected from a coastal marine sediment (Brandes et al., 2007). Covariance map is color-coded: Green represents P, blue represents Si, and red represents Na fluorescence signals. Regions 1, 3, and 5 have spectra consistent with organic phosphorus or polyphosphate, while regions 2 and 4 closely match calcium phosphate, specifically the mineral apatite.

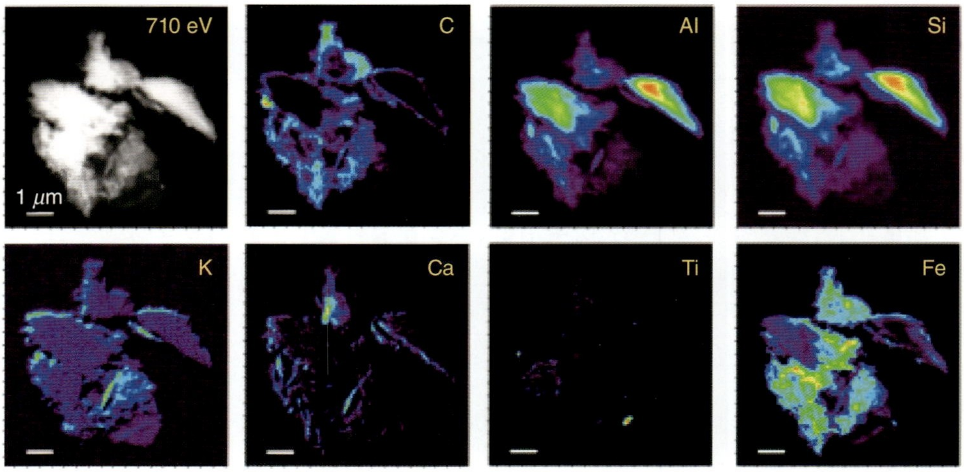

Figure 17.21. Distribution of carbon and mineral elements in an unsectioned micro-assemblage of a cambisol. The outline of the micro-assemblage is shown by the optical density map obtained at 710 eV (Wan et al., 2007). Observe the silicon- and aluminum-rich areas that could not be penetrated by the beam at the carbon edge in this unsectioned sample.

composition of the particles, their hygroscopic behavior may be quite different and thus affect aerosol CCN activity. In particular, organic compounds can introduce competing effects on the activation behavior of CCN by altering the surface tension of the particles, by contributing to the growing particle and affecting its growth kinetics (Kanakidou et al., 2005). Additionally, atmospheric aging of OAs seem also to be a key process determining the wet removal and hence the residence time of carbonaceous particles in the atmosphere. Recently, Stephanou (2005) discussed the potential of •OH-induced oxidative degradation of OAs, which leads to a rapid volatilization of the organic matter, as an important sink for OA removal from the atmosphere. As suggested by Stephanou (2005), the efficiency of this process could be comparable to precipitation in removing OAs from the atmosphere, which implies the need for further studies to improve the knowledge on the reactions and effects of OAs in the environment.

12.4. ORGANIC AEROSOLS: IMPACTS ON CLIMATE AND HUMAN HEALTH

In 2001, the IPCC recognized greenhouse gases (water vapor, CO_2, CH_4, N_2O, O_3, and halogenated hydrocarbons) and aerosols as the main drivers for climate changes (Penner et al., 2001). In the Fourth Assessment Report of the IPCC the importance of the long-lived greenhouse gases and aerosols as forcing agents is reassessed with updated estimates of their contribution to the energy budget changes of the climate system (Forster et al., 2007). However, climate forcing induced by greenhouse gases differs substantial from that of aerosols in several important ways. As recently discussed by Andreae et al. (2005), atmospheric aerosols counteract the warming effects of anthropogenic greenhouse gases by an uncertain, but potentially large, amount. Unlike greenhouse gases, atmospheric aerosols lifetimes in the troposphere are only a week or less (Horvath, 2000), resulting in a heterogeneous spatial and temporal distribution of aerosols with peak concentrations near the sources (Charlson et al., 1992; Ramanathan et al., 2001). Because of their long lifetimes, greenhouse gases are well mixed over the globe, thereby perturbing the global heat balance. Aerosols, on the other hand, have both regional and local impacts on the energy budget, leading to a differential spatial forcing with net heating in some areas and net cooling in others (Penner et al., 1994). Aerosol forcing is greatest in daytime and in summer, whereas greenhouse gases forcing acts over the full and seasonal cycles. Furthermore, aerosol phenomena depend nonlinearly on aerosol concentration, size, and composition. Such differences make a description of the aerosol influences on climate much more complex than a treatment of the radiative influences of greenhouse gases (Charlson et al., 1992). Radiative forcing is defined as the changes in the energy fluxes of solar radiation (maximum intensity in the spectral range of visible light) and terrestrial radiation (maximum intensity in the infrared spectral range) in the atmosphere induced by anthropogenic or natural changes in atmospheric composition, earth surface properties, or solar activity (Pöschl, 2005).

Radiative influences of aerosols on climate are generally distinguished as *direct*, referring to scattering and absorption of radiation by the aerosol particles themselves, and *indirect*, referring to the influence of aerosols on the radiative properties and lifetime of the clouds (Charlson et al., 1992). The direct radiative effect of aerosols is very sensitive to the composition and aerodynamic diameter (d_{ae}) of the particles, as well as to the radiation wavelength (λ) (Penner et al., 2001; Baltensperger et al., 2003). The absorption/scattering efficiency of aerosols is maximum for $\pi d_{ae}/\lambda \sim 1$. Therefore, accumulation mode particles ($d_{ae} = 0.1–1\,\mu m$) are most effective in the absorption and scattering of short-wave (solar, $\lambda \sim 0.4–0.7\,\mu m$) radiation. They also interact with the long-wave (infrared, $\lambda \sim 7\,\mu m$) radiation re-radiated by the earth's surface, but in a much lower degree (Baltensperger et al., 2003).

By intercepting incoming solar radiation, aerosols (e.g., SO_4^{2-}, organics, mineral dust, sea salt) reduce the energy flux arriving at the earth's surface, thus producing a cooling effect (Charlson et al., 1992). On the other hand, aerosols containing black graphitic and tarry carbon strongly absorb incoming sunlight. The effect of this type of aerosol are twofold, both warming the atmosphere and cooling the surface, thus reducing the atmosphere's vertical temperature gradient and causing a decline in evaporation and cloud formation (Kaufman et al., 2002). Studies conducted over the equatorial Indian Ocean during periods of heavy aerosol concentrations, containing sulfates, nitrates, organics, soot and fly ash, show that black graphite carbon warms the lowest 2–4 km of the atmosphere while reducing by 15% the amount of sunlight reaching the surface (Satheesh and Ramanathan, 2000; Ramanathan et al., 2001). Furthermore, the absorption of incoming solar radiation by BC is not only related to its concentration, but also depends on its location in the aerosol particle (Kaufman et al., 2002). Absorption can be two to three times stronger if the BC is located inside the scattering particle (Haywood and Boucher, 2000; Kaufman et al., 2002). With regard to the direct radiative forcing of OC, negative values where found to be associated with the biomass burning and fossil fuel OC (Haywood and Boucher, 2000), thus indicating that the earth–atmosphere system loses radiant energy, resulting in cooling (Kaufman et al., 2002).

Aerosols also indirectly influence climate through their important role as CCN and ice nuclei (IN). The concentration, size, and water solubility of the aerosol particles have an immediate effect on the concentration and size of cloud droplets (Charlson et al., 1992), which in turn affect cloud properties and rainfall generation (Kaufman et al., 2002). The increase in aerosols concentration produces more, but smaller, droplets in a given cloud, making it more reflective and leading to a climate cooling (Haywood and Boucher, 2000; Kaufman et al., 2002). This is called the "first" indirect radiative forcing (also known as Twomey effect). Smaller droplets are less likely to coalesce into raindrops, thus inhibiting precipitation development. This direct microphysical effect leads to an increase in cloud lifetime and in turn in the amount of clouds, which will lead to a further increase in the earth's albedo (Charlson et al., 1992; Ramanathan et al., 2001; Andreae et al., 2005). This is called the "second" indirect radiative forcing. Inhibited precipitation development might further modify the earth's hydrological cycle (Charlson et al., 1992).

However, not every aerosol particle serves as CCN. Accumulation mode aerosols provide the nuclei for most cloud drops (Penner et al., 2001). As in the case of anthropogenic and natural sulphate particles, OAs can also serve as CCN (Ramanathan et al., 2001). Additionally, the presence of water-soluble organic compounds in the particles and the presence of soluble gases (HNO_3) in the atmosphere can amplify the CCN activity of the aerosols and further increase the concentration of cloud droplets and the indirect forcing (Charlson et al., 2001). Also, biomass combustion aerosols can act as CCN (Andreae et al., 2004), mostly due to the presence of water-soluble organic species in smoke particles (Mayol-Bracero et al., 2002).

Several extensive investigations and coordinated field campaigns have been carried out to assess the impact of aerosols on climate. Nevertheless, the current knowledge about aerosol, cloud, and precipitation interactions is still highly uncertain (Lohmann and Feichter, 2005). It is well established that OAs serve as CCN, but their water solubility, hygroscopy, and surface tension properties are still poorly known. The actual climate responses to natural or anthropogenic perturbations are highly uncertain, not being clear whether a perturbation will be reinforced (positive feedback) or diminished (negative feedback) (Pöschl, 2005). More data regarding the radiative effects of natural and anthropogenic aerosols are discussed in the literature. We would like to direct the reader to the works of Haywood and Boucher (2000), Kaufman et al. (2002), Schwartz (2004), Lohmann and Feichter (2005) and Satheesh and Moorthy (2005) on this topic.

Several epidemiological studies show that fine and ultrafine (<0.1 μm) particulate matter and air pollution can pose adverse health effects including respiratory, cardiovascular, allergic, and carcinogenic diseases (Künzli et al., 2000; Donaldson et al., 2003; Bernstein et al., 2004). It appears also that ultrafine particles, after deposition in the lung and gain access to the pulmonary interstitium, can penetrate the systemic circulation and exert more toxicity than coarse and fine particles (Oberdörster, 2001; Bernstein et al., 2004).

Among the parameters and components potentially relevant for aerosol health effects are the particle size, structure, number, mass concentration, solubility, chemical composition, and individual components such as transition metals and organic compounds (Pöschl, 2005). However, the exact mechanisms by which air pollutants cause severe health effects have not yet been clearly identified. Bernstein et al. (2004) suggested a number of possibilities including pulmonary inflammation by particulate matter or ozone, free radical and oxidative stress generated by transition metals or organic compounds (e.g., polycyclic aromatic hydrocarbons (PAHs)), covalent modification of intracellular proteins (e.g., enzymes), inflammation induced by biological compounds such as endotoxins and glucans, and suppression of normal defense mechanisms (e.g., suppression of alveolar macrophage functions) (Bernstein et al., 2004).

12.5. CHEMICAL AND PHYSICAL PROPERTIES OF ORGANIC AEROSOLS

12.5.1. Chemical Characterization of Organic Aerosols (OAs) and Source Apportionment

The organic fraction present in atmospheric particles is a highly complex mixture, which makes the speciation of individual compounds a difficult task. The traditional analytical approach has usually been solvent extraction of aerosol particles collected in a filter followed by gas chromatographic separation coupled to mass spectrometry (GC-MS) detection for individual compound identification and quantification. Although a large number of compounds, sometimes in trace amounts, have been identified with this analytical methodology (see Rogge et al., 1993a), it is very likely that many more compounds did not even elute through the chromatographic column. In particular, a substantial portion of polar oxygenated organic compounds, the most water soluble ones, remains undetected (Facchini et al., 1999).

The prevailing resolved organic compounds by means of the traditional individual compound analysis are *n*-alkanes, fatty acids, aliphatic dicarboxylic acids, aromatic polycarboxylic acids, PAHs, polycyclic aromatic quinones, diterpenoid acids, lignin pyrolysis products, saccharides, levoglucosan and related anhydrosaccharides, and nitrogen-containing compounds. Some of these compounds are considered chemical fingerprints of numerous emissions sources. Most urban/industrial regions produce a similar set of organic compounds associated with atmospheric particulate matter, whereas regional vegetation cover produces distinct signatures and molecular markers (Simoneit, 2002). For example, levoglucosan and related anhydrosaccharides (e.g., mannosan, galactosan, and 1,6-anhydro-β-D-glucofuranose) are degradation products from cellulose and are used as key tracers for smoke particulate matter from biomass burning (Simoneit et al., 1999). PAHs, on the other hand, can be formed by anthropogenic processes (e.g., combustion processes) or be derived from natural sources (e.g., natural wildfires) (Rocha and Duarte, 1997; Rocha et al., 1999; Simoneit, 2002). Additional sources of the above-mentioned organic compounds include meat cooking operations (Rogge et al., 1991), road dust (Rogge et al., 1993b), vegetation (Rogge et al., 1993c), natural gas home appliances (Rogge et al., 1993d), cigarette smoke (Rogge et al., 1994; Kubátová et al., 2002), car exhausts (Rogge et al., 1993e), asphalt (Rogge et al., 1997a), and boilers (Rogge et al., 1997b).

Several compounds were also found to have a seasonal distribution. Kubátová et al. (2002) found that concentrations of lignin and cellulose pyrolysis products from wood burning were higher in aerosol samples collected during low-temperature conditions. On the other hand, concentrations of dicarboxylic acids and related products that are believed to be the oxidation products of hydrocarbons and fatty acids were highest in summer aerosols. PAHs, which are susceptible to atmospheric oxidation, were also more prevalent in winter than in summer. These results suggest that atmospheric oxidation of VOCs into secondary OAs and related oxidative degradation products are key factors in any OA mass closure, source identification, and source apportionment study. However, additional work is much desirable to assess the extent and seasonal variation of these processes.

Studies on detailed organic composition of OAs usually end up with a long list of individual compounds which together account only to few percent (typically ≤10%) of the OC composition (Rogge et al., 1993a; Kubátová et al., 2002). Clearly, only a small fraction of the organic matter is resolved in the form of specific compounds by the traditional individual compound approach (organic solvent extraction followed by GC-MS analysis). Experimental evidences that a highly variable fraction (12–95%) of the total organics is soluble in water have prompted the development of new analytical strategies to assess the chemical nature of this fraction (Facchini et al., 1999; Zappoli et al., 1999; Decesari et al., 2000; Duarte and Duarte, 2005).

Recently, carbon (1*s*) near-edge X-ray absorption fine structure (C (1*s*) NEXAFS) spectroscopy using synchrotron radiation has been applied to carbon (C) speciation and to investigate source apportionment of combustion derived particulate matter, as well as to predict a possible decay path for carboxylic groups in soot by photochemical processes (Braun, 2005; Braun et al., 2006). C (1*s*) NEXAFS spectra of woodsmoke deposits from different fireplaces and of a NIST standard urban particulate matter (SRM 1648) revealed multiple C (1*s*) electron transitions in the fine structure of C NEXAFS region (284–292 eV) indicating the presence of C=C π-bond systems (285 eV and 291 eV) (associated with graphite-like structures), aliphatic C (289 eV), carboxylic (288 eV), hydroxyl (287 eV), and carbonyl (286 eV) groups (Braun, 2005).

The application of C (1*s*) NEXAFS spectroscopy to C speciation in airborne particulate matter is still in its early stages, and the assignment of NEXAFS absorption peaks to particular molecular species is not an easy task. On the other hand, there are also experimental evidences that exposure to radiation can induce reactions and alter the sample, which implies the need to monitor radiation damages by performing more than one scan of a spectrum (Braun, 2005; Braun et al., 2006). Nevertheless, this analytical technique can be used not only to identify and fingerprint structural characteristics of OC but also to simulate the chemical and physical aging of airborne particulate matter (Braun et al., 2006).

12.5.2. Chemistry of Water-Soluble Organic Matter (WSOM)

12.5.2.1. Chemical Characteristics of WSOM. Currently, there is a consensus opinion about the importance of WSOM in particulate organics and how this affects the behavior of aerosol particles in different atmospheric processes. Due to their complexity, however, a good understanding of the chemical composition of the aerosol WSOM is still lacking. This current state-of-the-art significantly contributes to the uncertainties in describing the global radiative effects and public health problems associated with the atmospheric aerosols.

A theoretical study conducted by Saxena and Hildemann (1996) provided a list of candidate multifunctional compounds (e.g., diacids, polyols, amino acids) that may contribute to the WSOC of atmospheric particles. Facchini et al. (1999), using the traditional single compound speciation approach, identified about 120 individual compounds in a polluted area, including aliphatic dicarboxilic acids, sugars, aliphatic alcohols, and aliphatic carboxylic acids. However, these compounds only accounted for less than 5% of the total WSOC. From these results, it follows that the individual organic compounds identified so far cannot characterize the bulk of WSOM, suggesting that there are other classes of compounds whose occurrence has been neglected.

In a previous study conducted in a rural area of Japan (Mukai and Ambe, 1986), a brown substance having the solubility characteristics of humic acid was extracted from airborne particulate matter. Based on infrared and UV–vis spectroscopies plus elemental analysis, the authors suggested that this substance consists of polycyclic ring structures with hydrocarbon side chains, hydroxyl groups, carbonyl, and carboxyl groups. Twelve years later, Havers et al. (1998) found that 10% or more of aerosol OC can be attributed to macromolecular substances with spectroscopic characteristics resembling those of humic and fulvic acids. Since these studies, many different methods have been developed in an attempt to characterize these so-called humic-like substances (HULIS). However, one of the major differences between these studies and those undertaken since then has been the extraction methodology of the macromolecular substances. Havers et al. (1998) applied an aqueous alkaline extraction procedure similar to those commonly applied to extract humic acids from soil samples. The more recent works, on the other hand, have focused mainly on the aerosol WSOC fraction.

By definition, the WSOM present in the aqueous extract of aerosol particles interact with other dissolved species, both organic of low molecular weight and inorganic. Before their chemical and physical properties can be defined thoroughly, the WSOM must be isolated from other compounds, especially from inorganic species since they interfere with the application of advanced analytical techniques such as proton nuclear magnetic resonance (^{1}H-NMR), solid-state ^{13}C nuclear magnetic resonance (^{13}C-NMR) or Fourier transform infrared (FTIR) spectroscopies. Simultaneously, it is of great advantage for any laboratory study of WSOM if the isolation procedure produces unaltered material and ensures a representative mixture of the WSOC from the original atmospheric aerosol.

Accordingly, a variety of different off-line methods have been developed based on a combination of chromatographic separations, organic functional group analysis, and total OC analysis (Decesari et al., 2000; Krivácsy et al., 2001b; Kiss et al., 2002; Duarte et al., 2005; Sannigrahi et al., 2006). Decesari et al. (2000) used an anion exchange column to separate aerosol WSOC into three main classes of compounds: fraction 1 (neutral and/or basic compounds), mainly composed of polyols or polyethers; fraction 2 (monocarboxylic and dicarboxylic acids), mainly composed of hydroxylated aliphatic acids, and fraction 3 (polyacidic compounds), composed of highly unsaturated polyacidic compounds of predominantly aliphatic character, with a minor content of hydroxyl groups. On the basis of ^{1}H-NMR, the authors observed that fraction 3 had a more pronounced aromatic region than the other two fractions. A model structure consisting of an aromatic core bearing substituted aliphatic chains with $-COOH$, CH_2OH, $-COCH_3$, or CH_3 terminal groups was suggested to be consistent with the ^{1}H-NMR features observed in fraction 3. However, Chang et al. (2005) found that some compounds are not separated into their expected groups by this method. Decesari et al. (2001) verified in the Po Valley (Italy) that polycarboxylic acids were the most abundant class of WSOC in all seasons, while the monocarboxylic and dicarboxylic acids were dominant in summer.

On the basis of the above methodology, Fuzzi et al. (2001) have then proposed a conceptual approach to derive a set of a few model compounds whose chemical structure can simulate the chemical and physical properties of whole aerosol WSOM. These raw formulas do not represent any specific compound, but only define the percent distribution of the main identified organic functionalities within the specific classes of WSOC (Fuzzi et al., 2001). According to this conceptual model, aliphatic functional groups represent up to 90% of the total OC content of neutral compounds and mono-/di-acids, whereas the polyacidic compounds are much more aromatic (up to 50% of the total OC). The alcohols and ethers add up to 31% of the total OC of neutral compounds, while another 52% is represented by non-oxygenated aliphatic groups. The composition of aliphatic groups of mono-/di-acids is similar to that of polyacids, and carboxylic groups represent about one-fourth of the total aliphatic carbon (Fuzzi et al., 2001). Nevertheless, this hypothetical distribution of organic functionalities within each class of WSOC requires experimental validation.

Krivácsy et al. (2001b) applied a two-step solid-phase extraction (SPE) procedure for group separation of the WSOM from aerosol aqueous extracts. Overall, the

WSOM was found to be composed of two main fractions (each approximately 50% of WSOC): (i) highly polyconjugated weak polyacids with a varying degree of hydrophobicity (which authors identified as HULIS) and (ii) slightly polyconjugated, neutral, and very hydrophilic compounds. The average elemental composition of the HULIS fraction was found to be 52.3% carbon (C), 6.7% hydrogen (H), 38.5% oxygen (O), and 2.5% nitrogen (N). The reported O/C ratio of 0.55 indicates a predominance of oxygenated functional groups. This is in agreement with the FTIR spectrum, which exhibits spectral features common to humic materials such as oxygen-containing functional groups and aliphatic C–H groups. Studies on the acidic strength and electrophoretic behavior of the compounds suggested that phenolic OH groups are not typical for these organic substances, which are mainly composed of weak carboxylic acids.

Kiss et al. (2002), on the other hand, employed a one-step SPE procedure to separate the WSOM into moderately hydrophilic (retained on the column) and strongly hydrophilic (passed through the column) organic compounds. The isolated WSOM contained, on average, 52% C, 6.2% H, 39% O, and 2.5% N. These results of elemental analysis fit very well with those obtained by Krivácsy et al. (2001b). The average elemental composition indicated that the predominance of oxygenated functional groups, along with the low H/C ratio, implied the presence of unsaturated or polyconjugated structures. On the basis of the average elemental composition and spectroscopic features (UV–vis, fluorescence, and FTIR spectroscopies) of the isolated fraction, it was suggested that aerosol WSOM is mainly composed of short-chain carboxylic acids, hydroxy acids, and polyhydroxy compounds (e.g., carbohydrates).

Non-ionic macroporous XAD-8 and XAD-4 resins in tandem were used by Duarte and Duarte (2005) to isolate and fractionate the WSOM of summer and autumn aerosol samples from a rural location. This method was adapted from the methods traditionally applied to isolate and fractionate humic substances from aquatic environments (see Aiken et al., 1992). In this procedure, the XAD-8 resin retains a mixture representative of those WSOC that are highly conjugated compounds (60% of total WSOC). On the other hand, the most hydrophilic and of low-molecular-size compounds of the WSOC fraction are retained in the XAD-4 resin (9% of total WSOC). The composition of the WSOM fractions isolated by the XAD-8 resin was further investigated by means of UV–vis, synchronous fluorescence (with $\Delta\lambda = 20\,\text{nm}$), FTIR, and ^{13}C-NMR spectroscopies (Duarte et al., 2005). Typical FTIR and ^{13}C-NMR spectra of the WSOM fractions are shown in Figure 12.4 together with the spectra of the Suwannee River fulvic acid. The comparison of the spectroscopic features between the WSOM and the standard fulvic acid is notably striking in their similarities. Although the authors did not provide estimates for percentage of various functional groups, the ^{13}C-NMR spectra suggest that WSOM samples are dominated by a high content of aliphatic structures (10–50 ppm range), carboxyl groups (160–190 ppm) and aliphatic carbons single bonded to one oxygen or nitrogen atom (60–95 ppm). However, the autumn sample exhibits a higher aromatic content (110–160 ppm) than the summer sample, plus signals typical of lignin structural components—namely, (a) the resonance at ~55 ppm due to methoxyl groups like those of syringyl and guaiacyl units and (b) resonance at ~147 ppm due to oxygen-substituted aromatic ring carbons. These signals were attributed to lignin breakdown products which are likely to be released

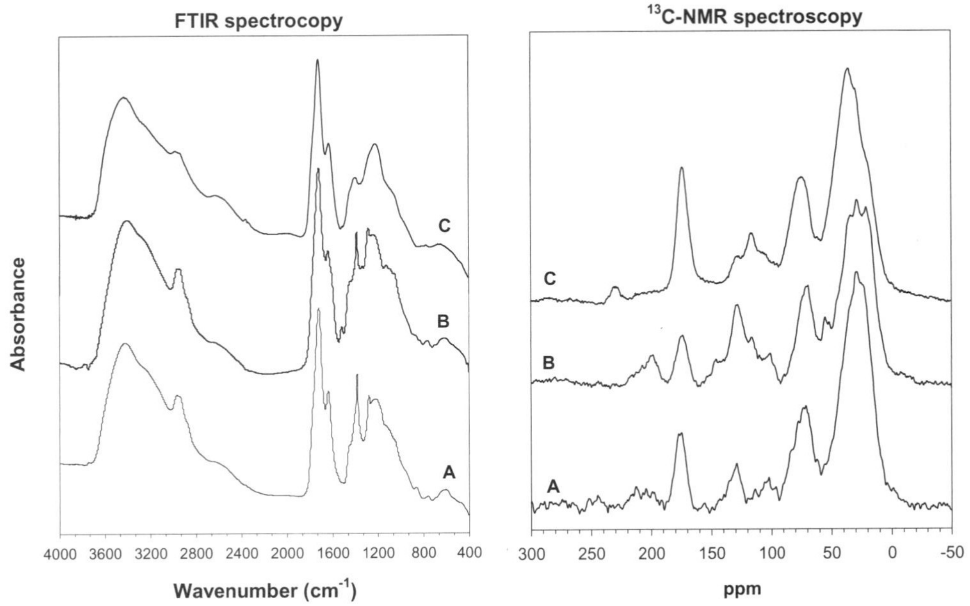

Figure 12.4. FTIR and ^{13}C-NMR spectra of WSOM isolated from the aqueous extracts of aerosol samples collected in summer (**A**) and autumn (**B**) from a rural area in Portugal [redrawn from Duarte et al. (2005)]. The spectra of the Suwannee River fulvic acid (**C**) are also shown for comparison.

during wood combustion processes in domestic fireplaces during low-temperature conditions.

The XAD-8 resin separation of hydrophobic and hydrophilic components of WSOM was also employed by Sannigrahi et al. (2006). The ^{13}C-NMR results indicated that WSOM in urban atmospheric particles is mostly aliphatic in nature (~95% C mass) with major contributions from alkyl and oxygenated alkyls (~80%), carboxylic acid (~10%), and aromatic functional groups (~4%). The authors also found that urban aerosol WSOC are only qualitatively similar to aqueous humic material in terms of functional group distribution.

Compared to the works of Krivácsy et al. (2001b) and Kiss et al. (2002), the studies of Duarte et al. (2005) and Sannigrahi et al. (2006) provided unique information on the structural characteristics and potential sources of aerosol WSOM. Furthermore, these studies have demonstrated that solid-state ^{13}C-NMR technique is a more suitable tool than solution ^{1}H-NMR for investigating the distribution of C functional groups. For example, carboxylic acids is a major class of WSOC that cannot be detected using solution ^{1}H-NMR due to the presence of rapidly exchangeable protons (Decesari et al., 2001).

Other nonspectroscopic characterization methods have been also employed in the search for aerosol WSOM structure. These methods include pyrolysis GC-MS (Gelencsér et al., 2000a; Subbalaksmi et al., 2000), capillary electrophoresis (Krivácsy et al., 2001b), elemental analysis (Krivácsy et al., 2001b; Kiss et al., 2002) and thermal profiling (Gelencsér et al., 2000b). Capillary electrophoresis and elemental analysis were used together with other spectroscopic techniques and their results were already mentioned above. With the pyrolysis GC-MS technique, only a small number of chromatographic peaks can be resolved by their characteristic mass spectra. This means that this technique is not suitable for drawing any quantitative conclusions since it yields structural information only on a few percentage of the total OC (Gelencsér et al., 2000a). Nevertheless, Gelencsér et al. (2000a) and Subbalakshmi et al. (2000) interpreted their results in terms of a polymeric material with a chemical structure similar to that of natural humic matter. The thermal profiling of fine continental aerosol (Gelencsér et al., 2000b) revealed the existence of two fractions: The first one consists of more volatile or easily oxidizable organic compounds, and the second one is attributed to the so-called "air polymers."

The studies of Krivácsy et al. (2001b), Kiss et al. (2002), Duarte et al. (2005) and Sannigrahi et al. (2006), which entail a separation protocol followed by spectroscopic characterization of aerosol WSOM, highlight the great advantage of group separation in providing unique insights into the bulk chemical nature and sources of OAs. In general, the bulk chemical characterization of aerosol WSOM indicated the presence of a complex mixture of compounds difficult to be characterized in unambiguous structural terms. It has been also suggested that these organic compounds have spectroscopic characteristics that resemble those of humic substances. As such, they share the same analytical difficulties of chemical characterization, both as a whole or as fractions separated from the whole. Moreover, this type of organic matter has been found in both urban and rural locations, suggesting that they are ubiquitous constituents of atmospheric aerosols.

12.5.2.2. Molecular Weight Distribution of WSOM. Currently, only a few papers reported estimative of the molecular-weight range of WSOM in atmospheric aerosols. Zappoli et al. (1999) estimated an upper-molecular-weight limit of WSOM of about 3000 Da using a model humic acid as standard. Recently, the studies of Kiss et al. (2003) and Samburova et al. (2005) have contradicted this idea that aerosol WSOM is mainly composed of high-molecular-weight structures.

For assessing the molecular weight distribution, Kiss et al. (2003) employed a liquid chromatography and concomitant pthodiode array detection and electrospray ionization (ESI) tandem quadropole mass spectrometer (MS) in negative ionization mode. This technique is prone to a variety of sources of errors of manifold origin including fragmentation in the ESI source, formation of multiply charged ions, and differing ionization and detection efficiencies for components of complex mixtures (Kiss et al., 2003). Nevertheless, the authors estimated the average molecular weight of the organic matter isolated from 15 aerosol samples in the range of 200–300 Da. In order to reduce uncertainties in the molecular weight estimates, another independent technique (vapor pressure osmometry) was also applied and resulted in number average molecular weight estimates in the range of 215–345 Da. These values compare fairly well with the estimates calculated from the mass spectra, although the latter are somewhat lower.

The results of Samburova et al. (2005) are in agreement with those of Kiss et al. (2003). Using a laser desorption ionization mass spectrometer (LDI-MS) in water extracts of urban particulate matter, the authors found a broad range of peaks between *m/z* 100–550. Signals with lower intensities are present up to *m/z* 700, but no peaks were observed at higher mass.

Other studies have also employed ESI-MS for characterizing *m/z* ion distributions of WSOC from fog (Kiss et al., 2001; Cappiello et al., 2003) and cloud waters (Feng and Möller, 2004). The reported values for bulk cloud waters samples were in the range of 100–500 Da, with the most intensive peaks in the 250- to 300-Da range. In fog waters, and due to the complexity of the organic fraction, an unresolved hump of ions has been recorded in the lower half of the mass spectrum, in the range of *m/z* 100–600, but with the most intense peaks being detected around *m/z* = 200–250. These findings compare fairly well with the estimates of *m/z* ion distributions reported for WSOM from fine atmospheric particles. Furthermore, it has been suggested that a significant fraction of the WSOC content in fog and cloud droplets is likely to be composed of a great number of acidic compounds with polar functional groups (Kiss et al., 2001; Feng and Möller, 2004).

In analogy to humic substances degradation in lake waters by UV radiation, O_3, and hydroxyl radical oxidation, it has been anticipated that conditions in the atmosphere are such that relatively larger macromolecules, even if formed, cannot be long-lived, thus explaining the relatively small molecular size of aerosol WSOM (Graber and Rudich, 2006). An alternative explanation is that aerosol WSOM consists mainly of a complex, unresolved mixture of relatively small molecules rather than a mixture of complex macromolecular compounds (Graber and Rudich, 2006). Furthermore, the average molecular weight of aerosol WSOM is markedly lower than those commonly reported for reference aquatic fulvic and humic acids under the same conditions (Gelencsér, 2004). This could be related to very different mechanisms of formation for WSOM as compared with humic substances (McDonald et al., 2004; Gelencsér, 2004).

12.5.2.3. On the Possible Origin of WSOM in Atmospheric Aerosols. Some authors have postulated that WSOM components may be derived primarily from biomass combustion (Facchini et al., 1999; Zappoli et al., 1999; Mayol-Bracero et al., 2002). Different mechanistic pathways for the formation of WSOM during biomass combustion have been anticipated, including: (a) soil-derived and/or decaying leaf litter humic matter lofted during combustion, (b) WSOM generation through chemical transformations of the biomass fuel and/or the initial volatile combustion products, as well as thermal breakdown of plant lignins and cellulose, and (c) recombination and condensation reactions of volatile, low-molecular-weight, primary products of combustion (Mayol-Bracero et al., 2002).

Recently, Simoneit et al. (2004) identified a group of saccharides including glucose, sucrose, trehalose, maltose, and iso-maltose, in the aqueous extracts of both aerosol and soil samples. In light of such results, the authors suggested that saccharides can be used as specific tracers of resuspension of soil from agricultural activities. Although considered as a major component of aerosol particles, soil dust are not assumed to be the origin of the bulk WSOM in atmospheric particles, mostly because soil-derived aerosols are classified in the coarse mode and the WSOM is predominantly found in fine particles. The results of Simoneit et al. (2004) are significant and add a new group of organic compounds to the list of WSOM components.

Another potential origin of WSOM in aerosols includes primary marine sources. Bubble bursting at the ocean surface has been suggested as a possible mechanistic pathway for the presence of this type of organic matter in marine aerosols (Graber and Rudich, 2006 and references therein).

Other researchers, however, have proposed various secondary formation pathways for aerosol WSOM. For example, Gelencsér et al. (2002) tentatively relate the formation of WSOM to the vast pool of soil organic matter via the evaporation, condensation, and aerosol-phase polymerization of low-molecular-weight polar degradation products of organic debris in the soil. This conceptual model also predicts that SVOCs emitted directly or formed as photooxidation products from anthropogenic and biogenic VOCs can participate as precursors. However, due to the complexity of the possible sources and the extent to which they are involved in the formation of atmospheric WSOM, the conceptual model proposed by Gelencsér et al. (2002) awaits direct experimental validation.

Decesari et al. (2002) suggested a secondary formation pathway for aerosol WSOM based on the reaction of soot particles with atmospheric oxidants. The authors found that oxidizing hexane-produced soot with O_3 produces water-soluble polycarboxylic compounds. The authors correlated this class of OC to the WSOM commonly found in atmospheric particles. However, and according to the same authors, the ^{1}H-NMR spectrum of the oxidized soot-derived polyacidic fraction differs substantially from that of the WSOM found in atmospheric particles, which means that this formation pathway alone cannot account for the variety of chemical constituents of aerosol WSOM (Decesari et al., 2002). Nevertheless, it could provide an explanation for the occurrence of polymeric UV-absorbing polar species (Havers et al., 1998) in urban aerosols.

A number of laboratory studies have also suggested the formation of aerosol WSOM through photochemical and oxidative processes. For example, Limbeck et al. (2003) presented evidences of secondary aerosol formation of atmospheric polymers by means of heterogeneous reaction of dienes (e.g., isoprene) in the presence of sulfuric acid. Competing oxidants such as ozone or the presence of humidity decreased the reaction yield, but the formation of polymeric organic matter was not disabled. Hoffer et al. (2004), on the other hand, studied the reaction between a representative lignin-type component (3,5-dihydroxybenzoic acid) from biomass burning aerosol and OH radicals in model cloud water. After a reaction time ranging from 1 day to 7 days, a light-absorbing material with features resembling those of humic materials was formed. The authors also concluded that the reaction products consist of a large number of species of different molecular weights well below 1000 Da and that the reaction itself may be oligomerization rather than polymerization. Kalberer et al. (2004) also demonstrated the occurrence of polymerization reactions during the photooxidation of 1,3,5-trimethylbenzene over the course of 20 h in a reaction chamber. The authors found that about 50% of the particle mass consists of polymers with a molecular mass up to 1000 Da. Other studies can be found in the review work of Graber and Rudich (2006).

Globally, and as put forward by Gelencsér (2004), it is possible that to a certain extent all pathways for secondary aerosol WSOM formation could occur; this would explain the ubiquitous nature and abundance of WSOM in atmospheric aerosol. What is known for certain is that biomass burning is a primary source of WSOM. Its secondary origin, however, still is under discussion and awaits proper experimental support.

12.5.3. Hygroscopic, Surface, and Colloidal Properties of Organic Aerosols

The affinity of various solutes in atmospheric particles for water uptake is a key factor controlling the mass and size of the particles—and consequently their scattering and absorption efficiencies, their acidity and rates of multiphase reactions, and more importantly their ability to act as CCN (Gelencsér, 2004). Regardless of their widespread occurrence, studies on the CCN properties of aerosol organic matter itself are scarce and address only the water-soluble fraction. Other studies are focused on CCN properties of model organic compounds, namely, aquatic fulvic and humic acids (Chan and Chan, 2003) and specific organic compounds known to be present in aerosol organic matter (e.g., adipic acid, glutaric acid, succinic acid, glycine, alanine, serine, glutamine, threonine, arginine, asparagine, levoglucosan, mannosan, and galactosan) (Cruz and Pandis, 1997; Bilde and Svenningsson, 2004; Chan et al., 2005).

Investigating the CCN activity of pure organic compounds surely helps to recognize the role of organic compounds in the activation of aerosol particles. However, model compounds represent only a few percent of the organic content in real atmospheric aerosols. Besides, physical chemistry of complex mixtures can differ considerably from that of the pure organic substances often used in laboratory experiments (Murphy, 2005). Likewise, aquatic fulvic and humic acids may not describe the activation of real aerosol particles properly (Kiss et al., 2005). Although sharing similar functionalities, aerosol WSOM differ from aquatic humic and fulvic acids in terms of their fine structure [e.g., different relative proportions of C functional groups (Sannigrahi et al., 2006)] and average molecular weight (Gelencsér, 2004). Such differences can lead to contradictory conclusions on the effect of organic components on activation. It becomes obvious, therefore, that the measurement of the hygroscopic properties of real atmospheric WSOM samples is of great importance for the study of aerosol CCN activity.

To date, just one study has examined the hygroscopic growth and deliquescence behavior of atmospheric WSOM. Gysel et al. (2004) examined hygroscopic properties of both water-soluble matter (WSM) and isolated organic matter extracts. The authors used a SPE extraction procedure to isolate the less hydrophilic fraction (ISOM) of the WSOM from remaining inorganic salts and "most" hydrophilic organic matter (MHOM). The authors reported that a particle's hygroscopic growth cannot be explained completely by the inorganic component solely or by the system "inorganic plus ISOM," but can only be explained by the complete mixed particle model "inorganic plus ISOM plus MHOM." The authors concluded that the deliquescence properties of WSM extracts are mainly determined by the inorganic salts (42–53% of WSM mass). However, at 90% of RH, about 80–62% of particulate water in the samples are associated with inorganic salts and about 20–38% with WSOM. The relative contributions of both distinguished WSOM fractions, ISOM and MHOM, remains uncertain since MHOM was not available in isolated form. These findings suggest that, apart from inorganic salts, organic compounds are also actively involved in atmospheric processes in which particle hygroscopic properties are crucial (Gysel et al., 2004).

Aerosol WSOM was also found to be surface-active (Facchini et al., 2000; Kiss et al., 2005; Dinar et al., 2006). Facchini et al. (2000) reported a decrease in surface tension as a function of increasing total OC concentration in both wet aerosol and cloud/fog samples. The same authors also investigated the characteristics of the compounds responsible for the decrease of surface tension in the fog droplets. The chosen fog sample was fractionated into three classes of compounds: neutral, mono-/dicarboxylic acids, and polycarboxylic acids. Polycarboxylic acids were found to be three times more surface-active than mono-/dicarboxylic acids and 10 times more active than neutral compounds. These results are in good agreement with those of Kiss et al. (2005) in their studies of the surface activity of WSOM fractions isolated from aqueous extracts of rural aerosol samples. The isolated organic matter decreased the surface tension of the aqueous solutions by 25–42% as compared to pure water. This effect was further enhanced when the WSOM was mixed with ammonium sulfate. A seasonal trend in the surface tension decreasing power was also observed with the most pronounced effect in the summer samples and the smallest in the winter samples. The surface activity of WSOM samples was also found to exceed that of standard humic and fulvic acids. This last finding is of particular interest and shows that standard humic substances are not good substitutes for atmospheric WSOM in every aspect (Kiss et al., 2005). This was also confirmed by Dinar et al. (2006), who found that WSOM extracted from three different aerosol sources (fresh smoke, aged smoke, and photochemical pollution) are more surface-active than standard fulvic acids.

The studies of Facchini et al. (2000), Gysel et al. (2004), Kiss et al. (2005) and Dinar et al. (2006) highlight the importance of organic aerosol components in the cloud-nucleating activities of atmospheric particles. However, a question still remains to be answered: What is the truly effect of organic components on aerosol CCN activity? Furthermore, does the WSOM enhance cloud droplet activation or, on the other hand, delay droplet activation?

Following the study of Ellison et al. (1999), Tabazadeh (2005) presented a quite interesting discussion on organic aggregate formation in aerosols and its potential impacts on atmospheric chemistry of aerosols. This discussion was based on the fact that when metals ions are added to aqueous solutions of humic and fulvic acids, micelle-like aggregates are formed by intermolecular coiling and/or intramolecular aggregation. Assuming that aerosol WSOM has similar colloidal properties as humic substances, the author suggested that in aerosols, which contain high amounts of inorganic ions, WSOM micelle-like aggregates are very likely to be present in large amounts in aqueous solutions. Once these pseudomicelles are formed, then the concentration of free surfactant molecules in solution decreases, making it difficult for a full coat to develop on the particle surface (Tabazadeh, 2005).

Aggregate formation in WSOM-containing atmospheric particles could also affect water uptake by the aerosol particle: If inorganic ions form stable complexes with organic-aggregates, then they are not free to uptake water from the atmosphere with increased humidity (Tabazadeh, 2005). Therefore, the water uptake reduction by some WSOM components may be due to removal of inorganic ions from the aqueous solution and not the physical attachment of the WSOM to the particle surface as currently hypothesized (Ellison et al., 1999; Decesari et al., 2003). On the basis of Tabazadeh (2005) assumptions, one may infer that WSOM could delay cloud droplet activation process. Besides, the lowered surface tension of cloud droplets may affect their coagulation rate and, consequently, rain formation. Furthermore, based on data for humic substances, WSOM pseudomicelles could also affect the ability of the aerosol as a whole to solubilize hydrophobic organic pollutants. Also, WSOM pseudomicelles could potentially increase light scattering by aerosol particles due to increased solution turbidity (Tabazadeh, 2005). Considering its possible importance to atmospheric chemistry, the colloidal nature of aerosol WSOM needs to be thoroughly explored by undertaking experimental studies in real atmospheric samples.

12.6. CONCLUSIONS: KNOWLEDGE GAPS AND RESEARCH NEEDED

Over the past decade, significant progress has been made to understand atmospheric organic matter and its role in climate and human health. Currently, there is a consensus opinion that the chemical and physical properties of OAs have strong implications on regional air quality and global climate. Therefore, the knowledge of OA chemical composition is the key to understanding many aerosol properties. However, our knowledge of OA composition, chemical and physical properties, sources, and transformations is far from being completed. In particular, the WSOM of atmospheric particles, which can be a large fraction of the total OC, is definitively not well understood. This fact also hampered reliable information on source apportionment of OAs. Understanding the sources implies more detailed studies with special emphasis on anthropogenic versus biogenic emissions, and primary versus secondary origin, including the formation mechanisms of secondary OAs and effects of seasonal cycles. A better understanding of the sources of OAs is critically needed in order to be able to predict its contribution to climate forcing.

Reviewing the data, we find several other open questions and research gaps needing further elucidation of aerosol organic matter effects on atmospheric chemistry, hydrological cycle, and human health. Among these, there is the quantification of OC and EC fractions in aerosol particles. That is to say, how much of the ambient aerosol is composed of organic material? This question highlights the need of achieving a standardized method for the determination of EC and OC in atmospheric particles. For example, during the 1990s a trend to build OC/EC instruments has emerged in Europe. However, because these instruments are operated under quite different conditions, considerable variations in OC/EC results are obtained (Schmid et al., 2001). Eventually, these uncertainties on the concentration of OC and EC could affect aerosol chemical mass closure.

What is known for certain is that WSOM components constitute a substantial fraction of OAs. To date, only a small number of aerosol WSOM samples have been studied and characterized, and most of them were collected in European locations. Besides, each collected sample was extracted and isolated by different procedures, which hinders the comparison of the fine physical and chemical properties of the isolated organic fractions. The establishment of a standardized analytical methodology for WSOM extraction and isolation seems to be an important prerequisite for efficient further characterization of aerosol WSOM samples from different areas with different degrees of pollution (urban versus rural versus remote areas in different regions of the world).

Furthermore, little is known about the alkaline-extractable and insoluble fractions of aerosol organic matter. On average these fractions together can account for 7–20% of the total aerosol mass (Zappoli et al., 1999; Krivácsy et al., 2001b). Thus, the characterization of this water-insoluble fraction is highly desirable, as well as a comparative study between this fraction and the WSOM fraction.

Likewise, further studies are needed regarding the structural composition and molecular weight distribution of the WSOM by advanced analytical techniques. Their elucidation will enable understanding the CCN properties of WSOM in atmospheric particles. Too often, these properties are simulated by model compounds, whose physical chemistry properties are far from the original organic matter contained in the real atmospheric aerosols. Consequently, little is still known about the hygroscopic, surface, colloidal, and pseudomicellular behavior of aerosol organic matter itself. Their elucidation, therefore, in real atmospheric samples, is highly required for understanding not only aerosol organic matter CCN activity but also its binding capacity toward metals and organic pollutants. Acknowledging these effects may provide insight on climate change and adverse health effects of anthropogenic atmospheric aerosols and, thus, may encourage the development of efficient strategies and guidelines for air quality control and medical treatment of airborne diseases.

REFERENCES

Aiken, G. R., McKnight, D. M., Thorn, K. A., and Thurman, E. M. (1992). Isolation of hydrophilic organic acids from water using non-ionic macroporous resins. *Org. Geochem.* **18**(4), 567–573.

Alves, C., Carvalho, A., and Pio, C. (2002). Mass balance of organic fractions in atmospheric aerosols. *J. Geophys. Res.* **107**, **D21**, 8345, doi: 10.1029/2001JD000616.

Alves, C., Pio, C., and Duarte, A. C. (2000). Particulate size distributed organic compounds in a forest atmosphere. *Environ. Sci. Technol.* **34**, 4287–4293.

Andreae, M. O., Jones, C. D., and Cox, P. M. (2005). Strong present-day aerosol cooling implies a hot future. *Nature* **435**, 1187–1190.

Andreae, M. O., and Crutzen, P. J. (1997). Atmospheric aerosols: Biogeochemical sources and role in atmospheric chemistry. *Science* **276**, 1052–1058.

Andreae, M. O., Rosenfeld, D., Artaxo, P., Costa, A. A., Frank, G. P., Longo, K. M., and Silva-Dias, M. A. F. (2004). Smoking rain clouds over the Amazon. *Science* **303**, 1337–1342.

Andrews, E., Saxena, P., Musarra, S., Hildemann, L. M., Koutrakis, P., McMurry, P. H., Olmez, I., and White, W. H. (2000). Concentration and composition of atmospheric aerosols from the 1995 SEAVS experiment and a review of the closure between chemical and gravimetric measurements. *J. Air Waste Manage. Assoc.* **50**, 648–664.

Baltensperger, U., Nyeki, S., and Kalberer, M. (2003). Atmospheric particulate matter. In *Handbook of Atmospheric Science. Principles and Applications*, Hewitt, C. N., and Jackson, A. V., eds., Blackwell Publishing, London, pp. 228–254.

Bauer, H., Kasper-Giebl, A., Löflund, M., Giebl, H., Hitzenberger, R., Zibuschka, F., and Puxbaum, H. (2002). The contribution of bacteria and fungal spores to the organic carbon content of cloud water, precipitation and aerosols. *Atmos. Res.* **64**, 109–119.

Berner, A., Galambos, Z., Ctyroky, P., Frühauf, P., Hitzenberger, R., Gomiščček, B., Hauck, H., Preining, O., and Puxbaum, H. (2004). On the correlation of atmospheric aerosol components of mass size distributions in the larger region of a central European city. *Atmos. Environ.* **38**, 3959–3970.

Bernstein, J. A., Alexis, N., Barnes, C., Bernstein, L., Nel, A., Peden, D., Diaz-Sanchez, D., Tarlo, S. M., and Williams, P. B. (2004). Health effects of air pollution. *J. Allergy Clin. Immunol.* **114**, 1116–1123.

Bilde, M., and Svenningsson, B. (2004). CCN activation of slightly soluble organics: the importance of small amounts of inorganic salt and particle phase. *Tellus* **56B**, 128–134.

Bond, T. C., Streets, D. G., Yarber, K. F., Nelson, S. M., Woo, J.-H., and Klimont, Z. (2004). A technology-based global inventory of black and organic carbon emissions from combustion. *J. Geophys. Res.* **109**, D14203, doi: 10.1029/2003JD003697.

Braun, A. (2005). Carbon speciation in airborne particulate matter with C (1s) NEXAFS spectroscopy. *J. Environ. Monit.* **7**, 1059–1065.

Braun, A., Wirick, S., Kubátová, A., Mun, B. S., and Huggins, F. E. (2006). Photochemically induced decarboxylation in diesel soot extracts. *Atmos. Environ.* **40**, 5837–5844.

Cappiello, A., Simoni, E., Fiorucci, C., Mangani, F., Palma, P., Trufelli, H., Decesari, S., Facchini, M. C., and Fuzzi, S. (2003). Molecular characterization of the water-soluble organic compounds in fogwater by ESIMS/MS. *Environ. Sci. Technol.* **37**, 1229–1240.

Carvalho, A., Pio, C., and Santos, C. (2003). Water-soluble hydroxylated organic compounds in German and Finnish aerosols. *Atmos. Environ.* **37**, 1775–1783.

Chan, M. N., and Chan, C. K. (2003). Hygroscopic properties of two model humic-like substances and their mixtures with inorganics of atmospheric importance. *Environ. Sci. Technol.* **37**, 5109–5115.

Chan, M. N., Choi, M. Y., Ng, N. L., and Chan, C. K. (2005). Hygroscopicity of water-soluble organic compounds in atmospheric aerosols: Amino acids and biomass burning derived organic species. *Environ. Sci. Technol.* **39**, 1555–1562.

Chang, H., Herckes, P., and Collett, J. L., Jr. (2005). On the use of anion exchange chromatography for the characterisation of water soluble organic carbon. *Geophys. Res. Lett.* **32**, doi: 10.1029/2004GL021322.

Charlson, R. J., Schwartz, S. E., Hales, J. M., Cess, R. D., Coakley, J. R., J. A., Hansen, J. E., and Hofmann, D. J. (1992). Climate forcing by anthropogenic aerosols. *Science* **255**, 423–430.

Charlson, R. J., Seinfeld, J. H., Nenes, A., Kulmala, M., Laaksonen, A., and Facchini, M. C. (2001). Reshaping the theory of cloud formation. *Science* **292**, 2025–2026.

Claeys, M., Wang, W., Ion, A. C., Kourtchev, I., Gelencsér, A., and Maenhaut, W. (2004). Formation of secondary organic aerosols from isoprene and its gas-phase oxidation products through reaction with hydrogen peroxide. *Atmos. Environ.* **38**, 4093–4098.

Cruz, C. N., and Pandis, S. N. (1997). A study of the ability of pure secondary organic aerosol to act as cloud condensation nuclei. *Atmos. Environ.* **31**, 2201–2214.

Decesari, S., Facchini, M. C., Fuzzi, S., and Tagliavini, E. (2000). Characterization of water-soluble organic compounds in atmospheric aerosols: A new approach. *J. Geophys. Res.* **105**, 1481–1489.

Decesari, S., Facchini, M. C., Matta, E., Lettini, F., Mircea, M., Fuzzi, S., Tagliavini, E., and Putaud, J.-P. (2001). Chemical features and seasonal variation of fine aerosol water-soluble organic compounds in the Po Valley, Italy. *Atmos. Environ.* **35**, 3691–3699.

Decesari, S., Facchini, M. C., Matta, E., Mircea, M., Fuzzi, S., Chughtai, A. R., and Smith, D. M. (2002). Water soluble organic compounds formed by oxidation of soot. *Atmos. Environ.* **36**, 1827–1832.

Decesari, S., Facchini, M. C., Mircea, M., Cavalli, F., and Fuzzi, S. (2003). Solubility properties of surfactants in atmospheric aerosol and cloud/fog water samples. *J. Geophys. Res.* **108**, D21, 4685, doi: 10.1029/2003JD003566.

Denman, K. L., Brasseur, G., Chidthaisong, A., Ciais, P., Cox, P. M., Dickinson, R. E., Hauglustaine, D., Heinze, C., Holland, E., Jacob, D., Lohmann, U., Ramachandran, S., Dias, P. L. S., Wofsy, S. C., and Zhang, X. (2007). Couplings between changes in the climate system and biogeochemistry. In *Climate Change 2007: The Physical Science Basis. Contribution of*

Working Group I to the Fourth Assessment Report of the Intergovernmental Panel on Climate Change, Solomon, S., Qin, D., Manning, M., Chen, Z., Marquis, M., Averyt, K. B., Tignor, M., and Miller, H. L., eds., Cambridge University Press, Cambridge, UK, pp. 499–587.

Dinar, E., Taraniuk, I., Graber, E. R., Katsman, S., Moise, T., Anttila, T., Mentel, T. F., and Rudich, Y. (2006). Cloud Condensation Nuclei properties of model and atmospheric HULIS, *Atmos. Chem. Phys.* **6**, 2465–2482.

Donaldson, K., Stone, V., Born, P. J. A., Jimenez, L. A., Gilmour, P. S., Schins, R. P. F., Knaapen, A. D. M., Rahman, I., Faux, S. P., Brown, D. M., and MacNee, W. (2003). Oxidative stress and calcium signaling in the adverse effects of environmental particles (PM_{10}). *Free Radical Biol. Med.* **34**, 1369–1382.

Duarte, R. M. B. O., and Duarte, A. C. (2005). Application of non-ionic solid sorbents (XAD resins) for the isolation and fractionation of water-soluble organic compounds from atmospheric aerosols. *J. Atmos. Chem.* **51**, 79–93.

Duarte, R. M. B. O., Pio, C. A., and Duarte, A. C. (2005). Spectroscopic study of the water-soluble organic matter isolated from atmospheric aerosols collected under different atmospheric conditions. *Anal. Chim. Acta* **530**, 7–14.

Ellison, G. B., Tuck, A. F., and Vaida, V. (1999). Atmospheric processing of organic aerosols. *J. Geophys. Res.* **104**, 11633–11641.

Facchini, M. C., Decesari, S., Mircea, M., Fuzzi, S., and Loglio, G. (2000). Surface tension of atmospheric wet aerosol and cloud/fog droplets in relation to their organic carbon content and chemical composition. *Atmos. Environ.* **34**, 4853–4857.

Facchini, M. C., Fuzzi, S., Zappoli, S., Andracchio, A., Gelencsér, A., Kiss, G., Krivácsy, Z., Mészáros, E., Hansson, H.-C., Alsberg, T., and Zebühr, Y. (1999). Partitioning of the organic aerosol component between fog droplets and interstitial air. *J. Geophys. Res.* **104**, D21, 26821–26832.

Feng, J. S., and Möller, D. (2004). Characterization of water-soluble macromolecular substances in cloud water. *J. Atmos. Chem.* **48**, 217–233.

Forster, P., Ramaswamy, V., Artaxo, P., Berntsen, T., Betts, R., Fahey, D. W., Haywood, J., Lean, J., Lowe, D. C., Myhre, G., Nganga, J., Prinn, R., Raga, G., Schulz, M., and Van Dorland, R. (2007). Changes in atmospheric constituents and in radiative forcing. In *Climate Change 2007: The Physical Science Basis. Contribution of Working Group I to the Fourth Assessment Report of the Intergovernmental Panel on Climate Change*, Solomon, S., Qin, D., Manning, M., Chen, Z., Marquis, M., Averyt, K. B., Tignor, M., and Miller, H. L., eds., Cambridge University Press, Cambridge, UK, pp. 129–234.

Fuzzi, S., Decesari, S., Facchini, M. C., Matta, E., Mircea, M., and Tagliavini, E. (2001). A simplified model of the water soluble organic component of atmospheric aerosols. *Geophys. Res. Lett.* **20**, 4079–4082.

Gajananda, K., Kuniyal, J. C., Momin, G. A., Rao, P. S. P., Safai, P. D., Tiwari, S., and Ali, K. (2005). Trend of atmospheric aerosols over the north western Himalayan region, India. *Atmos. Environ.* **39**, 4817–4825.

Gelencsér, A. (2004). *Carbonaceous Aerosol*, Atmospheric and Oceanographic Sciences Library, Springer, Dordrecht.

Gelencsér, A., Meszaros, T., Blazso, M., Kiss, G., Krivacsy, Z., Molnar, A., and Meszaros, E. (2000a). Structural characterisation of organic matter in fine tropospheric aerosol by pyrolysis–gas chromatography–mass spectrometry. *J. Atmos. Chem.* **37**, 173–183.

Gelencsér, A., Hoffer, A., Molnar, A., Krivacsy, Z., Kiss, G., and Meszaros, E. (2000b). Thermal behaviour of carbonaceous aerosol from a continental background site. *Atmos. Environ.* **34**, 823–831.

Gelencsér, A., Hoffer, A., Krivácsy, Z., Kiss, G., Molnár, A., and Mészáros, E. (2002). On the possible origin of humic matter in fine continental aerosol. *J. Geophys. Rese.* **107**, D12, 4137, doi: 10.1029/2001JD001299.

Graber, E. R., and Rudich, Y. (2006). Atmospheric HULIS: How humic-like are they? A comprehensive and critical review. *Atmos. Chem. Phys.* **6**, 729–753.

Gysel, M., Weingartner, E., Nyeki, S., Paulsen, D., Baltensperger, U., Galambos, I., and Kiss, G. (2004). Hygroscopic properties of water-soluble matter and humic-like organics in atmospheric fine aerosol. *Atmos. Chem. Phys.* **4**, 35–50.

Havers, N., Burba, P., Lambert, J., and Klockow, D. (1998). Spectroscopic characterization of humic-like substances in airborne particulate matter. *J. Atmos. Chem.* **29**, 45–54.

Haywood, J., and Boucher, O. (2000). Estimates of the direct and indirect radiative forcing due to tropospheric aerosols: a review. *Rev. Geophys.* **38**, 513–543.

Hidy, G., and Blanchard, C. L. (2005). The Midlatitude North American Background Aerosol and Global Aerosol Variation. *J. Air & Waste Manage. Assoc.* **55**, 1585–1599.

Hildemann, L. M., Markowski, G. R., and Cass, G. R. (1991). Chemical composition of emissions from urban sources of fine organic aerosol. *Environ. Sci. Technol.* **25**, 744–759.

Hoffer, A., Kiss, G., Blazso, M., and Gelencser, A. (2004). Chemical characterization of humic-like substances (HULIS) formed from a lignin-type precursor in model cloud water. *Geophys. Res. Lett.*, **31**, doi:10.1029/2003GL018962.

Horvath, H. (2000). Aerosols—An introduction. *J. Environ. Radioactivity* **51**, 5–25.

Hueglin, C., Gehrig, R., Baltensperger, U., Gysel, M., Monn, C., and Vonmont, H. (2005). Chemical characterization of PM2.5, PM10 and coarse particles at urban, near-city and rural sites in Switzerland. *Atmos. Environ.* **39**, 637–651.

Hughes, L. S., Cass, G. R., Gone, J., Ames, M., and Olmez, I. (1998). Physical and chemical characterization of atmospheric ultrafine particles in the Los Angeles area. *Environ. Sci. Technol.* **32**, 1153–1161.

Jacobson, M. C., Hanson, H.-C., Noone, K. J., and Charlson, R. J. (2000). Organic atmospheric aerosols: Review and state of the science. *Rev. Geophys.* **38**, 267–294.

Jaenicke, R. (1998). Atmospheric aerosol size distribution. In *Atmospheric Particles*, Harrison, R. M., and van Grieken, R. E., eds., John Wiley & Sons, Chichester, UK, pp. 1–28.

Kalberer, M., Paulsen, D., Sax, M., Steinbacher, M., Dommen, J. A., Prevot, S. H., Fisseha, R., Weingartner, E., Frankevich, V., Zenobi, R., and Baltensperger, U. (2004). Identification of Polymers as Major Components of Atmospheric Organic Aerosols. *Science* **303**, 1659–1662.

Kanakidou, M., Seinfeld, J. H., Pandis, S. N., Barnes, I., Dentener, F. J., Facchini, M. C., Van Dingenen, R., Ervens, B., Nenes, A., Nielsen, C. J., Swietlicki, E., Putaud, J. P., Balkanski, Y., Fuzzi, S., Horth, J., Moortgat, G. K., Winterhalter, R., Myhre, C. E. L., Tsigaridis, K., Vignati, E., Stephanou, E. G., and Wilson, J. (2005). Organic aerosol and global climate modelling: A review. *Atmos. Chem. Phys.* **5**, 1053–1123.

Kaufman, Y. J., Tanré, D., and Boucher, O. (2002). A satellite view of aerosols in the climate system. *Nature* **419**, 215–223.

Kiss, G., Varga, B., Gelencser, A., Krivacsy, Z., Molnar, A., Alsberg, T., Persson, L., Hansson, H. C., and Facchini, M. C. (2001). Characterisation of polar organic compounds in fog water. *Atmos. Environ.* **35**, 2193–2200.

Kiss, G., Varga, B., Galambos, I., and Ganszky, I. (2002). Characterisation of water-soluble organic matter isolated from atmospheric fine aerosol. *J. Geophys. Res.* **107**, D21, 8339, doi: 10.1029/2001JD000603.

Kiss, G., Tombácz, E., Varga, B., Alsberg, T., and Persson, L. (2003). Estimation of the average molecular weight of humic-like substances isolated from fine atmospheric aerosol. *Atmos. Environ.* **37**, 3783–3794.

Kiss, G., Tombacz, E., and Hansson, H. C. (2005). Surface tension effects of humic-like substances in the aqueous extract of troposphere fine aerosol. *J. Atmos. Chem.* **50**, 279–294, doi:10.1007/s10874-005-5079-5.

Krivácsy, Z., Hoffer, A., Sárvári, Z., Temesi, D., Baltensperger, U., Nyeki, S., Weingartner, E., Kleefeld, S., and Jennings, S. G. (2001a). Role of organic and black carbon in the chemical composition of atmospheric aerosol at European background sites. *Atmos. Environ.* **35**, 6231–6244.

Krivácsy, Z., Gelencsér, A., Kiss, G., Mészáros, E., Molnár, A., Hoffer, A., Mészáros, T., Sárvári, Z., Temesi, D., Varga, B., Baltensperger, U., Nyeki, S., and Weingartner, E. (2001b). Study on the chemical character of water soluble organic compounds in fine atmospheric aerosol at the Jungfraujoch. *J. Atmos. Chem.* **39**, 235–259.

Krivácsy, Z., and Molnár, Á. (1998). Size distribution of ions in atmospheric aerosols. *Atmos. Res.* **46**, 279–291.

Kubátová, A., Vermeylen, R., Claeys, M., Cafmeyer, J., and Maenhaut, W. (2002). Organic compounds in urban aerosols from Gent, Belgium: characterisation, sources and seasonal differences. *J. Geophys. Res.* **107**, D21, 8343, doi: 10.1029/2001JD000556.

Künzli, N., Kaiser, R., Medina, S., Studnicka, M., Chanel, O., Filliger, P., Herry, M., Horak, F., Puybonnieux-Texier, V., Quénel, P., Schneider, J., Seethaler, R., Vergnaud, J.-C., and Sommer, H. (2000). Public-health impact of outdoor and traffic-related air pollution: A European assessment. *Lancet* **356**, 795–801.

Lee, H. S., and Kang, B.-W. (2001). Chemical characteristics of principal $PM_{2.5}$ species in Chongju, South Korea. *Atmos. Environ.* **35**, 739–746.

Limbeck, A., Kulmala, M., and Puxbaum, H. (2003). Secondary organic aerosol formation in the atmosphere via heterogeneous reaction of gaseous isoprene on acidic particles. *Geophys. Res. Letters* **30**, 1996, doi: 10.1029/2003GL017738.

Lin, C.-C., Chen, S.-J., Huang, K.-L., Hwang, W.-I., Chang-Chien, G.-P., and Lin, W.-Y. (2005). Characteristics of metals in nano/ultrafine/fine/coarse particles collected beside a heavily trafficked road. *Environ. Sci. Technol.* **39**, 8113–8122.

Lohmann, U., and Feichter, J. (2005). Global indirect aerosol effects: A review. *Atmos. Chem. Phys.* **5**, 715–737.

Mader, B. T., Yu, J. Z., Xu, J. H., Li, Q. F., Wu, W. S., Flagan, R. C., and Seinfeld, J. H. (2004). Molecular composition of the water-soluble fraction of atmospheric carbonaceous aerosols collected during ACE-Asia. *J. Geophys. Res.* **109**, D06206, doi: 10.1029/2003JD004105.

Maenhaut, W., Cafmeyer, J., Dubtsov, S., and Chi, X. (2002). Detailed mass size distribution of elements and species, and aerosol mass closure during fall 1999 at Gent, Belgium. *Nucl. Instr. Meth. Phys. Res. B* **189**, 238–242.

Manoli, E., Voutsa, D., and Samara, C. (2002). Chemical characterization and source identification/apportionment of fine and coarse air particles in Thessaloniki, Greece. *Atmos. Environ.* **36**, 949–961.

Mark, D. (1998). Atmospheric aerosol sampling. In *Atmospheric Particles*, Harrison, R. M., and van Grieken, R. E., eds., John Wiley & Sons, Chichester, UK, 29–94.

Matsumoto, K., Nagao, I., Tanaka, H., Miyaji, H., Iida, T., and Ikebe, Y. (1998). Seasonal characteristics of organic and inorganic species and their size distributions in atmospheric aerosols over the northwest Pacific Ocean. *Atmos. Environ.* **32**, 1931–1946.

Mayol-Bracero, O. L., Guyon, P., Graham, B., Roberts, G. C., Andreae, M. O., Decesari, S., Facchini, M. C., Fuzzi, S., and Artaxo, P. (2002). Water-soluble organic compounds in biomass burning aerosols over Amazonia. 2. Apportionment of the chemical composition and importance of the polyacidic fraction. *J. Geophys. Res.* **107**, D20, 8091, doi: 10.1029/2001JD000522.

McDonald, S., Bishop, A. G. Prenzler, P. D., and Robards, K. (2004). Analytical chemistry of freshwater humic substances. *Anal. Chim. Acta* **527**, 105–124.

Mészáros, E., Bareza, T., Gelencsér, A., Hlavay, J., Kiss, G., Krivácsy, Z., Molnár, A., and Polyák, K. (1997). Size distributions of inorganic and organic species in the atmospheric aerosol in Hungary. *J. Aerosol Sci.* **28**, 1163–1175.

Molnár, A., Mészáros, E., Hansson, H. C., Karlsson, H., Gelencsér, A., Kiss, G., and Krivácsy, Z. (1999). The importance of organic and elemental carbon in the fine atmospheric aerosol particles. *Atmos. Environ.* **33**, 2745–2750.

Mukai, H., and Ambe, Y. (1986). Characterization of a humic acid-like brown substance in airborne particulate matter and tentative identification of its origin. *Atmos. Environ.* **5**, 813–819.

Murphy, D. M. (2005). Something in the air. *Science* **307**, 1888–1890.

Oberdörster, G. (2001). Pulmonary effects of inhaled ultrafine particles. *Int. Arch. Occup. Environ. Health* **74**, 1–8.

Penner, J. E., Charlson, R. J., Hales, J. M., Laulainen, N. S., Leifer, R., Novakov, T., Ogren, J., Radke, L. F., Schwartz, S. E., and Travis, L. (1994). Quantifying and minimizing uncertainty of climate forcing by anthropogenic aerosols. *Bull. Am. Meteorol. Soc.* **75**, 375–400.

Penner, J. E., Andreae, M., Annegarn, H., Barrie, L., Feitcher, J., Hegg, D., Jayaraman, A., Leaitch, R., Murphy, D., Nganga, J., and Pitari, G. (2001). Aerosols, their direct and indirect effects. In *Climate Change 2001: The Scientific Basis. Contribution of Working Group I to the Third Assessment Report of the Intergovernmental Panel on Climate Change*, Houghton, J. T., et al., eds., Cambridge University Press, Cambridge, UK, pp. 289–348.

Pillai, P. S., and Moorthy, K. K. (2001). Aerosol mass-size distributions at a tropical coastal environment:response to mesoscale and synoptic processes. *Atmos. Environ.* **35**, 4099–4112.

Pio, C. A., Castro, L. M., and Ramos, M. O. (1993). Differentiated determination of organic and elemental carbon in atmospheric aerosol particles by a thermal-optical method. In *Physico-Chemical Behaviour of Atmospheric Pollutants*, Report EUR 15609/2 EN, Angeletti, G., and Restelli, G., eds., Proceedings of the Sixth European Symposium, Vol. 2, Varese, 18–22 October, pp. 706–711.

Pöschl, U. (2005). Atmospheric aerosols: Composition, transformation, climate and health effects. *Angew. Chem. Int. Ed.* **44**, 7520–7540.

Putaud, J.-P., Raes, F., Van Dingenen, R., Brüggemann, E., Facchini, M. C., Decesari, S., Fuzzi, S., Gehrig, R., Hüglin, C., Laj, P., Lorbeer, G., Maenhaut, W., Mihalopoulos, N., Müller, K., Querol, X., Rodriguez, S., Schneider, J., Spindler, G., ten Brink, H., Tørseth, K., and Wiedensohler, A. (2004). A European aerosol phenomenology—2:chemical characteristics of particulate matter at kerbside, urban, rural and background sites in Europe. *Atmos. Environ.* **38**, 2579–2595.

Raes, F., Van Dingenen, R., Vignati, E., Wilson, J., Putaud, J.-P., Seinfeld, J. H., and Adams, P. (2000). Formation and cycling of aerosols in the global troposphere. *Atmos. Environ.* **34**, 4215–4240.

Ramanathan, V., Crutzen, P. J., Kielh, J. T., and Rosenfeld, D. (2001). Aerosols, climate, and the hydrological cycle. *Science* **294**, 2119–2124.

Rocha, T. A. P., and Duarte, A. C. (1997). Application of chemometrics to the identification of trends in polynuclear aromatic hydrocarbons and alkanes in air samples form Oporto. *Analyst* **122**, 1509–1515.

Rocha, T. A. P., Horvath, H., Oliveira, J. A. B. P., and Duarte, A. C. (1999). Trends in alkanes and PAHs in airborne particulate matter from Oporto and Vienna: Identification and comparison. *Sci. Total Environ.* **236**, 231–236.

Rogge, W. F., Hildemann, L. M., Mazurek, M. A., Cass, G. R., and Simoneit, B. R. T. (1991). Sources of fine aerosol. 1. Charbroilers and meat cooking operations. *Environ. Sci. Technol.* **25**, 1112–1125.

Rogge, W. F., Mazurek, M. A., Hildemann, L. M., and Cass, G. R. (1993a). Quantification of urban organic aerosols at a molecular level: Identification, abundance and seasonal variation. *Atmos. Environ.* **27A**, 1309–1330.

Rogge, W. F., Hildemann, L. M., Mazurek, M. A., Cass, G. R., and Simoneit, B. R. T. (1993b). Sources of fine aerosol. 3. Road dust, tire debris and organometallic brake lining dust/roads as sources and sinks. *Environ. Sci. Technol.* **27**, 1892–1904.

Rogge, W. F., Hildemann, L. M., Mazurek, M. A., Cass, G. R., and Simoneit, B. R. T. (1993c). Sources of fine aerosol. 4. Particulate abrasion products from leaf surfaces of urban plants. *Environ. Sci. Technol.* **27**, 2700–2711.

Rogge, W. F., Hildemann, L. M., Mazurek, M. A., Cass, G. R., and Simoneit, B. R. T. (1993d). Sources of fine aerosol. 5. Natural gas home appliances. *Environ. Sci. Technol.* **27**, 2736–2744.

Rogge, W. F., Hildemann, L. M., Mazurek, M. A., Cass, G. R., and Simoneit, B. R. T. (1993e). Sources of fine aerosol. 2. Noncatalyst and catalyst-equipped automobiles and heavy-duty diesel trucks. *Environ. Sci. Technol.* **27**, 636–651.

Rogge, W. F., Hildemann, L. M., Mazurek, M. A., Cass, G. R., and Simoneit, B. R. T. (1994). Sources of fine aerosol. 6. Cigarette smoke in the urban atmosphere. *Environ. Sci. Technol.* **28**, 1375–1388.

Rogge, W. F., Hildemann, L. M., Mazurek, M. A., Cass, G. R., and Simoneit, B. R. T. (1997a). Sources of fine aerosol. 7. Hot asphalt roofing tar pot fumes. *Environ. Sci. Technol.* **31**, 2726–2730.

Rogge, W. F., Hildemann, L. M., Mazurek, M. A., Cass, G. R., and Simoneit, B. R. T. (1997b). Sources of fine aerosol. 8. Boilers burning no. 2 distillate fuel oil. *Environ. Sci. Technol.* **31**, 2731–2737.

Rogge, W. F., Hildemann, L. M., Mazurek, M. A., Cass, G. R., and Simoneit, B. R. T. (1998). Sources of fine organic aerosol. 9. Pine, oak and synthetic log combustion in residential fireplaces. *Environ. Sci. Technol.* **32**, 13–22.

Ruellan, S., and Cachier, H. (2001). Characterisation of fresh particulate vehicular exhausts near a Paris high flow road. *Atmos. Environ.* **35**, 453–468.

Samara, C., and Voutsa, D. (2005). Size distribution of airborne particulate matter and associated heavy metals in the roadside environment. *Chemosphere* **59**, 1197–1206.

Samburova, V., Zenobi, R., and Kalberer, M. (2005). Characterization of high molecular weight compounds in urban atmospheric particles. *Atmos. Chem. Phys.* **5**, 2163–2170.

Sannigrahi, P., Sullivan, A. P., Weber, R. J., and Ingall, E. D. (2006). Characterization of water-soluble organic carbon in urban atmospheric aerosols using solid-state ^{13}C NMR spectroscopy. *Environ. Sci. Technol.* **40**, 666–672.

Satheesh, S. K., and Ramanathan, V. (2000). Large differences in tropical aerosol forcing at the top of the atmosphere and earth's surface. *Nature* **405**, 60–63.

Satheesh, S. K., and Moorthy, K. K. (2005). Radiative effects of natural aerosols: A review. *Atmos. Environ.* **39**, 2089–2110.

Saxena, P., and Hildemann, L. M. (1996). Water-soluble organics in atmospheric particles: A critical review of the literature and application of thermodynamics to identify candidate compounds. *J. Atmos. Chem.* **24**, 57–109.

Schmeling, M. (2003). Seasonal variations in diurnal concentrations of trace elements in atmospheric aerosols in Chicago. *Anal. Chim. Acta* **496**, 315–323.

Schmid, H., Laskus, L., Abraham, H. J., Baltensperger, U., Lavanchy, V., Bizjak, M., Burba, P., Cachier, H., Crow, D., Chow, J., Gnauk, T., Even, A., ten Brink, H. M., Giesen, K.-P., Hitzenberger, R., Hueglin, C., Maenhaut, W., Pio, C., Carvalho, A., Putaud, J.-P., Toom-Sauntry, D., and Puxbaum, H. (2001). Results of the carbon conference international aerosol carbon round robin test stage I. *Atmos. Environ.* **35**, 2111–2121.

Schwartz, S. E. (2004). Uncertainty requirements in radiative forcing of climate change. *J. Air Waste Manage. Assoc.* **54**, 1351–1359.

Sciare, J., Oikonomou, K., Cachier, H., Mihalopoulos, N., Andreae, M. O., Maenhaut, W., and Sarda-Estève, R. (2005). Aerosol mass closure and reconstruction of the light scattering coefficient over the Eastern Mediterranean Sea during the MINOS campaign. *Atmos. Chem. Phys.* **56**, 2253–2265.

Seinfeld, J. H. (1986). *Atmospheric Chemistry and Physics of Air Pollution*, John Wiley & Sons, New York.

Seinfeld, J. H., and Pandis, S. N. (1998). *Atmospheric Chemistry and Physics: From Air Pollution to Climate Changes*, John Wiley & Sons, New York.

Seinfeld, J. H., and Pankow, J. F. (2003). Organic atmospheric particulate material. *Annu. Rev. Phys. Chem.* **54**, 121–140.

Sempéré, R., and Kawamura, K. (1994). Comparative distributions of dicarboxylic acids and related polar compounds in snow, rain and aerosols from urban atmospheres. *Atmos. Environ.* **28**, 449–459.

Simoneit, B. R. T., Schauer, J. J., Nolte, C. G., Oros, D. R., Elias, V. O., Fraser, M. P., Rogge, W. F., and Cass, G. R. (1999). Levoglucosan, a tracer for cellulose in biomass burning and atmospheric particles. *Atmos. Environ.* **33**, 173–182.

Simoneit, B. R. T. (2002). Biomass burning—A review of organic tracers for smoke from incomplete combustion. *Applied Geochem.* **17**, 129–162.

Simoneit, B. R. T., Elias, V. O., Kobayashi, M., Kawamura, K., Rushdi, A. I., Medeiros, P. M., Rogge, W. F., and Didyk, B. M. (2004). Sugars—Dominant water-soluble organic compounds in soils and characterization as tracers in atmospheric particulate matter. *Environ. Sci. Technol.* **38**, 5939–5949.

Smith, D. M., and Chughtai, A. R. (1995). The surface structure and reactivity of black carbon. *Colloids Surfaces A: Physicochem. Eng. Aspects* **105**, 47–77.

Smolík, J., Ždímal, V., Schwarz, J., Lazarídis, M., Havránek, V., Eleftheriadis, K., Mihalopoulos, N., Bryant, C., and Colbeck, I., (2003). Size resolved mass concentration and elemental composition of atmospheric aerosols over the Eastern Mediterranean area. *Atmos. Chem. Phys.* **3**, 2207–2216.

Stephanou, E. G. (2005). The decay of organic aerosols. *Nature* **434**, 31.

Subbalakshmi, Y., Patti, A. F., Lee, G. S. H., and Hooper, M. A. (2000). Structural characterization of macromolecular organic material in air particulate matter using Py-GC-MS and solid state ^{13}C-NMR. *J. Environ. Monit.* **2**, 561–565.

Tabazadeh, A. (2005). Organic aggregate formation in aerosols and its impact on the physicochemical properties of atmospheric particles. *Atmos. Environ.* **39**, 5472–5480.

Temesi, D., Molnár, A., Mészáros, E., Feczkó, T., Gelencsér, A., Kiss, G., and Krivácsy, Z. (2001). Size resolved chemical mass balance of aerosol particles over rural Hungary. *Atmos. Environ.* **35**, 4347–4355.

ten Brink, H., Maenhaut, W., Hitzenberger, R., Gnauk, T., Spindler, G., Even, A., Chi, X., Bauer, H., Puxbaum, H., Putaud, J.-P., Tursic, J., and Berner, A. (2004). INTERCOMP2000:

The comparability of methods in use in Europe for measuring the carbon content of aerosol. *Atmos. Environ.* **38**, 6507–6519.

Valavanidis, A., Fiotakis, K., Vlahogianni, T., Bakeas, E. B., Triantafillaki, S., Paraskevopoulou, V., and Dassenakis, M. (2006). Characterization of atmospheric particulates, particle-bound transition metals and polycyclic aromatic hydrocarbons of urban air in the centre of Athens (Greece). *Chemosphere* **65**, 760–768.

Van Dingenen, R., Raes, F., Putaud, J.-P., Baltesnpergerb, U., Charron, A., Facchini, M. C., Decesari, S., Fuzzi, S., Gehrig, R., Hansson, H.-C. Harrison, R. M., Hüglin, C., Jones, A. M., Laj, P., Lorbeer, G., Maenhaut, W., Pamgren, F., Querol, X., Rodriguez, S., Schneider, J., ten Brink, H., Tunved, P., Tørseth, K., Wehnern, B., Weingartnerb, E., Wiedensohler, A., and Wåhlin, P. (2004). A European aerosol phenomenology—1: Physical characteristics of particulate matter at kerbside, urban, rural and background sites in Europe, *Atmos. Environ.* **38**, 2561–2577.

Viidanoja, J., Sillanpää, M., Laakia, J., Kerminen, V.-M., Hillamo, R., Aarnio, P., and Koskentali, T. (2002). Organic and black carbon in PM2.5 and PM10: 1 year of data from an urban site in Helsinki, Finland. *Atmos. Environ.* **36**, 3183–3193.

Wang, G., Wang, H., Yu, Y., Gao, S., Feng, J., Gao, S., and Wang, L. (2003). Chemical characterization of water-soluble components of $PM_{2.5}$ and PM_{10} atmospheric aerosols in five locations of Nanjing, China. *Atmos. Environ.* **37**, 2893–2902.

Xie, S., Zhang, Y., Qi, L., and Tang, X. Y. (2005). Characteristics of air pollution in Beijing during sand-dust storm periods. *Water, Air, Soil Pollution: Focus* **5**, 217–229.

Yang, H., Yu, J. Z., Ho, S. S. H., Xu, J., Wu, W.-S., Wan, C. H., Wang, X., and Wang, L. (2005). The chemical composition of inorganic and carbonaceous materials in $PM_{2.5}$ in Nanjing, China. *Atmos. Environ.* **39**, 3735–3749.

Zappoli, S., Andracchio, A., Fuzzi, S., Facchini, M. C., Gelencsér, A., Kiss, G., Krivácsy, Z., Molnár, Á., Mészáros, E., Hansson, H.-C., Rosman, K., and Zebühr, Y. (1999). Inorganic, organic and macromolecular components of fine aerosol in different areas of Europe on relation to their water solubility. *Atmos. Environ.* **33**, 2733–2743.

13

SEPARATION TECHNOLOGY AS A POWERFUL TOOL FOR UNFOLDING MOLECULAR COMPLEXITY OF NATURAL ORGANIC MATTER AND HUMIC SUBSTANCES

I. V. PERMINOVA, A. I. KONSTANTINOV, AND E. V. KUNENKOV

Department of Chemistry, Lomonosov Moscow State University, Moscow, Russia

A. GASPAR, P. SCHMITT-KOPPLIN, AND N. HERTKORN

Institute of Ecological Chemistry, GSF, National Research Center for Environment and Health, Neuherberg, Germany

N. A. KULIKOVA

Department of Soil Science, Lomonosov Moscow State University, Moscow, Russia

K. HATFIELD

Department of Civil and Coastal Engineering, University of Florida, Gainesville, Florida

Biophysico-Chemical Processes Involving Natural Nonliving Organic Matter in Environmental Systems, Edited by Nicola Senesi, Baoshan Xing, and Pan Ming Huang

13.1. INTRODUCTION

Various separation methods have been used to isolate, fractionate, and characterize humic materials. Originally it was fractionation, based on solubility differences of humic components in diluted alkalis and acids, which laid the ground work for the first classifications of humic substances (HS) in the 19th century (Mulder, 1861; Sprengel, 1837) and provided for operational definition of HS (Kononova, 1966). And now, alkali extraction is the method of choice for isolating HS from solid humus-containing substrates like soil, peat, coal, and so on (Swift, 1996), while hydrophobic resins (e.g., Amberlite XAD resins) are typically used to extract HS dissolved in natural waters (Aiken, 1985). Initial research on HS began with the used simple separation methods to prove, examine, and define characteristics of components of humic matter (Oden, 1919). Today, however, advances in HS research require ever more sophisticated techniques of separation combined with structural analysis (Orlov, 1990; Stevenson, 1994).

Because of polydisperse nature of HS, the importance of separation methods increased as the science evolved. Various separation methods were widely used for conventional fractionation and characterization of components based on differences in component solubility, charge, molecular weight, and/or size, polarity, hydrophobicity, and so on (Janos, 2003). More recent research focused on advanced molecular-level analyses of humic mixtures (Hertkorn and Schmitt-Kopplin, 2007), in which a combination of separation techniques, mostly, chromatography, or capillary electrophoresis) were coupled with high-resolution instrumental analysis [e.g., mass spectrometry (MS) or nuclear magnetic resonance (NMR) spectroscopy]. Several examples appeared in the literature, including those that used size exclusion chromatography (SEC) coupled to electrospray ionization mass spectrometry (ESI/MS) (Reemtsma and These, 2003), capillary zone electrophoresis (CZE) combined with ESI/MS (Schmitt-Kopplin and Kettrup, 2003), and liquid chromatography coupled to NMR (LC-NMR) (Simpson et al., 2004).

Janos (2003) reviewed various separation techniques but paid particular attention to the specific features of analytical setups used by different authors. Burba et al. (1998) published a concise overview of ultrafiltration and applications to fractionate aquatic HS. Piccolo (2001) summarized a topical selection of SEC studies on humics. Hassel'ov et al. (2007) discussed the use of field-flow fractionation to characterize aquatic colloids and macromolecules. Schmitt-Kopplin and Frommberger (2003) gave a broad overview of developments in capillary electrophoresis and mass spectrometry over the last 15 years including developments relevant to analyses of natural organic matter (NOM) and HS. Hertkorn et al. (2007) provided unique perspectives on the integrated use of high-performance separation technologies, ultra-high-resolution organic structural spectroscopy [i.e., Fourier transform ion cyclotron resonance mass spectrometry (FTICR MS) and NMR spectroscopy], and the mathematical treatment of data for the molecular-level analysis of NOM, HS, and other complex systems. They found that methods providing high-precision frequency measurements were most effective when studying complex materials. For example, NMR spectroscopy provided unsurpassed insights into close-range molecular order, while FTICR MS generated excellent resolution and resultant isotope-specific molecular signatures.

The goal of this chapter is to explore the molecular complexity of NOM and HS by examining first structural-molecular trends observed within operationally defined humic isolates and then link these trends to entropy-driven processes, which govern the evolution of humified matter in the environment. Through this in-depth analysis, NOM and HS molecular heterogeneity is interpreted in the context of viewing humification as a stochastic process to which insights can be derived on mechanisms of "combinatorial" information transfer within the nonliving organic matter.

To achieve this goal, highlights of contemporary debates pertinent to covalent and noncovalent interactions within HS and NOM systems are reviewed. Given first are theoretical underpinnings and results of separation technologies applied to fractionate NOM and HS to assist with analysis. The immediate focus is on separation techniques based on molecular size/weight and charge density. Related techniques are discussed next, and they include membrane filtration, chromatography, electrophoresis, and the less commonly employed method of field-flow fractionation (FFF). Application of paired techniques such as those based on coupling a given separation method to a specific high-resolution analytic technique are also relevant and are reviewed. The utility of combining offline separation and advanced organic structural analytics is examined in an effort to reconcile macroscopic and microscopic properties of humic matter. A new hypothesis is proposed that describes the evolution of a humic system in terms of an evolving distribution of organic components such that the state of humification is represented in the molecular weight distribution of these components. An analysis of sampled humics from a given location generates measures of the ensemble molecular distribution. It is assumed that this distribution is time invariant under local quasi-steady-state environmental conditions. Argued is the need for further developments in stochastic chemistry such as those set forth by Cabaniss et al. (2005) to further explain entropy-driven changes in covalently bonded chemical structures and the composition of noncovalently bonded molecular assemblies.

13.2. COVALENT AND NONCOVALENT INTERACTIONS WITHIN NOM AND HS

Covalent and noncovalent interactions within NOM and HS have been a subject of active debate in the literature over the last 10 years (Baigorri et al., 2007; Perminova, 1999; Piccolo, 2001; Simpson, 2002; Sutton and Sposito, 2005; Swift, 1999; Varga et al., 2000; Wershaw, 1999, 2004). The topic goes back to differences in humification models described in detail in monographs by Kononova (1966), Orlov (1990), Stevenson (1994), and others. The two most common hypotheses of humification are biopolymer degradation (BD) and abiotic condensation (AC) (Hatcher and Spiker, 1988; Hedges, 1988). The AC hypothesis assumes for the first stage of humification, complete disintegration of biogenic debris into smaller individual molecules followed by a combining/condensation of these smaller components into macromolecules (Hatcher and Spiker, 1988). In contrast, under the BD hypothesis, humification is a slow process involving abiotic oxidation, condensation, and decomposition reactions among plant and animal biopolymers; in addition, there is microbial activity that does not completely destroy the integrity of parent materials (Hedges, 1988; Hedges and Keil, 1999). As a result, the AC hypothesis considers fulvic acids (FA) to be the more oxidized, lower-molecular-weight precursors of humic acids (HA), whereas the BD hypothesis assumes HA are the precursors of FA on the path of humification. Despite these differences, both hypotheses recognize the predominantly macromolecular nature of HS, and both require the coexistence of high- and intermediate-molecular-weight molecules. Similar conceptual models were adopted by supporters of other humification paradigms as well (Alexandrova, 1977; Kononova, 1966).

Piccolo et al. (1996) and Piccolo (1997) presented results of HS fractionation using size exclusion chromatography (SEC), which fomented considerable debate. They proposed that HS could be viewed as micelles of low-molecular-weight acids held together by weak (hydrophobic, hydrogen, pi) bonds, and they contended that the micelle theory better explained SEC fractionation than the macromolecular theory. Recent SEC experiments, however, appeared to undermine the validity of the micell theory (De Nobili and Chen, 1999; Perminova, 1999; Swift, 1999; Varga et al., 2000). These experiments were among the first to characterize humic materials using FTICR MS, which provided unambiguous proof that the compositional space of HS was extremely complex (Kujawinski et al., 2002; Stenson et al., 2003). Application of this ultra-high-resolution technique to Suwannee River FA and HA provided molecular weight estimates from 400 to 600 Da, and recorded spectra gave no evidence that high-molecular-weight components were present. Data of DOSY NMR published by Simpson et al. (2001, 2002) were in sync with FTICR MS results and yielded low-molecular-weight estimates for humic materials of both aquatic and terrestrial origin. These results in turn engendered a critical review by Sutton and Sposito (2005), who hypothesized the structure of soil HS to be supramolecular associations of low molecular weight compounds. To the contrary, recent investigations on the aqueous aggregation of HS (Baigorri et al., 2007) indicated that bulk humic material was composed of both low-molecular-weight aggregates and a macromolecular fraction with a distinct macromolecular behavior in solution. This supported an emerging view that HS constituted a system of high- and low-molecular-weight components sharing common structural features defined by specific conditions of multiple humification processes.

To further understand the molecular organization in humic systems at the level of covalent and noncovalent interactions, additional developments in the high-resolution analytical tools are needed. To achieve this level of resolution, techniques used to explore the complexity of humic materials must be coupled to separation methods that facilitate substantial reductions in the molecular heterogeneity of studied systems.

The follow-up section will deal with separation methods based upon: (a) molecular size and related to it hydrodynamic volume (size-exclusion chromatography and ultrafiltration), (b) molecular size and related to it molecular diffusivity (field-flow fractionation), and (c) charge/size ratio and related to it molecular polarity (electrophoresis and mass spectrometry). Also reviewed will be hyphenated techniques or those that combine separation by chromatography or electrophoresis with spectral detection.

13.3. SEPARATION OF NOM AND HS BASED ON MOLECULAR SIZE

Molecular size or weight is an important parameter that determines the physical–chemical properties of a chemical and its behavior in the environment (Lead and Wilkinson, 2007). Unlike small molecules, the molecular weight of a macromolecule is not unique. Rather, a solution or mixture of a given macromolecular compound will contain a distribution of macromolecules with close, but not identical molecular weights. Thus, for macromolecular compounds an average molecular weight $<M>$ is used (Belenkii and Vilenchik, 1983). The same is true for complex mixtures of small molecules. Properties of a polymer or of a mixture of small molecules vary with the molecular weight distribution (MWD) function or average molecular weight.

There are several ways to calculate an average molecular weight. If we consider a property that is only sensitive to the number of molecules present—for example, the colligative properties of solutions such as boiling point elevation, freezing point depression, and osmotic pressure—the most relevant average molecular weight is the number-average molecular weight, M_n, which is defined as a polymer weight normalized to the molecule number:

$$M_n = \frac{\sum_{i=1}^{\infty} N_i \cdot M_i}{\sum_{i=1}^{\infty} N_i} = \sum_{i=1}^{\infty} X_i \cdot M_i \quad (13.1)$$

where N_i is the number of macromolecules with molecular weight M_i. The term $N_i \Big/ \sum_{i=1}^{\infty} N_i$, denoted as X_i, is a mole fraction. Physically, it means the number fraction of polymers with molecular weight M_i.

If we consider a property of a polymer solution which depends not just on the number of molecules but on the size or weight of each polymer molecule (e.g., light absorbance), then a weight average molecular weight, M_w, is needed. To calculate M_w, the number of polymers of molecular weight $i(N_i)$ in the formula (13.1) should be replaced with the weight of the polymer having molecular weight $i(N_iM_i)$. The result is

$$M_W = \frac{\sum_{i=1}^{\infty} N_i \cdot M_i^2}{\sum_{i=1}^{\infty} N_i \cdot M_i} = \sum_{i=1}^{\infty} w_i \cdot M_i \tag{13.2}$$

where $w_i = N_iM_i \Big/ \sum_{i=1}^{\infty} N_iM_i$ is the weight fraction of polymer with molecular weight i.

By generalizing the above process, other average molecular weights can be calculated by replacing N_i with $N_iM_i^k$—for example, $M_z(k = 2)$ and $M_{z+1}(k = 3)$. For any molecular weight distribution, the various average molecular weights always rank in the following order: $M_n < M_w < M_z < M_{z+1}$. (Belenkii and Vilenchik, 1983).

This fact is of particular importance for comparing average molecular weight measured with techniques sensitive to a number of molecules or to a weight of molecules. For example, signal intensity in SEC with DOC detectors is proportional to weight of the analyte, which yields M_w, whereas signal intensity in MS techniques is proportional to the abundances of ions and yields M_n. Consider the following numerical example of what number and weight average values reveal about the molecular weight distribution of a mixture. Assuming a mixture containing 12 molecules with nominal masses of 1000 Da (one molecule), 100 Da (one molecule), and 10 Da (10 molecules), the M_n of the mixture is 100 whereas M_w is 842.5. Hence, despite the presence of only one "heavy" molecule among 12 in the mixture, the value of M_w reflects most closely the value of the heaviest (most influential) molecule. Hence, a few large molecules in the mixture of very many small molecules can shift substantially a value of M_w which is often used to characterize molecular weight distributions of mixtures.

Figure 13.1 shows the size range of common materials and applicable methods for separating those materials. It also gives information on the interrelationship between sizes and weights of dissolved or suspended materials estimated on the dextran scale. HS and NOM are shown in Figure 13.1 in accordance with reported MW estimates (see Section 13.3.1. for details) to give an idea of how HS and NOM sizes are related to different chemicals and bacteria. For HS and NOM, molecular weights (kDa) and molecular sizes (nm) vary three orders of magnitude encompassing a range from 1 to 1000 nm. This enables us to speak of HS as of natural nanomaterials. Due to those large variations, separation by molecular size or weight is commonly used to fractionate HS and NOM (Janos, 2003). Both chromatographic and membrane techniques shown in Figure 13.1 are used for this purpose. Reverse osmosis is mostly used for the isolation of NOM and HS from aquatic environments (Perdue and Ritchie, 2003). Size exclusion chromatography (SEC) is frequently applied for molecular weight determinations of NOM and HS (Swift, 1999). Preparative applications are reported as well (Egeberg and Alberts, 2003; Mueller et al., 2000; Swift et al., 1992). Among membrane techniques, ultrafiltration is the most popular method of fractionating humic materials (Burba et al., 1998; Janos, 2003). Field-flow fractionation (FFF) is an emerging technique, which has found

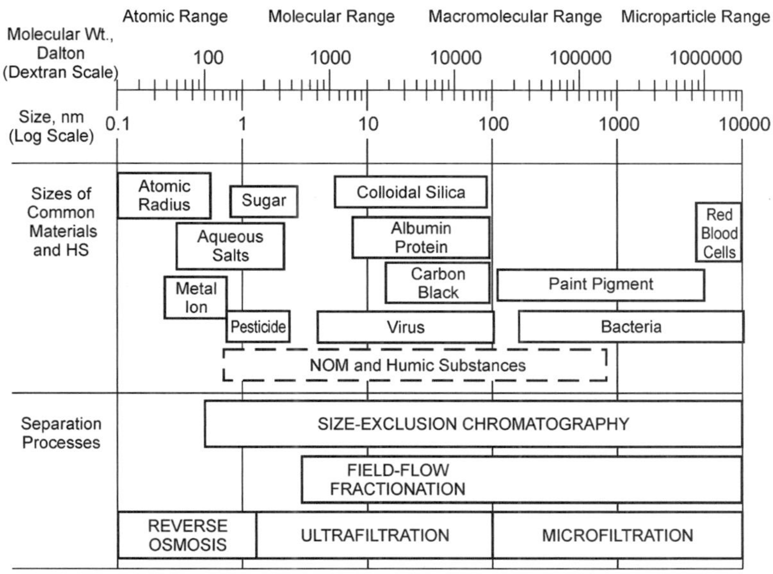

Figure 13.1. Size range of common materials, conversion scale between molecular size and weight based on the dextran scale, and separation range of different analytical methods.

broad applications in the fractionation of humic materials and aquatic colloids (Janos, 2003; Hassel'ov et al., 2007).

13.3.1. Size Exclusion Chromatography

13.3.1.1. Theory of Size-Exclusion Chromatography. Size-exclusion chromatography (SEC) is one of the most powerful techniques for separating mixture components according to molecular size and for determining the molecular weight (MW) distribution of macromolecular compounds (Belenkii and Vilenchik, 1983; Yau et al., 1979). The technique was developed by Porath and Flodin (1959). The underlying principle of SEC is shown in Figure 13.2, where it is implied that particles of different sizes will elute at different rates through a stationary phase that functions like a sieve: Larger molecules will take less time (or elution volume) to reach the outlet of an elution column as compared to the smaller ones. The prerequisite for a direct correlation between elution time and molecular size of the analyte is an assumption of no interactions between stationary phase and analyte. Otherwise, nonexclusion effects such as ionic exclusion and sorption will also govern constituent separation as shown in Figure 13.3 (Belenkii, 1998; Dubin, 1988).

Ionic exclusion arises from repulsive interactions between a partially charged gel matrix and an analyte possessing the same charge. As a result, the observed elution volume (V_e) for this analyte is lower (indicating incorrectly higher MWs) compared to the value expected from size exclusion alone. In the case of a highly charged analyte, V_e can be close to the void volume value of the column (V_0).

Sorption occurs and then ion-exchange or hydrophobic or hydrogen bonding occurs between an analyte and a gel matrix. Sorption retards analyte mobility. Hence, observed elution volumes increase for sorbed analytes (suggesting incorrectly lower MWs) as compared to those controlled by size exclusion alone. In the case of a

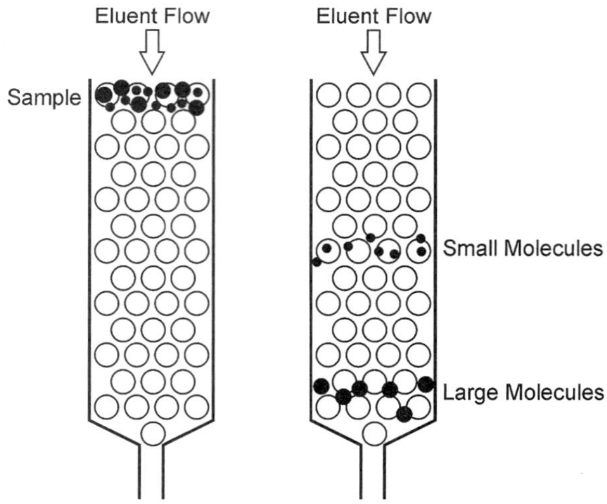

Figure 13.2. Illustration of molecular separation according to their size in a size-exclusion chromatography (SEC) column. The underlying principle of SEC is that particles of different size will elute through a stationary phase at different rates: Larger molecules will take less time (or elution volume) to reach outlet of the column as compared to the smaller ones. The prerequisite of a direct correlation between elution time and molecular size is an absence of interactions between the stationary phase and an analyte. Otherwise, nonexclusion effects such as electrostatic repulsion or sorption must be considered.

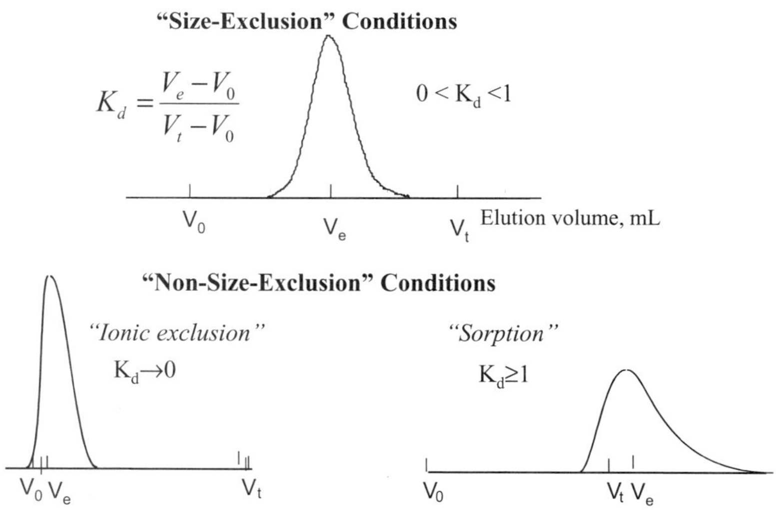

Figure 13.3. Appearance of "non-size-exclusion effects" on SEC-elution curves of polyelectrolytes and other charged analytes including low-molecular-weight organic acids. K_d is the distribution coefficient; and V_e, V_0, and V_t are elution volume of the analyte, column void volume, and total column volume, respectively.

strongly interacting analyte, V_e can reach values much larger than the total pore volume of the column (V_t). Given size exclusion effects take place only under conditions of $0 < K_d < 1$ (Yau et al., 1979), elution of the analyte at the volumes exceeding the total pore volume ($V_e > V_t$) unambiguously indicates the presence of sorption.

Eluents are often modified to avoid the effects of ionic exclusion and sorption (Belenkii and Vilenchik, 1983). For example, low-MW electrolytes or buffers are included in eluents to suppress ion-exclusion effects. The hydrophobic interactions are eliminated through the use of hydrophobic solvents or surfactants designed to occupy the hydrophobic sites of the gel, whereas the effects of hydrogen bonding are minimized by the addition of urea which competes for hydrogen bonding sites on the gel. However, the application of modified eluents can give rise to other kinds of artifacts, known as salt peaks, if target analytes are not in equilibrium with the mobile phase (De Nobili et al., 1989). Such artifacts evolve from concentration and charge gradients at the edges of the chromatographic zone. The position of the salt peaks on the SEC chromatogram depends on the elution volume of the low-MW electrolyte, which is smaller than the total pore volume of the column due to repulsive interactions between the electrolyte and the gel.

13.3.1.2. Application of SEC for HS Analysis and Fractionation. SEC was first applied to the analysis of HS by Posner in 1963 (Posner, 1963). Since then, a vast amount of experimental data has been gathered which show that elution conditions such as pH and ionic strength are crucial for the results of this analysis (De Haan et al., 1987; Frimmel et al., 1992; Gjessing, 1973; Mori et al., 1987; Pershina et al., 1989; Piccolo, 1997; Piccolo et al., 1996; Swift, 1999; Town and Powell, 1992; Varga et al., 2000). The acidic nature of HS gives rise to non-size-exclusion effects that depend not just on molecular size but also on electrostatic and/or sorptive interactions between ionogenic analytes and hydrophilic gel matrices (Belyaeva et al., 2006; De Nobili et al., 1989; Perminova, 1999). Hence, proper interpretation of the SEC results must consider possible non-size-exclusion effects not related to molecular size but intrinsic to experimental conditions.

Reported MW values of HS from different sources as determined by SEC are summarized in Table 13.1.

Molecular weights appear to range from 0.3 to 700 kDa; however, this range is more a reflection of differences among experimental conditions than the real differences in MW caused by different sources. Problems of SEC application for the analysis of HS were addressed in a special issue of *Soil Science* (1999, Vol. 164, No. 11). In addition to non-size exclusion effects, another source of MW variability is associated with a lack of proper standards (De Nobili et al., 1989). This problem was the subject of attention in Perminova et al. (1998).

Estimates of "true" MW variations in HS of different genesis can be derived from results reported by Perminova et al. (2003). Here a large suite of humic materials, isolated from different aquatic and terrestrial environments, were analyzed using standardized SEC conditions: the same column packing (Toyopearl HW50S) and mobile phase (0.028 mol liter^{-1} phosphate buffer at pH 6.8). Figure 13.4 summarizes results.

Observed trends show average MW values changed consistently with the source and fractional composition of HS known in the literature (Table 13.1): Aquatic humic materials and soil fulvic acids possessed the lowest MWs, whereas HS from peat and HA from soil had the largest. Results show that SEC gives meaningful estimates of trends in MW properties inherent within HS under conditions compensating for the strongest interactions between the humic analyte and gel matrix induced by extremely low values of ionic strength and pH for the mobile phase.

TABLE 13.1. SEC-Determined Molecular Weight Characteristics of Humic Substances from Different Sources as Reported in the Literature

	M_n (kDa)	M_w (kDa)	M_w/M_n
HS Sample			
Aquatic FA	0.64–5.4	1.0–16.73	1.6–2.87
Aquatic HA and HF	0.92–7.82	1.45–28.91	1.58–5.57
Aquatic DOM	0.55–7.13	0.85–17.77	1.5–7.1
Sediments FA	1.2–1.88	3.07–3.99	2.56–2.13
Aldrich HA[a]	1.63	4.1	2.5
		Range of Molecular Weights (kDa)	
Soil FA and HA			
Peat soil HA[c]		4–22	
Sod podzolic FA[d]		1–5	
Sod podzolic HA[d]		100–700	
Soil (n.d.)[e]		100–200	

[a]Polystyrenesulfonates.
[b]Dimethylformamide (DMFA) is used as a mobile phase.
[c]Proteins.
[d]Calibration standards are not known.
[e]Polydextrans are used as calibration standards.
Source: Based on Perminova et al. (2003).

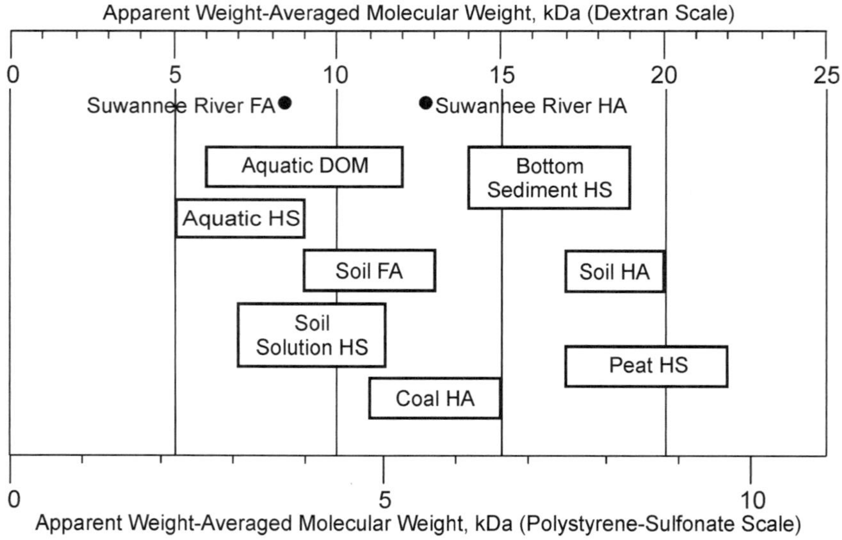

Figure 13.4. Apparent MWs of HS from different environments as determined by SEC [compiled from the data reported by Perminova et al. (2003)]. Bars represent a range of weight-averaged molecular weights determined for different groups of humic materials. Two black dots represent two IHSS standards: Suwannee River FA and HA. SEC chromatograms were obtained with column-packing Toyopearl HW-50S using 0.028 M phosphate buffer at pH 6.8 as a mobile phase. Two different sets of standards were used to calibrate the column: polystyrene sulfonates (PSS) and polydextrans; the upper MW scale corresponds to dextran calibration and the bottom one—to PSS calibration.

However, due to uncertainties among criteria used to define complete compensation for non-size-exclusion effects and inadequate calibration standards, molecular weight determinations can be inaccurate. Hence, estimates should be treated as apparent molecular weights.

SEC can be used both for analytical and preparative purposes. The latter has been demonstrated by Egeberg and Alberts (2003) and Mueller et al. (2000), who investigated the applicability of high-pressure SEC (HPSEC) for the preparative fractionation of NOM. The authors conducted five different experiments to examine the non-size-exclusion effects and the reproducibility of HPSEC system. Examined were various experimental conditions including (1) different flow rates, (2) different sample concentrations, (3) reinjection of collected fractions, (4) addition of acetonitrile to the mobile phase, and (5) reinjection of hydrophobic and hydrophilic subfractions. They concluded that gel–solute interactions were a minor problem, and that the selected HPSEC system separated NOM molecules mainly on the basis of molecular size. The primary distinguishing features of collected fractions were the relatively strong absorbance at shorter (<290 nm) wavelengths for smaller molecules (a structural indication of higher aromatic content), strong specific UV–vis absorbance for NOM molecules of intermediate molecular mass, and very low fluorescence of larger molecules. Similar trends were reported by Mueller et al. (2000) who used low-pressure SEC for fractionating NOM from brown water lake.

The present overview allows us to conclude that SEC can be considered an appropriate analytical tool for investigation of molecular weight distribution of HS as well as a preparative tool for large-scale fractionation of HS.

13.3.2. Application of Ultrafiltration for Fractionation of NOM and HS

Ultrafiltration (UF) is a technique that has been widely used for fractionation of HS on the basis of molecular size differences (Janos, 2003). Its application to quantify size and molecular weight characteristics of aquatic humic materials is reviewed in detail by Burba et al. (1998) and Guo and Santchi (2007).

13.3.2.1. Theory of Ultrafiltration. Ultrafiltration is a membrane process with the ability to separate molecules in solution on the basis of size (Ghosh, 2003). Particles are separated with the use of pressure and specially designed semipermeable membranes (Figure 13.5). An ultrafiltration membrane acts as a selective barrier. It

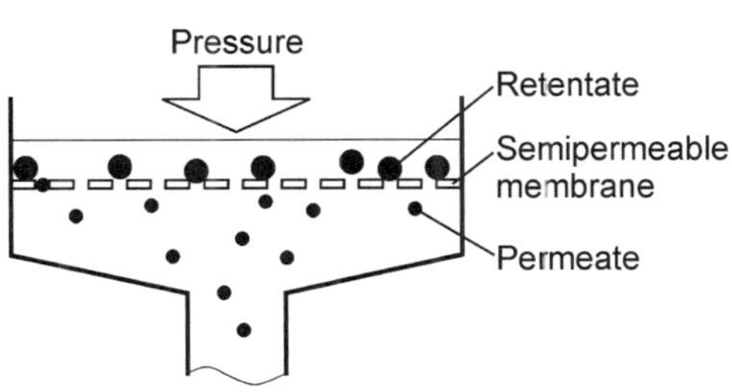

Figure 13.5. The separation principle of membrane processing is a sieving mechanism, and the driving force is a pressure range of 1–10 bars. Particles are separated on the basis of their molecular size and shape with the use of pressure and specially designed semipermeable membranes. Permeate designates the liquid passing through the membrane, while retentate (concentrate) designates the fraction not passing through the membrane.

retains species with molecular sizes larger than cutoff determined by pore size. The separation is achieved by concentrating the large molecules present in the feed on one side of the membrane (the retentate), while the solvent and microsolutes pass through the membrane and produce the permeate.

Under the influence of pressure, the membrane permits specific components to pass through (or permeate). The membrane also inhibits transport of some components. This selective transport forms the basis of the membrane separation process. Rejection is a bulk separation capability of the membrane. The observed solute rejection coefficient R_i for a given species i is given by

$$R_i = 1 - c_{ip}/c_{ir} \tag{13.3}$$

where i is the component, or collection of components, c_{ip} is the concentration in the permeate, and c_{ir} is the concentration in the retentate. The ratio of c_{ip} to c_{ir} is called concentration polarization. Concentration polarization is a common feature of all pressure-driven membrane processes. Fouling is a boundary layer phenomenon, caused by concentration polarization, in which solutes deposit on the membrane surface and reduce membrane flux and selectivity.

The most important factor in ultrafiltration is membrane selection. The membrane characteristics are surface pore size, pore size distribution, % porosity, rejection, flux, temperature stability, solvent resistance, and pressure resistance. Most UF membranes are made from polymers or ceramic materials—for example, polysulfone (PS), polyethersulfone (PES), sulfonated polysulfone (SPS), polyamide (PA), cellulose acetate (CA), zirconium oxide (inorganic), and alumina (inorganic) (Ghosh, 2003). There are several geometries of commercial membrane systems including plate and frame, tubular, spiral wound, and hollow fiber. The majority of these membranes (modified or unmodified) are able to withstand temperature greater than 100 °C and pH ranges of 1–14. The UF membranes are asymmetric porous and the pore sizes range from 0.05 μm to 1 nm. The separation principle for these membranes is a sieving mechanism and the driving force is a pressure range of 1–10 bars. Most UF processes operate in cross-flow mode. UF membranes are typically rated by molecular weight cutoff (MWCO), giving the molecular weight of a hypothetical macrosolute that the membrane will retain. UF generally has a MWCO of 1000–1,000,000 Å (see Figure 13.1).

13.3.2.2. UF Fractionation of Humic Materials. Conventional UF procedures have been studied since the early 1970s with respect to their fractionation performance for naturally occurring colloids such as HS in aquatic and terrestrial ecosystems (Burba et al., 1998) as well as in fouling membranes during drinking water treatments (Combe et al., 1999; Kennedy et al., 2005; Manttari et al., 2000). Concentration polarization and fouling as a boundary layer phenomenon play a substantial role in UF separation of NOM and humics (Combe et al., 1999; Costa and De Pinho, 2005; Kennedy et al., 2005; Lin et al., 2000; Manttari et al., 2000; Tang and Leckie, 2007). For example, for cellulose acetate (CA) membranes with pore sizes of 20, 54, and 62 Å, Costa and De Pinho (2005) have reported rejection coefficients close to 100% for all HS solutions tested. In contrast, membranes with pore sizes of 86 and 106 Å had lower rejection coefficients, which decreased when ionic strength increased. Different permeation patterns were correlated to membrane pore size

TABLE 13.2. Molecular Weight Cutoffs of Membranes Used for UF Fractionation of Humic Materials

Organic Matter	Molecular Weight (kD)	Reference
Aldrich HA	1, 10, 50, 100, 300	Shin et al. (1999)
Aldrich HA	3, 10, 30, 100	Hur and Schlautman (2003)
Aquatic HF	10, 30, 50, 100	Rocha et al. (2000)
Aquatic HF	30–100	Sargentini et al. (2001)
Dissolved organic matter	1–10	Chow et al. (2005)
Dissolved organic matter	1, 30, 100, 300	Kennedy et al. (2005)
Soil HS	1–100	Dick and Burba (1999)
Soil HA	1, 3, 10, 30, 50, 100, 300	Khalaf et al. (2003)
Soil HA	10, 30, 100, 300	Christl et al. (2000)
Peat HA	1, 3, 5, 10, 30, 100, 300	Li et al. (2004)

and to solution ionic strength through a control of mass transfer mechanisms, configurations of humic macromolecules, and membrane/humic macromolecules electrostatic interactions. It was concluded that the effect of ionic strength on humic acid deposition was enhanced in the more permeable membranes due to a higher contribution of convection to permeation fluxes. Some examples of molecular weight cutoffs of membranes used for UF fractionation of HS from different sources are given in Table 13.2.

From examples given in Table 13.2, the most common approach to fractionate HS is to use a series of Amicon membranes with reducing MWCOs. Of particular interest are reports of substantial humic material retention by membranes of the highest cutoff. For example, Li et al. (2004) reported 50% retention of TOC of bulk peat HA by a membrane with an MWCO of 300 K.

Hur and Schlautman (2003) reported 32.3% retention for the Aldrich HA UF size fraction in the range of 30–100 K. Khalaf et al. (2003) reported 33% retention of the mass of the whole HA by membranes up to 100 MWCO. Even aquatic NOM showed rejection with membranes of the highest MWCO (Kennedy et al., 2005). The supposed MW values of rejected fractions should be much higher as compared to MW estimates obtained for the same or similar materials fractionated with SEC (Figure 13.4). This might indicate that concentration polarization and membrane fouling have a substantial impact on HS fractionation. Hence, care must be exercised when interpreting MWCOs of membranes used as approximate estimates of molecular weights of the retained fractions. Systematic studies on the elimination of specific interactions between UF membranes and humic materials are needed as well as comparisons of molecular weight estimates acquired by alternative techniques (e.g., field-flow fractionation).

13.3.3. Field-Flow Fractionation (FFF)

Another technique widely used for size separation of humic materials is field-flow fractionation (FFF) (e.g., Baalousha et al., 2006; Boehme and Wells, 2006; Geckeis et al., 2003; Hassil'ov et al., 2007; Siripinyanond et al., 2005; Suteerapataranon et al., 2006; Zanardi-Lamardo et al., 2002). This technique was developed and introduced in 1966 by Giddings (1966) as a method for the separation and characterization of materials ranging in size from macromolecules to particulates. Similar to SEC, FFF is an elution-based separation technique. Unlike chromatography, however, FFF separations are carried out in a single phase. The separation of analytes by FFF is achieved in a laminar carrier flow under the action of applied flow field, consisting of a flow perpendicular to the carrier flow. The FFF separation mechanism does not rely on adsorption or partitioning; consequently, this technique is less likely to suffer sample loss. In principle, the particles of analyte do not experience intense contact with a separation medium. This makes FFF as a promising tool for separating the constituents of humic materials. An added benefit is that FFF instruments can be linked readily to analytical instruments to provide real-time analysis.

13.3.3.1. FFF Theory and Instrumentation. The principle of FFF is shown in Figure 13.6. Under laminar flow conditions, the flow velocity across the channel thickness (diameter of the capillary) w has a parabolic profile with the highest flow velocity at the center of the channel and slowest velocity at the walls. An external field is applied at the right angle to the direction of carrier flow. The external field interacts to different extents with different sample components, and the separation mechanism in FFF is based on the physical interaction of particles with an applied field and the subsequent migration down the channel caused by the carrier fluid. Molecules, depending on their size and diffusion coefficient, are distributed over different velocity lines of the flow in the axial flow, and separated accordingly. Larger particles that possess less diffusional motion and higher interaction with the applied field, will be caught up in the slower moving streams near the channel wall and elute later than smaller particles.

Different types of distributions correspond to different operating modes. The two most frequently used operating modes are normal mode and steric and hyperlayer mode (Reschiglian et al., 2005). The normal mode of separation is active for particles $<\sim 1\,\mu m$ and the steric and hyperlayer modes are applicable to particles $>\sim 1\,\mu m$. In the normal mode as macromolecules or particles that constitute the sample are driven by the field toward the accumulation wall, their concentration increases with decreasing distance from the wall. This creates a concentration gradient that causes sample diffusion away from the wall. Retention time in the normal FFF is therefore

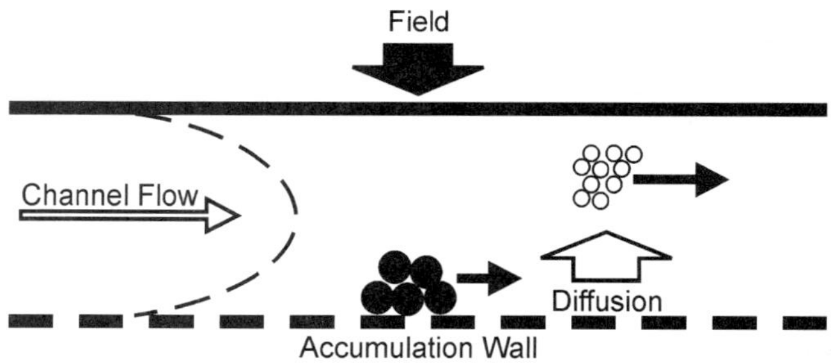

Figure 13.6. Separation principle of field-flow fractionation (FFF) is based on physical interactions of particles within an applied field and subsequent field-induced migration to the FFF channel wall ("accumulation wall"). Molecules, depending on their size and diffusion coefficient, are distributed over different velocity lines of axial flow, and they separate accordingly. Larger particles possess less diffusional motion and higher interaction with the applied field; hence, they will be caught up in slower-moving streams near the channel wall and elute later than smaller particles.

shorter for lower molar mass or size (Reschiglian et al., 2005). In the steric and hyperlayer mode, sample components are micron-sized particles, and their diffusion away from the wall is negligible. Particles are in fact driven by the field directly to the accumulation wall. Particles of a given size form a thin layer of a given thickness, hugging the wall. Larger particles form thicker layers that penetrate into faster streamlines of the parabolic flow profile, and they are eluted more rapidly than smaller particles. Hence, both the steric and hyperlayer modes are referred to as reversed modes. Retention in the steric FFF depends only on particle size. When particles are driven from the wall by a distance that is greater than their diameter, the retention mode is called hyperlayer; and in addition to size, it also depends on the various physical features of the particles (Reschiglian et al., 2005).

At present, FFF in the normal mode is mainly presented where the transport of each sample component is thus formed as a result of two opposing transport processes occurring continuously throughout the separation: field-induced migration to the FFF channel wall ("accumulation wall") and diffusion from regions of high concentration at the accumulation wall to regions of lower concentration away from the wall. At equilibrium, a steady state is established for each sample component at a unique distance l from the channel wall, depending on the magnitude of the diffusion coefficient and the strength of the interaction of the field with the component. The mean thickness of this sample equilibrium layer is related to the retention time and can be expressed by the formula (Giddings, 1993)

$$l = \frac{kT}{F} \tag{13.4}$$

where k is the Boltzmann constant, T is the absolute temperature, and F is the force exerted by the field on a single particle. The retention time t_R of each sample component in FFF analysis can be described as follows:

$$t_R = \frac{w}{6l} t^0 = \frac{Fw}{6kT} t^0 \tag{13.5}$$

where t^0 is void time (the emergence time for a nonretained tracer).

As it can be seen from the equation, t_R is roughly proportional to F. The magnitude of F, in turn, depends on particle properties, field strength, and type of field.

The family of FFF techniques includes (Schimpf et al., 2000)

- Centrifugal (giving rise to sedimentation FFF or SdFFF)
- Electrical (electrical FFF or ElFFF)
- Cross-flow (flow FFF or FlFFF)
- Temperature gradient (thermal FFF or ThFFF)

Although many other types of fields have been also used (e.g., magnetic, dielectric), equipment is commercially available only for these four. Equations have been derived for each specific field to relate the retention time to physicochemical properties of the sample and the experimental conditions. Among all those techniques, cross-flow FFF is most widely used for characterization of humic materials (Janos, 2003) and its principles will be considered below.

Sedimentation FFF implies application of the centrifugal field, which is produced by placing the channel in a centrifuge basket. SdFFF instruments can be linked readily to analytical instruments to provide analysis in real time. For the first time, Beckett (1991) introduced FFF–ICP–mass spectroscopy (MS) as a powerful analytical tool for characterizing macromolecules and particles. Taylor et al. (1992) illustrated the characterization of some inorganic colloidal particles and river-borne suspended particulate matter of size <1 μm using SdFFF and ICP-MS.

Other analytical tools have also been used as offline detectors for the FFF techniques. Blo et al. (1995) used for this purpose a graphite furnace atomic absorption spectrometer (GFAAS) to analyze colloidal kaolin particles. Contado et al. (1997) used same design to characterize river-suspended particulate matter of size <1 μm. However, the centrifuges available for SdFFF are only capable of separating particles with sizes down to 80 nm. Usage of SdFFF for characterization of HS is therefore restricted because of this limitation.

Cross-flow FFF (FlFFF) utilizes a second fluid flow to transport sample components across the channel thickness to the accumulation wall, and the position of individual species in the laminar carrier profile corresponds to their ordinary (Fick's) diffusion coefficient. As the particle size increases, the diffusion coefficient decreases until it becomes a relatively insignificant transport process. For micron-size particles, the extent of protrusion into the channel becomes the decisive factor in determining the order of elution.

For cross-flow FFF, the entire sample is displaced regardless of mass, density, size, charge, and so on. As a consequence, FlFFF is the most universally applicable FFF technique with applications encompassing macromolecules with molecular weight of 10^3–10^9 Da and particles 50 μm in diameter (Ratanathanawongs and Lee, 2006). For materials <1 μm, separation occurs in the normal mode and the retention time is inversely proportional to diffusion coefficient D and proportional to the hydrodynamic diameter. The order of sample elution for the normal mode of operation is high diffusion coefficient (small particles or low-molecular-weight molecules) followed by decreasing D (large particles or high-molecular-weight molecules).

Cross-flow FFF allows the determination of the diffusion coefficient and according to the Stoke's equation the hydrodynamic diameter is accessible without any calibration. However, molecular weight and molecular weight distributions require the use of calibration standards (Manh Thang et al., 2001) or a detector capable of measuring molecular weight of the eluting sample. The UV detector is the most commonly used in FlFFF. In the absence of molecular calibration standards, the multiangle light-scattering techniques (MALS) are used for this purpose (Becket and Hart, 1993; Ratanathanawongs and Lee, 2006).

13.3.3.2. Use of Field-Flow Fractionation for the Separation of Humic Materials. Among the FFF techniques available, cross-flow FFF is the most used to separate components of humic materials. Size characteristics were reported by several authors (Baalousha et al., 2006; Boehme and Wells, 2006; Geckeis et al., 2003; Siripinyanond et al., 2005; Suteerapataranon et al., 2006; Zanardi-Lamardo et al., 2002). Over the past decade, about 20 papers were published on FlFFF fractionation of humic materials. Overview of the data on molecular weights determined for HS by FlFFF technique are presented in Table 13.3.

The data reveal significant heterogeneity in measured sizes of HS. For example, the value of hydrodynamic diameter d_h for the Aldrich HA determined by different authors varies from 2 to 5.8 nm. This demonstrates clearly that FlFFF parameters can influence significantly the measured size of HS. Indeed, the need for optimizing FlFFF for relevant data acquisition was mentioned by Schimpf and Petteys (1997); Manh Thang et al. (2001), and Benincasa et al. (2003).

Schimpf and Petteys (1997) studied the effect of pH and salt concentration on hydrodynamic size of HA, FA, and DOC from compost. They generally found that the hydrodynamic size decreased along with a decrease in pH, but at a pH below 5, HA formed large aggregates. Small amounts of sodium chloride had little effect on the size distribution of HS, whereas calcium chloride reduced the hydrodynamic size of individual molecules inducing the formation of aggregates.

Benincasa et al. (2003) also studied the effect of ionic strength and electrolyte composition on hydrodynamic characteristics of HS. The author's reported that components of different HS fractions behave like organic acids, but that the retention level of fractions with larger components may not be accurately modulated by varying mobile phase properties as these species are either totally retained in acidic phases or released before the void peak at pH 4.2. Authors concluded that pronounced differences exist in the physicochemical properties of some HS components even when particle sizes were similar.

Manh Thang et al. (2001) studied another important factor influencing the measured size distribution of HS by FlFFF, namely, the interactions of HS with channel components and with the ultrafiltration membrane used as the accumulation wall.

TABLE 13.3. Peak (M_p), Number Averaged (M_n), and Weight Averaged (M_w) MWs and Hydrodynamic Diameters (d_h) for Natural Organic Matter of Different Origin as Determined by FlFFF

		Molecular Weight (kDa)			
Organic Matter	d_h (nm)	M_p	M_n	M_w	Reference
Aquatic DOM		2–2.4			Zanardi-Lamardo et al. (2002)
Aquatic DOM	<10				Baalousha et al. (2006)
Aquatic DOM		1–5 15–150			Boehme and Wells (2006)
Aquatic FA	1.5–2.5				Lead et al. (2000)
Aquatic FA	3.0	1.7	1.6	2.8	Manh Thang et al. (2001)
Aquatic HA	3.2	1.9	1.7	3.3	ibid.
Porewater DOM	1.8	1.0	1.1	1.8	ibid.
Groundwater FA	2.1	1.0	1.4	2.5	ibid.
Groundwater DOM	2.2	1.0	1.8	4.1	ibid.
Groundwater HS	<3				Geckeis et al. (2003)
Aldrich HA	2.6	1.3	1.7	3.3	Manh Thang et al. (2001)
Aldrich HA	3.0				Bouby et al. (2002)
Aldrich HA	2.9–5.8				Siripinyanond et al. (2005)
Aldrich HA	2				Suteerapataranon et al. (2006)
Sediment HS	2–6				Siripinyanond et al. (2002)

They concluded that the surface charge of HS controlled their recovery, and any carrier component decreasing the effective surface charge of HS led to a sorption increase. Furthermore, the cross-flow rate influenced recovery dramatically, which supports the idea that either sorption to membrane or permeation through it is responsible for losses. Finally, other channel components play a minor role. The authors recommendations were to use 0.005 M TRIS-buffer (pH 9.1) as a carrier solution (recovery of about 85–90%) and use a regenerated cellulose membrane with a 5-kDa cutoff as the accumulation wall.

As can be seen from Table 13.3, most FFFF studies give estimates of hydrodynamic diameter d_h of HS, whereas estimates of molecular weight characteristics are relatively scarce. This can be explained by a lack of adequate calibration standards required for determination of MW from hydrodynamic characteristics. Reported MWs are most often estimated using polystyrene sulfonates (PSS) and globular proteins as standards. At the same time, as in case of SEC determinations, there are substantial differences in surface charge properties of PSS and HS (Manh Thang et al., 2001). For PSS, a considerable dependence of the elution volume on the concentration was demonstrated: An increase in concentration led to a decrease in the elution volume. The concentration effect was not observed for the smallest PSS standard of 1.37 kDa, while it was very significant for the larger-size (3.8–46.4 kDa) standards. These findings were explained as a consequence of the high pH and low ionic strength of the selected carrier solution. Under those conditions, repulsive forces led to an increase in the apparent hydrodynamic radius of the polyelectrolytes. These forces weaken along with a decrease in PSS concentration and an increase in the ionic strength of the carrier. However, higher ionic strength promotes sorption and loss of humic analyte. To alleviate the problem, the authors offer to extrapolate measured PSS elution volumes to those obtained for an infinitely small PSS concentration.

Comparison of MW data determined with a use of FlFFF (Table 13.3) with those determined by SEC (Table 13.1) shows good consistency. For example, weight-averaged MW value for Aldrich HA were 4.1 and 3.3 kDa for SEC and FlFFF techniques, respectively. Of importance is that as in the case of SEC, the size distributions obtained by FlFFF were considerably smaller than expected from the stated pore size of the UF membranes. These results demonstrate that care must be taken in using UF to classify size distributions of humic materials.

13.4. SEPARATION OF COMPLEX NOM AND HS SYSTEMS BASED ON CHARGE DENSITY AND POLARITY: ELECTROPHORETIC TECHNIQUES

Electrophoresis plays a key role as an analytical or preparative technique in the characterization of natural organic matter because it gives information about the behavior of these molecular mixtures in controlled solution conditions, depending on both the size and the charge distribution frequency of the analytes in the complex mixture. Historically, the first electrophoretic separations were conducted with environmental colloids; and over the years all the techniques based on zone, gel electrophoresis, or isoelectric focusing were used in their different setups to analyze natural organic matter and environmental particles to a minor extent. The goal of most of the electrophoretic studies pertinent to NOM has been fractionation, often directed toward specific biomolecules (e.g., polysaccharides, N-containing compounds, enzymes) potentially present in mixtures. Methods primarily focused on the analysis of humic component of materials (alkali extracts of soils); and in most of the cases, electrophoresis was used as a fingerprinting method rather than as a method to study quantitatively the structural characteristics of NOM such as molecular size or charge. Free solution and paper electrophoresis, gel electrophoresis, and isoelectric focusing (electrofocusing) were the most frequently used techniques from 1950 to the late 1980s (Duxbury, 1989). Since the early years of electrophoretic separations of humic substances, the main research goal was to achieve a degree of resolution with respect to separation or the creation of bands consistent with that of biomolecules. These bands were in most of the cases attributed to humic fractions having similar electrophoretic mobility; furthermore, possible interactions of the humic substances with buffer components or separation matrices were known in the early 1950s to produce artefacts. For example, Stevenson stated in 1953 that associations of humic molecules could interfere with the separation process (Stevenson et al., 1953); however, these facts were often neglected.

Relevant literature on the overall topic of electrophoresis was reviewed by Deyl et al. (1979). For a review of capillary electrophoresis see Khaledi (1998), Kuhn and Hoffstetter-Kuhn (1993), Schmitt-Kopplin (2005), and Schmitt-Kopplin and Frommberger (2003). Classical electrophoretic techniques applied in the characterization of humic substances are discussed by Duxbury (1989) and Schmitt-Kopplin and Frommberger (2003).

13.4.1. Zone Electrophoresis

In a free solution, the electrophoretic mobility (i.e., μ_{elec}, the particle velocity per unit applied electric field) is a function of the net charge, the hydrodynamic drag on a molecule, and the properties of the solutions (viscosity; present ions—their concentration and mobility). It can be expressed as the ratio of its electric charge Z (Z = $q{\cdot}e$, with e the charge if an electron and q the valance) to its electrophoretic friction coefficient. Different predictive models have been demonstrated involving the size, flexibility, and permeability of the molecules or particles. Henry's theoretical model of μ_{elec} for colloids (Henry, 1931) can be combined with the Debye–Hückel theory predicting a linear relation between mobility and the charge Z:

$$\mu_{elec} = \frac{eZ}{6\pi\eta R_h} \frac{f_1(\kappa R_h)}{(1+\kappa R_h)} \tag{13.6}$$

R_h is the hydrodynamic radius of the analyte, κ is the inverse of the Debye length, η is the viscosity of the separation buffer, e is the fundamental unit of charge, and f_1 is a function that describes the effect of the molecule (or particle) on the electric field and is defined between two limits: (i) the Hückel limit, $f_1 = 1$ when $\kappa.R_h < 1$ (when the hydrodynamic radius is lower than the Debye length) and (ii) the Helmholtz–Smoluchovski limit, $f_1 = 3/2$ when $\kappa.R_h > 10$ (when the hydrodynamic radius is higher than the Debye length). Between the limits f_1 is calculated from the following equation:

$$f_1(\kappa R_h) = \left(1 + \frac{\kappa^2 R_h^2}{16} - \frac{5\kappa^3 R_h^3}{48} - \frac{\kappa^4 R_h^4}{96} + \frac{\kappa^5 R_h^5}{96} + -\frac{11}{96} e^{\kappa R_h} \int_{\infty}^{\kappa R_h} \frac{e^{-\kappa}}{r} dr\right) \quad (13.7)$$

For small molecules (metabolites, monomers, and small oligomers) the mobility equation may be empirically approached with the Offord model (linear relation to charge-to-size ratio, the charge being obtained directly from the ionization constants and the size being approached with the molecular mass exponent a factor α) [see Offord (1966)].

$$\mu_{\text{elec}} = a \cdot \frac{eZ}{M^{\alpha}} \quad (13.8)$$

M is the molecular mass, and a and α are two constants determined experimentally which can vary between classes of chemical compounds (DNA, peptides, organic acids) (Jalali-Heravi et al., 2005).

This empirical framework is well accepted and has been adapted to sufficiently simulate the separation of various types of analytes such as DNA fragments and peptides up to proteins (Carbeck and Negin, 2001), to enable separation optimization, or to assign chemical structures and properties. However, from oligomers to polymers, theoretical and empirical approaches often include the degree of polymerization (N, number of monomer units) in the models (Schmitt-Kopplin, 2005); this approach yields exact results but is not applicable in a general manner to polydisperse polyelectrolytes such as NOM. Cottet et al. (2000) developed a model for different polymeric materials dependent on the number of the monomers and their frictional characteristics (charge distribution as a function of chemical substitution of the monomers to which they attributed relative frictional coefficients). Fairly good data interpretation was achieved with benzene polycarboxylic acids, polyalanines, polyglycines, linear fatty acids, polystyrene sulfonates (PSS), and polycytidines. The same approach was effective with PSS of various sulfonation degrees (Cottet et al., 2000). Such electrophoretic models may describe structurally well-defined polymers but are more difficult to apply to NOM.

For colloids and particles, the electrophoretic mobility contains information on electric (surface charge) and hydrodynamic properties (size, conformation) of the analytes as well. In the lower nanometer range of polymers to the lower micrometer range for living organisms or their fractions (bacteria, mitochondria, cells, and cell wall fractions), the solution of the mobility of charged particles can also be approached as a solution of the Poisson–Boltzmann equation expressed in spherical coordinates for the distribution of ions around a sphere and the consequent potentials, coupled with the equations for force as a result electroosmotic flow (Stokes et al., 2004). Approximate solutions lead to a relatively complex picture of a balance of nonnegligible forces on a particle: the driving force of electrical field on the charged particle, frictional retarding force from moving a particle through a fluid, a retardation due to the diffuse double layer on the particle surface (equivalent to electroosmotic backflow from the charged particle surface), and a relaxation effect due to polarization of the diffuse double layer (Stokes et al., 2004). Many theoretical approaches were developed with different geometrical models and assumptions concerning the characteristics of the studied analytes [Hoagland et al. (1999) and references therein). Mixed models appeared recently that consider polymer-coated particles (Ohshima, 2002) or stabilized colloids (Fritz et al., 2002).

Hard-sphere or cylinder models (Avena et al., 1999; Benedetti et al., 1996; Carballeira et al., 1999; De Wit et al., 1993), permeable Donnan gel phases (Ephraim et al., 1986; Marinsky and Ephraim, 1986), and branched (Klein Wolterink et al., 1999) or linear (Gosh and Schnitzer, 1980) polyelectrolyte models were proposed for NOM. Here the various models must be differentiated in detail—that is, impermeable hard spheres, semipermeable spherical colloids (Marinsky and Ephraim, 1986; Kinniburgh et al., 1996), or fully permeable electrolytes. The latest new model applied to NOM (Duval et al., 2005) incorporates an electrokinetic component that allows a soft particle to include a hard (impermeable) core and a permeable diffuse polyelectrolyte layer. This model is the most appropriate for humic substances.

13.4.1.1. Capillary Zone Electrophoresis. The primary advantage of capillary electrophoresis can be found in the simplicity of the instrument. Basic experimental components include a high-voltage power supply, two buffer reservoirs, a fused silica capillary, and a detector. The basic setup is usually completed with enhanced features such as multiple injection devices, autosamplers, sample and capillary temperature controls, programmable power supplies, multiple detectors, fraction collection, and computer interfacing.

Capillary electrophoresis separation is performed in a flexible fused silica capillary tube that is filled with an appropriate buffer solution of defined pH and ionic strength (aqueous/nonaqueous). A small volume of sample (lower than 3–4% of the column volume) is needed to achieve efficient separation. This volume is introduced hydrodynamically (or less often electrokinetically) into the capillary to which an electrical potential is applied (Figure 13.7). Charged species of the sample exhibit

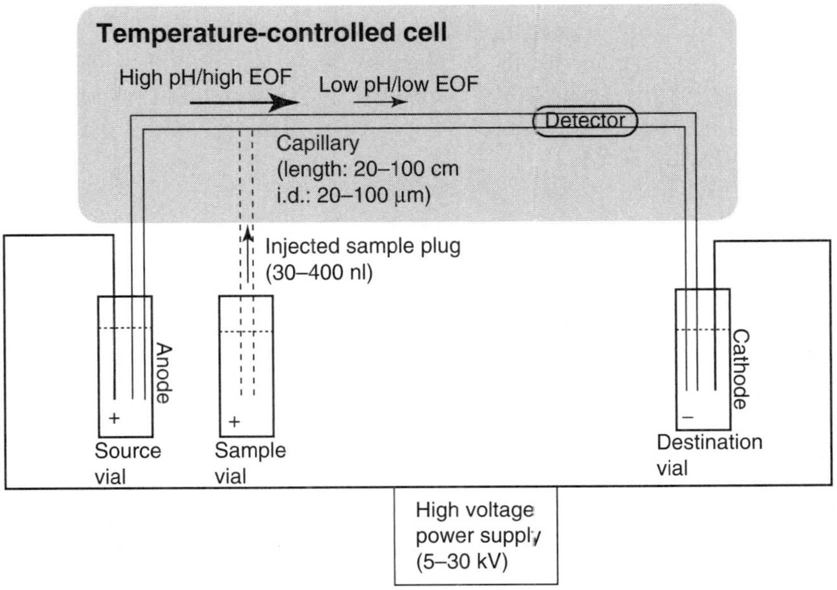

Figure 13.7. Simple setup of capillary electrophoresis (the potential setup is illustrated as generally used for NOM).

different effective electrophoretic mobilities (field strength reduced velocities) and are thereby separated. Detection may be possible with UV–vis, laser-induced fluorescence, electrochemistry, conductivity, or mass spectrometry. Different techniques are possible as a function of the type of capillary column relative to the buffer, allowing separation of charged, neutral, polar, or hydrophobic analytes. Compared to chromatography, the main differences in CE are the technique of injection into the capillary and the means by which the sample is driven through the column that is not by hydrodynamic pumping but instead by electroosmosis (Schmitt-Kopplin and Frommberger, 2003; Schmitt-Kopplin et al., 2001b). This elementary process occurs as a direct consequence of the surface charge on the walls of the uncoated fused silica capillary. The capillary surface carries silanol groups (pK_a 3–5) that ionize as a function of the pH of the separation buffer (Schmitt et al., 1997). The dissociation of SiOH groups to SiO^- produces a negatively charged surface on which an electrical double layer is established at the solid–liquid interface to preserve electroneutrality. The counterions and their associated solvating water molecules migrate in the electric field, producing solution flow toward or against the detector and in turn called *electroosmotic flow (EOF)*. This flow depends on the composition of the separation buffer, its viscosity η, and dielectric constant ε (especially important in the nonaqueous mode—NACE). For example, an increase in ionic strength can cause a decrease in the zeta potential, which will result in a decrease of the EOF (Figure 13.7). Choice of coatings (such as neutral or charged polymers) for the capillary surface can also modify the EOF.

Capillary zone electrophoresis (CZE) is the basis of separation within all above CE techniques. By changing the buffer, optimized interactions of the sample with buffer constituents will facilitate an increased selectivity in the separation of the charged or neutral analytes. CZE enables the separation of anions and cations based on their charge density (effective charge to size ratio) (Schmitt et al., 1997a,b). The addition of organic solvents to the running buffer (up to 100% in case of nonaqueous CE), can improve the solubility of some analytes. This approach is often used for separation of pharmaceuticals or plant secondary metabolites (Bianco et al., 2002). The selectivity is governed by their effective charge and thus by the separation buffer pH and the electroosmotic flow (EOF). This method was the one most frequently used with NOM.

13.4.1.2. Toward an Adequate Data Interpretation. The definitions of apparent and effective electrophoretic mobility need to be developed, since a conversion of electropherograms from the time scale into the effective mobility scale (μ-scale) improves the reproducibility of separation patterns. With the anode placed at the injection inlet and the cathode placed at the outlet, neutral samples will move toward the detector with the velocity of the EOF, while cations move toward the cathode with a higher apparent velocity (apparently faster) and anions will move against the EOF with a reduced apparent velocity (apparently slower). The peaks are detected with increasing times in the electropherogram. The electrophoretic mobility is defined as a coefficient of proportionality between particle velocity and electric field strength. With an applied field E across a capillary of total length L_t the field strength is E/L_t. After time t following the injection, the analytes (anions, neutrals, cations) will cross the detector situated at a length L_d from the inlet one after another; the observed velocity (v) of the analyte is thus equal to L_d/t. The electrophoretic mobility calculated from the observed velocity is called apparent electrophoretic mobility (μ_{ap}). The effective electrophoretic mobility (μ_{ef}) takes account of the velocity of the buffer toward the detector [EOF, (μ_{eo})] and is thus the EOF-normalized electrophoretic mobility of the ions ($\mu_{ef} = \mu_{ap} - \mu_{eo}$); μ_{ef} is equal to zero for neutral analytes, negative for anions and positive for cations. The effective electrophoretic mobility is an absolute parameter independent of the applied field or the column length and is only dependent on the charge and size of the analyte. A conversion of the electropherograms from the time scale into the effective mobility scale (μ-scale) improves the reproducibility of separation patterns and pertinent quantitative parameters.

In the case of NOM samples, the plot of UV absorbance versus effective mobility (μ_{eff}) shows a Gaussian-like distribution around the average electrophoretic mobility value (AEM). This representation of electrophoretic data using μ_{eff} values takes into account the changes in electroosmotic flow that can occur from one measurement to the other (depends on buffer chemistry: pH, ionic strength, type of buffer) (Schmitt-Kopplin et al., 2001b; Schmitt-Kopplin et al., 1999b; Schmitt-Kopplin et al., 1998; Menzinger et al., 2001). An electropherogram in this new scale can be considered as a frequency distribution of individual molecules (or "*molecular aggregates*" in these experimental conditions). From the relation between mobility with charge and size using model compounds (aliphatic and phenolic acids, polyacrilic acids, polystyrene sulfonates), charge density information can be derived directly from the mobility-scale electropherograms (Schmitt-Kopplin et al., 1999a). When the scale transformation is used to analyze single molecules such as exemplified with cationic and anionic analytes, better peak tracking is possible from real matrices (qualitative improvement) and an increased reproducibility in quantitative data is achieved (Schmitt-Kopplin et al., 2001a,b). In addition, a direct linear relation between the effective mobility (μ) and the charge-to-mass ratio can be verified (μ is the field-strength-reduced velocity of the ions and is independent of experimental setup such as voltage, capillary length, and EOF).

Mobility scaling (Figure 13.8) includes (i) the use of an internal standard (charged *p*-hydroxybenzoic acid, polystyrene sulfonate, or neutral EOF marker such as mesithyl oxide), (ii) a baseline correction, (iii) a scale transformation from migration time to effective mobility, and (iv) the deletion of an internal standard peak. The sign of the mobility scale is negative for anions and positive for cations. Changing the time scale to mobility can have effects on the electropherogram shape and resulting interpretations when samples are mixtures. Since data acquisition occurs with respect to the time (between 5 and 10 data points per second), when converting to a mobility scale (mobility = time^{-1}), the electropherogram's shape is "*compressed*." More data points describe peaks in the high-mobility region than near the EOF; consequently, high-resolution data acquisition is required for describing analyte migration near the EOF. Viewed on the mobility scale, data proffer a representation of the mobility distribution for sample constituents, and the time scale visually overestimates the contribution of high-mobility components.

The change in the mobility profile as a function of pH provides a valuable information on the changing charge density of the analyzed materials. Neihof and Loeb (1972) reported electrophoretic measurements on particulate and dissolved organic matter from seawater using the *microelectrophoresis* technique back to 1970s. These early studies reported on organic–mineral interactions based upon changes

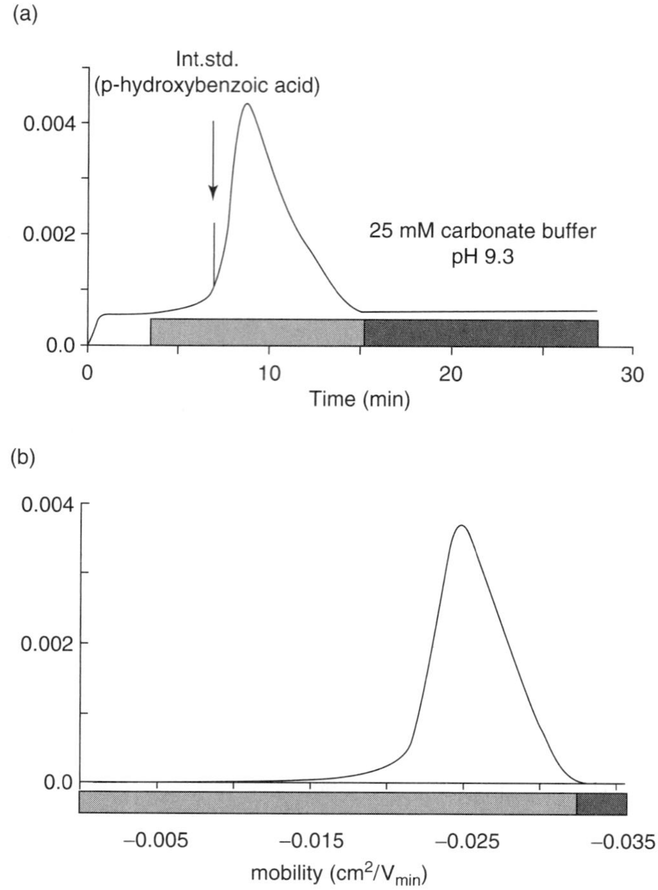

Figure 13.8. CZE electropherograms of an NOM measured with *p*-hydroxybenzoic acid as internal standard: **(a)** raw data; **(b)** mobility transformation.

in surface charge of the particulate matter in different water salinity (Neihof and Loeb, 1973). The method was further developed and adapted by Hunter (1980) to measure the pH-dependent electrophoretic mobility of organic coated minerals in marine systems. In that study the authors were able to determine the importance of COOH and OH groups with respect to the binding of different metal ions. Tipping (2002) used pH-dependent microelectrophoretic techniques to also study metal and mineral binding to NOM. Both average mobilities (as measured by taking the mobility value at the peak maximum) in addition to the mobility distribution profiles should be taken into consideration. For instance, Suwannee River humic acid (SRHA) and fulvic acid (SRFA) had been analyzed over the pH range of 2.75–11.5 (Figure 13.9).

All three components showed a very similar patter in their pH-dependent mobility. These patterns were also similar to the one found in previous CE studies (Hosse and Wilkinson, 2001; Schmitt-Kopplin et al., 1998). The SRFA were comparable to the SRNOM in terms of peak shape changes at lower pH. The two humic acids showed Gaussian mobility distributions and highly symmetric peaks over all the pH range. At pH < 3, the humic acids still showed very sharp and nicely shaped signals

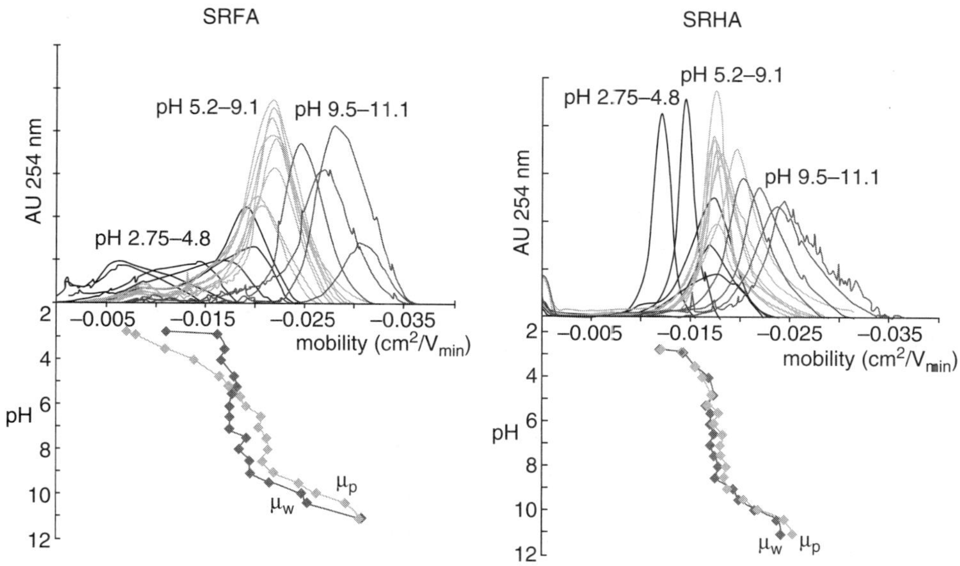

Figure 13.9. Suwannee River fulvic acid (SRFA) and humic acid (SRHA) analyzed over the pH range of 2.75–11.5.

(no signs of interaction with the wall), which were in a high charge-density region (close to the charge density of benzoic acid). This behavior could not be explained with a molecular model but suggests molecular associations and a free draining behavior as described with the partially charged PSS previously. It should be also noted that, the humic acid displayed a series of sharp peaks out of the hump at very high pH, where electrostatic charge repulsions, due to highly charged component mixtures, could be anticipated. The trend is that high mobility is found with highly charged and/or small molecules, but no linear correlation is found to effective charge as measured with potentiometric titration. This had been seen with polyelectrolytes (i.e., polystyrene sulfonates) in the free draining mode (Cottet and Gareil, 2000; Hosse and Wilkinson, 2001, 2002; Schmitt-Kopplin, 2002). The represented ionic strength- and pH-dependent CZE data (Schmitt-Kopplin, 2002) are consistent with a model of semipermeable spheres as demonstrated by Duval et al. (2005). Using a one-dimensional detection system such as UV–vis is thus not sufficient to distinguish subfractions: Only a combination of preparative approaches with orthogonal characterizations (see FFE) or a structure dependent detection such as in CE-ESI/MS might give further descriptive structural information about individual constituents in such a complex mixtures like humic substances.

13.4.2. Isoelectric Focusing

Isoelectric focusing (IEF) was introduced for biochemical separations in the 1960s (Righetti et al., 1978). The separation is carried out in a pH gradient, created by the addition of zwitterionic substances called ampholytes to the separation solution or integrated into the gels (analytical agarose, sephadex, or polyacrilamide gels); the analytes migrate in the solution or in the gel matrix to the pH where their positive charges balance their negative charges [isoelectric points (pI)]. IEF in gels of humic substances generates a series of regular bands in different isoelectric point domains, extending from a more diffuse area of increasing colour when going through lower pH (anode) to higher pH values (cathode). Although the results obtained with the IEF technique in gel with HS are extremely dependent on the experimental conditions (Govi et al., 1994), IEF in gels was one of the most frequently used electrophoretic techniques for the fingerprinting of NOM (for example, to follow the humification processes) (Kutsch and Schumacher, 1994). It was used to fingerprint HS from different soils, composts (Ciavatta et al., 1996; De Nobili, 1988; Govi et al., 1994), and fulvic and humic acids (Ceccanti et al., 1986) and to differentiate phenoloxydase products (Kutsch and Schumacher, 1994). It was also used to follow the humification processes (Ciavatta et al., 1996, 1997; De Nobili, 1988; Garcia et al., 1995; Govi et al., 1992; Kutsch and Schumacher, 1994; Petrussi et al., 1988).

Some authors assigned these bands to humic fractions with a specific isoelectric point; others attributed them to possible artifacts (Aak et al., 1984; Duxbury, 1989; Govi et al., 1994; Kutsch and Schleich, 1989). Different observations were described; for example, after the IEF-refocusing of one selected band, series of bands would appear around the position of the original band, especially toward lower pH ranges (Aak et al., 1984; Duxbury, 1989). In addition, the band pattern is strongly influenced by the prefocusing of the gel prior sample application and by the quality of the obtained pH gradient at the moment of the contact between the NOM and the ampholytes (especially for the bands near the application zone). The flocculation of acidic ampholytes in contact with NOM was often described (e.g., Aak et al., 1984; Gjessing and Gjerdahl, 1972; Kutsch and Schumacher, 1994). In some cases, up to 85% flocculation of the aquatic HS was observed in a solution IEF column (without gel) with *Ampholine* at low pH (pH 3). Comparable behavior in flocculation and focusing was observed in preparative solution IEF (BIORAD-Rotofor) by Schmitt et al. (1997a,b) with humic substances from soils, surface water, and groundwater.

Different commercially available ampholytes were tested with the same separation conditions and samples resulting different band positions (Govi et al., 1994), which were due to the interaction with ampholites in pH gradients neutral pH regions, furthermore several bands formed in the acidic and neutral pH (Ceccanti and Nannipieri, 1978 and Orioli et al., 1980, respectively). In the alkaline region, bands are mainly found at the position of the application confirming a sieving effect or an immobilization in relation to the polyacrilamide gel; the presence of HS with high ash content due to clay residues especially gives perturbed areas (precipitation) at the points where they are applied (cathode).

With a median nitrogen content around 2% (Perdue and Ritchie, 2003), NOM may contain only a limited amount of free positive charge (amino acid and peptide bonds play a minor role) which could be involved in charge neutralization and responsible for band formation. In solution IEF, only limited amount of carbon from the original sample will be found in the pI range >4; the flocculation of around 60% of the carbon occurred at pH value <4 (Schmitt et al., 1997a,b). Numerical simulations showed, in addition, that anionic molecules (such as organic acids) could not reach a steady state, but would approach asymptotically a mobility minimum with 2 pH units lower than their original pK_a values, resulting an accumulation of analytes and possible band formation without the necessity of zwitterions.

Capillary isoelectric focusing separates analytes based on differences in their isoelectric points (pI) using the same principles as in preparative solution IEF. After a focusing step, that builds up a linear pH gradient in the capillary (controlled with zwitterionic internal markers), the analytes move as a function of their respective charge until they reach a position of zero charge (isoelectric point). The solution is then mobilized in CIEF to the detector hydrodynamically.

Preparative solution isoelectric focusing can be performed in order to fractionate NOM (sample load up to 50 mg); the harvested 20 fractions were characterized with UV–vis spectroscopy, gel permeation chromatography, and capillary zone electrophoresis (CZE) and showed the distribution of the molecular weight of the focused fulvic fractions in the created pH gradient. CZE of the obtained fractions showed a series of sharp peaks that resulted from the disaggregation of the colloidal NOM after interaction with the ampholites; moreover, even some signals could be observed in the cationic range (Schmitt et al., 1997a,b; Schmitt-Kopplin et al., 2001b). In-column borate complexation was also used in CZE to compare the harvested fractions by fingerprinting. Only 32% of the total FA did not precipitate with the ampholites that were employed, and this FA fraction was mainly composed of low-molecular-weight compounds as was shown with CZE. Parallelisms were found in the distribution (at least three individualized fractions) of the low pH-focused NOM fractions in the preparative IEF compared to CIEF. Due to the interactions between the NOM and the ampholites, this method found only limited application, especially when the fractions needs to be analyzed with mass spectrometry or NMR spectrometry.

13.4.3. Capillary Gel Electrophoresis

Other efficient solutions were to use buffer and a supporting medium such as paper, starch, agarose, or polyacrilamide; not only was the diffusion reduced but also molecular sieving enabled an additional separation based on size (Block et al., 1958; Tiselius, 1953). The first paper electrophoresis systems of early 1940s could be used successfully on small molecules (amino acids, lipids, charged sugars). After the sample was applied to the middle of the paper strip, a potential difference of about $100\,V\,cm^{-1}$ was applied and the analytes were separated on the paper according to their charge and size. Using this technique on larger paper strips allowed the two-dimensional orthogonal separation of mixtures at two different pHs. Separations on agarose or polyacrilamide gels are also over 100 years old. Many interesting developments occurred about 30 years ago, when it was realized that these supports were ideal for the separation of proteins in denaturing agents such as urea or sodium dodecyl sulfate (SDS): polyacrilamide gel electrophoresis (PAGE) was born. Polyacrilamide gel electrophoresis (PAGE) was first reported as anticonvective matrix for humic materials in 1969 (Stepanov and Pakhonov, 1969). At the time it yielded comparable results to paper zone electrophoresis (at least visually in terms of the creation of "separation" bands). Because PAGE first found application with proteins in combination with denaturing agents [urea to reduce intermolecular H-bonds, sodium dodecyl sulfonic acid (SDS) to give the proteins a similar charge density and thus to create a separation in the gel based only on sieving], it was worthwhile to try the method with humic type of materials. SDS is not used with acidic proteins (low binding/charge repulsion) and its role has been previously described, and "any disaggregating effect of SDS on humic substances would probably be due to displacement of polyvalent cations from humic substances and would therefore be a function of the purification procedure undertaken prior electrophoresis" (Duxbury, 1989).

In terms of the separation of bands (fingerprinting without possible structure information), the PAGE approach certainly yields good results but a direct correlation in terms of structural characteristics is difficult because of specific interactions with the buffer/gel components. The separation is certainly caused by chemical interactions between fractions of the NOM and polyacrilamide, urea, SDS, or buffer constituents that depend on the sample structure. Numerous recent papers from Trubetskaya and Trubetskoj (Trubetskoj et al., 1997, 1998, 1999, 2001; Trubetskaya et al., 2001, 2002) have showed that this approach using complexing buffers may still be used; recent findings have shown that most fluorophores and a large proportion of photoinductive chromophores are located in the low mass fractions of soil humic substances, and such a distribution of photochemically active constituents may be characteristic across different soil types (Richard et al., 2004).

Capillary gel electrophoresis allows the analysis of molecules based on their size. The separation is done in entangled polymer solutions acting as replaceable physical gels (diluted polymer solutions such as methylcellulose, hydroxypropylcellulose, and polyethylene glycols) whose molecular size and concentration determine the selectivity by sieving effects. This technique set the basis of the gene sequencing techniques used in modern molecular biology and Genomics.

Capillary gel electrophoresis (CGE) has historically been developed for the separation of proteins and DNA fragments, and this technique is used to sequence the genomes. In the presence of anionic detergents such as sodium dodecylsulfate (SDS), proteins get a homogeneous charge density and in the presence of sieving medium the CGE migration times can directly be correlated to size. SDS, however, cannot be used with NOM because strong interactions alter the electropherograms significantly (Schmitt-Kopplin and Junkers, 2003). When comparing systems with different charge density, the linear relation between the mobility and size (De Nobili et al., 1999) is not verified anymore and only the variation in effective mobility in presence of the gel is correlated to the size. A physical gel composed of 0.3% methylcellulose in a 25 mM carbonate buffer pH 9.3 was shown as possible separation medium for NOM in CGE. From all tested gels, the polar polysaccharides were found the best for the separation of NOM; the interactions are minimized but can never be suppressed. Change in the mobility profile of a NOM in CGE (with gel) compare to CZE (without gel) (Figure 13.10) from which the PSS-mass equivalents and corresponding hydrodynamic ratios of the fraction can be calculated from the experimental linearization. In these conditions the average hydrodynamic radii of this NOM is in the range of 2–3 nm (values found with independent approach (Schmitt-Kopplin et al., 1999a) and also illustrated several times in this book in the chapter on fluorescence correlation spectroscopy).

The 2D-PAGE technique was recently modernized due to advances in proteomics. The new methodology combines the sieving effects and size selectivity of polyacrilamide gels with the pH-dependent charge selectivity of isoelectric focusing. For example, this approach could succesfully be used to fractionate Th-binding polysaccharidic fractions from colloidal marine organic matter (Quigley et al., 2002) and can thus be used as a fractionation tool for specific substance classes within NOM complex mixtures.

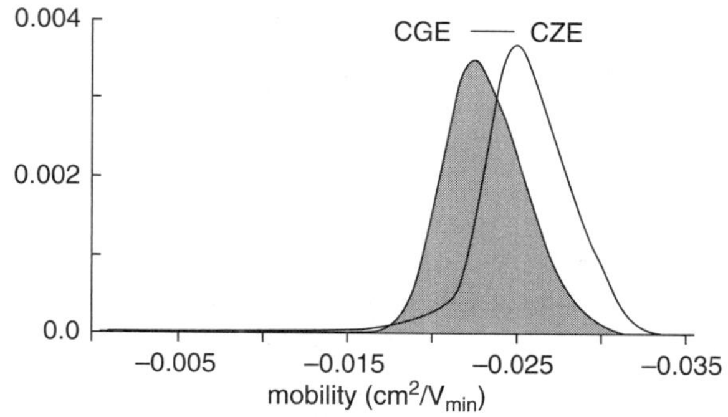

Figure 13.10. Changes in the electrophoretic mobility profile with 0.3% methylcellulose in the 25 mM carbonate buffer pH 9.3 (CGE) versus profile without methylcellulose (CZE).

13.5. HYPHENATED TECHNIQUES: TOWARD OFFLINE AND ONLINE MULTIDIMENSIONAL TECHNIQUES

13.5.1. Hyphenated Liquid Chromatography Techniques

SEC-ESI/MS. Mass spectrometry with electrospray ionisation (ESI/MS) combined with SEC is a promising technique for the molecular size determination and characterization of HS at the same time (Phillips and Olesik, 2003; Reemtsma and These, 2003, 2005). The SEC separation offers a possibility to interpret the highly complex spectra obtained from high resolution MS, equipped with soft ionization techniques (e.g., ESI). The apparent ability of HS to produce both positive and negative ions offers additional possibilities for structural determinations. More efficient methods to suppress the formation of multiply charged ions in favour of singly, or moderately, charged species are greatly needed to allow unequivocal molecular weight determinations (Persson et al., 2000). Investigations of humic (HA) and fulvic acid (FA) fractions with ESI-MS found lower mass distributions for humic acids (300–1200 Da) compared to size exclusion chromatography (1000–100,000) (Richardson, 2003). Therefore, new questions arise as to whether there are colloidal aggregates responsible for the higher masses found by previously applied methods like SEC, or whether there is a more selective ionization in ESI responsible for the recent findings. The latter would mean that only a specific portion of HS can be ionized by ESI. A reason for this may be the selective ionization due to the different ionization efficiencies of different compounds in ESI.

As in biological applications, ESI-MS may be suited to detect noncovalent interactions of dissolved species in aqueous samples. This type of ionization may assure the existence of certain aggregated supramolecular structures which was supported by SEC, but not detected in MS where the observed average masses support the low-molecular-weight molecules. It is interesting to note that ionization experiments on poly(carboxylic acids) indicated a low molecular weight of ESI-MS (about 330 in the positive ion mode and about 280 in the negative ion mode), although the observed mass distribution of this standard was 2000 Da (Leenheer et al., 2001). On the other hand, a first indication that ions generated by ESI are at least partially representative of the entire sample is that elemental composition values of combustion agree well with data obtained from ESI-MS. Ultrahigh-resolution electrospray ionization Fourier transform ion cyclotron resonance mass spectrometry (ESI-FT-ICR-MS) is a promising technique for advanced structural and chemical characterization of natural organic matter (Kujawinski et al., 2002; Stenson et al., 2002; Hertkorn et al., 2007).

LC-NMR. Separations using reverse-phase (RP) liquid chromatography are potentially more powerful because samples can be studied without derivatization. Numerous attempts have been made to separate NOM; and while most studies exhibit some degree of separation, to date the complete separation of a NOM sample has not been accomplished. Even only partial separation is possible, and it is worth to hyphenate a separation method with structure information-oriented analytical applications. Liquid chromatography combined with nuclear magnetic resonance and preliminary studies with solid-phase extraction were conducted on NOM isolated from freshwater and soil (Simpson et al., 2004).

13.5.2. Hyphenated Gas Chromatography Techniques

Py GC/MS involves chromatographic separation of pyrolysis products into single components so that mass spectral data can be obtained for each component. The interpretation of the data requires a detailed knowledge of the pyrolysis behavior of the compounds; since secondary reactions inevitably modify the original compound, the pyrolysis data are susceptible to bias. For example, pyrolysis of cellulose results in carbonyl compounds, acids, furans, pyranones, anhydrosugars, and phenols. Fatty acids may be decarboxylated under the pyrolysis procedure, especially in the presence of mineral soil that may have a catalytic effect on such reactions. Thus mainly alkanes and alkenes can be identified in the pyrolysates obtained from soils, with only minor occurrence of fatty acids (Saiz-Jimenez, 1994). Nitriles found in soil pyrolysis could originate from the reaction of long-chain fatty acid with some nitrogen derivatives present in the soil (van Bergen et al., 1998). The pyrolysis of polar macromolecular materials is well known to produce volatile polar products, only some of which can be chromatographed; very polar products remain attached to the column, undetected and unquantified. Pyrolysis with tetramethyl ammonium hydroxide (TMAH) derivatizes polar compounds to less polar products, which are more amenable to chromatographic separation. This procedure avoids decarboxylation and produces methyl esters of carboxylic acids and methyl ethers of hydroxyl groups (Martın et al., 1994; de Leeuw and Baas, 1993). This already multidimensional technique in combination with cross-polarization magic angle spinning (CP-MAS) ^{13}C-NMR provided chemical information on the chemical composition of soil organic matter (Dai et al., 2002).

The combination of pyrolysis in the presence of methylating agent (TMAH), nondestructive techniques, like XANES (X-ray absorption near-edge structure) and XPS (X-ray photoelectronon spectroscopy), allows us to characterize polar moieties and sulfur functionalities at the same sample. The combination of these different approaches allowed a more complete understanding of the organic sulfur structures in the leonardite coal. In agreement with previous studies, the results showed that oxidized sulfur functionalities, such as sulfonate and sulfate, represent the major forms of sulfur in leonardite coal. The organic matter in coals contains significant amounts of oxygen, sulfur, and nitrogen heteroatom incorporated in various functionalities (e.g., carboxylic group), which vary in abundance depending on the specific conditions of the coal beds. Nuclear microprobe analysis (NMA), which is a nondestructive technique, allows detection and quantification of all elements of the periodic table located on the solid surface with large spatial resolution. The technique provides an evidence of a strong affinity with the smallest HA colloids of numerous trace elements (Mercier et al., 2002).

13.5.3. Hyphenated Electrophoretic Techniques

Capillary electrophoresis (CE) is a relatively nonperturbing method that allows the electrophoretic separation of small molecules, polymers, or macromolecules in various modes of separation using aqueous or nonaqueous buffers with coated or uncoated capillary columns. The combination of CE online with mass spectrometry provides a very powerful analytical system, characterized by its high resolution capability and detection sensitivity (von Brocke et al., 2001). Additionally, the small volume of the analyte used in CE techniques allows the analysis of cellular components. For polydisperse mixtures, such as NOM, hypenation could provide insights into the distribution of charge density at specified pH and ionic strength. For correct interpretation of the CE_ESI/MS results of NOM, an independent understanding of the behavior in both CE and ESI/MS is necessary and be aware of any artifactual signals; for this reason the use of CE with NOM and humic substances were reviewed (Schmitt-Kopplin and Junkers, 2003).

A possible use of CZE-ESI/MS for the characterization of Suwannee River NOM was presented by Schmitt-Kopplin and Kettrup (2003). In a first step the ESI interface conditions were optimized with model compounds and the behavior of different oligomers were investigated. However, the used model compounds during the optimization (trimellitic acid and phthalic acid) are multiply charged in solution but are only single-charged after electrospray, testing the charge retention under the ionization conditions; higher molecular mass oligomers (polycarboxylic glycyrrhizic acid) were also injected, resulting in doubly charged detected peaks. High-molecular-mass humic substances presenting multiple charges in solution could thus also bear multiple charges after separation and ESI with a high probability. To test this hypothesis, several oligomers [polyelectrolytic polystyrene sulfonates (PSS) and polyacrylic acid (PAA)] were separated as multiply charged anions in CZE and detected in ESI-MS. The results showed that it is actually possible to detect multiply charged states with this approach. The analysis of NOM confirms that molecules smaller in average and more charged are present in the high-mobility region of the hump and that higher m/z are found in the low-mobility range of the colloidal charge density distribution, which was already suggested (Schmitt-Kopplin and Junkers, 2003).

13.5.4. Hyphenated Field-Flow Fractionation Techniques

Since the surface properties (charge density) of the NOM influence their environmental functions like buffering capacity or binding to pollutants in natural systems like groundwater or soil, separation and analysis based on these properties are highly favorable. Despite availability of the electrophoretic separation methods and existence of the powerful detection methods like MS, their offline coupling is not possible due to not sufficient sample amount. To overcome this problem, an upscale method, in the milliliter range, is available in electrophoretic separation, namely free-flow electrophoresis. The behavior of the fulvic acid were described by the CZE profiles and mobility distributions, obtained from the fractions separated by FFE (Whelan et al., 2005). The FFE approach enables new possibilities for the structural investigation of humic materials with other analytical characterization techniques (chromatographic and spectroscopic such as nuclear magnetic resonance spectroscopy). A further advantage of these obtained fractions is that they can be conserved, avoiding certain rearrangement by the time; and the different fractions, beside offline investigation, can be a base of different interaction studies, revealing the responsible reactive parts of these highly altered mixtures in nature like NOM. To follow the FFE separation, the fractions can be analyzed offline by FTICR MS, and the obtained spectra can be visualized by van Krevelen diagram to evaluate the NOM composition and structure. The continuous movement of the loci, which was observed from the relatively low O/C and high H/C ratio region toward the higher O/C and lower H/C determined region, is in good correlation with the fact that higher charge density is expected toward the anode; hence the increasing O/C elemental ratio may signify the carboxylic group moiety (Figure 13.11). This assumption was verified by the NMR analysis of the fractions, where the high mobility fractions indicated the occurrence of aromatic carboxylic groups. The necessity of parallel application of the MS and NMR techniques is supported by another fact that, for instance, aliphatic or carbohydrate derivatives are not or hardly revealed in ESI conditions but in NMR spectrum are clearly visible (Figure 13.12).

The reviewed results showed that online coupling of separation techniques to multiple analytical detectors can provide the most powerful tools for exploring multidimensional chemical space of NOM and HS. The huge potential of applying this approach to unfolding molecular complexity of natural materials is best of all demonstrated by the results of offline characterization of the fractionated humic materials which are discussed below.

13.6. RECONCILING MACROSCOPIC AND MICROSCOPIC PROPERTIES OF NOM AND HS

13.6.1. Exploring Molecular Heterogeneity within Bulk Humics

Separation techniques in conjunction with offline characterization of fractionated materials can be of particular value for both exploring the heterogeneity of macroscopic parameters within bulk NOM and HS and for reconciling macroscopic and microscopic properties of molecular assemblies of humic matter. To demonstrate this, we have examined relationships between molecular weights of size-fractionated humic materials and the aromatic carbon content in those fractions. The idea is to use molecular size or molecular weight as an "evolutionary index" of fractionated materials, and in turn their status or location along the path of humification. This interpretation of various molecular size parameters can be justified by the entropy-driven character of the humification process, since the natural course of events takes the system to a more disordered (higher entropy) state (Seddon and Gale, 2002). In the case of humification, such an increase in disorder could be provided by a

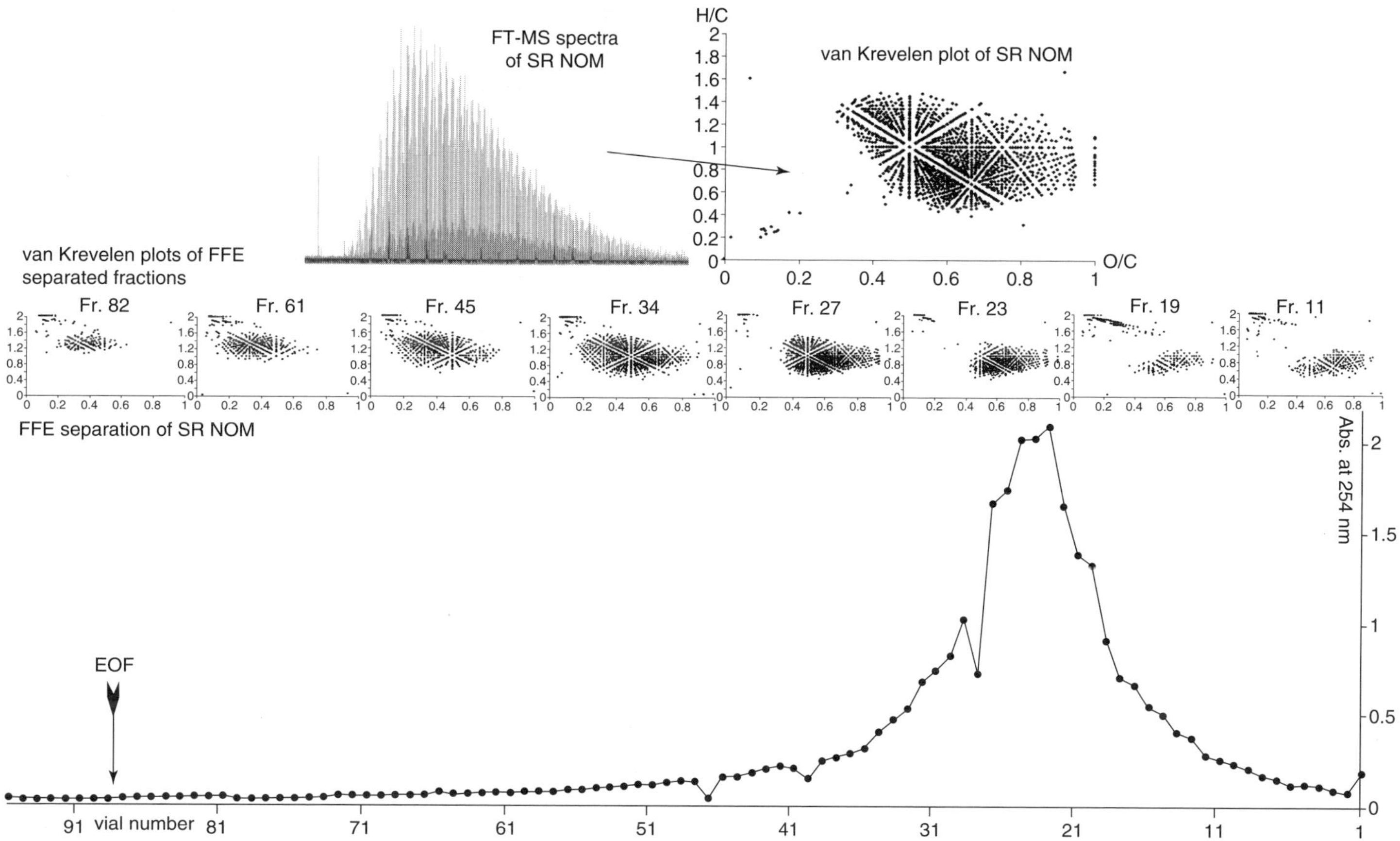

Figure 13.11. FFE separation of the SR DOM. FFE fractions were analyzed by FT/MS separately.

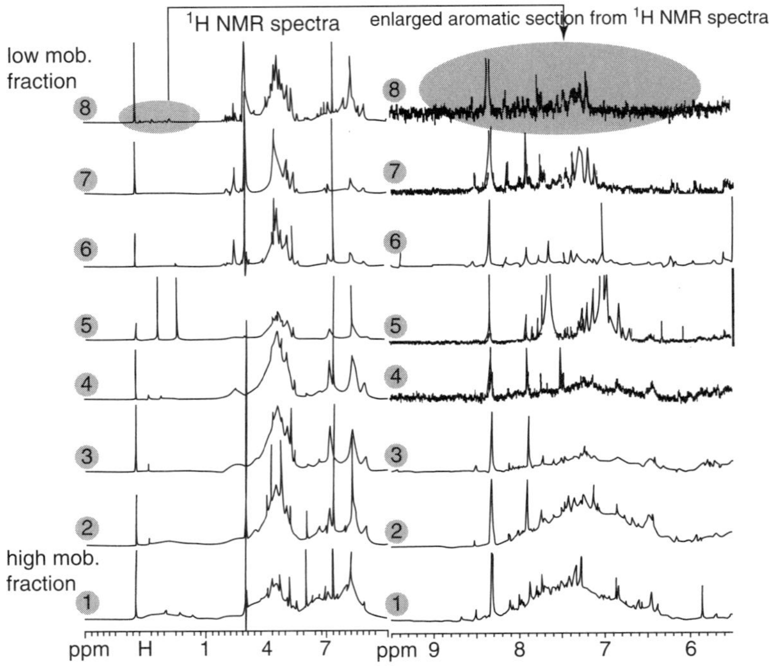

Figure 13.12. ^{1}H NMR spectra of pooled FFE fragments.

disintegration of covalently bonded biomolecules to noncovalently bonded oligomers and monomers of much smaller size accompanied by elimination of water and gaseous CO_2.

The aromatic carbon content, which for terrestrial NOM and HS contributes substantially to the content of double bound equivalents (DBE), was chosen as a "variability index." The role of DBE content as an indicator of the level of compositional variability of the molecular assembly within a given molecular weight was first revealed for marine NOM by Hertkorn et al. (2007). Authors showed that the major component of marine NOM, carboxyl-rich alicyclic molecules [CRAM ($C_nH_mO_q$)], occupied the section of the compositional space on the van Krevelen diagrams with the highest counts of feasible isomers (Figure 13.13). CRAM represented highly substituted alicyclic ring structures. The connection between the "isomer count" and the abundances of those structures in compositional space as determined by FTICR MS is demonstrated by Figure 13.13. This observation is consistent with the statistical definition of entropy, which is proportional to the natural logarithm of the number of microstates within the macroscopic state of the system (Seddon and Gale, 2002). For humic systems, the isomer count can serve as a measure of a number of microstates for the given macroscopic state defined by the elemental composition: The compositions that can be realized by the maximum amount of possible combinations will dominate the system.

The above considerations allow us to conclude that size fractionation can be considered as equivalent to sampling the states of a humic system along its evolution during humification whereby higher-molecular-weight fractions evolve toward lower-molecular-weight fractions with an inherent higher degree of disorder (or

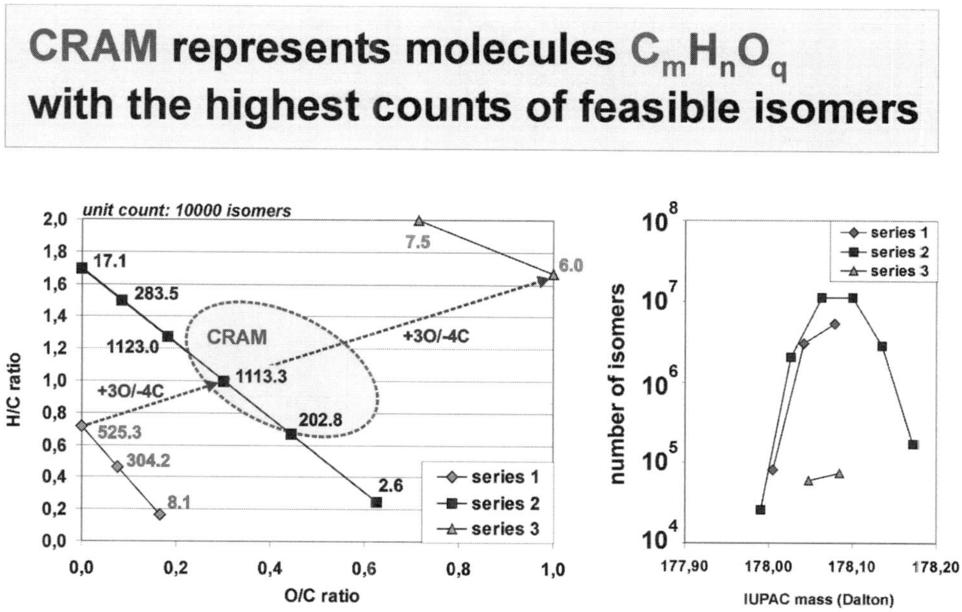

Figure 13.13. Van Krevelen diagram **(left)** of 11 molecular compositions of $C_nH_mO_q$ for marine NOM (UDOM) materials. Represented are three series of CH4/O isobaric molecules with an IUPAC nominal mass of 178 Da and numbers of feasible isomers given (unit count = 10,000 isomers). Certain restrictions have been applied to exclude mathematically possible, but chemically unlikely, structures. The dotted line relates isobaric molecules, in which 4 C atoms were exchanged against 3 O atoms. **Right:** Numbers of calculated isomers according to series 1–3 and IUPAC mass. [From Hertkorn (2006), with permission.]

entropy). If this is true, it can be expected that the averaged molecular weight of any NOM or HS size fraction will be inversely correlated with the content of DBE or aromaticity of the system. To confirm this hypothesis, we analyzed data reported in the literature for offline structural characterization of the size fractions of humic and NOM materials. Data analyses included humic materials isolated from major environmental sources. They are shown in Figure 13.14 for coal HA (A, B), soil HA (C, D), peat HA (E), and freshwater NOM (F). Fractionation conditions and descriptions of environmental sources are given in the corresponding legend to Figure 13.14.

It should be noticed here that the relationships shown in Figures 13.14A–E for Aldrich, soil, and peat humic acids were obtained for humic materials fractionated using ultrafiltration (UF), whereas brown water NOM (Figure 13.14F) was fractionated using SEC. Shin et al. (1999) and Khalaf et al. (2003) did not perform further MW characterization of the fractions using SEC; instead, corresponding MWs were given in the UF scale—according to cutoffs of the UF membranes used (Figures 13.14A and 13.14C). All data on fractions, which were further characterized using SEC, were displayed in SEC-MW scale. As already mentioned in Section 13.2.2, MW measures by SEC and the MW scale of UF differ substantially in estimating the absolute MW values of humics. This is reflected in data shown in Figures 13.14B, 13.14D, and 13.14E as well, which demonstrate much higher values of UF estimates

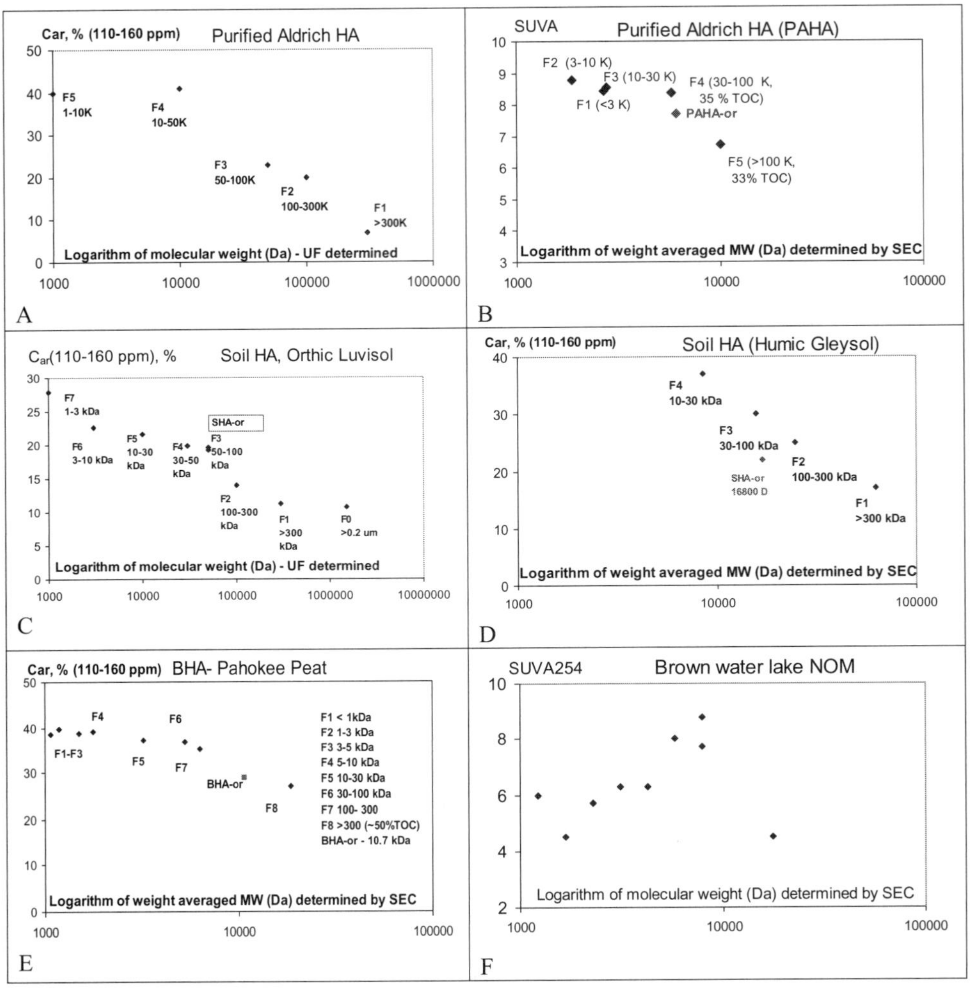

Figure 13.14. Aromaticity versus logarithm of molecular weight correlations plotted for size-fractionated humic materials from different sources. Aromaticity was estimated using two different indices: (1) percentage of aromatic carbon (C_{ar}, %) calculated from ^{13}C NMR spectrum as a partial integral of spectral density within the region from 110 to 160 ppm or (2) specific UV-light absorbance at $\lambda = 254$ nm (SUVA254) calculated as a ratio of optical density at 254 nm multiplied with a length of optical path to concentration of HS expressed on organic carbon basis mgC/liter). **(A)** Humic acid (Aldrich Chemical Company, 200 mg/liter in 0.1 M NaClO4 pH 7) was fractionated using ultrafiltration and membranes of different molecular weight cutoffs to acquire fractions with the nominal sizes (Da): >300 K (F1), 300–100 K (F2), 100–50 K (F3), 50–10 K (F4), and 10–1 K (F5) (Shin et al., 1999). **(B)** Humic acid (Aldrich Chemical Company, PAHA) was fractionated by ultrafiltration and membrane cartridges having nominal molecular weight cutoffs (MWCO) of 3 K, 10 K, 30 K, and 100 K. The starting concentration of PAHA was 200 mg C/L in a 0.01 M NaCl solution (Hur and Schlautman, 2003). **(C)** Soil was collected from the Ap horizon of the orthic luvisol, Merzenhausen, Germany. Eight fractions of isolated HA were obtained using filtration through a 0.2 um Nylon filter (retentate, F0) and a series of Amicon membranes with molecular weight cutoffs (kDa): >300, F1; >100, F2; >10, F3; >50, >30, F4; >10; F5; >3, F6; >1, F7. The fractions were characterized by CP-MAS ^{13}C-NMR spectroscopy (Khalaf et al., 2003).

expressed as cutoffs of the membranes used (about an order of magnitude higher) as compared to corresponding SEC estimates. At the same time, the both techniques are consistent with respect to trends observed within fractions.

As can be seen from Figures 13.14A–E for all data sets of humic acid samples isolated from different sources—namely, soil, peat, and coal—a decrease in molecular weight within fractions of bulk humic material was accompanied by an increase in the aromaticity of those fractions expressed either as content of aromatic carbon (C_{ar}, % determined from ^{13}C NMR spectrum as a partial integral of spectral density within 110 to 160-ppm region) or as specific ultraviolet absorption (SUVA, liter·mgC^{-1}·m^{-1} determined from UV absorbance spectrum as a ratio of optical density at 254 nm multiplied with a length of optical path to concentration of HS expressed on organic carbon basis mgC/liter). This trend was also consistent with data from Swift et al. (1992), who conducted SEC fractionation of soil humic acid, followed by NMR characterization, and then observed higher aromaticity within the fractions with smaller molecular weights. In contrast, it follows from Figure 13.14F that the opposite trend is observed for aquatic NOM, which is dominated by fulvic acid fractions: Aromaticity increases with an increase in molecular weight. This is in line with the data of Chin et al. (1997) and Cabaniss et al. (2000). Chin et al. (1997) reported direct correlations between molecular weight and aromaticity for a set of six aquatic HS samples. In this regard, data published by Perminova et al. (1999) are of particular importance. Here it was shown for sets of humic materials grouped by the similar environments and fractional compositions that correlations between molecular weight and partition coefficients (K_{oc}) for polycyclic aromatic hydrocarbons (PAHs) and aromaticity were positive for nonfractionated aquatic HS, peat HS, and soil FA and negative for soil HA. These opposing trends for molecular weight versus aromaticity can be explained assuming different vectors of evolution for humic and fulvic acid fractions: The humic acids evolve toward kerogen, which represents a condensed form of carbon, while fulvic acid evolve into aliphatic structures with formation of small carbonic acids and CO_2 as end product.

Hence, these results can be interpreted as supportive of the hypothesis that changes in molecular weight within bulk humic material are indicative of evolution along the path of humification. In other words, fractionation of operationally defined humic substances in accordance with size allows for separation of fractions with different degrees of humification: The smaller the fractions, the more they have

Figure 13.14. **(D)** Humic and fulvic acids were extracted from the surface horizon of a humic gleysol in northern Switzerland. Four size fractions were obtained using hollow fiber ultrafiltration cartridges with nominal molecular weight cutoffs (kDa) >300, >100, >30, and >10. The fractions were characterized by SEC and CPMAS ^{13}C NMR (Christl et al., 2000). **(E)** The BHA was base extracted from Pahokee peat and separated into eight fractions using ultrafiltration membranes with the following molecular weight cutoffs (kDa): <1, 1–3, 3–5, 5–10, 10–30, 30–100, 100–300, and >300. The molecular size distributions of fractions were further calibrated using HPSEC. Aromaticity was determined using solid-state ^{13}C NMR spectroscopy (Li et al., 2004). **(F)** A NOM sample was concentrated by membrane filtration and fractionated using SEC system with a Superdex 75 column. The fractions were analyzed using a LCDOC system with a TSK HW-50 (S) column. Peak molecular weights (M_p) were determined using polystyrene sulfonate (PSS) standards (Mueller et al., 2000).

undergone humification and vice versa; the larger the fractions, the less they are humified. Of particular importance is that the revealed trends were observed for datasets obtained by independent authors for the sources of humic materials covering all major environments including coal (Aldrich HA), peat, soil, and freshwater.

13.6.2. Connecting Evolution of Humic Matter in the Environment to Measurable Properties of Isolated Humic Samples

In nature, there exists a continuous exchange of mass and energy between any humic system and the environment. Large biomacromolecules (e.g., cellulose, lipids, proteins, and lignins) enter this system and undergo humification or transformations to produce disintegrated macromolecules, oligomers, a host of smaller organic fragments, and ultimately water and carbon dioxide which exit the system. The temporal scale of humification is on the order of 1–1000 years; and even under seasonal cycles, humification processes are repeated or continuous such that the distribution of remaining humic components approach quasi-steady-state conditions. The validity of this conceptualization in the context of modeling the synthesis and degradation of NOM was examined by Cabaniss et al. (2005).

Under conditions of seasonal cycles, where humification process is repeated and continuous, we can suppose that the humic system approaches steady state, which means that the average value of a property over humic ensemble is the same as the time average of that property for a single system. This statement is known as ergodic hypothesis (Seddon and Gale, 2002). In this case, upon extracting soil, peat, or other humics-containing sources, we isolate the humic system from energy and material exchange with its natural environment. This is equivalent to stopping time flow or freezing humification at the extraction point. The analysis of the extracted sample then gives us a statistical snapshot of humic ensemble. The ensemble is a collection of all possible systems that have different microscopic states but have an identical macroscopic state. Hence, the value of some measurable macroscopic property, P, of the system can be related to its microscopic properties using the following expression:

$$P_{\text{macro}} = \langle P \rangle_{\text{ensemble}} = \frac{1}{N_S}\left(\sum_{i=1}^{N_S} P_i\right) \tag{13.9}$$

where N_S is the number of microstates (i.e., components) consistent with the ensemble, P is the parameter of interest (i.e., molecular weight), and the subscript i represents the ith microstate (i.e., component). The above statement can be also rewritten in terms of probability:

$$P_{\text{macro}} = \langle P \rangle_{\text{ensemble}} = \sum_{i=1}^{N} \rho_i P_i \tag{13.10}$$

where ρ_i is the probability that the ith microstate or component occurs.

The probability of finding a system in a given macroscopic state (i.e., given molecular weight) depends upon the multiplicity of that state, which is proportional to the number of ways it can occur. For example, multiple isomers could have the same molecular weight. Thus, the molecular weight distribution of components found in a humic sample will be skewed toward values where multiplicity (or number of isomers) is greatest. A similar argument can be made for the elemental composition of an isolated humic fraction. The most probable elemental composition will correspond to molecular formula generating the highest multiplicity of structural isomers. It is of importance to note here that the multiplicity of a system is directly related to its entropy in accord with the following expression (Seddon and Gale, 2002):

$$S = k \ln W \tag{13.11}$$

where S is entropy, W is multiplicity, and k is Boltzmann's constant.

For a system of a large number of particles, like a mole of atoms, the most probable state will be overwhelmingly probable. Hence, it can be expected that the system at equilibrium will be found in the state of highest multiplicity since fluctuations from that state will usually be too small to measure. As a large system approaches equilibrium, its multiplicity (entropy) tends to increase. The validity of this postulate for the system of NOM or HS was demonstrated by the data of Hertkorn et al. (2006, 2007), who showed that CRAM represented $C_mH_nO_q$ molecules with the highest number of feasible isomers—or, in other words, the state of highest multiplicity. These data are shown in Figure 13.13 for $C_nH_mO_q$ compositions with nominal IUPAC mass of 178: eleven feasible molecular compositions were grouped into three series of CH_4/O molecules. From this graph, compositions with highest amount of feasible isomers produced the patterns on the van Krevelen diagram seen for the real CRAM. The highest multiplicity is provided by the presence of substituted aromatic or alicyclic rings, which provide clues as to the forces driving the evolution of chemical structures in humic matter. For example, for molecules with an IUPAC mass of 178 Da and an H/C ratio of 1.692, Hertkorn et al. (2007) have shown that the corresponding molecule $C_{13}H_{22}$ has three double-bond equivalents (DBE) and 1.7×10^5 isomers, whereas its fully saturated analogue $C_{13}H_{28}$ exhibits only 802 isomers. The maximum number of isomers (1.1×10^7 each) is reached for the molecules with 5 DBE ($C_{11}H_{14}O_2$) and 6 DBE ($C_{10}H_{10}O_3$), respectively.

Of interest is that the above consideration is consistent with the experimental spectrum shown in Figure 13.15 [details of acquisition are described in Kunenkov et al. (2009)]. It demonstrates a peak distribution around 177-Da $[M{-}H]^-$ ions, which corresponds to the above-mentioned 178-Da molecular peaks. The most intensive peaks are provided by $C_9H_5O_4^-$, $C_{10}H_9O_3^-$, and $C_{11}H_{13}O_2^-$, all of which have the highest feasible isomer counts based on their molecular formulas. The slight difference between number of feasible isomers shown in Figure 13.13 and observed peak intensities in the mass spectrum shown in Figure 13.15 can be explained by two factors. First, electrospray ionization efficiency depends on the molecular structure; so, different compounds with the same concentration yield peaks of different intensity in the mass spectrum. In addition, since humification conditions pose certain constraints on the kind of isomers formed, not all feasible isomers are observed. As a result, the total number of feasible isomers can be considered as good approximation but not as an exact number of actual microstates suited to the given macroscopic state defined by elemental composition of a humic molecule.

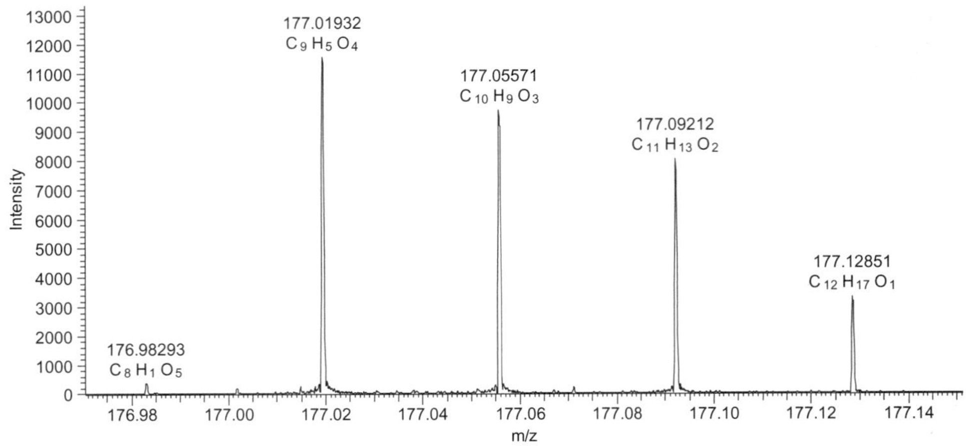

Figure 13.15. Expanded region at a nominal mass of 177 from FTICR mass spectrum of Suwannee River fulvic acid (100 mg/liter solution in acetonitrile). The spectrum was acquired on 7T Finnigan linear quadrupole ion trap–Fourier transform (LTQ FT) mass spectrometer (Thermo Electron Corp., Bremen, Germany) equipped with Ion Max electrospray ion source located at the facilities of the Emanuel Institute of Biochemical Physics RAS (Moscow, Russia). Instrument settings: selected ion monitoring, *m/z* range 90–190 amu, flow rate 1 μl/min, negative ion mode; needle voltage 3.0 kV; no sheath and auxiliary gas flow; capillary voltage –44 V; tube lens voltage 200 V; heated capillary temperature 240 °C.

These considerations allow us to link the time required for humification (always directed to an increase in entropy) to the type of chemical transformations in humic system, which best suit this demand: The system of NOM and HS should unavoidably evolve toward molecular compositions with the maximum number of isomers. Given that the overwhelming part of humic matter is being formed under oxic conditions, these structures are represented by low-molecular-weight aromatic and alicyclic acids. This suggests that under the same environmental constraints, the humification of NOM should lead to the formation of structures with an increased content of aromatic structures (or more precisely, the amount of DBE) and with a decrease in size similar to what was revealed by the results of data analysis on size-fractionated samples of humic materials shown in Figures 13.14A–D.

Given considerations show that further studies on compositional space of humic system are needed to reveal the mechanisms controlling humic system evolution. The role of advanced separation technique and high-resolution analytics in disclosing this mystery of nature will be critical.

13.7. CONCLUSIONS AND FUTURE PROSPECTS

Separation techniques are shown to be powerful tools for unfolding molecular complexity of natural nonliving organic matter. A separation approach is shown to be the most viable option for resolving the fundamental problem in molecular understanding of natural nonliving organic matter—that is, connecting macroscopic and microscopic properties of this complex chemical system. Along with the theoretical issues mentioned above, development of advanced separation techniques can be of particular value for drawing the attention of industrial chemists to conversion of huge resources of humified biomass to alternative feedstock for bio-based products. With oil prices steadily climbing, development of alternative feedstocks is critical to maintain the viability of the manufacturing industry. The 21st century will be a century of bioeconomy based on the use of biomass—that is, plants and plant-based materials, produced by photosynthesis within biological rather than geologic time. With sources encompassing different stages of biomass humification from mature lignites, peats, sapropels, and so on, to young composts, vermicomposts, activated sludges, and so forth, humic materials occupy a niche between fossil rocks and fresh biomass. The most striking feature of humic materials viewed in the context of a biobased economy is their unique constellation of properties: nontoxicity, biocompatibility, resistance to biodegradation, and polyfunctionality. These properties will allow humic-based materials to be competitive in the market of biobased products, remedial agents, bioplastics, and green specialty chemicals (dispersants, flocculants, chelators, etc.).

The single major obstacle to the successful industrial processing of crude humic materials is their immense complexity: A substantial reduction in molecular heterogeneity is needed to convert humic materials into competitive feedstock for the chemical industry. The boost to petroleum chemistry resulting from the invention of efficient methods for fractionating crude oil can serve as the closest historical analogy of how advanced separation technologies can further the developments in the industrial chemistry of humic materials. Unlike crude oil, highly oxygenated HS are not volatile and cannot be separated using distillation. So, the search for efficient fractionation methods applicable to complex mixtures of hydrophilic compounds is a real challenge of modern analytical and physical chemistry. In view of the high polarity of humic compounds, selective dissolution using ionic liquids could pose significant advantages.

The potential of sequential fractionation and field-flow fractionation should be thoroughly explored as well. Deep refining and modification of humic substances may launch a new chain of humic products with added value. Mechanochemical fractionation can be considered as very promising cost-effective approach for conditioning humic feedstock. Given that all raw humic materials are solid, the major advantage of mechanochemistry as compared to other techniques is its capability to separate solid organic materials. Transfer of recent successes in mechanochemical fractionation of biomass into cellulose, hemicellulose, and lignin onto humic materials can provide the chemical industry with conditioned humic feedstock already in the nearest future.

ACKNOWLEDGMENTS

The authors wish to acknowledge A. Kudryavtsev and E. Belyaeva (Lomonosov MSU) for technical assistance in preparing figures and references for the manuscript. This work was supported by the Interdisciplinary Scientific Program of Lomonosov MSU (grant MNP4-08), by Russian Foundation for Basic Research (grant 06-04-49017), and by NATO CLG (grant ESP.EAP.CLG 983197).

REFERENCES

Aak, O. V., Galynkin, V. A., Kashkin, A. P., and Yakovlev, V. I. (1984). Isoelectric focusing of high-molecular-weight shale and humic acids. *Appl. Biochem. Microbiol.* **20**, 290–294.

Aiken, G. R. (1985). Isolation and concentration techniques for aquatic humic substances. In *Humic Substances in Soil, Sediment, and Water: Geochemistry, Isolation and Characterization*, Aiken, G. R., et al., eds., Wiley-Interscience, New York, pp. 363–385.

Alexandrova, I. V. (1977). Soil organic matter and the nitrogen nutrition of plants. *Sov. Soil Sci.* **9**, 293–301.

Avena, M. J., Koopal, L. K., and van Riemsdijk, W. H. J. (1999). Proton binding to humic acids: Electrostatic and intrinsic interactions. *Colloid Interface Sci.* **217**, 37–48.

Baalousha, M., Kammer, F. V. D., Motelica-Heino, M., Baborowski, M., Hofmeister, C., and Le Coustumer, P. (2006). Size-based speciation of natural colloidal particles by flow field flow fractionation, inductively coupled plasma-mass spectroscopy, and transmission electron microscopy/x-ray energy dispersive spectroscopy: Colloids–trace element interaction. *Environ. Sci. Technol.* **40**(7), 2156–2162.

Baigorri, R., Fuentes, M., Gonzalez-Gaitano, G., and Garcia-Mina, J-M. (2007). Analysis of molecular aggregation in humic substances in solution. *Colloids Surf. A* **302**, 301–306.

Beckett, R. (1991). Field-flow fractionation-ICP-MS: A powerful new analytical tool for characterising macromolecules and particles. *At. Spectrosc.* **12**, 215–246.

Belenkii, B. G. (1998). Size exclusion chromatography. In *Handbook of Chemistry*, Vol. 5, Big Russian Encyclopaedia Publishers, Moscow, pp. 812–818.

Belenkii, B. G., and Vilenchik, L. Z. (1983). *Modern Liquid Chromatography of Macromolecules*, Elsevier Science, New York.

Belyaeva, E. Yu., Perminova, I. V., Kudryavtsev, A. V., and Novikov, A. P. (2006). Compension of non-exclusion effects during gel-permeation fractionation of humic substances. In *Humic Substances—Linking Structure to Functions*, Vol. **45-1**, Frimmel F. H., and Abbt-Braun G., eds., Book Series Division of Water Chemistry, Engler-Bunte Institute, University of Karlsruhe, Karlsruhe, pp. 217–220.

Benedetti, M. F., Van Riemsdijk, W. H., and Koopal, L. K. (1996). Humic substances considered as a heterogeneous Donnan gel phase. *Environ. Sci. Technol.* **30**, 1805–1813.

Benincasa, M.-A., Cartoni, G., and Imperia, N. (2003). HPLC of humic substances fractionated by flow FFF. *J. Sep. Sci.* **26**(5), 429–432.

Bianco, G., Schmitt-Kopplin, P., De Benedetto, G., Kettrup, A., and Cataldi, T. R. (2002). Determination of glycoalkaloids and relative aglycones by non-aqueous capillary electrophoresis coupled with electrospray ionization-ion trap mass spectrometry. *Electrophoresis* **23**, 2904–2912.

Blo, G., Contado, C., Fagioli, F., Bollain Rodgiguez, M. H., and Dondi, F. (1995). Analysis of kaolin by sedimentation field-flow fractionation and electrothermal atomic absorption spectrometry detection. *Chromatographia* **41**, 715–721.

Block, R. J., Durrum, E. L., and Zweig, G. (1958). *A Manual of Paper Chromatography and Paper Electrophoresis*, Academic Press, New York.

Boehme, J., and Wells, M. (2006). Fluorescence variability of marine and terrestrial colloids: Examining size fractions of chromophoric dissolved organic matter in the Damariscotta River estuary. *Mar. Chem.* **101**, 95–103.

Bouby, M., Manh, T. N., Geckeis, H., Scherbaum, F., and Kim, J. I. (2002). Characterization of aquatic colloids by a combination of LIBD and ICP-MS following the size fractionation. *Radiochim. Acta* **90**, 727–732.

Burba, P., Aster, B., Nifant'eva, T., Shkinev, V., and Spivakov, B. Y. (1998). Membrane filtration studies of aquatic humic substances and their metal species: A concise overview. Part 1. Analytical fractionation by means of sequential-stage ultrafiltration. *Talanta* **45**, 977–988.

Cabaniss, S., Zhou, Q., Maurice, P., Chin, Y.-P., and Aiken, G. (2000). A log-normal distribution model for the molecular weight of aquatic fulvic acids. *Environ. Sci. Technol.* **34**, 1103–1109.

Cabaniss, S., Madey, G., Leff, L., Maurice, P., and Wetzel, R. (2005). A stochastic model for the synthesis and degradation of natural organic matter. Part I. Data structures and reaction kinetics. *Biogeochemistry* **76**, 319–347.

Carballeira, J. L., Antelo, J. M., Rey, F., and Arce, F. (1999). Modeling the effects of ionic strength on ionization parameters for a soil fulvic acid at low concentrations. *Anal. Chim. Acta* **401**, 243–249.

Carbeck, J. D., and Negin, R. S. (2001). Measuring the size and charge of proteins using protein charge ladders, capillary electrophoresis, and electrokinetic models of colloids. *J. Am. Chem. Soc.* **123**, 1252–1253.

Ceccanti, B., and Nannipieri, P. (1978). In *Recent Developments in Chromatography and Electrophoresis*, Frigerio, A., and Renoz, L., eds., Amsterdam, pp. 225–232.

Ceccanti, B., Bertolucci, M. T., Rustighi, G., and Calcinai, M. (1986). Isoelectric focusing of soil humic substances in the presence of 8M urea. *Biol. Fertil. Soils* **2**, 71–75.

Chin, Y.-P., Aiken, G. R., and Danielsen, K. M. (1997). Binding of pyrene to aquatic and commercial humic substances: The role of molecular weight and aromaticity. *Environ. Sci. Technol.* **31**, 1630–1635.

Chow, A. T., Gao, S., and Dahlgren, R. A. (2005). Physical and chemical fractionation of dissolved organic matter and trihalomethane precursors: A review. *J. Water Supply: Res. Technol.—AQUA* **54**(8), 475–507.

Christl, I., Knicker, H., Koegel-Knabner, I., and Kretzschmar, R. (2000). Chemical heterogeneity of humic substances: Characterization of size fractions obtained by hollow-fibre ultrafiltration. *Euro J. Soil Sci.* **51**, 617–625.

Ciavatta, C., Govi, M., Bonoretti, G., and Gess, C. (1996). Identification of peat and leonardite using humification parameters and isoelectric focusing (IEF): A first approach. *Fertil. Res.* **44**, 225–230.

Ciavatta, C., Govi, M., Sitti, L., and Gessa, C. (1997). Influence of blood meal organic fertilizer on soil organic matter. *J. Plant Nutrition* **20**, 1573–1591.

Combe, C., Molis, E., Lucas, P., Riley, R., and Clarka, M. M. (1999). The effect of CA membrane properties on adsorptive fouling by humic acid. *J. Membr. Sci.* **154**, 73–87.

Contado, C., Blo, G., Fagioli, F., Dondi, F., and Beckett, R. (1997). Characterization of River Po particles by sedimentation field-flow fractionation coupled to GFAAS and ICP-MS. *Colloids Surf. A* **120**, 47–59.

Costa, A. R., and De Pinho, M. N. (2005) Effect of membrane pore size and solution chemistry on the ultrafiltration of humic substances solutions. *J. Membr. Sci.* **255**, 49–56.

Cottet, H., and Gareil, P. (2000). From small charged molecules to oligomers: A semiempirical approach to the modeling of actual mobility in free solution. *Electrophoresis* **21**, 1493–1504.

Dai, X. Y., Ping, C. L., and Michaelson, G. J. (2002). Characterizing soil organic matter in Arctic tundra soils by different analytical approaches. *Org. Geochem.* **33**, 407–419.

De Haan, H., Jones, R. I., and Salonen, K. (1987). Does ionic strength affect the configuration of aquatic humic substances, as indicated by gel filtration? *Freshwater Biol.* **17**, 453–459.

de Leeuw, J. W., and Baas, M. (1993). The behaviour of esters in the presence of tetramethylammonium salts at elevated temperatures; flash pyrolysis or flash chemolysis. *J. Anal. Appl. Pyrolysis* **26**, 175–184.

De Nobili, M. (1988). Electrophoretic evidence of the integrity of humic substances separated by means of electrofocusing. *J. Soil Sci.* **39**, 437–445.

De Nobili, M., and Chen, Y. (1999). Size exclusion chromatography of humic substances: Limits, perspectives and prospectives. *Soil Sci.* **164**, 803–811.

De Nobili, M., Bragato, G., and Mori, A. (1999). Capillary electrophoretic behaviour of humic substances in physical gels. *J. Chromatogr. A* **863**, 195–204.

De Nobili, M., Gjessing, E., and Sequi, P. (1989). Sizes and shapes of humic substances by gel chromatography. In *Humic Substances II: In Search of Structure*, Hayes, M. H. B., MacCarthy, P., Malcolm, R. L., and Swift, R. S., eds. John Wiley & Sons, Chichester, UK, pp. 561–592.

De Wit, J. C., van Riemsdijk, W. H., and Koopal, L. K. (1993). Proton binding to humic substances. 1. Electrostatic effects. *Environ. Sci. Technol.* **27**, 2005–2014.

Deyl, Z., Everaerts, F. M., Prusik, Z., and Svendsen, P. J. (1979). *Electrophoresis: A Survey of Techniques and Applications*, Elsevier, Amsterdam.

Dick, D. P., and Burba, P. (1999). Extraction kinetics and molecular size fractionation of humic substances from two Brazilian soils. *J. Braz. Chem. Soc.* **10**(2), 146–152.

Dubin, P. L. (1988). *Aqueous Size-Exclusion Chromatography*, Elsevier Science, Amsterdam.

Duval, J., Wilkinson, K. J., VanLeeuwen, H., and Buffle, J. (2005). Humic substances are soft and permeable: Evidence from their electrophoretic mobilities. *Environ. Sci. Technol.* **39**, 6435–6445.

Duxbury, J. M. (1989). In *Humic Substances II: In Serach of Structure*, Hayes, M. H. B., MacCarthy, P., Malcolm, R. L., and Swith, R. S., eds., John Wiley & Sons, New York, pp. 594–618.

Egeberg, P. K., and Alberts, J. J. (2003). HPSEC as a preparative fractionation technique for studies of natural organic matter (NOM). *Environ. Technol.* **24**, 309–318.

Ephraim, J. H., Alegret, S., Mathuthu, A., Bicking, M., Malcom, R.L., and Marinsky, J. A. (1986). A unified physicochemical description of the protonationandmetal ion complexation equilibria of natural organic acids (humic and fulvic acids). 2. Influence of polyelectrolyte properties and functional group heterogeneity on the protonation equilibria of fulvic acid. *Environ. Sci. Technol.* **20**, 354–366.

Frimmel, F. H., Gremm, T., and Huber, S. A. (1992). Liquid chromatographic characterization of refractory organic acids. *Sci. Total Environ.* **117/118**, 197–206.

Fritz, G., Schaedler, V., Willenbacher, N., and Wagner, N. J. (2002). Electrosteric stabilization of colloidal dispersions. *Langmuir* **18**, 6381–6390.

Garcia, C., Hernandez, T., and Ceccanti, B. (1995). Characterization by isoelectric focusing of the organic matter of a regenerated soil. *Commun. Soil Sci. Plant Anal.* **26**, 3033–3041.

Geckeis, H., Manh Thang, N. M., Bouby, M., and Kim, J. I. (2003). Aquatic colloids relevant to radionuclide migration: Characterization by size fractionation and ICP-mass spectrometric detection. *Colloids Surf. A* **217**(1–3), 101–108.

Ghosh, R. (2003). *Protein Bioseparation Using Ultrafiltration: Theory, Applications and New Developments*, Imperial College Press, London, **166** pp.

Giddings, J. C. (1966). A new separation concept based on a coupling of concentration and flow nonuniformities. *Sep. Sci.* **1**, 123–125.

Giddings, J. C. (1993). Micro-FFF: Theoretical and practical aspects of reducing the dimensions of field-flow fractionation channels. *J. Microcol.* **5**, 497–503.

Gjessing, E. T. (1973). Gel- and ultramembrane filtration of aquatic humus: A comparison of the two methods. *Schweiz. Z. Hydrol.* **35**, 286–294.

Gjessing, E. T., and Gjerdahl, T. (1972). Electromobility of aquatic humus: Fractionation by the use of the isoelectric focusing technique. In Proceedings, International Meeting on *Humic Substances*, Nieuwersluis, Pudoc, pp. 33–51.

Gosh, K., and Schnitzer, M. (1980). Macromolecular structures of humic substances. *Soil Sci.* **129**, 266–276.

Govi, M., Bonoretti, G., Ciavata, C., and Sequi, P. (1994). Characterization of soil organic matter using isoelectric focusing: a comparison of six commercial carrier ampholytes. *Soil Sci.* **157**, 91–96.

Govi, M., Ciavatta, C., and Gessa, C. (1994). Evaluation of the stability of the organic matter in slurries, sludges and composts using humification parameters and isoelectric focusing. In *Humic Substances in the Global Environment and Implications on Human Health*, Senesi, N., and Miano, T. M., eds., Amsterdam, 1311–1316.

Guo, L., and Santschi, P. (2007). Ultrafiltration and its applications to sampling and characterisation of aquatic Colloids. IUPAC Series on Analytical and Physical Chemistry of Environmental Systems, **10**, 159–222.

Hassel'ov, M., van der Kammer, F., and Becket, R. (2007). Characterisation of aquatic colloids and macromolecules by field-flow fractionation. IUPAC Series on Analytical and Physical Chemistry of Environmental Systems, **10**, 223–276.

Hatcher, P. G., and Spiker, E. C. (1988). Selective degradation of plant biomolecules. In *Humic Substances and Their Role in the Environment*, Frimmel, F. H., and Christman, R. F., eds., John Wiley & Sons, New York, pp. 45–58.

Hedges, J. I. (1988). Polymerization of humic substances in natural environments. In *Humic Substances and Their Role in the Environment*, Frimmel, F. H., and Christman, R. F., eds., John Wiley & Sons, New York, pp. 59–74.

Hedges, I. J., and Keil, R. G. (1999). Organic geochemical perspectives on estaruine processes: Sorption reactions and consequences. *Mar. Chem.* **65**, 55–65.

Henry, D. C. (1931). *Proc. R. Soc. London* **133**, 106–140.

Hertkorn, N. (2006). *Molecular level structural analysis of natural organic matter and of humic substances by NMR spectroscopy*. Habilitation thesis, Technical University of Munich, Munich, Germany, **285** pp.

Hertkorn, N., Benner, R., Frommberger, M., Schmitt-Kopplin, P., Witt, M., Kaiser, K., Kettrup, A., and Hedges, J. I. (2006). Characterization of a major refractory component of marine dissolved organic matter. *Geochim. Cosmochim. Acta* **70**, 2990–3010.

Hertkorn, N., Ruecker, C., Meringer, M., Gugisch, R., Frommberger, M., Perdue, E. M., Witt, M., and Schmitt-Kopplin, P. (2007). High-precision frequency measurements: Indispensable tools at the core of molecular-level analysis of complex systems. *Anal. Bioanal. Chem.* **389**, 1311–1327.

Hoagland, D. A., Arvanitidou, E., and Welch, C. (1999). Capillary electrophoresis measurements of the free solution mobility for several model polyelectrolyte systems. *Macromolecules* **32**, 6180–6190.

Hosse, M., and Wilkinson, K. J. (2001). Determination of electrophoretic mobilities and hydrodynamic radii of three humic substances as a function of pH and ionic strength. *Environ. Sci. Technol.* **35**, 4301–4306.

Hosse, M., and Wilkinson, A. E. (2002). Response to comment on "Determination of electrophoretic mobilities and hydrodynamic radii of three humic substances as a function of pH and ionic strength." *Environ. Sci. Technol.* **36**, 3043–3044.

Hunter, K. A. (1980). Microelectrophoretic properties of natural surface-active organic matter in coastal seawater. *Limnol. Oceanogr.* **25**, 807–822.

Hur, J. I. N., and Schlautman, M. A. (2003). Using selected operational descriptors to examine the heterogeneity within a bulk humic substance. *Environ. Sci. Technol.* **37**, 880–887.

Jalali-Heravi, M., Shen, Y., Hassanisadi, M., and Khaledi, M. G. (2005). Prediction of electrophoretic mobilities of peptides in capillary zone electrophoresis by quantitative structure–mobility relationships using the Offord model and artificial neural networks. *Electrophoresis* **26**, 1874–1885.

Janos, P. (2003). Separation methods in the chemistry of humic substances. *J. Chrom. A* **983**, 1–18.

Kennedy, M. D., Chun, H. K., Yangali, V. A. Q., Heijman, B. G. J., and Schippers, J. C. (2005). Natural organic matter (NOM) fouling of ultrafiltration membranes: Fractionation of NOM in surface water and characterization by LC-OCD. *Desalination* **178**(1–3), 73–83.

Khalaf, M., Kohl, S. D., Klumpp, E., Rice, J., and Tomacz, E. (2003). Comparison of sorption domains in molecular weight fractions of a soil humic acid using solid-state ^{19}F NMR. *Environ. Sci. Technol.* **37**, 2855–2860.

Khaledi, M. G. (1998). *High-Performance Capillary Electrophoresis: Theory Techniques, and Applications*, John Wiley & Sons, Chichester, UK.

Kinniburgh, D. G., Milne, C. J., Benedetti, M. F., Pinheiro, J. P., Filius, J., Koopal, L. K., and van Riemsdijk, W. H. (1996). Metal ion binding by humic acid: Application of the NICA model. *Environ. Sci. Technol.* **30**, 1687–1698.

Klein Wolterink, J., Leermakers, F. A. M., Fleer, G. J., Koopal, L. K., Zhulina, E. B., and Borisov, O. V. (1999). Screening in solutions of starbranched polyelectrolytes. *Macromolecules* **32**, 2365–2377.

Kononova, M. M. (1966). *Soil Organic Matter*, Pergamon Press, London.

Kuhn, R., and Hoffstetter-Kuhn, S. (1993). *Capillary Electrophoresis: Principles and Practise*, Springer-Verlag, Berlin.

Kujawinski, E. B., Hatcher, P. G., and Freitas, M. A. (2002). High-resolution Fourier transform ion cyclotron resonance mass spectrometry (FT-ICR MS) of humic and fulvic acids: Improvements and comparisons. *Anal. Chem.* **74**, 413–419.

Kunenkov, E. V., Kononikhin, A. S., Perminova, I. V., Garmash, A. V., Popov, I. A., and Nikolaev, E. N. (2009). In search for optimum conditions of electrospray ionization of humic substances from natural waters using Fourier transform ion cyclotron resonance mass spectrometry. *Moscow Univ. Chemi. Bull.* **50**(2) (in press).

Kutsch, H., and Schleich, C. (1989). Improved colorimetric determination of high-molecular weight hyaluronic acid from synovial fluids. *Fresenius J. Anal. Chem.* **333**, 810–817.

Kutsch, H., and Schumacher, B. (1994). Isoelectric focusing of humic substances on ultrathin polyacrylamide gels: Evidence of fingerprint performance. *Biol. Fertil. Soils* **18**, 163–167.

Lead, J. R., and Wilkinson, K. J. (2007). Environmental colloids and particles: Current knowledge and future developments. IUPAC Series on Analytical and Physical Chemistry of Environmental Systems, **10**, 1–16.

Lead, J. R., Wilkinson, K. J., Balnois, E., Cutak, B. J., Larive, C. K., Assemi, S., and Beckett, R. (2000). Diffusion coefficients and polydispersities of the Suwannee River Fulvic Acid: Comparison of fluorescence correlation spectroscopy, pulsed-field gradient nuclear magnetic resonance, and flow field-flow fractionation. *Environ. Sci. Technol.* **34**(16), 3508–3513.

Leenheer, J. A., Rostad, C. E., Gates, P. M., Furlong, E. T., and Ferrer, I. (2001). Molecular resolution and fragmentation of fulvic acid by electrospray ionization/multistage tandem mass spectrometry. *Anal. Chem.* **73**, 1461–1471.

Li, L., Zhao, Z. Y., Huang, W. L., Peng, P., Sheng, G. Y., and Fu, J. M. (2004). Characterization of humic acids fractionated by ultrafiltration. *Org. Geochem.* **35**(9), 1025–1037.

Lin, C.-F., Lin, T.-Y., and Hao, O. J. (2000). Effects of humic substance characteristics on UF performance. *Water Res.* **34**(4), 1097–1106.

Manh Thang, N., Geckeis, H., Kim, J. I., and Beck, H. P. (2001). Application of the flow field flow fractionation (FFFF) to the characterization of aquatic humic colloids: Evaluation and optimization of the method. *Colloids Surf. A* **181**(1–3), 289–301.

Manttari, M., Puro, L., Nuortila-Jokinen, J., and Nystrom, M. (2000). Fouling effects of polysaccharides and humic acid in nanofiltration. *J. Membr. Sci.* **165**, 1–17.

Marinsky, J. A., and Ephraim, J. (1986). A unified physicochemical description of the protonation and metal ion complexation equilibria of natural organic acids (humic and fulvic acids). 1. Analysis of the influence of polyelectrolyte properties on protonation equilibria in ionic media: Fundamental concepts. *Environ. Sci. Technol.* **20**, 349–354.

Martın, F., Gonzalez-Vila, F. J., del Rıo, J. C., and Verdejo, T. (1994). Pyrolysis derivatization of humic substances 1. Pyrolysis of fulvic acids in the presence of tetramethylammonium hydroxide. *J. Anal. Appl. Pyrolysis* **28**, 71–80.

Menzinger, F., Schmitt-Kopplin, P., Frommberger, M., Freitag, D., and Kettrup, A. (2001). Partial-filling micellar electrokinetic chromatography and non-aqueous capillary electrophoresis for the analysis of selected agrochemicals. *Fresenius J. Anal. Chem.* **371**, 25–34.

Mercier, F., Moulin, V., Guittet, M. J., Barre, N., Gautier-Soyer, M., Trocellier, P., and Toulhoat, P. (2002). Applications of new surface analysis techniques (NMA and XPS) to humic substances. *Org. Geochem.* **33**, 247–255.

Mori, S., Hiraide, M., and Mizuike, A. (1987). Aqueous size-exclusion chromatography of humic acids on a sephadex gel column with diluted phosphate buffers as eluents. *Anal. Chim. Acta* **193**, 231–238.

Mulder, G. J. (1861). *Die Chemie der Ackerkrume* Ueber, J. Mueller, Leipzig, two volumes.

Mueller, M. B., Schmitt, D., and Frimmel, F. H. (2000). Fractionation of natural organic matter by size exclusion chromatography—Properties and stability of fractions. *Environ. Sci. Technol.* **34**, 4867–4872.

Neihof, R. A., and Loeb, G. I. (1972). The surface of particulate matter in seawater. *Limnol. Oceanogr.* **17**, 7–16.

Neihof, R. A., and Loeb, G. I. (1973). Dissolved organic matter in seawater and the electric charge of immersed surfaces. *J. Mar. Res.* **32**, 5–11.

Oden, S. (1919). *Die Huminsaeuren*, Leipzig.

Offord, R. E. (1966). Electrophoretic mobilities of peptides on paper and their use in the determination of amide groups. *Nature* **211**, 591.

Ohshima, H. (2002). Electrophoretic mobility of a charged spherical colloidal particle covered with an uncharged polymer layer. *Electrophoresis* **23**, 1993–2000.

Orioli, G. A., and Curvetto, N. R. (1980). Evaluation of extractants for soil humic substances. I. Isotachophoretic studies. *Plant Soil.* **55**, 353–361.

Orlov, D. S. (1990). *Soil Humic Acids and General Theory of Humification*, Moscow State University Publisher, Moscow.

Perdue, E. M., and Ritchie, J. D. (2003). Dissolved organic matter in freshwaters. *Treatise on Geochemistry* **5**, 273–318.

Perminova, I. V. (1999). Size exclusion chromatography of humic substances: complexities of data interpretation. *Soil Sci.* **164**, 834–840.

Perminova, I. V., Frimmel, F. H., Kovalevskii, D., Abbt-Braun, G., Kudryavtsev, A., and Hesse, S. (1998). Development of a predictive model for calculation of molecular weight of humic substances. *Water Res.* **32**, 872–881.

Perminova, I. V., Frimmel, F. H., Kudryavtsev, A. V., Kulikova, N. A., Abbt-Braun, G., Hesse, S., and Petrosyan, V. S. (2003). Molecular weight characteristics of aquatic, soil, and peat humic substances as determined by size exclusion chromatography and their statistical evaluation. *Environ. Sci. Technol.* **37**, 2477–2485.

Perminova, I. V., Grechishcheva, N. Y., and Petrosyan, V. S. (1999). Relationships between structure and binding affinity of humic substances for polycyclic aromatic hydrocarbons: Relevance of molecular descriptors. *Environ. Sci. Technol.* **33**, 3781–3787.

Pershina (Perminova), I. V., Vermool, V. M., Polenova, T. V., and Ivanova, E. K. (1989). Study of molecular weight distribution and spectral parameters of the fulvic acids of natural water origin. I. Gel-permeation chromatography fractionation of fulvic acids. *Bulletin of Moscow University [Vestnik MGU], Series 2 (Chemistry)* **30**, 176–182.

Persson, L., Alsberg, T., Kiss, G., and Odham, G. (2000). On-line size-exclusion chromatography/electrospray ionisation mass spectrometry of aquatic humic and fulvic acids. *Rapid Commun. Mass Spectrom.* **14**, 286–292.

Petrussi, F., De Nobili, M., Viotto, M., and Sequi, P. (1988). Characterization of organic matter from animal manures after digestion by earthworms. *Plant Soil.* **105**, 41–46.

Phillips, S. L., and Olesik, S. V. (2003). Initial characterization of humic acids using liquid chromatography at the critical condition followed by size-exclusion chromatography and electrospray ionization mass spectrometry. *Anal. Chem.* **75**, 5544–5553.

Piccolo, A. (1997). New insights on the conformational structure of humic substances as revealed by size exclusion chromatography. In *The Role of Humic Substances in the Ecosystems and in Environmental Processes*, Drozd, J., Gonet, S. S., Senesi, N., and Weber, J., eds., IHSS-Polish Society of Humic Substances, Wroclaw, Poland, pp. 19–35.

Piccolo, A. (2001). The supramolecular structure of humic substances. *Soil Sci.* **166**, 810–832.

Piccolo, A., Nardi, S., and Concheri, G. (1996). Macromolecular changes of humic substances induced by interaction with organic acids. *European J. Soil Sci.* **47**, 319–328.

Porath, J., and Flodin, F. (1959). Gel filtration: A method for desalting and group separation. *Nature* **183**, 1657–1659.

Posner, A. M. (1963). Importance of electrolyte in the determination of molecular weights by Sephadex gel filtration, with especial reference to humic acid. *Nature* **196**, 1161–1163.

Quigley, M. S., Santschi, P. H., Hung, C. C., Guo, L., and Honeyman, B. D. (2002). Importance of acid polysaccharides for ^{234}Th complexation to marine organic matter. *Limnol. Oceanogr.* **47**, 367–377.

Ratanathanawongs, K. W.-S., and Lee, D. (2006). Field-flow fractionation of proteins, polysaccharides, synthetic polymers, and supramolecular assemblies. *J. Sep. Sci.* **29**, 1720–1732.

Reemtsma, T., and These, A. (2003). On-line coupling of size exclusion chromatography with electrospray ionization-tandem mass spectrometry for the analysis of aquatic fulvic and humic acids. *Anal. Chem.*, **75**, 1500–1507.

Reemtsma, T., and These, A. (2005). Comparative investigation of low-molecular-weight fulvic acids of different origin by SEC-Q-TOF-MS: New insights into structure and formation. *Environ. Sci. Technol.* **39**, 3507–3512.

Reschiglian, P., Zattoni, A., Roda, B., Michelini, E., and Roda, A. (2005). Field-flow fractionation and biotechnology. *Trends Biotechnol.* **23**(9), 475–483.

Richard, C., Trubetskaya, O., Trubetskoj, O., Reznikova, O., Afanaseva, G. V., Aguer, J. P., and Guyot, G. (2004). Key role of the low molecular size fraction of soil humic acids for fluorescence and photoinductive activity. *Environ. Sci. Technol.* **38**, 2052–2057.

Richardson, S. D. (2003). Water analysis: Emerging contaminants and current issues. *Anal. Chem.* **75**, 2831–2857.

Righetti, P.G., Brown, R. P. and Stone, A. L. (1978). Aggregation of ampholine on heparin and other acidic polysaccharides in isoelectric focussing. *Biochim. Biophys. Acta* **542**, 232–244.

Rocha, J. C., De Sene, J. J., Dos Santos, A., Toscano, I. A. S., and Zara, L. F. (2000). Aquatic humus from an unpolluted Brazilian dark-brown stream: General characterization and size fractionation of bound heavy metals. *J. Environ. Monit.* **2**(1), 39–44.

Saiz-Jimenez, C. (1994). Analytical pyrolysis of humic substances: Pitfalls, limitations, and possible solutions. *Environ. Sci. Technol.* **28**, 1773–1780.

Sargentini, E., Jr., Rocha, J. C., Rosa, A. H., Zara, L. F., and dos Santos, A. (2001). Aquatic humic substances: Molecular size fractionation and characterization of inner rearrangements after metal ions complexation. *Quimica Nova* **24**(3), 339–344.

Schimpf, M., Caldwell, K., and Giddings, J. C. (2000). *Field Flow Fractionation Handbook*, John Wiley & Sons, New York, **616** pages.

Schimpf, M. E., and Petteys, M. P. (1997). Characterization of humic materials by flow field-flow fractionation. *Colloids Surf. A* **120**, 87–100.

Schmitt, P., Garrison, A.W., Freitag, D., and Kettrup, A. (1997a). Capillary isoelectric focusing (CIEF) for the characterization of humic substances. *Water Res.* **31**, 2037–2049.

Schmitt, P., Poiger, T., Simon, R., Garrison, A. W., Freitag, D., and Kettrup, A. (1997b). Simultaneous ionization constants and isoelectric points determination of 12 hydroxy-*s*-triazines by capillary zone electrophoresis (CZE) and capillary electrophoresis isoelectric focusing (CIEF). *Anal. Chem.* **69**, 2559–2566.

Schmitt-Kopplin, P. (2002). Comment on "Determination of electrophoretic mobilities and hydrodynamic radii of three humic substances as a function of pH and ionic strength." *Environ. Sci.Technol.* **36**, 3041–3042.

Schmitt-Kopplin, P. (2005). *Capillary Electrophoresis; from Small Ions to Macromolecules*, Humana Press, Totowa, NJ.

Schmitt-Kopplin, P., and Frommberger, M. (2003). Capillary electrophoresis–mass spectrometry: 15 years of developments and applications. *Electrophoresis* **24**, 3837–3867.

Schmitt-Kopplin, P., and Junkers, J. (2003). Capillary zone electrophoresis of natural organic matter. *J. Chromatogr. A* **998**, 1–20.

Schmitt-Kopplin, P., and Kettrup, A. (2003). Capillary electrophoresis–electrospray ionization–mass spectrometry for the characterization of natural organic matter: An evaluation with free flow electrophoresis–off-line flow injection electrospray ionization mass spectrometry. *Electrophoresis* **24**, 3057–3066.

Schmitt-Kopplin, P., Freitag, D., Kettrup, A., Schoen, U., and Egeberg, P. (1999a). Capillary zone electrophoretic studies on Norwegian surface water natural organic matter. *Environ. Int.* **25**, 259–274.

Schmitt-Kopplin, P., Garmash, A. V., Kudryavtsev, A. V., Perminova, I. V., Hertkorn, N., Freitag, D., and Kettrup, A. (1999b). Mobility distribution description of synthetic and natural polyelectrolytes with capillary zone electrophoresis. *J. AOAC Int.* **82**, 1594–1603.

Schmitt-Kopplin, P., Garmash, A. V., Kudryavtsev, A. V., Menzinger, F., Perminova, I. V., Hertkorn, N., Freitag, D., Petrosyan, V. S., and Kettrup, A. (2001a). Quantitative and qualitative precision improvements by effective mobility-.scale data transformation in capillary electrophoresis analysis. *Electrophoresis* **22**, 77–87.

Schmitt-Kopplin, P., Garrison, A. W., Perdue, E. M., Freitag, D., and Kettrup, A. (1998). Capillary electrophoresis in the analysis of humic substances—Facts and artifacts. *J. Chromatogr. A* **807**, 101–109.

Schmitt-Kopplin, P., Menzinger, F., Freitag, D., and Kettrup, A. (2001b). Improving the use of CE in a chromatographer's world. *LC-GC Eur.* **14**, 284–388.

Seddon, J. M., and Gale, J. D. (2002). *Thermodynamics and Statistical Mechanics. Royal Society of Chemistry*, 166pp.

Shin, H.-S., Monsallier, J. M., and Choppin, G. R. (1999). Spectroscopic and chemical characterizations of molecular size fractionated humic acid. *Talanta* **50**, 641–647.

Simpson, A. J. (2002). Determining the molecular weight, aggregation, structures and interactions of natural organic matter using diffusion ordered spectroscopy. *Magn. Reson. Chem.* **40**, 72–82.

Simpson, A. J., Kingery, W. L., Spraul, M., Humpfer, E., Dvortsak, P., and Kerssebaum, R. (2001). Separation of structural components in soil organic matter by diffusion ordered spectroscopy. *Environ. Sci. Technol.* **35**, 4421–4425.

Simpson, A. J., Kingery, W. L., Hayes, M. H. B., Spraul, M., Humpfer, E., Dvortsak, P., Kerssebaum, R., Godejohann, M., and Hofmann, M. (2002). Molecular structures and associations of humic substances in the terrestrial environment. *Naturwissenschaften* **89**, 84–88.

Simpson, A. J., Tseng, L. H., Simpson, M. J., Spraul, M., Braumann, U., Kingery, W. L., Kelleher, B. P., and Hayes, M. H. B. (2004). The application of LC-NMR and LC-SPE-NMR to compositional studies of natural organic matter. *Analyst.* **129**, 1216–1222.

Siripinyanond, A., Barnes, R. M., and Amarasiriwardena, D. (2002). Flow field-flow fractionation–inductively coupled plasma mass spectrometry for sediment bound trace metal characterization. *J. Anal. At. Spectrom.* **17**(9), 1055–1064.

Siripinyanond, A., Worapanyanond, S., and Shiowatana, J. (2005). Field-flow fractionation–inductively coupled plasma mass spectrometry: An alternative approach to investigate metal–humic substances interaction. *Environ. Sci. Technol.* **39**(9), 3295–3301.

Sprengel, C. (1837). *Die Bodenkunde oder Die Lehre von Boden*, Leipzig.

Stenson, A., Landing, W. M., Marshall, A. G., and Cooper, W. T. (2002). Ionization and fragmentation of humic substances in electrospray ionization fourier transform-ion cyclotron resonance mass spectrometry. *Anal. Chem.* **74**, 4397–4409.

Stenson, A. C., Marshall, A. G., and Cooper, W. T. (2003). Exact masses and chemical formulas of individual Suwannee River fulvic acids from ultrahigh resolution electrospray ionization fourier transform ion cyclotron resonance mass spectra. *Anal. Chem.* **75**, 1275–1284.

Stepanov, V. V., and Pakhonov, A. N. (1969). Electrophoresis of humus substances in polyacrylamide gel. *Pochvovedenie*, **12**, 48–56.

Stevenson, F. J. (1994). *Humus Chemistry: Genesis, Composition, Reactions*, John Wiley & Sons, New York.

Stevenson, F. J., van Winkle, Q., and Martin, W. P. (1953). Physicochemical investigations of clay absorbed organic colloids. *Soil Sci. Soc. Am. Proce.* **17**, 31–34.

Stokes, J. C., and Johnson, M. J. (2004). Resolution in sub-micrometer particle separations by capillary electrophoresis. *Microchem. J.* **76**, 121–129.

Suteerapataranon, S., Bouby, M., Geckeis, H., Fanghaenel, T., and Grudpan, K. (2006). Interaction of trace elements in acid mine drainage solution with humic acid. *Water Res.* **40**(10), 2044–2054.

Sutton, R., and Sposito, G. (2005). Molecular structure in soil humic substances: The new view. *Environ. Sci. Technol.* **39**, 9009–9015.

Swift, R. S. (1996). Organic matter characterization. In: *Methods of Soil Analysis. Part 3: Chemical Methods*, Sparks, D. L., ed., SSSA Book Series No. 5, SSSA and ASA, Madison, WI, pp. 1011–1069.

Swift, R. S. (1999). Macromolecular properties of soil humic substances: Fact, fiction, and opinion. *Soil Sci.* **164**, 790–802.

Swift, R. S., Leonard, R. L., Newman, R. H., and Theng, B. K. G. (1992). Changes in humic acid composition with molecular weight as detected by ^{13}C-nuclear magnetic resonance spectroscopy. *Sci. Total Environ.* **117–118**, 53–61.

Tang, C. Y., and Leckie, J. O. (2007). Membrane independent limiting flux for RO and NF membranes fouled by humic acid. *Environ. Sci. Technol.* **41**, 4767–4773.

Taylor, H. E., Garbarino, J. R., Murphy, D. M., and Beckett, R. (1992). Inductively coupled-plasma-mass spectrometry as an element-specific detector for field-flow fractionation particle separation. *Anal. Chem.* **64**, 2036–2041.

Tipping, E. (2002). *Cation Binding by Humic Substances*, Cambridge Univeristy Press, Cambridge, UK.

Tiselius, A. (1953). Zone electrophoresis in filter paper and other media. *Dis. Faraday Soc.* **13**, 29–33.

Town, R. M., and Powell, H. K. (1992). Elimination of adsorption effects in gel permeation chromatography of humic substances. *Anal. Chim. Acta* **256**, 81–86.

Trubetskoj, O. A., Trubetskaya, O. E., Afanaseva, G. V., Reznikova, O. I., and Saiz-Jimenez, C. (1997). Polyacrylamide gel electrophoresis of soil humic acid fractionated by size-exclusion chromatography and ultrafiltration. *J. Chromatogr. A* **767**, 285–292.

Trubetskoj, O. A., Trubetskaya, O. E., Afanasieva, G. V., Reznikova, O. I., Hermosin, B., and Saiz-Jimenez, C. (1998). Tandem size exclusion chromatography polyacrylamide gel electrophoresis of humic acids. *J. Plant Nutr. Soil Sci.* **161**, 619–625.

Trubetskoj, O. A., Trubetskaya, O. E., Reznikova, O. I., and Afanas'eva, G. F. (1999). Weight and optical differences between soil humic acids fractions obtained by coupling SEC-PAGE. *Geoderma* **93**, 277–287.

Trubetskoj, O. A., Trubetskaya, O. E., Reznikova, O. I., Afanas'eva, G. V., and Saiz-Jimenez, C. (2001). Electrophoretic analysis of humic acids and the products of their acid hydrolysis. *Eurasian Soil Sci.* **34**, 1097–1100.

Trubetskaya, O. E., Trubetskoj, O. A., and Saiz-Jimenez, C. (2001). Polyacrylamide gel electrophoresis of humic and fulvic acids after acid hydrolysis. *Fresenius Environ. Bull.* **10**, 635–637.

Trubetskaya, O., Trubetskoj, O., Guyot, G., Andreux, F., and Richard, C. (2002). Fluorescence of soil humic acids and their fractions obtained by tandem size exclusion chromatography–polyacrylamide gel electrophoresis. *Org. Geochem.* **33**, 213–220.

van Bergen, P., Flannery, M. B., Poulton, P. R., and Evershed, R. P. (1998). Fate of N-containing macro-molecules. In *Biosphere and Geosphere*, Stankiewicz, B. A., van Bergen, P. F., eds., Symposium Series 707, American Chemical Society, Washington, D.C., pp. 321–338.

Varga, B., Kiss, G., Galambos, I., Gelencser, A., Hlavay, J., and Krivacsy, Z. (2000). Secondary structure of humic acids. Can micelle-like conformation be proved by aqueous size exclusion chromatography? *Environ. Sci. Technol.* **34**, 3303–3306.

von Brocke, A., Nicholson, G., and Bayer, E. (2001). Recent advances in capillary electrophoresis/electrospray–mass spectrometry. *Electrophoresis* **22**, 1251–1266.

Wershaw, R. L. (1999). Molecular aggregation of humic substances. *Soil Sci.* **164**, 803–811.

Wershaw, R. L. (2004). Evaluation of conceptual models of natural organic matter (humus) from a consideration of the chemical and biochemical processes of humification. Scientific Investigations Report 2004–5121, U.S. Geological Survey, Reston, VA.

Whelan, T. J., Shalliker, R. A., McIntyre, C., and Wilson, M. A. (2005). Development of a multidimensional high-performance liquid chromatography (HPLC) separation for bayer humic substances. *Ind. Eng. Chem. Res.* **44**, 3229–3237.

Yau, W. W., Kirkland, J. J., and Bly, D. D. (1979). *Modern Size-Exclusion Chromatography*, John Wiley & Sons, New York.

Zanardi-Lamardo, E., Clark, C. D., Moore, C. A., and Zika, R. G. (2002). Comparison of the molecular mass and optical properties of colored dissolved organic material in two rivers and coastal waters by flow field-flow fractionation. *Environ. Sci. Technol.* **36**(13), 2806–2814.

14

ANALYTICAL PYROLYSIS AND SOFT-IONIZATION MASS SPECTROMETRY

P. LEINWEBER, G. JANDL, K.-U. ECKHARDT, AND H.-R. SCHULTEN
Institute for Land Use, University of Rostock, Rostock, Germany

A. SCHLICHTING
Steinbeis-Transferzentrum Soil Biotechnology, Huckstorf, Germany

D. HOFMANN
Central Division of Analytical Chemistry/BioSpec, Research Centre Jülich, Jülich, Germany

Biophysico-Chemical Processes Involving Natural Nonliving Organic Matter in Environmental Systems, Edited by Nicola Senesi, Baoshan Xing, and Pan Ming Huang

14.1. INTRODUCTION

Analytical pyrolysis is defined as the characterization of a material or a chemical process by the instrumental analysis of its pyrolysis products (Ericsson and Lattimer, 1989). The most important analytical pyrolysis methods widely applied to environmental samples are Curie-point (flash) pyrolysis combined with electron impact (EI) ionization gas chromatography/mass spectrometry (Cp Py-GC/MS) and pyrolysis–field ionization mass spectrometry (Py-FIMS). In contrast to the fragmenting EI ionization, soft ionization methods, such as field ionization (FI) and field desorption (FD) each in combination with MS, result in the formation of molecule ions either without, or with only very low, fragmentation (Lehmann and Schulten, 1976; Schulten, 1987; Schulten and Leinweber, 1996; Schulten et al., 1998). The molecule ions are potentially similar to the original sample, which makes these methods particularly suitable to the investigation of complex environmental samples of unknown composition.

The broad application of complementary analytical pyrolysis methods in conjunction with wet-chemical and other spectroscopic methods initiated a breakthrough in the development of novel structural concepts of humic substances (Schnitzer and Schulten, 1995). Previous reviews demonstrated the great potential of analytical pyrolysis (especially in combination with soft ionization mass spectrometry) in structural, molecular-chemical based investigations not only of humic substances but also of soil fractions and whole soil organic matter (SOM) (Schulten, 1996; Leinweber and Schulten, 1998). The identification of nitrogen- and sulfur-containing organic compounds, the development of a better understanding of organic-mineral bonds and structural arrangements, and the disclosure of structure–property relationships (e.g., adsorption and bonding of contaminants) were emphasized as future applications of analytical pyrolysis and mass spectrometry (Schulten et al., 1998). Since these reviews in the last decade, great progress has been made in the development of new mass spectrometry techniques and in the broad application to fundamental and applied soil chemistry problems. Among others, analytical pyrolysis and mass spectrometry data were used in molecular-mechanics calculations and computational chemistry to explain the three-dimensional structure of nonliving organic matter and whole organic-mineral soil particles (Schulten and Leinweber, 1999).

This chapter intends to review advances in analytical pyrolysis and soft ionization mass spectrometric techniques and applications to the chemistry of nonliving organic matter that have been achieved in the last decade. Innovative soft ionization techniques such as liquid injection field desorption ionization (LIFDI) combined with MS (Linden, 2001, 2002, 2004; Qian et al., 2004; Schaub et al., 2004, 2005; Rodgers et al., 2005; Fu et al., 2006a,b; Gross et al., 2006), as well as the ultrahigh-resolution Fourier transform–ion cyclotron resonance mass spectrometry (FT-ICR MS) (e.g., Fievre et al., 1997; Marshall, 2000; Qian et al., 2001; Kujawinski et al., 2002; Llewelyn et al., 2002; Stenson et al., 2002, 2003; Fard et al., 2003), were recently introduced to applied environmental chemistry. These technical innovations will be described briefly, and a few examples of applications will be given to demonstrate the tremendous potential to disclose composition, structure, and functions of nonliving organic matter. Moreover, methodological innovations and the multiple application of Py-FIMS will be demonstrated with data compiled from studies of lipids, "unknown N" compounds, dissolved organic matter of very diverse origin, organic-mineral soil fractions, and nonfractionated whole soil samples.

14.2. OVERVIEW ON ANALYTICAL TECHNIQUES

14.2.1. Pyrolysis–Gas Chromatography/Electron Impact Mass Spectrometry

Curie-point pyrolysis–gas chromatography/electron impact mass spectrometry (Cp Py-GC/MS) is the most frequently used analytical pyrolysis method for the investigation of humic substances and soil organic matter (SOM). This widely applied method, which combines flash pyrolysis with electron impact, needs a short description despite its hard ionization technique and because of new instrumental developments in this field. The detailed analytical procedure of Cp Py-GC/MS was described previously and was used in recent research (Schulten, 1987; Schulten and Leinweber, 1996; Leinweber and Schulten, 1998; Schulten et al., 1998; Schulten et al., 2002). The dried samples (up to 11 mg) were pyrolyzed at 500 °C for 9.9 s in a Curie-point pyrolyzer (Fischer 1040 PSC, Germany). The pyrolysis products were separated on a gas chromatograph Varian 3800 (Varian, USA) equipped with a 25-m capillary column BPX 5 (SGE, Australia) that was coated with 0.25-μm film thickness and had an inner diameter of 0.32 mm. Following a splitless injection at 300 °C for 45 s, the split ratio was 1:100 from 45 s up to 90 s and was 1:5 from 90 s on. The flow rate of the helium carrier gas was adjusted to 2 ml min^{-1}. The starting temperature for the gas chromatographic program was 28 °C (5 min), followed heating at 5 °C min^{-1} to 280 °C (30 min). The gas chromatograph was connected to a double-focusing Finnigan MAT 212 mass spectrometer (Germany). Conditions for mass spectrometric detection in the electron impact mode were 3-kV acceleration voltage, 70-eV electron energy, 2.2-kV multiplier, 1.1-s (mass decade)$^{-1}$ scan speed, and m/z 48–450 mass range.

Furthermore, the technical development of analytical pyrolysis revealed the multi-step pyrolysis or double-shot technique, which can be used online in combination with GC/MS. This method performs the pyrolysis at two different temperatures. The stepped heating process allows the separation of individual substances at different decomposition temperatures. Thereby, the differentiation of thermal labile and stable tightly trapped biogenic compounds and comparison with classical Cp-Py results provided additional information on the composition, origin and nature of the insoluble, nonhydrolyzable organic fraction (Quénéa et al., 2006a). Furthermore, double-shot pyrolysis also showed that in volcanic soils a large amount of aliphatic compounds was stabilized by intense organic–mineral interactions affected by poorly crystalline materials like allophanes, imogolite, and other Fe and Al oxyhydroxides (González-Pérez et al., 2007). Double-shot pyrolysis is an approach to relate the thermal stability of compounds to their biological stability and resistance against biodegradation, a concept that will be explained in more detail later in this chapter.

Generally, the mass spectra corresponding to peaks in the gas chromatograms obtained in the EI mode at 70 eV are assigned by comparisons with mass spectral libraries. The fragments were assumed to be characteristic of the original structure (Galletti and Bocchini, 1995; Schulten, 1996). However, one often-described disadvantage of Curie-point pyrolysis is the formation of so-called "artifacts" during the pyrolysis. The high-energy impact during heating in milliseconds lead to thermally induced secondary reactions and fragmentation which can produce fragments different from the original structures (Schulten, 1996; Hatcher et al., 2001; Zang and Hatcher, 2002). Common examples are the formation of furanes from carbohydrates, alkylaromatics from cycled fatty acids (Saiz-Jimenez, 1994; Saiz-Jimenez et al., 1994), and N-heterocycles from peptides (Schulten et al., 1995a). However, this does not necessarily mean that these compounds cannot be genuine constituents of nonliving organic matter as will be shown for N-heterocycles (Section 14.3.1). Another difficulty in using GC is that some high-molecular-weight products may not be detected due to their poor volatility and thus retention on the GC column (Reeves and Francis, 1997). Nevertheless, under optimal analytical conditions, pyrolysis provided valuable information on nonliving organic matter as shown recently, for example, by Kögel-Knabner (2000), Schulten and Leinweber (2000) and Zang and Hatcher (2002). Furthermore, our general recommendation is to use at least two independent analytical methods for the determination of composition and structures of nonliving organic matter, as was done in a recent study of organic matter from municipal waste (Franke et al., 2006, 2007).

14.2.2. Pyrolysis–Field Ionization and Field Desorption Mass Spectrometry

The pyrolysis–field ionization (Py-FI) and field desorption (FD) mass spectrometry (MS) is a powerful analytical tool for the characterization of the molecular composition of SOM. Moreover, the temperature-resolved pyrolysis in combination with the soft field ionization offers the possibility to describe the thermal stability of biogenic marker substances and classes. The experimental setup for Py-FIMS has been described in detail by Schulten (1987) and Schulten et al. (1998). In brief, for temperature-resolved Py-FIMS, approximately 0.5 mg of organic material or 6 mg of soil were heated in a direct inlet system of the double-focusing mass spectrometer (Finnigan MAT 731, Germany) from 110 °C to 700 °C at heating steps of 10 K. During the analysis, 60 spectra were recorded in the mass range *m/z* 15–900. For most samples of nonliving organic matter, about 3–5 replicates were measured and the data averaged. In automated routine data analyses the total ion intensities (TII), normalized to 1 mg sample weight, are plotted against pyrolysis temperature, producing TII thermograms. Furthermore, averaged (for replicate measurements) and summed (over the whole temperature range) Py-FI mass spectra are calculated and plotted. A first valuable step is the multivariate statistical data evaluation, for example, by principal component analysis of this mass spectrometric "fingerprint" to derive similarity and dissimilarity among samples within sample sets under study. The *m/z* signals contributing to differences and arranged according to discrimination power can be assigned to chemical compounds. These assignments as well as the calculation of 10 important compound classes were based on an *in-house* spectra library including high-resolution spectra of numerous biogenic substances (Guggenberger et al., 1994; Schulten, 1996) and thermal properties (Schulten and Schnitzer, 1991, 1993).

Marker signals for these compound classes were published by Schnitzer and Schulten (1992), Schulten and Schnitzer (1992), and Schulten and Leinweber (1996)

TABLE 14.1. Marker Signals in Py-FIMS of Aquatic and Terrestrial Humic Substances, Soil Fractions, and Whole Soils

Compound Class	*m/z* of Marker Signals
Carbohydrates	60, 72, 82, 84, 96, 98, 110, 112, 114, 126, 132, 144, 162
Phenols + lignin monomers	94, 108, 110, 122, 124, 138, 140, 150, 152, 154, 164, 166, 168, 178, 180, 182, 194, 196, 208, 210, 212
Lignin dimers	246, 260, 270, 272, 274, 284, 286, 296, 298, 300, 310, 312, 314, 316, 326, 328, 330, 340, 342, 356
Lipids, alkanes, alkenes, fatty acids, *n*-alkyl esters	202, 216, 230, 244, 256, 258, 270, 272, 284, 286, 298, 300, 312, 314, 326, 328, 340, 342, 354, 368, 380, 382, 394, 396, 408, 410, 422, 424, 438, 452, 466, 480, 494, 508, 648, 662, 676, 704, 732
Alkylaromatics	92, 106, 120, 134, 142, 148, 156, 162, 170, 176, 184, 190, 192, 198, 204, 206, 218, 220, 232, 234, 246, 260, 274, 288, 302, 316, 330, 344, 358, 372, 386
N-containing compounds	59, 67, 79, 81, 95, 103, 109, 111, 123, 125, 137, 139, 153, 161, 167, 181, 183, 195, 203, 233, 245, 255, 257, 271, 285, 333, 359, 363, 393
Sterols	372, 386, 388, 390, 392, 394, 396, 398, 400, 402, 408, 410, 412, 414, 416, 426, 430
Peptides	57, 70, 73, 74, 75, 84, 87, 91, 97, 99, 115, 120, 129, 135
Suberin	432, 446, 460, 474, 488, 502, 516, 530
n-C_{16} to *n*-C_{34} free fatty acids	256, 270, 284, 298, 312, 326, 340, 354, 368, 382, 396, 410, 424, 438, 452, 466, 480, 494, 508
Low mass signals	15–56
[M+H] and isotope ^{13}C signals	58, 61, 71, 73, 76, 80, 85, 88, 97, 99, 104, 113, 116, 121, 127, 130, 133, 135, 136, 141, 143, 145, 149, 151, 155, 157, 163, 165, 169, 171, 177, 179, 185, 191, 193, 197, 199, 205, 207, 209, 211, 213, 217, 219, 221, 231, 235, 247, 259, 261, 273, 275, 287, 289, 297, 299, 301, 303, 311, 313, 315, 317, 327, 329, 331, 334, 341, 343, 355, 357, 360, 364, 369, 373, 381, 383, 387, 389, 391, 395, 397, 399, 401, 403, 409, 411, 413, 415, 417, 423, 425, 427, 431, 433, 439, 447, 453, 461, 467, 475, 481, 489, 495, 503, 509, 517, 531, 649, 663, 677, 705, 733

Source: Reprinted from Schulten, H.-R., and Leinweber, P. (1999). Thermal stability and composition of mineral-bound organic matter in density fractions of soil. *European Journal of Soil Science* **50**, 237–248, with permission from Blackwell.

and were applied since then (Table 14.1). These marker signals were derived for aquatic and terrestrial humic substances, soil fractions, and whole soils.

The marker signals for carbohydrates represent thermal fragments of pentose and hexose units, which were identified by high-resolution Py-FIMS (e.g., Schulten and Görtz, 1978). The quantitative evaluation of carbohydrates in DOM was confirmed by a strong correlation with ^{13}C NMR data (proportions of *O*-alkyl C) and wet-chemical carbohydrate analyses (Schulten et al., 2002). The signals for phenols and lignin monomers originate from the thermal fragmentation of lignin such as catechol (*m/z* 110), coniferyl (*m/z* 178/180), and sinapyl aldehyde/alcohol (*m/z* 208/210). The important *m/z* 110 originates from methylfuraldehyde (pyrolysis product of carbohydrate) and catechol (pyrolysis product of lignin), which both have the same elemental composition $C_6H_6O_2$. Therefore this prominent *m/z*, always appearing with high intensity, is assigned to both of these compound classes. Thermal fragmentation of lignin also results in dimers with phenlycoumaran, biphenyl-, diarylpropane- and resinol-type structures (Schulten and Schnitzer, 1991). Lipids, alkanes, alkenes, fatty acids, and *n*-alkyl esters were first derived from homologous series in supercritical carbon dioxide extracts (Schulten and Schnitzer, 1991). Alkylaromatics were identified as major pyrolysis products of humic acids and interpreted as backbone of humic substances (Schulten et al., 1991). These signals are generally abundant in the Py-FI mass spectra of nonextracted whole soil samples and may indicate humic substances. However, in plant litter or soil samples with large proportions of nondecomposed plant residues the *m/z* assigned to alkylaromatics probably originate from pyrolysis products of lignin or tannins (Sleutel et al., 2008). Signals assigned to soil organic nitrogen (N) were separated into two compound classes. Nonpeptidic N-compounds are discussed in detail in Section 14.3.2. The marker signals of peptides were derived from analyses of pure amino acids using Py-GC/MS and Py-FIMS (Sorge et al., 1993) and confirmed by correlations with α-amino N in DOM (Schulten et al., 2002) and whole soil samples (Schulten and Schnitzer, 1998). In this compound class *m/z* 84 is a pyrolysis product of glutamine and glutamic acid. However, origin of this nominal mass from carbohydrate was also reported (Schulten and Görtz, 1978). Sterols originated from plants (e.g., β-sitosterol, *m/z* 414) indicate fungal biomass (e.g., ergosterol, *m/z* 396) or feces of vertebrates (Jardé et al., 2007). A homologous series of aromatic esters was assigned to thermal degradation products of suberin, a major constituent of bark (Hempfling et al., 1991), also occurring in roots and litter-rich soil samples. The low masses *m/z* 15–56 are unspecific because they can be produced by various thermal decomposition reactions.

It must be emphasized that these marker signals can be used exclusively for the interpretation of Py-FI mass spectra, but not for the interpretation of conventional Cp Py-GC/MS because of the completely different heating and ionization conditions. The interpretation of the routine Py-FI mass spectra will be incrementally improved by the application of complementary techniques such as high-resolution Py-FIMS with a MAT 900 (see below) and synchrotron-based X-ray absorption near edge structure (XANES) spectroscopy (Section 14.3.2).

An example for the identification of organic matter constituents of plant materials by high-resolution MS is shown in Figure 14.1. Wheat straw was analyzed by Py-FIMS on a Finnigan MAT 900 sector field MS at a mass resolution of 4000. The spectra were recorded in the mass range *m/z* 50 to 450 with a scan rate of 20 s (mass decade)$^{-1}$. For temperature-resolved pyrolysis, the sample was heated with a direct inlet probe from 50 °C to 700 °C in steps of 20 °C. Figure 14.1 shows the thermograms in a mass window of ±30 mmu of palmitic acid (*m/z* 256.240), lignin dimer (phenylcoumaran type) $C_{16}H_{14}O_4$ (*m/z* 270.089), and its cleavage product $C_{16}H_{16}O_3$ (*m/z* 256.110). The high resolution of the nominal mass *m/z* 256 helped to distinguish between the palmitic acid and the lignin fragment. Furthermore, palmitic acid was volatilized at a much lower temperature maximum at approximately 230 °C than the lignin dimer and its cleavage product, which showed a similar thermogram with peak at 380 °C. In summary, the thermograms of high-resolved masses were used successfully to distinguish plant constituents of the same nominal mass.

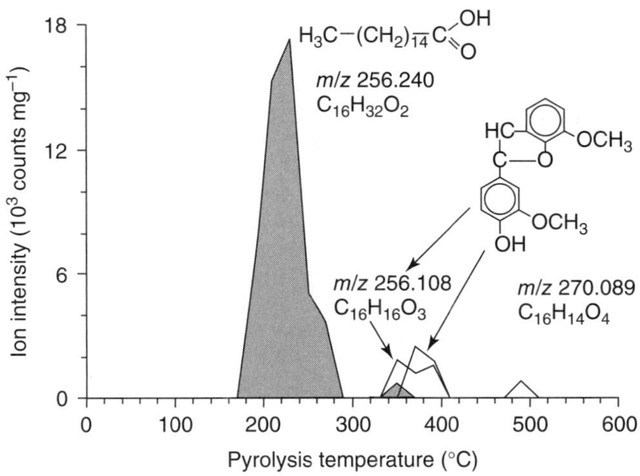

Figure 14.1. Temperature-resolved intensities of precise masses in a high-resolution Py-FIMS measurement of wheat straw. The thermograms of palmitic acid (*m/z* 256.240) and a thermal fragment (*m/z* 256.108) of the lignin dimer $C_{16}H_{14}O_4$ (*m/z* 270.089) are shown with a mass window of ±30 mmu.

Field desorption (FD) can be regarded as one of the softest ionization methods in mass spectrometry (Kane-Maguire et al., 1995; Guo et al., 1999). It is the only technique capable of successfully ionizing nonpolar as well as polar compounds in all three states of aggregation. Furthermore, FD ionization allowed the identification of long-chain aliphatics up to 2000 Da and higher in natural waxes (Murray and Schulten, 1981), in epicuticular waxes isolated from coniferous needles (Schulten et al., 1986), and in soil extracts (Schulten and Schnitzer, 1991). Thus, the softness of field desorption ionization makes it a powerful tool for analysis of fragile molecules, unless analytes can undergo immediate decomposition. Decomposition is possible by reaction with ambient air and/or water during conventional emitter loading by either the dipping (Beckey, 1969) or the syringe (Beckey et al., 1970) technique outside of the vacuum. Such problems can be avoided by using the improved *online* sample supply technique liquid injection field desorption ionization (LIFDI) as explained in the following chapter.

14.2.3. Liquid Injection Field Desorption Ionization Mass Spectrometry

Liquid injection field desorption ionization (LIFDI), first introduced as *in-source* liquid injection (ISLI) FD, is a quick technique enabling a high throughput of samples (Linden, 2001, 2002). Driven by the pressure gradient between atmosphere (outside) and the inner vacuum, a small volume of sample (40 nl) is transferred through a fused silica capillary column from a sample vial into the ion source. This sample volume enters the activated emitter (Figure 14.2). The gaps between the whiskers are filled by the sample solution. After complete sample transfer, the ionizing electric field is established by applying high voltage to the counterelectrode (negative) and the emitter (positive) in positive ion mode. The counterelectrode is located approximately 2 mm away from the emitter. Release and successive ioniza-

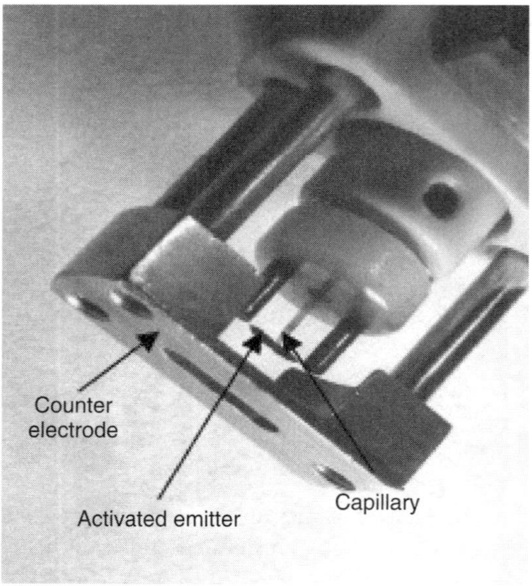

Figure 14.2. Arrangement of an activated emitter in a liquid injection field desorption ionization (LIFDI) system. Courtesy of Linden ChroMasSpec GmbH, Leeste, Germany.

tion of highly volatile compounds may start immediately after activation of high voltage. Subsequently, the emitter is heated stepwise with a defined increase of amperage (heating current in milliamperes). Scanning is synchronized with heating steps, and the volatilization maximum is defined by the limit of heating current of the emitter wire. LIFDI has some advantages: (1) There is no need to remove the FD probe from the vacuum between the samples, which minimizes the risk for incidental breakage of the emitter wire during sample transfer. (2) Readjustment of the ion optics is unnecessary because of the permanent location of the LIFDI probe in the ion source. (3) The decomposition of the analyte substances prior starting the measurement is prevented by handling of the sample solution under inert conditions until ionization. (4) The controlled supply of liquid analytes enables the quantification of compounds and higher yield of ion intensities through the continuous-flow technique. (5) Furthermore, LIFDI is well-suited for automation (Griep-Raming and Linden, 2005).

LIFDI has been successfully applied for analysis of petrochemical compounds (Schaub et al., 2004, 2005; Rodgers et al., 2005; Fu et al., 2006a,b), reactive transition metal complexes (Gross et al., 2006) and nonpolar hydrocarbons (Linden, 2004) in combination with TOF, FT-ICR and magnetic sector mass spectrometers (Linden, 2004; Qian et al., 2004).

For example, the LIFDI mass spectrum of biodiesel from oilseed rape revealed methyl esters of long-chain fatty acids as typical plant lipid constituents (Figure 14.3). The most prominent signal originated from the methyl ester of oleic acid ($C_{18:1}$, m/z 296.4), accounting to 42.6% of the TII, followed by the methyl esters of linoleic acid ($C_{18:2}$, m/z 294.4, 23.8%), linolenic acid ($C_{18:3}$, m/z 292.4, 4.4%), stearic acid ($C_{18:0}$, m/z 298.5, 2.8%), palmitic acid ($C_{16:0}$, m/z 270.4, 1.4%), and gondoic acid ($C_{20:1}$,

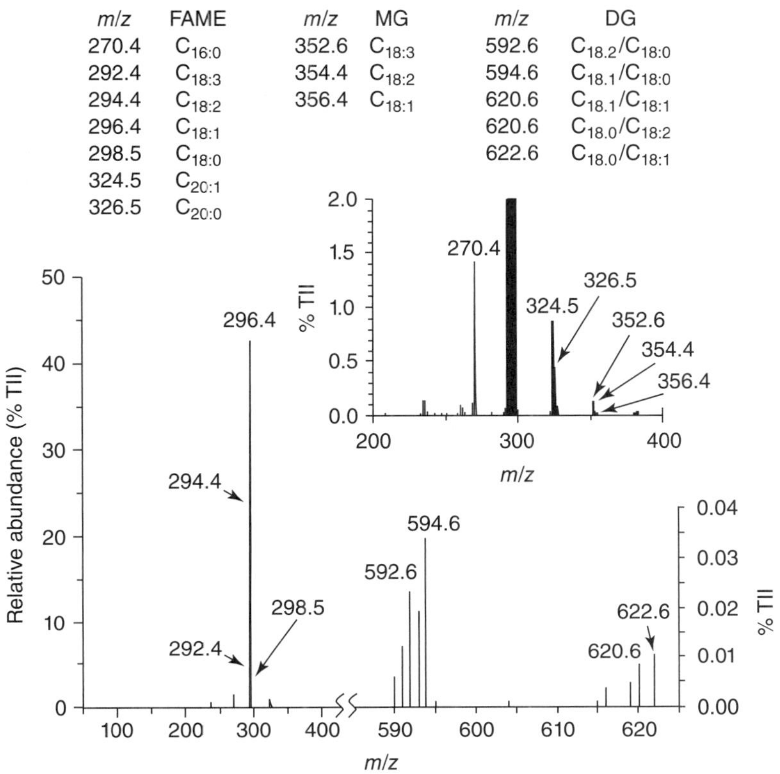

Figure 14.3. Distribution of fatty acid methyl esters (FAME), monoacylglycerols (MG), and diacylglycerols (DG) in a liquid injection field desorption ionization (LIFDI) mass spectrum of biodiesel (Schlichting et al., unpublished).

m/z 324.5, 0.9%). Furthermore, the $C_{18:1}$ (*m/z* 356.4), $C_{18:2}$ (*m/z* 354.4), and $C_{18:3}$ (*m/z* 352.6) monoglycerides and $C_{18:2}/C_{16:0}$ (*m/z* 592.6), $C_{18:1}/C_{16:0}$ (*m/z* 594.6), $C_{18:1}/C_{18:1}$ (*m/z* 620), $C_{18:0}/C_{18:2}$ (*m/z* 620.6), and $C_{18:0}/C_{18:1}$ (*m/z* 622.6) diglycerides were tentatively assigned in lower abundance. Thus, LIFDI MS offers a time efficiently determination of a wide range of lipid marker substances such as nonpolar glycerides, saponified or free derivatized fatty acids, and polar free fatty acids.

14.2.4. Ultrahigh-Resolution Mass Spectrometry

Technological advances of ion-trap mass spectrometers are the ultrahigh-resolution Fourier transform ion cyclotron resonance mass spectrometry (FT-ICR MS) and the recently released technique, the Orbitrap Fourier transform mass spectrometry (Hu et al., 2005), which enable the determination of molecular formulae with a high mass resolution and mass accuracy in mixtures. Today these ion-trap mass spectrometers are most frequently coupled with atmospheric pressure ionization (API) techniques such as electrospray ionization (ESI) (e.g., Fievre et al., 1997; Qian et al., 2001; Kujawinski et al., 2002; Llewelyn et al., 2002; Stenson et al., 2002, 2003; Fard et al., 2003) or matrix-assisted laser desorption/ionization (MALDI) (e.g., Solouki et al.,

1997; André et al., 2006; Taban et al., 2007). Furthermore, the use of alternative ionization methods to ESI, especially APCI and APPI for the determination of rather low-polar compounds, result in a more complex spectrum (Marshall, 2007; Schmitt-Kopplin, 2007; Witt et al., 2007). Mass resolution above 100,000 and mass accuracy from 2 ppm (Orbitrap, internal calibration) down to 0.5 ppm (FT-ICR in dependence on magnetic field strength, scan mode and calibration type) can be achieved and offer a great potential for the molecular-level characterization of very complex systems such as nonliving organic matter. Although an FT-ICR performs a 10- to 100-fold higher resolution than all other techniques including high-resolution sector field mass spectrometry, the additional combination with a chromatographic interface for sample purification, pre-concentration, and/or mixture separation (e.g., with regard to reduce quench effects under ESI conditions) is still desirable (Marshall et al., 1998). Typical successfully used combinations are the *online* coupling with liquid chromatography (Schrader and Klein, 2004) beside electrophoresis (Hofstadler et al., 1993)—for example, in the field of proteom research (e.g., tryptic digested proteins), combinatorial chemistry (e.g., analysis of compound libraries), or environmental research (e.g., metabolic investigations).

The principle of the FT-ICR MS and the Orbitrap MS is roughly the same. The ions circulate in orbits and their masses can be determined ultrahigh-resolved and ultra-precise out of the orbit frequency, based on the ability to measure frequencies more exactly than any other physical properties. In the ICR cell the ions are trapped within a static magnetic field. The ions describe under the influence of this homogenous magnetic field a circular motion, whose orbit frequency depends solely on their m/z and the constant magnetic field strength. In order to measure the cyclotron frequency, it is necessary to force the ions into coherent cyclotron motion by applying a frequency pulse through two opposite excitation electrodes, which is in resonance with the cyclotron frequency of the ions. After excitation, the increased radius of the circulating ion cloud induce an oscillating image current in two opposite detection plates whose frequency corresponds to the cyclotron frequency and hence the m/z of the single ions. The detection of ions differing in m/z needs a broadband rf pulse during the excitation, which induces superimpositions of many single frequencies. This so-called transient can be deconvoluted to single frequencies by Fourier transformation.

The ultrahigh mass resolution linearly increases with the magnitude of the magnetic field. Therefore, further efforts are made toward the application of stepwise higher-field magnets, such as superconductive magnets up to 20 Tesla and in the future possibly even hybrid magnets. Today, resolutions of more than 1 million can be achieved. This enables the determination of up to 20,000 different elemental compositions, which can be consequently resolved out of a single electrospray ionization mass spectrum, as shown for a petroleum sample, and even more become accessible by field desorption and/or atmospheric pressure photoionization (Marshall, 2007).

An entirely new principle of the ion-trap mass spectrometers is the Orbitrap with a coaxial inner spindle-shaped electrode in an outer barrel-like electrode. The peripheral injected ions move due to their electrostatic attraction to the inner electrode on orbits around and swing simultaneously along the electrode. The frequency of these harmonic oscillations is inversely proportional to the square root of m/z. The detected signals are induced by the frequency of these swings and resulted likewise in the mass spectra signals after their Fourier transformation. Since Orbitrap works with an electrostatic field instead of a magnetic field as used for FT-ICR, no cooling of the magnet with liquid helium and nitrogen is required. The specified resolution of approximately 100,000 is considerably lower in contrast to FT-ICR MS. The comparison of FT-ICR MS versus Orbitrap MS at its present stage of development shows that the high mass accuracy in the upper ppb range (Marshall, 2007) versus 2 ppm (Hu et al., 2005), a similar mass range up to *m/z* 4000 and a similar high dynamic range offers parameters, which possibly make the Orbitrap MS to a lower cost alternative in contrast to FT-ICR MS.

Ultrahigh-resolution and high precise mass spectrometry is used to understand the molecular basis in a wide range of analytical research fields, such as human diseases (Sihlbom et al., 2004), petroleomics (Marshall and Rodgers, 2004), the determination of dissolved organic carbon (Llewelyn et al., 2002), and so on. An important challenge is the investigation of natural organic matter (NOM) in terms of inherent complex mixtures of humic substances. The determination of the exact masses in the fully resolved spectra and, still more important, the assignments to unique molecular formulas are the fundamentals to indicate the dominant diagenetic processes of humic and fulvic acids in different environments (Kujawinski et al., 2002). So, the exact masses and chemical formulas of mostly degraded lignins have been assigned for approximately 5000 individual Suwannee River fulvic acids, which contained different homologous series (Stenson et al., 2003).

The evaluation of ultrahigh-resolved mass spectra of Suwannee River fulvic and humic acid revealed the presence of molecular families containing ions that differ from each other in the number of CH_2 groups, degree of saturation, and functional group substitution (Stenson et al., 2002). These findings were also valid for hot water extracted soil organic matter from a forest soil (Figure 14.4a). The spectrum was recorded by an ESI FT-ICR MS (LTQ FT Ultra, ThermoFisher Scientific, 7 Tesla) in the negative-ion mode. In the detail spectrum (Figure 14.4b) the asterisks and dots indicate peaks with a mass spacing of 14 Da, which is attributed to ions of two homologous series differing from each other in the number of their CH_2 groups. The interpretation of the 2 Da spacing is apparently due to, for example, the saturation degree of the ions with a formal loss of two hydrogen atoms (mass space 2.0156 Da) or the formal exchange of an oxygen atom by a CH_2 group (mass space 1.9793 Da) (Figure 14.4c). The expanded section (Figure 14.4c) show that each single odd-numbered mass peak has a corresponding lower abundant even-numbered mass peak higher in one *m/z*. The mass space of 1.0032 Da is thereby clearly induced by the [M + 1] peak attributed of the ^{13}C isotope (13.0034 Da – 12.0000 Da = 1.0034 Da). The difference of 0.9953 Da higher in *m/z* is assigned to the formal substitution of CH by N (14.0031 Da – 13.0078 Da = 0.9953 Da). Furthermore, the ions with the mass space of 0.0364 Da were derived from the mass difference between CH_4 and O, which represents the most important formal functional group substitution, giving the typical clusters within a nominal mass (Figure 14.4d). This is also similar but at an adequate lower intensity in the appropriated even-numbered mass peaks, which are more complex, because of further inherent peaks caused by N-containing compounds, as identified for peak X (Figure 14.4e).

This example shows that ultrahigh-resolution FT-ICR MS is a powerful tool for the characterization of the molecular constituents of nonliving organic substances.

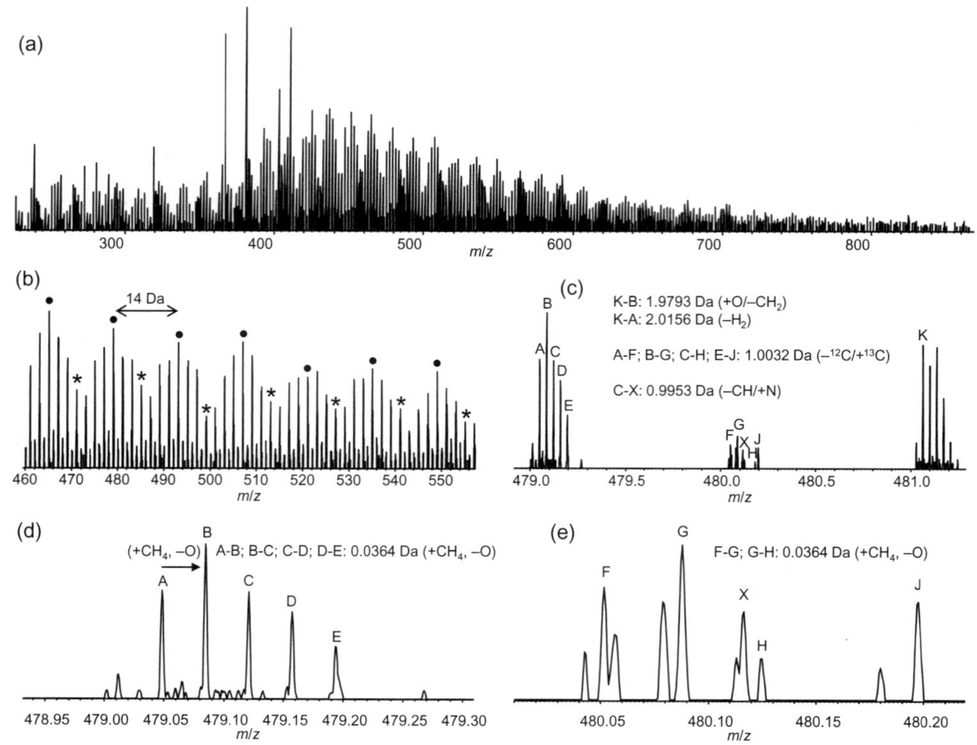

Figure 14.4. Negative-ion mode ESI FT-ICR (7 Tesla) mass spectrum of (**a**) a hot water extract from a forest soil. The expanded regions of the spectrum show (**b**) 14-Da spacing of two homologous series, indicated by asterisks and dots, (**c**) the 2-Da spacing, which is apparently due to a formal loss of two hydrogen atoms (mass space 2.0156 Da) or the formal exchange of an O atom by a CH_2 group (mass space 1.9793 Da), the predominance of each single odd- over even-numbered mass peak, the mass space of 1.0032 Da induced by the [M+1] of the ^{13}C isotope (13.0034 Da – 12.0000 Da = 1.0034 Da), and the difference of 0.9953 Da higher in *m/z* as assigned to the formal substitution of CH by N (14.0031 Da – 13.0078 Da = 0.9953 Da), (**d**) the mass space of 0.0364 Da, which represents the mass difference between CH_4 and O, and (**e**) a N-containing compound (peak X) at the even-numbered mass peak.

14.3. RECENT APPLICATIONS TO NATURAL NONLIVING ORGANIC MATTER COMPOSITION AND DYNAMICS

14.3.1. Extracted and Nonextracted Lipids

Lipids are an important class of organic soil substances, and they were studied in detail because they are biomarkers and can be used to trace the turnover of nonliving organic matter in soils (van Bergen et al., 1997; Amblès et al., 1998; Bull et al., 1998; Quénéa et al., 2006b). Lipids are extracted by organic solvents and are traditionally identified and quantified by gas chromatography coupled to MS (Jandl et al., 2002, 2004, 2005, 2007; Wiesenberg et al., 2004). An alternative approach using Py-FIMS was developed for the investigation of time series of archived soil samples from long-term agricultural experiments. Comparison of the fatty acid patterns from the same lipid extract shows that *in-source* Py-FIMS without derivatization yielded almost the same bimodal distribution and even-over-odd predominance as the conventional GC/MS (Figure 14.5a). However, whereas GC/MS can be applied to solvent extractable lipids only, Py-FIMS additionally released homologues of nonextracted fatty acids from the solid extraction residues as well. The quantity of registered molecules, expressed in arbitrary units in Figure 14.5b, indicates that extracted fatty acids occurred with roughly the same abundance as nonextracted but volatilized fatty acids. This nonextracted lipid pool seems to be an important

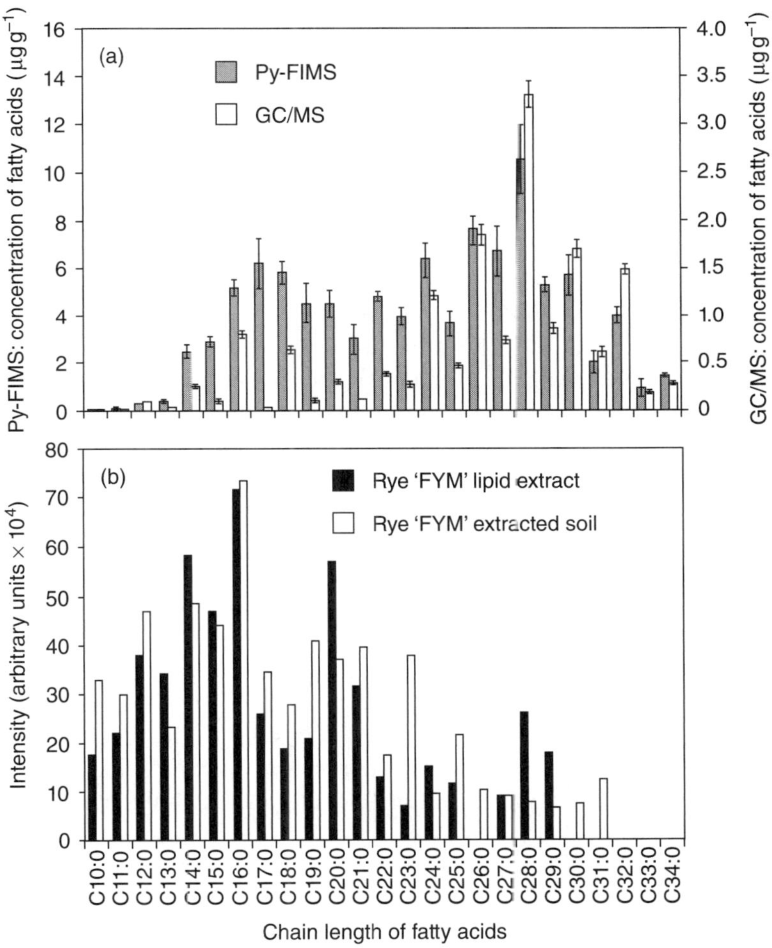

Figure 14.5. Fatty acids patterns of soils under long-term monoculture. (**a**) Lipid extract of soil under maize, unfertilized, after derivatization with tetramethylammonium hydroxide determined by conventional gas chromatography/mass spectrometry (GC/MS) in comparison to direct, *in-source* pyrolysis–field ionization mass spectrometry (Py-FIMS) without derivatization (Jandl et al., unpublished). (**b**) Py-FIMS of lipid extract of soil under rye, farmyard manure (FYM) treatment, compared to solid extraction residue, both directly measured without derivatization. Reprinted from Marschner, B., Brodowski, S., Dreves, A., et al. (2008). How relevant is recalcitrance for the stabilization of organic matter in soils? *Journal of Plant Nutrition and Soil Science* **171**, 91–110, with permission from Wiley-VCH.

recalcitrant fraction because it accounts for half of the total lipids in soil. However, this pool is largely unknown because of its insolubility and inaccessibility to the conventional GC/MS. Therefore, investigations into molecular composition and fate of the nonextracted lipids challenge future *in-source* Py-FIMS studies.

14.3.2. "Unknown" Organic Nitrogen

The organic N compounds of nonliving organic matter are an important pool in the global N cycle. They are the basis for the microbial formation of mineral NH_4^+ and NO_3^- which are essential major plant nutrients. Moreover, unbalanced N cycles in agricultural systems can also be a reason for serious environmental pollution such of groundwater contamination with NO_3^- and gaseous emissions of N_2O which contribute to climate change. Therefore, the chemical nature of soil N, especially of the "unknown N" (Schulten and Schnitzer, 1998), which may account for one-third to one-half of soil N (Stevenson, 1994), has been a challenge for analytical pyrolysis.

Direct Cp Py-GC/MS with N-selective detector of 6 *N* HCl hydrolysis residues of particularly N-rich soil clay fractions enabled the identification of 37 N-containing compounds such as aliphatic and aromatic nitriles, substituted five-membered (pyrroles, pyrazoles, imidazoles) and six-membered (pyridines, pyrazines) N-heterocycles (Leinweber and Schulten, 1998). However, it is well documented that flash pyrolysis and EI ionization MS of amino acids and peptides results in the formation of N-heterocycles (Schulten et al., 1995a). Since these consecutive reactions were less important for slow *in-source* heating and soft ionization, Py-FIMS of the same samples indicated the presence of peptides in the hydrolysis residue. A sequential wet-chemical extraction procedure was applied to remove dithionite-citrate-bicarbonate (DCB)-extractable, pedogenic Fe-oxides from clay and fine silt fractions and to determine hydrolyzable and nonhydrolyzable organic N compounds (Leinweber and Schulten, 2000). This pretreatment resulted in the release of additional 64–100% of hydrolyzable N in the freeze-dried DCB extracts and 12–66% of hydrolyzable N in the extracted residues with an amino acid composition typical for soils. In conclusion of this combined wet-chemical and complementary Py-FIMS of DCB extracts and extracted residues, approximately 25% of N in soils was assigned to heterocyclic N compounds. Although there is general agreement that proteinaceous materials constitute the majority of organic N in soils (Schulten and Schnitzer, 1998; Stevenson and Cole, 1999; Knicker, 2000), there is still a controversial discussion about the significance of heterocyclic N compounds in soils (Knicker, 2000; Burdon, 2001; Vairavamurthy and Wang, 2002; DiCosty et al., 2003; Sjöberg et al., 2004; Smernik and Baldock, 2006). Thus, the identification and quantification of the remaining proportions of "unknown N" is a current challenge for organic soil chemistry.

Recent research efforts and advances in the identification of "unknown N" were achieved by the complementary application of pyrolysis–mass spectrometry and synchrotron-based XANES at the N K-edge (N-XANES) (e.g., Jokic et al., 2004a).

There is no doubt that the *in-source* pyrolysis and soft ionization in the high electric field (anode voltage = 8 kV) results in much less fragmentation and formation of consecutive reaction products than flash pyrolysis-EI-GC/MS (Schulten, 1996). For this reason about 1–4% (DOM) and 1–5% (soils) of TII in the Py-FI mass spectra can be assigned to peptides (Tables 14.2 and 14.3). The validity of the signal assignments to peptides was proved by Py-FIMS of pure amino acids and peptides (Sorge et al., 1993) and by significant correlations to the concentrations of amino acids as determined by high performance liquid chromatography (HPLC) (Schulten et al., 2002). However, it appears that the proportions of peptides as percent of TII assigned to all N-containing compounds in the range of 13–45% (DOM) and 26–65% (soils) (calculated from data in Tables 14.2 and 14.3) are slight to small, if compared to wet-chemically determined amino acid concentrations by an improved hydrolysis and detection technique (Martens and Loeffelmann, 2003; Martens et al., 2006). One reason for the discrepancy in estimates of peptide-N quantity between wet-chemistry (35–76% of N after correction for NH_3, Martens and Loeffelmann, 2003; Martens et al., 2006) and the TII proportions of peptides in Py-FIMS (13–65% of N) could be the pyrolytic formation of N-heterocycles. Such a pyrolytic formation of pyrroles, pyridines, pyrazines, indoles, and carbazoles was also found for aquatic natural organic matter in a closed system, microscale sealed vessel pyrolysis followed by GC/MS analyses (Berwick et al., 2007).

Ongoing research is directed to systematically investigate the pyrolytic formation of heterocycles by (1) spiking of soil samples with increasing amounts of nonheterocyclic N compounds, (2) investigating the spiked samples with Py-FIMS and N K-edge XANES, and (3) investigating pyrolysis residues of the spiked soils removed from Py-FIMS after temperature steps of 100 °C from 300 °C to 700 °C. The first results were obtained by a spiking experiment with *N*-acetyl-D-(+)-glucosamine which was added to soil to increase the N content by factor 2 and 10. The N K-edge XANES spectra in Figure 14.6a,b indicate the pyrolytic formation of two structural families of N compounds with binding energies around 398.9 and 399.9 eV, respectively, in the temperature range 300–700 °C. According to our own N K-edge XANES results of a wide range of reference compounds (Leinweber et al., 2007) and a few literature data (Vairavamurthy and Wang, 2002; Jokic et al., 2004a,b), these structures probably comprise of pyridinic N in pyridines, imidazoles, pyrazoles, pyrazines, and pyrimidines (398.7–400 eV) and nitriles (399.9 eV). The latter feature was more pronounced in the lower spiked sample (Figure 14.6a) than in the higher spiked sample where only a shoulder appeared in this eV range (Figure 14.6b). During subsequent heating to 700 °C, the first peak (398.9 eV) disappeared in the lower spiked sample. This disappearance indicates intermediately formed products which were volatilized during heating from 500 °C to 700 °C. Consequently, these pyrolysis products then should be detected in the Py-FI mass spectra in the temperature range >500 °C. Furthermore, the peak around 401.4 eV, indicative of the spiked *N*-acetyl-D-(+)-glucosamine (Leinweber et al., 2007), gradually became smaller or was slightly shifted with increasing pyrolysis temperatures in both of the samples. This could be due to volatilization and then should be detected by Py-FIMS over the whole temperature range but should be most intensive up to 400 °C. For the evaluation of the Py-FI mass spectra, thermograms of *N*-acetylglucosamine in the spiked soil samples were calculated from the relevant marker signals published by Bahr and Schulten (1983). The thermograms for the sum of these marker signals in Figure 14.6c show distinct peaks of maximum volatilization at 240–250 °C. It can be calculated that about 55–68% of added *N*-acetyl-D-(+)-glucosamine was thermally volatilized up to 300 °C and 84–86% up to 400 °C. This conforms to the visible reduction of peak area in the N K-edge XANES spectra (Figure 14.6a,b).

TABLE 14.2. A Summary of the Abundance of Compound Classes Determined by Py-FIMS in Sets of Various Dissolved Organic Matter (DOM) Samples

Origin of DOM	Number	VM	CHYDR	PHLM	LDIM	LIPID	ALKYL	NCOMP	STEROL	PEPTI	SUBER	FATTY
Rhizodeposits (potato)[a]	20	26.9 ± 12.3	4.8 ± 0.8	5.9 ± 1.2	1.9 ± 0.9	5.7 ± 2.1	7.1 ± 2.0	5.7 ± 0.9	1.4 ± 0.8	3.7 ± 0.8	0.1 ± 0.1	1.9 ± 1.0
Cold water extracts (O, L)[b]	6	49.3 ± 5.7	3.1 ± 0.2	4.8 ± 0.3	4.5 ± 0.6	11.5 ± 0.5	7.8 ± 0.4	5.0 ± 0.2	5.1 ± 0.5	2.1 ± 0.4	0.7 ± 0.1	3.3 ± 0.4
Hot water extracts (O, L)[b]	6	56.0 ± 6.9	9.2 ± 3.2	9.6 ± 2.3	2.6 ± 0.7	9.2 ± 2.3	9.1 ± 0.6	6.9 ± 1.2	3.6 ± 1.4	3.5 ± 0.6	0.3 ± 0.1	3.6 ± 1.3
DOM fens (A, H)[c]	6	63.2 ± 5.4	2.8 ± 0.8	7.0 ± 1.6	3.4 ± 2.3	10.7 ± 0.9	11.0 ± 0.5	5.9 ± 0.5	1.4 ± 0.3	1.1 ± 0.2	0.1 ± 0	3.6 ± 0.9
DOM fens (groundwater)[c]	6	65.1 ± 5.2	1.7 ± 0.4	5.4 ± 0.8	3.9 ± 1.2	12.6 ± 0.7	11.8 ± 0.7	4.8 ± 0.2	1.5 ± 0.5	0.7 ± 0.2	0.1 ± 0.1	3.8 ± 0.7
DOM Siberian rivers[a]	8	49.3 ± 3.3	3.2 ± 1.3	5.8 ± 2.0	2.5 ± 1.3	6.0 ± 1.1	11.9 ± 1.2	3.0 ± 0.8	0.6 ± 0.1	2.5 ± 0.5	0 ± 0	1.5 ± 0.5
Municipal waste leachates[d]	7	40.5 ± 9.0	4.5 ± 1.1	7.4 ± 2.7	0.6 ± 0.6	3.0 ± 1.3	7.4 ± 2.7	8.4 ± 2.1	0.3 ± 0.1	3.3 ± 0.5	0 ± 0	0.8 ± 0.2

VM, percentage matter volatilized in pyrolysis; CHYDR, carbohydrates with pentose and hexose subunits; PHLM, phenols and lignin monomers; LDIM, lignin dimers; LIPID, lipids, alkanes, alkenes, bound fatty acids, and alkylmonoesters; ALKY, alkylaromatics; NCOMP, mainly heterocyclic N-containing compounds; STEROL, sterols; PEPTI, peptides; SUBER, suberin; FATTY, free fatty acids in % of total ion intensity.

[a]Leinweber et al. (unpublished).

[b]Landgraf et al. (2006).

[c]Leinweber et al. (2001).

[d]Franke et al. (2006, 2007).

TABLE 14.3. A Summary of the Abundance of Compound Classes Determined by Py-FIMS in Major Soil Units of the World and in Samples Showing the Influence of Primary Organic Matter Input Either from Vegetation or with Farmyard Manure (FYM) on the Composition of Soil Organic Matter

Major Soil Unit	Number	VM	CHYDR	PHLM	LDIM	LIPID	ALKYL	NCOMP	STEROL	PEPTI	SUBER	FATTY
Histosols (various)[a]	7	52.5 ± 15.9	4.6 ± 0.6	6.6 ± 2.6	1.8 ± 0.8	5.9 ± 2.3	7.6 ± 3.6	3.9 ± 1.1	1.3 ± 1.8	2.5 ± 0.8	0.6 ± 0.9	1.6 ± 1.6
Gelic histosols[b]	3	51.6 ± 4.2	7.4 ± 0.9	1.8 ± 0.2	0.8 ± 0.1	12.7 ± 3.7	1.6 ± 0.2	2.7 ± 0.1	13.8 ± 2.6	5.0 ± 0.5	0.7 ± 0.1	7.0 ± 1.5
Plaggic anthrosols[c]	7	7.8 ± 1.3	4.5 ± 0.9	4.5 ± 0.7	3.7 ± 0.7	11.6 ± 1.1	6.4 ± 0.5	4.1 ± 0.4	5.2 ± 0.9	2.8 ± 0.3	0.7 ± 0.2	4.6 ± 1.5
Terric anthrosols	6	n.d.	7.1 ± 0.8	7.8 ± 0.4	6.7 ± 1.1	9.8 ± 0.9	9.4 ± 1.2	6.9 ± 0.4	3.6 ± 0.6	3.1 ± 0.3	0.2 ± 0.1	0.9 ± 0.9
Vertisols (various)[d]	5	13.7 ± 2.5	9.5 ± 4.4	4.6 ± 3.5	0.2 ± 0.2	0.3 ± 0.4	3.7 ± 2.3	9.2 ± 4.3	0.1 ± 0.1	5.1 ± 1.8	0 ± 0	0 ± 0
Andosols[a,e]	4	16.3 ± 9.2	7.0 ± 0.7	6.8 ± 0.3	3.2 ± 0.6	7.5 ± 2.5	7.8 ± 0.7	5.1 ± 0.2	2.1 ± 0.7	3.5 ± 0.4	0.1 ± 0.1	2.0 ± 1.6
Podzols (Bh horizons)[f]	8	3.4 ± 1.1	3.2 ± 2.1	3.6 ± 2.7	9.2 ± 1.6	16.6 ± 2.9	9.0 ± 2.9	2.5 ± 1.2	12.0 ± 4.4	0.9 ± 0.6	1.1 ± 1.3	2.8 ± 1.8
Chernozems[a,g]	28	7.3 ± 2.6	8.7 ± 2.9	8.8 ± 1.9	3.4 ± 2.4	5.7 ± 3.7	9.6 ± 2.9	8.5 ± 2.5	1.6 ± 2.6	4.2 ± 1.6	0.1 ± 0.2	0.3 ± 0.6
Kastanozems[h]	2	5.1 ± 0.4	17.3 ± 0.4	14.3 ± 1.3	0.9 ± 1.1	1.8 ± 1.6	10.7 ± 0.1	13.0 ± 3.3	0 ± 0	4.6 ± 1.3	0 ± 0	0 ± 0
Luvisols[a]	72	5.9 ± 1.3	12.0 ± 1.8	18.1 ± 3.0	0.7 ± 0.8	2.2 ± 1.3	13.6 ± 2.1	8.5 ± 0.6	0.1 ± 0.5	4.2 ± 0.8	0.0 ± 0.1	0.0 ± 0.2
Luvisols (Bt horizons)[a]	4	6.3 ± 2.5	8.0 ± 0.8	5.8 ± 0.6	0.6 ± 0.4	2.0 ± 1.0	4.7 ± 0.4	9.8 ± 4.4	0.4 ± 0.2	4.4 ± 0.9	0.1 ± 0.2	0.3 ± 0.2
Stagnosols (tropic)[a]	3	21.2 ± 0.3	10.3 ± 1.8	8.5 ± 1.9	1.5 ± 0.9	2.2 ± 1.0	7.2 ± 1.3	8.1 ± 0.3	0.1 ± 0.2	5.3 ± 0.3	0 ± 0	0.1 ± 0.1
Phaeozems[a,i]	80	6.3 ± 2.0	5.3 ± 1.3	7.7 ± 0.8	7.9 ± 1.1	10.3 ± 1.3	12.5 ± 1.5	5.9 ± 0.7	2.2 ± 0.7	2.7 ± 0.6	0.1 ± 0.1	0.1 ± 0.1
Cambisols, regosols[j]	4	21.2 ± 5.4	6.0 ± 3.1	3.4 ± 1.7	4.9 ± 1.2	11.2 ± 0.4	5.4 ± 1.4	3.6 ± 0.9	7.1 ± 0.7	2.2 ± 0.4	2.3 ± 0.8	2.2 ± 1.3

For abbreviations see Table 14.2. n.d., not determined; origin of data and more detailed descriptions of sampling sites.

[a]Leinweber et al. (unpublished).
[b]Beyer et al. (1995).
[c]Blume and Leinweber (2004)
[d]Leinweber et al. (1999).
[e]Baglieri et al. (2007).
[f]Wilcken et al. (1997).
[g]Leinweber et al. (1994).
[h]Schulten et al. (1995b).
[i]Schmidt et al. (2000).
[j]Leinweber et al. (1996).

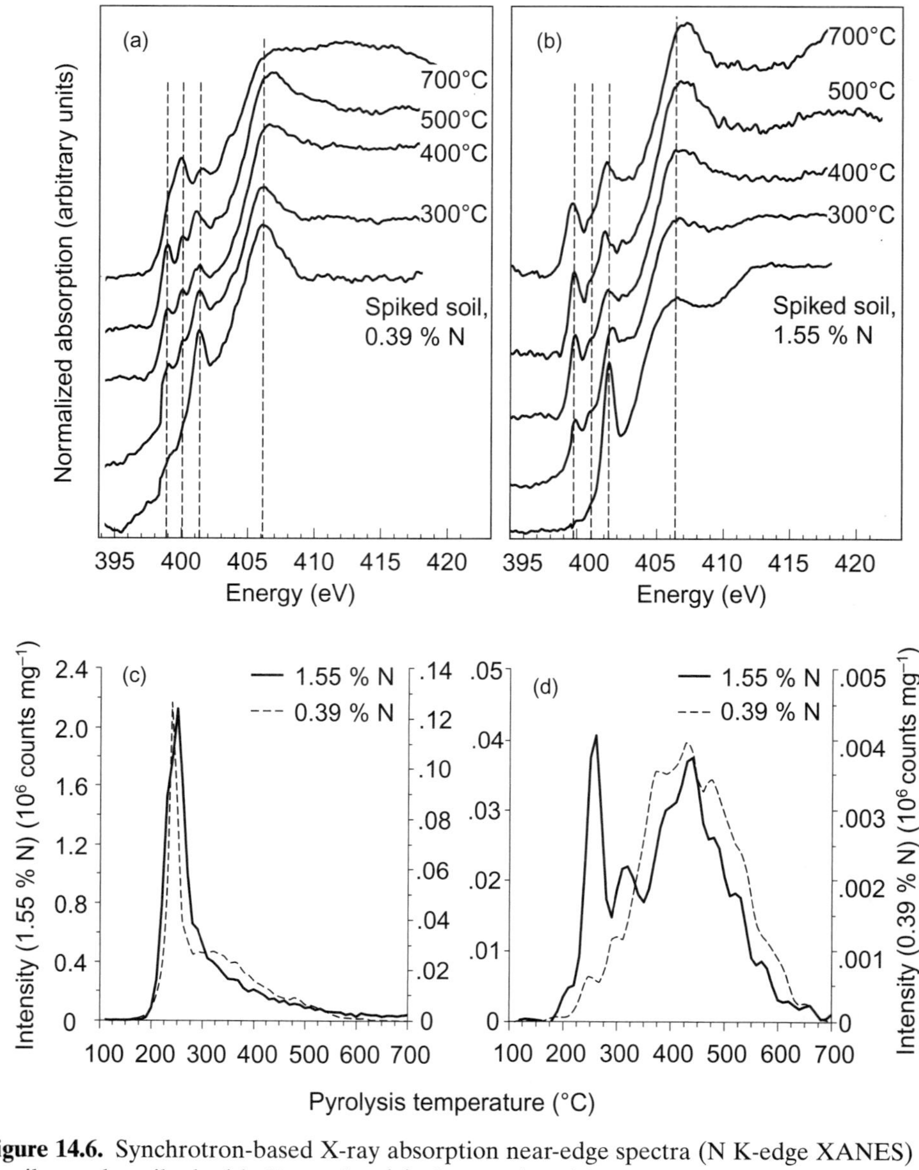

Figure 14.6. Synchrotron-based X-ray absorption near-edge spectra (N K-edge XANES) of a soil sample spiked with *N*-acetyl-D-(+)-glucosamine: (**a**) N content enriched by factor 2 (0.39% N), (**b**) N content enriched by factor 10 (1.55% N). Spiked soil and samples removed from *in-source* Py-FIMS after distinct pyrolysis temperatures. (Leinweber et al., unpublished). The XANES spectra were recorded on the 11ID-1, spherical grating monochromator (SGM) beamline at the Canadian Light Source Ltd., Saskatoon, Saskatchewan, Canada. Results of Py-FIMS of the spiked samples: (**c**) Thermograms for the volatilization of marker signals for *N*-acetyl-D-(+)-glucosamine (Bahr and Schulten, 1983) and (**d**) thermograms for the summed ion intensities of marker signals for heterocyclic N compounds that were indicated by the N XANES features in parts a and b (Leinweber et al., unpublished).

Furthermore, Figure 14.6d shows thermograms for pyridine, methylpyridine, alkylhydroxpyridine, methylpyrimidine, C_4-pyrazole, indolethanol, and propylchinoline, which represent basic structures showing N K-edge XANES peaks mostly around 399 eV. Both samples revealed volatilization maxima at 450 °C; however, only the sample with 1.55% N showed a prominent peak at about 250 °C. Figure 14.6c shows that this is the characteristic temperature of *N*-acetylglucosamine release from the sample. This indicates that the evaporation and thermal cleavage of high concentrations of *N*-acetylglucosamine in the sample led to the formation of *N*-heterocyclic compounds, mainly of methylpyridine. First cautious estimates indicate that about 5% (spiked soil 1.55% N) to 10% (spiked soil 0.39% N) of the added *N*-acetylglucosamine may have been transformed into heterocyclic and nitrile N. These preliminary data evaluations demonstrate the methodological approach by which possible pyrolytic formations of nonproteinaceous N compounds will be detected and quantified to validate the interpretation and evaluation of Py-FI mass spectra. Progress in this research will provide multi-methodological evidence for the organic N forms in nonliving organic matter.

In summary, this recent research shows that N heterocyclic compounds without a doubt form an important pool of soil organic N, because they enter soils in hundreds if not thousands of different substances from plant materials (Pedras et al., 2003; Somei and Yamada, 2003). Furthermore, they can be formed in soil by the Mn (IV) oxide-catalyzed Maillard reaction (Jokic et al., 2004a,b), as shown for substituted pyrroles, pyridines, and pyrazines by Cp Py-GC/MS, Py-FIMS, and N K-edge XANES (Jokic et al., 2004a). One major controversial argument—that is, the absence of distinct N-heterocyclics in ^{15}N NMR spectra—was recently explained by the invisibility of significant N proportions to solid-state ^{15}N NMR. Consequently, up to half or more of the N soil clay fractions was in a form, insensitive to NMR detection (Smernik and Baldock, 2006). The proportions of N-heterocyclic compounds may be smaller than previously derived from analytical pyrolysis studies; but application of pyrolysis soft ionization MS techniques, including the high-resolution mode, synchrotron-based N K-edge XANES, optimized wet-chemical amino acid extraction and quantification, and the use of ^{15}N-labeled compounds in conjunction, will result in valuable quantitative data about proportions and turnover of heterocyclic N in nonliving organic matter.

14.3.3. Dissolved Organic Matter (DOM): Origin, Composition, and Transformations

Dissolved organic matter (DOM) has received increasing attention in recent years because it is an important pool in the global organic matter cycles and a controlling factor in soil formation, mineral weathering, and pollutant transport (Kalbitz et al., 2000). The state of the art in DOM characterization by analytical pyrolysis and soft ionization mass spectrometry resulting from a Priority Program of the German Research Council (ROSIG; Frimmel et al., 2002) was summarized by Schulten et al. (2002). The DOM in soil can originate from rhizodeposition [i.e., the release of carbon (C) compounds from living plant roots into the surrounding soil] and from the dissolution of organics in soil, the latter sometimes simulated by mild extractions (e.g., boiling water). Furthermore, DOM can be leached out of organic layers covering mineral soils or out of waste piles, and so on, and may enter freshwater systems such as lakes and rivers. Table 14.2 gives an overview on the composition of DOM samples of widely different origin.

The proportion of volatile matter (VM) is generally a function of organic matter and moisture content in a sample (Leinweber and Schulten, 1998). Typically, DOM samples show a volatilization of more than 50% of the sample weight. Only rhizodeposits and leachates from municipal waste had less VM, probably due to higher proportions of mineral matter. Obviously, the DOM composition differs according to origin and mode of collection. Hot water extracted more carbohydrates, phenols, lignin monomers, peptides, and N-containing compounds than did cold water from the same organic layers in forest soils at the expense of more stable lignin dimers, lipids, and sterols. DOM collected in a fen landscape was particularly rich in lipids and alkylaromatics, indicating a residual enrichment of the most stable compounds, especially in the groundwater. DOM from a catchment in Siberia had high proportions of alkylaromatics, probably indicating a contamination of the river water, but lipids were in the lower range for DOM samples. Leachates from municipal waste had more N-containing compounds and peptides than did the other sample groups. This points to significant proportions of easily metabolized OM and intensive microbial decomposition, as also confirmed by the conventional sum parameters in wastewater [e.g., chemical and biological oxygen demand (Franke et al., 2006, 2007)]. Besides these obvious differences in the general DOM composition, the standard deviations and more detailed spectra evaluations demonstrate the analytical capabilities of Py-FIMS in the detection of even small differences according to physiological reactions of plants, microbial metabolism, distribution in landscapes, and technical processes.

Rhizodeposition is affected by many biotic and abiotic factors of plant and soil (Jones et al., 2004). A simple and efficient technical solution for the collection of rhizodeposits in a soil environment consists of small boxes in which plants are grown and from which the rhizodeposits are leached by a siphon–elution system (Kuzyakov and Siniakina, 2001). In the past five years, this was applied to a large number of rhizopeposits from maize (*Zea mais* L.) (e.g., Melnitchouck et al., 2005; Fischer et al., 2007) and potato (*Solanum tuberosum* L.) (e.g., Melnitchouck et al., 2006; Schlichting and Leinweber, 2008).

The TII thermogram in the summed Py-FI mass spectra of 54 potato rhizodeposits in Figure 14.7 shows at least three thermally different moieties with maximum volatilization at 220 °C, 360 °C, and 420 °C to 480 °C, the latter appearing rather as shoulders. Most prominent signals in the lower mass range were assigned to carbohydrates (e.g., *m/z* 60, 72, 84, 96, 98, 110, 112, 114, and 126), and phenols and lignin monomers (e.g., *m/z* 94, 108, 110, 122, 124, 150, 152, 164, 166, 168, 178, 180, 182, 194, 208, 210 and 212). Lignin dimers were indicated by signals in the medium mass range such as 246, 260, 270, 272, 274, 284, 286, 296, 298, 300, 310, 312, 314, 316, 326, 328, 330, 340, 342, and 356 but except for *m/z* 310 they were not very prominent. Intensive signals of lipids, alkanes, alkenes, bound fatty acids, and *n*-alkyl esters were *m/z* 216, 230, 256, 258, 270, 272, and 410. Alkylaromatics were indicated by intensive signals at *m/z* 206, 220, and 246. N-containing compounds were not very intensive. Some signals of sterols were detected at higher *m/z* 372, 386, 390, 392, and 410. Peptides showed signals in the lower mass range at *m/z* 70, 73, 74, 75, 84, 87, 97, 99, 115, 129 and 135. Signals for suberin were largely absent. Finally, a homologous series of saturated fatty acids could be traced from n-C_{16} (*m/z* 256) to n-C_{27} (*m/z* 410), and

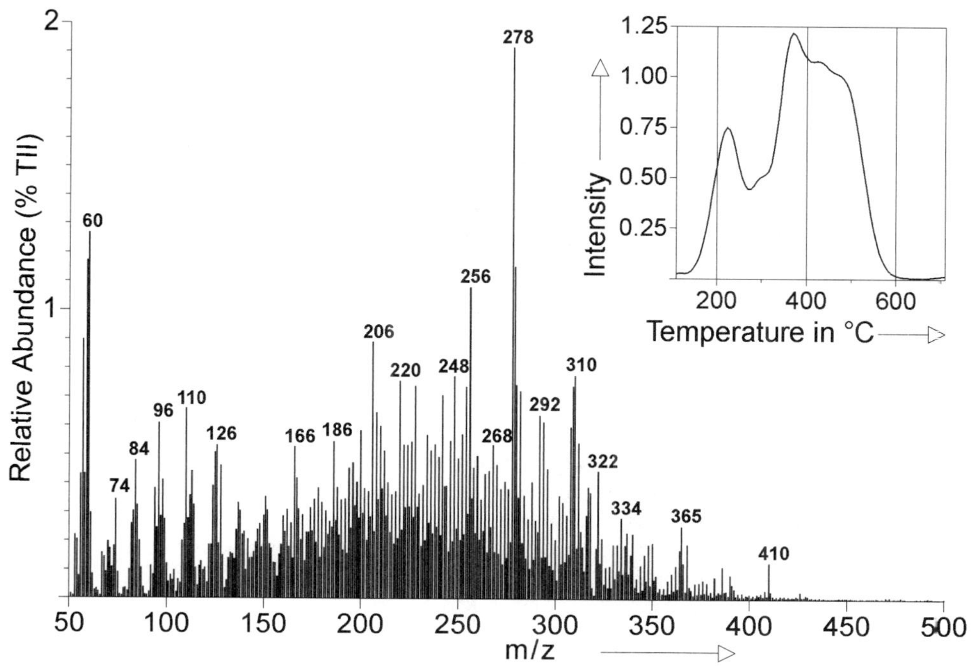

Figure 14.7. Summed and averaged ($n = 54$ samples) thermogram of total ion intensity (insert upper right) and Py-FI mass spectrum of potato rhizodeposits recorded at a Finnigan MAT 900 (Schlichting et al., unpublished).

the most intense peak (*m/z* 278) in the spectrum originated from the unsaturated linolenic acid.

In summary, the Py-FI mass spectrum shows a great diversity in the molecular rhizodeposit composition which could not be explained by previous chromatographic analyses of root exudates (e.g., Gransee and Wittenmayer, 2000). These focused mainly at the identification and quantification of *a priori* expected compounds (Fan et al., 2001). Therefore, Py-FIMS may contribute to the detection of previously unknown rhizodeposits and high-molecular-weight products of rhizodeposit interaction with genuine SOM compounds.

Freeze-dried DOM samples collected with the siphon–elution system (Kuzyakov and Siniakina, 2001) for the first time showed diurnal dynamics in the molecular-chemical composition of maize rhizodeposits (Kuzyakov et al., 2003). In a forthcoming study with maize, Melnitchouck et al. (2005) showed that amino acids, especially aspartic acid, asparagine, glutamic acid, phenylalanine, leucine and isoleucine contributed to the more intensive rhizodeposition during daytime than during nighttime. Furthermore, the maximum of thermal volatilization of peptides at low pyrolysis temperature in Figure 14.8 indicates the rhizodeposition or microbial formation of free amino acids rather than amino acids bound in peptides or trapped in soil humic substances.

A detailed analytical characterization of rhizodeposits also could be an important component of assessing possible environmental risks of genetically modified (GM) crops. It can be hypothesized that possible effects of GM crops on soil microorgan-

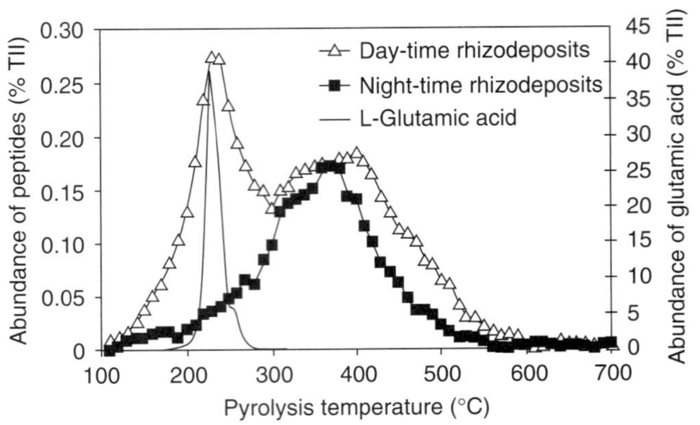

Figure 14.8. Thermograms for the volatilization of peptide-derived compounds in freeze-dried rhizodeposits leached from a soil cropped with maize after a daytime and a nighttime growth period and thermogram for the volatilization of L-glutamic acid. Reprinted from Leinweber, P., Jandl, G., Baum, C., Eckhardt, K.-U., and Kandeler, E. (2008). Stability and composition of soil organic matter control respiration and soil enzyme activities. *Soil Biology and Biochemistry* **40**, 1496–1505, with permission from Elsevier.

isms should be detected first in the rhizosphere which is a hot spot of microbial activity in soil. Potatoes from a nontransgenic line (*Solanum tuberosum L.* cv. Désirée) and from three GM lines, which expressed a gene for the resistance to kanamycin (DLH 9000) and a gene for T4 lysozyme (DL10 and DL12) (Heuer et al., 2002), were grown and rhizodeposits were leached by the siphon–elution system (Kuzyakov and Siniakina, 2001) after defined growth periods. The Py-FI mass spectra indicated that the potato growth generally altered the composition of soil solution, confirming all previous experiments (Kuzyakov et al., 2003; Melnitchouck et al., 2005). A principal component analysis of the Py-FI mass spectra showed differences between the leachates from the nontransgenic parent line and the GM potatoes as well as within the latter group (Melnitchouck et al., 2006). The signals in order of their importance for the discrimination between the four sample groups were *m/z* 189, 131, 202, 214, 226, 71, 84, 208, 125, 164, and 280. These indicate N-containing compounds (3-(3-indolyl)-propionic acid: *m/z* 189, methylindole: *m/z* 131, hydroxypropionitrile: *m/z* 71, 2(5*H*)-furanone: *m/z* 84; substituted pyrroles and pyridines: *m/z* 125) and lignin building blocks (*m/z* 202, 214, 226, 208, 164, 280). A principal component analysis unequivocally showed differences between the leachates from the same line and between leachates from the different lines. One principal component discriminated between the wild-type potato and all transgenic lines. The two signals with the highest negative loadings, *m/z* 189 (genuine indole derivative) and *m/z* 131 (indole derivate from pyrolysis of glutamine and/or glutamic acid), were more abundant in the leachates from the wild type than in the leachates from the GM plants. However, these differences in molecular composition could not be assigned to the release of T4-lysozyme into soil. Moreover, general parameters of microbial activity in soil such as dehydrogenase activity and substrate-induced soil respiration did not show any significant effects of the various potatoes grown. This agreed with the conclusions of many recent reviews and detailed studies on effects of GM crops on soil microorganisms (e.g., Bruinsma et al., 2003; Sessitsch et al., 2004; Brusetti et al., 2004). Thus, it is concluded that the

Py-FI mass spectrometric "fingerprint" can be developed as a fast, comprehensive, highly sensitive and reproducible analytical approach to discern any effects that GM crops may exert on soil ecological parameters (Melnitchouck et al., 2006).

One limitation in the development of Py-FIMS as a fast screening technique for rhizodeposition arises from the relatively large sample amount and time required for freeze-drying to obtain solid samples. Therefore, in ongoing methodological experiments we placed about 5 µl of rhizodeposit directly in the quartz tube of the Py-FI sample holder, evaporated the water over night in a desiccator and analyzed the solid residue directly by Py-FIMS without any further pretreatment. The principal component analysis in Figure 14.9 shows that rhizodeposits could be separated according to their origin either from the loamy soil or from the same soil diluted by additions of 50% w/w of quartz. Important *m/z* contributing to the discrimination were assigned to phenols and lignin monomers (not shown). This is interpreted as indication of some mobilization of these compounds from the genuine SOM by the impact of exudates or microbial metabolites (Leinweber et al., 2008a).

This example shows progress in the miniaturizing of the experimental approach and acceleration of rhizodeposit characterization, which is a first step toward time-resolved, high-throughput analyses to disclose plant physiological and microbial processes in the rhizosphere.

Microbial transformations of DOM were also studied by incubation experiments. The DOM extracted from maize straw and forest floors was incubated for 90 days and samples taken before and after incubation were analyzed by Py-FIMS and complementary UV absorbance, fluorescence emission spectroscopy, FTIR-spectroscopy, ^{1}H NMR spectroscopy, and ^{13}C natural abundance (Kalbitz et al., 2003). The Py-FI mass spectra showed increases in the proportions of phenols and lignin monomers at the expense of lignin dimers and alkylaromatics during

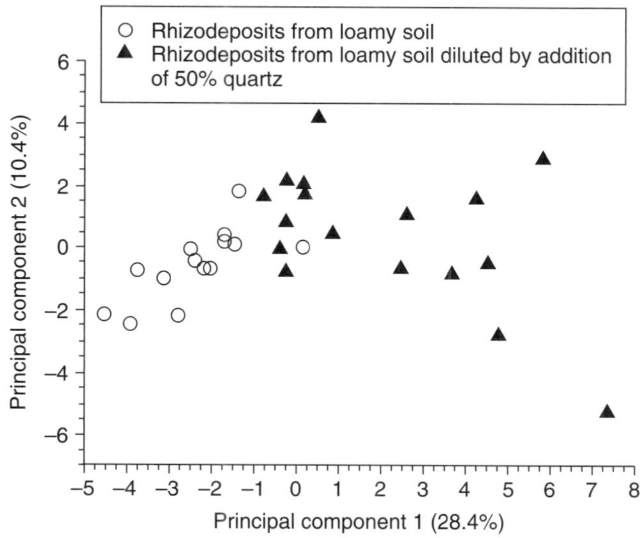

Figure 14.9. Principal component analysis of Py-FI mass spectra of rhizodeposits from a loamy soil and the same loamy soil diluted with 50% of quartz. The soils were grown with maize for 20 days and leached with distilled water. Py-FIMS was done with a 5-µl liquid rhizodeposit that was evaporated in the quartz sample tube (Leinweber et al., 2008a).

incubation. This partial degradation of higher-molecular, lignin-derived DOM compounds was accompanied by relative increases in the proportions of lower-molecular degradation products and microbial metabolites. Carbohydrates, especially when abundant at high initial contents, were the preferred substrate for microorganisms. However, all methods also suggested some microbial production of carbohydrates and peptides during DOM degradation.

Extraction of organic substrates by cold and hot water was another approach to study the decomposability of DOM from the forest floor (Oi, Oe, and Oa horizons) of a 170-year-old beech stand (*Fagus sylvatica*) in the Ore mountains, Germany (Landgraf et al., 2006). The C and N concentrations were always lower in the cold (C: 2.69–3.95 $g\,kg^{-1}$; N: 0.14–0.29 $g\,kg^{-1}$) than in the hot water extracts (C: 13.77–15.51 $g\,kg^{-1}$; N: 0.34–0.83 $g\,kg^{-1}$). By contrast to the solid soil samples, the C:N ratios of both extracts increased with increasing depth, indicating N inputs to the upper O horizons or N losses from the lower O horizons. The Py-FI mass spectra of the hot water extracts revealed more intensive signals of carbohydrates, phenols and lignin monomers. Additionally, the n-C_{28} fatty acid and the n-C_{38} to n-C_{52} alkyl monoesters clearly distinguished the hot from the cold water extract. A principal component (PC) analysis was carried out to visualize how the two extracts reflected the SOM decomposition in the sequence of O horizons (Figure 14.10). Although the aerobic incubation of solid soil samples altered the molecular-chemical composition of cold and hot water extracts, the plot of PC1 versus PC2 clearly gave the impression of gradual changes of DOM composition in the O horizons. However, in all samples, the data point triangles for the Oe horizon were aside a straight line from the Oa to the Oi. This indicated a discontinuity in organic matter changes which was reflected stronger by PCs 2 and 3 than by PC1, irrespectively on the extraction method. In the cold water extracts (Figure 14.10a), the organic matter decomposition in the profile was reflected by decreasing values of PC1, which explained about 36% of variance between the Py-FI mass spectra. The PC3, which explained about 8% of sample variance, contributed stronger to the separation of extracts from pre- and post-incubated samples. In the hot water extracts (Figure 14.10b), organic matter decomposition in the profiles was reflected by increasing values for PC1 of this sample set. The effects of the incubation were stronger reflected by PC2. This provided evidence that the macromorphological litter decomposition was reflected by the chemical composition of water extracts. Similar pictures were obtained for the evaluation of Py-FIMS data from daily sampled, short-time, composting experiments (Leinweber et al., 2001). This agreement indicates that Py-FIMS was sensitive enough to reflect differences in the composition of transforming solid organic matter (e.g., in composting experiments) and DOM extracted from solids by cold and hot water.

At a larger scale, environmental effects on DOM composition were studied in fen landscapes and in the Yenisei catchment in Siberia. The common topic of both studies was the influence of changing environmental conditions on SOM as a whole and on DOM which can be transferred to other ecosystem compartments.

Many fen areas in Central Europe were degraded by intensive agricultural use since the mid-18th century, and nowadays governmental programs intent to restore these unique habitats (Meissner and Leinweber, 2004). Since peat degradation and restoration may be accompanied by transformations of DOM, samples from differently degraded fen peat in the Droemling area (Saxony-Anhalt, Germany) were

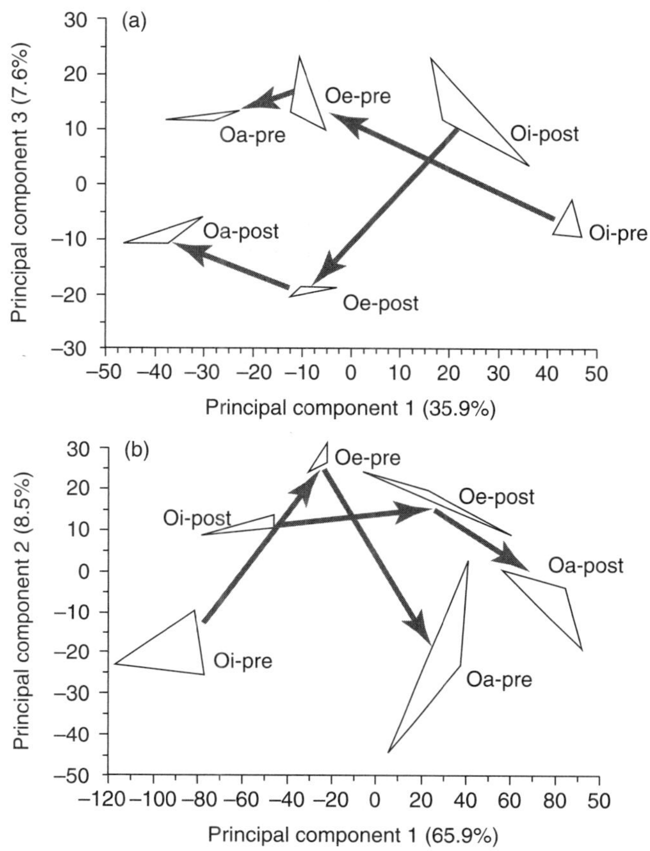

Figure 14.10. Principal component analysis of Py-FI mass spectra of (**a**) cold and (**b**) hot water extracts from the sequence of organic litter layers Oi–Oe–Oa in a beech stand (*Fagus sylvatica*) obtained before (-pre) and after (-post) aerobic incubation. The arrows indicate changes due to progressive decomposition top-down in the litter profile. Reprinted from Landgraf, D., Leinweber, P., and Makeschin, F. (2006). Cold and hot water extractable organic matter as indicators of litter decomposition in forest soils. *Journal of Plant Nutrition and Soil Science* **169**, 76–82, with permission of Wiley-VCH.

extracted with water, and the extracts were adsorbed on XAD-8 resin, purified (Kalbitz et al., 1999) and analyzed by Py-FIMS. The Py-FI mass spectra of DOM from the severely degraded topsoils (Ap/a/h and Hn horizons) had more signals in the lower mass range (assigned to carbohydrates, phenols, lignin monomers, and peptides) than in DOM samples from subsoils (Leinweber et al., 2001). This indicated that DOM with increasing proportions of microbial metabolites and decomposition products was formed with increasing intensity of soil tillage, aeration, and peat degradation.

Correspondingly, the thermograms in Figure 14.11a show that the curve for the summed ion intensities of carbohydrates, phenols, lignin monomers, and peptides from the arable Ap horizon (AR-Ap) peak at relatively low temperature and exceed the corresponding curves from the other horizons. The abundance of these compounds was greatly reduced in the groundwater of this field. In conserved extensive

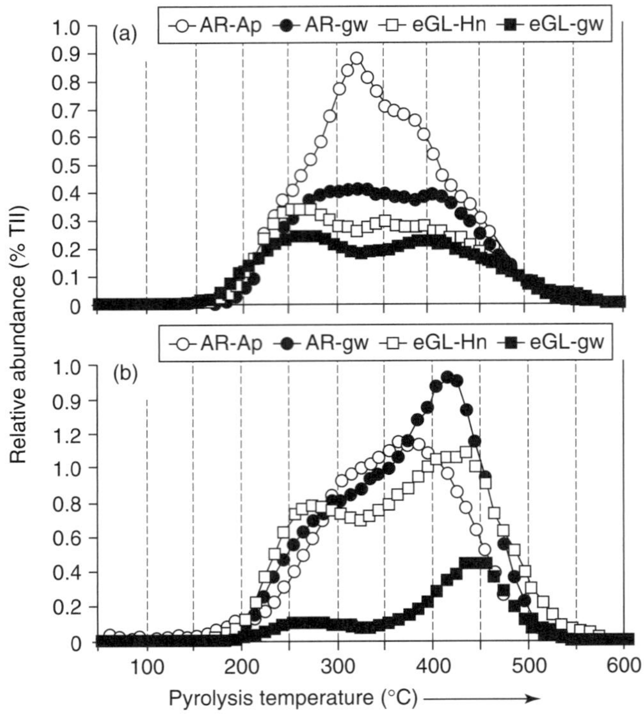

Figure 14.11. Thermograms for the volatilization of (**a**) carbohydrates, phenols, lignin monomers, and peptides and (**b**) lipids and alkylaromatics in dissolved organic matter from a highly degraded fen peat, extracted from the arable ploughed topsoil (AR-Ap) and the corresponding groundwater (AR-gw) in comparison to extracts from a conserved, less degraded peat under extensive grassland (eGl-Hn) and the corresponding groundwater (eGl-gw) in the Drömling nature reserve.

grassland the thermograms for carbohydrates, phenols, lignin monomers, and peptides show pronounced bimodal shape with peaks at lower (270 °C) and higher (350–400 °C) pyrolysis temperature, along with only minor differences within the soil profile indicating small alterations in the topsoil. The summed curves for lipids and alkylaromatics (Figure 14.11b) show great enrichments of a thermally stable fraction in AR compared to eGL, and in AR-Ap compared to AR-gw, which indicate the relative enrichment of these molecules with advanced peat decomposition. Under extensive grassland, which preserves the fen peat, a thermally labile fraction of lipids and alkylaromatics is indicated by the peak at about 270 °C in the thermogram, which perhaps originates from recent nonbound rhizodeposits or microbial biomass pyrolysis products.

The composition and fate of DOM at even larger scale was studied in the Yenisei catchment, Siberia (Kawahigashi et al., 2004). Water samples were collected from eight tributaries along the Yenisei between 67 °30′ N and 65 °49′ N latitude and analyzed for chemical compounds, DOM fractions, and mineralizable DOC. Results of wet-chemical analyses of the water samples were compared to compound classes derived from Py-FIMS of freeze-dried samples. Figure 14.12 shows excellent agreement of the results for contents of carbohydrates and peptide-derived N when based on the total organic matter of samples. This confirms previously published correla-

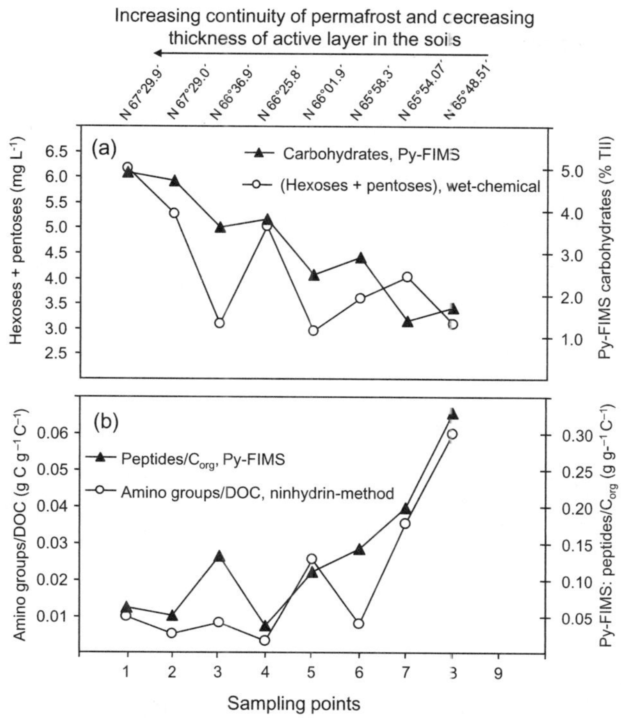

Figure 14.12. Quantification of carbohydrates and peptides in dissolved organic matter in tributaries of the Yenisei river (Siberia) by conventional wet-chemical methods (Kawahigashi et al., 2004) and Py-FIMS of freeze-dried samples.

tions between proportions of compound classes or C functions as quantified by Py-FIMS, ^{13}C NMR and wet-chemical methods (Schulten et al., 2002). Moreover, molecular weight characteristics and the thermal properties of whole DOM, compound classes, or single molecules can be determined from the Py-FIMS data sets and can be evaluated to explain differences according to permafrost proportions, microbial decomposition, and stabilization by soil minerals.

14.3.4. Organic-Mineral Particle Size, Density, and Aggregate Fractions

The majority of SOM is bound to soil minerals, forming organic–mineral particles and fractions. Physically, these fractions can be separated according to particle size, particle density, and aggregate-size fractions or location in the inter-aggregate and intra-aggregate space. The latter is restricted to specific light, particulate organic matter. State of the art in the composition, properties, and models of molecular structure of organic–mineral soil particles was reviewed by Schulten and Leinweber (2000). Py-FIMS was used to investigate the chemical composition and stability of organic matter associated with size and density fractions. Figure 14.13 shows decreasing proportions of carbohydrates and N-containing compounds (N in heterocycles, nitriles, amino acids and peptides) from fine clay to medium silt with increasing spherical equivalence diameter of the size fractions. The proportions of lignin dimers increase in this direction, and phenols and lignin monomers show no distinct trend with particle size in this sample set, although in other samples sometimes decreasing proportions with increases in particle size were observed (Schulten and Leinweber,

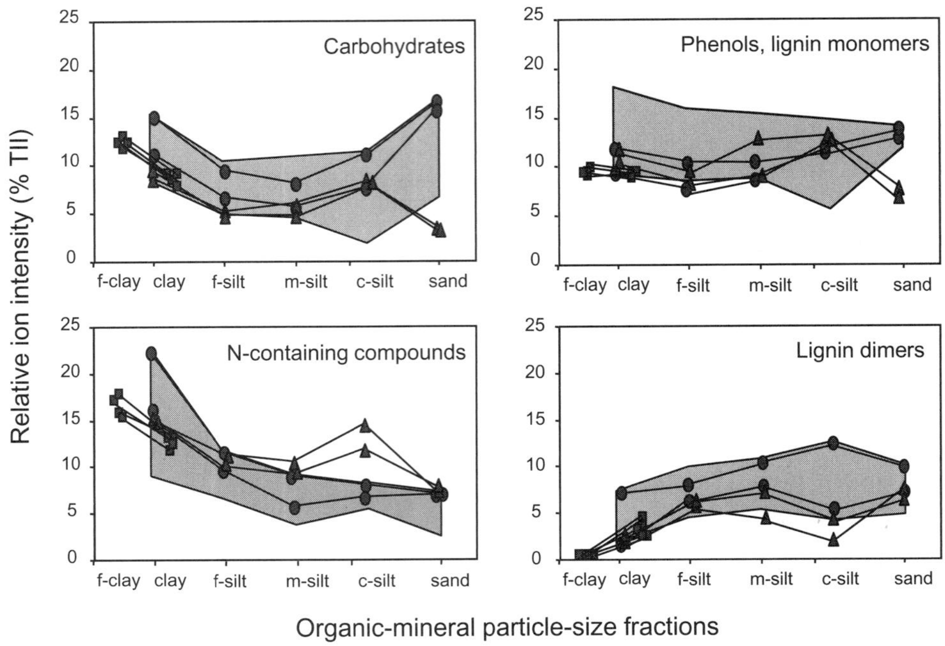

Figure 14.13. Proportions of compound classes in organic–mineral particle-size fractions of long-term fertilization experiments in Germany. The gray areas indicate the largest and smallest proportions of these compound classes found by Py-FIMS of other particle-size fractions.

2000). Lipids (not shown) were often enriched in clay fractions (Jandl et al., 2004). These general trends agree well with findings of ^{13}C NMR spectroscopy, if we compare lipids with alkyl C, carbohydrates with *O*-alkyl C, and lignin dimers with aromatic C (Schulten and Leinweber, 2000). Differences among SOM associated with size fractions were explained by (1) the progressive decomposition of primary plant materials and successive transfer of the transformation products from coarser to finer size fractions and (2) by the binding of colloidal and soluble organic matter like DOM, rhizodeposits and cell lysates by the fractions with the largest specific surface area, and enrichments of pedogenic oxides. This general concept, proposed on the basis of extensive Py-FIMS and complementary wet-chemical analyses (Leinweber, 1995), was partially confirmed by more recent studies (e.g., Mikutta et al., 2006; Kögel-Knabner et al., 2008).

Py-FIMS of density fractions showed the predominance of carbohydrates and lignin building blocks in specific light (<2 g cm^{-3}) particulate organic matter (Schulten and Leinweber, 1999), similar to the SOM isolated with sand-size fractions (Figure 14.13). Recently, the density fractionation was applied to isolate straw that was incubated with lignocellulytic fungi such as *Trichoderma saturnisporum* (Wiedow et al., 2007). In the Py-FI mass spectrum of the wheat straw before incubation, marker signals for carbohydrates, phenols, and lignin monomers were most prominent. The clearest impact of inoculation on compound classes was the loss of lignin dimers and the increase of the ratio of phenols + lignin monomers:lignin dimers after 4 weeks. This indicates accelerated lignin decomposition after the inoculation with the fungi. After 13 weeks this effect of the inoculation was no longer detectable.

One explanation for this could be an additional decomposition of higher polymer lignin to lignin dimers in the inoculated treatment. The straw decomposition was reflected in the thermograms by reduced energy required for the volatilization of lignin dimers (peak shift by –50 °C) and phenols and lignin monomers (–70 °C). Accordingly, the inoculation effects were most obvious in the range of 350–400 °C (lignin dimers) and 300–350 °C (phenols and lignin monomers) (Figure 14.14). The higher range of relevant temperatures for inoculation effects on lignin dimers agreed with a larger thermal stability of lignin dimers compared to lignin monomers as shown for a light soil fraction (Schulten and Leinweber, 1999) and macroaggregates of gleysols (Monreal et al., 1997). This confirmed that the thermal properties as reflected by Py-FIMS were a good indicator for the decomposability of plant materials in soil (Wiedow et al., 2007).

Storage in aggregates can be the reason for the inaccessibility of SOM constituents to microbial decomposition, and in turn for the stabilization of otherwise

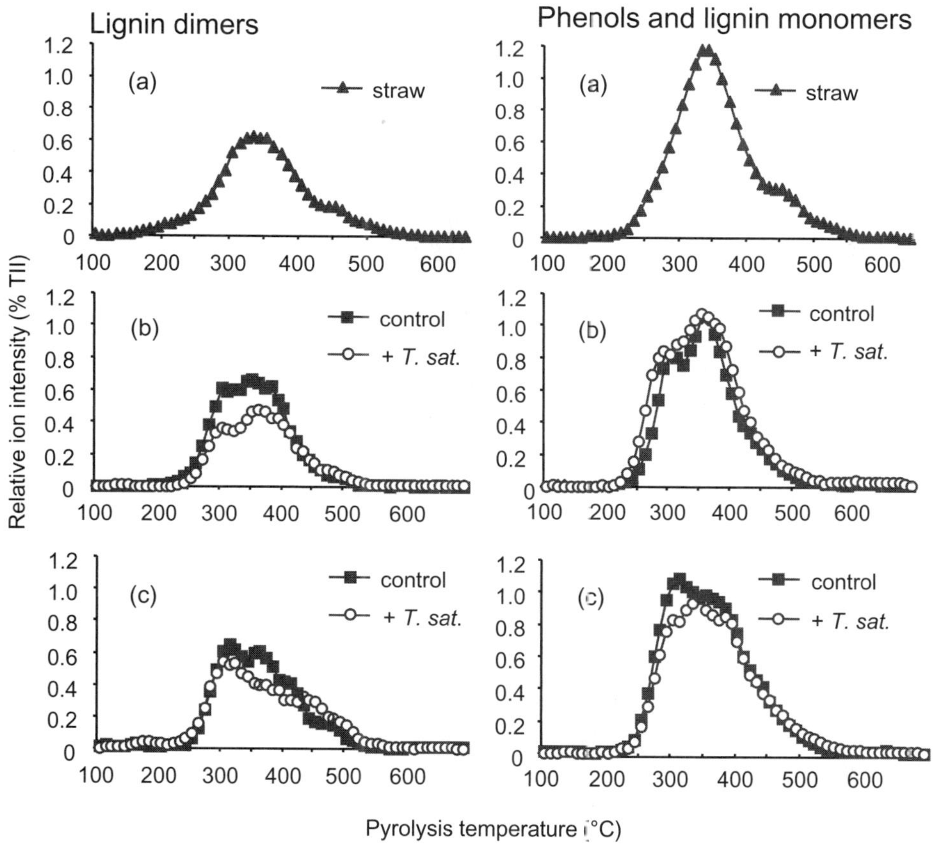

Figure 14.14. Thermograms for volatilization of lignin dimers and lignin monomers of wheat straw (**a**) before incubation, without incubation (control), and with inoculation with *Trichoderma saturnisporum* (+T. sat.), (**b**) after 4 weeks, and (**c**) after 13 weeks of incubation. Reprinted from Wiedow, D., Baum, C., and Leinweber, P. (2007). Inoculation with *Trichoderma saturnisporum* accelerates wheat straw decomposition on soil. *Archives of Agronomy and Soil Science* **53**, 1–12, with permission from Taylor & Francis.

decomposable substrates. This can be explained by (1) the inaccessibility of intra-aggregate space for microorganisms and reduced diffusion of their enzymes and (2) the restricted aerobic decomposition due to a reduced oxygen availability. Relationships between soil aggregation and SOM turnover were studied by Py-FIMS. For instance, Monreal et al. (1995, 1997) observed a small difference in SOM composition in different aggregate-size fractions of forest soils, but found more pronounced differences under arable use. Organic matter in macroaggregates (>250 µm) was characterized by strong volatilization of phenols and lignin monomers at 300–400 °C. The peak temperatures of thermograms for microaggregates were mostly shifted to higher temperatures, pointing to a stronger degree of cross-linking and organic–mineral interactions in microaggregates. Irrespective of soil management, proportions of lignin dimers decreased and those of N-containing compounds increased with a decrease in aggregate-size from >250 µm to <50 µm. This corresponds to observations made for particle-size fractions (Figure 14.13) and reflects the degree of decomposition and transformation of plant materials.

More recently, functionally different SOM pools such as free particulate organic matter (POM) and microaggregate (50–250 µm) occluded POM were separated (Six et al., 2002) and subjected to chemical analyses. Solid-state ^{13}C-NMR revealed that free POM contained more *O*-alkyl C and less alkyl C than the occluded POM (Golchin et al., 1994; Kölbl and Kögel-Knabner, 2004). Recent Py-FIMS of similar size-, aggregate- and density-fractionated samples showed that free POM had more volatile matter, lignin dimers, lipids, sterols, and free fatty acids than the other corresponding fractions. The occluded POM showed changes in the compound classes toward the organic matter composition of the whole soil such as decreasing volatile matter and proportions of lignin dimers, lipids, and sterols, as well as increased proportions of carbohydrates, phenols, lignin monomers, peptides, and fatty acids. The latter compounds indicate the impact of microbial decomposition. Reduced versus conventional tillage was reflected stronger by occluded than by free POM and organic–mineral silt- or clay-size fractions (Sleutel et al., 2007).

14.3.5. Nonfractionated Whole Soil Organic Matter: Factors Influencing Its Composition and Turnover

Probably one of the greatest advantages of Py-FIMS is the possibility to analyze chemically unaltered, nonextracted or chemically enriched, simply dried and milled whole soil samples. Table 14.3 compiles proportions of SOM compound classes in major soil groups of the world. The data sets for individual soil groups sometimes included different geographical origins, vegetation, land uses or management regimes, which could account for the standard deviations. However, besides these variations, some soil-group-specific proportions became obvious.

Gelic histosols were characterized by high contents of volatile matter and proportions of lipids and free fatty acids. This and an additional high content of sterols was distinct for gelic histosols from the Antarctica (Beyer et al., 1995). Plaggic anthrosols from Northwest Germany also showed high proportions of lipids, alkylaromatics, sterols, and free fatty acids. This can be explained by the former heather vegetation that was used for "Plaggen" and its enrichment with aliphatic compounds from feces during use as bedding material (Blume and Leinweber, 2004). The terric anthrosols were formed by erosion of humus-rich soil at top positions and accumulation downslope and in depressions. They also had relatively high proportions of lipids and alkylaromatics, but in the proportions of N-containing compounds and peptides they were more similar to luvisols because these two groups of soil are associated in the Weichselian-pleistocene landscape in Northern Germany. The specificity of vertisols, sampled in various parts of the world, were high proportions of N-containing compounds, as shown by Py-FIMS and complementary Cp Py-GC/MS (Leinweber et al., 1999). The volatile matter contents of Andosols indicated high contents of organic matter and bound moisture. The compound classes showed no particular enrichment. González-Pérez et al. (2007) found large yields of aliphatic (both alkyl- and carbohydrate-derived) pyrolysis compounds in soils with andic horizons by double-shot Py-GC/MS. Podzol Bh horizons were formed by leaching of dissolved and colloidal organic matter that originates from undecomposed litter layers, often from heathland vegetation, and the precipitation of these organics in subsoils (Wilcken et al., 1997). This SOM was characterized by the highest proportions of lipids and sterols among all mineral soils. Recent work on SOM stabilization in subsoils (Kaiser et al., 2002; Schmidt and Kögel-Knabner, 2002; Rumpel et al., 2004a,b) confirmed the described adsorption of DOM of a podzol Bh horizon (Rumpel et al., 2004a) and the obtained presence of aliphatic C, most probably derived from ester-bound moieties, in sandy subsoil horizons of a dystric cambisol (Rumpel et al., 2004b). The sterols in the podzol Bh horizons unequivocally originate from the vegetation (phytosterols), so that (for instance) comparison of the sterol patterns between podzol-Bh horizons and the plaggic anthrosols would indicate possible enrichments of feces-derived sterols in the latter. Chernozems belong to the most fertile soils worldwide. Their relatively large proportions of carbohydrates, phenols, lignin monomers, N-containing compounds, and peptides indicate a high level of primary organic matter decomposition and biological activity. Kastanozems, worldwide adjacent to chernozems, but under drier and warmer climatic conditions, had still more carbohydrates, phenols, lignin monomers, and N-containing compounds. The large number of luvisols originated from arable soils grown with potato in Northern Germany. They revealed the largest proportions for phenols and lignin monomers and alkylaromatics as compared to the other major soil units. Together with the relatively high proportion of carbohydrates, this seems to be an indicator for intensive arable land use. Py-FI mass spectra for some argillic subsoil (Bt) horizons were generally low in intensity but showed all compound classes and relatively high proportions of N-containing compounds. The few stagnosols investigated originate from tropical regions in Cuba which naturally should show a vertisol dynamics. However, due to irrigation under intensive sugarcane production, these soil were practically saturated in most parts of the profile and for large periods of the year, and thus showed more stagnic than vertic properties. Nevertheless, they resembled vertisols in similar high proportions of carbohydrates and N-containing compounds. Finally, most cambisols and regosols originated from Pleistocene moraines in the Bolivian Andes (Leinweber et al., 1996). Due to incomplete decomposition of plant residues and young soil ages, they showed relatively high proportions of aliphatics, although less than podzol Bh horizons (Table 14.3).

In summary, this set of 264 nonfractionated whole soil samples, which could be grouped according to major soil units of the World Reference Base soil classification, is possibly the largest sample set that has been characterized by an extremely sensitive, versatile analytical method under identical experimental conditions.

The standard deviations of means in Table 14.3 indicated that there were significant variations of SOM composition within major soil units. These differences can be due to the well-known factors that influence content and composition of SOM: (1) effects of vegetation that provides litter as a source of SOM formation, (2) action of soil organisms that decompose litter and release metabolites into soil, and (3) climate that generally governs input from vegetation and organic matter decomposition as well as weather that may cause annual variations in the input-decomposition dynamics. The following examples demonstrate how Py-FIMS was used to investigate the action of these factors on SOM composition.

The influence of native vegetation on the composition of SOM was studied in well-dated volcanic soils at the Mount Etna region, Sicily, Italy. Whole soil samples were taken under chestnut and oak stands, and their humic acids (Baglieri et al., 2007) and nonfractionated whole soil samples were analyzed by Py-FIMS. The proportions of compound classes in Table 14.4 show that the roughly 2000-year-old soil under an oak stand was significantly richer in lipids, sterols, and free fatty acids than was the soil under chestnut. In another pair of soil samples with a longer period of soil formation (about 7000–9000 years), the difference between oak and chestnut was +4.2% for lipids, +2.6% for sterols and +2.5% for free fatty acids (not shown). This predominance of aliphatics in the soils under oak confirms Nierop et al. (2003), who also found the release of aliphatics from oak stands into soil by conventional lipid extractions and GC/MS.

The recent discussion about limiting climate change by reducing atmospheric CO_2 output was accompanied by research on renewable sources and energy. Among other high-biomass crops, *Miscanthus × giganteus* (Greef et Deu.) was grown experimentally in Europe, and its effects on soil properties were studied (Kahle et al., 2001). Py-FIMS of a field with Miscanthus showed increased proportions of alkanes, alkenes, sterols, and free fatty acids, compared to an adjacent plot without Miscanthus. Subsequently we studied effects of Miscanthus on the depth distribution of SOM compound classes from litter layers to the argillic Bt horizons in the subsoils. Figure 14.15 shows that in the organic litter horizons (L and Of), both groups of compounds, carbohydrates, and all N-containing compounds on the one hand side and lipids, sterols, and free fatty acids on the other were abundant in roughly the same proportions. With increasing soil depth toward the Ah and Bt horizon, the proportions of lipids, sterols, and fatty acids decreased whereas the proportions of carbohydrates and all N-containing compounds had a maximum in the Ah horizon but remained at a high level in the Bt horizon. Generally, the distribution of compound classes in soil horizons is a function of the input-turnover dynamics and transport processes. The increasing proportions of carbohydrates and N compounds in the Ah and Bt horizon are assumed to be an combined effect of Miscanthus rooting, resulting rhizodeposits and microbial transformation processes, and larger leaching inputs than for aliphatics with lower solubility and larger hydrophobicity. In the clay-rich argillic Bt horizon, close and stabile organic matter–clay bonds and, perhaps, restrictions in the thermally induced diffusion of large molecules, such as long-chain aliphatics, sterols, and fatty acids, in the quartz sample tube of Py-FIMS also could contribute to disproportional large detection of lower *m/z* signals (carbohydrates and N-containing compounds).

Besides the organic matter input with native vegetation or crops cultivated, manuring is a traditional measure to amend soils, in particular by direct inputs of

TABLE 14.4. A Summary of the Abundance of Compound Classes Determined by Py-FIMS in Sample Pairs Showing the Influence of Primary Organic Matter Input Either from Vegetation or with Farmyard Manure (FYM) on the Composition of Soil Organic Matter

Origin of Samples	Number	VM	CHYDR	PHLM	LDIM	LIPID	ALKYL	NCOMP	STEROL	PEPTI	SUBER	FATTY
Etna soil, chestnut	1	11.6	7.8	6.4	3.2	5.2	7.3	4.9	1.6	4.0	0.0	1.0
Etna soil, oak	1	11.5	7.4	6.7	3.5	9.7	8.0	5.3	2.2	3.3	0.1	3.8
"Eternal Rye" NPK	22	6.0 ± 1.4	4.9 ± 0.7	7.1 ± 0.4	8.1 ± 0.6	10.5 ± 0.7	12.0 ± 0.7	5.8 ± 0.3	2.4 ± 0.4	2.5 ± 0.3	0.1 ± 0	0.2 ± 0.1
"Eternal Rye" FYM	5	5.1 ± 2.3	5.8 ± 0.7	7.4 ± 0.8	6.7 ± 0.9	9.3 ± 1.0	9.0 ± 0.3	6.4 ± 0.6	3.3 ± 0.5	3.2 ± 0.3	0.2 ± 0.1	0.6 ± 0.2
Static experiment, U	4	6.6 ± 1.1	8.7 ± 0.9	16.1 ± 1.7	0.6 ± 0.1	2.1 ± 0.4	15.1 ± 0.8	9.3 ± 0.6	<0.1	4.0 ± 0.5	<0.1	<0.1
Static experiment, NPK + FYM	3	6.5 ± 0.7	7.7 ± 0.3	9.3 ± 0.3	3.2 ± 0.4	4.3 ± 0.5	10.4 ± 0.2	8.0 ± 0.3	0.2 ± 0.1	4.3 ± 0.3	<0.1	<0.1

For abbreviations see Table 14.2. U, unfertilized; NPK, mineral nitrogen + phosphate + potassium fertilizer.

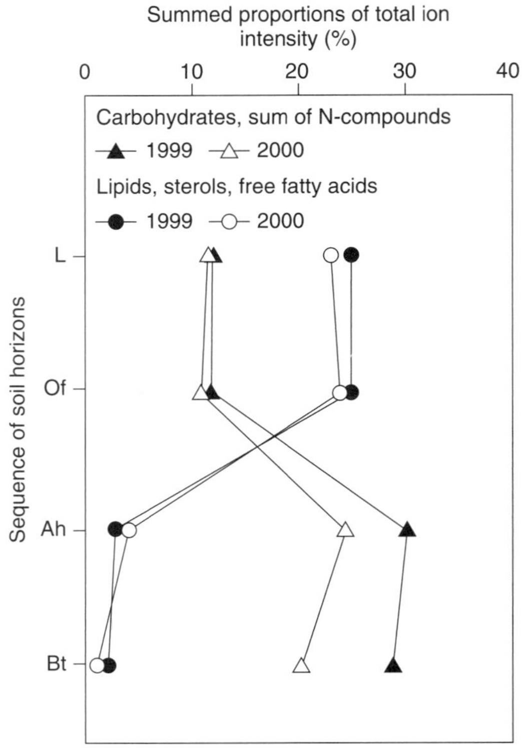

Figure 14.15. Depth distribution of summed compound classes in soil profiles under Miscanthus stands. L and Of describe organic litter layers mainly composed of the Miscanthus residues, Ah the humic mineral topsoil, and Bh an argillic subsoil horizon. Samples were taken in the years 1999 and 2000.

available plant nutrients and by the buildup of stable SOM, which then indirectly acts as buffer for storage and supply of nutrients, as aggregate stabilizing agent, and, thus, as physical soil conditioner. Beneficial effects of manuring were documented in many studies, but Py-FIMS and multivariate statistical data evaluation for the first time visualized the long-lasting effect of manuring on SOM quality. The data pairs in Table 14.4 show that the manured soils at both sites had more sterols and peptides. This is not surprising since sterols are important constituents of feces (Jardé et al., 2007) and peptides indicate larger proportions of microbial biomass. For other compound classes the effect of manuring was inconsistent among experiments. For the "Eternal Rye" cultivation the relative enrichments of the manured soil in phenols and lignin monomers and N-containing compounds confirm Schmidt et al. (2000). In the Static Experiment, the relative enrichment of the manured soil in lignin dimers and lipids reflects the input relatively stable primary organic matter with FYM. Recent evaluations of Py-FIMS data from time series of differently managed experimental soils enabled us to distinguish pools of different stability and turnover among individual compound classes.

Without any doubt, the composition of nonliving organic matter is largely influenced by the action of soil organisms. Particulate vegetation residues are first consumed by macro- and mesofauna organisms and subsequently by microorganisms. Martius (1992) and Amelung et al. (2002) studied feeding strategies and food residues in termite nests in the Amazonian rain forest. Py-FIMS of nest materials and spectra evaluation by principal component analyses clearly separated soil, microepiphytes, and wood as principal food sources (Amelung et al., 2002) and the nest materials of the various termite genera and species (Figure 14.16). The nests of *Nasutitermes* sp. were similar to wood, confirming their assignment to wood feeders as derived from density fractionation and lignin analyses of density separates (Amelung et al., 2002). Furthermore, the shift from wood to soil food incorporation in the sequence *Nasutitermes* > *Cornitermes* > *Termes* > *Emibratermes*, *Anoplotermes* as proposed by Amelung et al. (2002) was confirmed. However, the Py-FIMS data indicated that microepiphytes possibly also contributed to the food and nest chemical composition of *Cornitermes, Termes, Anoplotermes* and *Emibratermes*, although to a lesser extent than to the food of *Constrictotermes*. This hypothesis will be tested by more detailed data evaluations.

Traditionally, the bulk decomposability of primary OM and SOM is studied in incubation experiments, and the CO_2 evolution is considered as direct measure for the microbial decomposition. Such an incubation experiment was done with an unfertilized soil of a long-term experiment under rye (Leinweber et al., 2008b). The cumulative CO_2–C flux increased from 154 mg kg^{-1} (day 7) to over 375 mg kg^{-1} (day 21) and reached 669 mg kg^{-1} after 63 days. More interesting, the incubation changed the chemical and thermal properties of SOM and affected the microbial decomposer population and activities. Data from this study, compiled in Figure 14.17 show relationships between the thermal stability of compound classes as determined by temperature-resolved Py-FIMS. The ion intensities of monomeric and dimeric lignin building blocks were separated into labile and stable proportions. The temperature limits for this separation were set up according to experience from previous studies (e.g., Leinweber, 1995; Leinweber and Schulten, 1998; Kalbitz et al., 2003) and visual

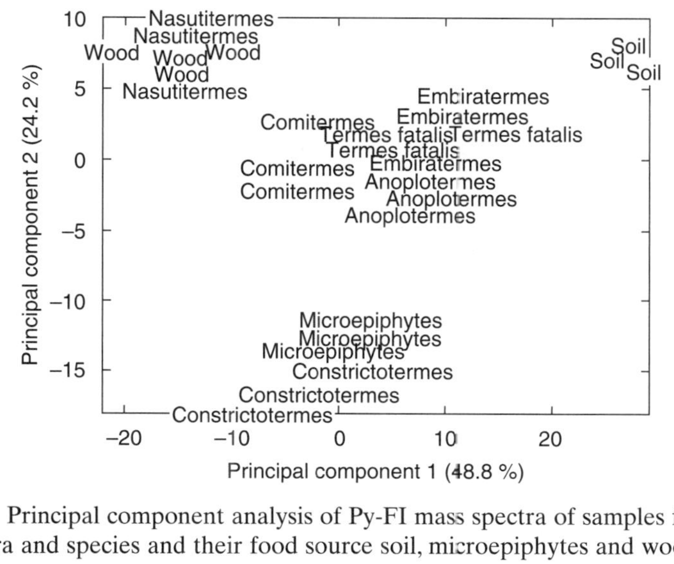

Figure 14.16. Principal component analysis of Py-FI mass spectra of samples from different termite genera and species and their food source soil, microepiphytes and wood.

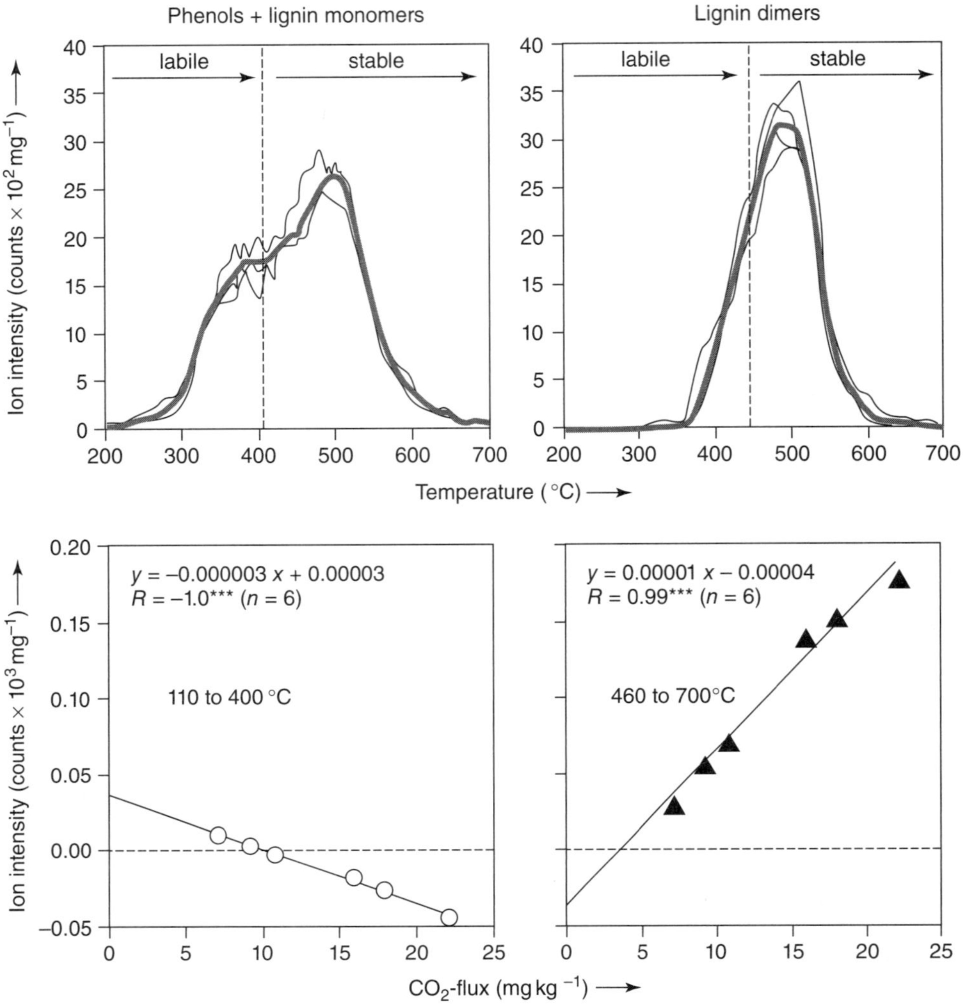

Figure 14.17. Thermograms of phenols + lignin monomers and lignin dimers released during an incubation experiment with a soil under rye from the unfertilized treatment of the "Eternal Rye Cultivation" experiment at Halle, Saxony-Anhalt, Germany. The bold gray line is the average of the single thermograms from samples after different incubation periods. The dotted vertical line symbolizes the temperature derived to distinguish thermally labile and stable compounds (**upper graphs**). **Lower graphs:** Linear regression functions showing correlations between differences in the ion intensities of marker signals for lignin-derived compounds in characteristic temperature intervals of Py-FIMS and the corresponding differences in CO_2–C flux for time intervals. Positive intensity differences indicate relative gains and negative relative losses of these compound classes (Leinweber et al., 2008b).

detection of large differences according to incubation. The temperature thresholds of 405 °C (phenols and lignin monomers) and 455 °C (lignin dimers) were derived and used in correlation and regression analyses. For phenols and lignin monomers, this temperature indicates a shoulder in the averaged thermogram which may originate from the overlay of two Gaussian-like curves (thick gray curve in Figure 14.17, upper left). Lignin dimers generally show Gaussian-like thermograms at high pyrolysis temperature which made it difficult to derive thermally labile and stable proportions (Figure 14.17, upper right). Application of these labile and stable proportions of monomeric and dimeric lignin building block resulted in highly significant correlations with the CO_2–C flux. The differences in ion intensity of labile phenols and lignin monomers were positively, and those in stable lignin dimers negatively, correlated with the differences in CO_2–C flux (Figure 14.17, lower). This indicates that thermally labile phenols and lignin monomers were decomposed and contributed to the CO_2 release. In contrast, thermally stable lignin dimers could not be decomposed and, thus, were relatively enriched during incubation. Similar relationships were also derived for other compound classes. Furthermore, the intensities of single Py-FIMS signals also explained changes in the enzyme activities during SOM decomposition (Kögel-Knabner et al., 2008; Leinweber et al., 2008b). This contributed to compelling evidence for the unique value of the temperature-resolved Py-FIMS data as universal indicators of the stability and microbial decomposability of nonliving organic matter.

Impact of the climate on the composition of SOM is partly reflected by the standard deviations of means in Table 14.3. More detailed insights can be obtained when soils along climatic gradients under similar native vegetation or management are studied (Amelung et al., 1998, 1999). Py-FIMS of pairs of native and cultivated prairie soils in Canada (Schulten et al., 1995b; Schnitzer et al., 2005) were compiled, accomplished, and re-evaluated for the abundance and thermal stability of various N-containing compounds (Leinweber et al., 2009). The difference thermograms in Figure 14.18 upper) clearly show that the soils at all three sites have lost thermally labile organic nitrogen compounds. This was valid for peptides as well as for nonproteinaceous mainly heterocyclic compounds, and the extent of losses followed the order Lethbridge > Macklin > St. Denis. On the other hand, the negative intensity differences indicate relative enrichments of thermally stable compounds, particularly pronounced for heterocyclic N compounds at Lethbridge. The cultivation-induced changes in the chemical composition in organic N compounds were also confirmed by N K-edge XANES (Figure 14.18, lower). The positive intensity differences with peak around 401 eV at Lethbridge indicate losses in amino N and peptides, although an overlap with pyrrolic N cannot be excluded in this range of the XANES spectrum (Leinweber et al., 2007). The sharp negative peak at 399.8 eV indicates a residual enrichment in pyridinic N and nitrile N; and the second negative peak at 403–404.5 eV is assigned to nitro-N, which is the only N function showing features in this range. Features in the N-XANES spectra >405 eV reflect $1s \rightarrow \sigma^*$ transitions that are not specific for inorganic or organic N functions (Leinweber et al., 2007). For the St. Denis soil pair the broad positive intensity difference indicates losses of various N compounds with any specific enrichments, which is in good agreement with the Py-FI mass spectra as visualized by the difference thermograms. The order Lethbridge > Macklin > St. Denis reflects a gradient in soil conditions from warmer to colder conditions (Figure 14.18). The small alterations in the proportions of all N-containing compounds in the St. Denise soil, which is located in a transitional zone straddling the Dark Brown and Black soil zones in Saskatchewan, are explained by the largest organic N pool, the greatest thermal stability of peptides (volatilization at +50–100 °C higher pyrolysis temperature compared to Macklin and Lethbridge), and, perhaps, the shortest time of cultivation.

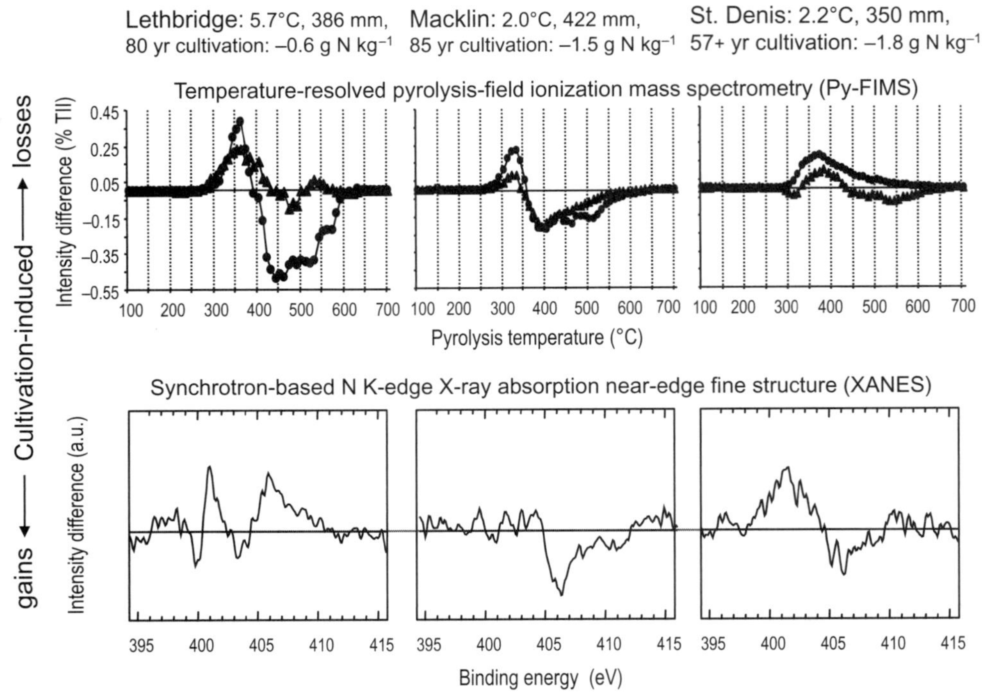

Figure 14.18. Difference thermograms showing changes in the volatilization of N-containing compounds (-●-) and peptides (-▲-) in Py-FIMS (**upper**) and the corresponding difference N K-edge XANES spectra of samples taken at native and long-term cultivated sites along a climate gradient in Canada. Data for native soils minus data for cultivated soils.

This complements to various examples in the present review (Figures 14.6, 14.8, 14.11, 14.14, and 14.17) as well as in other papers (Kalbitz et al., 2003; Marschner et al., 2008) and confirms that the thermal properties as determined by Py-FIMS are a universal indicator of resistance to microbial decomposition. Furthermore, it could be concluded that heterocyclic N compounds form an important pool in Canadian prairie soil that was enriched by pedogenesis and cultivation and that may have a currently unknown contribution to the N nutrition of plants (Leinweber et al., 2008b).

In addition to long-term and large-scale climatic effects, the annual variation in weather conditions also affected the amount and turnover of plant residues and thus the chemical composition SOM sampled at a certain date (Leinweber et al., 1994, 1995). Figure 14.19 shows the ion intensities for carbohydrates in two plots of the "Eternal Rye Cultivation" experiment at Halle, Saxony-Anhalt, Germany. In 1958, parts of the former rye monoculture were shifted into maize monoculture. This resulted in decreases in the ion intensities for carbohydrates. In the plot that remained under rye monoculture the ion intensities for carbohydrates increased initially. For several time periods, the changes in ion intensities were roughly parallel for the two plots, which is considered as evidence for annual variations in the decomposition dynamics. For some time (1993 to 1997), there was no significant difference between the rye and maize plots in the ion intensities for carbohydrates, but the

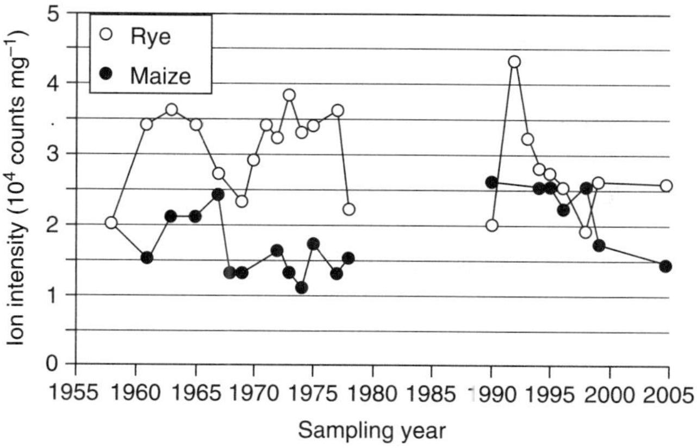

Figure 14.19. Long-term changes and annual variation in the proportions of carbohydrates of the unfertilized plot grown with rye and maize after previously rye from the "Eternal Rye Cultivation" experiment at Halle/Saxony-Anhalt, Germany.

most recent sampling indicated larger carbohydrate proportions in soils under rye. Similar annual variations were also found for other compound classes (not shown).

14.4. CONCLUSION AND OUTLOOK

The intention of this chapter was to review advances in analytical pyrolysis and soft ionization mass spectrometric techniques as applied to the molecular-chemical characterization of nonliving organic matter.

Technical innovations introduced to the chemistry of nonliving organic matter in the past decade included liquid injection field desorption ionization (LIFDI)-MS, ultrahigh-resolution FT-ICR MS, and Orbitrap MS. These techniques can be applied to solvent extracts and liquid samples. Their broad application will reveal new insights into the molecular and structural diversity of nonliving organic matter. This new analytical possibilities will provide a much better basis for building models of the molecular structures of dissolved organic matter and modeling its interactions with natural molecules and xenobiotics.

For solid samples, Cp Py-GC/MS, Py-FIMS and recently developed synchrotron-based X-ray absorption fine-structure analysis (XANES) at the K edges of C, N, and S in organic matter are the methods of choice if applied in conjunction. Important trends to be highlighted are high-resolution measurements with Py-FIMS, and the combination of Py-FIMS with N-XANES of pyrolysis residues. This will result in improved qualitative and quantitative evaluations of the Py-FI mass spectra, along with new evidence for the occurrence and transformations of heterocyclic N-containing compounds in nonliving organic matter. Furthermore, first experience with Cp Py-GC/MS of ^{15}N-labeled hay, feces, and soil along with the high-resolution mass spectrometry techniques open up the door for much more detailed studies of the fate of agricultural N in the environment.

Besides these new experimental approaches, which so far could be only demonstrated on a few examples, fundamental new evidence on organic matter composition in various groups of samples was obtained by the accumulation, storage, and evaluation of a huge number of Py-FIMS mass spectra in the past decade. The environmental samples can be grouped into (1) dissolved organic matter, (2) organic matter in physical fractions such as particle-size, density, and aggregate fractions and (3) whole, nonfractionated samples from plant materials, waste, manure, or soils. Although not complete, the data sets compiled for these three groups of samples (except for solid waste) allow, by comparison, interesting conclusions about the impact of origin and environmental factors on the molecular composition. The multipurpose evaluation opportunities and, thus, the value of this spectra library increases with each new set of analyses saved. This philosophy is the basis for the ongoing build-up of a rhizodeposit database of Py-FI mass spectra of a wide range of nontransgenic cultivates of agricultural crops and their corresponding transgenic lines.

Two unique scientific outputs of the Py-FI mass spectra are emphasized:

1. The mass-spectrometric "fingerprint"—that is, the abundance of up to 900 single properties in the form of *m/z* with specific intensities—was shown to be most sensitive to detect, prove, and visualize even minor differences between samples, by the use of appropriate statistical procedures. This is independent on the specific sample properties (dissolved/solid, fractionated/nonfractionated) and was shown to disclose agronomic (fertilizer, manure, or crop-specific impacts on SOM quality), and ecological (parent material–metabolite, consumer–food, plant–soil) interrelationships.
2. The thermal properties of individual nominal and high-resolution masses, of compound classes or even of the whole sample, recorded during temperature-resolved Py-FIMS were shown to be universal markers of the stability in the sense of resistance to microbial decomposition. This was proved among all groups of samples (DOM, organic-mineral fractions, whole samples) and by various experimental approaches (comparison of the abundance along gradients of environmental impacts, incubation of liquid and solid samples in the laboratory and in the field).

These two scientific outputs distinguish Py-FIMS from all other chemical-analytical methods widely applied to nonliving organic matter. Therefore, irrespective of the limitations mentioned in the previous reviews (Schulten, 1996; Schulten et al., 1998; Schulten, 2002), diversification in Py-FIMS applications to other fractions of nonliving organic matter, a wider range of major soil units, or agronomic problems and more complex and complicated ecological problems can be foreseen. This also includes product development and success control of new agrobio-technologies such as diverse inoculates for pest control, stimulation beneficial transformation processes, or waste management.

ACKNOWLEDGMENTS

The mass-spectrometric investigations of the University of Rostock group was supported financially by the European Community (Projects 2002-1345/001-001 CPT

CPTACN and FOOD-CT-514082), the Deutsche Forschungsgemeinschaft (projects Lei 903/3-1 to 3-3, Lei 903/4-1 to 4-2, Schu 416/18-5/6), the Federal Ministry of Education and Research (Projects 03WKS04B and WTTR02058406), the German Academic Exchange Service (project D/05/50492), and Ministry of Education of Mecklenburg-Western Pomerania (HSPIII project 4200/0037 5001, project UR 07 079.). The authors are very grateful to the technical co-workers of the research group, especially Rolf Beese. The close and constructive collaboration with Sören Thiele-Bruhn (Chair of Soil Science, University of Trier/Germany), Steven Sleutel (Department of Soil Management and Soil Care, Ghent University/ Belgium), Fran Walley, P. Ming Huang, and Adam Gillespie (Department of Soil Science, University of Saskatchewan, Saskatoon/Canada) is gratefully acknowledged. We also thank Wolfgang Metelmann-Strupat (ThermoFisher Scientific, Bremen/Germany) for the acquisition of two soil extract mass spectra with the LTQ-FT Ultra.

REFERENCES

Amblès, A., Colina-Tejada, A., Jambu, P., Lemee, L., and Parlanti, E. (1998). Experimental leaching of podzol soil lipids. Nature and biological origin of water soluble components. *Agrochimica* **42**, 158–171.

Amelung, W., Zech, W., Zhang, X., Follett, R. F., Tiessen, H., Knox, E., and Flach, K.-W. (1998). Carbon, nitrogen and sulfur pools in particle-size fractions as influenced by climate. *Soil Sci. Soc. Am. J.* **62**, 172–181.

Amelung, W., Zhang, X., Flach, K.-W., and Zech, W. (1999). Amino sugars in native grassland soils along a climosequence in North America. *Soil Sci. Soc. Am. J.* **63**, 86–92.

Amelung, W., Martius, C., Bandeira, A. G., Garcia, M. V. B., and Zech, W. (2002). Lignin characteristics and density fractions of termite nests in an Amazonian rain forest—indicators of termite feeding guilds? *Soil Biol. Biochem.* **34**, 367–372.

André, M., Le Caer, J. P., Greco, C., Planchon, S., El Nemer, W., Boucheix, C., Rubinstein, E., Chamot-Rooke, J., and Le Naour, F. (2006). Proteomic analysis of the tetraspanin web using LC-ESI-MS/MS and MALDI-FTICR-MS. *Proteomics* **6**, 1437–1449.

Baglieri, A., Ioppolo, A., Negre, M., and Gennari, M. (2007). A method for isolating soil organic matter after the extraction of humic and fulvic acids. *Org. Geochem.* **38**, 140–150.

Bahr, U., and Schulten, H.-R. (1983). Pyrolysis field ionization mass spectrometry of cell wall components and bacterial cell walls. *J. Anal. Appl. Pyrolysis* **5**, 27–37.

Beckey, H. D. (1969). Field desorption mass spectrometry, a technique for the study of thermally unstable substances of low volatility. *Int. J. Mass Spectrom. Ion Phys.* **2**, 500–503.

Beckey, H. D., Heindrichs, A., and Winkler, H. U. (1970). New field desorption techniques. *Int. J. Mass Spectrom. Ion Phys* **3**, A9–A11.

Berwick, L., Greenwood, P., Kagi, R., and Croue, J. P. (2007). Thermal release of nitrogen organics from natural organic matter using micro scale sealed vessel pyrolysis. *Org. Geochem.* **38**, 1073–1090.

Beyer, L., Sorge, C., Blume, H.-P., and Schulten, H.-R. (1995). Soil organic matter composition and transformation in gelic histosols of coastal continental Antarctica. *Soil Biol Biochem.* **27**, 1279–1288.

Blume, H.-P., and Leinweber, P. (2004). Plaggen soils: Landscape history, properties, and classification. *J. Plant Nutr. Soil Sci.* **167**, 319–327.

Bruinsma, M., Kowalchuk, G. A., and van Veen, J. A. (2003). Effects of genetically modified plants on microbial communities and processes in soil. *Biol. Fertil. Soils* **37**, 329–337.

Brusetti, L., Francia, P., Bertolini, C., Pagliuca, A., Borin, S., Sorlini, C., Abruzzese, A., Sacchi, G., Viti, C., Giovannetti, L., Giuntini, E., Bazzicalupo, M., and Daffonchio, D. (2004). Bacterial communities associated with the rhizosphere of transgenic Bt 176 maize (*Zea mays*) and its non transgenic counterpart. *Plant Soil* **266**, 11–21.

Bull, I. D., Van Bergen, P. F., Poulton, P. R., and Evershed, R. P. (1998). Organic geochemical studies of soils from the Rothamsted Classical Experiments—II, Soils from the Hoosfield Spring Barley Experiment treated with different quantities of manure. *Org. Geochem.* **28**, 11–26.

Burdon, J. (2001). Are the traditional concepts of the structures of humic substances realistic? *Soil Sci.* **166**, 752–769.

DiCosty, R. J., Weliky, D. P., Anderson, S. J., and Paul, E. A. (2003). ^{15}N-CPMAS nuclear magnetic resonance spectroscopy and biological stability of soil organic nitrogen in whole soil and particle-size fraction. *Org. Geochem.* **34**, 1635–1650.

Ericsson, I., and Lattimer, R. P. (1989). Pyrolysis nomenclature. *J. Anal. Appl. Pyrolysis* **14**, 219–221.

Fan, T. W. M., Lane, A. N., Shenker, M., Bartley, J. P., Crowley, D., and Higashi, R. M. (2001). Comprehensive chemical profiling of gramineous plant root exudates using high-resolution NMR and MS. *Phytochemistry* **57**, 209–221.

Fard, A. M., Turner, A. G., and Willett, G. D. (2003). High-resolution electrospray-ionization Fourier-transform ion cyclotron resonance and gas chromatography-mass spectrometry of Macadamia nut oil. *Aust. J. Chem.* **56**, 499–508.

Fievre, A., Solouki, T., Marshall, A., and Cooper, W. T. (1997). High-resolution Fourier transform ion cyclotron resonance mass spectrometry of humic and fulvic acids using laser desorption and electrospray ionization. *Energy Fuels* **11**, 554–560.

Fischer, H., Meyer, A., Fischer, K., and Kuzyakov, Y. (2007). Carbohydrate and amino acid composition of dissolved organic matter leached from soil. *Soil Biol. Biochem.* **39**, 2926–2935.

Franke, M., Jandl, G., and Leinweber, P. (2006). Characterization of organic compounds in aerobic biological treated municipal solid waste and re-circulated leachates. *Biodegradation* **17**, 473–485.

Franke, M., Jandl, G., and Leinweber, P. (2007). Analytical pyrolysis of re-circulated leachates: Towards an improved municipal waste treatment. *J. Anal. Appl. Pyrolysis* **79**, 16–23.

Frimmel, F. H., Abbt-Braun, G., Heumann, K. G., Hock, B., Lüdemann, H.-D., and Spiteller, M. (2002). *Refractory Organic Substances in the Environment*, Wiley-VCH, Heidelberg.

Fu, J., Klein, G. C., Smith, D. F., Kim, S., Rodgers, R. P., Hendrickson, C. L., and Marshall, A. G. (2006a). Comprehensive compositional analysis of hydrotreated and untreated nitrogen-concentrated fractions from syncrude oil by electron ionization, field desorption ionization, and electrospray ionization ultrahigh-resolution FT-ICR mass spectrometry. *Energy Fuels* **20**, 1235–1241.

Fu, J., Kim, S., Rodgers, R. P., Hendrickson, C. L., Marshall, A. G., and Qian, K. (2006b). Nonpolar compositional analysis of vacuum gas oil distillation fractions by electron ionization Fourier transform ion cyclotron resonance mass spectrometry. *Energy Fuels* **20**, 661–666.

Galletti, G. C., and Bocchini, P. (1995). Pyrolysis/gas chromatography/mass spectrometry of lignocellulose. *Rapid Comm. Mass Spectrom.* **9**, 815–826.

Golchin, A., Oades, J. M., Skjemstad, J. O., and Clarke, P. (1994). Study of free and occluded particulate organic matter in soils by solid state ^{13}C Cp/MAS NMR spectroscopy and scanning electron microscopy. *Aus. J. Soil Res.* **32**, 285–309.

González-Pérez, J. A., Arbelo, C. D., González-Vila, F. J., Rodríguez, A. R., Almendros, G., Armas, C. M., and Polvillo, O. (2007). Molecular features of organic matter in diagnostic horizons from andosols as seen by analytical pyrolysis. *J. Anal. Appl. Pyrolysis* **80**, 369–382.

Gransee, A., and Wittenmayer, L. (2000). Qualitative and quantitative analysis of water-soluble root exudates in relation to plant species and development. *J. Plant Nutr. Soil Sci.* **163**, 381–385.

Griep-Raming, J., and Linden, H. B. (2005). Fully automated liquid injection field desorption/ionization (LIFDI) mass spectrometry for high throughput screening of compounds with various polarity. 38. Annual DGMS Conference, Rostock, Germany.

Gross, J. H., Nieth, N., Linden, H. B., Blumbach, U., Richter, F. J., Tauchert, M. E., Tompers, R., and Hofmann, P. (2006). Liquid injection field desorption/ionization of reactive transition metal complexes. *Anal. Bioanal. Chem.* **386**, 52–58.

Guggenberger, G., Zech, W., and Schulten, H.-R. (1994). Formation and mobilization pathways of dissolved organic matter: Evidence from chemical structural studies of organic matter fractions in acid forest floor solutions. *Org. Geochem.* **21**, 51–66.

Guo, X., Fokkens, R. H., Peeters, H. J. W., Nibbering, N. M. M., and de Koster, C. G. (1999). Multiple cationization of polyethylene glycols in field desorption mass spectrometry: A new approach to extend the mass scale on sector mass spectrometers. *Rapid Commun. Mass Spectrom.* **13**, 2223–2226.

Hatcher, P. G., Dria, K. J., Kim, S., and Frazier, S. W. (2001). Modern analytical studies of humic substances. *Soil Sci.* **166**, 770–794.

Hempfling, R., Simmleit, N., and Schulten, H.-R. (1991). Characterization and chemodynamics of plant constituents during maturation, senescence and humus genesis in spruce ecosystems. *Biogeochemistry* **13**, 27–60.

Heuer, H., Kroppenstedt, R. M., Lottmann, J., and Smalla, K. (2002). Effects of T4-lysozyme release from transgenic potato roots on bacterial rhizosphere communities are negligible relative to natural factors. *Appl. Environ. Microb.* **68**, 1325–1335.

Hofstadler, S. A., Wahl, J. H., Bruce, J. E., and Smith, R. D. (1993). On-line capillary electrophoresis with Fourier transform ion cyclotron resonance mass spectrometry. *J. Am. Chem. Soc.* **115**, 6983–6984.

Hu, Q., Noll, R. J., Li, H., Makarov, A., Hardman, M., and Cooks, R. G. (2005). The Orbitrap: A new mass spectrometer. *J. Mass Spectrom.* **40**, 430–443.

Jandl, G., Schulten, H.-R., and Leinweber, P. (2002). Quantification of long-chain fatty acids in dissolved organic matter and soils. *J. Plant Nutr. Soil Sci.* **165**, 133–139.

Jandl, G., Leinweber, P., Schulten, H.-R., and Eusterhues, K. (2004). The concentrations of fatty acids in organo-mineral particle-size fractions of a Chernozem. *Eur. J. Soil Sci.* **55**, 459–469.

Jandl, G., Leinweber, P., Schulten, H.-R., and Ekschmitt, K. (2005). Contribution of primary organic matter to the fatty acid pool in agricultural soils. *Soil Biol. Biochem.* **37**, 1033–1041.

Jandl, G., Leinweber, P., and Schulten, H.-R. (2007). Origin and fate of soil lipids in a Phaeozem under rye and maize monoculture in Central Germany. *Biol. Fertil. Soils* **43**, 321–332.

Jardé, E., Gruau, G., Mansuy-Huault, L., Peu, P., and Martinez, J. (2007). Using sterols to detect pig slurry contribution to soil organic matter. *Water Air Soil Pollut.* **178**, 169–178.

Jokic, A., Schulten, H.-R., Cutler, J. N., Schnitzer, M., and Huang, P. M. (2004a). A significant abiotic pathway for the formation of unknown nitrogen in nature. *Geophys. Res. Let.* **31**, L05502.

Jokic, A., Cutler, J. N., Anderson, D. W., and Walley, F. (2004b). Detection of heterocyclic N compounds in whole soils using N-XANES spectroscopy. *Can. J. Soil Sci.* **84**, 291–293.

Jones, D. L., Hodge, A., and Kuzyakov, Y. (2004). Plant and mycorrhizal regulation of rhizodeposition. *New Phytologist* **163**, 459–480.

Kahle, P., Beuch, S., Boelke, B., Leinweber, P., and Schulten, H.-R. (2001). Cropping of Miscanthus in Central Europe: Biomass accumulation and influence on soil organic matter and nutrient cycling. *Eur. J. Agron.* **15**, 171–184.

Kaiser, K., Eusterhues, K., Rumpel, C., Guggenberger, G., and Kögel-Knabner, I. (2002). Stabilisation of organic matter in the subsoil of sandy, acidic forest soils. *J. Plant Nutr. Soil Sci.* **165**, 451–459.

Kalbitz, K., Geyer, W., and Geyer, S. (1999). Spectroscopic properties of dissolved humic substances—a reflection of land use history in a fen area. *Biogeochemistry* **47**, 219–238.

Kalbitz, K., Solinger, S., Park, J.-H., Michalzik, B., and Matzner, E. (2000). Controls on the dynamics of dissolved organic matter in soils: A review. *Soil Sci.* **165**, 277–304.

Kalbitz, K., Schwesig, D., Schmerwitz, J., Kaiser, K., Haumeier, L., Glaser, B., Ellerbrock, R., and Leinweber, P. (2003). Changes in properties of soil-derived dissolved organic matter induced by biodegradation. *Soil Biol. Biochem.* **35**, 1129–1142.

Kane-Maguire, L. A. P., Kanitz, R., and Sheil, M. M. (1995). Comparison of electrospray mass spectrometry with other soft ionization techniques for the characterization of cationic α-hydrocarbon organometallic complexes. *J. Organomet. Chem.* **486**, 243–248.

Kawahigashi, M., Kaiser, K., Kalbitz, K., Rodionov, A., and Guggenberger, G. (2004). Dissolved organic matter in small streams along a gradient from discontinuous to continuous permafrost. *Glob. Change Biol.* **10**, 1576–1586.

Knicker, H. (2000). Biogenic nitrogen in soils as revealed by solid-state ^{13}C and ^{15}N NMR spectroscopy. *J. Environ. Qual.* **29**, 715–723.

Kögel-Knabner, I. (2000). Analytical approaches for characterizing soil organic matter. *Org. Geochem.* **31**, 609–625.

Kögel-Knabner, I., Guggenberger, G., Kleber, M., Kandeler, E., Kalbitz, K., Scheu, S., Eusterhues, K., and Leinweber, P. (2008). Organo-mineral associations in temperate soils: Integrating biology, mineralogy and organic matter chemistry. *J. Plant Nutr. Soil Sci.* **171**, 61–82.

Kölbl, A., and Kögel-Knabner, I. (2004). Content and composition of free and occluded particulate organic matter in a differently textured arable Cambisol as revealed by solid-state ^{13}C NMR spectroscopy. *J. Plant Nutr. Soil Sci.* **167**, 45–53.

Kujawinski, E. B., Hatcher, P. G., and Freitas, M. A. (2002). High-resolution Fourier transform ion cyclotron resonance mass spectrometry of humic and fulvic acids: Improvements and comparisons. *Anal. Chem.* **74**, 413–419.

Kuzyakov, Y., and Siniakina, S. V. (2001). Siphon method of separating root-derived organic compounds from root respiration in non-sterile soil. *J. Plant Nutr. Soil Sci.* **164**, 511–517.

Kuzyakov, Y., Leinweber, P., Sapronov, D., and Eckhardt, K.-U. (2003). Quantitative assessment of root exudates in non-sterile soil by analytical pyrolysis. *J. Plant Nutr. Soil Sci.* **166**, 719–723.

Landgraf, D., Leinweber, P., and Makeschin, F. (2006). Cold and hot water extractable organic matter as indicators of litter decomposition in forest soils. *J. Plant Nutr. Soil Sci.* **169**, 76–82.

Lehmann, W. D., and Schulten, H.-R. (1976). Physikalische Methoden in der Chemie: Massenspektrometrie II. Chemische Ionisations-, Feldionisations- und Felddesorptions-Massenspektrometrie. *Chemie in unserer Zeit* **10**, 163–174.

Leinweber, P. (1995). Organische Substanzen in Partikelgrößenfraktionen: Zusammensetzung, Dynamik und Einfluß auf Bodeneigenschaften. In *Vechtaer Studien zur Angewandten Geographie und Regionalwissenschaft*, Band 15, Vechtaer Druckerei und Verlag, Vechta, p. 148.

Leinweber, P., and Schulten, H.-R. (1998). Advances in analytical pyrolysis of soil organic matter. *J. Anal. Appl. Pyrolysis* **47**, 165–189.

Leinweber, P., and Schulten, H.-R. (2000). Nonhydrolyzable forms of soil organic nitrogen: Extractability and composition. *J. Plant Nutr. Soil Sci.* **163**, 433–439.

Leinweber, P., Schulten, H.-R., and Körschens, M. (1994). Seasonal variations of soil organic matter in a long-term agricultural experiment. *Plant Soil* **160**, 225–235.

Leinweber, P., Schulten, H.-R., and Körschens, M. (1995). Hot water extracted organic matter: Chemical composition and temporal variations in a long-term field experiment. *Biol. Fertil. Soils* **20**, 17–23.

Leinweber, P., Jordan, E., and Schulten, H.-R. (1996). Molecular characterization of soil organic matter in pleistocene moraines from the Bolivian Andes. *Geoderma* **72**, 133–148.

Leinweber, P., Schulten, H.-R., and Jancke, H. (1999). New evidence for the molecular composition of soil organic matter in Vertisols. *Soil Sci.* **164**, 857–870.

Leinweber, P., Schulten, H.-R., Kalbitz, K., Meissner, R., and Jancke, H. (2001). Fulvic acid composition in degraded fenlands. *J. Plant Nutr. Soil Sci.* **164**, 371–379.

Leinweber, P., Kruse, J., Walley, F. L., Gillespie, A., Blyth, R. I. R., and Regier, T. (2007). N K-edge XANES—An overview of reference compounds used to identify "unknown" organic nitrogen in environmental samples. *J. Synchrotron Rad.* **14**, 500–511.

Leinweber, P., Eckhardt, K.-U., Fischer, H., and Kuzyakov, Y. (2008a). A new rapid micromethod for the molecular-chemical characterization of rhizodeposits by soft-ionization mass spectrometry. *Rapid Comm. Mass Spectom.* **22**, 1230–1234.

Leinweber, P., Jandl, G., Baum, C., Eckhardt, K.-U., and Kandeler, E. (2008b). Stability and composition of soil organic matter control respiration and soil enzyme activities. *Soil Biol. Biochem.* **40**, 1496–1505.

Leinweber, P., Walley, F. L., Kruse, J., Jandl, G., Eckhardt, K.-U., Blyth, R., and Regier, T. (2009). Cultivation affects soil organic nitrogen: Pyrolysis–mass spectrometry and N K-edge XANES evidence. *Soil Sci. Soc. Am. J.* **73**, 82–92.

Linden, H. B. (2001). Sensitivity improvement by 2–3 orders of magnitude and significantly raised sample throughput, even under inert conditions, by In-Source Liquid Injection FD. *Proceedings of the 49th Annual Conference on Mass Spectrometry and Allied Topics, Chicago*, MPA 024.

Linden, H. B. (2002). Quick soft analysis of sensitive samples under inert conditions by in-source liquid injection FD. In *Proceedings of the 50th Annual Conference on Mass Spectrometry and Allied Topics*, Orlando 2002, MPL 373.

Linden, H. B. (2004). Liquid injection field desorption ionization: A new tool for soft ionization of samples including air sensitive catalysts and non-polar hydrocarbons. *Eur. Mass Spectrom.* **10**, 459–468.

Llewelyn, J. M., Landing, W. M., Marshall, A. G., and Cooper, W. T. (2002). Electrospray ionization Fourier transform ion cyclotron resonance mass spectrometry of dissolved organic phosphorus species in a treatment wetland after selective isolation and concentration. *Anal. Chem.* **74**, 600–606.

Marschner, B., Brodowski, S., Dreves, A., Gleixner, G., Grootes, P. M., Hamer, U., Heim, A., Jandl, G., Ji, R., Kaiser, K., Kalbitz, K., Kramer, C., Leinweber, P., Rethemeyer, J., Schmidt, M. W. I., Schwark, L., and Wiesenberg, G. L. B. (2008). How relevant is recalcitrance for the stabilization of organic matter in soils? *J. Plant Nutr. Soil Sci.* **171**, 91–110.

Marshall, A. G., Hendrickson, C. L., and Jackson, G. S. (1998). Fourier transform ion cyclotron resonance mass spectrometry: A primer. *Mass Spectrum Rev.* **17**, 1–35.

Marshall, A. G. (2000). Milestones in Fourier transform ion cyclotron resonance mass spectrometry technique development. *Int. J. Mass Spectrom.* **200**, 331–356.

Marshall, A. G., and Rodgers, R. P. (2004). Petroleomics: The next grand challenge for chemical analysis. *Acc. Chem. Res.* **37**, 53–59.

Marshall, A. G. (2007). Petroleomics: Chemistry of the underworld. 8. European FTMS conference, Moscow, Russia.

Martens, D. A., and Loeffelmann, K. L. (2003). Soil amino acid composition quantified by acid hydrolysis and anion chromotography-pulsed amperometry. *J. Agric. Food Chem.* **51**, 6521–6529.

Martens, D. A., Jaynes, D. B., Colvin, T. S., Kaspar, T. C., and Karlen, D. L. (2006). Soil organic nitrogen enrichment following soybean in an Iowa corn–soybean rotation. *Soil Sci. Soc. Am. J.* **70**, 382–392.

Martius, C. (1992) Food provisioon storing by xylophagous termites in Amazonia (Isoptera: Nasutitermitidae). *Entomol. Gener.* **17**, 269–276.

Meissner, R., and Leinweber, P. (2004). PROWATER: Prevention of diffuse water pollution with phosphorus from degraded and re-wetted peat soils—Final Report of an European Research Project. ed. UFZ Halle-Leipzig.

Melnitchouck, A., Leinweber, P., Eckhardt, K.-U., and Beese, R. (2005). Qualitative differences between day- and nighttime rhizodeposition in maize *(Zea mays L.)* as investigated by pyrolysis-field ionization mass spectrometry. *Soil Biol. Biochem.* **37**, 155–162.

Melnitchouck, A., Leinweber, P., Broer, I., and Eckhardt, K.-U. (2006). Pyrolysis-field ionization mass spectrometry of rhizodeposits—A new approach in the risk assessment of genetically modified plants. *Environ. Biosafety Res.* **5**, 37–46.

Mikutta, R., Kleber, M., Torn, M., and Jahn, R. (2006). Stabilization of soil organic matter: Association with minerals or chemical recalcitrance. *Biogeochemistry* **77**, 25–56.

Monreal, C. M., Schnitzer, M., Schulten, H.-R., Campbell, C. A., and Anderson, D. W. (1995). Soil organic structures in macro- and microaggregates of a cultivated Brown Chernozem. *Soil Biol. Biochem.* **27**, 845–853.

Monreal, C. M., Schulten, H.-R., and Kodama, H. (1997). Age, turnover and molecular diversity of soil organic matter in aggregates of a Gleysol. *Can. J. Soil Sci.* **77**, 379–388.

Murray, K. E., and Schulten, H.-R. (1981). Field desorption mass spectrometry of lipids. I. The application of field desorption mass spectrometry to the investigation of natural waxes. *Chem. Phys. Lipids.* **29**, 11–21.

Nierop, K. G. J., Naafs, D. F. W., and Verstraten, J. M. (2003). Occurrence and distribution of ester-bound lipids in Dutch coastal dune soils along a pH gradient. *Org. Geochem.* **34**, 719–729.

Pedras, M. S. C., Jha, M., and Ahiahonu, P. W. K. (2003). The synthesis and biosynthesis of phytoalexins produced by cruciferous plants. *Curr. Org. Chem.* **7**, 1635–1647.

Qian, K., Rodgers, R. P., Hendrickson, C. L., Emmett, M. R., and Marshall, A. G. (2001). Reading chemical fine print: Resolution and identification of 3000 nitrogen-containing aromatic compounds from a single electrospray ionization Fourier transform ion cyclotron resonance mass spectrum of heavy petroleum crude oil. *Energy Fuels* **15**, 492–498.

Qian, K., Green, L. A., Olmstead, W. N., Schaub, T. M., Hendrickson, C. L., Rodgers, R. P., and Marshall, A. G. (2004). Petroleum characterization by field desorption mass spectrometry: A comparison of magnetic sector, time-of-flight and Fourier transform ion cyclotron resonance. In *Proceedings of the 52nd Annual Conference on Mass Spectrometry and Allied Topics, Nashville.*

Quénéa, K., Derenne, S., González-Vila, F. J., González-Pérez, J. A., Mariotti, A., Largeau, C. (2006a). Double-shot pyrolysis of the non-hydrolysable organic fraction isolated from a sandy forest soils (Landes de Gascogne, South-West France)—Comparison with classical Curie-point pyrolysis. *J. Anal. Appl. Pyrolysis* **76**, 271–279.

Quénéa, K., Largeau, C., Derenne, S., Spaccini, R., Bardoux, G., and Mariotti, A. (2006b). Molecular and isotopic study of lipids in particle size fractions of a sandy cultivated soil (Cestas cultivation sequence, southwest France): Sources, degradation, and comparison with Cestas forest soil. *Org. Geochem.* **37**, 20–44.

Reeves, J. B., and Francis, B. A. (1997). Pyrolysis–gas-chromatography–mass-spectrometry for the analysis of forages and by-products. *J. Anal. Appl. Pyrolysis* **40–41**, 243–265.

Rodgers, R. P., Schaub, T. M., and Marshall, A. G. (2005). Petroleomics: MS returns to its roots. *Anal. Chem. A-Pages* **77**, 20A–27A.

Rumpel, C., Seraphin, A., Goebel, M.-O., Wiesenberg, G., González-Vila, F. J., Bachmann, J., Schwark, L., Michaelis, W., Mariotti, A., and Kögel-Knabner, I. (2004a). Alkyl C in B and C horizons of an acid forest soil. *J. Plant Nutr. Soil Sci.* **167**, 685–692.

Rumpel, C., Eusterhues, K., and Kögel-Knabner, I. (2004b). Location and chemical composition of stabilized organic carbon in topsoil and subsoil horizons of two acid forest soils. *Soil Biol. Biochem.* **36**, 177–190.

Saiz-Jimenez, C. (1994). Analytical pyrolysis of humic substances: Pitfalls, limitations, and possible solutions. *Environ. Sci. Tech.* **28**, 1773–1780.

Saiz-Jimenez, C., Ortega-Calvo, J. J., and Hermosin, B. (1994). Conventional pyrolysis: A biased technique for providing structural informations on humic substances? *Naturwissenschaften* **81**, 28–29.

Schaub, T. M., Linden, H. B., Hendrickson, C. L., and Marshall, A. G. (2004). Continuous flow sample introduction for field desorption/ionization mass spectrometry. *Rapid Commun. Mass Spectrom.* **18**, 1641–1644.

Schaub, T. M., Rodgers, R. P., Marshall, A. G., Qian, K., Green, L. A., and Olmstead, W. N. (2005). Speciation of aromatic compounds in petroleum refinery streams by continuous flow field desorption ionization FTICR mass spectrometry. *Energy Fuels* **19**, 1566–1573.

Schlichting, A., and Leinweber, P. (2009). New evidence for the molecular-chemical diversity of potato plant rhizodeposits obtained by pyrolysis–field ionisation mass spectrometry. *Phytochem. Anal.* **20**, 1–13.

Schmidt, L., Warnstorff, K., Dörfel, H., Leinweber, P., Lange, H., and Merbach, W. (2000). The influence of fertilization and rotation on soil organic matter and plant yields in the long-term Eternal Rye trial in Halle (Saale), Germany. *J. Plant Nutr. Soil Sci.* **163**, 639–648.

Schmidt, M. W. I., and Kögel-Knabner, I. (2002). Organic matter in particle-size fractions from A and B horizons of a Haplic Alisol. *Eur. J. Soil Sci.* **53**, 383–391.

Schmitt-Kopplin, P. (2007). Analysis of the unresolved organic fraction in aerosols with ultra-high resolution ICR-FT/MS. 8. In *European FTMS Conference, Moscow, Russia.*

Schnitzer, M., and Schulten, H.-R. (1992). The analysis of soil organic matter by pyrolysis–field ionization mass spectrometry. *Soil Sci. Soc. Am. J.* **56**, 1811–1817.

Schnitzer, M., and Schulten, H.-R. (1995). The analysis of organic matter in soil extracts and whole soils by pyrolysis–mass spectrometry [Review]. *Advan. Agron.* **55**, 167–217.

Schnitzer, M., McArthur, D. F. E., Schulten, H.-R., Kozak, L. M., and Huang, P. M. (2005). Long-term cultivation effects on the quantity and quality of organic matter in selected Canadian prairie soils. *Geoderma* **130**, 141–156.

Schrader, S., and Klein, H.-W. (2004). Liquid chromatography/Fourier transform ion cyclotron resonance mass spectrometry (LC-FTICR MS): An early overview. *Anal. Bioanal. Chem.* **379**, 1013–1024.

Schulten, H.-R. (1987). Pyrolysis and soft ionization mass spectrometry of aquatic/terrestrial humic substances and soils. *J. Appl. Anal. Pyrolysis* **12**, 149–186.

Schulten, H.-R. (1996). Direct pyrolysis–mass spectrometry of soils: A novel tool in agriculture, ecology, forestry, and soil science. In *Mass Spectrometry of Soils*, Boutton, T. W., and Yamasaki, S., eds., Marcel Dekker, New York, pp. 373–436.

Schulten, H.-R. (2002). New approaches to the molecular structure and properties of soil organic matter: Humic-, xenobiotic-, biological-, and mineral-bonds. In *Soil Mineral–Organic Matter–Microorganism Interactions and Ecosystem Health. Developments in Soil Science*, Vol. **28A**, Violante, A., Huang, P. M., Bollag, J.-M., and Gianfreda, L., eds., Elsevier B. V., Amsterdam, pp. 351–381.

Schulten, H.-R., and Görtz, W. (1978). Curie-point pyrolysis and field ionization mass spectrometry of polysaccharides. *Anal. Chem.* **50**, 428–433.

Schulten, H.-R., and Leinweber, P. (1996). Characterization of humic and soil particles by analytical pyrolysis and computer modeling. *J. Appl. Anal. Pyrolysis* **38**, 1–53.

Schulten, H.-R., and Schnitzer, M. (1998). The chemistry of soil organic nitrogen: A review. *Biol. Fertil. Soils* **26**, 1–15.

Schulten, H.-R., and Leinweber, P. (1999). Thermal stability and composition of mineral-bound organic matter in density fractions of soil. *Eur. J. Soil Sci.* **50**, 237–248.

Schulten, H.-R., and Leinweber, P. (2000). New insights into organic–mineral particles: Composition, properties and models of molecular structure. *Biol. Fertil. Soils* **30**, 399–432.

Schulten, H.-R., and Schnitzer, M. (1991). Supercritical carbon dioxide extraction of long-chain aliphatics from two soils. *Soil Sci. Soc. Am. J.* **55**, 1603–1611.

Schulten, H.-R., and Schnitzer, M. (1992). A contribution to solving the puzzle of the chemical structure of humic substances: Pyrolysis–soft ionization mass spectrometry. *Sci. Total Environ.* **117/118**, 27–39.

Schulten, H.-R., and Schnitzer, M. (1993). Temperature-resolved in-source pyrolysis–soft ionization mass spectrometry of soil humic acids. *Org. Geochem.* **20**, 17–25.

Schulten, H.-R., Simmleit, N., and Rump, H. H. (1986). Soft ionization mass spectrometry of epicuticular waxes isolated from coniferous needles. *Chem. Phys. Lipids* **41**, 209–224.

Schulten, H.-R., Plage, B., and Schnitzer, M. (1991). A chemical structure for humic substances. *Naturwissenschaften* **78**, 311–312.

Schulten, H.-R., Sorge, C., and Schnitzer, M. (1995a) Structural studies on soil nitrogen by Curie-point pyrolysis–gas chromatography/mass spectrometry with nitrogen-selective detection. *Biol. Fertil. Soils.* **20**, 174–184.

Schulten, H.-R., Monreal, C. M., and Schnitzer, M. (1995b). Effects of long-term cultivation on the chemical structure of soil organic matter. *Naturwissenschaften* **82**, 42–44.

Schulten, H.-R., Leinweber, P., and Schnitzer, M. (1998). Analytical pyrolysis and computer modelling of humic and soil particles. In *IUPAC Environmental Analytical and Physical Chemistry Series, Vol. 4: Environmental Particles: Structure and Surface Reactions of Soil Particles*, Huang, P. M., Senesi, N., and Buffle, J., eds., John Wiley & Sons, Chichester, England, Chapter 8, pp. 281–324.

Schulten, H.-R., Leinweber, P., and Jandl, G. (2002). Analytical pyrolysis of humic substances and dissolved organic matter in water. In *Refractory Organic Substances in the Environment*, Frimmel, F. H., Abbt-Braun, G., Heumann, K. G., Hock, B., Lüdemann, H.-D., and Spiteller, M., eds., Wiley-VCH, Heidelberg, Section 2.8, pp. 163–187.

Sessitsch, A., Gyamfi, S., Tscherko, D., Gerzabek, M. H., and Kandeler, E. (2004). Activity of microorganisms in the rhizosphere of herbicide treated and untreated transgenic glufosinate-tolerant and wildtype oilseed rape grown in containment. *Plant Soil* **266**, 105–116.

Sihlbom, C., Davidsson, P., Emmett, M. R., Marshall, A. G., and Nilsson, C. L. (2004). Glycoproteomics of cerebrospinal fluid in neurodegenerative disease. *Int. J. Mass Spectrom.* **234**, 145–152.

Six, J., Conant, R. T., Paul, E. A., and Paustian, K. (2002). Stabilization mechanisms of soil organic matter: Implications for C-saturation of soils. *Plant Soil* **241**, 155–176.

Sjöberg, G., Knicker, H., Nilsson, S. I., and Berggren, D. (2004). Impact of long-term N fertilization on the structural composition of spruce litter and mor humus. *Soil Biol. Biochem.* **36**, 609–618.

Sleutel, S., Abdul Kader, M., Leinweber, P., D'Haene, K., and De Neve, S. (2007). Effect of tillage management on the composition of soil organic matter in different soil fractions as examined by Py-FIMS. *Soil Sci. Soc. Am. J.* **71**, 1620–1628.

Sleutel, S., Leinweber, P., Begum, S. A., Vandenbruwane, J., Abdul Kader, M., and De Neve, S. (2008). Composition of organic matter in sandy relict and cultivated heathlands as examined by pyrolysis–field ionization MS. *Biogeochemistry* **89**, 253–271.

Smernik, R. J., and Baldock, J. A. (2006). Does solid-state 15N NMR detect all soil organic nitrogen? *Biogeochemistry* **75**, 507–528.

Solouki, T., Emmett, M. R., Guan, S., and Marshall, A. G. (1997). Detection, number, and sequence location of sulfur-containing amino acids and disulfide bridges in peptides by ultrahigh-resolution MALDI FTICR mass spectrometry. *Anal. Chem.* **69**, 1163–1168.

Somei, M., and Yamada, F. (2003). Simple indole alkaloids and those with nonarranged monoterpenoid unit. *Nat. Prod. Rep.* **20**, 171–211.

Sorge, C., Schnitzer, M., and Schulten, H.-R. (1993). In-source pyrolysis–field ionization mass spectrometry and Curie-point pyrolysis–gas chromatography/mass spectrometry of amino acids in humic substances and soils. *Biol. Fertil. Soils* **16**, 100–110.

Stenson, A. C., Landing, W. M., Marshall, A. G., and Cooper, W. T. (2002). Ionization and fragmentation of humic substances in electrospray ionization Fourier transform ion cyclotron resonance mass spectrometry. *Anal. Chem.* **74**, 4397–4409.

Stenson, A. C., Marshall, A. G., and Cooper, W. T. (2003). Exact masses and chemical formulas of individual Suwannee River fulvic acids from ultrahigh resolution electrospray ionization Fourier transform ion cyclotron resonance mass spectra. *Anal. Chem.* **75**, 1275–1284.

Stevenson, F. J. (1994). *Humus Chemistry, Genesis, Composition, Reactions*, 2nd edition, John Wiley & Sons, New York.

Stevenson, F. J., and Cole, M. A. (1999). *Cycles of Soil. Carbon, Nitrogen, Phosphorus, Sulfur, Micronutrients*, 2nd edition, John Wiley & Sons, New York.

Taban, I. M., Altelaar, A. F. M., van der Burgt, Y. E. M., McDonnell, L. A., Heeren, R. M. A., Fuchser, J., and Baykut, G. (2007). Imaging of peptides in the rat brain using MALDI-FTICR mass spectrometry. *J. Am. Soc. Mass Spectrom.* **18**, 145–151.

Vairavamurthy, A., and Wang, S. (2002). Organic nitrogen in geomacromolecules: Insights on speciation and transformation with K-edge XANES spectroscopy. *Environ. Sci. Technol.* **36**, 3050–3056.

van Bergen, P. F., Bull, I. D., Poulton, P. R., and Evershed, R. P. (1997). Organic geochemical studies of soils from the Rothamsted Classical Experiments—I. Total lipid extracts, solvent insoluble residues and humic acids from Broadbalk Wilderness. *Org. Geochem.* **26**, 117–135.

Wiedow, D., Baum, C., and Leinweber, P. (2007). Inoculation with *Trichoderma saturnisporum* accelerates wheat straw decomposition on soil. *Arch. Agron. Soil Sci.* **53**, 1–12.

Wiesenberg, G. L. B., Schwarzbauer, J., Schmidt, M. W. I., and Schwark, L. (2004). Source and turnover of organic matter in agricultural soils derived from *n*-alkane/*n*-carboxylic acid compositions and C-isotope signatures. *Org. Geochem.* **35**, 1371–1393.

Wilcken, H., Sorge, C., and Schulten, H.-R. (1997). Molecular composition and chemometric differentiation and classification of soil organic matter in Podzol B-horizons. *Geoderma* **76**, 193–219.

Witt, M., and Fuchser, J. (2007). Crude oil analysis by Fourier transform ion cyclotron resonance mass spectrometry using different ionization techniques. 8. In *European FTMS Conference, Moscow, Russia*.

Zang, X., and Hatcher, P. G. (2002). A Py-GC-MS and NMR spectroscopy study of organic nitrogen in Mangrove Lake sediments. *Org. Geochem.* **33**, 201–211.

15

NUCLEAR MAGNETIC RESONANCE ANALYSIS OF NATURAL ORGANIC MATTER

A. J. SIMPSON AND M. J. SIMPSON
Department of Chemistry, University of Toronto, Toronto, Canada

Biophysico-Chemical Processes Involving Natural Nonliving Organic Matter in Environmental Systems, Edited by Nicola Senesi, Baoshan Xing, and Pan Ming Huang

15.1. INTRODUCTION

Nuclear magnetic resonance (NMR) spectroscopy is the single most powerful analytical technique for the analysis of organic structures. This has been recently realized in the biomedical arena, and in 2002 the Nobel Prize in chemistry was awarded for the applications of NMR in protein structure determination. To date, six recipients have received the Nobel Prize in the field of NMR spectroscopy. Today, hundreds of instruments are dedicated to the determination of biological structures globally. In environmental applications, NMR can not only provide information as to the basic chemical structures present in a mixture, but can also potentially provide information as to the self-associations of molecules (aggregation and flocculation processes), their mechanistic interactions with xenobiotics (transport of contaminants), and the direct connection between molecular scale processes (environments of individual nuclei) and macroscopic scale, via NMR micro-imaging. Furthermore, NMR is unique in its ability to provide comprehensive molecular and structural information *in vivo*, permitting rare opportunities to study the role of natural organic matter (NOM) indirectly or directly in the life cycle of animals, plants, and humans.

This chapter aims to outline the key role that NMR has played in understanding the structure, reactivity, and preservation of NOM in the environment. It will highlight the most informative experiments and techniques, as well as demonstrate some of the key results that have provided an unparalleled insight into complex systems such as soils and sediments. This chapter will conclude with a treatise of emerging techniques and their role in answering the key scientific questions both now and in the future. This chapter will not attempt to address the theory behind NMR mainly because there are numerous excellent resources already available in the literature. Wilson covers applications in geochemistry and soil chemistry thoroughly (Wilson, 1987), while Keeler covers NMR in general (Keeler, 2005). Preston specifically deals with the correct implementation of solid-state NMR spectroscopy for the study of soil organic matter (SOM) (Preston, 2001) while others (Simpson, 2001; Cardoza et al., 2004; Cook, 2004) have produced various treatise covering the practical and theoretical aspects of solution-state NMR studies to NOM.

15.1.1. What Can Be Studied by NMR?

15.1.1.1. Commonly Studied NMR Nuclei. NMR provides information as to the chemical environment of nuclei within a molecule. A simple one-dimensional (1D) spectrum has one frequency axis and analysis relies upon the relative frequency shifts between chemically inequivalent nuclei, combined with difference in the relative integrated intensities of the peaks. By far the two most common nuclei studied in NOM are protons (^{1}H) and carbons (^{13}C). Unfortunately, the most common carbon isotope ^{12}C (98.93%) is not directly observable by NMR while 99.99% of protons are detectable. Considering this and numerous other theoretical concepts, beyond the scope of this chapter, if the absolute receptivity of a proton is defined as 1, then the relative receptivity of a ^{13}C nuclei (at natural abundance) will be 1.7×10^{-4}, the important message being that ^{13}C is much less sensitive than ^{1}H NMR spectroscopy. Luckily as discussed later there are some techniques that can be used to increase the sensitivity of ^{13}C detected NMR spectroscopy. In addition to ^{1}H and ^{13}C NMR, other commonly studied nuclei include ^{31}P, ^{15}N, ^{29}Si, and ^{27}Al, which are all present in soils and sediments. Many other nuclei for example ^{19}F, ^{113}Cd and ^{199}Hg are also studied as they themselves are either environmentally important contaminants or are constituents of contaminants (i.e., ^{19}F is a very sensitive NMR nucleus and is common in many pharmaceuticals, pesticides and herbicides).

15.1.1.2. The Sample for NMR Spectroscopy. NMR is a very versatile technique that can be performed on solids or solutions. In the case of solid samples, whole soils or extracted organic matter can be studied. For traditional solution-state NMR, the sample must be dissolved in an appropriate solvent. Recently, a newly emerging technique, high-resolution magic angle spinning (HR-MAS) NMR, permits the analysis of semi-solid (swellable) materials as well as samples with multiple phases. This has important implications because components of soils and sediments can be studied *in situ* without any pretreatment, not even drying. Thus information as to the components in hydrophobic domains and those at the soil–water interface can be discerned. Complementary to this NMR, micro-imaging can be employed to study unaltered samples such as sediment cores, or the crucial root–soil interface. Methodologies and sample preparation can vary considerably, depending on the exact NMR analysis chosen. Thus the main four types of NMR spectroscopy—namely solid-state HR-MAS, solution-state, and micro-imaging—will be introduced separately in the next section.

15.2. THE BASIC NMR TECHNIQUES

15.2.1. Solid-State NMR

Solid-state NMR spectroscopy is arguably the most commonly applied NMR technique in the study of NOM structure because "whole" soil or sediment analysis can be performed without sample extraction. The ^{13}C nucleus is typically the focus of NOM solid-state NMR studies because strong ^{1}H–^{1}H dipolar interactions (which cannot be easily overcome experimentally) in the solid state result in extremely broad lines. However, because the natural abundance of the ^{13}C isotope is only ~1.13% of the total carbon present, observing ^{13}C signals directly is often difficult.

The development of cross-polarization (CP) improved the sensitivity of ^{13}C detection (Pines et al., 1972). The CP technique transfers the polarization from the abundant ^{1}H nuclei to the less abundant ^{13}C nucleus, which enhances the ^{13}C signal by up to a factor of ~4. Another advantage of the CP method is that the relaxation of the ^{1}H nuclei determines the repetition rate, which is much shorter than that of ^{13}C nuclei. In turn, this reduces the time required between pulses and the overall time of the experiment. Spinning samples at the "magic angle" of 54.7° (Andrew et al., 1958, 1959) reduces broadening from dipolar interactions and chemical shift anisotropy. Chemical shift anisotropy results from the various orientations of molecules in the solid phase; however, these orientations are partially averaged by spinning the sample at the magic angle. Low spinning speeds (for example: <10 kHz at a ^{1}H frequency of 300 MHz) may result in spinning side bands (resulting from residual chemical shift anisotropy and dipolar interactions) that fall within the spectral window. Spinning side bands can interfere with both quantitative and qualitative analysis. With some commercial solid-state NMR probes, it is possible to spin the sample such that the spinning side bands fall outside the window of interest; for NOM this corresponds to the chemical shift range of 0–250 ppm. Another option is to apply experiments that suppress the formation of sidebands such as total suppression of side bands (TOSS) (Dixon et al., 1982) and sideband elimination by temporary interruption of the chemical shift (SELTICS) (Hong and Harbison, 1993). High-power decoupling is also applied to eliminate the broadening due to heteronuclear interactions between ^{13}C and ^{1}H nuclei. Quantification of ^{13}C signals in the solid state has been an active area of debate amongst NOM researchers because indirect methods may underestimate ^{13}C nuclei that are not within the vicinity of ^{1}H nuclei (Mao et al., 2000, 2002a; Hatcher et al., 2001; Baldock and Smernik, 2002; Smernik et al., 2002; Keeler and Maciel, 2003; Simpson and Hatcher, 2004; Smernik, 2005; Keeler et al., 2006). Consequently, CP methods are considered to be semiquantitative and are suitable for relative comparisons amongst samples analyzed using identical NMR parameters and acquisition times (Preston, 1996, 2001; Simpson and Hatcher, 2004).

Solid-state NMR methods have been used to study NOM structure more than other available NMR methods. Consequently, the body of literature on solid-state NMR applications is extensive and, for brevity, will not be repeated here. However, readers are encouraged to refer to the following publications: a comprehensive text on solid-state NMR and applications to geochemistry and soil chemistry (Wilson, 1987); practical guides on the use of solid-state NMR to study soil organic matter (Preston, 1996, 2001; Simpson and Preston, 2007); information on using NMR as a tool in the study of soil organic matter structure (Kögel-Knabner, 1997; Hatcher et al., 2001; Smernik and Oades, 2003) and the applicability of different solid-state NMR methods to the study of NOM (Mao et al., 2002a). This chapter aims to highlight some important aspects of solid-state NMR applications in NOM but due to the vastness of the topic readers are encouraged to refer to other published works on this topic. Applications of solid-state NMR to various NOM samples including whole soils, sediments, and humin are covered in Sections 15.3.2 and 15.3.3.

15.2.1.1. Sample Preparation for Solid-State NMR. The amount of sample needed for solid-state NMR analysis depends on the size of the rotor used. Between 75 and 100 mg of sample is needed to fill a 4-mm rotor whereas 7-mm rotors require

~300–400 mg of sample. Samples will need to be packed tightly into the rotor to avoid problems with spinning; therefore, the samples should be ground to be less than 100 μm (if possible) in size before packing the sample. Certain samples such as plant litters can be very difficult to grind by hand, and the use of a commercial grinder/ball mill may be required. Samples should also be dried prior to analysis. A recent study by Smernik (2006) showed that ^{13}C sensitivity decreased with increasing water content but the observed chemical shifts of the two soils studied did not change (Smernik, 2006). Therefore, whenever possible, freeze-drying the sample is preferable. Furthermore, it is easier to pack the sample into the NMR rotor when the sample is dry.

Soils and sediments that are low in organic matter will challenge the detection limits of any NMR technique. Analysis of whole soils and sediments is often additionally confounded by the presence of paramagnetic species, such as iron, in the mineral component of the sample. In 2003 Keeler performed an excellent study examining the influence of paramagnetic species on the ^{13}C NMR of whole soils; any reader intending to attempt NMR studies on whole soils or sediments is strongly encouraged to refer to this article (Keeler and Maciel, 2003). To overcome the influence of paramagnetic species, many researchers have proposed the use of hydrofluoric acid (HF) to remove the mineral component through successive dissolution steps (Preston et al., 1989; Skjemstad et al., 1994b; Schmidt et al., 1997a; Gelinas et al., 2001; Smernik and Oades, 2002; Gonçalves et al., 2003). This procedure also increases the concentration of NOM in the sample, thus improving the signal-to-noise ratio (and resolution of the spectrum), and decreases the acquisition time of the analysis. Samples rich in carbonates should be pretreated with hydrochloric acid (HCl) to increase the relative concentration of organic carbon in the sample. Different concentrations of HF and HCl have been used in the past to reduce the iron content and increase the amount of organic carbon in a sample (Preston et al., 1989; Skjemstad et al., 1994a; Schmidt et al., 1997a; Gelinas et al., 2001; Gonçalves et al., 2003). Schmidt et al. (1997b) did not detect a change in the organic carbon distribution after samples were treated with 10% (v/v) HF. Gonçalves et al. (2003) reported carbon losses in B horizons when using 10% (v/v) HF but did not observe any change in the distribution of carbon functional groups. Gelinas et al. (2001) reported only small changes in the $(C/N)_a$ and $(H/C)_a$ ratios of recovered sedimentary NOM after treatment with HF and HCl and suggested that the molecular composition was not altered significantly. Despite concerns of NOM structure changes during demineralization, pretreating samples using HF and/or HCl is considered to be the most advantageous approach because of increased spectral resolution and decreases in interferences from paramagnetic minerals (Skjemstad et al., 1994a; Gelinas et al., 2001; Gonçalves et al., 2003; Keeler and Maciel, 2003).

15.2.2. Solution-State NMR

In solution-state NMR the tumbling of the individual molecules reduces anisotropic interactions resulting in high-resolution spectra without interference from side bands. Additionally, both proton (this is difficult in solids because strong dipolar interactions produce broad bands) and carbon experiments can be employed to provide quantitative information on NOM mixtures. Furthermore, a range of multidimensional experiments can provide unparalleled information as to the structure and reactivity of NOM. The main drawback of solution-state NMR spectroscopy is that the NOM sample must be soluble. For samples such as dissolved organic matter (DOM), solution-state NMR is ideal, because a great wealth of information can be obtained on samples close to their natural state (dissolved in an aqueous system). In the case of NOM from soils and sediments, extracts are often prepared and are amenable to study using solution-state techniques but must be prepared appropriately.

15.2.2.1. Sample Preparation for Solution-State NMR. Sample preparation is perhaps the single most important step for obtaining high-quality solution-state NMR spectra of NOM. Often, the use of a cation exchange resin such as Amberlite IR-120 or IR-1200+ can remove paramagnetic metals that adversely effect the quality of NMR spectra. The use of a cation exchange resin has the added benefit that NOM is converted to its H^+-exchanged form. Dimethyl sulfoxide (DMSO) is an excellent solvent for NMR (Simpson, 2001; and see later); however, DMSO is only an efficient solvent for cations, thus only effective for NOM in its H^+-exchanged form. For DOM it has been shown that the use of C_{18} solid-phase extraction disks can remove many of the inorganic impurities and results in increased spectral quality (Kim et al., 2003). Additionally, the pre-treatment of solid samples (i.e., soils sediments, etc.) with hydrofluoric acid (Schmidt et al., 1997b) before extraction (Simpson et al., 2006a) or the use of chelating agents can also be beneficial (Fan et al., 2000).

The solvent employed for NMR studies is of fundamental importance for solution-state studies. The most obvious choices for studies of NOM are DMSO-d_6 or D_2O (or D_2O/NaOD, depending on the solubility of the NOM). D_2O and D_2O/NaOD have been employed in numerous NOM studies (Buddrus et al., 1989; Larive et al., 1996; Dixon et al., 1999; Fan et al., 2000; Simpson et al., 2001c; Wang et al., 2003; Hertkorn et al., 2004; Peuravuori, 2005) while others utilize DMSO-d_6 (Wais et al., 1996; Simpson, 2001; Simpson et al., 2003a; Uyguner et al., 2004; Hertkorn et al., 2006; Kelleher and Simpson, 2006). D_2O and D_2O/NaOD have advantages in that the solution conformation adopted by the solute are likely to be more predictable than those adopted in DMSO. For example, when investigating contaminant interactions with humic material, it is important to reproduce conditions that are relevant to the natural soil or water environment (Hinedi et al., 1997; Nanny et al., 1997; Dixon et al., 1999). Under these circumstances, D_2O would be the most appropriate solvent. However, it is practically impossible to prepare a humic solution in D_2O without obtaining some water contamination because the solvent will exchange with protons in the sample or in the water in the atmosphere. Such water peaks lead to “t1 noise” (a problem whereby a streak runs vertically down the spectrum centered around a specific chemical shift, in this instance water). Note that t1 noise is in no way related to T_1 relaxation (which is often capitalized), and in many cases it is necessary to employ solvent suppression techniques in 2D experiments in order to reduce the water signal. Furthermore, signals from exchangeable functionalities such as amide and hydroxyl groups will disappear in the presence of deuterated aqueous solvents. This will result in the loss of important information from components such as sugars and amino acids. Therefore, should the purpose of the NMR study be to access compositional and structural information only, then the use of DMSO-d_6 may be more appropriate. Although it is important to point out that DMSO-d_6 does contribute a solvent peak in a ^{13}C spectra (not the case with aqueous solvents) and recovering a sample from DMSO can be challenging. Samples when properly dried (the authors recommend freeze-drying the sample and then further drying over P_2O_5 for an additional 48 hr) will dissolve in DMSO-d_6 to produce a solution with little or no water signal. In addition, exchangeable signals such as amide can be easily identified by the addition of a small volume of D_2O. Under these circumstances, amide signals that are fully visible in DMSO-d_6 exchange and disappear from the spectrum when deuterium is added, which aids in their identification (Kingery et al., 2000).

The amount of sample used is also an important consideration in preparing samples. At very high concentrations, the sample can become very viscous and spectral resolution will diminish. The authors have found that a concentration of ~50–100 mg ml^{-1} provides good sensitivity and resolution for 1D and 2D NMR experiments carried out using 5-mm tubes and field strengths of 300–900 MHz. However, should the amount of sample be restricted, useful information can still be obtained from 2D NMR experiments, as demonstrated by Haiber et al. (1999), who obtained HMQC (^{1}H–^{13}C one bond correlation) data using sample concentrations less than 30 mg ml^{-1}.

15.2.2.2. Key Experiments. Solution-state NMR is an extremely versatile tool for understanding the structure and reactivity of NOM. There are literally hundreds, arguably thousands of potential solution-state NMR experiments that can be applied to NOM, and limitless opportunities to devise novel experimental approaches. However, there are a handful of key experiments that provide the most amount of information in the shortest amount of time. Here we will refer to these as the "Top 10" which are listed in Table 15.1. Figure 15.1 highlights the main classes of 2D experiments and how information can be extracted from their data sets. Simpson (2001) provides an in-depth overview as to the optimized parameters used for acquiring and processing these types of data sets (Simpson, 2001). Applications of solution-state NMR to the study of various forms of NOM will be covered in Section 15.3.

15.2.3. HR-MAS NMR

Using conventional solid- or solution-state NMR techniques, it is not possible to assess which organic components are accessible to soil microbes, which are physically protected, and which are chemically recalcitrant. For solid-state NMR, samples must be dried such that they are no longer in their most biologically active state, and key interfaces (for example, the soil–water interface) or potential hydrophobic domains can no longer be studied. Furthermore, advanced multidimensional NMR of whole soils is challenging because researchers often have to detect the less sensitive ^{13}C nuclei because the more sensitive proton signal is significantly broadened due to strong ^{1}H–^{1}H dipolar interactions in solids. Conversely, solution-state NMR provides excellent spectral resolution, but information as to the conformation and arrangement of the organic matter in the whole soil would be lost during extraction.

Luckily due to recent developments in NMR probe design, there is now a third option, which can, to some extent, be used to address key questions as to the

TABLE 15.1. The "Top Ten" NMR Experiments for the Study of NOM in Solution

NMR Experiment	Acronym	Information	Interpretation	Comments
1D ^{1}H NMR	N/A	Quantitative overview as to the distribution of protons in a sample	In the case of NOM, often 2D NMR is central to the interpretation of the 1D NMR, which often contains considerable overlap.	^{1}H NMR, care must be taken to avoid water contamination in the sample. This is especially important for samples run in DMSO-d6, which is very hygroscopic. In aqueous solvents a "solvent suppression" technique is often required.
1D ^{13}C NMR	N/A	Can provide a quantitative overview as to the carbon distribution.	In the case of NOM, often 2D NMR is central to the interpretation of the 1D NMR, which often contains considerable overlap.	For quantitative data, the recycle delay (d1) should be $\geq 5 \times T1$ for the slowest relaxing component in the sample and inverse gated decoupling should be carried out, to prevent the transfer of ^{1}H NOE to the ^{13}C nuclei.
Diffusion edited ^{1}H NMR spectroscopy	DE	Only large molecules, macromolecular domains, and stable aggregates are observed.	Often simplifies the 1D ^{1}H as signals from small molecules (including solvents and water) are removed.	A very powerful approach that is underutilized. Diffusion editing is relates to molecular self-diffusion in solution. It should not be confused with spin diffusion in solids, which is completely unrelated.
Correlation spectroscopy	COSY	Connectivity information of protons on adjacent carbons.	Cross-peaks connect the chemical shifts of protons that are coupled; they are symmetrically placed around a central diagonal (see Figure 15.1).	Specifically a ^{1}H–H 2–3J coupling through bonds. Can provide useful information. However, TOCSY tends to be more sensitive than COSY for NOM. With a short mixing time (~30 ms), COSY-type correlations can be observed using TOCSY.
Total correlation spectroscopy	TOCSY	Long range proton–proton couplings.	As for COSY, in addition long range correlations form additional horizontal or vertical lines or cross-peaks (see Figure 15.1).	Very sensitive experiment for the study of DOM. The mixing time used in the experiment is critical; for most studies a mixing time of 60–80 ms is acceptable. For sample with extremely fast relaxation (for example a whole soil swollen in a solvent see later), short mixing times of ~30 ms are recommended.

Heteronuclear single quantum coherence or heteronuclear multiple quantum coherence	HSQC/ HMQC	^{1}H–^{13}C 1 bond correlation.	A cross-peaks represents in one dimension the units carbon chemical shift and in the other dimension the proton chemical shift (see Figure 15.1).	HSQC and HMQC essentially provide similar information. HSQC theoretically provides higher sensitivity than HSQC. However, HMQC is less prone to instrumental miscalibrations and, due to the shorter pulse sequence, is often more effective in samples with fast relaxation.
Heteronuclear multiple bond correlation	HMBC	^{1}H–^{13}C 2–4 bond correlations. Quaternary carbon are observed	Read information as vertical lines (see Figure 15.1).	Arguably the most powerful NMR experiment. However, it is difficult to perform due to the relatively long delays during which components can relax and the relatively low sensitivity of the technique.
Nuclear Overhauser effect spectroscopy	NOESY	Interactions through space/chemical exchange	As for COSY and TOCSY.	For small molecules, NOE is positive; for large molecules and chemical exchange, peaks appear negative.
Rotational nuclear Overhauser effect spectroscopy	ROESY	Interactions through space/chemical exchange.	As for COSY and TOCSY.	For all molecular weights, interactions through space are "negative" while those from chemical exchange are positive. ROESY and NOESY are best used as complementary techniques; when used together, peaks from exchange and those from spatial interactions can be discerned.
Heteronuclear single quantum coherence—total correlation spectroscopy	HSQC-TOCSY	Long-range ^{1}H–^{13}C correlations. Quaternary carbons not observed.	Similar to HMBC, although the ^{1}H–^{13}C 1 bond correlations are also present.	Can be a full 3D technique, or a 2D technique. Provides information on ^{1}H–^{13}C framework of a molecule. Provides information complementary to that from HMBC. However, quaternary carbon are not detected using HSQC-TOCSY.

Note there are hundreds of other experiments that can provide key information on NOM; some others are discussed later in this chapter. The "Top Ten" are among the easiest to apply and interpret, giving the greatest amount of information in the shortest amount of time.

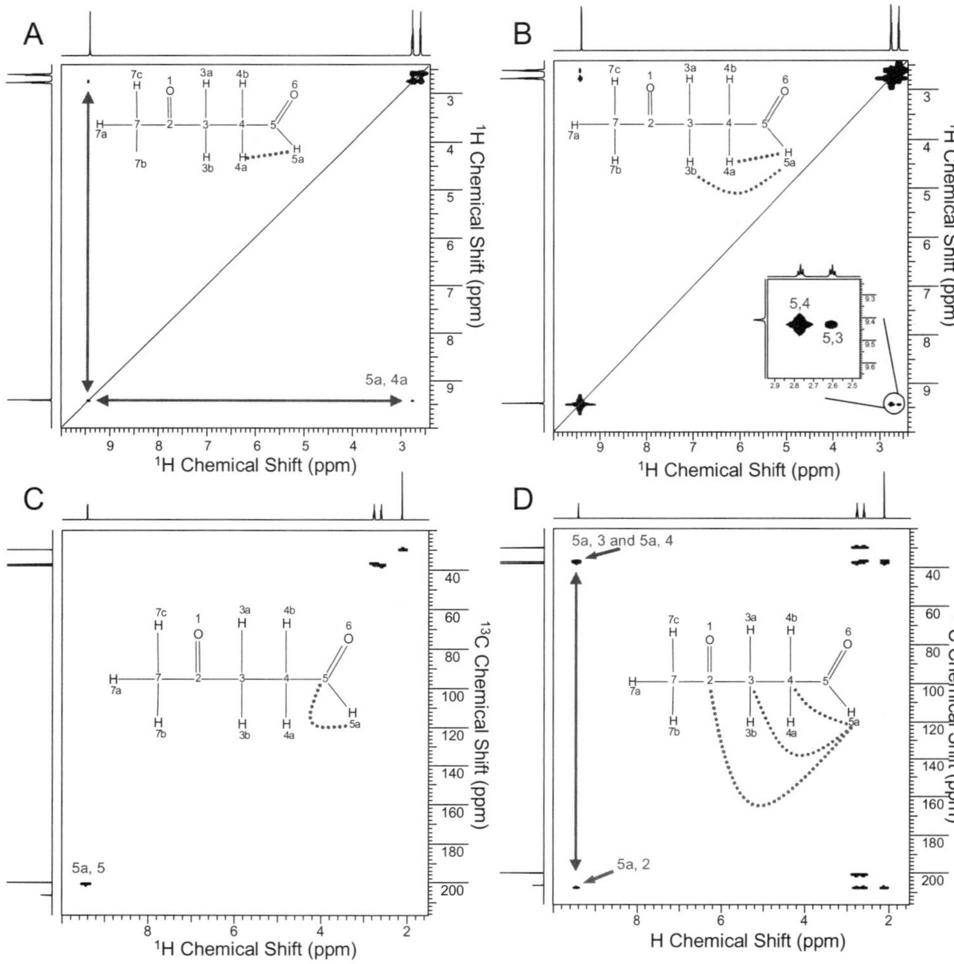

Figure 15.1. (**A**) ^{1}H–^{1}H COSY, (**B**) ^{1}H–^{1}H TOCSY, (**C**) ^{1}H–^{13}C HSQC or HMQC, (**D**) ^{1}H–^{13}C HMBC, for 4-oxopentanal. For clarity, only key assignments have been given as an example. Note that the double-ended arrows indicate how to "interpret" the spectra. In the case of COSY and TOCSY the information is represented as cross-peaks that are symmetrically oriented with respect to the central diagonal. In the single-bond correlation (HSQC/HMQC) a cross-peak represents in one dimension the carbon chemical shift and in the other dimension the proton chemical shift. Note there is no diagonal in heteronuclear NMR experiments. In the HMBC, lines are drawn vertically to connect the cross-peaks. In HMBC 2–4 bonds, ^{1}H–^{13}C correlations are often observed. Note that the 4-bond correlation is less common in NMR but has been included here as an example, and 1-bond correlation is commonly filtered from the HMBC experiment to improve detection limits for the weaker 2–4 bond correlations.

arrangement and conformation of organic matter *in situ*. The recently developed technique of HR-MAS NMR spectroscopy allows the analysis of semi-solid and swellable materials with resolution close to that afforded by true solution-state NMR (Keifer et al., 1996). Recently, HR-MAS has been applied in structural studies of plant polymers (Stark et al., 2000; Deshmukh et al., 2003; Simpson et al., 2003a,b) and the degradation of plant materials in the soil environment (Kelleher et al., 2006). Investigations have also assessed the organic components at the whole-soil aqueous–solid interface (Simpson et al., 2001b) as well at the organo-mineral interface (Simpson et al., 2006b). In HR-MAS NMR, deuterated solvents are used to swell the sample. Dipolar interactions that are predominant in solid materials are decreased by the addition of solvent and are averaged by magic angle spinning, allowing solution-state experiments to be carried out on samples that are not soluble. As dipolar interactions are reduced by the addition of solvent and magic angle spinning, ^1H line widths are reduced to close to that observed in solution. Thus by using HR-MAS NMR it is possible to study a sample "as is" and using the full suite of powerful solution-state NMR experiments that are available. At present, many of the NMR approaches used to study NOM by HR-MAS are similar or identical to those used in solution-state NMR. The "Top 10" solution-based NMR techniques (see Table 15.1 and Figure 15.1) will also be applicable in HR-MAS NMR spectroscopy. Applications of HR-MAS NMR to the study of organic matter will be considered in Section 15.3.

15.2.4. NMR Micro-imaging

NMR micro-imaging of complex porous systems such as soils and sediments is highly beneficial for water and contaminant distribution and flow studies (Simpson et al., 2007b) but can also be applied to study tissues, bioreactors, biofilms, plants, wood, and food products (Van As and Van Dusschoten, 1997; Humbert, 2001). NMR micro-imaging, also referred to as magnetic resonance imaging (MRI), has great potential as a tool to study environmental processes *in situ*, in real time and also results in a spatially resolved image of the nuclei of interest. NMR micro-imaging measures a NMR signal at different locations within a sample and then reconstructs this information to create an image. This technology is nondestructive and noninvasive and produces images of both dynamic and static phenomena in porous media (Amin et al., 1996; Van As and Van Dusschoten, 1997; Lens and Hemminga, 1998; Humbert, 2001; Nestle et al., 2002). MRI studies mostly focus on abundant nuclei such as ^1H because water is prolific in most natural systems. ^1H methods are also routinely used in medical studies; however, medical applications utilize a horizontal magnet with a bore large enough to accommodate a human patient whereas NMR micro-imaging is performed using vertical magnets, with bores of only a few centimeters in diameter. Although this limits the size of the sample, higher-resolution images can be obtained and with greater sensitivity for smaller specimens, such as small animals, soil, and sediment cores. The use of NMR micro-imaging in understanding the role of NOM in environmental processes will be discussed further in Sections 15.3 and 15.4.

15.3. STRUCTURAL STUDIES OF NOM

It is important to note that the application of 1D and 2D NMR to the study of NOM has been extensive, with hundreds of published works utilizing the different available approaches. Given the broad scope of this chapter it is not possible to summarize all the findings here. This section simply aims to provide the reader with an overview as to the most powerful NMR approaches that are available and highlight where possible the major findings/breakthroughs that have been made in the field in general.

15.3.1. Extractable NOM from Soils

15.3.1.1. A Brief History. NMR has been used to study humic substances for over 40 years. To the best of our knowledge, the first application was published in *Nature* in 1963 (Barton and Schnitzer, 1963). Since then, NMR studies of NOM has progressed rapidly, with the first applications of multidimensional solution-state NMR to study NOM occurring in 1989 (Buddrus et al., 1989). In the last 10 years the development of NMR experiments to study complex systems has expanded exponentially, mainly as a result of advances in the biomedical arena. Similarly, improvements in NMR hardware, stability, and sensitivity have kept pace in both solid-state and solution-state NMR. Thus, it is not surprising that in the last decade the application of multidimensional NMR has grown exponentially (Simpson et al., 1997, 2001a,c, 2002a, 2003a; Haiber et al., 1999, 2001a,b; Fan et al., 2000; Kingery et al., 2000; Knicker, 2000a; Simpson, 2001; Hertkorn et al., 2002b, 2006; Cook et al., 2003; Kaiser et al., 2003; Kim et al., 2003; Cardoza et al., 2004; Kelleher and Simpson, 2006; Mao and Schmidt-Rohr, 2006).

In 1989, the first applications of multidimensional NMR were applied to humic substances (Buddrus et al., 1989). This study involved the application of ^{13}C detected J-resolved (J-Res) spectroscopy. The study was successful in that it showed multidimensional NMR was applicable to the study of humic substances. However, in 1989 the lack of various modern experiments and the corresponding hardware (mainly probes fitted with pulse field gradients) made applying NMR to humic materials very challenging. In 1997, Simpson et al. demonstrated that the more sensitive inverse-detected NMR experiments were applicable to NOM (Simpson et al., 1997). In this manuscript COSY, TOCSY and HMQC were applied (Simpson et al., 1997) and in 1998 Schmitt-Kopplin et al. used HMQC and COSY to monitor the change in dissolved humic acid during photo-degradation (Schmitt-Kopplin et al., 1998). In 1999, Haiber et al. demonstrated that HMQC could be used to study size fractionated humic structures (Haiber et al., 1999) and Morris et al. showed diffusion ordered spectroscopy (DOSY) to be a useful tool (Morris et al., 1999). Diffusion order spectroscopy will be discussed later in Section 15.4. In 2000, studies showed that humic acid had a considerable peptide content (Fan et al., 2000), and that 2D NMR was applicable to the study of the IHSS standard soil humic acid (Kingery et al., 2000). In the following year, Simpson et al. applied a full suite of NMR experiments, including COSY, TOCSY, HSQC, HMQC-TOCSY, HMBC, and NOESY, to the study of a charge-fractionated fulvic acid and attempted detailed interpretation of the data (Simpson et al., 2001a). Furthermore, the same group offering a practical guide and review (Simpson, 2001) demonstrated the applicability of HR-MAS NMR (Simpson et al., 2001b) and the application of DOSY to separate NOM components (Simpson et al., 2001c). In the same year, Haiber et al. demonstrated the quantification of carbohydrate components in humic acid using 2D NMR (Haiber et al., 2001a), and they also reported studies of the IHSS Suwannee River reference humic and fulvic acids (Haiber et al., 2001b). During 2002, Simpson et al. summarized a range of NMR data to explain the major structural components in soil NOM

(Simpson et al., 2002a), offered an in-depth study of diffusion based techniques (Simpson, 2002), and compared some basic 2D experiments (Simpson et al., 2002b). In the same year, Hertkorn utilized 2D NMR to understand the utilization of aquatic humics by microorganisms (Hertkorn et al., 2002a) and studied peat humic and fulvic acids (Hertkorn et al., 2002b). The year 2003 saw the introduction of 3D NMR to study NOM (Simpson et al., 2003a) and studies of a Laurentian fulvic acid (Cook et al., 2003). Online hyphenated NMR was introduced in the form of HPLC-NMR and HPLC-SPE-NMR in 2004 (Simpson et al., 2004b), as well as the use of 2D spectral simulations (Simpson et al., 2004a), and a review of the field (Cook, 2004). In 2005 and 2006 some key studies included, understanding the preservation of peptides in humic materials (Hsu and Hatcher, 2005), a look at the biopolymeric contribution to extractable soil organic matter (Kelleher and Simpson, 2006), studying the domains in solid humic substances (Mao and Schmidt-Rohr, 2006), and following the conversion of labeled plant material through humification (Kelleher et al., 2006).

15.3.1.2. Understanding Solution-State 1D and 2D Data of NOM. 1D NMR provides an overview as to a samples composition, while multidimensional NMR helps assign the broader resonances in the 1D spectrum, ultimately providing spectral dispersion and connectivity information from which more detailed structural information can be discerned. As an example, Figure 15.2 shows the 1D NMR spectrum of a typical NOM soil extract. Figure 15.2A shows generic assignments that could be made on the basis of the 1D NMR alone. 1D NMR data can provide a useful overview of some components in the sample; but when 1D and 2D NMR are combined, a detailed overview of the sample is obtained. Figure 15.2B depicts the more detailed assignments that can be defined when 1D and 2D spectra are combined. It should be noted that only the major components are highlighted in Figure 15.2B; structures from minor components can also be identified in 2D NMR but cannot be labeled on the 1D spectrum because they are masked by overlap from more abundant species.

Arguably, the most important experiment for the study of humic materials is the HMQC or HSQC experiment. The ^{1}H–^{13}C HSQC or HMQC experiment detects the H–C bonds in an organic structure. Note that HMQC results are very similar to those from HSQC and are considered together for the purpose of this chapter. A cross peak in an HSQC spectrum represents the chemical shift of both carbon and proton atoms in a C–H unit. When considered together, the cross peaks form a specific pattern that can be thought of as the "molecular fingerprint" of a specific structure or class of structures. In Figure 15.3 the ^{1}H–^{13}C HSQC of the International Humic Substances Society (IHSS) peat humic acid is compared to that of categories of biopolymers (protein, carbohydrate, cuticular (aliphatic), and lignin) that are common in soil. The detailed interpretation of the 2D is well beyond the scope of this chapter, although the color overlays permit nonspecialists to visualize how detailed information can be easily extracted from multidimensional NMR data sets when appropriate reference materials are studied. Readers are encouraged to refer to the original manuscript for full details (Kelleher et al., 2006). After HSQC (and HMQC), TOCSY is probably the most useful form of 2D NMR for NOM studies. Figure 15.4 compares the TOCSY spectra of the same biopolymers to that of the IHSS peat humic acid standard.

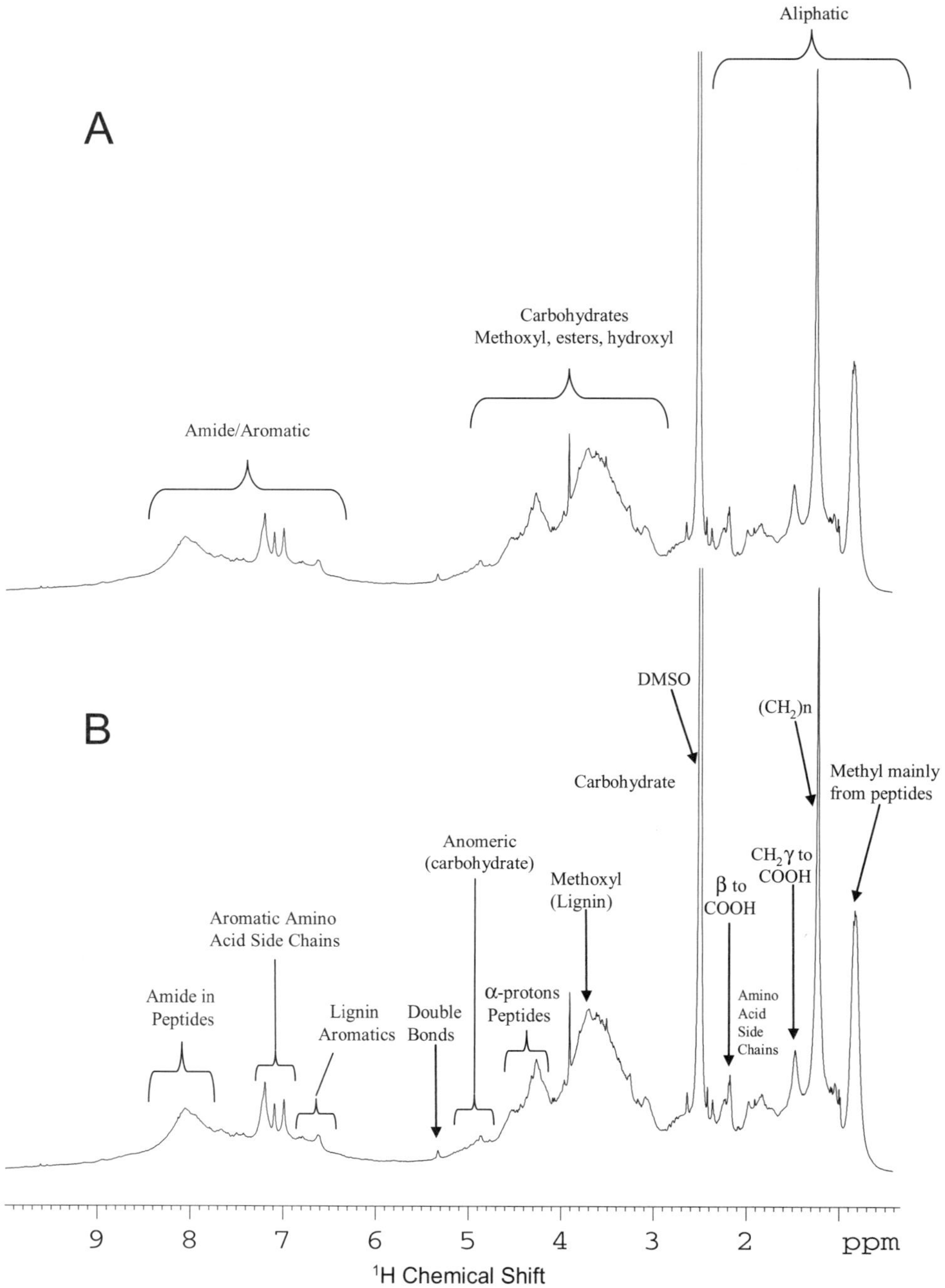

Figure 15.2. An ^{1}H solution-state spectrum of a total alkaline extract from a grassland soil. (**A**) Generic assignments that can be assigned from the 1D data alone. (**B**) Detailed assignments that can be made when information from numerous 1D and 2D NMR techniques are combined. Note that only the major components are highlighted, structures from minor components can be identified in 2D NMR but cannot be labeled on the 1D spectrum due to overlap.

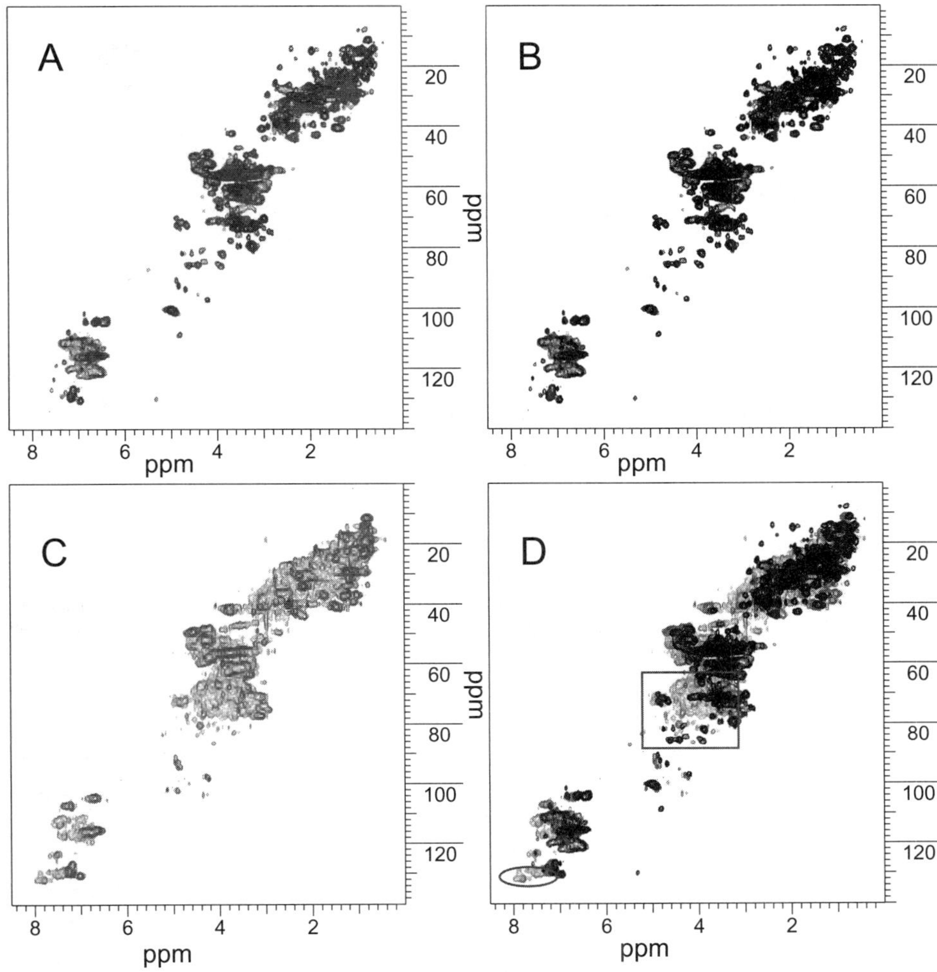

Figure 15.3. Overlaid HSQC spectra of biopolymers on IHSS peat. (**A**) Biopolymers; lignin (gray), amylopectin (red), albumin (blue) and cuticle (green) overlaid on each other. (**B**) All biopolymers are illustrated in black. (**C**) IHSS humic acid extract from peat. (**D**) Biopolymers (black) overlaid on IHSS peat (green). The highlighted areas in 2D are those not well represented by biopolymers in the HA, namely complex carbohydrates and *p*-hydroxybenzoates from lignin [see Kelleher and Simpson (2006) for more details]. See color insert. Reprinted from Kelleher, B. P., and Simpson, A. J. (2006). Humic substances in soils: Are they really chemically distinct? *Environ. Sci. Technol.* **40**, 4605–4611, with the permission of the American Chemical Society.

Like HSQC, TOCSY produces a molecular fingerprint of a molecule or mixture; but in TOCSY, peaks arise from the interactions of protons over numerous bonds. In simple terms, HSQC describes the H–C units in a mixture and TOCSY describes how these units are linked together. Considered together, the two experiments can be used to describe the complete H–C framework of any organic structure. This concept can be combined to create the HSQC (or HMQC)-TOCSY experiment, and this can be either a 2D or full 3D version, both of which contain information as to long-range ^{1}H–^{13}C couplings (Simpson et al., 2003a) (Figure 15.5). The full 3D

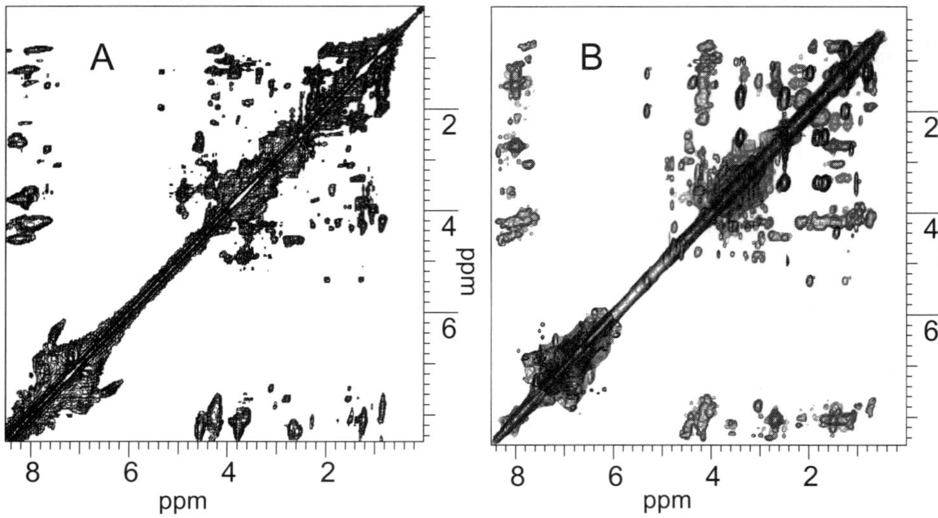

Figure 15.4. TOCSY spectra of the IHSS peat humic acid (**A**) and biopolymers (**B**). The four biopolymers are lignin (gray), albumin (blue), amylopectin (red), and cuticle (green). See color insert. Reprinted from Kelleher, B. P., and Simpson, A. J. (2006). Humic substances in soils: Are they really chemically distinct? *Environ. Sci. Technol.* **40**, 4605–4611, with the permission of the American Chemical Society.

version spreads the chemical shift information into three dimensions providing excellent spectral dispersion even for very complex mixtures such as NOM. However, it should be noted that the acquisition time required to collect a full 3D dataset is extensive sometimes requiring a week or more of NMR time. Thus while 3D NMR is extremely powerful, it may never become a routine tool for the study of NOM, mainly because in most cases similar information can be gained in a fraction of the time if a series of carefully planned 2D experiments are performed (Simpson, 2001).

Multidimensional solution-state NMR has played a key role thus far in understanding the main components in extractable humic substances. Combined, information from the above-mentioned studies indicates that a large proportion of NOM can be assigned to intact or degrading plant or microbial structures. The major structures include cuticular components (species derived from the cuticular coating of plant leaves) peptide/proteins, lignin, and carbohydrate (Kelleher and Simpson, 2006). This is consistent with other NMR studies that identify these components (Almendros et al., 1991a; Kögel-Knabner et al., 1992; Zech et al., 1992; Knicker and Ludemann, 1995; Rumpel et al., 1998; Fan et al., 2000; Haiber et al., 2001a; Chefetz et al., 2002; Simpson, 2002; Simpson et al., 2002a) in abundance in soil organic matter. This is logical considering that soluble NOM from soils is operationally defined and at any one time microbial and plant residues will be the major contributors to soil biomass (Kögel-Knabner, 2002).

15.3.2. Whole Soils and Sediments

Structural studies of whole soils and sediments by NMR have been limited to solid-state and HR-MAS NMR techniques due to the limited solubility of whole samples. Although solid-state spectra are typically less resolved than those acquired using

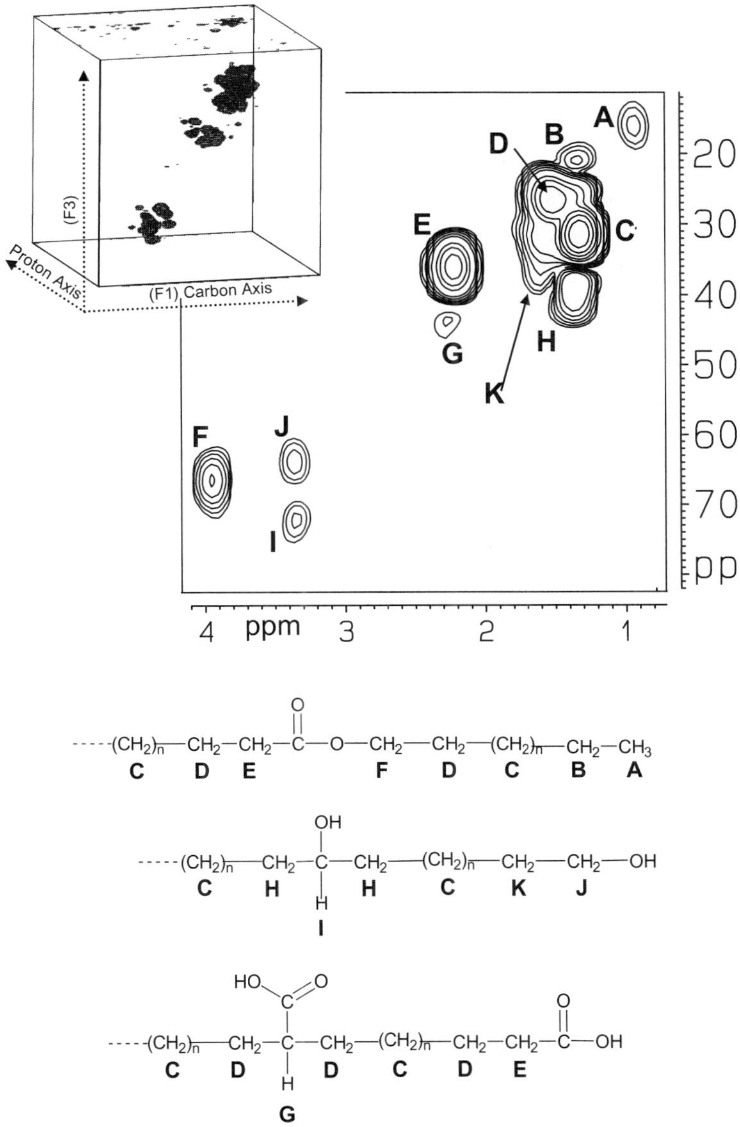

Figure 15.5. Two-dimensional spectrum produced from an F1–F2 slice through the 3-D HMQC-TOCSY spectrum of a pine forest soil fulvic acid at 1.3 ppm on the F3 (proton) axis (Figure 15.2). Labels on cross-peaks correspond to the C–H structures in the aliphatic structures shown. The full 3D cube is superimposed onto the example slice. Reprinted from Simpson, A. J., Kingery, W. L., and Hatcher, P. G. (2003a). The identification of plant derived structures in humic materials using three-dimensional NMR spectroscopy. *Environ. Sci. Technol.* **37**, 337–342, with permission from the American Chemical Society.

liquid-state techniques, solid-state methods are advantageous because the spectra are representative of the whole soil or sediment and little sample pretreatment is necessary. Furthermore, solid-state NMR is nondestructive and therefore, after analysis, samples can be used for other types of structural investigations. The earliest application demonstrated the applicability of the technique (Hatcher et al., 1980;

Barron and Wilson, 1981; Wilson, 1981; Wilson et al., 1981a–c; Preston and Ripmeester, 1982). More recently, solid-state NMR techniques have been used extensively to gain insight into the structural components of NOM in whole soils and sediments (Wilson, 1987; Knicker et al., 1993; Kögel-Knabner and Ziegler, 1993; Preston, 1996; Kögel-Knabner, 1997, 2000; Preston et al., 1997; Mahieu et al., 1999; Nierop et al., 1999; Knicker, 2000a, 2004; Spaccini et al., 2000; Hatcher et al., 2001; Keeler and Maciel, 2003; Smernik and Baldock, 2005; Xing et al., 2005; Dickens et al., 2006; Keeler et al., 2006). Solid-state NMR studies mostly examine the distribution of carbon (^{13}C nuclei) within a whole soil or sediment; however, there have also been applications of ^{15}N solid-state NMR techniques (Knicker et al., 1993; Guggenberger et al., 1994; Knicker, 2000a, 2004; Smernik and Baldock, 2005). NMR has been used to study the structure of soil NOM for close to 40 years (Preston, 1996) and has improved the understanding of the chemical structure and reactivity of NOM in soils and sediments. Several bodies of literature have been written regarding the use of solid-sate NMR techniques in the study of NOM (Wilson, 1981, 1987; Preston, 1996, 2001; Kögel-Knabner, 1997, 2000; Conte et al., 2004) as well as reviews on practical aspects of solid-state NMR (Bryce et al., 2001; Simpson and Preston, 2007). Thus, this section aims to highlight some of the important aspects of the techniques, and readers are referred to the previously mentioned publications for further information.

Solid-state ^{13}C NMR studies have been used extensively in the structural investigations of NOM in soils and sediments (see references above and references therein). The application of solid-state ^{13}C NMR has undoubtedly advanced the understanding of NOM structure and environmental reactivity in the last two decades. The most common means by which solid-state techniques are applied involves the structural assignments and quantification of the major functional groups in NOM. Figure 15.6, from Keeler et al. (2006), nicely depicts the structural information that can be gained from using one-dimensional solid-state ^{13}C NMR methods allowing one to differentiate between aliphatic, substituted aliphatic, aromatic, phenolic, carboxylic, and carbonyl carbon forms. The ^{13}C chemical shift ranges for structures often found in NOM are listed in Table 15.2. These chemical shifts are useful for studying soil and sediment NOM processes such as plant degradation in soil. For example, cellulose and lignin degradation in soil are often the focus of studies that monitor the rate of plant residue turnover in soil. In these types of experiments, specific resonances of plant biopolymers can be monitored. Lignin moieties can be distinguished by the presence of characteristic methoxyl (56 ppm) and phenolic (~150 ppm) resonances, whereas carbohydrate components have a number of chemical shifts (Figure 15.6; Table 15.2); 62–66 ppm for C-6 carbons, 65–85 ppm for ring carbons, and 105 ppm for the anomeric carbon (Preston et al., 1997; Keeler et al., 2006). Polymethylene domains have also been identified in whole soils and sediments (Dria et al., 2002; Smernik, 2005); however, the source of methylene carbon in soil and sedimentary environments differs. In terrestrial environments, it is believed that plant waxes from cutin and suberin (Nierop, 1998; Derenne and Largeau, 2001; Kögel-Knabner, 2002; Buurman et al., 2007) are preserved through mineral binding to soil particles whereas methylene carbon found in sedimentary environments is most likely derived from the preservation of recalcitrant algal material referred to as algaenan (Lichtfouse et al., 1994; Derenne et al., 1997).

The issue of quantitative reliability of solid-state ^{13}C NMR methods has been investigated by a number of researchers (Mao et al., 2000, 2002a; Hatcher et al.,

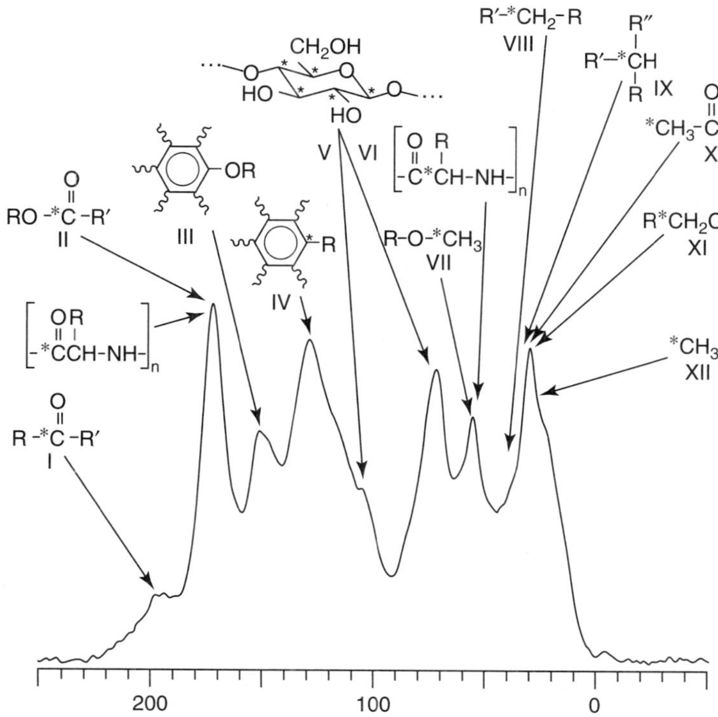

Figure 15.6. Example of ^{13}C chemical shift assignments of structural groups found in NOM. The asterisk marks the C atom which is found in the corresponding chemical shift region. Reprinted from Keeler, C., Kelly, E. F., and Maciel, G. E. (2006). Chemical-structural information from solid-state C-13 NMR studies of a suite of humic materials from a lower montane forest soil, Colorado, USA. *Geoderma* **130**, 124–140, with permission from Elsevier.

TABLE 15.2. ^{13}C Chemical Shift Ranges Used to Identify NOM Constituents in Soils and Sediments from Solid-State NMR Spectra

^{13}C Chemical Shift Range (ppm)	Types of Structures Present
0–45 ppm	Unsubstituted alkyl carbon: includes straight-chain methylene carbon (30–34 ppm) and terminal methyl groups (15 ppm). Branched methylene carbon is found more downfield (35–45 ppm)
45–65 ppm	Substituted alkyl carbon such as that found in amines (45–46 ppm) and methoxyl groups (56 ppm)
65–95 ppm	Oxygen-substituted carbon, ring carbons in carbohydrates, and carbons in ethers
95–110 ppm	Di-oxygen-substituted aliphatic carbon and anomeric carbon in carbohydrates (105 ppm)
110–145 ppm	Aromatic carbon
145–160 ppm	Phenolic carbon
160–190 ppm	Carboxylic, amide and ester carbon
190–220 ppm	Carbonyl carbon

Source: Adapted from Malcolm (1989) and Baldock and Skjemstad (2000).

2001; Baldock and Smernik, 2002; Smernik et al., 2002; Keeler and Maciel, 2003; Simpson and Hatcher, 2004; Smernik, 2005; Keeler et al., 2006). The emphasis of this research has been on comparisons between single pulse excitation methods that directly polarize ^{13}C nuclei (referred to as direct polarization, DP or Bloch decay) versus cross-polarization (referred to as CP) methods that polarize ^{13}C nuclei via nearby ^{1}H nuclei. It should be noted that both ^{13}C DP and CP are used with magic angle spinning (MAS). Furthermore, depending on the field strength, the rate of MAS may not be sufficient and spinning side bands may result due to chemical shift anisotropy. Consequently, some researchers use TOSS in addition to CP to overcome quantification interferences due to the presence of spinning side bands. At higher MAS spinning speeds, the CP efficiency can be compromised and can be overcome by applying the ramped amplitude (RAMP) CP pulse program (Cook and Langford, 1998, 1999), which matches the Hartman–Hahn condition necessary for efficient CP. Studies comparing DP and CP methods have collectively demonstrated that DP methods are more quantitative and that CP methods may underestimate carbons that are not in the vicinity of protons (Conte et al., 1997, 2004; Mao et al., 2000; Baldock and Smernik, 2002; Smernik et al., 2002; Keeler and Maciel, 2003; Simpson and Hatcher, 2004). For example, Mao et al. compared DP and CP techniques for humic materials from 14 different NOM samples of varying origins (Mao et al., 2000). These results indicate that carbonyl groups in NOM are underestimated in some cases by a factor of 2. This is likely due to the remoteness of carbonyl groups from protons, and thus these groups are not cross-polarized as readily as carbons near protons. In addition, Keeler and Maciel (2003) suggest that some of the "missing carbon" may be related to the presence of paramagnetic Fe^{3+}. Other studies have focused on differences between CP and DP of charred NOM (i.e., black carbon). Baldock and Smernik (2002) reported that the amounts of observable carbon from charred vegetation formed at high temperatures (>250 °C) varied greatly between CP and DP MAS methods. The relative amounts of aromatic carbon, however, did not differ significantly between the two techniques. Smernik et al. (2002) reported that the CP observable carbon for three charred woods varied between 69% and 82% in comparison to observing 100% of the carbon in uncharred wood. Simpson and Hatcher (2004) found that the difference between ramp-CP and DP MAS methods was less than 5% for an ancient charred wood sample, and the difference was mostly in the carboxylic region. Dria et al. (2002) and Smernik (2005) have also explored the use of higher field strengths (400 MHz) and increased spinning speeds on CP efficiency. Their results agree with that of Cook et al. (2002) in that ramp-CPMAS is more quantitative than CPMAS. However, it is generally agreed that DPMAS is more quantitative than all CPMAS methods. Despite the disadvantages that impact the quantitative reliability of CPMAS techniques, CPMAS remains the most commonly used method for the analysis of NOM in whole soils and sediments mostly because it is significantly faster (DPMAS is limited by the long relaxation times of carbon nuclei) and requires less instrument time.

Unlike ^{13}C, ^{15}N analysis in the solid state using direct polarization is especially challenging due to the low natural abundance of ^{15}N, which is <0.4% (Knicker et al., 1993), and thus CPMAS methods are often applied (Kögel-Knabner, 1997; Hatcher et al., 2001). Knicker and co-workers have performed a number of ^{15}N NMR studies investigating the forms of nitrogen in soil and sediment NOM (Knicker et al., 1993, 1996, 2002; Knicker, 2000a,b, 2001, 2002, 2004; Knicker and Hatcher,

2001; Maie et al., 2006a). These studies have shown that amide and amino nitrogen are the primary forms of organic nitrogen detected by ^{15}N CPMAS NMR techniques and suggest that proteinaceous material may be preserved in soil and sedimentary NOM (Knicker, 2004). Similarly, other researchers have also reported the dominance of amide-nitrogen in soil NOM (Clinton et al., 1995; Mahieu et al., 2000). A recent study investigated the intriguing finding that most of the detectable ^{15}N in soils has been in the form of amide-nitrogen (Smernik and Baldock, 2005). Smernik and Baldock (2005) measured ^{15}N in a number of soil clay-size fractions and, when compared to a wheat protein (gliadin), concluded that half or even more of the organic nitrogen may be in a form that is insensitive to ^{15}N NMR detection. Based on the ^{15}N and ^{13}C NMR measurements of the soil fractions, they further surmised that nitrogen in clay fractions may be in a heterocyclic form (Smernik and Baldock, 2005). Future studies will undoubtedly focus on the detection of different organic nitrogen forms in soils and sediments.

Advanced solid-state NMR techniques have been emerging more recently. Smernik and Oades (2000a,b, 2003) have reported the use of spin counting techniques that allows one to assess the quantitative reliability of carbon detection in different samples (Smernik and Oades 2000a,b, 2003). Spin counting is useful for delineating the degree of carbon that is underestimated by CP methods due to rapid $T_{1\rho}H$ relaxation and/or inefficient cross-polarization (Smernik and Oades, 2003). Spectral editing techniques, such as RESTORE (Restoration of Spectra via TCH and $T_{1\rho}H$ editing), produce sub-spectra that can then be summed to reconstruct a corrected CP spectrum (Smernik and Oades, 2003). Smernik and Oades (2003) reported that the RESTORE spectra for eight different soil samples closely resembled the DP spectra suggesting that the RESTORE technique may be more quantitative than conventional CP methods. Additional spectral editing techniques have recently been reported (Mao and Schmidt-Rohr 2003, 2004a,b, 2005). These sophisticated methods have the power to isolate signals from fused aromatic rings (Mao and Schmidt-Rohr, 2003, 2004a,b), which is beneficial for black carbon NOM studies and avoids the requirement for removing non-black carbon moieties prior to analysis by NMR (Skjemstad et al., 1999; Simpson and Hatcher, 2004). Mao and Schmidt-Rohr (2005) have also devised a method to isolate methylene (CH_2) carbons in humic materials using spectral editing methods. Two-dimensional solid-state NMR techniques have also been reported (Knicker, 2000b, 2002; Mao et al., 2001); however, are not used as routinely as 1D methods.

High-resolution magic angle spinning (HR-MAS) NMR is a recent NMR method that can be used in combination with solid-state NMR methods to study semi-soluble components of NOM in whole soils and sediments. HR-MAS NMR can be thought of as a "hybrid" of both liquid- and solid-state techniques. In HR-MAS NMR, various deuterated solvents can be used to swell the sample. Any dipolar interactions that arise from the insoluble portion are averaged by MAS, and thus the "soluble" components that are in contact with the solvent can be observed using liquid-state NMR methods. This enables researchers to take advantage of the abundance of ^1H nuclei in the sample and also facilitates the application of 2D methods that are more difficult to perform in the solid state. Simpson et al. (2001b) reported the first application of ^1H HR-MAS NMR techniques to whole soil samples. Soils were swollen with two contrasting solvents: D_2O, which would allow the analysis of NOM structures at the soil–water interface; and DMSO-d_6, a more penetrating solvent that is capable of breaking hydrogen bonds (Simpson et al., 2001b). [1]H HR-MAS spectra are shown in Figure 15.7. When swollen in D_2O, the soil–water interface is dominated by polar structures such as sugars and amino acids, but long-chain fatty acids are also visible. When DMSO-d_6 is used as the primary solvent, the

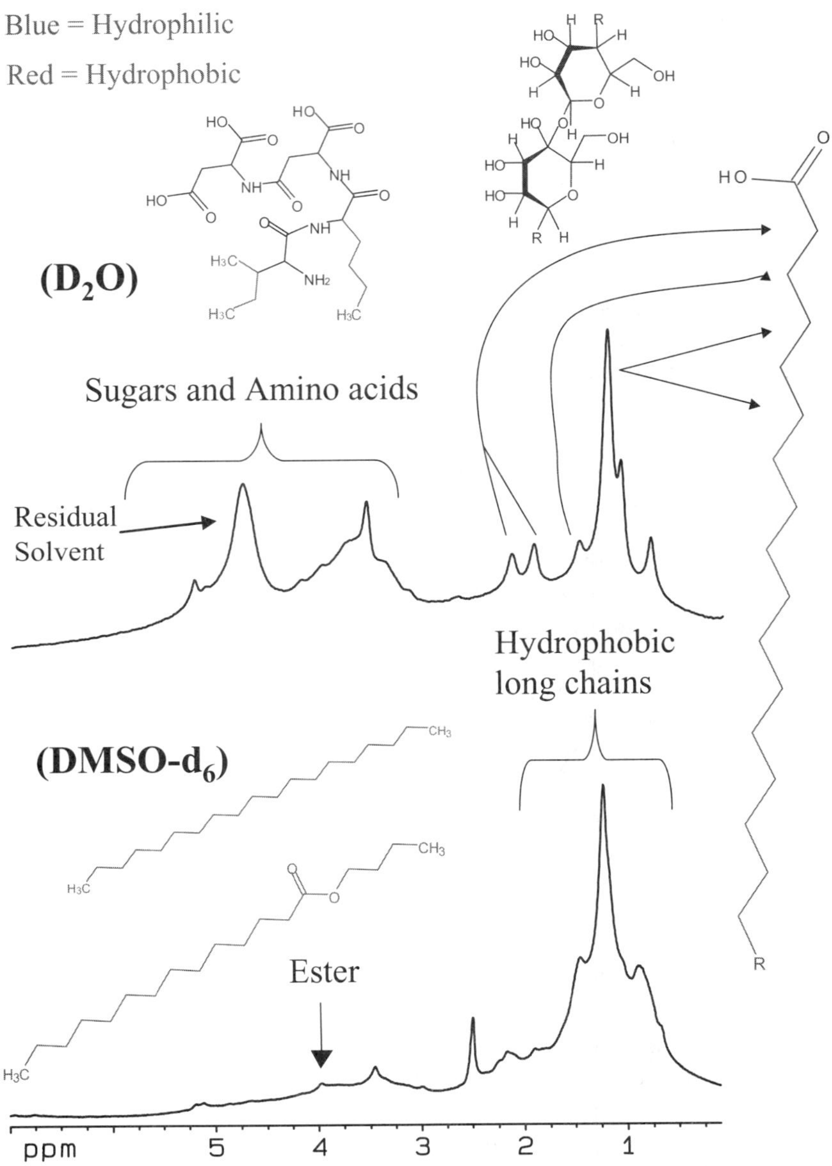

Figure 15.7. ^{1}H HR-MAS NMR of a forest soil. (**Top**) Sampled and analyzed "as is" after the addition of 10 μl of D_2O as a lock signal. Resonances in the top spectrum are those that are in contact with water, and thus at the soil–water interface. (**Bottom**) Same sample as top, but freeze-dried and swollen in DMSO-d_6. Note that DMSO is an excellent swelling solvent and penetrates into both the polar and hydrophobic domains in NOM (Simpson et al., 2001b). See color insert.

proportion of nonpolar compounds increases (as evident from the large peak corresponding to long-chain fatty acids in Figure 15.7) and dominates the spectrum. Signals from ester linkages are also visible. Signals from polar functionalities such as sugars and amino acids are not as prominent compared to the spectrum acquired when D_2O is the solvent. Also notable is the lack of aromatic signals (Figure 15 7), which suggests that aromatic components may be buried within the soil matrix and may not exist at the soil–water interface. HR-MAS NMR can therefore complement other NMR methods, namely solid-state methods, by providing some insight into organic matter components at the soil–water interface.

15.3.3. Nonextractable Soil Organic Matter (Humin)

Nonextractable NOM (referred to as the humin fraction) can represent more than 50% of the total carbon in NOM found in soils and sediments (Rice, 2001). However, because it is insoluble, most researchers opt to use solid-state ^{13}C NMR methods to analyze the structure of humin (Preston et al., 1989; Almendros et al., 1996; Nierop et al., 1999; Kang et al., 2003; Guignard et al., 2005; Xing et al., 2005; Keeler et al., 2006; Simpson and Johnson, 2006). Traditionally, there have been fewer NMR studies of soil humin than other NOM fractions (humic and fulvic acids). More recently, humin characterization by solid-state ^{13}C NMR methods has become more prevalent due to the prominent role of humin in the sequestration of problematic organic contaminants (Kohl and Rice, 1998; Chefetz et al., 2000; Kang and Xing, 2005; Simpson and Johnson, 2006). For example, several studies have shown that the humin fraction yields higher sorption values than does its corresponding source material (Nearpass, 1976; Garbarini and Lion, 1986; Chiou et al., 2000; Salloum et al., 2001; Kang and Xing, 2005; Bonin and Simpson, 2007). Solid-state ^{13}C NMR studies of humin have demonstrated that humin has functional groups similar to those of the whole soil or sediment from which it was isolated; however, the humin fraction has been observed to be higher in aliphatic and substituted aliphatic carbon but lower in aromatic carbon than corresponding humic acids (Almendros et al., 1991b, 1996; Nierop et al., 1999; Salloum et al., 2001; Kang et al., 2003; Guignard et al., 2005; Wang and Xing, 2005b; Keeler et al., 2006; Simpson and Johnson, 2006; Bonin and Simpson, 2007). A comparison of the ^{13}C CPMAS NMR spectra from two types of soil and corresponding humin isolates are shown in Figure 15.8. The peat and peat humin have similar characteristics and their integration results are comparable and do not vary more than 2% (Simpson and Johnson, 2006), suggesting that the peat soil and humin contain similar carbon structures. In contrast, the grassland soil and humin display differences in their distribution of carbon. The humin sample contains more unsubstituted aliphatic carbon (0–50 ppm), less substituted aliphatic carbon (50–110 ppm), less aromatic carbon (110–165 ppm), and less carboxylic and carbonyl carbon (165–215 ppm) than its corresponding soil sample.

The concentration of unsubstituted aliphatic carbon in soil humin has also been reported by a number of other researchers (Almendros et al., 1991b; Nierop et al., 1999; Salloum et al., 2001; Kang et al., 2003; Guignard et al., 2005; Wang and Xing, 2005b; Keeler et al., 2006; Bonin and Simpson, 2007) and has garnered attention. This region can contain signals from terminal methyl groups (10–20 ppm), methylene carbon from alkyl chains (15–50 ppm), methylene carbon in branched chains (35–50 ppm), and methine carbon in alkyl chains (25-50 ppm). Amorphous (29-

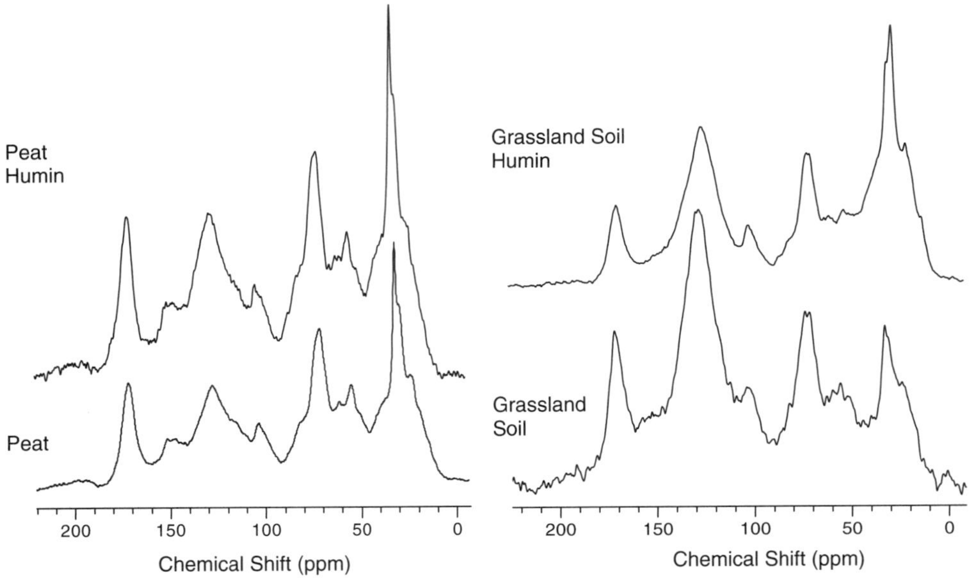

Figure 15.8. ^{13}C CPMAS NMR spectra of the IHSS Pahokee Peat and a Canadian Grassland (black chernozem) soil and their corresponding humin samples. Reprinted from Simpson, M. J., and Johnson, P. C. E. (2006). Identification of mobile aliphatic sorptive domains in soil humin by solid-state ^{13}C nuclear magnetic resonance. *Environ. Toxi. Chem.* **25**, 52–57, with permission from the Society of Environmental Toxicology and Chemistry.

30 ppm) and crystalline (32-33 ppm) methylene carbon has been identified in humic substances (Hu et al., 2000) and is believed to arise from the recalcitrance and accumulation of plant cuticles (Chefetz et al., 2000; Salloum et al., 2002). Signals from amorphous (mobile) and crystalline (rigid) methylene carbon have been identified in humin samples (Mao et al., 2002b; Simpson and Johnson, 2006), and an example of the doublet resulting from both amorphous and crystalline methylene domains is shown in Figure 15.9. These results suggest that plant cuticular material is preserved in soil humin potentially through the preferential binding to clay mineral surfaces. Other chemical studies of soil humin have shown that its chemical nature is highly aliphatic (Almendros et al., 1991b, 1996; Nierop et al., 1999; Kang et al., 2003; Feng et al., 2005; Guignard et al., 2005; Xing et al., 2005; Keeler et al., 2006); however, one of the disadvantages of using solid-state ^{13}C NMR techniques is that the broad lines in other regions of the spectra do not enable the identification and source apportionment of the other structures within humin.

While the main applications of NMR to study humin have been restricted to the solid state, it has recently been shown that >70% of the traditional humin fraction can be isolated using concentrated urea and $DMSO/H_2SO_4$ (Simpson et al., 2007a). These extracts, soluble in DMSO-d_6, are amenable to solution-state studies. Figure 15.10 shows the solution-state HMQC spectrum for humin, along with some of the detailed assignments that can be made using solution-state NMR approaches. This solution-state study identified strong contributions from five main categories of components—namely peptides, aliphatic species, carbohydrates, peptidoglycan, and lignin—to be the major constituents in the humin samples studied. The components

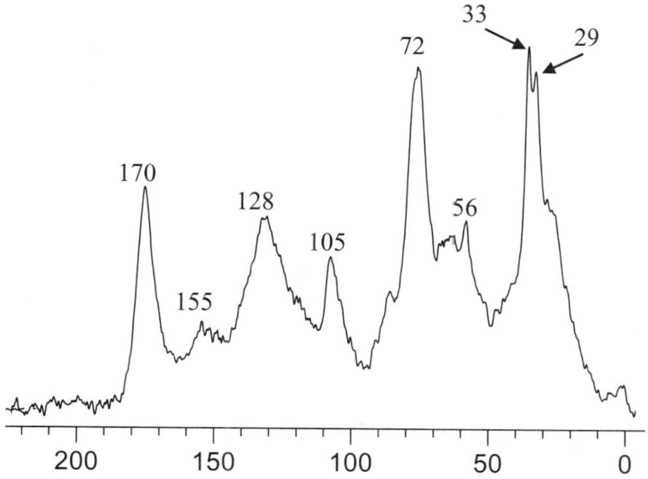

Figure 15.9. ^{13}C CPMAS NMR spectrum of humin extracted from a brown chernozem soil from Western Canada. The characteristic doublet in the unsubstituted aliphatic region is characteristic of methylene carbon (28–34 ppm) and shows the presence of both amorphous (soft) domains at 29 ppm and crystalline (rigid) domains at 33 ppm in soil humin. Reprinted from Simpson, M. J., and Johnson, P. C. E. (2006). Identification of mobile aliphatic sorptive domains in soil humin by solid-state ^{13}C nuclear magnetic resonance. *Environ. Toxi. Chem.* **25**, 52–57, with permission from the Society of Environmental Toxicology and Chemistry.

found in humin are generally found to be similar to those in traditional humic and fulvic fractions, with the exception that the components are larger (macromolecular), and a significant proportion of peptidoglycan (from microbial cells walls) is also present. This initial study clearly demonstrates that solution-state techniques will play a key role in understanding the structural domains in humin. However, extracting humin with harsh solvents (for example DMSO/H_2SO_4) can potentially functionalize and modify the humin structure. While it is clear the main categories of biopolymers identified in this study are major constituents in humin, it is not clear as to the exact state in which they exist (for example, are they highly oxidized and, if so, is this from humification processes or modification during extraction?). Furthermore, extraction disrupts organo-mineral associations, producing material with different reactivities to those found in nature. However, assignments of the major components in solution-state NMR spectra of the extractable components are critical. Once assigned, this information can be used, in combination with state-of-the-art high-resolution (HR-MAS) NMR (which utilizes solution-state experiments to study "swellable" materials), to study humin *in situ*. HR-MAS NMR studies have yet to be applied to humin and will be essential to further evaluate this most recalcitrant and least understood fraction of soil organic matter.

15.3.4. Dissolved Organic Matter

Dissolved organic matter (DOM) is ideally suited to solution-state NMR studies because, by definition, DOM is soluble. Considering that modern solution-state NMR generally produces higher resolution spectra than does solid-state NMR and

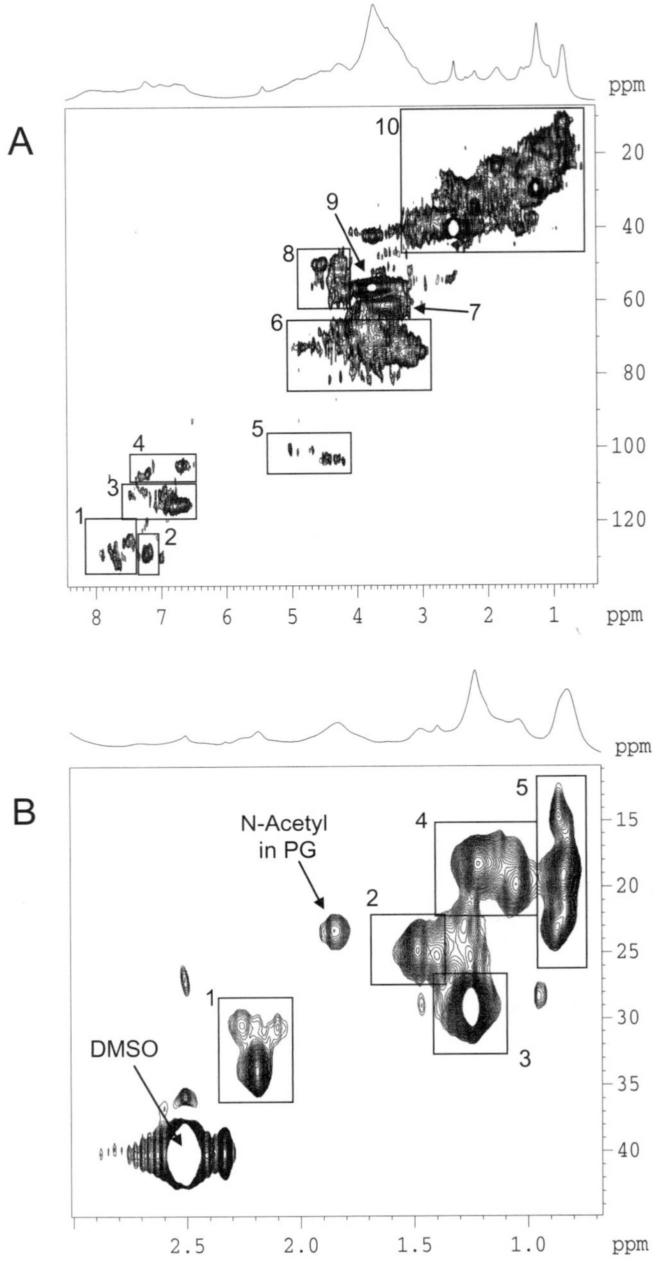

Figure 15.10. HMQC of a DMSO soluble humin fraction. (**A**) Complete spectrum. (**B**) Expansion of the aliphatic region. Assignments in part A are as follows: 1, aromatic protons in *p*-hydroxybenzoates (lignin); 2, phenylalanine (in peptide); 3, mainly aromatic protons adjacent to an Ar-OR functionality in lignin; 4, units in syringyl units (lignin); 5, anomeric protons (carbohydrates); 6, other CH in carbohydrates; 7, CH_2 in carbohydrates; 8, α-protons in peptides and proteins; 9, methoxyl in lignin; 10, aliphatic linkages including signals from various lipids, and side-chain protons in peptides. Assignments in part B are as follows: 1, R–OCO–$\underline{CH_2}$–R methylene unit adjacent to the carbonyl in lipids (including lipoproteins and cutins); 2, methylene units in an aliphatic chains β to an acid or ester; 3, methylene $(CH_2)_n$ in aliphatic chains, 4, aliphatic methylene γ to an acid or ester; 5, CH_3 (a small contribution in this region will be from terminal CH_3 from lipids; however, the majority of signals are from peptides (indicated by the distribution of ^{13}C shifts common in proteins). Reprinted from Simpson, A. J., Song, G. X., Smith, E., et al. (2007a). Unraveling the structural components of soil humin by use of solution-state nuclear magnetic resonance spectroscopy. *Environ. Sci. Technol.* **41**, 876–883, with permission from the American Chemical Society.

that there is a wealth of 1D and 2D approaches that can be applied, solution-state NMR is the likely choice for the study of DOM. Both ^{1}H and ^{13}C 1D NMR has been applied extensively to the studies of DOM (Leenheer et al., 1991, 2003, 2004; Wershaw et al., 1993, 1996a; Chen et al., 1994; Hejzlar et al., 1994; McCarthy et al., 1997; McKnight et al., 1997; Aluwihare and Repeta, 1999; Hernes et al., 2001; Kolowith et al., 2001; Sannigrahi et al., 2005; Hwang et al., 2006; Maie et al., 2006b). A number of key studies, published in *Science* and *Nature*, have utilized NMR to offer novel insights into the structure of DOM (Benner et al., 1992; Aluwihare et al., 1997, 2005; McCarthy et al., 1997, 1998). Benner showed using solid-state NMR how DOM changes with depth in the ocean (Benner et al., 1992). A key finding was that while the surface DOM contained nearly ~50% carbohydrate, contributions in samples from greater depths were reduced (Benner et al., 1992). Aluwihare et al. (1997) showed that a biosynthetically derived acetylated carbohydrate polymer contributes significantly to the DOM in ocean water. In the same year, McCarthy et al. (1997) showed that the majority of organic nitrogen in the oceans is present as amide, and in the following year they showed that a proportion of this amide is likely to be in the form of microbial peptidoglycan (McCarthy et al., 1998). In 2005 a study demonstrated that there may be two distinct pools of organic nitrogen in the ocean: one in the form of *N*-acetyl amino polysaccharides (this fraction will include peptidoglycan) and the other being amide nitrogen (Aluwihare et al., 2005). However, despite the great wealth of information that 1D NMR studies have provided, fewer studies have applied multidimensional NMR to the structural studies of DOM (Haiber et al., 2001b; Kaiser et al., 2003; Kim et al., 2003; Hertkorn et al., 2006). This is likely in large part due to a relatively large amount of DOM needed for 2D solution-state NMR (ideally >50 mg), which is time-consuming and expensive to isolate. Haiber et al. (2001b) may have been the first to collect 2D NMR data of DOM. In this study the authors showed that ^{13}C detected ^{1}H–^{13}C correlation experiments were useful for understanding the fate of lignin residues in aquatic environment through the study of the Suwannee River fulvic and humic acids. Kaiser et al. (2003) showed that 2D NMR was a useful tool for structural studies of solid phase and ultrafiltered DOM, and Kim et al. (2003) showed the characterization of DOM isolated on C-18 solid phase extraction disks. Simpson et al. (2004b) showed that hyphenated NMR (HPLC-NMR) was applicable to the study of DOM isolated from Lake Ontario and Hertkorn et al. (2006) concluded an extensive study of DOM from the Pacific Ocean and was able to highlight some of the major components present in oceanic DOM. In this study, Hertkorn et al. demonstrated that solution-state NMR holds great promise in unraveling the structural components in DOM. Figure 15.11 shows an expansion from an HSQC spectrum of DOM from the Pacific Ocean. The key structural components are identified as heteropolysaccharides, carboxyl-rich alicyclic molecules (CRAM), aliphatics, and peptides. While previous studies based on 1D solution- and solid-state NMR have argued the presence of some of these structures, (Benner et al., 1992; Aluwihare et al., 1997; Leenheer et al., 2003, 2004), the multidimensional studies by Hertkorn et al. demonstrate clearly that multidimensional NMR has an unrivaled capacity to identify components expected to be present and elucidate novel structures (carboxyl-rich alicyclic molecules, CRAM) in very complex mixtures (Hertkorn et al., 2006). Future applications of multidimensional solution-state NMR hold great promise in further refining the structural components in DOM. Hedges argued that if oceanic dissolved

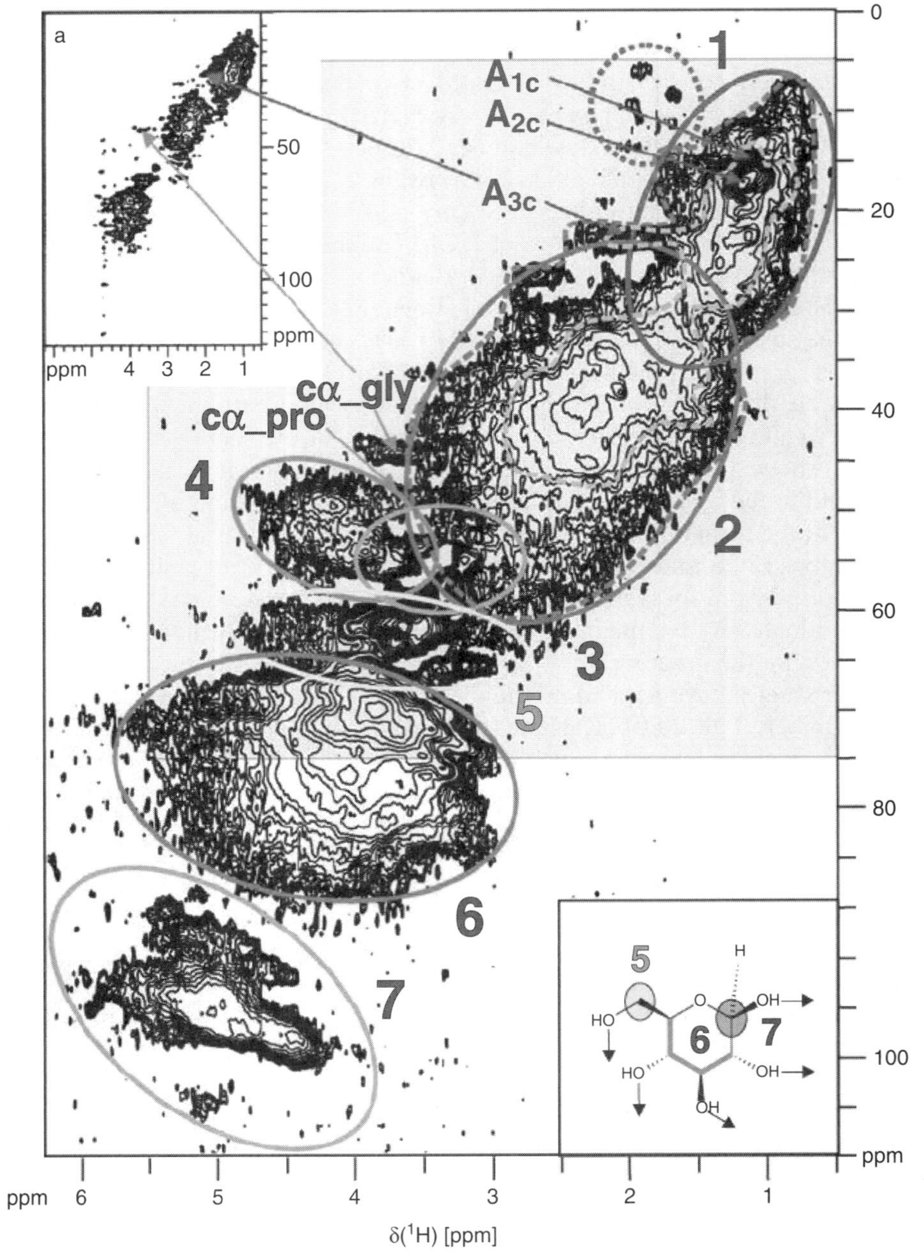

Figure 15.11. Two-dimensional ^{1}H–^{13}C HSQC NMR spectra of ultrafiltered DOM isolated from Pacific Ocean water. For the purposes of this chapter, main assignments can be summarized as follows: 1, methyl bound to carbon and sulfur (dotted circle); and in its lower left corner, branched purely aliphatic CH pairs and polymethylene; 2, methylene and methine cross-peaks without direct bonds to heteroatoms; 3, low-intensity cross-peaks from methoxyl; 4, cross-peaks representative mainly of α-CH in proteins and vicinal dicarboxylic acids; 5, carbohydrate methylene cross-peaks, 6, carbohydrate methine cross-peaks; 7, anomeric units in carbohydrates. Reprinted from Hertkorn, N., Benner, R., Frommberger, M., et al. (2006). Characterization of a major refractory component of marine dissolved organic matter. *Geochim. Cosmochim. Acta* **70**, 2990–3010, with permission from Elsevier.

organic carbon were to experience a net annual decomposition of 1%, it would create a CO_2 flux larger than that created by human fossil fuel usage (Hedges, 2002). Furthermore, Repeta et al. (2002) summarized "Biogeochemical processes that produce, accumulate, and recycle DOM may share important similarities and be broadly comparable across a range of environmental settings." Therefore, it is possible that DOM from many environments may be structurally similar and that these structures form a key link in the global carbon cycle. Thus there is an urgent need to further our understanding of the structural components and how these vary on a global scale. Considering that solution-state NMR is a powerful tool for the analysis of dissolved organic structures, it likely that future multidimensional studies will significantly contribute to our understanding of DOM in the environment.

15.3.5. NMR of Atmospheric NOM

Other potentially important sources of NOM, include that in the atmosphere (air particulates, dissolved materials in rainwater), and that deposited on surfaces (chemical films). Havers et al. (1998) applied ^{1}H NMR to study the alkaline extract from air particles. The material termed "humic-like substances" provided a very interesting NMR profile dominated by relatively sharp lines, suggesting that the mixture contained a large portion of relatively small molecules. However, no detailed assignments were offered. Suzuki et al. (2001) carried out an extensive study using 1D ^{1}H NMR to study water-soluble compounds in air particulates. In this study, numerous small-molecular-weight species were identified (see Figure 15.12), which were superimposed upon broad resonances from larger-molecular-weight species in the mixture. A recent study performed ^{1}H NMR on similar water-soluble materials (Decesari et al., 2005) obtaining a spectral profile broadly analogous to that of Suzuki et al. (2001). In addition, Decesari et al. also studied NOM in drizzle and rainwater (collected as "wet deposits"). As with the air particulate material, the ^{1}H NMR spectra contained a complex overlapping profile (Decesari et al., 2005). Interestingly, the aliphatic content of the "wet deposit" samples was considerably higher than that obtained from the air particles. At present, the reason for this along with the majority of structures in air particulate samples are not fully understood. There remains great potential for the applications of NMR to study species in the atmosphere. To the authors' knowledge, the only application of multidimensional NMR to the study of atmospherically derived material was in 2006 (Simpson et al., 2006a). In this study, urban films were collected from various urban surfaces. The material was sequentially extracted and subject to 1D and 2D solution-state NMR. In addition, the intact whole films were analyzed by HR-MAS NMR (with and without HF pretreatment) and by solid-state NMR. This study was able to (a) provide a semi-quantitative overview as to the main components in the sample and (b) identify specific polymers such as a styrene-derived polymer and polybutadiene (artificial rubber), as well as numerous other components including aliphatic acids, alcohols, alkanes, alkenes, esters, carbohydrates, and hydroxylated PAHs. To the authors' knowledge, this is the first example in the literature where all three main types of NMR (solid-state, solution, and HR-MAS) have been combined synergistically to understand a very complex medium. The synergistic use of NMR approaches will be discussed in Section 15.5.

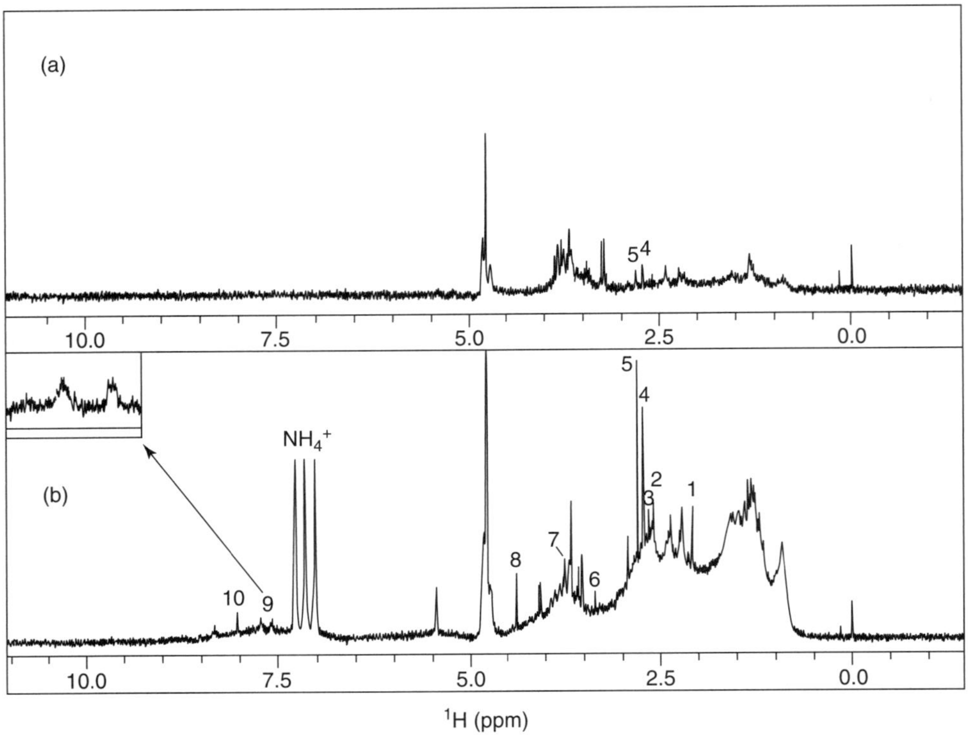

Figure 15.12. ^1H NMR spectra of aqueous soluble species from: (**A**) coarse air particles with diameters of 3.3–4.7 mm (pH 6.55); (**B**) fine air particles with diameters of 0.65–1.1 mm (pH 3.92). Compounds identified in the spectra include: 1, acetic acid; 2, monomethylamine; 3, succinic acid; 4, dimethylamine; 5, methanesulfonic acid; 6, methanol; 7, monomethyl hydrogen sulfate; 8, hydroxymethanesulfonic acid; 9, phthalic acid; 10, terephthalic acid. Reprinted from Suzuki, Y., Kawakami, M., and Akasaka, K. (2001). ^1H NMR application for characterizing water-soluble organic compounds in urban atmospheric particles. *Environ. Sci. Technol.* **35**, 2656–2664, with permission from the American Chemical Society.

15.4. INTERACTIONS AND ASSOCIATIONS OF NOM

15.4.1. Self-Association and Aggregation of NOM

Solution-state NMR is extremely versatile, and many possibilities exist for studying the aggregation and/or self-association of NOM in solution. The application of advanced techniques has largely been hampered in the past due to the lack of detailed understanding of the individual components in NOM. This is changing slowly with recent studies describing the major components present in extractable soil NOM (Kelleher and Simpson, 2006) and aquatic DOM (Hertkorn et al., 2006) and is reflected in the "new view" of NOM, which depicts NOM as a complex mixture of species that can aggregate in solution (Sutton and Sposito, 2005). Thus, in the future, numerous NMR studies aimed at understanding the aggregation of humic substances seem likely (Peuravuori, 2005).

Some powerful NMR approaches have already been applied to the study of NOM and have in some cases been vital in describing NOM as an aggregated system. Wershaw (1999) provided an excellent review and discussion of aggregation for humic materials. A few papers cited by Wershaw (1999) used NMR, but mainly in the context of understanding the structural components present and not specifically to study aggregation processes. For example, Piccolo et al. (1999) used ^{13}C CPMAS NMR to characterize humic fractions separated after disaggregation by gel permeation chromatography (GPC) and organic acids. Tombacz (1999) used various chemical and physical measurements along with ^{13}C CPMAS NMR to provide some insights into aggregation processes. Morris et al. (1999) published the first work dealing with 2D diffusion ordered spectroscopy (DOSY) which has become a very powerful tool for studying NOM aggregation. In this key paper, Morris et al. (1999) demonstrated that diffusion coefficients could be measured for NOM in solution by NMR. Lead et al. (2000) compared numerous different approaches to measure diffusion coefficients. These preliminary diffusion-based studies conclude that DOSY NMR was especially powerful because it not only measured the diffusivities but also correlates, in a two-dimensional plot, the diffusivities with chemical shift information (structural information). Thus, DOSY NMR can theoretically be used to understand which components are, for example, the largest components in a mixture but also describe what these larger components are (i.e., carbohydrates, peptides, etc.). Simpson et al. (2001c) applied DOSY NMR to separate species in a fractionated fulvic acid and a whole soil extract. The study demonstrated that components in NOM were separable based on their diffusion coefficients; and components such as peptides, carbohydrates, and aliphatic and aromatic species were observed. In 2002, a detailed study was carried out exploring the role of DOSY NMR in the study of NOM and its interactions (Simpson, 2002). In this paper, the author showed that NOM aggregation increased with NOM concentration and could be disaggregated by the addition of simple organic acids (Piccolo et al., 1999). Figure 15.13 shows the diffusivities of a fulvic and humic acid at different concentrations obtained by DOSY NMR. At all concentrations, the fulvic acid behaves similarly to a maltodextrin (MD) of ~1300 Das. The slight decrease in diffusion coefficient with concentration results from increased viscosity of the solution. The humic acid behaves very differently, and at low concentrations (1 mg/ml) it exhibits a diffusion in between that of 1300 to 6100-Da maltodextrans. At higher concentrations the humic acid aggregates, indicated by a rapid decrease in the diffusion coefficient, and the humic acid behaves as a colloid greater than 100 kDa (Simpson, 2002). This is strong evidence that humic substances are aggregate systems. Furthermore, DOSY NMR can be used to study disaggregation via the addition of acetic acid (Piccolo et al., 1999). Figure 15.14 shows the DOSY NMR spectra before and after the addition of acetic acid. Prior to the addition of acetic acid, all the NOM components diffuse with approximately the same diffusion coefficient, indicating that the species are aggregated. After the addition of the acetic acid, the main components in the mixture (peptide, carbohydrates, and lignin) are separable on basis of their diffusion coefficients and have been disaggregated. In the same article the author introduced 3D DOSY NMR-based approaches and demonstrated that while at low concentrations, the humic acid components behave *on average* like smaller moieties, very large components, including 12-kDa peptides/proteins, were also present in the humic mixtures. The study concluded that "It is logical that an operationally defined extract of soils will result in a mixture of plant components at various stages of humification with a range of molecular sizes and structures rather than macromolecules with

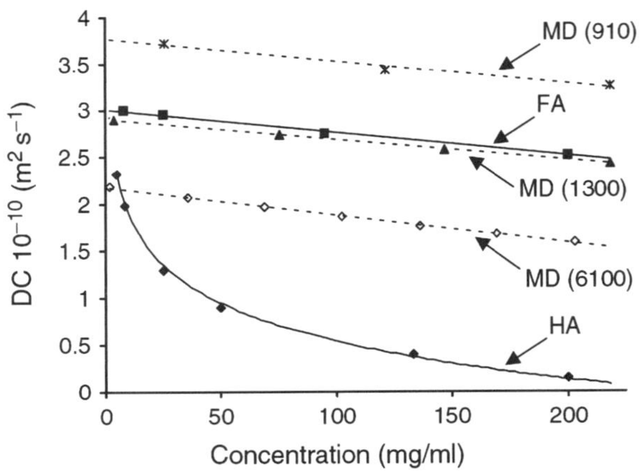

Figure 15.13. Diffusivities of a fulvic and humic acid at different concentrations obtained by DOSY NMR. At all concentrations the fulvic acid behaves similarly to a maltodextrin (MD) of ~1300 Da. The slight decrease in diffusion coefficient with concentration results from increased viscosity of the solution. The humic acid behaves very differently, and at low concentrations (1 mg/ml) it exhibits diffusion between that of 1300- and 6100-Da maltodextrans. At higher concentrations the humic acid aggregates, indicated by a rapid decrease in the diffusion coefficient; at higher concentrations, the humic acid behaves as a colloid of >100,000 Da. Reprinted from Simpson, A. J. (2002). Determining the molecular weight, aggregation, structures and interactions of natural organic matter using diffusion ordered spectroscopy. *Magn. Reson. Chem.* **40**, S72–S82, with permission from John Wiley & Sons, Ltd.

undetermined structures." (Simpson, 2002). The same year, a study combined information from DOSY NMR and nuclear Overhauser enhancements to conclude that some humic materials had relatively low average molecular weights (Simpson et al., 2002a). Additional studies have shown some surfactants can form stable ions with humic substances and may alter the way in which humic acids aggregate (Otto et al., 2003) while Wang et al. (2003) showed that relaxation measurements in various solvents could be a useful tool for understanding NOM associations. Recently, Kazpard et al. (2006) showed that solid-state ^{13}C CPMAS and ^{27}Al MAS NMR were useful tools to study the aggregation of synthetic NOM in the presence of aluminum. The authors demonstrated that at pH 6, carboxylic groups from the DOM bind selectively to the aluminum, while at pH 8 the phenolic groups are more influential in the aggregation process.

In summary, the application of advanced NMR approaches to study the aggregation, associations, and ultimately the environmental reactivity of NOM are in their infancy. Various parameters that can be measured by NMR, including relaxation rates, diffusion coefficients, dipolar interactions, and saturation transfer, can be exploited in 1D to nD NMR experiments. These experiments should theoretically be able to describe which components in humic materials associate with each other, as well as describe their organization in solution to form various "domains " that may be key in understanding the interactions of NOM with many organic contaminants (Weber et al., 1999). Finally, it is important to point out that it is also possible to study the associations and organization of organic matter in its natural swollen state in whole soils, as well as study organo-mineral associations by employing

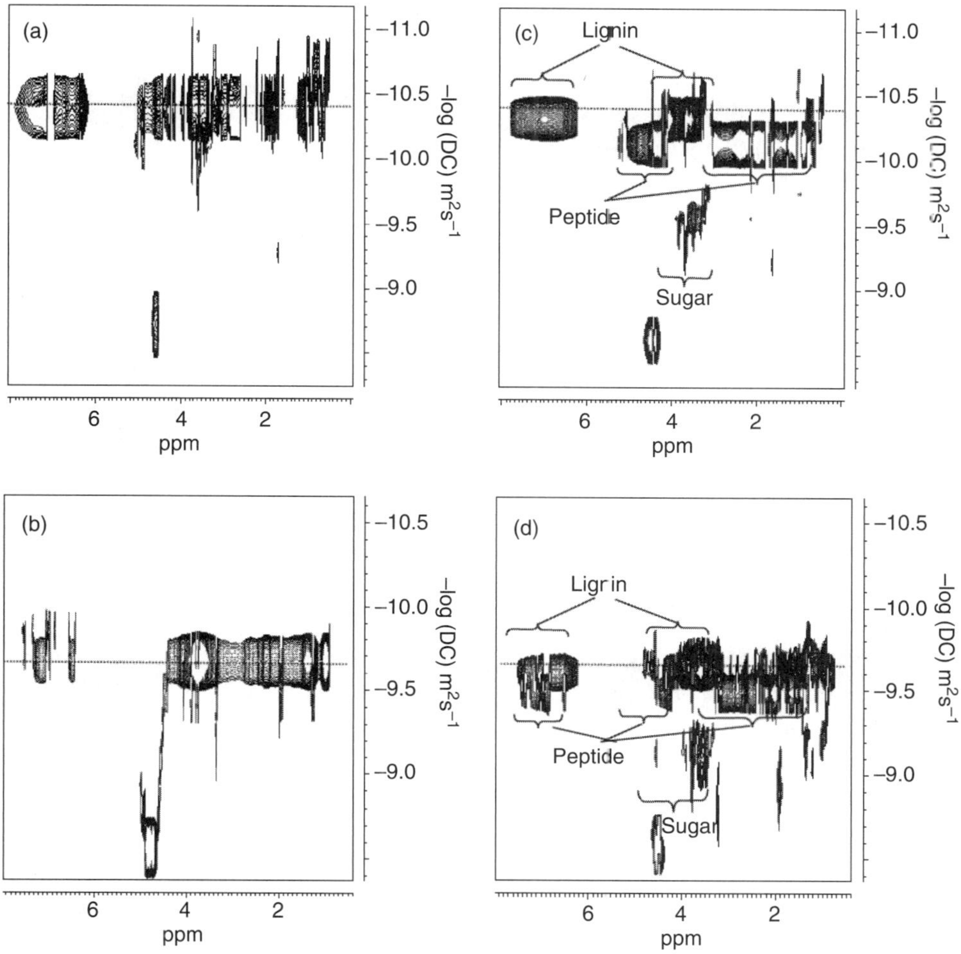

Figure 15.14. DOSY spectra of the IHSS peat humic acid in D_2O at 5 mg/ml (**A**) and 133 mg/ml (**B**) and after the addition of 5 µl of acetic acid (**C**, **D**). Prior to the addition of acetic acid, all the NOM components diffuse with approximately the same diffusion coefficient, indicating that the species are aggregated. After the addition of the acetic acid, the main components in the mixture (peptide, carbohydrates, and lignin) are separable on the basis of their diffusion coefficients, and have been disaggregated. Reprinted from Simpson, A. J. (2002). Determining the molecular weight, aggregation, structures and interactions of natural organic matter using diffusion ordered spectroscopy. *Magn. Reson. Chem.* **40**, S72–S82, with permission from John Wiley & Sons, Ltd.

HR-MAS NMR spectroscopy. Applications to whole soils (Simpson et al., 2001b) are discussed in Section 15.3.3, and applications to organo-mineral interactions (Simpson et al., 2006b) are covered in Section 15.4.3.

15.4.2. Contaminant Interactions

The applications of NMR to the study of NOM interactions have been extensive, with hundreds of reports in the literature applying NMR to some degree. Several recent reviews have summarized the use of NMR in the study of organic contaminants and NOM components (Cardoza et al., 2004; Delort et al., 2004; Simpson, 2006). Some studies will be highlighted in this section; however, this topic will not be covered in detail because of existing recent reviews; for more information, readers are encouraged to refer to the review articles by Cardoza et al. (2004), Delort et al. (2004), and Simpson (2006) written on this topic. In general, contaminants can be categorized as either organic or metallic in nature. The first application of NMR to study contaminant interactions with DOM was probably that of Lindman and Lindqvis (1969), who looked at the interaction of rubidium with humic acid. A number of studies of metallic nuclei have examined: aluminum (Howe et al., 1997; Lookman et al., 1997; Matthias et al., 2003), cesium (Xu et al., 2006), vanadium (Lu et al., 1998), europium (Shin et al., 1996), and lithium (Nielsen et al., 2005). However, the most widely applied NMR studies of heavy metals have been the study of cadmium (Chung et al., 1996; Larive et al., 1996, 1997; Li et al., 1998; Otto et al., 2001a,b; Simpson, 2002; Hertkorn et al., 2004; Grassi and Daquino, 2005; Perdue et al., 2005). Cadmium NMR studies have been used to quantify ^{113}Cd binding (Chung et al., 1996; Larive et al., 1996; Otto et al., 2001b), evaluate chemical exchange (Larive et al., 1996), identify the groups (mainly oxygen, but also nitrogen and sulfur at certain pH values) responsible for binding (Hertkorn et al., 2004), and identify specific (inner/outer sphere) complexes (Grassi and Daquino, 2005).

Similarly, in the area of organic contaminant interactions, a wide array of studies have been performed (Thorn et al., 1996a,b, 1997; Bortiatynski et al., 1997; Hinedi et al., 1997; Jayasundera et al., 1997; Nanny et al., 1997; Achtnich et al., 1999; Dixon et al., 1999; Green et al., 1999; Knicker et al., 1999; Nanny, 1999; Xiong et al., 1999; Bruns-Nagel et al., 2000; Kohl et al., 2000; Emery et al., 2001; Smernik and Oades, 2001; Kacker et al., 2002; Thorn and Kennedy, 2002; Strynar et al., 2004; Smernik et al., 2006). Using NMR, it has been possible to determine the amount of organic contaminants bound (Simpson et al., 2004c), evaluate binding type (covalent and noncovalent) (Nanny et al., 1997; Achtnich et al., 1999; Nanny and Maza, 2001), extract mechanistic information (Wais et al., 1996; Hinedi et al., 1997), and even focus on the structural components of NOM involved (Dixon et al., 1999). It is important to note that numerous studies in the literature have examined the transformation of contaminant species in the environment, but these topics will not been covered here either because the NOM is not directly involved or because covalent interactions are formed between the NOM and the contaminant species. Generally speaking, in the case of covalent bond formation the same structural tools as outlined in Sections 15.1 and 15.2 would be appropriate. The aim of this section is to introduce the main types of information that can be obtained from NMR to noncovalent interactions, which can differ somewhat from the techniques already outlined in Sections 15.2 and 15.3.

15.4.2.1. The Chemical Shift. The most accessible information in relation to contaminant interactions can be simply inferred from a change in chemical shift of the contaminant species. In the case of metal NMR studies, the observed nucleus is most commonly the metal itself—for example, ^{113}Cd studies (Dehorter et al., 1992; Larive et al., 1996; Li et al., 1998; Otto et al., 2001a,b; Hertkorn et al., 2004; Grassi and Daquino, 2005). In the case of an organic species, many different contaminant nuclei have been studied, including: ^{1}H (Dixon et al., 1999; Simpson et al., 2004c), ^{2}H

(Nanny, 1999; Nanny and Maza, 2001), ^{13}C (Hatcher et al., 1993; Wais et al., 1996), ^{15}N (Thorn et al., 1996a,b, 1997; Weber et al., 1996), and ^{19}F (Kohl et al., 2000; Khalaf et al., 2003; Strynar et al., 2004). Figure 15.15 shows an example where the chemical shift of the contaminant can change drastically when interacting with NOM in a whole soil. The aromatic protons (IV) are shifted drastically to a higher chemical shift. It is hypothesized that the aromatic nitro groups of the contaminant are interacting strongly with NOM carboxyl groups that are abundant at the NOM soil–water interface (Figure 15.15). This draws the aromatic protons into an environment that is strongly electron-withdrawing and, in turn, leads to de-shielding of the aromatic protons, and a downfield shift for these protons is observed (Simpson et al., 2001b).

15.4.2.2. Relaxation. In addition to changes in chemical shift, changes in lineshape are often also observed. The lineshape of an NMR signal is directly correlated to relaxation. Relaxation is the time taken for nuclei to return to equilibrium after equilibrium has been perturbed by a radio-frequency pulse. In NMR, there are two main types of relaxation, namely T_1 and T_2. The theoretical details are beyond the scope of this chapter; for an excellent discussion, readers should refer to Keeler (2005). It is simply enough for the reader to understand that both T_1 (also called

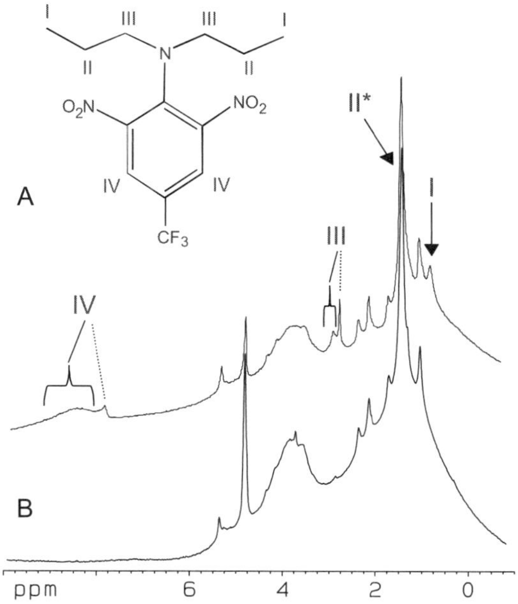

Figure 15.15. ^1H HR-MAS NMR spectra of a whole soil swollen in D_2O and doped with trifluralin (**A**). ^1H HR-MAS spectra whole soil swollen in D_2O (**B**). For C-III and C-IV, dashed lines indicate the "unbound trfluralin" while solid lines with brackets indicate the "bound" trifluralin. *protons II are masked by the aliphatic signals from the soil. Reprinted from Simpson, A. J., Kingery, W. L., Shaw, D. R., et al. (2001b). The application of ^1H HR-MAS NMR spectroscopy for the study of structures and associations of organic components at the solid—Aqueous interface of a whole soil. *Environ. Sci. Technol.* **35**, 3321–3325, with permission from the American Chemical Society.

longitudinal relaxation or spin-lattice relaxation time) and T_2 (also called transverse relaxation or spin–spin relaxation time) are easily measured and can be used to quantify contaminant interactions. Figure 15.16 shows the change in relaxation of 1-naphthol protons as the concentration of humic acid is increased. If relaxation rates differ significantly for protons at different positions in the contaminant structure, it is possible to obtain information as to the mechanism through which the contaminant binds. However, in the case shown in Figure 15.16 the protons relax at a similar rate, indicating that the contaminant does not show a specific binding mechanism (or at least one that can be elucidated using relaxation studies) and may be undergoing partitioning with the NOM. Furthermore, using the simple relationship below the percentage of associated versus free aromatic compound can be calculated:

$$A = \frac{T_{1\text{Obs}} - T_{1\text{Free}}}{T_{1\text{Assoc}} - T_{1\text{Free}}}$$

where A represents the fraction of humic-associated contaminant, T_1Obs is the measured T_1 value that represents an average T_1 (in seconds) of both free and humic-associated aromatic compound, T_1Free is the measured T_1 in the absence of humic acid, and T_1Assoc is the T_1 value when the aromatic compound is fully associated with the humic acid (the aromatic compound takes on the T_1 value of the humic acid and the T_1 versus humic acid concentration curve plateaus). The example shown

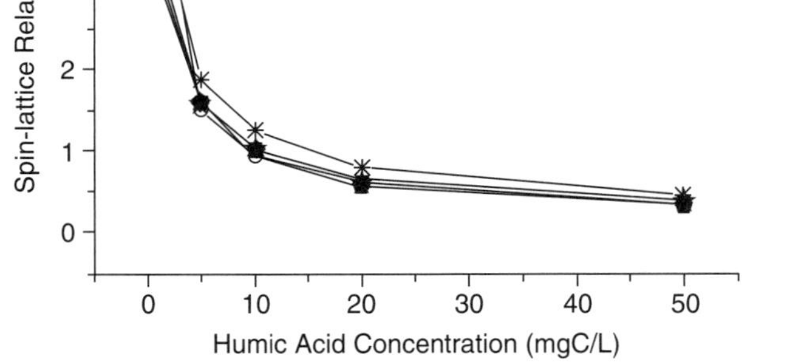

Figure 15.16. ^{1}H Relaxation of 1-naphthol protons with increasing humic acid concentration at pH 7. All protons are observed to relax at a similar rate, suggesting a nonselective interaction between the protons of 1-naphthol and humic acid. Reprinted from Simpson, M. J., Simpson, A. J., and Hatcher, P. G. (2004). Noncovalent interactions between aromatic compounds and dissolved humic acid examined by nuclear magnetic resonance spectroscopy. *Environ. Toxi. Chem.* **23**, 355–362, with permission from the Society of Environmental Toxicology and Chemistry.

here demonstrates the very simple use of T_1 relaxation to probe the interactions of an organic contaminant. It is possible to use T_2 relaxation in a similar way, calculate correlation times (providing information as to the rate of molecular tumbling) (Carper, 1999), and even extract relaxation parameters directly from multidimensional NMR spectra (Leisen et al., 1993; Yang et al., 1997). Many of these techniques have not been exhaustively applied to the study of NOM–contaminant interactions, and there is great potential for future studies. This said, numerous excellent studies have been carried out to study NOM interactions (Lindman and Lindqvis, 1969, 1971; Lindqvis and Lindman, 1970; Andrasko et al., 1972; Chien et al., 1997; Jayasundera et al., 1997; Nanny et al., 1997; Carper, 1999; Culligan et al., 2001; Nanny and Maza, 2001; Otto et al., 2001a; Wang et al., 2003; Jaeger et al., 2006; Smernik, 2006). Combined, these studies have utilized relaxation to provide qualitative and quantitative information regarding the interactions of a range of organic contaminants in the presence of humic and fulvic acid.

15.4.2.3. Molecular Diffusion. Diffusion measurements can be easily measured using DOSY NMR. In a NOM/contaminant mixture, if the contaminant is not interacting with NOM, it will display the same diffusion coefficient as the free molecule (measured without any NOM present). If the contaminant binds permanently to the NOM, then the contaminant takes on the same diffusion coefficient as the NOM itself and a contaminant in exchange with the NOM will exhibit a diffusion coefficient somewhere between that of the bound and free states. A simple example of how DOSY can be used to monitor the interactions of cadmium with a soil fulvic acid is shown in Figure 15.17. Parts A and B of the figure demonstrate the broadening in the ^{113}Cd chemical shift that occurs after the addition of the fulvic acid, indicating that the cadmium interacts strongly with the fulvic acid. This is also reflected in the diffusion coefficients of the $CdCl_2$ before and after the addition of the fulvic acid (see Figures 15.17, parts C, D, E, and F). In this example, the fraction of cadmium interacting with the DOM was found to display diffusion coefficients in between that of the free cadmium and that of the fulvic acid itself, indicating the $CdCl_2$ to be in exchange with the fulvic acid (Simpson, 2002), which is consistent with other studies (Larive et al., 1996; Otto et al., 2001a,b).

DOSY NMR can also be used to study organic contaminants. Thus far only a few applications have been published, the best examples being Dixon et al. (1999), Simpson et al. (2002a), and Otto et al. (2003). Dixon et al. looked at the interactions between fluoro-acetonaphthone and the Suwannee River fulvic acid, Simpson et al. looked at competitive binding between a herbicide and MTBE (a petroleum additive) with NOM, and Otto et al. investigated the interactions between humic substances and surfactants.

15.4.2.4. Nuclear Overhauser Effects. Nuclear Overhauser effects (NOEs) can be used to measure both interactions through space, and chemical exchange (Neuhaus and Williamson, 2000). In a system where a contaminant interacts strongly with NOM, NOEs should be measurable between the NOM and the contaminant. In theory, using such an approach should provide information as to which components in DOM the contaminant is associated, as well as possible information on exchange rates, molecular dynamics, and strength of the interactions. Relatively few studies have used NOEs extensively to study NOM–contaminant interactions directly. The

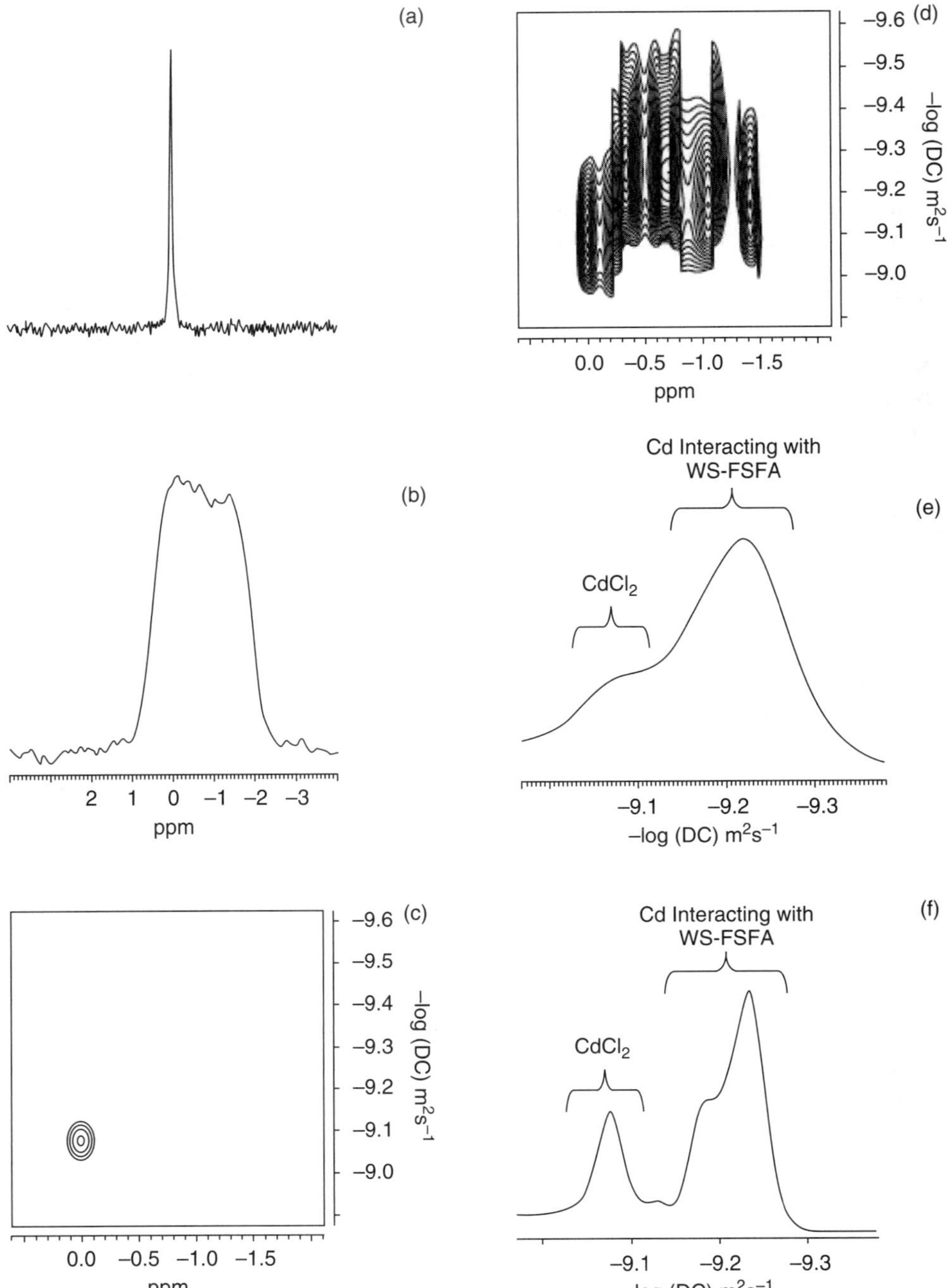

Figure 15.17. ^{113}Cd NMR spectra of $CdCl_2$ at a concentration of 50 mg/ml metal ions (**A**, 8 scans); after the addition of 5 mg of a water soluble forest soil fulvic acid (**B**, 2048 scans); DOSY of $CDCl_2$ (linewidth factor 1) (**C**); DOSY after the addition of the fulvic acid (linewidth factor 2.5) (**D**); *F*1 diffusion projection (**E**); and *F*1 diffusion projection with linewidth factor 0.3 (**F**). Reprinted from Simpson, A. J. (2002). Determining the molecular weight, aggregation, structures and interactions of natural organic matter using diffusion ordered spectroscopy. *Magn. Reson. Chem.* **40**, S72–S82, with permission from John Wiley & Sons, Ltd.

main reason for this is likely due to the overlap in ^{1}H NMR spectra, making the implementation and interpretation of 1D and 2D NOE experiments quite challenging. The most notable study has been that of Dixon et al. (1999), who used the ^{19}F–^{1}H heteronuclear NOE effect to study the interactions of fluoro-acetonaphthone and the Suwannee River fulvic acid. In this study, because the ^{19}F signal from the contaminant did not overlap with the ^{1}H signals from the NOM mixture, the most sensitive 1D ^{19}F–^{1}H NOE difference approach could be implemented. The authors detected a significant NOE indicating that the F-acetonaphthone molecules were strongly associated with Suwannee River fulvic acid for a period of time sufficient for the NOE to develop. Furthermore, they were able to determine that both the aliphatic and aromatic components of the NOM were interacting with contaminant. The measurement of the NOE is a very powerful tool in modern NMR spectroscopy, and it can be implemented in 1D to nD NMR experiments. Future studies taking advantages of this phenomenon will likely provide key information as to the interactions of NOM with contaminant species.

15.4.2.5. NMR Micro-imaging. Recently, applications of NMR micro-imaging that focus on contaminant processes in porous media have emerged (Reeves and Chudek, 2001; Chen et al., 2002; Zhang et al., 2002; Chu et al., 2004). Reeves and Chudek (2001) used this technique to study diesel oil migration in sediment cores. Similarly, Chen et al. (2002) measured the oil distribution in aquifer columns and Zhang et al. (2002) studied the movement of dense nonaqueous phase liquid (NAPL) in silica gel columns. Chu et al. (2004) evaluated the efficiency of NAPL removal from silica gel columns during soil vapor extraction (SVE). NMR micro-imaging techniques are beneficial because contaminant movement or flow through a column can be monitored in real time and images can be acquired that depict the location of the contaminant of interest. Monitoring nuclei that are not naturally abundant in soil or sediment is also advantageous because interference from soil or sediment materials is negligible. For example, fluorine is an ideal nucleus for contaminant micro-imaging studies because background levels of ^{19}F in soils and sediments is typically negligible, thereby minimizing any interference from natural sources (Bondar et al., 1998; Simpson, 2006). Furthermore, there is only one fluorine isotope (^{19}F represents 100% of the total abundance); thus the sensitivity of ^{19}F by NMR is greater than most other nuclei such as ^{13}C and ^{15}N at natural abundance (Bondar et al., 1998; Simpson, 2006).

Recently, Simpson et al. (2007b) explored the use of ^{19}F NMR micro-imaging techniques for use in contaminant-soil studies. The fate of hexafluorobenzene (F_6C_6) added to a miniature soil core was monitored over a 24-h period. Select images from the study are shown in Figure 15.18. At the start of the experiment (time = 0 h, Figure 15.18A), the hexafluorobenzene is visible at the top part of the soil column. Over time, the hexafluorobenzene appears in localized regions throughout the soil matrix (Figures 15.18B–D). After 1.5 h (Figure 15.18B), evidence of hexafluorobenzene transport is visible in the form of small, filled pores. After 8 h (Figure 15.18C), the ^{19}F signal intensity in the same regions identified after 1.5 h has intensified, indicating that these areas represent fast-filling pores or preferential flow channels. This observation is more apparent after 16 h (Figure 15.18D) when signal intensity in addition to the size of the filled area increases. The appearance of the hexafluorobenzene in localized positions provides direct evidence for preferential flow models and dem-

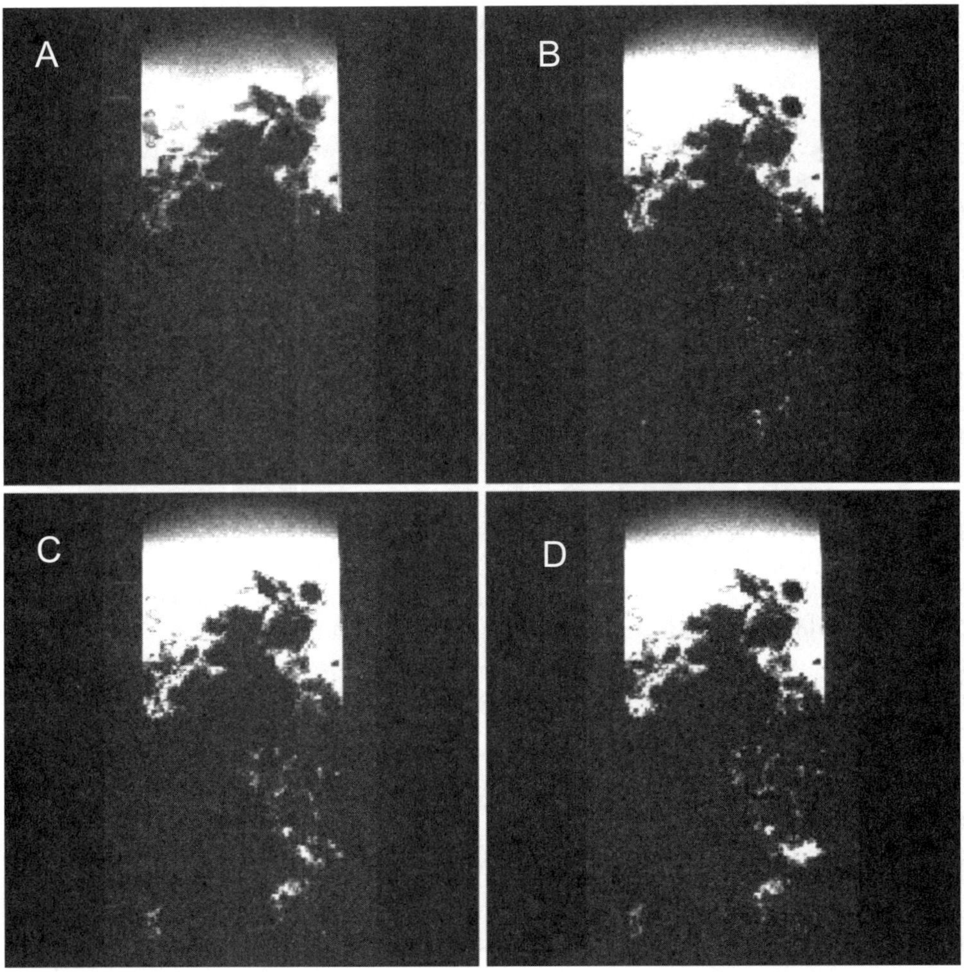

Figure 15.18. A series of 15-mm × 30-mm images (acquired at a ^{1}H frequency of 400-MHz) taken at different times showing hexafluorobenzene penetration into an organic matter rich soil. The brighter the region, the higher the concentration of hexafluorobenzene present at that location. The four images were taken at time = 0 hours (**A**), 1.5 hours (**B**), 8 hours (**C**) and 16 hours (**D**). Reprinted from Simpson, M. J., Simpson, A. J., Gross, D., et al. (2007). ^{1}H and ^{19}F nuclear magnetic resonance microimaging of water and chemical distribution in soil columns. *Environ. Toxi. Chem.* **26**, 1340–1348, with permission from the Society of Environmental Toxicology and Chemistry.

onstrates that contaminants selectively move though soil channels and pores. The process also appears to be relatively slow because after 16 h only a small fraction of the hexafluorobenzene has entered the soil matrix (Figure 15.18). Although only a preliminary study, this work demonstrates the potential of NMR micro-imaging techniques for studying contaminant movement in soil.

15.4.2.6. Other Approaches. There are numerous other NMR approaches that can be applied to probe NOM–contaminant interactions. One possibility is to take advantage of dipolar interactions in solid-state NMR to probe the proximity of specific structures. Recently, Sachleben et al. (2004) showed that it was possible to use 2D solid-state NMR to show interactions between pyrene and cuticular species, and Smernik et al. (2006) used various parameters to probe the mobility of ^{13}C-labeled benzene sorbed onto charcoal. Contaminants that contain quadrapolar nuclei such as deuterium are also well-suited to NMR studies. Nanny has demonstrated that noncovalent interactions between NOM and deuterated contaminants can be carried out in solution (Nanny, 1999; Nanny and Maza, 2001), while other studies have monitored the quadrapolar lineshape of deuterated contaminants as a sensitive probe to their environment in solid-state NMR (Xiong et al., 1999; Emery et al., 2001). Paramagnetic species can induce fast relaxation of species in close proximity. In 1997 a study demonstrated that hydrophilic and hydrophobic paramagnetic probes could be used to investigate the formation of hydrophobic domains in NOM solutions and demonstrated that atrazine sorption likely takes place within these domains (Chien et al., 1997). Recently, Xu et al. (2006) showed that relatively uncommon NMR nuclei ^{133}Cs and ^{35}Cl could be used to understand their complexation by NOM (Xu et al., 2006). Finally, it is important to note that there are numerous other studies in the literature that look at the transformation of contaminant species in the environment, but these are not covered here in detail for brevity.

15.4.3. Organo-Mineral Interactions

Interactions between NOM and the mineral fraction are important in the stabilization and/or preservation of NOM components in the environment. The investigation of mechanisms of NOM sorption to mineral surfaces is an active area of research because these processes are important in regulating the movement and/or preservation of carbon in both aquatic and terrestrial systems (Mayer, 1994; Collins et al., 1995; Wershaw et al., 1996b; Arnarson and Keil, 2000; Chorover and Amistadi, 2001; Guggenberger and Kaiser, 2003; Feng et al., 2005, 2006; Wang and Xing, 2005a; Simpson et al., 2006b). NOM can be sorbed to mineral surfaces via six mechanisms: ligand exchange, cation bridging, anion and cation exchange, van der Waals interactions, and hydrophobic bonding (Arnarson and Keil, 2000; Chorover and Amistadi, 2001; Feng et al., 2005). The majority of prior research has indicated that the sorption of soluble organic matter to the clay surface is competitive and higher-molecular-weight compounds are preferentially sorbed over lower-molecular-weight species, hydrophobic structures are preferentially adsorbed over hydrophilic structures, adsorption of organic matter increases inversely with pH, and the thickness of organic matter coatings varies with concentration. However, the majority of these conclusions are based on macroscopic observations and only recently has NMR been used as a tool to study NOM–mineral interactions (Wershaw et al., 1996b; Wattel-Koekkoek et al., 2001; Feng et al., 2005, 2006; Wang and Xing, 2005a; Simpson et al., 2006b).

NMR can provide detail regarding the types of NOM structures that are preferentially sorbed to mineral surfaces in soils and sediments. Simpson et al. (2006b) used ^1H liquid-state and HR-MAS NMR methods to study the sorption of model compound mixtures to calcium-saturated montmorillonite. The model compound mixture included one representative compound from each of the following structural classes: sugars, lignin, peptides, and long-chain aliphatics. After sorption, the supernatant was analyzed by liquid-state NMR and the organo-mineral complex was analyzed by HR-MAS NMR. Of the four compounds in the mixture, aliphatic structures were observed to be preferentially sorbed to the montmorillonite surface and the majority of lignin, peptide, and sugar remained in the supernatant (Simpson et al., 2006b). Simpson et al. (2006b) also applied the same NMR approach to study the sorption of isolated soil NOM to montmorillonite. As with the model compound mixtures, aliphatic structures were the predominant compounds on the mineral surface, but signals from carbohydrate structures were also evident although to a lesser extent. Feng et al. (2005, 2006) studied the sorption of peat humic acid (PHA) to both kaolinite and montmorillonite and employed both liquid-state and HR-MAS NMR techniques. Kaolinite and montmorillonite were found to preferentially sorb different types of NOM structures. Figure 15.19 shows the ^1H NMR spectrum of the unsorbed compounds (i.e., those remaining in the supernatant after batch equilibrium sorption) and the ^1H HR-MAS spectrum of the components sorbed to the mineral surface (Feng et al., 2006). The kaolinite surface contains appreciable signals from CH_2 and CH_3, suggesting the sorption of aliphatic chains. Very little signals from other NOM structures are observed on the kaolinite surface, suggesting that sorption of peptides, aromatic compounds, and sugars are not as dominant. This is confirmed by the presence of these compounds in the ^1H liquid-state spectrum of the supernatant (Figure 15.19). Conversely, the ^1H HR-MAS NMR spectrum of the montmorillonite surface contains signals from both peptide and aliphatic materials. This is evident from the ratio of CH_3 and CH_2 signals as well as the presence of amide protons in the spectrum (Feng et al., 2005). Similarly, Wang and Xing (2005a) observed that aliphatic components of humic acids were preferentially sorbed by

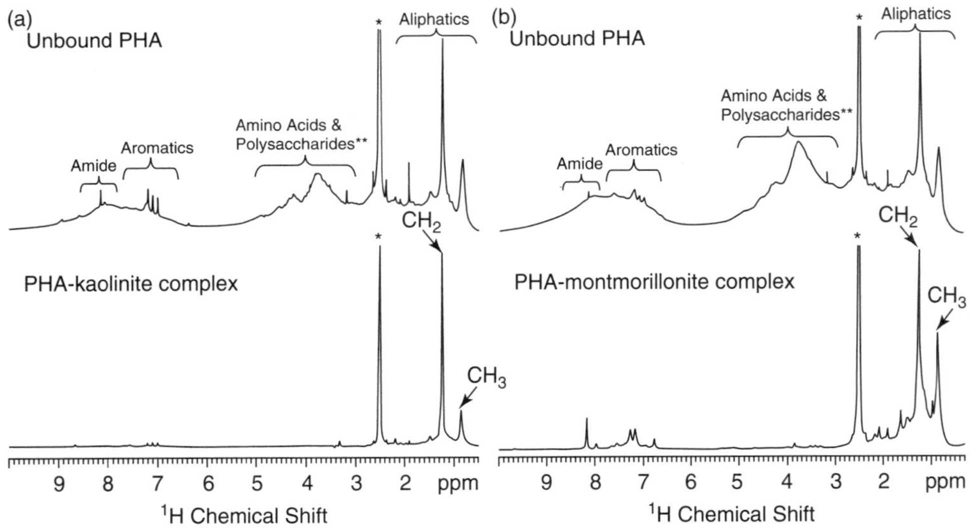

Figure 15.19. ^1H liquid-state NMR spectra of peat humic acid (PHA) that did not sorb to clay minerals (material remaining in the supernatant) and ^1H HR-MAS NMR spectra of sorbed peat humic material to kaolinite and montmorillonite. The signal from DMSO-d_6 is labeled with an asterisk. Reprinted from Feng, X. J., Simpson, A. J., and Simpson, M. J. (2006). Investigating the role of mineral-bound humic acid in phenanthrene sorption. *Environ. Sci. Technol.* **40**, 3260–3266, with permission from the American Chemical Society.

montmorillonite and kaolinite. Subsequent contaminant sorption studies suggested that kaolinite sorbed more aliphatic NOM than did montmorillonite; but due to the insensitivity of ^{13}C NMR, the authors could not confirm this using the NMR technique employed (Wang and Xing, 2005a).

Other studies have used solid-state ^{13}C NMR techniques to study the composition of NOM in clay fractions of soils (Baldock et al., 1992; Baldock and Skjemstad, 2000; Wattel-Koekkoek et al., 2001). Wattel-Koekkoek et al. (2001) used ^{13}C NMR in addition to pyrolysis GC-MS to characterize the NOM in the clay fraction of kaolinitic and smectitic soils. The NOM in the clay fraction was extracted and the isolated NOM analyzed by ^{13}C CPMAS NMR. The study found that NOM associated with kaolinite was more enriched with polysaccharide structures whereas the smectite-associated NOM was more aromatic in nature. Their findings suggest that mineral type and content are important for regulating NOM structure in organo-mineral complexes in soils. Baldock et al. (1992) found that the fine fraction ($<2\,\mu m$) of some soils contained signals predominantly from alkyl structures and suggested that these structures may arise from the preservation of plant biomolecules by association with mineral surfaces as well as the production of new compounds *in situ* by soil microbes. The study concluded that soil type and particle size of the soil greatly impacted the nature of NOM in different size fractions of soil (Baldock and Skjemstad, 2000). Mineral components may also protect NOM from biological degradation and may increase the residence time of some NOM compounds in the environment (Baldock and Skjemstad, 2000). The use of NMR in studies of organo-mineral associations has aided in the development of hypotheses regarding NOM preservation and degradation in the environment (Simpson and Johnson, 2006) as well as provided detailed, mechanistic information about NOM sorption to minerals (Feng et al., 2005, 2006; Wang and Xing, 2005a; Simpson et al., 2006b). NMR is an invaluable tool for the study of NOM composition and structure in organo-mineral complexes and will likely be used more frequently in the future.

15.5. ADVANCED AND EMERGING AREAS IN RELATION TO NOM

NMR spectroscopy is a vibrant and exciting area of research due to its versatility and is evolving at an extremely fast pace. There are numerous "standard" experiments that are supplied with spectrometers in the experiment library which have not yet been applied to the study of NOM rigorously. But more importantly, with a basic knowledge of the experimental details, hundreds and even thousands of novel experiments can be designed specifically for the study of NOM. As new techniques, discoveries, and improvements are made, the potential for new exciting and informative applications grows exponentially. However, over time the greatest breakthroughs in NOM research are not likely to come from a single cutting edge technology but by the rigorous implementation of all modern NMR techniques in combination.

15.5.1. Synergistic Use of Modern NMR Approaches

In a recent special issue of *Science*, Young and Crawford (2004) stated: "There is mounting evidence that the essential features of soil will emerge only when the relevant physical and biochemical approaches are integrated." Indeed if we are to truly understand the chemistry and function of NOM, future studies will need to focus more on NOM in its natural state. For example, if solid-state NMR is used as the only tool for analysis, samples such as soils and sediments must be dried before analysis making it impossible to study: hydrophobic domains, the reactive soil–water interface, or microbial species in their most biologically active state. On the other hand, solution-state NMR provides highly resolved structural information but only for the soluble fractions, and it can be argued that NOM may be chemically altered during any isolation procedure. Ideally, the goal of future research should be to gain as much information as possible while still studying NOM as close to its natural state as possible. In simple terms, this translates to tailoring the NMR method to suit the sample, rather than changing the sample to suit the analysis. For example, if the goal is to unravel as much information as possible about whole soil NOM structure and function, the first key step would be to study the sample by NMR micro-imaging which is completely nondestructive and would probe information as to water distribution pore size and even some information regarding the chemical composition of the larger components (roots, leaves, etc.). Next, the sample could be studied in its naturally swollen state by HR-MAS NMR, which could provide information as to the chemistry of the components at the soil–water interface. A water extraction would permit detailed solution-state NMR studies of the water-soluble soil components. If required, additional HR-MAS NMR studies could be carried out to study the components hidden in hydrophobic domains (through the use of a penetrating solvents such as DMSO-d_6 (Simpson et al., 2001b), or even study components that become HR-MAS NMR observable during demineralization, thus gaining some information as to components at the organo-mineral interface. Next, solid-state NMR could be used to obtain an overview as to the carbon distributions in the sample, and advanced solid-state techniques could be used to study specific domains and structures (Mao and Schmidt-Rohr, 2006). Finally, the sample could be further fractionated into extractable humics, ideal for high-resolution solution-state studies, and humin for further HR-MAS and solid-state NMR studies. The focus here is not the experimental protocols but to demonstrate the versatility of NMR spectroscopy and demonstrate how NMR can be used to study structures and processes from the macroscopic scale (cm-µm scale with micro-imaging and some solid-state approaches) to the molecular level (individual bond distances in some solution and HR-MAS approaches). In certain cases, the techniques can be employed with little, if any, sample treatment, and if employed synergistically, hold the potential to provide an unrivaled overview of a complex system well beyond that of any other analytical approach. Two prohibiting factors for such "synergistic" studies will likely be financial (costs in upgrading a standard spectrometer to run, solution-state, solid-state, HR-MAS, and micro-imaging NMR can be costly) and the intellectual hurdles of being an expert in all four areas of NMR spectroscopy. Thus, interdisciplinary and inter-university collaborations will likely become even more valuable in the future. To our knowledge, there has only been one NOM study that employed solution-state, HR-MAS, and solid-state NMR in tandem. In this study the authors extracted a large amount of information as to the chemical make-up and physical layering in organic-rich films deposited on urban surfaces (Simpson et al., 2006a). The addition of NMR micro-imaging could have further improved the study. It is important to note that the interplay between solution-state, HR-MAS, solid-state, and micro-imaging NMR has been the focus here, because this is where future studies will likely benefit significantly. Finally, while NMR is an extremely versatile and a very powerful analytical approach, the utilization of NMR in combination with other techniques is more desirable. One key direction in which this is being carried out directly is in the area of " hyphenated" NMR approaches.

15.5.2. Hyphenated NMR

Hyphenated or "linked" NMR approaches have existed for decades (Buddrus and Herzog, 1980, 1981; Buddrus et al., 1981); however, recently, due to gains in sensitivity, improved solvent suppression, and an increased interest in mixture analysis, hyphenated NMR has become an increasingly powerful tool. The primarily application for the study of NOM is the hyphenation of NMR spectroscopy with high-performance liquid chromatography (HPLC). Other techniques such as capillary electrophoresis and gas chromatography have been hyphenated with NMR, but at present the detection limit of NMR is hindering their widespread use (Buddrus and Herzog, 1981; Wu et al., 1994). Using a relatively standard but modern HPLC-NMR probe, $\sim\geq 1\,\mu g$ of compound can be detected using 1D ^1H NMR while $\sim\geq 50\,\mu g$ is required for 2D analysis. Using standard analytical columns (4.6-mm diameter), it has been shown that LC-NMR is a potentially very powerful tool for the analysis of NOM (Simpson et al., 2004b). There are, however, a considerable number of drawbacks of directly hyphenated HPLC to NMR: Expensive deuterated solvents are required, suppression of signal from nondeuterated additives (for example, ion pair reagents, or a nondeuterated solvent) are challenging and can suppress signals from the sample, the volume of the NMR detection cell cannot be optimized for samples with both sharp and broad LC peaks, and further hyphenation with mass spectrometry is limited as the deuterated solvents exchange with the sample interfering with mass determination and library searches. LC-SPE-NMR uses solid-phase extraction (SPE) cartridges to trap components prior to NMR analysis and offers many advantages over LC-NMR. Primarily, chromatographic separation can be carried out more economically with the use of nondeuterated solvents and additives that are not compatible with NMR spectroscopy (such as ion pair reagents). Secondly, after trapping on the SPE cartridge, the sample is eluted with a small volume (~25 µl for 2-mm cartridges). This sample concentration provides a substantial increase in sensitivity, especially for broader peaks. Thirdly, multiple "trapping" from subsequent chromatographic separations of the same sample can be carried out on different SPE phases. Thus, even trace components in a mixture can be concentrated until they are amenable to NMR analysis. If required, repeated preparation and sample pre-concentration can be carried out offline such that the NMR spectrometer is available for other experiments. Finally, the deuterated solvent that is used for the elution and transfer is independent of the chromatographic conditions and can be selected to improve spectral quality or make exchangeable protons observable in NMR and facilitates the hyphenation of mass spectrometry. Thus far the LC-SPE-NMR has been successfully applied to study a soil fulvic acid, and it was possible to identify a number of lignin-derived aromatic species (Simpson et al., 2004b). Applications in the future integrating LC-SPE-NMR-MS and other hyphenated forms of NMR will likely be central in identifying specific structures in

NOM and in turn provide key information as to the transformation and reactivity of NOM in the environment. Increased detection limits afforded by cryogenically cooled probes will be especially beneficial in this area.

15.5.3. Cryogenically Cooled Probes

Cryogenically cooled probes (referred to as cryo-probes) are analogous to conventional room-temperature NMR probes, with the exception that the receiver coil and electronics are cooled to temperatures at which they become superconducting. At present, commercial probes are restricted to solution-state NMR, although cryogenically cooled HR-MAS and solid-state NMR probes are in development. In an organic solvent system, cryogenically cooled probes provide a three- to fourfold enhancement in sensitivity (Kim et al., 1995; Odoj et al., 1998; Voehler et al., 2006). This advantage decreases as the salinity of the sample increases. It has been recently shown that with samples that have >4 M NaCl, the use of a smaller diameter temperature probe (1–4 mm) can be advantageous over a cryo-probe, especially for "mass limited" samples (Voehler et al., 2006). However, in NOM research, the sample is often not limited (i.e., at least 50 mg of sample is available). Furthermore, even if samples are measured in an solvent such as D_2O/NaOD, the salt concentration can be carefully controlled such that considerable sensitivity gains in 1D and 2D NMR are observed (Hertkorn et al., 2006). With NOM dissolved in DMSO, excellent receptivity is observed (Simpson et al., 2001c, 2002a; Hertkorn et al., 2006). Thus the applications of cryoprobes for the study of NOM structure is particularly attractive and still in its infancy. The largest drawback of a dedicated cryoprobe for the study of environmental samples is the cost of maintenance (the compressor head needs to be replaced every year at a cost of $20,000 US). With environmental samples, analysis can take many days or weeks; thus recovering costs can be much more challenging than, for example, in the medical arena where high-throughput analysis combined with high user fees allow cryo-probe maintenance to be manageable. However, with this in mind, and considering that a cryogenically cooled probe, under ideal conditions, can permit the study of samples with concentrations ~4 times less, or in one-sixteenth of the time, than that of an analogous room temperature probe, their future role as a key tool in NOM research is highly likely.

15.6. CONCLUSIONS AND FUTURE PROSPECTS

The future of NMR spectroscopy is extremely bright. In the last few years the development of novel techniques has been increasing rapidly. It is now possible to acquire a 2D NMR spectrum of a protein in less than 1 second (Schanda and Brutscher, 2006) or enhance the signals in a carbon spectrum by a factor >44,000 (Ardenkjaer-Larsen et al., 2003) by dynamic nuclear polarization. These techniques, as well as countless others, are continuously evolving and potential applications to NOM are practically unlimited. It is essential that NOM researchers embrace current and future NMR technologies and apply them to their full potential. The key aspect here will be interdisciplinary training of future scientists. Preston (2001) wrote "However, among those investigating SOM the level of understanding of the technique generally remains low, and few graduate students acquire the crucial hands-on experience or adequate understanding of the basic theory. On the other hand, few operators in NMR centers have much understanding of SOM, or familiarity with these types of samples." Unfortunately, this is often still the case and it is critical to stress that a strong foundation in both NMR spectroscopy and NOM research is essential. NMR spectroscopy can be involved, highly theoretical, and thus extremely daunting to a student lacking a strong background in physics, mathematics and chemistry. However, with appropriate training that stresses the information that can be gained and limitations of various techniques, along with hands-on experience, a researcher with little or no previous NMR knowledge can be acquiring and interpreting state-of-the-art NMR experiments in a matter of weeks. It is imperative for professors, lecturers, and facility managers to dig through the reams of theory and select only that essential for students from different disciplines to grasp the fundamentals while still providing the key knowledge required to apply modern NMR effectively to complex samples. Consequently, students that would normally not be exposed to NMR spectroscopy become immersed in the techniques, and overtime can develop into leaders in a field that is undoubtedly essential to the future of NOM research.

REFERENCES

Achtnich, C., Fernandes, E., Bollag, J. M., Knackmuss, H. J., and Lenke, H. (1999). Covalent binding of reduced metabolites of TNT to soil organic matter during a bioremediation process analyzed by ^{15}N NMR spectroscopy. *Environ. Sci. Technol.* **33**, 4448–4456.

Almendros, G., Frund, R., Gonzalezvila, F. J., Haider, K. M., Knicker, H., and Ludemann, H. D. (1991a). Analysis of ^{13}C and ^{15}N CP-MAS NMR-spectra of soil organic-matter and composts. *FEBS Lett.* **282**, 119–121.

Almendros, G., Guadalix, M. E., GonzalezVila, F. J., and Martin, F. (1996). Preservation of aliphatic macromolecules in soil humins. *Org. Geochem.* **24**, 651–659.

Almendros, G., Sanz, J., Gonzalezvila, F. J., and Martin, F. (1991b). Evidence for a polyalkyl nature of soil humin. *Naturwissenschaften* **78**, 359–362.

Aluwihare, L. I., and Repeta, D. J. (1999). A comparison of the chemical characteristics of oceanic DOM and extracellular DOM produced by marine algae. *Mar. Ecol. Prog. Ser.* **186**, 105–117.

Aluwihare, L. I., Repeta, D. J., and Chen, R. F. (1997). A major biopolymeric component to dissolved organic carbon in surface sea water. *Nature* **387**, 166–169.

Aluwihare, L. I., Repeta, D. J., Pantoja, S., and Johnson, C. G. (2005). Two chemically distinct pools of organic nitrogen accumulate in the ocean. *Science* **308**, 1007–1010.

Amin, M. H., Richards, K. S., Chorley, R. J., Gibbs, S. J., Carpenter, T. A., and Hall, L. D. (1996). Studies of soil–water transport by MRI. *Magn. Reson. Imag.* **14**, 879–882.

Andrasko, J., Bull, T. E., and Lindqvis, I. (1972). Quadrupole relaxation in solutions of humic acids. *Chem. Scripta* **2**, 93–95.

Andrew, E. R., Bradbury, A., and Eades, R. G. (1958). Nuclear magnetic resonance spectra from a crystal rotated at high speed. *Nature* **182**, 1659–1659.

Andrew, E. R., Bradbury, A., and Eades, R. G. (1959). Removal of dipolar broadening of nuclear magnetic resonance spectra of solids by specimen rotation. *Nature* **183**, 1802–1803.

Ardenkjaer-Larsen, J. H., Fridlund, B., Gram, A., Hansson, G., Hansson, L., Lerche, M. H., Servin, R., Thaning, M., and Golman, K. (2003). Increase in signal-to-noise ratio of >10,000 times in liquid-state NMR. *Proc. Natl. Acad. Sci. USA* **100**, 10158–10163.

Arnarson, T. S., and Keil, R. G. (2000). Mechanisms of pore water organic matter adsorption to montmorillonite. *Marine Chem.* **71**, 309–320.

Baldock, J. A., Oades, J. M., Waters, A. G., Peng, X., Vassallo, A. M., and Wilson, M. A. (1992). Aspects of the chemical-structure of soil organic materials as revealed by solid-state ^{13}C NMR spectroscopy. *Biogeochemistry* **16**, 1–42.

Baldock, J. A., and Skjemstad, J. O. (2000). Role of the soil matrix and minerals in protecting natural organic materials against biological attack. *Org. Geochem.* **31**, 697–710.

Baldock, J. A., and Smernik, R. J. (2002). Chemical composition and bioavailability of thermally altered Pinus resinosa (red pine) wood. *Organic Geochemistry* **33**, 1093–1109.

Barron, P. F., and Wilson, M. A. (1981). Humic soil and coal structure study with magic-angle spinning ^{13}C CP NMR. *Nature* **289**, 275–276.

Barton, D. H. R., and Schnitzer, M. (1963). A new experimental approach to humic acid problem. *Nature* **198**, 217–218.

Benner, R., Pakulski, J. D., McCarthy, M., Hedges, J. I., and Hatcher, P. G. (1992). Bulk chemical characteristics of dissolved organic-matter in the ocean. *Science* **255**, 1561–1564.

Bondar, V. S., Boersma, M. G., Golovlev, E. L., Vervoort, J., Van Berkel, W. J. H., Finkelstein, Z. I., Solyanikova, I. P., Golovleva, L. A., and Rietjens, I. M. C. M. (1998). ^{19}F NMR study on the biodegradation of fluorophenols by various Rhodococcus species. *Biodegradation* **9**, 475–486.

Bonin, J. L., and Simpson, M. J. (2007). Variation in phenanthrene sorption coefficients with soil organic matter fractionation: The result of structure or conformation? *Environ. Sci. Technol.* **41**, 153–159.

Bortiatynski, J. M., Hatcher, P. G., and Minard, R. D. (1997). The development of ^{13}C labeling and ^{13}C NMR spectroscopy techniques to study the interaction of pollutants with humic substances. In *Nuclear Magnetic Resonance Spectroscopy in Environmental Chemistry*. Nanny, M. A., Minear, R. A., and Leenheer, J. A., eds., Oxford University Press, New York, pp. 26–50.

Bruns-Nagel, D., Knicker, H., Drzyzga, O., Butehorn, U., Steinbach, K., Gemsa, D., and Von Low, E. (2000). Characterization of ^{15}N TNT residues after an anaerobic/aerobic treatment of soil/molasses mixtures by solid state ^{15}N NMR spectroscopy. 2. Systematic investigation of whole soil and different humic fractions. *Environ. Sci. Technol.* **34**, 1549–1556.

Bryce, D. L., Bernard, G. M., Gee, M., Lumsden, M. D., Eichele, K., and Wasylishen, R. (2001). Practical aspects of modern routine solid-state multinuclear magnetic resonance spectroscopy: One-dimensional experiments. *Can. J. Anal. Sci. Spectrosc.* **46**, 46–82.

Buddrus, J., Burba, P., Herzog, H., and Lambert, J. (1989). Quantitation of partial structures of aquatic humic substances by one-dimensional and two-dimensional solution ^{13}C nuclear magnetic-resonance spectroscopy. *Anal. Chem.* **61**, 628–631.

Buddrus, J., and Herzog, H. (1980). Coupling of HPLC and NMR. 1. Analysis of flowing liquid-chromatographic fractions by proton magnetic-resonance. *Org. Magn. Res.* **13**, 153–155.

Buddrus, J., and Herzog, H. (1981). Coupling of chromatography and NMR. 3. Study of flowing gas-chromatographic fractions by proton magnetic-resonance. *Org. Magn. Res.* **15**, 211–213.

Buddrus, J., Herzog, H., and Cooper, J. W. (1981). Coupling of HPLC and NMR. 2. Analysis of flowing liquid-chromatographic fractions by proton magnetic-resonance in the presence of hydrogen-containing eluting solvents. *J. Magn. Reson.* **42**, 453–459.

Buurman, P., Schellekens, J., Fritze, H., and Nierop, K. G. J. (2007). Selective depletion of organic matter in mottled podzol horizons. Soil Biol. *Biochem.* **39**, 607–621.

Cardoza, L. A., Korir, A. K., Otto, W. H., Wurrey, C. J., and Larive, C. K. (2004). Applications of NMR spectroscopy in environmental science. *Prog. Nucl. Magn. Res. Sp.* **45**, 209–238.

Carper, W. R. (1999). Direct determination of quadrupolar and dipolar NMR correlation times from spin–lattice and spin–spin relaxation rates. *Concept Mag. Res.* **11**, 51–60.

Chefetz, B., Deshmukh, A. P., Hatcher, P. G., and Guthrie, E. A. (2000). Pyrene sorption by natural organic matter. *Environ. Sci. Technol.* **34**, 2925–2930.

Chefetz, B., Salloum, M. J., Deshmukh, A. P., and Hatcher, P. G. (2002). Structural components of humic acids as determined by chemical modifications and ^{13}C NMR, pyrolysis-, and thermochemolysis-gas chromatography/mass spectrometry. *Soil Sci. Soc. Am. J.* **66**, 1159–1171.

Chen, Q., Kinzelbach, W., and Oswald, S. (2002). Nuclear magnetic resonance imaging for studies of flow and transport in porous media. *J. Environ. Qual.* **31**, 477–486.

Chen, R. F., Repeta, D. J., and Eglinton, T. I. (1994). Characterization of dissolved organic-matter in seawater by ^{1}H NMR and analytical pyrolysis. *Abs. Pap. Am. Chem. Soc.* **207**, 121-GEOC.

Chien, Y. Y., Kim, E. G., and Bleam, W. F. (1997). Paramagnetic relaxation of atrazine solubilized by humic micellar solutions. *Environ. Sci. Technol.* **31**, 3204–3208.

Chiou, C. T., Kile, D. E., Rutherford, D. W., Sheng, G., and Boyd, S. A. (2000). Sorption of selected organic compounds from water to a peat soil and its humic-acid and humin fractions: Potential sources of the sorption nonlinearity. *Environ. Sci. Technol.* **34**, 1254–1258.

Chorover, J., and Amistadi, M. K. (2001). Reaction of forest floor organic matter at goethite, birnessite and smectite surfaces. *Geochim. Cosmochim. Acta* **65**, 95–109.

Chu, Y. J., Werth, C. J., Valocchi, A. J., Yoon, H., and Webb, A. G. (2004). Magnetic resonance imaging of nonaqueous phase liquid during soil vapor extraction in heterogeneous porous media. *J.Cont. Hydrol.* **73**, 15–37.

Chung, K. H., Rhee, S. W., Shin, H. S., and Moon, C. H. (1996). Probe of cadmium(II) binding on soil fulvic acid investigated by ^{113}Cd NMR spectroscopy. *Can. J. Chem.* **74**, 1360–1365.

Clinton, P. W., Newman, R. H., and Allen, R. B. (1995). Immobilization of ^{15}N in forest litter studied by ^{15}N CPMAS NMR spectroscopy. *Eur. J. Soil Sci.* **46**, 551–556.

Collins, M. J., Bishop, A. N., and Farrimond, P. (1995). Sorption by mineral surfaces: rebirth of the classical condensation pathway for kerogen formation? *Geochim. Cosmochim. Acta* **59**, 2387–2391.

Conte, P., Piccolo, A., Van Lagen, B., Buurman, P., and de Jager, P. A. (1997). Quantitative aspects of solid-state C-13-NMR spectra of humic substances from soils of volcanic systems. *Geoderma* **80**, 327–338.

Conte, P., Spaccini, R., and Piccolo, A. (2004). State of the art of CPMAS ^{13}C NMR spectroscopy applied to natural organic matter. *Prog. Nucl. Mag. Res. Sp.* **44**, 215–223.

Cook, R., and Langford, C. (1998). Structural characterization of a fulvic acid and a humic acid using solid-state ramp-CP-MAS ^{13}C nuclear magnetic resonance. *Environ. Sci. Technol.* **32**, 719–725.

Cook, R., and Langford, C. (1999). Ramped amplitude cross polarization magic angle spinning NMR (Ramp-CP-MAS-NMR): A technique for quantitative study of the composition of solid state polymers. *Polymer News* **24**, 6–15.

Cook, R. L. (2004). Coupling NMR to NOM. *Anal. Bioanal. Chem.* **378**, 1484–1503.

Cook, R. L., Langford, C. H., et al. (1996). A modified cross-polarization magic angle spinning C-13 NMR procedure for the study of humic materials. *Anal. Chem.* **68**, 3979–3986.

Cook, R. L., McIntyre, D. D., Langford, C. H., and Vogel, H. J. (2003). A comprehensive liquid-state heteronuclear and multidimensional NMR study of Laurentian fulvic acid. *Environ. Sci. Technol.* **37**, 3935–3944.

Culligan, P. J., Sinfield, J. V., Maas, W. E., and Cory, D. G. (2001). Use of NMR relaxation times to differentiate mobile and immobile pore fractions in a wetland soil. *Water Resour. Res.* **37**, 837–842.

Decesari, S., Facchini, M. C., Fuzzi, S., McFiggans, G. B., Coe, H., and Bower, K. N. (2005). The water-soluble organic component of size-segregated aerosol, cloud water and wet depositions from Jeju Island during ACE-Asia. *Atmos. Environ.* **39**, 211–222.

Dehorter, B., Kontchou, C. Y., and Blondeau, R. (1992). ^{13}C NMR Spectroscopic Analysis of soil humic acids recovered after incubation with some white rot fungi and Actinomycetes. *Soil Biol. Biochem.* **24**, 667–673.

Delort, A.-M., Combourieu, B., and Haroune, N. (2004). Nuclear magnetic resonance studies of interactions between organic pollutants and soils components, a review. *Environ. Chem. Lett.* **1**, 209–213.

Derenne, S., and Largeau, C. (2001). A review of some important families of refractory macromolecules: Composition, origin, and fate in soils and sediments. *Soil Sci.* **166**, 833–847.

Derenne, S., Largeau, C., Hetenyi, M., Brukner-Wein, A., Connan, J., and Lugardon, B. (1997). Chemical structure of the organic matter in a Pliocene maar-type shale: implicated Botryococcus race strains and formation pathways. *Geochim. Cosmochim. Acta* **61**, 1879–1889.

Deshmukh, A. P., Simpson, A. J., and Hatcher, P. G. (2003). Evidence for cross-linking in tomato cutin using HR-MAS NMR spectroscopy. *Phytochemistry* **64**, 1163–1170.

Dickens, A. F., Baldock, J. A., Smernik, R. J., Wakeham, S. G., Arnarson, T. S., Gelinas, Y., and Hedges, J. I. (2006). Solid-state ^{13}C NMR analysis of size and density fractions of marine sediments: Insight into organic carbon sources and preservation mechanisms. *Geochim. Cosmochim. Acta* **70**, 666–686.

Dixon, A. M., Mai, M. A., and Larive, C. K. (1999). NMR investigation of the interactions between 4-fluoro-1-acetonaphthone and the Suwannee river fulvic acid. *Environ. Sci. Technol.* **33**, 958–964.

Dixon, W. T., Schaefer, J., Sefcik, M. D., Stejskal, E. O., and Mckay, R. A. (1982). Total Suppression of Sidebands in CP-MAS ^{13}C NMR. *J. Magn. Reson.* **49**, 341–345.

Dria, K. J., Sachleben, J. R., and Hatcher, P. G. (2002). Solid-state ^{13}C nuclear magnetic resonance of humic acids at high magnetic field strengths. *J. Environ. Qual.* **31**, 393–401.

Emery, E. F., Junk, T., Ferrell, R. E., De Hon, R., and Butler, L. G. (2001). Solid-state ^{2}H MAS NMR studies of TNT absorption in soil and clays. *Environ. Sci. Technol.* **35**, 2973–2978.

Fan, T. W. M., Higashi, R. M., and Lane, A. N. (2000). Chemical characterization of a chelator-treated soil humate by solution-state multinuclear two-dimensional NMR with FTIR and pyrolysis-GCMS. *Environ. Sci. Technol.* **34**, 1636–1646.

Feng, X., Simpson, A. J., and Simpson, M. J. (2005). Chemical and mineralogical controls of humic acid sorption to clay mineral surfaces. *Org. Geochem.* **36**, 1553–1566.

Feng, X. J., Simpson, A. J., and Simpson, M. J. (2006). Investigating the role of mineral-bound humic acid in phenanthrene sorption. *Environ. Sci. Technol.* **40**, 3260–3266.

Garbarini, D. R., and Lion, L. W. (1986). Influence of the nature of soil organics on the sorption of toluene and trichloroethylene. *Environ. Sci. Technol.* **20**, 1263–1269.

Gelinas, Y., Baldock, J. A., and Hedges, J. I. (2001). Demineralization of marine and freshwater sediments for CP-MAS ^{13}C NMR analysis. *Org. Geochem.* **32**, 677–693.

Gonçalves, C. N., Dalmolin, R. S. D., Dick, D. P., Knicker, H., Klamt, E., and Kögel-Knabner, I. (2003). The effect of 10% HF treatment on the resolution of CPMAS 13C NMR spectra and on the quality of organic matter in ferralsols. *Geoderma* **116**, 373–392.

Grassi, M., and Daquino, V. (2005). ^{113}Cd NMR and fluorescence studies of multiple binding mechanisms of Cd(II) by the Suwannee River fulvic acid. *Ann. Chim-Rome* **95**, 579–591.

Green, N. A., Meharg, A. A., Till, C., Troke, J., and Nicholson, J. K. (1999). Degradation of 4-fluorobiphenyl in soil investigated by ^{19}F NMR spectroscopy and ^{14}C radiolabelling analysis. *Chemosphere* **38**, 1085–1101.

Guggenberger, G., and Kaiser, K. (2003). Dissolved organic matter in soil: Challenging the paradigm of sorptive preservation. *Geoderma* **113**, 293–310.

Guggenberger, G., Zech, W., and Schulten, H. R. (1994). Formation and mobilization pathways of dissolved organic-matter—Evidence from chemical structural studies of organic-matter fractions in acid forest floor solutions. *Org. Geochem.* **21**, 51–66.

Guignard, C., Lemee, L., and Ambles, A. (2005). Lipid constituents of peat humic acids and humin. Distinction from directly extractable bitumen components using TMAH and TEAAc thermochemolysis. *Org. Geochem.* **36**, 287–297.

Haiber, S., Burba, P., Herzog, H., and Lambert, J. (1999). Elucidation of aquatic humic partial structures by multistage ultrafiltration and two-dimensional nuclear magnetic resonance spectrometry. *Fresenius J. Anal Chem.* **364**, 215–218.

Haiber, S., Herzog, H., Burba, P., Gosciniak, B., and Lambert, J. (2001a). Quantification of carbohydrate structures in size fractionated aquatic humic substances by two-dimensional nuclear magnetic resonance. *Fresenius J. Anal Chem.* **369**, 457–460.

Haiber, S., Herzog, H., Burba, P., Gosciniak, B., and Lambert, J. (2001b). Two-dimensional NMR studies of size fractionated Suwannee River fulvic and humic acid reference. *Environ. Sci. Technol.* **35**, 4289–4294.

Hatcher, P. G., Bortiatynski, J. M., Minard, R. D., Dec, J., and Bollag, J. M. (1993). Use of high-resolution ^{13}C NMR to examine the enzymatic covalent binding of ^{13}C-labeled 2,4-dichlorophenol to humic substances. *Environ. Sci. Technol.* **27**, 2098–2103.

Hatcher, P. G., Breger, I. A., and Mattingly, M. A. (1980). Structural characteristics of fulvic-acids from continental-shelf sediments. *Nature* **285**, 560–562.

Hatcher, P. G., Dria, K. J., Kim, S., and Frazier, S. W. (2001). Modern analytical studies of humic substances. *Soil Sci.* **166**, 770–794.

Havers, N., Burba, P., Lambert, J., and Klockow, D. (1998). Spectroscopic characterization of humic-like substances in airborne particulate matter. *J. Atmos. Chem.* **29**, 45–54.

Hedges, J. I. (2002). Why dissolved organics matter? In *Biogeochemistry of Marine Dissolved Organic Matter*, Hansell, D. A., and Carlson, C. A., eds., Academic Press, New York, pp. 1–33.

Hejzlar, J., Szpakowska, B., and Wershaw, R. L. (1994). Comparison of humic substances isolated from peatbog water by sorption on Deae-cellulose and Amberlite XAD-2. *Water Res.* **28**, 1961–1970.

Hernes, P. J., Benner, R., Cowie, G. L., Goni, M. A., Bergamaschi B. A., and Hedges, J. I. (2001). Tannin diagenesis in mangrove leaves from a tropical estuary: A novel molecular approach. *Geochim. Cosmochim. Acta* **65**, 3109–3122.

Hertkorn, N., Benner, R., Frommberger, M., Schmitt-Kopplin, P., Witt, M., Kaiser, K., Kettrup, A., and Hedges, J. I. (2006). Characterization of a major refractory component of marine dissolved organic matter. *Geochim. Cosmochim. Acta* **70**, 2990–3010.

Hertkorn, N., Claus, H., Schmitt-Kopplin, P. H., Perdue, E. M., and Filip, Z. (2002a). Utilization and transformation of aquatic humic substances by autochthonous microorganisms. *Environ. Sci. Technol.* **36**, 4334–4345.

Hertkorn, N., Perdue, E. M., and Kettrup, A. (2004). A potentiometric and ^{113}Cd NMR study of cadmium complexation by natural organic matter at two different magnetic field strengths. *Anal. Chem.* **76**, 6327–6341.

Hertkorn, N., Permin, A., Perminova, I., Kovalevskii, D., Yudov, M., Petrosyan, V., and Kettrup, A. (2002b). Comparative analysis of partial structures of a peat humic and fulvic acid using one- and two-dimensional nuclear magnetic resonance spectroscopy. *J. Environ. Qual.* **31**, 375–387.

Hinedi, Z. R., Chang, A. C., and Borchardt, D. B. (1997). Probing the association of fluorobenzene with dissolved organic matter using NMR spectroscopy. *Water Res.* **31**, 877–883.

Hong, J., and Harbison, G. S. (1993). Magic-angle-spinning side-band elimination by temporary interruption of the chemical-shift. *J. Magn. Reson. Ser. A* **105**, 128–136.

Howe, R. F., Lu, X. Q., Hook, J., and Johnson, W. D. (1997). Reaction of aquatic humic substances with aluminium: A ^{27}Al NMR study. *Mar. Freshwater Res.* **48**, 377–383.

Hsu, P. H., and Hatcher, P. G. (2005). New evidence for covalent coupling of peptides to humic acids based on 2D NMR spectroscopy: A means for preservation. *Geochim. Cosmochim. Acta* **69**, 4521–4533.

Hu, W. G., Mao, J. D., Xing, B. S., and Schmidt-Rohr, K. (2000). Polymethylene crystallites in humic substances detected by nuclear magnetic resonance. *Environ. Sci. Technol.* **34**, 530–534.

Humbert, F. (2001). Potentials of radio-frequency field gradient NMR microscopy in environmental science. *J. Indust. Microbio. Biotech.* **26**, 53–61.

Hwang, J., Druffel, E. R. M., Eglinton, T. I., and Repeta, D. J. (2006). Source(s) and cycling of the nonhydrolyzable organic fraction of oceanic particles. *Geochim. Cosmochim. Acta* **70**, 5162–5168.

Jaeger, F., Grohmann, E., and Schaumann, G. E. (2006). ^{1}H NMR relaxometry in natural humous soil samples: Insights in microbial effects on relaxation time distributions. *Plant Soil* **280**, 209–222.

Jayasundera, S., Torrents, A., and Schmidt, W. J. (1997). Noncovalent interactions between acenaphthenone and dissolved fulvic acid as determined by ^{13}C NMR T_1 relaxation measurements—Comment. *Environ. Sci. Technol.* **31**, 3742–3743.

Kacker, T., Haupt, E. T. K., Garms, C., Francke, W., and Steinhart, H. (2002). Structural characterisation of humic acid-bound PAH residues in soil by ^{13}C CPMAS-NMR-spectroscopy: evidence of covalent bonds. *Chemosphere* **48**, 117–131.

Kaiser, E., Simpson, A. J., Dria, K. J., Sulzberger, B., and Hatcher, P. G. (2003). Solid-state and multidimensional solution-state NMR of solid phase extracted and ultrafiltered riverine dissolved organic matter. *Environ. Sci. Technol.* **37**, 2929–2935.

Kang, S. H., Amarasiriwardena, D., Veneman, P., and Xing, B. S. (2003). Characterization of ten sequentially extracted humic acids and a humin from a soil in western Massachusetts. *Soil Sci.* **168**, 880–887.

Kang, S. H., and Xing, B. S. (2005). Phenanthrene sorption to sequentially extracted soil humic acids and humins. *Environ. Sci. Technol.* **39**, 134–140.

Kazpard, V., Lartiges, B. S., Frochot, C., de la Caillerie, J. B. D., Viriot, M. L., Portal, J. M., Gorner, T., and Bersillon, J. L. (2006). Fate of coagulant species and conformational effects during the aggregation of a model of a humic substance with Al-31 polycations. *Water Res.* **40**, 1965–1974.

Keeler, C., Kelly, E. F., and Maciel, G. E. (2006). Chemical–structural information from solid-state C-13 NMR studies of a suite of humic materials from a lower montane forest soil, Colorado, USA. *Geoderma* **130**, 124–140.

Keeler, C., and Maciel, G. E. (2003). Quantitation in the solid-state ^{13}C NMR analysis of soil and organic soil fractions. *Anal. Chem.* **75**, 2421–2432.

Keeler, J. (2005). *Understanding NMR Spectroscopy*, John Wiley & Sons, Chichester, UK.

Keifer, P. A., Baltusis, L., Rice, D. M., Tymiak, A. A., and Shoolery, J. N. (1996). A comparison of NMR spectra obtained for solid-phase-synthesis resins using conventional high-resolution, magic-angle-spinning, and high-resolution magic-angle-spinning probes. *J. Magn. Reson. Ser. A* **119**, 65–75.

Kelleher, B. P., and Simpson, A. J. (2006). Humic substances in soils: Are they really chemically distinct? *Environ. Sci. Technol.* **40**, 4605–4611.

Kelleher, B. P., Simpson, M. J., and Simpson, A. J. (2006). Assessing the fate and transformation of plant residues in the terrestrial environment using HR-MAS NMR spectroscopy. *Geochim. Cosmochim. Acta* **70**, 4080–4094.

Khalaf, M., Kohl, S. D., Klumpp, E., Rice, J. A., and Tombacz, E. (2003). Comparison of sorption domains in molecular weight fractions of a soil humic acid using solid-state ^{19}F NMR. *Environ. Sci. Technol.* **37**, 2855–2860.

Kim, S., Simpson, A. J., Kujawinski, E. B., Freitas, M. A., and Hatcher, P. G. (2003). High resolution electrospray ionization mass spectrometry and 2D solution NMR for the analysis of DOM extracted by C_{18} solid phase disk. *Org. Geochem.* **34**, 1325–1335.

Kim, Y. W., Earl, W. L., and Norberg, R. E. (1995). Cryogenic probe with low-loss transmission-line for nuclear-magnetic-resonance. *J. Magn. Reson. Ser. A* **116**, 139–144.

Kingery, W. L., Simpson, A. J., Hayes, M. H. B., Locke, M. A., and Hicks, R. P. (2000). The application of multidimensional NMR to the study of soil humic substances. *Soil Sci.* **165**, 483–494.

Knicker, H. (2000a). Biogenic nitrogen in soils as revealed by solid-state ^{13}C and ^{15}N nuclear magnetic resonance spectroscopy. *J. Environ. Qual.* **29**, 715–723.

Knicker, H. (2000b). Solid-state 2-D double cross polarization magic angle spinning ^{15}N ^{13}C NMR spectroscopy on degraded algal residues. *Org. Geochem.* **31**, 337–340.

Knicker, H. (2001). Incorporation of inorganic nitrogen into humified organic matter as revealed by solid-state 2-D ^{15}N ^{13}C NMR spectroscopy. *Abstr. Pap. Am. Chem. Soc.* **221**, U517–U517.

Knicker, H. (2002). The feasibility of using DCPMAS ^{15}N ^{13}C NMR spectroscopy for a better characterization of immobilized N-15 during incubation of ^{13}C and ^{15}N enriched plant material. *Org. Geochem.* **33**, 237–246.

Knicker, H. (2004). Stabilization of N-compounds in soil and organic-matter-rich sediments—What is the difference? *Marine Chem.* **92**, 167–195.

Knicker, H., Bruns-Nagel, D., Drzyzga, O., Von Low, E., and Steinbach, K. (1999). Characterization of N-15-TNT residues after an anaerobic/aerobic treatment of soil/molasses mixtures by solid state N-15 NMR spectroscopy. 1. Determination and optimization of relevant NMR spectroscopic parameters. *Environ. Sci. Technol.* **33**, 343–349.

Knicker, H., Frund, R., and Ludemann, H. D. (1993). The chemical nature of nitrogen in native soil organic-matter. *Naturwissenschaften* **80**, 219–221.

Knicker, H., and Hatcher, P. G. (2001). Sequestration of organic nitrogen in the sapropel from Mangrove Lake, Bermuda. *Org. Geochem.* **32**, 733–744.

Knicker, H., Hatcher, P. G., and Gonzalez-Vila, F. J. (2002). Formation of heteroaromatic nitrogen after prolonged humification of vascular plant remains as revealed by nuclear resonance spectroscopy. *J. Environ. Qual.* **31**, 444–449.

Knicker, H., Hatcher, P. G., and Scaroni, A. W. (1996). A solid-state ^{15}N NMR spectroscopic investigation of the origin of nitrogen structures in coal. *Int. J.* Coal Geol. **32**, 255–278.

Knicker, H., and Ludemann, H. D. (1995). ^{15}N and ^{13}C CP-MAS and solution NMR-studies of ^{15}N enriched plant-material during 600 days of microbial-degradation. *Org. Geochem.* **23**, 329–341.

Kögel-Knabner, I. (1997). ^{13}C and ^{15}N NMR spectroscopy as a tool in soil organic matter studies. *Geoderma* **80**, 243–270.

Kögel-Knabner, I. (2000). Analytical approaches for characterizing soil organic matter. *Org. Geochem.* **31**, 609–625.

Kögel-Knabner, I. (2002). The macromolecular organic composition of plant and microbial residues as inputs to soil organic matter. *Soil Biol. Biochem.* **34**, 139–162.

Kögel-Knabner, I., Hatcher, P. G., Tegelaar, E. W., and Deleeuw, J. W. (1992). Aliphatic components of forest soil organic-matter as determined by solid-state ^{13}C NMR and analytical pyrolysis. *Sci. Total Environ.* **113**, 89–106.

Kögel-Knabner, I., and Ziegler, F. (1993). Carbon distribution in different compartments of forest soils. *Geoderma* **56**, 515–525.

Kohl, S. D., and Rice, J. A. (1998). The binding of contaminants to humin: a mass balance. *Chemosphere* **36**, 251–261.

Kohl, S. D., Toscano, P. J., Hou, W. H., and Rice, J. A. (2000). Solid-state ^{19}F NMR investigation of hexafluorobenzene sorption to soil organic matter. *Environ. Sci. Technol.* **34**, 204–210.

Kolowith, L. C., Ingall, E. D., and Benner, R. (2001). Composition and cycling of marine organic phosphorus. *Limnol. Oceanogr.* **46**, 309–320.

Larive, C. K., Rogers, A., Morton, M., and Carper, W. R. (1996). ^{113}Cd NMR binding studies of Cd–Fulvic acid complexes: Evidence of fast exchange. *Environ. Sci. Technol.* **30**, 2828–2831.

Larive, C. K., Rogers, A., Morton, M., and Carper, W. R. (1997). ^{113}Cd NMR binding studies of Cd-fulvic acid complexes: Evidence of fast exchange. *Environ. Sci. Technol.* **31**, 948.

Lead, J. R., Wilkinson, K. J., Balnois, E., Cutak, B. J., Larive, C. K., Assemi, S., and Beckett, R. (2000). Diffusion coefficients and polydispersities of the Suwannee River fulvic acid: Comparison of fluorescence correlation spectroscopy, pulsed-field gradient nuclear magnetic resonance, and flow field-flow fractionation. *Environ. Sci. Tech.* **34**, 3508–3513.

Leenheer, J. A., Nanny, M. A., and McIntyre, C. (2003). Terpenoids as major precursors of dissolved organic matter in landfill leachates, surface water, and groundwater. *Environ. Sci. Technol.* **37**, 2323–2331.

Leenheer, J. A., Noyes, T. I., Rostad, C. E., and Davisson, M. L. (2004). Characterization and origin of polar dissolved organic matter from the Great Salt Lake. *Biogeochemistry* **69**, 125–141.

Leenheer, J. A., Wershaw, R. L., Brown, P. A., and Noyes, T. I. (1991). Detection of poly(ethylene glycol) residues from nonionic surfactants in surface-water by ^{1}H and ^{13}C nuclear-magnetic-resonance spectrometry. *Environ. Sci. Technol.* **25**, 161–168.

Leisen, J., Schmidtrohr, K., and Spiess, H. W. (1993). Nonexponential relaxation functions above T(G) analyzed by multidimensional NMR and novel spin-echo decay techniques. *Physica A* **201**, 79–87.

Lens, P. N. L., and Hemminga, M. A. (1998). Nuclear magnetic resonance in environmental engineering: Principles and applications. *Biodegradation* **9**, 393–409.

Li, J., Perdue, E. M., and Gelbaum, L. T. (1998). Using ^{113}Cd NMR spectrometry to study metal complexation by natural organic matter. *Environ. Sci. Technol.* **32**, 483–487.

Lichtfouse, E., Derenne, S., Mariotti, A., and Largeau, C. (1994). Possible algal origin of long chain odd *n*-alkanes in immature sediments as revealed by distributions and carbon isotope ratios. *Org. Geochem.* **22**, 1023–1027.

Lindman, B., and Lindqvis, I. (1969). Nuclear quadrupole relaxation of ^{85}RB in an aqueous solution of a humic acid. *Acta Chem. Scand.* **23**, 2215–2216.

Lindqvis, I., and Lindman, B. (1970). Quadrupole relaxation studies of humic acids. *Acta Chem. Scand.* **24**, 1097–1098.

Lindman, B., and Lindqvis, I. (1971). Quadrupole relaxation studies of metal-ion bonding to humic acids. *Chem. Scripta* **1**, 195–200.

Lookman, R., Grobet, P., Merckx, R., and Van Riemsdijk, W. H. (1997). Application of P-31 and Al-27 MAS NMR for phosphate speciation studies in soil and aluminium hydroxides: Promises and constraints. *Geoderma* **80**, 369–388.

Lu, X. Q., Johnson, W. D., and Hook, J. (1998). Reaction of vanadate with aquatic humic substances: An ESR and V-51 NMR study. *Environ. Sci Technol.* **32**, 2257–2263.

Mahieu, N., Olk, D. C., and Randall, E. W. (2000). Accumulation of heterocyclic nitrogen in humified organic matter: A N-15-NMR study of lowland rice soils. *Eur. J. Soil Sci.* **51**, 379–389.

Mahieu, N., Powlson, D. S., and Randall, E. W. (1999). Statistical analysis of published ^{13}C CPMAS NMR spectra of soil organic matter. *Soil Sci. Soc. Am. J.* **63**, 307–319.

Maie, N., Knicker, H., Watanabe, A., and Kimura, M. (2006a). Heterocyclic N in the highly humified humic acids extracted from the subsoil of paddy fields and surface ando soils. *Org. Geochem.* **37**, 12–19.

Maie, N., Parish, K. J., Watanabe, A., Knicker, H., Benner, R., Abe, T., Kaiser, K., and Jaffe, R. (2006b). Chemical characteristics of dissolved organic nitrogen in an oligotrophic subtropical coastal ecosystem. *Geochim. Cosmochim. Acta* **70**, 4491–4506.

Malcolm, R. L. (1989). Applications of solid-state ^{13}C NMR spectroscopy to geochemical studies of humic substances. In *Humic Substances II. In Search of Structure*, Hayes, M. H. B., MacCarthy, P., Malcolm, R. L., and Swift, R. S., John Wiley and Sons Ltd., New York, 340–372.

Mao, J. D., Hu, W. G., Schmidt-Rohr, K., Davies, G., Ghabbour, E. A., and Xing, B. S. (2000). Quantitative characterization of humic substances by solid-state ^{13}C nuclear magnetic resonance. *Soil Sci. Soc. Am. J.* **64**, 873–884.

Mao, J. D., Hub, W. G., Ding, G. W., Schmidt-Rohr, K., Davies, G., Ghabbour, E. A., and Xing, B. S. (2002a). Suitability of different ^{13}C solid-state NMR techniques in the characterization of humic acids. *Int. J. Environ. Anal. Chem.* **82**, 183–196.

Mao, J.-D., Hundal, L. S., Thompson, M. L., and Scmidt-Rohr, K. (2002b). Correlation of polymethylene-rich amorphous aliphatic domains in humic substances with sorption of a nonpolar organic contaminant, phenanthrene. *Environ. Sci. Technol.* **36**, 929–936.

Mao, J. D., and Schmidt-Rohr, K. (2003). Recoupled long-range C-H dipolar dephasing in solid-state NMR, and its use for spectral selection of fused aromatic rings. *J. Magn. Reson.* **162**, 217–227.

Mao, J. D., and Schmidt-Rohr, K. (2004a). Accurate quantification of aromaticity and nonprotonated aromatic carbon fraction in natural organic matter by ^{13}C solid-state nuclear magnetic resonance. *Environ. Sci. Technol.* **38**, 2680–2684.

Mao, J. D., and Schmidt-Rohr, K. (2004b). Separation of aromatic-carbon ^{13}C NMR signals from di-oxygenated alkyl bands by a chemical-shift-anisotropy filter. *Solid State Nucl. Magn. Reson.* **26**, 36–45.

Mao, J. D., and Schmidt-Rohr, K. (2005). Methylene spectral editing in solid-state ^{13}C NMR by three-spin coherence selection. *J. Magn. Reson.* **176**, 1–6.

Mao, J. D., and Schmidt-Rohr, K. (2006). Absence of mobile carbohydrate domains in dry humic substances proven by NMR, and implications for organic-contaminant sorption models. *Environ. Sci. Technol.* **40**, 1751–1756.

Mao, J.-D., Xing, B., and Schmidt-Rohr, K. (2001). New structural information on a humic acid from two-dimensional 1H-13C correlation solid-state nuclear magnetic resonance. *Environ. Sci. Technol.* **35**, 1928–1934.

Matthias, A., Maurer, M., and Parlar, H. (2003). Comparative aluminium speciation and quantification in soil solutions of two different forest ecosystems by 27Al NMR. *Fresenius Environ. Bull.* **12**, 1263–1275.

Mayer, L. M. (1994). Surface area control of organic carbon accumulation in continental shelf sediments. *Geochim. Cosmochim. Acta* **58**, 1271–1284.

McCarthy, M., Pratum, T., Hedges, J., and Benner, R. (1997). Chemical composition of dissolved organic nitrogen in the ocean. *Nature* **390**, 150–154.

McCarthy, M. D., Hedges, J. I., and Benner, R. (1998). Major bacterial contribution to marine dissolved organic nitrogen. *Science* **281**, 231–234.

McKnight, D. M., Harnish, R., Wershaw, R. L., Baron, J. S., and Schiff, S. (1997). Chemical characteristics of particulate, colloidal, and dissolved organic material in Loch Vale Watershed, Rocky Mountain National Park. *Biogeochemistry* **36**, 99–124.

Morris, K. F., Cutak, B. J., Dixon, A. M., and Larive, C. K. (1999). Analysis of diffusion coefficient distributions in humic and fulvic acids by means of diffusion ordered NMR spectroscopy. *Anal. Chem.* **71**, 5315–5321.

Nanny, M. A. (1999). Deuterium NMR characterization of noncovalent interactions between monoaromatic compounds and fulvic acids. *Org. Geochem.* **30**, 901–909.

Nanny, M. A., Bortiatynski, J. M., and Hatcher, P. G. (1997). Noncovalent interactions between acenaphthenone and dissolved fulvic acid as determined by ^{13}C NMR T_1 relaxation measurements—Response. *Environ. Sci. Technol.* **31**, 3744–3745.

Nanny, M. A., and Maza, J. P. (2001). Noncovalent interactions between monoaromatic compounds and dissolved humic acids: A deuterium NMR T_1 relaxation study. *Environ. Sci. Technol.* **35**, 379–384.

Nearpass, D. C. (1976). Absorption of picloram by humic acids and humin. *Soil Sci.* **121**, 272–277.

Nestle, N., Baumann, T., and Niessner, R. (2002). Magnetic resonance imaging in environmental science. *Environ. Sci. Technol.* **36**, 154A–160A.

Neuhaus, D., and Williamson, M. P. (2000). *The Nuclear Overhauser Effect in Structural and Conformational Analysis*, Wiley-VCH, New York.

Nielsen, U. G., Paik, Y., Julmis, K., Schoonen, M. A. A., Reeder, R. J., and Grey, C. P. (2005). Investigating sorption on iron-oxyhydroxide soil minerals by solid-state NMR spectroscopy: A ^{6}Li MAS NMR study of adsorption and absorption on goethite. *J. Phys. Chem. B* **109**, 18310–18315.

Nierop, K. G. J. (1998). Origin of aliphatic compounds in a forest soil. *Org. Geochem.* **29**, 1009–1016.

Nierop, K. G. J., Buurman, P., and de Leeuw, J. W. (1999). Effect of vegetation on chemical composition of H horizons in incipient podzols as characterized by C-13 NMR and pyrolysis-GC/MS. *Geoderma* **90**, 111–129.

Odoj, F., Rommel, E., von Kienlin, M., and Haase, A. (1998). A superconducting probehead applicable for nuclear magnetic resonance microscopy at 7T. *Rev. Sci. Instrum.* **69**, 2708–2712.

Otto, W. H., Britten, D. J., and Larive, C. K. (2003). NMR diffusion analysis of surfactant–humic substance interactions. *J. Colloid Interface Sci.* **261**, 508–513.

Otto, W. H., Burton, S. D., Carper, W. R., and Larive, C. K. (2001a). Examination of Cadmium(II) complexation by the Suwannee River fulvic acid using ^{113}Cd NMR relaxation measurements. *Environ. Sci. Technol.* **35**, 4900–4904.

Otto, W. H., Carper, W. R., and Larive, C. K. (2001b). Measurement of cadmium(II) and calcium(II) complexation by fulvic acids using ^{113}Cd NMR. *Environ. Sci. Technol.* **35**, 1463–1468.

Perdue, E. M., Hertkorn, N., and Kettrup, A. (2005). A potentiometric and ^{113}Cd NMR study of cadmium complexation by natural organic matter at two different magnetic field strengths. *Geochim. Cosmochim. Acta* **69**, A539–A539.

Peuravuori, J. (2005). NMR spectroscopy study of freshwater humic material in light of supramolecular assembly. *Environ. Sci. Technol.* **39**, 5541–5549.

Piccolo, A., Conte, P., and Cozzolino, A. (1999). Effects of mineral and monocarboxylic acids on the molecular association of dissolved humic substances. *Eur. J. Soil Sci.* **50**, 687–694.

Pines, A., Gibby, M. G., and Waugh, J. A. (1972). Proton-enhanced nuclear induction spectroscopy. A method for high sesolution NMR of dilute spins in solids. *J. Chem. Phys.* **56**, 1776–1777.

Preston, C., Trofymow, J., Sayer, B., and Niu, J. (1997). ^{13}C nuclear magnetic resonance spectroscopy with cross-polarization and magic-angle spinning investigation of the proximate-analysis fractions used to assess litter quality in decomposition studies. *Can. J. Bot.* **75**, 1601–1613.

Preston, C. M. (1996). Applications of NMR to soil organic matter analysis: History and prospects. *Soil Sci.* **161**, 144–166.

Preston, C. M. (2001). Carbon-13 solid-state NMR of soil organic matter—Using the technique effectively. *Can. J. Soil Sci.* **81**, 255–270.

Preston, C. M., and Ripmeester, J. A. (1982). Application of solution and solid-state ^{13}C NMR to 4 organic soils, their humic acids, fulvic-acids, humins and hydrolysis residues. *Can. J. Spectrosc.* **27**, 99–105.

Preston, C. M., Schnitzer, M., and Ripmeester, J. A. (1989). A spectroscopic and chemical investigation on the de-ashing of a humin. *Soil Sci. Soc. Am. J.* **53**, 1442–1447.

Reeves, A. D., and Chudek, J. A. (2001). Nuclear magnetic resonance imaging (MRI) of diesel oil migration in estuarine sediment samples. *J. Indt. Microbiol. Biotechnol.* **26**, 77–82.

Repeta, D. J., Quan, T. M., Aluwihare, L. I., and Accardi, A. M. (2002). Chemical characterization of high molecular weight dissolved organic matter in fresh and marine waters. *Geochim. Cosmochim. Acta* **66**, 955–962.

Rice, J. A. (2001). Humin. *Soil Sci.* **166**, 848–857.

Rumpel, C., Knicker, H., Kögel-Knabner, I., Skjemstad, J. O., and Huttl, R. F. (1998). Types and chemical composition of organic matter in reforested lignite-rich mine soils. *Geoderma* **86**, 123–142.

Sachleben, J. R., Chefetz, B., Deshmukh, A., and Hatcher, P. G. (2004). Solid-state NMR characterization of pyrene—Cuticular matter interactions. *Environ. Sci. Technol.* **38**, 4369–4376.

Salloum, M. J., Chefetz, B., and Hatcher, P. G. (2002). Phenanthrene sorption to aliphatic-rich natural organic matter. *Environ Sci Technol* **36**, 1953–1958.

Salloum, M. J., Dudas, M. J., and McGill, W. B. (2001). Variation of 1-naphthol sorption with organic matter fractionation: The role of physical conformation. *Org. Geochem.* **32**, 709–719.

Sannigrahi, P., Ingall, E. D., and Benner, R. (2005). Cycling of dissolved and particulate organic matter at station Aloha: Insights from ^{13}C NMR spectroscopy coupled with elemental, isotopic and molecular analyses. *Deep-Sea Res. Pt. I* **52**, 1429–1444.

Schanda, P., and Brutscher, B. (2006). Hadamard frequency-encoded SOFAST-HMQC for ultrafast two-dimensional protein NMR. *J. Magn. Reson.* **178**, 334–339.

Schmidt, M., Knicker, H., Hatcher, P., and Kögel-Knabner, I. (1997a). Improvement of ^{13}C and ^{15}N CPMAS NMR spectra of bulk soils, particle size fractions and organic material by treatment with 10% hydrofluoric acid. *Eur. J. Soil Sci.* **48**, 319–328.

Schmidt, M. W. I., Knicker, H., Hatcher, P. G., and KögelKnabner, I. (1997b). Improvement of C-13 and N-15 CPMAS NMR spectra of bulk soils, particle size fractions and organic material by treatment with 10% hydrofluoric acid. *Eur. J. Soil Sci.* **48**, 319–328.

Schmitt-Kopplin, P., Hertkorn, N., Schulten, H. R., and Kettrup, A. (1998). Structural changes in a dissolved soil humic acid during photochemical degradation processes under O_2 and N_2 atmosphere. *Environ. Sci. Technol.* **32**, 2531–2541.

Shin, H. S., Rhee, S. W., Lee, B. H., and Moon, C. H. (1996). Metal binding sites and partial structures of soil fulvic and humic acids compared: Aided by Eu(III) luminescence spectroscopy and DEPT/QUAT ^{13}C NMR pulse techniques. *Org. Geochem.* **24**, 523–529.

Simpson, A. (2001). Multidimensional solution state NMR of humic substances: A practical guide and review. *Soil Sci.* **166**, 795–809.

Simpson, A. J. (2002). Determining the molecular weight, aggregation, structures and interactions of natural organic matter using diffusion ordered spectroscopy. *Magn. Reson. Chem.* **40**, S72–S82.

Simpson, A. J., Boersma, R. E., Kingery, W. L., Hicks, R. P., and Hayes, M. H. B. (1997). Applications of NMR spectroscopy for studies of the molecular compositions of humic substances. In *Humic Substances in Soils, Peats, and Waters: Environmental and Health Aspects*, Special Publication, Royal Society of Chemistry, vol. **172**, pp. 46–62.

Simpson, A. J., Burdon, J., Graham, C. L., Hayes, M. H. B., Spencer, N., and Kingery, W. L. (2001a). Interpretation of heteronuclear and multidimensional NMR spectroscopy of humic substances. *Eur. J. Soil Sci.* **52**, 495–509.

Simpson, A. J., Kingery, W. L., and Hatcher, P. G. (2003a). The identification of plant derived structures in humic materials using three-dimensional NMR spectroscopy. *Environ. Sci. Technol.* **37**, 337–342.

Simpson, A. J., Kingery, W. L., Hayes, M. H. B., Spraul, M., Humpfer, E., Dvortsak, P., Kerssebaum, R., Godejohann, M., and Hofmann, M. (2002a). Molecular structures and associations of humic substances in the terrestrial environment. *Naturwissenschaften* **89**, 84–88.

Simpson, A. J., Kingery, W. L., Shaw, D. R., Spraul, M., Humpfer, E., and Dvortsak, P. (2001b). The application of ^1H HR-MAS NMR spectroscopy for the study of structures and associations of organic components at the solid—Aqueous interface of a whole soil. *Environ. Sci. Technol.* **35**, 3321–3325.

Simpson, A. J., Kingery, W. L., Spraul, M., Humpfer, E., Dvortsak, P., and Kerssebaum, R. (2001c). Separation of structural components in soil organic matter by diffusion ordered spectroscopy. *Environ. Sci. Technol.* **35**, 4421–4425.

Simpson, A. J., Lam, B., Diamond, M. L., Donaldson, D. J., Lefebvre, B. A., Moser, A. Q., Williams, A. J., Larin, N. I., and Kvasha, M. P. (2006a). Assessing the organic composition of urban surface films using nuclear magnetic resonance spectroscopy. *Chemosphere* **63**, 142–152.

Simpson, A. J., Lefebvre, B., Moser, A., Williams, A., Larin, N., Kvasha, M., Kingery, W. L., and Kelleher, B. (2004a). Identifying residues in natural organic matter through spectral prediction and pattern matching of 2D NMR datasets. *Magn. Reson. Chem.* **42**, 14–22.

Simpson, A. J., Salloum, M. J., Kingery, W. L., and Hatcher, P. G. (2002b). Improvements in the two-dimensional nuclear magnetic resonance spectroscopy of humic substances. *J. Environ. Qual.* **31**, 388–392.

Simpson, A. J., Simpson, M. J., Kingery, W. L., Lefebvre, B. A., Moser, A., Williams, A. J., Kvasha, M., and Kelleher, B. P. (2006b). The application of ^{1}H high-resolution magic-angle spinning NMR for the study of clay–organic associations in natural and synthetic complexes. *Langmuir* **22**, 4498–4503.

Simpson, A. J., Song, G. X., Smith, E., Lam, B., Novotny, E. H., and Hayes, M. H. B. (2007a). Unraveling the structural components of soil humin by use of solution-state nuclear magnetic resonance spectroscopy. *Environ. Sci. Technol.* **41**, 876–883.

Simpson, A. J., Tseng, L. H., Simpson, M. J., Spraul, M., Braumann, U., Kingery, W. L., Kelleher, B. P., and Hayes, M. H. B. (2004b). The application of LC-NMR and LC-SPE-NMR to compositional studies of natural organic matter. *Analyst* **129**, 1216–1222.

Simpson, A. J., Zang, X., Kramer, R., and Hatcher, P. G. (2003b). New insights on the structure of algaenan from *Botryococcus braunii* race A and its hexane insoluble botryals based on multidimensional NMR spectroscopy and electrospray–mass spectrometry techniques. *Phytochemistry* **62**, 783–796.

Simpson, M. J. (2006). Nuclear magnetic resonance based investigations of contaminant interactions with soil organic matter. *Soil Sci. Soc. Am. J.* **70**, 995–1004.

Simpson, M. J., and Hatcher, P. G. (2004). Determination of black carbon in natural organic matter by chemical oxidation and solid-state ^{13}C nuclear magnetic resonance spectroscopy. *Org. Geochem.* **35**, 923–935.

Simpson, M. J., and Johnson, P. C. E. (2006). Identification of mobile aliphatic sorptive domains in soil humin by solid-state ^{13}C nuclear magnetic resonance. *Environ. Toxicol Chem.* **25**, 52–57.

Simpson, M. J., and Preston, C. M. (2007). Soil organic matter analysis by solid-state ^{13}C nuclear magnetic resonance spectroscopy. In *Soil Sampling and Methods of Analysis*. CRC Press, Boca Raton, FL, pp. 195–200.

Simpson, M. J., Simpson, A. J., Gross, D., Spraul, M., and Kingery, W. L. (2007b). ^{1}H and ^{19}F nuclear magnetic resonance microimaging of water and chemical distribution in soil columns. *Environ. Toxicol. Chem.* **26**, 1340–1348.

Simpson, M. J., Simpson, A. J., and Hatcher, P. G. (2004c). Noncovalent interactions between aromatic compounds and dissolved humic acid examined by nuclear magnetic resonance spectroscopy. *Environ. Toxicol. Chem.* **23**, 355–362.

Skjemstad, J. O., Clarke, P., Taylor, J. A., Oades, J. M., and Newman, R. H. (1994a). The removal of magnetic materials from surface soils. A solid state 13C CP/MAS NMR study. *Aust. J. Soil Res.* **32**, 1215–1229.

Skjemstad, J. O., Clarke, P., Taylor, J. A., Oades, J. M., and Newman, R. H. (1994b). The removal of magnetic materials from surface soils. A solid state ^{13}C CP-MAS NMR study. *Aust. J. Soil Res.* **32**, 1215–1229.

Skjemstad, J. O., Taylor, J. A., and Smernik, R. J. (1999). Estimation of charcoal (char) in soils. *Commun. Soil Sci. Plant Anal.* **30**, 2283–2298.

Smernik, R. J. (2005). Solid-state C-13 NMR spectroscopic studies of soil organic matter at two magnetic field strengths. *Geoderma* **125**, 249–271.

Smernik, R. J. (2006). The effect of water content on solid-state ^{13}C NMR quantitation and relaxation rates of soil organic matter. *Eur. J. Soil Sci.* **57**, 665–676.

Smernik, R. J., and Baldock, J. A. (2005). Does solid-state N-15 NMR spectroscopy detect all soil organic nitrogen? *Biogeochemistry* **75**, 507–528.

Smernik, R. J., Baldock, J. A., and Oades, J. M. (2002). Impact of remote protonation on ^{13}C CPMAS NMR quantitation of charred and uncharred wood. *Solid State Nucl. Magn. Reson.* **22**, 71–82.

Smernik, R. J., Kookana, R. S., and Skjemstad, J. O. (2006). NMR characterization of ^{13}C benzene sorbed to natural and prepared charcoals. *Environ. Sci. Technol.* **40**, 1764–1769.

Smernik, R. J., and Oades, J. M. (2000a). The use of spin counting for determining quantitation in solid state C-13 NMR spectra of natural organic matter 1. Model systems and the effects of paramagnetic impurities. *Geoderma* **96**, 101–129.

Smernik, R. J., and Oades, J. M. (2000b). The use of spin counting for determining quantitation in solid state C-13 NMR spectra of natural organic matter 2. HF-treated soil fractions. *Geoderma* **96**, 159–171.

Smernik, R. J., and Oades, J. M. (2001). Solid-state ^{13}C NMR dipolar dephasing experiments for quantifying protonated and non-protonated carbon in soil organic matter and model systems. *Eur. J. Soil Sci.* **52**, 103–120.

Smernik, R. J., and Oades, J. M. (2002). Paramagnetic effects on solid state ^{13}C nuclear magnetic resonance spectra of soil organic matter. *J. Environ. Qual.* **31**, 414–420.

Smernik, R. J., and Oades, J. M. (2003). Spin accounting and RESTORE—two new methods to improve quantitation in solid-state ^{13}C NMR analysis of soil organic matter. *Eur. J. Soil Sci.* **54**, 103–116.

Spaccini, R., Piccolo, A., Haberhauer, G., and Gerzabek, M. H. (2000). Transformation of organic matter from maize residues into labile and humic fractions of three European soils as revealed by C-13 distribution and CPMAS-NMR spectra. *Eur. J. Soil Sci.* **51**, 583–594.

Stark, R. E., Yan, B., Ray, A. K., Chen, Z., Fang, X., and Garbow, J. R. (2000). NMR studies of structure and dynamics in fruit cuticle polyesters. *Solid State Nucl. Magn. Reson.* **16**, 37–45.

Strynar, M., Dec, J., Benesi, A., Jones, A. D., Fry, R. A., and Bollag, J. M. (2004). Using ^{19}F NMR spectroscopy to determine trifluralin binding to soil. *Environ. Sci. Technol.* **38**, 6645–6655.

Sutton, R., and Sposito, G. (2005). Molecular structure in soil humic substances: The new view. *Environ. Sci. Technol.* **39**, 9009–9015.

Suzuki, Y., Kawakami, M., and Akasaka, K. (2001). ^{1}H NMR application for characterizing water-soluble organic compounds in urban atmospheric particles. *Environ. Sci. Technol.* **35**, 2656–2664.

Thorn, K. A. (1997). Covalent binding of the reductive degradation products of TNT to humic substances examined by ^{15}N NMR. *Abs. Pap. Am. Chem.Soc.* **213**, 107-ENVR.

Thorn, K. A., Goldenberg, W. S., Younger, S. J., and Weber, E. J. (1996a). Covalent binding of aniline to humic substances—Comparison of nucleophilic addition, enzyme-, and metal-catalyzed reactions by N-15 NMR. *ACS Symp. Ser.* **651**, 299–326.

Thorn, K. A., and Kennedy, K. R. (2002). N-15 NMR investination of the covalent binding of reduced TNT amines to soil humic acid, model compounds, and lignocellulose. *Environ. Sci. Technol.* **36**, 3787–3796.

Thorn, K. A., Pettigrew, P. J., and Goldenberg, W. S. (1996b). Covalent binding of aniline to humic substances. 2. N-15 NMR studies of nucleophilic addition reactions. *Environ. Sci. Technol.* **30**, 2764–2775.

Tombacz, E. (1999). Colloidal properties of humic acids and spontaneous changes of their colloidal state under variable solution conditions. *Soil Sci.* **164**, 814–824.

Uyguner, C. S., Hellriegel, C., Otto, W., and Larive, C. K. (2004). Characterization of humic substances: Implications for trihalomethane formation. *Anal. Bioanal. Chem.* **378**, 1579–1586.

Van As, H., and Van Dusschoten, D. (1997). NMR methods for imaging of transport processes in micro-porous systems. *Geoderma* **80**, 389–403.

Voehler, M. W., Collier, G., Young, J. K., Stone, M. P., and Germann, M. W. (2006). Performance of cryogenic probes as a function of ionic strength and sample tube geometry. *J. Magn. Reson.* **183**, 102–109.

Wais, A., Burauel, P., deGraaf, A. A., Haider, K., and Fuhr. F. (1996). Using ^{13}C NMR spectroscopy to evaluate the binding mechanism of bound pesticide residues in soils. 2. Investigations of model humic acids. *J. Environ. Sci. Health B* **31**, 1–24.

Wang, K., and Xing, B. (2005a). Structural and sorption characteristics of adsorbed humic acid on clay minerals. *J. Environ. Qual.* **34**, 342–349.

Wang, K. J., Dickinson, L. C., Ghabbour, E. A., Davies, G., and Xing, B. S. (2003). Proton spin-lattice relaxation times of humic acids as determined by solution NMR. *Soil Sci* **168**, 128–136.

Wang, K. J., and Xing, B. S. (2005b). Chemical extractions affect the structure and phenanthrene sorption of soil humin. *Environ. Sci. Technol.* **39**, 8333–8340.

Wattel-Koekkoek, E. J. W., van Genuchten, P. P. L., Buurman, P., and van Lagen, B. (2001). Amount and composition of clay associated soil organic matter in a range of kaolinitic and smectitic soils. *Geoderma* **99**, 27–49.

Weber, E. J., Spidle, D. L., and Thorn, K. A. (1996). Covalent binding of aniline to humic substances. 1. Kinetic studies. *Environ. Sci. Technol.* **30**, 2755–2763.

Weber, W. J., Huang, W. L., and LeBoeuf, E. J. (1999). Geosorbent organic matter and its relationship to the binding and sequestration of organic contaminants. *Colloid Surf. A* **151**, 167–179.

Wershaw, R. L. (1999). Molecular aggregation of humic substances. *Soil Sci* **164**, 803–813.

Wershaw, R. L., Leenheer, J. A., Kennedy, K. R., and Noyes, T. I. (1996a). Use of ^{13}C NMR and FTIR for elucidation of degradation pathways during natural litter decomposition and composting. 1. Early stage leaf degradation. *Soil Sci.* **161**, 667–679.

Wershaw, R. L., Llaguno, E. C., and Leenheer, J. A. (1996b). Mechanism of formation of humus coatings on mineral surfaces. 3. Composition of adsorbed organic acids from compost leachate on alumina by solid-state ^{13}C NMR. *Colloid Surface A* **108**, 213–223.

Wershaw, R. L., Melvin, R. L., and Rigatti, G. P. (1993). Characterization of a Compost Leachate by ^{13}C NMR Spectroscopy. *Abs. Pap. Am. Chem. Soc.* **205**, 40-Envr.

Wilson, M. A. (1981). Applications of nuclear magnetic resonance spectroscopy to the study of the structure of soil organic matter. *Eur. J. Soil Sci.* **32**, 167–186.

Wilson, M. A. (1987). *N.M.R. Techinques and Applications in Geochemistry and Soil Chemistry*, Pergamon Press, Oxford.

Wilson, M. A., Barron, P. F., and Gillam, A. H. (1981a). The structure of fresh-water humic substances as revealed by ^{13}C NMR spectroscopy. *Geochim. Cosmochim. Acta* **45**, 1743–1750.

Wilson, M. A., Barron, P. F., and Goh, K. M. (1981b). Differences in structure of organic-matter in 2 soils as demonstrated by ^{13}C cross polarization Nuclear magnetic-resonance spectroscopy with magic angle spinning. *Geoderma* **26**, 323–327.

Wilson, M. A., Pugmire, R. J., Zilm, K. W., Goh, K. M., Heng, S., and Grant, D. M. (1981c). Cross-polarization ^{13}C NMR spectroscopy with magic angle spinning characterizes organic-matter in whole soils. *Nature* **294**, 648–650.

Wu, N., Peck, T. L., Webb, A. G., Magin, R. L., and Sweedler, J. V. (1994). Nanoliter volume sample cells for ^{1}H NMR application to online detection in capillary electrophoresis. *J. Am. Chem. Soc.* **116**, 7929–7930.

Xing, B. S., Liu, J. D., Liu, X. B., and Han, X. Z. (2005). Extraction and characterization of humic acids and humin fractions from a black soil of China. *Pedosphere* **15**, 1–8.

Xiong, J. C., Lock, H., Chuang, I. S., Keeler, C., and Maciel, G. E. (1999). Local motions of organic pollutants in soil components, as studied by ^{2}H NMR. *Environ. Sci. Technol.* **33**, 2224–2233.

Xu, X., Kalinichev, A. G., and Kirkpatrick, R. J. (2006). ^{133}Cs and ^{35}Cl NMR spectroscopy and molecular dynamics modeling of Cs^{+} and Cl^{-} complexation with natural organic matter. *Geochim. Cosmochim. Acta* **70**, 4319–4331.

Yang, D. W., Konrat, R., and Kay, L. E. (1997). A multidimensional NMR experiment for measurement of the protein dihedral angle psi based on cross-correlated relaxation between (H alpha-^{13}C alpha) ^{1}H dipolar and ^{13}C (carbonyl) chemical shift anisotropy mechanisms. *J. Am. Chem. Soc.* **119**, 11938–11940.

Young, I. M., and Crawford, J. W. (2004). Interactions and self-organization in the soil–microbe complex. *Science* **304**, 1634–1637.

Zech, W., Ziegler, F., Kögel-Knabner, I., and Haumaier, L. (1992). Humic substances distribution and transformation in forest soils. *Sci. Total Environ.* **118**, 155–174.

Zhang, C., Werth, C. J., and Webb, A. G. (2002). DNAPL distribution in silica gel columns. *Environ. Sci. Technol.* **36**, 3310–3317.

16

EPR, FTIR, RAMAN, UV–VISIBLE ABSORPTION, AND FLUORESCENCE SPECTROSCOPIES IN STUDIES OF NOM

L. Martin-Neto, D. M. B. P. Milori, W. T. L. Da Silva, and M. L. Simões

Brazilian Agricultural Research Corporation (EMBRAPA), National Center for Agricultural Instrumentation, São Carlos, Brazil

Biophysico-Chemical Processes Involving Natural Nonliving Organic Matter in Environmental Systems, Edited by Nicola Senesi, Baoshan Xing, and Pan Ming Huang

16.1. INTRODUCTION

Questions associated with natural nonliving organic matter (NOM) are of immense importance due to several functions of NOM in different terrestrial, aquatic, and aerial environments. Nowadays, this importance has been greatly intensified by their association with the C cycle that has a direct relationship with climate changes in the planet. However, to advance in this NOM research agenda, it is fundamentally pivotal to include analytical tools that can provide precise insights on the dynamics and reactivities of NOM in different ecosystems, including interfaces and interconnections. The scenario makes clear that in order to advance in NOM research, more than relevant quantitative data on C levels are required. Qualitative information that can be provided by spectroscopic methods are also necessary—for example, electron paramagnetic resonance (EPR), Fourier transform infrared (FTIR), Raman, ultraviolet–visible absorption (UV–vis), and fluorescence. Therefore, in this chapter the principles of such techniques and the results obtained in different ecosystems will be presented by the authors of the chapter, as well as from current literature. Additionally, in closing this chapter, conclusions and some suggestions on research opportunities and needs are expounded.

16.2. ELECTRON PARAMAGNETIC RESONANCE (EPR)

16.2.1. Principles of EPR

Electron paramagnetic resonance (EPR) or electron spin resonance (ESR) is a spectroscopy that detects species (atoms, ions, or molecules) with, at least, one unpaired electron. Although most species have all electrons paired, there are several detectable EPR systems: free radicals in the solid, liquid, or gaseous phases; systems with more than one unpaired electron (biradicals and triplet-state systems); point defects in solids or localized crystal imperfections; systems with conducting electrons (semiconductors and metals); transition-metal ions (Fe^{3+}, Cu^{2+}, Mn^{2+}, VO^{2+}, Mo^{5+}, Cr^{3+}, and others) and some rare-earth ions. This spectroscopy is extremely sensitive, and under favorable circumstances the minimum detectable can be of 10^{11} spins g^{-1}, 10^{13} spins ml^{-1}, or 10^{-8} mol l^{-1} (Knowles et al., 1976; Parish, 1990; Goodman and Hall, 1994; Weil et al., 1994).

The electron possesses a magnetic moment (μ) with spin $S = 1/2$. In the presence of an external magnetic field (H), μ has two allowed orientations. The two spin states differ in energy (ΔE) owing to the orientation differences in relation to H. Resonant absorption occurs if the frequency (ν) is adjusted so that

$$\Delta E = h\nu \tag{16.1}$$

The amount of energy E required to misalign μ from H is calculated by the equation

$$E = -\mu H \tag{16.2}$$

where

$$\mu = -g\frac{eh}{4\pi mc}M_S = -g\beta M_S \tag{16.3}$$

where g is the electron g-factor, e is the charge on the electron (-1.6×10^{-19} C), h is Planck's constant (6.63×10^{-34} J s), m is the mass of the electron (9.1×10^{-31} g), c the speed of light (3×10^{8} m s^{-1}), β is the Bohr magneton (0.92×10^{-23} JT), and $M_S = \pm 1/2$ is the spin quantum number. Hence a combination of Eqs. (16.1)–(16.3) yields

$$\Delta E = h\nu = g\beta H \tag{16.4}$$

Transitions between the two spin states (+1/2 and −1/2) can be induced by oscillating electromagnetic radiation (ν in the microwave region) applied perpendicularly to H. The energy-level splitting is referred to as the Zeeman effect, illustrated in Figure 16.1. Normally in the EPR measurements, ν is maintained at a fixed value and H is permitted to vary until the resonance is matched.

16.2.2. *g* Values

The g value is used to characterize the position of a resonance line. It is a measure of the local magnetic field experienced by the electron in the orbital. This parameter can be considered as a quantity characteristic of the molecule in which the unpaired electron is located, reflecting the nature of this orbital. According to Eq. (16.4), the g-factor can be obtained from

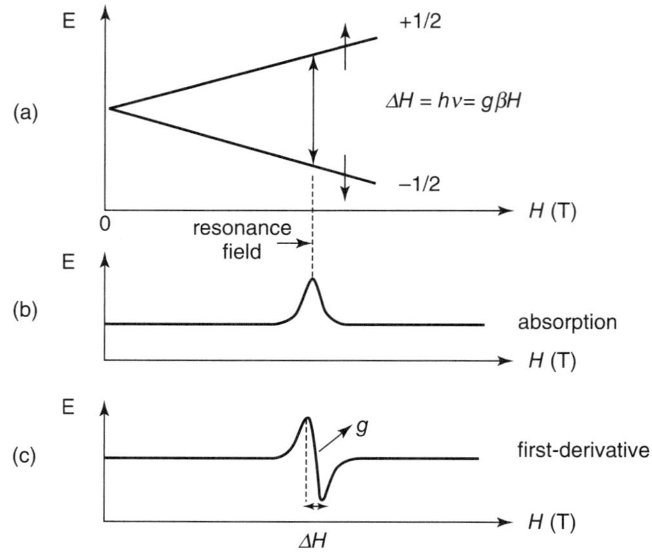

Figure 16.1. (**a**) Simplified scheme of EPR phenomenon, showing the energy-level splitting (Zeeman effect) for the electron spin $S = 1/2$ ($M_S = \pm 1/2$) as a function of applied magnetic field (H), (**b**) the EPR absorption line, and (**c**) first derivative of absorption line, indicating the g value and line width (ΔH), normally detected in the EPR spectra.

$$g = \frac{h\nu}{\beta H} = 0.714487 \frac{\nu(\text{MHz})}{H(\text{Gauss})} \tag{16.5}$$

The g value for a free electron is 2.0023. The principal source of the local magnetic fields, which causes the g factor to deviate from the free electron g value, is an orbital magnetic moment introduced by a mixing of excited states into the ground state.

In almost all cases the admixture of excited states is anisotropic; that is, the observed g value varies according to the orientation of the paramagnetic species in relation to the applied magnetic field (orientation-dependent). The g-factor anisotropy is characterized by three principal g values, namely, g_{xx}, g_{yy}, and g_{zz}. When these three values are different, the symmetry is defined as rhombic; and in the case of axial symmetry, $g_{xx} = g_{yy} \neq g_{zz}$. In the orientation-independent (isotropic) situation the g factor is represented by a single value. This is also true if the species paramagnetic is in a solution of low viscosity (water) where the molecular tumbling causes all the g factor anisotropy to be averaged out (Knowles et al., 1976; Campbell and Dwek, 1984).

16.2.3. Mechanisms of Relaxation and Line Width

The speed by which the spin in the upper state loses energy is known as spin-lattice relaxation time, T_1. The magnitude of T_1 depends on the molecular environment, that is, how strongly the spin is coupled to the lattice. For example, in solid material, T_1 is short, but in liquids it is relatively long. Thus, the rate of excitation from the lower to the upper state must not exceed the rate of relaxation back to the ground state in order to avoid saturation of the system (Knowles et al., 1976).

From the Boltzmann function, the relative population of the upper state, N^+, and the lower state, N^-, energy levels at thermal equilibrium is given by

$$N^+/N^- = \exp(-h\nu/kT) = 1 - g\beta H/kT \tag{16.6}$$

where k is Boltzmann's constant ($1.381 \times 10^{-23}\,\mathrm{J\,K^{-1}}$) and T is the absolute temperature (given in K). Besides the influence in the relative population of the spin states, T affects T_1, which almost always decreases as the T increases.

Another mechanism of relaxation is associated with the magnetic interaction between nuclei and paramagnetic electrons (the so-called magnetic dipole interactions). This process is known as spin–spin relaxation time (T_2).

The line width is influenced by secular broadening, which is caused by processes that generate varying local magnetic fields, and lifetime broadening. The latter is associated with the Heisenberg uncertainty principle (Levine, 1974), establishing that there is an uncertainty in the spin state lifetime (Δt), as well as an uncertainty in the energy state (ΔE), namely,

$$\Delta E \times \Delta t \sim \frac{h}{2\pi} \tag{16.7}$$

Equations (16.1) and (16.4) yield

$$\Delta H \sim \frac{h}{g\beta}\frac{1}{\Delta t} \tag{16.8}$$

From Eq. (16.8), it is noted that the line width (ΔH) is inversely proportional to the relaxation time. It should be emphasized that under normal circumstances (system not saturated), it is T_2 and not T_1 which determines the intrinsic line width. Just like spin-lattice relaxation time (T_1), T_2 is also influenced by temperature, because decreasing the temperature results in larger T_2 and, consequently, smaller line-width values. This means, in several situations, that EPR spectra obtained at lower temperatures are better resolved, for example, at N_2 (77 K) or He (4 K) liquid temperatures.

16.2.4. Nuclear Hyperfine Interactions

The lines in an EPR spectrum can be split by interaction of the electron spin with the nuclear magnetic moment of atoms on which the unpaired electron is located (Parish, 1990). Only atoms with nuclear spin (I) nonzero exhibit this type of interaction, which can be of two types: (1) contact interaction that is isotropic and results from the delocalization of the unpaired electron onto the nucleus and (2) dipolar interaction between electron spin and the nucleus. In the second case, the interaction is dependent on orientation and, therefore, anisotropic (Campbell and Dwek, 1984).

The number of isotropic hyperfine lines from a particular nucleus depends on the nuclear spin, I, and the line multiplicity is $2I + 1$. For n equivalent nuclei, the EPR spectrum consists of $2nI + 1$ lines whose relative intensities are given by binomial coefficients obtained in the expansion of $(1 + x)^n$ (Knowles et al., 1976). When nuclear hyperfine interactions occur, Eq. (16.4) becomes

$$h\nu = g\beta H + AM_I \tag{16.9}$$

where A is the hyperfine splitting constant and M_I is the nuclear quantum number.

The allowed transitions, according to selection rules, correspond to $\Delta M_S = \pm 1$ and $\Delta M_I = 0$. Hence, for $S = 1/2$ and $I = 1/2$, M_S and M_I are $+1/2$ or $-1/2$. The energy-level for this situation is illustrated in Figure 16.2.

The magnitude of the splitting between the lines defined as hyperfine splitting constant (A^*) can be given, for practical reasons, in magnetic field units in gauss (G) or tesla (T) and can be obtained directly from EPR spectra. The A value depends on the interaction intensity that occurs between the unpaired electron and the nucleus and can also be orientation-dependent depending on the charge distribution symmetry of the system, similar to the g-factor symmetry dependence.

Another type of splitting of the EPR spectrum can occur when an unpaired electron interacts with the nuclei having nonzero I on adjacent atoms. This type of interaction is known as superhyperfine splitting; and in an analogous way to hyperfine splitting, the magnitude of the interaction depends on the extent of delocalization of the unpaired electron on the adjacent atoms and the number of bonds involved (Parish, 1990).

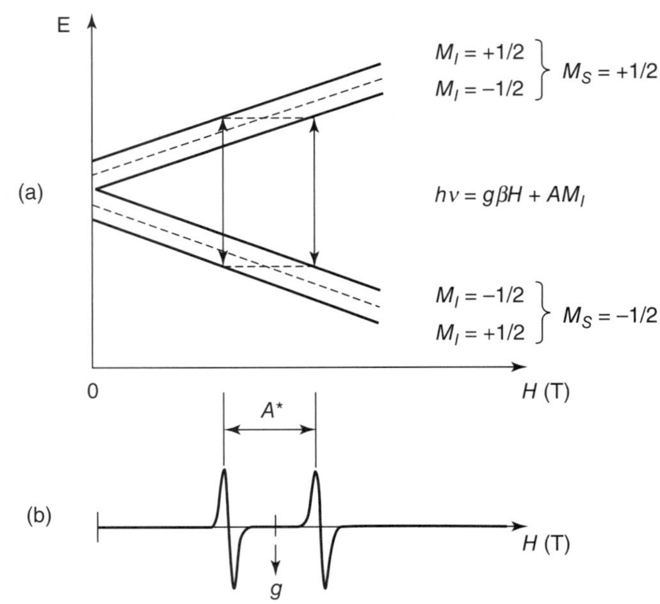

Figure 16.2. (**a**) Energy-level splitting as a function of applied magnetic field (H) for electron spin $S = 1/2$ ($M_S = \pm 1/2$) and nuclear spin $I = 1/2$ ($M_I = \pm 1/2$); (**b**) first derivative of EPR spectra showing the g value and the hyperfine splitting constant (A^*).

There are yet other interactions that can be detected by EPR: the dipolar interaction that arises from spin–spin coupling; zero-field splitting that occurs in systems which have more than one unpaired electron, such as some transition-metal ions, caused by the surrounding ligand ions (and also by electron–electron interactions in triplet-sate molecules); the quadrupole interaction that arises from nuclear quadrupole moment and electric field gradient; and nuclear Zeeman interaction that arises from the nuclear magnetic moment and external magnetic field (Knowles et al., 1976; Senesi, 1992). These interactions are not presented in this chapter for the sake of simplicity and also because the main examples and results presented can be supported by those given principles.

16.2.5. Applications

16.2.5.1. Determination of Humification Degree of Soil Organic Matter and Humic Substances in Different Ecosystems, Including Areas Under Carbon Sequestration. Humification is defined as the transformation of macromorphologically identifiable matter into amorphous compounds, as a rule involving the changes that occur in vegetal residues or soil organic matter (SOM) during the humification process (Zech et al., 1997). According to Plaza et al. (2003) a fundamental requirement for any organic material to be safely, conveniently, and efficiently used as soil amendment—that is, avoiding adverse effects of amendment on soil properties—is to have achieved a biological and chemical stability or maturity, which are associated with the humification degree.

Several transformation processes of terrestrial and aquatic organic matter in the environment are connected with reactions of organic free radicals. Complex aromatic structures are believed to stabilize semiquinone free radicals (Figure 16.3a) in humic substances (HS) (Riffaldi and Schnitzer, 1972; Senesi, 1990a; Stevenson, 1994) in coexistence with carbon-centered "aromatic" radicals (Paul et al., 2006), although contributions from methoxybenzene and nitrogen-associated radicals cannot be excluded (Senesi, 1990a). The semiquinone-type free radical (SFR) EPR signal, detected in HS and NOM in solid state, is normally characterized by a narrow single line, although hyperfine lines of HS in solution have been detected (Cheshire and McPhail, 1996; Watanabe et al., 2005). For soil humic acids (HA) and fulvic acids (FA) the g value is in the range of 2.0032 to 20051, and ΔH at around 3.9 to 7.5 G (Senesi, 1990a). For NOM the g value is in the range of 2.0031–2.0045 and ΔH is in the range of 4.4–8.0 G (Paul et al., 2006). Watanabe et al. (2005), analyzing HA from a wide range of soil types, reported that ΔH ranged between 3.2 and 5.5 G and showed an inverse correlation with the free radicals concentration, suggesting a decrease in the heterogeneity of free radicals with their concentration increase.

A spectroscopically demonstrated molecular property of SOM relating to the degree of humification is the SFR concentration, as measured by EPR (Riffaldi and Schnitzer, 1972; Schnitzer and Levesque, 1979; Martin-Neto et al., 1991; Senesi et al., 1996; Jerzykiewicz et al., 1999; Watanabe et al., 2005). The content of paramagnetic species is proportional to the EPR spectrum area that can be obtained by double integration of the first derivative EPR spectrum, which is normally detected. An approximation commonly used to obtain the relative area of free radicals is the

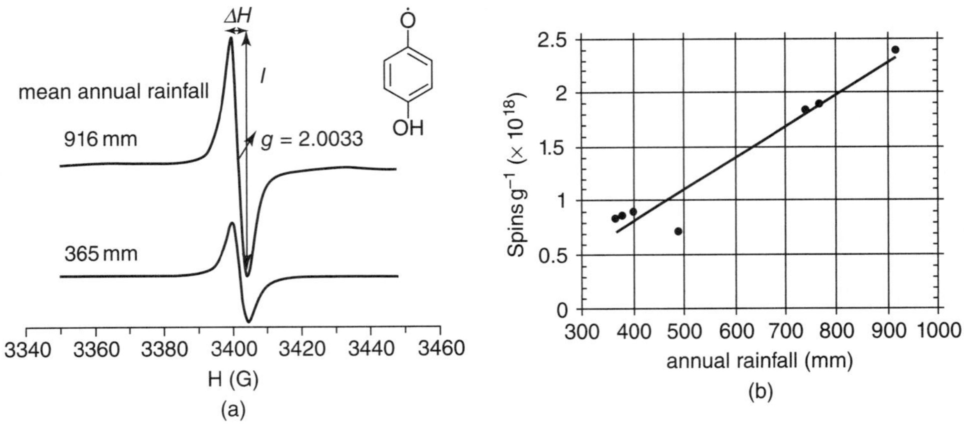

Figure 16.3. (**a**) Representative EPR spectra of the HA extracted from soil under differing mean annual rainfall, showing the peak-to-peak line width (ΔH), the intensity of signal (I), and the g value. The inset in the figure shows the chemical structure of the SFR. (**b**) Graph of SFR concentration, in spins g^{-1} of HA sample, against mean annual rainfall. The line through the data points is a least-squares fit. Adapted from Martin-Neto et al. (1998).

approximation $I \times \Delta H^2$ (Poole, 1967), where I is the first derivative EPR signal intensity and ΔH is the peak-to-peak line width (Figure 16.3a). For conversion of the EPR spectrum area in spin concentration (spins g^{-1} or spin $(g\ C)^{-1}$, for solid samples) a standard sample is used (strong pitch or weak pitch, generally provided by EPR spectrometer factories) of known free radicals content (Martin-Neto et al., 1994a). For more accurate free radicals quantification a secondary standard is used during the acquisition of EPR spectra, in agreement with Singer's method (Singer, 1959; Martin-Neto et al., 1994a), to detect possible alterations in the quality factor of the EPR cavity, known as Q factor (Weil et al., 1994). These possible alterations in the Q factor are due to the differences in the sample characteristics, such as residual humidity.

Applying this spectroscopic information to HA samples extracted from a climosequence of temperate grassland soils under different mean annual rainfall, Martin-Neto et al. (1998) demonstrated significant positive linear correlation ($R = 0.969$) between the humification degree (inferred by SFR concentration) and a pluviometric index increase (Figure 16.3b), hence showing the impact of rainfall, and indirectly from microbial activity, on SOM. Figure 16.3a illustrates typical, well-resolved (first-derivative) EPR spectra for HA samples, representing two extremes of the soil-sampling transect.

Generally, spectroscopy analyses of HS are made only in some of its fractions, usually HA and FA. The difficulty in studying humin fraction is its strong association with soil mineral fraction, mainly those containing Fe^{3+} ions, which interfere in the EPR analysis. However, Saab and Martin-Neto (2004) obtained spectroscopic data from all HS fractions and whole soil samples (Gley soil). The results indicated that for this type of soil, the humin was the fraction with the highest SFR concentration, followed by HA and FA, with the lowest concentration. Also, Martin-Neto et al. (1994a) showed the possibility of using EPR to determine the humification degree (through quantification of SFR) of physical fractions (organo-mineral aggregates) of a Mollisol, from Argentina.

Investigations of the OM dynamics of agricultural tropical soils can provide valuable information on how to manage such soils to increase C stocks and promote C sequestration, seeing that molecular recalcitrance and organo-mineral interaction are factors that determine SOM stability (Sollins et al., 1996; Baldock and Skjemstad, 2000).

Molecular recalcitrance of SOM can be inferred through the SFR concentration, since the latter is supposed to derive from and be stabilized by recalcitrant aromatic structures (Riffaldi and Schnitzer, 1972; Senesi, 1990a; Stevenson, 1994), and are also believed to relate to the recalcitrance imparted by disordered structural conformation of HS (Almendros and Dorado, 1999). The data obtained using this spectroscopic information showed that the SFR concentration is generally higher in HA and physical fractions of soils under conventional tillage (Figure 16.4) (Bayer et al., 2002b) or under low-input cropping systems (Bayer et al., 2000) than in soils under no-tillage and with addition of higher crop residues (Bayer et al., 2002a) (Figure 16.5). This indicates that with higher tillage intensity and lower residue addition, only the most recalcitrant structures of SOM tend to remain. Therefore, the concept of the degree of humification obtained by SFR concentration needs careful interpretation, depending on specific conditions (e.g., soil management) and the characteristics of ecosystems and their evolution. According to Bayer et al. (2006b), no-tillage management increased the C stock in Brazilian savanna (Cerrado) oxisols compared with conventional tillage, and this is in agreement with previous results found in Brazilian subtropical areas from the country's southern region (Bayer et al., 2000).

Organo-mineral interaction can be assessed through power saturation experiments of EPR (Weil et al., 1994). Bayer et al. (2006a), using data analysis from power

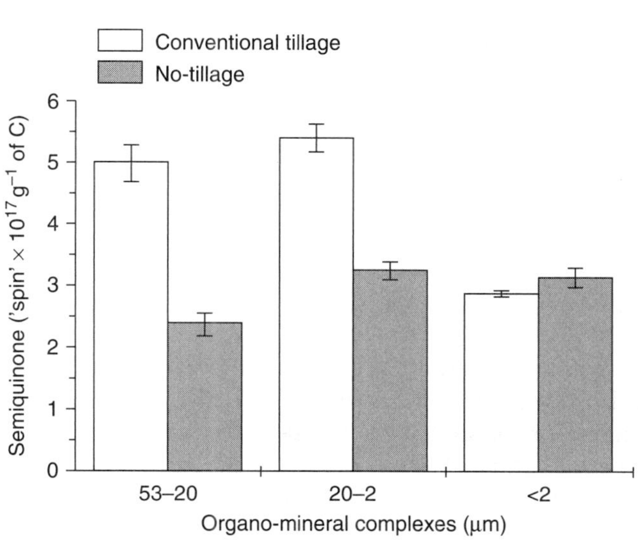

Figure 16.4. SFR concentration of SOM in the 53–20-, 20–2- and <2-μm-sized organo-mineral complexes from a 0- to 25-mm layer of an acrisol under conventional and no-tillage systems. Reprinted from Bayer et al. (2002b).

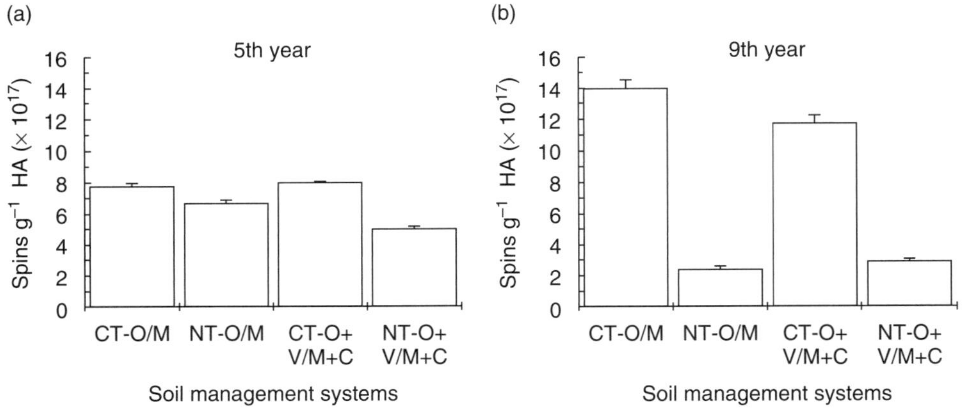

Figure 16.5. SFR concentration in soil HA samples extracted from different tillage practices, cropping systems, and period of experiments: (**a**) 5 years and (**b**) 9 years. NT, no tillage; CT, conventional tillage; O/M, oat/maize; O + V/M + C, oat + vetch/maize + cowpea. Reprinted from Bayer et al. (2002a).

saturation experiments, obtained an experimental evidence of organo-mineral interactions. They observed that microwave power saturation easily occurred in the coarse silt fraction, required more power to occur in fine silt, and was not observed in the clay fraction, within the employed power range. This result evidenced the existence of stronger organo-mineral interactions in the clay-size fraction (which contains the highest iron oxides and kaolinite concentrations) than in fine and coarse silt-size fractions.

More recently, Canellas et al. (2008) evaluated the chemical characteristics of HA isolated from tropical soils under different weathering degrees and analyzed a possible relationship between the degree of humification obtained by chemical and spectroscopic analyses (among them the SFR concentration) and biological activity, evaluated by the activity of the plasma membrane H^+-ATPase of these HA on maize root growth. Results showed that all HA were able to stimulate the plasma membrane H^+-ATPase activity and root growth. However, those HA with higher degree of humification were more effective in the H^+-ATPase stimulation than were the HA with lower degree of humification. Additional studies must be conducted to adequately understand the mechanisms of reaction in organic agriculture. Nevertheless, the result of Canellas et al. (2008) is a very interesting example of the importance of the degree of humification of humic materials in association with plant biological activity.

Relevant studies showed the electron accepting capacity of HS from a variety of environments (Lovley et al., 1996; Scott et al., 1998). The mechanism is based on Fe^{3+}-reducing microorganisms that can transfer electrons derived from the oxidation of organic compounds and/or H_2 to HS (Lovely et al., 1996), given that this electron transfer yields energy to support the growth of microorganisms. Scott et al. (1998) observed that quinones are important electron-accepting groups in HS, showing that the latter, which were traditionally thought to be microbially inert in anoxic environments, participate in microbial reactions. Quinone content in HS can be estimated by measuring the SFR (Senesi and Steelink, 1989). Hence, it is possible to evaluate the electron-accepting capacity of the HS through EPR. Using the electron-accepting property of HS, Perminova et al. (2005) developed quinonoid-enriched humic materials with enhanced redox properties to be used as potentially effective redox mediators and reducing agents for *in situ* remediation of soil and aquatic environments.

Despite the useful information to study biophysico-chemical processes involving NOM in environmental systems, the SFR concentration depends on various laboratory conditions such as pH and irradiation (Senesi, 1990a; Paul et al., 2006), redox conditions, acid hydrolysis, methylation, and temperature (Senesi, 1990a). Moreover, ΔH and SFR concentration can also be affected by sample humidity and the presence of paramagnetic ions (Novotny and Martin-Neto, 2002; Novotny et al., 2006).

16.2.5.2. Metal Ions Complexation with Humic Materials in Different Environments. Complexation mechanisms involving metal ions and NOM play a vital role in establishing the behavior of the former, influencing the bioavailability of metal nutrients to plants and soil microorganisms, toxicity hazard of potentially toxic metals, migration–accumulation phenomenon of metals in the soil–water–sediment system, pedogenic processes, and geochemical transfer and mobility pathways (Senesi, 1992). Some metal ions are paramagnetic (Fe^{3+}, Cu^{2+}, Mn^{2+}, VO^{2+}, Mo^{5+}, and Cr^{3+}), hence detectable by EPR (Weil et al., 1994), which is almost unique among laboratory methods for investigating metal ions–HS complexes in providing structural information without artifacts or restrictive experimental conditions (Senesi et al., 1985). In the specific case of Cr^{3+}, only unresolved EPR spectra were obtained for a peat HA doped with either Cr^{3+} or diamagnetic dichromate, which is apparently reduced to Cr^{3+} by the HA (Lakatos et al., 1977).

Several papers in the literature have shown the HS complexes and their respective spectroscopic data, and some examples will be presented here. Customarily, these studies demonstrate that paramagnetic metal ions complexed with HS can be identified, generally by g values and A^* values, with type and symmetry of coordination sites eventually also determined (Lakatos et al., 1977; McBride, 1978; Boyd et al., 1983; Senesi, 1990a, 1992; Martin-Neto et al., 1991; Jezierski et al., 1998; Flogeac et al., 2004). It is important to keep under scrutiny that depending on the metal ions and the origin of the sample, some factors can alter specific spectroscopic parameters—for example, type of ligand atoms, metal ions concentration, humidity, temperature, and pH (McBride, 1982a; Schnitzer and Ghosh, 1982; Senesi et al., 1985; Martin-Neto et al., 1991; Mangrich et al., 1998; Novotny and Martin-Neto, 2002; Flogeac et al., 2004; Novotny et al., 2006).

Recent papers have used spectroscopy analysis in the study of metal ions complexation in different environments. For example, González-Pérez et al. (2006a), based on the spectroscopic data from literature cited above, observed the complexation between soil HA and some paramagnetic metal ions—for example, Cu^{2+} and VO^{2+} (Figure 16.6). In Figure 16.6b, HA was extracted from soil under sewage sludge application, showing that the sewage sludge, besides being a source of OM, nitrogen, and phosphorous, can be a source of micronutrients essential for plant growth. Zelano et al. (2006), based on the fact that natural sediments show sequestering properties that can lead to a self-purification process of aquatic environment from metal pollution, used EPR to better characterize the Cu^{2+}-sediment interaction and to obtain further information concerning the nature of ligand site(s) involved.

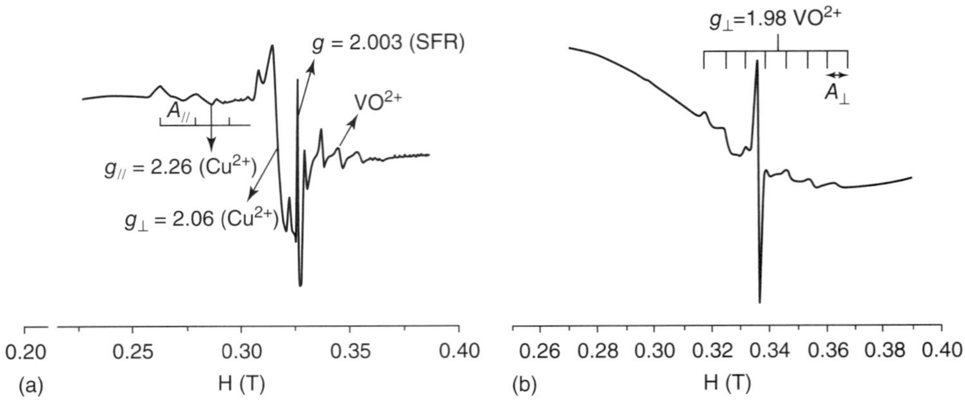

Figure 16.6. EPR spectra of the HA extracted from soil under no-tillage and sewage sludge application, showing several resonance lines (measured at −160 °C): (**a**) In $g_{//} = 2.26$ and $g_{\perp} = 2.06$ with $A_{//} \approx 17\,mT$ associated with Cu^{2+} complexed with oxygen ligands; $g = 2.003$ due to SFR. (**b**) In $g_{\perp} = 1.98$ and $A_{\perp} \approx 7.1–7.3\,mT$ assigned to VO^{2+} complexed with oxygen ligands with axial symmetry. Adapted from González-Pérez et al. (2004, 2006a).

Besides using the EPR to evaluate the binding capacity of NOM for metal ions, it is also possible to use the latter as spin probe (Senesi, 1990a). Analyses of spectroscopic data comprise studying the adsorptive properties, surface interactions, and structural chemistry of soil, synthetic metal oxides, hydrous oxides, and clay minerals (McBride et al., 1984; Coyne and Banin, 1986; Senesi et al., 1991a; Spagnuolo et al., 2004).

16.2.5.3. Studies of Reaction Mechanisms of Pesticides with Humic Substances. EPR has been used in studies of pesticide reaction mechanisms with HS. Spectroscopic data from HS, as SFR concentration and line width (ΔH) can be used to obtain information about the reaction mechanisms of pesticides-HS (Senesi, 1990a).

A reaction mechanism between pesticides *s*-triazines and HS, which has been discussed in the literature, is the charge transfer (Senesi et al., 1987; Martin-Neto et al., 1994b, 2001; Sposito et al., 1996). This relatively strong reaction mechanism is based on the possibility that electron-deficient, quinone-like structures in the HS can remove electrons from donating amine and/or N atoms of the triazine ring by single-electron donor–acceptor processes that involve semiquinone free radical intermediates (Senesi, 1990a). This reaction mechanism is shown in Figure 16.7.

Senesi et al. (1987) observed significant increase in SFR concentration on reaction with HA for some *s*-triazines (e.g., promotone, ametryne, desmetryne, and methoprotyne), assigning the spectroscopic alteration to charge-transfer mechanism. However, Martin-Neto et al. (1994b, 2001) showed that the absence of changes in the SFR concentration of samples of atrazine (6-chloro-N^2-ethyl-N^4-isopropyl-1,3,5-triazine-2,4-diamine)-HA complexes when compared with HA samples was an indication that the electron-donating capability of the atrazine was not sufficient to engage in an electron-transfer mechanism. This was in agreement with theoretical studies realized by Welhouse and Bleam (1993a,b). Moreover, Martin-Neto et al. (1994b, 2001), considering the data from UV–vis absorption, FTIR, and EPR spec-

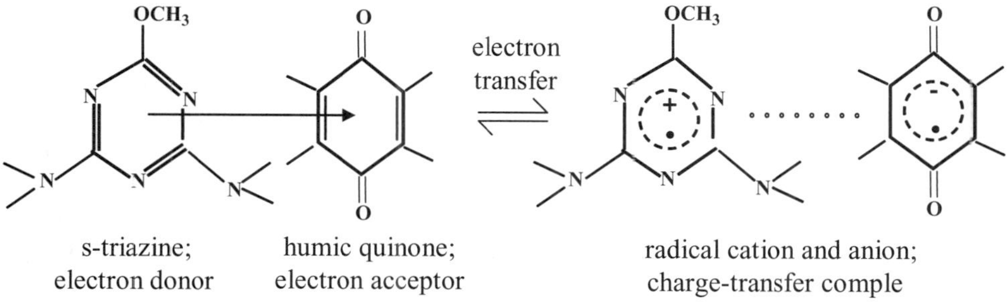

Figure 16.7. Scheme of charge-transfer reaction mechanism between *s*-triazine molecule and humic quinone structure, showing formation of SFR that can be detected by EPR measurement (Senesi, 1990a).

troscopies, observed the weak sorption mechanisms of atrazine by HA, involving hydrogen bonding, proton transfer (at low pH), and possibly hydrophobic interaction. The reason that atrazine was not involved in the charge-transfer reaction is probably due to the fact that the Cl atom at the 6-position on the triazine ring of atrazine molecule is sufficiently electron-withdrawn to inhibit its electron-donating capacity and thus is able to prevent the formation of this kind of charge-transfer complex (Welhouse and Bleam, 1993a,b).

Also, Martin-Neto et al. (2001), by using different spectroscopies, among which were EPR to monitor the SFR concentration, demonstrated that hydroxyatrazine (6-hydroxy-N^2-ethyl-N^4-isopropyl-1,3,5-triazine-2,4-diamine), which is an atrazine by-product and whose transformation to the hydroxy form is facile, readily forms electron-transfer complexes with HS. The occurrence of the electron-transfer mechanism was attributed to the high basicity of hydroxyatrazine ($pK_a = 4.28$), with the presence of hydroxy group in the 6-position on the triazine derivative ring, and is different from atrazine ($pK_a = 1.68$), which has the presence of Cl atom at 6-position on the triazine ring. These complexes probably are the cause of the well-known strong sorption by HA and they may be the undetected cause of apparent electron-transfer complexes between SOM and atrazine. When in fact uncontrolled hydroxyatrazine formation occurs, this generates misunderstanding in identifying the correct reaction mechanism of atrazine. Ferreira et al. (2002) obtained similar results using the pesticide imazaquin (2-[4,5-dydro-4-methyl-4-(1-methylethyl)-5-oxo-1*H*-imidazol-2-yl]-3-quinoline-carboxylic acid); that is, weak binding forces played a prominent role in the imazaquin sorption on the soil and on its HA, involving hydrogen bonding, proton transfer, cation exchange contribution (at low pH), and mainly hydrophobic interactions.

Hence, through EPR, it is possible to obtain information at a molecular level regarding the chemistry of pesticides and other xenobiotic interactions with soil constituents or any of the transporting agents that may carry xenobiotics through a soil profile to groundwater.

16.2.5.4. Spin-Label Methodology Applied to Hydrophobic Interaction Studies of Humic Substances. Many molecules have no intrinsic paramagnetic centers, but EPR can provide useful information about these molecules if a structurally sensitive spin label is attached (Knowles et al., 1976). Spin-label molecules are stable free radicals usable as reporter groups or probes. Spin probes have been used in environmental systems—for example, clay minerals (McBride, 1986; Dumestre et al., 2006) and aluminum oxides (McBride, 1982b). The most commonly used spin labels contain the relatively inert nitroxide group (Ohnishi and McConnell, 1965), where the unpaired electron is principally localized in a 2pπ-orbital of the nitrogen–oxygen bond (Figure 16.8a) interacting with the nitrogen nuclear spin ($I = 1$), giving rise to the characteristic three isotropic narrow hyperfine lines of the spin label EPR spectrum due to the three nuclear spin projections ($M_I = +1$, 0, and −1), as shown in Figure 16.8b.

The anisotropy of the spin-label hyperfine splitting is a consequence of the particular charge distribution symmetry within the molecule. The analysis of spectral anisotropy provides the immobilization extent of the spin label and is sensitive to the molecular conformation. Moreover, the spin-label molecules have their charge distribution distorted by the polarity of their environment. The isotropic hyperfine splitting constant (a_0) provides a relative polarity indicator (Knowles et al., 1976; Campbell and Dwek, 1984).

Recently, Bonin and Simpson (2007) showed in their study that polycyclic aromatic hydrocarbons' sorptive behavior could not be solely attributed to a specific SOM chemical characteristic (i.e., aliphaticity or aromaticity) and concluded that both structure and OM physical conformation are important in sorption processes. Martin-Neto et al. (1994b) used analysis of microwave power saturation curves (Weil et al., 1994) to obtain information about HA conformation. They observed that at pH > 3.5 a typical curve of homogeneous saturation was obtained for the oxisol HA, whereas at pH 2.3 inhomogeneous saturation occurred. The peat HA showed only homogeneous saturation. However, a similar "inhomogeneous trend"

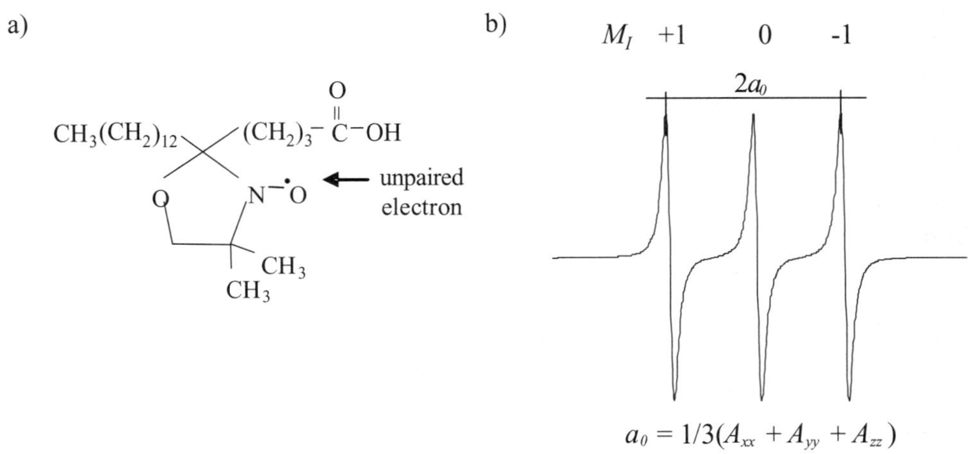

Figure 16.8. (**a**) Chemical structure of typical spin-label 5-SASL containing the nitroxide group. Shown in the figure is a nitrogen–oxygen bond and the unpaired electron. (**b**) Characteristic three isotropic narrow hyperfine EPR lines of the spin-label spectra due to the three nuclear spin projections of nitrogen atom ($M_I = +1, 0$, and -1). The term a_0 is isotropic hyperfine splitting constant, and A_{xx}, A_{yy}, A_{zz} are the hyperfine constants. In nonviscous solution (e.g., water) all anisotropic effects are lost and a_0 is the average of all hyperfine constants (Knowles et al., 1976).

with pH was apparent. At low pH (inhomogeneous saturation), the dissipation of microwave energy is a slow process because of a less effective interaction between the SFR and its molecular environment. This could be the result of a "globular" HA conformation that produces less interaction among constituent molecules and those surrounding them. At higher pH, the relaxation process is more rapid (homogeneous saturation), and an HA conformation with more effective interactions may have occurred.

Ferreira et al. (2001), through the use of the spin-label methodology, observed substantial differences in the 5-SASL spin label (5-(4,4-dimethyloxazolidine-*N*-oxyl)stearic acid) EPR spectra when in aqueous solution (Figure 16.9a) and when in the presence of HA suspensions (Figures 16.9b and 16.9c). It was observed that below pH 5, the typical three isotropic narrow hyperfine lines of 5-SASL (typical of free spin label) were less evident due to formation of additional broad lines in the 5-SASL EPR spectrum (Figure 16.9b). On the other hand, these broad lines in the 5-SASL EPR spectrum were not observed in pH 6.5 (Figure 16.9c). From these results, the existence of at least two kinds of hydrophobic sites at HA molecules were proposed. First, inner hydrophobic sites (probably water-protected) at pH below 5, with a very strong affinity with nonpolar chemical compounds, possibly including nonpolar pesticides. These inner hydrophobic sites, however, would be partially or totally destroyed at higher pH due to conformational HA changes. Second, surface hydrophobic sites at pH above 5, for example, formed by hydrophobic moieties such as the poly(methylene) groups (Hu et al., 2000), which are exposed to water but keep their capacity to bind nonpolar chemical compounds. These less protected (not "closed") hydrophobic sites certainly have a lower affinity than those inner sites existing at low pH, thus explaining the increase of HA sorption capacity at low pH of several nonpolar compounds (Li et al., 1992; Martin-Neto et al., 2001). However, these HA surface hydrophobic moieties could be very important in the pH above 5, which is normally observed in soils and natural water systems.

According to Ferreira et al. (2001), the formation of inner hydrophobic sites (mainly at pH below 5) is due to H-bonding association of mainly carboxylic groups from different small molecules generating an expanded (or globular) HA structure with inner hydrophobic sites. Simões (2005), using different spin-label molecules with different hydrophobicity—that is, TEMPO (2,2,6,6-tetramethyl-4-aminopiperidine-1-oxyl; 5-SASL 5-(4,4-dimethyloxazolidine-*N*-oxyl)stearic acid;

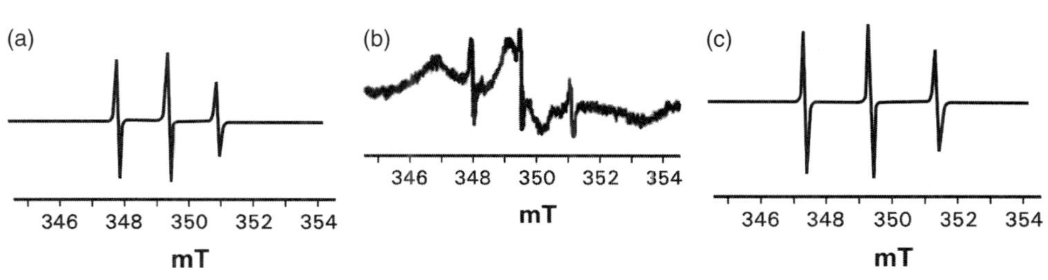

Figure 16.9. EPR spectra of 5-SASL spin label in aqueous solutions: (**a**) At pH 3.3 in aqueous solution. (**b**) At pH 3.3 in HA presence. (**c**) At pH 6.5 in HA presence. Adapted from Ferreira et al. (2001).

16-SASL (2-(14-carboxytetradecyl)-2-ethyl-4,4-dimethyl-3-oxazolidinyloxy) and 5-MSSL (2-(4-methoxy-4-oxobuthyl)-4,4-dimethyl-2-tridecyl-3-oxazolidinyloxy)—confirmed that the interactions established with HA are mainly hydrophobic.

Thus, spin-label methodology can provide useful data for better understanding the structure of HS, independent of an existing consensus regarding molecular structures of HS (Burdon, 2001; Hayes and Clapp, 2001; Piccolo, 2001; MacCarthy, 2001; Sutton and Sposito, 2005).

16.2.5.5. Spin-Trapping Technique Applied to Photoreaction Studies of Humic Substances. Several highly reactive species—for example, singlet oxygen, hydrated electrons, superoxide, hydrogen peroxide, and hydroxyl radicals (·OH)—are formed through solar irradiation of NOM in environmental waters (Paul et al., 2005; Neamtu and Frimmel, 2006; Hassett, 2006; Latch and McNeill, 2006; Wang et al., 2007). Moreover, it is known that HS can exhibit photocatalytic effects to pesticide residues in aqueous environment (Kamiya and Kameyama, 1998). However, HS can also reduce the pesticide photolysis rate (Si et al., 2004). Since they present high absorption bands (especially in the ultraviolet range), they undergo photolysis under incident ultraviolet and visible light and they can also quickly react with ·OH (Lindsey and Tarr, 2000).

EPR spin-trapping technique can be used to detect free radical intermediates, which are free radicals that have a short lifetime to be detected by EPR—for example, ·OH (Janzen, 1980). The spin-trapping technique is based on the fact that during certain reactions in solution a transient radical will interact with a diamagnetic reagent to form a more persistent radical. The radical product accumulates to a concentration where detection and, frequently, identification are possible by EPR. The key reaction is usually one of attachment. The diamagnetic reagent is said to be a "spin trap" and the persistent radical product is then the "spin-adduct" (IUPAC, 1997).

Garbin et al. (2007) used the spin-trapping technique to evaluate the photogeneration and scavenging of free radicals by HS during irradiation time. Using the DMPO (5,5-dimethyl-1-pirrolyne-*N*-oxide) spin-trap molecule, the spin-adduct formed a signal such as a 1:2:2:1 quartet, with 14.8 G hyperfine splitting (Figure 16.10), where signal intensity provided a valuation of the number of ·OH photogen-

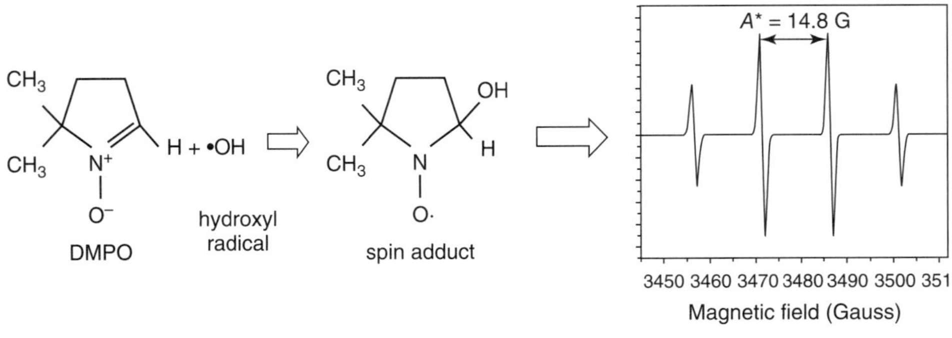

Figure 16.10. Chemical structure of spin-trap DMPO (diamagnetic), reacting with hydroxyl radical (·OH), showing the formation of spin-adduct DMPO-hydroxyl radical (·OH) and typical DMPO-OH adduct EPR signal.

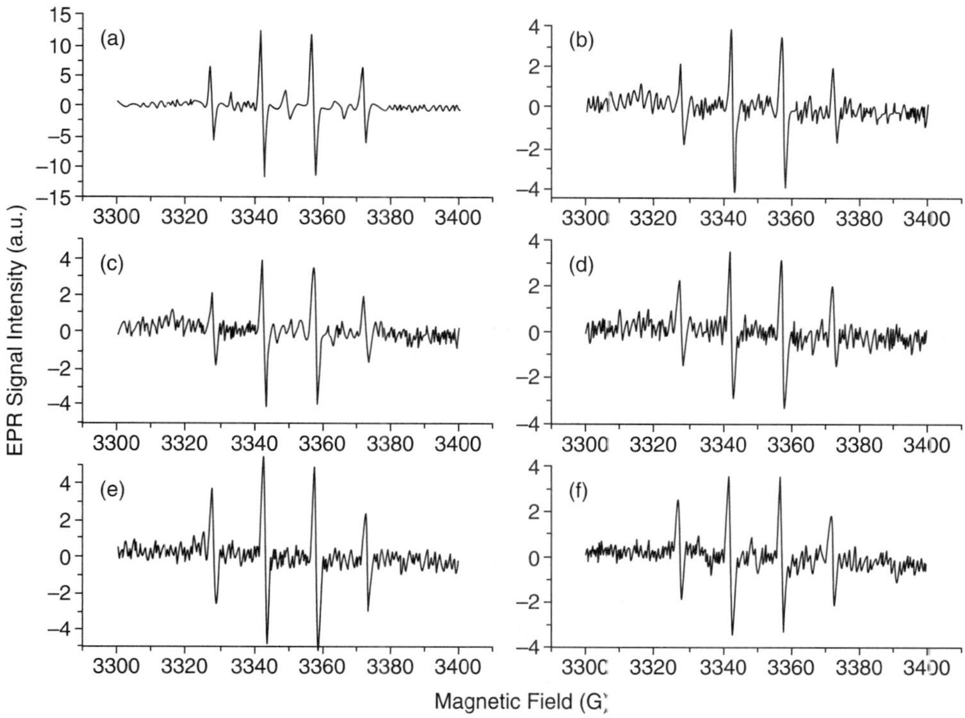

Figure 16.11. DMPO-OH adduct EPR signal for following samples: (**a**) TiO_2, (**b**) soil HA, (**c**) peat HA, (**d**) aquatic HA, (**e**) peat FA, and (**f**) aquatic FA. All HS were purchased from IHSS and used at 50 mg liter^{-1} of initial concentration. DMPO concentration at 20 mmol liter^{-1} and 3 min of irradiation at 880 mW cm^{-2} of light intensity in the UV–vis region. Reprinted from Garbin et al. (2007).

erated (Figure 16.11). Tests were performed to confirm that the spin-adduct signal detected by EPR is due to ·OH capture by DMPO, and that the sources of ·OH are TiO_2 and HS molecules.

From these results and others from UV–vis absorption spectroscopy and differential pulse polarography (data not shown), Garbin et al. (2007) concluded that HS can act as photocatalyst to pesticide photolysis in aqueous solution only for specific ranges of concentration (as seen in Figure 16.12), which in turn depended on the HS and pesticide chemical characteristics. Under ultraviolet and visible radiation, this photocatalysis is based on photogeneration of ·OH radicals, and the susceptibility of pesticide molecules to ·OH attacks defines the efficiency of the photocatalysis.

16.3. FOURIER-TRANSFORM INFRARED (FTIR)

16.3.1. Principles and Equipment

Infrared (IR) radiation refers broadly to that part of the electromagnetic spectrum between visible and microwave regions. IR spectroscopy is one of the most powerful

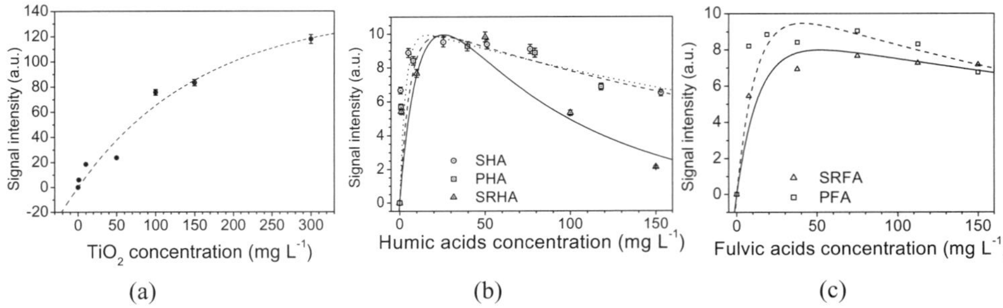

Figure 16.12. DMPO-OH adduct EPR signal intensity versus (**a**) titanium dioxide (TiO_2), (**b**) different HA, and (**c**) different FA concentrations. All HS were purchased from IHSS, being SHA standard soil HA, PHA standard peat HA, SRHA standard Suwannee River HA, PFA standard peat FA, and SRFA standard Suwannee River FA, with same light irradiation conditions, as given in Figure 16.11 Reprinted from Garbin et al. (2007).

tools available to the chemist for identifying pure organic and inorganic compounds because, with the exception of a few homomolecular molecules such as O_2, N_2, and Cl_2, all molecular species absorb infrared radiation. Of greatest practical use to the organic chemist is a limited range of wavenumber between 4000 and 400 cm^{-1}. There has been also some interest in the near-infrared (NIR) (14,290–4000 cm^{-1}) and the far-infrared regions (700–200 cm^{-1}).

Infrared radiation of frequencies less than about 100 cm^{-1} are absorbed and converted by an organic molecule into energy of molecular rotations. This absorption is quantized; thus a molecular rotation spectrum consists of discrete lines. IR radiation ranging from about 10,000 to 100 cm^{-1} is absorbed and converted by an organic molecule into energy of molecular vibrations. There is a selection rule that allows transitions only between adjacent levels. For a vibration of a diatomic molecule to be spectroscopically active, the dipole moment must change during the vibration. For example, atomic molecules such as O_2, N_2, and Cl_2, where the changes in dipole moment are equal to zero, are inactive in the infrared spectrum. In a diatomic molecule there is only one mode of vibration—that is, the stretching of the bond. In polyatomic molecules, it has to allow for bonds to stretch and bend as the molecule distorts. For a molecule that consists of N atoms, there are $(3N - 6)$ ways in which the molecule can vibrate, or $(3N - 5)$ if the molecule is linear. For this reason, in order to describe the motions of a molecule of N atoms, a total of $3N$ coordinates is required (x-, y-, and z-coordinates for each atom). Three of these are required to specify the position of the mass center of the molecule. Three more are required to specify the rotational motions (only two are required to specify rotation in a linear molecule). This means that relative atomic positions at a given time can be fully specified by $(3N - 6)$ coordinates, or $(3N - 5)$ in a linear molecule. Any overall motion of the molecule can be represented by a superposition (or combination) of these fundamental modes of vibration.

Figure 16.13 shows an energy-level scheme for a diatomic molecule, showing the rotational energy transition, the vibrational energy transition in the ground state, and electronic energy transition from ground state to the excited state. The frequency of absorption wavelength depends on the reduced masses of the atoms, the

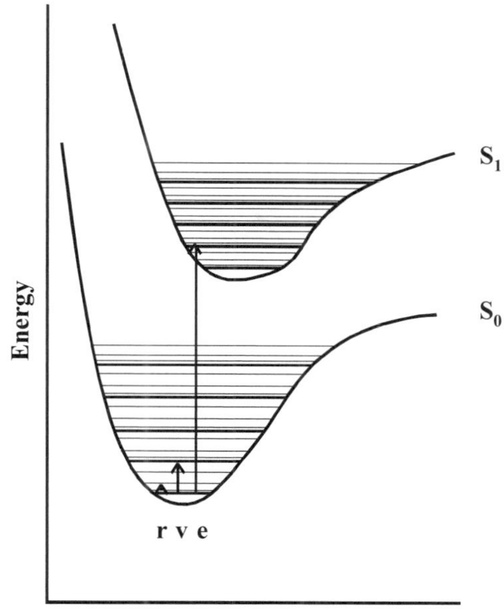

Figure 16.13. Energy-level scheme for a diatomic molecule, showing the rotational energy transition (r), the vibrational energy transition (v) in the ground state (S_0), and electronic energy transition (e) from S_0 to the excited state (S_1).

force constants of the bonds, and the geometry of the atoms. Assignments for stretching frequencies can be approximated by the application of Hooke's Law:

$$\bar{\nu} = \frac{1}{2\pi c}\sqrt{\frac{f}{(MxMy)/(Mx+My)}} \tag{16.10}$$

where $\bar{\nu}$ is the vibrational frequency (cm^{-1}), c is the velocity of light ($cm\ s^{-1}$), f is the force constant of bond ($dyne\ cm^{-1}$), and Mx and My represent the mass (g) of atom x and atom y, respectively.

Although the infrared spectrum is characteristic of the entire molecule, it is true that certain groups of atoms give rise to bands at or near the same frequency regardless of the structure of the rest of the molecule.

Fourier transform infrared (FTIR) spectroscopy has been extensively developed over the past decade and provides a number of advantages. The main part of FTIR spectrophotometer is the Michelson interferometer. Radiation containing all IR wavelengths (e.g., 4000–400 cm^{-1}) is emitted by source of infrared radiation (Globar) and is split into two beams. One beam is of fixed length, and the other is of variable length (movable mirror).

The varying distance between two pathlengths results in a sequence of constructive and destructive interferences and hence variations in intensities, as shown in an interferogram. Fourier transformation converts this interferogram from the time domain into one spectral point on the more familiar form of the frequency domain. Smooth and continuous variation of the piston lengths adjusts the position of mirror B and varies the length of beam B (Figure 16.14); Fourier transformation, at suc-

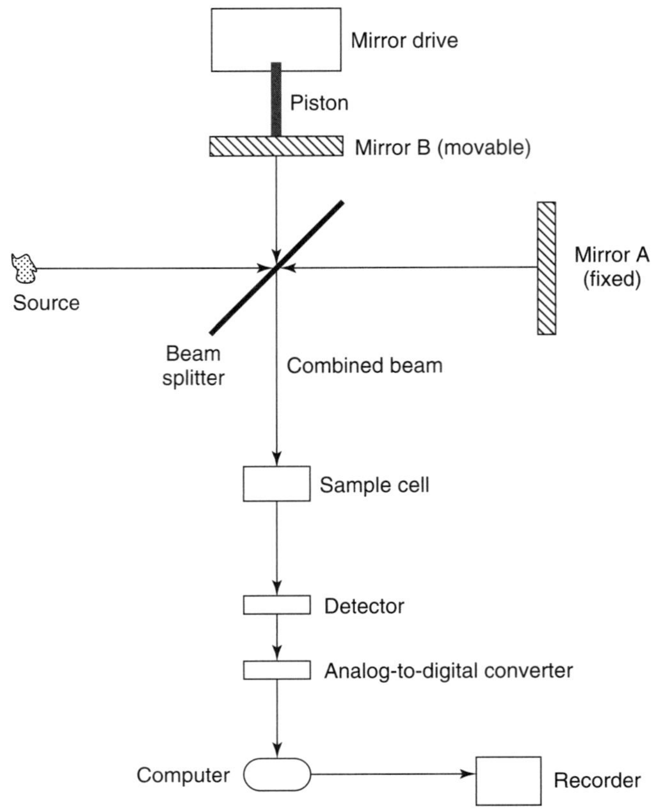

Figure 16.14. Schematic representation of a FTIR spectrometer (Silverstein et al., 2005).

cessive points throughout this variation, gives rise to the complete IR spectrum. Passage of this radiation through a sample subjects the compound to a broad band of energies. In principle, the analysis of one broad-banded pass of radiation through the sample will result in a complete IR spectrum.

There are many advantages for using the FTIR methods. Since a monochromator is not used, the entire radiation range is passed through the sample simultaneously and much time is saved (Felgett's advantage). FTIR instruments can have very high resolution ($\leq$0.001 cm^{-1}). Moreover, since the data undergo analog-to-digital conversion, IR results are easily manipulated. By combining the results of several scans to average out random absorption artifacts, excellent spectra from very small samples can be obtained.

Band position in IR spectra is shown as wavenumbers (ν) whose unit is reciprocal centimeter (cm^{-1}); this unity is proportional to the energy of vibration and modern instruments are linear in reciprocal centimeters.

One of the advantages of the FTIR spectrometer is the facility of sample preparation procedure. It is possible to press powder into pellet. In this method, different diluents (matrixes) can be selected for several applications. For mid-IR frequency range, KBr, KCl, or diamond dust can be used. For far-Infrared testing, high-density polyethylene (HDPE) or diamond dust is suitable. For near-infrared analysis, CsI or KBr can be selected or mulls may be used as alternatives to pellets. The sample

(1–5 mg) is ground with a mulling agent (1–2 drops) to give a two-phase mixture that has a consistency similar to toothpaste. The mull is pressed between two IR-transmitting plates to form a thin film. Common mulling agents include mineral oil or Nujol (a high-boiling hydrocarbon oil), Fluorolube (a chlorfluorocarbon polymer), and hexachlorobutadiene. To obtain a full IR spectrum that is free of mulling agent bands, the use of multiple mulls (such as Nujol and Fluorolube) are generally required.

16.3.2. Functional Groups Detection in Soil Organic Matter

Infrared photon energy associated with a 4000–400 cm^{-1} interval (1–15 kcal mol^{-1}) is not sufficient to excite electrons. However, it can produce vibrational changes between atoms and chemical groups that are covalently bound. They can produce a number of movements, depending on their characteristics and all the atoms present.

In general, molecular vibrations can be classified in two ways: axial deformation (stretching) and angular deformation (bending). Stretchings are radial oscillations of distances between the same axis bounding nucleus, and bendings show angle changes between bounds and they take place either in asymmetric deformation out of plan or in the plan which contains the chemical bounds in a referenced plan (Silverstein et al., 2005). The specific frequency in which such vibration occurs is established by bond forces and mass of respective atoms. The required energy to stretch a chemical bound is higher in comparison to bending it (Silverstein et al., 2005).

Identification of organic substance from a FTIR spectrum starts by determining which functional groups are the most probable to be present by considering the observed frequency region, followed by a detailed comparison of spectrum with correlation tables. The exact interpretation of spectrum is not always possible due to their complex characteristics. In fact, this characteristic makes individualization and consequent utility for the identification possible, if the sample has a high grade of purity.

A characteristic FTIR spectrum of HS shows overlapping bands as a result of its complex mixture of organic molecules, which indicates the diversity of functional groups present in the structure (Stevenson, 1994). Figure 16.15 shows a typical spectrum of HA. Table 16.1 shows typical bands found in HS spectra.

In general, a FTIR spectrum in environmental studies is not conclusive due to overlapping spectral bands, which, in turn, makes unambiguous and complete interpretation difficult. The use of several chemical and spectroscopic methods, in addition to FTIR, is largely applied to confirm the information obtained by FTIR and vice versa.

FTIR is also used to study oxidation and mineralization processes, revealing the dynamics of HS as a function of several environmental variables, such as pH, ionic strength, metal concentration, presence of pollutants, radiation, and so on.

Fu and Quan (2006) observed the complexes of FA derived from leonardite on the surface of iron oxides (hematite, goethite, and akaganeite) by FTIR studies at several pH values. The FTIR data provided the consistent evidence that the interaction mechanism was the ligand-exchange involving carboxylic functional groups of the FA and the surfaces sites of both hematite and goethite, while no complexation

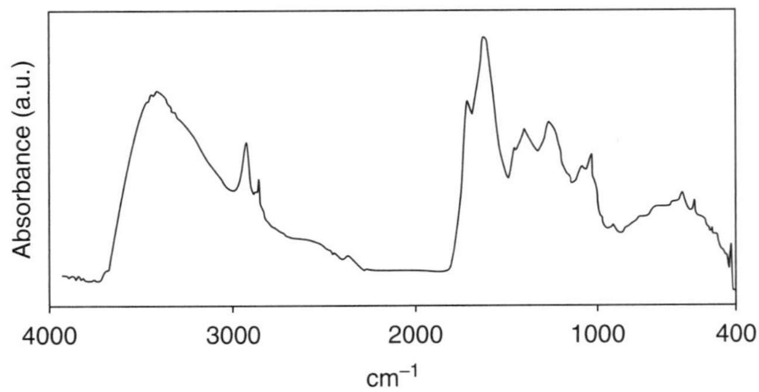

Figure 16.15. Typical FTIR HA spectrum.

TABLE 16.1. Infrared Bands Detected in HA Spectra (Niemeyer et al., 1992)

Spectrum Region (cm^{-1})	Signal Origin
3380	Phenolic O–H Stretching (contribution of aliphatic OH and water). N–H Stretching.
3030	Aromatic C–H stretching
2930	C–H asymmetric stretching
2840	C–H symmetric stretching
2600	O–H stretching of –COOH
1720	–C=O stretching of COOH
1610	Aromatic C=C stretching and/or COO^- asymmetric stretching
1520–1525	Aromatic C=C stretching and also ascribed to N–H deformation and C=N stretching of amides
1450	CH_2 in-plane bending or scissoring
1350	$-COO^-$ symmetric stretching and carboxylic O–H bending
1270	Phenolic C–O stretching
1225	C–O stretching and carboxylic O–H bending
1170	Aliphatic C–OH stretching
1070	C–C stretching of aliphatic groups
1050 and 1040	C–O stretching of polysaccharides or polysaccharide-like substances and Si–O of silicate impurities
830	Aromatic CH out-of-plane bending. Nontronite, Muscovite impurities.
775	Aromatic CH out-of-plane bending

can be evidenced in the case of akaganeite. In general, the binding affinities of the iron oxides with the FA were on the order of hematite > goethite > akaganeite.

Guan et al. (2006) investigated the role of phenolic groups in the interaction of some NOM model [dihydroxybenzoic acids (DHBAs)] compounds with aluminum hydroxide, by using attenuated total reflectance–FTIR (ATR–FTIR). Carboxylic groups governed the complexation of DHBAs with aluminum hydroxide at low pH or in cases when the two hydroxyl groups were not adjacent to each other, and neither of them was ortho to the carboxylic group. The involvement of the phenolic groups, ortho to another phenolic group or ortho to the carboxylic groups, in the complexation increased with increasing pH because the deprotonation of phenolic groups was easier at higher pH. The presence of phenolic groups increased the electron density of the carboxylic groups and facilitated the inner-sphere complexation of the carboxylic groups with metal hydroxide.

FTIR can be also coupled to gas systems like gas chromatography (GC) and pyrolysis. Kuckuk et al. (1994) showed by pyrolysis–GC–FTIR that some principal structures were present in aquatic HS. Many of the pyrolysis products (e.g., methanol, acetone, alkylbenzenes, cyclopentane, aliphatic and aromatic organic acids, acetamide, pyrrole, and phenols) could be identified by their FTIR spectra using a digital library for automatic comparison. Some of the compounds are related to lignin fragments, which form a large part of the HS investigated. Geyer et al. (2000) investigated soil HS from different environments by using thermogravimetric equipment coupled to FTIR, in addition to multivariate data analysis. They observed the degree of soil pollution to a different extent by difference in FTIR spectra obtained after combustion at several temperatures.

HS are a mixture of several different structures, and also the same functional group can be bound to different parts of organic chain or metal ions. Therefore it is not possible to obtain a "pure" HS. The International Humic Substances Society (IHSS, www.ihss.gatech.edu) has several reference samples that are well-characterized, and this is a good way to compare results of HS spectra. Depending on the chemical environment, the position and shape of each band can be modified. The relative intensities of the band at $1700\,cm^{-1}$ (C=O stretching of COOH) and $1600\,cm^{-1}$ (COO^- asymmetric stretching) are strongly pH-dependent and do not necessarily represent any structural modification. However, the carboxylate band will be more pronounced if metallic–carboxylic complexes are present. Finally, if KBr pellets (pressed-disk) sampling are not done well, several problems like light scattering, opacity, and bad distribution of the sample in the pellet are produced, promoting deformed baselines and poor spectra resolution. Sample and KBr previously dried must be mixed in a mortar and pressed correctly. The use of FTIR for quantitative analysis must be well-evaluated, and the use of mathematical tools is necessary (Small, 2006).

16.3.3. Detection of Soil Tillage Effects on Humic Substances Characteristics

FTIR is a very useful tool to observe structural variations of HS due to environmental changes resulting from soil tillage. After the "green revolution" in agriculture, between 1940s and 1960s, characterized by mechanization and better productivity control, the use of inorganic fertilizers, liming and pesticides, has proved to ameliorate productivity in the field. However, the impact of the continuous use of these materials in soil, in addition to the systematic use of conventional tillage systems, produced variations in the quantity, composition, reactivity, and stability of SOM.

Conventional tillage exposes the OM, which are naturally stabilized in the soil, to the action of sunlight and oxidation conditions that favor microbial activity. Consequently, only the very stable fraction of SOM generally remains. The importance of SOM functions is well known, but structural information, chemical composition, and changes induced by anthropogenic factors, such as tillage practices, are still under research.

González-Pérez et al. (2004) characterized Brazilian HA from an oxisol under different treatments: conventional tillage/maize-bare fallow; conventional tillage/maize rotation with soybean-bare fallow; no-till/maize-bare fallow; no-till/maize rotation with soybean-bare fallow; no-till/maize-cajanus and noncultivated soil under natural vegetation. Soil HA samples were analyzed by FTIR, EPR, ^{13}C NMR, and UV–vis fluorescence spectroscopies and elemental analysis (CHNS). The FTIR spectra of the HA were similar for all treatments. HA from treatments on noncultivated and no-till/maize-cajanus were less aromatic, as shown by the correlation among obtained ^{13}C NMR, EPR, and fluorescence data. However, no important effect due to tillage system was observed in these areas after 5 years of cultivation. Probably, the studied oxisol has a high clay content that offers protection to the clay–Fe–OM complex against strong structural alterations.

Chang Chien et al. (2006) studied the qualitative and quantitative characteristics of soil organic carbon (SOC) and related HA extracted from the soils of field plots which had undergone 8 years of annual paddy (*Oryza sativa* L.) and upland maize (*Zea mays* L.) rotation with different fertilizations. Samples collected were subjected to soil characterizations and extraction of HS, which were used for chemical, spectroscopic (FTIR, ^{13}C NMR, EPR, X-ray diffractometry), ^{13}C, and ^{14}C dating analyses. The relative intensities of FTIR absorption bands of the seven composite HA from the seven treatments did not differ much in wavenumber range of 4000–2000 cm^{-1}. The spectra revealed main changes in the regions at 1240–1245 cm^{-1} (C–O stretching vibrations of esters, ethers, and phenols), 1036–1038 cm^{-1} (alcohols and carbohydrates), and at 1135–1140 cm^{-1} (C–O stretching vibrations of carbohydrates). These results, in addition to other spectroscopic analyses, indicated the input of fresh C in the composition of SOM, as well as an increase of the SOC turnover rate.

The labile mobile humic acid (MHA) and the more recalcitrant calcium humate (CaHU) in double-cropped lowland rice soils were analyzed by chemical and spectroscopical methods by Olk et al. (1999). Compared with CaHU fractions, the MHA generally featured more intense absorption in six bands: 1650 and 1510 cm^{-1} (C=O stretching of amide I band, and N–H deformation and C=N stretching of amide group); 1460 and 1450 cm^{-1} (aliphatic C–H); 1420 and 1415 cm^{-1} (OH deformation and C–O stretching of phenolic OH); 1125 cm^{-1} (C–O of alcoholic and ether groups); and 1036 cm^{-1} (C–O stretching of polysaccharide-like components or Si–O silicate impurities). These results suggested incomplete humification of both fractions and indicated that MHA has lower humification than CaHU.

Soil amendment with animal manures is a common practice for either increasing SOM and nutrient content or disposing of wastes from intensive animal farms. However, the application of organic amendments that are not sufficiently mature and stable may adversely affect soil properties, especially the content and quality of SOM pools.

Francioso et al. (2000) used spectroscopic methods to investigate molecular changes in SOM treated with different residues. The experiment consisted of soil treated over a 22-year period with different amendments: cattle manure, cow slurries, and crop residues. The presence of a new band at 1640 cm^{-1} (SOM from cattle manure amended soil), described as either NH_2 bending or amide 1 motions, was supported by an increase of total organic N concentration. The increase of absorption at 1409 cm^{-1} (phenolic components) was more intense in the SOM sampled from soil amended with cattle manure. These results, in addition of other spectroscopic analyses, demonstrated that the composition of the soil amended with cattle manure varied significantly in relation to other amendments given that some specific aromatic and aliphatic moieties are resistant to the degradation.

Plaza et al. (2003) studied the effect of the consecutive annual additions of pig slurry (PS) on the soil FA fraction of SOM on a calcic luvisol, using FTIR and several other chemical and spectroscopical FA characterizations. Changes in the band intensities centered at about 2900, 1520, 1230, and 1040 cm^{-1} showed that PS-FA was characterized by a prevalent aliphatic character, large contents of acidic functional groups, S– and N–containing groups, and polysaccharide components. In addition, extended molecular heterogeneity, small SFR contents, and low degrees of aromatic ring polycondensation, polymerization, and humification were observed by other techniques. Statistical analysis of experimental data showed that, with some exceptions, these effects generally increased with increasing cumulative amount of PS applied to soil over time. Thus, this material should not be considered as a mature organic amendment and should be treated appropriately before being applied to the soil, so as to increase the degree of humification and thus enhance its potential as a soil organic fertilizer.

FTIR spectroscopy can also be used to monitor the sewage sludge-based compost, evaluate the degradation rate, and thus determine the maturity (Grube et al., 2006). Although the composition of the input mixture strongly affects the shape of the infrared spectra, typical bands of components can be selected and used to follow the composting process. The appearance, shape, and intensity of the nitrate band at 1384 cm^{-1} was well-pronounced and evident for a sewage sludge-based compost maturity. An increase of the peak ratios 1384/2925 and decrease of 2925/1034, 1034/1384 correlated with the degree of decomposition. For the composting mixture under study, the peak ratios 1034/1384 and 1384/2925 were more demonstrative. Considering the influence of the composting mixture (components and their ratio) on the shape of the FTIR spectra, the nitrate band at 1384 cm^{-1} can be overlapped by other absorption bands (e.g., lignin bands in 1300- to 1400-cm^{-1} region) thus appearing in the spectra as a shoulder, and therefore the ratios 1384/2925 and 1034/1384 become unusable for maturity evaluation.

16.3.4. Determination of Reaction Mechanisms Between Humic Substances and Pesticides

Soil sorption of most hydrophobic organic compounds (e.g., nonpolar pesticides) is directly related to SOM content. HS are the major SOM components (Ferreira et al., 2002). FTIR spectroscopy makes possible the observation of how some chemical functions, present in humic structures, are involved in the sorption process.

Sorption mechanism of atrazine by SOM has been a subject of controversy. The early works (Weber et al., 1969; Hayes, 1970) showed that the sorption process is inhibited due to the low pK_a value of herbicide, along with the proton transfer between carboxylic groups as well as the charger transfer at low pH values. These were discussed as probable retention mechanisms by organic colloids. However, Martin-Neto et al. (1994b, 2001) observed by FTIR (Figure 16.16) and UV–vis spectra that a charge-transfer mechanism was not operative in the HA–atrazine (HA–AT) interaction. FTIR showed that in pH <4, the carboxylate band (1610 cm^{-1}) was observed in HA–AT spectrum, but a decrease in the wavenumber of C–H

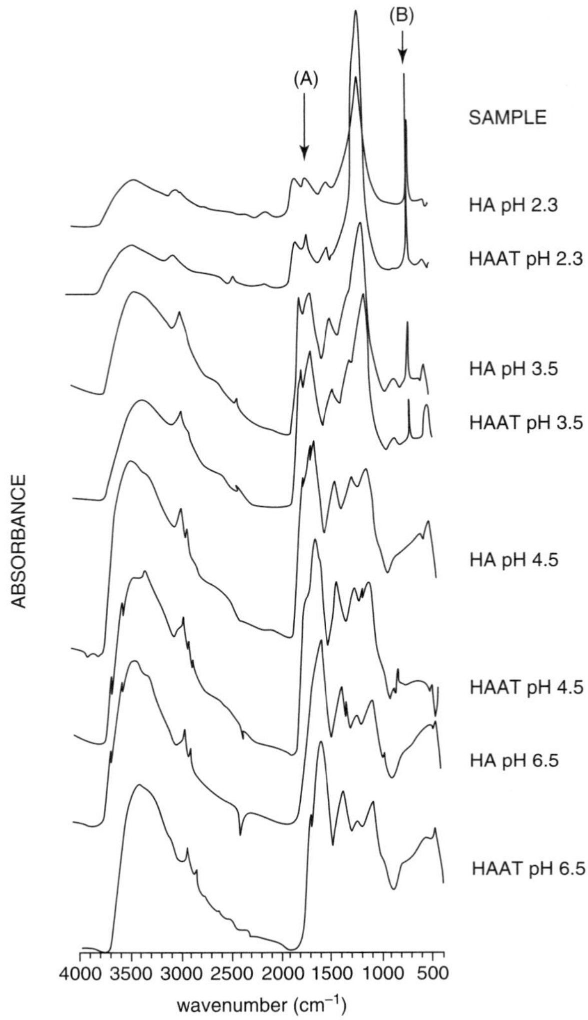

Figure 16.16. FTIR spectra of oxisol HA and HA-AT at pH 2–6 (Martin-Neto et al., 1994b). Bands (A) and (B) represent the 1610-cm^{-1} and 800-cm^{-1} wavenumbers, respectively, associated with carboxylate group of HA and C–H stretching of atrazine.

stretching of atrazine (800 cm^{-1}) was not observed, indicating that only proton transfer was operating between both HA and AT. Martin-Neto et al. (2001) showed, by EPR, that hydroxyatrazine readily forms electron-transfer complexes with HS. These complexes probably are the cause of the well-known strong adsorption by HA, and they may be the undetected cause of apparent electron-transfer complexes between SOM and atrazine, whose transformation to the hydroxy form is facile (Martin-Neto et al., 2001).

In addition to polarography to quantify the sorption process, sorption interactions between the herbicide imazaquin and a HA (extracted from a Brazilian oxisol) were observed by Ferreira et al. (2002) by EPR and FTIR spectroscopies. The imazaquin amount sorbed on the HA was much higher than that on the whole soil within the pH range studied, emphasizing the prominent role played by SOM on imazaquin sorption. Moreover, the sorption increased as the soil–solution pH decreased. This enhancement in sorption was attributed to the hydrophobic affinity of the herbicide by the HA and to the electrostatic interactions between the protonated quinoline group of imazaquin and the negative sites of the HA. The FTIR and EPR spectroscopies and polarography data indicated weak interactions between the herbicide and the soil and its HA, involving hydrogen bonding, proton transfer, and cation exchange (at low pH), and that mainly hydrophobic interactions were involved.

Humic acids, isolated from vermicompost (VHA), were compared with the humic acid samples extracted from peat (PHA) in respect of its chemical interactions and sorption with metribuzin herbicide (4-amino-6 *t*-butyl-3-methylthio)-1,2,4-triazine-5(4*H*)-one). The results showed that metribuzin was degraded by the HA in the 5.8–7.12 pH range. The nature and arrangement of functional groups in both kinds of HA are discussed on the basis of the FTIR spectroscopy. The VHA spectrum shows a very strong band at $1400\,cm^{-1}$. This band should have a different origin than the one in the same position present in the PHA spectrum, since it is not pH-dependent. In the PHA spectrum the band at $1400\,cm^{-1}$, assigned to symmetric stretching of carboxylate group, is strongly pH-dependent. Since VHA shows a higher concentration of nitrogen, the band could be assigned mainly to C–N stretching of amine groups. The metribuzin degradation mechanism involves the removing of aminogroup bounded at position 4 triazine nitrogen, catalyzed either by carboxylic acid, present in the PHA, or by aliphatic –OH, which is present in the VHA (Landgraf et al., 1998).

16.3.5. Carbon Quantification by Near-Infrared Spectroscopy (NIRS)

Near-infrared spectroscopy (NIRS) has become a very popular technique for a wide range of analysis in various industries. The usefulness of this technique is mainly attributed to allowing the rapid analysis of bulk materials with a simple manipulation procedure. On the other hand, improvements of instrumentation, and specially the development of chemometric software, have contributed to the tremendous expansion of and to the current state of popularity of this technique (Bokobza, 2002).

In many research endeavors, large numbers of samples are either required or generated. Therefore, these samples most likely need to be analyzed for one or more analytes of interest, which is often time-consuming and expensive. An important example for the need of large-scale analyses would be soil C inventory assessments that become mandated under international climate change agreements, including accounting for soil C sequestration, as may occur under future enforcement periods as described in the Kyoto Protocol. As a result of this, the ability to measure C sequestration in soils and to assess the role of soil in C fluxes is becoming more relevant (Madari et al., 2006).

While NIRS has come to dominate many areas of analysis in agriculture, the analysis of soils using NIRS has only relatively recently been examined for quantitative analysis (Reeves et al., 1999; Reeves and McCarty, 2001). In many areas of agriculture, both commercial and research spectroscopic methods based on NIRS have come to be used in reducing the need for chemical base and even other instrumentally based methods of sample analysis (Roberts et al., 2004; Madari et al., 2006). In NIRS, calibrations are developed using a representative sample set to relate spectral information to the analyte in question (calibrations)—for example, protein in wheat—using a process known as chemometrics (Siesler et al., 2002; Roberts et al., 2004).

Madari et al. (2006) showed that both diffuse reflectance infrared Fourier transform (DRIFT) and NIRS, when combined with chemometric methods, have great potential to quantify total C, N, sand, and clay in soil samples. It was also shown that these techniques could also be calibrated to estimate parameters such as soil aggregation indices, which are not directly measurable by other means. Bruun et al. (2005) described the determination of C and N mineralization patterns of a wide range of plant materials using the richness of information in the NIRS to improve predictions compared to traditional stepwise chemical digestion (SCD) or C/N ratios. The predictions with the NIRS spectra were only slightly better for C but worse for N mineralization. When compared to SCD fractions, NIRS still holds advantages, because it is a much less laborious and cheaper analytical method. However, the authors state that applications of NIRS spectroscopy in decomposition studies have only just begun and that they offer new ways to gain insights into the decomposition process. McCarty et al. (2002) report that the spectral signature of inorganic C was very strong relative to soil organic C. The presence of CO_3^- reduced the ability to quantify organic C using middle infrared (MIR) as indicated by improved ability to measure organic C in acidified soil samples (Figure 16.17). Small (2006) shows that the use of chemometrics in quantitative NIRS is necessary to avoid pitfalls that may lead to misleading or overly optimistic results.

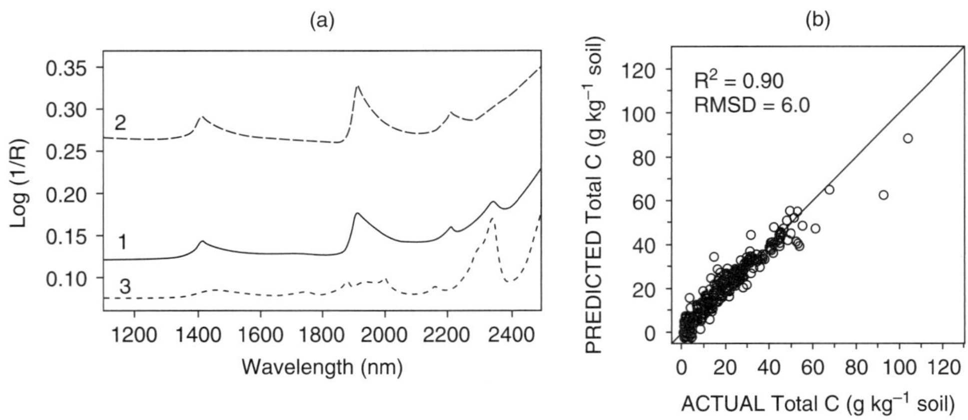

Figure 16.17. (**a**) NIR spectra of a highly calcareous soil (1) before and (2) after treatment with acid for removal of carbonates. The carbonate (i.e., $CaCO_3$) spectrum is included for additional comparison (3). (**b**) Calibration for near-infrared spectroscopy based on total soil C measured by dry combustion (actual). Deviation is represented by RMSD (McCarty et al., 2002).

16.4. RAMAN SPECTROSCOPY

16.4.1. Principles of Raman Spectroscopy

Raman spectroscopy can be used in qualitative and quantitative measurements of both organic and inorganic materials, and it is successfully employed to solve complex analytical problems such as determining chemical structures. Gases, vapors, aerosols, liquids, and solids can be analyzed by spectroscopy. As well as room temperature observations, cryogenic and high-temperature measurements can be made, including *in situ* identification and quantification of combustion products in flames and plasmas (Laserna, 1996).

When light interacts with matter, the photons can be absorbed or scattered, or will not interact with the material and may pass straight through it. In the case of scattering, most of the photons are elastically scattered, a process called Rayleigh scattering, but when incident photons are scattered in an inelastic process (i.e., in frequencies different from incident photons), the process is known as Raman[1] scattering. It is inherently a weak process since only one in every 10^6–10^8 photons that scatters is Raman scattered (Smith and Dent, 2005).

Raman scattering occurs with a change in a molecule's vibrational energy, and the difference in energy between the incident photons and the Raman scattered photons is equal to the vibrational energies of the molecule. In classical terms, the interaction can be viewed as a perturbation of the molecule's electric field. In essence, the transition induced by the electromagnetic field from the vibrational state in the electronic ground state to an excited state occurs, and the subsequent transition from the excited state to the vibrational state in the electronic ground state is accompanied by spontaneous emission of Raman scattered photons (Suëtaka, 1995), as shown in Figure 16.18.

The energy variation of the system to either lower or higher energy depends on the start of the process. When it starts with a molecule in the ground state, it is the so-called Stokes process; and if the molecule is in a vibrational excited state, it is the so-called anti-Stokes process. Experimentally, it is found that Rayleigh scattering is about three orders of magnitude higher than Raman scattering. Classic theory predicts that Raman scattering by a Stokes process (gain of vibrational energy) or an anti-Stokes process (loss of energy) is equally probable; however, quantum mechanics predicts that anti-Stokes vibrational Raman scattering is not possible when the molecule is in the lowest vibrational level. Considering that, at room temperature, most of the molecules are predicted to be in the lowest vibrational state, Stokes Raman scattering will be stronger than anti-Stokes Raman scattering (Stevenson and Vo-Dinih, 1996). A plot of intensity of scattered light versus energy difference from incident radiation, normally given in wavenumber (cm^{-1}), is a Raman spectrum.

Although FTIR and Raman spectroscopies are similar in that both provide information on vibrational levels (spectral region range of 10^4–$10^2\,cm^{-1}$), in the former the absorption of infrared light is measured, while in the latter the scattered light absorption is measured, which is usually observed in the direction perpendicular to the incident light. There are many advantages and disadvantages unique to each spectroscopy (Ferraro and Nakamoto, 1994). For example, Raman spectroscopy is

[1]The scattering was discovered by the Indian physicist C. V. Raman in 1928; and as a result of his studies about inelastic scattering phenomenon, he was awarded the Nobel Prize for Physics in 1930.

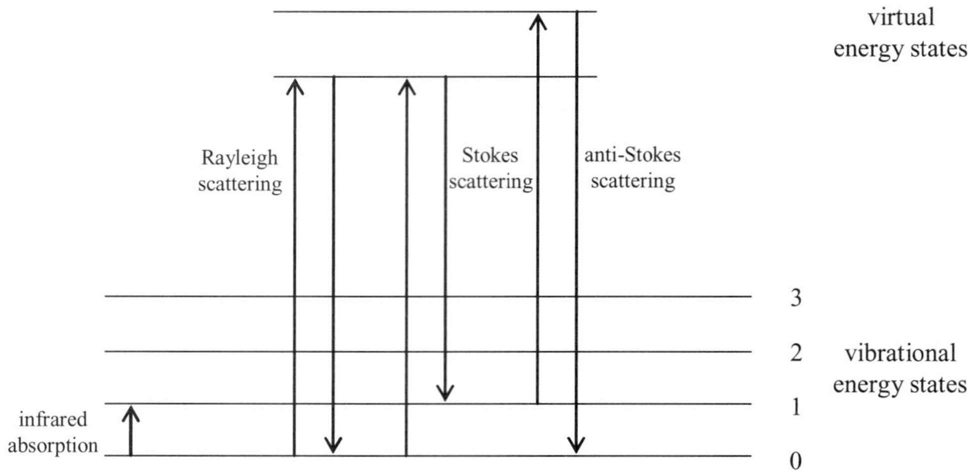

Figure 16.18. Energy-level scheme showing the infrared absorption, Rayleigh scattering and Raman scatterings (Stokes and anti-Stokes). Virtual states are not real states of the molecule but are states created when the photons interact with the electrons and cause polarization, being that the energy of these states is determined by the frequency of light.

less sensitive than FTIR spectroscopy. However, the Raman scattering of solvent (mainly water) brings far less serious problems than infrared absorption of the solvent. However, each of these spectroscopies provide complementary spectra, because the mutual exclusion rule is generally applicable to molecules having a center of symmetry (Wilson et al., 1955); that is, infrared active vibrations are Raman inactive and vice versa (Suëtaka, 1995). Thus, some vibrations are inherently weak in FTIR and strong in Raman spectra—for example, stretching vibrations of the C=C, C=C, P=S, S–S and C–S bonds. In general, vibrations are strong in Raman if the bond is covalent, at which the ratio of relative intensities of the C≡C, C=C, and C–C bond stretching vibrations is 3:2:1. This occurs because both FTIR and Raman spectroscopies measure the vibrational energies of molecules, but these methods only rely on different selection rules. For a vibrational motion to be infrared active, the dipole moment of the molecule must change. This is different from Raman spectroscopy, which depends upon the polarizability of the molecules, which have no net dipole moment.

The polarizability (α) describes the ease with which molecular orbitals are deformed by the presence of an external electric field (E) (Stevenson and Vo-Dinih, 1996). The dipole moment induced (P) on a molecule by the presence of E is most generally described by the following equation:

$$P = \alpha E \tag{16.11}$$

The electric field strength of the electromagnetic wave fluctuates with time (t) as shown by Eq. (16.12).

$$E = E_0 \cos 2\pi \nu_0 t \tag{16.12}$$

where E_0 is the vibrational amplitude and ν_0 is the frequency of the electromagnetic wave.

Hence a combination of Eqs. (16.11) and (16.12) yields

$$P = \alpha E_0 \cos 2\pi\nu_0 t \tag{16.13}$$

If the molecule is vibrating with the frequency ν_m, the nuclear displacement q is given by

$$q = q_0 \cos 2\pi\nu_m t \tag{16.14}$$

where q_0 is the amplitude of vibration. For small amplitude of vibration, α is a linear function of q and can be expressed by

$$\alpha = \alpha_0 + \left(\frac{\partial\alpha}{\partial q}\right)_0 q + \cdots \tag{16.15}$$

Here, α_0 is the polarizability at the equilibrium position, and $(\partial\alpha/\partial q)_0$ is the change rate of α with respect to the change in q, evaluated at the equilibrium position. Combining Eqs. (16.13)–(16.15) and the trigonometric identity, $\cos\alpha * \cos\beta = \frac{1}{2}[\cos(\alpha-\beta)+\cos(\alpha+\beta)]$, yields

$$P = \alpha_0 E_0 \cos 2\pi\nu_0 t + \frac{1}{2}\left(\frac{\partial\alpha}{\partial q}\right)_0 q_0 E_0 [\cos\{2\pi(\nu_0+\nu_m)t\} + \cos\{2\pi(\nu_0-\nu_m)t\}] \tag{16.16}$$

According to classical theory, in Eq. (16.16), the first term represents an oscillating dipole that radiates light of frequency ν_0, that is, Rayleigh scattering. The second term is associated with the Raman scattering of frequency $\nu_0 - \nu_m$ (Stokes process) and $\nu_0 + \nu_m$ (anti-Stokes process). If $(\partial\alpha/\partial q)_0$ is zero, the vibration is not Raman active (Ferraro and Nakamoto, 1994).

Basically, the main components utilized in the Raman spectroscopy are the excitation source, sample illumination and collection system, wavelength selector, and detection and computer control/processing systems (Ferraro and Nakamoto, 1994). In this spectroscopy the sample is illuminated by a laser beam and the light from the illuminated spot is collected by a lens and sent through a monochromator. The scattered photons with wavelengths near the laser line (due to elastic Rayleigh scattering) are filtered out, and those in a certain spectral window away from the laser line (due to inelastic Raman scattering) are dispersed onto a detector. The Raman scattering energy is typically very weak, and as a result the main difficulty of Raman spectroscopy is separating the weak inelastically scattered photons from the intense Rayleigh scattered laser light. Raman spectrophotometers typically use holographic diffraction gratings and multiple dispersion stages to achieve a high degree of laser rejection. A photon-counting photomultiplier or, more commonly, a charge-coupled device (CCD) detector is used to detect the Raman scattered light.

Several variations of Raman spectroscopy have been developed with the purpose of enhancing the sensitivity [surface-enhanced Raman spectroscopy (SERS)], improving the spatial resolution (micro-Raman spectroscopy), or acquiring very specific information (resonance Raman spectroscopy) (Laserna, 1996). Specifically, SERS (Fleischmann et al., 1974) is normally done in a silver or gold colloid or a substrate containing silver or gold. Surface plasmons of silver and gold are easily excited by the laser, and the resulting electric fields cause other nearby molecules to become Raman active. The result is amplification of the Raman signal by factors of up to 10^6 (Rupérez and Laserna, 1996). If the substrate is coated with a target compound or the molecules are adsorbed on rough metallic substrate, the scattered light contains specific information on compounds or molecules, which are useful for both qualitative and quantitative chemical analysis.

In relation to sample preparation, Raman spectra can be obtained from pure complexes in the bulk state, seeing that for better performance the careful grinding of samples is required. Contrary to FTIR spectroscopy, where samples are mixed with mineral oil (Nujol) or KBr pellets, in Raman spectroscopy a pure substance is used. For this reason, the Raman spectroscopy is called a nondestructive measurement method. Additionally, analysis can be carried out through many containers such as glass, Pyrex reaction vessels, plastic containers, and so on.

16.4.2. Applications in Studies of Humic Substances

Raman spectroscopy is widely used for the investigation of the structure of molecules, crystal, and polymer materials (Suëtaka, 1995). In relation to HS, the development of new techniques has contributed to improve the application of Raman spectroscopy in this area. The main problem arises from inherent intense fluorescence emitted by these substances. If the sample absorbs the laser beam and emits it as fluorescence, the Raman spectra are obscured by a broad and strong fluorescence band. Normally, to minimize the fluorescence interference, FT (Fourier transform)–Raman spectroscopy and SERS (Ferraro and Nakamoto, 1994; Suëtaka, 1995; Laserna, 1996) are used. In the FT (Fourier transform)–Raman spectroscopy, the exciting wavelength is changed and by shifting it to a longer wavelength, fluorescence may be greatly reduced. FT–Raman spectroscopy is ideal since it employs an exciting line in the infrared region where electronic transitions are rare (Ferraro and Nakamoto, 1994). An advantage of SERS over normal Raman techniques in the characterization of HS is that no interference due to intense fluorescence emitted by the HS is observed. Besides the smaller sample size required for SERS measurements, the application of this Raman technique also produces an intense signal from the large amount of aromatic or aliphatic groups contained in HS. This is in contrast to IR techniques, which is more sensitive to more polar groups (Francioso et al., 2002).

In HS studies, the Raman spectroscopy has been used, for example, to obtain structural and conformational information on different HS (Francioso et al., 1996, 2001; Yang and Wang, 1997; Vogel et al., 1999); to investigate HA fractionated into different nominal molecular weights (Francioso et al., 2002); to characterize thin solids films of FA (Avarez-Puebla et al., 2004; dos Santos et al., 2005); to analyze the ability of HS to bind environmental analytes, specifically metallic ions (Liang et al., 1999; Alvarez-Puebla et al., 2006); to study on the interaction between HS and conducting polymers for sensor application (Venancio et al., 2005); to observe the size and shape of gold nanoparticles in FA colloidal solutions (dos Santo et al., 2005); to analyze the influence of HS on the corrosion of steel (Dick and Rodrigues, 2006); and so on.

The most characteristic group frequencies in the Raman spectra of HS reported in the literature is shown in Table 16.2.

An important property of HS is its ability to interact with metal ions (Weber, 1988). Moreover, the functional groups responsible for the affinity of HS toward metallic ions are also involved in the strong interaction observed between these organic macromolecules and mineral surfaces occurring in soils, sediments and water (Schnitzer and Khan, 1972). Sanchez-Cortes et al. (1998) analyzed the adsorption of several fractions of HA and FA on silver colloids by SERS, with detection of HS chemical groups directly attached to the metallic surface. The results obtained showed a strong correlation among the structure, size, and electric charge of HS and their mode of adsorption on the surface. The aromatic to aliphatic ratio is higher in HA, while FA was found to have a higher content of carboxylic acids. The pH dependence in the SERS spectra of HS was attributed to an expansion of the humic matrices on increasing the pH and a possible reorientation of the chain on the surface. From this behavior they were able to identify different structural domains in these macromolecules: (a) a more hydrophobic domain placed inside the HS structure and (b) a more hydrophilic one placed outside the humic matrix. The interactions with the metal surfaces occurred through COO^- bands at acidic pH. At alkaline pH, the aliphatic and phenols groups, which are likely placed inside the

TABLE 16.2. Characteristic Raman Band Frequencies (in cm^{-1}) for HS, Including the Assignments and the Vibrational Modes

Wavenumber (cm^{-1})	Assignments
3240	ν(N–H) of bonded amines
2956, 2885	ν(C–H)
2135	ν(C≡C) or accumulated ν(C=C) or a vibrational mode of $-NH_3^+$
1700–1650	ν(C=O)
1618, 1590	$\nu_A(COO^-)$ and benzene ring stretching
~1580	Aromatic ring stretching vibrations in plane stretching vibrations of highly substituted phenols
1574, 1475, 498	Polymeric benzene rings
1540	Ring stretching vibrations of aromatic moieties
1450	$\delta(C–H_2)$ aliphatic
1379	$\nu_S(COO^-)$
1315	$\nu_S(COO^-)$ and benzene substituted ring
1300–1000	Stretching modes of the C–C, C–O or C–N bonds and/or rocking and wagging modes of the C–H and N–H units of the molecule
1211	ν(C–O) ether
1171	δ(C–H)
1165	ν(C–O) alcohol and aliphatic ethers
1060	ν(C–C)
700–400	In-plane deformation of the $-COO^-$ group and torsional motion of a $-NH_3^+$ moiety
365	Skeletal deformation mode

Source: Adapted from Alvarez-Puebla et al. (2004).

humic matrix, were more strongly attached to the metal surface, indicating that hydrophobic associations and, eventually, hydrogen bonds are very important mechanisms of adsorption (Sanchez-Cortes et al., 1998).

Francioso et al. (2002), using SERS, obtained information regarding the conformational structure and the chemical groups of peat HA separated into different nominal weight fractions. Using catechol and gallic acid as a model molecule, they observed that many bands detected in the HS fraction (Figure 16.19) could be correlated to SERS of the model molecules, indicating the existence of polyphenols with a large number of substituents. Therefore, the SERS spectra obtained showed a difference in the content of poorly substituted aromatic groups among HS fractions and also detected structural and conformational modifications caused by the extraction method of HS.

Considering the importance that knowledge of the structure of HS can contribute toward understanding the role of their interactions with other elements and compounds in the environment, Alvarez-Puebla et al. (2004) used thin solid FA films on gold islands to study the structure and organization of the aggregation by using the Raman and FTIR spectroscopies and atomic force microscopy (AFM). From SERS studies it was observed that at pH 5, FA form small aggregates, and at pH 8, FA

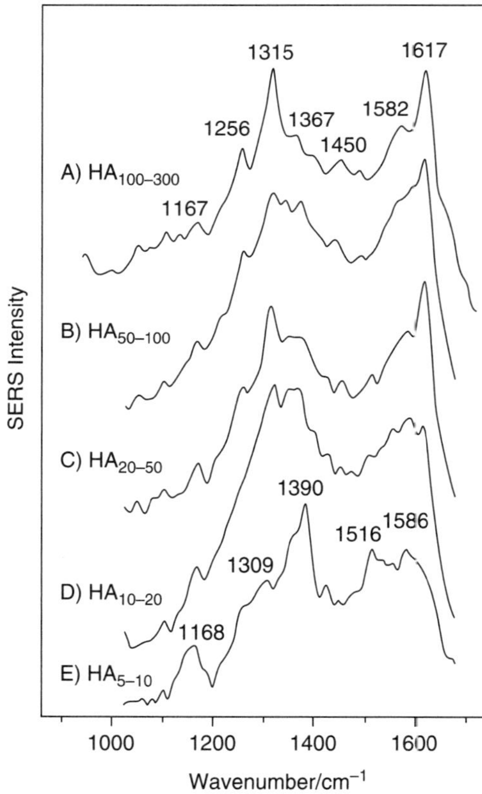

Figure 16.19. SERS spectra of different nominal molecular weight of HA, showing the band position (in cm^{-1}). The number nearby of HA abbreviation in the legend is the nominal molecular weight given in kilodaltons. Reprinted from Francioso et al. (2002).

tend to expand due to electrostatic repulsion, thus giving rise to a fractal surface (with spatial resolution of ~1 μm^2) composed of different domains. SERS studies of these domains revealed that the most polar molecules (carboxylic groups) are located on the external faces, according to Sanchez-Cortes et al. (1998). At pH 11, fractal conformations are even more pronounced and give rise to radial patterned structures. At this pH, the position of FA molecules in the fractal micelles is the same as that observed at pH 8 (Figure 16.20). Thus, they concluded that SERS can be viewed as a powerful tool for the analysis of the composition, surface functional groups, and building blocks (i.e., catechol, gallic, salicylic, or ftalic acids) in the structures of these materials.

Venancio et al. (2005) used Raman spectroscopy associated UV–vis absorption and AFM to investigate the interaction between HS and poly(*o*-ethoxyaniline)

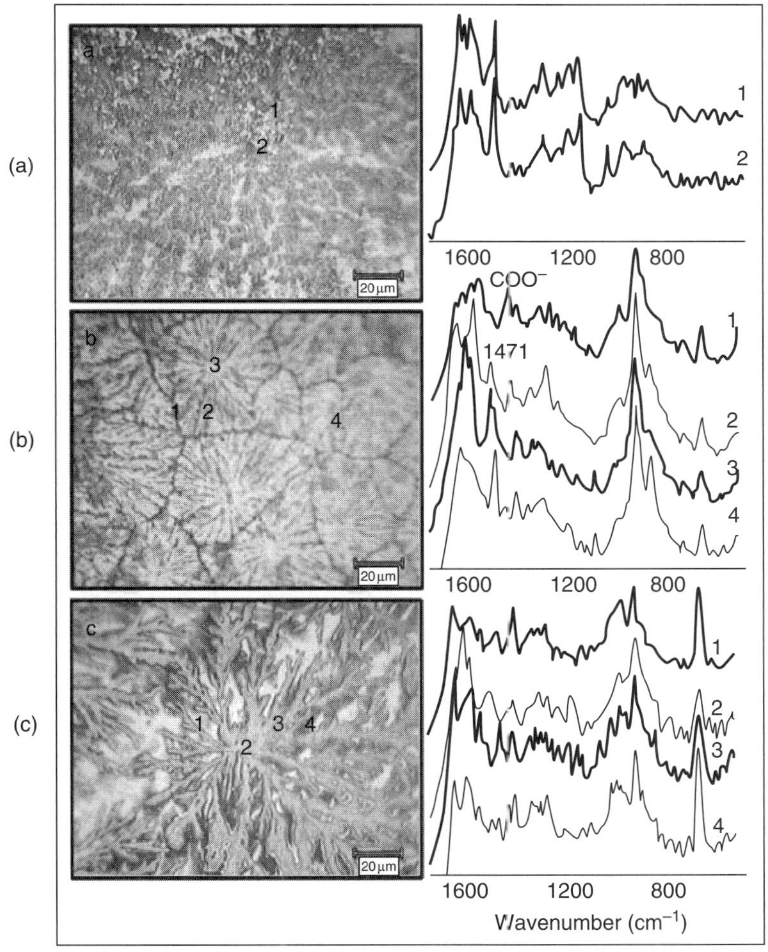

Figure 16.20. Micro-SERS spectra of FA films cast from solutions at (**a**) pH 5, (**b**) pH 8, and (**c**) pH 11 on 9-nm gold island films. Laser excitation at 785 nm and spectra were collected from different spots as marked on the optical image. Reprinted from Alvarez-Puebla et al. (2004).

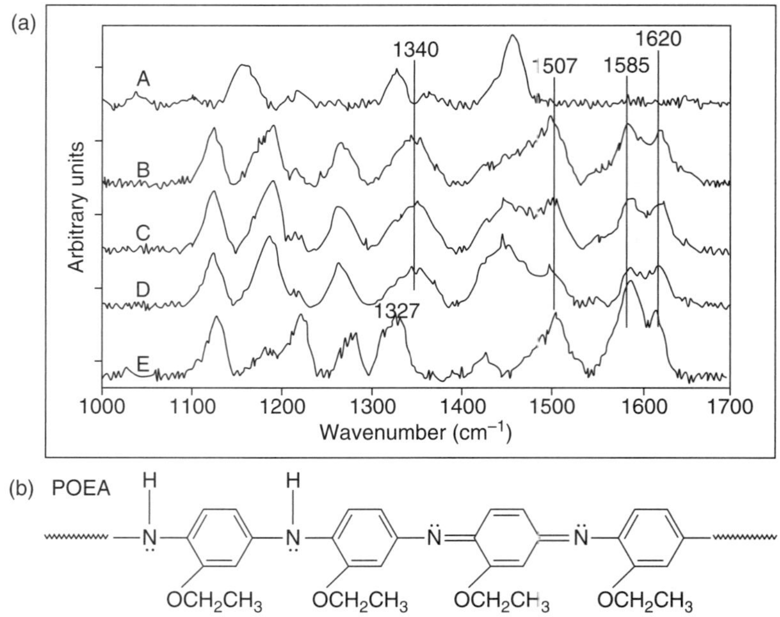

Figure 16.21. (**a**) Raman spectra recorded with the 633-nm laser line for aqueous solution at pH 5.0 containing: (A) FA 5 mg liter^{-1}, (B) POEA + 30 mg liter^{-1} FA, (C) POEA + 10 mg liter^{-1} FA, (D) POEA + 5 mg liter^{-1} FA, (E) POEA powder doped with HCl (1 mol liter^{-1}). Reprinted from Venancio et al. (2005). (**b**) Chemical structure of poly(*o*-ethoxyaniline) polymer (POEA).

(POEA), a conducting polymer. The measurements were carried out using both solution and polymer films deposited by the self-assembly technique (Paterno and Mattoso, 2001). Figure 16.21 presents the Raman spectra of FA, POEA, and POEA/FA complexes in different concentrations of FA.

According to Venancio et al. (2005) the difference in the Raman spectra indicated an interaction between POEA and HS. The UV–vis absorption spectra from POEA aqueous solution showed a shift in the POEA polaronic band to higher wavelength (data not shown), indicating an increase of the doping level of POEA due to the action of the HS, whose reasoning is supported by Raman spectroscopy, even for a constant pH. A POEA/HS complex, in which the carboxylic and phenolic acid groups of the HS induce a primary doping by protonation of the imine groups of the POEA (Figure 16.21b), may be formed. These POEA self-assembled films allowed the development of a sensing array capable of detecting and distinguishing HS in aqueous solutions (EMBRAPA, 2002).

16.5. ULTRAVIOLET AND VISIBLE ABSORPTION (UV–VIS)

16.5.1. Principles and Equipment

Absorption of ultraviolet and visible radiation by molecules generally occurs in one or more electronic absorption bands, each of which is made up of numerous closely packed but discrete lines. The electromagnetic radiation induces the oscillation of the electrons as a response to the incident magnetic field. If the induced frequency of oscillation coincides with the difference of energy between different electronic states, the probability for a transition is high.

The energy involved in electronic transitions corresponds to photon absorptions in visible region (400–750 nm) and ultraviolet (200–400 nm) from the electromagnetic spectrum. Each line arises from the transition of an electron from the ground state to one of the many vibrational and rotational energy states associated with each excited electronic energy state. Figure 16.22 shows the relative position of different energy levels of bonding, antibonding, and nonbonding orbitals, as well as the possible transitions that can occur in organic molecules. The transitions with most probability are those which require low energy for their occurrence ($n \rightarrow \pi^*$) and ($\pi \rightarrow \pi^*$) (Rohatgi-Mukherjee, 1992). Because there are so many of these vibrational and rotational states and because their energies differ only slightly, the number of lines contained in the typical band is large and their displacement from one another minute (Skoog et al., 1992).

Two types of electrons are responsible for the absorption of ultraviolet and visible radiation by organic molecules: (1) shared electrons that participate directly in bond formation and are thus associated with more than one atom and (2) unshared outer electrons that are largely localized about such atoms as oxygen, the halogens, sulfur, and nitrogen (Skoog et al., 1992).

The wavelengths at which an organic molecule absorbs depend upon how tightly its various electrons are bound. Organic compounds containing double or triple bonds generally exhibit useful absorption peaks in the readily accessible ultraviolet region because the electrons in unsaturated bonds are relatively loosely held and thus easily excited. Unsaturated organic functional groups that absorb in the ultraviolet and visible region are termed chromophores (Skoog, 1992). Single-bound spectra have not been widely exploited because the excitation required energies corresponding to wavelengths in the vacuum ultraviolet region (below 180 nm). Through the register of both wavelength absorption and adsorbed light intensity, the adsorption spectra represent a direct value of quantity and kind of molecule that possesses π systems able to adsorb in ultraviolet region. Table 16.3 describes some chromophores and their correspondent wavelengths.

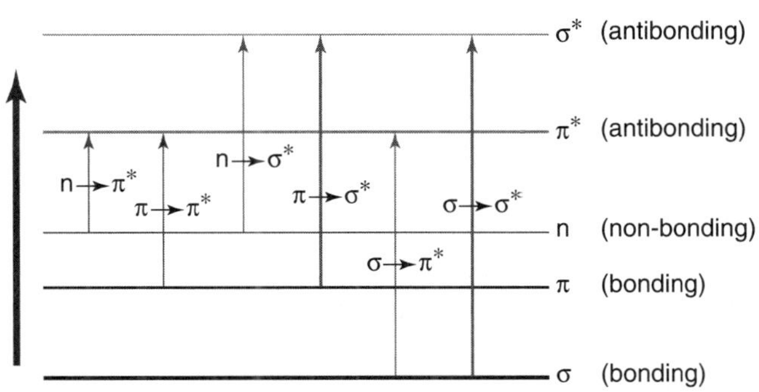

Figure 16.22. Permitted electronic transitions which occur during a molecule excitation (Reusch, 1999).

TABLE 16.3. Chromophores Structures of Organic Molecules and Their Respective Electronic Transitions

Wavelength (nm)	Chromophores	Electronic Transition
177	–C=C–	$\pi \rightarrow \pi^*$
280	–C=O	$\pi \rightarrow \pi^*$
204	–COOH	$n \rightarrow \sigma^*, n \rightarrow \pi^*$
214	–CNO	$n \rightarrow \pi^*$
339	–N=N–	$n \rightarrow \pi^*$
280	$–NO_2$	$n \rightarrow \pi^*$
270	$–NO_3$	$n \rightarrow \pi^*$

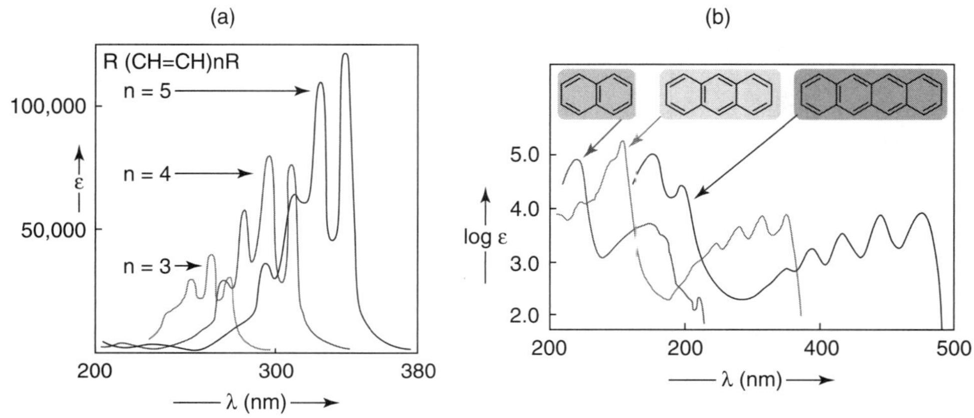

Figure 16.23. (**a**) Bathochromic shift by conjugation of polyenes, and (**b**) conjugation of aromatic systems (naphthalene, anthracene, and tetracene) (Reusch, 1999).

The absorption wavelength of each chromophore cannot be evaluated without observing the conjugation of π system in the structure. The conjugation of molecular systems ($\pi - \pi^*$) generally shifts the absorption to visible region, a phenomenon known in spectroscopy as bathochromic shift (Pavia et al., 2001). Figures 16.23a and 16.23b show the bathochromic shift in polyene systems and aromatic systems, respectively. Conjugation of double bounds diminishes the energetic difference of orbitals, shifting the characteristic spectrum to longer wavelength. Analogously, the addition of aromatic rings in the chemical structure also produces bathochromic shifts. In the spectra of naphthalene, anthracene and tetracene, the bands exhibited shifts in different proportions to the red region, in accordance with the increase of conjugated aromatic rings in the chain (Korshin et al., 1997).

An UV–vis spectrophotometer is the most popular spectroscopic equipment due to facility of use, robustness, and affordable cost. Figure 16.24 shows the optics of single- and double-beam spectrophotometers. The instrument is equipped with interchangeable deuterium/tungsten sources, a reflection grating monochromator, and photomultiplier detector. The beam splitter in general is half mirrors or motor-driven circular disk that is divided into three segments, one of which is transparent, the second reflecting, and the third opaque. With each rotation, the detector receives

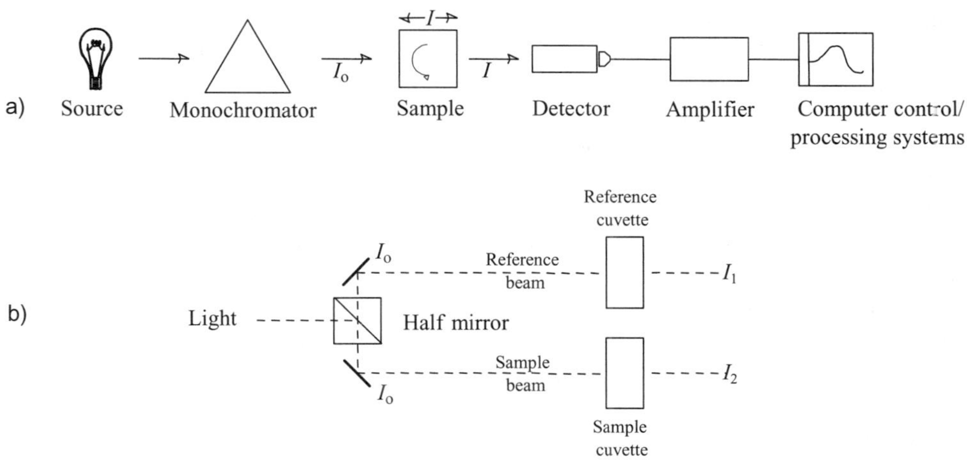

Figure 16.24. (**a**) Schematic representation of a single-beam spectrophotometer and (**b**) a beam splitter in a double-beam spectrophotometer.

three signals, the first corresponding to P_0, the second to P, and the third to the dark current. The resulting electrical signals are then processed electronically to give the transmittance or absorbance on a readout device.

The incident light has intensity I_0 in a wavelength λ and reaches perpendicularly to both cuvettes, is generally produced in quartz, which contains either sample or reference, and generally has length of 1 cm. The not absorbed light passes through the cell and reaches to the detector with intensity I.

Absorbance (A) is defined by the Lambert–Beer law as $\log(I/I_0)$ and is equal to the product of sample concentration C (mol liter^{-1}), molar absorbance ε characteristic of the sample (liter mol^{-1} m^{-1}), and the optical length L (m) of the sample:

$$A = \log(I/I_0) = C \cdot \varepsilon \cdot L \tag{16.17}$$

If the absorbance value is known, this law allows us to determine the concentration of the solution. The UV–vis spectrophotometry is a very easy and cheap technique useful to quantify the concentration of several organic and inorganic solutions.

However, for NOM investigation, most interest in UV–vis absorption is related to qualitative analysis. Relative intensity and change of position of bands indicate structural differences as chain breakdown, formation or cleavage of inter- and intramolecular bounds, and interactions of OM with xenobiotics such as pesticides. The necessity to dissolve the analyte makes the analysis of the whole SOM difficult. Thus, in general, HS dissolved in water are the usual way to obtain an UV–vis spectrum.

16.5.2. Spectral Parameters and Characterization of Humic Substances

Because of the large quantity of aromatic condensations and other conjugated π systems present in the very complex HS structure, the UV–vis absorption spectra of HS do not present particular bands, and thus it is not possible to quantify or characterize a particular chromophore. The typical spectra show an overlapping of bands without a defined band (Stevenson, 1994), followed by a format with gradual decay in function of the increase of wavelength (Figure 16.25). Humic-like substances obtained from composting can show spectra with some characteristic shoulders at 280 nm, which is mainly due to the presence of well defined extractable lignin aromatic rings in the composition (Fuentes et al., 2006).

Absorption of light by HA in the ultraviolet range reflects aromatic and carboxylic electron systems as well as their conjugates. On the other hand, functional groups with quinoide structures and keto–enol systems are more responsible for the absorption in the visible range (Uyguner and Bekbolet, 2005a,b). Principal chromophore structures identified in the HS act to form the color that varies from yellow-brown to black, as shown in Figure 16.26 (Baes and Bloom, 1990; Korshin et al., 1997).

Despite some authors' opinions that the absorption spectra of HS contain little important information, a number of research articles showed that an adequate

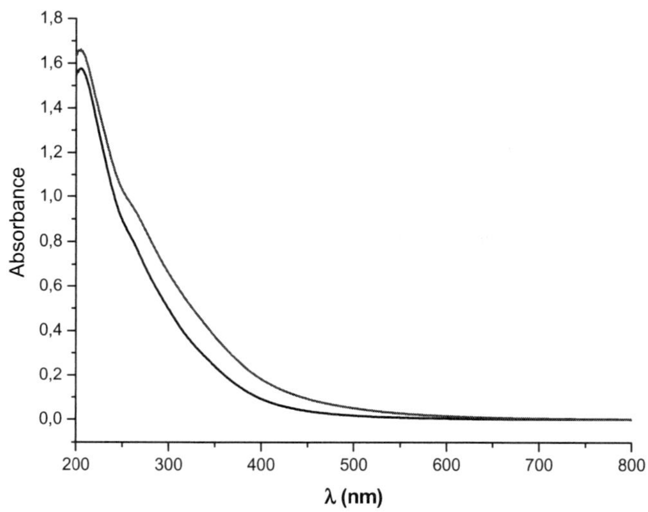

Figure 16.25. UV–vis spectra of IHSS Suwanee River HA (gray line) and FA (black line). 50 mg liter^{-1} of HS in 0.01 mol liter^{-1} sodium acetate solution (pH 6.8). Adapted from Simões et al. (2006).

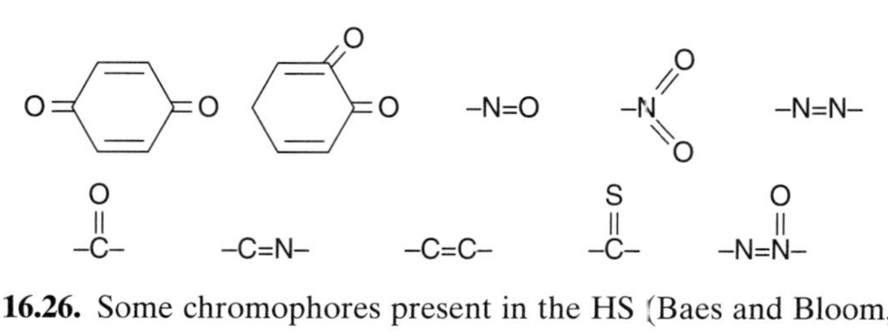

Figure 16.26. Some chromophores present in the HS (Baes and Bloom, 1990).

analysis can help elucidate HS chemical structures (Kulovaara et al., 1996; Peuravuori and Pihlaja, 1997).

A number of UV–vis absorption ratios have been measured to provide information about the state of humification and content of humic material in the dissolved organic carbon (DOC). For example, the ratio of the ultraviolet absorbance at 254 nm (UV_{254}) with the DOC content provides an estimate of the abundance of ultraviolet absorbing species and may also be used for comparison of the aromaticity of various humic materials (McDonald et al., 2004).

The assumption that HS are polymers has promulgated the use of simple physical–chemical measurements to characterize HS, such as the E_4/E_6 (absorbance at 465 nm and 665 nm) ratio. However, the use of E_4/E_6 is controversial, and frequently the E_4/E_6 and E_2/E_3 (absorbance at 250 nm and 365 nm) ratios are used to indicate an inverse relationship with progressive humification and increased condensation or molecular weight (McDonald et al., 2004). The correlation of the E_2/E_3 and E_4/E_6 ratios with the condensed aromatic C content of a sample has been shown to be poor and also has been repeatedly shown not to hold the predicted relationship with molecular weight (Piccolo, 2001). Results obtained by Saab and Martin-Neto (2007), using ^{13}C NMR CP-MAS [with dipolar decoupling (DD)] and EPR techniques, indicated that the E_4/E_6 ratio identifies mainly the degree of aromatic rings condensation. Piccolo (1988) compared E_4/E_6 ratios of various HS with gel permeation chromatograms and found that the results were comparable only when the HS had been subjected to extensive purification. Summers et al. (1987) demonstrated that the E_4/E_6 ratio values varied considerably with concentration of HS. E_4/E_6 ratio is also dependent of pH, ionic strength, and presence and quantity of metals in the structure (Stevenson, 1994).

Compost producers are increasingly subject to quality standards, and the users of compost require precise knowledge of the product used and hope to be able to have confidence in the quality of the product (Domeizel et al., 2004). Jerzykiewicz et al. (1999) and Fuentes et al. (2006) observed an apparently incoherent increase of E_4/E_6 ratio in function of the composting, possibly indicating that the composting process may be associated with a fragmentation of the basic structures (mainly lignin) to form fresh materials. This result shows the need for the use of procedures other than E_4/E_6 to determine the degree of humification of humic-like substances (Fuentes et al., 2006).

Kulovaara et al. (1996) associated the absorbance in wavelengths close to 254 nm to π–π^* transitions found in benzene substituted structures and in the most conjugated polyenes. Fuentes et al. (2006) used the ratio between the absorption at 253 (E_{253}) and 203 nm, corresponding to the electron-transfer band (ET) and the benzenoid band (Bz) of benzene ultraviolet light absorption respectively, as described by Korshin et al. (1997). Fuentes et al. (2006) also used E_{280}—that is, molar absorptivity at 280 nm—in order to estimate the relative aromaticity of the products. This method was used to estimate the humification degree of HS from several composts produced from agricultural and urban residues. These results were correlated with fluorescence technique and presented a good correlation to the procedure described by Milori et al. (2002). Domeizel et al. (2004) proposed a new method to evaluate the state of humification based on a spectrum of total HS using the method of ultraviolet spectral deconvolution. The main band centered at 280 nm was one of the most important parameters observed.

16.5.3. Mechanisms of Reactions Between Pesticides and Humic Substances

Absorption by organic matter is a key factor in the behavior of many pesticides in soil, including bioactivity, persistence, biodegradability, leachability, and volatility (Stevenson, 1994). HS have an ability to interact, immobilize, and degrade pesticides and organic xenobiotics. The resultant effect is dependent on both HS and xenobiotic chemical characteristics and is also dependent on external effects like moisture, pH, ionic strength, interaction time, temperature, wavelength, intensity of light exposure, and so on. The use of UV–vis spectroscopy is one of the tools for this study which, in general, uses other techniques like chromatography, NMR, and infrared spectroscopies.

Effect of HS on the electrochemical reduction of *p*-nitrophenol (PNP) was studied by Simões et al. (2006). PNP is the main hydrolysis product of methylparathion (MP), one of the most commonly used organophosphate insecticides in the world. The study was conducted using electroanalytical and UV–vis techniques, to understand how the HS can influence PNP degradation in the environment. Electroanalytical results showed that the HS experience the reduction of the nitro group of PNP by electrocatalysis (Figure 16.27a).

Changes in the reduction peak potentials to less negative potentials indicate an electrocatalysis phenomenon that could mean a catalytic degradation of the PNP due to interactions with the HS. This result shows either an alteration of the electronic environment in the electroactive substance or that HS can act like a proton donor, favoring the reduction of nitro-group reduction to –NO, –NHOH, or $-NH_2$ by H^+ equilibrium displacement. PNP presents two absorbance bands at approximately 320 nm and 400 nm, and the HS an intense band at about 210 nm with a slow decrease in absorbance, which is characteristic of the complex constitution of the HS (Figure 16.27b). When the PNP are added to the HS solution, there is an increase in the peak at 320 nm, compared with that at 400 nm, characteristic of acidic medium as verified with the spectrum of the PNP at pH 2.3 (Figure 16.27c). According to Fourage et al. (1999), the absorbance bands at 320 and 400 nm correspond to nitrophenol and nitrophenolate forms, respectively. UV–vis absorption results confirm the hypothesis that HS acted as H^+ donor favoring the reduction of PNP.

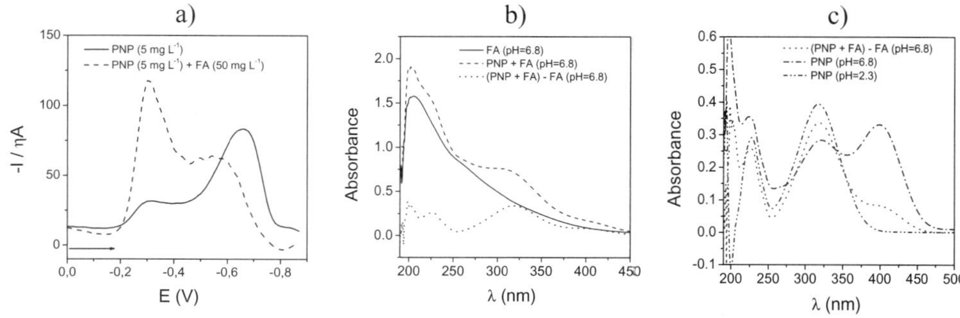

Figure 16.27. (**a**) Differential pulse voltammograms of PNP (5 mg liter^{-1}) and PNP (5 mg liter^{-1}) + FA (50 mg liter^{-1}) (pH 6.8). Scan rate 20 mV s^{-1} and pulse height 50 mV. (**b**) UV–vis spectra of PNP (5 mg liter^{-1}) + FA (50 mg liter^{-1}), pH = 6.8; FA (50 mg liter^{-1}) and (PNP + FA)-FA; (**c**) UV–vis spectra of (PNP + FA)-FA, PNP (5 mg liter^{-1}), pH = 6.8 and PNP (5 mg liter^{-1}), pH = 2.3 (Simões et al., 2006).

Interactions between HS and pesticides can be observed by using of several techniques and statistics to obtain more trustworthy results. Martin-Neto et al. (2001) observed by UV–vis spectra and polarography that the interactions of HS with the herbicide atrazine are strongly pH-dependent (Figure 16.28). Clearly, it was possible to observe an increase of AT sorption on HA as pH decreased, indicating hydrophobic sorption due to formation of more globular structure of HA (Ferreira et al., 2001). However, at pH below 3, an unexpected decrease of AT sorption was detected. The explanation to this was that the AT molecule becomes more protonated at very low pH (pK_a 1.68) and its water solubility increases, thus reducing the AT hydrophobic character and consequently decreasing the hydrophobic sorption on HA (Martin-Neto et al., 2001). This phenomenon is clearly observed by difference and derivative UV–vis spectra, as shown by Martin-Neto et al. (1994b). Lucio and Schmitt-Kopplin (2006) modeled the binding of triazine herbicides to HS using capillary electrophoresis, spectroscopic measurements (UV–vis, FTIR, NMR), elemental analysis, and potentiometric titration. The study involved four *s*-triazine herbicides and metabolites, (ameline, hydroxyatrazine, atraton, and ametryn) and 12 structurally different HS. Principal component analysis and partial least-squares analysis clearly showed the importance of carboxylic acidity and aromaticity of the humic ligands in relation to the partial positive charge and relative hydrophobicity of the pesticides.

Trubetskoj et al. (2008) studied humic-like acids (HLA) extracted from compost at the beginning and after 70, 130, and 730 days of maturation in order to be investigated for their ability to induce the transformation of 2,4,6-trimethylphenol under irradiation at 365 nm. The reaction was accompanied by absorbance at 365 nm. The rate of 2,4,6-trimethylphenol phototransformation in the presence of HLA (25 mg liter^{-1}) varied within HLA0 << HLA70 ~ HLA130 ~ HLA730. The changes of photoinductive capacity paralleled the changes of HLAs absorptivity, indicating that the formation of photoinductive constituents is related to that of the colored moieties.

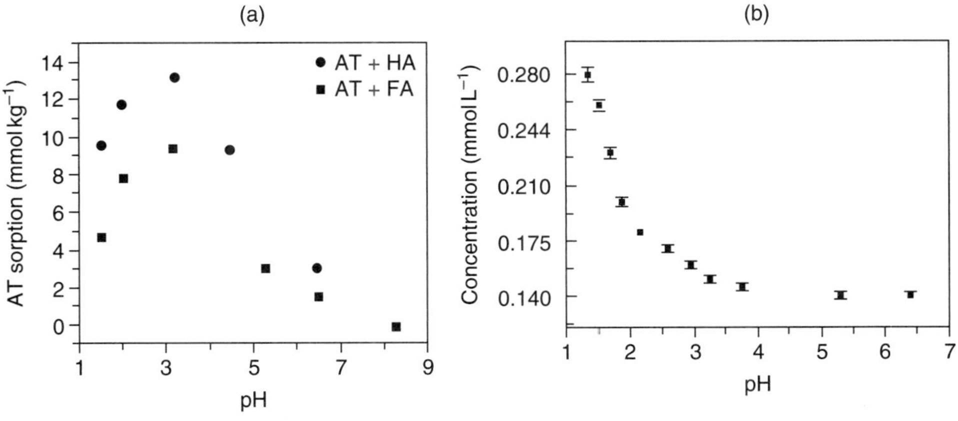

Figure 16.28. (**a**) Sorption envelopes for atrazine (AT; 0.14 mmol liter^{-1}) on oxisol HA and FA, both at a concentration of 600 mg liter^{-1}, and (**b**) pH dependence of AT solubility in water as detected by UV–vis spectroscopy, monitoring the absorption band at 223 nm. The initial concentration of the saturated aqueous solution of atrazine was 0.28 mmol liter^{-1} (Martin-Neto et al., 2001).

16.5.4. Analysis of Photoreactions of Humic Substances

The photocatalytic removal of model HA and FA of different origins (terrestrial and aquatic) was investigated by Uyguner and Bekbolet (2005a). After each run, the absorption of the supernatant was determined using a double-beam spectrophotometer at 436 nm, 280 nm, and 254 nm for the evaluation of the kinetics of the photocatalytic degradation of HS. The absorbance values measured at 254 nm and 280 nm are known to explain aromaticity removal, whereas 436 nm is used to monitor decolorization. The reaction rates calculated for the UV–vis parameters showed that soil HA reacts more rapidly than FA. However, the rate constants showed differences depending on the UV–vis absorption parameters. Furthermore, the data related to the molecular size distribution profiles of raw and treated HA suggested that the photocatalytic degradation occurs irrespective of the molecular size fractions and the light action is dependent on which kind of chromophores are present in the structure.

Garbin et al. (2007) investigated the direct and indirect photolysis of pesticide residues atrazine, imazaquin, and iprodione (3-(3,5-dichlorophenyl)-*N*-(1-methylethyl)2,4-dioxo-1-imidazoline-carboxamide) in aqueous solutions in the presence and absence of HS and under ultraviolet and visible radiation (280–480 nm) (Figure 16.29). All pesticides showed a fast direct photolysis following a first-order kinetics. HS were added to the pesticide solutions in concentrations from 1 to 100 mg liter^{-1} by means of HS mixture and pesticide stock solutions. HS only exhibited photocatalytic effect within specific concentration ranges—that is, about 30 mg liter^{-1} for atrazine and below 10 mg liter^{-1} for iprodione. For imazaquin, only a decrease was observed in the photolysis rate with HS addition.

16.5.5. Reactions of Chlorine and Chlorine Dioxide with Humic Substances

During the last 30 years, extensive research has been carried out to understand the speciation, yield, and kinetics of the formation of chlorine-, bromine- and, less fre-

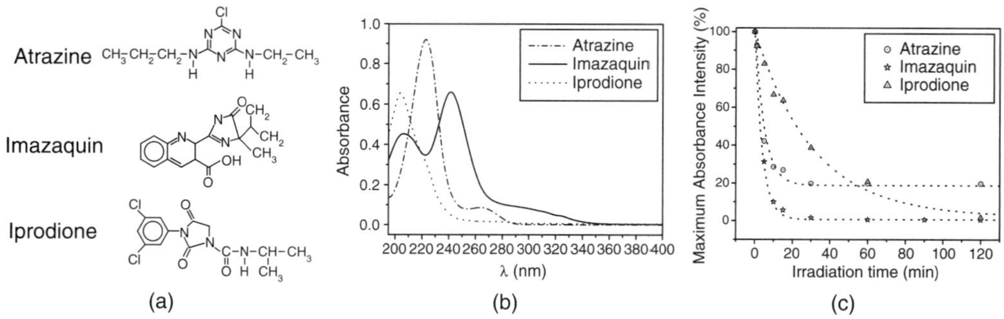

Figure 16.29. (**a**) Chemical structures of pesticides atrazine, imazaquin and iprodione. (**b**) UV–vis absorption spectra, obtained in samples with 10 mg liter^{-1} of pesticide concentrations. (**c**) Direct photolysis curves obtained by means of monitoring the changes in pesticide absorbance intensity during irradiation time. The emission range was 280–480 nm and incidence light intensity was 880 mW cm^{-2}. Initial concentration of pesticides was 10 mg liter^{-1} (Garbin et al., 2007).

quently, iodine-containing disinfection by-products (DBPs) formed during the treatment of terrestrial waters and seawater. One of the earliest articles that described possible chlorination pathways and products was written by Hook (1977). In drinking water, the identified DBPs are predominated by chloroform and mixed trihalomethanes (THMs) such as $CHCl_2Br$ and $CHClBr_2$ and by nine chlorine- and bromine-containing haloacetic acids (HAAs) (Fabbricino and Korshin, 2005). The kinetics of the formation of DBPs can be different for the different categories or species of compounds, depending also on the chlorine dose, OM content, and the presence of bromide ion (Nikolaou et al., 2004).

According to Korshin et al. (1997), the decrease in ultraviolet absorbance caused by chlorination of NOM correlated linearly with the amount of THM formed, for a remarkably wide range of water quality conditions and reaction times. They used the decrease of ultraviolet absorbance as an indicator of THM formation and it also correlated with ΔUV, but the relationship is more sensitive to solution pH and is probably not linear over the entire range of conditions/times relevant for water treatment. It appeared that all chlorine–DOC reactions that destroy ultraviolet absorbance generate organic halogens, but substantial amounts of THMs are formed only after a significant amount of reaction has occurred. They also reported a strong correlation (R^2 = 0.95) between $CHCl_3$ formation and the decrease of ultraviolet absorbance at 254 nm (ΔUV_{254}) after chlorination of water from a single source, for a range of reaction times, DOC concentrations, and chlorine doses.

Fabbricino and Korshin (2005) reported that by using UV–vis spectrophotometric measurements in addition to chromatography, chlorination caused the absorbance of deep ocean (sampled at a 1500-m depth) seawater (DO) and chlorinated coastal seawater (CS) to decrease at all wavelengths >250 nm. The main features of the corresponding differential (Figure 16.30) absorbance spectra were remarkably close to that reported for chlorinated drinking water, but their intensity was lower. The concentrations of $CHBr_3$, $CHBr_2Cl$, and $CHBrCl_2$ found in CS and DO chlorinated at varying chlorine doses and reaction times were strongly correlated with the corresponding $-\Delta A_{272}$ and $-\Delta A_{405}$ values. The examination of chlorination reactions by conventional and stop-flow differential absorbance spectroscopy were also presented by the same research group (Korshin et al., 2002, 2007; Li et al., 2000). Carvalho (2003) and Carvalho et al. (2004) observed that the chlorination using different reagents may produce different effects in a Brazilian FA. Absorbance of DOM decreased after chlorination and, by differential absorbance, the authors remarked that the use of Cl_2 produced greater degradation of HS than ClO_2, mainly in the region between 200 and 400 with a maximum at 272 nm, corresponding to benzene and low conjugated structures. The existence of these correlations indicates that the nature of chlorine attack sites in marine and terrestrial NOM is very likely to be similar. Kinetics of phenol chlorination and of THM formation are described by Gallard and von Gunter (2002a,b).

16.6. ULTRAVIOLET AND VISIBLE FLUORESCENCE

16.6.1. Basic Concepts of Fluorescence

Presently, several organic substances are recognized for emitting light when activated by different kinds of radiation (luminescence). Depending on the type of

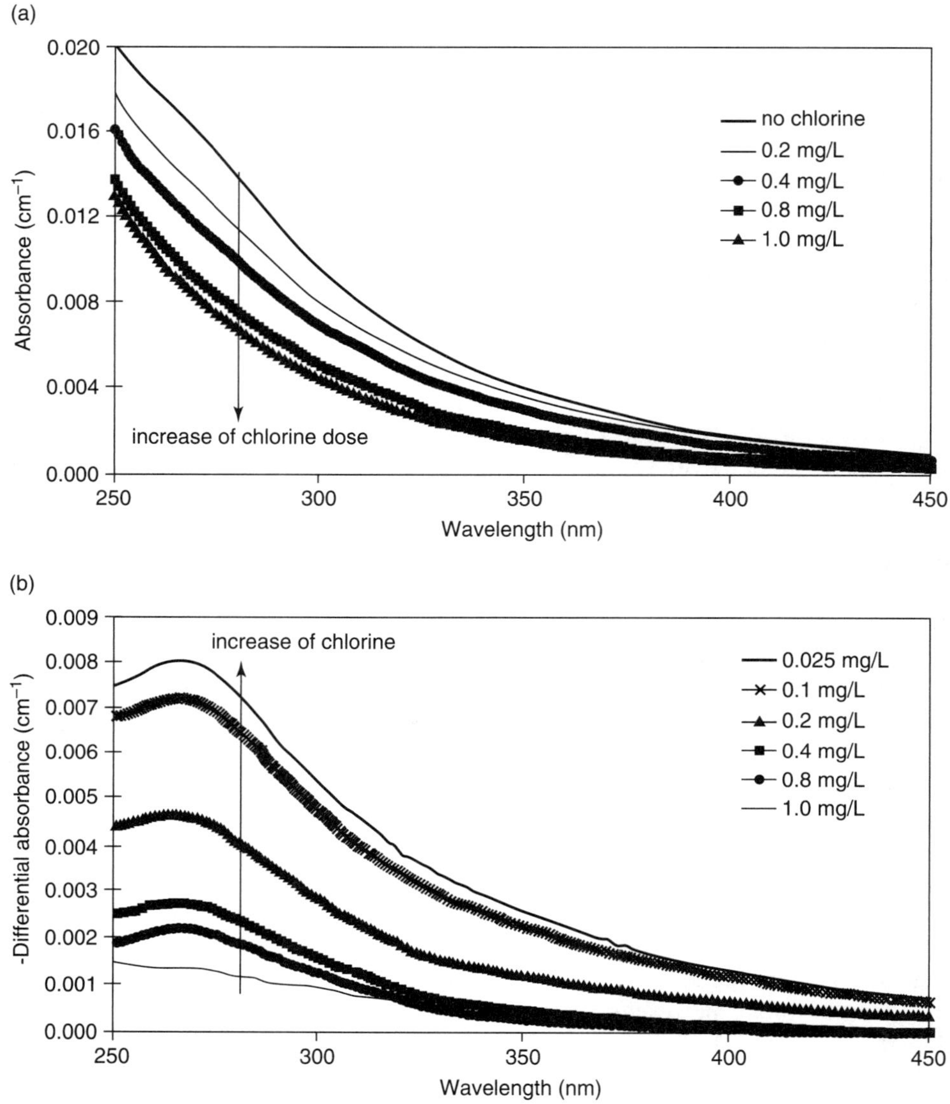

Figure 16.30. (**a**) Absorbance spectra of coastal seawater chlorinated at varying chlorine doses. (**b**) Differential absorbance spectra of coastal seawater chlorinated at varying reaction chlorine doses. Reaction time: 168h (Fabbricino and Korshin, 2005).

excitation energy, a distinction of the process is performed, such as photoluminescence, which is stimulated by visible or ultraviolet radiation; radioluminescence (scintillation), excited by radioactive substances; cathodoluminescence, caused by high-velocity electron bombardment; X-ray luminescence, triggered by X-ray; chemiluminescence and electrochemiluminescence, resulting from some chemical end electrochemical reactions, and so on.

All the substances that exhibit luminescence in response to various factors are known as luminescent materials or luminophors. Two basic kinds of luminescent materials are known: organics (organoluminophors) and inorganics (phosphors). The lumininescence of inorganic materials is determined by electronic transitions of one determined atom and its interaction with lattice structure. For example, when doped inorganic crystals are melted, their luminescence is altered or disappears altogether. In the case of organic luminophor, it is the structure of an individual molecule that is responsible for luminescence. Therefore, when a substance passes from a solid state into a melt or vapor, or is dissolved, its luminescence persists.

Molecular total energy is a sum of its kinetic (E_{kin}), potential (E_p), electronic (E_{elect}), vibrational (E_{vibr}), and rotational energy (E_{rot}). Usually, the magnitude of these energies decreases in the following sequence: $E_{\text{kin}} > E_p > E_{\text{elect}} > E_{\text{vibr}} > E_{\text{rot}}$.

Energy absorbed at the visible–ultraviolet region produces changes in molecular electronic energy, causing a transition of valence electrons. Such transition consists of excitation of one electron from an occupied molecular orbital (usually a nonbonding n or bonding π) to the next more energetic orbital (an antibonding π^* or σ^*). The antibonding orbitals are indicated by an asterisk. Thus, the absorption of a photon with appropriate energy that can promote one of π electron to an antibonding orbital π^* is denoted as $\pi \rightarrow \pi^*$. Molecules that have this kind of transition show double bonds and conjugated systems. Figure 16.31 illustrates each kind of transition with their respective absorption at electromagnetic spectrum.

The relationship between absorbed energy in an electronic transition and the radiation wavelength (λ) is expressed as

$$\Delta E = \frac{hc}{\lambda} \tag{16.18}$$

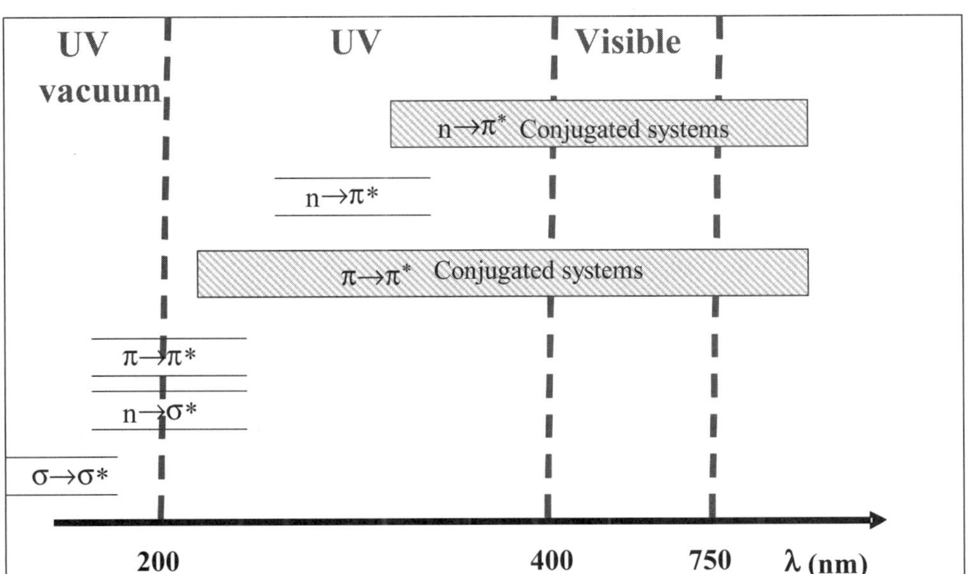

Figure 16.31. Energy range corresponding to each type of molecular transition.

where h is Plank's constant and c is the speed of light. The absorbed energy is dependent on the energy difference between the ground state and the excited state. The smaller the energy difference, the higher is the wavelength absorption.

If the absorbed energy by the molecule is not sufficient for ionization or disassociation of the molecule, it will remain at an excited state for a certain amount of time, which is described as excited-state lifetime. Afterwards, it reemits as heat and/or light. The light emitted by this process is what we call fluorescence or phosphorescence, depending on the lifetime of the excited state. Short lifetime of excited states (10^{-7}–10^{-9} s) lead to processes called fluorescent, hence long lifetime (>10^{-6} s) of excited states lead to processes called phosphorescent.

Since the molecule's electronic states are quantized, the absorption spectrum originating from a single electronic transition consists of discrete lines. However, in the case of molecular composites, this discrete line is not obtained, because the electronic absorption is overlapped with vibrational sublevels. In more complex molecules, the multiplicity of vibrational sublevels and proximity of its spacing causes colligated discrete bands, generating large absorption bands or "envelope bands." Under solution, the loss of well-defined peak structures is even more significant, especially if interaction between molecules and/or solvent occurs. This also causes minor displacements of energy levels in relation to those of the individual molecule. Such characteristics are also observed in the fluorescence emission spectra.

The vibrational energy of excited molecules is quickly dissipated by means of collisions or by other processes that are quicker than the emission of a photon at the excited state. Consequently, within a short period of time (<10^{-10} s) the molecule decays to the lowest vibrational level of the excited state ($\nu' = 0$). Such processes are called nonradioactive.

Most molecules have a ground state with the valence electrons occupying the same orbital that have anti-paralleled spins. The resultant spin S of the electrons is zero, and the level multiciplicity, given by the quantity $|2S| + 1$, is equal to 1. Thus, generally the molecules have a ground-state-type singlet S_0. If the excited state allows for one of the electron's spin inversion, the total spin will be unitary, and the multiplicity of the state will be equal to 3. Such states are called triplet T_n^*. If the absorption of a light quantum does not bring about any change in multiplicity, the corresponding excited state will be a singlet S_n^*. Figure 16.32 illustrates a graph of possible processes that can occur in a molecule excited by light.

The emission intensity ($S_n^* \rightarrow S_0$ or $T_n^* \rightarrow S_0$) is dependent on the transition probability in the energy levels involved in the process. The higher the transition probability, the more intense is an emission peak.

The energy of transitions $\pi \rightarrow \pi^*$ is considerably dependent on the size of the π system. Since the chain of conjugated bonding increases in size, for instance, in the condensation of the benzene rings in aromatic compounds, such energy decreases (the spectrum displaces toward larger wavelengths). The energy of transitions $n \rightarrow \pi^*$ does not decrease with increase of the system π, but can sometimes also become bigger.

The molecule passes from the excited state S^* either into the ground state S_0, having emitted a quantum light, or to the triplet T_1, having undergone intersystem crossing (transition between states with different multiplicities) (Figure 16.32). Since the lifetime of an excited molecule in triplet state is much longer than in the singlet state ($S^* \rightarrow S_0$ has a $\tau = 10^{-9}$–10^{-7} s, while $S^* \rightarrow T_1$ has a $\tau = 10^{-4}$–10^{-2} s), the probability

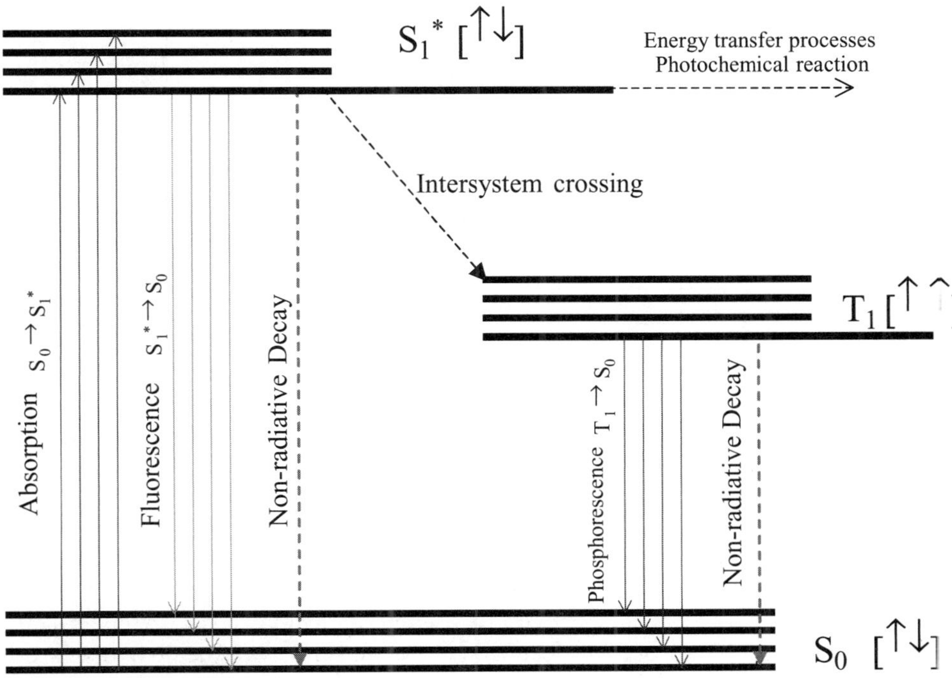

Figure 16.32. Vibrational and electronic levels of a polyatomic molecule and paths of radioactive and nonradioactive decay (Krasovitskii and Bolotin, 1988).

for nonradioactive losses is much higher in a triplet state (Krasovitskii and Bolotin, 1988).

The singlet–singlet transitions associated with the absorption of a light quantum can result in the transfer of an electron of valence π of the molecule to an antibonding orbital π^* (transition $S_0 \rightarrow S_{\pi\pi^*}$), or of transference to the same orbital of an electron n pertaining to a heteroatom, if the molecule has a heteroatom (transition $S_0 \rightarrow S_{n\pi^*}$). A transition of this type is possible if the molecule contains groups type C=C, –N=N–, C N, $-NO_2$, among others.

If the molecule absorbs ultraviolet radiation and does not fluoresce, it probably possesses another nonradioactive mechanism to return to the ground state. Nonradioactive decay from the excited state to ground state can occur in two ways:

1. An intramolecular redistribution of the energy between the electronic states and vibrational states can occur. This takes place in two stages. First, the system goes through an internal energy conversion and later dissipates the energy through a process of vibrational relaxation.
2. A combination of the redistribution of intra- and intermolecular energy can occur. Such process also takes place in two stages. The first one is an energy transference to another level of molecular energy by a crossing between systems. At this point, losses can take place by vibrational relaxation and ultimately an external suppression, which can take place by resonant energy transference with another system, or simply by external quenching.

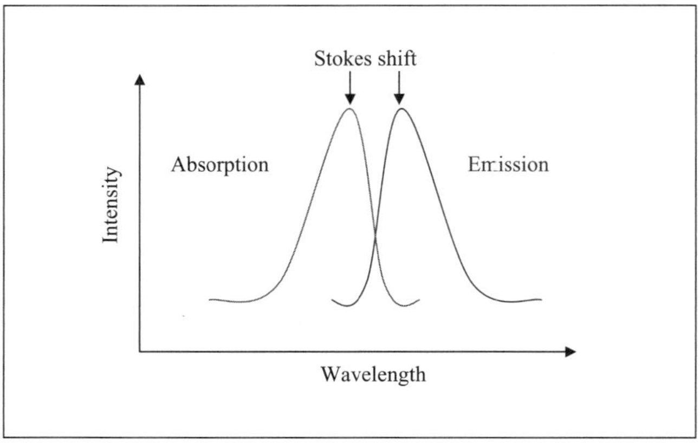

Figure 16.33. Typical emission and absorption spectra for molecular compounds.

Another process that can take place is decomposition of the molecule if the energy received is comparable to the energy of the molecule's chemical disassociation. Both mechanisms are illustrated in Figure 16.32.

Generally, the fluorescence spectrum of organic molecules has a specular symmetry in relation to the absorption spectrum (Levshin rule), with the peak displaced toward bigger wavelengths (Stokes shift) (Figure 16.33). Usually, this displacement is at around 50–70 nm, representing nonradioactive losses in luminescence. When the molecule's light absorption is accompanied by structural changes, such as angular changes or at interatomic distances, Stokes shift can take place at the interval of 150–250 nm.

When speaking of fluorescence spectroscopy, an important parameter is the quantum efficiency of the process or *fluorescence efficiency* ϕ_F, defined as

$$\phi_F = \frac{n^\circ \text{ of emitted photons}}{n^\circ \text{ of absorbed photons}} = \frac{\text{fluorescence intensity}}{\text{absorption intensity}} \tag{16.19}$$

The values of ϕ_F (range 0–1) are inherent to the molecule and depend on their structure. A high value of ϕ_F is generally associated with the molecules that have a broad and delocalized system of double conjugated bondings, which results in a relatively rigid structure, and such is the case for molecules of fluorescein, anthracene, and other condensed aromatic structures. To understand this behavior, one must assess the entire system—that is, both the electronic and vibrational levels. The conversion of electronic energy to vibrational energy and its subsequent dissipation is much easier if the molecules are loose and floppy because they can reorient themselves in order to help promote the internal energy transfer. It is behavior similar to that of a piece of soft rubber or plastic which absorbs the shock waves of a sudden impact such as a blow with a hammer and then distributes the energy throughout its entire volume. On the other hand, a rigid block of metal or stone merely transmits the shock waves to its surroundings. This behavior is similar to that of a rigid molecule, which does not efficiently bring about the internal conversion of energy to return to its ground state. Therefore, it is more probable that it will emit a photon. However, not every rigid molecule is fluorescent, since in certain structures there is the possibility of an intersystems crossing taking place (Rendell, 1987). Such route to the ground state becomes particularly effective when the excited state of least energy can be reached by a transition $n \rightarrow \pi^*$ from the ground state. The excited state $n \rightarrow \pi^*$ has a much longer lifetime than the excited state $\pi \rightarrow \pi^*$. Therefore, there are more probabilities for a crossing between systems to take place. Once the crossing takes place, the vibrational relaxations take the molecule to level $\nu' = 0$, where it remains trapped for a long period of time (transition $T_1 \rightarrow S_0$ is prohibited), long enough to be deactivated by collisions or other mechanisms of internal suppression (quenching). This explains the low fluorescence efficiency of compounds such as nitrogen ketones and heterocycles. Intersystems crossing are also very efficient in structures containing heavy metals. Other environmental factors such as temperature, nature of solvent, pH, and other solute can influence the fluorescence of a compound in solution by either influencing the rates of one or more processes involving excited state or by perturbing one or more of the energy levels involved in fluorescence. In the case of pure solid samples, these environmental problems are minimized.

16.6.2. Fluorescence Measurements and Instrumentation

The simplest geometry to measure fluorescence of transparent samples is instrumentation of excitation at 90°. In this case, it can be considered that the intensity of absorbed light (I_{abs}) is equal to the intensity of incident light (I_0) minus the intensity of transmitted light (I_{trans}). The transmitted light is expressed by Lambert–Beer Law:

$$I_{trans} = I_0 e^{-\varepsilon_f C d} \tag{16.20}$$

where ε_f is the of the molar absorption coefficient (absorptivity) of the fluorescent material, expressed in liter mol^{-1} cm^{-1}; C is the concentration of the molecules in solution, expressed in mol liter^{-1}; and d is the distance that the light travels through the material (the optical path length), expressed in centimeters. Thus, the intensity of absorbed light by the sample is given by

$$I_{abs} = I_0 - I_0 e^{-\varepsilon_f C d} = I_0 \left(1 - e^{-\varepsilon_f C d}\right) \tag{16.21}$$

On the other hand, by Eq. (16.12) there is the quantum efficiency of the fluorescence process (ϕ) given by the ratio between intensities of fluorescence and absorbed light; therefore,

$$I_F = \phi I_{abs} = \phi I_0 \left(1 - e^{-\varepsilon_f C d}\right) \tag{16.22}$$

By Eq. (16.22), one can observe that the intensity of fluorescence for transparent samples is directly proportional to the intensity of incident light in the sample, depending on the absorptivity of the sample in the wavelength of excitation and concentration of chromospheres. Figure 16.34a shows a typical curve for the fluorescence intensity as a function of fluorophore concentration.

However, there is a probability that the sample absorbs the wavelength of emitted light, thus inducing fluorescence reduction (inner filter effect). In this case, fluorescence, as a function of concentration, presents another behavior. The effective fluo-

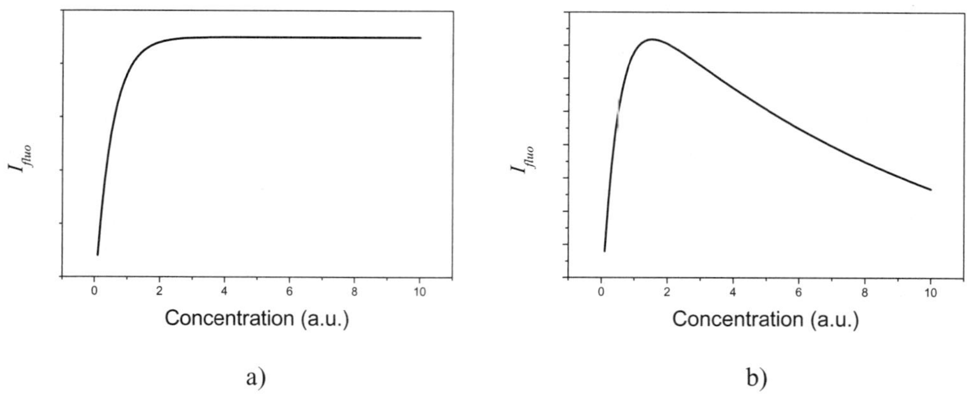

Figure 16.34. (**a**) Intensity of fluorescence emission as a function of concentration for samples without superimposition between emission and absorption spectra. (**b**) Intensity of fluorescence emission as a function of concentration for the case where the samples present absorption in the region of emission.

rescence (I_{EF}) will be the emitted fluorescence minus what was absorbed by the sample. At a first approximation, if we consider that the sample absorbs at a single wavelength the emission spectrum, it can be expressed that

$$I_{EF} = I_{fluo} - I_{fluo}\left(1 - e^{-\varepsilon Cd}\right) \tag{16.23}$$

where ε is the absorptivity in the wavelength of emission.

$$I_{EF} = I_{fluo}e^{-\varepsilon Cd} = \phi I_0\left(1 - e^{-\varepsilon_f Cd}\right)e^{-\varepsilon Cd} \tag{16.24}$$

In Figure 16.34b there is a typical curve that describes Eq. (16.24). At high concentrations, even a complete extinction of the fluorescence sign can be observed. A similar behavior can occur when the sample contains chromospheres that compete for the absorption of light excitation.

16.6.2.1. Instrumentation for Fluorescence Spectroscopy

16.6.2.1.1. Conventional Fluorescence Spectroscopy. Figure 16.35 shows the typical instrumental requirements for fluorescence spectroscopy. Basically they are:

- (i) A source of ultraviolet–visible radiation for excitation (mercury lamp, xenon arc).
- (ii) A wavelength selector to choose the excitation. When excitation is done through a lamp which presents a broad emission spectrum, a monochromator or a filter is necessary.
- (iii) Optical system to direct the exciting radiation onto the sample and collect the emitted radiation onto detection system. The best geometry depends on the sample. For transparent samples, excitation and emission at 90° is commonly used. However, for opaque samples the optical system must collect the scattered emission from the sample surface.

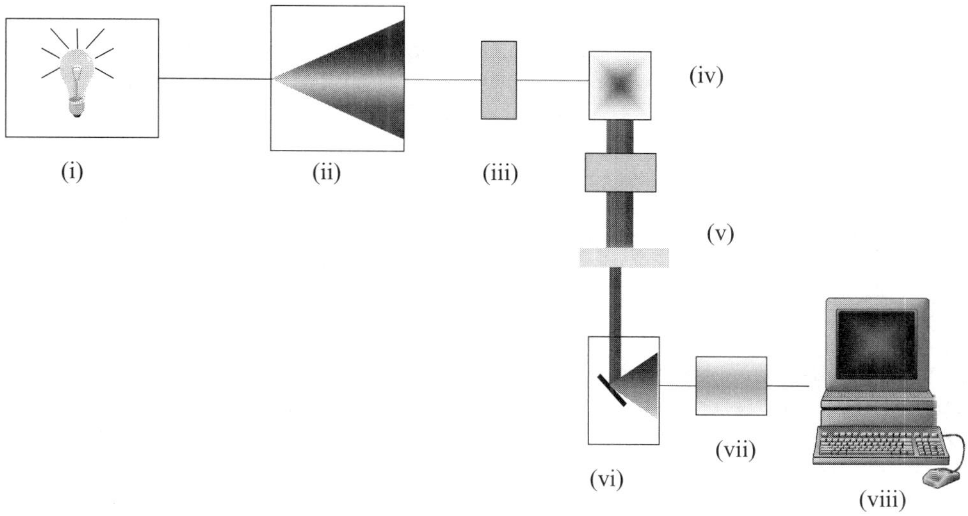

Figure 16.35. Layout of a typical fluorescence spectrometer. (i) Source, (ii) excitation monochromator, (iii) optical system, (iv) sample, (v) filter, (vi) emission monochromator, (vii) detector, and (viii) data acquisition system.

(iv) Sample holder (cuvette, cell, flowcell).
(v) A filter to eliminate the excitation light. Harmonics of excitation wavelength produced inside the monochromator usually disturb the emission spectrum.
(vi) A second wavelength selector (emission monochromator).
(vii) A sensitive detector to respond to the emitted radiation. Usually a photomultiplier or photodiodes is used.
(viii) A data acquisition system to record the intensity of the fluorescence emission at each wavelength. This system must also control the monochromator scan.

Two-dimensional fluorescence spectra (2D) can be measured in three different ways:

1. Emission Spectrum. Excitation monochromator is maintained in a specific wavelength, and the data acquisition system scans the emission monochromator measuring all wavelengths that the sample emits.
2. Excitation Spectrum. Emission monochromator is maintained in a specific wavelength, and the data acquisition system scans the excitation monochromator measuring all wavelengths that the sample may excite.
3. Synchronous Scan. Emission and excitation monochromator scan at the same speed with a determined wavelength difference. This kind of scan is usually applied to study complex materials with several fluorophors or in mixtures of several fluorescent substances. The result of a synchronous scan is a product of excitation and emission spectra. This tool produces more defined peaks for interpreting the behavior of chemical structures during a dynamical process.

4. Time-Resolved Fluorescence. Emission and excitation monochromators are maintained in a specific wavelength, but the excitation is chopped off and fluorescence decay is measured as a function of time. This kind of spectroscopy is interesting for studying structural changes or different complexation sites.

Another useful tool in fluorescence spectroscopy is three-dimensional (3D) fluorescence spectra. In this case, the graph is composed of emission wavelength, excitation wavelength, and intensity axes. To construct this kind of graph, several emission spectra are recorded in different excitations. The presentation of this kind of spectrum can be topographic or contour curves.

16.6.2.1.2. Laser-Induced Fluorescence Spectroscopy. The laser-induced fluorescence (LIF) spectroscopy technique is widely used in research for a variety of analytical applications, from interrogation of plasma plumes in laser-induced breakdown spectroscopy (LIBS) for elemental analysis (Martin et al., 2007) to diagnosis of cancerous tissues using fluorescence spectroscopy of single molecules (Li and Xie, 2005). LIF is one of the most sensitive approaches available for analytical purposes. It is relatively easy to implement, phenomenologically straightforward and well-investigated, and largely noninvasive, so that it can be useful for environmental applications.

All of the concepts outlined the previous session regarding the basic principles are valid for LIF spectroscopy. The only difference is the fact that excitation of the fluorescence material is performed by a laser. The advantages of using a laser are high intensity, coherence, and polarization. The light intensity is higher in a wavelength of specific excitation. Hence, it produces a good signal–noise ratio when compared to fluorescence induced by a lamp. Besides, it gives greater selectivity in excitation, and it causes a decrease in interference factors in the fluorescent signal. Furthermore, the possibilities of working with polarized and high-potency coherent excitations are important tools in some characterization studies.

Figure 16.36 shows a basic layout of the experimental setup to carry out analyses of LIF spectroscopy of opaque samples using an argon laser (Ar) as excitation source. This laser is an interesting choice because it possesses lines at ultraviolet and visible spectrum (351, 454.6, 457.9, 465.8, 476.5, 488.0, 496.5, 501.7, 514.5, and 528.7 nm), thus enabling a certain degree of freedom in selecting the excitation wavelength.

The prism at the outlet of the laser serves to separate the laser emission of the gas fluorescence and allows for a "clean" excitation of the sample. For excitation using solid-state lasers, this element is dispensable. The lens (element 5) collects the fluorescent signal and focuses on the aperture of the monochromator. The filter is used to eliminate excitation that is spread over the surface of the sample. The optical chopper serves to modulate the light at a defined frequency, which serves as reference for the lock-in amplifier. A data acquisition system controls the pace of the monochromator and reads the signal of the lock-in, generating the sample's emission spectrum.

16.6.3. Fluorescence Analysis of Humic Substances

The fluorescence phenomenon is highly probable in molecular systems containing atoms with lone pairs of electrons such as C=O, aromatic, phenolic, quinone, and

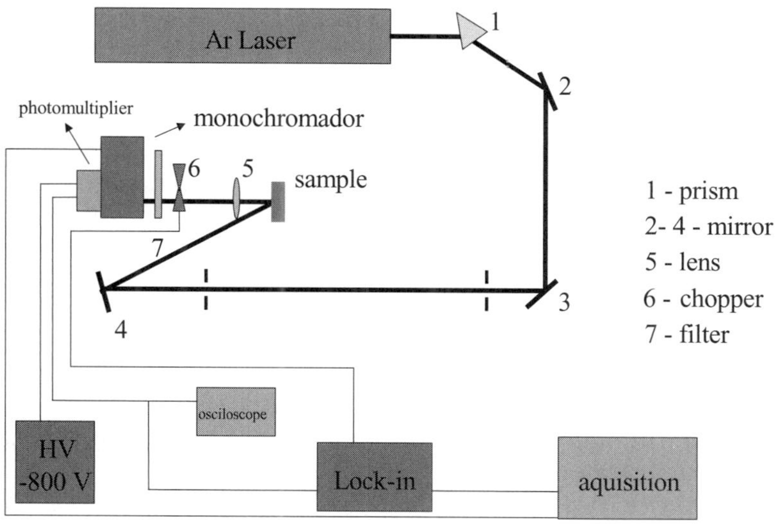

Figure 16.36. Experimental layout to perform analyses of laser-induced fluorescence (LIF) spectroscopy of opaque samples.

more rigid unsaturated conjugated aliphatic systems. These functional groups are present in HS (Senesi, 1990b; Senesi et al., 1991b).

The extension of π-electron systems, the level of heteroatomic substitution, and type and number of substituting groups on the aromatic rings strongly affect the intensity and wavelength of the molecular fluorescence. Increasing the extension of the π-electron system has the effect of increasing fluorescence and lowering the energy difference between the ground state and the first excited state, thus shifting the emission wavelengths toward higher values. Therefore, the behavior of the molecule's total fluorescence results in an accumulative effect contingent on, primarily, various structural components of the molecule, and the spectrum seen in any of the three fluorescence modes consists of the sum of individual spectra of different emitter focus in the molecule. The well-known heterogeneous characteristics of HS therefore provides the fundamental reason for using fluorescence analysis, since the fluorescence properties are considered as criterion for valid diagnosis to distinguish these materials based on their nature, genesis, and origin. On the other hand, the structural and chemical complexity of the humic molecules represents an obstacle for the identification of individual structural components responsible for the fluorescence of HS.

Although the problem of identifying the molecular components responsible for the fluorescence of HA and FA is a complex task, some hypotheses can be suggested concerning the possible chemical nature of the fluorescent structure in several humic materials already examined.

The fluorescence of HS is already a widely used technique in environmental samples (Miano et al., 1992; Senesi et al., 1995, 2003; Olk et al., 2000; Li and Korshin, 2002; Carvalho et al., 2004; Cilenti et al., 2005; Rosa et al., 2005; Li et al., 2006; Plaza et al., 2006; Saadi et al., 2006; Rodríguez-Zúñiga et al., 2008). Depending on the origin of the sample, the 2D-fluorescence spectrum presents a characteristic profile. Usually, the fluorescence spectra of HS are composed of broad bands. Senesi (1990b) compared excitation and emission spectra of different FA from terrestrial (soil, peat, sludge) and aquatic systems (freshwater, estuary, sea, and ocean). Table 16.4 shows ranges of fluorescence wavelengths measured for emission and excitation bands of FA from various sources. Maximum fluorescence emission for soil samples occurs at bigger wavelengths than that of peat or sludge, indicating that the structure of FA from soil is more complex with bigger π-electron system.

Humic substances are among the most reactive structures of SOM strongly interacting with metal ions (Stevenson, 1994). Consequently, the speciation of metal ions in soil and natural water is dependent on the quantity and quality of HS. Molecular Fluorescence spectroscopy is a particularly useful technique to study complexation phenomena of metal and HS because it is a highly sensitive and nondestructive analytical technique (Saar and Weber, 1980; Ryan and Weber, 1982; Cabaniss and Shumam, 1986; Senesi, 1990b; Pullin and Cabaniss, 1995). When metal ions are brought into contact with HS, the fluorescence of HS can be changed because the ions can produce a more rigid or more floppy structure, or they can even change the rate of intersystem crossing. If the complexation produces a more rigid structure, the quantum yield of fluorescence increases by decreasing the probability of competing with nonradiative transitions (Lakowicz, 1983). Thus, an enhancement of fluorescence emission can be observed. On the other hand, if the complexation produces a more floppy structure, nonradiative processes are more probable and a quenching of fluorescence can be observed. If the complexation increases the rate of intersystem crossing, the quantum yield of fluorescence decreases, resulting again in a fluorescence quenching. However, fluorescence spectroscopy provides only a lower bound on the number of HS binding sites, because it cannot observe all the sites that might be present. Aliphatic sites are less likely to fluoresce than aromatic sites, and ^{13}C NMR and elemental composition studies indicate that over 50% of the carboxylate groups in FA may be aliphatic (Wilson et al., 1987). A site present in very low relative concentration will not be detected, nor will two sites filled in the same proportions be detected as separate sites (Cabaniss, 1992). Nevertheless, changes in the fluorescence emission of HS, as a consequence of the formation of

TABLE 16.4 Ranges of Fluorescence Wavelengths (nm) Measured for Emission Band (λ_{exc} = 320–379 nm) and for the Excitation Peaks (λ_{em} = 520–570 nm) of Fulvic Acids from Several Sources

Sample Origin	λ_{em} (nm)	λ_{exc} (nm)
Terrestrial		
Soil	500–520	465–470
Peat	455–461	390
Sludge	435	390
Aquatic		
Freshwater	410–450	325–360
Estuary	430–450	350
Sea	420	325
Ocean	488	385

Source: Adapted from Senesi (1990b).

stable complex species, can be significant and, in these cases, they are used as quantitative information about complexation (Ryan and Weber, 1982; Cabaniss, 1992; Machado et al., 1994; Geipel et al., 2004; Provenzano et al., 2004; da Silva et al., 2006; Plaza et al., 2006).

16.6.3.1. Degree of Humification of Humic Substances. Figure 16.37 shows typical examples of 2D-fluorescence spectra of HA from Brazilian soil (Hapludox) under different management systems (Milori et al., 2002). Significant differences were observed between the fluorescence spectra for the HA extracted from the forest, conventional tillage, and no-tillage soils. When excited with ultraviolet radiation (240 nm), the emission spectra of forest soil samples showed a peak in the blue region (~450 nm), while the spectra for HA from soils in conventional tillage were broader and the peak was seen in the green region (~507 nm). The fluorescence syncronous-scan excitation showed an interesting inversion of intensity at 465- and 399-nm peaks when we compared HA from native forest with conventional tillage

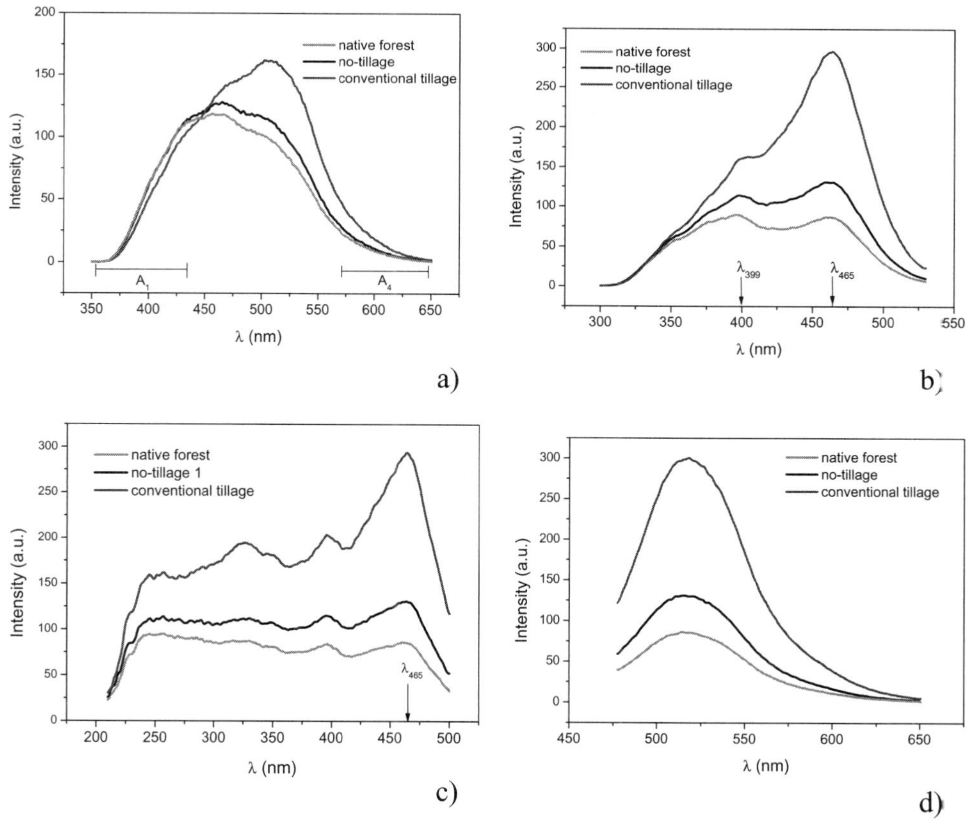

Figure 16.37. Typical examples of fluorescence spectra of humic acids from Brazilian soil (Hapludox) under different tillage systems. The samples were prepared in aqueous solutions (20 mg liter^{-1}, pH 8). (**a**) Fluorescence emission ($\lambda_{exc} = 240$ nm). (**b**) Fluorescence synchronous-scan excitation spectra ($\Delta\lambda = 55$ nm). (**c**) Fluorescence excitation spectra ($\lambda_{em} = 517$ nm). (**d**) Fluorescence emission spectra ($\lambda_{exc} = 465$ nm).

soils. It seemed that emission above 500 nm and excitation at 465 nm were related with humified structures. A similar conclusion could be reached from the excitation spectra. When HA from native forest was compared with conventional tillage soil, the biggest variations occurred at 465 nm. With respect to the emission spectra with excitation at 465 nm, the spectral profiles remained the same, but significant variations of intensity were observed.

In the literature, techniques to assess the humification degree of HS in solution based on photoluminescence were recently published. All techniques are based on the fact that complex structures usually have absorption and emission spectra shifted to red. Zsolnay et al. (1999) proposed to use the ratio between areas (a quarter more to red/a quarter more to blue) of an emission spectrum with excitation at the ultraviolet (240 nm) to assess humification of dissolved organic matter (DOM). This index was called A_4/A_1. Kalbitz et al. (1999) employed the ratio between peaks in the spectrum of synchronous scan (peak intensity more to the red/peak intensity more to the blue) to define the humification degree of FA. This index was called I_{465}/I_{399}. Milori et al. (2002) demonstrated high correlation between the intensity of fluorescence emission with excitation in the blue (465 nm) and SFR concentration, which was determined by EPR. Figure 16.38 shows a typical result of analyses performed with soils from the Brazilian savanna. Complex aromatic structures are believed to stabilize semiquinone free radicals in HS (Riffaldi and Schnitzer, 1972; Senesi, 1990a; Stevenson, 1994). According to Martin-Neto et al. (1998), the increase of complex structures that originated from humification processes can be used as a humification index, and this parameter is proportional to SFR concentration. On the other hand, the blue light excites mainly those complex molecules, such as condensed or substituted aromatic rings with large π-electronic system. Therefore, in this case the fluorescent intensity can be considered to be directly proportional to the humification degree of HA. The area under fluorescence emission when excitation is 465 nm is defined as a humification index called A_{465}. Figure 16.39 shows a

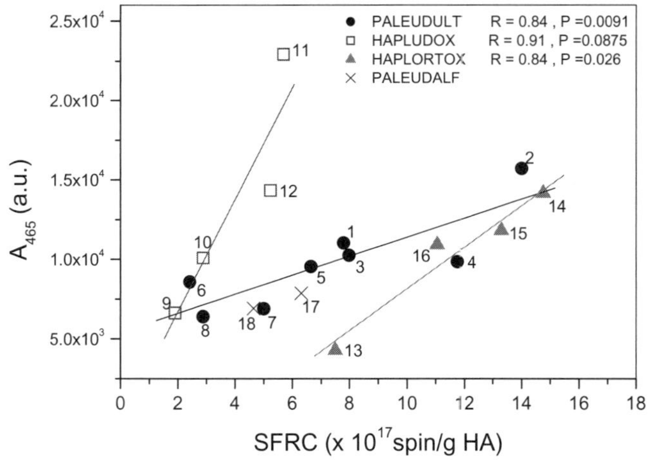

Figure 16.38. Correlation between humification index, obtained by fluorescence (A_{465}), as proposed by Milori et al. (2002), and the concentration of SFR, determined by EPR, found in HA extracted from soils of the Brazilian savanna region (Cerrado).

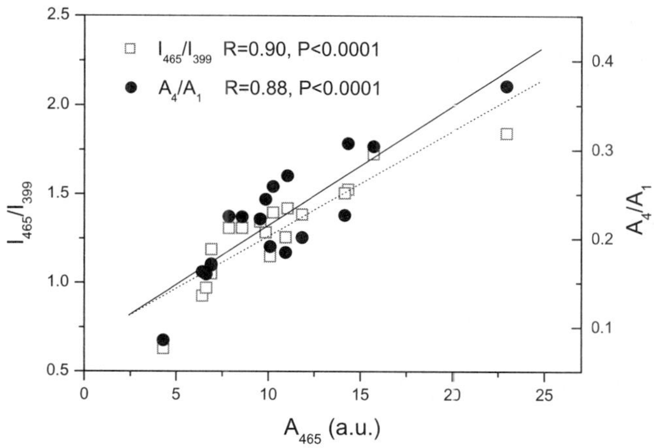

Figure 16.39. Data showing correlation among fluorescence methods to determine humification degree of HA. A_4/A_1 is the humification index proposed by Zsolnay et al. (1999); it is calculated through the ratio between areas of the upper quarter of emission spectra (435–480 nm) and the lower quarter (300–445 nm) when excitation is made at 240 nm. I_{465}/I_{399} is the humification index proposed by Kalbitz et al. (1999); it is calculated through the ratio of peak intensities in 465 and 399 nm measured in fluorescence synchronous-scan excitation spectra. A_{465} is the humification index proposed by Milori et al. (2002); it is calculated by fluorescence area of emission spectra when the excitation is made at 465 nm.

comparison among the three proposed indexes for HA from Brazilian soils (Milori et al., 2002).

These methods for analysis of humification degree have been used to assess soil management. Management of SOM is essential to sustaining the quality and productivity of soils around the globe. This appears to be particularly true in the tropics where there is a greater proportion of nutrient-poor, highly weathered soils that are more susceptible to losses of SOM. Developing management practices that promote the maintenance and storage of SOM in the tropics depends on understanding the factors that control SOM dynamics (Feller and Beare, 1997). Bayer et al. (2002a) studied long-term effects of tillage and cropping systems on characteristics of HA from surface layer (0–25 mm) of a subtropical Brazilian paleudult soil. Fluorescence and EPR results were compared. Both methodologies were consistent ($R = 0.84$, $P < 0.01$), showing positive effects of conservation management systems on soil humus characteristics. On the other hand, González-Pérez et al. (2004) studied Brazilian HA from an oxisol under different treatments using several spectroscopic techniques, including fluorescence. They reported that no important effect due to tillage system was observed in these areas after 5 years of cultivation. They concluded that, probably, the studied oxisol had a high clay content that offered protection to the clay–Fe–OM complex against strong structural alterations. The above were two interesting applications of fluorescence spectroscopy in soil management analysis.

16.6.3.2. Structural and Interaction Studies of Humic Substances. Changes in structural aspects and interactions with other compounds usually affect fluorescence properties of HS. This kind of information is important to understand structure and reactivity with metal ions and hydrophobic organic compounds.

Wandruszka et al. (1997) studied the role of selected cations in the formation of pseudomicelles in aqueous HA using fluorescence spectroscopy. Ragle et al. (1997) applied fluorescence techniques to study mechanisms for the interactions of dissolved HA with nonpolar micropollutants.

Reactions between aquatic HS and halogen-based oxidants, during water chlorination process, can lead to the formation of trihalomethanes, with potential carcinogenic effects. Carvalho et al. (2004) investigated reactions of tropical aquatic fulvic acids (AFA) with chlorine and formation of trihalomethanes through fluorescence spectroscopy.

3D-Fluorescence spectroscopic analysis has also been used for analysis of terrestrial and aquatic HS. Figure 16.40 shows an example of topographic and contour plot of 3D-fluorescence spectrum. In this case, the Fluorescence spectroscopy involved scanning and recording 17 individual emission spectra (260–700 nm) at sequential 10-nm increments of excitation wavelength between 250 and 410 nm (Parlanti et al., 2002). The authors used this technique to obtain structural information about HS and also used it in studies concerning their transformation processes. They reported that there were five major fluorescent components in bulk seawater based on 3D-fluorescence spectroscopy. They defined α and α′ (excitation at 330–350 nm and emission at 420–480 nm; excitation at 250–260 nm and emission at

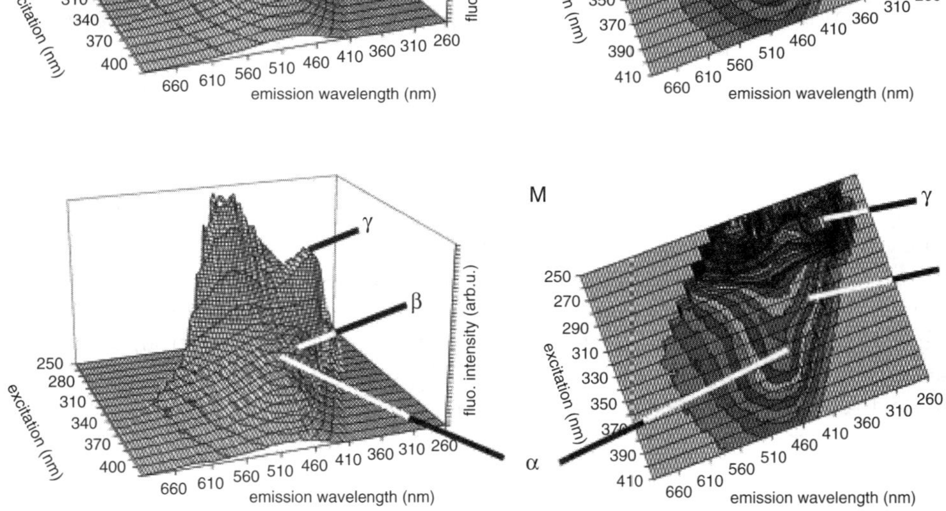

Figure 16.40. Topographic and contour 3D-fluorescence plots for fresh water (F) and marine water (M) samples (Parlanti et al., 2002).

380–480 nm, respectively) as peaks related to the component type of humic-like substances; β (excitation at 310–320 nm and emission at 380–420 nm) as a peak related to the component type of marine humic-like substances; γ (excitation at 270–280 nm and emission at 300–320 nm) as a peak related to the component type of tyrosine-like and/or protein-like; δ (excitation at 270–280 nm and emission at 320–350 nm) as a peak related to the component type of tryptophane-like, protein-like, and/or phenol-like substances. Parlanti et al. (2002) used the β and γ peaks as markers to estimate the biological activity in coastal zones and the different stages of the biological production.

Many metal ions can quench the fluorescence of HS by increasing the rate of intersystem crossing and/or increasing nonradiative processes. At low concentration, these two quenching mechanisms may only occur if the metal ion is associated with the fluorophore. Plaza et al. (2006) showed that titration of HA with Cu^{2+}, Zn^{2+}, Cd^{2+}, or Pb^{2+} ions causes significant modification of fluorescence EEM spectra, which depends both on the metal ion and on the origin of the HA sample (Figure 16.41). The different extents of the modifications observed can be ascribed to (a) the different strengths of the bonding between Cu^{2+}, Zn^{2+}, Cd^{2+}, or Pb^{2+} and the various HA and (b) conformational changes of the HA as the complexing process occurs.

16.6.4. Laser-Induced Fluorescence of Whole Soils

Generally, spectroscopic techniques require extraction and chemical fractioning of HS from the soil, which results in production of chemical residues, hence making soil analysis a slow and laborious task. Moreover, the products from this treatment (HA, FA, and humin) can undergo modifications in relation to its *in situ* form (Feller and Beare, 1997). In order to render SOM analysis viable at the closest to natural state possible, laser-induced fluorescence spectroscopy (LIF) has been used. Such methodology applied to soils is recent (Milori et al., 2006) and has evidenced very interesting results in assessing the humification degree of OM in soils under different managements and depths.

Fluorescence of soils originates from OM, since it has several functional groups containing unsaturated bondings in rigid systems. Such groups are present mainly in more humified OM, such as aromatic rings, phenolics, quinone structures, and carboxylic groups. These are the main fluorophors of SOM. The extension of the π-electronic system, the level of heteroatomic substitution and type and number of substituting groups under the aromatic rings strongly affect the intensity and wavelength of molecular fluorescence.

Figure 16.42 illustrates the spectra of a whole soil sample and of the same calcinated soil. As seen, the soil fluorescent emission is basically due to organic matter (OM) (Milori et al., 2006). As the soil samples are illuminated with near-ultraviolet or blue light, mainly the more complex structures are excited, and the area under the normalized curve (divided by C content) provides a parameter proportional to the humification degree of SOM.

To simplify the measurements of LIF spectroscopy of whole soil, pellets were usually prepared. Soil sample aliquots of approximately 0.50 g, after being further grinded to pass through a 250-μm mesh, were pressed into pellets of 1-cm diameter and 2-mm thickness, which were then inserted into an apparatus like the one

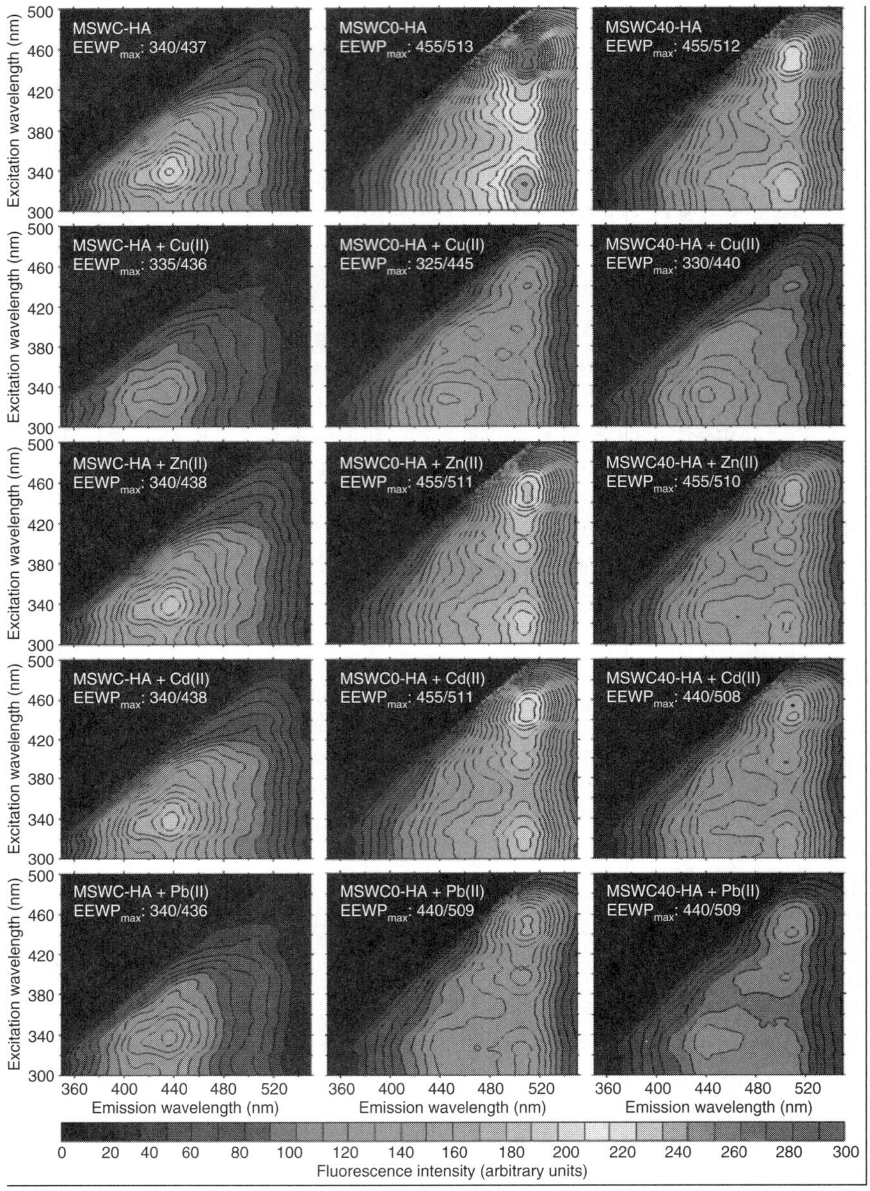

Figure 16.41. Fluorescence excitation–emission matrix spectra of humic acids (HA) isolated from municipal solid waste compost (MSWC), from soil amended with MSWC at 40 t ha^{-1} yr^{-1} (MSWC40), and from the corresponding unamended control soil (MSWC0), in the absence and presence of Cu^{2+}, Zn^{2+}, Cd^{2+}, and Pb^{2+} ions at a total concentration of 40 μmol $liter^{-1}$. $EEWP_{max}$ denotes the excitation/emission wavelength pairs at maximum fluorescence intensity (Plaza et al., 2006).

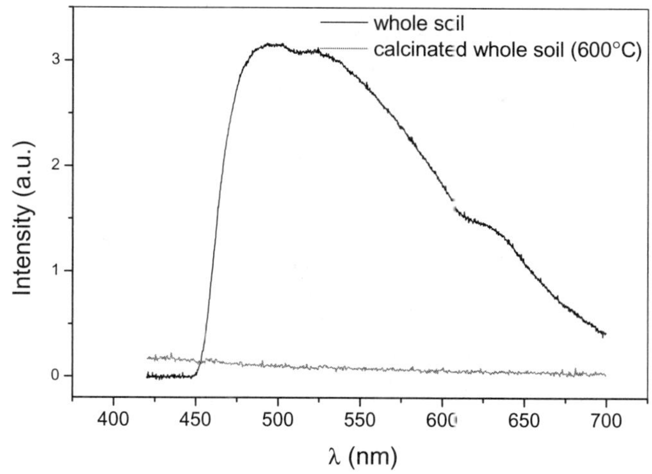

Figure 16.42. Spectra typical of LIF emission of whole soils. **Top curve:** Untreated soil. **Bottom curve:** Calcinated soil, at 600 °C.

described in Figure 16.36. Figure 16.43a illustrates a photo of the experimental mounting at the Optic and Laser Laboratory from EMBRAPA Agricultural Instrumentation Center. Figure 16.43b shows a detail of the soil pellets used in this technique.

Based on the results of Milori et al. (2002), excitation at near-ultraviolet or blue wavelengths interacts mainly with structures present in more humified substances. Therefore, LIF of whole soil with this range of excitation provides information about more humified structures of OM. In this case, the area under the fluorescence emission normalized by C content of soil was defined as a humification index called H_{LIF}. Figure 16.44 presents the results of the correlation between the humification degree of SOM measured using LIF(H_{LIF}) and the humification degree of HA in aqueous solution extracted from the same soils using conventional fluorescence (Milori's method (2006). Generally, the correlation between the techniques is high and significant, thus supporting the application of LIF for SOM analyses.

Figure 16.45 shows an example of LIF application for soil management analysis. The graph shows the behavior of the SOM humification degree in different layers and how it is affected by land use from samples of the Brazilian savanna (Cerrado) region. In conventional tillage treatment, H_{LIF} was rather uniform along the profile. This humification uniformity is consistent with the homogeneity imparted by tillage disturbances on the 0- to 20-cm layer of these soils. The H_{LIF} for the native Cerrado and no-tillage soils showed a tendency to increase from top to deeper layers. This is a consistent result, considering that the proportion of particulate OM fraction, and thus of labile compounds like carbohydrates and peptides, decreases in depth. Similar results were observed by González-Perez et al. (2007) for subtropical Brazilian soils.

Besides applications with whole soils, LIF can be applied to study physical and insoluble chemical fractions of soil. González-Pérez et al. (2006b) carried out studies about fractions of SOM under sewage sludge application. In this work, the high fluorescence contribution of humin fraction to the fluorescence of whole soils was

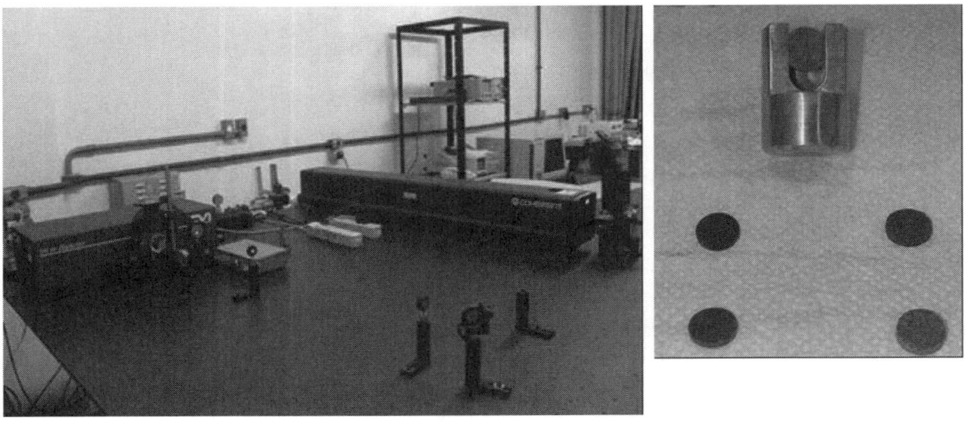

a) b)

Figure 16.43. (**a**) Experimental apparatus to carry out LIF spectroscopy. (**b**) Details of the soil pellets.

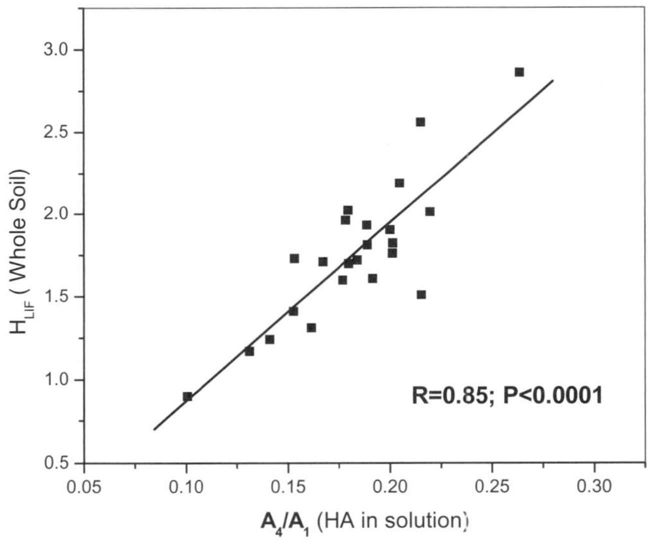

Figure 16.44. Correlation between the humification index determined by LIF spectroscopy (H_{LIF}) in whole soil samples and the humification index A_{465} (Milori et al., 2002) determined by conventional fluorescence in the corresponding humic acid samples. The indexes are expressed as arbitrary units (a.u.) (Milori et al., 2006).

shown, stressing the importance of studying OM associated with the mineral matrix of soil.

16.7. CONCLUSIONS AND PERSPECTIVES

This chapter presented technical principles and some relevant results from NOM using EPR, FTIR, Raman, UV–vis absorption, and fluorescence spectroscopies.

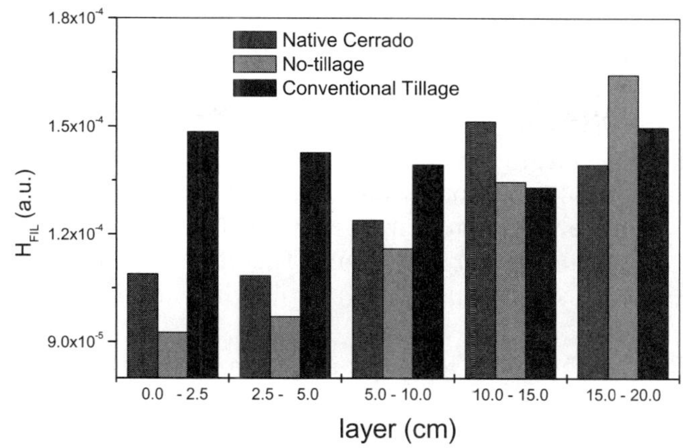

Figure 16.45. SOM humification degree obtained through LIF spectroscopy (H_{LIF}) as affected by land use and soil management system in the experiment at Brazilian Cerrado (Milori et al., 2006).

Each methodology has given its specific contribution and the combined use of these methodologies with other analytical tools will certainly help toward the development of this very exciting NOM research field. Also, many other new opportunities and research needs are identifiable for the future. Therefore, briefly, some suggestions are made, as possible future research opportunities:

1. Development and application of portable equipment systems, based on spectroscopic methods, for *in situ* measurements of NOM content in different ecosystems, including areas with potential to C sequestration, such as conservative soil tillage (for example, no-till), reforestation, and others. These apparatus could provide faster and lower cost measurements to improve data basis of C content in soils and their compartments. These data are essential to be used in C balance models and reduce extrapolation risks from reduced number of C analysis, especially with regard to identifying C sequestration, mitigating increase of greenhouse effect, and, consequently, climate changes. A real possibility is the use of LIBS (laser-induced breakdown spectroscopy), which enables multi-elemental analysis, including C, and must be validated in soil with different characteristics. Another opportunity is the use of LIF (laser-induced fluorescence) technique in NOM studies, as an excellent alternative with which to obtain qualitative information (mainly chemical stability) of organic material in soils, without the use of any chemical and/or physical treatments to realize measurements.
2. Research advances in the role of free radicals and phenolic groups in the structure and reactivity of HS. The EPR technique permits monitoring of the generation and suppression of stable free radicals, as semiquinone, or transients, as hydroxyl (which can be captured and quantified by spin-trap methodology), and this is relevant molecular information that helps to boost the advancement of structural and reactivity aspects of HS, including their eventual antioxidants properties. The combined use of EPR and composition of functional groups, such as phenolic groups, quinones, carboxylic acids, and

others, could be of help in understanding the reaction mechanisms of HS and NOM in different ecosystems, including aquatics and aerial.

3. Research on pyrogenic organic materials, also called black carbon, have recently increased their standing and have aroused the interest of many researchers and institutions. For example, the so-called "Terra Preta do Índio," from the Amazonian region, has not yet been fully understood but is considered as a challenge when combined with an opportunity for sustainable tillage of soils in different ecosystems. Therefore, the simultaneous applications of spectroscopies could bring new insights into this challenging subject matter.
4. In aquatic environments, reactions are, in many situations, driven by DOC and their interactions with natural compounds and xenobiotics. The more intensive use of spectroscopic tools could help in making headway toward understanding the fate of metals, pesticides, and other xenobiotics in the environment.
5. Reactions of DOC and chlorine compounds, in the water disinfection process, generate the formation of carcinogenic compounds, such as trihalomethanes, which is mainly chloroform. Recent studies using spectroscopic tools, such as light absorption and fluorescence, together with chromatographic methods, allowed the identification of additional steps of reaction mechanisms of DOC and chlorine compounds and thus help in identifying alternative methods of water disinfection for human consumption. These spectroscopic tools could have also been used in situations of water reuse, a practice that is increasingly required in the world.
6. The challenge remains for consolidating a fully acceptable structural model of HS. The macromolecular model, until recently well-accepted, has been questioned and a supramolecular model has been proposed. However, additional efforts must be devoted to have a more definitive structural model for HS and spectroscopic methods combined with other analytical tools would be essential in this endeavor.
7. The undertaking of substituting petroleum derivatives, by biofuels and other biomaterials, using products from biomass and different organic residues, such as urban and sewage sludge residues, demands the characterization and processing of organic constituents, and the spectroscopic techniques can most certainly help in this very important and challenging issue.
8. Scientific basis for organic agriculture is missing, and spectroscopic methods can help in the understanding of the mechanisms of plant growth, disease control, and others.
9. With the advance of nanoscience and nanotechnology, several important challenges exist on the use and impact of nanomaterials in the environment and human and animal health. For example, slow release of fertilizers and pesticides could be established through complexation with nano soil particles, such as clays and HA, and spectroscopic methods could be used in the sorption–desorption analysis of chemical compounds with nano soil particles. Also, the potential generation of free radicals resulting from the use of nanoparticles in the environment and post ingestion by humans and animals, including potential molecular damage and the generation of cancer disease,

can be monitored by EPR methods, with direct free radical detection and/or use of spin-trap methods.

10. Heightened interest on air reactions that can be associated with C cycle and climate changes include NOM compounds. Therefore, there are demands for characterization and reaction mechanisms of NOM in aerosols from different origins, namely, urban, rural, from biomass burning, and others. On that account, spectroscopic techniques, combined with adequate sample preparation methods, could bring additional insights into aerosol research studies.

REFERENCES

Almendros, G., and Dorado, J. (1999). Molecular characteristics related to the biodegradability of humic acids preparations. *Eur. J. Soil Sci.* **50**, 227–236.

Alvarez-Puebla, A. R., Garrido, J. J., and Aroca, R. F. (2004). Surface-enhanced vibrational microspectroscopy of fulvic acid micelles. *Anal. Chem.* **76**, 7118–7125.

Alvarez-Puebla, R. A., Aroca R. F., Valenzuela-Calahorro, C., and Garrido, J. J. (2006). Retention of cobalt on a humin derived from brown coal. *J. Hazard. Mater.* **135**, 122–128.

Baes, A. U., and Bloom, P. R. (1990). Fulvic acid ultraviolet–visible spectra: Influence of solvent and pH. *Soil Sci. Soc. Am. J.* **54**, 1248–1254.

Baldock, J. A., and Skjemstad, J. O. (2000). Role of the soil matrix and minerals in protecting organic materials against biological attack. *Org. Geochem.* **31**, 697–710.

Bayer, C., Martin-Neto, L., Mielniczuk, J., and Ceretta, C. A. (2000). Effect of no-till cropping systems on soil organic matter in a sandy clay loam Acrisol from Southern Brazil monitored by electron spin resonance and nuclear magnetic resonance. *Soil Til. Res.* **53**, 95–104.

Bayer, C., Martin-Neto, L., Mielniczuk, J., Saab, S. C., Milori, D. M. B. P., and Bagnato, V. S. (2002a). Tillage and cropping system effects on soil humic acid characteristics as determined by electron spin resonance and fluorescence spectroscopies. *Geoderma* **105**(1–2), 81–92.

Bayer, C., Mielniczuk, J., Martin-Neto, L., and Ernani, P. R. (2002b). Stocks and humification degree of organic matter fractions as affected by no-tillage on a subtropical soil. *Plant and Soil* **238**(1), 133–140.

Bayer, C., Martin-Neto, L., Mielniczuk, J., Dieckow, J., and Amado, T. J. C. (2006a). C and N stocks and the role of molecular recalcitrance and organomineral interaction in stabilizing soil organic matter in a subtropical Acrisol managed under no-tillage. *Geoderma* **133**(3–4), 258–268.

Bayer, C., Martin-Neto, L., Mielniczuk, J., Pavinato, A., and Dieckow, J. (2006b). Carbon sequestration in two Brazilian Cerrado soils under no-till. *Soil Till. Res.* **86**, 237–245.

Bokobza, L. (2002). *Origin of near-infrared absorption bands*. In *Near-Infrared Spectroscopy: Principles, Instruments, Applications*, Siesler, H. W., Ozaki, Y., Kawata, S., and Heise, H M., eds., Wiley-VCH, Weinheim, Germany, pp. 11–39.

Bonin, J. J., and Simpson, M. J. (2007). Variation in phenanthrene sorption coefficients with soil organic matter fractionation: The result of structure or conformation? *Environ. Sci. Technol.* **41**, 153–159.

Boyd, S. A., Sommers, L. E., Nelson, D. W., and West, D. X. (1983). Copper(II) binding by humic acid extracted from sewage sludge: An electron spin resonance study. *Soil Sci. Soc. Am. J.* **47**, 43–46.

Bruun, S., Stenberg, B., Breland, T. A., Gudmundsson, J., Henriksen, T. M., Jensen, L. S., Korsæth, A., Luxhøi, J., Pálmason, F., Pedersen, A., and Salo, T. (2005). Empirical predictions of plant material C and N mineralization patterns from near infrared spectroscopy, stepwise chemical digestion and C/N ratios. *Soil Biol. Biochem.* **37**, 2283–2296.

Burdon, J. (2001). Are the traditional concepts of structures of humic substances realistic? *Soil Sci.* **166**, 11, 752–769.

Cabaniss S. (1992). Synchronous fluorescence spectra of metal–fulvic acid complexes. *Environ. Sci. Technol.* **26**, 6, 1133–1139.

Cabaniss, S. E., and Shuman, M. S. (1986). Combined ion-selective electrode and fluorescence quenching detection for copper-dissolved organic-matter titrations. *Anal. Chem.* **58**(2), 398–401.

Campbell, I. D., and Dwek, R. A. (1984). *Biological Spectroscopy*, Benjamin/Cumming, Menlo Park, CA, p. 404.

Canellas, L. P., Zandonadi, D. B., Busato, J. G., Baldotto, M. A., Simões, M. L., Martin-Neto, L., Façanha, A. R., Spaccini, R., and Piccolo, A. (2008). Bioactivity and chemical characteristics of humic acids from tropical soils sequence. *Soil Sci.* **173**(9), 624–637.

Carvalho, E. R. (2003). Study of trihalomethanes formation from aquatic humic substances treated by chlorine and chlorine dioxide. Ph.D. thesis, University of São Paulo-USP, Brazil.

Carvalho, E. R., Martin-Neto, L., Milori, D. M. B. P., Rocha, J. C., and Rosa, A. H. (2004). Interactions of clorine with tropical aquatic fulvic acids and formation of intermediates observed by fluorescence spectroscopy. *J. Brazilian Chem. Soc.* **15**(3), 421–426.

Chang Chien, S. W., Wang, M. C., Hsu, J. H., and Seshaiah, K. (2006). Influence of fertilizers applied to a paddy-upland rotation on characteristics of soil organic carbon and humic acids. *J. Agric. Food Chem.* **54**, 6790–6799.

Cheshire, M. V., and McPhail, D. B. (1996). Hyperfine splitting in the electron spin resonance solution spectra of humic substances. *Eur. J. Soil.* **47**, 205–213.

Cilenti, A., Provenzano, M. R., and Senesi, N. (2005). Characterization of dissolved organic matter from saline soils by fluorescence spectroscopy. *Environ. Chem. Lett.* **3**(2), 53–56.

Clapp, C. E., and Hayes, M. H. B. (1999). Sizes and shapes of humic substances. *Soil Sci.* **164**, 777–789.

Coyne, L. M., and Banin, A. (1986). Effect of adsorbed iron on thermo-luminescence and electron-spin resonance-spectra of Ca-Fe-exchanged montmorillonite. *Clays Clay Miner.* **34**(6), 645–650.

da Silva, J. C. G. E., Tavares, M. J. C. G., and Tauler, R. (2006). Multivariate curve resolution of multidimensional excitation-emission quenching matrices of a Laurentian soil fulvic acid. *Chemosphere*, **64**(11), 1939–1948.

Dick, L. F. P., and Rodrigues, L. M. (2006). Influence of humic substances on the corrosion of the API 5LX65 steel. *Corrosion*, **62**, 35–43.

Domeizel, M., Khalil, A., and Prudent, P. (2004). UV spectroscopy: a tool for monitoring humification and for proposing an index of the maturity of compost. *Bioresour. Technol.* **94**, 177–184.

Dos Santos, D. S., Alvarez-Puebla, R. A., Oliveira, O. N., and Aroca, R. F. (2005). Controlling the size and shape of gold nanoparticles in fulvic acid colloidal solutions and their optical characterization using SERS. *J. Mater. Chem.* **15**(29), 3045–3049.

Dumestre, A., Spagnuolo, M., Bladon, R., Berthelin, J., and Baveye, P. (2006). EPR monitoring of the bioavailability of an organic xenobiotic (4-hydroxy-TEMPO) in model clay suspensions and pastes. *Environ. Pollut.* **143**, 73–80.

Empresa Brasileira de Pesquisa Agropecuária (BR/DF). (EMBRAPA, 2002) Venâcio, E. C., Neto, L. M., Júnior, A. R., Fernando Josepetti Fonseca, F. J., Capparelli Mattoso, L. H., Mello, S. V., and Taylor, D. M. Sensor for analysis of mixtures by global selectivity and its use in sensor system. BR no. PI 02 00 409-7, Jan. 30, 2002. US no. PI PCT BR 02/00028 PI 02 00 409-7, Feb. 21, 2002.

Fabbricino, M., and Korshin, G. V. (2005). Formation of disinfection by-products and applicability of differential absorbance spectroscopy to monitor halogenation inchlorinated coastal and deep ocean seawater. *Desalination* **176**, 57–69.

Feller, C., and Beare, M. H. (1997). Physical control of soil organic matter dynamics in the tropics. *Geoderma.* **79**, 69–116.

Ferraro, J. R., and Nakamoto, K. (1994). *Introductory Raman Spectroscopy*, Academic Press, New York, p. 370.

Ferreira, J. A., Nascimento, O. R., and Martin-Neto, L. (2001). Hydrophobic interactions between spin-label 5-SASL and humic acid as revealed by ESR spectroscopy. *Environ. Sci. Technol.* **35**, 761–765.

Ferreira, J. A., Martin-Neto, L., Vaz, C. M. P., and Regitano, J. B. (2002). Sorption interactions between imazaquin and a humic acid extracted from a typical Brazilian oxisol. *J. Environ. Qual.* **31**, 1665–1670.

Fleischmann, M. J., Hendra, P. J., McQuillan, A. J. (1974). Raman spectra of pyridine adsorbed at a silver electrode. *Chem. Phys. Lett.* **26**, 163–166.

Flogeac, K., Guillon, E., and Aplincourt, M. (2004). Surface complexation of copper(II) on soil particles: EPR and XAFS studies. *Environ. Sci. Technol.* **38**, 3098–3103.

Fourage, L., Helbert, M., Nicolet, P., and Colas, B. (1999). Temperature dependence of the ultraviolet–visible spectra of ionized and un-ionized forms of nitrophenol: consequence for the determination of enzymatic activities using nitrophenyl derivatives—A warning. *Anal. Biochem.* **270**, 184.

Francioso, O., Sánchez-Cortés, S., Tugnoli, V., Ciavatta, C., Sitti, L., and Gessa, C. (1996). Infrared, raman, and nuclear magnetic resonance (^{1}H, ^{13}C, and ^{31}P) spectroscopy in the study of fractions of peat humic acids. *Appl. Spectrosc* **50**(9), 1165–1174.

Francioso, O., Ciavatta, C., Sanches-Cortez, S., Tugnoli, V., Sitti, L., and Gessa, C. (2000). Spectroscopic characterization of soil organic matter in long term amended trials. *Soil Sci.* **165**, 495–504.

Francioso, O., Sánchez-Cortéz, S., Tugnoli, V., Marzadori, C., and Ciavatta, C. (2001). Spectroscopic study (DRIFT, SERS and ^{1}H NMR) of peat, leonardite and lignite humic substances. *J. Mol. Struct.* **565–566**, 481–485.

Francioso, O., Sánchez-Cortés, S, Casarini, D., Garcia-Ramos, J. V., Ciavatta, C., and Gessa, C. (2002). Spectroscopy study of humic acids fractionated by means of tangential ultrafiltration. *J. Mol. Struct.* **609**, 137–147.

Fu, H., and Quan, X. (2006). Complexes of fulvic acid on the surface of hematite, goethite, and akaganeite: FTIR observation. *Chemosphere.* **63**, 403–410.

Fuentes, M., González-Gaitano, G., and García-Mina, J. M. (2006). The usefulness of UV–visible and fluorescence spectroscopies to study the chemical nature of humic substances from soils and composts. *Org. Geochem.* **37**, 1949–1959.

Gallard, H., and von Gunter, U. (2002a). Chlorination of Phenols: Kinetics and Formation of Chloroform. *Environ. Sci. Technol.* **36**, 884–890.

Gallard, H., and von Gunter, U. (2002b). Chlorination of natural organic matter: Kinetics of chlorination and of THM formation. *Water Res.* **36**, 65–74.

Garbin, J. R., Milori, D. M. B. P., Simões, M. L., da Silva, W. T. L., and Martin-Neto, L. (2007). Influence of humic substances on the photolysis of aqueous pesticide residues. *Chemosphere* **66**, 1692–1698.

Geipel, G., Acker, M., Vulpius, D., Bernhard, G., Nitsche, H., and Fanghanel, T. (2004). An ultrafast time-resolved fluorescence spectroscopy system for metal ion complexation studies with organic ligands. *Spectrochim. Acta Part A—Mol. Biol. Spectrosc.* **60**(1–2), 417–424.

Geyer, W., Hemidi, F. A.-H., Brüggemann, L., and Hanschmann, G. (2000). Investigation of soil humic substances from different environments using TG-FTIR and multivariate data analysis. *Thermochim. Acta* **361**, 139–146.

González-Pérez, M., Martin-Neto, L., Saab, S. C., Novotny, E. H., Milori, D. M. B. P., Bagnato, V. S., Colnago, L. A., Melo, W. J., and Knicker, H. (2004). Characterization of humic acids from a Brazilian Oxisol under different tillage systems by EPR, ^{13}C-NMR, FTIR and fluorescence spectroscopy. *Geoderma* **118**(3–4), 181–190.

González-Pérez, M., Martin-Neto, L., Colnago, L. A., Milori, D. M. B. P., De Camargo, O. A., Berton, R., and Bettiol, W. (2006a). Characterization of humic acids extracted from sewage sludge-amended oxisols by electron paramagnetic resonance. *Soil Til. Res.* **91**(1–2), 95–100.

González-Pérez, M., Milori, D. M. B. P., Martin-Neto, L., Colnago, L. A., De Camargo, O. A., Berton, R., and Bettiol, W. (2006b). Laser-induced fluorescence of organic matter from a Brazilian oxisol under sewage-sludge applications. *Sci. Agric.* **63**(3), 269–275.

González-Pérez, M., Milori, D. M. B. P., Colnago, L. A., Martin-Neto, L. A., and Melo, W. J. (2007). Study of organic matter in Brazilian Oxisol under different tillage systems by laser induced fluorescence spectroscopy. *Geoderma* **138**, 20–24.

Goodman, B. A., and Hall, P. L. (1994). Electron paramagnetic resonance spectroscopy. In *Clay Mineralogy Spectroscopy and Chemical Determinative Methods,* Wilson, M. J., ed., *Chapman & Hall*, London, pp. 173–225.

Grube, M., Lin, J. G., Lee, P. H., and Kokorevicha, S. (2006). Evaluation of sewage sludge-based compost by FT-IR spectroscopy. *Geoderma* **130**, 324–333.

Guan, X. H., Shang, C., and Chen, G. H. (2006). ATR-FTIR investigation of the role of phenolic groups in the interaction of some NOM model compounds with aluminum hydroxide. *Chemosphere* **65**, 2074–2081.

Hassett, J. P. (2006). Dissolved natural organic matter as a microreactor. *Science* **311**, 1723–1724.

Hayes, M. H. B. (1970). Adsorption of triazine herbicides on soil organic matter, including a short review on organic matter chemistry. *Res. Rev.* **32**, 131–174.

Hayes, M. H. B., and Clapp, C. E. (2001). Humic substances: considerations of compositions, aspects of structures, and environmental influences. *Soil Sci.* **166**, 723–737.

Hook, J. J. (1977). Chlorination reactions of fulvic acids in natural waters. *Environ. Sci. Technol.* **11**, 478–482.

Hu, W. G., Mao, J., Xing, B., and Schmidt-Rohr, K. (2000). Poly(methylene) crystallites in humic substances detected by nuclear magnetic resonance. *Environ. Sci. Technol.* **34**, 530–534.

IUPAC (1997). *Gold Book Compendium of Chemical Terminology*, 2nd edition, McNaught, A. D., and Wilkinson, A. eds., Blackwell Scientific Publications, London.

Janzen, E. G. (1980). A critical review of spin trapping in biological systems. In *Free Radical in Biology*, Pryor, W. A., ed., Academic Press, New York, pp. 115–154.

Jerzykiewicz, M., Drozd, J., and Jezierski, A. (1999). Organic radicals and paramagnetic metal complexes in municipal solid waste composts. An EPR and chemical study. *Chemosphere* **39**, 253–268.

Jezierski, A., Drozd, J., Jerzykiewicz, M., Chen, Y., and Kaye, K. J. (1998). EPR in the environmental control: Copper complexes and free radicals in soil and municipal solid waste compost. *Appl. Magn. Reson.* **14**, 275–282.

Kalbitz, K., Geyer, W., and Geyer, S. (1999). Spectroscopic properties of dissolved humic substances—A reflection of land use history in a fen area. *Biogeochemistry* **47**, 219–238.

Kamiya, M., and Kameyama, K. (1998). Photochemical effects of humic substances on the degradation of organophosphorus pesticides. *Chemosphere* **36**(10), 2337–2344.

Knowles, P. F., Marsh, D., and Rattle, H. W. E. (1976). *Magnetic Resonance of Biomolecules*, John Wiley & Sons, New York, p. 343.

Korshin, G. V., Li, C. W., and Benjamin, M. M. (1997). Monitoring the properties of natural organic matter through UV spectroscopy: A consistent theory. *Water Res.* **31**, 1787–1795.

Korshin, G. V., Wu, W. W., Benjamin, M. M., and Hemingway, O. (2002). Correlations between differential absorbance and the formation of individual DBPs. *Water Res.* **36**, 3273–3282.

Korshin, G. V., Benjamin, M. M., Chang, H.-S., and Gallard, H. (2007). Examination of NOM Chlorination reactions by conventional and stop-flow differential absorbance spectroscopy. *Environ. Sci. Technol.* **41**, 2776–2781.

Krasovitskii, B., and Bolotin, B. M. (1988). *Organic Luminescent Materials*, VCH Publishers, New York.

Kuckuk, R., Hill, W., Burba, P., and Davies, A. N. (1994). Pyrolysis-GC-FTIR for structural elucidation of aquatic humic substances. *Fresenius J. Anal. Chem.* **350**, 528–532.

Kulovaara, M., Corin, N., Backlund, P., and Tervo, J. (1996). Impact of UV_{254}-radiation on aquatic humic substances. *Chemosphere* **33**, 783–790.

Lakatos, B., Tibai, T., and Meisel, J. (1977). ESR spectra of humic acids and their metal complexes. *Geoderma* **19**, 319–338.

Lakowicz, J. R. (1983). *Principles of Fluorescence Spectroscopy*; Plenum Press: New York. **496** pp.

Landgraf, M. D., da Silva, S. C., and Rezende, M. O. O. (1998). Mechanism of metribuzin herbicide sorption by humic acid samples from peat and vermicompost. *Anal. Chim. Acta.* **368**, 155–164.

Laserna, J. J. (1996). *Modern Techniques in Raman Spectroscopy*, John Wiley & Sons, New York, p. 427.

Latch, D. E., and McNeill, K. (2006). Microheterogeneity of singlet oxygen distributions in irradiated humic acid solutions. *Science* **311**, 1743–1746.

Levine, I. N. (1974). *Quantum Chemistry*, 2nd edition, Allyn and Bacon, Boston, p. 506.

Li, B. H., and Xie, S. S. (2005). Fluorescence spectroscopy and imaging for optical biopsy. *Spectrosc. Spectral Anal.* **25**(7), 1083–1087.

Li, C. W., and Korshin, G. V. (2002). Studies of metal-binding sites in natural organic matter and their role in the generation of disinfection by-products using lanthanide ion probes. *Chemosphere* **49**(6), 629–636.

Li, J. H., Gamble, D. S., Pant, B. C., and Langford, C. H. (1992). Interaction of lindane with laurentian fulvic-acid—a complexation model. *Environ. Technol.* **13**(8), 739–749.

Li, C. W., Benjamin, M. M., and Korshin, G. V. (2000). Use of UV spectroscopy to characterize the reaction between NOM and free chlorine. *Environ. Sci. Technol.* **34**, 2570–2575.

Li, C. W., Benjamin, M. M., and Korshin, G. V. (2006). Characterization of NOM and its adsorption by iron oxide coated sand (IOCS) using UV and fluorescence spectroscopy. *J. Environ. Eng. Sci.* **5**(6), 467–472.

Liang, E. J., Yang, Y., and Kiefer, W. (1999). Surface-enhanced Raman spectra of fulvic and humic acids adsorbed on copper electrodes. *Spectrosc. Lett.* **32**, 689–701.

Lindsey, M. E., and Tarr, M. A. (2000). Inhibition of hydroxyl radical reaction with aromatics by dissolved natural organic matter. *Environ. Sci. Technol.* **34**, 444–449.

Lovley, D., Coates, J., Blunt-Harris, E., Phillips, E., and Woodward, J. (1996). Humic substances as electron acceptors for microbial respiration. *Nature* **382**, 445.

Lucio, M., and Schmitt-Kopplin, P. (2006). Modelling the binding of triazine herbicides to humic substances using capillary electrophoresis. *Environ. Chem. Lett.* **4**, 15–21.

MacCarthy, P. (2001). The principles of humic substances. *Soil Sci.* **166**(11), 738–751.

Machado, A. A. S. C., da Silva, J. C. G. E., and Maia, J. A. C. (1994). Multiwavelength analysis of synchronous fluorescence-spectra of the complexes between a soil fulvic-acid and Cu(II). *Anal. Chim. Acta*, **292**(1–2), 121–132.

Madari, B. E., Reeves, J. B., III, Machado, P. L. O. A., Guimarães, C. M., Torres, E., and McCarty, G. W. (2006). Mid- and near-infrared spectroscopic assessment of soil compositional parameters and structural indices in two ferralsols. *Geoderma* **136**, 245–259.

Mangrich, A. S., Lermen, A. W., Santos, E. J., Gomes, R. C., Coelho, R. R. R., Linhares, L. F., and Senesi, N. (1998). Electron paramagnetic resonace and ultraviolet–visible spectroscopic evidence for copper porphyrin presence in actinomycete melanims from soils. *Biol. Fertil. Soils* **26**, 341–345.

Martin, M. Z., Labbe, N., Andre, N., Harris, R., Ebinger, M., Wullschleger, S. D., and Vass, A. A. (2007). High resolution applications of laser-induced breakdown spectroscopy for environmental and forensic applications. Spectrochim. *Acta Part B—Atomic Spectrosc.* **62**(12), 1426–1432.

Martin-Neto, L., Nascimento, O. R., Talamoni, J., and Poppi, N. R. (1991). EPR of micronutrients–humic substance complexes extracted from Brazilian soil. *Soil Sci.* **151**, 369–376.

Martin-Neto, L., Andriulo, A. E., and Traghetta, D. G. (1994a). Effects of cultivation on ESR spectrum of organic matter from soil size fractions of a Mollisol. *Soil Sci.* **157**, 365–372.

Martin-Neto, L., Vieira, E. M., and Sposito, G. (1994b). Mechanism of atrazine sorption by humic acid: A spectroscopic study. *Environ. Sci. Technol.* **28**, 1067–1073.

Martin-Neto, L., Rossel, R., and Sposito, G. (1998). Correlation of spectroscopic indicators of humification with mean annual rainfall along a temperate grassland climosequence. *Geoderma* **81**, 305–311.

Martin-Neto, L., Traghetta, D. G., Vaz, C. M. P., Crestana, S., and Sposito, G. (2001). On the interaction mechanisms of atrazine and hydroxyatrazine with humic substances. *J. Environ. Qual.* **30**, 520–525.

McBride, M. (1978). Transition metal bonding in humic acid: an ESR study. *Soil Sci.* **126**, 200–209.

McBride, M. B. (1982a). Electron spin resonance investigation of Mn^{2+} complexation in natural and synthetic organics. *Soil Sci. Soc. Am. J.* **46**, 1137–1143.

McBride, M. B. (1982b). Organic anion adsorption on aluminum hydroxides: Spin probe studies. *Clays Clay Miner.* **30**, 438–444.

McBride, M. B. (1986). *Paramagnetic probes of layer silicate surfaces*. In *Geochemical Processes at Mineral Surfaces*, Davis, J. A, and Hayes, K. F., eds., American Chemical Society, Washington, D.C., pp. 362–388.

McBride, M. B., Fraser, A. R., and McHardy, W. J. (1984). Cu^{2+} interaction with microcrystalline gibbsite—Evidence for oriented chemisorbed copper ions. *Clays Clay Miner.* **32**(1), 12–18.

McCarty, G. W., Reeves, J. B., III, Reeves, V. B., Follett, R. F., and Kimble, J. M. (2002). Mid-infrared and near-infrared diffuse reflectance spectroscopy for soil carbon measurement. *Soil Sci. Soc. Am. J.* **66**, 640–646.

McDonald, S., Bishop, A. G., Prenzler, P. D., and Robards, K. (2004). Analytical chemistry of freshwater humic substances. *Anal. Chim. Acta* **527**, 105–124.

Miano, T. M., Piccolo, A., Celano, G., and Senesi, N. (1992). Infrared and fluorescence spectroscopy of glyphosate–humic acid complexes. *Sci. Total Environ.* **123**, 83–92.

Milori, D. M. B. P., Martin-Neto, L., Bayer, C., Mielniczuk, J., and Bagnato, V. S. (2002). Humification degree of soil humic acid determined by fluorescence spectroscopy. *Soil Sci.* **167**(11), 739–749.

Milori, D. M. B. P., Galeti, H. V. A., Martin-Neto, L., Dieckow, J., González-Pérez, M., Bayer, C., and Salton, J. (2006). Organic matter study of whole soil samples using laser-induced fluorescence spectroscopy. *Soil Sci. Soc. Am. J.* **70**, 57–63.

Neamtu, M., and Frimmel, F. H. (2006). Photodegradation of endocrine disrupting chemical nonylphenol by simulated solar UV-irradiation. *Sci. Total Environ.* **369**, 295–306.

Niemeyer, J., Chen, Y., and Bollag, J. M. (1992). Characterization of humic acids, composts, and peat by diffuse reflectance Fourier-transform infrared spectroscopy. *Soil Sci. Soc. Am. J.* **56**, 135–140.

Nikolaou, A. D., Lekkas, T. D., and Golfinopoulos, S. K. (2004). Kinetics of the formation and decomposition of chlorination by-products in surface waters. *Chem. Eng. J.* **100**, 139–148.

Novotny, E. H., and Martin-Neto, L. (2002). Effects of humidity and metal ions on the free radicals analysis of peat humus. *Geoderma* **106**, 305–317.

Novotny, E. H., Knicker, H., Colnago, L. A., and Martin-Neto, L. (2006). Effect of residual vanadyl on the spectroscopic analysis of humic acids. *Org. Geochem.* **37**, 1562–1572.

Ohnishi, S., and McConnell, H. M. (1965). Interaction of the radical ion of chlorpromazine with deoxyribonucleic acid. *J. Am. Chem. Soc.* **87**, 2293.

Olk, D. C., Brunetti, G., and Senesi, N. (1999). Organic matter in double-cropped lowland rice soils: Chemical and spectroscopic properties. *Soil Sci.* **164**, 633–649.

Olk, D. C., Brunetti, G., and Senesi, N. (2000). Decrease in humification of organic matter with intensified lowland rice cropping: A wet chemical and spectroscopic investigation. *Soil Sci. Soc. Am. J.* **64**(4), 1337–1347.

Parish, R. V. (1990). *NMR, NQR, EPR and Mössbauer Spectroscopy in Inorganic Chemistry*. Ellis Horwood, New York, pp. 168–201.

Parlanti, E., Morin, B., and Vacher, L. (2002). Combined 3D-spectrofluorometry, high performance liquid chromatography and capillary electrophoresis for the characterization of dissolved organic matter in natural waters. *Org. Geochem.* **33**(3), 221–236.

Paterno, L. G., and Mattoso, L. H. C. (2001). Effect of pH on the preparation of self-assembled films of poly(o-ethoxyaniline) and sulfonated lignin. *Polymer* **42**, 5239–5245.

Paul, A., Hackbarth, S., Zwirnmann, E., Röder, B., and Steinberg, C. E. W. (2005). In *Humic Substances: Molecular Details and Applications in Land and Water Conservation*, Ghabbour, E. A., and Davies, G., eds., Taylor & Francis, New York, pp. 99–116.

Paul, A., Stösser, R., Zehl, A., Zwirnmann, E., Vogt, R. D., and Steinberg, C. W. (2006). Nature and abundance of organic radicals in natural organic matter: Effect of pH and irradiation. *Environ. Sci. Technol.* **40**, 5897–5903.

Pavia, D. L., Lampman, G. M., and Kriz, G. S. (2001). *Introduction to Spectroscopy: A Guide for Students of Organic Chemistry*, 3th edition. Brooks/Cole, Australia.

Perminova, I., Kovalenko, A. N., Schmitt-Kopplin, P., Hatfield, K., Hertkorn, N., Belyaeva, E. Y., and Petrosyan, V. S. (2005). Design of quinonoid-enriched humic materials with enhanced redox properties. *Environ. Sci. Technol.* **39**, 8518–8524.

Peuravuori, J., and Pihlaja, K. (1997). Molecular size distribution and spectroscopic properties of aquatic humic substances. *Anal. Chim. Acta* **337**, 133–149.

Piccolo, A. (1988). Characteristics of soil humic substances extracted with some organic and inorganic solvents and purified by HCl–HF treatment. *Soil Sci.* **146**, 418–426.

Piccolo, A. (2001). The supramolecular structure of humic substances. *Soil Sci.* **166**, 810–832.

Plaza, C., Senesi, N., Polo, A., Brunetti, G., Garcia-Gil, J. C., and D'Orazio, V. (2003). Soil fulvic acid properties as a means to assess the use of pig slurry amendment. *Soil Till. Res.* **74**, 179–190.

Plaza, C., Brunetti, G., Senesi, N., and Polo, A. (2006). Fluorescence characterization of metal ion-humic acid interactions in soils amended with composted municipal solid wastes. *Anal. Bioanal. Chem.* **386**(7–8), 2133–2140.

Poole, C. P., Jr. (1967). *Electron Spin Resonance: A Comprehensive Treatise on Experimental Techniques*, Wiley-Interscience, New York, p. 921.

Provenzano, M. R., D'Orazio, V., Jerzykiewicz, M., and Senesi, N. (2004). Fluorescence behaviour of Zn and Ni complexes of humic acids from different sources. *Chemosphere* **55**(6), 885–892.

Pullin, M. J., and Cabaniss, S. E. (1995). Dissolved organic-matter fluorescence as a mixing tracer—Quantitation of interferences, method sensitivity, and preliminary data. *Abstr. Papers Am. Chem. Soc.* **210**, 37-IEC Part 1.

Ragle, C., Engebretson, R. R., and Wandruszka, R. von. (1997). The sequestration of hydrophobic micropollutants by dissolved humic acids. *Soil Sci.* **162**, 106.

Reeves, J. B., McCarty, G. W., and Meisinger, J. J. (1999). Near infrared reflectance spectroscopy for the analysis of agricultural soils. *J. Near Infrared Spectrosc.* **7**, 179–193.

Reeves, J. B., III, McCarty, G. W., and Reeves, V. B. (2001). Mid-infrared diffuse reflectance spectroscopy for the quantitative analysis of agricultural soils. *J. Agric. Food Chem.* **49**, 766–772.

Rendell, D. (1987). *Fluorescence and Phosphorescence*, John Wiley & Sons, New York.

Reusch, W. (1999). *Virtual Textbook of Organic Chemistry: Visible and Ultraviolet Spectroscopy.* http://www.cem.msu.edu/~reusch/VirtualText/Spectrpy/UV-Vis/spectrum.htm#uv1. Accessed April 22, 2009.

Riffaldi, and Schnitzer, M. (1972). Electron spin resonance spectrometry of humic substances. *Soil Sci. Soc. Am.* **36**, 301–305.

Roberts, C. A., Workman, J., Jr., Reeves, J. B., III. (2004). *NIR in Agriculture*, Agronomy Series, Vol. 44, American Society of Agronomy, Inc., Publishers, Madison, WI.

Rodríguez-Zúñiga, U. F., Milori, D. M. B. P., da Silva, W. T. L., Martin-Neto, L., Oliveira, L. C., and Rocha, J. C. (2008). Changes in optical properties caused by UV-irradiation of aquatic humic substances from the Amazon River basin: Seasonal variability evaluation. *Environ. Sci. Technol.* **42**(6), 1948–1953.

Rohatgi-Mukherjee, K. K. (1992). *Fundamentals of Photochemistry*, Wiley Eastern, New Delhi, India.

Rosa, A. H., Simões, M. L., De Oliveira, L. C., Rocha, J. C., Martin-Neto, L., and Milori, D. M. B. P. (2005). Multimethod study of degree humification of humic substances extracted from different tropic soils profiles in Brazil's Amazonian region. *Geoderma* **127**(1–2), 1–10.

Rupérez, A., and Laserna, J. J. (1996). Surface-enhanced raman spectroscopy. In *Modern Techniques in Raman Spectroscopy*, Laserna, J. J., ed., John Wiley & Sons, New York, pp. 227–264.

Ryan, D. K., and Weber, J. H. (1982). Fluorescence quenching titration for determination of complexing capacities and stability-constants of fulvic-acid. *Anal. Chem.* **54**(6), 986–990.

Saab, S. C., and Martin-Neto, L. (2004). Studies of semiquinone free radicals by ESR in the whole soil, HA, FA and humin substances. *J. Braz. Chem. Soc.* **15**, 34–37.

Saab, S. C., and Martin-Neto, L. (2007). Anéis aromáticos condensados e relação E4/E6: estudo de ácidos húmicos de gleissolos por RMN de ^{13}C no estado sólido utilizando a técnica de CP/MAS desacoplamento defasado. *Quim. Nova.* **30**, 260–263.

Saadi, I., Borisover, M., Armon, R., and Laor, Y. (2006). Monitoring of effluent DOM biodegradation using fluorescence, UV and DOC measurements. *Chemosphere* **63**(3), 530–539.

Saar, R., and Weber, J. H. (1980). Metal-ion complexation of fulvic-acids studied by fluorescence. *Abstr. Papers Am. Chem. Soc.* **180**, 62-envr.

Sanchez-Cortes, S., Francioso, O., Ciavatta, C., Garcia-Ramos J. V., and Gessa, C. (1998). pH-dependent adsorption of fractionated peat humic substances on different silver colloids studied by surface-enhanced Raman spectroscopy. *J. Colloid Interface Sci.* **198**(2), 308–318.

Schnitzer, M., and Ghosh, K. (1982). Characteristics of water-soluble fulvic acid-copper and fulvic acid–iron complexes. *Soil Sci.* **134**, 354–363.

Schnitzer, M., and Khan, S. U. (1972). *Humic Substances in the Environment*, Marcel Dekker, New York.

Schnitzer, M., and Levesque, M. (1979). Electron spin resonance as a guide to the degree of humification of peats. *Soil Sci.* **127**(3), 140–145.

Scott, D. T., McKnight, D. M., Blunt-Harris, E. L., Kolesar, S. E., and Lovley, D. R. (1998). Quinone moieties act as electron acceptors in the reduction of humic substances by humics-reducing microorganisms. *Environ. Sci. Technol.* **32**, 2984–2989.

Senesi, N. (1990a). Application of electron spin resonance (ESR) spectroscopy in soil chemistry. *Adv. Soil. Sci.* **14**, 77–130.

Senesi, N. (1990b). Molecular and quantitative aspects of the chemistry of fulvic acids and its interactions with metal ions and organic chemicals. Part II. The fluorescence spectroscopy approach. *Anal. Chim. Acta* **232**, 77–106.

Senesi, N. (1992). Metal humic substances complexes in the environment. Molecular and mechanistic aspects by multiple spectroscopic approach. In *Biogeochemistry of Trace Metals*, Adriano, D. C., ed., CRC Press, Boca Raton, FL, pp. 429–496.

Senesi, N., and Steelink, C. (1989). Application of ESR Spectroscopy to the study of humic substances. In *Humic Substances II In Search of Structure*, Hayes, M., MacCarthy, P., Malcolm, R., and Swift, R., eds., John Wiley & Sons, New York, pp. 373–408.

Senesi, N., Bocian, D. F., and Sposito, G. (1985). Electron spin resonance investigation of copper(II) complexation by soil fúlvico acid. *Soil Sci. Soc. Am. J.* **49**, 114–119.

Senesi, N., Testini, C., and Miano, T. M. (1987). Physico-chemical investigations of interaction mechanisms between *s*-triazine herbicides and soil humic acids. *Geoderma* **28**, 129–146.

Senesi, N., Sposito, G., Bradford, G. R., and Holtzclaw K. M. (1991a). Residual metal reactivity of humic acids extracted from soil amended with sewage-sludge. *Water Air Soil Pollut.* **55**(3–4), 409–425.

Senesi, N., Miano, T. M., Provenzano, M. R., and Brunetti, G. (1991b). Characterization, differentiation, and classification of humic substances by fluorescence spectroscopy. *Soil Sci.* **152**(4), 259–271.

Senesi, N., Dorazio, V. D., and Miano, T. M. (1995). Adsorption mechanisms of *s*-triazine and bipyridylium herbicides on humic acids from hop field soils. *Geoderma* **66**(3–4), 273–283.

Senesi, N., Miano, T. M., and Brunetti, G. (1996). Humic-like substances in organic amendments and effects on native soil humic substances. In *Humic Substances in Terrestrial Ecosystems*, Piccolo, A., ed., Elsevier, Amsterdam, pp. 531–593.

Senesi, N., D'orazio, V., and Ricca, G. (2003). Humic acids in the first generation of EUROSOILS. *Geoderma* **116**(3–4), 325–344.

Si, Y., Zhou, J., Chen, H., Zhou, D., and Yue, Y. (2004). Effects of humic substances on photodegradation of bensulfuron-methyl on dry soil surfaces. *Chemosphere* **56**, 967–972.

Siesler, H. W., Ozaki, Y., Kawata, S., and Heise, H. M., editors. (2002). *Near-Infrared Spectroscopy: Principles, Instruments, Applications*, Wiley-VCH, Weinheim.

Silverstein, R. M., Bassler, G. C., and Morril, T. C. (1991). *Spectrometric Indentification of Organic Compounds*, 5th edition, John Wiley & Sons, New York.

Silverstein, R. M., Webster, F. X., and Kiemle, D. J. (2005). *Spectrometric Indentification of Organic Compounds*, John Wiley & Sons, Hoboken, NJ, p. 502.

Simões, F. R., da Silva, W. T. L., Simões, M. L., Martin-Neto, L., and Vaz, C. M. P. (2006). Effect of humic substances on the electrochemical reduction of p-nitrophenol. *Int. J. Environ. Anal. Chem.* **86**, 779–787.

Simões, M. L. (2005). Study of hydrophobic interactions in humic substances and soil components using spectroscopic methodologies. Ph.D. thesis, University of São Paulo-USP, Brazil.

Singer, L. S. (1959). Synthetic ruby as a secondary standard for the measurement of intensities in electron paramagnetic resonance. *J. Appl. Phys.* **30**, 1463–1464.

Skoog, D. A., West, D. M., and Holler, F. J. (1992). *Fundamentals of Analytical Chemistry*, 6th edition, WB Saunders, Fort Worth, TX, p. 892.

Small, G. W. (2006). Chemometrics and near-infrared spectroscopy: Avoiding the pitfalls. *Trends Anal. Chem.* **25**, 11.

Smith, E., and Dent, G. (2005). *Modern Raman Spectroscopy: A Practical Approach*. John Wiley & Sons, Hoboken, NJ.

Sollins, P., Homann, P., and Caldwell, B. A. (1996). Stabilization and destabilization of soil organic matter: Mechanisms and controls. *Geoderma* **74**, 65–105.

Spagnuolo, M., Martínez, C. E., Jacobson, A. R., Baveye, P., McBride, M. B., and Newton, J. (2004). Coprecipitation of trace metal ions during the synthesis of hectorite. *Appl. Clay Sci.* **27**, 129–140.

Sposito, G., Martin-Neto, L., and Yang, A. (1996). Atrazine complexation by soil humic acids. *J. Environ. Qual.* **25**, 1203–1209.

Stevenson, C. L., and Vo-Dinh, T. (1996). Signal expressions in raman spectroscopy. In *Modern Techniques in Raman Spectroscopy*, Laserna, J. J., ed., John Wiley & Sons, New York, pp. 1–39.

Stevenson, F. J. (1994). *Humus Chemistry. Genesis, Composition, Reactions*, 2nd edition, John Wiley & Sons, New York, **496** pages.

Suëtaka, W. (1995). *Surface Infrared and Raman spectroscopy. Methods and Applications*, vol. 3, Plenum Press, New York, p. 270.

Summers, R. S., Cornel, P. K., and Roberts, P. V. (1987). Molecular size distribution and spectroscopic characterization of humic substances. Sci. *Total Environ.* **65**, 27–37.

Sutton, R., and Sposito, G. (2005). Molecular structure in soil humic substances: The new view. *Environ. Sci. Technol.* **39**(23), 9009–9015.

Trubetskoj, O., Richard, C., Grigatti, M., Ciavatta, C., and Trubetskaya, O. (2007). Evaluation of photochemical properties of compost humic-like material. *Bioresour. Technol.* **99**(11), 5090–5093.

Uyguner, C. S., and Bekbolet, M. (2005a). A comparative study on the photocatalytic degradation of humic substances of various origins. *Desalination* **176**, 167–176.

Uyguner, C. S., and Bekbolet, M. (2005b). Implementation of spectroscopic parameters for practical monitoring of natural organic matter. *Desalination* **176**, 47–55.

Venancio, E. C., Consolin-Filho, N, Constantino, C. J. L., Martin-Neto, L., and Mattoso, L. H. C. (2005). Studies on the interaction between humic substance and conducting polymers for sensor application. *J. Braz. Chem. Soc.* **16**(1), 24–30.

Vogel, E., Gessner, R., Hayes, M. H. B., and Kiefer, W. (1999). Characterisation of humic acid by means of SERS. *J. Mol. Struct.* **483**, 195–199.

Wandruszka, R., von., Ragle, C., and Engebretson, R. R. (1997). The role of selected cations in the formation of pseudomicelles in aqueous humic acid. *Talanta* **44**, 805.

Wang, W., Zafiriou, O. C., Chan, I., Zepp, R. G., and Blough, N. V. (2007). Production of hydrated electrons from photoionization of dissolved organic matter in natural waters. *Environ. Sci. Technol.* **41**, 1601–1607.

Watanabe, A., McPhail, D. B., Maie, N., Kawasaki, S., Anderson. H. A., and Cheshire, M. V. (2005). Electron spin resonance characteristics of humic acids from a wide range of soil types. *Org. Geochem.* **36**, 981–990.

Weber, J. B., Weed, S. B., and Ward, T. M. (1969). Adsorption of *s*-triazines by soil organic matter. *Weed Sci.* **17**, 417–421.

Weber, J. H. (1988). Binding and transport of metals by humic materials. In *Humic Substances and Their Role in the Environmen*, Frimmel, F. H., and Christman, R. F., eds., John Wiley & Sons, Chichester, pp. 165–178.

Weil, J. A., Bolton, J. R., and Wertz, J. E. (1994). *Electron paramagnetic resonance: Elementary Theory and Practical Applications*, John Wiley & Sons, New York, p. 568.

Welhouse, G. J., and Bleam, W. F. (1993a). Atrazine hydrogen-bonding potentials. *Environ. Sci. Technol.* **27**, 494–500.

Welhouse, G. J., and Bleam, W. F. (1993b). Cooperative hydrogen bonding of atrazine. *Environ. Sci. Technol.* **27**, 500–505.

Wilson, E. B., Jr., Decius, J. C., and Cross, P. C. (1955). *Molecular vibrations: The Theory of Infrared and Raman Vibrational Spectra*, McGraw-Hill, New York, p. 388.

Wilson, M. A., Vasallo, A. M., Perdue, E. M., and Reuter, J. H. (1987). Compositional and solid-state nuclear-magnetic-resonance study of humic and fulvic-acid fractions of soil organic-matter. *Anal. Chem.* **59**, 4, 551–558.

Yang, Y., and Wang, T. (1997). Fourier transform raman spectroscopi characterization of humic substances. *Vib. Spectrosc.* **14**, 105–112.

Zech, W., Senesi, N., Guggenberger, G., Kaiser, K., Lehmann, J., Miano, T. M., Miltner, A., and Schroth, G. (1997). Factors controlling humification and mineralization of soil organic matter in the tropics. *Geoderma* **79**, 117–161.

Zelano, V., Daniele, P. G., Berto, S., Ginepro, M., Laurenti, E., and Prenesti, E. (2006). Metal ion distribution between water and river sediment: Speciation model and spectroscopic validation. *Anal. Chim.* **96**(1–2), 1–11.

Zsolnay, A., Baigar, E., Jimenez, M., Steinweg, B., and Saccomandi, F. (1999). Differentiating with fluorescence spectroscopy the sources of dissolved organic matter in soils subjected to drying. *Chemosphere* **38**, 45–50.

17

SYNCHROTRON-BASED NEAR-EDGE X-RAY SPECTROSCOPY OF NATURAL ORGANIC MATTER IN SOILS AND SEDIMENTS

J. Lehmann and D. Solomon

Department of Crop and Soil Sciences, Cornell University, Ithaca, New York

J. Brandes

Skidaway Institute of Oceanography, University of Georgia, Athens, Georgia

H. Fleckenstein and C. Jacobson

Brookhaven National Laboratory, SUNY Stony Brook, Stony Brook, New York

J. Thieme

Institut Für Röntgenphysik, University of Göttingen, Göttingen, Germany

Biophysico-Chemical Processes Involving Natural Nonliving Organic Matter in Environmental Systems, Edited by Nicola Senesi, Baoshan Xing, and Pan Ming Huang

17.1. INTRODUCTION

Spectroscopic techniques have received increased attention for the study of natural organic matter (NOM) over the past decades (Hatcher et al., 2001; Abbt-Braun et al., 2004). Such techniques allow the determination of molecular speciation in many cases without the need for extractions, derivatization, or hydrolysis. Spectroscopy is generally less selective in nature than for example chemical extraction techniques, even of chemically or thermally recalcitrant compounds (Frimmel et al., 2002; Haberstroh et al., 2006), though important restrictions for specific bonds apply for some spectroscopic techniques. Equally important are the potentials to investigate the spatial relationships between NOM and mineral phases, surface properties and alteration, and micro-scale heterogeneity within NOM. With improved capabilities and access to synchrotron facilities, worldwide efforts in applying an entire range of powerful spectroscopic tools have proliferated in all areas of science.

In this chapter, we discuss near-edge X-ray fine structure (NEXAFS) for the study of elemental speciation and distribution that are most relevant to the study of NOM in soils and sediments (carbon, nitrogen, oxygen, phosphorus, and sulfur with brief mentioning of metals) and will use the term NEXAFS rather than X-ray absorption near-edge structure (XANES), which is more often used for hard X-ray (e.g., sulfur and phosphorus). NEXAFS (Stöhr, 1992) involves excitation of core-level electrons to unoccupied or partially occupied molecular orbitals and uses a variety of resulting phenomena such as absorption of energy, fluorescence, or emission of photons for the assessment of the bonding environment of a specific element (Hitchcock et al., 2008). NEXAFS spectroscopy, microspectroscopy and the development of NEXAFS spectromicroscopy (explained in Section 2.3) have been one of the most promising tools for the study of soils and sediments since its development in the 1980s (Kirz et al., 1995; Ade and Urquhart, 2002), but was only more widely used for investigations of NOM since the mid-1990s. The scope of studies has included contaminant and metal speciation, uptake, and distributions (Schulze et al., 1995), model humic and natural substances (Vairavamurthy et al., 1997; Plaschke et al., 2004 2005; Rothe et al., 2004), microorganisms (Thieme et al., 1994; Lawrence et al., 2003; Liang et al., 2006), soils (Schmidt et al., 2003; Kinyangi et al., 2006; Schumacher et al., 2005; Lehmann et al., 2007; Wan et al., 2007) and soil extracts (Solomon et al., 2005a 2007a, b), dissolved organic matter (DOM) (Schumacher et al., 2006), sediments (Schäfer et al., 2005; Benzerara et al., 2006; Haberstroh et al., 2006, Brandes et al., 2007) or aquatic colloids (Niemeyer et al., 1994; Thieme et al., 1998; Rothe et al., 2000; Brandes et al., 2004), and DNA (Ade et al., 1992). NEXAFS has also been successfully used for the environmental study of coal (Cody et al., 1998), plant fossils (Boyce et al., 2002) or soot (Braun et al., 2007), which are not the subject of this contribution. Other synchrotron-based techniques for the study of NOM chemistry such as Fourier-transform infrared spectroscopy (FTIR) (Solomon et al., 2005a; Lehmann et al., 2007) have also been used with great success but will not be discussed in this chapter.

NEXAFS methods can be applied to nearly all elements, is not restricted to certain isotopes, does not typically exhibit interferences (e.g., paramagnetic iron with nuclear magnetic resonance (NMR) spectroscopy), and does not necessarily require sample drying (Williams et al., 1993; Kirz et al., 1995). An additional advantage is the ability of NEXAFS to directly interrogate the composition of minor or trace elements in NOM without extraction or chemical modifications (Brandes et al., 2007). For example, NEXAFS has provided an entirely new approach to the study of organic sulfur compounds in soils that was not possible before (Xia et al., 1998; Prietzel et al., 2003; Solomon et al., 2003).

What has proved especially useful is the coupling of NEXAFS with imaging in an approach called spectromicroscopy in X-ray imaging (Ade et al., 1992; Jacobsen et al., 2000; Hitchcock et al., 2008; detailed description in Section 2.3) or spectrum imaging in electron microscopy (Jeanguillaume and Colliex, 1989; Hunt and Williamds, 1991). With spectromicroscopy, entire areas of soils and sediments can be measured by NEXAFS with very high spatial resolution (Jacobsen et al., 2000). Both soils and sediments are complex assemblages of minerals, living and dead organic matter, water, and pore space filled with various gases. The notion that the spatial organization of all these components holds the key to an understanding of processes controlling NOM quantity and quality is gaining more and more attention (Young and Crawford, 2004). The spatial information obtained by scanning entire areas within organo-mineral assemblages of soils (Kinyangi et al., 2006; Lehmann et al., 2007), sediments (Neuhäusler, 1999), and suspended colloids (Rothe et al., 2000; Schumacher et al., 2005) affords an important step forward in gaining critical insight into the biogeochemistry of NOM in the environment.

In the following sections, we explain the basic principles of NEXAFS, methods of data acquisition and analysis, and its application to NOM.

17.2. PRINCIPLES

17.2.1. Synchrotron Facilities

Synchrotron radiation is produced by relativistic electron or positron beams being forced to radiate when traveling through strong magnetic fields. Beginning in the late 1970s, a growing number of facilities dedicated to the production of synchrotron radiation have been developed worldwide. A comprehensive listing is provided by www.lightsources.org. Compared to laboratory X-ray sources such as X-ray tubes, synchrotron radiation is (by many orders of magnitude) brighter and more readily tunable. Along with developments in high spatial resolution X-ray optics, this has enabled the development of X-ray microscopes with spatial resolution in the 30- to 300-nm range and with spectromicroscopy capabilities as noted above (detailed description in Section 2.3). Most spectromicroscopy studies are carried out using scanning microscopes or microprobes, where the X-ray beam is focused down to a small spot through which the sample is scanned at each desired photon energy (see, e.g., Kirz et al., 1995). In these scanning microscopes, fields of up to several millimeters can be examined, though high resolution scans are typically carried out on field sizes of several micrometers.

17.2.2. NEXAFS

In the historical Bohr (Niels Bohr, 1885–1962) model the atom is pictured as having electrons in circular orbits surrounding a small, positively charged nucleus, to which the electrons are attracted by the electrostatic Coulomb force. These orbits have discrete radii indexed by a principal quantum number $n = 1,2,3 \ldots$, and discrete binding energies for single-electron atoms of $E_n = -E_0Z^2/n^2$, where Z is the atomic number and $E_0 = 13.6\,\text{eV}$ (see Figure 17.1). When the energy of incident photons is increased to match the binding energy of an electron, the photon absorption probability suddenly increases in what is called an X-ray absorption edge (step function in Figure 17.2, which is at about 290 eV for carbon), so that the electron is completely removed from the atom in an ionization event (absorption edges are also labeled

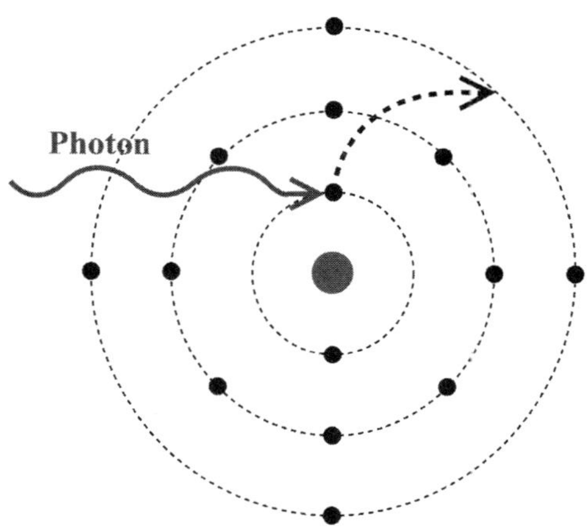

Figure 17.1. Atom in the Bohr model. The electrons surround the nucleus. An incoming photon can lift an electron to a not fully occupied, higher orbital or remove it entirely from the atom.

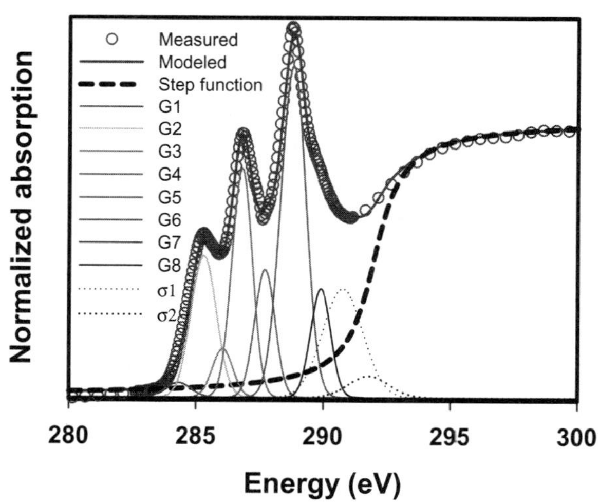

Figure 17.2. Carbon NEXAFS spectrum of NOM from the Suwannee River (IHSS standard humic acid mounted on indium foil; total electron yield using a dwell time of 200 msec and an exit slit of 50 µm, calibrated to CO at 287.38 eV, Canadian Light Source SGM beamline 11-ID.1) to show pre-edge features and the so-called "edge". The spectrum is deconvoluted using a series of Gaussian curves (G) at energy positions of known transitions, along with a step function at the edge as described by Solomon et al. (2005). See color insert.

as K edges when the $n = 1$ electron is excited, or L edges when the $n = 2$ electron is excited, and so on). At an energy just below the absorption edge, however, the energy is not sufficient to remove the electron from the atom entirely but instead it may be promoted to a weakly bound orbital. The probability for this will depend on the degree of occupancy of this orbital. Since chemical binding energies are in the range of about 2–10 eV, the occupancy of orbitals within ~2–10 eV of the absorption edge is in turn modified by the chemical binding state of the atom. These pre-edge resonances (peaks between 284 and 290 eV in Figure 17.2) are known as near-edge X-ray absorption fine structure (NEXAFS) or X-ray absorption near-edge structure (XANES). Their narrow width of ~0.1–0.2 eV in carbon is determined by a combination of the relatively long core hole lifetime and the shorter lifetime of the excited state that it is coupled to. Finally, in addition to these pre-edge resonances from coupling to bound states, the inner-shell electron can also be promoted to classically unbound states (such as the $1s \rightarrow \sigma^*$ transition that will be noted in carbon NEXAFS spectra shown below) that have broad energy widths of several electron volts due to their short lifetime.

The Bohr model of the atom is an oversimplification that has been supplanted by Schrödinger equation solutions for electron wavefunctions in the presence both of the Coulomb potential of the nucleus, and the distributed charge of other electron wavefunctions in the atom or molecule. Once wavefunction solutions are known, the energy position and strength of NEXAFS resonances can in principle be calculated from the overlap of wavefunctions in the presence of an incident photon of a particular energy. In practice, calculations of this sort are guided by observing the trend of the appearance of NEXAFS resonances in simple molecules with various chemical functional groups (Stöhr, 1992). In complex molecules both multiple, partially overlapping NEXAFS resonances and even a slight shift in the ionization potential can be observed as the configuration of outer electrons affect the charge seen by inner-shell electrons. The challenge in NEXAFS analysis is to interpret the observed absorption spectrum as a sum of all resonances plus the ionization step of the atom, as is shown in Figure 17.2.

At photon energies above the absorption edge, the liberated electron has a de Broglie wavelength given by its kinetic energy (the photon energy minus the ionization potential). This electron can undergo partial reflection from the charge distribution of other nearby atoms and can then interfere with itself. This effect, called extended X-ray absorption fine structure (EXAFS) (Teo, 1986), manifests itself in subtle "wiggles" in the absorption spectrum at energies of tens to hundreds of electron volts above the absorption edge due to alternating constructive and destructive self-interference of the ejected electron. EXAFS spectroscopy is typically used to study minority atoms at higher energies where one can better separate EXAFS and NEXAFS features, and it has been used for sulfur. It is of less use in the case of organic molecules (Teo, 1986) and will not be discussed further here.

The discussion above has considered only what happens to the electron that is removed from an inner shell. An atom with an inner-shell electron removed is unstable, and an electron from a higher orbital will soon fill the inner-shell void. Therefore, the energy difference between these two orbitals must be emitted by the atom (called a secondary process). This can occur by two competing processes: (a) emission of a fluorescent photon or by (b) Auger electron emission (Stöhr, 1992: p116). If the atom in question is at a surface, the energy of the Auger electron can be directly measured in an electron spectrometer (this is done in scanning photoelectron microscopy (SPEM) (Ade, 1998), in X-ray photoelectron spectroscopy (XPS) (Gerin et al., 2003), and in some photoelectron emission microscopy systems (PEEM)). If the atom is more deeply buried (about 10 nm or more, depending on density of the material and elemental content), the Auger electron will undergo multiple inelastic scattering and information on its initial energy will be lost; however, one can still measure the total electron yield which is proportional to X-ray absorption. The relative probability of the fluorescence versus Auger process is given by the fluorescence yield of the atom in question. This quantity has been tabulated (Krause, 1979) and is shown in Figure 17.3. As can be seen, for low-Z atoms such as carbon, nitrogen, or oxygen, the fluorescence yield is exceedingly low, but is more frequently used for phosphorus and sulfur.

Radiation damage is possible for some compounds in NOM and requires preliminary tests where no prior experience for a beamline and a certain NOM fraction is available. The hierarchy of radiation induced modifications includes: (i) incapacitation of biological function; (ii) organizational changes that affect inter-molecular organization and orientation; (iii) chemical transformation such as bond breaking, formation, and reorganization; and (iv) mass loss (Ade and Hitchcock, 2008). Amino groups appear to be especially sensitive to beam damage, and appearance of spectral features of heterocyclic nitrogen forms were reported by repeated measurement of pure amino acid standards (Zubavichus et al., 2004; Leinweber et al., 2007). Interestingly, such beam damage caused by repeated measurement is less apparent in NOM–mineral mixtures such as soils than in pure substances (Leinweber et al., 2007). In general, beam damage is less of a problem in NEXAFS than, for example, in electron microscopy (Rightor et al., 1997, Hitchcock et al., 2008). The level of radiation that does not cause damage also depends on whether the specimen is dry or wet (Kirz et al., 1995) and increases in a wet stage due to the diffusion of radicals (Williams et al., 1993). Measurements under cryo conditions decreases morphological changes and mass loss, but has little effect on reducing beam damage of the

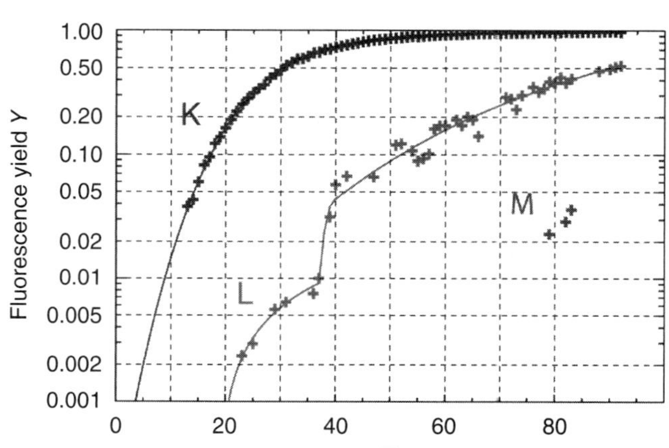

Figure 17.3. Fluorescence yield, or fraction of core-level ionization events resulting in emission of a fluorescent photon, for all elements as a function of their atomic number (Z). K, L and M refer to the atomic shells. Data from Krause (1979).

functional group chemistry (Beetz and Jacobsen, 2003). Beam damage can be reduced for example by decreasing the dwell time or by defocusing the beam when measuring in transmission. In spectromicroscopy, sequential images at different energy levels also pose lower risks of beam damage than line scans or point spectra, due to lower exposure times at similar spectral quality (Ade and Hitchock, 2008).

17.2.3. NEXAFS Techniques and Instrumentation

NEXAFS experiments on NOM can be conducted in several modes that differ in the type of detected particle and objectives of the experiment: transmission (X rays transmitted through the sample), fluorescence (fluorescent X rays due to absorption of the X-ray beam), or electron yield (photo-emitted electron) (Sparks, 2003). Alternatively, the techniques can be divided into full-field applications such as transmission X-ray microscopy (TXM) and X-ray photoemission electron microscopy (PEEM), in comparison to scanning techniques such as scanning transmission X-ray microscopy (STXM) and scanning photoemission microscopy (SPEM) that provide spatial information of elemental forms.

Low-Z elements such as carbon have low fluorescence yields (Figure 17.3) as mentioned above and the spectral resolution and therefore quality of the spectra therefore can but does not have to be limited (Stöhr, 1992). These fluorescence experiments at the carbon K edge are often carried out under high vacuum. A possible challenge may pose evaporation losses, if volatile compounds are to be measured.

Compared with total electron yield, which is a fairly surface-sensitive (~10 nm) technique, fluorescence yield spectroscopy is much more bulk sensitive and can be used to probe deeper (~100 nm for C 1s excitation and proportional to photon energy and elemental content) into the sample. The low depth of excitation is a limitation when information about bulk properties of a sample is desired. Thorough grinding of the sample prior to measurement is then important. Yet, in some instances, surface properties may be of interest, such as in aggregate studies. In these cases, intact aggregates can be compared with ground samples to obtain information about aggregate surfaces.

In a transmission experiment using STXM, the X-ray intensities incident on the sample (I_0) and transmitted through the sample (I) are recorded as a function of the incident X-ray energy E to yield an absorption spectrum. Because of the Lambert–Beer Law for X-ray absorption of $I = I_0e^{-\mu t}$ where μ is a linear absorption coefficient (e.g., in cm^{-1}) and t is the specimen thickness (e.g., in cm), it is common to express the result of the transmission measurement as an optical density $OD = \mu t = -\ln(I/I_0)$. This optical density as a function of energy OD(E) is what is usually plotted for absorption spectra such as in Figure 17.2. The development of X-ray spectromicroscopy (such as scanning transmission X-ray microscopy, STXM) means that from a series of images at different X-ray energies one can obtain a measurement of the absorption spectrum at each pixel in the image set. Therefore, entire regions can be mapped for their NEXAFS properties (as shown below and in Section 17.4.3). At present the highest-quality spectra are delivered by scanning (STXM) microscopes. State-of-the-art scanning transmission X-ray microscopes (STXMs) for soft X rays are, for example, found at the National Synchrotron Light Source (NSLS) (Feser et al., 1998), the Advanced Light Source (ALS) (Kilcoyne et al., 2003), the Advanced Photon Source (APS) (Bluhm et al., 2006), BESSY (Wiesemann et al., 2003), the Canadian Light Source (CLS) (Hitchcock et al., 2008), or the Swiss Light Source (Flechsig et al., 2007) (see Table 17.1 for a list of beam lines often used for NOM research). In soft X-ray STXMs up to 2 keV, wet specimens with a thickness up to a few micrometer can be examined for elements such as carbon, nitrogen, oxygen, iron, calcium, aluminum, silicon, or potassium at a spatial resolution of about 35 nm at present (with advances being currently made), with an energy resolution of about 0.1 eV. In the hard X-ray range above 2 keV, only a few beamlines offer STXM capabilities for measurement of, for example, phosphorus and sulfur (e.g., ID-21 at the European Synchrotron Radiation Facility (ESRF)), while others operate with fluorescence to obtain spatial data (e.g., APS XOR-2-ID-B; Table 17.1).

Figure 17.4 shows an example layout of a beamline equipped for NEXAFS, here depicted in combination with STXM. At beamline X1A of the National Synchrotron Light Source (NSLS) an undulator (a device made up of pairs of permanent magnets stacked with interchanging polarity, located in a straight section of the storage ring) can be tuned to generate a bright beam of soft X rays between 250 and 800 eV. The spherical grating monochromator in the following beamline section disperses the X rays into a horizontal rainbow of "colors," from which a set of entrance and exit slits sorts out partially spatially coherent X rays of a small energy range (Winn et al., 2000). A Fresnel zone plate then focuses the X rays onto the sample. Zone plates are circular diffraction gratings. Modern "phase zone plates" have transparent zones alternating with zones of a phase shifting material (typically nickel or gold) to improve efficiency compared to "amplitude zone plates" (alternating transparent and opaque zones). The zone plates for soft X rays commonly used in STXM have diameters of 80, 160, or 240 μm and an outermost zone width of 30–40 nm and are produced with a central stop. If a pinhole [called the order sorting aperture (OSA)] is matched to the central stop in size and a position is placed between zone plate and sample, only the first diffraction order can pass onto the sample.

The sample is then raster scanned through the focus spot, and the transmission in each image pixel is recorded. Among the variety of soft X-ray detectors are proportional counters and phosphors with photomultipliers. A recent development is a silicon detector with a segmented chip (Feser et al., 2003) allowing for dark and bright field as well as differential phase contrast imaging.

For the precise positioning and scanning of the sample, a combination of motorized stages is used. The NSLS STXM employs a stack of three stepping motor stages for the selection of sample areas and coarse overview scans. The stepping motors have a sufficiently large travel range and a step size of 1 μm in the transverse direction (imaging plane) and below 0.1 μm in the longitudinal (focusing) direction. Since the focal length of a zone plate grows linearly with the X-ray energy, it is necessary to re-position the sample relative to the zone plate at each new wavelength with sufficiently high step resolution. On top of the stepping motors sits a two-axis piezo stage for high-resolution imaging of 10 μm and below. The travel range of the piezo stage in each transverse direction is limited to about 80–100 μm.

The most advanced STXMs presently use laser interferometers to reduce the effect of imperfections in sample positioning during NEXAFS scans (Kilcoyne et al., 2003). High-precision mirrors attached to the sample and the zone plate

TABLE 17.1. Selection of Synchrotron X-ray Beamlines Commonly Used or Projected to Be Used in the Study of NOM as of 2007

Beamline	Energy Range	Microscopy Supported and Spatial Resolution (Y: Yes; N: No)
ALS 5.3.2	250–700 eV	Y, 35 nm
ALS 10.3.1	3–20 KeV	Y, 1 μm
ALS 10.3.2	2.5–17 KeV	Y, 5 μm
ALS 11.0.2	90–2150 eV	Y, 35 nm
APS XOR-2-ID-B	1–4 KeV	Y, 50 nm
APS XOR-2-ID-D,E	1–30 KeV	Y, 200 nm
APS XOR/PNC-20-BM-B	2.7–25 KeV	N
APS XOR/PNC-20-ID-B,C	11–50 KeV	Y, 5 μm
APS MR-CAT-10-ID-B	15–90 KeV	N
APS GSECARS-13-ID-C,D	4–45 KeV	N
AS BL5	5–45 KeV	N, under construction
AS BL6	90–2000 eV	N, under construction
AS BL9	4–12 KeV	Y, under construction
CLS SGM-11ID-1	240–2000 eV	N
CLS SM-10ID-1	100–2500 eV	Y, 35 nm
CLS SXRBM- 06B1-1	1.7–10 KeV	under construction
DIA I18	2–20 KeV	Y, 1 μm
ESRF ID-21	2–8 KeV	Y, 0.3–1 μm
NSLS X1A	250–800 eV	Y, 40 nm
NSLS X15B	1–10 KeV	N
NSLS X26A	3–30 KeV	Y, 10 μm
NSLS X27A	4.5–20 KeV	Y, 10 μm
NSLS X11	4.5–35 KeV	N
NSLS X18B	4.8–40 KeV	N
NSLS X19A	2.1–17 KeV	N
SLS X03MA	400–1800 eV	N
SLS X07DA	200–1200 eV	Y, 35 nm
SLS X07MA	0.8–8 KeV	Y, 1 μm
SLS X11MA	90–2000 eV	Y, 100 nm
SPRING-8 BL17SU	300–1800 eV	under construction
SPRING-8 BL47XU	5–37 KeV	Y, 1 μm, under construction
SOLEIL Antares	10–1000 eV	under construction
SOLEIL Lucia	0.8–8 KeV	Y, 5 μm
SOLEIL Galaxies	2–12 KeV	N
BESSY	various	various
SSRL BL 2-3	2.4–30 KeV	N
SSRL BL 6-2	2–5 KeV	Y, 0.5 μm

ALS: Advanced Light Source at Lawrence Berkeley National Laboratories (www.als.lbl.gov)
APS: Advanced Photon Source at Argonne National Laboratories (www.aps.anl.gov)
AS: Australian Synchrotron (www.synchrotron.vic.gov.au)
CLS: Canadian Light Source (www.lightsource.ca)
DIA: Diamond Synchrotron Facility (www.diamond.ac.uk/)
ESRF: European Synchrotron Radiation Facility (www.esrf.eu)
NSLS: National Synchrotron Light Source at Brookhaven National Laboratory (www.nsls.bnl.gov)
SLS: Swiss Light Source (sls.web.psi.ch/view.php/about/index.html)
SOLEIL: SOLEIL Synchrotron (www.synchrotronhyphen;soleil.fr)
SSRL: Stanford Synchrotron Radiation Laboratory (wwwhyphen;ssrl.slac.stanford.edu/)

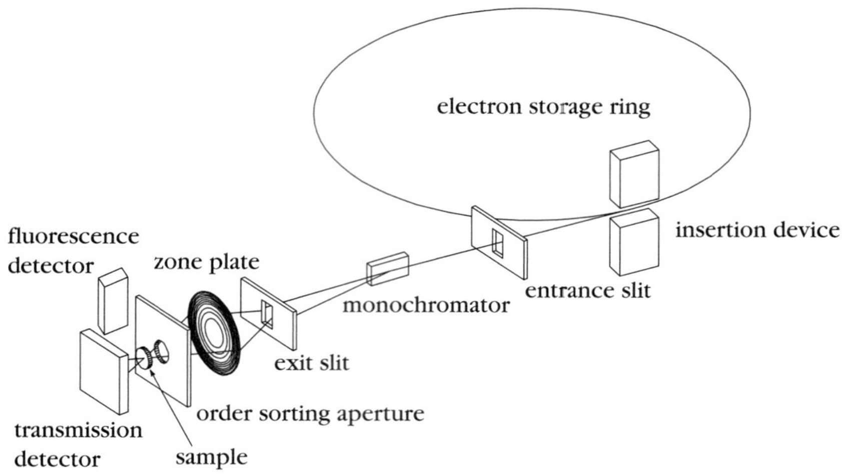

Figure 17.4. Schematic of a scanning transmission X-ray microscope. This setup uses soft X rays (250–800 eV) from an undulator on beamline X1A of the National Synchrotron Light Source in Brookhaven National Laboratory, Upton, NY.

mount reflect parts of a divided laser beam. The interference information from the reflections is used to correct the sample position relative to the zone plate.

In imaging mode the sample is scanned in the two transverse directions, whereas in spectrum mode the monochromator is moved to scan the X-ray energy while the sample simultaneously follows the focus position in the beam direction. The combination of the two modes is referred to as spectromicroscopy. Typically, a series of images of the same specimen area is acquired over a range of energies. The resulting data set with a full spectrum in each image pixel can be analyzed to show a map of the sample's chemical composition. Important here is the alignment between images that may be necessary due to small shifts between images at different energy levels. The spatial resolution can be adjusted, and beamlines offering the highest resolution reach about 35 nm. For overviews to identify regions of interest, a lower spatial resolution can be chosen, which increases the step size between measurement spots, but not the size of the focal spot. In some cases, defocusing to a spot size of a few micrometers helps to obtain information for a larger area at the expense of spatial resolution, but this is a more representative measurement for heterogeneous samples such as NOM.

Spatial information of elemental distribution can be obtained not only by transmission experiments but also by using fluorescence. This technique is mainly used for high-Z elements such as phosphorus and sulfur due to the low fluorescence of low-Z elements. However, at this point of the development of the technique, spectral features can not be obtained for entire maps, since the acquisition time for each energy level is too long to allow stacks to be measured.

17.2.4. Sample Preparation

The preparation of samples can be the most daunting step in obtaining NEXAFS measurements. The precise technique to be used will vary significantly between samples, at different beamlines, the NEXAFS experiment (i.e., fluorescence, total electron yield, or transmission) and for the element(s) and spatial resolution to be examined. Some general differences exist between analyses of bulk samples and spatially explicit analyses. Sample carriers should be very low in the element(s) to be analyzed, especially when fluorescence or total electron yield is used. In transmission, carriers may contain the investigated element in a homogenous and stable form, which can be deducted by choosing an appropriate I_0, and with an optical density that still allows the sample to be clearly detected. This is typically verified in advance. If bulk properties of a sample are measured using fluorescence or total electron yield, finely ground samples are often mounted onto adhesive material or contained within a thin foil to keep the sample in place. For example, adhesive material such as carbon tape or indium is often used where ultrahigh vacuum is applied, whereas mylar or regular Scotch tape is used when the chamber is flushed with helium. However, transmission experiments (for example) with STXM requires a sample holder that the X rays can penetrate with as little loss as possible, and often silicon nitride or silicon monoxide are used.

Sample thickness plays a critical role for obtaining good spectra. This applies to fluorescence or total electron yield as well as to absorption using X-ray transmission. Humic extracts or standard materials such as glucose measured in transmission have to be sufficiently dilute to be penetrated by the X ray, and background measurements need to be collected next to the sample. Microorganisms can very often be measured without sectioning in their hydrated stage (Thieme et al., 1994). Individual colloids or soil particles can in some instances also be measured by transmission without pretreatment (Prietzel et al., 2003). More difficult is the measurement by transmission of mixtures of NOM and minerals or particulate NOM in their natural assemblage. Either sufficiently thin particles have to be chosen or the sample has to be thin-sectioned. In many instances, sectioning is required to obtain information about the spatial distribution in micro-assemblages because unsectioned specimen invariably contain areas that cannot be penetrated by the beam (Wan et al., 2007). Sectioning also allows interior regions of aggregates and pores to be examined and reduces the bias created by variable thickness and complexity. The thickness to which a sample has to be sectioned depends on its optical density, X-ray flux, detector sensitivity and headroom, and the concentration of the element measured. A general challenge is posed by the heterogeneity of environmental materials (Hitchcock et al., 2005a), which often contain not only biological but also mineral matter. Ade et al. (1992) recommended a sample thickness between 40 and 800 nm for biological materials. When minerals are present or when carbon is present in high concentrations, samples should preferably have a thickness of less than 200 nm, whereas analyses of the mineral elements (Al, Si) may require slightly thicker samples. Successful measurements were conducted with a thickness ranging from 80 nm for sediments (Benzerara et al., 2005), 100–200 nm for black carbon (Lehmann et al., 2005; Liang et al., 2006), to 300–800 nm for soils (Kinyangi et al., 2006; Lehmann et al., 2007). However, it is important to realize that sectioning, even with a diamond knife, does not yield samples of uniform thickness. Both sample deformation and sample density differences can generate such variations.

If samples have to be sectioned, for transmission experiments the type of embedding medium is important. Epoxy embedding media traditionally used in soil or sediment investigations (Tippkotter and Ritz, 1996) contain carbon and nitrogen compounds and are therefore not suitable for organic matter studies. Lehmann et al. (2005) introduced an embedding approach for black carbon particles using elemental sulfur based on prior research that was done on interplanetary dust (Bradley et al., 1993; Flynn et al., 2004). The spatial integrity of entire aggregates is better preserved, however, when they are misted, frozen, and cryosectioned either with (Kinyangi et al., 2006) or without (Lehmann et al., 2007) additional embedding in elemental sulfur. Cryosectioning was also successfully applied to black carbon particles (Ghosh et al., 2000). Key to the success of cryosectioning is rapid freezing of the moistened samples to prevent the build-up of ice crystals. Liquid N_2 has been used in several studies (Ghosh et al., 2000; Kinyangi et al., 2006), but liquid ethane is preferable due to its higher heat capacity despite the higher boiling point. The temperature at which the specimen is cut varied from −150 °C (Ghosh et al., 2000) to −55 °C (Kinyangi et al., 2006). Another recently developed method is the focused ion beam (FIB) technique, which allows very thin sections to be obtained and does not require embedding (Langford, 2006).

Rapid freezing is also important for preserving the spatial integrity of samples. If the sample is mounted on a rotating stage, three-dimensional measurements can be taken (Johansson et al., 2007). Such spatially explicit analyses of both colloids and aggregates provide exciting perspectives for understanding the biogeochemistry of NOM.

The necessity of keeping strict anoxic conditions during sampling of anoxic soil horizons is seen for example by comparing two NEXAFS spectra taken at the K-absorption edge of sulfur from a haplosaprist soil horizon, as seen in Figure 17.5 (Thieme et al., 2006). The two peaks representing reduced and oxidized sulfur species are visible. However, for the sample taken without maintaining anoxic conditions, their ratio is shifted, favoring the oxidized species at higher energy levels as expected. A significant error would have occurred when extracting the relative amount of each sulfur species using only the oxidized sampling procedure.

As the sample has to be supported with thin carrier foils, the question arises whether they can act sufficiently as a barrier for oxygen. The time lag t for the first

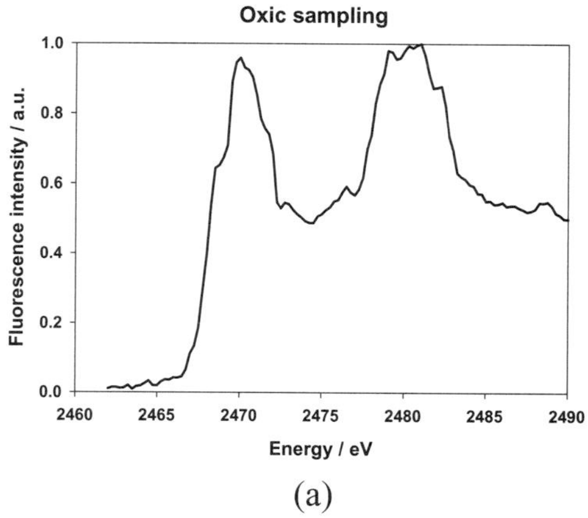

Figure 17.5. Spectra of the same sampling position, but taken under oxic **(a)** or anoxic **(b)** conditions (data from Thieme et al., 2006).

TABLE 17.2. Time Lag for two Polymers and Silicon (Kjeldsen, 1993)

Material	Diffusion Coefficient ($cm^2 s^{-1}$)	Thickness (μm)	Time Lag (s)
Polyethylene	0.30×10^{-7}	4	0.89
Polypropylene	0.29×10^{-7}	4	0.92
Silicon	0.15×10^{-30}	0.1	1.11×10^{20}

molecule to travel a specified distance through a barrier with a thickness d is described as

$$t = \frac{1}{6} \cdot \frac{d^2}{D} \tag{17.1}$$

where D is the diffusion coefficient. In Table 17.2 the results for two polymers and for silicon, which was used to approximate Si_3N_4, are listed. The thicknesses were chosen according to normal experimental conditions. The table explains clearly that polymer foils are not efficient to keep anoxic conditions and Si_3N_4 is required.

17.3. DATA ANALYSES

17.3.1. Spectral Features and Peak Assignments

One of the current challenges in the application of synchrotron-based NEXAFS techniques in bioorganic objects is the complexity of the molecular composition of the target object. Unlike studies of polymers and other man-made materials, the structural complexity of NOM makes it impossible to use spectral modeling approaches to determine absolute molecular structure. Nonliving organic matter is a heterogeneous mixture composed of organic molecules representing both compounds released from living plant and microbial cells (e.g., extracellular enzymes, surface-active proteins, chelating compounds, etc.) to complex plant, animal and microbial residues ranging in size and complexity from simple monomers or organic acids to mixtures of complex biopolymers. The chemical composition of NOM can be transformed by degradation reactions into different groups of complex organic compounds. The amount, chemical composition, and polyelectrolytic characteristics of the resulting organic C compounds may also vary considerably, depending on the origin, age of the organic material, and the environment under which it is found (Chen et al., 2002). It is these changes and heterogeneities along the molecular structure and size continuum that create significant analytical problems and that have made structural and functional characterization of NOM challenging.

Nonetheless, the ability of combined spectromicroscopic NEXAFS studies to elucidate composition and structure in NOM is unparalleled as pointed out in Sections 17.3.2 and 17.4.3. These sections illustrate one significant fact: NOM is often heterogeneous at the submicron level. Thus all other spectroscopic techniques of lower spatial resolution will "spatially average" NOM composition.

Simple model compounds provide first insights into peak assignments as demonstrated for carbon in Figure 17.6. Citric acid shows only one sharp peak at around

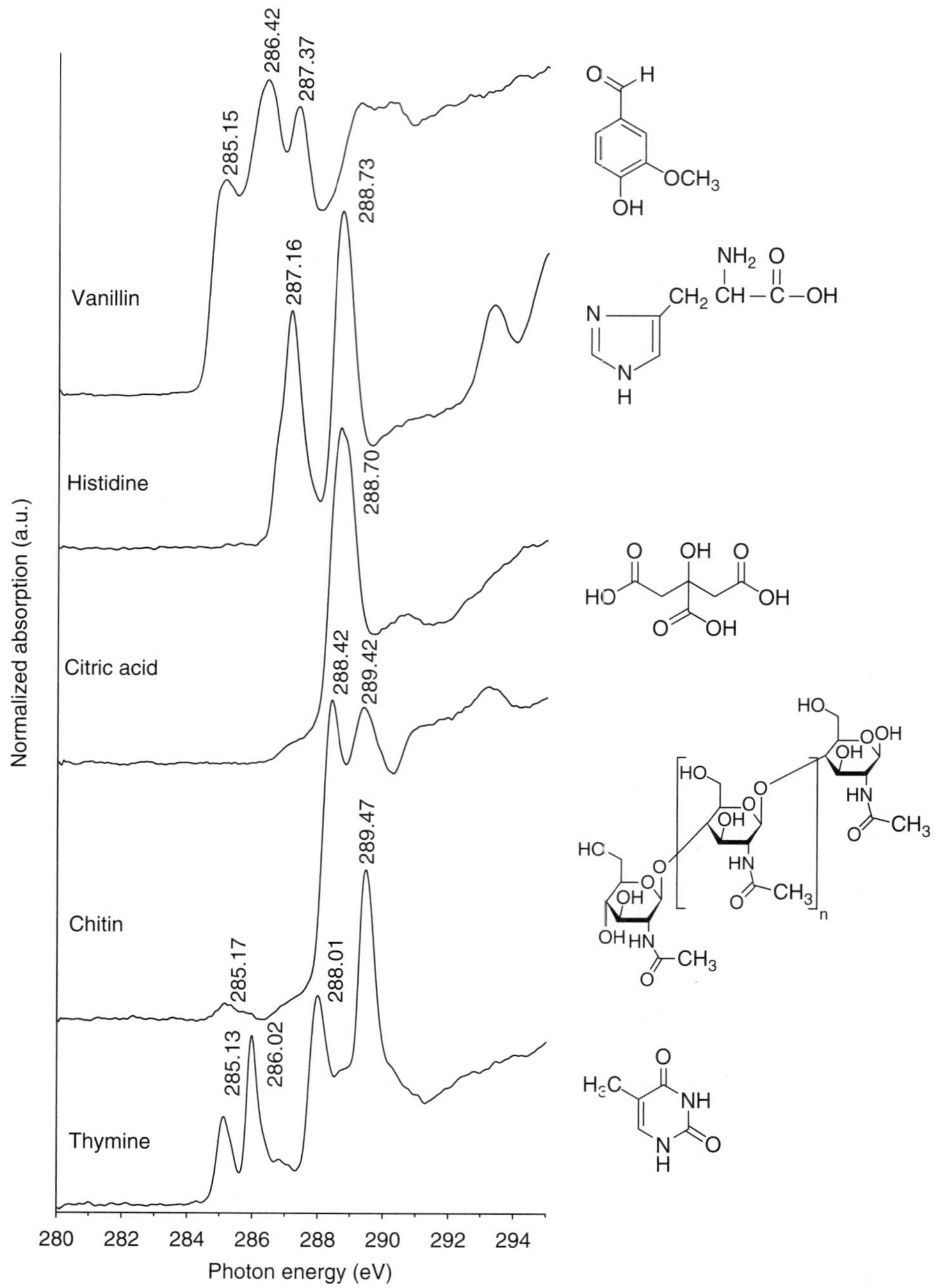

Figure 17.6. Experimental C (1*s*) NEXAFS spectra of reference organic C compounds using STXM (Solomon et al., 2009; a more comprehensive spectral library for biogeochemically relevant model organic C compounds is provided by these authors at http://knb.ecoinformatics.org/knb/metacat/datastar.50.3/knb). Energy levels are indicated above major peaks. Measurements were made at beamline X1A1 of the National Synchrotron Light Source, by adding dissolved standards to 100 nm thick silicon nitride windows, defocusing the beam to 10 µm, and moving the grating from 280 to 310 eV on a single spot with a 0.1 eV energy step and 120 msec dwell time. The monochromator was calibrated using the CO_2 adsorption band at 290.74 eV.

288.7 eV most likely derived from carboxyl carbons (lower energy levels indicating N-substitution for example at 288.2 eV). Histidine additionally shows a well-developed peak at 287.2 eV which may indicate aliphatic carbon but also π* features shifted to higher energy by nitrogen. Aromatic carbon around 285.0 eV typically has strong bands as shown for vanillin and thymine, as does oxygen- and nitrogen-substituted aromatic carbon at 286.6 eV, often attributed in NOM studies to lignin-derived phenol groups but possibly representing other carbon bonds in aromatic rings. Peak attribution is challenged by the complexity of NOM and overlapping energy regions of different bonds. Few model compounds relevant to NOM have been published to date, while much more information is available about biopolymers (e.g., database for gas-phase molecules by A. Hitchcock: http://unicorn.mcmaster.ca/corex/cedb-title.html). The study of such model compounds enables us to develop peak assignments and relate those to the chemical environment of the excited atom with its neighboring atoms (Tables 17.3–17.7). Assignments for peak positions for carbon are listed in Table 17.3. These are based on a range of studies and provide guidelines for interpretation of NEXAFS spectra. From this list, it can be seen that these broad peak assignments are not unique and still comprise very different types of compounds, depending for example on the chemical neighborhood of a specific bond (Hitchcock and Ishii, 1987). For example, De Stasio et al. (2005) showed a significant shift of the carboxyl peak of a peptide from 288.4 eV to 288.6 eV by binding to calcium. With more research and combination of carbon with oxygen and nitrogen or metal NEXAFS, it will be possible to obtain more refined information.

In Table 17.4, we present the relative energy position and structure of representative sulfur compounds relevant to NOM studies. Based on information gathered from the literature, the different oxidation states in NOM and the organic sulfur functional moieties associated with them fall in to three major groups: (i) C-bonded organic sulfur in its strongly reduced state (e.g., polysulfides, disulfides, thiols, mono-sulfides, and thiophenes), (ii) C-bonded organic sulfur in its intermediate oxidation state (e.g., sulfoxides and sulfonates), and (iii) organic sulfur in its strongly oxidized state (predominantly ester sulfates).

Tables 17.5–17.7 show literature assigned NEXAFS peak positions for nitrogen, oxygen, and phosphorus, respectively. The theory and application of NEXAFS measurements to studies of these three elements lags considerably behind that of carbon and sulfur. There have been a few notable studies of N-NEXAFS in NOM (Vairavamurthy and Wang, 2002; Jokic et al., 2004a b; Leinweber et al., 2007), and this technique represents perhaps in principle the best method for examining pyridine, pyridone, pyrrazole, or pyrrole nitrogen in NOM, because NMR is relatively insensitive to heterocyclic nitrogen (Smernik and Baldock, 2005). However, energy ranges for amide nitrogen significantly overlap with those of most nitrogen heterocycles, making a full distinction between amino groups and nitrogen-substituted heterocycles difficult without additional information about the sample or complementary analyses by other techniques. Presently, only spectral features with peak positions below 400.0 eV can be unambiguously interpreted as non-amide nitrogen such as pyridine nitrogen (six-ring heterocycles with one nitrogen atom). Possibly pyrrole nitrogen (five-ring heterocycles with one nitrogen) may be separately quantified once more information about NOM is available.

Oxygen is the least well explored common element in NOM, and O-NEXAFS studies suffer from a lack of a consistent effort in developing both a standards

TABLE 17.3. C (1s) NEXAFS Approximate Transition Energy Ranges and Assignments of Primary Absorption Peaks

Organic C Forms	Bond	Examples	Transition	Peak Energy (eV)
Aromatic C and quinone C	C=O	Quinone-type C Protonated and alkylated aromatic C Heteroatom substituted aromatic C	$1s$–π^*	283.0–284.5[g,k,q,t,v]
Aromatic C and double-bonded alkyl C	C=C	Protonated and alkylated aromatic C Carbonyl substituted aryl C Alkene C	$1s$–π^*	284.9–285.5[b,f,k,l,m,n,r,t,v]
Aromatic C with side chain hydroxylation and N-substituted aromatic C	C=C–OH C=O R–(C=O)–R′ C=N, C–N	Carbonyl C in aromatic ring Aromatic C attached to amide group Phenol C Aryl ethers and ketones Carbonyl C Pyrimidine C	$1s$–π^*	286.0–287.4[h,j,k,o,n,u,s,t,v]
Aliphatic C	C–H	Aliphatic C of CH_3, CH_2, and CH nature	$1s$–$3p/\sigma^*$	287.0–288.5[a,c,i,o,k,q,v]
Carboxylic C	R–COOH COO C=O (NH2)–C–O R–(NH2)–R′	Carboxylic C Carboxyamide C Carbonyl C	$1s$–π^*	288.0–288.7[c,d,k,o,q,s,u,t,v]
O-alkyl C	C–OH	Polysaccharides, alcohols, and other hydroxylated- and ether-linked C	$1s$–$\pi^*/1s3p/\sigma^*$	289.2–289.5[e,i,k,l,q,v]

Peak assignments were based on data collected by [a]Hitchcock et al. (1986), [b]Hitchcock et al. (1987), [c]Hitchcock and Ishii (1987), [d]Ishii and Hitchcock (1987), [e]Ishii and Hitchcock (1988), [f]Robin et al. (1988), [g]Francis and Hitchcock (1992), [h]Hitchcock et al. (1992), [i]Hitchcock and Mancini, (1994), [j]Boese et al. (1997), [k]Cody et al. (1996, 1998), [l]Scheinost et al. (2001), [m]Stöhr et al. (2001), [n]Ade and Urquhart (2002), [o]Kaznacheyev et al. (2002), [p]Jokic et al. (2003), [q]Schäfer et al. (2003, 2005), [r]Cooney and Urquhart (2004), [s]Hitchcock et al. (2005b), [t]Lehmann et al. (2005), [u]Schumacher et al. (2005), and [v]Solomon et al. (2005a, 2007, 2009).

TABLE 17.4. Relative Energy Position, Predicted Oxidation States and Structures of Representative Organic Sulfur Compound in NOM

Organic Sulfur Forms	Bond	Examples	Transition	Peak Energy (eV)
Most Reduced Sulfur				
Polysulfides	R–S–S–S–R′	Polyisoprene	$1s$–σ^*	2472.3–2476.0[b,c,d,e,f,g,h,i,j]
Disulfides	R–S–S–R′	Cystine, lipoic acid	$1s$–σ^*	
Thiols	R–S–H	Cysteine, coenzyme A	$1s$–σ^*	
Monosulfides	R–S–R′	Methionine	$1s$–σ^*	
Thiophenes	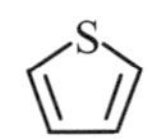	Benzothiophene	$1s$–π^*	2473.5–2473.8[e,g,h,i]
Intermediate Sulfur				
Sulfoxides	R-S(=O)-R'	DMSO, diethyl-sulfoxide, thiophene sulfoxide	$1s$–π^*	2477.0–2478.0[a,b,d,e,g,h,i]
Sulfonates	H-O-S(=O)(=O)-R	Alkylbenzene sulfonate, benzene sulfonate, sulfolipids, sulfoquinovose	$1s$–π^*	2479.0–2483.0[b,e,g,h,i]
Highly Oxidized Sulfur				
Sulfates	H-O-S(=O)(=O)-O-R	Chitin sulfate, arylsulfate esters, aromatic sulfate esters	$1s$–π^*	2481.5–2482.2[a,b,e,g,h,i]

Peak assignments were based on data collected by [a]Waldo et al. (1991a,b), [b]Vairavamurthy et al. (1993, 1997), [c]Morra et al. (1997), [d]Rompel et al. (1998), [e]Xia et al. (1998), [f]Prange et al. (1999), [g]Sarret et al. (1999), [h]Prietzel et al. (2003), [i]Solomon et al. (2003, 2005b), and [j]Szilagyi and Schwab (2005).

TABLE 17.5. Characteristic Energies of the Main Peaks in Nitrogen NEXAFS and Peak Assignments

Nitrogen Forms	Structure	Example	Transition	Peak Energy (eV)
Pyridine	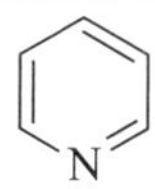	Quinoline 6-carboxylic acid, 6-hydroxyquinoline, acridine	$1s$–π^*	398.6–399.9[d,f,j,k]
Oxidized pyridine derivatives	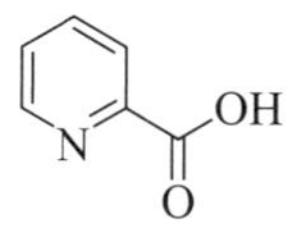	Pyridine 2–5-carboxylic acid, 2,3-pyridinedicarboxylic acid	$1s$–π^*	398.7–401.4[f,j]
Pyridone	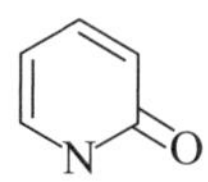	2-Hydroxypyridone	$1s$–π^*	401.0–401.9[f,j]
Nitrogen gas	N_2		$1s$–π^*	401.10[a]
Indol	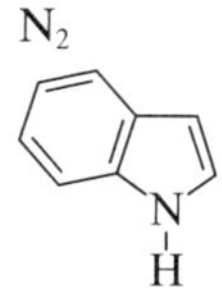	Tryptophane	$1s$–π^*	401.7[h]
Pyrrole		2-Phenylindole, pyrrole-2-carboxylic acid, carbazole	$1s$–π^*	401.1–403.2[d,e,f,j]
Imidazole	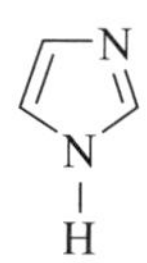	Histidine, imidazolelactic acid	$1s$–π^*	400.0–402.5[c,d,j]
Amide/amine	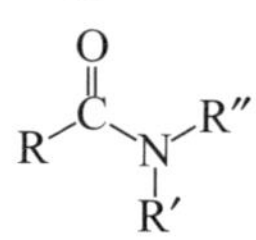 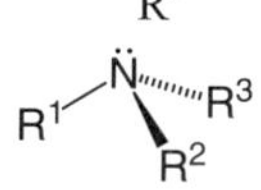	Glucosamine, arginine, histidine, lysine, alanine, proline, chitine	$1s$–π^*	400.7–402.5[b,f,i,j]
Pyrazine	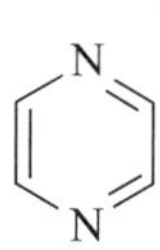	Pyrazinecarboxylic acid, pyrazinecarboxamide	$1s$–π^*	398.7–398.8[j,k]
Pyrimidine	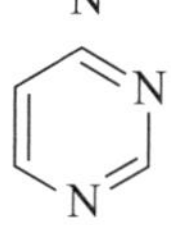	Pyrimidine-carbonitrile, cytosine, thymine, histidine	$1s$–π^*	398.8–402.2[j,k]
Imine	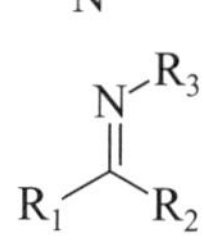	Arginine	$1s$–π^*	401.7[j]
Saturated Amine	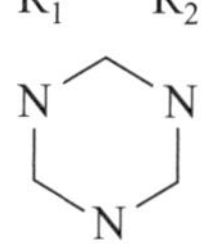	Tri-benzyl-hexa-hydro-triazine	$1s$–π^*	407.1[d]

TABLE 17.5. *Continued*

Nitrogen Forms	Structure	Example	Transition	Peak Energy (eV)
Nitrile	C≡N	Carbonitriles	$1s$–π^*	399.4–399.9[c,d,j]
Nitro compounds	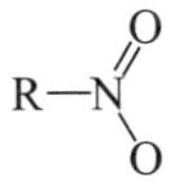	4-Nitrophenyl acetic acid, nitrophtalic acid	$1s$–π^*	403.8–404.8[f,j]
All amino compounds		Alanine, histidine, phenylalanine	$1s$–σ^*	405.5–409.7[g,i,j]

Peak positions were compiled from literature data collected by [a]Sodhi and Brion (1984), [b]Mitra-Kirtley et al. (1992), [c]Apen et al. (1993), [d]Mitra-Kirtley et al. (1993), [e]Hennig et al. (1996), [f]Vairavamurthy and Wang (2002), [g]Zubavichus et al. (2004), [h]Zubavichus et al. (2005), [i]Otero and Urquhart (2006), [j]Leinweber et al. (2007), and [k]Vall-Ilosera et al. (2008).

TABLE 17.6. O (1s) NEXAFS Approximate Transition Energy Ranges and Assignments of Primary Absorption Peaks

Organic O Forms	Bond	Examples	Transition	Peak Energy (eV)
Ketones, aldehydes	C=O, HC=O	Vanillin	$2s$–π^*	530.6–531.3[b]
Carboxylic acids, carbox-amides	COOH, CONH	Alanine, cysteine, hydroxamic acid	$2s$–π^*	532.0–532.7[a,b,c,e]
Alcohols	C–OH, C–N–O	Ethanol	$1s$–σ^*	534.1[e,g]
Ethers	C–O–C, O–C–N	Diethylether	$1s$–σ^*	535.4–535.6[e,g]
Water	H_2O		$1s$–σ^*	535,537–542[d,f]

Peak assignments were based on data collected by [a]Myneni (2002), [b]Urquhart and Ade (2002), [c]Zubavichus et al. (2004), [d]Cappa et al. (2005), [e]Edwards and Myneni (2006), [f]Takahashi et al. (2006), and [g]Cody et al. (2008).

TABLE 17.7. Peak Energies for P (1s) NEXAFS Approximate Transition Energy Ranges and Assignments of Primary Absorption Peaks

Organic P Forms	Bond	Examples	Transition	Peak Energy (eV)
Phosphines	(C)3–P, $R_{1\text{-}3}$–P	Tri-butyl phosphine, tri-naphtyl phosphine	$1s$–π^*	2145[b]
Phosphinic acids	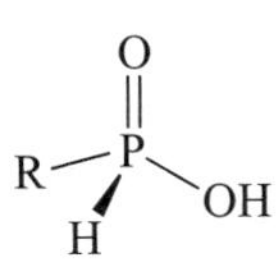	Phenylphosphinic acid	$1s$–π^*	2148.8[b]
Phosphonate	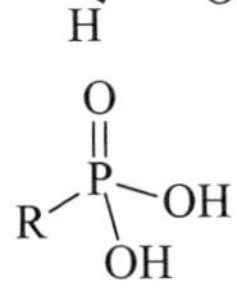	α-Aminoethylphosphonate	$1s$–π^*	2150–2151[a,b]
Polyphosphate	C–O–P–O–P–	ATP, ADP	$1s$–π^*	2152[b]
Phosphates	C–O–PO_3^{3-}	Phospholipids, RNA, DNA	$1s$–π^*	2152[b]

Peak assignments were based on data collected by [a]Myneni (2002) and [b]Brandes et al. (2007).

database and examples of standard NOM spectra. The best constrained peak in Table 17.6 is that for carboxylic and peptide oxygen, at 532 eV (Urquhart and Ade, 2002; Figure 17.7). The other peak assignments primarily come from data presented in Myneni (2002), and they remain unconfirmed and are contradicted in some cases. For example, Lawrence et al. (2003) list a lipid O-NEXAFS peak at 534 eV, which should according to Table 17.6 be due to alcohol groups. The presence of large water peaks on NOM can obscure the presence of ether moieties and can also preclude the use of combined carbon, nitrogen, and oxygen NEXAFS to measure NOM elemental ratios. However, the potential ability to characterize oxygen speciation in NOM at submicron scales, in tandem with carbon and other element NEXAFS studies on the same material, remains a strong incentive to develop this field. As noted above, carbon and sulfur NEXAFS typically have distinctive and well-spaced spectral features that aid in identification of NOM composition as shown for the Suwannee River NOM in Figure 17.7. Fewer peaks are present with oxygen and nitrogen NEXAFS because these elements are generally associated with carboxylic (O) and amide (N) moieties. However, note that in the N-NEXAFS spectra shown in Figure 17.7 a small peak at 399.1 eV is due to the presence of aromatic nitrogen.

Phosphorus NEXAFS has been used primarily to examine organic phosphorus in amended soils and animal manure (Peak et al., 2002; Beauchemin et al., 2003; Shober et al., 2006). The high phosphorus content of these samples allowed for the use of NEXAFS techniques, but phosphorus typically exists in very low concentrations within natural soils and sediments. With the advent of scanning fluorescence X-ray microscopes capable of conducting fluorescence (functionally equivalent to absorbance) spectroscopy, low phosphorus abundance samples can be examined (Brandes et al., 2007). Organic phosphorus X-ray spectra are very similar among different compounds, exhibiting a single peak at 2152 eV and a broader resonance in the 2165- to 2175-eV region (Beauchemin et al., 2003). There is some evidence to suggest that differences exist between mono- and poly-phosphorus compounds in the higher 2160-eV region (Brandes et al., 2007); but if spectral differences exist between phospholipids and other mono-esters, they will require instruments with greater energy resolution and stability than currently present. Currently, the most promising use for P-NEXAFS and fluorescence spectroscopy is in separating organic phosphorus from inorganic phases and in identifying the interactions of phosphorus with other elements. Phosphorus minerals and salts have distinctive secondary peak features that can identify elemental interactions on submicron scales. For example, the presence of small features at lower energy levels (to the left of the primary peak in Figure 17.7) are indicative for the presence of Fe- and Ca-associated phosphorus (see below). In the future, phosphorus L-edge in addition to K-edge NEXAFS may be useful to explore in NOM studies with high P concentration, because it is superior in differentiating chemical environments and may offer prospects for organic phosphorus speciation.

17.3.2. Quantification of Bonds and Compounds

The functional group chemistry of samples with unknown composition have been inferred from the spectral features of well-characterized reference compounds such as the ones shown in Figure 17.6 by comparing band height and shapes in a qualita-

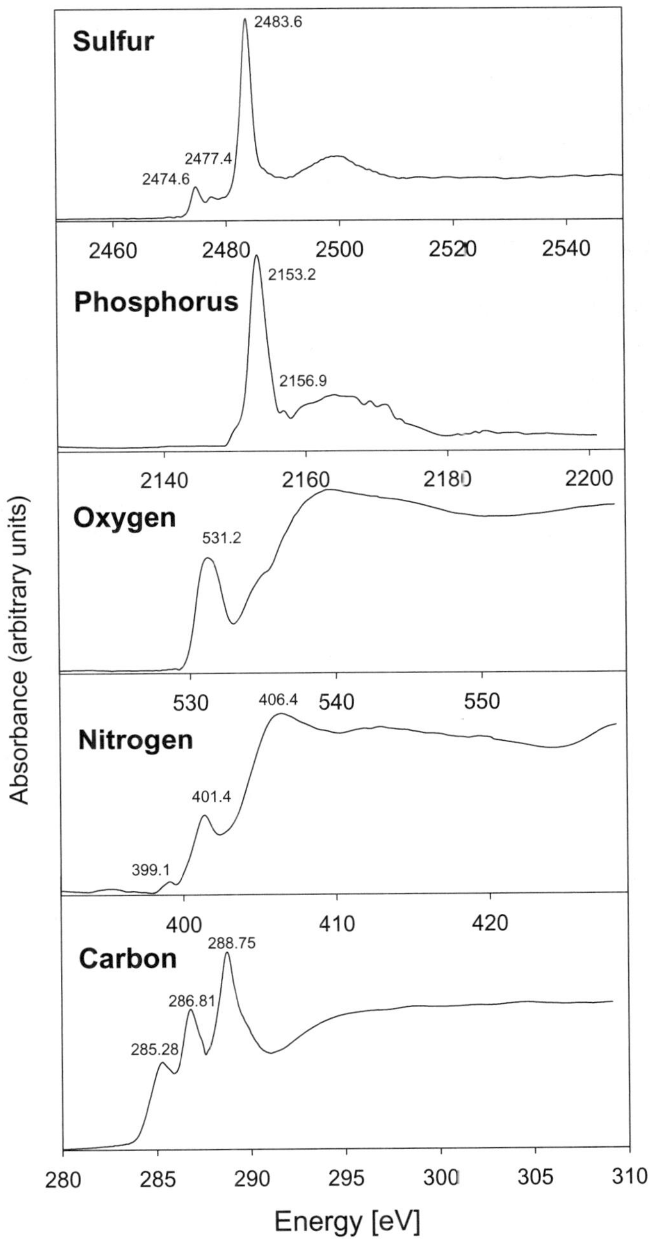

Figure 17.7. Carbon, nitrogen, oxygen, phosphorus, and sulfur near-edge fine structure of a standard humic substance obtained from the Suwannee River (IHSS) measured by fluorescence (carbon and oxygen, mounted on indium; phosphorus and sulfur, mounted in mylar bags) or by total electron yield (nitrogen, mounted on indium). Carbon, nitrogen, and oxygen measured using total electron yield with a dwell time of 200 msec and an exit slit of 50 µm, Canadian Light Source SGM beamline 11-ID.1. Phosphorus and sulfur measured by fluorescence yield at beamline X15B at the National Synchrotron Light Source (see Solomon et al. (2003) and Sato et al. (2005) for information on experimental details).

tive way. In addition, synchrotron-based NEXAFS spectroscopy has also been effectively employed to study structural composition of NOM, including semiquantitative approaches to speciate NOM composition (Scheinost et al., 2001; Gleber et al., 2003; Jokic et al., 2003; Schäfer et al., 2003, 2005; Solomon et al., 2003, 2005a, b, 2007a, b). Such approaches involve principal component analysis (Beauchemin et al., 2002), deconvolution (Scheinost et al., 2001), or least-squares linear combination fitting (Hutchison et al., 2001).

Spectral deconvolution uses a series of Gaussian or Lorentzian functions to describe pre-edge peaks and arctangents functions to describe the edge (Figure 17.2). Good knowledge about peak positions and identity of peaks are required to successfully deconvolute spectra. This has mostly been done for carbon and sulfur NEXAFS data in NOM studies (Scheinost et al., 2001; Solomon et al., 2003, 2005a, b). Least-squares linear combination fitting is often done for phosphorus NEXAFS spectra using a range of standards spectra that are fitted to the spectrum of the sample (Hutchison et al., 2001; Sato et al., 2005). Programs that can be used for quantification and for spectral transformation include Athena (Ravel and Newville, 2005), Peakfit (Systat Software Inc., San Jose, CA), Kaleidagraph (Synergy Software Co., Reading, PA), WinXAS (WinXAS Software, Hamburg, Germany), aXis2000 (http:\\unicorn.mcmaster.ca\axis2000.html), or Microsoft Excel Solver.

Energy calibration is crucial to obtaining reliable data for quantification. Slight shifts in peak positions can also be used for identification of chemical forms and are an important tool in interpreting ecological changes of NOM. For example, shifts in peak position were used to discern between oxidized black carbon and organic matter adsorbed to black carbon surfaces (Lehmann et al., 2005; Liang et al., 2006). Energy calibration is, for example, done using CO_2 for carbon [with characteristic π^*, $3s$, and $3p$ peaks at 290.7, 292.74, and 294.96 eV, respectively (Sham et al., 1989; Ma et al., 1991)] and oxygen [with characteristic π^* and $3s$ peaks at 535.0 and 539.3 eV, respectively (Sham et al., 1989), and 535.4 and 539.0 eV (Hitchcock and Ishii, 1987)], using N_2 for nitrogen [with a characteristic $1s \rightarrow \pi^*(v = 1)$ peak (second large peak) at 401.10 eV (Sodhi and Brion 1984)], using variscite [2149 eV (Beauchemin et al., 2003)] for phosphorus, and elemental sulfur [2872 eV (Hutchison et al., 2001; Solomon et al., 2003)] or calcium sulfate (2482.5 eV) for sulfur NEXAFS. Careful measurement of standard substances, explicit mentioning of the energy position to which a standard was calibrated is essential to working with and reporting NEXAFS results.

17.3.3. Spatial Analyses

In transmission experiments using STXM but also with fluorescence experiments, entire maps from sample regions can be obtained. Such maps can be taken at one or multiple energy levels, and several different opportunities exist to analyze these data. As described in Section 17.2.2 the absorption coefficient varies considerably between energy levels in the proximity of absorption edges. A simple way to probe for a known element—for example, carbon—is to take images below and above its K-edge. In Figures 17.8a and 17.8b the image was taken at an energy below and at an energy above, respectively, the carbon K-edge at 284.2 eV. The carbon-rich

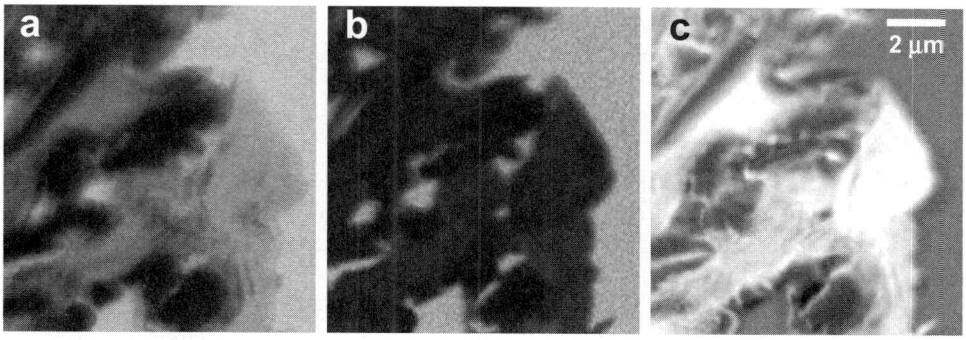

Figure 17.8. Optical density maps of a thin section from a topsoil obtained from the Arnot forest, Upstate New York. **(a)** STXM image below the carbon K-edge at 280 eV. **(b)** STXM image above the carbon K-edge at 310 eV. **(c)** Difference map (subtracting images at energy levels of 280–282 eV from images at energy levels of 308–310 eV). White color in (c) indicates areas of high carbon concentration. Black color in (a) indicates areas of minerals (J. Lehmann, unpublished data 2006, measured as described in Lehmann et al., 2007).

areas, which are little absorptive at 280 eV in image (a), show up dark at 310 eV in image (b). The carbon distribution in the sample can be visualized by forming a difference map. Image (c) in Figure 17.8 shows the difference between images from 308–310 eV and images from 280–282 eV. Carbon-rich areas show up as white in this image.

We can collect STXM images at many energy levels $n = 1, \ldots, N$ across the K-edge of a specific element with $p = 1, \ldots, P$ pixels each. In simple cases an open area in the sample can be chosen to extract the flux without sample during data acquisition and calculate the optical densities according to Eq. (17.3). In organo-mineral specimens, where open areas are rarely found, it may be appropriate to choose areas dominated by minerals instead. Kinyangi et al. (2006) found that by choosing mineral matter as I_0, features hidden in the original analysis appeared in the spectra as well as peaks that were better expressed.

With the appropriate incident flux, we can process our recorded data to form a matrix of optical densities $OD_{N\times P}$. In this matrix, the individual images of our data set are reformed in single lines P (while preserving the overall order of pixels in the images). The spectra at each of the pixels form rows N in the same matrix. If all spectra in the matrix $\mu_{N\times S}$ are known, the thickness maps $t_{S\times P}$ can be calculated via

$$t_{S\times P} = (\mu_{N\times S})^{-1} \cdot OD_{N\times P} \tag{17.2}$$

This is commonly not true for complex specimens. In the case of environmental samples, we often deal with mixtures of many NOM types, for which the complete set of spectra is a priori unknown. So other approaches have to be used, such as principal component analysis (PCA).

PCA finds a representation of a data set according to its most significant variations (Osanna and Jacobsen, 2000). The matrix of optical densities OD is factorized into a product of two matrices with a number of $m = 1, \ldots, M$ ($M \leq N$) abstract components:

$$\mathrm{OD}_{N\times P} = C_{N\times M} \cdot R_{M\times P} \tag{17.3}$$

where the columns of $C_{N\times M}$ are so-called eigenspectra and the rows of $R_{M\times P}$ eigenimages of our M components. The spectra in the columns of $C_{N\times M}$ are the eigenvectors and the eigenvalues found in the same eigenvalue equation are measures for the significance of the individual components.

It is important to note that the abstract components are not physical spectra present in our data set and therefore cannot be used to identify chemical components in the sample and the eigenimages are not their thickness maps. In fact the principal components may contain parts of various spectral signatures. Components with lower eigenvalues usually contain noise, which we eliminate by forming a reduced data set.

The following data were obtained from a soil sample collected from a Dystrochrept on bedrock and glacial till parent material in the Arnot Forest near Ithaca, New York. Figure 17.9 shows the eigenvalues of all abstract components. This plot suggests reducing our data set to the first five components, since their eigenvalues are significantly higher than those of the other components. Whether the eigenimages and eigenspectra of individual components contain valuable information or appear to be noise helps in deciding which subset of components to keep. Figure 17.10 shows eigenimages and eigenspectra of the first five components and the component number 20. It is obvious in this case that we want to keep the first five components, but not the 20th. The first component is an average of all spectra present and is typically a function of the sample thickness, which is of little interest

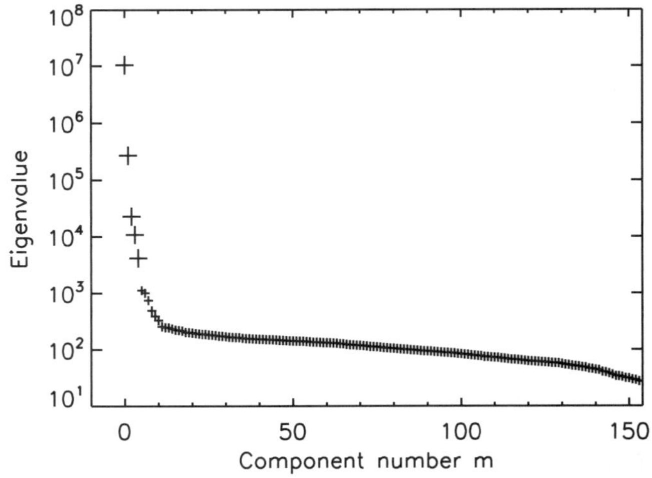

Figure 17.9. Eigenvalues for the Arnot forest data set. The first five eigenvalues (marked by larger symbols) are significantly higher than the rest. This suggests reducing our data set to five principal components (J. Lehmann, unpublished data 2006, measured as described in Lehmann et al., 2007).

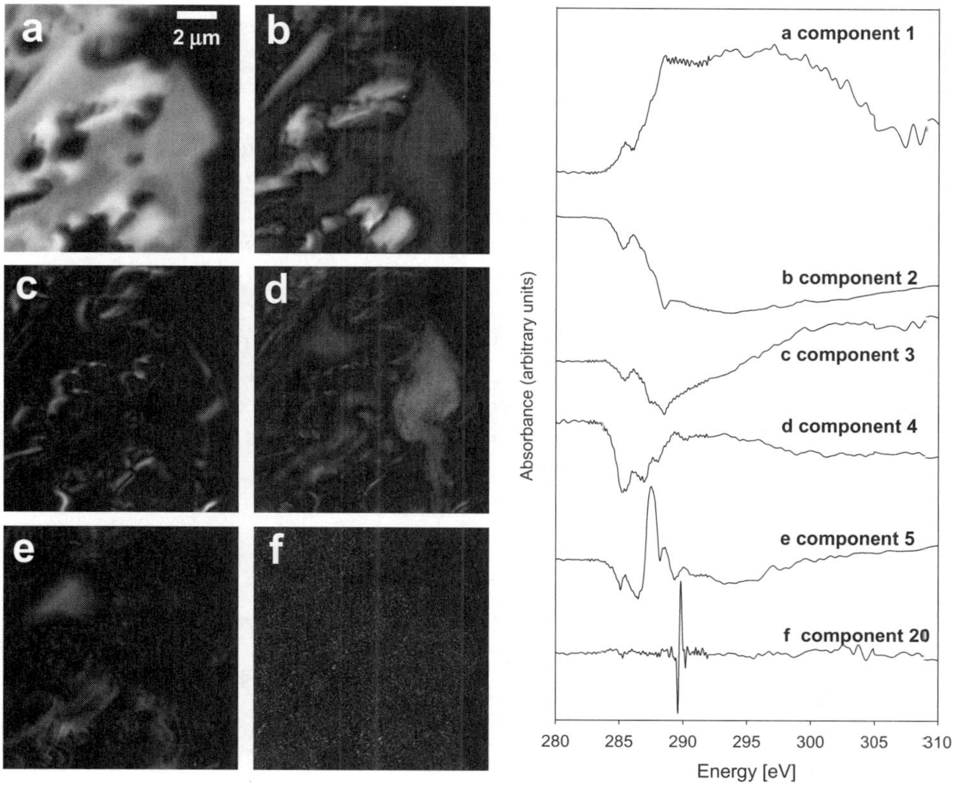

Figure 17.10. Eigenimages and eigenspectra for the first five components and component 20 of the Arnot Forest data set. Unlike the other shown components, image and spectrum of component 20 appears to be composed of noise (J. Lehmann, unpublished data 2006, measured as described in Lehmann et al., 2007).

for the analysis of environmental samples that have been sectioned. So it may be advisable to discard the first component as well.

Cluster analysis as an approach to find spatial groupings of the real absorption spectra has been used by Lerotic et al. (2004). Since computing time still is an issue in this method, a reduced data set from PCA is used. All data points are fanned out in a multidimensional space, with one dimension for each principal component. The algorithm then throws in a number of cluster centers at random positions in this multidimensional space and iteratively moves them closer to neighboring data points with similar spectral characteristics, until they finally mark an average of data points that have a similar component coefficients and therefore often similar chemical composition. The groups of data points can be shown in a color-coded map along with their average spectra as shown in Section 17.4.3. These maps then allow spatial associations to be discussed for example in the context of questions about organo-mineral interactions or the exudation and habitat of microorganisms in soils and sediments. In most analyses, the number of clusters would then be further reduced to decrease redundancies while taking care that minor but ecologically important clusters (such as a single microorganism in a larger aggregate section) are not grouped together in larger clusters.

One shortcoming of these cluster maps is that every image point is identified with one spectrum only—the average spectrum of all pixels in the cluster. An approach to overcome this limitation is to map thickness variations as shown by Lerotic et al. (2004). In this approach, the cluster spectra—and other spectra if known—can be used as target spectra to a singular value decomposition (SVD) calculation to find thickness maps (see Figure 17.11). These target maps show the distribution of organic species better than cluster maps, because they allow for characterization of the same areas by different spectral signatures. Since NOM in mineral assemblages is commonly present as both organic matter adsorbed to minerals as well as organic matter occluded in pore space, multiple characterization of small areas are advantageous. However, the set of target spectra will generally not be usable in a direct calculation of thicknesses as given in Eq. (17.2). The maps from our approach contain negative thicknesses (shown in red). It is therefore not possible to interpret these pseudo-thicknesses directly as physical thicknesses.

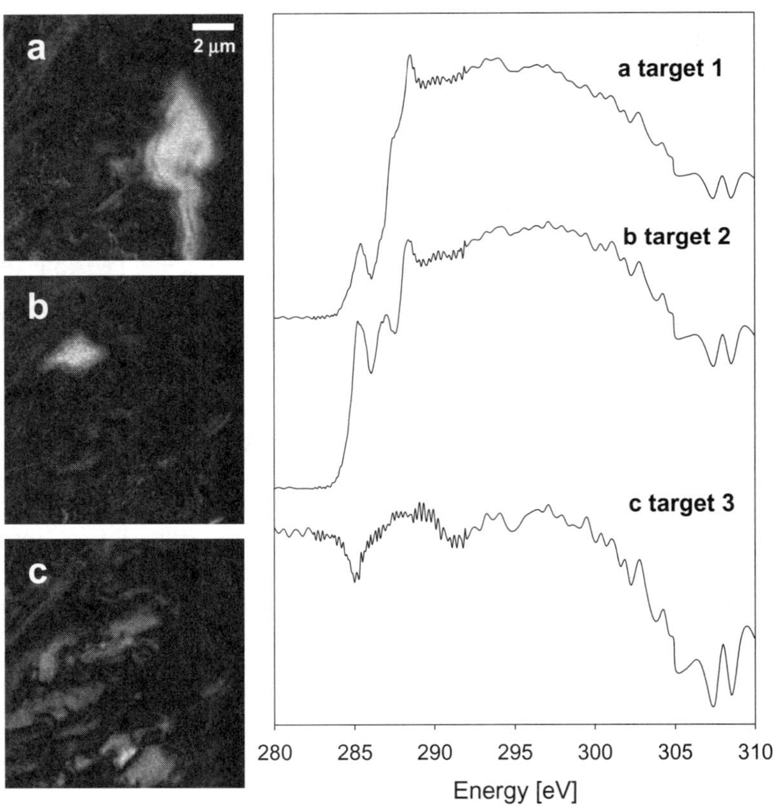

Figure 17.11. Pseudo-thickness (target) maps and associated spectra from singular value decomposition (J. Lehmann, unpublished data 2006, measured as described in Lehmann et al., 2007).

Another powerful application of SVD is that known spectra of measured compounds can be fed into the calculation to trace these compounds in the complex mixture. This procedure requires knowledge about the sample and its composition, but can in many cases be useful for identifying location of defined substrates.

As described above, the factorization of the data matrix by PCA does not lead to physical spectra or thickness maps. The abstract spectra and eigenimages for example can have negative values, while real spectra and thicknesses are strictly positive. And while the results of clustering are averages of measured absorption spectra, they generally arise from mixtures of several chemical species. What we are actually interested in is the factorization of our data into real spectra and thickness maps of individual chemical components. So-called non-negative matrix factorization (NMF) is an iterative algorithm to break up a matrix into a product of two factors under non-negativity constraint (Lee and Seung, 1999). The iteration is driven toward optimizing the value of a function that describes how far the computed factor product is from the actual data (Lee and Seung, 2001).

The correct number of chemical components in the sample is generally unknown, so that the iteration has to be started with a sufficiently large number of components which is reduced as computation progresses. Spectral components that are close to identical can be merged and their thicknesses are added, while components that are obviously noise can ultimately be ignored in further calculations.

Another improvement to the algorithm is curve smoothing, since absorption spectra of real chemical components generally have a smooth moving trend. Smoothing at each iteration reduces variations in the spectra, but a number of cycles without smoothing can be added to the end of a calculation to restore peaks to their full height. Measurement and photon noise in the data set can make it hard for the algorithm to converge. Using PCA to find a noise reduced version of our data can help to find a better factorization.

In Figure 17.12 we see four NMF thickness maps and spectra found in a highly weathered soil from a forest site in Kenya. While especially the regions found in the first shown component seem to be of special interest, the spectra need to be evaluated with caution. It is often unclear if the NMF algorithm found the global extremum of the cost function or did not move beyond a local extremum. The spectra found may not be close to reality. While the results can be better in many other cases, it is always worthwhile to compare them with the results of other methods or verify against additional information one might have.

17.4. COMPOSITION OF NATURAL ORGANIC MATTER IN THE ENVIRONMENT

17.4.1. Natural Organic Matter Properties in the Environment

NEXAFS is a useful tool for comparative studies of NOM in the environment. Spectral properties show characteristic fingerprints of organic matter types under investigation. For example, spectra of black carbon exhibit characteristic aromatic

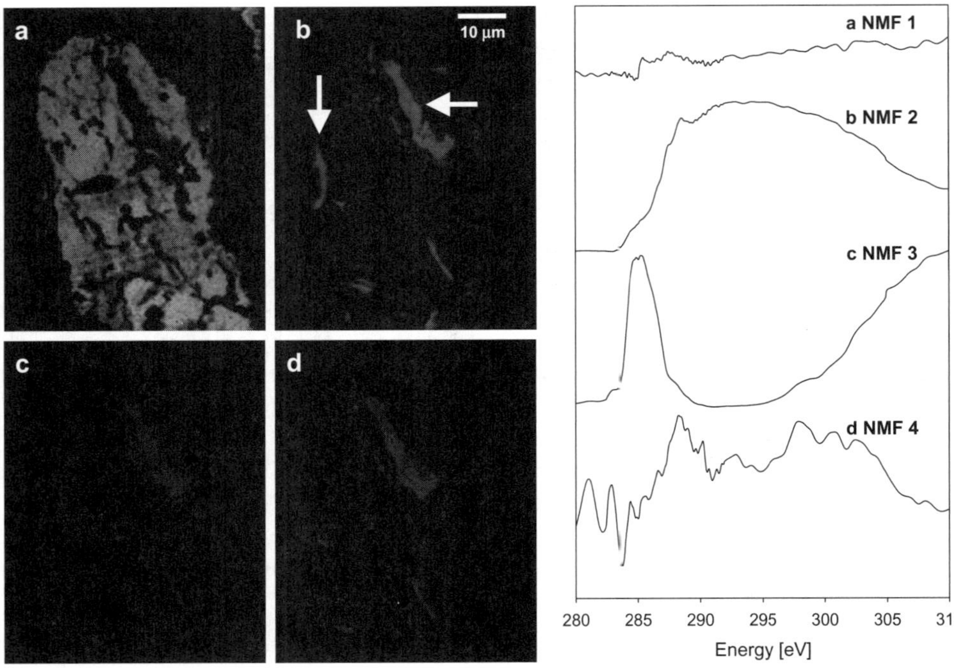

Figure 17.12. Four components found by NMF in an oxisol obtained from a forest site in Western Kenya (J. Lehmann, unpublished data 2006, for site description see Kinyangi et al., 2006). **(a)** Mineral matter with low contents of organic carbon; **(b)** organic carbon dominated by aliphatic and carboxylic forms; **(c)** organic carbon dominated by aromatic forms; **(d)** organic carbon dominated by carboxylic forms. Arrows in map (b) point to carbon features that share structures characterized by spectrum (b), and the feature at the horizontal arrow also contains aromatic carbon in contrast to the feature at vertical arrow.

carbon features with peak positions below 285 eV (Figure 17.13). In comparison, peak height in the aromatic region obtained from a leaf litter sample containing lignin structures was much lower and at slightly higher energy level (Figure 17.13). Spectral signatures of bacteria are dominated by carboxyl (possibly lipid-rich) or carboxamide carbon at energy levels of 288.6 and 288.2 eV (Hitchcock et al., 2005a,b), respectively, which is also characteristic for bacterial exudates (Lawrence et al., 2003). Spectral features vary significantly between these components of NOM and can be used not only in quantification of NOM properties, but also in identification of different regions within the small-scale assemblage of NOM when spectromicroscopy is used in conjunction with NEXAFS.

Despite these large differences in spectral properties between components of NOM, the spectral properties of C (1*s*) NEXAFS for humic substances extracts from different grasslands and forests soils were remarkably similar (Figure 17.14). They showed peaks at 285.2 eV indicating aromatic carbon, at 286.7 eV indicating phenolic carbon, and at 288.6 eV indicating carboxyl C. These peak positions were also found in the river NOM of the IHSS standard (Figure 17.7 in Section 17.3.1) and have been reported for a wide variety of NOM (Rothe et al., 2000; Scheinost et al., 2001; Schäfer et al., 2003). Also, a significant decline in total soil organic matter

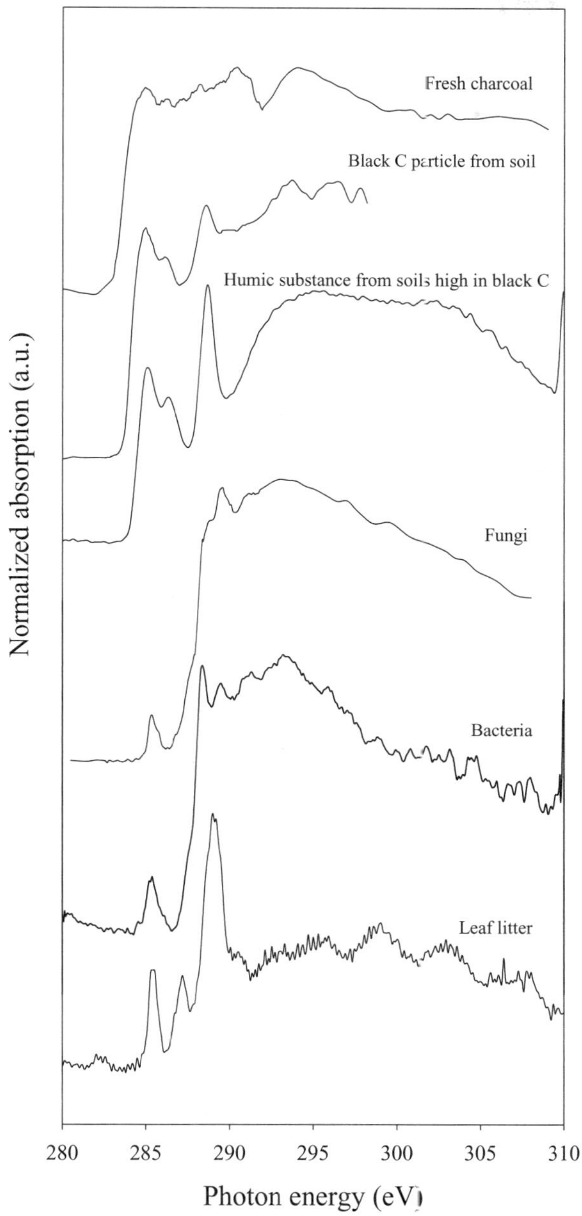

Figure 17.13. Stacked C (1*s*) NEXAFS spectra of NOM from various sources and environments showing molecular-level structural and compositional heterogeneity. The spectra were recorded in transmission mode (fungi, bacteria, fresh charcoal, black C particle from Liang et al., 2006; black-C rich humic substance from Solomon et al., 2007b; litter: J. Lehmann, unpubl. data).

from more than 10% to about 1% did not change the peak positions in a chronosequence study in Western Kenya (Solomon et al., 2007a). The positions of these three dominant peaks appear to be constant across a wide climatic, edaphic, and degradational gradient.

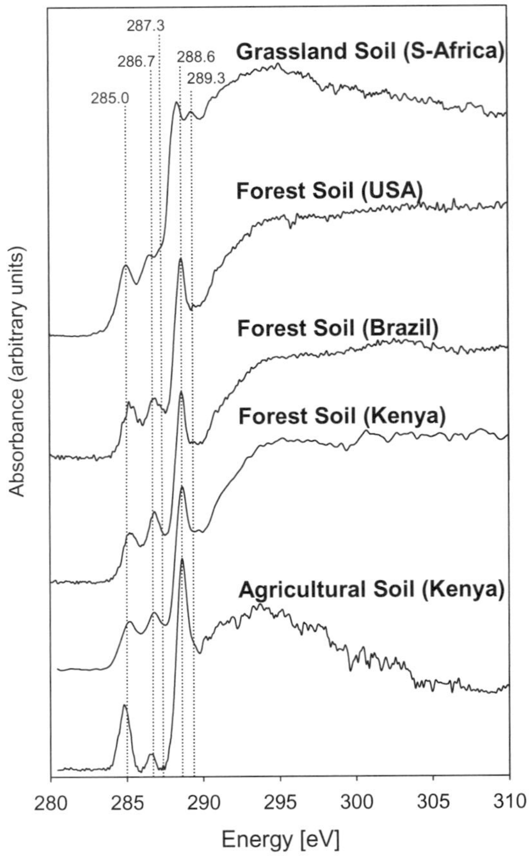

Figure 17.14. Carbon (1*s*) NEXAFS from STXM of humic substance extracts obtained from different soils. Kenyan and South African soils are from Solomon et al. (2007a); soils from the United States and Brazil are described in Lehmann et al. (2007).

The relative peak heights were almost identical between different forest soils (Figure 17.14). The available data from grassland soils may indicate a slightly lower proportion of phenolic carbon compounds (Figure 17.14). With progressive degradation, Solomon et al. (2007a) found a clear relative decrease in phenolic carbon and increase in aromatic carbon as well as the most oxidized carbon forms (compare agricultural and forest soil from Kenya in Figure 17.14). A greater oxidation through longer agricultural activity was also observed for organic sulfur forms in grassland soils from South Africa (Solomon et al., 2005b). The increase in aromatic features at the K-edge of carbon is related to its recalcitrance and could indicate black carbon residues.

Extraction of humic substances has been a useful way of characterizing NOM properties by NEXAFS. In those cases where the element studied also has significant inorganic fractions such as sulfur in submerged soils and sediments, or phosphorus in most mineral soils and sediments, an extract appears to be the only approach to obtain meaningful information about NOM. Often, it was found that the spectra have a lower background and can be more easily analyzed quantitatively for example by deconvolution (Solomon et al., 2003). It has to be recognized, however, that humic substances extractions such as those using NaOH typically extract only a portion of total NOM, with values ranging widely from 20–80%. An extraction may also not necessarily capture a representative portion of total NOM, because stable compounds such as strongly adsorbed NOM or black carbon may not dissolve to the same extent as more readily available NOM forms. It can therefore be desirable to analyze the soil without prior extraction. However, direct measurements of the entire soil will necessarily capture inorganic species in addition to carbon and nutrients in NOM. This is not a constraint for carbon in noncalcareous soils and for nitrogen or sulfur in well-drained soils, but will include inorganic species of phosphorus in all mineral soils and sediments and of sulfur in sediments and water-logged or submerged soils. In such cases, measurements of humic substances extracts may in fact be preferable to obtain information about organic forms only.

Terrestrial humic substances showed greater proportions of reduced organic sulfur forms compared to humic substance extracted from mineral and organic soils (Xia et al., 1998). A generally greater amount of reduced organic sulfur was also found in temperate forest soils (Prietzel et al., 2003) or wetlands (Jokic et al., 2003) than in well-drained agricultural soil (Solomon et al., 2003, 2005b). Forms of organic sulfur appear to react significantly to the oxygen environment, with the lowest degree of oxidation in submerged soils and the highest in agricultural soils (Zhao et al., 2006). Yet these changes do not seem to occur over very short time frames as shown by aeration experiments (Hutchison et al., 2001). It should be noted, however, that a significant amount of ester sulfate can even be found in marine sediments (Vairavamurthy et al., 1997).

Different physical fractions of soils showed clear trends in NOM speciation. Clay size separates ($<2\,\mu m$) had more reduced organic sulfur (12–28% of its total organic sulfur contents) than did silt size fractions (2–20 μm) (7–15%) (Solomon et al., 2003) and had more aromatic carbon (11–16% of total organic carbon) than did silt size separates (6–8%) (Solomon et al., 2007a) regardless of vegetation cover. Density separates obtained from a Kenyan Hapludox showed little difference in peak heights at the carbon K-edge between free light, intra-aggregate light, and organo-mineral fractions (NaI at $1.8\,g\,cm^{-3}$; Lehmann et al., unpublished data 2007). However, we observed a significant peak shift from 286.7 eV to 287.3 eV from heavy to light fractions. Such a shift may indicate a greater contribution of aliphatic carbon in plant litter and more phenolic carbon in organic matter associated with mineral surfaces. Such spectral differences between density fractions were much less pronounced than typically shown by NMR (Sohi et al., 2001).

Sulfur K-edge NEXAFS also detected significant differences between oxidation states in hydrophobic and hydrophilic organic matter (Hundal et al., 2000). Oxidized organic sulfur forms were dominant in the hydrophilic fraction, whereas reduced sulfur forms were dominant in the hydrophobic fraction.

Information about nitrogen is still very scarce, and only a few works have been published so far on NOM (Vairavamurthy and Wang, 2002; Jokic et al., 2004a,b). The primary challenges here are the lack of a standards database and the low abundance of nitrogen in NOM, which results in poor signal-to-noise ratios of spectra. Research on nitrogen using NEXAFS is at present mainly concerned with characterization of organic nitrogen. The incentive for employing nitrogen K-edge spectroscopy is the opportunity to investigate the proportion of heterocyclic organic nitrogen, which is difficult to assess with other spectroscopic techniques such as NMR (Smernik and Baldock, 2005).

Similarly, few studies of oxygen NEXAFS in natural samples have been conducted. Here there have been several high-quality standards papers (Ade et al., 1997) as well as a limited amount of studies of biofilms and extraterrestrial organic matter (Flynn et al., 2003; Lawrence et al., 2003). For NOM, one difficulty with oxygen NEXAFS studies stems from the presence of water, which generates a large peak in the 534- to 544-eV region. This water peak can obscure features in this region, especially those associated with carbonates, and precludes the use of NEXAFS to generate C:N:O elemental ratios. However, several diagnostic features in the 530- to 540-eV range have been identified (Ade et al., 1997) and can be used to verify features observed in carbon K-edge NEXAFS spectra that may have ambiguous interpretations. For example, a sharp peak at 286.5 eV can be either phenolic carbon, ketone, aldehyde, or cyano group (Lawrence et al., 2003).

Phosphorous NEXAFS studies have concentrated on soil and animal litter samples with high levels of phosphorus (Peak et al., 2002). While organic and phosphate phosphorus (V) compounds have very similar spectra and are thus difficult to differentiate, P NEXAFS of mineral phases have distinctive secondary features and reduced phosphorus compounds such as phosphonates have 1- to 5-eV shifts in primary peak positions (Figure 17.15). Additionally, iron, magnesium, and several other alkaline earth or transition metal–phosphorus interactions generate a distinctive pre-edge feature in NOM (see Figure 17.7), and calcium phosphates exhibit a strong shoulder above the primary absorption or fluorescence peak (Figure 17.15).

17.4.2. Relationship with Other Methods

In addition to NEXAFS, several approaches using modern and fairly effective analytical techniques have been available for the characterization of NOM. For organic carbon, the techniques involve the use of (i) chemolysis (Kögel-Knabner, 1995; Zhang and Amelung, 1996; Solomon et al., 2002), (ii) thermochemolysis (del Rio et al., 1998; Chefetz et al., 2002), and (iii) pyrolysis (Saiz-Jimenez, 1994; Leinweber and Schulten, 1998), as well as advanced *ex situ* invasive spectroscopy techniques such as X-ray photoelectron (XPS), AES (Auger electron spectroscopy), and SIMS (secondary ion mass) spectroscopy (McKeague and Wang, 1980; McHardy and Robertson, 1983; Yuan et al., 1998) to obtain information about the chemical composition, microheterogeneity, and physical location of organic and mineral materials in soils (Kögel-Knabner, 2000; Scheinost et al., 2001). The disadvantage of these methods is that they often involve invasive oxidative treatments or other degradative wet-chemical techniques or the experiments must be performed under adverse experimental conditions (e.g., sample drying, ultrahigh vacuum, heating, or particle bombardment). Such conditions do not simulate most natural conditions in the environment. They can also alter the nature of samples, yielding misleading data as a result of experimental artifacts (Sparks, 2003). Such limitations have led to the use of noninvasive *in situ* magnetic [solid-state nuclear magnetic resonance (NMR)] and vibrational [conventional and synchrotron-based Fourier transform infrared (FTIR)]

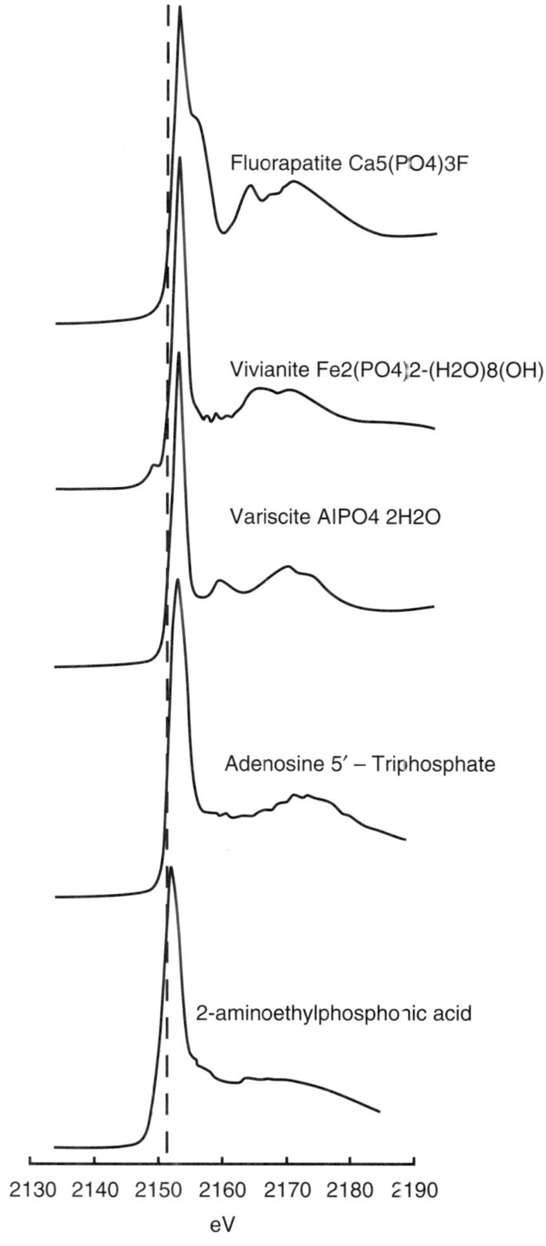

Figure 17.15. Phosphorus (1*s*) fluorescence spectra from chemical and mineral standards. The dashed line represents the edge position (2152 eV) of inorganic phosphorus (V) compounds (redrawn after Brandes et al., 2007).

spectroscopy techniques. However, while the rich spectral features from these techniques compared to NEXAFS enabled researchers to get detailed information about the chemical composition of NOM, they offer no information about the micro-scale lateral distribution, heterogeneity, and physical allocation of NOM on soil aggregates, mineral surfaces, or sediments. Thus it is not possible to decide whether the organic compounds are in a continuous layer of dispersed coatings or accumulated in distinct patches or microsites (Ransom et al., 1997; Mayer, 1999). Moreover, while these techniques provide atomic-level information, they do not always provide precise information about the local structure of a sorbed species. Synchrotron-based STXM can circumvent these problems.

Despite the wide range of origin and chemical heterogeneity of the samples investigated, several studies (Scheinost et al., 2001; Schäfer et al., 2003; Solomon et al., 2005a, 2007a) showed that the semiquantitative results generated from C (1*s*) NEXAFS spectra compared very well with the results of the more established NOM characterization techniques such as ^{13}C NMR (Figure 17.16) and FTIR spectroscopy. Despite clear evidence that the different spectroscopic techniques used in these investigations have variable degrees of sensitivity for the different organic carbon functionalities, Figure 17.16 shows that the characteristic band positions of NEXAFS spectra were sufficiently separated to allow discrimination and quantification of organic carbon functional groups from heterogeneous NOM sources. The goodness of fit shown in Figure 17.16 are lower than the values reported for aliphatic C (correlation coefficient $r = 0.99$), aromatic C ($r = 0.95$), phenolic C ($r = 0.99$), and carbonyl C ($r = 0.98$) by Schäfer et al. (2003) for four fulvic acid samples extracted from ground water, but compare very well with the values reported from six humic acid, fulvic acid and NOM samples by Schumacher et al. (2005). Similarly, Solomon et al. (2007a) observed good correlation coefficients between the organic C functional groups identified by C (1*s*) NEXAFS and Sr-FTIR-ATR spectroscopy ($r = 0.70$ for aromatic C, $r = 0.45$ for phenolic C, $r = 0.42$ for aliphatic C, $r = 0.67$ for carboxylic C, and $r = 0.62$ for *O*-alkyl C, respectively, for humic substances extracted from soil samples collected from undisturbed tropical forest and subtropical grassland ecosystems; $N = 17$), indicating the complementary nature of these spectroscopic techniques in investigations involving the speciation and structural chemistry of NOM and their usefulness in providing process-oriented data for soil carbon and ecosystem models.

Although both HI-reduction (Freney et al., 1975; Neptune et al., 1975) and S NEXAFS spectroscopy (Morra et al., 1997; Xia et al., 1998) were used in the past to speciate sulfur in environmental and geochemical samples, there are only few studies that attempt to compare the conventional wet-chemical technique with NEXAFS spectroscopy in soils and humic substances. Recently, Hundal et al. (2000) found that the relative proportions of reduced and oxidized sulfur species determined by selective dissolution and by NEXAFS spectroscopy were of similar magnitude in biosolid-derived fulvic acid, and they concluded that the two sulfur-speciation methods are in close agreement. Prietzel et al. (2003) reported that the average proportion of reduced organic sulfur analyzed by NEXAFS agreed with that obtained by wet-chemical analysis for organic horizons collected from Schluchsee, Germany. However, these authors did not observe this similarity for organic horizons sampled from Rotherdbach, Germany. Similarly, Solomon et al. (2003, 2005b) reported a poor relationship between the relative proportions of organic sulfur species identified by NEXAFS spectroscopy in the humic substances and by degradative wet-chemical fractionation techniques of the bulk soils (ester SO_4–S as revealed by NEXAFS versus HI fractionation, $r = 0.27$; $N = 9$). Solomon et al. (2005b) showed that the proportion of ester SO_4–S measured by wet-chemical

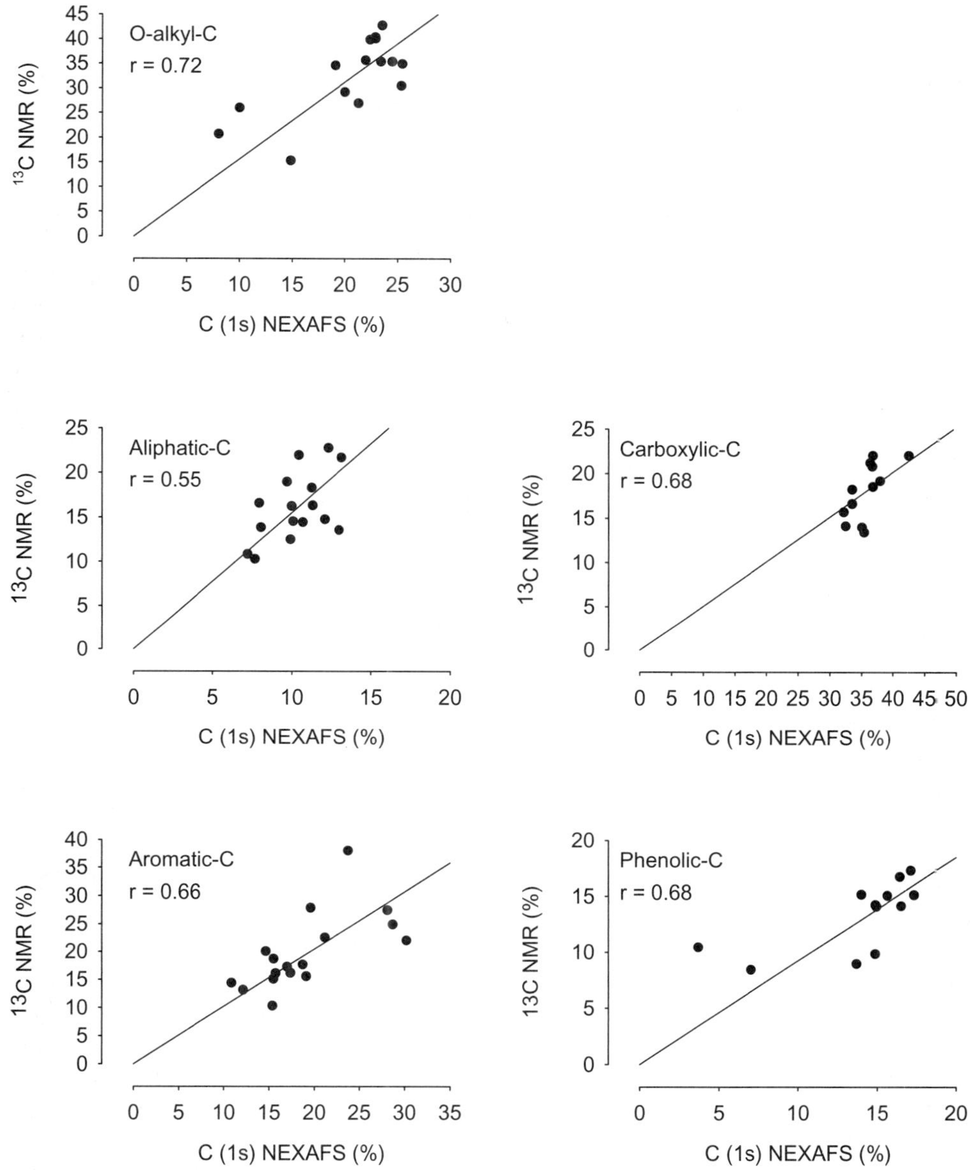

Figure 17.16. Correlation plots of organic C functional groups as a fraction of total carbon (in %) identified by C (1s) NEXAFS and ^{13}C NMR spectroscopy of humic substances from native and less disturbed tropical and subtropical agroecosystems [N = 17, P < 0.05; data recalculated from Solomon et al. (2005a, 2007a) and D. Solomon et al. (unpublished data 2007)].

analysis from the bulk soils (11–30%) was generally less than the proportion of ester SO_4–S determined by NEXAFS spectroscopy (39–55%). In order to compare the two sulfur speciation techniques on a similar matrix, these authors also determined the proportions of ester SO_4–S and C-bonded sulfur in the humic substances extracted from the bulk soils using the degradative wet-chemical technique. The results showed that C-bonded sulfur was the major form of organic sulfur associated with the humic substance extracts, representing 58–72% of the total organic sulfur. The proportion of ester SO_4–S accounted for 28–42% of the total organic sulfur. As for the bulk soils, the proportion of ester SO_4–S measured by wet-chemical reduction from the humic substances did not correlate significantly with the proportion of ester SO_4–S identified by S NEXAFS spectroscopy (ester SO_4–S as revealed by NEXAFS spectroscopy versus HI fractionation, $r = 0.39$). These results agree, in turn, with our previous findings using both differential sulfur reduction and NEXAFS spectroscopy (Solomon et al., 2003), where the solid-state spectroscopic technique determined a larger proportion of strongly oxidized sulfur species compared to wet-chemical fractionation. The discrepancy in these sets of values could be caused by limitations in the comparability of the two sulfur speciation techniques for different subgroups of organic sulfur fractions. Prietzel et al. (2003) and Solomon et al. (2003) suggested that the wet-chemical technique relies on the differential reduction of organic sulfur compounds to H_2S and thus it might not recover all the organic sulfur fractions compared to NEXAFS, which is a more direct and nondestructive technique. The shortcoming of sulfur K-edge NEXAFS spectroscopy, however, is its inability to discriminate precisely between soil organic sulfur forms with an oxidation state of 0 to +1 (polysulfides, disulfides, thiols, monosulfides, and thiophenes). Sulfur L-edge NEXAFS spectroscopy could provide complementary information to identify these sulfur moieties (Kasrai et al., 1996), but this technique only works well for samples that have large amounts of sulfur such as in coal (M. Kasrai, personal communication), and our attempt to resolve the organic sulfur species in humic substance extracts using S L-edge NEXAFS spectroscopy did not provide satisfactory results (D. Solomon, unpublished data 2005).

While NEXAFS studies of organic phosphorus are rare (Toor et al., 2005; Sato et al., 2005), Beauchemin et al. (2003) and Lombi et al. (2006) demonstrated that P K-edge spectroscopy has great potential to investigate inorganic phosphorus species in long-term-fertilized, P-rich soils differing in pH, clay, and organic matter. In an experiment conducted in conjunction with wet chemical techniques to characterize the dominant solid-phase species of phosphorus, Beauchemin et al. (2003) also showed that the results obtained by NEXAFS spectroscopy for calcium ($r = 0.87$)-related and aluminum or iron ($r = 0.99$)-related phosphorus species correlated very well with the HCl extractable and NaOH extractable inorganic phosphorus species obtained from sequential phosphorus fractionation technique, respectively. Although chemical fractionation results indicated that some soil samples contained up to 26% of total phosphorus as organic phosphorus ($NaHCO_3$–Po and NaOH–Po), none of the organic phosphorus standards included in the fitting of NEXAFS spectra were able to explain the variation in the spectra recorded by these authors. This result may be partly explained by the absence of strong and unique spectral features in the spectrum of certain organic phosphorus species such as inositol monophosphate, which is considered the most important fraction of organic phosphorus in soils (Harrison, 1987), or by the presence of organic phosphorus species at concentrations below detection, despite a unique spectral feature. The detection limit of the technique was not yet tested using carefully controlled standard mixtures. Beauchemin et al. (2003) suggested that organic phosphorus species can be detected if they represent more than 10–15% of total phosphorus and if they additionally have a spectrum that is unique from other standards. For this reason, these authors suggested that other complementary techniques such as NMR spectroscopy (in iron depleted samples) might prove to be better suited for direct soil organic phosphorus speciation than NEXAFS spectroscopy.

Nitrogen NEXAFS spectroscopy method avoids the use of more traditional wet-chemistry methods for organic nitrogen speciation in soils where the material is treated with hot mineral acids/bases to liberate nitrogenous constituents, thereby primarily identifying the amino compounds. These compounds have been grouped as acid-insoluble nitrogen, NH_3–N, amino acid nitrogen, amino sugar nitrogen, and hydrolyzable unknown nitrogen (Sowden et al., 1977). NEXAFS is able to minimize the possible artifactual formation of heterocyclic organic nitrogen at high temperature in methods such as analytical pyrolysis by dissociation or rearrangement of the original structures (Vairavamurthy and Wang, 2002). In addition, NEXAFS is expected to provide information complementary to that provided by nondestructive techniques such as NMR which have inadequate sensitivity due to low abundance of ^{15}N in NOM. At the present stage, no data have been published which directly relate NEXAFS with other techniques for NOM. Relationship between oxygen spectra and other techniques have also not been attempted.

17.4.3. Spatial Distribution of Natural Organic Matter in Aggregates and Colloids

The combination of scanning X-ray microscopy with NEXAFS offers the possibility to map the spatial association between NOM and mineral matter and between NOM compounds on a very fine spatial scale. This requires that the preparation of the specimen preserves the spatial assemblage. Suitable methods for sample preparation are described in Section 17.2.4.

Organic matter is generated with nanoscale heterogeneity by bacteria, autotrophs, and heterotrophs. As diagenesis progresses, evidence suggests that such heterogeneity is not lost and may in fact play a role in NOM preservation (Lee et al., 2004). Consider the example of North Sea suspended particulate marine NOM shown in Figure 17.17. Although there are no intact organisms visible, NOM in the sample has a high amount of structure and chemical heterogeneity. Domains containing both protein-like (in yellow), lipid-like (red), and carbohydrate-like OM are present and distinctive. Both associations with mineral matter and encapsulation of hydrophobic mineral phases are observable in this sample. This high degree of heterogeneity has also been observed in deep sea NOM (Brandes et al., 2004).

Similarly, organic carbon in an acid forest soil from upstate New York (Figure 17.18A) was unevenly distributed throughout an approximately 30-µm microaggregate. This confirms earlier results for three forest soils from the United States, Kenya, and Brazil (Lehmann et al., 2007). A detailed observation of a portion of the microaggregate showed an even greater variation in NOM forms which was visualized by principal component and cluster analysis (Figure 17.18D). The carbon properties ranged from highly oxidized (e.g., cluster 8) or aliphatic (e.g., cluster 2) to highly aromatic (clusters 14 and 19, Figure 17.19).

The number of clusters obtained depends on the inherent variability. In the example shown here, a relatively uncommon spectral property still provided ecologically useful information as seen from cluster 19.

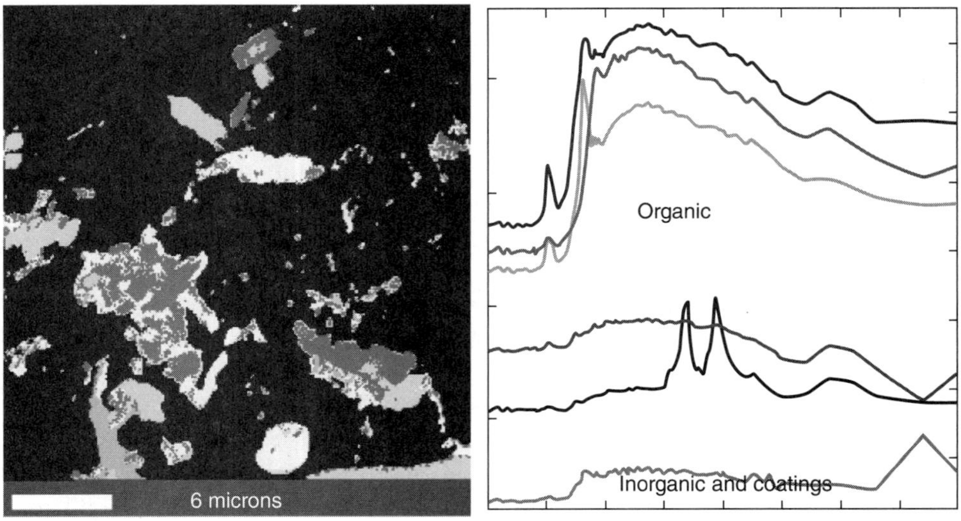

Figure 17.17. Principal component analysis map of sample **(left)** and color-coded spectra **(right)** from a sample of marine suspended particulate matter. The lower three spectra are characteristic of low organic mineral phases, while the upper three organic phases have distinctively different C-NEXAFS spectra. Background regions are shown in black (J. Brandes, unpublished data 2007). See color insert.

Typical acquisition areas range from ~1 to ~$10^4\,\mu m^2$. Such areas are limited by (i) the microscope stage, (ii) the chosen spatial and energy resolution, (iii) the dwell time for each pixel, and (iv) the time available for measurement. Individual STXM images at one single energy level can take from 3–60 s (80 × 80 pixel) to 0.5–10 min (250 × 250 pixel) or more, depending on the beamline design and capabilities. Because stack data sets contain 50 to >100 individual STXM images, a detailed analysis can require several hours. Fluorescence images take significantly longer because of the time needed to collect enough detector counts (1–5 s a pixel, 8 h for a 100 × 100 pixel map) but have the advantage of mapping several elements at once. Thus there is a tradeoff between sample detail and sample area (or number of samples) which must be considered when analyzing NOM. The newest generation of STXM has a much shorter acquisition time, which is expected to decrease even further. If the instrument allows it, individual point spectra or spectra collected along 1-D lines can significantly accelerate the exploration of sample composition. Coarse or low dwell time overview scans can also be conducted to determine regions where more detailed analyses are warranted. For example, the entire aggregate in Figure 17.18A was measured with a step size of 500 nm, whereas the detailed investigation of a portion of the aggregate in Figure 17.18B was done using a spatial resolution of 50 nm. This strategy allows us to economize valuable measurement time and to maximize the success to answer specific questions that require finding certain features that can typically not be discerned from a map at one energy level or from difference maps. For example, microbes or their metabolites are difficult to discern in complex assemblages unless both spectral and morphological features are available.

Because fluorescence analyses require significantly more time than transmission, the approach taken by the investigator in such studies needs to be modified. Data

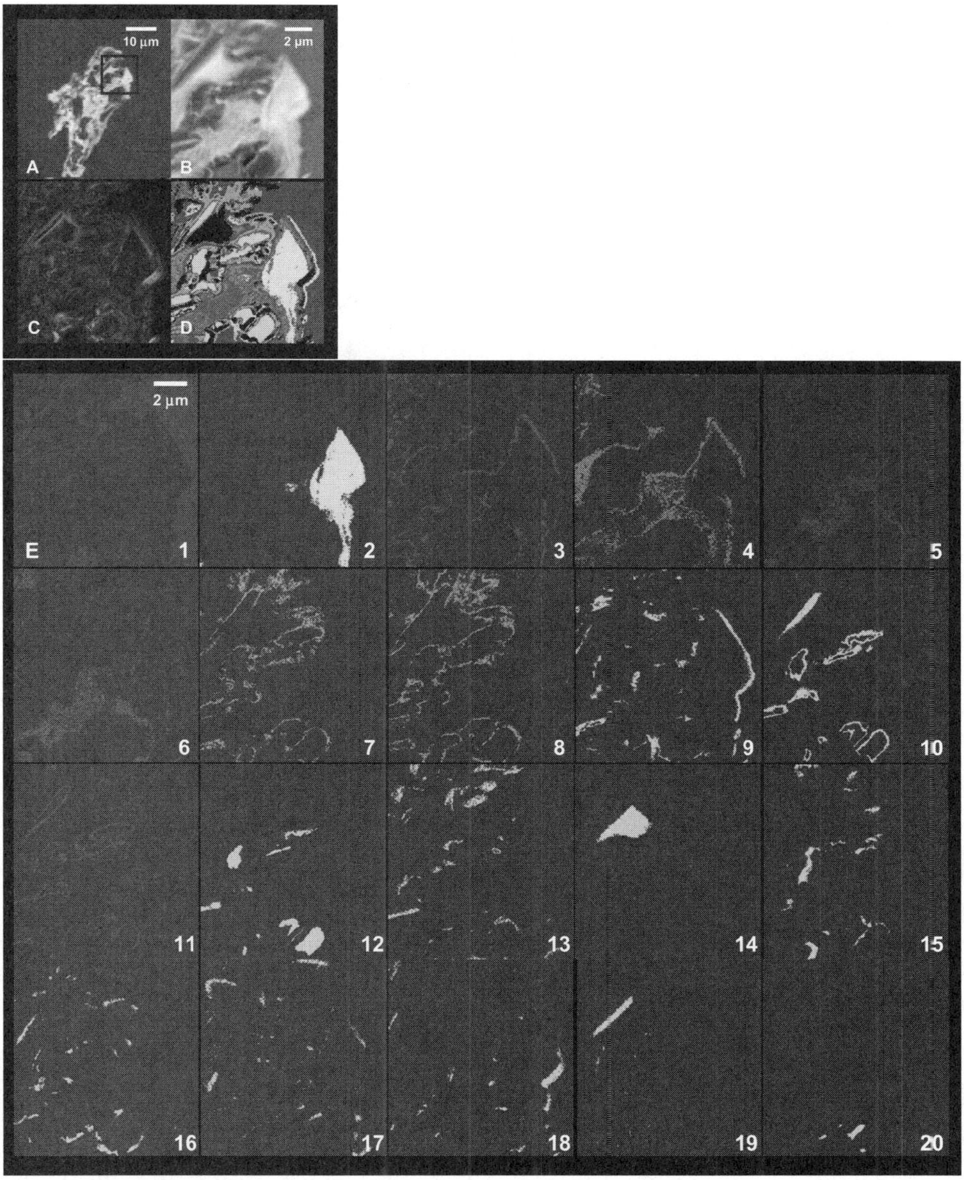

Figure 17.18. **(A)** Carbon contents in an entire microaggregate from an alfisol at Arnot Forest in Upstate New York (500-nm resolution). **(B)** Detail of the microaggregate (red box in A) (50-nm resolution). **(C)** X-ray map of B. **(D)** Cluster map [3 components, 20 clusters, without first component; PCA_GUI 1.0 developed by Lerotic et al. (2004)]. **(E)** Individual clusters from D; numbers in each cluster map correspond to spectra shown in Figure 17.19 (J. Lehmann, unpublished data 2006, measured as described in Lehmann et al., 2007). See color insert.

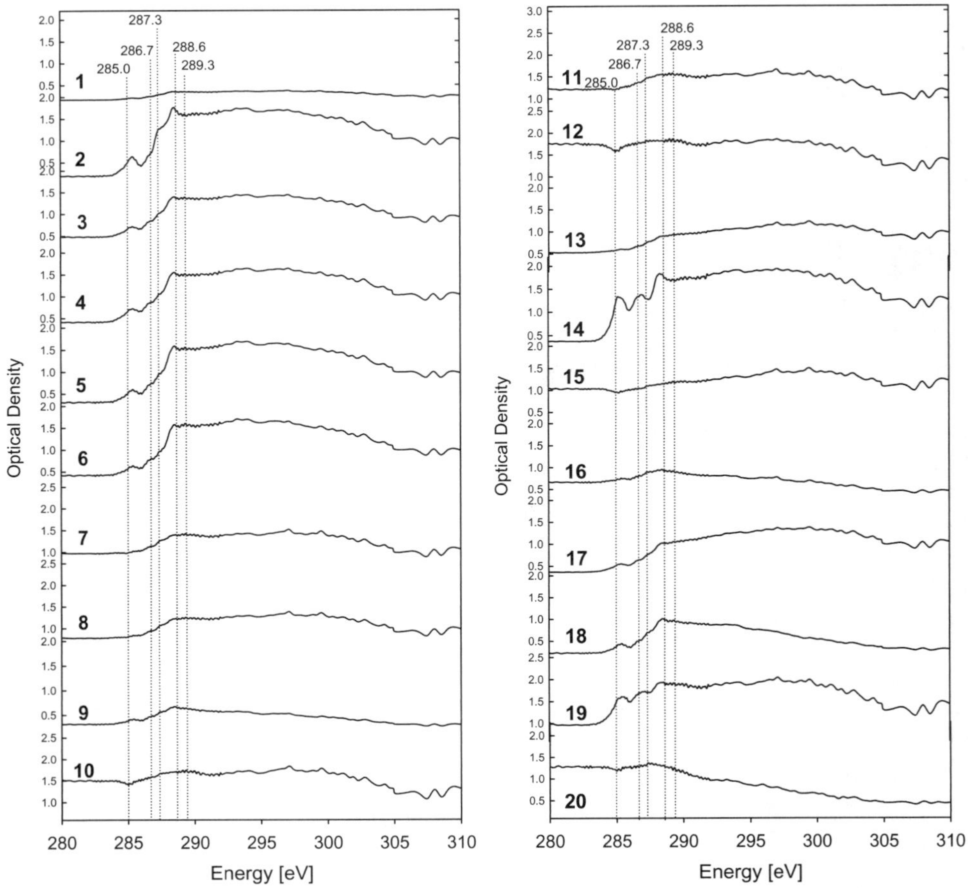

Figure 17.19. Carbon (1*s*) NEXAFS spectra of clusters obtained from principal component analysis. Cluster maps are shown in Figure 17.18. Further reduction in the number of clusters will reduce redundancy, but can mask minor features such as those in cluster 19 (J. Lehmann, unpublished data 2006, measured as described in Lehmann et al., 2007).

stack collection is impractical because individual fluorescence images can require 8 or more hours to collect. One approach is to conduct an overview scan at moderate resolution (e.g., 1 or $4\,\mu m^2$ pixels) followed by detailed small area scans and spectra collected at individual points in the image. Ideally X-ray focus should be adjusted so that the spot size illuminated is roughly equal to the pixel size. Figure 17.20 shows a fluorescence elemental covariance map of a marine sediment sample (Brandes et al., 2007). It is easily apparent that phosphorus is heterogeneously distributed within the sample and is not covariant with silicon or sodium. Follow-up P-fluorescence spectra collected on individual regions identified them as being either calcium phosphate (specifically the mineral apatite) or organic (possibly polyphosphates given the size and density in an aquatic medium) (Figure 17.20). This approach has also been used successfully for sulfur studies. As with the other spectromicroscopy techniques discussed in this chapter, phosphorus spectromicros-

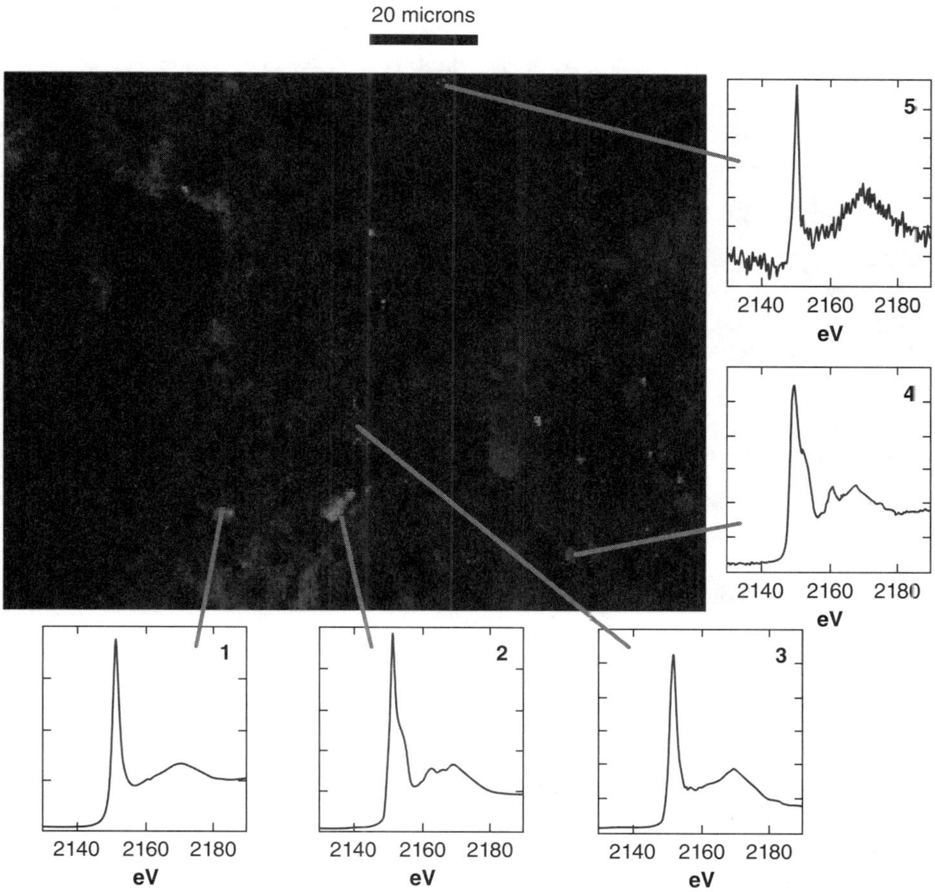

Figure 17.20. Phosphorus covariance map and fluorescence P-NEXAFS spectra collected from a coastal marine sediment (Brandes et al., 2007). Covariance map is color-coded: Green represents P, blue represents Si, and red represents Na fluorescence signals. Regions 1, 3, and 5 have spectra consistent with organic phosphorus or polyphosphate, while regions 2 and 4 closely match calcium phosphate, specifically the mineral apatite. See color insert.

copy represents a unique window into the biogeochemistry of this element in natural systems.

Not only the spatial distribution of NOM between pores or surfaces is important, but also the specific association with certain minerals. Some minerals such as short-range order aluminum oxides are known for their ability to stabilize large amounts of organic matter (Torn et al., 1997). NEXAFS spectroscopy may be able to provide clues as to the binding mechanisms between certain minerals and NOM. Similar to organic carbon, also the distribution of iron, aluminum, silicon, calcium, or potassium are very heterogeneous (Figure 17.21). In this example from a cambisol (Wan et al., 2007), carbon appears to be spatially associated with iron rather than aluminum or silicon. Further research is needed to more clearly investigate which NOM forms are spatially associated with different mineral phases.

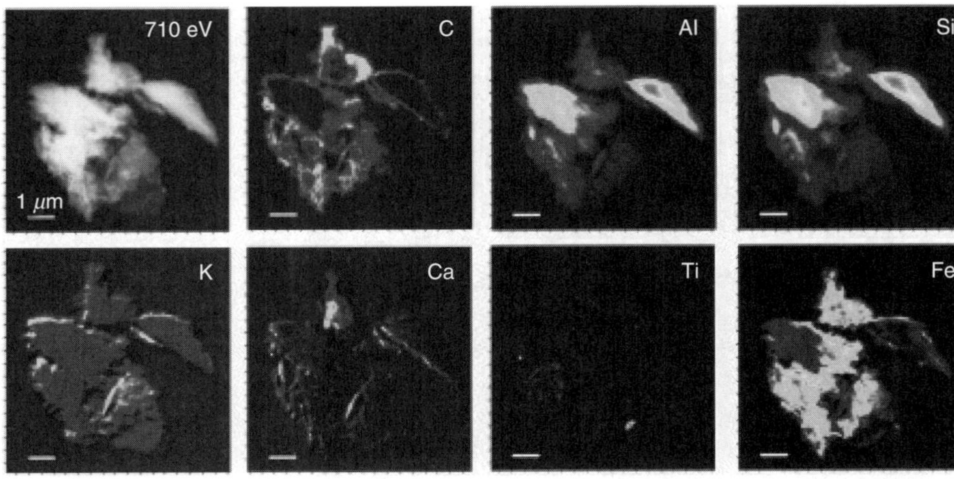

Figure 17.21. Distribution of carbon and mineral elements in an unsectioned micro-assemblage of a cambisol. The outline of the micro-assemblage is shown by the optical density map obtained at 710 eV (Wan et al., 2007). Observe the silicon- and aluminum-rich areas that could not be penetrated by the beam at the carbon edge in this unsectioned sample. See color insert.

The next stage in the development of spatially explicit analyses of NOM is to move from two- to three-dimensional mapping. Three-dimensional NEXAFS chemical information can be obtained by serial sectioning, measurement of two-dimensional maps, and computer reconstruction (Hitchcock et al., 2003). This requires multiple sections to be obtained and measured posing a significant challenge to widespread application, but is most likely the only solution for NOM–mineral mixtures. NOM samples with lower optical density such as suspended organic colloids with little mineral admixture may be measured by tomography using STXM, which was done for low-density latex particles at the oxygen K-edge (Johansson et al., 2007). The sample stage is rotated and multiple two-dimensional images are obtained and subsequently combined using suitable software application. This procedure may offer exciting perspectives for chemical speciation of the three-dimensional nature of suspended NOM and should be explored in the future.

17.5. CONCLUSIONS

NEXAFS and spectromicroscopy are relatively new techniques for the study of environmental samples and NOM. While still in early stages of application, these methods have unique advantages over other spectroscopic techniques such as NMR, FTIR, Raman–IR, or mass spectrometry. Sample preparation can be very simple and does not require extraction, although thin sectioning and embedding are often needed for spatially explicit analyses of light elements. Interferences are quite rare and do not lead to signal loss as in the case of NMR and paramagnetic elements. The opportunity to map and chemically characterize very fine-scale spatial distribution of a wide variety of elements is unparalleled and is a tremendous asset for environmental science studies. For many elements there are no alternative methods that can provide speciation information in context with the sample matrix. This is particularly true for organic matter studies, where NEXAFS can provide detailed information on functional group abundance distributions with 35 nm (e.g., a few hundred atoms wide) detail. Finally, conducting multi-element and possibly three-dimensional NEXAFS mapping of NOM holds the promise of deconvolving complex and challenging NOM compositions.

No analytical method is perfect. Spectral interpretation is still difficult, and standard spectra databases are scarce. The issues of quantification, comparison with data collected by other methods, and "scale up" are important, especially in spectromicroscopy studies. Radiation damage and sectioning artifacts can make analysis of susceptible samples difficult. The biggest obstacle to widespread use of NEXAFS spectroscopy, microscopy, and spectromicroscopy in environmental studies remains the extremely limited number of such instruments. Typically, each beamline allocation committee receives 2 or 3 times as many requests for time as is available. Studies, when granted, are usually for 2–5 days every 4–6 months. Thus, scientists have to be very selective about the types of questions and samples that they choose to examine using these techniques. Continued pressure and education from the scientific community will be needed to increase the number of beamlines suitable for NOM studies in the future, even as new synchrotron facilities are planned or built.

REFERENCES

Abbt-Braun, G., Lankes, U., and Frimmel, F. H. (2004). Structural characterization of aquatic humic substances—The need for a multiple method approach. *Aquatic Sci.* **66**, 151–170.

Ade, H. (1998). X-ray spectromicroscopy. In *Experimental Methods in the Physical Science*, Vol. 32, Samson, J. A. R., and Ederer, D. L., eds., Academic Press, San Diego, CA, pp. 225–261.

Ade, H., and Urquhart, S. G. (2002). NEXAFS Spectroscopy and microscopy of natural and synthetic polymers. In *Chemical Applications of Synchrotron Radiation, Advanced Series in Physical Chemistry*, Vol. 12, Sham, T. K., ed., World Scientific Publishing, River Edge, NJ, pp. 285–355.

Ade, H., and Hitchcock, A. P. (2008). NEXAFS microscopy and resonant scattering: composition and orientation probed in real and reciprocal space. *Polymer* **49**, 643–675.

Ade, H., Zhang, X., Cameron, S., Costello, C., Kirz, J., and Williams, S. (1992). Chemical contrast in X-ray microscopy and spatially resolved XANES spectroscopy of organic specimens. *Science* **258**, 972–975.

Ade, H., Smith, A. P., Zhang, H., Zhuang, G. R., Kirz, J., Rightor, E., and Hitchcock, A. (1997). X-ray spectromicroscopy of polymers and tribological surfaces at beamline X1A at the NSLS. *J. Electron. Spectrosc. Relat. Phen.* **84**, 53–72.

Apen, E., Hitchcock, A. P., and Gland, J. L. (1993). Experimental studies of the core excitation of imidazole, 4,5-dicyanoimidazole, and s-triazine. *J. Phys. Chem.* **97**, 6859–6866.

Beauchemin, S., Hesterberg, D., and Beauchemin, M. (2002). Principal component analysis approach for modeling sulfur K-XANES spectra of humic acids. *Soil Sci. Soc. Am. J.* **66**, 83–91.

Beauchemin, S., Hesterberg, D., Chou, J., Beauchemin, M., Simard, R. S., and Sayers, D. E. (2003). Speciation of phosphorus in phosphorus-enriched agricultural soils using X-ray absorption near-edge structure spectroscopy and chemical fractionation. *J. Environ. Qual.* **32**, 1809–1819.

Beetz, T., and Jacobsen, C. (2003). Soft x-ray radiation-damage studies in PMMA using a cryo-STXM. *J. Synchr. Rad.* **10**, 280–283.

Benzerara, K., Yoon, T.-H., Menguy, N., Tyliszczak, T., and Brown, G. E. (2005). Nanoscale environments associated with bioweathering of a Mg-Fe-pyroxene. *Proc. Natl. Acad. Sci.* **102**, 979–982.

Benzerara, K., Menguy, N., López-Garcia, P., Yoon, T.-H., Kazmierczak, J., Tyliszczak, T., Guyot, F., and Brown, G. E. (2006). Nanoscale detection of organic signatures in carbonate microbialites. *Proc. Natl. Acad. Sci.* **103**, 9440–9445.

Bluhm, H., Andersson, K., Araki, T., Benzerara, K., Brown, G. E., Dynes, J. J., Ghosal, S., Gilles, M. K., Hansen, H.-C., Hemminger, J. C., Hitchcock, A. P., Ketteler, G., Kilcoyne, A. L. D., Kneedler, E., Lawrence, J. R., Leppard, G. G., Majzlam, J., Mun, B. S., Myneni, S. C. B., Nilsson, A., Ogasawara, H., Ogletree, D. F., Pecher, K., Salmeron, M., Shuh, D. K., Tonner, B., Tyliszczak, T., Warwick, T., and Yoon, T. H. (2006). Soft X-ray microscopy and spectroscopy at the molecular environmental science beamline at the advanced light source. *J. Electr. Spectr. Rel. Phen.* **150**, 86–104.

Boese, J., Osanna, A., Jacobsen, C., and Kirz J. (1997). Carbon edge XANES spectroscopy of amino acids and peptides. *J. Electr. Spectr. Rel. Phen.* **85**, 9–15.

Boyce, C. K., Cody, G. D., Feser, M., Jacobsen, C., Knoll, A. H., and Wirick, S. (2002). Organic chemical differentiation within fossil plant cell walls detected with X-ray spectroscopy. *Geol. Soc. Am.* **30**, 1039–1042.

Bradley, J. P., Keller, L., Thomas, K. L., Vander Wood, T. B., and Brownlee, D. E. (1993). Carbon analyses of IDPs sectioned in sulfur and supported on beryllium films. In *24th Lunar and Planetary Sciences Conference*, Blanchard, D., and Black, D., eds., Houston, TX, pp. 173–174.

Brandes, J. A., Lee, C., Wakeham, S., Peterson, M., Jacobsen, C., Wirick, S., and Cody, G. (2004). Examining marine particulate organic matter at sub-micron scales using scanning transmission X-ray microscopy and carbon X-ray absorption near edge structure spectroscopy. *Marine Chem.* **92**, 107–121.

Brandes, J. A., Ingall, E., and Paterson, D. (2007). Characterization of minerals and organic phosphorus species in marine sediments using soft X-ray fluorescence spectromicroscopy. *Marine Chem.* **103**, 250–265.

Braun, A., Mun, B. S., Huggins, F. E., and Huffman, G. P. (2007). Carbon speciation of diesel exhaust and urban particulate matter NIST standard reference materials with C(1s) NEXAFS spectroscopy. *Environ. Sci. Technol.* **41**, 173–178.

Cappa, C. D., Smith, J. D., Wilson, K. R., Messer, B. M., Gilles, M. K., Cohen, R. C., and Saykally, R. J. (2005). Effects of alkali metal halide salts on the hydrogen bond network of liquid water. *J. Phys. Chem. B* **109**, 7046–7052.

Chefetz, B., Salloum, M. J., Deshmukh, A. P., and Hatcher, P. G. (2002). Structural components of humic acids as determined by chemical modifications and carbon-13 NMR, pyrolysis–, and thermochemolysis–gas chromatography/mass spectrometry. *Soil Sci. Soc. Am. J.* **66**, 1159–1171.

Chen, J., Gu, B., LeBoeuf, E. J., Pan, H., and Dai, S. (2002). Spectroscopic characterization of the structural and functional properties of natural organic matter fractions. *Chemosphere* **48**, 59–68.

Cody, G. D., Botto, R. E., Ade, H., and Wirick, S. (1996). The application of soft X-ray microscopy to the in-situ analysis of sporinite in coal. *Int. J. Coal Geol.* **32**, 69–86.

Cody, G. D., Ade, H., Wirick, S., Mitchell, G. D., and Davis, A. (1998). Determination of chemical–structural changes in vitrinite accompaying luminescence alteration using C-NEXAFS analysis. *Org. Geochem.* **28**, 441–455.

Cody, G. D., Ade, H., Alexander, C. M.O'D., Araki, T., Butterworth, A., Fleckenstein, H., Flynn, G., Gilles, M. K., Jacobsen, C., Kilcoyne, A. L. D., Messenger, K., Sandford, S. A., Tyliszczak, T., Westphal, A. J., Wirick, S., and Yabuta, H. (2008). Quantitative organic and light-element analysis of comet 81P/Wild 2 particles using C- N-, and O-mu-XANES. *Meteoritics and Planetary Science*, **43**, 353–365.

Cooney, R. R., and Urquhart, S. G. (2004). Chemical trends in the near-edge X-ray absorption fine structure of monosubstituted and para-bisubstituted benzenes. *J. Phys. Chem. B* **108**, 18185–18191.

De Stasio, G., Schmitt, M. A., and Gellmann, S. H. (2005). Spectroscopy at the organic–inorganic interface in biominerals. *Am. J. Sci.* **305**, 673–686.

del Rio, J. C., McKinney, D. E., Knicker, H., Nanny, M. A., Minard, R. D., and Hatcher, P. G. (1998). Structural characterization of bio- and geo-macromolecules by off-line thermochemolysis with tetramethylammonium hydroxide. *J. Chrom.* **823**, 433–448.

Edwards, D. C., and Myneni, C. B. (2006). Near-edge x-ray absorption fine structure spectroscopy of bacterial hydroxamate siderophores in aqueous solutions. *J. Phys. Chem. A* **110**, 11809–11818.

Feser, M., Carlucci-Dayton, M., Jacobsen, C., Kirz, J., Neuhäusler, U., Smith, G., and Yu, B. (1998). Applications and instrumentation advances with the Stony Brook scanning transmission x-ray microscope. In *X-Ray Microfocusing: Applications and Techniques*, McNulty, I., ed., Proceedings of Society of Photo-optical Instrumentation Engineers, San Diego, CA, Vol. 3449, pp. 19–29.

Feser, M., Jacobsen, C., Rehak, P., and DeGeronimo, G. (2003). Scanning transmission x-ray microscopy with a segmented detector. *J. Phys. IV* **104**, 529–534.

Flechsig, U., Quitmann, C., Raabe, J., Böge, M., Fink, R., and Ade, H. (2007). The PolLux microspectroscopy beam line at the Swiss light source. *Conf. Proc. Am. Inst. Phys.* **870**, 505–508.

Flynn, G. J., Keller, L. P., Wirick, S., Jacobsen, C., and Sutton, S. R. (2003). Analysis of interplanetary dust particles by soft and hard X-ray microscopy. *J. Phys. IV* **104**, 367–372.

Flynn, G. F., Keller, L. P., Jacobsen, C., and Wirick, S. (2004). An assessment of the amount and types of organic matter contributed to the earth by interplanetary dust. *Adv. Space Res.* **33**, 57–66.

Francis, J. T., and Hitchcock, A. P. (1992). Inner-shell spectroscopy of *p*-benzoquinone, hydroquinone, and phenol: Distinguishing quinoid and benzenoid structures. *J. Phys. Chem.* **96**, 6598–6610.

Freney, J. R., Melville, G. E., and Williams, C. H. (1975). Soil organic matter fractions as sources of plant-available sulfur. *Soil Biol. Biochem.* **22**, 1163–1165.

Frimmel, F. H., Abbt-Braun, G., Heumann, K. G., Hock, B., Luedemann, H.-D., and Spiteller, M. (2002). *Refractory Organic Substances in the Environment*, Wiley-VCH, Weinheim.

Gerin, A., Genet, M. J., Herbillon, A. J., and Delvaux, B. (2003). Surface analysis of soil material by X-ray photoelectron spectroscopy. *Eur. J. Soil Sci.* **54**, 589–604.

Ghosh, U., Gilette, J. S., Luthy, R. G., and Zare, R. N. (2000). Microscale location, characterization, and association of polycyclic aromatic hydrocarbons on harbor sediment particles. *Environ. Sci. Technol.* **34**, 1729–1736.

Gleber, G., Thieme, J., Niemeyer, J., and Feser, M. (2003). Interaction of organic substances with iron studied by O1s spectroscopy—Development of an analysis program. *J. Phys. IV France* **104**: 429–432.

Haberstroh, P. R., Brandes, J. A., Gelinas, Y., Dickens, A. F., Wirick, S., and Cody, G. (2006). Chemical composition of the graphitic black carbon fraction in riverine and marine sediments at submicron scales using carbon X-ray spectromicroscopy. *Geochim. Cosmochim. Acta* **70**, 1483–1494.

Harrison, A. F. (1987). *Soil Organic Phosphorus. A Review of World Literature*, CABI Publishers, Wallingford, UK.

Hatcher, P. G., Dria, K. J., Kim, S., and Frazier, S. W. (2001). Modern analytical approaches of humic substances. *Soil Sci.* **166**, 770–794.

Hennig, C., Hallmeier, K. H., Bach, A., Bender, S., Franke, R., Hormes, J., and Szargan, R. (1996). Influence of substituents on the N K X-ray absorption near-edge structure of pyrrole derivatives. *Spectrochim. Acta* **52**, 1079–1083.

Hitchcock, A., and Ishii, I. (1987). Carbon K-shell excitation spectra of linear and branched alkanes. *J. Electron Spectrosc. Relat. Phen.* **42**, 11–26.

Hitchcock, A. P., and Mancini, D. C. (1994). Bibliography of atomic and molecular inner-shell excitation studies. *J. Electron Spectrosc. Relat. Phen.* **67**, 1–132.

Hitchcock, A. P., Newbury, D. C., Ishii, I., Stohr, J., Horsley, J. A., Redwing, R. D., Johnson, A. L., and Sette, F. (1986). Carbon K-shell excitation of gaseous and condensed cyclic hydrocarbons: C_3H_6, C_4H_8, C_5H_8, C_5H_{10}, C_6H_{10}, C_6H_{12}, and C_8H_8. *J. Chem. Phys.* **85**, 4849–4862.

Hitchcock, A. P., Fischer, P., Gedanken, A., and Robin, M. B. (1987). Antibonding σ^* valence MOs in the inner shell and outer shell spectra of the fluorobenzenes. *J. Phys. Chem.* **91**, 531–540.

Hitchcock, A. P., Urquhart, S. G., and Rightor, E. G. (1992). Inner shell spectroscopy of benzaldehyde, terephthalaldehyde, ethylbenzoate, terephthaloyl chloride, and phosgene: Models for core excitation of poly(ethylene terephthalate). *J. Phys. Chem.* **96**, 8736–8750.

Hitchcock, A. P., Araki, T., Ikeura-Sekiguchi, H., Iwata, N., and Tani, K. (2003). 3*d* chemical mapping of toners by serial section scanning transmission X-ray microscopy. *J. Phys. IV* 104, 509–512.

Hitchcock, A. P., Morin, C., Zhang, X., Araki, T., Dynes, J., Stöver, H., Brash, J., Lawrence, J. R., and Leppard, G. G. (2005a). Soft X-ray spectromicroscopy of biological and synthetic polymer systems. *J. Electr. Spectr. Rel. Phen.* **144–147**, 259–269.

Hitchcock, A. P., Stöver, H. D., Croll, L. M., and Childs, R. F. (2005b). Chemical mapping of polymer microstructure using soft X-ray spectromicroscopy. *Austr. J. Chem.* **58**, 423–432.

Hitchcock, A. P., Dynes, J. J., Johansson, G., Wang, J., and Botton, G. (2008). Comparison of NEXAFS microscopy and TEM-EELS for studies of soft matter. *Micron* **39**, 311–319.

Hundal, L. S., Carmo, A. M., Bleam, W. L., and Thompson, M. L. (2000). Sulfur in biosolids-derived fulvic acid: Characterization by ZXANES spectroscopy and selective dissolution approaches. *Environ. Sci. Technol* **34**, 5184–5188.

Hunt, J. A., and Williamds, D. B. (1991). Electron energy-loss spectrum-imaging. *Ultramicr.* **38**, 47–73.

Hutchison, K. J., Hesterberg, D., and Chou, J. W. (2001). Stability of reduced organic sulfur in humic acids as affected by aeration and pH. *Soil Sci. Soc. Am. J.* **65**, 704–709.

Ishii, I., and Hitchcock, A. P. (1987). A quantitative experimental study of the core excited electronic states of foramide, formic acid, and formyl fuoride. *J. Chem. Phys.* **87**, 830–839.

Ishii, I., and Hitchcock, A. P. (1988). The oscillator strengths for C1s and O1s excitation of some saturated and unsaturated organic alcohols, acids, and esters. *J. Electr. Spectr. Rel. Phen.* **46**, 55–84.

Jacobsen, C., Wirick, S., Flynn, G., and Zimba, C. (2000). Soft X-ray spectroscopy from image sequences with sub-100 nm spatial resolution. *J. Microsc.* **197**, 173–184.

Jeanguillaume, C., and Colliex, C. (1989). Spectrum-image: The next step in EELS digital acquisition and processing. *Ultramicr.* **28**, 252–257.

Johansson, G. A., Tyliszczak, T., Mitchell, G., Keefe, M., and Hitchcock, A. P. (2007). Three dimensional chemical mapping by scanning transmission X-ray spectromicroscopy. *J. Synchr. Rad.* **14**, 395–402.

Jokic, A., Cutler, J. N., Ponomarenko, E., van der Kamp, G., and Anderson, D. W. (2003). Organic carbon and sulfur compounds in wetland soils: Insights on structure and transformation processes using K-edge XANES and NMR spectroscopy. *Geochim. Cosmochim. Acta.* **67**, 2585–2597.

Jokic, A., Cutler, J. N., Anderson, D. W., and Walley, F. L. (2004a). Detection of heterocyclic N compounds in whole soils using N-XANES spectroscopy. *Can. J. Soil Sci.* **84**, 291–293.

Jokic, A., Schulten, H. R., Cutler, J. N., Schnitzer, M., and Huang, P. M. (2004b). A significant abiotic pathway for the formation of unknown nitrogen in nature. *Geophys. Res. Letters* **31**, L05502.

Kasrai, M., Brown, J. R., Bancroft, G. M., Yin, Z., and Tan, K. H. (1996). Sulphur characterization in coal from X-ray absorption near edge spectroscopy. *Int. J. Coal Geol.* **32**, 107–135.

Kaznacheyev, K., Osanna, A., Jacobsen, C., Plashkevych, O., Vahtras, O., Ågren, H., Carravetta, V., and Hitchcock, A. P. (2002). Innershell Absorption Spectroscopy of Amino Acids. *J. Phys. Chem. A*, **106**, 3153–3168.

Kilcoyne, A. L. D., Tylisczcak, T., Steele, W. F., Fakra, S., Hitchcock, P., Franck, K., Anderson, E., Harteneck, B., Rightor, E. G., Mitchell, G. E., Hitchcock, A. P., Yang, L., Warwick, T., and Ade, H. (2003). Interferometer-controlled scanning transmission X-ray microscopes at the advanced light source. *J. Synchrotron Rad.* **10**, 125–163.

Kinyangi, J., Solomon, D., Liang, B., Lerotic, M., Wirick, S., and Lehmann, J. (2006). Nanoscale biogeocomplexity of the organo-mineral assemblage in soil: Application of STXM microscopy and C 1s-NEXAFS spectroscopy. *Soil Sc. Soc. Am. J.* **70**, 1708–1718.

Kirz, J., Jacobsen, C., and Howells, M. (1995). Soft x-ray microscopes and their biological applications. *Q. Rev. Biophys.* **28**, 33–130.

Kjeldsen, P. (1993). Evaluation of gas diffusion through plastic materials used in experimental and sampling equipment. *Water Res.* **27**, 121–131.

Kögel-Knabner, I. (1995). Composition of soil organic matter. In *Methods in Applied Soil Microbiology and Biochemistry*, Nannipieri, P., and Alef, K., eds., Academic Press, London, pp. 66–78.

Kögel-Knabner, I. (2000). Analytical approaches for characterizing soil organic matter. *Org. Geochem.* **31**, 609–625.

Krause, M. O. (1979). X-ray fluorescence cross-sections for *K* and *L* X-rays of the elements. *J. Phys. Chem. Ref. Data* **8**, 307–327.

Langford, R. M. (2006). Focused ion beam techniques for nanomaterials characterization. *Microsc. Res. Techn.* **69**, 538–549.

Lawrence, J. R., Swerhone, G. D. W., Leppard, G. G., Araki, T., Zhang, X., West, M. M., and Hitchcock, A. P. (2003). *Scanning* transmission X-ray, laser scanning, and transmission electron microscopy mapping of the exopolymeric matrix of microbial biofilms. *Appl. Environ. Microbiol.* **69**, 5543–5554.

Lee, D. D., and Seung S. (1999). Learning the parts of objects by non-negative matrix factorization. *Nature* **401**, 788–791.

Lee, D. D., and Seung S. (2001). Algorithms for non-negative matrix factorization. In *Advances in Neural Information Processing Systems 13*, Proceedings of the 2000 Conference, pp. 556–562.

Lee, C., Wakeham, S., and Arnosti, C. (2004). Particulate organic matter in the sea: the composition conundrum. *Ambio* **33**, 565–575.

Lehmann, J., Liang, B., Solomon, D., Lerotic, M., Luizão, F., Kinyangi, F., Schäfer, T., Wirick, S., and Jacobsen, C. (2005). Near-edge X-ray absorption fine structure (NEXAFS) spectroscopy for mapping nano-scale distribution of organic carbon forms in soil: Application to black carbon particles. *Global Biogeochem. Cycles* **19**, GB1013.

Lehmann, J., Kinyangi, J., and Solomon, D. (2007). Organic matter stabilization in soil microaggregates: Implications from spatial heterogeneity of organic carbon contents and carbon forms. *Biogeochemistry.* **85**, 45–57.

Leinweber, P., and Schulten, H. R. (1998). Advances in analytical pyrolysis of soil organic matter. *J. Anal. Appl. Pyrol.* **47**, 165–189.

Leinweber, P., Kruse, J., Walley, F. L., Gillespie, A., Eckardt, K.-U., Blyth, R., and Regier, T. (2007). Nitrogen K-edge XANES—An overview of reference compounds used to identify 'unknown' organic nitrogen in environmental samples. *J. Synchrotron Rad.* **14**, 500–511.

Lerotic, M., Jacobsen, C., Schäfer, T., and Vogt, S. (2004). Cluster analysis of soft x-ray spectromicroscopy data. *Ultramicroscopy* **100**, 35–57.

Liang, B., Lehmann, J., Solomon, D., Kinyangi, J., Grossman, J., O'Neill, B., Skjemstad, J. O., Thies, J., Luizão, F. J., Petersen, J., and Neves, E. G. (2006). Black carbon increases cation exchange capacity in soils. *Soil Sci. Soc. Am. J.* **70**, 1719–1730.

Lombi, E., Scheckel, K. G., Armstrong, R. D., Forrester, S., Cutler, J. N., and Paterson, D. (2006). Speciation and distribution of phosphorus in a fertilized soil: a synchrotron-based investigation. *Soil Sci. Soc. Am. J.* **70**, 2038–2048.

Ma, Y., Chen, C. T., Meigs, G., Randall, K., and Sette, F. (1991). High-resolution K-shell photoabsorption measurements of simple molecules. *Phys. Rev. A* **44**, 1848–1858.

Mayer, L. M. (1999). Extent of coverage of mineral surfaces by organic matter in marine sediments. *Geochim. Cosmochim. Acta* **63**, 207–215.

McHardy, W. J., and Robertson, L. (1983). An optical scanning electron microscopic and microanalytical study of cementation in some podzols. *Geoderma* **30**, 161–170.

McKeague, J. A., and Wang, C. (1980). Micromorphology and energy dispersive analysis of ortstein horizons of podzolic soils from New Brunswick and Nova Scotia, Canada. *Can. J. Soil Sci.* **60**, 9–21.

Mitra-Kirtley, S., Mullins, O. C., Chen, J., van Elp, J., George, S. J., Chen, C. T., O'Halloran, T., and Cramer, S. P. (1992). Nitrogen chemical structure in DNA and related molecules by X-ray absorption spectroscopy. *Biochim. Biophys. Acta* **1132**, 249–254.

Mitra-Kirtley, S., Mullins, O. C., van Elp, J., George, S. J., Chen, J., and Cramer, S. P. (1993). Determination of the nitrogen chemical structures in petroleum asphaltenes using XANES spectroscopy. *J. Am. Chem. Soc.* **115**, 252–258.

Morra, M. J., Fendorf, S. E., and Brown, P. D. (1997). Speciation of S in humic and fulvic acids using X-ray absorbtion near-edge structure (XANES) spectroscopy. *Geochim. Cosmochim. Acta* **61**, 683–688.

Myneni, S. C. B. (2002). Soft X-ray spectroscopy and spectromicroscopy studies of organic molecules in the environment. In *Applications of Synchrotron Radiation in Low-Temperature Geochemistry and Environmental Sciences*, Fenter, P., Rivers, M., Sturchio, N., and Sutton, S., eds., *Rev. Mineralogy Geochem.* **49**, 485–579.

Neptune, A. M. L., Tabatabai, M. A., and Hanway, J. J. (1975). Sulfur fractions and carbon–nitrogen–phosphorus–sulfur relationship in some Brazilian and Iowa soils. *Soil Sci. Soc. Am. Proc.* **39**, 51–55.

Neuhäusler, U. (1999). Soft x-ray spectromicroscopy on hydrated colloidal and environmental samples. Ph.D. thesis, University of Göttingen.

Osanna, A., and Jacobsen, C. (2000). Principle component analysis for soft X-ray spectromicroscopy. In *X-Ray Microscopy: Proceedings of the Sixth International Conference*, Meyer-Ilse, W., Warwick, T., and Attwood, D., eds., American Institute of Physics, Melville, New York, pp. 350–357.

Otero, E., and Urquhart, S. G. (2006). Nitrogen 1*s* near-edge x-ray absorption fine structure spectroscopy of amino acids: Resolving zwitteronic effects. *J. Phys. Chem. A* **110**, 12121–12128.

Peak, D., Sims, J. T., and Sparks D. L. (2002). Solid-state speciation of natural and alum-amended poultry litter using XANES spectroscopy. *Environ. Sci. Technol.* **36**, 4253–4261.

Plaschke, M., Rothe, J., Denecke, M. A., and Fanghänel, T. (2004). Soft X-ray spectromicroscopy of humic acid europium(III) complexation by comparison to model substances. *J. Electron Spectrosc. Relat. Phenom.* **135**, 53–62.

Plaschke, M., Rothe, J., Altmaier, M., Denecke, M. A., and Fanghänel, T. (2005). Near edge X-ray absorption fine structure (NEXAFS) of model compounds for the humic acid/actinide ion interaction. *J. Electron Spectrosc. Relat. Phenom.* **148**, 151–157.

Prange, A., Arzberger, I., Engemann, C., Modrow, H., Schumann, O., Trüper, H. G., Steudel, R., Dahl, C., and Hormes, J. (1999). In situ analysis of sulfur in the sulfur globules of phototrophic sulfur bacteria by X-ray absorption near edge spectroscopy. *Biochim. Biophys. Acta* **1428**, 446–454.

Prietzel, J., Thieme, J., Neuhäusler, U., Susini, J., and Kögel-Knabner, I. (2003). Speciation of sulphur in soils and soil particles by X-ray spectroscopy. *Eur. J. Soil Sci.* **54**, 423–433.

Ransom, B., Bennett, R. J., Baerwald, R., and Shea, K. (1997). TEM study of in situ organic matter on continental shelf margins: Occurrence and the "monolayer" hypothesis. *Marine Geol.* **138**, 1–9.

Ravel, B., and Newville, M. (2005). ATHENA and ARTEMIS: interactive graphical data analysis using IFEFFIT. *Phys. Scripta* **115**, 1007–1010.

Rightor, E. G., Hitchcock, A. P., Ade, H., Leapman, R. D., Urquhart, S. G., Smith, A. P., Mitchell, G., Fischer, D., Shin, H. J., and Warwick, T. (1997). Spectromicroscopy of poly(ethylene terephthalate): Comparison of spectra and radiation damage rates in x-ray absorption and electron energy loss. *J. Phys. Chem. B* **101**, 1950–1960.

Robin, M. B., Ishii, I., McLaren, R., and Hitchcock, A. P. (1988). Fluorination effects on the inner shell spectra of unsaturated molecules. *J. Electron Spectrosc. Relat. Phen.* **47**, 53–92.

Rompel, A., Cinco, R. M., Latimer, M. J., McDermott, A. E., Guiles, R. D., Quintanilha, A., Krauss, R. M., Sauer, K., Yachandra, V. K., and Klein, M. P. (1998). Sulfur K-edge X-ray absorption spectroscopy: a spectroscopic tool to examine the redox state of S-containing metabolites in vivo. *Proc. Nat. Ac. Sci.* **95**, 6122–6127.

Rothe, J., Denecke, M. A., and Dardenne, K. (2000). Soft X-ray spectromicroscopy investigation of the interaction of aquatic humic acid and clay colloids. *J. Coll. Interface Sci.* **231**, 91–97.

Rothe, J., Plaschke, M., and Denecke, M. A. (2004). Scanning transmission X-ray microscopy as a speciation tool for natural organic molecules. *Radiochim. Acta* **92**, 711–715.

Saiz-Jimenez, C. (1994). Analytical pyrolysis of humic substances: Pitfalls, limitations and possible solutions. *Environ. Sci. Technol.* **28**, 1773–1780.

Sarret, G., Connan, J., Kasrai, M., Bancroft, G. M., Charrié-Duhaut, A., Lemoine, S., Adam, P., Albrecht, P., and Eybert-Bérard, L. (1999). Chemical forms of sulfur in geological and archaeological asphaltenes from Middle East, France, and Spain determined by sulfur K- and L-edge X-ray absorption near-edge structure spectroscopy. *Geochim. Cosmochim. Acta* **63**, 3767–3779.

Sato, S., Solomon, D., Hyland, C., Ketterings, Q. M., and Lehmann, J. (2005). Phosphorus speciation in manure and manure-amended soils using XANES spectroscopy. *Environ. Sci. Technol.* **39**,7485–7491.

Schäfer, T., Hertkorn, N., Artinger, R., Claret, F., and Bauer, A. (2003). Functional group analysis of natural organic colloids and clay association kinetics using C(1*s*) spectromicroscopy. *J. Phys. IV* **104**, 409–412.

Schäfer, T., Buckau, G., Artinger, R., Kim, J. I., Geyer, S., Wolf, M., Bleam, W. F., Wirick, S., and Jacobsen, C. (2005). Origin and mobility of fulvic acids in the Gorleben aquifer system: Implications from isotopic data and carbon/sulfur XANES. *Org. Geochem.* **36**, 567–582.

Scheinost, A. C., Kretzschmar, R., Christl, I., and Jacobsen, C. (2001). Carbon group chemistry of humic and fulvic acid: A comparison of C-1*s* NEXAFS and ^{13}C-NMR spectroscopies. In *Humic Substances: Structures, Models and Functions*, Ghabbour, E. A., and Davies, G., eds., Royal Society of Chemistry, Gateshead, UK, pp. 39–47.

Schmidt, C., Thieme, J., Neuhäusler, U., Jacobsen, C., Kaulich, B., Salome, M., and Susini, J. (2003). Spectromicroscopy of soil colloids. *J. Phys. IV* **104**, 405–408.

Schulze, D. G., McCay-Buis, T., Sutton, S. R., and Huber, D. H. (1995). Manganese oxidation states in *Gaeumannomyces*-infested wheat rhizospheres probed by micro-XANES spectroscopy. *Phytopathology* **85**, 990–994.

Schumacher, M., Christl, I., Scheinost, A. C., Jacobsen, C., and Kretzschmar, R. (2005). Chemical heterogeneity of organic soil colloids investigated by scanning transmission X-ray microscopy and C-1s NEXAFS microspectroscopy. *Environ. Sci. Technol.* **39**, 9094–9100.

Schumacher, M., Christl, I., Vogt, R. D., Barmettler, K., Jacobsen, C., and Kretzschmar, R. (2006). Chemical composition of aquatic dissolved organic matter in five boreal forest catchments sampled in spring and fall seasons. *Biogeochemistry* **80**, 263–275.

Sham, T. K., Yang, B. X., Kirz, J., and Tse, J. S. (1989). K-edge near-edge x-ray-absorption fine structure of oxygen- and carbon-containing molecules in the gas phase. *Phys. Rev. A* **40**, 652–669.

Shober, A. L., Hesterberg, D. L., Sims, J. T., and Gardner, S. (2006). Characterization of phosphorus species in biosolids and manures using XANES spectroscopy. *J. Environ. Qual.* **35**, 1983–1993.

Smernik, R. J., and Baldock, J. A. (2005). Does solid-state ^{15}N NMR spectroscopy detect all soil organic nitrogen? *Biogeochemistry* **75**, 507–528.

Sodhi, R. N. S., and Brion, C. E. (1984). Reference energies for inner shell electron energy-loss spectroscopy. *J. Electr. Spectr. Rel. Phen.* **34**, 363–372.

Sohi, S. P., Mahieu, N., Arah, J. R. M., Powlson, D. S., Madari, B., and Gaunt, J. L. (2001). A procedure for isolating soil organic matter fractions suitable for modeling. *Soil Sci. Soc. Am. J.* **65**, 1121–1128.

Solomon, D., Fritzsche, F., Tekalign, M., Lehmann, J., and Zech, W. (2002). Soil organic matter composition in the sub-humid Ethiopian highlands as influenced by deforestation and agricultural management. *Soil Sci. Soc. Am. J.* **66**, 68–82.

Solomon, D., Lehmann, J., and Martinez, C. E. (2003). Sulfur K-edge XANES spectroscopy as a tool for understanding sulfur dynamics in soil organic matter. *Soil Sci. Soc. Am. J.* **67**, 1721–1731.

Solomon, D., Lehmann, J., Kinyangi, J., Liang, B., and Schäfer, T. (2005a). Carbon K-edge NEXAFS and FTIR-ATR spectroscopic investigation of organic carbon speciation in soils. *Soil Sci. Soc. Am. J.* **69**, 107–119.

Solomon, D., Lehmann, J., Lobe, I., Martinez, C. E., Tveitnes, S., Du Preez, C. C., and Amelung, W. (2005b). Sulphur speciation and biogeochemical cycling in long-term arable cropping of subtropical soils: Evidence from wet-chemical reduction and S K-edge XANES spectroscopy. *Eur. J. Soil Sci.* **56**, 621–634.

Solomon, D., Lehmann, J., Kinyangi, J., Amelung, W., Lobe, I., Ngoze, S., Riha, S., Pell, A., Verchot, L., Mbugua, D., Skjemstad, J., and Schäfer, T. (2007a). Long-term impacts of anthropogenic perturbations on dynamics and speciation of organic carbon in tropical forest and subtropical grassland ecosystems. *Global Change Biol.* **13**, 511–530.

Solomon, D., Lehmann, J., Thies, J., Schäfer, T., Liang, B., Kinyangi, J., Neves, E., Petersen, J., Luizão, F., and Skjemstad, J. (2007b). Molecular signature and sources of biochemical recalcitrance of organic C in Amazonian Dark Earths. *Geochim. Cosmochim. Acta* **71**, 2285–2298.

Solomon, D., Lehmann, J., Kinyangi, J., Liang, B., Hanley, K., Heymann, K., Wirick, S., and Jacobsen, C. (2009). Carbon (1s) NEXAFS spectroscopy of biogeochemically relevant organic reference compounds. *Soil Sci. Soc. Am. J.*, in press.

Sowden, F. J., Chen, Y., and Schnitzer, M. (1977). The nitrogen distribution of soils formed under widely different climatic conditions. *Geochim. Cosmochim. Acta* **41**, 1524–1526.

Sparks, D. L. (2003). Environmental Soil Chemistry, 2nd edition, Academic Press, San Diego.

Stöhr, J. (1992). *NEXAFS Spectroscopy*, Springer Series in Surface Sciences, Vol. 25, Springer, Berlin.

Stöhr, J., Samant, M. G., Lüning, J., Callegari, A. C., Chaudhari, P., Doyle, J. P., Lacey, J. A., Lien S. A., Purushothaman, S., and Speidell, J. L. (2001). Liquid crystal alignment on carbonaceous surfaces with orientational order. *Science* **292**, 2299–2302.

Szilagyi, R. K., and Schwab, D. E. (2005). Sulfur K-edge X-ray absorption spectroscopy as an experimental probe for S-nitroso proteins. *Biochem. Biophys. Res. Comm.* **330**, 60–63

Takahasi, O., Yamanouchi, S., Yamamoto, K., and Tabayashi, K. (2006). Theoretical study of the x-ray absorption spectra of small formic acid clusters. *Chem. Phys. Letters* **419**, 501–505.

Teo, B. K. (1986). *EXAFS: Basic Principles and Data Analyses*, Inorganic Chemistry Concepts, Vol. 9, Springer, Berlin.

Thieme, J., Wilhein, T., Guttmann, P., Niemeyer, J., Jacob, K.-H., and Dietrich, S. (1994). Direct visualization of iron and manganese accumulating microorganisms by x-ray microscopy. In *Proceedings of the 4th International Conference*, Aristov, V. V., and Erko, A. I., eds., Chernogolovka, Russia, September 20–24, 1993, pp. 152–156.

Thieme, J., Niemeyer, J., Machulla, G., and Schulte-Ebbert, U. (1998). Aggregation of colloids observed by X-ray microscopy. In *X-Ray Microscopy and Spectromicroscopy*, Thieme, J., Schmahl, G., Rudolph, D., and Umbach, E., eds., Springer, Heidelberg, pp. 11–20.

Thieme, J., Prietzel, J., Tyufekchieva, N., Paterson, D., and McNulty, I. (2006). Speciation of sulfur in oxic and anoxic soils using X-ray spectromicroscopy. In *Proceedings of the 8th International Conference on X-Ray Microscopy*, IPAP Conference Series 7, pp. 318–320.

Tippkotter, R., and Ritz, K. (1996). Evaluation of polyester, epoxy and acrylic resins for suitability in preparation of soil thin sections for in situ biological studies. *Geoderma* **69**, 31–57.

Toor, G. S., Peak, J. D., and Sims, J. T. (2005). Phosphorus speciation in broiler litter and turkey manure produced from modified diets. *J. Environ. Qual.* **34**, 687–697.

Torn, M. S., Trumbore, S. E., Chadwick, O. A., Vitousek, P. M., and Hendricks, D. M. (1997). Mineral control of soil organic carbon storage and turnover. *Nature* **389**, 170–173.

Urquhart, S. G., and Ade, H. (2002). Trends in the carbonyl core (C 1S, O 1S) $\rightarrow$ pi* C=O transition in the near-edge X-ray absorption fine structure spectra of organic molecules. *J. Phys. Chem. B* **106**, 8531–8538.

Vairavamurthy, A., Manowitz, B., Luther, G. W., III, and Jeon, Y. (1993). Oxidation state of sulfur in thiosulfate and implications for anaerobic energy metabolism. *Geochim. Cosmochim. Acta* **57**, 1619–1623.

Vairavamurthy, A., Maletic, D., Wang, S., Manowitz, B., Eglinton, T., and Lyons, T. (1997). Characterization of sulfur-containing functional groups in sedimentary humic substances by X-ray absorption near-edge spectroscopy. *Energy Fuels* **11**, 546–553.

Vairavamurthy, A., and Wang, S. (2002). Organic nitrogen in geomacromolecules: Insights on speciation and transformationwith K-edge XANES spectroscopy. *Environ. Sci. Techn.* **36**, 3050–3056.

Vall-Ilosera, G., Gao, B., Kivimaki, A., Coreno, M., Alvarez Ruiz, J., de Simone, M., Agren, H., and Rachlew, E. (2008). The C 1*s* and N 1*s* near edge x-ray absorption fine structure spectra of five azabenzenes in the gas phase. *J. Chem. Phys.* **128**, 044316.

Waldo, G. S., Carlson, R. M. K., Moldowan, J. M., Petters, K. E., and Penner-Hahn, J. E. (1991a). Sulfur speciation in heavy petroleum: Information from x-ray absorption near edge structure. *Geochim. Cosmochim. Acta* **55**, 801–814.

Waldo, G. S., Mullins, O. C., Penner-Hahn, J. E., and Cramer, S. E. (1991b). Determination of the chemical environment of sulfur in petroleum asphaltenes by x-ray absorption spectroscopy. *Fuel* **70**, 549–559.

Wan, J., Tyliszczak, T., and Tokunaga, T. K. (2007). Organic carbon distribution, speciation, and elemental correlation within soil microaggregates: Application of STXM and NEXAFS spectroscopy. *Geochim. Cosmochim. Acta* **71**, 5439–5449.

Wiesemann, U., Thieme, J., Guttmann, P., Fruke, R., Rehbein, S., Niemann, B., Rudolph, D., and Schmahl, G. (2003). First results of the new scanning transmission X-ray microscope at BESSY-II. *J. Phys. IV* **104**, 95–98.

Williams, S., Zhang, X., Jacobsen, C., Kirz, J., Lindaas, S., van't Hof, J., and Lamm, S. S. (1993). Measurements of wet metaphase chromosomes in the scanning transmission x-ray microscope. *J. Microsc.* **170**, 155–165.

Winn, B., Ade, H., Buckley, C., Feser, M., Howells, M., Hulbert, S., Jacobsen, C., Kaznacheyev, K., Kirz, J., Osanna, A., Maser, J., McNulty, I., Miao, J., Oversluizen, T., Spector, S., Sullivan, B., Wang, Y., Wirick, S., and Zhang, H. (2000). Illumination for coherent soft X-ray applications: The new X1A beamline at the NSLS. *J. Synchrotron Rad.* **7**, 395–404.

Xia, K., Weesner, F., Bleam, W. F., Bloom, P. R., Skyllberg, U. L., and Helmke, P. A. (1998). XANES studies of oxidation states in aquatic and soil humic substances. *Soil Sci. Soc. Am. J.* **62**, 1240–1246.

Young, I. M., and Crawford, J. W. (2004). Interactions and self-organizations in the soil-microbe complex. *Science* **304**, 1634–1637.

Yuan, G., Soma, M., Seyama, H., Theng, B. K. G., Lavkulich, L. M., and Takamatsu, T. (1998). Assessing the surface composition of soil particles from some podzolic soils by X-ray photoelectron spectroscopy. *Geoderma* **86**, 169–181.

Zhang, X., and Amelung, W. (1996). Gas chromatographic determination of muramic acid, glucosamine, galactosamine and mannosamine in soils. *Soil Biol. Biochem.* **28**, 1201–1206.

Zhao, F. J., Lehmann, J., Solomon, D., Fox, M. A., and McGrath, S. P. (2006). Sulphur speciation and turnover in soils: evidence from sulphur K-edge XANES spectroscopy and isotope dilution studies. *Soil Biol. Biochem.* **38**, 1000–1007.

Zubavichus, Y., Fuchs, O., Weinhardt, L., Heske, C., Umbach, E., Denlinger, J. D., and Grunze, M. (2004). Soft X-ray-induced decomposition of amino acids: An XPS, mass spectrometry, and NEXAFS study. *Rad. Res.* **161**, 346–358.

Zubavichus, Y., Shaporenko, A., Grunze, M., and Zharnikov, M. (2005). Innershell absorption spectroscopy of amino acids at all relevant absorption edges. *J. Phys. Chem. A.* **109**, 6998–7000.

18

THERMAL ANALYSIS FOR ADVANCED CHARACTERIZATION OF NATURAL NONLIVING ORGANIC MATERIALS

E. J. LEBOEUF AND L. ZHANG

Department of Civil and Environmental Engineering, Vanderbilt University, Nashville, Tennessee

Biophysico-Chemical Processes Involving Natural Nonliving Organic Matter in Environmental Systems, Edited by Nicola Senesi, Baoshan Xing, and Pan Ming Huang

18.1. INTRODUCTION

Used widely in synthetic macromolecular and natural biopolymer fields to evaluate structural and thermodynamic properties of macromolecular materials, thermal analytical methods have been applied to assist in the characterization of natural organic matter (NOM). Originally applied to whole soils, early thermal studies focused on qualitative and quantitative examination of soil constituents. Information derived from such analyses included water, organic matter, and mineral contents (Matejka, 1922; Tan and Hajek, 1977), composition of organic matter (Tan and Clark, 1969), and type of minerals (Matejka, 1922; Hendricks and Alexander, 1940). Additional early studies applied thermal analyses in a focused effort for NOM characterization, including structure (Turner and Schnitzer, 1962; Ishiwata, 1969) and NOM–metal complexes (e.g., Schnitzer and Kodama, 1972; Jambu et al., 1975a,b; Tan, 1978). Summaries of early thermal analytical methods for soils and humic substances may be found in Tan and Hajek (1977) and Schnitzer (1972), respectively, while more current reviews of thermal techniques are provided by Senesi and Loffredo (1999) and Barros et al. (2006).

Recently, thermal analytical methods for NOM characterization have been extended to include the material properties of specific heat capacity, thermal expansion coefficient, a variety of thermal transition phenomena including beta and alpha transitions (e.g., glass transitions), so-called step transitions, and NOM–solvent/mineral interactions. Methods reviewed in this chapter include thermal gravimetric analysis, differential thermal analysis, differential scanning calorimetry, temperature-modulated differential scanning calorimetry, thermal mechanical analysis, dynamic mechanical thermal analysis, and dielectric thermal analysis. Each analytical technique is presented in terms of a state-of-the-art literature review (including recent technological advances), followed by discussion of the potential for use as a tool to provide advanced characterization of NOM. We begin our discussion by first exploring environmentally relevant characteristics of NOM that may be conducive to application of thermal analysis techniques.

18.1.1. NOM and Environmental Relevance

NOM in the environment varies widely in composition, from newly deposited bits and pieces of biopolymers, to moderately aged humin, humic acid, and fulvic acid, to well-aged kerogen/coal formed over millions of years, and naturally and anthropogenically altered black carbon (e.g., charcoals and soots). NOM, though typically comprising only a small mass fraction of soils and sediments, widely exists in the environment and plays important roles in many environmentally related processes. For example, proper management of humic substances has been recognized as the heart of sustainable agriculture, and humic materials contribute to plant growth through their effects on the physical, chemical, and biological properties of soil (Stevenson, 1994). NOM may also serve as important sources of energy [e.g., kerogen (Benyamna et al., 1991)] and impact climatic conditions through absorption and scattering of solar radiation [e.g., black carbon (Goldberg, 1985)]. The role of NOM also possesses a significant influence on the transport and bioavailability of contaminants. NOM tends to increase the retention, inactivation, and persistence of contaminants in the environment by sorption/desorption behavior (Riise and Salbu, 1986). As such, significant efforts have been invested to better elucidate NOM's physical and chemical structures, functionality, and reactivity responsible for observed behaviors. A number of chapters in this book are devoted to illustrating the application of various physicochemical techniques to the characterization of NOM.

18.1.2. Motivation for Use of Thermal Analytical Techniques

While physicochemical and spectroscopic techniques elucidate valuable physical and structural information, thermal analysis techniques offer an additional approach to characterize NOM with respect to thermal stability, thermal transitions, and even interactions with solvents. Information such as thermal degradation temperature (or peak temperature), glass transition temperature, heat capacity, thermal expansion coefficient, and enthalpy can be readily obtained from thermal analysis; these properties, when correlated with structural information, may serve to provide additional insights into NOM's environmental reactivity.

Such thermodynamic properties may also be used as constraints to structural modeling of NOM. Spectroscopic methods have been employed to examine the compositional, structural, and conformational nature of NOM, facilitating the development of NOM models through adjustment of chemical functional group composition and their connectivities. In spite of the increased sensitivity and resolution of spectroscopic methods, along with improvements in computational capabilities, the heterogeneous nature of NOM continues to pose significant difficulties in providing a complete depiction of molecular structure. Efforts in comparing phenomenological experimental studies with simulated macroscopic properties, including thermal properties, provide an additional means to screen/select appropriate NOM structural models (Diallo et al., 2003). The main objectives of this chapter are to explore a number of thermal analysis techniques that have been, or may be, employed in the advanced characterization of NOM. We begin our discussion with an exploration of the material properties of NOM that make them amendable to thermal characterization.

18.2. MATERIAL PROPERTIES OF MACROMOLECULES

Studies supporting a macromolecular perspective for NOM are typically based upon analogies between NOM and synthetic and natural polymers in terms of thermal transition behavior (LeBoeuf and Weber, 1997; DeLapp and LeBoeuf, 2004; Schaumann et al., 2005) and sorption phenomena (e.g., nonlinear sorption and desorption hysteresis) described by a polymer-based dual domain sorption model (Vieth et al., 1976; Huang et al., 1997; LeBoeuf and Weber, 1997, 2000a,b; Xing and Pignatello, 1997; Xia and Ball, 1999; Xia and Pignatello, 2001; Gunasekara et al., 2003). [A more detailed discussion of the supramolecular and macromolecular views of NOM, while beyond the scope of this chapter, may be found in a recent critical review of this topic (Schaumann, 2006), as well as in Chapter 1 of this book).] Despite the ongoing debate of the supramolecular/macromolecular nature of NOM, it is generally accepted that NOM is predominantly amorphous in nature with an occasional presence of crystalline or microcrystalline regions. Thermal behaviors such as glass transition, crystallization, and crystalline melt may thus be expected to be observed in NOM as well. Since each of these thermal behaviors are widely observed in polymers, it may be beneficial to examine polymer physicochemical characteristics believed to be responsible for these behaviors, thus enabling improved interpretation of the similar behaviors observed in NOM. As such, we continue with a discussion of material properties shared by NOM and polymers while we treat them both as macromolecules.

For the purpose of clarity, the remainder of this chapter defines macromolecules as molecules in which at least 500 atoms are linked together by covalent bonds to form linear or three-dimensional networks. The term *polymer* is referenced as a natural or synthetic macromolecule possessing a relatively simple chemical structure consisting of identical repeating units. No further distinction between polymer and macromolecule is made.

18.2.1. Structure of Synthetic Organic Macromolecules

Phenomenologically, there are two types of polymer structures: thermoplastics and thermosets. Thermoplastics are characterized by (a) their ability to soften when heated and (b) their capacity to flow upon application of stress. Upon cooling, they reversibly regain their original rubbery behavior. Thermosets are similar in their ability to soften when heated, but are differentiated from thermoplastics by their inability to return to their original state upon cooling. Often, heating of thermosets results in a curing reaction, and continuation of heating results in thermal degradation. Rosen (1993) notes that rubber provides a classic illustration of the two types of polymer structures. Natural rubber is a thermoplastic that becomes soft and sticky at temperatures slightly greater than ambient. However, heating rubber in the presence of sulfur results in a curing reaction that converts the natural polymer into a thermoset upon cooling. This process, called vulcanization by its inventor, Charles Goodyear, allows the new rubber to perform at significantly elevated temperatures while maintaining most of the desirable properties of a natural rubber.

Polymers are more fundamentally categorized as linear, branched, and cross-linked structures. Linear polymers (e.g., polyesters) are constructed from strictly difunctional monomers, while branched polymers are produced from the introduction of tri- or higher-functional monomers within linear chains. Linear and branched polymer chains are held together by hydrogen bonds, dipole interactions, van der Waals forces, or ionic bonds. Addition of longer and more frequent multifunctional monomers allows for greater interaction between individual polymer chains, along with the eventual formation of a fully connected, three-dimensional polymer network bonded by covalent cross-links, or a so-called cross-linked polymer. This cross-linked polymer is normally formed either from chemically created cross-links between previously formed linear or branched chains (e.g., vulcanization) or by addition of sufficient quantities of multifunctional monomers during the original polymerization process. In terms of natural systems, we can think of diagenesis as the process of chemically creating additional cross-links between relatively homogeneous biochemical monomer or biopolymer chains, resulting in an everincreasingly complex cross-linked structure.

In general, exposing a polymer to increasing thermal energy can cause dissociation of interchain bonds. Linear and branched polymers, with relatively low-energy interchain bonds, can undergo dissociation with relatively little energy input, thus allowing the chains to slide freely past one another and to flow under an applied stress. Cross-linked polymers, with high-energy covalent bonding providing linkages to both main-chain molecules and interchain links (cross-links), require much larger energy input to pull the bonds apart. These polymers can undergo dissociation of both cross-link and main-chain bonds at the same time, resulting in little or no flow and ultimate degradation of the polymer.

Exposing polymers to "good" solvents (i.e., solvents that can form noncovalent bonds with polymer chains) can cause replacement of interchain bonds with solvent molecules, resulting in dissolution of linear and branched polymers. The stronger, covalent interchain bonds of cross-linked polymers are not affected by these solvent interactions and thus normally do not dissolve. However, the un-cross-linked portion of the polymer can undergo solvation, resulting in swelling of the polymer matrix about the cross-linkages. Thus, increased chain length, or molecular weight, between cross-links can result in increased swelling, whereas increased cross-linking—and thus shorter chain lengths—can result in reduced polymer swelling.

18.2.2. Molecular Weight

Molecular weight constitutes a basic physical property of NOM. Unlike a pure compound, which possesses a single molecular weight, NOM is heterogeneous and polydispersed and thereby may exhibit a wide range of molecular weights, ranging from several hundred to several hundred thousand daltons depending on origin and diagenetic history (Behar and Vandenbroucke, 1987; Stevenson, 1994; Gaffney et al., 1996; Geoffrey and Ghabbour, 1998). Notably, reported molecular weights can vary considerably, even for identical NOM. These discrepancies may arise in part from application of different experimental techniques and treatment conditions, but they may also be complicated by the possibility that NOM molecules can aggregate to form micellar structures or be bridged by multivalent cations, suggesting that select techniques can contribute to specific experimental artifacts and lead to overestimation of molecular weight (Guetzloff and Rice, 1996, Piccolo et al., 1996; Conte and Piccolo, 1999).

Polymers normally consist of a number of different chain lengths characteristic of the degree of polymerization, with each chain length comprising different molecular weights. Given a range of molecular weights, the polymer is usually characterized in terms of a number-average, M_n, or weight-average, M_w, molecular weight. Number-average molecular weight is calculated by

$$M_n = \frac{W}{N} = \sum_{x=1}^{\infty}\left(\frac{n_x}{N}\right)M_x \tag{18.1}$$

where W is the total weight of the sample [M], N is the total number of moles (of all chain length sizes) in the sample [moles], n_x is the number of moles of x-mer [moles], and M_x is the molecular weight of x-mer [M/mole]. Weight-average molecular weight is determined by

$$M_n = \frac{W}{N} = \frac{\sum_{x=1}^{\infty} w_x M_x}{\sum_{x=1}^{\infty} w_x} = \sum_{x=1}^{\infty}\left(\frac{w_x}{W}\right)M_x = \frac{\sum_{x=1}^{\infty} n_x M_x^2}{\sum_{x=1}^{\infty} n_x M_x} \tag{18.2}$$

where w_x is the total weight of x-mer [M].

Several analytical procedures for determination of molecular weights exist, including membrane and vapor pressure osmometry, end-group determination or identification of number-average molecular weight, and light-scattering or ultracentrifugation for weight-average molecular weight calculations (Kroschwitz, 1990; Rosen, 1993). Size-exclusion chromatography and ultrafiltration are primarily used to determine the molecular weight distribution (Provder, 1984; Kilduff and Weber, 1992). The ratio of weight-average molecular weight to number-average molecular weight, referred to as the polydispersivity index (M_w/M_n), also gives an indication of breadth of the molecular weight distribution. The closer M_w/M_n is to unity, the greater similarity (with respect to total volume and chemical composition) of the individual polymer chains. Polymers normally span a polydispersivity index of 1.5–2.0 or greater, depending on the polymerization process (Kroschwitz, 1990).

Molecular weight information is especially useful since variances in molecular weight are shown to impact adhesion and brittleness properties (Kroschwitz, 1990), solubility (Flory, 1970), and sorption and mass transfer characteristics of organic solutes (Kilduff and Weber, 1992). The impacts of varying molecular weight on polymer crystallinity, solubility, and, to a lesser extent, the glass transition temperature are discussed in the following sections.

18.2.3. Crystallinity

Crystallinity refers to a regular, ordered, three-dimensional crystal lattice portion of a polymer (Kroschwitz, 1990) where the polymer chains align themselves in perfect parallel array. There are no purely crystalline polymers, as all polymers, even so-called crystalline polymers, have some portion of amorphous content. Nonetheless, crystalline regions of polymers can form up to 98% of the polymer structure (Rosen, 1993) and have consequent large impacts on mechanical behavior.

Crystalline regions are formed by noncovalent bonds between polymer chains of similar structure which are strong enough to overcome the disordering effects of thermal energy. Hence, polymers with a high crystalline melting point (T_m) (i.e., that temperature at which the crystalline regions begin to come apart) generally are bound together by stronger interactions (such as hydrogen or dipole bonds) than are those polymers with lower T_m's. Variances in the degree of crystallinity also obtains from differential thermal histories of samples. For example, a normally semicrystalline polymer can be formed in a totally amorphous state by rapid quenching of the sample from a temperature above the T_m. This rapid cooling causes the polymer chains to "freeze" in an expanded, amorphous position by not providing sufficient time or thermal energy at the lower temperature to allow polymer chain reallignment into crystalline regions.

Polymers with relatively homogeneous chain compositions are also more likely to form crystalline regions than polymers with irregular chain structures that may have a number of protruding side functional groups (Rosen, 1993). Given the relative heterogeneity of soil or sediment organic macromolecules, one would thus expect the occurrence of crystallinity in these natural systems to be rare, with only very small or so-called microcrystalline regions present. However, for more homogeneous biopolymers or largely diagenetically altered natural organic matter (e.g., coals), one may expect to find significantly larger regions of crystallinity.

The physical properties of semicrystalline polymers have been described using a number of models, including the fringed micelle model (Elias, 1977), the (more recent) folded-chain model (represented in Figure 18.1), and reptation theory. In the folded-chain model, the crystalline regions (or so-called crystallites) are represented by portions of polymer chains tightly folded onto themselves in parallel array of nanometer dimensions. Since the polymer chains are on the order of micrometers in length, they tend to extend from one crystalline region to another through less dense amorphous regions. The crystalline regions of polymers are similar to crosslinks in that the polymer chains are interconnected by the crystallites, but differ in that exposure to increased thermal energy or a "good" solvent will tend to separate and dissolve the polymer chains contained in crystallites, while the polymer itself will degrade at the higher temperatures required for disintegration of covalently bonded cross-links.

Reptation theory description of polymer structure is analogous to a bowl of live snakes (Teraoka et al., 1992). In this "bowl" reside a mesh of entangled, linear flexible polymer chains that continue to wriggle within a minimal range, effectively forming a tube-like structure. It is within this tube that the polymer chains move back and forth; and over sufficient periods of time, the polymer chain can actually move along the tube to new interaction sites with fellow polymer chains or other media.

As one might imagine, an increasing presence of crystallinity imparts a rigidity or stiffness to a polymer structure. In semicrystalline polymers significantly below their melting point, crystallinity imparts an inaccessibility to a portion of the polymer (Barton, 1990). Considering the polymer as a whole, this results in a lower average solute concentration at equilibrium during hydrophobic organic compound (HOC) sorption compared to the actual solute concentration in the amorphous regions. This, of course can have profound influences on the rate of uptake of solutes within polymeric matrices.

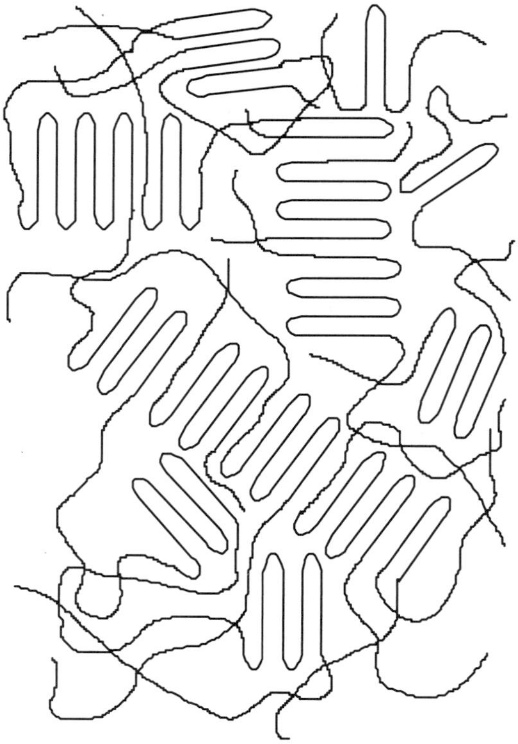

Figure 18.1. Folded-chain model for crystallinity. Adapted from Rosen (1993). Reproduced by permission of John Wiley & Sons, Ltd.

The degree or amount of crystallinity is determined by a number of different methods, including small-angle X-ray diffraction. In this technique, the scattering of X rays in the small angle (so-called Bragg angle) region by crystalline sections is differentiated from the broad diffraction pattern characteristic of amorphous regions (Kroschwitz, 1990).

18.2.4. Thermodynamic States of Polymers

Amorphous polymers are known to exhibit two distinct forms of mechanical behavior, depending on whether they exist in a glassy or rubbery so-called "thermodynamic state." (*Note*: The phrase "thermodynamic state" appears in quotes because, as will be explained in the next section, there is not a complete theoretical understanding of the glassy state, since its existence appears to rely on both thermodynamics and kinetics.) The existence of a macromolecule in either of these states can result in widely varying HOC sorption behavior, where nonlinear sorption, slow, non-Fickian diffusion, and competitive multi-solute sorption are attributed to glassy regions; and linear, partition-like sorption, relatively fast, Fickian-type diffusion, and no multi-solute competitive sorption are attributed to rubbery regions (LeBoeuf and Weber, 1997).

Glassy polymers, or glasses, are differentiated from rubbery polymers by their hard, rigid, glass-like structure [e.g., poly(methyl methacrylate) (PMMA) (i.e., Lucite®, Plexiglas®)] at room temperature. Rubbery polymers (e.g., polyethylene, rubber) are soft, and flexible at room temperature. This flexible nature of rubbery polymers can be attributed to additional long-range (multi-molecule) motions of polymer chains not fully present in glassy polymers. For example, heating a glassy polymer such as PMMA significantly above room temperature (to approximately 120 °C) will result in rubber-like behavior. This is a result of the additional thermal energy available at elevated temperatures that allow an increasing number and magnitude of molecular motions, with resulting increases in volume (larger molecular motions creating greater free volume) and heat capacity (larger molecular motions are better able to dissipate energy) of the polymer. In similar fashion, cooling a rubbery polymer results in less long-range molecular motion, along with consequent stiffness and rigidity of the polymer, such as what may occur to a less expensive garden hose on a frosty morning (with consequent frustration of the individual attempting to roll the garden hose into a tight bundle). The temperature (or more accurately, range of temperatures) at which increased molecular motions lead to rubbery behavior is referred to as the glass transition temperature, or T_g. Glass transitions have been observed in a number of different materials besides high-molecular-weight organic polymers, including silica-based glasses (SiO_2), specially prepared salts ($ZnCl_2$) and metals, and even low-molecular-weight species such as 2-methyl pentane (Mark et al., 1984).

As previously noted, a material's T_g is manifested in its ability to move large molecular segments. Table 18.1 provides a summary of the expected molecular motions of polymers in order of increasing activation energy. Rubbery polymers normally possess all of the above motions, while glassy polymers possess motions only up to small groups of five or six atoms, although larger motions can manifest themselves in glassy polymers over extended periods of time. To appreciate how a reduction in thermal energy restricts volumetric relaxation, Eisenberg (1993) indicates that the volume relaxation time for polystyrene, with a T_g of 100 °C, is on the order of 10^{-2} s; at 95 °C it is about 1 s; at 85 °C, 5 h; at 79 °C, 60 h; and at 77 °C, about 1 y.

TABLE 18.1. Molecular Motions of Polymers with Increasing Thermal Energy

Increasing activation energy ↓	
	1. Wriggling, or vibrations of atoms about equilibrium positions. For crystalline regions of polymers, this vibration is about a fixed position, while for amorphous polymers the motion is about a less-ordered atomic center.
	2. Wriggling of side functional groups on the main chain (backbone) of a polymer, or small groups of atoms (5 or 6) along the backbone.
	3. Segmental movement of polymer chains 40–50 carbon atoms in length, permitting flexing of individual chains, leading to elasticity.
	4. Flow due to translational motion of entire molecules.

Source: After Rosen (1993).

As noted above, cooling a rubbery polymer more rapidly than the polymer chains can reconfigure themselves leads to quenching, or temporary "freezing" of the polymer chains in a more expanded configuration. Over time, however, the continual wriggling of glassy segments leads to a lower energy configuration, and ultimate return of the polymer segments to their less expanded, lower energy state. This process is known as relaxation and is a direct result of the nonequilibrium state in which many glassy polymers reside. Hence, during relaxation, the mechanical properties of a glassy polymer may change with time, and this is often referred to as physical aging (Struik, 1978). With respect to transitions from the glassy state to the rubbery state, the thermal, configurational, and temporal histories directly influence the activation energy required to cause an abrupt transition in physical behavior, as measured by the T_g.

18.2.4.1. Thermodynamic and Kinetic Basis of the T_g. Given the rather straightforward definition of T_g as the demarcation between glassy and rubbery states, it appears that measuring this transition region can be accomplished fairly easy by plotting the volume change with temperature, or the heat capacity change with temperature. As shown in Figures 18.2 and 18.3, this is indeed the case. Figures 18.2 and 18.3 support a thermodynamic conclusion about the glass transition temperature—that is, that it manifests itself in an abrupt, discontinuous change in heat capacity (and also coefficient of thermal expansion) as a function of temperature. This is described thermodynamically as a second-order transition and is illustrated by addressing the Gibbs free energy equation

$$G = H - TS \tag{18.3}$$

where enthalpy (H) is defined as

$$H = U + pV \tag{18.4}$$

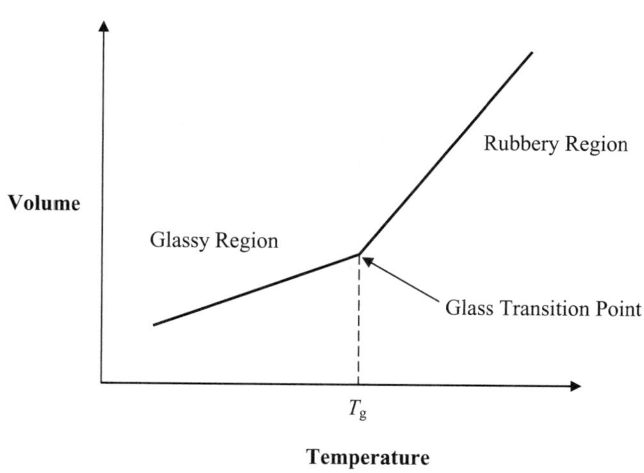

Figure 18.2. Total volume versus absolute temperature.

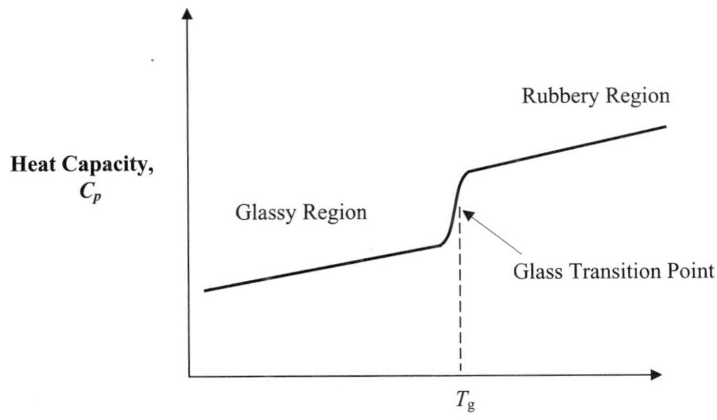

Figure 18.3. Constant pressure specific heat capacity, C_p, versus absolute temperature.

The Gibbs free energy may then be expressed as

$$G = U + pV - TS \tag{18.5}$$

So-called first-order transitions include evaporation or fusion, where volume (V), entropy (S), and enthalpy (H) all exhibit a discontinuity upon differentiation of Eq. (18.5) with respect to state variables pressure (p), or temperature (T).

$$\left(\frac{\partial G}{\partial T}\right)_p = -S \tag{18.6}$$

$$\left(\frac{\partial G}{\partial p}\right)_T = V \tag{18.7}$$

$$\left(\frac{1}{T}\right)G = \left(\frac{1}{T}\right)U + \left(\frac{1}{T}\right)pV - S \tag{18.8}$$

$$\left(\frac{\partial (G/T)}{\partial (1/T)}\right)_p = H \tag{18.9}$$

The glass transition is characterized in part by an observed second-order transition distinguished by a discontinuity of the Gibbs free energy with respect to the aforementioned state variables, but by continuity of entropy, volume, and enthalpy. Hence, heat capacity, C_p, as well as the thermal expansion coefficient, α, as defined below, both exhibit a discontinuity at the glass transition temperature (McKenna, 1989).

$$\frac{\partial}{\partial p}\left(\left(\frac{\partial G}{\partial p}\right)_T\right)_T = \left(\frac{\partial V}{\partial p}\right)_T = \alpha V \tag{18.10}$$

where

$$\alpha = \frac{1}{V}\left(\frac{\partial V}{\partial p}\right)_T \tag{18.11}$$

and

$$\frac{\partial}{\partial T}\left(\left(\frac{\partial(G/T)}{\partial(1/T)}\right)_p\right)_T = \left(\frac{\partial H}{\partial T}\right)_p = C_p \tag{18.12}$$

Normally, true thermodynamic, second-order transitions are found to have no dependence on time. However, as shown in Figure 18.4, this is not necessarily the case. Phenomenologically, the glass transition temperature is specific to thermal histories, with T_g varying as a function of sample heating or cooling rates. Recall that the time scale of structural response (i.e., relaxation) is highly dependent on temperature. When the sample temperature is well above T_g, the relaxation rate of the macromolecule is generally much faster than experimental heating/cooling rates, resulting in samples being able to quickly assimilate added/subtracted heat energy, enabling quick attainment of equilibrium between sample structural configuration and temperature. If the sample temperature is lowered to near T_g, however, the relaxation rate may be of the same order of magnitude as the experimental time scale. At sample temperatures below T_g, the relaxation rate is typically much slower than the heating/cooling rate such that materials simply do not have sufficient time

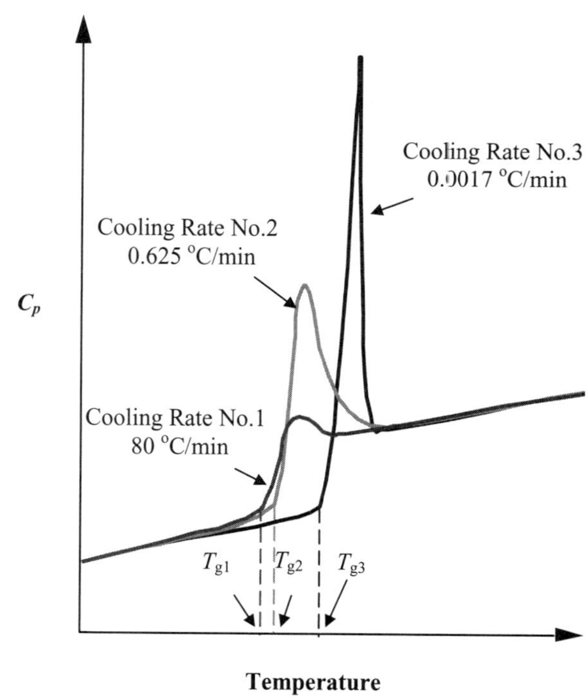

Figure 18.4. Heat capacity curves for poly(styrene) cooled at different rates. Adapted from Richardson (1989). Reproduced by permission of Elsevier, Ltd.

to reach equilibrium between temperature and structural relaxation. Similar behaviors have been observed for NOM subject to differential heating rates (LeBoeuf and Weber, 2000a).

One theory that describes the temperature dependence of relaxation time and structural recovery is the Tool–Narayanaswamy–Moynihan (TNM) model developed to describe the often nonlinear relationship between heating rate and T_g. In this model, the structural relaxation time, τ, is referenced as a function of temperature (T), activation enthalpy (Δh^*), universal gas constant (R), fictive temperature (T_f), and nonlinearity factor (x) (Tool, 1946; Narayanaswamy, 1971; Moynihan et al., 1976):

$$\tau = \tau_0 \exp\left[\frac{x\Delta h^*}{RT}\right]\exp\left[\frac{(1-x)\Delta h^*}{RT_f}\right] \tag{18.13}$$

Here, T_f is used to characterize the entropy change of the cooperative motions of the macromolecule as a function of temperature. For melts, T_f is equivalent to the temperature, T, of the thermal bath (Schawe, 1998). The activation enthalpy can be obtained directly from observations in the change of T_g with changes in cooling rate, q (Moynihan et al., 1974)

$$\frac{d\ln(q)}{d(1/T_g)} = -\frac{\Delta h^*}{R} \tag{18.14}$$

and the non-linearity factor, x ($0 \le x \le 1$), can be obtained through experimental fitting (note that for $x = 1$, Eq. (18.13) is transformed to an Arrehenius expression). The TNM model has been found to be a successful predictor of material behavior in the T_g region (Schawe, 1998; Hodge and Huvard, 1983; Privalko et al., 1986).

Although the T_g may shift due to different thermal histories, the actual change in heat capacity or thermal expansion coefficient does not (McKenna, 1989). It is worthwhile at this point to also note the large endothermic response for the less rapidly cooled polymer. This "enthalpic overshoot" is attributed to an overrelaxation of polymer chains and is characteristic of (a) the thermal history of the polymer and (b) the heating rate of the experiment. If the experiment was conducted at a rate similar to the relaxation rate of the polymer, then little to no enthalpic overshoot would be expected. The following two sections explain the glass transition in terms of the Gibbs–DiMarzio configurational statistical thermodynamics model and the free volume theory model.

18.2.4.2. Gibbs-DiMarzio Theory. Gibbs–DiMarzio theory predicts a true second-order transition to occur at one temperature (at constant pressure), T_2 where the configurational entropy contribution to free energy approaches zero. Gibbs–DiMarzio assume that each polymer chain has a lowest energy shape, and that configurational deviations from this lowest energy state increases the internal energy of the molecule. This internal energy is composed of the energy associated with the flexing of polymer chains out of their lowest energy state, as well as "hole" energy attributed to the number of intermolecular bonds broken by introduction of vacancies, or "holes" into a polymer lattice due to increased flexing or wriggling action. At temperature T_2, there is not enough thermal energy to cause the larger scale

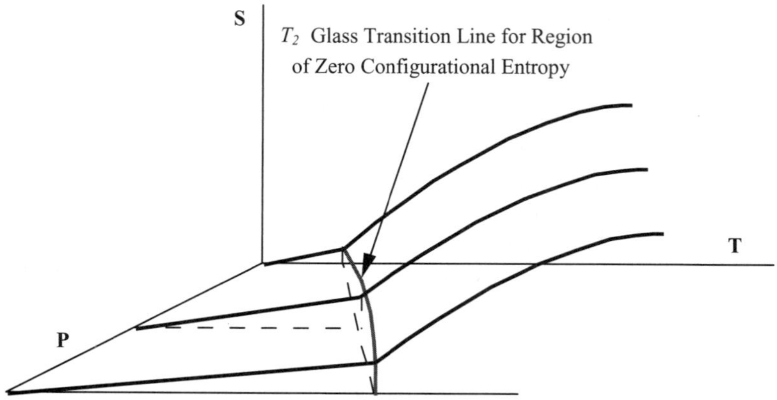

Figure 18.5. Gibbs–DiMarzio configurational entropy glass transition model representation of T_2. Reproduced from DiMarzio (1981), by permission of John Wiley & Sons, Ltd.

wriggling, and hence below this temperature, the molecules remain in their low energy configuration. The theoretical transition temperature, T_2, is illustrated on an entropy–pressure–temperature (SPT) plot in Figure 18.5.

The Gibbs–DiMarzio theory is very important for two reasons. First, although the temperature, T_2, never obtains (Eisenberg, 1993), the theory allows for an explanation of why one observes glass to rubber transitions at different temperatures for the same polymer if the polymer is aged in different manners or is subjected to different thermal histories. For example, in Figure 18.6, the raising of the observed glass transition temperature as the rate of initial cooling decreases can be attributed to the ability of the less quickly cooled polymer to have more time to move more closely to equilibrium configurations (and lower total volumes). This results in a movement in the glass transition to values closer to the true thermodynamic transition, T_g. In similar fashion, an "aged" polymer will have a higher T_g than a younger polymer (given similar thermal histories). In addition, the rate at which the glass transition experiment takes place (i.e., duration) has an impact on the observed values of T_g. In this case, a polymer heated at a slower rate will have a lower T_g than its more quickly heated counterpart (given similar "ages" and previous thermal history) because the relaxation time of the polymer heated at the faster rate is much longer than the time scale of the shorter experiment (McKenna, 1989).

Second, Gibbs–DiMarzio theory allows for an explanation for behavior of atactic polymers (of which most natural organic polymers are composed), which cannot crystallize due to an irregular backbone structure. Essentially, this theory predicts a "fourth thermodynamic state of matter" (besides gas, liquid, and solid) for glassy polymers that cannot form crystalline structures, the lowest-energy state for solids (McKenna, 1989). Basically, "aged," atactic glassy polymers exist in a metastable, lowest-energy state.

18.2.4.3. Free Volume Theory. Free volume theory suggests that the glass transition temperature is observed for polymers when their viscosity approaches that of their liquid state. Following a derivation based on the Doolittle expression for polymer viscosity (η) as a function of free volume (Eisenberg, 1984)

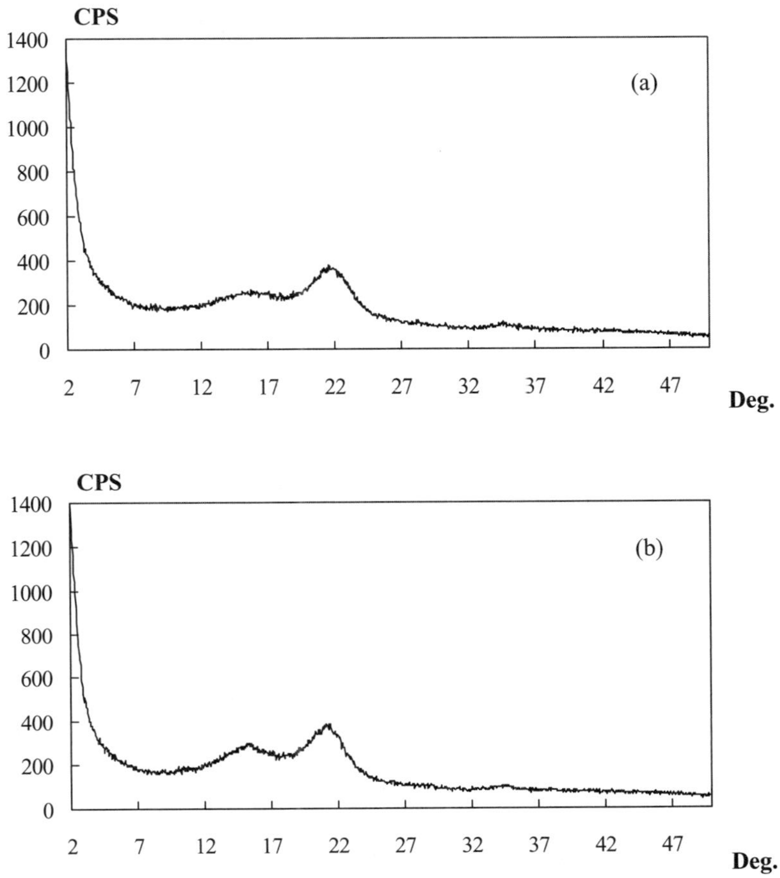

Figure 18.6. X-ray diffraction spectrum of poplar wood at (**a**) 20 °C and (**b**) 180 °C.

$$\eta = Ae^{\left(B\frac{Vp}{Vf}\right)} \tag{18.15}$$

where A and B are constants, V_p is the volume of the polymer itself, and V_f is the free volume. This theory suggests that substantially less viscous behavior occurs at a free volume fraction, f,

$$f = \frac{Vf}{Vp + Vf} \cong 0.025 \tag{18.16}$$

for a wide range of materials. [Although the derivation of this theory is rather straightforward, it is beyond the scope of this chapter, and the reader is referred to Eisenberg (1993).] The advantage of this theory is its rather simple relationship to polymer free volume, along with the ability to easily visualize the impacts of a number of variables affecting free volume on determination of the glass transition temperature.

18.2.4.4. Variables Influencing T_g. Since T_g is a function of macromolecular mobility, any changes to the macromolecular structure that increases or decreases this mobility will have an effect upon T_g. The following discussion highlights several variables that can impact the observed glass transition temperature.

Molecular Weight. Shorter polymer chains give rise to an increased number of wriggling polymer end groups (which generally have larger motions than mid-chain molecules). This increased wriggling consequently results in an increase in free volume, and a reduction in the glass transition temperature. Fox and Flory (1948, 1951, 1954) developed an empirical correlation where T_g varies inversely with the number-average molecular weight, M_n:

$$T_g = T_g^{\infty} - \frac{B}{M_n} \tag{18.17}$$

where T_g^{∞} is the glass transition for a polymer of infinite chain length, and B is a constant characteristic of the individual polymer. Examination of this equation shows that little reduction in T_g is realized for polymers with relatively large molecular weights, as is the case with most synthetic polymers (Rosen, 1993).

Cross-Linking. Increased cross-linking restricts chain mobility of larger segments. Additionally, crosslinks of shorter covalent bonds replace van der Waals bonds between adjacent polymer chains, resulting in a decrease in the total volume occupied by the polymer (McKenna, 1989). The reduction in chain mobility and free volume give rise to increased glass transition temperatures. However, Eisenberg (1993) notes that addition of a crosslinking agent, such as divinyl benzene, into a polymer matrix can also give rise to the same effect that the addition of a monomer or other polymer, forming a copolymer-type matrix (i.e., the polymer may swell due to the presence of monomer, resulting in a lower T_g).

Solubility Parameter. Increased attractive forces between molecules, as measured by a larger solubility parameter, requires more thermal energy to produce molecular motion. This results in an increase in the T_g with increasing σ_p. Eisenberg (1993) provides an equation that relates the molar cohesive energy density, σ_p^2, to the glass transition by

$$\sigma_p^2 = 0.5(mRT_g) - 25\,m \tag{18.18}$$

where m is a parameter whose value is a function of the ability of the polymer chains to rotate and R is the universal gas constant (unfortunately, units of the variables used in the above equation were not identified by the author). Nonetheless, the above relationship shows a linear dependence of the cohesive energy density with the glass transition temperature (Barton, 1983).

Pressure. Isothermally increasing the pressure applied to a polymer also reduces molecular mobility, and consequently, free volume. Thus, the T_g will generally increase with increased pressure [approximately 20 °C increase in T_g per thousand atmosphere increase in pressure (Eisenberg, 1984)].

Stiffness of Polymer Chains and Impact of Side-Chain Functional Groups. The internal mobility of polymer chains is primarily affected by the size of the side chain, or substituent functional groups attached to one of the carbon–carbon bonds of the backbone (Tobolsky, 1960). In general, the larger the side-chain group, the greater the activation energy required to move or rotate the chain. For example, poly(propylene) with a methyl group attached to every other carbon atom has a T_g of −10 °C, while poly(styrene), with a phenyl group attached in the same location, has a T_g of about 100 °C (Eisenberg, 1993). Polymer chains with aromatic backbones or parallel bonds in their backbone have extremely stiff bonds, with resulting reduction in molecular mobility and increased glass transition temperature (Rosen, 1993).

Free Volume. An increase in the free volume of the macromolecule (or that volume not occupied by the molecules themselves) allows more room for the molecules to move around, and thus an accompanying reduction in the T_g. Therefore, swelling of a macromolecule by a thermodynamically compatible solute (i.e., possessing similar σ_p values as the sorbent) will tend to increase the free volume and lower the T_g (Kelley and Bueche, 1961; Haward, 1973; Barton, 1983). Lucht et al. (1987) developed a relationship between the glass transition temperature of several coals and the weight fraction of pyridine within the coal network,

$$T_g = 6121 - 998.5\, w_p + 1260\, w_p^2 \qquad (18.19)$$

where w_p is the mass fraction of pyridine

$$w_p = \frac{\left(\dfrac{M_p}{M_c}\right)}{\left(\dfrac{M_p}{M_c}+1\right)} \qquad (18.20)$$

and where M_p and M_c represent the mass of pyridine and coal, respectively. This relationship was further developed to express the amount of solute, on a mass fraction basis, that is required to depress the glass transition temperature so that the matrix becomes rubbery at the experimental temperature (Lucht et al., 1987),

$$\frac{M_p}{M_c} = \frac{(T_g - T_{\text{experiment}})}{\beta/\alpha_t} \qquad (18.21)$$

where β represents the free volume contribution constant of the solute in the coal network, and α_t represents the free volume constant of a solute-free coal. Values for β of 0.15 (Fujita and Kishimoto, 1958) and α_t of 3.7×10^{-4} K (Ferry, 1980) are found to hold for most polymers.

Copolymerization. Introducing a comonomer (one of the compounds comprising the monomer) with known glass transition temperature, T_{g1}, into a polymer with known T_{g2} to form a copolymer (a polymer comprised of two or more monomers) can affect the glass transition temperature. If the two homopolymers (polymer formed from a single monomer) are introduced in random order and are compatible, one intermediate glass transition temperature will be observed according to the empirical relation (Elias, 1977; Eisenberg, 1993; Rosen, 1993)

$$\frac{1}{T_g} = \frac{W_1}{T_{g1}} + \frac{W_2}{T_{g2}} \tag{18.22}$$

where W_1 and W_2 represent the mass fraction of homopolymer 1 and 2, respectively, within the copolymer. However, if the two homopolymers are not compatible, or are formed in a block fashion where segments or blocks of polymer A are attached to segments of polymer B (e.g., AAAABBBBAAA versus random polymerization, ABAABABABABABBABBBAAB), then two distinct glass transition temperatures may be observed.

18.2.4.5. Natural Polymers and T_g. With respect to the glass transition in natural macrolecular systems, the process of diagenesis can be viewed as the action of converting relatively young, expanded, lightly cross-linked organic matter into a more condensed, highly cross-linked, more aromatic (more "stiff"), glassy structure with consequent increasing glass transition temperatures. Following this view, it is expected that "younger" soil organic matter such as humic and fulvic acids, although polar in nature (and thus likely possessing large σ_p values), may possess lower glass transition temperatures than do much more diagenetically advanced, or "aged" NOM such as kerogen. Coal, a highly aromatic material derived from kerogen, has been shown to undergo a glass transition ranging from 307 °C to 359 °C (Lucht et al., 1987), with increasing glass transition temperature shown to be a function of increasing carbon presence in the coal structure. Biopolymers, such as cellulose, have also exhibited glass transition temperatures, and have shown a remarkable reduction in their T_g in the presence of water (Akim, 1978), a "good" swelling solvent for cellulose. Many dry lignins have been observed to undergo a glass transition near 200 °C (Bouajilae et al., 2005), while lignins in aqueous solution typically have observed glass transitions between 80 °C and 100 °C (Salmen, 1984). In similar fashion, reductions in the glass transition temperature of other polar polymers such as humic and fulvic acids in aqueous solution is also expected. Furthermore, in most natural systems composed of several different components of organic macromolecules of varying cross-link densities and molecular weight, there likely exists a composition or range of glass transition temperatures. If the polymers are relatively compatible with one another (i.e., have similar chemical structure), then one may observe one relatively broad glass transition. However, given the relatively heterogeneous nature of NOM, as well as the probable block-type polymerization process of natural systems, one may expect to find a range of glass transition temperatures existing within a soil particle at any one time. This, unfortunately, can significantly increase the difficulty in actually measuring T_g of heterogeneous natural systems. It is also quite possible, as is the case for cellulose, that the observed T_g in the dry state of some components of NOM may be above the beginning of thermal degradation. This normally does not lend itself to ease of measurement. However, in the case of cellulose, subjecting it to compatible, swelling solvents allows sufficient depression of the T_g below the point of thermal degradation, and subsequent measurement of the "wet" T_g, that extrapolation to the dry T_g is possible (Akim, 1978). In addition, the attachment of NOM to mineral surfaces may impose sufficient cross-links to cause a large increase in the T_g; however, once removed from these mineral bonds (as may be the case in soil solvent extraction or demineralization), the polymer segments may revert to more rubbery behavior (Teraoka et al., 1992).

18.3. THERMAL BEHAVIORS OF NOM

18.3.1. Thermal Degradation and Moisture Loss

Similar to synthetic polymers, NOM will also experience thermal degradation upon thermal treatment when the temperature is sufficiently high to cause the scission of molecular segments and evolution of light compounds with consequent mass loss. Compared with the relatively abrupt weight losses observed in synthetic polymers at certain temperatures, NOM, as a heterogeneous mixture of many constituents, typically displays a somewhat more continuous mass loss pattern. Nevertheless, NOM is expected to share some similarities in mass evolution and degradation patterns upon heating. For example, NOM in the natural environment typically contains a fair amount of moisture, where mass loss occurring below 110 °C can generally be attributed to evolution of physisorbed water. Mass loss may also be associated with thermal degradation of aliphatic and aromatic moieties of NOM, with degradation of aliphatic components typically occurring at lower temperatures relative to degradation of the aromatic constituents. Measurement of moisture loss and thermal degradation finds its importance in interpretation of NOM thermal stability, but it has also been extended to assist in the interpretation of NOM maturation (Dell'Abate et al., 2000) and elucidation of NOM–solvent interactions, such as water/NOM wetting/drying cycles (McBrierty et al., 1996) and their influence on the overall physicochemical properties of NOM (Schaumann and LeBoeuf, 2005). Sections 18.4.2 (TGA) and 18.4.3 (DSC) of this chapter further elucidate the experimental protocols and applications of thermal degradation and NOM-solvent interaction studies.

18.3.2. Thermal Transitions

Upon heating or cooling, a material may experience one or several thermal transitions, such as crystallization, crystalline melt, or glass transition, depending on the amorphous character of the material (i.e., fully amorphous versus semicrystalline) and thermal history (e.g., heating rate). As discussed from the perspective of thermodynamic theories in the preceding sections, crystallization, crystalline melts, and glass transitions display different patterns in terms of plots of thermodynamic properties as a function of temperature. Crystallization and crystalline melts are categorized as first-order thermal transitions, which are characterized by a discontinuity in the first partial derivatives of the Gibbs free energy, G (e.g., volume V and enthalpy H). Consequently, volume change and latent heat are expected to accompany a first-order thermal transition, although temperature remains constant in the transition. Albeit a first-order thermal transition, crystallization differs from crystalline melts in that crystallization is an exothermic process manifesting an exothermic peak as exemplified in a typical differential scanning calorimetry (DSC) plot (see, for example, Figure 18.15 in Section 18.4.2), whereas crystalline melting is an endothermic process, with corresponding observation of endothermic peaks. The integrated areas under the exothermic and endothermic peaks represent the latent heats associated with each thermal behavior.

Glass transitions represent second-order transitions, demonstrating discontinuities in second partial derivatives of the Gibbs free energy (i.e., specific heat capacity C_p and thermal expansion coefficient α) with respect to relevant state variables (i.e., P and T), but by a continuity in both Gibbs free energy G and its first partial derivatives (i.e., enthalpy H, volume V, and entropy S). Therefore, an abrupt change in C_p and α can be expected in a glass transition. Macromolecules displaying both first-order and second-order thermal transitions suggests the presence of both amorphous and semicrystalline phases, with poly(ethylene) representing a typical example. Nevertheless, this may not be the case in synthetic polymers alone, because crystallinity can also exist in soil and sediment organic matter.

Given the amorphous nature of soil and sediment organic matter, the occurrence of large crystalline components in these materials is less likely relative to that in homogeneous synthetic polymers. Nevertheless, smaller, more localized regions of crystallinity, or microcrystallinity, may exist in soil and sediment organic matter, especially for highly diagenetically altered materials such as kerogens (e.g., Aouad et al., 2002; Bhargava et al., 2005; Oh et al., 1990). A recent NMR study on humins, surface soil, and humic acids also detected crystalline domains composed of poly(methylene) chains (Hu et al., 2000). Characterization of crystalline domains normally involves use of X-ray diffraction (XRD), especially so-called small-angle XRD. Crystalline domains provide sharp Bragg reflection peaks due to orderly arrangement/packing of molecular chains, while these peaks are absent in amorphous domains because of irregular chain structure. Figure 18.6 illustrates example XRD scans of a poplar wood sample at 20 °C and 180 °C, where poplar wood demonstrates its amorphous character with broad maximum intensity (i.e., amorphous halo), with crystallinity not being detected in this case. However, XRD scans of Ohio Shale kerogen at 20 °C and 175 °C (Figure 18.7) manifests many sharp Bragg reflections, indicating significant presence of microcrystalline regions. This behavior may be reasonably attributed to high ash content in the kerogen sample. Comparison of XRD patterns of each sample at the higher and lower temperatures suggests no change in crystalline content over the examined temperature range. Use of temperature scanning XRD assists in the confirmation and differentiation of glass transitions from crystalline melts (DeLapp et al., 2004), since enthalpic overshoot behavior accompanying glass transitions bears resemblance to crystalline melt endothermic peaks (LeBoeuf and Weber, 2000a).

Amorphous macromolecules may possess several second-order thermal transitions, and it is customary to reference the highest temperature thermal transition as an α transition (i.e., glass transition) and assign other lower-temperature transitions β, γ, and so on, as they appear in the order of descending temperature. Multiple thermal transition processes have been reported in numerous prior studies in synthetic macromolecules (Ma et al., 1995), biopolymers (Kelley et al., 1987), and humic substances (DeLapp et al., 2004; DeLapp and LeBoeuf, 2004). Because the greatest amount of thermal energy is required for main polymer chain relaxation, the highest temperature transition is generally regarded as the glass transition. Other sub-glass

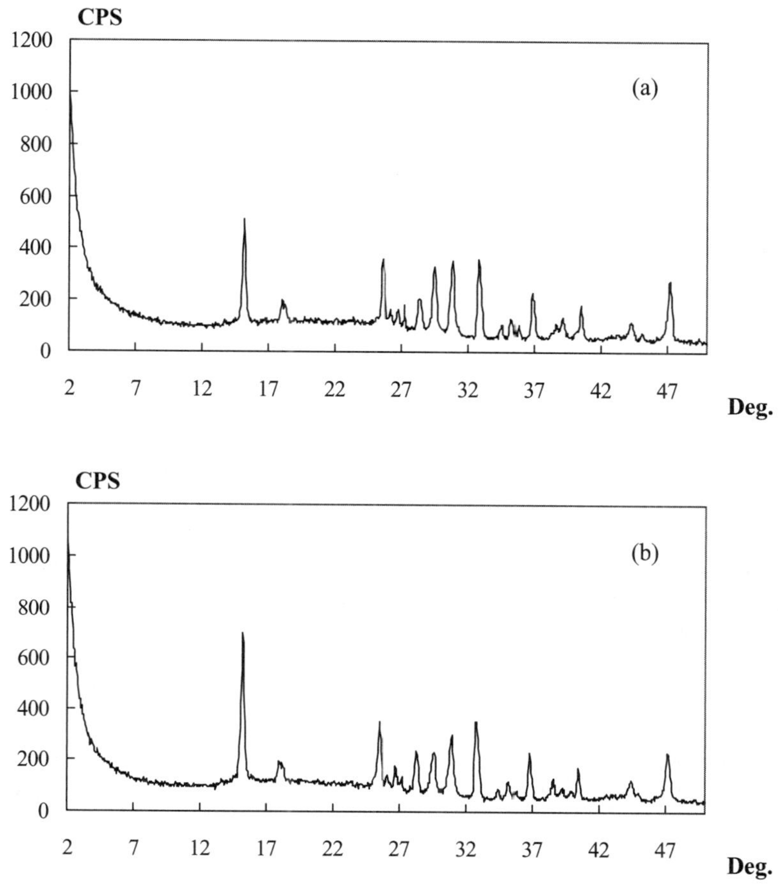

Figure 18.7. X-ray diffraction spectrum of Ohio Shale kerogen at (**a**) 20 °C and (**b**) 175 °C.

transitions are usually associated with the relaxation of side chains or functional groups, normally evidenced by reduced magnitude or "weaker" transitions. A second explanation for multiple transition behavior, however, is the presence of distinct regions of macromolecules manifesting different mobility. For example, Wunderlich (1990) detected multiple distinct thermal transitions in copolymers associated with the coexistence of various macromolecules (i.e., the copolymer phases did not sufficiently mix, thus retaining their individual polymer character). It is possible that this explanation may be especially applicable to soil and sediment organic matter based on their composition and method of formation/degradation.

18.4. THERMAL ANALYTICAL CHARACTERIZATION TECHNIQUES

Thermal analysis comprises a spectrum of techniques in which properties of a material are measured as a function of temperature, time, and other variables, while the material is subjected to a controlled temperature program. Analytical techniques include (i) gravimetric-based systems (thermal gravimetric analysis) for evaluation

TABLE 18.2. Thermal Analytical Techniques for the Determination of Selected Thermal Transitions

Thermal Transition	Applicable Thermal Analytical Technique
α, β, γ Transitions	DSC, DTA, TMA
Crystallization and crystalline melt	DSC, DTA
Specific heat capacity	DSC, DTA
Heat of vaporization	DSC, DTA
Coefficient of thermal expansion	TMA, DMA
Softening point, mechanical modulus	TMA, DMA
Damping factor, and stress–strain relationship	DMA
Thermal degradation temperature	TGA

of sample mass loss as a function of temperature, (ii) heat-flow based systems (differential thermal analysis and differential scanning calorimetry) for quantification of sample response to heating and cooling trends, (iii) mechanical based systems (thermal mechanical analysis and dynamic mechanical analysis) to examine changes in mechanical properties of materials as a function of temperature, and (iv) electrical based systems to evaluate temperature-dependent material dielectric properties. The following sections provide an overview of each of these techniques, including technological advances, general experimental protocols, example results, and potential for use as a tool for thermal analysis of natural organic materials. Additional detailed information on thermal transitions and general experimental protocols may be found in a number of references including Höhne et al. (1996), Hatakeyama and Hatakeyama (2004), and Wunderlich (2005).

Table 18.2 summarizes the variety of thermal analytical techniques that may be employed to measure α, β, and γ transitions, crystallization, crystalline melts, and thermal degradation temperatures, including thermodynamic properties specific heat capacity (C_p) and thermal expansion coefficient. Such methods include differential thermal analysis (DTA) (measuring changes in temperature between two sample cells as a function of increasing energy input), differential scanning calorimetry (DSC) (measuring heat capacity changes as a function of temperature), dilatometry and thermal mechanical analysis (TMA) (measuring total volume changes as a function of temperature or pressure), dynamic mechanical analysis (DMA) (measuring creep and relaxation as a function of temperature), dielectric relaxation (measuring changes in the dielectric constant with temperature), positron annihilation lifetime spectroscopy (PALS) (measuring micropore-size hole volume as a function of temperature), and thermal gravimetric analysis (measuring changes in sample mass with temperature). Additional information on the use of TGA, DTA, DSC, TMA, DMA, and dielectric relaxation of natural and synthetic samples is detailed in subsequent sections of this chapter.

18.4.1. Thermal Gravimetric Analysis

TGA is a technique in which a change in sample mass is recorded as a function of temperature and/or time. A TGA instrument comprises three major components: microbalance, furnace, and thermocouples. In a TGA experiment, a sample is placed into a sample pan (usually comprised of platinum or ceramic), which is attached to a microbalance, allowing the mass of the sample to be measured throughout the course of an experiment. A movable furnace positioned around the sample serves to heat and cool the sample to required temperatures under a controlled temperature program, and modern TGA instruments can heat samples in excess of 1000 °C (e.g., Q500 TGA from TA Instruments, Inc., New Castle, Delaware or TG from Perkin Elmer, Waltham, Massachusettes). Thermocouples provide precise measurement of the system temperature. As such, real-time mass and temperature of the sample can be determined, and the results can be presented by plotting mass or derivative of mass (rate of mass change) against temperature.

TGA is often employed as a means to derive moisture content of soils and sediments, with typical heating protocols including equilibration of samples at 105 °C to 110 °C for 24 or more hours, often under an inert atmosphere. Numerous other applications have included use of TGA to examine sample thermal stability following evolution of physisorbed water [e.g., Kolokassidou et al. (2007)]. Coupling of TGA with additional analytical instruments includes use of various evolved gas analysis (EGA) techniques, such as gas chromatography/mass spectrometry (GC/MS) and gas chromatography/Fourier transform infrared spectrometry (GS/FTIR). Identification of evolved gases serves to reveal underlying chemical reactions associated with consequent weight losses. Similarly, other techniques, such as X-ray diffraction, can assist in the quantification of solid reaction products left after thermal decomposition. For example, Cuypers et al. (2002) examined the amorphous and condensed domains of soil and sediment organic matter through use of TGA, pyrolysis–GC/MS, and ^{13}C NMR, noting that the composition of amorphous and condensed domains were highly dependent on the origin and nature of the organic matter. In particular, condensed domains present in relatively non–diagenetically altered organic matter were rich in aliphatic carbons, while more diagenetically altered organic matter possessed greater aromatic character.

18.4.1.1. Technological Advances. In a typical TGA experiment, a sample is often heated in a linear fashion. Modern TGA instruments, however, can now superimpose a sinusoidal temperature program on top of the linear heating rate, known as modulated TGA (Blaine and Hahn, 1998). This produces an oscillatory rate of weight loss from which kinetic information, such as activation energy and pre-exponential factor, can be derived. In addition, advances have been made to experimental protocols to achieve higher resolution, which serves to further separate/resolve thermal decompositions in more narrow temperature ranges.

Despite these advancements, TGA alone fails in its ability to provide detailed information with respect to the type of decomposition products and residues during the course of a thermal decomposition experiment. For this reason, TGA is often combined with other analytical techniques, such as DSC, which is especially useful in the study of energies associated with loss of moisture or volatile compounds upon heating. A typical application for TGA-DSC would be the determination of the heat of vaporization of a liquid in an open system. Weight loss associated with vaporization can be recorded by TGA, while latent heat of vaporization can be easily determined by integrating the accompanying endothermic peak provided in a DSC signal. Therefore, heat of vaporization can be obtained by dividing latent heat by weight loss.

18.4.1.2. General Experimental Protocols. Similar to other thermal analytical instruments, a TGA instrument requires regular calibration to maintain accuracy of measurements. Calibration of a TGA instrument includes taring, weighing, and temperature calibrations, with temperature calibrations typically derived from analysis of a high-purity metal in which the Curie point temperature is recorded and compared to the correct value. Following the instrument calibration, samples are added to the sample pan in preparation for analysis. Finely ground samples are preferred in TGA since sample size and form do affect TGA results. For example, a large sample size can develop thermal gradients, contributing to deviations in the measured thermal decomposition temperature. Subsequent experimental protocol development includes the selection of purge gas (generally employing relatively inert gases such as He or N_2), heating mode (e.g., isothermal model versus dynamic mode), and heating rate.

18.4.1.3. Example Results. As mentioned in the section of technological advances, TGA has been coupled with other techniques for more detailed analysis of thermal decomposition processes. Figure 18.8 illustrates the usefulness of the TGA-MS in the characterization of poly(vinylchloride) (PVC) in a helium atmosphere (TA Instruments, 1995a). The TGA plot indicates a two-stage decomposition process: a major decomposition around 309 °C and another, weaker decomposition process at a higher temperature of 468 °C. Mass spectrometer data further identifies evolved gases associated with each decomposition process. HCl, benzene, and vinyl chloride are the primary decomposition products in the first stage of weight loss, while

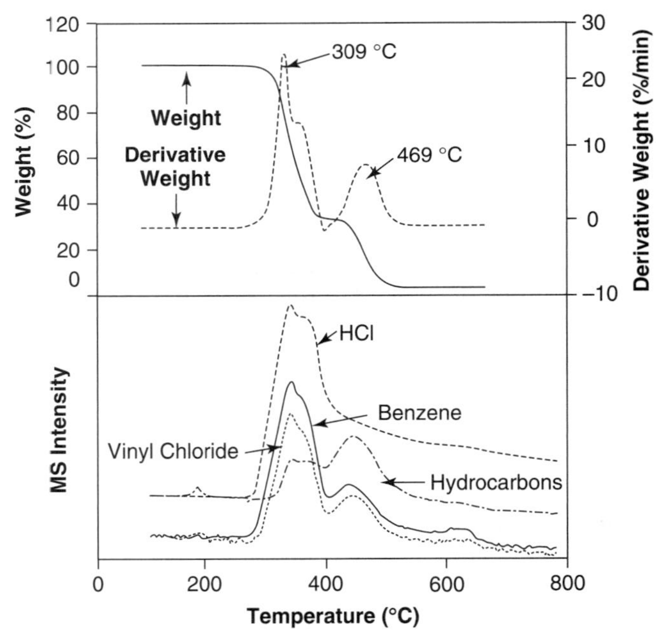

Figure 18.8. Thermal gravimetric analysis-mass spectroscopy characterization of poly(vinyl chloride). Reproduced from TA Instruments, Inc. (1995a), by permission of TA Instruments, Inc.

higher-molecular-weight hydrocarbon is the major decomposition product accompanying the second stage of weight loss.

TGA has also been widely used in the study of the kinetics of chemical reactions. The rate of a chemical reaction can be expressed in the following form:

$$\frac{da}{dt} = Zf(a)\exp\left(\frac{-E_a}{RT}\right) \tag{18.23}$$

where da/dt is the rate of reaction, Z is the pre-exponential factor, $f(a)$ is the reaction kinetics, R is the gas constant, T is the absolute temperature, and E_a is the activation energy. Determination of activation energy and pre-exponential factor of a chemical reaction is of great importance. Exploration of differing heating rates and temperatures through numerous experiments allows quantification of chemical reaction activation energy and pre-exponential factor using conventional TGA. Modulated TGA, however was developed to allow determination of activation energy and pre-exponential factor in a single experiment. Demonstration of the capability of modulated TGA is illustrated by the study of the kinetics of the thermal decomposition of high-density polyethylene, HDPE (Figure 18.9) (Aubuchon and Blaine, 1998). The solid line indicates thermal decomposition of HPDE near 450 °C, while the dashed line represents the oscillating rate of weight loss in response to modulation of the temperature and heating rate of the experiment. Subsequent employment of real-time discreet Fourier transform (DFT) can deconvolute oscillatory temperature and rate of weight loss, allowing accurate quantification of activation energy and pre-exponential factor.

18.4.1.4. Potential as a Tool for Thermal Analysis of NOM. TGA typically is employed to quantify both moisture content and thermal stability of soil and sediment organic matter samples (e.g., Tan and Hajek, 1977; and Schnitzer, 1972), including identification of structural components (e.g., Schnitzer et al., 1964). Examination of thermal stability prior to evaluation in thermal analysis experiments enables

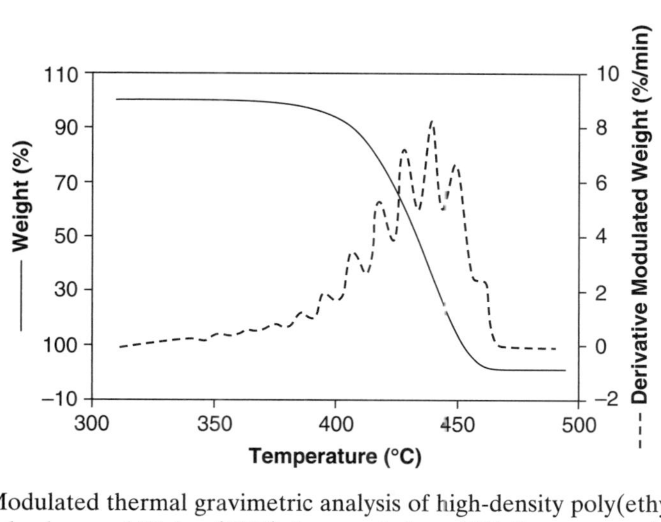

Figure 18.9. Modulated thermal gravimetric analysis of high-density poly(ethylene). Reproduced from Aubuchon and Blaine (1998), by permission of TA Instruments, Inc.

improved confidence in the temperatures employed in experimental protocols to ensure that thermal degradation temperatures are not exceeded, thus avoiding irreversible chemical and structural changes to samples (DeLapp et al., 2004; Young and LeBoeuf, 2000). Figure 18.10 illustrates the application of TGA to an Amherst humic acid. Note the onset of large mass losses near 100 °C, with the peak mass loss rate centered near 300 °C. Recently, Buurman et al. (2002) examined the thermal stability of humic substances saturated with various cations (e.g., calcium, sodium, and aluminum), suggesting the role of cations in bridging humic molecules and changing molecular mobility. In addition, TGA has been shown as a rapid and reliable tool for simultaneous quantification of organic carbon, total nitrogen, carbonate carbon, and clay contents in various mineral soils (Siewert, 2004). While TGA alone is useful in providing the aforementioned information, more often it is coupled with other analytical techniques, such as mass spectroscopy and differential scanning calorimetry. For example, Kolokassidou et al. (2007) studied the thermal stability of a soil humic acid employing TGA, temperature-programmed desorption coupled with mass spectrometry (TPD/MS), and *in situ* diffuse reflectance infrared Fourier transformed spectroscopy (*in situ* DRIFTS). Reversible water loss was identified below 70 °C, above which irreversible water loss took effect. In addition, gradual decomposition of the humic material resulted in the evolution of carbon dioxide between 110 °C and 240 °C and carbon monoxide between 140 and 240 °C, possibly as a result of decarboxylation. Dell'Abate et al. (2003) combined DSC with TGA for characterization of soil humic substances under both oxidizing and inert conditions, where derived thermograms allowed calculation of the index of thermal recalcitrance and the index of thermal decomposability, further assisting in the correlation of compositional changes in humic substances under different humification processes. Numerous other studies have employed TGA as a means to assist in the thermal character of soils and soil and sediment organic matter (Cuypers et al., 2002; Davies and Jabeen, 2002; Davies and Jabeen, 2003; Kumar et al., 2004; Wollenweber et al., 2006; Kolokassidou et al., 2007; Liebich et al., 2007; Novotny et al., 2007;

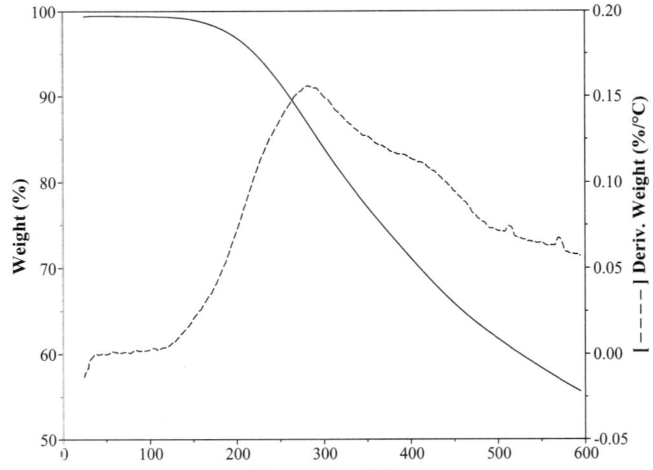

Figure 18.10. Thermal gravimetric analysis of Amherst humic acid.

Yamamoto et al., 2007), including evaluation of the level of maturation of organic matter in composts (Dell'Abate et al., 2000) and evaluation of relative soil "health" (Nunez-Regueira et al., 2006) through use of thermal analysis to evaluate relative microbial activity. Table 18.3 summarizes materials, methods employed, and relevant citations.

In addition to the ability of TGA to quantify the thermal character of soil and sediment organic matter, it has been employed to study the adsorption of volatile organic compounds onto soil and sediment organic matter. For example, Boussehain et al. (1986) used TGA to study alcohol adsorption on charcoals, from which adsorption/desorption isotherms were developed at various temperatures and adsorption models were used to fit experimental data. In addition, Risoul et al. (2002) performed laboratory pilot studies on thermal desorption of polychlorinated biphenyls (PCBs) from a contaminated soil. This study indicated that TGA appears to be a promising tool to determine optimum operating thermal extraction conditions.

18.4.2. Differential Thermal Analysis and Differential Scanning Calorimetry

Differential thermal analysis (DTA) measures changes in temperature between two sample cells (sample cell and reference cell) as a function of increasing energy input (Figure 18.11A). The platinum sensors detect the temperature in each cell, while energy input to the connected heaters is recorded as the instrument provides equivalent energy input into each cell while scanning a preset temperature range. The

TABLE 18.3. Select NOM Characterization Studies by Thermal Gravimetric Analysis (TGA), Differential Scanning Calorimetry (DSC), Differential Thermal Analysis (DTA), and Evolved Gas Analysis (EGA)

Materials Studied	Methods Employed	Reference
Soil organic matter	TGA	Schnitzer et al. (1964)
Humic acid, coal, peat	TGA and DTA	Buurman et al. (2002)
		Golebiowska et al. (1996)
		Francioso et al. (2003)
Soil and sediment organic matter	TGA	Cuypers et al. (2002)
Humic acids	TGA	Kolokassidou et al. (2007)
Humic and fulvic acids	TGA	LeBoeuf and Weber (2000a)
		DeLapp et al. (2004)
Composts	TGA and DSC	Dell'Abate et al. (2000)
Coal	TGA	Biswas et al. (2006)
		Wollenweber et al. (2006)
Soil humic substances	TGA, DSC, and EGA	Dell'Abate et al. (2003)
	TGA	Novotny et al. (2007)
Whole soil	TGA	Siewert (2004)
		Nunez-Regueira et al. (2006)
		Davies and Jabeen (2002)
		Davies and Jabeen (2003)
		Kumar et al. (2004)
Kerogen and humic substances	TGA	Yamamoto et al. (2007)
Maize straw	TGA	Liebich et al. (2007)

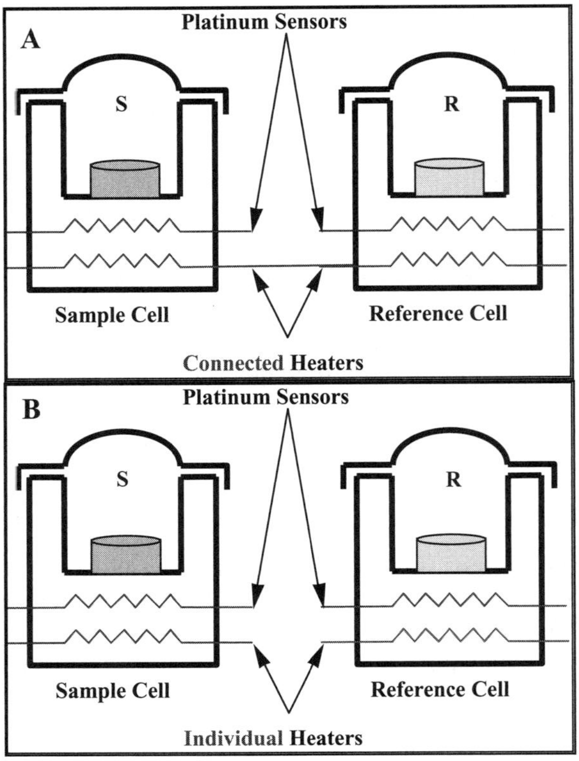

Figure 18.11. Schematic of (**A**) a differential thermal analyzer and (**B**) differential scanning calorimeter for a TA Instruments, Inc.-type configuration. Modified from Richardson (1989). Reproduced by permission of Elsevier, Ltd.

detector thus is able to directly measure the "difference" in temperature between the sample cell and the reference cell. Sample heat capacity can then be determined through difference in temperatures between sample cell and reference cell.

Similar to the DTA system noted above, differential scanning calorimetry (DSC) consists of a sample cell and reference cell, where platinum sensors detect the temperature in each cell. In the case of DSC, however, each cell possesses an individual heater (Figure 18.11B) so that energy input to the individual heaters is recorded as the instrument attempts to maintain equivalent temperatures in each cell while scanning a preset temperature range. The detector thus is able to directly measure the "difference" in heat capacity between the sample cell and the reference cell.

18.4.2.1. Technological Advances. DTA has been generally regarded as a qualitative thermal analysis technique since it does not possess quantitative calorimetric measurement capabilities of DSC. As such, there is a general trend for increased use of DSC relative to DTA. However, DTA still finds use in characterization of high-temperature behaviors (>750 °C), which impose design constrains for many DSC instruments. Conventional DTA instruments typically employ metallic thermocouples to measure system temperature, which limits the application of DTA to temperatures above 1600 °C and to noncorrosive atmospheres. Recently developed optical differential thermal analysis (ODTA) (Caslavsky, 1988) employing an infrared pyrometer, however, established DTA capabilities up to 3600°C. A second advancement for DTA involves the combination of DTA with other thermal analytical techniques, such as TGA, which offers valuable information that DTA alone cannot reveal. For example, an endothermic process revealed by DTA may be derived from many thermal behaviors, such as melting and sublimation. Weight loss information provided by TGA can further confirm the mechanism associated with the endothermic process.

Because DSC is dependent on thermal behavior of the sample cell relative to a reference cell, most technical improvements have focused on increasing the sensitivity of heat flow measurements. Such efforts include improving the heat flow and thus thermal response time of the sample cell through improved design of sample pans to enhance thermal contact between the pan and cell, as well as by providing improved thermal stability of the analysis and reference cells. Inherent in this endeavor for increased sensitivity are the efforts to reduce interference or noise that may be reflected in the instrument baseline (or heat flow between an empty sample pan analysis cell and the reference cell). Recent advancements in DSC technology by thermal analysis instrument manufacturers have resulted in nearly an order of magnitude increased sensitivity (to 0.2 μW) due to higher signal-to-noise ratios and improved baseline performance (reduced baseline curvature to less than 10 μW) (TA Instruments, Inc., 2006).

A second advancement has focused on the decoupling of the kinetic and thermodynamic basis of glass transitions, enabling, in many instances, improved sensitivity to so-called "weak" transitions, where the change in specific heat capacity upon transition from glassy to rubbery states is small (typically present when only a small portion of the sample undergoes the thermal transition). Separation of kinetic and thermodynamic components into reversible and nonreversible elements is achieved through the superposition of a sinusoidal modulation over a standard DSC heating ramp in which the temperature increases with time (referenced as temperature-modulated DSC, or TMDSC). The resulting heat flow information is then deconvoluted by discrete Fourier transformation for determination of C_p, T_g, and T_m (Hutchinson, 1998). An example TMDSC heating profile is illustrated in Figure 18.12.

18.4.2.2. General Experimental Protocols. Experimental protocols for both DSC and TMDSC involve sample preparation and selection of optimum heating/cooling rates. Although sample pans are limited in size (typically less than 0.5 ml in volume, although larger pans are available), users typically desire to maximize the amount of sample (e.g., 10–20 mg) in the pan to increase heat flow relative to the reference cell, thus increasing the signal-to-noise ratio. Sample pans typically are comprised of aluminum, and samples are pressed into the pan to optimize thermal contact between the sample and pan bottom. Covers are placed over the pans to minimize sample disturbance; and where loss of water moisture or other solvent is of concern, the pans may be hermetically sealed.

Heating protocols for DSC typically range from 2°C/min to 50°C/min, where the optimum heating rate is achieved at the balance between increased sensitivity to thermal transition (larger heating rates typically provide greater thermal response) and precision of the transition temperature. Heating protocols for TMDSC include reduced heating rates (1°C/minute to 5°C/minute, with a 50- to 100-s period of

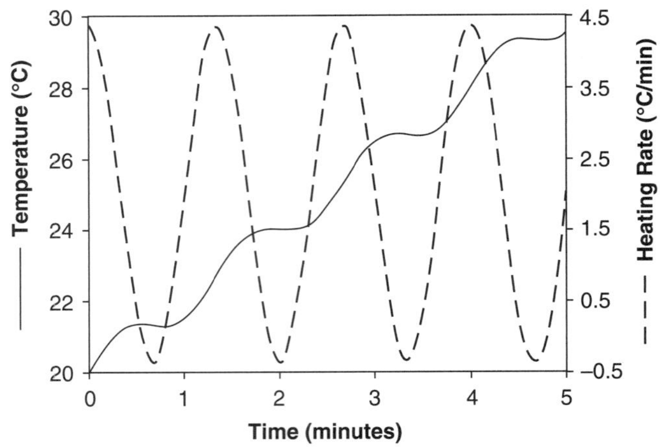

Figure 18.12. Example heat flow behavior of temperature-modulated differential scanning calorimetry. Reproduced from Young and LeBoeuf (2000), by permission of the American Chemical Society.

modulation and +0.5 °C temperature amplitude). An example heating protocol for TMDSC is provided in Figure 18.12. Detailed reviews of experimental protocols for DSC and TMDSC are provided in LeBoeuf and Weber (2000a) and Young and LeBoeuf (2000), respectively. Several heating/cooling cycles are typically used to evaluate the repeatability of the thermal transitions, where cooling rates usually range from 5 °C/minute to 10 °C/minute.

18.4.2.3. Example Results. Many inorganic materials such as metals and minerals possess very high transition temperatures, which severely limits the application of DSC to the study of these materials. In this case, DTA can be an alternative technique for high-temperature thermal studies. Figure 18.13 illustrates a DTA analysis of a titanium sample (TA Instruments, 1995b). The endothermic peak in the heating cycle and the exothermic peak in the cooling cycle indicate a transition temperature near 900 °C. Coupling X-ray diffraction with DTA serves to correlate exothermic processes with crystallization. Figure 18.14 demonstrates the simultaneous DTA and X-ray analyses of a poly(amide) sample heated from 250 °C to 600 °C (Androsch et al., 1996). The upper figure shows a sharp endothermic peak at approximately 500 °C due to polyamide melting. Before this melting temperature is reached, a small exothermic peak around 450 °C is visible, interpreted as the recrystallization of the γ phase into α phase. At the same temperature, the XRD pattern shows a small peak at 22 deg 2θ, confirming the exothermal process due to crystallization.

An example thermogram for a standard DSC experiment is illustrated in Figure 18.15. Beginning in a glassy state, samples first progress through a glass transition (an endothermic process), which, depending on the heating rate of the experiment and the relaxation rate of the sample, may exhibit enthalpic relaxations. Following the jump in C_p, the material resides in a rubbery state. Given additional thermal energy and sample homogeneity, the material may possess adequate mobility for the polymer chains to align themselves into parallel arrays, resulting in crystallization (an exothermic process). For crystalline materials, application of additional thermal energy eventually results in crystalline melts (endothermic process),

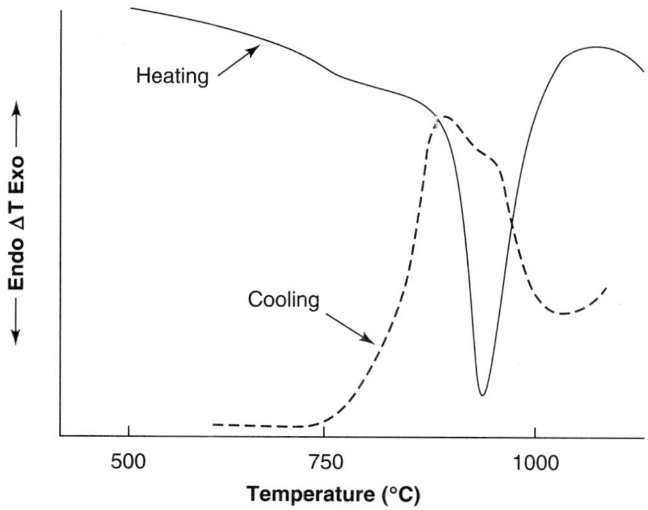

Figure 18.13. Differential thermal analysis of titanium. Reproduced from TA Instruments, Inc. (1995b), by permission of TA Instruments, Inc.

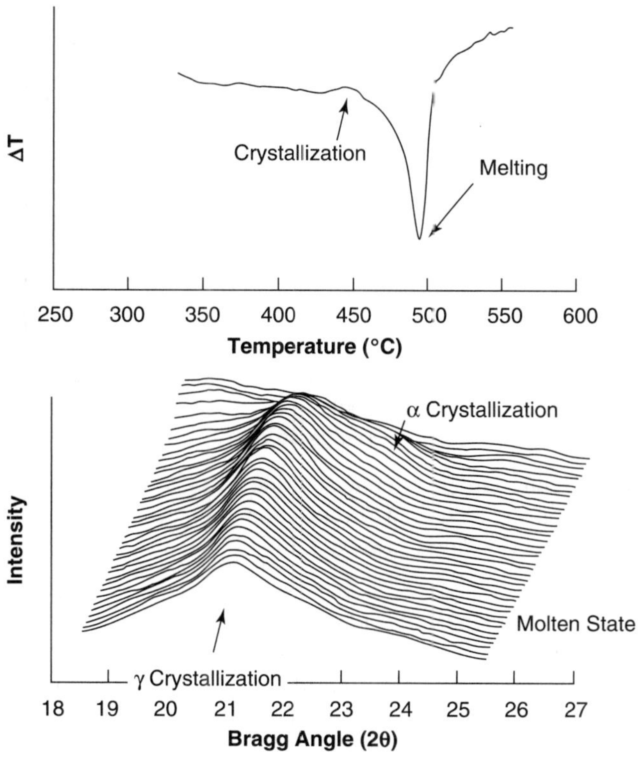

Figure 18.14. Simultaneous X-ray diffraction and differential thermal analysis of polyamide. Reproduced from Androsch et al. (1996), by permission of Elsevier, Ltd.

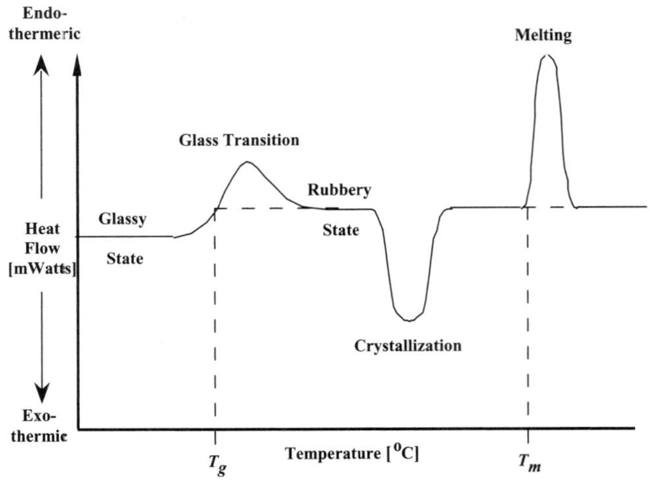

Figure 18.15. Demonstrated use of thermal analytical techniques to evaluate structural and thermodynamic properties of macromolecular materials.

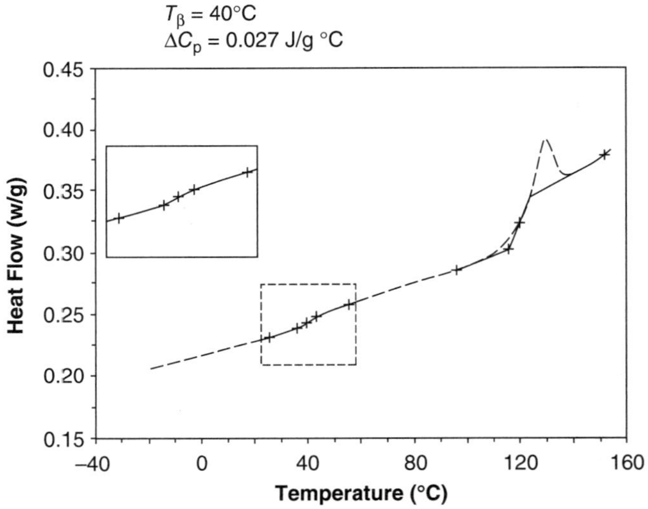

Figure 18.16. Standard differential scanning calorimetry scan of poly(methyl methacrylate) indicating β relaxation near 40 °C and α-relaxation (T_g) near 119 °C. Reproduced from DeLapp et al. (2004), by permission of Elsevier, Ltd.

followed eventually by thermal degradation and sample decomposition. Example thermograms for standard DSC and TMDSC analyses of poly(methyl methacrylate) (PMMA) are provided in Figures 18.16 and 18.17, respectively. Note the presence of both β and α transitions, plus the absence of enthalpic relaxations for TMDSC due to use of temperature modulation.

18.4.2.4. Potential as a Tool for Thermal Analysis of NOM. DTA studies of NOM are relatively scarce, due primarily to DTA's inability to provide quantitative

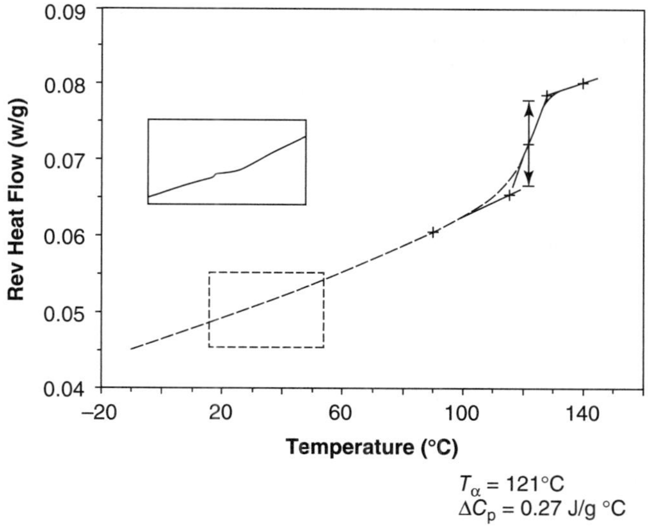

Figure 18.17. Temperature-modulated differential scanning calorimetry scan of poly(methyl methacrylate) indicating β-relaxation near 32 °C and α relaxation (T_g) near 121 °C. Reproduced from DeLapp et al. (2004), by permission of Elsevier, Ltd.

calorimetric information (i.e., C_p), limiting its application to determination of glass transition temperatures, crystallization, and crystalline melts. Example studies include Turner and Schnitzer (1962) and Ishiwata (1969), who used DTA to examine NOM structure and aromaticity, and Schnitzer and Kodama (1972), Jambu et al. (1975a, 1975b), and Tan (1978), who examined NOM–metal complexes. In addition, Shurygina et al. (1971), Francioso et al. (2003), and Golebiowska et al. (1996) utilized DTA in combination with TGA and spectrometric techniques to analyze humic acids, brown coal, and peat, from which information on thermal stability, exothermic/endothermic effects, evolved compounds, and thermal decomposition mechanisms (e.g., decarboxylation) were derived.

In contrast to DTA, DSC has become the most widely used thermal analytical technique for the characterization of NOM. Provenzano and co-workers have used DSC extensively to explore thermograms of soil, sediment, and aquatic organic matter and their fractions, relating the observed patterns to sample nature (i.e., composition) and their origin (Provenzano and Senesi, 1998, 1999; Provenzano et al., 2006). Additional analyses of compost-derived organic matter yielded information related to both composition (e.g., lignin content) and maturation (Provenzano and Senesi, 1998; Provenzano et al., 1998a, 1998b, 2000, 2005; Ouatmane, 2000). Another significant application of DSC for NOM characterization is the identification of glass transition behavior. Glass transition temperatures, T_gs, have been quantified for numerous NOM samples, including fulvic acid, humic acid, kerogen/coal, charcoal, peat, and whole soil and sediment samples. Table 18.4 summarizes materials, methods employed, and relevant citations for select studies. In a similar manner as that observed in some synthetic polymers, NOM may also undergo multiple thermal transitions. Figures 18.18 and 18.19 illustrate multiple thermal transitions

TABLE 18.4. Select NOM Characterization Studies by Differential Scanning Calorimetry (DSC) and Temperature-Modulated Differential Scanning Calorimetry (TMDSC)

Materials Studied	Methods Employed	Reference
Coal	DSC	Lucht et al. (1987) Yun and Suuberg (1992) Mackinnon et al. (1994)
Aldrich humic acid	DSC	LeBoeuf and Weber (1997)
River aquatic fulvic acids	DSC	Provenzano and Senesi (1998)
IHSS standard and reference Fulvic and humic acids	DSC	Provenzano and Senesi (1999)
Composted materials	DSC	Provenzano et al. (1998a) Provenzano et al. (1998b) Provenzano et al. (2000) Ouatmane et al. (2000) Provenzano et al. (2005)
Leonardite humic acid	DSC	LeBoeuf and Weber (2000a)
Peat humic acid Suwannee river fulvic acid	DSC, TMDSC	Young and LeBoeuf (2000)
Leonardite humic acid Soil humic acid Peat humic acid Peat fulvic acid Nordic aquatic humic acid Nordic aquatic fulvic acid Suwannee river fulvic acid	DSC, TMDSC	Zhang and LeBoeuf (2009) DeLapp et al. (2004)
Harpeth humin Harpeth humic acid Harpeth fulvic acid Harpeth carbohydrate Georgetown aquatic humic acid Georgetown aquatic fulvic acid Georgetown aquatic carbohydrate	DSC, TMDSC	DeLapp et al. (2005)
Peat	DSC	Schaumann (2005) Schaumann and LeBoeuf (2005)
Whole soil and sediment	DSC, TMDSC	DeLapp and LeBoeuf (2004) Hurrass and Schaumann (2007)
Dissolved organic matter	DSC	Provenzano et al. (2006)
Black carbon	Oxidative DSC	Leifeld (2007)
Kerogen and charcoal	DSC, TMDSC	Zhang et al. (2007)

for a Georgetown carbohydrate fraction and a Harpeth humin. As described by DeLapp et al. (2004), multiple transition behavior may be similar to that exhibited by block copolymers, which Ma et al. (1995) attributed to coexisting macromolecules. Given the heterogeneous nature of NOM, more than one T_g or a range of T_g values may result. Similar to that observed for synthetic macromoleclues, multiple thermal transitions may also indicate the presence of γ, β, or α transitions. Although it is difficult to differentiate the type of transitions for these two NOMs, the lower

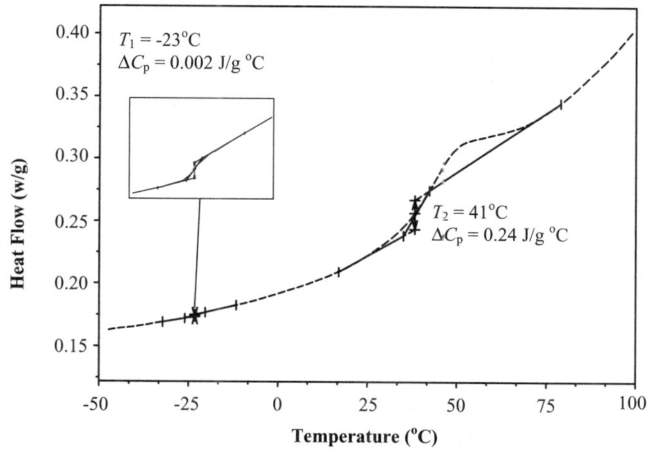

Figure 18.18. Standard differential scanning calorimetry scan of Georgetown carbohydrate fraction indicating multiple transitions: T_1 near −23 °C and T_2 near 41 °C. Reproduced from DeLapp et al. (2005), by permission of the Soil Science Society of America.

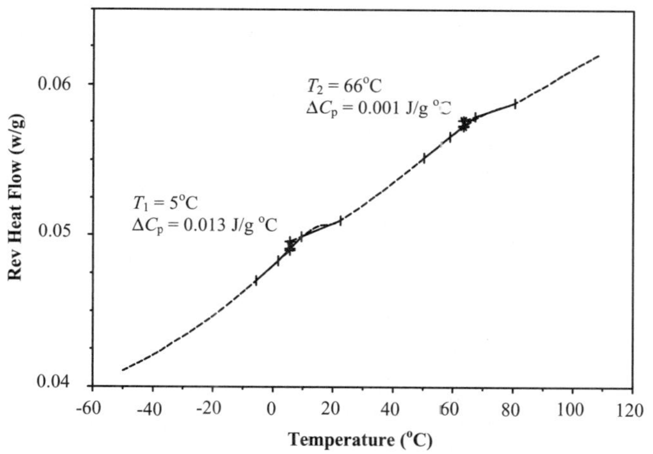

Figure 18.19. Temperature-modulated differential scanning calorimetry scan of Harpeth humin indicating multiple transitions: T_1 near 5 °C and T_2 near 66 °C. Reproduced from DeLapp et al. (2005), by permission of the Soil Science Society of America.

temperature transitions may be associated with either (a) noncooperative local mobility in the side chain (γ transition) or (b) cooperative side-chain mobility associated with β transitions, both of which occur below the main chain or α transitions. An additional explanation has recently been forwarded by Schaumann and LeBoeuf (2005) in terms of so-called "step transitions," where the step change in specific heat capacity at the thermal transitions may be associated with a combination of β transitions and influences of water molecule–NOM or metal–NOM complexes.

Detection of glass transition temperatures in several NOMs has further supported the concept of dual mode sorption, a phenomenon occurring in synthetic macromolecules advanced to explain nonideal sorption behaviors in NOM. For example, sorption isotherm nonlinearity was found to be associated with glassy domains of NOM, while exposing the same materials to increased temperature or sufficiently large concentration (i.e., chemical potential) of solvents resulted in increased molecular motions and matrix swelling (and thus increased macromolecular mobility), resulting in linear sorption behavior in the more rubber-like domains (LeBoeuf and Weber, 1997; Bouchard, 2002; Gunasekara and Xing, 2003).

While many DSC analyses were performed on NOM samples with thermal pretreatment (i.e., on "dry" samples), application of DSC to "wet" NOM should receive additional attention because (i) NOM in natural systems are, more or less, associated with various solvents, especially water, and (ii) physicochemical properties of NOM (e.g., flexibility and consequent sorption behavior) can change significantly due to the relative presence of solvents. Solvents are well known to swell NOM and result in increases in free volume with consequent increase in macromolecular mobility. As such, solvents have been generally regarded as plasticizers, implying that they serve to soften the macromolecular network with reduction in glass transition temperature (T_g), modulus of elasticity, and tensile strength. For example, DSC analysis on a wet Aldrich humic acid revealed a decrease of T_g by 15–20 °C relative to that of a "dry" sample (LeBoeuf and Weber, 2000a). Since water evaporation, depicted as a broad endotherm in a DSC thermogram, can interfere with interpretation and detection of glass transition behaviors, DSC analysis on hermetically sealed samples has been approached. Recent DSC work on hermetically sealed wet peat samples (Schaumann, 2005; Schaumann and LeBoeuf, 2005) detected both anti-plasticization and plasticization behaviors, where T_g increased with increasing water contents to a mass fraction of 0.12, followed by decreased T_g with water mass fractions exceeding 0.12. The detected glass transition appears to be "irreversible" because it cannot be detected in a second thermal analysis (cycle), but reappears again following storage for several days. These works suggested that NOM components may experience a rearrangement (i.e., structural relaxation) as a result of heating/cooling/resting cycles, resulting in changes to NOM–water interactions, especially hydrogen bonds.

Additional studies employing DSC and NMR explored the relative mobility of water molecules in NOM. For example, Norinaga et al. (1998) used DSC and H ^{1}NMR on a spectrum of coals of different ranks; based on freezing enthalpy and relaxation time, they were able to distinguish and quantify three types of water: free water, freezable bound water, and nonfreezable bound water. Similar work was also performed on peat (McBrierty et al., 1996) and soil samples (Hurrass and Schaumann, 2007), suggesting that NOM swelling was influenced by the type of water present and the apparently slow binding process of water to NOM.

Recent work by Zhang and LeBoeuf (in review) examined the effects of the presence of three solvents—water, acetone, and benzene—on the molecular mobility and structural relaxation of a humic acid through DSC analysis combined with molecular dynamics. Again, antiplasticization behavior was observed in two of the three systems (i.e., HA–water and HA–acetone) where solvents present in relatively low concentrations exhibited potential to form hydrogen bonds with the humic acid. Antiplasticization and plasticization behaviors were further interpreted from the perspective of hydrogen bonding analysis and free volume theory.

18.4.3. Thermal Mechanical Analysis and Dynamic Mechanical Thermal Analysis

Thermal mechanical analysis involves applying a stress to a sample while its deformation is measured as a function of temperature. Sample thermal expansion characteristics and or resistance to deformation are measured against temperature, thus yielding information regarding thermal transitions, including glass transitions and material softening points. Thermodynamic properties that may be derived from TMA include the coefficient of thermal expansion and deflection temperature under load. Modulating the temperature in similar fashion to TMDSC enables detection of more subtle glass transitions, while application of a sinusoidal force to samples in dynamic TMA mode enables measurement of storage modulus and loss modulus. Dynamic mechanical thermal analysis (DMTA or DMA) represents an extension of dynamic TMA because the specimen is also subjected to an oscillating stress instead of a constant stress, but DMA also provides measurement of additional mechanical properties (e.g., mechanical modulus, mechanical damping factor, and stress–strain relationships) as a function of temperature, loading frequency, and time. Principles underlying the mechanical behavior of materials are reviewed below.

Upon application of stress (σ), a specimen responds to produce a deformation or strain (ε). In the case of an elastic material, the strain is proportional to the stress described by Hooke's Law:

$$\sigma = E\varepsilon = E\frac{\Delta L}{L_0} \qquad (18.24)$$

where E is the elastic modulus (i.e., Young's modulus) and ε is the strain [ratio of dimension change (ΔL) to original dimension (L_0)]. Glassy amorphous polymers tend to behave like solids and deform in an approximately elastic matter. Un-cross-linked amorphous polymers, however, tend to behave like liquids above T_g. In this case, Newton's law holds in that the stress is proportional to the rate of strain:

$$d\sigma = \eta\frac{d\varepsilon}{dt} \qquad (18.25)$$

where η is the viscosity of the specimen and $d\varepsilon/dt$ is the rate of strain. Most materials nevertheless possess both solid and liquid characteristics when stress is simultaneously dependent on strain and rate of strain. By recording the stress–strain relationship for a material, DMA enables quantification of the complex modulus to account for viscoelastic behavior.

In the case of DMA, the applied oscillating stress, σ, usually follows a sinusoidal waveform:

$$\sigma = \sigma_0 \sin(\varpi t) \qquad (18.26)$$

where σ is the stress at time t, σ_0 is the peak stress, ω is the angular frequency of oscillation ($\omega = 2\pi f$), and f is the frequency in hertz. A sinusoidal stress will correspondingly produce a sinusoidal strain, and the shape of the strain wave will depend on both elastic and viscous behaviors of the materials. If the material is perfectly elastic, the strain wave has no phase difference relative to the stress wave. On the other hand, there is a phase difference of $\pi/2$ for perfectly viscous materials. For materials exhibiting viscoelastic behavior, the strain wave can be expressed as

$$\varepsilon = \varepsilon_0 \sin(\varpi t + \delta) \tag{18.27}$$

where ε_0 is the peak strain and δ is the phase difference between stress wave and strain wave. A schematic representation of stress and strain waves is illustrated in Figure 18.20. The ratio of peak stress to peak strain defines a complex modulus, E^*, as provided in Eq. (18.19). Similar expressions can be written for both shear modulus and bulk modulus:

$$E^* = \frac{\sigma_0}{\varepsilon_0} = E' + iE'' \tag{18.28}$$

where E', termed storage modulus, represents the real part of the complex modulus, while E'', termed loss modulus, represents the imaginary part of the complex modulus. Storage modulus and loss modulus are, respectively, reflections of elastic and viscous nature of a material. The ratio of loss modulus to storage modulus defines the mechanical damping factor ($\tan\delta$).

$$E' = \frac{\sigma_0}{\varepsilon_0}\cos\delta \tag{18.29}$$

$$E'' = \frac{\sigma_0}{\varepsilon_0}\cos\delta \tag{18.30}$$

$$\tan\delta = \frac{E''}{E'} \tag{18.31}$$

18.4.3.1. Technological Advances. TMA and DMA are both widely employed in the characterization of viscoelastic behavior of polymers, composites, and other materials. Notably, TMA and DMA are particularly useful in identifying glass transitions and other low energy-associated sub-glass transitions, which may not be easily

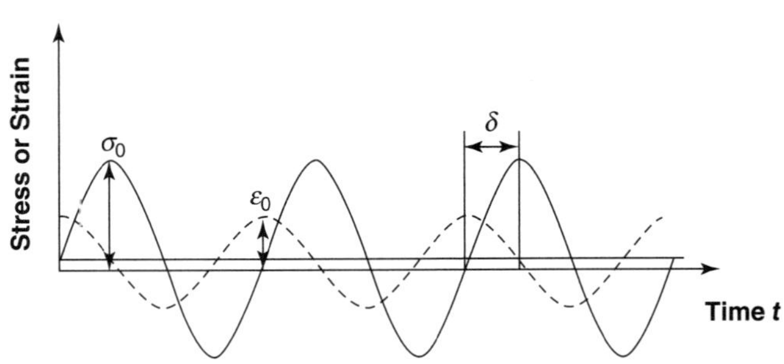

Figure 18.20. An illustrative relationship between sinusoidal stress and strain waves. Reproduced from Wetton et al. (1991), by permission of Elsevier, Ltd.

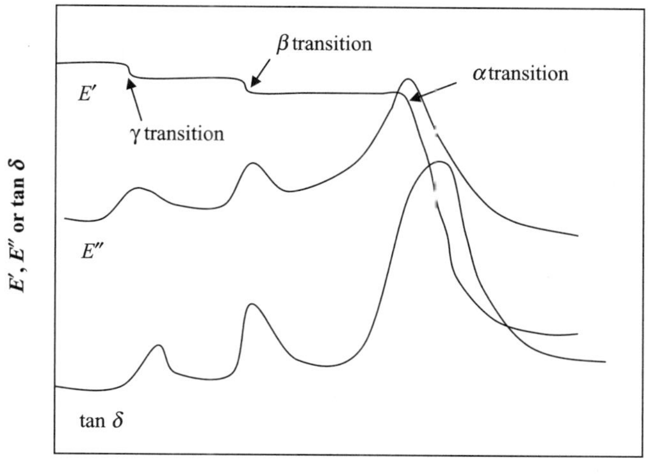

Figure 18.21. Plots of E', E'', and $\tan\delta$ as a function of temperature for a material showing multi-thermal transitions. Adapted from Wetton et al. (1991). Reproduced by permission of Elsevier, Ltd.

detected by aforementioned thermal analytical techniques (e.g., DSC). In addition, changes in storage modulus, loss modulus, and damping factor can each indicate secondary relaxation behaviors in different patterns as illustrated in Figure 18.21.

As temperature increases, materials may experience a transition from a frozen, glass-like state to a soft, rubber-like state with increased mobility. Accompanying this transition, E' of an amorphous material may be reduced up to three orders of magnitude, indicating loss of elasticity. Simultaneously, peaks in E'' and $\tan\delta$ appear in the same temperature region. The smaller peaks in E'' and $\tan\delta$, and reduced decrease in E' below T_g may suggest the presence of sub-glass transitions (e.g., β and γ transitions) associated with side group and/or segmental motions.

Besides quantifying thermal and mechanical behaviors of "dry" materials, recent developments have enabled samples to be immersed in liquid and vapor media (e.g., Bao and Bagga, 1993; Price, 1997). This advancement enables characterization of actual material performance under environmental conditions (e.g., dyeing operations) when materials are in contact with various solvents.

18.4.3.2. General Experimental Protocols. As noted above, thermal mechanical analysis may be conducted in three separate modes: standard, temperature-modulated, and force-modulated. Sample preparation requires dimensional stability, typically including either placement of the sample into a receptacle (useful for powders) or pressing into pellets or tablets.

Dynamic mechanical analyzers can further be divided into free resonance analyzers and forced resonance analyzers. In the case of free resonance analyzers, samples are allowed to oscillate freely (e.g., at their natural frequency) until oscillations cease. Forced resonance analyzers are nevertheless more frequently used. They are designed to apply oscillating stress at a fixed frequency and are ideal for scanning material performance over a wide temperature range (Menard, 1999). Sample preparation includes selection of an appropriate clamp system, which is a function of sample geometry (i.e., length, width, and thickness) and modulus. For example, shear plates are used for circular, thick, and soft materials. Instrument calibrations are then performed, which generally include clamp, electronic, force, and temperature calibrations. Temperature calibration is particularly important since material properties are strongly correlated to temperature. Experimental protocols are then established, including temperature range, frequency (typically between 0.01 Hz and 200 Hz), and heating rate (typically between 1 °C/min and 10 °C/min).

18.4.3.3. Example Results. The application of DMA to the study of the glass transition and sub-glass transitions of poly(benzoxazine) is presented in Figure 18.22 (Ishida and Lee, 2001). The figure illustrates the change of G' and $\tan\delta$ as a function of temperature. A glass transition (i.e., α transition) is observed around 170 °C, with a significant drop of G' and a strong peak of $\tan\delta$. Besides the glass transition, the plot of $\tan\delta$ versus temperature also manifests a β transition (the broad peak around 70 °C) and a γ transition (the peak around −80 °C), indicating the existence of local relaxations associated with small segments and/or side functional groups.

Figure 18.23 shows the effects of oscillation frequency on mechanical response of amorphous anhydrous sorbitol (Talja and Roos, 2001), where E' and E'' are plotted against temperature with oscillation frequencies between 1 and 20 Hz. In addition to the clear glass transition located around 0 °C, it can be seen that the peak in E'' and the onset temperature marking the decrease in E' progresses to higher temperatures as the frequency is increased. Similar to the kinetic behavior of the glass transition observed in conventional DSC experiments (i.e., shift of glass transition influenced by thermal history), the peak in E'' represents the temperature (i.e., T_g) at which the relaxation rate of the material is on the order of the mechanical oscillation rate. Below T_g, materials are retarded in their response to external perturbation (i.e., stress) while equilibrium can be reached instantly above T_g.

18.4.3.4. Potential as a Tool for Thermal Analysis of NOM. As mentioned above, TMA and DMA represent more sensitive thermal analytical techniques relative to

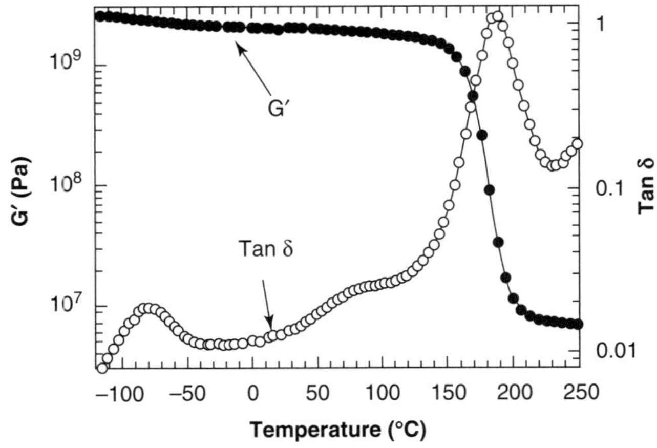

Figure 18.22. Differential mechanical analysis scans of polybenzoxazine. Reproduced from Ishida and Lee (2001), by permission of Elsevier, Ltd.

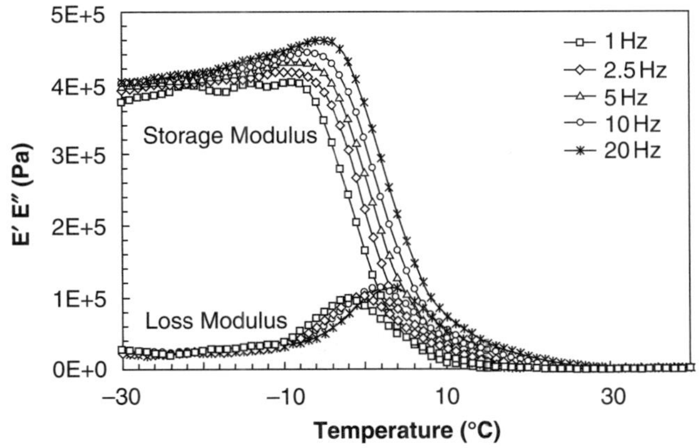

Figure 18.23. Storage modulus and loss modulus spectra for amorphous sorbitol at various frequencies. Reproduced from Talja and Roos (2001), by permission of Elsevier, Ltd.

DSC and DTA. They can be used to confirm suspicious glass transitions revealed by DSC; and most important, they can further quantify molecular mobility associated with sub-glass transitions. For example, DSC analysis of poly(ethylene 2,6-naphthalene dicarboxylate) (PEN) only revealed the presence of a glass transition around 112 °C (Hardy et al., 2001). DMA analysis of the same sample, however, revealed two secondary relaxations below this glass transition (Hardy et al., 2001). In the case of humic materials, it is not uncommon for DSC to fail to detect clear thermal transitions due to their heterogeneous nature, which contributes to overlap/broadening or "washout" of thermal transitions. As such, TMA and DMA represent powerful, complementary tools to DSC.

Several recent studies have illustrated the usefulness of TMA in detecting thermal transitions of humic materials (Young and LeBoeuf, 2000; DeLapp and LeBoeuf, 2004; DeLapp et al., 2004, 2005) and that of peats (Schaumann and LeBoeuf, 2005) and whole soils (Schaumann et al., 2005; Schaumann, 2006). Figure 18.24 illustrates a TMA scan of Leonardite humic acid, which corresponds well with the DSC-detected T_g (DeLapp and LeBoeuf, 2004). Other samples, such as the Nordic aquatic fulvic and humic acids illustrated in Figure 18.25, may not undergo a traditional transition, but may instead experience a softening point, where increased free volume caused by thermal expansion results in reduced sample viscosity. Forces applied to the sample in excess of the resistive forces may thus result in a collapse of the matrix, identified as the material's softening point (Eisenberg, 1993).

With the capability to operate with samples immersed in a liquid or vapor medium, DMA possesses the advantage of examining molecular mobility of humic materials under "actual" environmental conditions. The correlation of corresponding responses (i.e., shift of modulus curve) to solvent content may help identify specific chemical interactions between the sample and the solvent. For example, DMA studies on polysaccharides of various water contents observed changes in the tan δ curve: (a) increases in the higher temperature peak with water loss and (b) the appearance of a lower temperature peak at lower water contents (Craig and Johnson, 1995). The authors attributed the lower temperature peak to intermolecular hydrogen bonding between polysaccharides and water and attributed the higher tempera-

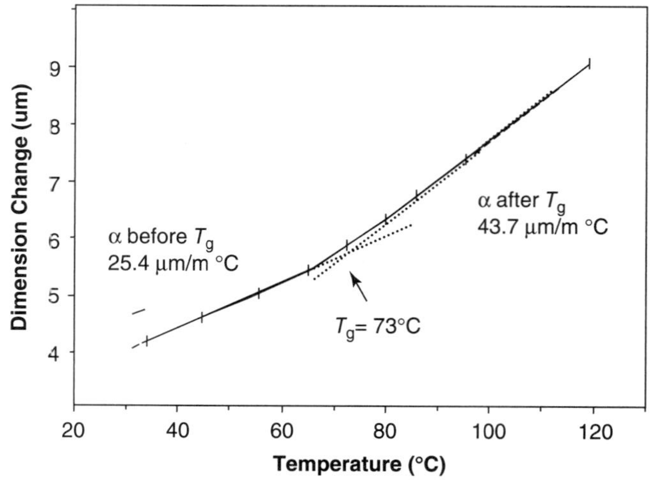

Figure 18.24. Thermal mechanical analysis scan of Leonardite humic acid at a heating rate of 2.5 °C min^{-1}.

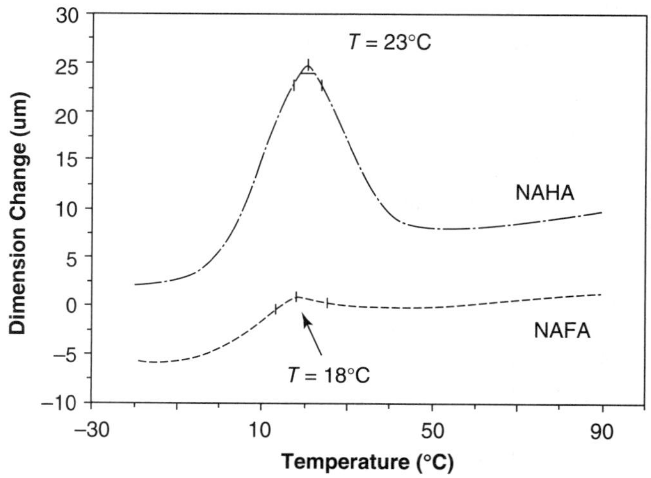

Figure 18.25. Thermal mechanical analysis scans of Nordic aquatic fulvic acid (NAFA) and Nordic aquatic humic acid (NAHA) indicating transitions near 18 °C and 23 °C, respectively, associated with the collapse of the structures following thermal transitions. Reproduced from DeLapp and LeBoeuf (2004), by permission of the Soil Science Society of America.

ture peak to free water at higher hydration levels. Interactions in the form of ionic bonding and hydrogen bonding are generally expected between humic materials and solvents such as water; DMA may thus offer improved mechanistic understanding of humic–solvent systems based on subtle changes in mechanical properties of humic materials.

Despite the potential of DMA as a tool to characterize humic materials, challenges remain with respect to sample preparation and data interpretation. Unlike conventional DSC experiments, samples in DMA experiments require particular shape and stiffness. In addition, the contact between the sample and instrument clamp may greatly influence the reproducibility and accuracy of results. In the case of data interpretation, assigning observed thermal transitions to glass transitions may be complicated by sample heterogeneity, and it is even more challenging to quantitatively relate observed thermal transitions (e.g., sub-glass transitions) to specific chemical functional groups or molecular segments in such complex molecular systems.

18.4.4. Dielectric Thermal Analysis

In an analogous manner to DMA, dielectric thermal analysis (DETA) represents a technique to apply an alternating electric field across the tested sample, which contributes to a polarization of the material with consequent current flow. DETA enables measurement of dielectric properties, which can further be related to material properties and thermal transitions as described below.

A material possesses both capacitive and conductive characteristics. The former represents the material's capability to store electrical charges, while the latter indicates its capability to pass electrical charges. In a DETA experiment, a sinusoidal voltage (V) is applied such that dipoles in the material align with the electric field and ions move to the oppositely charged electrode, producing a sinusoidal current flow (I). For a material exhibiting only conductive behavior, the current flow will always be in phase with the applied electric field, whereas current flow will lag the electric field by 90° for a material behaving completely like a capacitor (i.e., capacitive behavior). Because materials exhibit capacitive and conductive behaviors simultaneously, a phase angle (δ) between voltage and current is expected.

$$V = V_0 \sin \omega t \tag{18.32}$$

$$I = I_0 \sin(\omega t + \delta) \tag{18.33}$$

where V_0 and I_0 are, respectively, the amplitude of voltage and current, t is the time, and ω is the angular frequency ($\omega = 2\pi f$, f is the frequency in hertz). As such, the capacitance (C) and conductance (G) of the material can be expressed as

$$C = \frac{I_0}{V_0} \frac{\sin \delta}{\omega} \tag{18.34}$$

$$G = \frac{I_0}{V_0} \cos \delta \tag{18.35}$$

These two quantities can be converted to dimensionless quantities; and therefore, relative permittivity (ε') and dielectric loss factor (ε'') are more frequently used:

$$\varepsilon' = \frac{C}{\varepsilon_0} \frac{A}{D} \tag{18.36}$$

$$\varepsilon'' = \frac{G}{\omega \varepsilon_0} \frac{A}{D} \tag{18.37}$$

where C and G are capacitance and conductance defined above, ε_0 is the absolute permittivity of free space (8.85×10^{-12} F/m), and A and D represent electrode area and electrode separation distance (i.e., sample thickness), respectively. ε' represents the degree of alignment of dipoles to an electric field. This quantity is expected to increase with increasing temperature, particularly following thermal transitions because molecules will possess greater mobility at higher temperatures. ε'' represents energy loss due to alignment of dipoles and ions, where peaks in ε'' represent individual thermal transitions. In addition, the ratio $\varepsilon''/\varepsilon'$ (or $\tan\delta$) is referenced as the dissipation factor, displaying similar patterns to ε'' through thermal transitions.

$$\tan\delta = \frac{\varepsilon''}{\varepsilon'} \tag{18.38}$$

18.4.4.1. Technological Advances. DETA complements other thermal analytical methods by identifying thermal transitions, rheological phenomena, and polymerization and cure behaviors from measured dielectric properties of the material. DEA can evaluate samples of various forms, such as solids, liquids, gels, and thin films. One major advantage of DETA over other techniques is that DETA can span a wide frequency range (e.g., 0.003–100 kHz for TA Instruments DEA 2970), enabling resolution of relaxation behaviors at low frequencies and evaluation of material properties end-use conditions (high-frequency conditions). DETA is also very sensitive to local motions associated with dipole orientation, which may not be readily captured by other techniques. For example, it has been shown that an impurity concentration as low as 1 ppm can induce sufficient ion conductivity to enable detection (ion conductivity is positively correlated to dielectric loss factor ε'') (Senturia and Sheppard, 1986). Finally, DETA equipped with disposable micro-dielectric sensors may be applied *in situ*, enabling dielectric measurements during manufacturing processes (e.g., vulcanization) (Sircar, 1997).

18.4.4.2. General Experimental Protocols. Similar to other thermal techniques, DETA analyses require exacting experimental protocols to ensure accuracy and reproducibility of results. Though detailed protocols will vary depending on the instrument, general experimental preparation steps include sample preparation, mode/sensor selection, instrument calibration, and establishment of heating rate, temperature range, and electrical frequency. Samples are typically molded into solid disks or thin films of viscous liquid with thickness typically no more than 1–2 mm. Four types of sensors are available: ceramic parallel plate, sputter coated, ceramic single surface, and remote single surface, each corresponding to particular modes of operation. The selection of sensors depends on the properties of interest. For example, ceramic parallel plates are used to evaluate bulk dielectric properties, while ceramic single surfaces are used to evaluate surface and curing properties. Experimental parameters include heating rate, electrical frequency, and explored temperature range (typically between −150 °C and 500 °C).

18.4.4.3. Example Results. As a material passes its T_g, static dipoles will gain sufficient mobility to align with the electric field. This generally causes an increase in ε'' with a corresponding ε'' peak. Figure 18.26 illustrates the DETA analysis on polycarbonate (Foreman et al., 1995), where ε' and ε'' are plotted as a function of

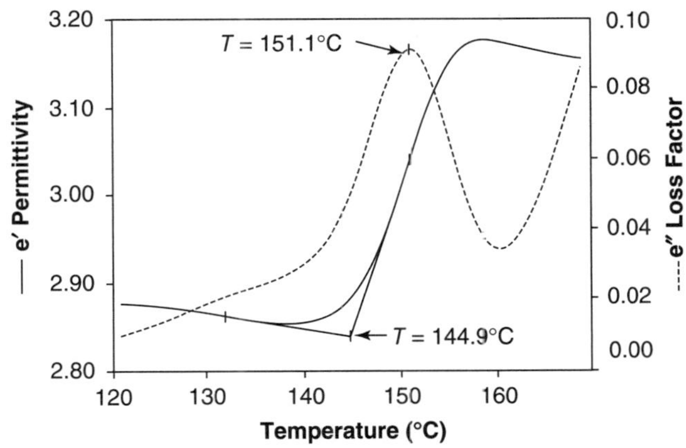

Figure 18.26. Dielectric thermal analysis scans of polycarbonate at 1 Hz and at a heating rate of 3 °C min^{-1}. Reproduced from Foreman et al. (1995), by permission of TA Instruments, Inc.

temperature. T_g of poly(carbonate) is interpreted from the onset of the increase in ε' (144.9 °C), and the peak (151.1 °C) in ε''.

DETA has been gaining popularity as a method for cure monitoring, and it can be used to examine the cure of elastomers via ionic conductivity measurements. Ionic conductivity is related to ε'' by the following equation:

$$\varepsilon'' = \varepsilon''_d + \frac{\sigma}{\varpi\varepsilon_0} \qquad (18.39)$$

where ε_d'' is the energy loss due to dipole orientation (dipolar loss factor), σ is ionic conductivity, and $\sigma/\omega\varepsilon_0$ represents ionic conductance. As ions move toward the opposite electrode, they will encounter resistance (a viscous force), and thus ionic conductivity is inversely related to viscosity. At low temperatures, ionic conductivity is insignificant and dipoles dominate, whereas ionic conductance is dominant at high temperatures. Figure 18.27 shows the online cure monitoring of a fiber-reinforced composite system (McIlhagger et al., 2000), where ionic conductivity and temperature are plotted versus cure time. Initially, the ionic conductivity increases as the temperature is raised toward the cure temperature and the sample becomes less viscous. Subsequent reaction (mobile ions acting as crosslinking agents) leads to increased viscosity, and therefore, drop in ionic conductivity. The two stages contribute to the observed ionic conductivity peak, marking the minimum viscosity.

18.4.4.4. Potential as a Tool for Thermal Analysis of NOM. Although there have been no published reported uses of DETA for characterization of NOM, the sensitivity of DETA may prove beneficial for examination of subtle thermal transitions. DETA identifies thermal transitions based on the strength of permanent dipoles and free ions in the material, and thus reasonable levels of dipolar groups and ions are required for the application of this technique. Humic materials, although heterogeneous in nature, are comprised of a variety of functional groups. Upon exposure to an electrical field, functional groups will be easily polarized, particularly for polar groups (e.g., carboxylate and hydroxyl groups) and cations. Depending on the chemical composition of the humic material, DETA may elucidate strong (e.g.,

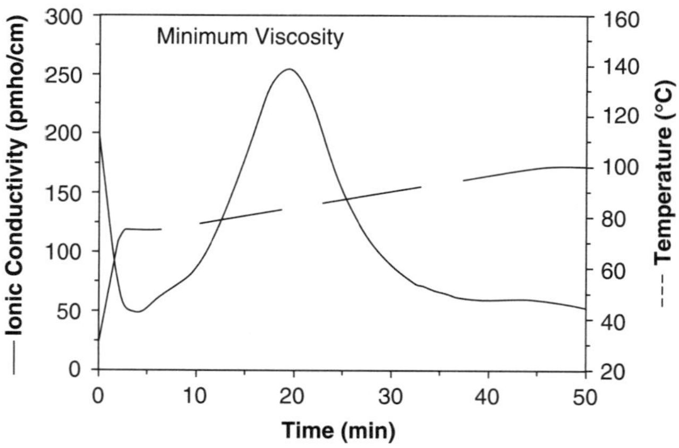

Figure 18.27. Dielectric thermal analysis online monitoring of a resin transfer molding process at 1000 Hz. Reproduced from McIlhagger et al. (2000), by permission of Elsevier, Ltd.

humic and fulvic acids) to weak (soot with very small amount of polar groups) thermal transition signals.

18.5. SUMMARY AND CONCLUSION

Current research needs require an improved understanding of the role of NOM in various environmental processes. The development of advanced thermal characterization techniques assists in this effort by improving our understanding of NOM structure and environmental reactivity through (i) improved accuracy of predictive tools such as molecular simulations of NOM structures, (ii) improved understanding of the influence of varying water and cation content on NOM structure, and (iii) improved identification of NOM–contaminant interactions through quantification of contaminant binding energies and influences of varying contaminant concentrations on NOM structural properties and consequent sorption/desorption mechanisms. Thermal analysis techniques provide a means to quantify thermodynamic properties of heat capacity, thermal expansion coefficient, glass transition temperatures, and binding energies. Combined, these parameters formulate both substantive new constraints for existing molecular simulations models and information sources to further elucidate NOM structure reactivity most responsible for observed contaminant sorption/desorption behaviors. The availability of thermal analytical data for NOM, although growing, is fairly scarce. The establishment of a database of thermodynamic properties for numerous NOM samples as proposed by DeLapp and LeBoeuf (2003), including standardized experimental protocols as described in this work, may thus be beneficial in filling this information void.

REFERENCES

Akim, E. L. (1978). Cellulose—Bellwether or old hat. *Chemtech* **8**(11), 676–682.

Androsch, R., Stolp, M., and Radusch, H.-J. (1996). Simultaneous X-ray diffraction and differential thermal analysis of polymers. *Thermochim. Acta Dev. Calorim. 1995*, **271**, 1–8.

Aouad, A., Benchanaa, M., Mokhlisse, A., and Arafan, A. (2002). Study of thermal behaviour of organic matter from natural phosphates (Youssoufia-Morocco). *J. Thermal Anal. Calorim.* **70**(2), 593–603.

Aubuchon, S. R., and Blaine, R. L. (1998). Thermal analysis & rheology: Recent development in the application of thermal analysis to polyolefins. TA Instruments, Inc., New Castle, DE.

Bao, Q.-B., and Bagga, C. S. (1993). The dynamic mechanical analysis of hydrogel elastomers. *Thermochim. Acta* **226**, 107–113.

Barros, N., Salgado, J., and Feijoo, S. (2006). Calorimetry and soil. *Thermochim. Acta* **458**(1–2), 11–17.

Barton, A. F. M. (1983). *CRC Handbook of Solubility Parameters and Other Cohesion Parameters*, CRC Press, Boca Raton, FL.

Barton, A. F. M. (1990). *CRC Handbook of Polymer-Liquid Interaction Parameters and Solubility Parameters*, CRC Press, Ann Arbor, MI.

Behar, F., and Vandenbroucke, M. (1987). Chemical modelling of kerogens. *Org. Geochem.* **11**(1), 15–24.

Benyamna, A., Bennouna, C., Moreau, C., and Geneste, P. (1991). Upgrading of distillate fractions of Timahdit Moroccan shale oil over a sulphided $NiO–MoO_3/\gamma–Al_2O_3$ catalyst. *Fuel* **70**(7), 845–848.

Bhargava, S., Awaja, F., and Subasinghe, N. D. (2005). Characterisation of some Australian oil shale using thermal, X-ray and IR techniques. *Fuel* **84**(6), 707–715.

Biswas, S., Choudhury, N., Sarkar, P., Mukherjee, A., Sahu, S. G., Boral, P., and Choudhury, A. (2006). Studies on the combustion behaviour of blends of Indian coals by TGA and Drop Tube Furnace. *Fuel Processing Technology* **87**, 191–199.

Blaine, R. L., and Hahn, B. K. (1998). Kinetic parameters obtained by modulated thermogravimetry. *J. Therm. Anal. Calorime.* **54**, 695–704.

Bouajila, J., Limare, A., Joly, C., and Dole, P. (2005). Lignin plasticization to improve binderless fiberboard mechanical properties. *Polym. Eng. Sci.* **45**(6), 809–816.

Bouchard, D. C. (2002). Cosolvent effects on sorption isotherm linearity. *J. Contam. Hydrol.* **56**(3–4), 159–174.

Boussehain, R., Feidt, M. L., and Balesdent, D. (1986). Thermogravimetric study of alcohol adsorption on active charcoal. *Thermochim. Acta* **103**(1), 113–122.

Buurman, P., van Lagen, B., and Piccolo, A. (2002). Increase in stability against thermal oxidation of soil humic substances as a result of self association. *Org. Geochem.* **33**(3), 367–381.

Caslavsky, J. L. (1988). Principles of the optical differential thermal analysis (ODTA). *Thermochim. Acta* **134**, 371–376.

Conte, P., and Piccolo, A. (1999). Conformational arrangement of dissolved humic substances. Influence of solution composition on association of humic molecules. *Environ. Sci. Technol.* **33**(10), 1682–1690.

Craig, D. Q. M., and Johnson, F. A. (1995). Pharmaceutical applications of dynamic mechanical thermal analysis. *Thermochim. Acta Pharmaceut. Therm. Anal.* **248**, 97–115.

Cuypers, C., Grotenhuis, T., Nierop, K. G. J., Franco, E. M., de Jager, A., and Rulkens, W. (2002). Amorphous and condensed organic matter domains: the effect of persulfate oxidation on the composition of soil/sediment organic matter. *Chemosphere* **48**(9), 919–931.

Davies, J. E. D., and Jabeen, N. (2002). The adsorption of herbicides and pesticides on clay minerals and soils. Part 1. Isoproturon. *J. Inclusion Phenom. Macrocyclic Chem.* **43**(3–4), 329–336.

Davies, J. E. D., and Jabeen, N. (2003). The adsorption of herbicides and pesticides on clay minerals and soils. Part 2. Atrazine. *J. Inclusion Phenom. Macrocyclic Chem.* **46**(1–2), 57–64.

DeLapp, R. C., and LeBoeuf, E. J. (2003). A proposal for the establishment of a database of thermodynamic properties of natural organic matter. In *Humic Substances: Nature's Most Versatile Materials*, Ghabbour, E. A., and Davies, G., eds., Taylor and Francis, New York, pp. 61–81.

DeLapp, R. C., and LeBoeuf, E. J. (2004). Thermal analysis of whole soils and sediment. *J. Environ. Quality* **33**(1), 330–337.

DeLapp, R. C., LeBoeuf, E. J., and Bell, K. D. (2004). Thermodynamic properties of several soil- and sediment-derived natural organic materials. *Chemosphere* **54**(4), 527–539.

DeLapp, R. C., LeBoeuf, E. J., Chen, J., and Gu, B. (2005). Advanced thermal characterization of fractionated natural organic materials. *J. Environ. Qual.* **34**(4), 842–853.

Dell'Abate, M., Benedetti, A., and Sequi, P. (2000). Thermal methods of organic matter maturation monitoring during a composting process. *J. Therm. Anal. Calorim.* **61**(2), 389–396.

Dell'Abate, M. T., Benedetti, A., and Brookes, P. C. (2003). Hyphenated techniques of thermal analysis for characterisation of soil humic substances. *J. Sep. Sci.* **26**(5), 433–440.

Diallo, M. S., Simpson, A., Gassman, P., Faulon, J. L., Johnson, J. H., Hatcher P. G., and Goddard, W. A. (2003). 3-D structural modeling of humic acids through experimental characterization, computer assisted structure elucidation and atomistic simulations. 1. Chelsea soil humic acid. *Environ. Sci. Technol.* **37**(9), 1783–1793.

DiMarzio, E. A. (1981). Equilibrium theory of glasses. In *Structure and Mobility in Molecular and Atomic Glasses*, O'Reilly, J. M., and Goldstein, M., eds., Annals of the New York Academy of Sciences, New York, pp. 1–20.

Eisenberg, A. (1984). The glassy state and the glass transition. In *Physical Properties of Polymers*, Mark, J. E., Eisenberg, A., Graessley, W. W., Mandelkern, L., and Koenig, J. L., eds., American Chemical Society, Washington, D.C., pp. 55–95.

Eisenberg, A. (1993). The glassy state and the glass transition. In *Physical Properties of Polymers*, Mark, J. E., Eisenberg, A., Graessley, W. W., Mandelkern, L., Samulski, E. T., Koenig, J. L., and Wignall, G. D., eds., American Chemical Society, Washington, D.C., pp. 61–95.

Elias, H. G. (1977). *Macromolecules 1: Structure and Properties*, translated from German by J. W. Stafford, Plenum Press, New York.

Ferry, J. D. (1980). *Viscoelastic Properties of Polymers*, Elsevier Scientific Publishing Company, New York.

Flory, P. J. (1970). Thermodynamics of polymer solutions, *Disc. Faraday Soc.* **49**, 7–29.

Foreman, J., Sauerbrunn, S. R., and Marcozzi, C. L. (1995). Thermal analysis & rheology: Exploring the sensitivity of thermal analysis techniques to the glass transition. TA Instruments, Inc., New Castle, DE.

Fox, T. G., Jr., and Flory, P. J. (1948). Viscosity–molecular weight and viscosity–temperature relationships for polystyrene and polyisobutylene, *J. Am. Chem. Soc.* **70**, 2384–2395.

Fox, T. G., Jr., and Flory, P. J. (1951). Further studies on the melt viscosity of polyisobutylene. *J. Phys. Colloid Chem.* **55**, 221–235.

Fox, T. G., Jr., and Flory, P. J. (1954). The glass temperature and related properties of polystyrene. influence of molecular weight, *J. Polym. Sci.* **14**, 315–319.

Francioso, O., Ciavatta, C., Montecchio, D., Tugnoli, V., Sanchez-Cortes, S., and Gessa, C. (2003). Quantitative estimation of peat, brown coal and lignite humic acids using chemical parameters, 1H-NMR and DTA analyses. *Bioresour. Technol.* **88**(3), 189–195.

Fujita, H., and Kishimoto, A. J. (1958). Diffusion-controlled stress relaxation in polymers. II. stress relaxation in swollen polymers, *J. Polym. Sci.* **28**, 547–567.

Gaffney, J. S., Marley, N. A., and Clark, S. B. (1996). Humic and fulvic acids and organic colloidal materials in the environment. *Humic and fulvic acids: Isolation, Structure, and Environmental Role*, American Chemical Society, Washington, D.C.

Geoffrey, D., and Ghabbour, E. A. (1998). *Humic Substances: Structures, Properties and Uses*, Royal Society of Chemistry, Cambridge, UK.

Goldberg, E. D. (1985). *Black Carbon in the Environment: Properties and Distribution*, John Wiley & Sons, New York.

Golebiowska, D., Mielnik, L., and Gonet, S. (1996). Characteristics of humic acids in bottom sediments of lobelia lakes. *Environ. Int.: The HUMEX/HUMOR Project and Humic Substances* **22**(5), 571–578.

Guetzloff, T. F., and Rice, J. A. (1996). Micellar nature of humic substances. *Humic and Fulvic Acids: Isolation, Structure, and Environmental Roles*, Clark, S. B., ed., American Chemical Society, Washington, D.C., p. 18.

Gunasekara, A. S., Simpson, M. J., and Xing, B. (2003). Identification and characterization of sorption domains in soil organic matter using structurally modified humic acids. *Environ. Sci. Technol.* **37**(5), 852–858.

Gunasekara, A. S., and Xing, B. (2003). Sorption and desorption of naphthalene by soil organic matter: Importance of aromatic and aliphatic components. *J. Environ. Qual.* **32**(1), 240–246.

Hardy, L., Stevenson, I., Boiteux, G., Seytre, G., and Schonhals, A. (2001). Dielectric and dynamic mechanical relaxation behaviour of poly(ethylene 2,6 naphthalene dicarboxylate). I. Amorphous films. *Polymer* **42**(13), 5679–5687.

Hatakeyama, T., and Hatakeyama, H. (2004). *Thermal Properties of Green Polymers and Biocomposites*, Kluwer Academic Publishers, Dordrecht.

Haward, R. N. (1973). *The Physics of Glassy Polymers*, John Wiley & Sons, New York.

Hendricks, S. B., and Alexander, L. T. (1940). Semiquantitative estimation of montmorillonite and clays. *Soil Sci. Soc. Am. Proc.* **5**, 95–99.

Hodge, I. M., and Huvard, G. S. (1983). Effects of annealing and prior history on enthalpy relaxation in glassy polymers. 3. Experimental and modeling studies of polystyrene. *Macromolecules* **16**(3), 371–375.

Höhne, G. W. H., Hemminger, W., and Flammersheim, H.-J. (1996). *Differential Scanning Calorimetry. An Introduction for Practitioners*. Springer, Berlin.

Hu, W.-G., Mao, J., Xing, B., and Schmidt-Rohr, K. (2000). Poly(methylene) crystallites in humic substances detected by nuclear magnetic resonance. *Environ. Sci. Technol.* **34**(3), 530–534.

Huang, W. L., Young, T. M., Schlautman, M. A., Yu, H., and Weber W. J., Jr. (1997). A distributed reactivity model for sorption by soils and sediments. 9. General isotherm nonlinearity and applicability of the dual reactive domain model. *Environ. Sci. Technol.* **31**(6), 1703–1710.

Hurrass, J., and Schaumann, G. E. (2007). Hydration kinetics of wettable and water-repellent soils. *Soil Sci. Soc. Am. J.* **71**(2), 280–288.

Hutchinson, J. M. (1998). Characterising the glass transition and relaxation kinetics by conventional and temperature-modulated differential scanning calorimetry. *Thermochimica Acta* **324**(1–2), 165–174.

Ishida, H., and Lee, Y.-H. (2001). Synergism observed in polybenzoxazine and poly(ε-caprolactone) blends by dynamic mechanical and thermogravimetric analysis. *Polymer* **42**(16), 6971–6979.

Ishiwata, R. (1969). An estimation of aromaticity of a lake sediment humic acid by air oxidation and evaluation of it. *Soil Sci.* **107**(1), 53–57.

Jambu, P., Dupuis, T., and Garais, M. (1975a). Use of differential thermal-analysis to characterize fulvic acid metal complexes and humic acid metal complexes. 2. Application to natural organometallic complexes. *J. Therm. Anal.* **8**(2), 231–237.

Jambu, P., Dupuis, T., and Garais, M. (1975b). Use of differential thermal-analysis to characterize humic acid metal complexes and rulvic acid metal complexes. 1. Characterization of complexes prepared in laboratory. *J. Therm. Anal.* **8**(1), 87–97.

Kelley, F. N., and Bueche, F. (1961). Viscosity and glasstemperaturerelations for polymer-diluent systems, *J. Polym. Sci.* **50**, 549–556.

Kelley, S. S., Rials, T. G., and Glasser, W. G. (1987). Relaxation behaviour of the amorphous behavior the amorphous components of wood. *J. Mater. Sci.* **22**(2), 617–624.

Kilduff, J. E., and Weber, W. J., Jr. (1992). Transport and separation of organic macromolecules in ultrafiltration processes, *Environ. Sci. Technol.* **26**(3), 569–577.

Kolokassidou, C., Pashalidis, I., Costa, C. N., Efstathiou, A. M., and Buckau, G. (2007). Thermal stability of solid and aqueous solutions of humic acid. *Thermochim. Acta* **454**(2), 78–83.

Kroschwitz, J. I. (1990). *Polymers: Polymer Characterization and Analysis*, Wiley-Interscience, New York.

Kumar, M., Ghosh, S. K., and Manjaiah, K. (2004). Organo-mineral complexes in some inceptisols of India: Physicochemical and thermal properties. *Commun. Soil Sci. Plant Anal.* **35**(13–14), 1865–1889.

LeBoeuf, E. J., and Weber, W. J., Jr. (1997). A distributed reactivity model for sorption by soils and sediments. 9. Sorbent organic domains: Discovery of a humic acid glass transition and an argument for a polymer-based model. *Environ. Sci. Technol.* **31**(6), 1697–1702.

LeBoeuf, E. J., and Weber, W. J., Jr. (2000a). Macromolecular characteristics of natural organic matter. 1. Insights from glass transition and enthalpic relaxation behavior. *Environ. Sci. Technol.* **34**(17), 3623–3631.

LeBoeuf, E. J., and Weber, W. J., Jr. (2000b). Macromolecular characteristics of natural organic matter. 2. Sorption and desorption behavior. *Environ. Sci. Technol.* **34**(17), 3632–3640.

Leifeld, J. (2007). Thermal stability of black carbon characterised by oxidative differential scanning calorimetry. *Org. Geochem.* **38**(1), 112–127.

Liebich, J., Schloter, M., Schaffer, A., Vereecken, H., and Burauel, P. (2007). Degradation and humification of maize straw in soil microcosms inoculated with simple and complex microbial communities. *Eur. J. Soil Sci.* **58**(1), 141–151.

Lucht, L. M., Larson, J. M., and Peppas, N. A. (1987). Macromolecular structure of coals. 9. Molecular structure and glass transition temperature. *Energy Fuels* **1**, 56–58.

Ma, X., Sauer, J. A., and Hara, M. (1995). Poly(methyl methacrylate) based ionomers. 1. dynamic mechanical properties and morphology. *Macromolecules* **28**, 3953–3962.

Mackinnon, A. J., Antxustegi, M. M., and Hall, P. J. (1994). Glass transitions and enthalpy relaxation in coals. *Fuel* **73**(1), 113–115.

Matejka, J. (1922). Thermal analysis as a means of detecting kaolinite in soils. *Chem. Listy* **16**, 8–14.

Mark, J. E., Eisenberg, A., Graessley, W. W., Mandelkern, L., and Koenig, J. L. (1984). The glassy state and the flass transition. In *Physical Properties of Polymers*, American Chemical Society, Washington, D.C., pp. 55–96.

McBrierty, V. J., Wardell, G. E., Keely, C. M., ONeill, E. P., and Prasad, M. (1996). The characterization of water in peat. *Soil Sci. Soc. Am. J.* **60**(4), 991–1000.

McIlhagger, A., Brown, D., and Hill, B. (2000). The development of a dielectric system for the on-line cure monitoring of the resin transfer moulding process. *Composites Part A: Appl. Sci. Manuf.* **31**(12), 1373–1381.

McKenna, G. B. (1989). Glass formation and glassy behavior. In *Comprehensive Polymer Science: The Synthesis, Characterization, Reactions, & Applications of Polymers*, Vol. **II**: *Polymer Properties*, G., Allen, and Bevington, J. C., eds., Pergamon Press, New York, pp. 311–362.

Menard, K. P. (1999). *Dynamic Mechanical Analysis: A Practical Introduction*, CRC Press, Boca Raton, FL.

Moynihan, C. T., Easteal, A. J., Bolt, M. A., and Tucker, J. (1976). Dependence of the fictive temperature of glass on cooling rate. *J. Am. Ceram. Soc.* **59**(1–2), 12–16.

Moynihan, C. T., Easteal, A. J., Wilder, J., and Tucker, J. (1974). Dependence of the glass transition temperature on heating and cooling rate. *J. Phys. Chem.* **78**(26), 2673–2677.

Narayanaswamy, O. S. (1971). A model of structural relaxation in glass. *J. Am. Ceram. Soc.* **54**(10), 491–498.

Norinaga, K., Kumagai, H., Hayashi, J.-I., and Chiba, T. (1998). Classification of water sorbed in coal on the basis of congelation characteristics. *Energy Fuels* **12**(3), 574–579.

Novotny, E. H., Deazevedo, E. R., Bonagamba, T. J., Cunha, T. J. F., Madari, B. E., Benites, V. D., and Hayes, M. H. B. (2007). Studies of the compositions of humic acids from Amazonian dark earth soils. *Environ. Sci. Technol.* **41**(2), 400–405.

Nunez-Regueira, L., Proupin-Castineiras, J., Rodriguez-Anon, J. A., Villanueva-Lopez, M., and Nunez-Fernandez, O. (2006). Design of an experimental procedure to assess soil health state. *J. Therm. Anal. Calorim.* **85**(2), 271–277.

Oh, C. H., Trichet, J., Oh, J. H., and Rouzaud, J. N. (1990). Relation between the thermal-behavior and the structure of kerogen in the Goesan uranium deposit, Korea. *Org. Geochem.* **16**(1–3), 609–619.

Ouatmane, A., Provenzano, M. R., Hafidi, M., and Senesi, N. (2000). Compost maturity assessment using calorimetry, spectroscopy and chemical analysis. *Compost Sci. Util.* **8**(2), 124–134.

Piccolo, A., Nardi, S., and Concheri, G. (1996). Micelle-1ike conformation of humic substances as revealed by size exclusion chromatography. *Chemosphere* **33**(4), 595–602.

Price, D. M. (1997). Thermal analysis of fibres and films in solvents and vapours. *Thermochim. Acta Curr. Res. Dev. Therm. Anal. Calorim.* **294**(1), 127–131.

Privalko, V. P., Demchenko, S. S., and Lipatov, Y. S. (1986). Structure-dependent enthalpy relaxation at the glass transition of polystyrenes. *Macromolecules* **19**(3), 901–904.

Provder, T., ed. (1984). Size exclusion chromatography: Methodology and characterization of polymers and related materials. In *ACS Symposium Series No. 245*, American Chemical Society, Washington, D.C.

Provenzano, M. R., Albuzio, A., and D'Orazio, V. (2005). Spectroscopic and thermal investigations on compost improved by iron salt addition. *J. Agric. Food Chem.* **53**(2), 374–382.

Provenzano, M. R., Gigliotti, G., Cilenti, A., Erriquens, F., and Senesi, N. (2006). Spectroscopic and thermal investigation of hydrophobic and hydrophilic fractions of dissolved organic matter. *Compost Sci. Util.* **14**(3), 191–200.

Provenzano, M. R., Ouatmane, A., Hafidi, M., and Senesi, N. (2000). Differential scanning calorimetric analysis of composted materials from different sources. *J. Therm. Anal. Calorim.* **61**(2), 607–614.

Provenzano, M. R., and Senesi, N. (1998). Differential scanning calorimetry of river aquatic fulvic acids and their metal complexes. *Fresenius Environ. Bull.* **7**(7–8), 423–428.

Provenzano, M. R., and Senesi, N. (1999). Thermal properties of standard and reference humic substances by differential scanning calorimetry. *J. Therm. Anal. Calorim.* **57**(2), 517–526.

Provenzano, M. R., Senesi, N., and Miikki, V. (1998a). Characterization of composts and humic acids from pulp and paper mill biosludges by DSC in association with FT-IR spectroscopy. *J. Therm. Anal. Calorim.* **52**(3), 1037–1046.

Provenzano, M. R., Senesi, N., and Piccone, G. (1998b). Thermal and spectroscopic characterization of composts from municipal solid wastes. *Compost Sci. Util.* **6**(3), 67–73.

Richardson, M. J. (1989). Thermal analysis. In *Comprehensive Polymer Science: The Synthesis, Characterization, Reactions, & Applications of Polymers, Volume I: Polymer Properties*, Allen, G., and Bevington, J. C., eds., Pergamon Press, New York, pp. 867–909.

Riise, G., and Salbu, B. (1986). The influence of aquatic humic substances on the octanol/water partition coeffficient of pesticides and trace elements at different pH values. In *Humic Substances in the Aquatic and Terrestrial Environment*, Grimvall, A., eds., Springer-Verlag, Linkoping, Sweden, pp. 357–365.

Risoul, V., Renauld, V., Trouve, G., and Gilot, P. (2002). A laboratory pilot study of thermal decontamination of soils polluted by PCBs. Comparison with thermogravimetric analysis. *Waste Manage.* **22**(1), 61–72.

Rosen, S. L. (1993). *Fundamental Principles of Polymeric Materials*, John Wiley & Sons, New York.

Salmen, L. (1984). Viscoelastic properties of in situ lignin under water-saturated conditions. *J. Mater. Sci.* **19**, 3090–3096.

Schaumann, G. E. (2005). Matrix relaxation and change of water state during hydration of peat. In *A Selection of Papers from the Third International Conference "Interfaces against Pollutions" (IAP 2004)*, May 24–27, Julich, Germany. *Colloids Surf. A: Physicochem. Eng. Aspects* **265**(1–3), 163–170.

Schaumann, G. E. (2006). Soil organic matter beyond molecular structure Part I: Macromolecular and supramolecular characteristics. *J. Plant Nutr. Soil Sci.* **169**(2), 145–156.

Schaumann, G. E., and LeBoeuf, E. J. (2005). Glass transitions in peat: Their relevance and the impact of water. *Environ. Sci. Technol.* **39**(3), 800–806.

Schaumann, G. E., LeBoeuf, E. J., DeLapp, R., and Hurrass, J. (2005). Thermomechanical analysis of air-dried whole soil samples. *Thermochim. Acta* **436**(1–2), 83–89.

Schawe, J. E. K. (1998). A description of the glass transition measured by temperature modulated differential scanning calorimetry. *Colloid Polym. Sci.* **276**(7), 565–569.

Schnitzer M. (1972). Chemical, Spectroscopic, and thermal methods for the classification of humic substances. In *Proceedings International Meetings on Humic Substances, Pudoc, Wageningen*, pp. 293–310.

Schnitzer, M., and Kodama, H. (1972). Differential thermal-analysis of metal–fulvic acid salts and complexes. *Geoderma* **7**(1–2), 93–103.

Schnitzer, M., Turner, R. C., and Hoffman, I. (1964). A thermogravimetric study of organic matter of representative Canadian podzol soils. *Can. J. Soil Sci.* **44**, 7–13.

Senesi, N., and Loffredo, E. (1999). The chemistry of soil organic matter. In *Soil Physical Chemistry*, Sparks, D. S., eds., CRC Press, Boca Raton, FL, pp. 317–319.

Senturia, S. D., and Sheppard, N. F. (1986). Dielectric analysis of thermoset cure. In *Advances in Polymer Science*, Springer, Berlin, Germany, pp. 1–47.

Shurygina, E. A., Larina, N. K., Chubarov, M. A., and Kononova, M. M. (1971). Differential thermal analysis (Dta) and thermogravimetry (Tg) of soil humus substances. *Geoderma* **6**(3), 169–177.

Siewert, C. (2004). Rapid screening of soil properties using thermogravimetry. *Soil Sci. Soc. Am. J.* **68**(5), 1656–1661.

Sircar, A. K. (1997). Elastomers. In *Thermal Characterization of Polymeric Materials*, Turi, E. A., ed., Academic Press, San Diego, CA.

Stevenson, F. J. (1994). *Humus Chemistry: Genesis, Composition, Reaction*, John Wiley & Sons, New York.

Struik, L. C. E. (1978). *Physical Aging in Amorphous Polymers and Other Materials*, Elsevier Scientific Publishing Company, New York.

TA Instruments, Inc. (1995a). Thermal Analysis and Rheology: Analysis of Polymer Decomposition by TGA-Mass Spectrometry, New Castle, DE.

TA Instruments, Inc. (1995b). *Thermal analysis and theology: Detection of beta transus in titanium alloys by differential thermal analysis.* New Castle, DE.

TA Instruments, Inc. (2006). *Q1000 Differential Scanning Calorimeter Product Information*, New Castle, DE.

Talja, R. A., and Roos, Y. H. (2001). Phase and state transition effects on dielectric, mechanical, and thermal properties of polyols. *Thermochim. Acta* **380**(2), 109–121.

Tan, K. H. (1978). Formation of metal humic acid complexes by titration and their characterization by differential thermal-analysis and infrared spectroscopy. *Soil Biol. Biochem.* **10**(2), 123–129.

Tan, K. H., and Clark, F. E. (1969). Polysaccharide constituents in fulvic and humic acids extracted from soil. *Geoderma* **2**(3), 245–255.

Tan, K. H., and Hajek, B. F. (1977). Thermal analysis of soils. In *Minerals in Soils Environments*, Dixon, J. B., Weed, S. B., Kittrick, J. A., Milford, M. H., and White, J. L., eds., Soil Science Society of America, Madison, WI, pp. 865–884.

Teraoka, I., Langley, K. H., and Karasz, F. E. (1992). Reptation dynamics of semirigid polymers in porous media. *Macromolecules* **25**(23), 6106–6112.

Tobolsky, A. V. (1960). *Properties and Structure of Polymers*, John Wiley & Sons, New York.

Tool, A. Q. (1946). Relation between inelastic deformability and thermal expansion of glass in its annealing range. *J. Am. Ceram. Soc.* **29**(9), 240–253.

Turner, R. C., and Schnitzer, M. (1962). Thermogravimetry of organic matter of a podzol. *Soil Sci.* **93**, 225–232.

Vieth, W. R., Howell, J. M., and Hsieh, H. H. (1976). Dual sorption theory. *J. Membr. Sci.* **1**, 177–220.

Wetton, R. E., Marsh, R. D. L., and Van-de-Velde, J. G. (1991). Theory and application of dynamic mechanical thermal analysis. *Thermochimica Acta* **175**, 1–11.

Wollenweber, J., Schwarzbauer, J., Littke, R., Wilkes, H., Armstroff, A., and Horsfield, B. (2006). Characterisation of non-extractable macromolecular organic matter in Palaeozoic coals. *Palaeogeogr. Palaeoclimatol. Palaeoecol.* **240**(1–2), 275–304.

Wunderlich, B. (1990). *Thermal Analysis*, Academic Press, Boston.

Wunderlich, B. (2005). *Thermal Analysis of Polymeric Materials*. Springer, New York.

Xia, G., and Ball, W. P. (1999). Adsorption-partitioning uptake of nine low-polarity organic chemicals on a natural sorbent. *Environ. Sci. Technol.* **33**(2), 262–269.

Xia, G., and Pignatello, J. J. (2001). Detailed sorption isotherms of polar and apolar compounds in a high-organic soil. *Environ. Sci. Technol.* **35**(1), 84–94.

Xing, B. S., and Pignatello, J. J. (1997). Dual-mode sorption of low-polarity compounds in glassy poly(vinyl chloride) and soil organic matter. *Environ. Sci. Technol.* **31**(3), 792–799.

Yamamoto, S., Yoshioka, H., and Ishiwatari, R. (2007). Pyrolysis– and chemical degradation–GC/MS analyses of environmental kerogen and humic substances and their applications to geochemistry. *Bunseki Kagaku* **56**(2), 71–91.

Young, K. D., and LeBoeuf, E. J. (2000). Glass transition behavior in a peat humic acid and an aquatic fulvic acid. *Environ. Sci. Technol.* **34**(21), 4549–4553.

Yun, Y., and Suuberg, E. M. (1992). Structural changes in high-rank bituminous coals caused by solvent swelling and heat. *Division of Fuel Chemistry, American Chemical Society Preprints of Papers* **37**(2), 856–865.

Zhang, L., LeBoeuf, E. J., and Xing, B. (2007). Thermal analytical investigation of biopolymers and humic- and carbonaceous-based soil and sediment organic matter. *Environ. Sci. Technol.* **41**(14), 4888–4894.

Zhang, L., and LeBoeuf, E. J. (2009). Effects of solvents on thermal and structural properties of humic acids: An integrated thermal analytical and molecular dynamics study, *Environ. Sci. Technol.* (in review).

INDEX

Biophysico-Chemical Processes Involving Natural Nonliving Organic Matter in Environmental Systems, Edited by Nicola Senesi, Baoshan Xing, and Pan Ming Huang